CHEMISTRY, MAN AND SOCIETY SECOND EDITION

MARK M. JONES
Chairman of the Department of Chemistry, Vanderbilt University, Nashville, Tennessee

JOHN T. NETTERVILLE
Chairman, Department of Chemistry, David Lipscomb College, Nashville, Tennessee

DAVID O. JOHNSTON
Professor of Chemistry, David Lipscomb College, Nashville, Tennessee

JAMES L. WOOD
Professor of Chemistry, David Lipscomb College, Nashville, Tennessee

SAUNDERS GOLDEN SUNBURST SERIES

W. B. SAUNDERS COMPANY / Philadelphia / London / Toronto

W. B. Saunders Company: West Washington Square
Philadelphia, Pa. 19105

1 St. Anne's Road
Eastbourne, East Sussex BN21 3UN, England

833 Oxford Street
Toronto, M8Z 5T9, Canada

Library of Congress Cataloging in Publication Data

Main entry under title:

Chemistry, man, and society.

(Saunders golden sunburst series)
Includes bibliographies.

1. Chemistry. I. Jones, Mark Martin, 1928–
QD31.2.C43 1976 540 75-22733
ISBN 0-7216-5220-4

Cover artwork supplied through the courtesy of Diamond Shamrock Corporation

Chemistry, Man and Society ISBN 0-7216-5220-4

Last digit is the print number: 9 8 7 6 5 4 3 2

PREFACE

Like the first edition of this text, this second edition has been prepared for those students wishing a one-semester or two-semester course in college chemistry. Flexibility in the use of this text is built into its design. As was widely demonstrated in the use of the first edition, this fundamental approach will serve well those who wish to examine the beginnings and growth of the science, giving critical consideration to the discoveries and thought processes that have brought us to the current state in chemistry. It will serve equally well those who wish to build on the consequences of modern chemistry, spending most of the study time in the examination of the man-made chemical effects that are of concern in our society. Obviously, intermediate choices in this teaching spectrum are also available. Regardless of the choice made, however, we have worked consistently on the belief that fundamental facts should be related to fundamental theory in dealing with chemical effects.

Certainly, there is mystery in chemistry, but to leave the workings of the chemist as a mystery is inevitably to argue that the liberally educated man must be dependent upon the chemist for the chemical decisions that affect society as a whole. This text is based on the belief that the liberal arts student can see and appreciate the chain of events leading from chemical fact to chemical theory and the ingenuous manipulation of materials based on the chemical theories. He will then see that the intellectual struggles in chemistry are closely akin to his own intellectual pursuits and will feel that he can be partner to the ongoing chemical manipulation of our environment.

The entire presentation in this book is based on stated student needs. Over the years we, through innumerable discussions with our students, have observed an intense interest in the following areas:

1. There is considerable pleasure in understanding a sensible, theoretical explanation of an observed natural phenomenon. Such curiosity appears innate in every person.

2. There is a desire to make highly personal choices as rational as possible. The role of fluoride in tooth structure, for example, is appreciated when understood. A class always perks up when the role of aluminum chloride in a body deodorant is explained, realizing the personal choice that is to be made in the matter.

3. Chemical choices to be made in the immediate environment, in the home, in the automobile, and in the work area are of vital interest. For example, the choice of fuel additives in gasoline proves to be interesting to almost every student.

4. The great environmental questions of a chemical nature also have a compelling interest. The liberal arts student appreciates a sense of awareness in the dilemmas presented in gross pollution problems and in the depletion of natural resources.

Although an effort has been made to tailor the text to suit the student with a minimum of scientific training, the approach is sufficiently different to challenge and interest the student with a high school chemistry background. Chemical theory is shown as a continuing development of conceptual ideas about the submicroscopic world of atoms and molecules rather than as a set of "facts" to be learned. Most of the topics covered in the last half of the text are not presented in high school texts.

Numerous study aids for the benefit of the student have been included in this second edition. Marginal notes have been added for the purposes of highlighting various ideas presented in the text and adding enrichment. Self-tests have been interspersed throughout the text to give the student an opportunity to see whether or not he has a comprehension of the material presented. Matching sets to help build and maintain the chemical vocabulary and numerous questions have been included at the end of each chapter. Though some of these questions are based on simple recall, most of them are intended to provoke thought about significant chemical problems, and some of them are open-ended in the sense that they call for a highly critical evaluation of a chemical issue. Important words in the text are presented in bold face type to aid the student in developing the necessary vocabulary. The references for further reading have been expanded with recent material pertaining to current problems. Considerable attention has been given to enriching the illustrative material in order to help the student grasp as quickly as possible the ideas presented.

The organization of this text allows a unique development of the energy question—perhaps the most fundamental question in man's future. The problem is dealt with from many different points of view. The student can readily see the energy that can be available to man, its channels both physical and chemical through our environment, and the choices that must be made in energy management.

As in previous texts, we dedicate this effort to our wives and gratefully acknowledge their support and understanding during the preparation of this manuscript.

MARK M. JONES

JOHN T. NETTERVILLE

DAVID O. JOHNSTON

JAMES L. WOOD

CONTENTS

CHAPTER 1

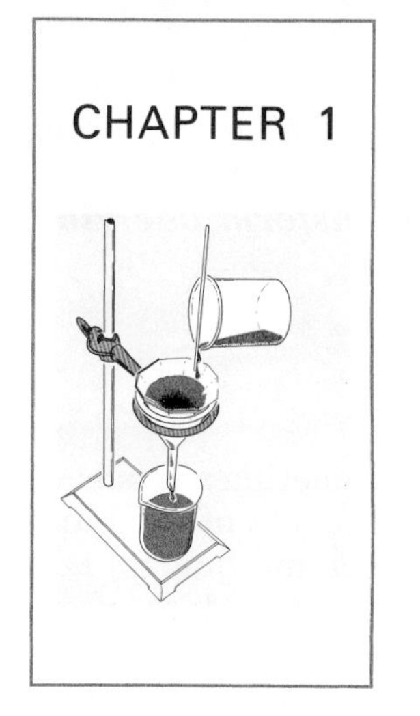

THE CHEMICAL VIEW OF MATTER

Look around! Most of the things we use in our daily life are very different from the materials which are an obvious part of our ***natural surroundings.*** Practically everything we use has been transformed from a natural state of little or no utility to one of very different appearance and much greater utility. The processes by which the materials of nature can be transformed and a detailed description of such changes are highly intriguing. Wherever we look we see matter undergoing change. This is what chemistry is all about: ***matter,*** and the ***changes*** it undergoes.

An understanding of our environment and the changes which are possible in it is necessary if we are to know the consequences of our acts on our environment and even more critical if we are to undertake the successful repair of previous damage. Since life itself involves very intricate sequences of changes, each of us has a more personal reason for valuing knowledge of such processes—a desire to gain some insight into the manner in which chemicals in our food, air, and water can influence our individual health, happiness, and behavior.

While all changes in matter are of concern, our attention here will be focused on a particular kind of change called ***chemical change.*** In any chemical change the starting material is changed into a different ***kind*** of matter. Matter may also undergo another kind of change which does not produce a new kind of matter, but simply results in a new form of the same material. This kind of change is a ***physical change.***

Chemical change alters the kinds of matter present.

Since the feature used to characterize chemical changes is the production of a different kind of matter, it is necessary to be able to classify matter into different types. In a natural source the types of matter are usually mixed together and the separation of such mixtures has to precede their systematic classification. After an examination of the methods of separating such mixtures into their components, some of the problems involved in an accurate definition of the terms "kinds of matter" and "chemical change" can be appreciated.

In spite of the difficulties experienced in establishing the exact bounds of chemistry, it will be helpful to think of ***chemistry*** as ***the study of the kinds of matter and the changes that transform one kind of matter into another.***

PURE SUBSTANCES

Most samples of matter are complex mixtures. Sometimes it is easy to see the various ingredients in a mixture, such as the bits of sand and clay in a sample of soil, or the glittering crystals (mica and quartz) and dark areas (feldspar and magnetite) in a piece of granite. More often, however, it is not apparent to the casual observer that a mixture is made up of distinctly separate ingredients. For example, it is only by an experiment that we can establish that the air we breathe is a mixture.

Bits of iron and sulfur mixed

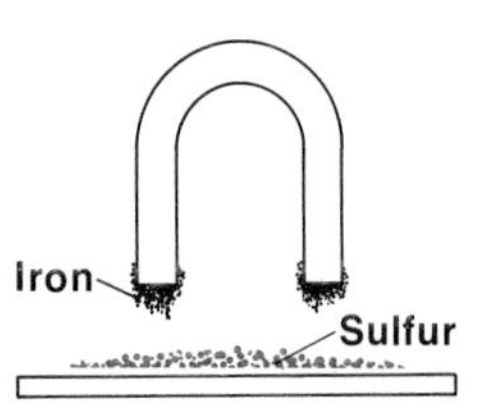

When a mixture is separated into its components, the components of the mixture are said to be purified. However, most efforts at separations are incomplete in a single operation or step, and repetition of the purification process results in a better separation. Ultimately in such a process, the experimenter arrives at pure substances, samples of matter that cannot be further purified. For example, if sulfur and iron powder are ground together to form a mixture, the iron can be separated from the sulfur by repeated stirrings of the mixture with a magnet. When the mixture is stirred the first time and the magnet removed, much of the iron is removed with it, leaving the sulfur in a higher state of purity. However, after just one stirring the sulfur may still have a dirty appearance due to a small amount of iron that remains. Repeated stirring with the magnet, or perhaps the use of a very strong magnet, will finally leave a bright yellow sample of sulfur that apparently cannot be purified further by this technique. In this purification process a property of the mixture, its color, is a measure of the extent of purification. After the bright yellow color is obtained, it could be assumed that the sulfur has been purified. Drawing a conclusion based on one property of the mixture is dangerous, however, because other methods of purification might change some other properties of the sample. It is only safe to call the sulfur a pure substance when all possible methods of purification fail to change its properties. This assumes that all pure substances have a set of properties by which they can be recognized just as a person can be recognized by a set of characteristics. ***A pure substance, then, is a kind of matter with properties that cannot be changed by further purification.***

Purification separates the kinds of matter.

There are some naturally occurring pure substances. Rain is very nearly pure water except for small amounts of dust and air. Gold, diamonds, and sulfur are also found in very pure form. These substances are special cases. Man, a complex assemblage of mixtures, lives in a world of mixtures—eating them, wearing them, living in houses made of them, and making most of his tools out of them.

Most materials in nature are mixtures; a few are relatively pure substances.

Although naturally occurring pure substances are not common, it has been possible to produce many pure substances from natural mixtures. A number of ingenious methods have been designed to separate mixtures into their component pure substances. Furthermore, processes have been developed to change such pure substances into a multitude of other pure substances, many of which do not occur in nature.

Relatively pure substances are now very common as a consequence of the development of modern purification techniques. Common examples are sugar, table salt (sodium chloride), copper, sodium bicarbonate, nitrogen, dextrose, ammonia, uranium, and carbon dioxide—to mention just a few. In all, about two million pure substances have been identified.

DEFINITIONS: OPERATIONAL AND THEORETICAL

Chemistry begins with observations and experiments.

The preceding definition given for a pure substance is an ***operational*** definition, or a definition in terms of specific experiments or operations. That is, if any further purification effort is unsuccessful in changing the properties of a substance, it is said to be a pure substance. It is evident then that operational definitions result from performing operations or tests on matter and summarizing the results in a statement. For example, pure sulfur is a yellow substance that boils at 833°F and has a density of 129 pounds per cubic foot. When all the properties of pure sulfur have been listed, we find that the pure substance has been characterized in a way that distinguishes it from any other pure substance.

A pure substance also can be defined in theoretical terms; that is, in terms of the molecules, atoms, and subatomic particles that compose it. Both types of definitions are important in the study of chemistry and both will be used in the presentation of this text. The theoretical definition will follow the development of the theory on which it is based.

SELF-TEST 1-A*

1. Four common materials that could not be pure substances are: a. ____________, b. ____________, c. ____________, and d. ____________.

2. Which must come first, the operational definition or the theoretical definition?____________________.

3. Four common materials that are very nearly pure substances are: a. ____________, b. ____________, c. ____________, and d. ____________.

4. Two different pure substances could have all properties identical. True ( ) or False ( )

SEPARATION OF MIXTURES INTO PURE SUBSTANCES

The separation of mixtures is usually more difficult than the magnetic separation of iron and sulfur described previously. Most beginning chemistry students would find it a bewildering task to separate a piece of granite into pure substances. Indeed, a chemist would find this a difficult assignment. Since each of the pure substances in the granite has a set of properties unlike those of any other pure substance, it should be

*Use these self-tests as a measure of how well you have learned the material. Take a test only after careful reading of the section which it follows. Do not return to the text during the self-test, but reread entire sections carefully if you do poorly on the self-test on those sections. The answers to the self-tests are at the end of the text.

possible to use these properties to separate the pure substances, just as the attraction of iron to a magnet is used to separate it from sulfur.

The numerous methods that have been devised for the ***physical*** separation of mixtures exploit a difference in a given property (such as boiling point or solubility) of the pure substances involved. Of equal importance are the ***chemical*** methods which have been developed for separations. They are based upon changing one substance into another one with different properties. Chemical methods of separation are discussed later in this text. At this point, an examination of some of the physical separation methods will be used to convey an understanding of their usefulness and their limitations.

Solubility is the extent to which a substance dissolves in a solvent (a solvent is often a pure liquid substance).

Filtration

Often two solids which are intimately mixed together can be separated by making use of the fact that one of them will dissolve in a certain liquid, while the other will not. The solid that will not dissolve can then be caught on a filter (Figure 1–1) while the one that dissolved will pass through the filter with the liquid. Salt mixed with sand can be separated by stirring the mixture in water. The salt will dissolve, leaving the sand as the only remaining solid. The sand then can be removed by filtration. Another mixture which can be separated by filtration is the mixture of iron and sulfur mentioned earlier. In this case the sulfur can be dissolved in a liquid, carbon disulfide, in which iron will not dissolve at all. Filtration can then be used to separate the iron powder from the sulfur

Filtration is a separation technique based on differences in solubility in a given solvent.

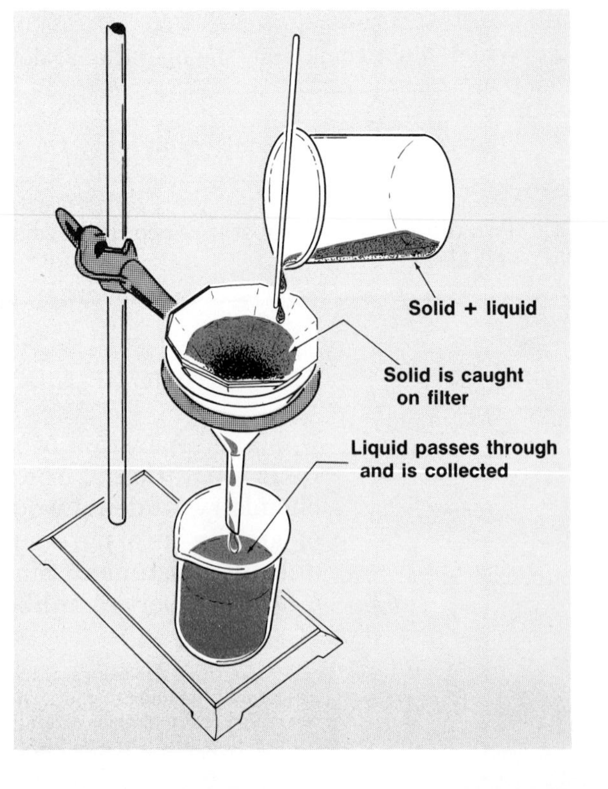
Solid + liquid
Solid is caught on filter
Liquid passes through and is collected

Figure 1–1 The separation of a solid from its mixture with a liquid by filtration. The solid particles are large enough to be held back by the filter paper in the filter.

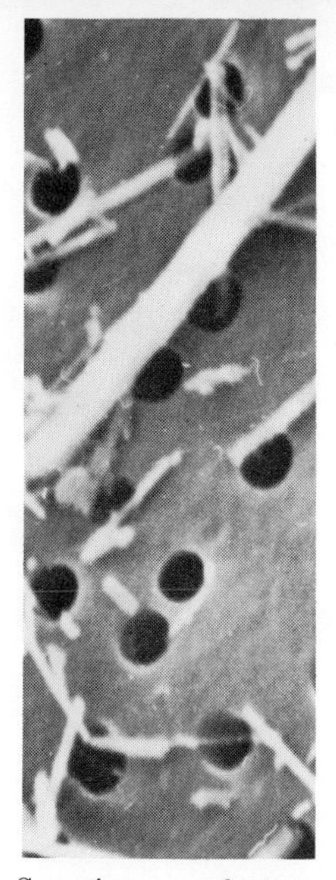

Scanning electron micrograph of particles of asbestos filtered from a sample of air by a small pore filter. (Courtesy of Dr. Spurny, Institut fur Aerobiologie.)

In distillation a mixture of substances is separated by changing one or more of its components from a liquid to a gas or vapor, and then back to a liquid. All liquid substances (except some which decompose when heated) boil at sufficiently high temperatures to allow for this separation.

solution. Most of the solid material in river water that is to be used for drinking is removed in water purification plants by passing the mixture through a filter bed composed in part of fine sand.

Distillation

The boiling point of a liquid is the lowest temperature at which bubbles of its vapor first form throughout it. Usually the components in a mixture will have different boiling points. If this is the case, a separation can be based on this difference. In a simple example, common table salt can be separated from its solution in water because the boiling point of water is very much lower than that of salt. The water can simply be boiled away, leaving the dry salt (Figure 1-2). To reclaim the water as well as the salt, the gaseous water can be passed through a cool tube where it will condense. The process of boiling followed by the condensation of the vapor into a separate vessel is called ***distillation.***

In some mixtures the boiling points are relatively close together and a single distillation will only partially separate them. In such cases, repeated distillations yield relatively pure substances. Such a purification can commonly be performed in a single step in a specially designed distillation apparatus; this process, which has the same effect as several repeated distillations, is called ***fractional distillation.*** The components of crude petroleum are partially (in some special cases completely) purified by fractional distillation.

Air is a mixture of nitrogen and oxygen along with small amounts of several other gases. If air is liquefied by cooling, the nitrogen, oxygen, and other gases can be separated from each other by taking advantage of their different boiling points. As the liquid air is warmed, the nitrogen, which has a lower boiling point than oxygen, will tend to boil off, leaving the oxygen in a higher state of purity. This process can be repeated until the nitrogen and oxygen are obtained in a very pure form.

A similar type of separation is based on a rather unusual property of some solids, i.e., they sublime (pass directly into the gaseous state

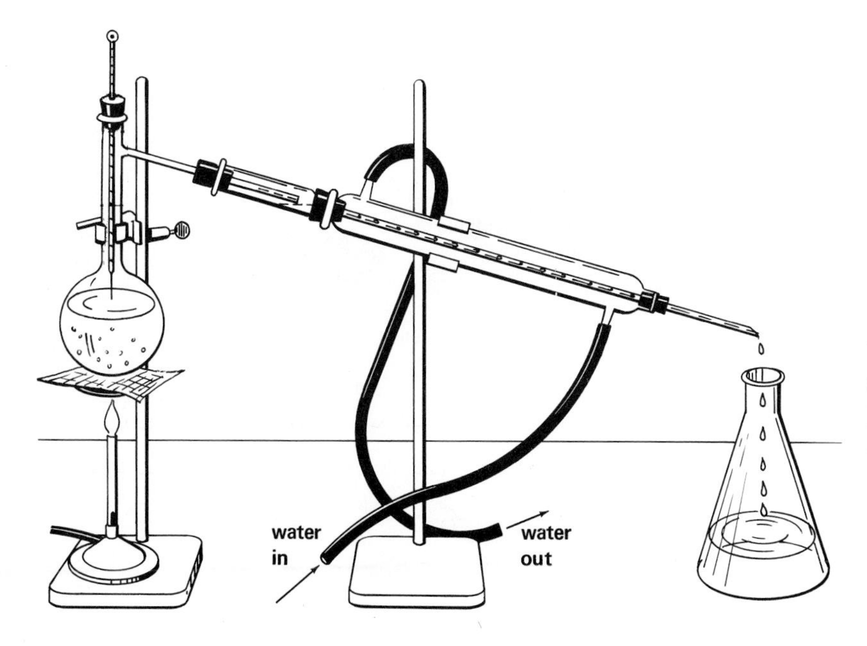

Figure 1-2 Distillation. Sodium chloride dissolves in water to form a colorless solution. When heated above the boiling point (indicated by thermometer), water will change into vapor and pass into the condenser. Cool water injected into the glass jacket of the condenser circulates over the inner tube causing the steam to liquefy and collect in the flask. In this simple example pure water collects in the receiving flask while the salt remains in the boiling flask.

Distillation

without melting). A solid that will sublime mixed with one that will not can be removed from a mixture of the two by simply heating the mixture. Dry ice, which is solid carbon dioxide, and iodine are familiar examples of solids that will sublime. The process of purification by sublimation

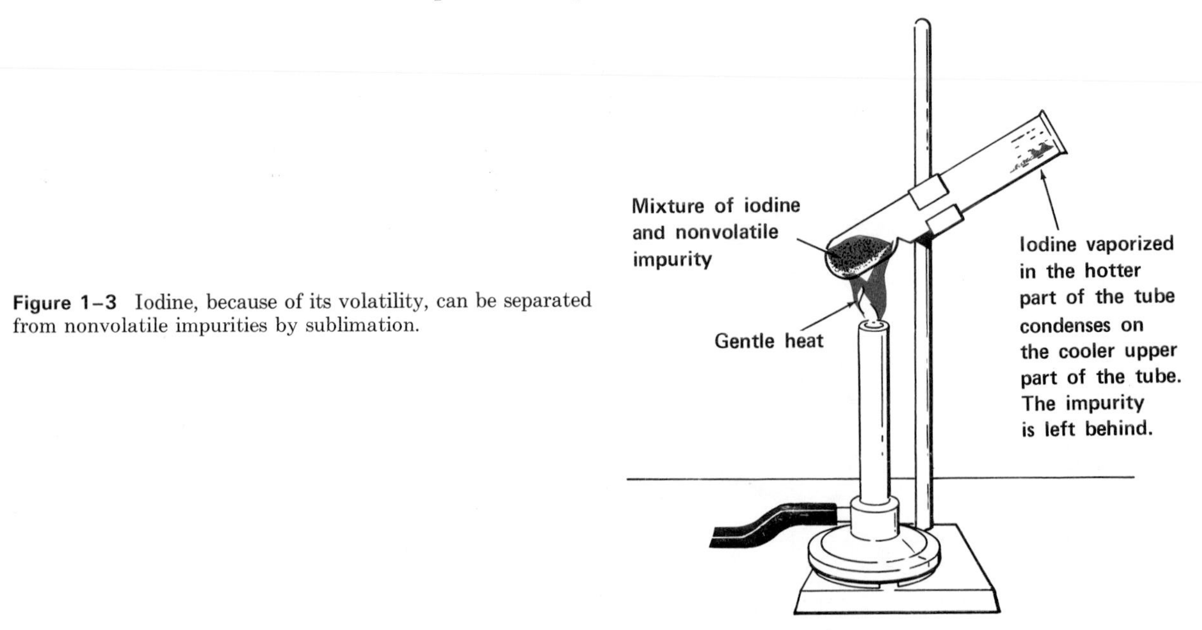
Mixture of iodine and nonvolatile impurity
Gentle heat
Iodine vaporized in the hotter part of the tube condenses on the cooler upper part of the tube. The impurity is left behind.

Figure 1–3 Iodine, because of its volatility, can be separated from nonvolatile impurities by sublimation.

is illustrated in Figure 1-3, and requires that the material which has passed off as a vapor be resolidified via cooling.

Extraction

"Miscibility" is the term used to designate the solubility of one liquid in another.

Extraction is based on difference in solubility between two immiscible solvents.

Some pairs of liquid, such as oil and water, will separate into two layers after being shaken together and allowed to stand. Such a pair of liquids is said to be ***immiscible.*** This is in contrast to the pair gasoline and kerosene, which will mix in all proportions, forming a single layer; gasoline and kerosene are said to be ***miscible.*** Solid iodine will dissolve in both carbon tetrachloride and water, two solvents which are immiscible with each other. If water, carbon tetrachloride, and a small amount of iodine are shaken together, one finds that the iodine is much more soluble in the carbon tetrachloride (85 times as much) than in the water. Iodine dissolved in water can be effectively removed by shaking the mixture with carbon tetrachloride. If this process is repeated a few times, essentially all of the iodine can be removed from the water.

Extraction is used to make caffeine-free coffee.

This technique of removing a substance from one liquid by making use of its greater tendency to dissolve in another liquid is called ***extraction,*** and it can be used to separate mixtures. For example, if one has a water solution containing both sodium chloride (table salt) and iodine, it would be easy to extract the iodine with carbon tetrachloride, leaving the salt in the water, since the salt is not soluble in carbon tetrachloride. The technique used in extraction is shown in Figure 1-4.

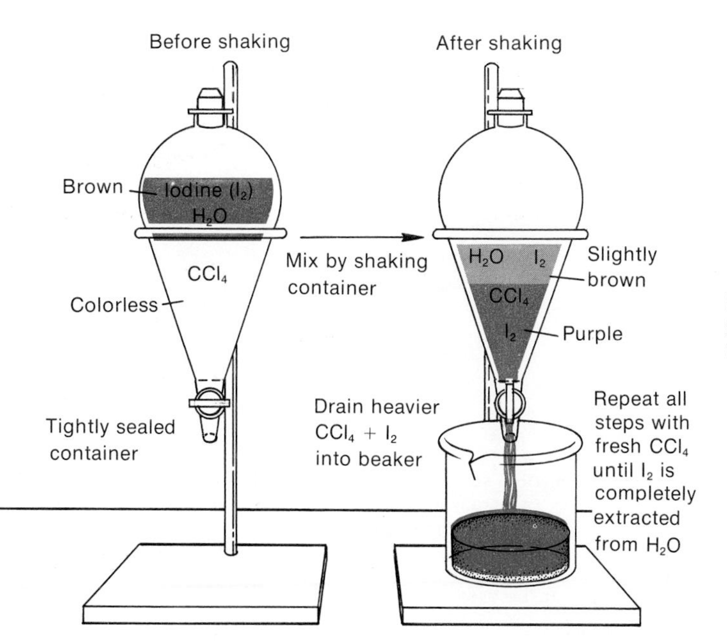

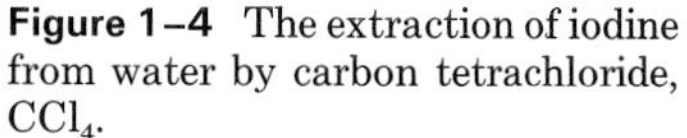
Figure 1-4 The extraction of iodine from water by carbon tetrachloride, CCl_4.

Recrystallization

Solid substances typically are more soluble in a hot liquid than in the same liquid when cold. In these cases a hot solution which contains the maximum amount of substance that will dissolve at that temperature will form crystals as the solution is cooled, since the cooler liquid cannot hold as much of the pure substance in solution.

It is often possible to take advantage of this behavior to separate the components of a solid mixture. Consider the case in which two solid substances are present in a mixture; both of the pure substances are very soluble in a hot liquid, and only one of the two is very soluble in the cold liquid. It is evident, then, that the two can be separated by dissolving the mixture in a minimum amount of the hot liquid, allowing the solution to cool (crystals of the less soluble pure substance form), and filtering the newly formed crystals of the less soluble material from the liquid. This process is called ***recrystallization*** (Figure 1-5). By its use, impure crystals often can be dissolved and reformed in a higher state of purity. The purification process is assisted by the fact that crystal formation often selectively removes only one substance from solution and rejects others.

Recrystallization depends on differences in solubility in hot and cold solvents and selectivity in forming the crystalline solids.

Recrystallization is used in the purification of sodium bicarbonate. In an important commercial process, the Solvay process, crystals of sodium bicarbonate are formed in a solution containing large amounts of another substance, ammonium chloride. As a result, the sodium bicarbonate is contaminated with ammonium chloride. Recrystallization

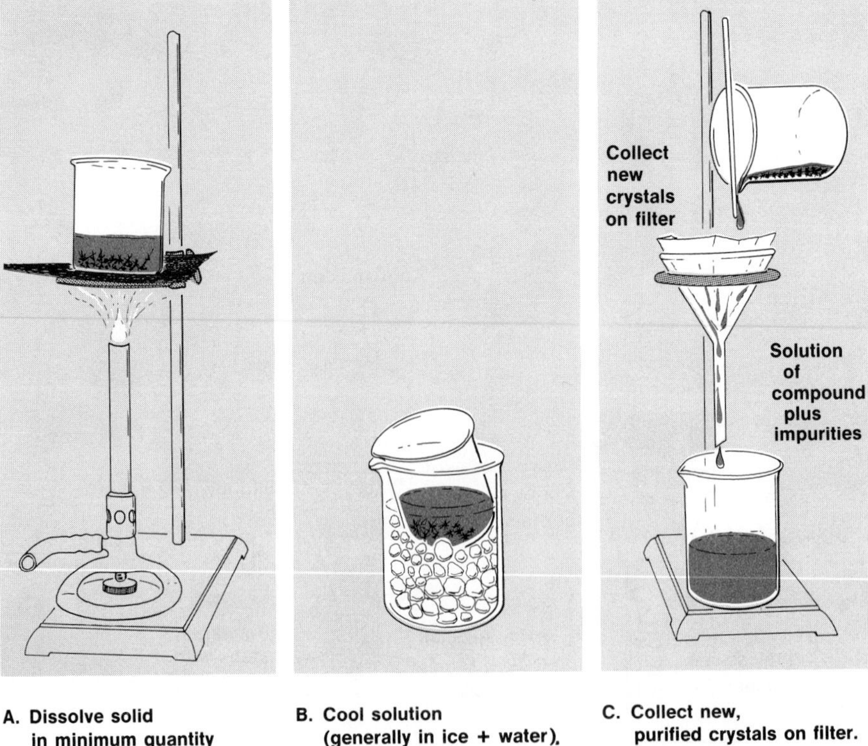
Collect
new
crystals
on filter
Solution
of
compound
plus
impurities
A. Dissolve solid in minimum quantity of hot solvent.
B. Cool solution (generally in ice + water). New crystals form.
C. Collect new, purified crystals on filter.
D. Repeat process if necessary.

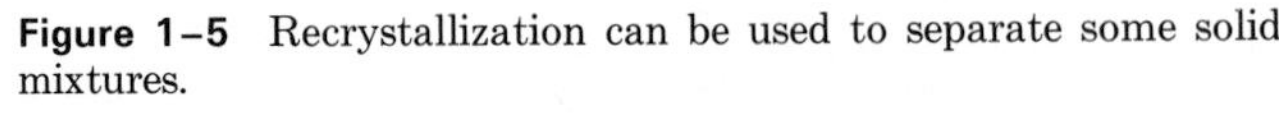

Figure 1-5 Recrystallization can be used to separate some solid mixtures.

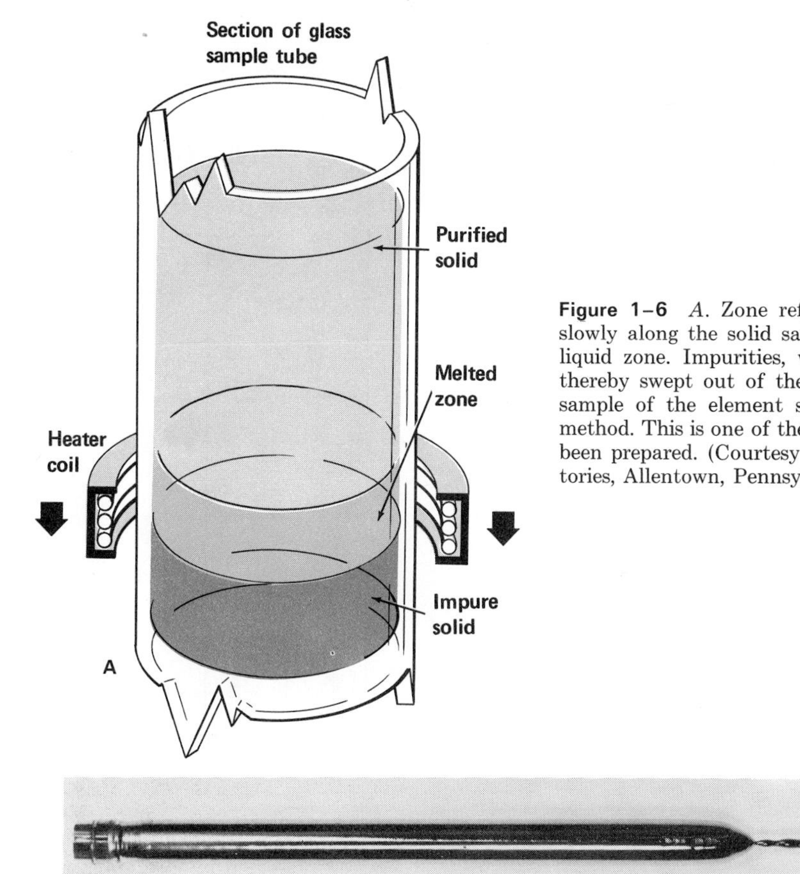

Figure 1–6 *A*. Zone refining purification. A heater coil is moved slowly along the solid sample to be purified, resulting in a moving liquid zone. Impurities, which are more soluble in the liquid, are thereby swept out of the sample. *B*. A 1-inch diameter crystalline sample of the element silicon (Si) prepared by the zone melting method. This is one of the purest samples of a solid element that has been prepared. (Courtesy of K. E. Benson, Bell Telephone Laboratories, Allentown, Pennsylvania.)

is an ideal method for the purification of sodium bicarbonate in this case, since it is sparingly soluble in cold water but more soluble in hot water, whereas the ammonium chloride is quite soluble even in cold water.

Zone refining is an interesting form of recrystallization that has recently been developed and is capable of producing substances in an extremely high state of purity. Pure substances with less than one part per billion of an impurity can be obtained using this technique. This degree of purity is required of materials used in the components of solid-state electronic devices such as the transistor. In zone refining, a heater ring around a tube is slowly moved along the tube containing the material to be purified (Figure 1-6). The material melts in the region of the heater, only to solidify again after the heater has moved to an adjacent region. If, as it generally happens, the impurity is more soluble in the molten (liquid) region, the impurity will be partially carried along with the moving molten region. Hence, the material left behind will be in a higher state of purity. Many passes of the heater over the same sample result in very high purity of the solidified material; the impurities are concentrated in the moving molten zone. At the end of the process, the impure zone is finally allowed to solidify and is then cut off the bar.

1 part per billion (ppb) = 1 part of impurity for every billion (10^9) parts. Few chemicals are this pure in practice. One ppb of the earth's population in 1975 represented four people.

Chromatography

Absorbent papers, such as paper towels, attract water to the extent that water will flow against the pull of gravity to wet the paper. If a

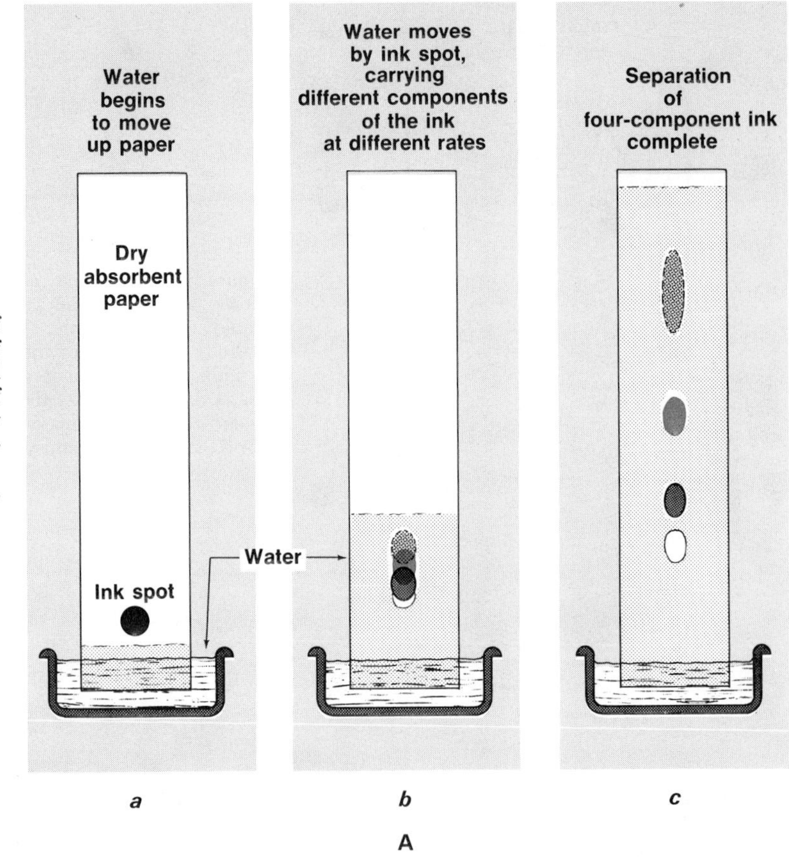

Figure 1–7 (*A*) Paper chromatography of ink. Owing to the absorbent character of paper, water moves against gravity, *a*, and carries the ink dyes along its path, *b*. If the ink dyes move at different rates, they will be separated in the developed chromatogram, *c*.
Illustration continued on opposite page.

spot of ordinary washable ink is placed in the path of the moving water, the ink will be extracted by and will move along with the water. However, some substances in the ink will move faster than others since they are more easily extracted by the water. Bands of color will develop on the absorbent paper as the ink mixture separates. This technique of separation is called ***paper chromatography*** (Figure 1–7). The term chromatography is taken from the Greek word "chromos," which means color. The original experiments involved the separation of colored substances from plants. However, this method of separation has been widely extended to include colorless mixtures, such as the amino acids from egg white protein, so that now the idea of color is not directly connected with the concept.

Chromatography is based on differences in mobility of dissolved substances on a supporting medium.

The scope of chromatography has been extended in a general way to any two phases of matter that are moving relative to each other. The term "phase" is used to refer to a homogeneous mass of matter with distinct boundaries which separate it from other phases. Thus water exists in a solid phase (ice), a liquid phase (water), and a gaseous phase (steam). A mixture of gases can be separated by passing it in a stream of another gas over a solid, if the solid tends to adsorb one gas more than the other. The solid, in the form of small particles, is packed in a tube and the mixture of gases is passed through the tube. Under ideal conditions the weakly adsorbed gas will pass out of the end of the tube completely before the strongly adsorbed gas begins to come out (Figure 1–7B). Similar separations can be made by passing mixtures of gases over liquid surfaces.

Chromatography requires:
1. stationary phase
2. moving phase
3. mixture attracted to both

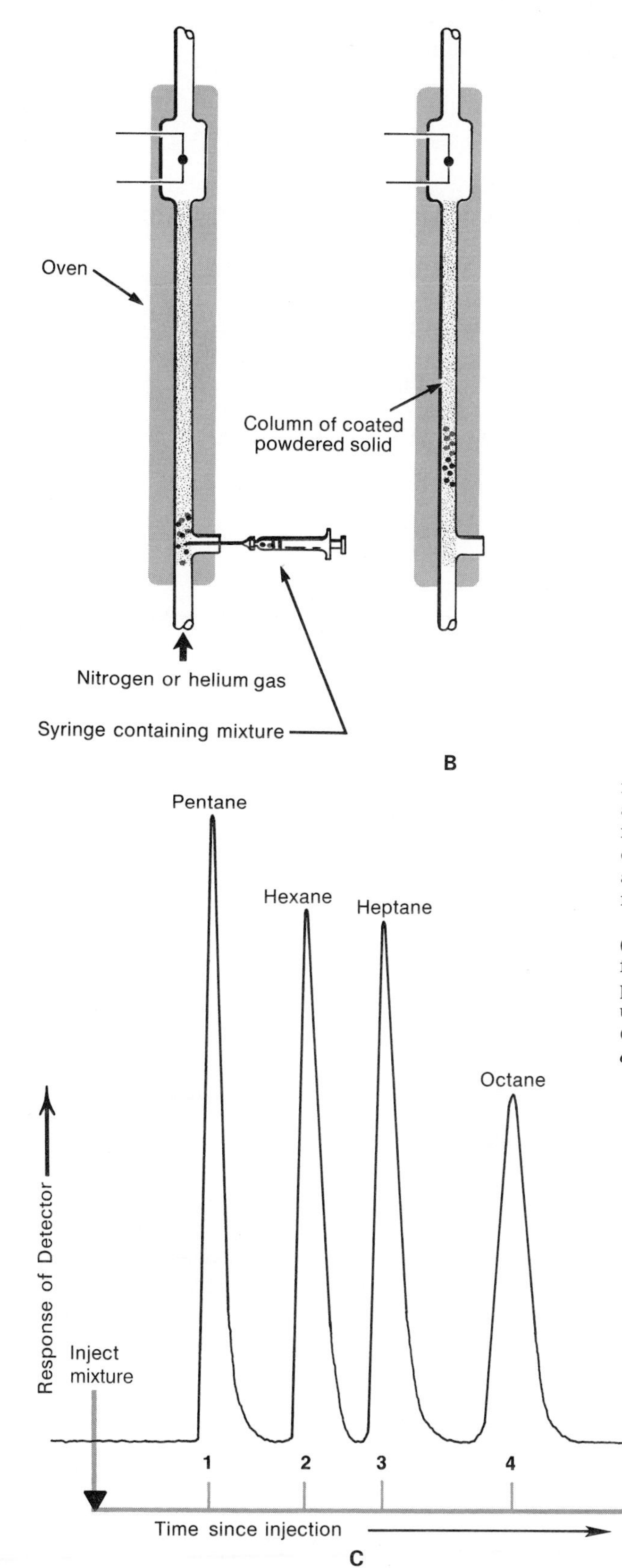

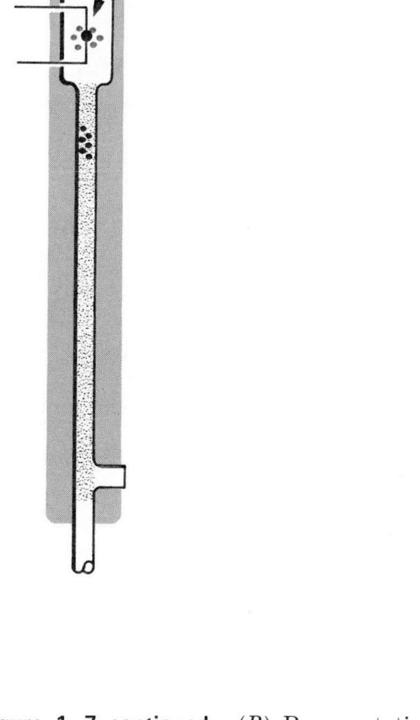

Figure 1–7 continued. (*B*) Representation of gas chromatography. The temperature of the column is controlled so as to keep all components vaporized. The components of the mixture separate owing to differences in adsorption on the column packing. The detector electrically senses (detects) and then draws, on a roll of paper, a peak for each component detected.

(*C*) A typical result is illustrated, showing the separation of a four-component mixture of highly flammable liquids. Each peak corresponds to one substance. The greater the area under a peak, the greater the amount of component. (*B* and *C* are from Breschia, F., et al., *Chemistry—A Modern Introduction.* W. B. Saunders Co., Philadelphia, 1974.)

SELF-TEST 1-B

1. If gasoline can be distilled out of crude oil leaving a heavy oil and tar, which fraction is said to be vaporized: gasoline ( ) or oil ( )?

2. Two common materials that will sublime are: a. __________, b. __________.

3. Wet clothes hung in dry, subfreezing weather will first become stiff and then soft and dry. To what purification technique would you compare this process?

   ____

4. Two liquids that will not mix are said to be __________.

5. Chromatography literally means graphing __________.

6. Zone melting is a special case of what purification technique? __________ __________.

7. In gas-liquid chromatography, the moving phase would likely be the __________ moving over a __________ surface.

ELEMENTS AND COMPOUNDS

These are operational definitions of "element" and "compound."

Experimentally, pure substances can be classified into two categories: those that can be broken down by chemical change into simpler pure substances and those that cannot. Table sugar (sucrose), a pure substance, will decompose when heated in the oven, leaving carbon,

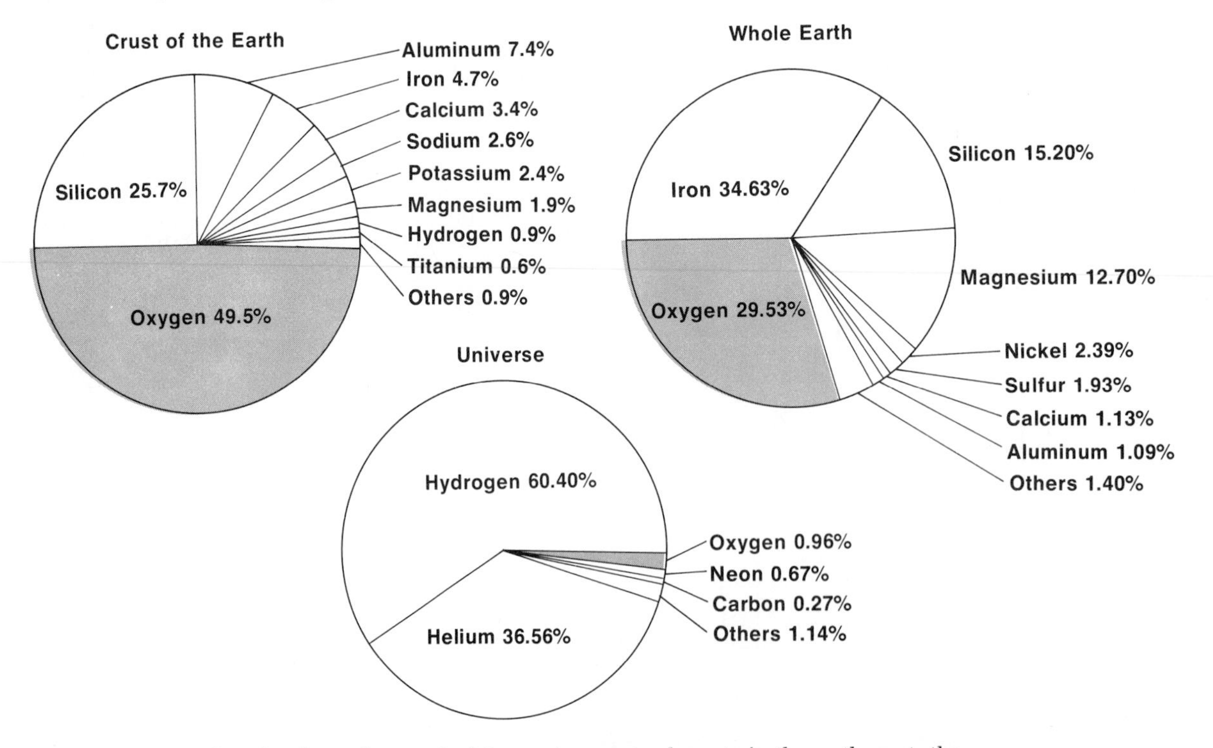

Figure 1–8 Relative abundance (by mass) of the most common elements in the earth crust, the earth, and the universe. Note that the earth crust environment is a very special case of the cosmic array of elements.

another pure substance, and evolving water. No chemical operation has ever been devised that will decompose carbon into simpler pure substances. Obviously sucrose and carbon belong to two different categories of pure substances. Only 89 substances that cannot be reduced chemically to simpler substances are found in nature; 17 others are available artificially. These 106 substances are called ***elements.*** Pure substances that can be decomposed into two or more different pure substances are referred to as ***compounds.*** Even though there are presently only 106 known elements, there appears to be no practical limit to the number of compounds that can be made.

Figure 1-8 lists the most common elements in the universe and on earth.

Elements are the basic building blocks of the universe and the world in which we live. Table 1-1 lists the properties of some of the common elements; a complete list of the elements is found inside the back cover of this text. A number of elements are found as the elementary substance in nature; examples include gold, silver, oxygen, nitrogen, carbon (graphite and diamond), copper, platinum, sulfur, and the noble gases (helium, neon, argon, krypton, xenon, and radon). Many more elements, however, are found chemically combined with other elements in the form of compounds.

Elements in compounds no longer show their characteristic physical

TABLE 1–1 Some Common Elements*

Metals

(A metal is a good conductor of electricity, can have a shiny or lustrous surface, and in the solid form can be deformed without breaking.)

Name	*Symbol*	*Properties of Pure Element*
Iron Latin, *ferrum*	Fe	strong, malleable, corrodes
Copper Latin, *cuprum*	Cu	soft, reddish-colored, ductile
Sodium Latin, *natrium*	Na	light silvery metal, very reactive, soft, low melting point
Silver Latin, *argentum*	Ag	shiny white metal, relatively unreactive, best conductor of electricity
Gold Latin, *aurum*	Au	heavy yellow metal, not very reactive, ductile
Chromium	Cr	resistant to corrosion, hard grayish-white

Nonmetals

(A nonmetal is often a poor conductor of electricity, normally lacks a shiny surface, and is brittle in crystal-solid form.)

Hydrogen	H	colorless, odorless, occurs as a very light gas (H_2), burns in air
Oxygen	O	colorless, odorless gas (O_2), reactive, constituent of air
Sulfur	S	yellow solid (S_8), low melting point, burns in air
Nitrogen	N	colorless, odorless gas (N_2), rather unreactive
Chlorine	Cl	greenish yellow gas (Cl_2), very sharp choking odor, poisonous
Iodine	I	dark purple solid (I_2), sublimes easily

*Chemists usually use the symbol rather than the name of the element. In addition to denoting the element, the chemical symbol has a very specialized meaning which is described later in this chapter. A complete list of the elements with the symbols can be found inside the back cover of this book.

References explaining the historical development of chemical symbols are given in the suggestions for further reading at the end of Chapter 1.

properties, such as color, hardness, and melting point. Consider sucrose, $C_{12}H_{22}O_{11}$, as an example. It is made up of three elements: carbon (which is usually a black powder), hydrogen (the lightest gas known), and oxygen (a gas necessary for respiration). The compound, sucrose, is completely unlike any of the three elements; it is a white crystalline powder which, unlike solid carbon, is readily soluble in water.

Compounds have a fixed composition in terms of the elements they contain.

A careful distinction should be made between a compound of two or more elements and a mixture of the same elements. The two gases, hydrogen and oxygen, can be mixed in all proportions, forming various mixtures of these elements. However, these two elements can and do react chemically to form the compound water. Not only does water exhibit properties peculiar to itself and different from hydrogen and oxygen, but it also has a definite percentage composition by weight (88.8 per cent oxygen and 11.2 per cent hydrogen). This is a second distinct difference between compounds and mixtures: ***compounds have a definite percentage composition by weight of the combining elements.***

CHEMICAL CHANGES AND PHYSICAL CHANGES

Chemical change results in the appearance of new substances.

The changes in matter can be distinguished as chemical or physical changes. A chemical change occurs when substances are used up in the production of different ones. For example, the burning of carbon (as charcoal) to form carbon dioxide is a chemical change. A physical change produces no new substances; it only alters the original state of the elements or compounds. An example is the boiling of water, a change of liquid water into gaseous water. Some additional examples of chemical and physical changes are the following:

Chemical Changes	Physical Changes
Rusting of iron	Evaporation of a liquid
Burning gasoline in an automobile engine	Melting of iron
Preparation of caramel by heating sugar	Freezing of water
Preparation of iron from its ores	Grinding or pulverizing of a solid
Solution of copper in nitric acid	

The properties of a new substance are essential since they identify it. There are physical properties and there are chemical properties. A chemical property of an element or compound describes its ability to undergo chemical change. It follows, then, that to determine a chemical property we must attempt a chemical change. Physical properties, on the other hand, are characteristics of a substance which can be determined without chemically changing the element or compound into some other element or compound. Physical properties of a pure substance include such things as melting temperature, boiling temperature, density (weight of the substance in a given volume), color, physical state (whether gas, liquid, or solid, at a specified temperature), crystalline form, and magnetic properties.

It should be emphasized that not all of the properties of a sample need to change in order to establish that a chemical change has occurred.

For example, if table salt is treated with sulfuric acid and heated strongly, the solid product of this reaction is much like table salt in its appearance, but it has a different melting point and different chemical properties. A change in only one significant property usually indicates that a newly formed substance has resulted from a chemical change.

While most changes can easily be classified on the basis of whether a different substance is produced or not, some changes are difficult to put into such neat categories. Examples include those which accompany the dissolving of a solid in a liquid. We often apply a simple test to determine the type of change. If the solid can be recovered unchanged by the evaporation of the liquid, the solution process was a physical change. Thus, table salt or sugar dissolves in water, but either solid can be recovered by careful evaporation, so such solutions are said to be formed by physical changes although the interaction between constituents is considerable. An example of a chemical change resulting from solution is observed when hydrogen chloride dissolves in water. In this case, the properties of the solution are very different from those of the constituents, and evaporation of the solution is not a feasible way to recover the starting materials completely. In general, however, solution processes are of such complexity that their detailed discussion is inappropriate at this point.

Nuclear changes which are basic to radioactivity, atomic bombs, hydrogen bombs and solar energy are a unique type of change because of the relatively large amounts of energy involved and the parts of the atoms which undergo change. (See Chaps. 6, 24 and 25.)

The classification of solution changes illustrates some difficulties about exact definitions as used in chemistry. Like the stamp collector arranging stamps and the cook arranging foods in the kitchen, we will find our work easier if we group similar things and phenomena into categories. This makes it easier to organize knowledge. Of course, we try to define the limits of our categories so that there will be no question about what fits into each classification, but nature sometimes presents situations that are very hard to classify. These are borderline cases, and all efforts to be specific and limiting will usually meet with an item or idea that defies exact classification. There are some changes in matter that are difficult to classify as either physical or chemical, and all of our definitions are subject to this kind of limitation. However, a few exceptions do not stop us from originating simple categories for the things we observe.

WHY STUDY PURE SUBSTANCES AND THEIR CHEMICAL CHANGES?

A knowledge of chemical changes allows us to carry them out to achieve desired results.

Perhaps by now you are wondering why we should be interested in elements and compounds and their chemical properties. There are two reasons. The first is the belief that the knowledge of chemical substances and chemical changes will allow us to bring about desired changes in the nature of everyday life. Two hundred years ago most of the materials surrounding a normal person were only physically changed from the way the material occurred in nature. Only a few of his useful materials, such as iron, were the product of man's control over chemical change. By contrast, synthetic fibers, plastics, drugs, latex paints, detergents, new and better fuels, and photographic films are but a few of the materials produced by controlled chemical change. (We will return to examine the chemistry of many of these later.) You will discover that it is difficult to find more than a few objects in your home that have not been altered by a desirable chemical change. The second reason for chemical studies is even more important—the curiosity of man. Chemicals and chemical change are a part of nature that is open to investigation, and, like the mountain climber, we will find this task both interesting and challenging. If we hope to understand matter, it is obviously necessary to discover the simplest forms of matter and to study their interactions.

STRUCTURE OF MATTER AND RESULTING PROPERTIES

For reasons which are partly theoretical and partly practical, we are deeply interested in the structure of matter—that is, the minute parts of matter and how these parts are fitted together to make larger units. Why does an element or compound have the properties it has? Why does one element or compound undergo a change that another element or compound will not undergo? Inanimate matter is the way it is because of the nature of its parts. A watch is what it is because of the nature of its individual parts. So is a car, a refrigerator, and the salt in your salt shaker. The individual parts (smaller than the whole) are what determine the nature (actions and properties) of the whole. The most basic parts of matter, as we shall see, are very, very small. If we even hope to understand the nature of matter, it is absolutely necessary that we have some understanding of these minute parts and how they are related to each other. A very large portion of the research in chemistry being done today is aimed at sorting out and elucidating the structure of matter. Indeed, the basic theme of this text is the relationship between the structure of matter and its properties. This theme of structure and related properties is of great interest because if we know exactly how and with what strength the minute parts of matter are put together, we can discover exact relationships between structure and properties. Armed with this understanding, we can make changes that result in new substances and predict the properties of these substances. Such knowledge can save many months of trial and error which otherwise may be required to prepare a product with the desired qualities. While this day of predicting chemical changes based on structural characteristics has not completely arrived, such significant advances have been made that the practice of modern chemistry would not be possible without such knowledge.

Samples of matter large enough for ordinary laboratory experiments are called ***macroscopic*** samples in contrast to ***microscopic*** samples, which are so small that they have to be viewed with the aid of a microscope. The structure of matter that really interests us, however, is at the ***submicroscopic*** level. Our senses have no direct access into this small world of structure, and any conclusions about it will have to be based on circumstantial evidence gathered in the macroscopic and microscopic worlds.

Submicroscopic structures help to explain chemistry.

We can now extend our concept of the science of chemistry. It is that science which investigates the properties and changes of pure substances. Chemistry is also deeply concerned with structure, both ***macrostructure*** and ***submicrostructure,*** in an effort to give plausible reasons for properties and change, with emphasis on chemical change.

FACTS, LAWS, AND THEORIES

A scientific fact can be verified independently of any particular observer.

A ***scientific fact*** is an observation about nature that can usually be reproduced at will. For example, carbon in some forms will burn in the presence of air. If you have any doubt about this fact it is easy enough to set up an experiment that will readily demonstrate the fact anew. You would only need some carbon, air, and a source of heat. The repeatability of a scientific fact distinguishes it from a historical fact, which obviously cannot be reproduced. Of course, some scientific facts are also historical facts—such as the movement of heavenly bodies—and are not repeatable at will.

A scientific law summarizes a large number of related facts.

Often a large number of related scientific facts can be summarized into broad, sweeping statements called natural ***laws.*** The law of gravity is a classic example of a natural law. This law, that all bodies in the universe have an attraction for all other bodies which is directly proportional to the product of their masses and inversely related to the square of their separation distance, summarizes in one sweeping statement an enormous number of facts. It implies that any heavy object lifted a short distance from the surface of the earth will fall back if released. Such a natural law can only be established in our minds by inductive reasoning; that is, you conclude that the law applies to all possible cases, since it applies in all of the cases studied. A well-established law allows us to predict future events. When convinced of the generality of a scientific law, we may reason deductively, based on our belief that if the law holds for all situations, it will surely hold for the events in question.

Theories are ideas or models used to explain facts and laws.

The same procedure is used in the establishment of chemical laws, as can be seen from the following example. Suppose an experimenter carried out hundreds of different chemical changes in closed, leakproof containers, and that he weighed the containers and their contents before and after each of the chemical changes. Also, suppose that in every case he found that the container and its contents weighed exactly the same before and after the chemical change occurred. Finally, suppose that he repeated the same experiments over and over again, obtaining the same results each time, until he was absolutely sure that he was dealing in reproducible facts. It can be understood then that the experimenter would reasonably conclude: ***"All chemical changes occur without any detectable loss or gain in weight."*** This is indeed a basic chemical law and serves as one of the foundations of modern scientific theory.

After a natural law has been firmly established, its explanation must be sought. Chemists are not satisfied until they have explained chemical laws logically in terms of the submicroscopic structure of matter. This is indeed a difficult process, and the progress, until recent time, has been painfully slow because of our lack of direct access into the submicroscopic structure of matter with our physical senses. All we can do is to collect information in the macroscopic world in which we live and try, by circumstantial reasoning, to visualize what the submicroscopic world must be like in order to explain our macroscopic world. Such a visualization of the submicroscopic world is called a ***theoretical*** model. If the theoretical model is successful in explaining a number of chemical laws, a major scientific theory is built around it. The atomic theory and the electron theory of chemical bonding are two such major theories and both will be discussed in relation to chemical laws in later chapters.

Consider again the chemical law concerning the conservation of

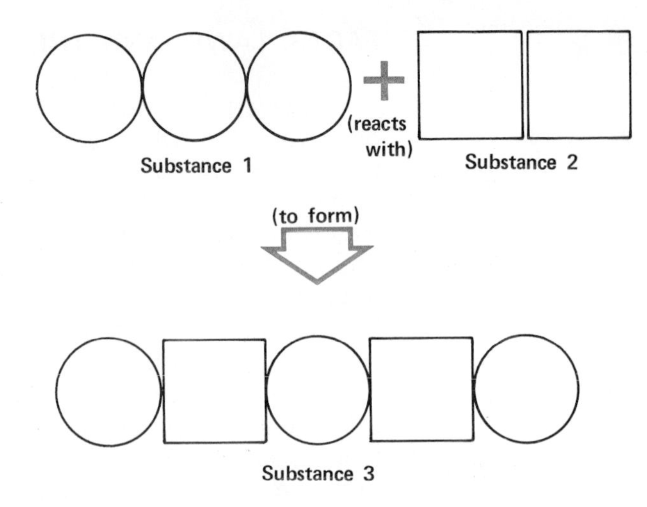

Figure 1–9 Atomic explanation for weight conservation in chemical changes. If substances are composed of atoms, if chemical changes are atomic rearrangements, and if atoms have fixed weights, the products in a reaction must weigh the same as the starting materials; hence, the law of conservation of weight is explained in terms of a theoretical model.

weight in chemical changes. What is a possible theoretical model that could explain this law? If we assume that matter is made up of atoms and that these atoms are grouped in a particular way in a given pure substance, we can reason that a chemical change is simply the rearrangement of these atoms into new groupings, and, consequently, into new substances. If the same atoms are still there, they should have the same weight, and the law of conservation of weight is explained (Figure 1–9).

Chemical theories use the concepts of atoms to explain chemical observations.

For a scientific theory to have much value, it must not only explain the pertinent facts and laws at hand, but it must also be able to explain new facts and laws that are obviously related. If the theory cannot consistently perform in this manner, it must be revised until it is consistent, or, if this is not possible, it must be completely discarded. You must not allow yourself to think that this process of trying to understand nature's secrets is nearing completion. The process is a continuing one.

The word theory is often used in a different sense from the one discussed above. If a student is absent from the chemistry class, his neighbor may say, "I do not know why he is absent, but my theory is that he is sick and unable to come to class." This use of the word theory is similar to the concept of the general scientific theory in that there is an element of speculation in both uses. But the speculative guess of the student about his absent friend is vastly different from the broad theoretical picture that is able to explain a number of laws. The reader should be alert for the considerable amount of confusion that has resulted from the different meanings associated with this word. In this book the word ***hypothesis*** is used when speaking of a speculation about a particular set of data, reserving the word ***theory*** for the broad imaginative concepts that have gained wide acceptance.

THE LANGUAGE OF CHEMISTRY

Symbols, formulas, and equations are used in chemistry to convey ideas quickly and concisely. These shorthand notations are merely a convenience, and contain no mysterious concepts that cannot be expressed in words. Certain characters are used often, and a general familiarity with them is desirable.

Figure 1–10 Direct observation stops at the microscopic level. Convinced of structure beyond the microscopic level, the chemist employs circumstantial evidence to construct the world of molecules, atoms, and subatomic parts in his mind's eye.

Chemical symbols are abbreviations for the different elements.

A *mole* is a *very* large number; it is 6.023×10^{23} particles.

A ***chemical symbol*** for an element is a one- or two-letter term, the first letter a capital and the second a lower case letter. The symbol is a sign for three concepts. First, the symbol stands for the element in general. H, O, N, Cl, Fe, Pt are shorthand notations for the elements hydrogen, oxygen, nitrogen, chlorine, iron, and platinum, respectively, and it is customary to substitute these symbols for the words themselves in describing chemical changes. Some symbols originate from Latin words, such as Fe, from *ferrum,* the Latin word for iron. Second, the chemical symbol stands for a single atom of the element. The ***atom*** is the smallest particle of the element that can enter into chemical combinations. Third, the elemental symbol stands for a mole of the atoms of the element. The ***mole*** is a term that has evolved in chemical usage (it is derived from the Latin for "a pile of" or "a quantity of") and has come to mean a particular number. Just as a dozen apples would be 12 apples, a mole of atoms would be 602,300,000,000,000,000,000,000, atoms or 6.023×10^{23} atoms. It turns out that a mole of atoms is usually a convenient amount for laboratory work. Thus, the symbol Ca can stand for the element calcium, a single calcium atom, or a mole of calcium atoms. It will be evident from the context which of these meanings is implied.

Atoms can unite (bond together) to form molecules. A ***molecule*** is the smallest particle of an element or a compound that can have a stable existence in the close presence of like molecules. One or more of the same kind of atom can make a molecule of an element. For example, two atoms of hydrogen will form a molecule of ordinary hydrogen, and eight sulfur atoms will form a single molecule. Subscripts in a chemical formula show the number of atoms involved: H_2 means a hydrogen molecule is composed of two atoms, and S_8 means a sulfur molecule is composed of eight atoms. The noble gases, such as helium, He, have monatomic molecules (monatomic—one atom).

When unlike atoms combine, as in the case of water (H_2O) or sulfuric acid (H_2SO_4) the formulas tell what atoms and how many of each are present. For example, H_2SO_4 molecules are composed of two hydrogen atoms, one sulfur atom, and four oxygen atoms. A formula can stand not only for the molecule itself, it also can stand for a mole of such molecules, or for the substance in general, depending on the context.

H_2SO_4
4 atoms of oxygen
1 atom of sulfur
2 atoms of hydrogen

When elements or compounds undergo a chemical change, the formulas, arranged in the form of a ***chemical equation,*** can present the information in a very concise fashion. For example, carbon can react with oxygen to form carbon monoxide. Like most solid elements, carbon is written as though it had one atom per molecule; oxygen exists as diatomic (two-atom) molecules, and carbon monoxide molecules contain two atoms, one each of carbon and oxygen. Furthermore, one oxygen molecule will combine with two carbon atoms to form two carbon monoxide molecules. All of this information is contained in the equation

$$2C + O_2 \longrightarrow 2CO$$

Chemical equations summarize information on chemical reactions in a concise fashion.

The arrow should be read "yields"; the equation then states the following information:

(a) carbon plus oxygen yields carbon monoxide
(b) two monatomic molecules of carbon plus one diatomic molecule of oxygen yield two molecules of carbon monoxide
(c) two moles of carbon atoms plus one mole of diatomic oxygen molecules yield two moles of carbon monoxide

The number written before a formula, the ***coefficient,*** gives the amount of the substance involved while the ***subscript*** is a part of the definition of the pure substance itself. Changing the coefficient only changes the amount of the element or compound involved, whereas changing the subscript would necessarily involve changing from one substance to another. For example, 2CO means either two molecules of carbon monoxide or two moles of these molecules.

MEASUREMENT

The heart and soul of reliable scientific facts, and, consequently, the basis for laws and theories, is the accurate measurement of weights, volumes, times, lengths, and other quantities. Perhaps more than any other person, Antoine Lavoisier, who lived in the late 1700's, put chemistry on a quantitative basis by very accurately measuring the weights of materials used in chemical reactions. Because of the significance of his work, Lavoisier, a French chemist, has been called the Father of Modern Chemistry. The reproducible data which he obtained led directly to the formulation of several laws concerning chemical change.

Since 1960, a coherent system of units known as the International System of Units, abbreviated SI, and bearing the authority of the International Bureau of Weights and Measures, has been in effect and is gaining acceptance among scientists. It is an extension of the metric system which began in 1790, and assigns each physical quantity an SI unit uniquely. We use the more familiar metric system in most of our discussions; many metric units are identical to SI units.

The metric system has one great advantage over older systems: it is simple. Its simplicity lies in its organized structure and its decimal counting which require relatively few terms to be remembered. For example, units such as those of length are defined in such a way that a larger unit is ten (or some power of ten) times larger than a smaller unit. A ***meter,*** a unit of length equal to 39.4 inches, is 100 centimeters and a centimeter is 10 millimeters. To make a conversion from one unit of length to another in the metric system merely involves shifting the decimal point. Compare this to the complexity of the conversion of miles to inches wherein one would probably employ the factors 12 and 5280. Instead of having a number of different root words for length as in the English system (league, mile, rod, yard, foot, inch, mil), the metric system has just one, the meter. With suitable prefixes, the meter can be expressive of a unit useful to the watchmaker (millimeter = 0.001 meter) or to the distance runner (kilometer = 1000 meters).

Accuracy of English and metric systems of measurement is the same, but metric is easier to use.

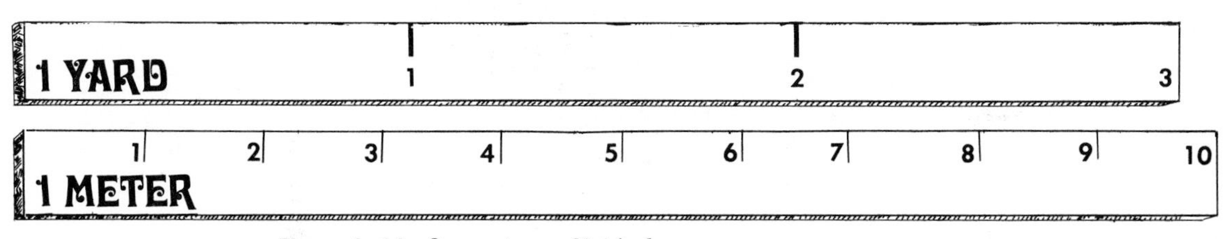

Figure 1-11 One meter = 39.4 inches.

Figure 1-12 The volume of a liter is almost that of a quart. One liter = 1.06 quarts.

While the convenience of the metric system can scarcely be overstated, it should be emphasized that the science of chemistry need not rest on this system of measurement. Measurements made in the English system can be and are just as accurate. However, in the metric system only four basic units will be required in this course, a unit of length, the ***meter;*** a unit of volume, the ***liter;*** a unit of mass (weight), the ***gram;*** and a temperature unit, the ***centigrade degree.*** English equivalents of the first three are:

Mass is a measure of the amount of matter in a body, whereas weight is a measure of the attraction of the body for the earth. The weight of the body varies with its distance from the earth, whereas the mass remains constant. Unless otherwise stated, the term "weight" refers to the earth-surface weight.

1 meter = 39.4 inches (Figure 1-11)
1 liter = 1.06 quarts (Figure 1-12)
1 gram = 0.0352 ounce

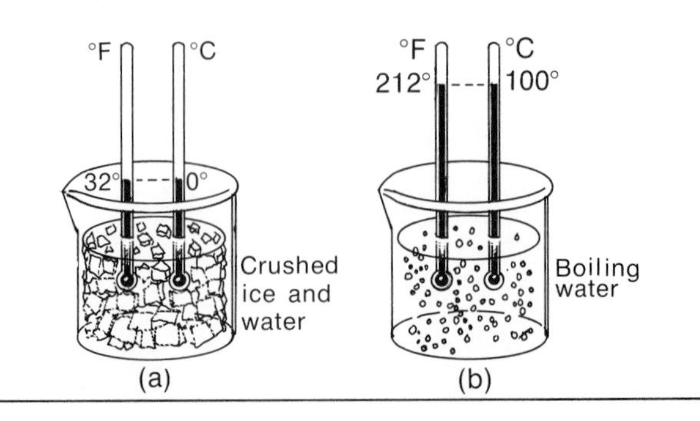

Figure 1-13 Thermometer scales are defined in terms of the expansion of common materials such as mercury and reference points such as the change of state of water and other common materials. It is only a matter of preference and convenience whether one number or another is used.

Figure 1–14 Units of measure are selected for convenience; the millimeter for the small parts of this tiny radio (above), and the kilometer for measuring intercontinental distances. The metric system simply adds the ease of using multiples of 10.

Temperature scales are defined in terms of the behavior of samples of matter. For example, the familiar Fahrenheit scale defines the temperature at which water freezes to be 32°F and that at which it boils to be 212°F. Thus there are 180 (212 − 32) Fahrenheit degrees between the freezing and boiling points of water. Most scientists prefer the centigrade scale which defines the freezing point of water to be 0°C and the boiling point, 100°C. There are 100 centigrade degrees in this same temperature range (Figure 1–13).

For the interested reader and those who need to understand additional aspects of the metric system to support laboratory work, more information is presented in Appendices A, B, and C.

SELF-TEST 1-C

1. The two large divisions into which elements can be divided are:

   a. ______________________ and b. ______________________

   A half dozen examples of each are:

   a. ______________________ a. ______________________

   b. ______________________ b. ______________________

   c. ______________________ c. ______________________

d. ______________________ d. ______________________

e. ______________________ e. ______________________

f. ______________________ f. ______________________

2. How many elements are presently listed in the periodic table? ________

3. A compound has properties that are combinations of the elemental properties. True( ) or False( )

4. Four chemical changes not listed in the text are:

   a. ______________________________

   b. ______________________________

   c. ______________________________

   d. ______________________________

5. Four physical changes not listed in the text are:

   a. ______________________________

   b. ______________________________

   c. ______________________________

   d. ______________________________

6. A chemical change always produces a new ____________.

7. ______________________ structures explain chemical properties.

8. Put in order of decreasing size: (1) microscopic, (2) molecular and (3) macroscopic.

   a. ____________, b. ____________, c. ____________

9. Arrange from most abstract to general to specific: facts, laws, and theories.

   a. ____________, b. ____________, c. ____________

10. Consider the equation: $2Na + 2HCl \longrightarrow H_2 + 2NaCl$. Explain what is meant by the symbols:

    1. Na ______________________________

    2. 2Na ______________________________

    3. HCl ______________________________

    4. $\rightarrow$ ______________________________

    5. H_2 ______________________________

MATCHING SET

______ 1.	produces a new type of matter	a.	filtration
______ 2.	air	b.	fractional distillation
______ 3.	unchanged by further purification	c.	miscible
______ 4.	used to separate a solid from a liquid	d.	zone refining
______ 5.	separation via "repeated" distillations	e.	chemical change
______ 6.	direct passage of a solid into gaseous state	f.	element
______ 7.	liquids which readily dissolve in each other	g.	properties of pure substance
______ 8.	process for purification of solids to obtain extremely pure substances	h.	mixture
______ 9.	cannot be reduced to simpler substances	i.	used to explain facts and laws
______ 10.	scientific theory	j.	sublimation
______ 11.	symbol for iron	k.	O_3
______ 12.	mole of atoms	l.	CO
______ 13.	molecule containing three oxygen atoms	m.	Fe
______ 14.	carbon monoxide	n.	one meter
______ 15.	100 centimeters	o.	6.023×10^{23} atoms

QUESTIONS

1. Name as many materials as you can that you have used during the past day that were not in any way the result of a man-made chemical change.

2. Identify the following as physical or chemical changes. Justify in terms of the operational definitions for these types of changes.

   a. formation of snow flakes
   b. rusting of a piece of iron
   c. ripening of fruit
   d. fashioning a table leg from a piece of wood
   e. fermenting grapes

3. Chemical changes can be both useful and destructive to man's purposes. Cite a few examples of each kind with which you have had personal experience. Also give observed evidence that each is indeed a chemical change and not a physical change.

4. Classify each of the following as a physical property or a chemical property. Justify in terms of operational definitions.

   a. density
   b. melting temperature
   c. substance decomposes into two elements upon heating
   d. electrical conductivity of a solid
   e. the substance does not react with sulfur
   f. ignition temperature of a piece of paper

5. Classify the following as element, compound, or mixture. Justify each answer.

   a. mercury
   b. milk
   c. pure water
   d. a tree
   e. ink
   f. iced tea
   g. pure ice
   h. carbon

6. Which of the materials listed in Question 5 can be pure substances?

7. Explain how the operational definition of a pure substance allows for the possibility that it is not actually pure.

8. Suggest a method for purifying water slightly contaminated with a dissolved solid.

9. Distinguish between theory and law as used in chemistry.

   a. Which has a better chance of being true? Why?
   b. Which summarizes? Which explains?

10. Given the sentence below, express a chemical reaction in terms of symbols that the chemist uses to convey the same information. "One nitrogen molecule containing two nitrogen atoms per molecule reacts with three hydrogen molecules containing two hydrogen atoms per molecule, to produce two ammonia molecules containing one nitrogen and three hydrogen atoms per molecule."

11. Aspirin is a pure substance, a compound of carbon, hydrogen, and oxygen. If two manufacturers produce equally pure aspirin samples, what can be said of the relative worth of the two products?

12. How is the salt content of the sea related to the purity of rain water? What method of purification does nature employ in the purification of rain water?

13. Suggest a hypothesis whereby recrystallization might explain the occurrence in nature of relatively pure salt deposits.

14. Two boys begin in front of a marching column of boy scouts. At the end of the march the playful fellow brings up the rear, having taken time to observe many of his surroundings, while the dedicated marcher remains in front. How is this analogous to chromatographic separations?

15. Do you think the fact that there are only 89 naturally occurring elements argues for the "simplicity idea" concerning matter? Why?

16. How many times do you think a given experiment should give a result before a scientific fact is established? How many failures would you require before rejecting the "fact?"

17. Suppose a wife and children discover the family car missing on returning from an afternoon ball game, even though the father rode to work with a neighbor that morning. The mother says to the children, "Don't worry, Dad must have had an unexpected need for the car and got it after lunch." Would you call the statement made by the mother a theory or a hypothesis? Why?

18. From the molecular formulas given below, tell what kind of atoms and how many

of each kind are present in a molecule. (You may look up the names of any elements whose symbols you do not know in the list printed inside the cover.)

$$SO_3, HCl, NH_3, H_2S$$

19. Describe in words, the chemical process which is summarized in the following equation:

$$2Na + Cl_2 \longrightarrow 2NaCl$$

20. How many *atoms* are present in each of the following:

    a. one mole of He
    b. one mole of Cl_2
    c. one mole of O_3

21. Name as many types of purification as you can. Be able to recite a list of these without resorting to the text.

22. Do the purification techniques described in this chapter involve physical or chemical changes? Defend your answer with a specific example.

23. Would you think that tea in tea bags is a pure substance? Use the process of making tea to make an argument for your answer. How would your argument apply to instant tea?

24. Crystals will almost always grow when a hot saturated solution is cooled. What does this say about the relative solubilities of most materials in hot and cold solvents? Do you think a carbonated beverage would be an exception to this generalization?

25. Why is zone melting more effective than a single recrystallization in purifying metals?

26. Do elements sometimes retain their physical properties in the formation of compounds? Give two examples to support your argument.

27. What is the most fundamental assumption relative to structure and properties in chemical theory?

28. Consider the equation:

$$2C_7H_{14} + 21O_2 \longrightarrow 14CO_2 + 14H_2O$$

$2C_7H_{14}$	$21O_2$	$14CO_2$	$14H_2O$
Heptane (a component of gasoline)	*Oxygen*	*Carbon dioxide*	*Water*

Write out in words as much of the information presented in this equation as you can decipher.

PROBLEMS

NOTE: The following problems should be attempted after a more thorough study of the metric system as presented in Appendices A, B, and C.

1. Determine:

    a. the mass in grams of a quantity of water weighing 8 ounces.
    b. the mass in kilograms of a 1 pound loaf of bread.
    c. the length in centimeters of a pencil 9 inches long.
    d. the volume in gallons of 10 liters of cider.
    e. the speed in kilometers per hour of a car going 45 miles per hour.

    Ans. (a) 227 g (b) 0.454 kg (c) 22.8 cm (d) 2.65 gal (e) 72.3 km/hr

2. If you prefer 70°F for your living quarters, what would be your preference on the centigrade scale?

    Ans. 21.1°C

3. While traveling in Europe, you have antifreeze put into your car radiator. The attendant tells you that your car will now be safe at temperatures down to −10°C. When you return to the United States, you hear on the radio that a low temperature of 20°F is expected for the coming evening. Do you need to worry about your car? Why?

   Ans. No, 20°F is only −6.6°C.

4. A doctor in the United States expects the body temperature of a normal person to be 98.6°F.

   a. If the doctor uses a European clinical thermometer with a centigrade scale, what temperature is normal?
   b. Does a person with a body temperature of 40°C have a fever?

   Ans. (a) 37°C (b) Yes.

5. Is the numerical value of the temperature in Fahrenheit degrees always larger than the numerical value in centigrade degrees?

   Ans. No. Below −40°F (−40°C) Farenheit values are more negative (or smaller) than the corresponding Centigrade values.

6. At what temperature is the numerical value on the Fahrenheit scale exactly twice the numerical value on the Centigrade scale?

   Ans. 160°C = 320°F

SUGGESTIONS FOR FURTHER READING

Benjamin, A. C., "Science, Technology, and Human Values," University of Missouri Press, Columbia, Mo., 1965.

Carter, Luther J., "Industrial Minerals: New Study of How to Avoid a Supply Crisis," *Science,* Vol. 170, p. 147 (1970).

"Chemistry in the Economy," American Chemical Society, 1973.

Garrett, A. B., "The Discovery Process and the Creative Mind," *Journal of Chemical Education,* Vol. 41, p. 479 (1964).

Hildebrand, J. H., "It Ain't Necessarily So," *Chemistry,* Vol. 40, No. 10, p. 19 (1967).

Ihde, A. J., "The Development of Modern Chemistry," Harper and Row, New York, 1964, pp. 3–31.

Kesselman, B., "The Skeptical Chemist," *Chemistry,* Vol. 40, No. 1, p. 9 (1967).

McKelvey, V. E., "Mineral Resources Estimates and Public Policy," *American Scientist,* Vol. 60, pp. 32–40 (1972).

Pratt, Christopher J., "Sulfur," *Scientific American,* Vol. 222, No. 5, May, 1970, pp. 62–72.

CHAPTER 2

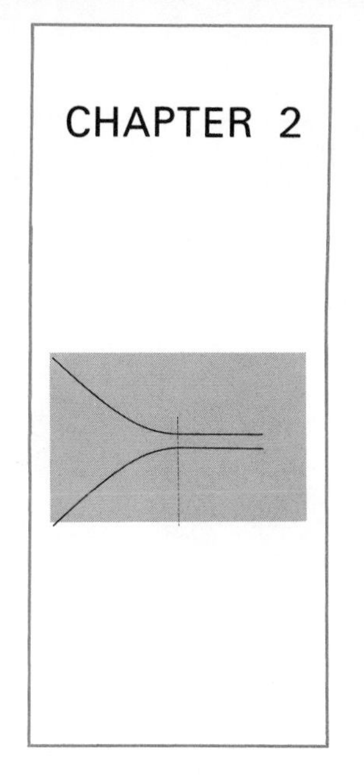

SOME PRINCIPLES OF CHEMICAL REACTIVITY

There are 106 different known chemical elements, each with its own characteristics. In addition to the broad classifications of metals and nonmetals mentioned in Chapter 1, smaller groups or families of elements can be recognized on the basis of similar chemical behavior. In this chapter we shall observe some general characteristics of chemical reactions and note some groupings of reactions that help to classify the elements.

CHARACTERISTICS OF CHEMICAL REACTIONS

All chemical reactions have common features.

A thorough consideration of chemical change involves more than is immediately evident in a cursory survey. In addition to the obvious production of new chemical substances, other phenomena are also present which should be studied and understood. ***Energy*** changes are as definite and reproducible as substance changes in a chemical reaction. The rate at which a chemical change occurs is affected by several factors which can be controlled. A chemical reaction is a two-way street: many chemical reactions are ***reversible*** in practice, and all reactions are reversible in theory. To an observer, a reversible reaction appears to proceed to a point at which both starting materials and product materials are present. Finally, most chemical reactions are not unique. For example, if substance A is chemically similar to substance B, then A and B will enter into similar chemical reactions. We shall now consider in detail some of these characteristics of reactions.

Figure 2–1 Chemical changes produce new substances, often with properties very different from the starting material. Bright, shiny nuts and bolts are changed to crumbly, dull rust by oxidation.

Reactants Become Products

In all chemical reactions some pure substances disappear and others appear. A familiar example is the change of shiny steel to a red-brown iron rust (Figure 2-1). This is a very important reaction since it has been estimated that the dollar loss from corrosion of iron and steel in the United States is slightly over $30 per person per year.

Some other chemical changes are given in the following equations:

REACTANTS				PRODUCTS	HEAT EFFECT*
CaO CALCIUM OXIDE (QUICKLIME)	+	H_2O WATER	$\longrightarrow$	$Ca(OH)_2$ CALCIUM HYDROXIDE (SLAKED LIME)	+ Heat [15.6 kcal per mole of $Ca(OH)_2$]
$2Na$ SODIUM	+	$Cl_2(gas)$ CHLORINE	$\longrightarrow$	$2NaCl$ SODIUM CHLORIDE (TABLE SALT)	+ Heat [98.2 kcal per mole of NaCl]
$H_2(gas)$ HYDROGEN	+	$I_2(gas)$ IODINE	$\longrightarrow$	$2HI(gas)$ HYDROGEN IODIDE	− Heat [6.20 kcal per mole of HI]

+Heat means heat is liberated.

—Heat means heat is required.

*Heat energy can be measured in calories (cal). A kilocalorie (kcal) is 1000 calories. A calorie is approximately the amount of heat required to raise the temperature of 1 gram of water 1 degree centigrade.

In chemical reactions new substances are formed and old ones disappear.

From these reactions we can note our first important point concerning chemical reactions:

> **In chemical reactions, reactants become products; some substances are consumed and new ones appear.**

Quantitative Energy Changes

Again consider the three reactions cited above.

In the first reaction, calcium oxide (quicklime) reacts with water to give calcium hydroxide (slaked lime) with the evolution of heat. In the second reaction, metallic sodium reacts with the greenish yellow gas, chlorine, to give sodium chloride or table salt. If a piece of hot sodium is put into a flask containing chlorine, the sodium burns, liberating a great deal of heat and light, to produce white crystals of sodium chloride. In the last reaction, gaseous hydrogen reacts with gaseous iodine to produce gaseous hydrogen iodide, with the absorption of heat. These facts, then, lead to a second point:

> **A given amount of a particular chemical change corresponds to an energy change of a fixed amount.**

Energy changes in chemical reactions are proportional to the amount of reactant (or product).

For example, the preparation of one mole of $Ca(OH)_2$ from CaO and H_2O releases 15.6 kcal; for two moles of $Ca(OH)_2$, 2×15.6 or 31.2 kcal of heat is released.

Figure 2–2 shows photographs of three chemical reactions, two of which release heat energy, while the third absorbs energy. The reaction of sodium with chlorine, *B*, is a spectacular reaction releasing energy in the forms of heat, light, and sound. In the electrolysis of water, *C*, water is decomposed; electrical energy is supplied to make the reaction proceed.

Sometimes energy changes in reactions are difficult to observe because of the very slow rate of reaction. An example is the rusting of iron. Here the reaction involved is complicated, but we can represent it by the simplified equation:

$$\underset{\text{Iron}}{4Fe} + \underset{\text{Moist air}}{3O_2 + 6H_2O} \longrightarrow \underset{\text{Iron Hydroxide (rust)}}{4Fe(OH)_3} + \underset{\text{heat}}{788\ \text{kcal}}$$

Ordinarily, the rusting of iron occurs so slowly that the liberation of heat is perceptible only with the aid of special instruments. The total amount of heat evolved in rusting is considerable, but it typically takes place over a long period of time.

Temperature Effect on Reaction Rate

We noted earlier that sodium reacts more rapidly with chlorine than iron does with oxygen of the atmosphere. The whole notion of how fast or how slow chemical reactions proceed is one which can be put on a quantitative basis by the concept of the ***reaction rate.*** The rate of a reaction is always defined in terms of the changes in the amounts of

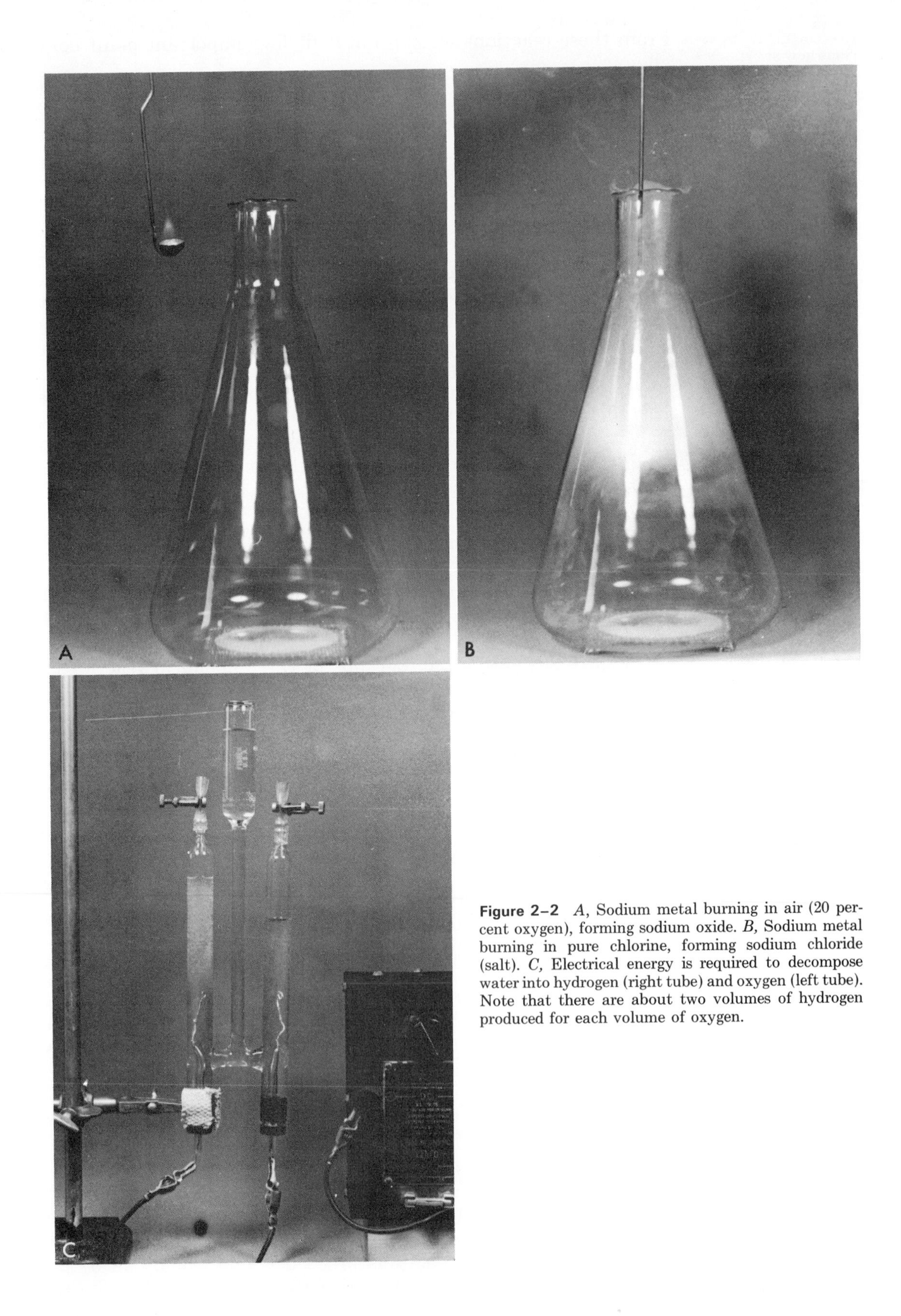

Figure 2–2 *A,* Sodium metal burning in air (20 percent oxygen), forming sodium oxide. *B,* Sodium metal burning in pure chlorine, forming sodium chloride (salt). *C,* Electrical energy is required to decompose water into hydrogen (right tube) and oxygen (left tube). Note that there are about two volumes of hydrogen produced for each volume of oxygen.

Figure 2–3 The biochemical processes of decomposition occur more rapidly at higher temperatures. Half the peach shown in this photograph was refrigerated, while the other half was kept warm. The refrigerated half on the right shows little discoloration, while the other one shows the typical signs of decay.

chemical substances present per unit of time. Thus, if we consider the burning of sulfur to produce sulfur dioxide,

$$S + O_2 \longrightarrow SO_2$$

we can discuss the rate of the reaction in terms of the amount of SO_2 formed per minute or of the amount of S or O_2 consumed per minute.

It is possible to alter the rate of a chemical reaction by changing the temperature. If the temperature is raised, the rates of chemical reactions are increased; if the temperature is reduced, they are decreased. We make use of this principle in cooking foods (a roast will cook at a faster rate at a higher temperature) and in preserving foods (foods spoil less quickly if refrigerated). Figure 2–3 illustrates the effect of temperature on the reactions that take place in a slice of fruit exposed to air.

Raising the temperature speeds up chemical reactions.

We conclude that another general characteristic of chemical reactions is:

The rate of a typical chemical reaction is increased if the temperature is increased.

Effect of Concentration on Reaction Rate

It is also possible to alter the rate of a reaction by changing the concentration of the reactants. For example, in the reaction of sulfur and oxygen given earlier, if air replaces the oxygen, the reaction will proceed at a slower rate, since air is a mixture of oxygen and nitrogen. The rusting of iron can be decreased by painting or coating the surface of the metal to cut down on the concentration of the oxygen and moisture at the surface. Figure 2–4 contrasts the oxidation of iron when oxygen is in relatively small or in relatively large concentrations. These facts illustrate still another characteristic of chemical reactions:

Increasing the concentration of reactants speeds up a reaction.

The rate of a chemical reaction depends on the concentrations of the reactants (Figure 2–4).

A molecular explanation can be given for the observed or empirical results for the reaction between oxygen and sulfur. The reaction proceeds more slowly in air because there are fewer oxygen molecules per volume of space to react with the sulfur. This example illustrates the useful technique of reducing the rate of a reaction by reducing the concentration of the reactants. A theoretical (molecular) explanation for both

Empirical results are those observed directly, as from an experiment.

Figure 2–4 Effect of concentration on reaction rate. Steel wool held in the flame of a gas burner is oxidized. It is in contact with air, which is 20 percent oxygen. When the red hot metal is placed in pure oxygen in the flask, it oxidizes much more rapidly.

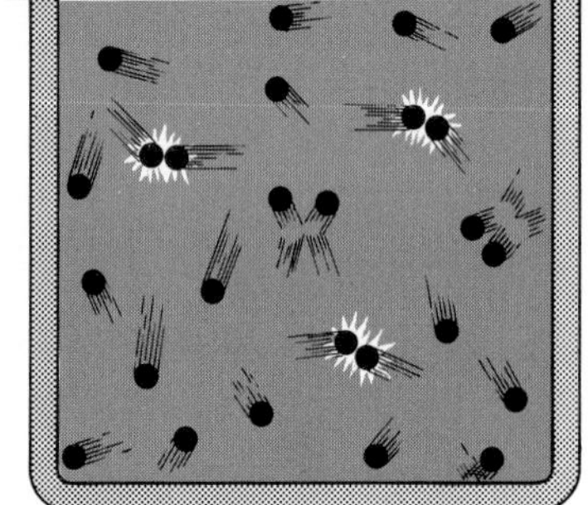

Figure 2–5 Effects of temperature and concentration on rates of chemical reactions. At the higher temperature, more collisions occur between molecules, and a greater percentage of the collisions produce a chemical reaction. At the higher concentration (no temperature change), more collisions occur, but the percentage of effective collisions remains the same.

Temperature increase

Concentration increase

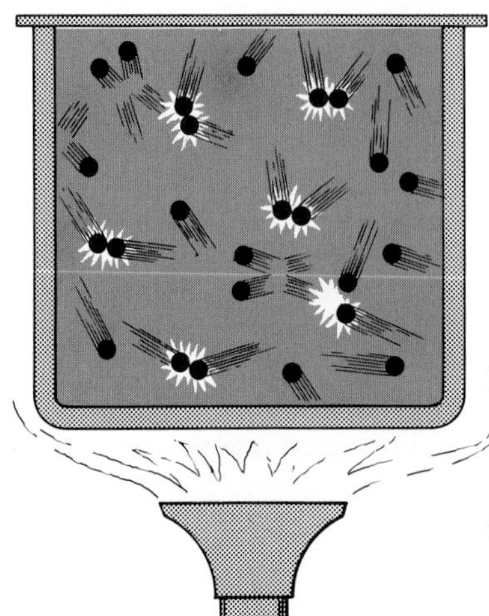

c

COLLISION RATE INCREASES IN BOTH CASES

the concentration effect and the temperature effect on the rate of chemical reactions is illustrated in Figure 2-5.

Reversibility of Chemical Reactions

Most chemical processes are capable of being reversed under suitable conditions. When a chemical reaction is reversed, some of the products are converted back into reactants. For example, heating calcium hydroxide will drive off some of the water. This process is the reverse of adding water to quicklime.

$$Ca(OH)_2 + \text{Heat} \longrightarrow CaO + H_2O$$

Chemical reactions are capable of going forward or backward.

Other methods can be used to reverse chemical reactions. Thus, if we put calcium hydroxide in a vacuum, there will soon be water vapor in the space around the solid. It is sometimes possible to reverse chemical reactions when they are associated with small heat changes. For the sodium-chlorine reaction, it is possible to break NaCl molecules apart into atoms,

$$NaCl(\text{gas}) + \text{Heat} \longrightarrow Na + Cl$$

but only at the very elevated temperatures at which the compound NaCl is vaporized. Since sodium chloride is very stable compared to sodium and chlorine, the reaction forming sodium chloride is reversed only under conditions similar to those in which water is decomposed into hydrogen

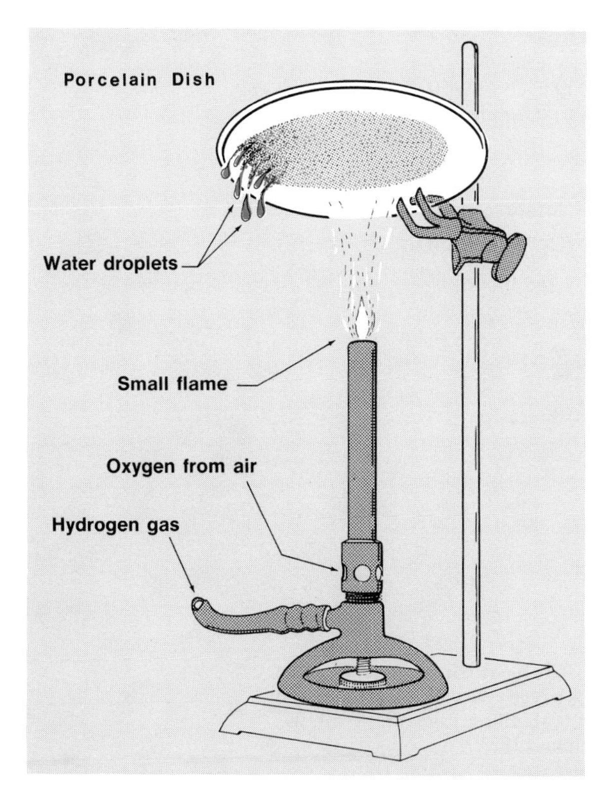

Figure 2–6 Hydrogen and oxygen burning to produce water in the gaseous state. The water is condensed on the cooler porcelain dish.

and oxygen using electrical energy (Figure 2–2). The fact that many reactions can be approached from either direction leads to the conclusion:

Chemical reactions are generally reversible.

The reverse reaction to the electrolytic decomposition of water is shown in Figure 2–6. Hydrogen when burned in air produces water vapor, which can be condensed readily.

If atoms in one arrangement—as in the case of water—can be decomposed into another arrangement, i.e., hydrogen and oxygen, there would be every reason to believe that the atoms could be rearranged into the original structure. After all, a set of building blocks can be used to

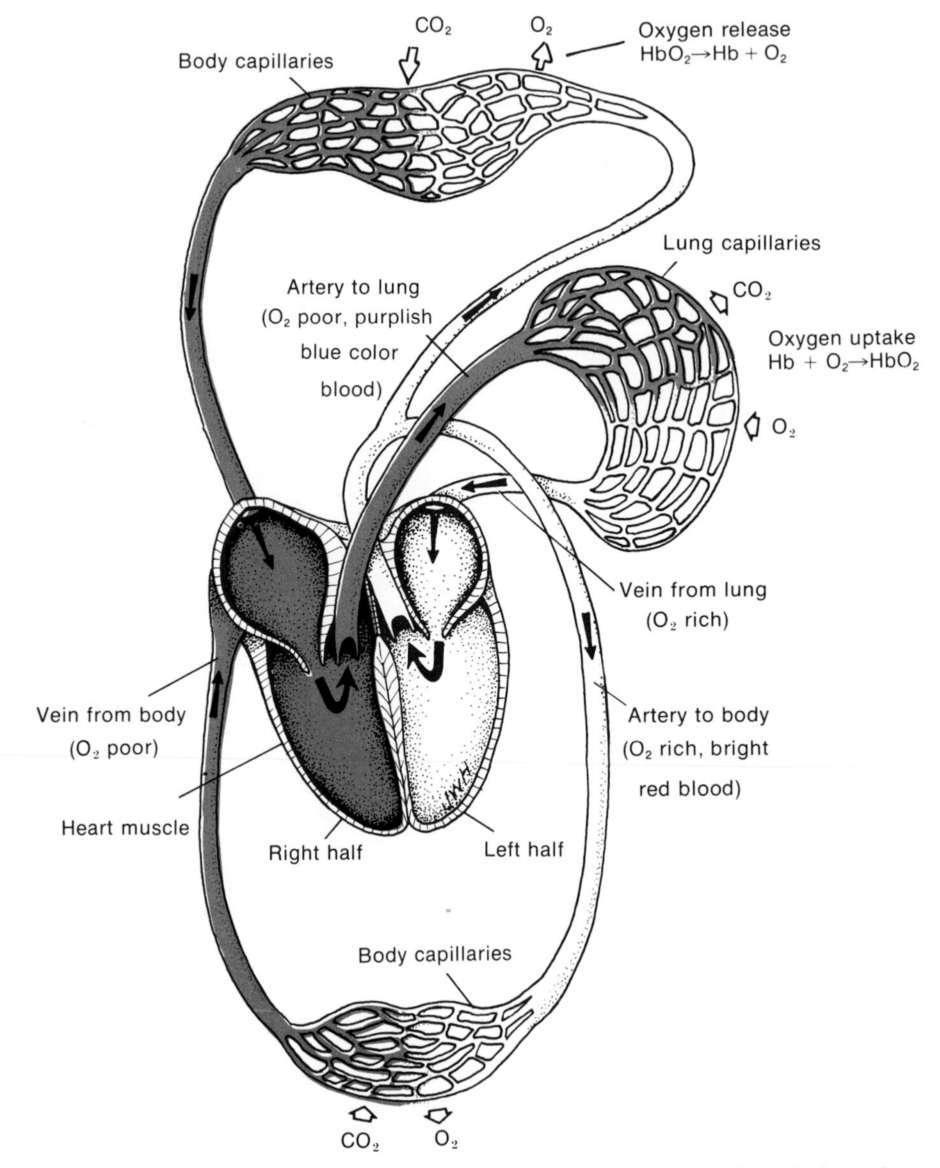

Figure 2–7 Simplified diagram of human circulation. The heart (shown in front view) is divided into two parallel halves. The right half pumps oxygen poor blood to the lungs; the left half pumps oxygen rich blood to the body. (Modified from Clark, M. E., *Contemporary Biology,* W. B. Saunders Co., Philadelphia, 1973.)

construct a wall, then a bridge, and then recombined into the wall again.

There are many reversible chemical reactions which are important to human life. One of these is involved in the transport of atmospheric oxygen from the lungs to the various parts of the body. This task is carried out by hemoglobin, a complex molecule which is found in the blood. This molecule takes up oxygen while in the lungs to form oxyhemoglobin.

The double arrows $\rightleftharpoons$, indicate a reversible reaction.

$$\text{Hemoglobin} + O_2 \rightleftharpoons \text{Oxyhemoglobin}$$

The oxyhemoglobin is then carried by the bloodstream to the various parts of the body, where it releases the oxygen for use in metabolic processes, as shown in Figure 2–7.

While these have been examples of reversible chemical reactions, it is appropriate to add here that many physical changes are also reversible. The physical change, ice to liquid water, is reversible and can be written as:

This reversible physical change involving forms of water is very important in the Earth's weather.

$$\text{ice} \longrightarrow \text{liquid} - \text{energy}$$

or

$$\text{liquid} \longrightarrow \text{ice} + \text{energy}$$

This process, and many other physical processes, can be reversed an indefinite number of times by the repeated addition or removal of energy.

Reactions by Chemical Groups

When a survey is made of the known types of chemical reactions, a diversity is found which is illustrated by the following examples:

$$\underset{\textit{Carbon}}{C} + \underset{\textit{Fluorine}}{2F_2} \longrightarrow \underset{\textit{Carbon tetrafluoride}}{CF_4}$$

$$\underset{\textit{Phosphorus trichloride}}{PCl_3} + \underset{\textit{Water}}{3H_2O} \longrightarrow \underset{\textit{Phosphorous acid}}{H_3PO_3} + \underset{\textit{Hydrogen chloride}}{3HCl}$$

$$\underset{\textit{Sulfur trioxide}}{SO_3} + \underset{\textit{Water}}{H_2O} \longrightarrow \underset{\textit{Sulfuric acid}}{H_2SO_4}$$

$$\underset{\textit{Carbon}}{C} + \underset{\textit{Sulfur vapor}}{2S} \xrightarrow{\text{Heat}} \underset{\textit{Carbon disulfide}}{CS_2}$$

$$\underset{\textit{Iron (III) oxide}}{Fe_2O_3} + \underset{\textit{Carbon monoxide}}{3CO} \xrightarrow{\text{Heat}} \underset{\textit{Iron}}{2Fe} + \underset{\textit{Carbon dioxide}}{3CO_2}$$

These reactions are not presented for you to learn at this point but only to show some of the wide variety of reactions and substances encountered in chemistry. Indeed, the number of known compounds runs into the millions and the number of reactions by which they are produced into the multimillions. However, becoming familiar with a large number of chemical reactions is not as hopeless as it may seem at first. There is much order to be found in the apparent chaos. For example, consider the following reactions:

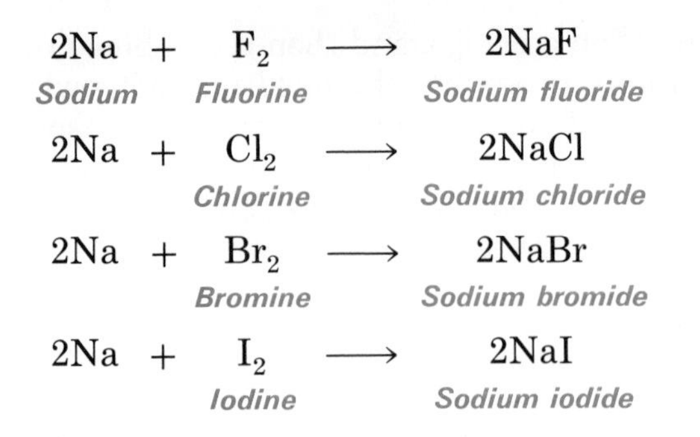

Many chemical reactions follow similar patterns.

Note that the group of elements, fluorine, chlorine, bromine, and iodine, all react with sodium in much the same way; that is, products with similar formulas are produced. Once we have learned one reaction in such a group of elements, we have some idea about what is likely to happen to other members in the group. Even the names of the compounds involved become easier to remember, because there is also a system of names.

There are elements that undergo similar chemical reactions. Elements showing such similarities are identified as a chemical group or family. The elements fluorine, chlorine, bromine, and iodine form one such family.

Another such family includes the group of elements oxygen, sulfur, and selenium, all of which react with calcium in much the same way.

$$2Ca + O_2 \longrightarrow 2CaO$$

CALCIUM + OXYGEN ⟶ CALCIUM OXIDE

$$Ca + S \longrightarrow CaS$$

SULFUR ⟶ CALCIUM SULFIDE

$$Ca + Se \longrightarrow CaSe$$

SELENIUM ⟶ CALCIUM SELENIDE

Elements occur in families; the members of a given family have similar chemical properties.

Not only do elements react as members of groups but groups of atoms may act as a single unit. Consider the following:

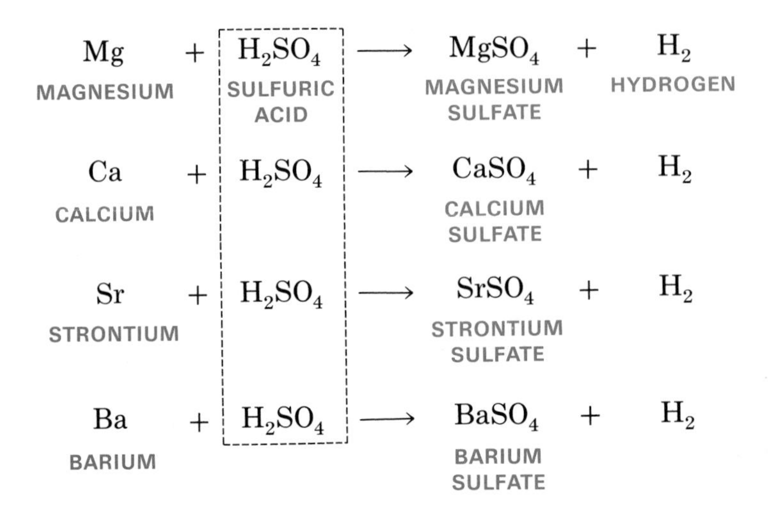

The sulfate group, SO_4, in this series of reactions acts very much like the oxygen, sulfur, and selenium in the group of reactions with calcium given previously.

Two major topics growing out of this discussion of the general characteristics of the chemical reactions are the periodic classification of the elements and chemical equilibrium. Usually these topics are developed much later in a chemistry course because of their complexity. However, these ideas in their simpler forms go back to the early history of chemistry, and we feel they should be carried along and expanded as an overview of the science is obtained.

The Periodic Chart

The periodic chart organizes the chemical elements into a basic pattern of chemical families.

Many of the chemical and physical properties of the elements can be correlated and presented in a periodic chart of the elements (Figure 2–8). This particular arrangement of the elemental symbols is dictated by nature, but it is also the result of the combination of factual information and theoretical models in an organized way. Interpretation of the periodic chart can be carried out at various levels. You can expect this chart to be meaningful from both a practical and a theoretical point of view.

Elements with similar chemical properties are placed in vertical ***groups*** (also known as ***families***). For example, Group IA, the Alkali Metal Group, is composed of lithium, sodium, potassium, rubidium, cesium, and francium. All of these elements are metals with general characteristics that distinguish them from all the other elements. For example, each of the alkali metals reacts with chlorine to form chlorides with the general formula MCl, where M stands for the metal atom. The formulas for the chlorides are LiCl, NaCl, KCl, RbCl, CsCl, and FrCl.

It is obvious that the elements within a group are not the same in all respects or they would be the same element. To illustrate, sodium reacts more vigorously with chlorine than does lithium. Note that sodium is found under lithium in this table. Also, potassium reacts more vigorously than sodium. In fact, there is actually a trend of reactivity from the top of this group to the bottom. For the alkali metals, the trend is to react more vigorously with chlorine as one moves down the group. Such trends of reactivity exist in all of the chemical groups.

In the periodic chart elements are arranged in the order of their atomic numbers. A *period* is a horizontal row in the periodic chart.

In the periodic chart, you will note that each element is assigned a number. This is the order number of the element and is based, from a historical point of view, on all the known chemical and physical information about the element. Later we shall find that the order number of an element is identical to its ***atomic number,*** which is a fundamental concept derived from atomic structure.

The chart is "***periodic***" because when the elements are arranged in the order of their atomic numbers, periodically we encounter an alklai metal and periodically we encounter members of each of the other elemental groups. Observe that the first period has only two elements, hydrogen and helium. Hydrogen is unique in that it can fit into two groups of the periodic table. The second and third periods have eight elements:

Li	Be	B	C	N	O	F	Ne
Na	Mg	Al	Si	P	S	Cl	Ar

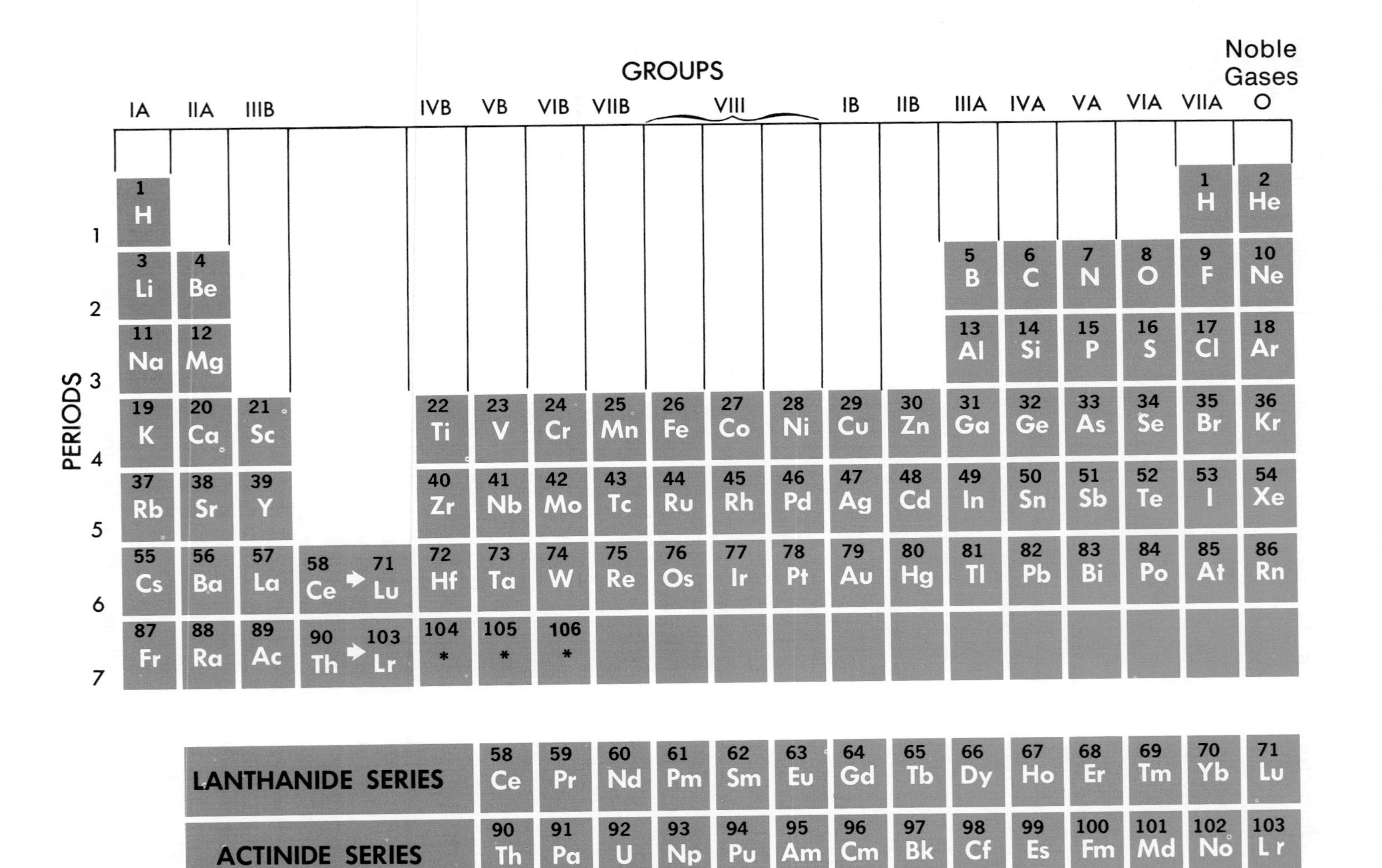

GROUPS

PERIODS	IA	IIA	IIIB		IVB	VB	VIB	VIIB	VIII	VIII	VIII	IB	IIB	IIIA	IVA	VA	VIA	VIIA	Noble Gases O
1	1 H																	1 H	2 He
2	3 Li	4 Be												5 B	6 C	7 N	8 O	9 F	10 Ne
3	11 Na	12 Mg												13 Al	14 Si	15 P	16 S	17 Cl	18 Ar
4	19 K	20 Ca	21 Sc		22 Ti	23 V	24 Cr	25 Mn	26 Fe	27 Co	28 Ni	29 Cu	30 Zn	31 Ga	32 Ge	33 As	34 Se	35 Br	36 Kr
5	37 Rb	38 Sr	39 Y		40 Zr	41 Nb	42 Mo	43 Tc	44 Ru	45 Rh	46 Pd	47 Ag	48 Cd	49 In	50 Sn	51 Sb	52 Te	53 I	54 Xe
6	55 Cs	56 Ba	57 La	58 Ce → 71 Lu	72 Hf	73 Ta	74 W	75 Re	76 Os	77 Ir	78 Pt	79 Au	80 Hg	81 Tl	82 Pb	83 Bi	84 Po	85 At	86 Rn
7	87 Fr	88 Ra	89 Ac	90 Th → 103 Lr	104 *	105 *	106 *												

LANTHANIDE SERIES	58 Ce	59 Pr	60 Nd	61 Pm	62 Sm	63 Eu	64 Gd	65 Tb	66 Dy	67 Ho	68 Er	69 Tm	70 Yb	71 Lu
ACTINIDE SERIES	90 Th	91 Pa	92 U	93 Np	94 Pu	95 Am	96 Cm	97 Bk	98 Cf	99 Es	100 Fm	101 Md	102 No	103 Lr

Figure 2–8 The periodic chart of the elements. Each block contains the symbol for the element and its atomic number.
*No official names yet accepted for these elements.

The fourth and fifth periods have 18 elements each, and the situation becomes more complex as we proceed. But the point to be made is that there are periods of such length that similar elements do fall beneath one another in the chart.

The periodic table is thus the most convenient and useful form of classification of the elements.

Chemical Equilibrium

At chemical equilibrium, both reactants and products are present.

When a chemical reaction proceeds to a point at which the amounts of the reactants and products do not change any further, the reaction has reached a state of ***chemical equilibrium.*** At the beginning of a reaction, we usually are dealing only with reactants. As the reaction proceeds, the amounts of reactants decrease while the amounts of products increase until the reaction arrives at equilibrium. At equilibrium there are still reactants remaining in the presence of products. The amounts of products may be greater than, equal to, or less than the amounts of reactants.

As an example of chemical equilibrium, consider a laboratory experiment in which N_2O_4 (dinitrogen tetroxide) is the reactant and NO_2 (nitrogen dioxide) is the product, according to the equation:

$$N_2O_4 \rightleftharpoons 2NO_2 - 13.9\ \text{kcal}$$

(13.9 kcal of heat is absorbed when 1 mole of N_2O_4 forms 2 moles of NO_2.)

Let us examine what happens if the reaction is carried out in a suitable container in which 1 mole of N_2O_4 is present initially, as shown in Figure 2–9. We can follow the progress of the reaction by measuring the color of the system, since NO_2 is brown and N_2O_4 is colorless. The intensity of the brown color of the system is thus related to the amount

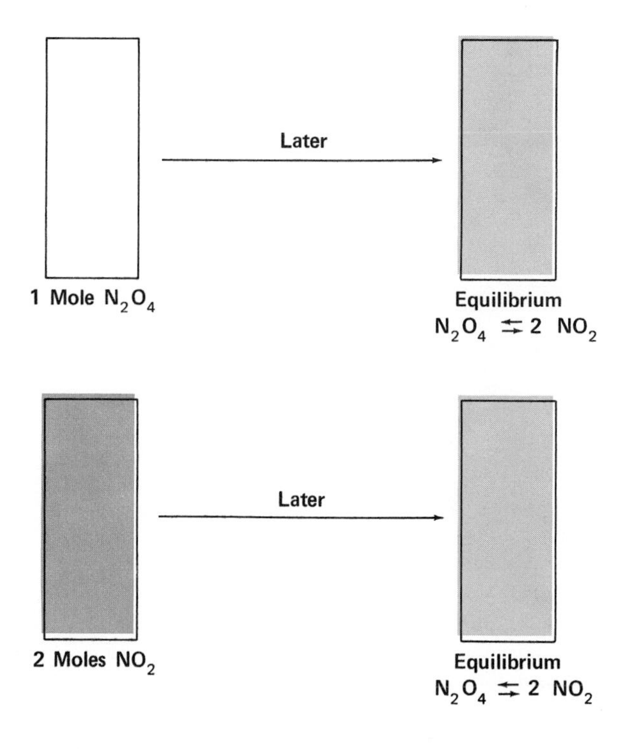

Figure 2–9 The equilibrium between N_2O_4 and NO_2 can be approached from either direction because the reactions are easily reversible. The intensity of the color indicates the amount of NO_2 present in the system.

of NO_2 present. As time passes, the brown color develops and then reaches a steady intensity. Since the intensity does not change as long as the container remains leakproof and is held at a constant temperature, you might be inclined to believe that the reaction had gone to completion. However, if you test the system, you will find that the vessel contains a mixture of NO_2 and N_2O_4. (As a simple test, heating the system will result in a darker brown color. This indicates that some of the N_2O_4 present at the lower temperature has been converted to brown NO_2.) Since some N_2O_4 remains, and since the amounts of N_2O_4 and NO_2 do not change with time, we say that the reaction has reached a state of equilibrium.

If the previous experiment is changed by starting with 2 moles of NO_2, rather than 1 mole of N_2O_4, the same shade of brown will eventually appear, provided, of course, the conditions of container size and temperature are the same (Figure 2–9). This emphasizes the fact that the same position of chemical equilibrium can be attained by approaching from either direction.

In a chemical reaction mixture at equilibrium, the rate of the forward reaction is equal to the rate of the reverse reaction.

Further investigations indicate that chemical equilibrium is ***dynamic*** in the sense that molecular transformations occur continuously when a system is at equilibrium. Although the relative amounts of reactants and products do not change with time, experimental studies indicate that product molecules are reverting to reactants and that reactant molecules are becoming products. ***However, both processes occur at the same rate, so the overall amounts of products and reactants remain the same*** (Figure 2–10).

For example, consider the reaction, $N_2 + 3H_2 \rightleftharpoons 2NH_3$. If we allow nitrogen, hydrogen, and ammonia to come to equilibrium, then remove some of the hydrogen (H_2) and replace it with exactly the same number of molecules of deuterium (D_2), we can show the dynamic nature of equilibrium. (Deuterium is a heavier form of hydrogen with similar properties, distinguishable from ordinary hydrogen by its mass.) After

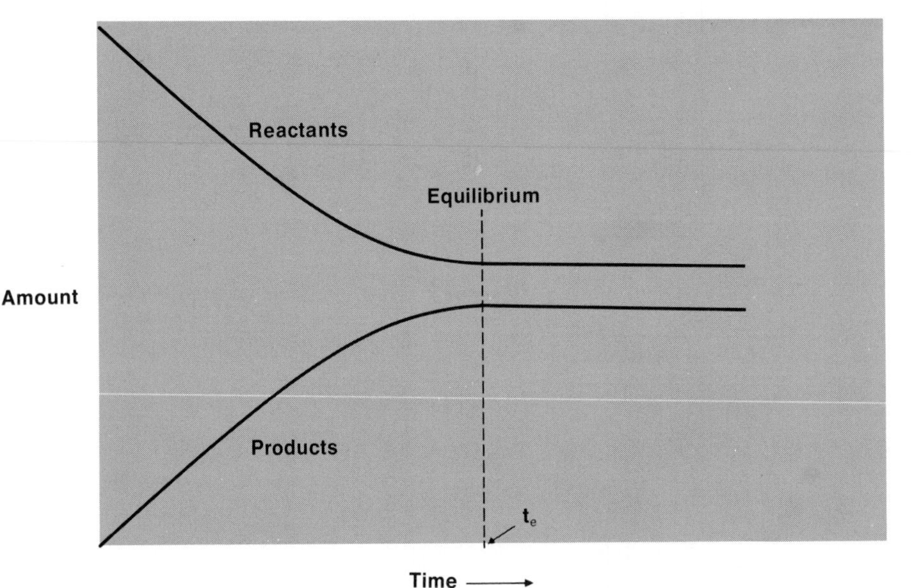

Figure 2–10 Changes in relative amounts of reactants and products with time. At time t_e, equilibrium has been established and relative amounts do not change.

At chemical equilibrium reactants and products are continually changing into each other but the amounts of each remain fixed.

a few moments, we analyze the reaction mixture, using a mass spectrograph (Chapter 6). Perhaps, to our surprise, we find the species HD, NH_2D, NHD_2, and ND_3, as well as the H_2, N_2, and NH_3 present in the original equilibrium mixture. The only way to explain this assortment of molecular species is to assume that hydrogen (D_2 and H_2) is reacting with nitrogen to form ammonia simultaneously with its decomposition into nitrogen and hydrogen. This can be generalized into a fundamental characteristic of chemical equilibrium.

At equilibrium, the rate of formation of products from reactants is equal to the rate at which products revert to reactants.

SELF-TEST 2-A

1. Balance the following equations:

   a. _______ Mg + _______ O_2 $\longrightarrow$ _______ MgO

   b. _______ Si + _______ Cl_2 $\longrightarrow$ _______ $SiCl_4$

   c. _______ Al + _______ O_2 $\longrightarrow$ _______ Al_2O_3

2. Selenium is directly below sulfur in the periodic table. Given the sulfur compounds below, write formulas that are expected for the analogous selenium compounds.

   a. Oxides
      Sulfur: SO_2
      SO_3

      Selenium: _______

      ____

   b. Fluorides
      Sulfur: SF_4
      SF_6

      Selenium: _______

      ____

   c. Oxyacids
      Sulfur: H_2SO_3
      H_2SO_4

      Selenium: _______

      ____

3. Describe how each of the following factors affects the rate of a chemical reaction.

   a. increase in temperature _______________

   b. decrease in temperature _______________

   c. increase in concentrations of reactants _______________

4. When 36 g of liquid water is prepared by the reaction

$$2H_2\,(g) + O_2\,(g) \longrightarrow 2H_2O\,(\ell)$$

   136.6 kilocalories of energy is released. How much energy is released if only 1 gram of liquid water is made by this reaction?

   _______________ kilocalories

5. Deuterium exchanges with the hydrogen of liquid water. What species will be found if we mix equal amounts of D_2O and H_2O?

   _______________, _______________, and _______________.

6. Explain briefly what is meant by the statement "chemical reactions are generally reversible." ____________

____

7. What meaning is attached to $\rightleftarrows$ in the equation below?

$$2H_2 + S_2 \rightleftarrows 2H_2S?$$

____

MATCHING SET

____ 1. reversible reaction
____ 2. corrosion
____ 3. refrigeration
____ 4. chemical family
____ 5. Mg and Ca
____ 6. periodic chart
____ 7. equilibrium
____ 8. Li

a. relative amounts of reactants and products are constant
b. F, Cl, Br, I, At
c. react similarly with oxygen
d. atomic number ordering
e. retards food spoilage
f. formation of rust
g. would form an oxide similar to K_2O
h. hemoglobin-oxygen reaction

QUESTIONS

1. Based on information presented in this chapter on likenesses of elements in groups, predict the formulas of the products of the following reactions.

$Rb + Cl_2 \longrightarrow$	$K + Cl_2 \longrightarrow$
$Ba + O_2 \longrightarrow$	$Mg + Se \longrightarrow$
$Na + Br_2 \longrightarrow$	$Mg + S \longrightarrow$
$Sr + S \longrightarrow$	$Be + S \longrightarrow$

2. In the periodic chart the elements are arranged in the order of their ________.

3. If you were arranging the elements in a row, A, B, C, D, . . . , and came upon an element, G, which had properties similar to A, where would you place it?

4. When a chemical reaction reaches equilibrium, what does this mean with respect to:

   a. the relative amounts of reactants and products?
   b. the cessation of chemical reaction?

5. Name four ways energy plays an important role in chemical reactions that we use in our daily lives.

6. Give an example of a chemical reaction whose rate is fast and one whose rate is slow.

7. List three characteristics of all chemical reactions and give an example of each.

8. Using the periodic chart, select elements which can be expected to have chemical properties similar to:

   a. Ca (calcium), atomic number, 20
   b. Fe (iron), atomic number, 26
   c. Sn (tin), atomic number, 50
   d. S (sulfur), atomic number, 16

9. In 1968 an Apollo spacecraft cabin fire killed three astronauts. The fact that pure oxygen was used as the cabin atmosphere contributed to the fire. How?

10. The element helium is very unreactive chemically. What type of behavior would you expect for argon? Study their relationship on the periodic chart.

11. The recycling of many by-products in our society, such as paper and glass, involves the principles of reversibility. Outline how paper and glass may be recycled for further use.

12. Consider what would be the relative rates of rusting of iron in a dry climate as opposed to a damp climate. What principle(s) of reaction rate is(are) involved?

13. Copper forms a chloride salt, CuCl. Is this consistent with the group number into which copper is placed in the periodic chart?

14. Fires have been started by water seeping into bags in which quicklime was stored. Why would this produce a fire?

SUGGESTIONS FOR FURTHER READING

Campbell, J. A., "Why Do Chemical Reactions Occur?," Prentice-Hall, Englewood Cliffs, N. J., 1965.

Howard, R. A. and Manck, W. A., "The Science of Chemistry: Periodic Properties and Chemical Behavior," Macmillan Co., New York, 1971.

Kieffer, W. F., "Chemistry: A Cultural Approach," Harper and Row, New York, 1971.

Pauling, Linus, "Chemistry," *Scientific American,* Vol. 183, No. 3, p. 32 (1950).

Schaff, J. F. and Westmeyer, P., "Dynamic Nature of Chemical Equilibrium," *Chemistry,* Vol. 41, No. 7, p. 48 (1968).

Vaczek, Louis, "The Enjoyment of Chemistry," Viking Press, New York, 1968.

Wettack, F. S., "A Photometric Study of the N_2O_4-NO_2 Equilibrium," *Journal of Chemical Education,* Vol. 49, p. 556 (1972).

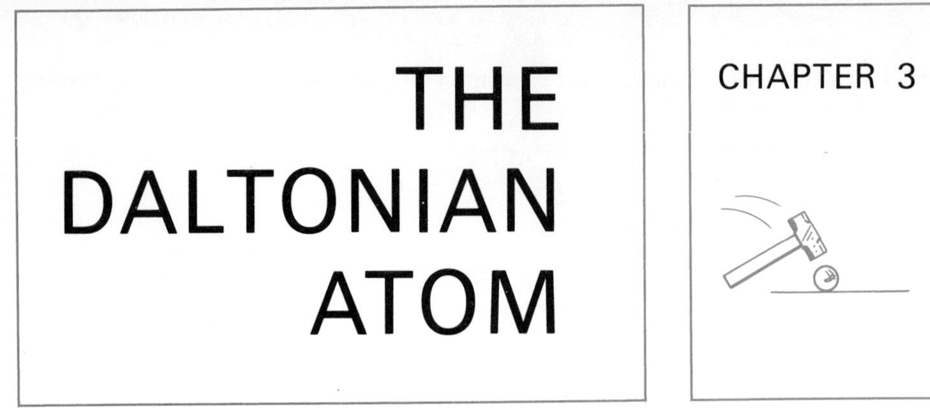

MATTER: CONTINUOUS OR DISCONTINUOUS?

A basic question that has concerned man from antiquity is: Can a sample of matter be subdivided indefinitely without losing its identity? For example, would you expect the division of gold into smaller and smaller particles to leave always a sample of gold, or would you expect to arrive ultimately at a tiny particle of gold which, if further subdivided, would yield something other than gold? Although most students would probably agree with the particle or atomistic nature of matter, direct observation with one or more of their physical senses is not sufficient to demonstrate the discontinuous nature of matter beyond all reasonable doubt.

THEORETICAL ATOMS VISUALIZED: THE GREEK INFLUENCE

The earliest known advocate of the atomic (discontinuous) nature of matter was a Greek philosopher, Leucippus, who lived in the fifth

Figure 3–1 Is matter continuous or discontinuous?

Democritus

century B.C. However, it was his student, Democritus (460–370 B.C.), who apparently aroused the interest of the natural philosophers of his own and of later times in this concept of matter. Democritus used the word ***atom,*** meaning "that which cannot be further divided," to describe the ultimate particles of matter. According to Democritus there was no limit to the kinds of atoms, and all of them were made of the same basic material. He also believed that the various kinds of atoms differed only in their shapes and sizes. Democritus even sought to explain the properties of substances like lead and iron in terms of the way in which the different kinds of atoms associated in forming the macrostructure. He reasoned that the shape of the lead atom enables the close packing for the formation of a very dense but soft material, whereas iron atoms fit together in a rigid network pattern, thus accounting for iron's lower density and greater strength.

It would be wrong to imply that the atomic concepts of the Greeks were based purely on intuitive feeling rather than on scientific observation. A physical theory of matter has value even if it can explain some, although not all, of the pertinent macroscopic facts in terms of submicroscopic models. Although the atomic theory of Democritus was limited in its scope and application, it did indeed explain in simple terms some well-known phenomena such as evaporation (the drying of clothes), condensation (the moisture which appears on the outside of a glass of ice water), diffusion (the movement of an odor through a room), and the growth of new material (crystal growth from a solution).

Based solely on philosophical thought, Aristotle (384–322 B.C.), Plato (427–327 B.C.), and consequently those in the mainstream of enlightened thought rejected the atomistic theories of Democritus. More than two thousand years were to pass before John Dalton (1766–1844), an English schoolteacher, forcefully reintroduced the idea of the atom. Later Dalton's ideas captured the imagination of the scientific community and provided a solid basis for modern atomic theory. Prior to Dalton, however, the intuitive feeling that matter was somehow composed of particles had showed up again and again in scientific thought. Two centuries before Dalton, Galileo reasoned that the appearance of a new substance through chemical change could be explained in terms of a rearrangement of parts

John Dalton (1766–1844), a self-taught English school teacher, moved to Manchester in 1793 and devoted the rest of his life to scientific investigations. His presentation of atomic theory in the early part of the 19th century served as the basis from which modern chemical theories have grown.

too small to be seen. Robert Boyle (1627–1691) and Isaac Newton (1642–1727) used atomic concepts to interpret certain chemical and physical laws. Even before this, Francis Bacon (1561–1626) had speculated that heat might be a form of motion by submicroscopic matter.

SELECTED PHYSICAL AND CHEMICAL INFORMATION ACCUMULATED BY 1810

What were the reasons that prompted Dalton to advance his new ideas of the atomic nature of matter? We shall look at these in some detail for two reasons. First, the chemical and physical laws explained by Dalton are valid today with minor refinements. Second, this offers an opportunity to gain insight into the most fundamental process in man's quest for knowledge and understanding of matter—the interpretation of reproducible facts in terms of the nature of the parts of matter.

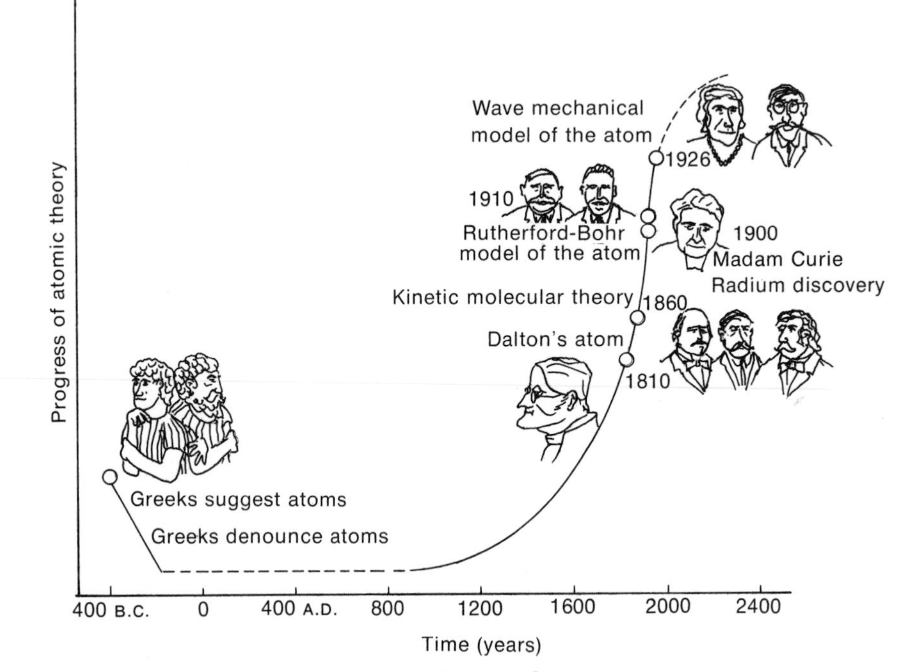

Figure 3–2 Highlights of the progress of atomic theory.

While Dalton's atomic theory primarily sought to explain the chemical laws that will be discussed in this chapter, there were certain physical properties of matter that undoubtedly affected his thinking. It is worth noting that a successful theory in the natural sciences must be able not only to explain laws in one science, but it must also, at the same time, be consistent with laws and well-established theories of interrelated sciences.

States of Matter

What other substances are you familiar with that can exist as solid, liquid, and gas?

Matter occurs in three well-defined forms called ***states:*** solid, liquid, and gas. A solid has a definite shape and volume, a liquid has a definite volume but no definite shape, and a gas has neither definite shape nor definite volume. Under ordinary conditions substances generally exist in just one of the three states. However, by proper manipulation of pressure (force per unit area) and temperature, substances can be made to change state. For example, oxygen at room temperature and atmospheric pressure is a gas, but at −200°C, oxygen is a liquid. At temperature and pressure conditions that are easily attained, water can exist as solid ice, liquid water, or gaseous steam (Figure 3–3).

Long before Dalton's time, scientists knew of the states of matter and understood, at least partly, how to bring about the physical changes from one state to another. Not understood, however, was exactly what was happening within the sample of matter as it changed from one state to another. This was particularly true in the case of gaseous matter, and it was the study of gases that, in a large measure, led Dalton to his understanding of the structure of matter.

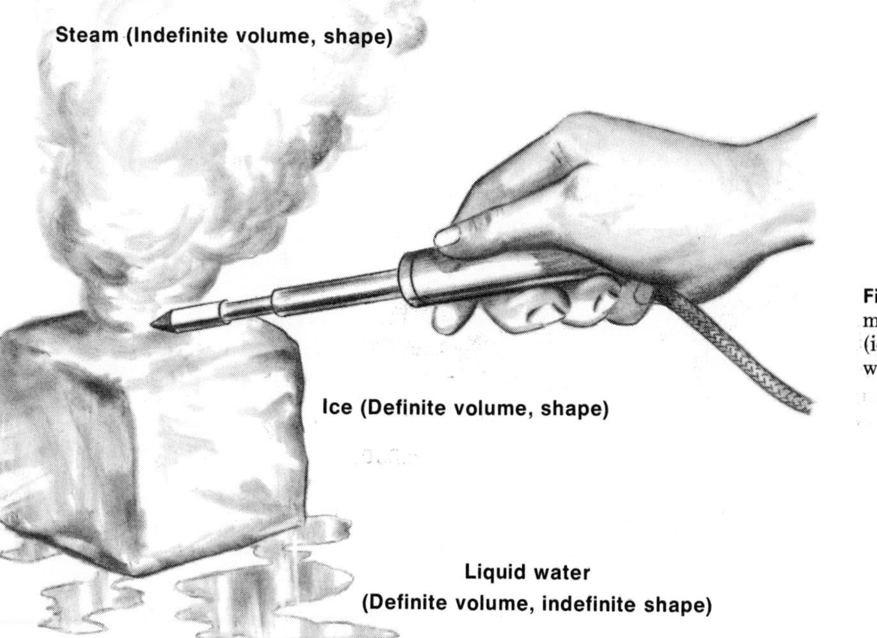

Figure 3–3 The three states of matter illustrated by solid water (ice), liquid water, and gaseous water (steam).

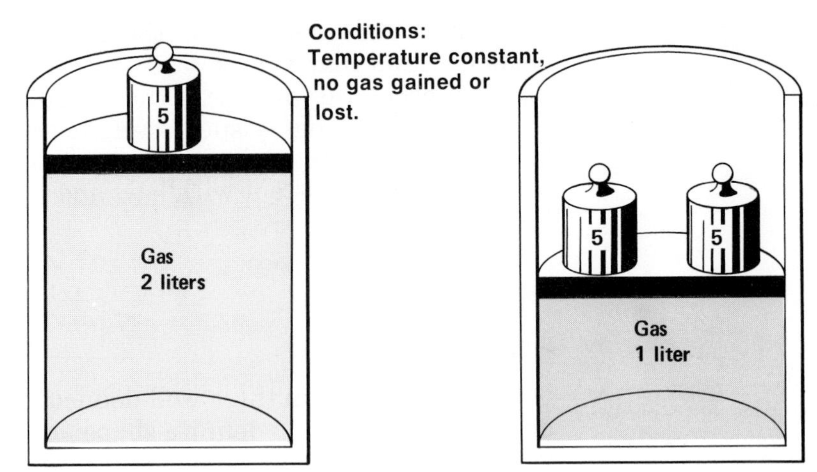

Figure 3–4 Illustration of Boyle's law. If the pressure is doubled, the volume is halved at constant temperature and amount of gas.

Boyle's Law

Robert Boyle, an Irish physicist of the 17th century, studied the relationship between the pressure of a gas and its volume at constant temperature. A simple way to measure the change in volume accompanying a change in pressure is to use an apparatus like that shown in Figure 3-4. The sample of gas trapped in the cylinder is subjected to various pressures by changing the weights. For each pressure, the piston will adjust until the pressure exerted by the weights is equal to the pressure exerted by the entrapped gas. The volume is then measured for each pressure. Some typical data emerging from this kind of apparatus could be as follows:

Pressure (*Pounds/square inch*)	Volume (*Cubic inches*)	Pressure × Volume
20	100	2000
25	80	2000
30	67	2010
40	50	2000

Is there any regular change in the volume as the pressure is changed? How can this relationship be expressed? It appears from these data that the product of the pressure times the volume is approximately constant (within the accuracy of the measurements) and that doubling the pressure halves the volume. In essence, these are statements of ***Boyle's law,*** which can be stated in several alternate equivalent ways (see also Figure 3–5).

For a series of pressure and volume measurements on the same sample of gas at constant temperature, the volume of a gas varies inversely with the pressure on that sample of gas, or

$$\text{pressure} \times \text{volume} = \text{a constant}$$
$$PV = \text{constant}$$

For any two sets of measurements P_1,V_1, and P_2,V_2, on the same sample of gas,

$$P_1V_1 = \text{constant} = P_2V_2$$

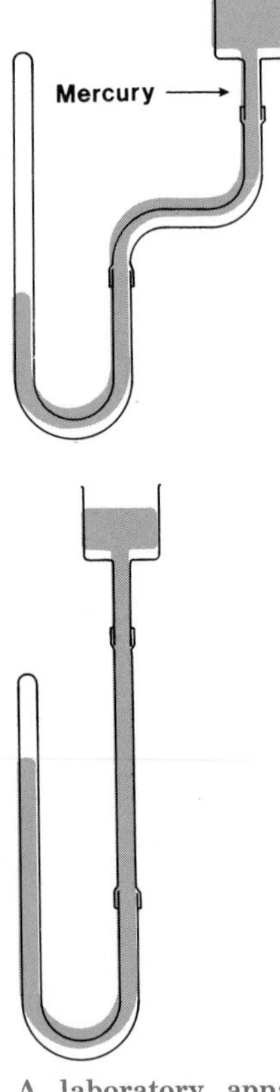

A laboratory apparatus of this type would involve a gas sample in a U-shaped tube with mercury to produce the pressure.

Boyle's Law: Pressure goes up, volume goes down, proportionately with temperature constant.

Boyle's observations and subsequent studies have shown that this relationship is a good approximation for describing the behavior of most gases. From the fact that gases can be squeezed into a smaller space, Boyle theorized that gases are made of particles much like submicroscopic threads of wool or small coiled springs, and that the pressure is due to these particles being whirled about so vigorously by the ether (a hypothetical medium proposed to support the particles of a gas) that each particle endeavors to beat off all others from coming within its own little sphere. The interpretation of the properties of matter in terms of small particles becomes more and more attractive as the knowledge of matter becomes more detailed.

Dalton's Law of Partial Pressures

Many of Dalton's experiments involved measurements on air. Scientists of his time believed that air was a compound composed of nitrogen, oxygen, and water. Dalton's own analyses of air samples gathered at different locations showed that in 100 volume units of air there were always about 21 parts oxygen and 79 parts nitrogen, with a trace of water which increased with temperature. His experiments showed that the pressure exerted by a sample of dry air increases when water vapor is added to it, and the increase is by an amount equal to the pressure exerted by the water vapor alone at that temperature.

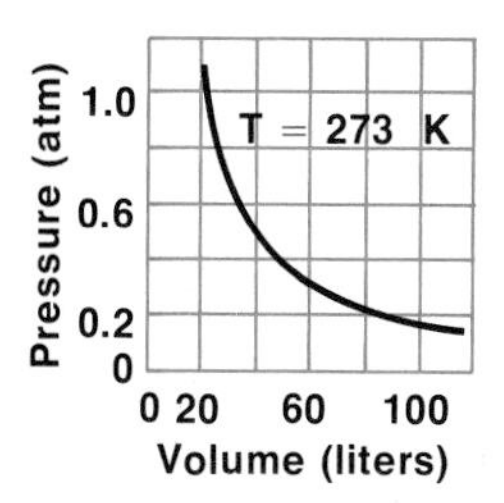

Figure 3–5 A graphical illustration of Boyle's Law—the inverse proportion between the volume and pressure of a gas at constant temperature.

This type of behavior is observed for other gaseous mixtures as well. As a typical example, 32 grams of oxygen at 0°C exerts 14.7 pounds per square inch (1 atmosphere) pressure when confined in 22.4 liters, and 28 grams of nitrogen exerts the same pressure (1 atmosphere) in the same volume at the same temperature. When the 32 grams of oxygen and the 28 grams of nitrogen are placed together in a 22.4 liter container, the pressure becomes 2 atmospheres (Figure 3-6). This is typical of many gaseous mixtures and is known as ***Dalton's law of partial pressures*** for a two-component mixture:

$$\text{Pressure of gas 1} + \text{Pressure of gas 2} = \text{Total pressure}$$
$$(P_1 + P_2 = P_{total})$$

which may be stated as:

The pressure exerted by a gaseous mixture is the sum of the pressures exerted by the individual gases when each is confined separately in the same volume.

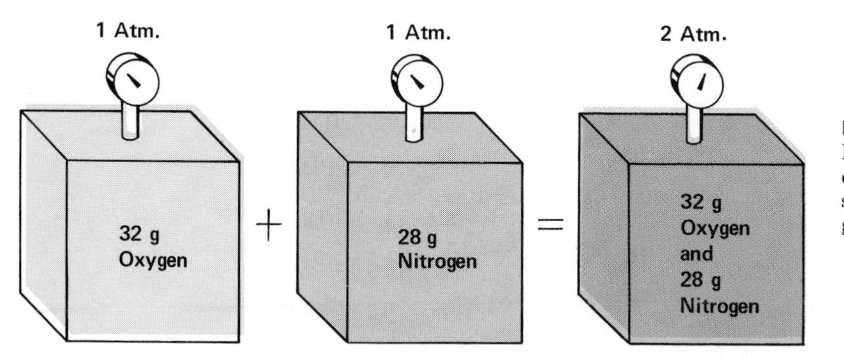

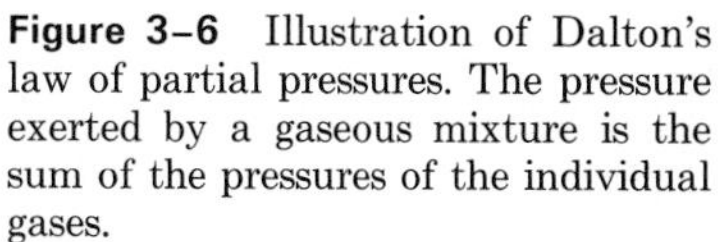

Figure 3–6 Illustration of Dalton's law of partial pressures. The pressure exerted by a gaseous mixture is the sum of the pressures of the individual gases.

Studies Dalton performed in 1801 led him to believe that gases must be somehow composed of particles, capable of diffusing through a volume already occupied by another gas, with little regard for the gas already there. This is a reasonable idea if the particles of the gas are extremely small relative to the great distances between them.

Henry's Law

In addition to studying the pressures of mixed gases, Dalton also investigated solutions of gases in liquids. The amounts of nitrogen that will dissolve in 100 liters of water at 20°C at various pressures of nitrogen are as follows:

1.9 g of nitrogen with a nitrogen pressure of 1 atmosphere
3.8 g — 2 atmospheres
5.7 g — 3 atmospheres
7.6 g — 4 atmospheres

In 1803, William Henry, a close friend of Dalton, summarized observations like these on the solution of several gases in liquids in a statement known as ***Henry's law.***

The weight of a gas that dissolves in a definite volume of a liquid is directly proportional to the partial pressure of the gas above the liquid at a constant temperature.

The behavior of carbonated beverages illustrates an effect of pressure on the solution of gases. When the container is opened gas is evolved because, as the pressure on the gas is reduced, the liquid can dissolve less gas.

When the pressure is released on a bottle of carbonated beverage, carbon dioxide bubbles out, illustrating Henry's law.

Charles' Law

The relationship between the volume of a sample of gas at various temperatures and at constant pressure can be illustrated by an apparatus such as that shown in Figure 3–7. Typical data that might be obtained with such apparatus are:

Temperature (°C)	Volume (ml)
27	600
54	654
127	800
227	1000
327	1200

Note from the data and Figure 3–3 that as the temperature is increased, the volume is increased. This generalization is true for all gases. However, a simple relationship between volume and centigrade temperature is not evident from these data. The data show that doubling the centigrade temperature from 27 to 54 degrees (100 percent increase) results in only an 8.8 percent $\left(\frac{654 - 600}{600} \times 100\%\right)$ increase in volume.

PLATE I

Lavoisier, the "father of modern chemistry," and his wife. His quantitative measurements provided the basis for our understanding of chemistry. (Courtesy of The Rockefeller University. Private Collection, New York City.)

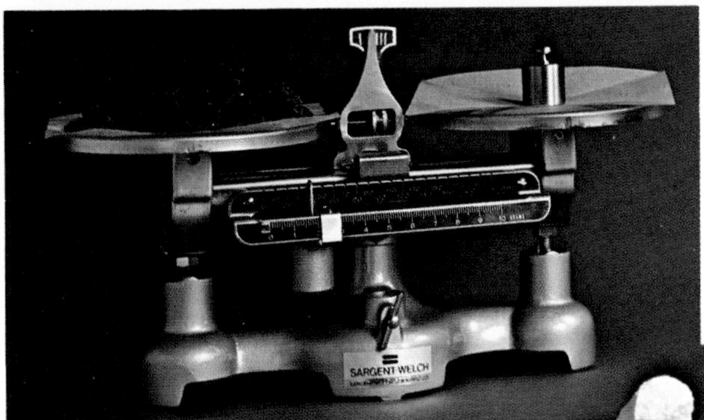

PLATE II

Mineral samples are, at times, pure substances, as pictured below. Most of the time, however, they are apt to be as complex as dirt or granite.

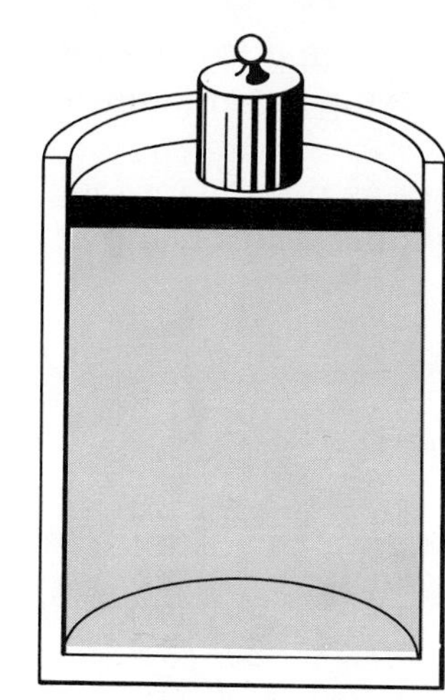

Figure 3–7 Illustration of Charles' law. If the absolute temperature is doubled, the volume is doubled at constant pressure and amount of gas.

Often, in the study of natural science, a simple relationship exists but is not immediately obvious. Further observations concerning the temperature-volume relationship for a gas led to an interestingly simple result. However, before this result can be appreciated fully, it will be necessary to introduce a new temperature scale.

It is observed that a sample of gas at 0°C decreases by $^1/_{273}$ of its original volume when cooled 1°C. If the gas sample is cooled 10 centigrade degrees, it decreases by $^{10}/_{273}$ of its starting volume. It appears then that a gas sample would lose all ($^{273}/_{273}$) of its volume if it were cooled from 0°C to −273°C (Figure 3-8).

In reality, this zero volume is never observed because all real gases first liquefy and most solidify before reaching −273°C. However, on the basis of this behavior of gases, and for other reasons as well, −273°C is termed ***absolute zero,*** and a new temperature scale measured in degrees absolute has been established. Temperatures measured on this scale are often called Kelvin temperatures (degrees K), named after the originator of the scale, Lord Kelvin. On the Kelvin or absolute temperature scale,

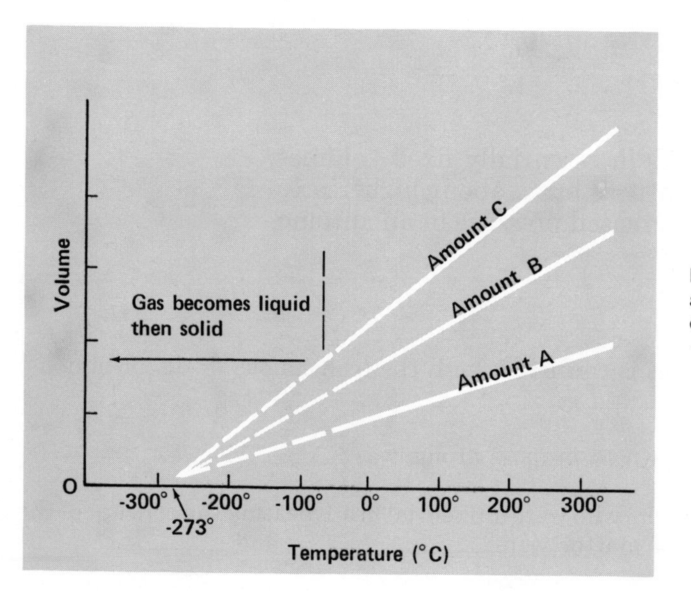

Figure 3–8 The variance of the volume of different amounts of a gas with temperature. The temperature corresponding to a hypothetical zero volume is − 273°C, absolute zero.

the freezing point of water is 273 K and its boiling point is 373 K. The relationship between centigrade temperature and Kelvin temperature is K = °C + 273.

Now, if the same set of data as used before is expressed in degrees Kelvin, a simple temperature-volume relationship at constant pressure is immediately obvious:

Normal body temperature is 98.6°F, 37.0°C, or 310 K.

Temperature		Volume
(°C)	(K)	(ml)
27	300	600
54	327	654
127	400	800
227	500	1000
327	600	1200

It is evident now that the volume of the sample is directly proportional to the temperature when the temperature is expressed in Kelvin degrees, because when the temperature is increased from 300 K to 600 K (doubled), the volume is doubled from 600 ml to 1200 ml.

In 1787, Jacques Charles, a French scientist, studied the temperature-volume relationship for gases and, in 1801, another French scientist, Joseph Gay-Lussac, published his findings, which were similar to Charles' but more refined. The result, named after the first discoverer, is known as ***Charles' law:***

At constant pressure, the volume of a gas is directly proportional to the Kelvin temperature of the gas.

That is,

$$\text{Volume} = \text{constant} \times \text{Kelvin temperature, or}$$
$$V = kT$$

For any two sets of measurements, V_1, T_1, and V_2, T_2, on the same sample of gas, held at constant pressure,

$$\frac{V_1}{T_1} = \frac{V_2}{T_2}$$

More frequently, we heat gas samples in essentially fixed volumes; under these conditions increased temperatures bring about higher pressures. A common example of this is the increased pressure in an automobile tire after a long, fast trip.

SELF-TEST 3-A

1. The concept of atoms is consistent with the (continuous or discontinuous) nature of matter. ____________

2. The first person known to propose atoms was ____________.

3. Two Greek philosophers who were influential in advocating the concept of the continuous nature of matter were ____________ and ____________.

4. The Greek's approach to the "discovery" of atoms can be best described by
   a. experimentation.
   b. philosophy (play on logic).
   c. direct observation of atoms.
   d. consistent explanation of well-known, established laws of nature.
   e. deductive reasoning.

5. If 2 liters of a gas at room temperature and 1 atmosphere of pressure is compressed to a 1-liter volume, what will be the new pressure? ____________ ____________

6. Consider the numbers in Figure 3–6 and determine what the pressure would be in the same box if an additional 32 grams of oxygen were added to the box on the right. ____________

7. At 20°C, how many grams of nitrogen will dissolve in 100 liters of water if the pressure of nitrogen above the liquid is 2.5 atmospheres? (Data are given in the text.) ____________

8. Convert the following temperatures:

a.	30°C to ________ K	d.	250 K to ________ °C
b.	−25°C to ________ K	e.	600 K to ________ °C
c.	100°C to ________ K	f.	4 K to ________ °C

9. If a sample of gas occupies 2 liters at 25°C and 1 atmosphere pressure, what volume will it occupy at 174°C and 1 atmosphere pressure? (Hint: Convert to degrees Kelvin first.) ____________

While the physical laws of gases provided some basis for the logical assumption of the particle nature of matter, the chemical laws of nature were the primary evidence for Dalton's concept of the atom. It is to the chemical laws known during Dalton's time that we now turn.

The Law of Conservation of Matter

While many had suspected that atoms merely became rearranged in a chemical change but were never destroyed (see Figure 1–9, page 18), it remained for Antoine Lavoisier, a French scientist, to establish firmly in 1785 that matter is neither created nor destroyed in chemical transformations.

Color Plate I shows a painting of Lavoisier and his wife.

Lavoisier discovered that mercury (Hg) combines with oxygen in the air to form mercuric oxide. This reaction is expressed by the equation:

$$2Hg + O_2 \longrightarrow 2HgO$$
$$\text{mercury} + \text{oxygen} \longrightarrow \text{mercuric oxide}$$

Upon heating mercuric oxide, an orange powder, Lavoisier recovered exactly, within the limits of his ability to measure weights, the same amount of mercury used to prepare the mercuric oxide originally.

$$2HgO \xrightarrow{\text{Heat}} 2Hg + O_2$$

$$\text{mercuric oxide} \xrightarrow{\text{Heat}} \text{mercury} + \text{oxygen}$$

Furthermore, he observed that the amount of gas consumed in the production of a sample of mercuric oxide was exactly equal to the amount produced in the destruction of the sample.

An examination of all chemical reactions that lend themselves to the kind of quantitative study carried out by Lavoisier leads to the generalization known as the ***law of conservation of matter:***

In a chemical reaction, the sum of all of the weights of the products is exactly equal to the sum of all of the weights of the substances that enter into the reaction.

It has been observed that substances can be created or destroyed in a chemical process but matter cannot. As a further example, consider Figure 3-9.

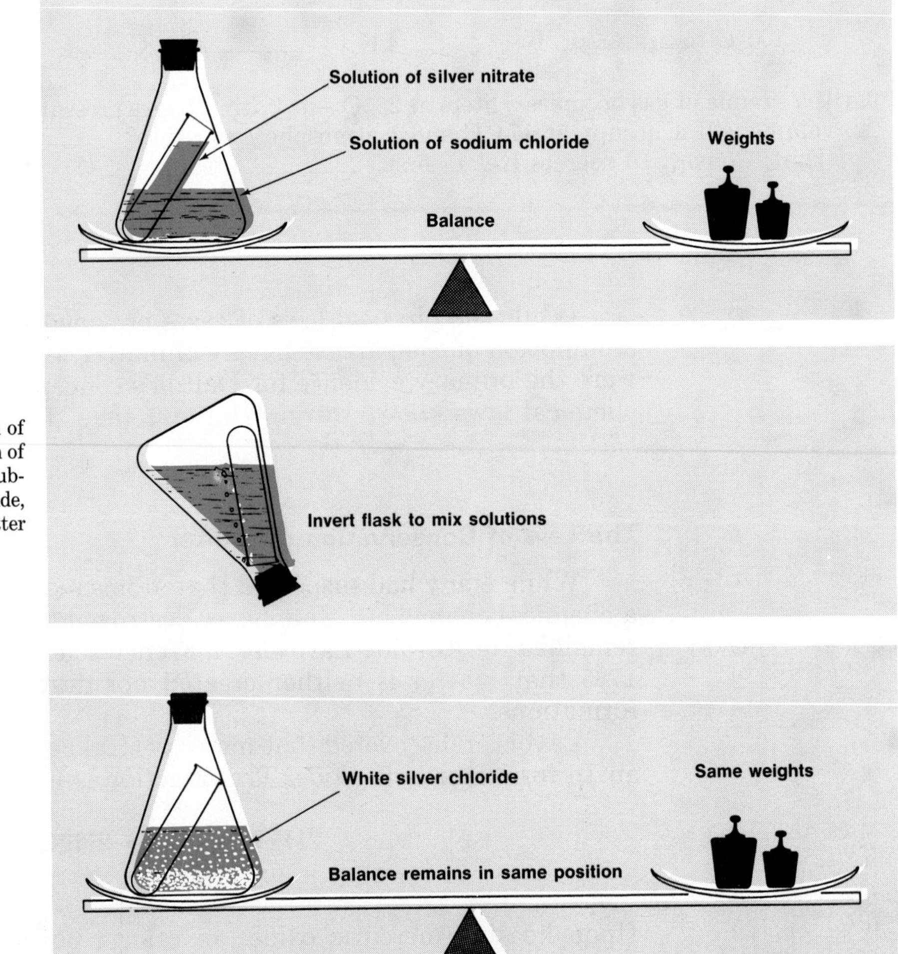

Figure 3-9 Mixing a solution of sodium chloride with a solution of silver nitrate produces a new substance, insoluble silver chloride, but the total weight of the matter remains the same.

The Law of Constant Composition

Sodium chloride (common table salt is a slightly impure form of this compound) has been found always to contain 39.4 percent by weight sodium and 60.6 percent by weight chlorine. Sucrose or table sugar, another common compound in our daily environment, is always found to be 42.1 percent by weight carbon, 6.4 percent by weight hydrogen, and 51.5 percent by weight oxygen.

Information of this type concerning the compositions of various compounds led to the ***law of constant composition:***

> **When two or more elements combine to form a given compound, the ratio of the weights of the elements involved is always the same.***

If you wish to prepare pure water, a compound that is 88.8 percent by weight oxygen and 11.2 percent by weight hydrogen, you will find that, if 11.2 grams of hydrogen react with 88.8 grams of oxygen, you will obtain 100 grams of water with no appreciable amount of hydrogen or oxygen left over. If you use 15.2 grams of hydrogen to react with the 88.8 grams of oxygen, you will not obtain more water because the 88.8 grams of oxygen will make only 100 grams of water—4 grams of hydrogen will be left over. This further illustrates the fact that a pure compound always has the same chemical composition.

The percent composition alone is not sufficient for positive identification. Physical properties such as melting point, boiling point, solubility, and so forth are also needed.

Since elements combine in fixed proportions to form compounds, it necessarily follows that compounds have characteristic compositions. One can readily see that the percent composition of a compound can be useful in identifying that compound.

Let us look at some simple examples of the usefulness of the law of constant composition.

EXAMPLE PROBLEM 1

Carbon dioxide, the gaseous product of respiration, is 27 percent carbon and 73 percent oxygen. This compound can be produced by burning pure carbon in air. What weight of oxygen will be needed to combine with 54 g of carbon to produce carbon dioxide?

ANSWER

73 grams of oxygen react with 27 grams of carbon to produce 100 grams of carbon dioxide. Therefore, since 54 grams of carbon is 2×27 grams, 146 grams of oxygen (that is, 2×73 grams) will be needed to combine with the 54 grams of carbon in order to keep the same ratio of weight of carbon to weight of oxygen.

---

*Over the years, chemists have encountered substances with varying compositions, substances that they choose to call compounds. Ordinary rust is an example. Such substances resemble compounds more than mixtures; as with many definitions of categories in nature, terms become difficult to apply in borderline cases.

EXAMPLE PROBLEM 2

Sodium chloride, 61 percent chlorine, is a commercial source of chlorine. What weight of sodium chloride would be required to produce 183 pounds of chlorine?

ANSWER

100 pounds of table salt produces 61 pounds of chlorine. Since 183 pounds of chlorine is desired (that is, 3×61 pounds of chlorine), it will take 3×100 pounds or 300 pounds of table salt.

EXAMPLE PROBLEM 3

Charcoal is an impure form of carbon made from wood. Why is it not possible to calculate the weight of charcoal obtainable from 100 g of wood?

ANSWER

Since wood and charcoal are mixtures, different samples of wood may contain different amounts of carbon.

Sources of Percentages by Weight

You may ask at this point, "How did early chemists establish that water is 88.8 percent oxygen, carbon dioxide is 27 percent carbon, and sodium chloride is 61 percent chlorine?" The answers are found through quantitative analyses. Two possible methods of arriving at percent composition—synthesis and decomposition—will now be discussed to answer this question.

Quantitative analysis is the determination of how much of each element is present in a given weight of a compound.

Percent Composition by Synthesis

When exactly the correct weighed amounts of two elements, A and B, are combined to form a compound, the weights of the two parts (wt. A and wt. B) are known, and their sum (wt.A + wt.B) would be the weight of the compound produced. Since percent is the $\frac{\text{part}}{\text{whole}} \times 100$, the percent of each element in the compound (the percent composition by weight) can be calculated in a straightforward manner.

$$\left(\frac{\text{wt. A}}{\text{wt. A + wt. B}}\right) \times 100\% = \left(\frac{\text{wt. A}}{\text{wt. compound}}\right) \times 100\% = \%\text{A (by weight)}$$

$$\left(\frac{\text{wt. B}}{\text{wt. A + wt. B}}\right) \times 100\% = \left(\frac{\text{wt. B}}{\text{wt. compound}}\right) \times 100\% = \%\text{B (by weight)}$$

With a slight modification, the analysis of carbon dioxide can be achieved in a similar way. A very pure sample of the element carbon, which is a black solid in its usual form, can be obtained easily and then weighed accurately on the analytical balance. Though conceivably possible, it would be extremely difficult to select just the right amount of

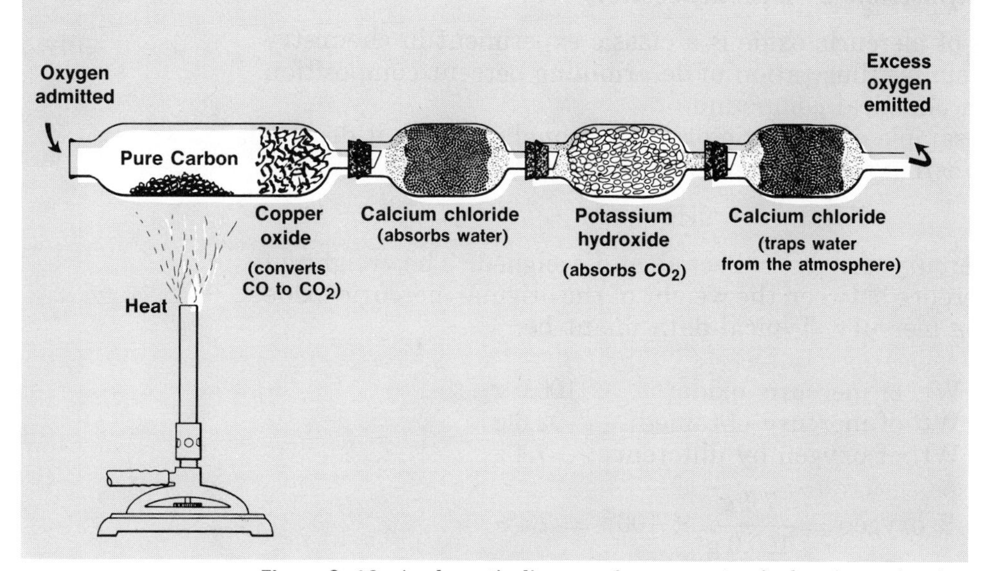

Figure 3–10 A schematic diagram of an apparatus for burning carbon in oxygen and trapping the carbon dioxide produced.

oxygen for a fixed amount of carbon and then weigh this sample of gas. It would be much easier to burn the weighed sample of carbon in excess oxygen, weigh the gaseous product, carbon dioxide, and, from the difference, obtain the weight of oxygen.

Pure oxygen is passed into the series of glass tubes, as shown in Figure 3–10. A weighed sample of carbon is heated until it is completely reacted. Because there is a tendency for some of the carbon to be converted only to carbon monoxide, CO, instead of carbon dioxide, CO_2, the solid copper oxide is heated to aid the complete conversion of carbon monoxide to carbon dioxide. Next in line is a tube filled with a drying agent, such as calcium chloride, which selectively absorbs water vapor but allows excess oxygen and the newly formed carbon dioxide to pass through. This step is necessary since it is difficult to keep the chemicals and apparatus perfectly dry. The water must be removed first or it will be absorbed along with the carbon dioxide. Potassium hydroxide in the next tube absorbs the carbon dioxide and allows the excess oxygen to pass. The potassium hydroxide is weighed before and after the experiment; the gain in weight is equal to the weight of carbon dioxide produced from the original amount of carbon. The last tube is filled with a drying agent to prevent water in the atmosphere from entering the tube containing the potassium hydroxide. Connections are placed between the tubes so they can be separated and closed easily. A set of data for this experiment might appear as follows:

Wt. of carbon:	0.105 g
Wt. of carbon dioxide produced:	0.385 g
Wt. of oxygen (by difference, 0.385 g − 0.105 g):	0.280 g

$$\%\ \text{carbon} = \frac{\text{wt. carbon}}{\text{wt. compound, } CO_2} \times 100\% = \frac{0.105\ \text{g}}{0.385\ \text{g}} \times 100\% = 27.3\%$$

$$\%\ \text{oxygen} = \frac{\text{wt. oxygen}}{\text{wt. compound, } CO_2} \times 100\% = \frac{0.280\ \text{g}}{0.385\ \text{g}} \times 100\% = 72.7\%$$

Percent Composition by Decomposition

The analysis of mercuric oxide is a classic experiment in chemistry and offers a very simple illustration of determining percent composition by decomposing a chemical compound.

The Greek letter Δ indicates heat is applied to the reactants.

If a weighed sample of mercuric oxide is strongly heated, it decomposes, liberating mercury and gaseous oxygen:

$$2HgO \xrightarrow{\Delta} 2Hg + O_2$$

The resulting mercury can be collected and weighed. The weight of oxygen is the difference between the weight of the original mercuric oxide and the remaining mercury. Typical data might be:

Wt. of mercuric oxide:	100.0 g
Wt. of mercury obtained:	92.6 g
Wt. of oxygen by difference:	7.4 g

$$\% \text{ oxygen} = \frac{7.4 \text{ g}}{100.0 \text{ g}} \times 100\% = 7.4\%$$

$$\% \text{ mercury} = \frac{92.6 \text{ g}}{100.0 \text{ g}} \times 100\% = 92.6\%$$

These calculations are presented to show that by simple experiments and the ability to recognize a pure substance, the early chemists could obtain the percentage composition for compounds even before they had definite ideas about atoms, molecules, or molecular formulas.

The Law of Multiple Proportions

It is often observed that two elements can form more than one compound. For example, carbon and oxygen can form carbon monoxide, a poisonous gas, and carbon dioxide, a gas produced in respiration. Hydrogen and oxygen form both water and hydrogen peroxide, and nitrogen and oxygen combine to form several different compounds. In an extreme case, hydrogen and carbon form compounds, called hydrocarbons, which number into the hundreds of thousands. In every case in which the compounds are composed of the same elements, each compound has a characteristic set of properties which distinguishes it from other compounds of these elements.

Carbon monoxide is CO, carbon dioxide is CO_2.

You might quickly guess that the weight of oxygen needed to convert a fixed weight of carbon, say 10 g, into carbon monoxide would be different from the amount needed to convert the 10 g of carbon to carbon dioxide. You would reason that since carbon monoxide and carbon dioxide are different compounds, each with its own fixed composition, the weight of oxygen for a fixed weight of carbon in the two compounds would necessarily have to be different.

A set of experimental data for carbon monoxide and carbon dioxide is the following:

0.75 g carbon and 1.00 g oxygen form 1.75 g carbon monoxide;
0.75 g carbon and 2.00 g oxygen form 2.75 g carbon dioxide.

Note that for a fixed weight of carbon (0.75 g) used in the formation of each compound, different amounts of oxygen are required. In fact, the ratio of the weights of oxygen is 2:1, a ratio of small whole numbers.

In 1804 Dalton recognized this as a general phenomenon, and a statement summarizing these facts is known as the ***law of multiple proportions:***

Many elements form compounds which obey the law of multiple proportions.
Fe: $FeCl_2$ and $FeCl_3$
N: N_2O, NO, and NO_2
S: SO_2 and SO_3

> In the formation of two or more compounds from the same elements, the weights of one element that combine with a fixed weight of a second element are in a ratio of small whole numbers (integers) such as 2:1, 3:1, 3:2, or 4:3.

Figure 3–11 is an illustration of this law as it applies to the two compounds of carbon with oxygen. If Dalton was to advance a satisfactory theory of matter in terms of atoms, he had to be able to explain the law of multiple proportions.

Law of Combining Volumes

Gay-Lussac carried his study of the volume of gas samples a step further than did Charles in that he studied the volumes of gases involved in chemical reactions. Gay-Lussac measured the amounts of gaseous samples in terms of volumes, being careful to measure these volumes at constant temperature and pressure, or to make suitable calculations according to Boyle's and Charles' laws for variation in these factors.

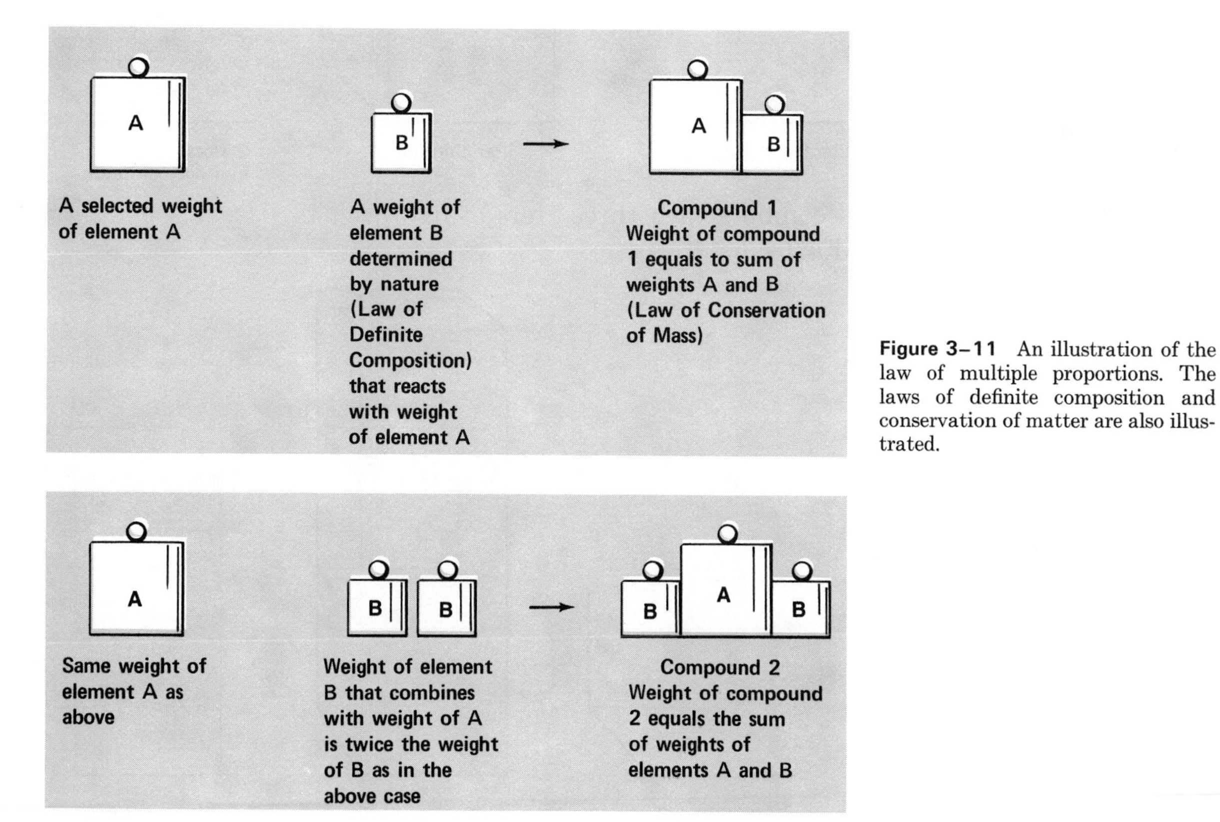
A
A selected weight of element A
B
A weight of element B determined by nature (Law of Definite Composition) that reacts with weight of element A
A
B
Compound 1
Weight of compound 1 equals to sum of weights A and B (Law of Conservation of Mass)
A
Same weight of element A as above
B
B
Weight of element B that combines with weight of A is twice the weight of B as in the above case
B
A
B
Compound 2
Weight of compound 2 equals the sum of weights of elements A and B

Figure 3–11 An illustration of the law of multiple proportions. The laws of definite composition and conservation of matter are also illustrated.

Some examples will illustrate the type of data obtained by Gay-Lussac.

Two liters of hydrogen react with one liter of oxygen to form two liters of steam (Figure 3–12).

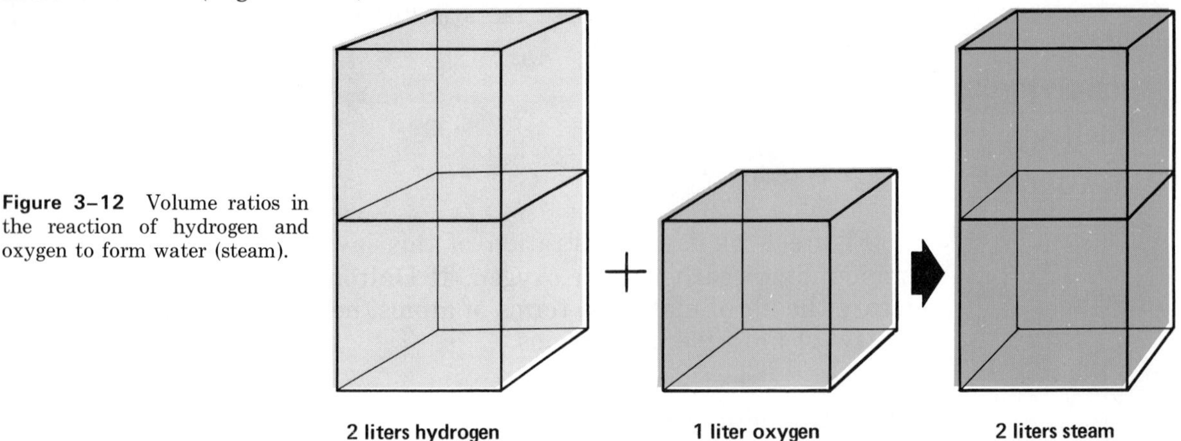

Figure 3–12 Volume ratios in the reaction of hydrogen and oxygen to form water (steam).

One liter of hydrogen will combine with one liter of chlorine gas to form two liters of hydrogen chloride gas (Figure 3–13).

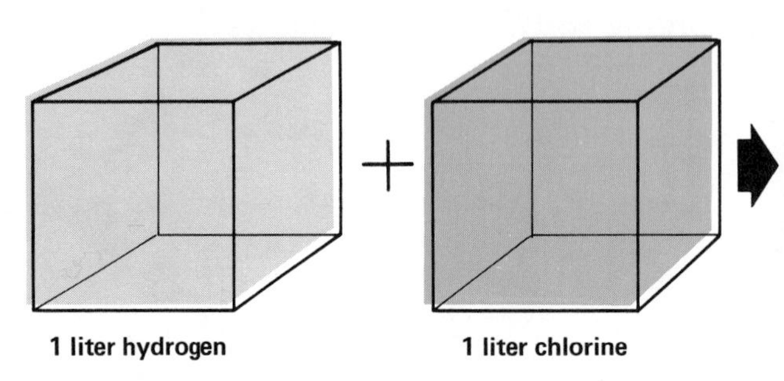

Figure 3–13 Volume ratios in the reaction of hydrogen and chlorine to form hydrogen chloride.

One liter of nitrogen will combine with three liters of hydrogen to form two liters of ammonia (Figure 3–14).

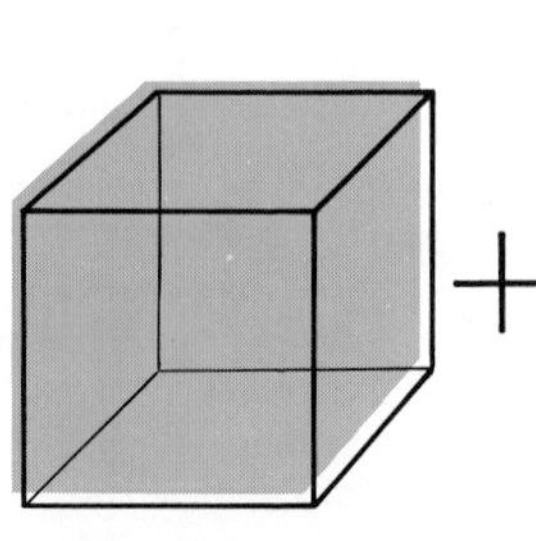

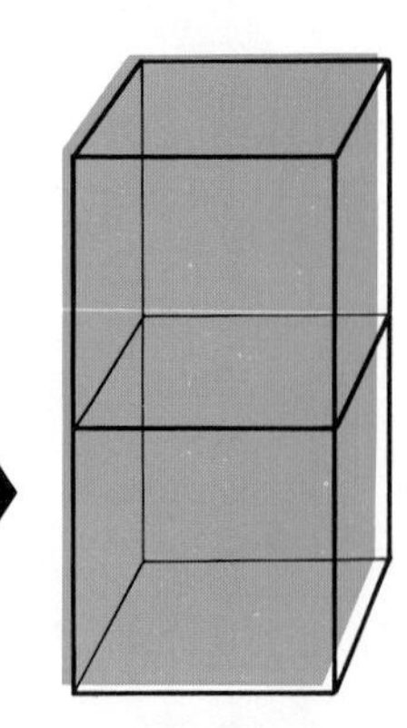

Figure 3–14 Volume ratios in the reaction of nitrogen and hydrogen to form ammonia.

Observations of this type led Gay-Lussac to state the rather simple relationship known as the ***law of combining volumes of gases:***

When measured at constant temperature and pressure, the volumes of gases involved in a chemical reaction are always in a ratio of small whole numbers.

In the previous examples, the ratios of the volumes of gases are 2:1:2 for the formation of steam, 1:1:2 for the formation of hydrogen chloride, and 1:3:2 for the formation of ammonia.

All of the reactants or products do not have to be gases. A certain weight of carbon, a solid, will combine with one liter of oxygen gas to form one liter of carbon dioxide gas (Figure 3–15).

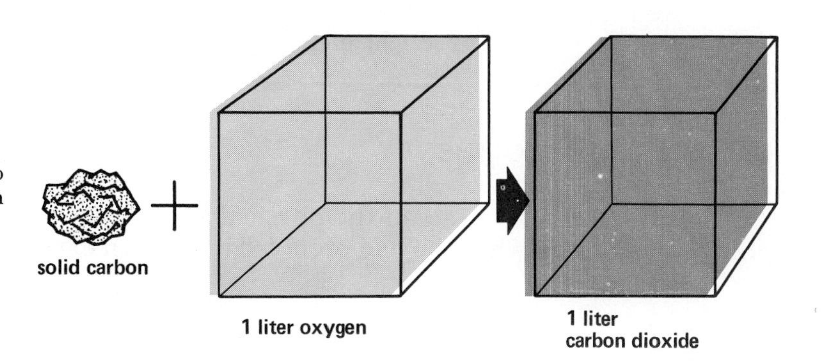

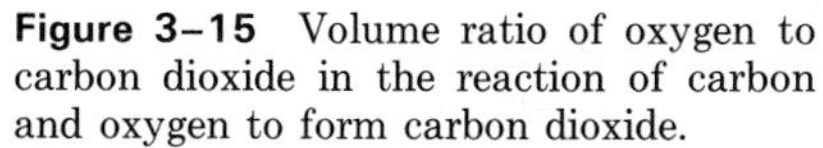

Figure 3–15 Volume ratio of oxygen to carbon dioxide in the reaction of carbon and oxygen to form carbon dioxide.

For the two substances that are gases, the volume ratio is 1:1, in agreement with the law of combining volumes. It should be noted from these examples that the volume of gaseous products in a reaction may be quite different from the volume of gaseous reactants, as volume is not conserved in chemical reactions as is mass.

SELF-TEST 3-B

1. The law of conservation of matter states that matter is neither created nor ______________ in a ______________ reaction.

2. Who discovered the law of conservation of matter? ______________

3. The law of constant composition is really the definition of a ______________.

4. Fifty-three and one-half grams (53.50 g) of silver will combine with 17.75 g of chlorine to form silver chloride. What is the percentage of silver and chlorine in this compound?

weight of Ag 53.50 g

weight of Cl 17.75 g

weight of compound ______ g

$$\% \text{ Ag} = \frac{53.50}{\_\_\_\_} \times 100\% = \_\_\_\_\_ \%$$

$$\% \text{ Cl} = \frac{17.75}{\_\_\_\_} \times 100\% = \_\_\_\_\_ \%$$

5. a. Assume that you are a chemist of many years ago. Your field of study is oxides of nitrogen. There are several you know about. One contains 16 g for every 14 g N, while another contains 32 g O for every 14 g N. Your assistant discovers what he claims is a new oxide which, upon analysis, contains 8 g O for every 14 g N. Has your assistant discovered a new oxide of nitrogen or is it one of the others?______. Explain.

   b. What is the ratio by weight of oxygen in these compounds for a given weight of N? _____ : _____ : _____

DALTON'S ATOMIC THEORY

John Dalton was aware of the physical and chemical laws described previously, some of which he had observed in his own experimental work.

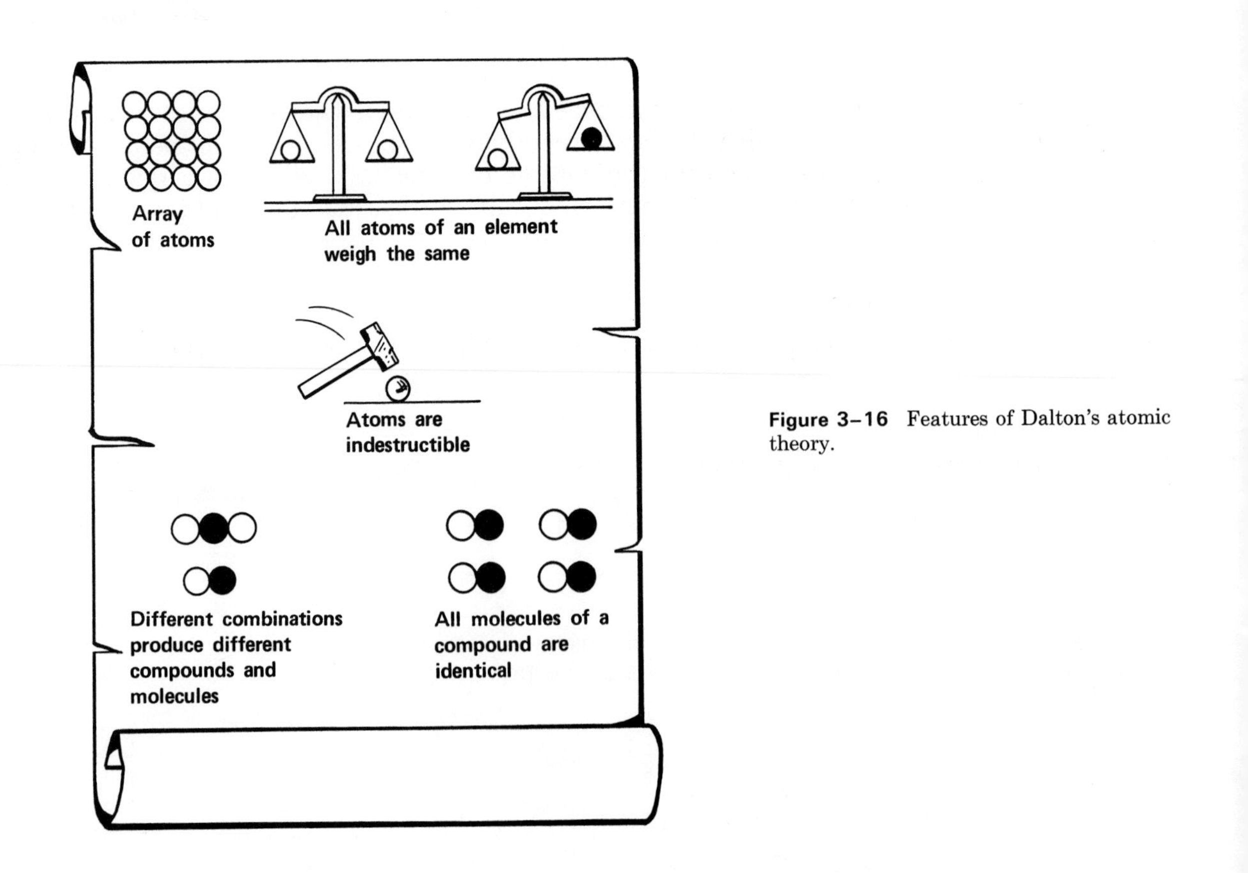

Figure 3–16 Features of Dalton's atomic theory.

Desiring to offer an explanation for these chemical laws, he postulated the following points about the submicroscopic world of atoms and molecules (Figure 3–16). His conclusions about elements and atoms were:

ELEMENTS

Hydrogen	1	Strontian	46
Azote	5	Barytes	68
Carbon	54	Iron	50
Oxygen	7	Zinc	56
Phosphorus	9	Copper	56
Sulphur	13	Lead	90
Magnesia	20	Silver	190
Lime	24	Gold	190
Soda	28	Platina	190
Potash	42	Mercury	167

1. A sample of an element is composed of an array of identical particles called ***atoms.***
2. The atom of each element is different in weight from the atoms of all other elements.
3. An elemental sample can be subdivided only to the point of yielding individual atoms. Further division of the element is impossible.
4. It is impossible to create or destroy an atom of an element.

About compounds and molecules:

1. The ultimate particle of a compound is a ***molecule**** which is made up of one or more atoms of at least two elements.
2. When atoms combine to form molecules of a given compound, they form identical molecules, each having the same ratio of the combining atoms.
3. Two or more kinds of atoms may combine in different ways to form more than one kind of molecule.
4. Since atoms are permanent unchanging bodies, atoms combine in simple numerical ratios, such as one to one, one to two, two to three, in the formation of molecules.
5. The most stable and abundant compound of two elements consists of molecules made up of one atom of each element.

The significantly new concept of the atom as proposed by Dalton is the definite distinguishing weight attached to atoms of a given element. However, in the light of present chemical knowledge and theory, most of the original statements of Dalton require some modification. The most obvious example is the statement made by Dalton that atoms cannot be created or destroyed. It is now well known that atoms of various elements can be fused together to create new elements (fusion), and that certain atoms can be split apart (fission) to produce other elements. However, all of Dalton's statements except the last one concerning compounds and molecules approached more closely the realities of nature than did previous concepts. In fact we now know that many compounds do not consist of simple molecules at all, but of more extensive solid structures. This is precisely the way an understanding of nature grows.

New experiments are suggested, new facts are gathered and recorded, and theoretical concepts are modified in keeping with the new findings.

Indeed, it would be truly surprising if Dalton had been able to define the theoretical atom in such a way that his statements would be in complete agreement with all later discoveries. His claim that the most stable molecules are made of atoms in a 1:1 ratio serves as an excellent example of the frailties of theoretical concepts. Not only are theories limited by insufficient data, but errors in human judgment also are often a significant limitation. For example, because of the great abundance of water, Dalton firmly believed that it was composed of molecules with the simplest possible formula, HO. His reasoning dismissed the possibility that H_2O might be the formula.

*For want of a better term Dalton called these molecules "atoms" but indicated these "atoms" to be compounded of elemental atoms. You should understand that in this presentation the thoughts of early workers are presented in modern terms.

DALTON'S THEORY EXPLAINS THREE OF FOUR CHEMICAL LAWS

The conservation of matter in chemical transformations and the composition of a compound were well established by Lavoisier and others prior to the major contributions of Dalton. Soon thereafter, the law of multiple proportions was clearly stated, based on laboratory experiments. All three of these laws were beautifully explained by Dalton's theory. However, Gay-Lussac's law of combining volumes of gases was at first not even considered by Dalton, nor was it ever explained consistently by him. Indeed, Gay-Lussac's law was in conflict with Dalton's idea that the most stable molecules of compounds containing only two elements were made up of just two atoms, one from each element.

Law of Conservation of Matter

If atoms are indestructible, as Dalton suggested, it follows that the appearance of new substances in chemical changes is simply the result of a new arrangement of the same atoms already present. The total weight of the products would have to be exactly the same as the weight of the reactants, since the very same atoms are involved (Figure 3–17).

If a child used all of his blocks in building first a fort and then a bridge, the two displays would necessarily weigh the same.

Law of Constant Composition

Carbon monoxide is a chemical compound composed of carbon and oxygen. Dalton explained the fact that carbon monoxide always gives the same analysis, 42.9 percent carbon and 57.1 percent oxygen by weight, by stating that a sample of carbon monoxide is simply an array of carbon monoxide molecules each of which is 42.9 percent carbon and 57.1 percent oxygen by weight. Since all carbon monoxide molecules would have the same number of carbon and oxygen atoms, any group of carbon monoxide molecules would have to yield the same analysis as would the single molecule (Figure 3–18).

If a party is made up only of couples, each of which is composed of a boy and a girl, the ratio of boys to girls at the party would have to be 1:1 regardless of the size of the party.

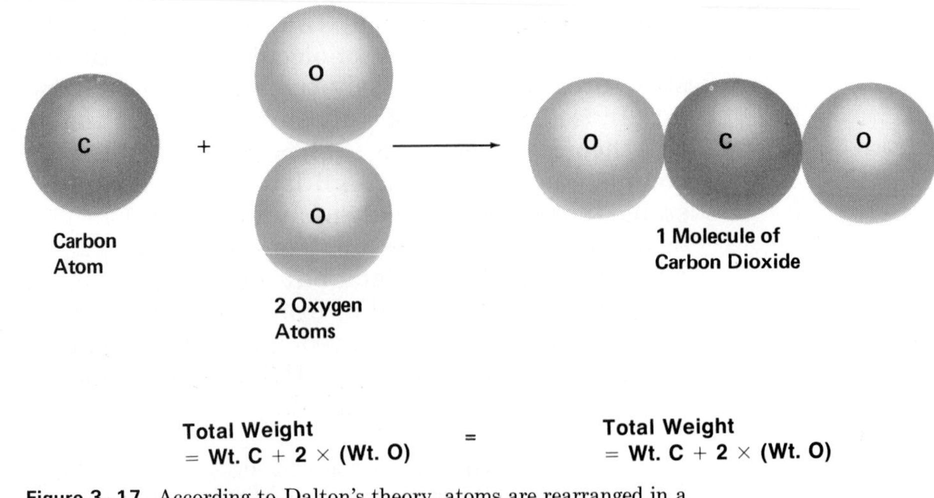

Figure 3–17 According to Dalton's theory, atoms are rearranged in a chemical reaction. Matter is conserved.

Carbon monoxide

One molecule:

$$\%C = \frac{\text{wt. C atom}}{\text{wt. C atom} + \text{wt. O atom}} \times 100\%$$

10 molecules:

$$\%C = \frac{10\ (\text{wt. C atom})}{10\ (\text{wt. C atom}) + 10\ (\text{wt. O atom})} \times 100\%$$

$$\%C = \frac{\text{wt. C atom}}{\text{wt. C atom} + \text{wt. O atom}} \times 100\%$$

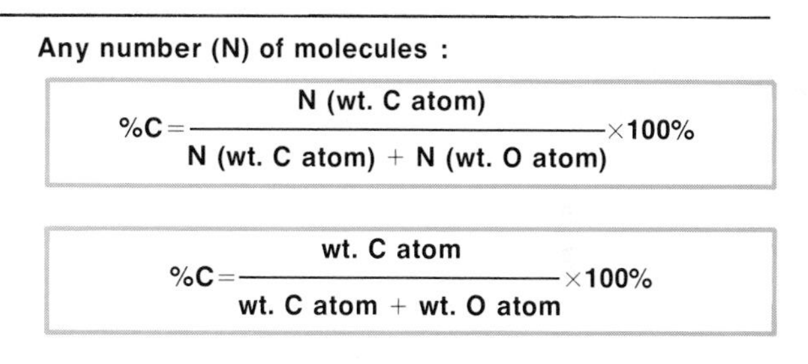

Any number (N) of molecules :
%C = N (wt. C atom) / N (wt. C atom) + N (wt. O atom) × 100%
%C = wt. C atom / wt. C atom + wt. O atom × 100%

Figure 3–18 All molecules of a given compound contain the same numbers and kinds of atoms. According to Dalton, all atoms of a given element have the same weight. Therefore, the percentage by weight of a given kind of atom in a molecule is the same as the percentage by weight of the element in a macroscopic sample of the compound.

Law of Multiple Proportions

Dalton's atomic theory stands as a monumental starting point for the modern concept of the atom.

In the earlier discussion of this law we stated that 0.75 g of carbon combines with 1.00 g of oxygen to form 1.75 g of carbon monoxide and that 0.75 g of carbon combines with 2.00 g of oxygen to form 2.75 g of carbon dioxide. A Daltonian explanation would say this is a simple matter if one considers that a carbon atom, weighing 0.75 unit of weight, combines with an oxygen atom, 1.00 unit of weight, to form a carbon monoxide molecule containing one carbon atom and one oxygen atom, together having a combined weight of 1.75 units. According to Dalton, all oxygen atoms have the same weight, all carbon atoms have the same weight, and whole atoms are combined to form molecules. Therefore, consistent with the units of weight assumed for the carbon monoxide molecule, the 2.75 units for a carbon dioxide molecule would necessarily require two atoms of oxygen, each weighing the same 1.00 unit, and one atom of carbon, weighing 0.75 unit of weight. In order to satisfy the weight requirements of 2.75 units for carbon dioxide and 1.75 units for carbon monoxide, it is obvious that the weight of oxygen in a carbon dioxide molecule, CO_2, is twice (2:1) the weight of oxygen in a carbon monoxide molecule, CO. Since this small whole number ratio holds at the molecular level, one would expect the same ratio to exist between macroscopic samples of these compounds, which are simply collections of large numbers of the two kinds of molecules (Figure 3–19).

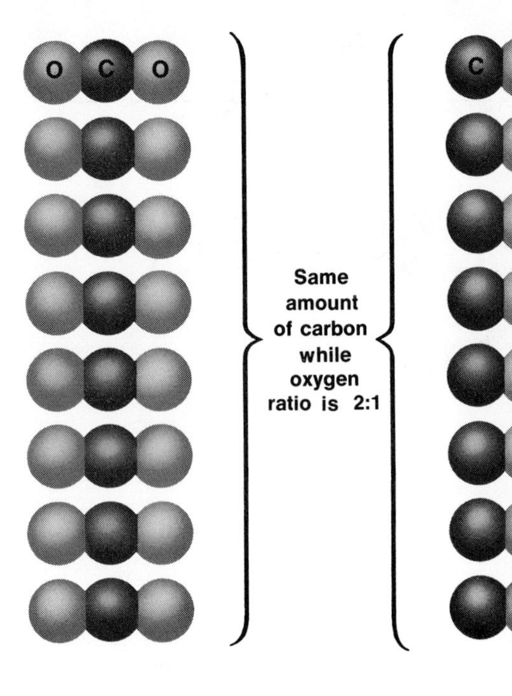

Figure 3–19 Illustration of the law of multiple proportions. Eight molecules of carbon dioxide contain the same number of carbon atoms as do eight molecules of carbon monoxide whereas the number of oxygen atoms is twice as great (2:1) in the carbon dioxide. Thus, we have an atomic explanation for the 2:1 weight relationships that holds on the macroscopic level.

AVOGADRO'S HYPOTHESIS RESOLVES A DIFFICULTY

An empirical law as simple as Gay-Lussac's law of combining volumes called for a simple theoretical explanation, but Dalton's atomic theory was inadequate primarily because it formulated water as HO and ammonia as NH. This discrepancy was so disturbing to Dalton that he even suggested that Gay-Lussac's data were incorrect. "The truth is," Dalton maintained, "that gases do not unite in equal or exact measures in any one instance; when they appear to do so, it is owing to the inaccuracy of our experiments." While other contemporary experimenters proved that Gay-Lussac's law of combining volumes was correct, and Jöns J. Berzelius, a most eminent and influential chemist of the time, wrote Dalton that he needed to alter his thinking on Gay-Lussac's data, Dalton was not convinced and the result was utter confusion.

In 1811, Amedeo Avogadro, an Italian physicist, proposed hypotheses that were capable of resolving the dilemma by adequately explaining Gay-Lussac's law of combining volumes while retaining Dalton's concept of unbreakable atoms. Avogadro's hypothesis can be stated as:

> Equal volumes of all gases under the same conditions of temperature and pressure contain the same number of molecules. Molecules of some elements are diatomic; that is, the molecule is composed of two atoms.

A diatomic molecule is composed of two atoms.

The first hypothesis means that regardless of the size of the molecules of gases, and regardless of the number of atoms per molecule, equal volumes contain equal numbers of molecules. Large molecules have very slightly less free space between them than the space between smaller molecules (Figure 3-20). The second hypothesis means that certain

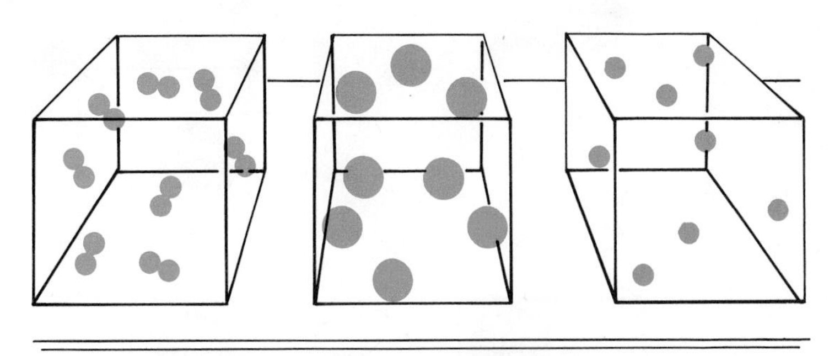

Figure 3–20 Avogadro's hypothesis. Some elements have diatomic molecules. Equal volumes of gases have equal numbers of molecules.

Some elements consist of diatomic molecules. These include
hydrogen, H_2
fluorine, F_2
chlorine, Cl_2
nitrogen, N_2
oxygen, O_2
and others.

elements which are gases, such as hydrogen, chlorine, nitrogen, oxygen, and fluorine, are packaged two atoms per molecule.

What evidence did Avogadro have for these bold assumptions? In essence, he had none. He could not count molecules; neither could he weigh them. It remained for Jean Baptiste Perrin, Robert Andrews Millikan, and other, later experimenters to devise methods to count the number of molecules in a sample of gas, and for Irving Langmuir, over a century later, to show that the hydrogen molecule is composed of two atoms. Although Avogadro did not have experimental evidence to support his hypotheses, he had a very consistent and workable explanation for Gay-Lussac's law of combining volumes. The following examples will illustrate this.

EXAMPLE 1

What are the formulas for hydrogen and chlorine molecules based on the experimental results shown in Figure 3-21? Since the volume ratios are believed to be the same as the molecular ratios (equal volumes contain equal numbers of molecules), the results indicate that one molecule of hydrogen reacts with one molecule of chlorine to form two molecules of hydrogen chloride. Now each of the two molecules of hydrogen chloride must contain at least one atom of hydrogen. This dictates that the molecule of hydrogen must contain at least two atoms, in which case the

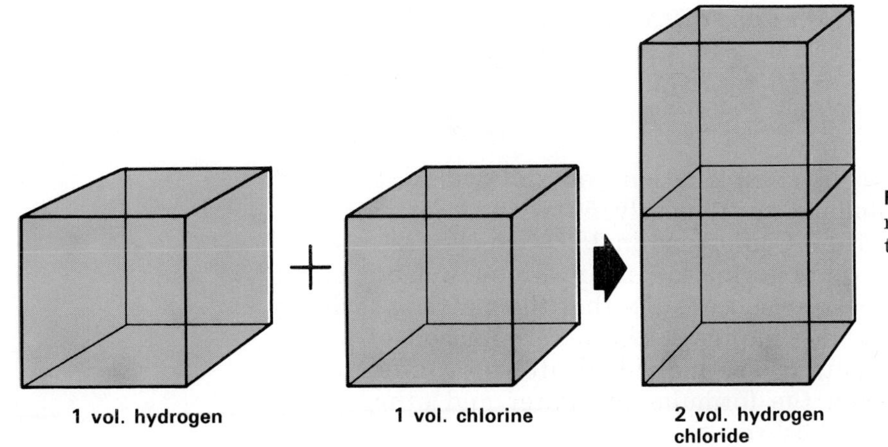

Figure 3–21 Volume ratios in reaction of hydrogen and chlorine to form hydrogen chloride.

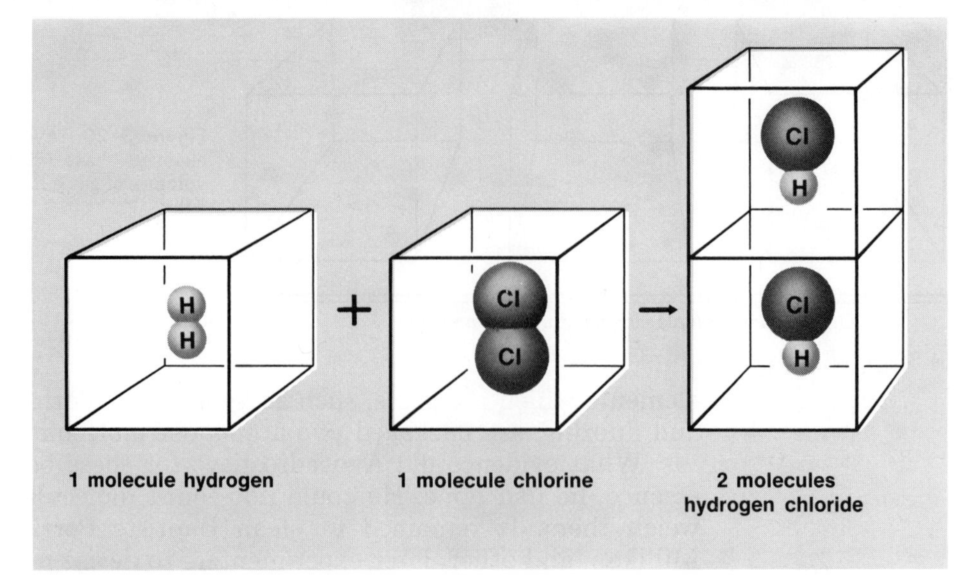

Figure 3–22 Molecular ratios are the same as volume ratios in the reaction of hydrogen and chlorine to form hydrogen chloride.

formula is H_2. Even though later evidence discounts the possibility, one has to admit that the above argument allows for the possibility that hydrogen molecules could be even larger than H_2, perhaps H_4 or H_6, and so forth, but *not* smaller—that is, not H. Believing in the inherent simplicity of nature, Avogadro chose the simplest and presently accepted possibility. The same argument can be made to show that chlorine molecules are diatomic. A molecular representation of the reaction might appear as shown in Figure 3–22.

EXAMPLE 2

Assuming the formula H_2 for hydrogen molecules, what is the formula for water if 2 volumes of hydrogen react with 1 volume of oxygen to produce 2 volumes of steam? (See Figure 3–23.) Two hydrogen molecules (4 atoms) plus one oxygen molecule (2 atoms) yield two water molecules, each of which would have to contain at least one oxygen atom. The simplest way to explain the volume ratios is to assume that the formula of water is H_2O.

Unfortunately, Avogadro's explanation was not widely accepted, because of Dalton's influence, until nearly 50 years later. Stanislao Cannizzaro, at the Karlsruhe Conference in 1860, which was called in an effort to clarify conflicting information about atomic weights, championed Avogadro's hypothesis so successfully that the matter was finally settled. As will be shown in Chapter 4, the determination of atomic weights (the relative weights of the individual atoms of different elements) could proceed when the formulas for water and other simple compounds were firmly established.

Dalton erroneously rejected Avogadro's hypothesis.

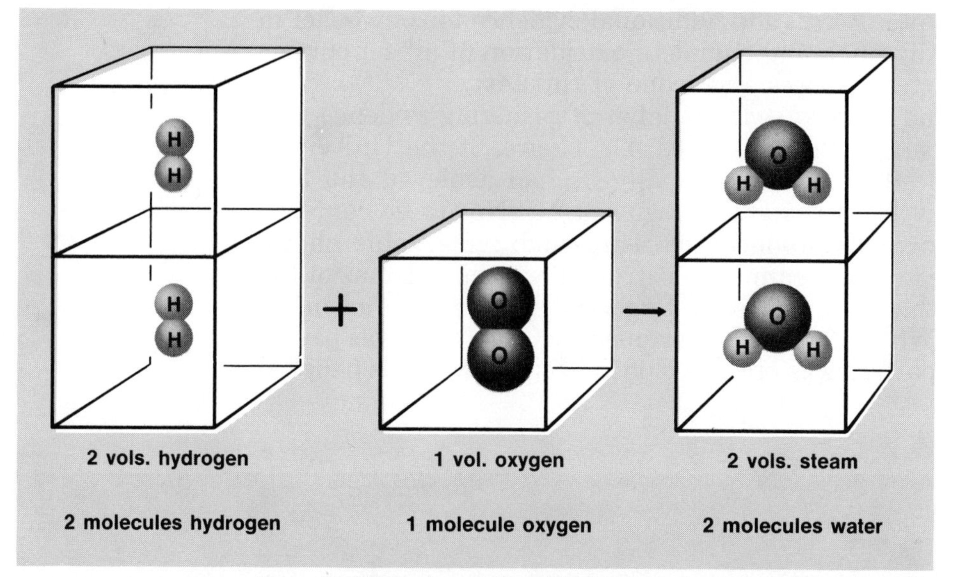

Figure 3–23 In the reaction of hydrogen and oxygen to form steam, the ratio of two volumes of hydrogen to one volume of oxygen to two volumes of steam suggests that two molecules of hydrogen react with one molecule of oxygen to form two molecules of water (steam).

SELF-TEST 3-C

1. According to Dalton's atomic theory, what happens to atoms during a chemical change? Select one.

   a. Atoms are made into new and different kinds of atoms.
   b. Atoms are destroyed.
   c. Atoms are created.
   d. Atoms are rearranged in the way they are combined with each other.

2. According to Dalton's atomic theory, a compound has a definite percent by weight of each element because

   a. all atoms of a given element weigh ____________________.
   b. all molecules of a given compound contain a definite number and kind of ________________.

3. Dalton didn't know it, but fluorine exists as a diatomic molecule, F_2. When fluorine reacts with elemental phosphorus, the volume ratio is 1 liter of phosphorus to 6 liters of fluorine to 4 liters of phosphorus fluoride. What is the formula for

   a. phosphorus vapor ____________________

   b. phosphorus fluoride ________________

MODERN EVIDENCE

Modern evidence in support of the atomic theory is massive and varied. Much of it is too complex for presentation at this point. Where it seems appropriate, brief presentations, such as the one concerning mass spectroscopy in Chapter 6, will add additional evidence for our belief in atoms. However, a thorough and complete elucidation of all supporting evidence is beyond the purposes and scope of this text.

For emphasis and to show that the labor of gathering evidence goes on, the work of Albert V. Crewe is cited. Dr. Crewe, of the University of Chicago, announced in 1970 that finally he had achieved the long sought-for "photograph" of the atom (Figure 3–24). Not to be confused with ordinary photographs produced by light, such remarkable photographs were achieved with a scanning electron microscope. Atoms of the element thorium, which are relatively large and heavy, were adsorbed on a very thin film. When a beam of electrons (see Chapter 6) was passed through this film, a pattern was projected onto a screen, which is believed

Figure 3–24 Chains of thorium atoms separated by an organic molecule. These chains were placed on a thin carbon film a tenth of a millionth of an inch thick. In each chain the smallest white dots represent single thorium atoms. The larger white dots are probably aggregates of a few thorium atoms very close together. (Courtesy of Professor Albert V. Crewe, Department of Physics, Enrico Fermi Institute, University of Chicago.)

to show the actual position of the thorium atoms. Figure 3–24 is an ordinary photograph of the image on the screen.

MATCHING SET

______ 1.	discontinuous matter	a.	law of constant composition
______ 2.	state of matter	b.	Kelvin temperature
______ 3.	experiment	c.	combination of separate substances
______ 4.	PV = constant	d.	atoms
______ 5.	(°C) + 273	e.	law of multiple proportions
______ 6.	conservation	f.	solid
______ 7.	definition of compound	g.	preservation
______ 8.	synthesis	h.	nature proceeding under controlled conditions
______ 9.	Cl_2O, ClO_2, Cl_2O_7	i.	hydrogen
______10.	diatomic molecule	j.	Boyle's law

QUESTIONS

1. What additional kinds of evidence did Dalton have for atoms that the early Greeks (Democritus, Leucippus) did not have?

2. Which gas laws most closely relate to the following effects?

   a. A soft drink fizzes when the top is removed.
   b. A pressure cooker blows up if it is heated too hot and the pressure valve stops up.
   c. If you hold your finger over the end of a bicycle pump, a rapid push on the plunger raises your finger off the outlet.
   d. As a balloon rises into the air, the volume becomes greater.
   e. If a jar is inverted over a burning candle arranged on a wooden block floating on water, the water rises into the bottle.

3. How does Dalton's atomic theory explain:

   a. the law of conservation of matter?
   b. the law of constant composition?
   c. the law of multiple proportions?

4. For the reaction of hydrogen with fluorine, one volume of hydrogen gas combines with one volume of fluorine gas, yielding two volumes of hydrogen fluoride gas.

   a. Use Avogadro's hypotheses to explain these observations.
   b. Why would Dalton's ideas be incapable of explaining these observations?

5. List one specific contribution of each man to atomic theory: Democritus, Boyle, Dalton, Avogrado.

6. What are two ways in which the percentage composition of compounds may be experimentally determined?

7. The laws of gases and the laws of chemical change presented in this chapter are

often referred to as empirical laws. What does "empirical" mean? How does empirical differ from theoretical?

8. Although there may not be a very reliable way to check the conservation of matter in a large explosion of dynamite, what leads us to believe that the law of conservation of matter is obeyed?

9. What is the basic assumption made when it is argued that one molecule of nitrogen reacts with three molecules of hydrogen to form two molecules of ammonia because, in the laboratory, one volume of nitrogen reacts with three volumes of hydrogen to form two volumes of ammonia? The volumes are measured at constant temperature and pressure.

10. A hydrogen atom has a weight of 1.0 unit and an oxygen atom has a weight of 16.0 units. Water is composed of molecules having two atoms of hydrogen and one atom of oxygen per molecule. Hydrogen peroxide, another compound of hydrogen and oxygen, is composed of molecules having two atoms of hydrogen and two atoms of oxygen in each molecule. If these statements can be assumed to be true, how do they illustrate the Law of Multiple Proportions?

PROBLEMS

1. In the chemical reaction, $2Ag_2O \longrightarrow 4Ag + O_2$, 10.7 g of silver oxide (Ag_2O) will decompose into 10.0 g of silver metal (Ag). According to the law of conservation of matter, how much oxygen must be formed?

   Ans. 0.7 g O_2

2. Twelve grams of carbon are burned in a closed container which originally had 80 grams of gaseous oxygen present. Forty-four grams of carbon dioxide are formed. How much oxygen remains unreacted in the container?

   Ans. 48 g O_2

3. In the preparation of sulfur trioxide by the oxidation of sulfur, the following data were obtained in a series of experiments:

	Wt. of sulfur used	Wt. of oxygen used	Wt. of sulfur trioxide obtained
Experiment 1	1.0 g	1.5 g	2.5 g
Experiment 2	2.0	3.0	5.0
Experiment 3	3.0	4.5	7.5

   Determine if these data are in accord with the law of definite composition.

   Ans. Ratio of wt. of O to wt. of S is same for all experiments: 1.5:1

4. Change to degrees Kelvin (K):
   a. 50°C
   b. −30°C
   c. 77°F

   Ans. 323 K
   243 K
   298 K

5. The pressure on 6 liters of gas is 2 atmospheres. What will be the volume (in liters) when the pressure is doubled without changing the temperature?

   Ans. 3 liters

6. The pressure on 2.00 liters of gas is 0.91 atmosphere. If the pressure is raised to 1.45 atmospheres on this sample without changing the temperature, what will the new volume have to be?

   Ans. 1.26 liters

7. What will be the new volume of a 5.00-liter sample of a gas if it is cooled from 80.0°C to 0.0°C at constant pressure?

   Ans. 3.87 liters

8. How much does the volume of a gas change if the temperature is changed from 200 K to 600 K at a constant pressure of 1 atmosphere?

   Ans. increases threefold

9. An experiment in which volumes of gases (all under standard conditions) were allowed to react with each other showed that 3 volumes of gaseous H_2 reacted with 1 volume of a gas X to produce 3 volumes of gaseous H_2X. How many atoms are present in a molecule of gaseous X?

   Ans. 3, X_3

10. The density (weight per unit volume) of carbon dioxide, CO_2, is 1.96 g per liter and the density of ammonia, NH_3, is 0.76 g per liter at 0°C and 1 atmosphere pressure. Hence, the density of carbon dioxide is 1.96/0.76 (or about 2.6 times) that of ammonia. How many times heavier is a carbon dioxide molecule than an ammonia molecule? Explain.

Ans. 2.6

SUGGESTIONS FOR FURTHER READING

"Atomic Theory in the Ancient World," *Chemistry,* Vol. 44, No. 4, p. 17 (1971).
Ihde, A. J., "The Development of Modern Chemistry," Harper and Row, New York, 1964.
Jaffe, B., "Crucibles: The Story of Chemistry," Fawcett World Library, New York, 1957.
Kesselman, B., "The Skeptical Chemist (Robert Boyle)," *Chemistry,* Vol. 39, No. 1, p. 9 (1966).
Lucretius, "The Nature of the Universe," Penguin Books, Inc., New York, 1959.
Patterson, Elizabeth C., "John Dalton and the Atomic Theory," Doubleday, New York, 1970.
Paul, Martin A., "International System of Units (SI)," *Chemistry,* Vol. 45, No. 9, p. 14 (1972).
Spritzer, M. S., "Testing Boyle's Law," *Chemistry,* Vol. 43, No. 9, p. 29 (1970).
Spritzer, M. S., and Markham, J., "Charles' Law: Estimating Absolute Zero," *Chemistry,* Vol. 42, No. 8, p. 24 (1969).
Szabadvary, F., "Great Moments in Chemistry. Part I: A Visit with Antoine Lavoisier," *Chemistry,* Vol. 42, No. 4, p. 14 (1969).
Szabadvary, F., "Great Moments in Chemistry. Part II: From Thales to Bohr," *Chemistry,* Vol. 42, (No. 11, p. 6 (1969).

ATOMIC WEIGHTS AND STOICHIOMETRY

CHAPTER 4

A major contribution of Dalton's atomic theory was the assumption that atoms of a given element have a unique weight. This assumption set into motion an enthusiastic search for the weights of the atoms. However, the ensuing 40 years following the proposal of Dalton's theory brought more confusion over atomic weights than agreement about them. The primary hindrances to accepting a consistent scale of atomic weights were the lack of acceptance of Avogadro's hypothesis (it was practically unnoticed even by the time of his death in 1856), and the confusion between the terms "atom" and "molecule." Dalton used these terms interchangeably and believed that one atom (molecule) of oxygen and one atom (molecule) of hydrogen would combine to form one molecule of water:

Dalton was confused over the difference between atoms and molecules.

$$H + O \longrightarrow HO$$

In contrast, Avogadro, as a result of his understanding of Gay-Lussac's law of combining volumes, believed that oxygen and hydrogen molecules are composed of two atoms each, O_2 and H_2. Furthermore, he believed that one molecule of oxygen reacts with two molecules of hydrogen to form two molecules of water, the molecular ratio being exactly the same as the ratio of combining volumes of gases:

$$2H_2 + O_2 \longrightarrow 2H_2O$$

As will be shown in the following section, Dalton's hypothesis led to the conclusion that an oxygen atom is eight times as heavy as a hydrogen atom, whereas Avogadro's hypothesis required an oxygen atom to be 16 times as heavy. Obviously, such a major point of confusion had to be resolved before an acceptable scale of atomic weights could be generally accepted.

RELATIVE WEIGHTS OF HYDROGEN AND OXYGEN ATOMS—WHY THE DILEMMA?

By 1860 experimental methods had been developed by which the weights of each element in a sample of a compound could be calculated. Two examples were given in Chapter 3. However, to state that water is 88.8 percent oxygen and 11.2 percent hydrogen by weight tells nothing concerning the relative weights of the oxygen and hydrogen atoms. However, if one added to this belief, as did Dalton, that the molecule of water has the formula HO, it would follow that 88.8 percent of the weight of each molecule is due to the oxygen atom and 11.2 percent due to the hydrogen atom. The ratio of weight of the oxygen atom to the hydrogen atom would then be 88.8:11.2. In other words, the oxygen atom would be $\frac{88.8}{11.2}$, or 7.9 times as heavy as the hydrogen atom. Avogadro believed the formula for water was not HO, but H_2O, and the thought process involved is a bit more complex. If two hydrogen atoms are 11.2 percent of the weight of the molecule, one hydrogen atom would weigh one half of 11.2, or 5.6 percent of the total weight of a molecule. Since the one oxygen atom in the molecule is 88.8 percent of the weight, the ratio of the weight of the oxygen atom to the weight of the hydrogen atom would be $\frac{88.8}{5.6}$, or 15.9, the approximate value which is accepted today.

In 1860 it was apparent that if an atomic weight scale were to be firmly established, it would be necessary to know the molecular formulas for compounds as well as to know the percent composition of the compounds. The latter was readily available but the former remained in doubt, in spite of Avogadro's efforts.

DILEMMA RESOLVED—THE KARLSRUHE CONFERENCE

In September, 1860, many of the most brilliant minds in chemistry met in Karlsruhe, Germany, to settle the atomic weight crisis. After several of the 140 chemists present had spoken and debated the distinction between the molecule and the atom, Stanislao Cannizzaro raised his voice to dispute their ideas and to revive clearly the ideas of the dead and forgotten Avogadro. The young orator described how Avogadro's molecules consistently explained Gay-Lussac's law of combining volumes and how the acceptance of molecules of hydrogen composed of two atoms would clarify the atomic weight dilemma. In spite of his arguments the confusion on atomic weights was not resolved during the conference. Fortunately, at the close of the meeting, Angelo Pavesi, a friend of Cannizzaro, distributed copies of a paper written by Cannizzaro two years earlier. After several years the chemical community finally recognized the great contribution made by Avogadro and, with a consistent set of atomic weights, surged forward in the use of atomic theory.

Cannizzaro revived the correct but forgotten ideas of Avogadro.

Stanislao Cannizzaro (1826–1910), Italian chemist whose work resulted in the clarification of the atomic weight scale. (From Ihde, A. J., *The Development of Modern Chemistry,* Harper and Row, New York, 1964.)

It now could be said with general agreement that the oxygen atom weighs approximately 16 times as much as the hydrogen atom. Evidence developed later in this text leaves little room to doubt this assertion. Man's ability to make this statement is truly one of his most notable accomplishments, for he is describing the relative weights of particles so small that it is impossible for him to visualize their size and weight in terms of the world that he experiences with his physical senses.

THE ATOMIC WEIGHT SCALE—AT LAST

Using an approach similar to that described previously for hydrogen and oxygen, we can learn that a carbon atom weighs 12 times as much as a hydrogen atom, nitrogen 14 times as much, chlorine 35.5 times as much, and sulfur 32 times as much. Even though these numbers tell us the relative weights of the atoms, it is important to realize that they are not the actual weights of atoms in any previously known units. If a specific number is assigned as the weight of any particular atom, this immediately fixes relative numbers to the weights of all other atoms. Since no atom has been found to be lighter than the hydrogen atom, a satisfactory system of numbers could be obtained based on the assignment of the number 1 as the weight of the hydrogen atom. This would result in all other atomic weights being greater than 1. On this scale the atomic weight of oxygen would be slightly less than 16, since the ratio of the weight of oxygen to hydrogen is not 16 to 1 but 15.873 to 1. A more popular scale in chemistry for many years placed oxygen at exactly 16.000, which resulted in hydrogen having the value of 1.008.

The lightest atom is the hydrogen atom.

In 1961, a new atomic weight scale was adopted for use in both chemistry and physics based upon the carbon-12 isotope (isotopes are atoms of the same element having different weights; see Chapter 6) having a value of exactly 12 ***amu (atomic mass unit).*** The amu is a unit of mass just as the gram of kilogram is a unit of mass. In fact, one amu is the same as 0.0000000000000000000000016605 gram

The atomic mass unit (amu) is a very small unit of mass. In recent years new experimental techniques have been developed to measure the actual atomic weights of the elements very accurately.

(1.6605×10^{-24} gram). The extremely small size of the amu is convenient for the description of the weights* of atoms just as, in another instance, inches are more convenient than miles for measuring household articles. For example, the weight of one hydrogen atom is 1.67×10^{-24} gram or 1.008 amu. In this case it is obvious that the amu is the more convenient unit to manipulate mathematically.

A complete listing of the atomic weights based on carbon-12 is shown in Table 4–1.

The consequences of atomic weights are far-reaching in the field of chemistry. Indeed, modern chemistry could not have developed without this cornerstone, and the characterization of every new substance depends on the atomic weights of the atoms involved.

SELF-TEST 4-A

1. The letters amu stand for ____________ ____________ ____________.
2. An amu is (1) larger than, (2) smaller than, (3) the same size as a gram.
3. If, as Dalton suggested, oxygen had an atomic weight of 8 amu, what would be the atomic weight of hydrogen?________ Of calcium?________
4. What is the current reference point in atomic weights? ____________ ____________

AVOGADRO'S NUMBER

When the number of grams of an element equal to its atomic weight is taken, this is a ***gram atomic weight*** of the element. The average atomic weight of carbon atoms is 12.011. A quantity of 12.011 grams of carbon is a gram atomic weight of carbon. Similarly, the gram atomic weight of magnesium (Mg) is 24.305 grams.

Names for numbers:
2—pair
12—dozen
144—gross
6.02×10^{23}—mole

The 12.011 grams of carbon is the mass of a certain quantity of carbon atoms, namely ***Avogadro's number*** of atoms. Avogadro's number is $\mathbf{6.02 \times 10^{23}}$—that is, 602,000,000,000,000,000,000,000. Recall that this number was also defined in Chapter 1 as the ***mole.*** Therefore, 12.011 grams of carbon is the mass of 6.02×10^{23} atoms of carbon.

Since the atomic weight of a magnesium atom is about twice that of a carbon atom (24.305 to 12.011), it takes about twice the weight of Mg to have the same number of atoms of Mg as the number of atoms of C in 12.011 grams of carbon. Thus, 24.305 grams of Mg is the mass of exactly the same number of atoms as 12.011 grams of C—one mole (or Avogadro's number) of atoms. In fact, *the gram atomic weight of any element is the mass of 6.02×10^{23} atoms of that element.*

---

*It is a common and accepted error to refer to the weights of atoms when actually we are speaking of atomic masses. See footnote in Table A–1 in Appendix A.

TABLE 4–1 Table of Atomic Weights
(Based on Carbon-12 = 12 amu)*

	Symbol	Atomic No.	Atomic Weight
Actinium	Ac	89	(227)
Aluminum	Al	13	26.98154
Americium	Am	95	(243)
Antimony	Sb	51	121.75
Argon	Ar	18	39.948
Arsenic	As	33	74.9216
Astatine	At	85	(210)
Barium	Ba	56	137.34
Berkelium	Bk	97	(247)
Beryllium	Be	4	9.01218
Bismuth	Bi	83	208.9804
Boron	B	5	10.81
Bromine	Br	35	79.904
Cadmium	Cd	48	112.40
Calcium	Ca	20	40.08
Californium	Cf	98	(251)
Carbon	C	6	12.011
Cerium	Ce	58	140.12
Cesium	Cs	55	132.9055
Chlorine	Cl	17	35.453
Chromium	Cr	24	51.996
Cobalt	Co	27	58.9332
Copper	Cu	29	63.546
Curium	Cm	96	(247)
Dysprosium	Dy	66	162.50
Einsteinium	Es	99	(254)
Erbium	Er	68	167.26
Europium	Eu	63	151.96
Fermium	Fm	100	(253)
Fluorine	F	9	18.99840
Francium	Fr	87	(223)
Gadolinium	Gd	64	157.25
Gallium	Ga	31	69.72
Germanium	Ge	32	72.59
Gold	Au	79	196.9665
Hafnium	Hf	72	178.49
Helium	He	2	4.00260
Holmium	Ho	67	164.9340
Hydrogen	H	1	1.0079
Indium	In	49	114.82
Iodine	I	53	126.9045
Iridium	Ir	77	192.22
Iron	Fe	26	55.847
Krypton	Kr	36	83.80
Lanthanum	La	57	138.9055
Lawrencium	Lr	103	(257)
Lead	Pb	82	207.2
Lithium	Li	3	6.941
Lutetium	Lu	71	174.97
Magnesium	Mg	12	24.305
Manganese	Mn	25	54.9380
Mendelevium	Md	101	(256)
Mercury	Hg	80	200.59
Molybdenum	Mo	42	95.94
Neodymium	Nd	60	144.24
Neon	Ne	10	20.179
Neptunium	Np	93	237.0482
Nickel	Ni	28	58.71
Niobium	Nb	41	92.9064
Nitrogen	N	7	14.0067
Nobelium	No	102	(254)

TABLE 4–1 (cont'd.)

	Symbol	Atomic No.	Atomic Weight
Osmium	Os	76	190.2
Oxygen	O	8	15.9994
Palladium	Pd	46	106.4
Phosphorus	P	15	30.97376
Platinum	Pt	78	195.09
Plutonium	Pu	94	(242)
Polonium	Po	84	(210)
Potassium	K	19	39.098
Praseodymium	Pr	59	140.9077
Promethium	Pm	61	(147)
Protactinium	Pa	91	231.0359
Radium	Ra	88	226.0254
Radon	Rn	86	(222)
Rhenium	Re	75	186.207
Rhodium	Rh	45	102.9055
Rubidium	Rb	37	85.4678
Ruthenium	Ru	44	101.07
Samarium	Sm	62	150.4
Scandium	Sc	21	44.9559
Selenium	Se	34	78.96
Silicon	Si	14	28.086
Silver	Ag	47	107.868
Sodium	Na	11	22.98977
Strontium	Sr	38	87.62
Sulfur	S	16	32.06
Tantalum	Ta	73	180.947
Technetium	Tc	43	(99)
Tellurium	Te	52	127.60
Terbium	Tb	65	158.9254
Thallium	Tl	81	204.37
Thorium	Th	90	232.0381
Thulium	Tm	69	168.9342
Tin	Sn	50	118.69
Titanium	Ti	22	47.90
Tungsten	W	74	183.85
Uranium	U	92	238.029
Vanadium	V	23	50.9414
Xenon	Xe	54	131.30
Ytterbium	Yb	70	173.04
Yttrium	Y	39	88.9059
Zinc	Zn	30	65.38
Zirconium	Zr	40	91.22

*Values in parentheses are estimates and denote, in most cases, the most stable isotopes.

Note: There is general agreement that elements 104, 105, and 106 have been prepared but there is disagreement concerning who prepared them first and what the names should be. See *Chemical and Engineering News,* Sept. 16, 1974, p. 4.

Since Avogadro's time a number of indirect methods have been developed that actually determine the number of atoms in one gram atomic weight of any element. All the methods give the same value, within the range of experimental error. One method, explained in Chapter 6, involves measuring the volume of helium produced in a radioactive decay reaction while counting the actual number of atoms of helium producing this volume.

Avogadro's number is a very large number. Compared with the number of atoms in 12.011 grams of carbon (one gram atomic weight of carbon), the earth's population, about four billion (4×10^9) people, is

Figure 4–1 Modern chemical theory tells us that water is made up of molecules and that there are a large number of molecules in a very small volume of water. In fact, 18 grams of water or one mole of water contains 6.02×10^{23} molecules. In order to dramatize the smallness of the molecule, consider this fact. It would take 126,000 years for one mole of water drops to pass over Niagara Falls at a flow rate of 120,000,000 gallons/minute (Courtesy of the Power Authority of the State of New York.)

exceedingly small. Indeed, it would require 150 trillion planets with the earth's population to have a total human population of 6×10^{23}. Since the mole is a number, it is perfectly consistent to speak of moles (or Avogadro's number) of atoms, moles of groups of atoms (molecules), and moles of subatomic particles.

HOW ARE FORMULAS DERIVED FROM ATOMIC WEIGHTS?

Once the atomic weight scale was firmly established, it was then possible to determine the simplest formula for any compound, provided the elemental percent analysis had been determined. Obviously, this is of major importance in the study of new compounds. To illustrate, let us determine the formula of silica, the principal compound in ordinary sand. Quantitative analyses of silica show it to have 46.7 percent silicon and 53.3 percent oxygen by weight. The percent analysis tells us the ratio of the weight of the two elements in the compound; for each 100 grams of silica, we expect to find 46.7 grams of silicon and 53.3 grams of oxygen. Other units of weight would do as well; 46.7 amu of silicon

The formula of a compound can be derived from its percentage composition and an accurate table of atomic weights.

would combine with 53.3 amu of oxygen. Since the oxygen and silicon atoms do not have the same weight, we will have to consider this fact when determining the atomic ratio (for the chemical formula of the compound) from the weight ratio (expressed by the percent composition). This can be done as follows:

1. Divide the weight of each element per 100 amu of compound by the corresponding relative atomic weight to obtain a ratio of atoms in the molecule.

$$\text{Si: } 46.7\,\cancel{\text{amu Si}} \times \frac{1 \text{ atom Si}}{28.09\,\cancel{\text{amu Si}}} = 1.66 \text{ atoms Si}$$

$$\text{O: } 53.3\,\cancel{\text{amu O}} \times \frac{1 \text{ atom O}}{16.0\,\cancel{\text{amu O}}} = 3.34 \text{ atoms O}$$

Note that the ratio of 46.7 to 53.3 is a weight ratio of silicon to oxygen in silica, whereas the ratio of 1.66 to 3.34 refers to the number of silicon atoms to oxygen atoms. Since the experimental error in determining percentages by weight of elements in compounds is generally less than one percent error, we are justified in rounding off to whole numbers only when the numbers are, for example, 1.01, 2.02, 3.03. Therefore, the numbers 1.66 and 3.34 cannot be rounded off to whole numbers in order to determine a formula. Instead the ratio of 1.66 to 3.34 must be changed to its equivalent form in whole numbers.

2. Convert to a whole number ratio by:
   a. dividing each number in the ratio by the smaller number, and
   b. (if necessary) converting the resulting decimal fractions to whole numbers.

$$\frac{1.66 \text{ atoms Si}}{1.66} = 1 \text{ atom Si}$$

$$\frac{3.34 \text{ atoms O}}{1.66} \cong 2 \text{ atoms O}$$

3. Write the formula.

$$SiO_2$$

In summary, divide the percentage for each element by the respective atomic weight. The resulting numbers are proportional to the number of atoms involved and a simple whole number ratio can be obtained by multiplication and/or division by a constant factor.

The simplest empirical formula shows the smallest whole number ratio of atoms.

If the ratio of atoms is reduced to the smallest whole number ratio possible, the simplest formula (the ***empirical formula***) for the compound is thus obtained. It should be carefully understood that the simplest formula is not necessarily the molecular formula. The simplest formula tells only the atom ratios. It does not tell how many of each kind of atom are in the molecule (molecular formula), or even whether or not molecules exist at all (there are none in SiO_2). While it is often the case that the simplest formula is the molecular formula (H_2O, HCl, CO, CO_2, NO), there are numerous examples in which this is not the case. Hydrogen peroxide has the simplest formula, HO, but there is ample evidence to show that its molecules are composed of two atoms of hydrogen and two atoms of oxygen. Its molecular formula is H_2O_2. Likewise, benzene is C_6H_6 (not CH), and ethane is C_2H_6 (not CH_3).

FORMULAS AND GRAM FORMULA WEIGHTS

The relative ***molecular weights*** (gram formula weight*) for the species in Table 4–2 can be obtained simply by adding up the relative weights of the atoms in the molecules. For example, the relative molecular weight of the ammonia molecule, NH_3, is the sum of the weights of one nitrogen atom and three hydrogen atoms (14 amu + 3 × 1 amu = 17 amu).

One molecule of ammonia is a submicroscopic particle which has a weight of 17 amu. The much larger quantity of 17 grams of ammonia is very real to man's senses and is a convenient amount of ammonia for small-scale laboratory experiments.

Formula weight of H_2S:
2H: 1 amu × 2 = 2 amu
S: 32 amu × 1 = 32 amu
TOTAL = 34 amu

The gram formula weight of an element or a compound is obtained by taking the numerical value of the relative formula weight and attaching the unit of grams to it.

A very interesting relationship develops when a study of the volumes of gram formula weights of different gases is made. Two grams of hydrogen occupy 22.4 liters when measured at ***0°C and 1 atm pressure*** (standard temperature and pressure or ***STP***), and 71 grams of chlorine occupy 22.4 liters at these same conditions. Furthermore, the gram formula weights of most gases occupy this same volume, 22.4 liters, at standard conditions. This volume is the ***gram molecular volume*** (Figure 4–2).

The concept of the gram molecular volume gives us an easy method for determining the molecular weight of a new gaseous substance. We simply find out how much 22.4 liters of the gas weighs at standard conditions. For example, natural gas contains several gaseous compounds, one

*If a chemical substance does not exist as molecules, it does not have a molecular weight, but rather a gram formula weight, which is a more general term.

TABLE 4–2 Formula and Gram Formula Weights for Some Elements and Compounds

Substance	Molecular Formula	Formula Weight	Gram Formula Weight
Hydrogen	H_2	2 amu	2 g
Chlorine	Cl_2	71 amu	71 g
Hydrogen chloride	HCl	36.5 amu	36.5 g
Nitrogen	N_2	28 amu	28 g
Ammonia	NH_3	17 amu	17 g
Oxygen	O_2	32 amu	32 g
Nitric oxide	NO	30 amu	30 g
Water	H_2O	18 amu	18 g

Figure 4–2 Gram molecular volume. *STP = standard temperature and pressure (0°C and 1 atmosphere pressure).

of which is ethane. Percent analysis and relative atomic weights indicate the simplest formula for ethane is CH_3. But the formula could be CH_3 or C_2H_6 or C_3H_9, etc., and we would still have the same ratio of carbon to hydrogen. However, since 22.4 liters of ethane has a weight of 30 grams, the molecular weight must be 30 amu, and the molecular formula must be C_2H_6[(2 × 12 amu) + (6 × 1 amu) = 30 amu].

SELF-TEST 4-B

1. The name for the number, 6.02×10^{23}, is ________ or ________ ________.

2. Complete the table:

Substance	Gram Formula Weight	Number of Units
K	39.1 g	________
Kr	________	1 mole of atoms
SO_2	64.1 g	________
NO_2	________	6.02×10^{23} molecules

3. What is the formula weight of the sugar, dextrose (alias glucose) $C_6H_{12}O_6$?

C: 12 amu × ________ = ________ amu

H: 1 amu × ________ = ________ amu

O: 16 amu × ________ = ________ amu

____

Formula Weight = ________ amu

4. How much does a mole of Ne atoms weigh? ________________

5. An acid component of vinegar has the composition: 40.0% carbon, 6.7% hydrogen, and 53.5% oxygen.

   a. What is the empirical (or simplest) formula of the acid?

$$\text{C}\ \frac{40.0\ \cancel{\text{amu}}}{\_\_\_\ \cancel{\text{amu}}/\text{atom C}} = \_\_\_\_\_\ \text{atoms C}$$

$$\text{H}\ \frac{6.7\ \cancel{\text{amu}}}{\_\_\_\ \cancel{\text{amu}}/\text{atom H}} = \_\_\_\_\_\ \text{atoms H}$$

$$\text{O}\ \frac{53.3\ \cancel{\text{amu}}}{\_\_\_\ \cancel{\text{amu}}/\text{atom O}} = \_\_\_\_\_\ \text{atoms O}$$

   This is the same as the whole number ratio of ________ C, to ________ H, to ________ O.

   b. What is the molecular formula of the acid if its molecular weight is 60 amu? ________________

6. What temperature and pressure are represented by the letters STP?

   Temperature: ________ Pressure: ________

7. How many liters would 16 grams of methane gas (formula: CH_4) occupy at STP? ________ liters

---

STOICHIOMETRY

The word ***stoichiometry*** comes from the Greek *stoicheion* (element) and literally means the measurement of amounts of elements. In its chemical usage, stoichiometry refers to the weights of elements and compounds that are consumed or produced in a chemical reaction. For example, an important stoichiometric question is: How much aluminum can be produced from one ton of aluminum oxide? Once the atomic weight scale was established, such important calculations could readily be accomplished. The stoichiometric problems presented in the following paragraphs involve simple number relationships and illustrate the considerations involved in such calculations. More challenging problems are presented in Appendix D.

Since the law of conservation of matter states that matter is neither lost nor gained in a chemical reaction, the weight of the reactants in a chemical reaction must be the same as the weight of the products. For example, according to the law of conservation of matter, when hydrogen reacts with chlorine to form hydrogen chloride, the weight of the reactants, hydrogen and chlorine, used in the reaction must equal the weight of the product, hydrogen chloride. The reaction can be written more concisely by using the symbolism of a chemical equation.

$$H_2 + Cl_2 \longrightarrow 2HCl$$

Accounting for atoms requires balancing an equation. To balance a chemical equation:

1. Place numbers (coefficients) only *before* formulae. ($2H_2$ means 2 molecules of H_2 *and* $2 \times 2 = 4$ atoms of H.)
2. Have same number of each kind of atom on each side of the arrow ($\longrightarrow$).
3. Do not change subscripts. Each symbol represents one atom. H_2 means a molecule composed of two atoms.

Note that a coefficient, 2, has been placed in front of the hydrogen chloride so there will be 2 atoms of hydrogen and 2 atoms of chlorine represented in both the reactants and the products; that is, none will be gained or lost. What is the meaning of the symbolism of the equation? Here are two alternative but equally meaningful ways to express its significance:

> 1 molecule of hydrogen reacts with 1 molecule of chlorine to form 2 molecules of hydrogen chloride;

or,

> 1 mole of hydrogen molecules reacts with 1 mole of chlorine molecules to form 2 moles of hydrogen chloride molecules.

Once the equation is balanced, the relative number of moles for each substance involved is given by the respective coefficients. Furthermore, since 1 mole weighs 1 gram formula weight, a set of weights for all substances involved in the reaction can easily by obtained by adding up the atomic weights in each formula.

These facts allow several types of calculations to be made for chemical reactions. The following examples will illustrate some of these.

Ammonia is used as a crop fertilizer.

EXAMPLE 1

How many moles of nitrogen (N_2) are required to react with 6 moles of hydrogen (H_2) in the formation of ammonia (NH_3)?

a. Write and balance the equation:

$$N_2 + 3H_2 \longrightarrow 2NH_3$$

b. Since 1 mole of N_2 reacts with 3 moles of H_2, how many moles of N_2 will react with 6 moles of H_2? The answer is 2 moles of N_2. If the number of moles of H_2 is doubled, then the number of moles of N_2 must be doubled to keep the same ratio of nitrogen and hydrogen that react with each other.

EXAMPLE 2

How many moles of nitrogen dioxide (NO_2) will be produced by 4 moles of oxygen reacting with sufficient nitric oxide (NO)?

$$NO + O_2 \longrightarrow NO_2$$

a. Balance the equation.

$$2NO + O_2 \longrightarrow 2NO_2$$

Nitrogen dioxide is a major air pollutant.

b. From the balanced equation we see that the number of moles of NO_2 produced is twice that of the oxygen reacting. Therefore 4 moles of oxygen would produce 8 moles of NO_2.

Carbon monoxide is also a major air pollutant in areas where there are many automobiles.

EXAMPLE 3

How many grams of carbon monoxide (CO) can be produced by burning 2650 grams of gasoline (C_8H_{18}), assuming the gasoline burns according to the following reaction?

$$C_8H_{18} + O_2 \longrightarrow CO + H_2O$$

a. Balance the equation.

$$2C_8H_{18} + 17O_2 \longrightarrow 16CO + 18H_2O$$

b. The balanced equation states that 2 moles of gasoline produces 16 moles of CO. Since the molecular weight of C_8H_{18} is 114, i.e., $(8 \times 12) + (18 \times 1)$, 2 moles would weigh 228 grams. Since the molecular weight of CO is 28, 16 moles would weigh $16 \times 28 = 448$ grams. Thus, 228 grams of gasoline would produce 448 grams of CO, or the weight of CO produced is about twice the weight of gasoline burned. This means our 2650 grams of gasoline should produce about 5300 grams of CO. To be more exact:

$$\text{grams of CO} = 2650\ \cancel{\text{g } C_8H_{18}} \times \frac{448 \text{ g CO}}{228\ \cancel{\text{g } C_8H_{18}}}$$

$$= 5210 \text{ g CO}$$

EXAMPLE 4

What volume of hydrogen gas would be required to react with 10 liters of oxygen gas at the same temperature and pressure in the formation of water?

$$H_2 + O_2 \longrightarrow H_2O$$

a. Balance the equation.

$$2H_2 + O_2 \longrightarrow 2H_2O$$

b. Avogadro's hypothesis tells us that the volume ratios are the same as the molecular-number ratios if pressure and temperature are held constant. Since 2 molecules of hydrogen react with one molecule of oxygen, it follows that 2 liters of hydrogen react with 1 liter of oxygen and 20 liters of hydrogen are required to react with 10 liters of oxygen.

After working through these examples, you should find it obvious that a consistent set of atomic weights, backed up by the law of conservation of matter, makes possible the calculation of amounts used and produced in a chemical reaction. Obviously, such calculations are much easier and less expensive than actually doing the experiments to gain such information. Several other applications of stoichiometry are shown in Appendix D.

MATCHING SET

______	1. Avogadro's hypothesis	a.	percentage composition and atomic weights
______	2. 6.02×10^{23}	b.	18 grams
______	3. basis for formula ratio	c.	20
______	4. gram formula weight of P_2O_5	d.	10
______	5. weight of one mole of H_2O molecules	e.	one mole
______	6. conserved during chemical reaction	f.	hydrogen
______	7. number of atoms in the formula: $(NH_4)_3PO_4$	g.	isotope of carbon
______	8. number of gram formula weights of H_2 in 20 grams of hydrogen	h.	gas volume is proportional to number of molecules
______	9. atom with smallest mass	i.	142 grams
______	10. reference point in atomic weights	j.	mass

QUESTIONS

1. Suppose two kinds of corn (A and B) have seeds that are essentially the same size, and that they pack equally well in a bushel container. If a bushel of seeds of corn A weighs only ¾ as much as a bushel of seeds of corn B, what can be said about the relative weights of a typical seed of each kind of corn? How is this analogous to the determination of the molecular weights of gases?

2. Why is it necessary to balance a chemical equation before it can serve as the basis of a stoichiometric calculation?

3. How is it possible for 1 mole of the larger propane molecules (C_3H_8, molecular weight 44) to occupy the same space as 1 mole of the much smaller hydrogen molecules (H_2, molecular weight 2) at STP? Both substances are gases at the same temperature and pressure.

4. Cite other pivotal moments in United States and world history similar to the Karlsruhe Conference in which one man came forth and solved a dilemma that had baffled mankind and stifled progress for many years.

5. Complete the following table for the reaction indicated. Be consistent with the one value given.

	NaH	+ H_2O	⟶	$NaOH$	+ H_2
Moles:	4 moles	+ ______	⟶	______	+ ______
Grams:	24 grams	+ ______	⟶	______	+ ______
Atoms:	1.02×10^{23} atoms	+ ______	⟶	______	+ ______

6. If one mole of boron atoms weighs 10.811 grams, by what number would you divide to obtain the weight of a single boron atom?

7. Suppose you wish to establish a new atomic weight scale and you decide the atomic weight of oxygen should be 1.00. Which atomic weight is consistent with your new scale? Why? (a) H, atomic weight: 1; (b) Ca, atomic weight: 2.5; (c) S, atomic weight: 64; (d) Sr, atomic weight: 4.476.

PROBLEMS

1. What are the molecular weights (formula weights) of CH_4 (methane), $C_{12}H_{22}O_{11}$ (sucrose), $Fe_2(SO_4)_3$ (ferric sulfate), $(NH_4)_3PO_4$ (ammonium phosphate)?

   Ans. 16, 342, 400, 149

2. Determine the correct simplest formulas for compounds with the following percentages by weight:

   a. 64.86% C, 13.51% H, and 21.63% O.
   b. 27.48% Mg, 23.66% P, and 48.85% O.
   c. 38.79% Cl and 61.21% O.
   d. 22.9% Na, 21.5% B, and 55.7% O.

   Ans. a. $C_4H_{10}O$ b. $Mg_3P_2O_8$ or $Mg_3(PO_4)_2$ c. Cl_2O_7 d. $Na_2B_4O_7$

3. How many atomic mass units are there in one gram?

   Ans. 6.02×10^{23} amu/gram

4. Use the conversion unit: 1 amu/1.6605×10^{-24} g and calculate the weight in grams of a C atom (atomic weight 12.011 amu).

   Ans. 19.93×10^{-24} g/C atom (or 1.99×10^{-23})

5. An important source of hydrogen, used as a rocket fuel, is the decomposition of water by electrical energy. The reaction is:

$$H_2O \xrightarrow{\text{Electrical energy}} H_2 + O_2$$

   a. Balance the equation.
   b. What weight of water is necessary to produce 2 grams of hydrogen?
   c. How many grams of oxygen would be produced as a by-product?
   d. How much water would be necessary to produce 2 tons of hydrogen?

   Ans. a. $2H_2O \longrightarrow 2H_2 + O_2$ b. 18 g c. 16 g d. 18 tons

NOTE: Problems 6 to 11 are somewhat more difficult than those above.

6. A major source of sulfur dioxide (SO_2) in the air is burning coal and oil that contain sulfur. If a shipment of coal contains 3% sulfur by weight, what weight of SO_2 is available to pollute the air if one ton of this coal is burned? The sulfur burns according to the following equation.

   Ans. 120 lbs. SO_2

$$S + O_2 \longrightarrow SO_2$$

7. The formula for vitamin C is $C_6H_8O_6$. What is the percentage of carbon in this compound?

   Ans. 41% C

8. What is the percent of carbon in morphine, $C_{17}H_{19}NO_3 \cdot H_2O$?

   Ans. 67.3% C

9. Naturally occurring carbon, as in soot, is 98.89 percent carbon atoms with atomic weight of 12.00000 amu and 1.11 percent carbon atoms with atomic weight of 13.00335 amu. Calculate the average atomic weight of naturally occurring carbon. Compare your value with the one given in Table 4–1.

   Ans. 12.0111

10. The Atlantic Ocean covers 16% of the 196,950,000 square miles of the earth's surface. The average depth of the ocean is about 12,900 feet. Is the volume of the water in the Atlantic Ocean more or less than the volume of 1 mole of sugar cubes that are 1 centimeter on each edge?

   Ans. Less. Atlantic Ocean: 3×10^{23} cm^3 sugar cubes: 6×10^{23} cm^3

11. If 11 liters of gas A at 25°C and 1 atmosphere pressure weighs 2.00 grams and 11 liters of gas B at 25°C and 1 atmosphere weighs 18.00 grams,
    a. What is the relative weight of molecule A to molecule B?
    b. If the atomic weight of A is assumed to be 6, what is the atomic weight of B on the same scale?

Ans. a. 1:9 b. 54

SUGGESTIONS FOR FURTHER READING

Coward, H. F., "John Dalton (b. 1766, d. 1844)," *Journal of Chemical Education,* Vol. 4, p. 22 (1927).

Douville, J. A., "Determining the Molar Volume: A Safe Method," *Chemistry,* Vol. 43, No. 1, p. 25 (1970).

Guggenheim, E. A., "The Mole and Related Quantities," *Journal of Chemical Education,* Vol. 38, p. 86 (1961).

Hawthorne, R. M., Jr., "Avogadro's Number: Early Values of Loschmidt and Others," *Journal of Chemical Education,* Vol. 47, p. 751 (1970).

"Hydrogen Trioxide," *Chemistry,* Vol. 43, No. 6, p. 20 (1970).

Ihde, A. J., "The Karlsruhe Congress: A Centennial Retrospect," *Journal of Chemical Education,* Vol. 38, p. 83 (1961).

Kieffer, W. F., *The Mole Concept in Chemistry,* Reinhold Publishing, New York, 1964.

Labbauf, A., "The Carbon-12 Scale of Atomic Masses," *Journal of Chemical Education,* Vol. 39, p. 282 (1962).

MacNevin, W. M., "Berzelius—Pioneer Atomic Weight Chemist," *Journal of Chemical Education,* Vol. 31, p. 207 (1954).

Moynihan, C. T., and H. Goldwhite, "Determining Avogadro's Number from the Volume of a Monolayer," *Journal of Chemical Education,* Vol. 46, p. 779 (1969).

Sunier, A. A., "Some Methods of Determining Avogadro's Number," *Journal of Chemical Education,* Vol. 6, p. 299 (1929).

Wichers, E., "Report of the International Commission on Atomic Weights—1961," *Journal of the American Chemical Society,* Vol. 84, p. 4175 (1962).

Zink, G. E., "Aspects of Nonstoichiometry," *Chemistry,* Vol. 41, No. 11, p. 13 (1968).

ENERGETICS OF PARTICLES AND PURE SUBSTANCES

CHAPTER 5

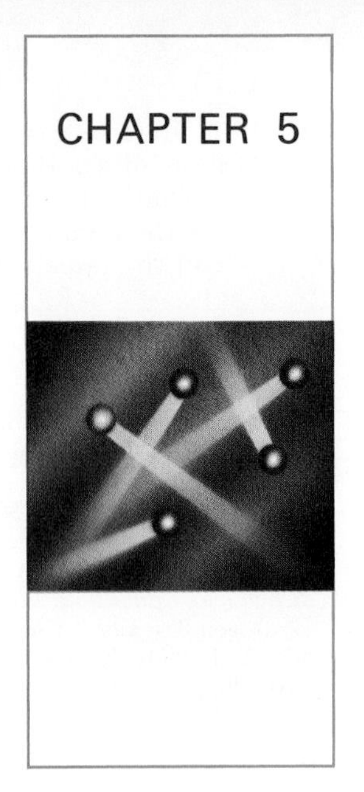

For many years it has been stated that matter "occupies space and has weight" and that energy is the "ability to do work." Since 1945 it has been widely recognized that matter and energy are actually two forms of the same reality that are interchangeable according to Einstein's famous equation, $E = mc^2$. No one needs to be told that energy transfer occurs when matter is changed, because we all have felt the warmth of the burning fire, the cold feeling when water evaporates from our skin, and many other indications that a storage or a depletion of energy is associated with changes in matter. The charged battery, for example, has a measure of energy that the same battery does not have when "dead," and, similarly, the water above the electrical generator of a dam has a measure of energy that the same water does not have when it is below the dam.

In understanding the properties of substances and the changes that transform one substance into another kind of substance (or into another form of the same substance), we must take into account many associated energy effects. The modern chemist seeks to explain macroscopic changes in matter and associated energy changes in terms of structural and energy changes in submicroscopic matter. Since submicroscopic particles have not been directly observed, the most obvious way to describe the interactions of these particles is to treat them analogously to macroscopic matter which we can observe and measure.

Macroscopic energetics can be explained by particle energetics.

In this chapter we will develop an understanding of two important kinds of energy which objects can have. These are kinetic and potential energy. These energy concepts are fundamental to the kinetic molecular theory, which helps to explain the behavior of atoms and molecules.

KINETIC ENERGY

Matter in motion has kinetic energy.

Our everyday experiences offer ample evidence that there is energy possessed by a moving piece of matter which is not possessed by matter at rest. ***Kinetic energy results from matter being in motion.*** While the motion is necessary, the mass of the matter is an important part of the kinetic energy of the object. A table tennis ball and a golf ball traveling at the same speed could have decidedly different effects on a plate glass window.

Elastic collisions conserve kinetic energy.

The batted ball illustrates, as do many other ordinary experiences, that the kinetic energy of one object can be transferred to another object through a collision. A careful study of such collisions indicates that in the absence of friction and other energy losses, the total kinetic energy of the colliding bodies is the same after the collision as it was before. This describes a perfectly elastic collision (Figure 5-1). As early as the 17th century, it was recognized that when two bodies collide in an elastic collision, the sum total of mass times the velocity squared values (mv^2) for the bodies is the same after the collision as it was before, although each body may be moving with a new velocity. Later the term *kinetic energy* was introduced for the term, $\frac{mv^2}{2}$. Thus, the kinetic energy for any particle in motion is given by the equation

$$\text{kinetic energy} = \frac{mv^2}{2}$$

Inelastic collisions convert kinetic energy into heat.

Actually there are no perfectly elastic collisions in the macroscopic world even though this ideal can be closely approached. The concept is introduced at this point for two reasons: first, the loss of some kinetic energy in inelastic collisions and the resulting change in the matter, such as a change in temperature or a change in structure, indicate that matter can contain energy in ways other than by motion; and second, the idea of the perfectly elastic collision, with absolutely no loss in kinetic energy, is assumed at the molecular level in the theoretical discussion of the structure of matter.

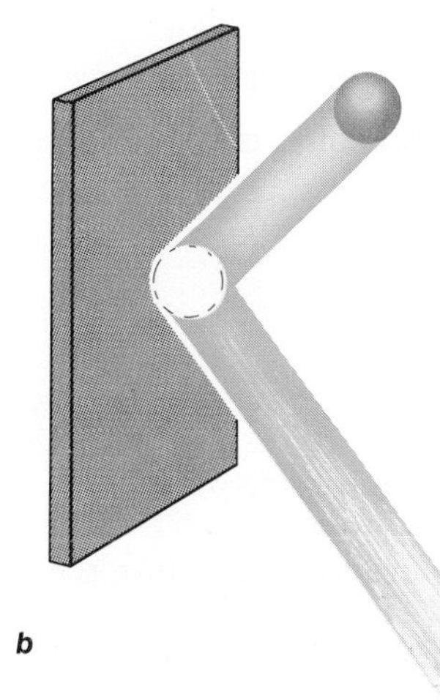

Figure 5–1 A diagrammatic representation of perfectly elastic collisions (*a*) between two similar bodies and (*b*) between a small body and a stationary object. Kinetic energy is conserved in both instances.

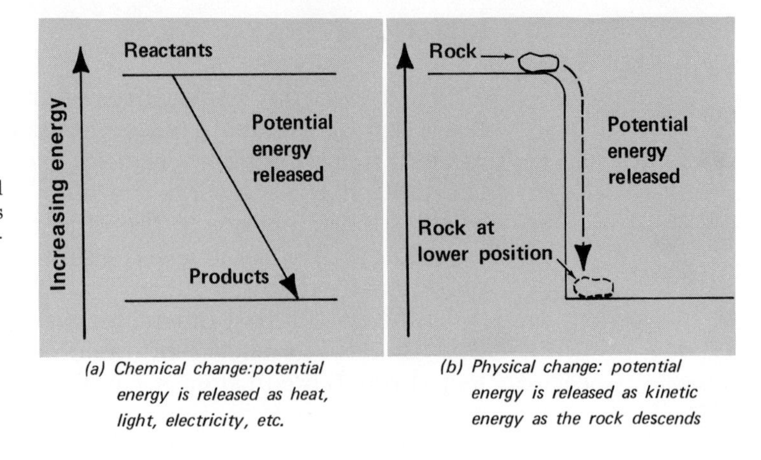

(a) Chemical change: potential energy is released as heat, light, electricity, etc.

(b) Physical change: potential energy is released as kinetic energy as the rock descends

Figure 5–2 The similarity between chemical and physical changes and the energy involved is illustrated in two examples. What happens, energetically, when these processes are reversed?

POTENTIAL ENERGY

Potential energy can be stored in the composition and structure of matter.

An object may contain energy by virtue of its ***position*** or ***composition.*** Such energy is called ***potential energy.*** A rock balanced on a mountain ledge has the potential energy to harm a mountain climber, although the rock would pose no threat if lying on the valley floor below. A piece of wood and the oxygen around it may release energy in the form of heat and light when they combine in a process called combustion. New substances—ashes and gases of combustion that weigh the same as the destroyed wood and oxygen but have less energy of composition—are left. It seems reasonable to believe that the rock's change in position and the chemical change are similar (Figure 5–2). Just as the rock has less energy after it has fallen, so the molecular structure of the wood and oxygen rearrange, through combustion, into products of less energy. Both of these processes result in a loss of potential (or stored) energy, which is the energy contained in a sample of matter due to its position or composition that can be released by means of a physical or chemical change.

SELF-TEST 5-A

1. Two kinds of energy that matter can have are:

   a. ____________

   b. ____________

2. Energy due to motion is ____________ energy.

3. Energy due to structure is ____________ energy.

4. Two factors determining the energy due to motion are:

   a. ____________

   b. ____________

CONSERVATION OF ENERGY

Energy is indestructible.

Kinetic energy can be converted into potential energy. A boy may give kinetic energy to his favorite ball only to find this energy converted into potential energy as the ball resides high in a tree. Likewise potential energy can be converted into kinetic energy as a squirrel nudges the ball from its lodgment in the tree. Many energy transformations are possible. However, there is no loss or gain in the amount of energy in any of these transformations. If energy appears in one form it has to come from an equal amount of energy in another form or the destruction of an equivalent amount of matter. These changes of energy into alternative forms without loss are generalized into the ***law of conservation of energy:***

The total energy of a closed system remains constant for all processes except nuclear processes.

ELECTRICAL CHARGE

In order to understand how energy can be stored in the structure of matter, it is necessary to understand something about the ability of matter to interact with other matter at a distance. Of special interest in chemistry are the forces that are exerted through space between electrical charges. Indeed, the forces that appear to hold matter together can be explained in terms of these forces between electrical charges. Macroscopic amounts of electrical charge are not at all uncommon in ordinary experience. The bolt of lightning, the spark between comb and hair in dry weather, the shock on touching the door handle when certain clothes are brushed across the seat covers in a car, are all the result of electrical charge.

Experiments indicate that charges either attract or repel each other. The fact that there are two and only two effects suggested to Ben Franklin that there are only two kinds of electrical charge, positive and negative. Figure 5-3 illustrates a very sensitive apparatus, which can be

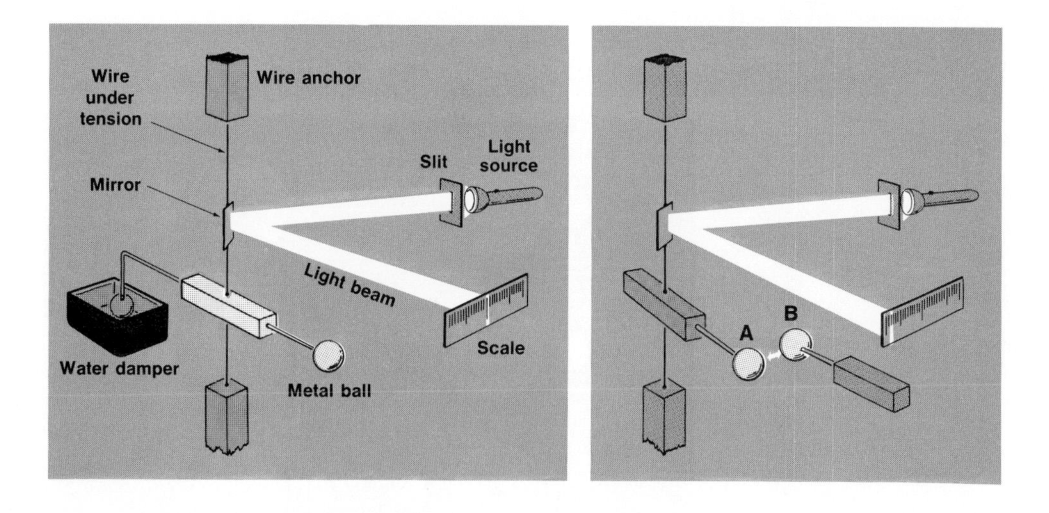

Figure 5-3 Apparatus for the demonstration of electrical charges and their interaction.

used to show the two types of charges. A wire under tension supports a small mirror and an arm connected both to a metal ball through an electrical insulator and a damping device. A bright line of light, produced by a slit in front of a parallel light beam, shines on the mirror and is reflected to a screen marked with a linear scale so that a very slight twist in the wire produces a much larger change in the position of the light on the screen. An electrical charge can be produced on a hard rubber rod by rubbing it vigorously for a few moments with fur. Part of the charge on the rubber rod can be transferred to metal ball *A* simply by touching the two together. Similarly, the same charge can be placed on ball *B*. Now, if *B* is placed near *A*, the balls will move away from each other. The resulting twist of the wire is clearly evident by the movement of the light on the screen. Exactly the same visible results can be obtained if a glass rod is rubbed with silk. The charge transferred from the glass rod to the metal balls causes them to repel each other. But, if the charge from the rubber rod is transferred to one of the metal balls and the charge from the glass rod to the other, the two balls attract and move closer together. Furthermore, any source of electrical charge will do one of two things when brought close to the charge from the rubber rod: (a) It will repel the rubber-rod charge just as this charge repels itself; or (b) it will attract the rubber-rod charge just as the glass-rod charge attracts it. Since all charges behave exactly the same as the rubber-rod charge or the glass-rod charge, we conclude that there are but two kinds of electrical charge and name them negative (rubber rod) and positive (glass rod). Also, we can conclude from these experiments that a positive charge attracts a negative charge, and that if both charges are positive or both negative, they will repel each other (Figure 5–4).

Two kinds of charge—by experience, not by theory.

Like charges repel. Unlike charges attract.

The attraction between unlike charges or the repulsion between like charges is directly proportional to the magnitude of each charge and inversely proportional to the square of their separation distance. This is summarized by ***Coulomb's law,*** which is

$$F = \frac{q_1 \times q_2}{d^2}$$

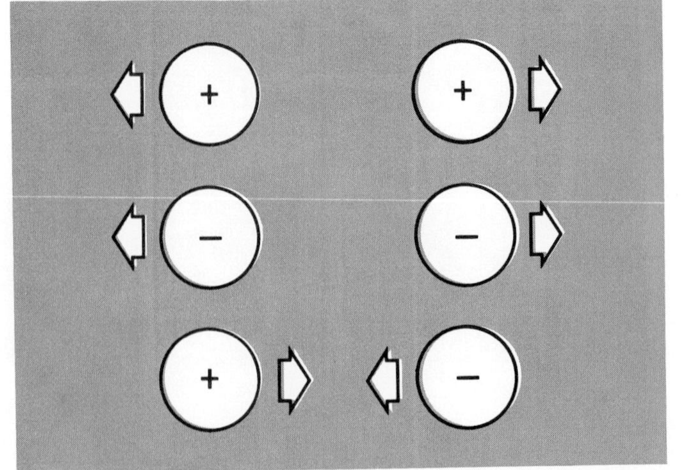

Figure 5–4 Like charges repel and unlike charges attract.

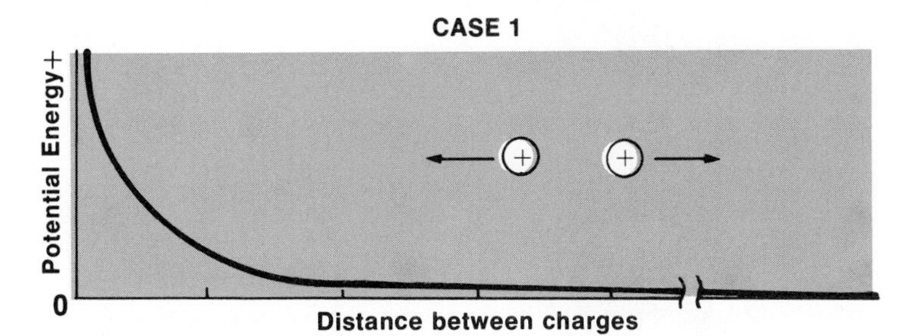

Figure 5–5 Potential energy of a system composed of two positive charges at different distances. Similar results are observed for two negative charges.

where F is the force of attraction or repulsion, q_1 and q_2 are the sizes of the two charges, and d is the distance between the charges.

ENERGY OF CHARGED SYSTEMS

Just as it requires an effort to separate small magnets, a similar effort is required to separate a positive and a negative charge. The energy that is used to separate the unlike charges is then stored in the system as potential energy. This energy, stored by virtue of the structure of the system, is potential energy because it can and will be released if the two charges are allowed to come together. The bolt of lightning is a display of the energy released when opposite charges come together.

Potential energy diagrams are very useful in understanding the energy stored in atomic and molecular systems; Figures 5–5 and 5–6 are potential energy diagrams.

A situation in which the distance of separation between opposite charges is reduced to zero is impossible in any real situation. The force of attraction between the charges is infinite for a zero value of distance. It would take infinite energy to separate the charges when together, and an infinite amount of energy would be released if the charges came together. Since this would involve the release of more energy than the universe contains, it becomes necessary to postulate a repulsive force between unlike charges at very small distances. The attractive force is still present, but at a given distance it is counterbalanced by a repulsive force. A macroscopic, physical system that is somewhat analogous will help to clarify this point. A steel cannon ball elevated 10 feet above the floor has considerable energy due to the arrangement of the system. However, all of this energy cannot be released from the system if the

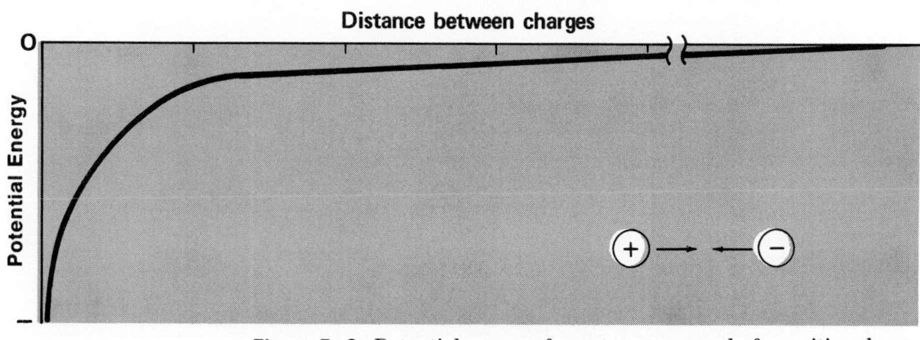

Figure 5–6 Potential energy of a system composed of a positive charge and a negative charge at different distances.

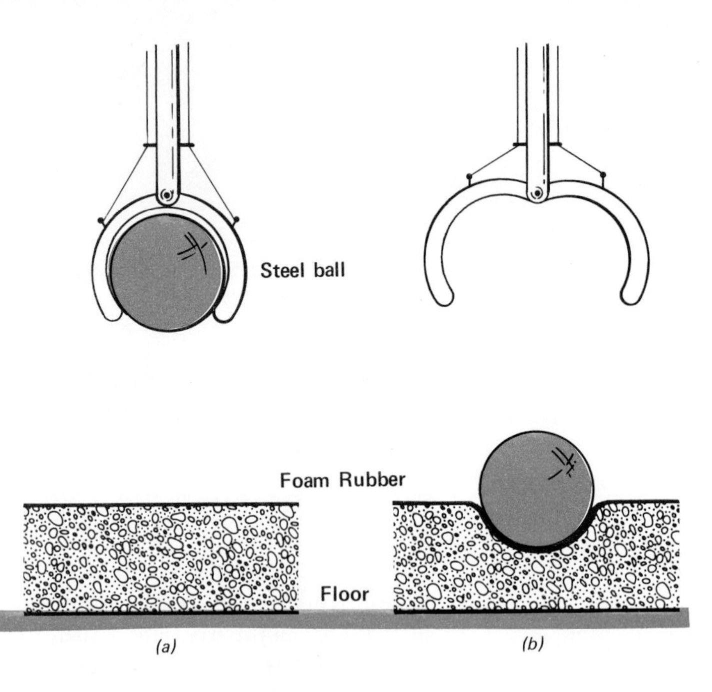

Figure 5–7 Resting on the floor would represent a lower potential energy for the steel ball, but the presence of the foam rubber pad prevents this from happening. A new position (*b*) is now a minimum potential energy position.

floor is covered with a 1-foot-thick pad of foam rubber. The steel ball, when dropped, will bounce a while and come to rest at a distance somewhat less than 1 foot above the floor (Figure 5–7). An amount of potential energy has been released, but the short-range upward force exerted by the compressed rubber prevents a closer approach and the release of even more energy.

Figure 5–8 illustrates the potential energy changes that occur as two oppositely charged particles approach one another and experience a repulsive force. At Point A, when the charges are far apart, there is a relatively large amount of potential energy in the system. The system contains less energy at B as the charges are closer together. Point C represents a separation distance for the system when it is at its lowest possible energy level (sometimes called the bottom of the energy well). At Point C any effort to increase the distance between the charges will require energy to overcome the electrostatic force of attraction, and any effort to reduce the distance will require energy to overcome the short-

Short-range repulsive forces were invented to make the theory agree with facts.

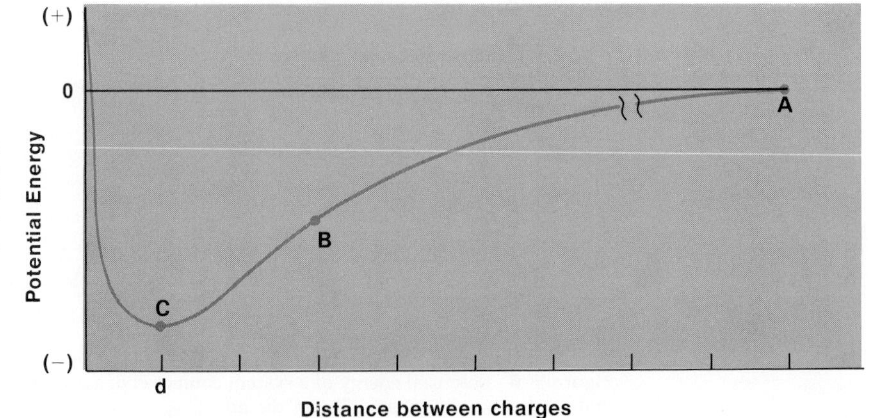

Figure 5–8 Potential energy of a system composed of a positive charge and a negative charge at different distances when short-range repulsive forces are present.

range force of repulsion. Unless energy is supplied from some outside source, the two unlike charges will lose all energy possible and will exist as a pair separated by the distance, d.

SELF-TEST 5-B

1. Kinetic energy can be converted into ____________ energy.
2. The force between two charges is given by ____________ law.
3. The total energy of a closed system remains ____________ for all processes except nuclear processes.
4. When a rocket is shot into orbit around the earth the potential energy of the rocket fuel is converted into the ____________ energy of the rocket.
5. The force of repulsion between two +3 charges at 3 cm will be (larger than, equal to, less than) that between two +2 charges at 2 cm.

   ____

EFFECT OF THE ADDITION OF ENERGY TO MATTER

Addition of energy into matter can:
1. raise temperature;
2. produce physical change;
3. produce chemical change.

Energy acting upon matter can be used to alter its structure. This sometimes brings about a chemical change and at other times causes a physical change. Most commonly the addition of energy will distort or break the bonds between the particles.

As an illustration of how added energy can produce either physical or chemical changes, consider the addition of energy to a pure substance composed of molecules at some low temperature. At this point of very low energy the molecules are very close together. The addition of energy spreads the molecules farther apart. The continued addition of energy eventually melts the solid, heats the liquid, and vaporizes it into a gas. Each process results in a further separation of the particles which attract each other. All of these changes are physical changes, since the same pure substance survived all of the additions of energy. However, if energy continues to be added, the gas is heated to the point where the molecules break apart into atoms or molecular fragments. At this point, a chemical change occurs, since the original substance composed of molecules is changed into another substance composed of atoms. Continued additions of energy can cause atoms to be separated into their component parts, producing additional chemical changes. Throughout the process of change from molecules at a low temperature to subatomic particles at higher temperatures, the addition of energy causes a separation of particles which attract each other. This absorbed energy is stored in the system by means of its expanded structure. At any point, the total energy contained in a sample of matter is called its ***internal energy.***

Internal energy results from:
1. Particle motion (kinetic) and
2. Particle structure (potential)

ENERGY CHANGES IN A PURE SUBSTANCE

The total energy contained in an isolated sample of matter such as a sample of a pure substance can be increased by heating the sample or by doing work on it. Heating can be accomplished by placing the sample in contact with other matter which is at a higher temperature. Work can be done on the sample by beating on it, stretching it, compressing it, or in any way forcing the parts of the sample to move against forces that tend to hold these parts in place. Since internal energy can be increased by either heat or work, we might suspect that heat and work are related in some way. Indeed, two of the signal advancements in all of physical science were the demonstration by Count Rumford (Benjamin Thompson), in about 1800, that heat and work are actually two expressions of energy, and the verification of James Prescott Joule, about 40 years later, that a given quantity of work is equivalent to a given amount of heat.

Heat and work are interconvertible.

Change in state: melting boiling

What are some of the results of increasing the internal energy of a pure substance? ***Either the temperature will increase or there will be a change in state.***

In order to deal with specific changes in a familiar substance in terms of given energy changes, it will be necessary to become acquainted with a unit used to measure heat or work. There are a number of such defined energy units: erg, joule, foot-pound, British thermal unit, kilowatt-hour, and calorie. Any of these units can be used to measure quantities of energy just as units of length (foot, inch, yard, meter) are used to measure distance. In this text we will use the calorie unit exclusively. Although defined in terms of the joule, the calorie is approximately the amount of heat necessary to raise the temperature of 1 gram of water 1 centigrade degree at a starting temperature of about 15°C. This calorie (cal) is one-thousandth of the large calorie (Cal, really a kilocalorie), which is commonly used to measure the potential energy available in foods.

It is also important to understand the difference between heat and temperature. Without being concerned with definitive statements, we can say simply that ***heat*** is energy and is measured in any of the units given in the last paragraph, whereas ***temperature*** is a measure of the relative hotness or coldness of a given substance and is measured in degrees. (More accurately, temperature is a measure of the driving force behind heat transfer.) The heat from a burning match will raise the temperature of an iceberg only a very small part of one degree. Although the burning match has a much higher temperature, it furnishes very little total heat compared to the large iceberg's capacity to store heat. Heat will flow from a body of higher temperature into a body of lower temperature (Figure 5–9) regardless of the amount of heat already present in each body. You should also keep in mind the fact that substances differ from one another in their capacity to store heat. While it takes 1 calorie to heat 1 gram of water 1 degree, it only takes one-half of a calorie to heat 1 gram of ice 1 degree. Iron is even more responsive in its temperature changes when energy is applied. It takes only 0.12 calorie to raise the temperature of 1 gram of iron 1 degree. These numbers are called the ***specific heats*** for these materials; the specific heat of iron, $0.12\ \frac{\text{cal}}{\text{g deg}}$, is read 0.12 calorie per gram per degree centigrade.

Temperature measures the intensity of heat.

When a pure substance undergoes a change of state, there is a

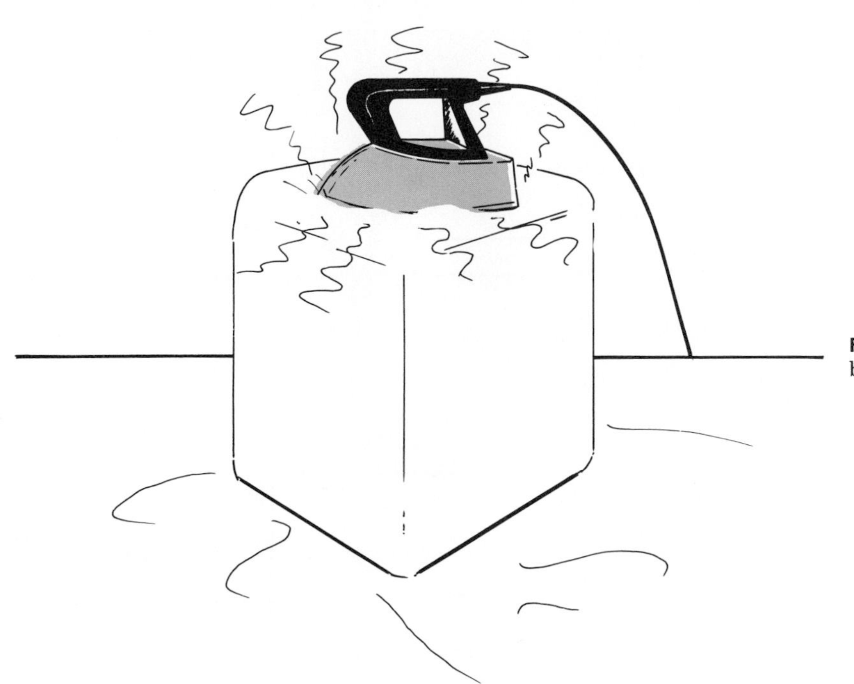

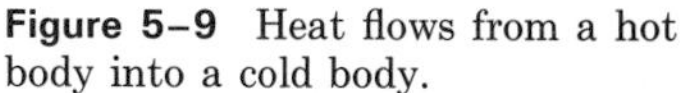

Figure 5–9 Heat flows from a hot body into a cold body.

relatively large amount of energy involved. It requires about 80 calories of heat to melt one gram of ice at 0°C. Similarly, about 80 calories of heat have to be removed from one gram of water at 0°C to convert it into 1 gram of ice at 0°C. The 80 cal/g is called the ***heat of fusion*** for ice, and each pure substance can be expected to have its own characteristic heat of fusion. The ***heat of vaporization*** of a liquid is the amount of heat necessary to convert one gram of the liquid to one gram of the gas at the same temperature. The heat of vaporization of water is about 540 cal/g.

Heats and temperatures of fusion and vaporization help in characterizing pure substances.

In order to clarify the points just made, consider Figure 5-10. Here, the interaction of energy with 1 gram of ice at 0°C is shown in such a way that one can easily see the results of fusion (melting of the ice), further heating (0°C ⟶ 100°C), and vaporization (boiling the water). An application of the law of conservation of energy leads to the conclusion that 1 gram of water at 0°C has more energy than 1 gram of ice

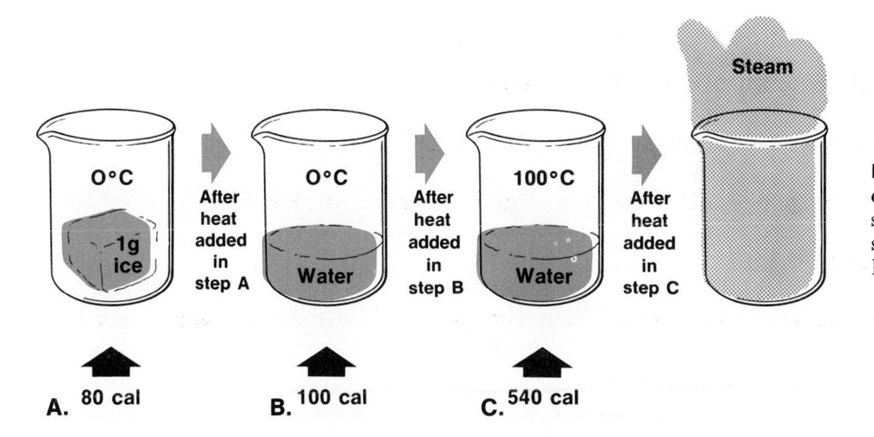

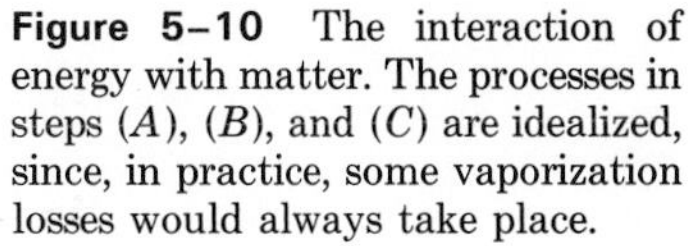

Figure 5–10 The interaction of energy with matter. The processes in steps (*A*), (*B*), and (*C*) are idealized, since, in practice, some vaporization losses would always take place.

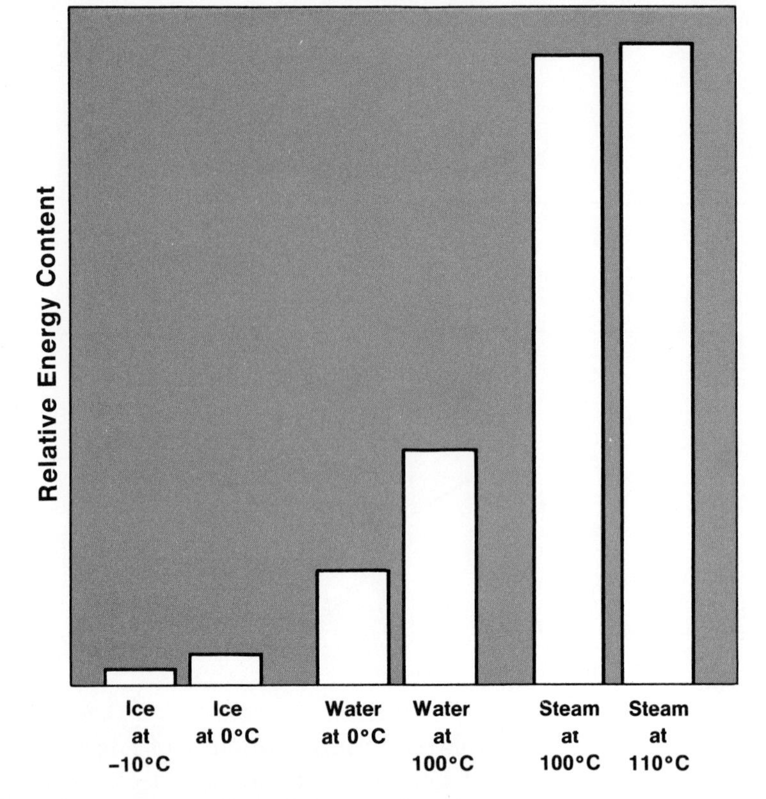

Figure 5–11 A graphic representation of the relative internal energies of equal weight samples of the compound H_2O at various temperatures.

at 0°C, 1 gram of water at 100°C has more energy than 1 gram of water at 0°C, and so on. These energy differences are illustrated for 1 gram of water (as ice, liquid, and steam) in Figure 5–11.

All pure substances have characteristic heats of fusion and vaporization, and specific heats, which are dependent on their submicroscopic properties. Each can be expected to exhibit behavior similar to that of water at characteristic temperatures.

SELF-TEST 5-C

1. Name three kinds of changes which can occur in pure substances upon adding energy if the pure substance is not changed into another pure substance.

   ____

   ____

   ____

2. Eighty calories of energy are added to 1.0 g of ice at 0°C and cause the ice to melt. This amount of energy is referred to as the ____________________.

3. The energy needed to vaporize 1 gram of water at 100°C is referred to as its heat of ____________________.

4. When a substance passes from the gaseous state to the liquid state, it ____________________ heat.

THE KINETIC-MOLECULAR THEORY EXPLAINS ENERGY CHANGES IN A PURE SUBSTANCE

The kinetic-molecular theory describes molecules in motion.

The kinetic-molecular theory developed rapidly after 1860 as a sophisticated model of the structure of gases. Later, when some of the theoretical predictions (such as molecular sizes) were found to agree with independent experimental results, the evidence was strong that scientists were on the right track in sorting out the submicroscopic picture of matter.

Although we will deal with the theory as a picture model, it can be expressed in mathematical terms. Not only does it offer a qualitative explanation for the gas laws as presented in Chapter 3 in mathematical form, but the theory can also be used to derive the gas laws from basic theoretical concepts. This helps to confirm our belief in molecules. The mathematical part of the theory allows one to calculate the velocities of molecules, the frequency of molecular collisions, and the distances traveled by molecules between collisions. Although it is beyond the scope of this presentation to develop the mathematics involved, perhaps the nonmathematical, qualitative aspects of the theory will help you to see the power of the kinetic theory. It provides a molecular explanation of existing gas laws and changes of state.

The kinetic-molecular theory visualizes a gas in the following way (only those points that are important to this discussion are included):

1. A gas is made up of molecules (monatomic molecules in the case of gases like helium).
2. The molecules are very small relative to the normally great distance between them.
3. The molecules have no appreciable attraction for each other at these great distances.
4. The molecules are moving with very high velocities, causing many intermolecular collisions and collisions with the walls of the container in each second.
5. Collisions at the molecular level are perfectly elastic. There is, therefore, no total loss of energy resulting from these collisions, and no tendency for the molecules to slow down and stop, as do billiard balls on a table.
6. Molecules move faster as the temperature is increased.

These assumptions are quite effective in explaining the gas laws that were discussed in Chapter 3. The following are examples:

Boyle's Law

A smaller volume causes more gas-container collisions.

Boyle observed that the pressure exerted by a gas decreases as the volume increases at constant temperature. This is readily understood if one conceives of the pressure on the walls of a balloon resulting from the molecules beating against the walls. If the balloon were suddenly made larger, the same number of molecules moving at the same speed would hit a given area of the walls less often, resulting in a *reduced* pressure.

Dalton's Law of Partial Pressures

The additivity of gas pressures is easily understood when one notes that gas molecules have little attraction for one another and that the space between molecules is rather large. Since the molecules do not interact they act independently upon the walls of the container and the pressure increases accordingly.

A gas is mostly empty space.

Henry's Law

The increased solubility of a gas in a liquid can be explained in terms of Boyle's law and the collisions that take place between the surface of the liquid and the gas molecules. As the pressure of a gas increases, the number of molecules in a given volume increases. This results in an increased number of collisions of gas molecules with the liquid surface, some of which lead to gas molecules going into the liquid (dissolving). The greater the pressure, the more collisions, and the greater the solubility.

Pressure crowds more molecules into open spaces within the liquid.

Charles' Law

Charles' law states that the volume of a gas is increased as the temperature is increased, if the pressure is held constant. If increased temperature causes molecules to move faster, as is asserted by the kinetic-molecular theory, an increase in the temperature of a gas would cause the molecules to hit the walls of the container more often. This would result in an increased pressure. If the pressure is to be held constant, it follows that the container will have to be made larger so that the total number of collisions from faster-moving molecules will equal the pressure caused by the collisions of the slower-moving ones in the small volume.

Faster moving molecules collide more frequently.

Liquids and Solids

The kinetic-molecular theory of ***liquids*** includes the following points:

1. Molecules are close-packed, essentially in contact with each other.

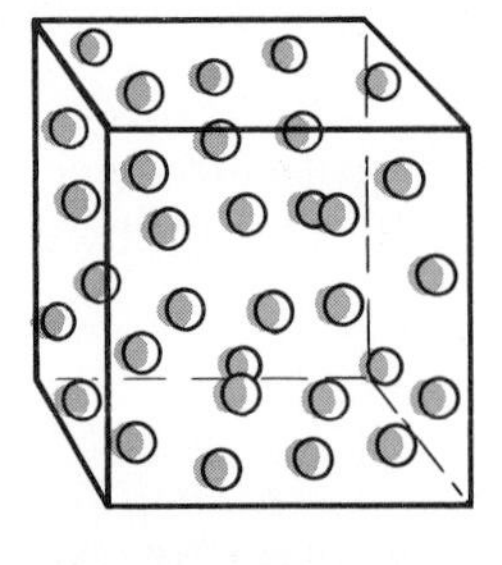

Some particles in gaseous state (not drawn to scale); particles far apart, completely fill container.

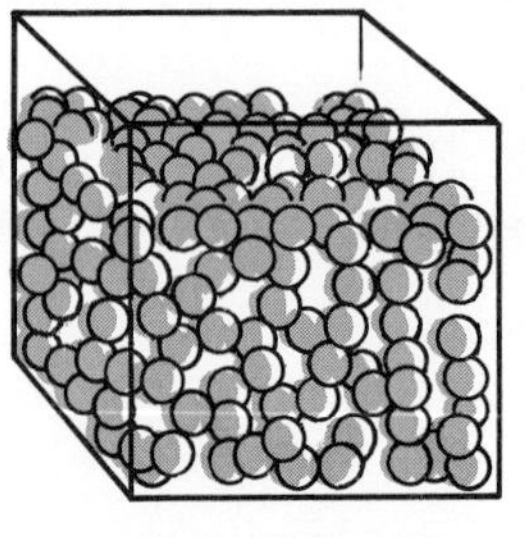

Some particles in a liquid state; not in fixed positions (liquid flows), definite volume, usually somewhat larger than when in solid state.

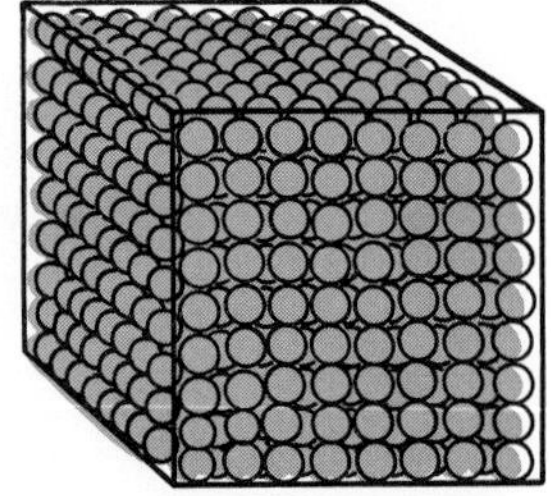

Particles in a solid; fixed positions, definite volume.

Figure 5–12 An illustration of gas, liquid, and solid states of matter. Molecules move more rapidly with increased temperature. This usually results in a liquid being less dense than the solid it was obtained from as the liquid has more space not occupied by molecules.

Liquid molecules are close together without rigid structure.

2. There are many solid-like crystal structures within the body of the liquid. These are constantly changing in size and in the particular molecules involved.
3. Molecules move more rapidly with increased temperature. This usually results in a liquid being less dense than the solid it was obtained from as it has more space not occupied by molecules.

The molecules of the solid hold their position rather well relative to the molecules around them. This is true in spite of the fact that the molecules are vibrating about fixed positions (see Figure 5–13), and some even have enough kinetic energy to escape the binding intermolecular forces, moving from place to place within the solid or even out of the solid.

Solids have the particles close to each other and arranged in a definite structure.

Using the assumptions of the kinetic-molecular theory, the behavior of a pure substance as energy is added can be explained in terms of the kinetic and potential energy of the individual molecules. As a pure substance is heated below the point at which it changes state, the increased internal energy can be stored in one of two ways. Primarily, the increased energy is stored as kinetic energy in the faster-moving molecules. However, some expansion usually occurs when a substance is heated. This expansion results in a movement of the molecules away from the restraining intermolecular forces, resulting in increased potential energy. In the change of state of a pure substance there is usually a dramatic change in volume. In going from a liquid to a gas the increase in volume is most pronounced. There is usually a relatively small increase in volume in going from the solid to the liquid phase. (Water is an exception to this rule.) Since the temperature does not change during a change of state, it is reasonable to conclude that heats of fusion and heats of vaporization are stored as potential energy in the molecular system. It then becomes easy to understand why heats of vaporization are larger than the corresponding heats of fusion. It takes much more energy to move the molecules through the relatively larger distances required to achieve the gas structure, while changes in distances that occur in a solid to liquid transition are relatively small.

The structure of water is discussed in Chapter 9.

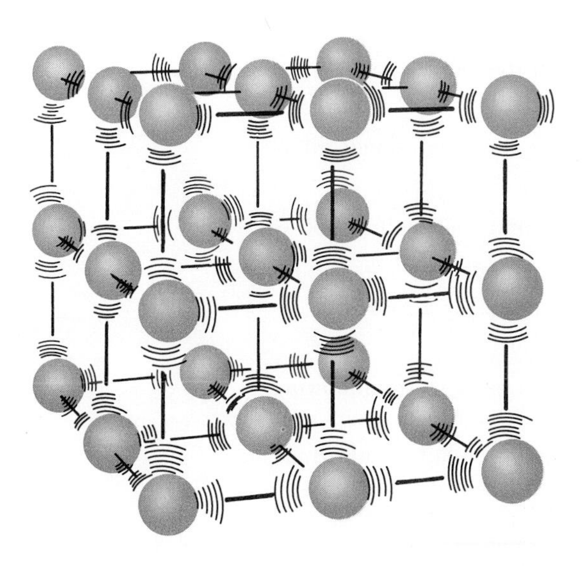

Figure 5–13 Vibrations of particles in their fixed positions in a solid lattice.

MATCHING SET

______ 1.	energy	a.	true of all energy changes
______ 2.	potential energy	b.	measured in energy units
______ 3.	conservation	c.	an energy unit
______ 4.	kinetic energy	d.	requires energy
______ 5.	unlike electrical charges	e.	always from hot to cold
______ 6.	heat and work	f.	energy due to motion
______ 7.	melting solid	g.	assumes molecular motion
______ 8.	calorie	h.	far apart
______ 9.	heat flow	i.	attract
______ 10.	internal energy	j.	increases with temperature
______ 11.	kinetic-molecular theory	k.	energy due to position
______ 12.	molecules of gas	l.	the ability to do work
______ 13.	molecules of solid	m.	vibrate

QUESTIONS

1. Which of the following are exhibiting primarily potential energy and which kinetic energy?
   a. wound clock spring
   b. a moving automobile
   c. water stored in a water tower
   d. a rock just beginning its tumble down a mountainside

2. Most steel bridges have expansion joints in them to allow for variations in the weather. Why?

3. Distinguish between gas, liquid, and solid with respect to the differences between the kinetic and potential energies of the particles in the three states.

4. Gravitational interactions involve only attractions. How do gravitational interactions differ from electrical and magnetic interactions?

5. Explain why water in a pot (at sea level) will heat up and then boil at 100°C until all of it is vaporized, as heat is added to it.

6. Describe the potential energy and kinetic energy relationships as a rock tumbles off a cliff.

7. Explain in detail how the law of conservation of energy applies to the following processes:
   a. Water flows from the storage pool behind a dam through conduits to a turbine which turns a generator; the energy produced turns a motor, furnishes a light, rings an electrical bell, and heats a house, in addition to sustaining frictional losses.
   b. A piece of wood is burned.

8. Draw a diagram showing how the potential energy of a positive particle and a negative particle changes as the two particles approach each other from a great distance apart, touch, and because of outside force, are caused to dent each other.

9. How would the kinetic molecular theory explain (in terms of molecules):

   a. the absorption of the heat of fusion in melting a solid without changing the temperature?
   b. the absorption of the heat of vaporization in vaporizing a liquid without changing the temperature?
   c. why a liquid expands as the temperature is increased?

10. Could two containers of water have the same temperature but contain different quantities of heat? Explain.

11. Two slowly moving automobiles crash into each other head-on. They recoil only slightly. What happened to the rest of the kinetic energy which they possessed prior to their crash?

12. Mercury has a melting point of −38.87°C and a boiling point of 356.58°C. What changes would you expect to find in the internal energy of a sample of mercury as you heated it from −50°C to 400°C?

13. Name some materials which are commonly used to provide energy because of the potential energy stored in their molecular structure.

14. Design a system that you could use to obtain potential energy from the following sources of kinetic energy.

    a. the tidal motion of the ocean
    b. wind

PROBLEMS

1. How many calories are required to melt 10 grams of ice at 0°C and raise the temperature of the liquid water to 40°C?

   Ans. 1200 calories

2. Which has greater kinetic energy: a freight train that weighs 2000 tons, which is moving at 40 miles per hour, or an airplane that weighs 20 tons and is moving at 1000 miles per hour?

   Ans. airplane

SUGGESTIONS FOR FURTHER READING

"A Sun Powered Furnace," *Chemistry,* Vol. 43, No. 10, p. 23 (1970).
Alder, B. J., and Wainwright, T. E., "Molecular Motions," *Scientific American,* Vol. 201, p. 113, 1959.
Angrist, S. W., and Hepler, L. G., "Order and Chaos; Laws of Energy and Entropy," Basic Books, New York, 1967.
Farber, E., "The Evolution of Chemistry," Ronald Press Co., New York, 1952, p. 204.
Hubbert, M. K., "The Energy Resources of the Earth," *Scientific American,* Vol. 225, p. 60, 1971.
Ihde, A. J., "The Development of Modern Chemistry," Harper and Row, New York, 1964, p. 395.
Lonsdale, K., "Disorder in Solids," *Chemistry,* Vol. 38, No. 12, p. 14 (1965).
Peisen, H. S., and Torgesen, J. L., "Crystal Growth as Chemical Research," *Chemistry,* Vol. 38, No. 9, p. 15 (1965).
Sabine, D. B., "Count Rumford," *Chemistry,* Vol. 39, No. 8, p. 18 (1966).
Tabor, D., "Gases, Liquids and Solids," Penguin Press, Baltimore, 1969.
"The Geometry of Liquids," *Chemistry,* Vol. 38, No. 12, p. 20 (1965).

Most elementary physics texts have a discussion of charged particles and their interactions.

ATOMIC THEORY—DALTON TO RUTHERFORD

CHAPTER 6

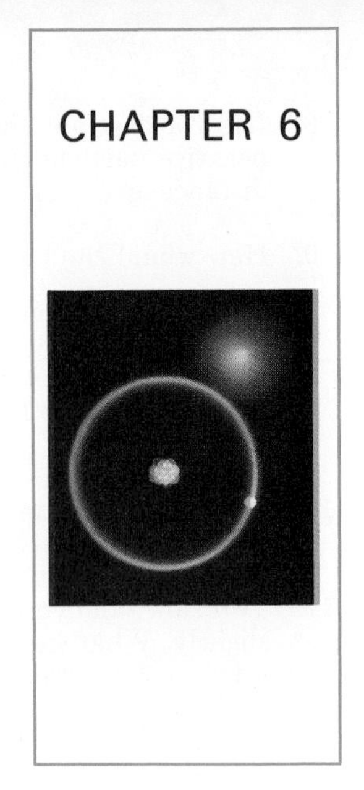

Up to the beginning of the 20th century, the Daltonian concept of the atom was used largely in its original form. The model was successful because it explained most of the chemical and many of the physical laws of interest during this period. About the end of the 19th century, however, certain experimental facts, which were in direct conflict with the atomic model presented by Dalton, became generally known. These facts necessitated a rather severe modification in atomic theory, but many of Dalton's ideas were retained and incorporated into the more elegant theory. The resulting atomic theory, as refined in 1913 by Niels Bohr, was not only able to explain the laws that occupied the attention of Dalton, but was also able to explain many other experimental facts as well. It is the purpose of this chapter to look at some of these experimental facts in detail and see how they led scientists to a more realistic model of the atom.

New experimental facts that are not compatible with a theory require a new theory.

ELECTRICAL DISCOVERIES

The fact that matter can be charged electrically has been observed since ancient times, but it was not until 1600 that William Gilbert coined the word electric (from the Greek *elektron,* for amber, fossilized pine gum). It was Benjamin Franklin, in 1747, who set out present definitions —that glass rubbed with silk becomes *positive* and that amber or rubber rubbed with fur becomes *negative*. In 1791, Luigi Galvani, a physician, noted the convulsions in a partially dissected frog leg when a static electric charge was applied. Subsequent experiments revealed that the static charge was unnecessary and that the mere contact of nerves with pieces of metals produced the same effect. Although it was not realized at first, Galvani was successful in passing a current through an electrical circuit (the network of nerves) by making use of the energy derived from a chemical change. Galvani concluded that he was observing "animal electricity" and that it was characteristic only of living tissues. A few years later, in 1800, Alessandro Volta made the first battery, a discovery that captured the imagination of many scientists and led to widespread investigations into electrical phenomena. Using an electric current furnished by a battery, Nicholson and Carlisle were successful in that same year in decomposing water into hydrogen and oxygen. Sir Humphry Davy and also Michael Faraday soon followed suit with studies of the electrical decomposition ***(electrolysis)*** of numerous compounds. These and other similar experiments strongly suggested that a fundamental relationship exists between electricity and the structure of matter. However, Dalton's atomic theory did not encompass this relationship and, as a result, atomic and electrical theory developed with little interaction until about 1900.

The electrical properties of matter are not explained by Dalton's atomic model.

CATHODE RAYS AND THE ELECTRON

If a high voltage is applied to metal discs (electrodes) connected through the ends of a partially evacuated glass tube (Figure 6–1), streaks of light can be observed streaming between the electrodes. Interesting and changing patterns of light are observed as the amount of gas in the tube is varied. Changing from air to other gases results in light of different colors. If the gas is almost completely removed, the colored light practically disappears. However, if a sample of solid matter is placed between the electrodes in such an evacuated tube, there is clear evidence that

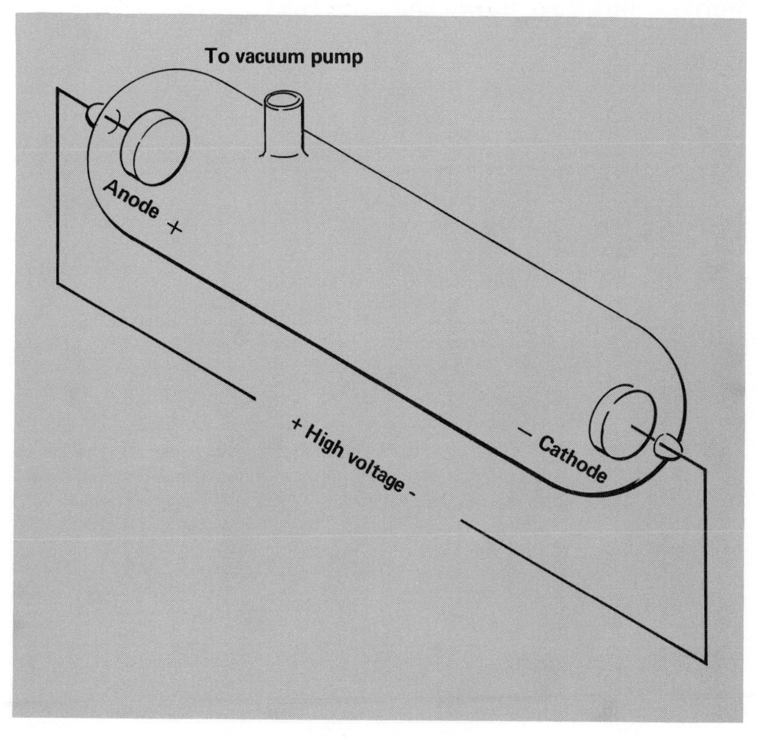

Figure 6–1 Simple discharge tube.

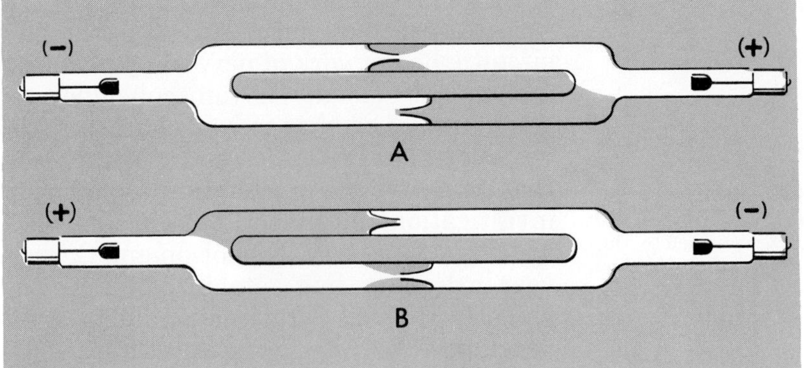

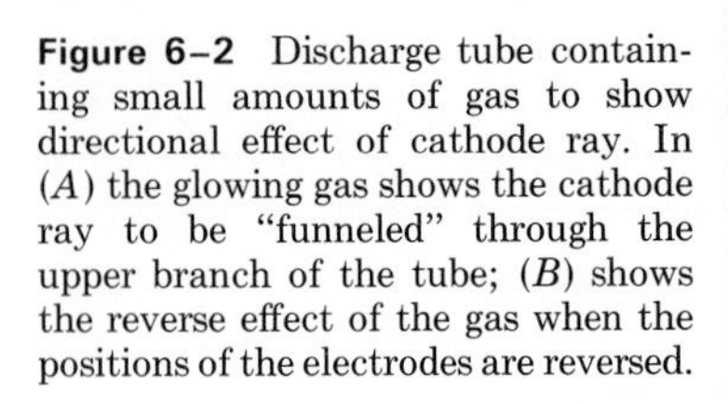
Figure 6–2 Discharge tube containing small amounts of gas to show directional effect of cathode ray. In (*A*) the glowing gas shows the cathode ray to be "funneled" through the upper branch of the tube; (*B*) shows the reverse effect of the gas when the positions of the electrodes are reversed.

an invisible ray is passing between the electrodes. For example, a thin piece of metal foil will become red hot when subjected to the ray, and a phosphor, such as zinc sulfide, will glow with light, as do the gas samples, when exposed to the ray. These results clearly indicate that considerable energy is transmitted in the ray.

These rays are called ***cathode rays.*** They always pass from the negative electrode to the positive electrode (Figure 6–2). (In discharge tubes, the negative electrode is the ***cathode,*** and the positive electrode is the ***anode.*** As illustrated in Figure 6–3, the path of the ray can be blocked partially with a metal plate containing a narrow slit. A metallic screen covered with phosphor is placed at an oblique angle beyond the slit. The ray that passes through the slit then strikes the screen in a straight line, and this results in a straight line of bright light on the phosphor coating. If the other electrode is made negative, this line does not appear. In this case, the entire screen glows, indicating that the ray always comes from the negative electrode. This experiment not only shows that the rays are negative in origin, but also shows that the rays

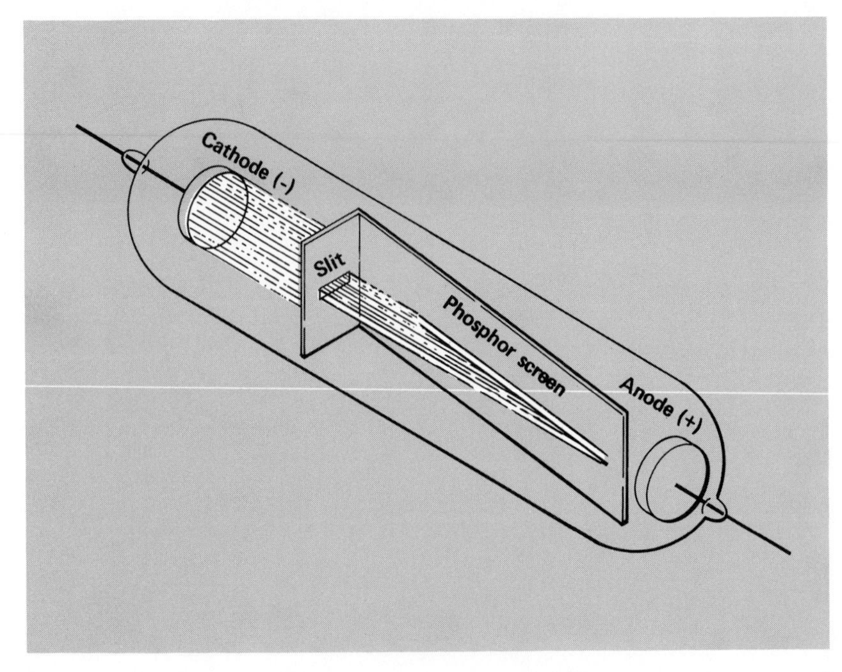

Figure 6–3 Discharge tube with slit and phosphor screen to show path of cathode ray.

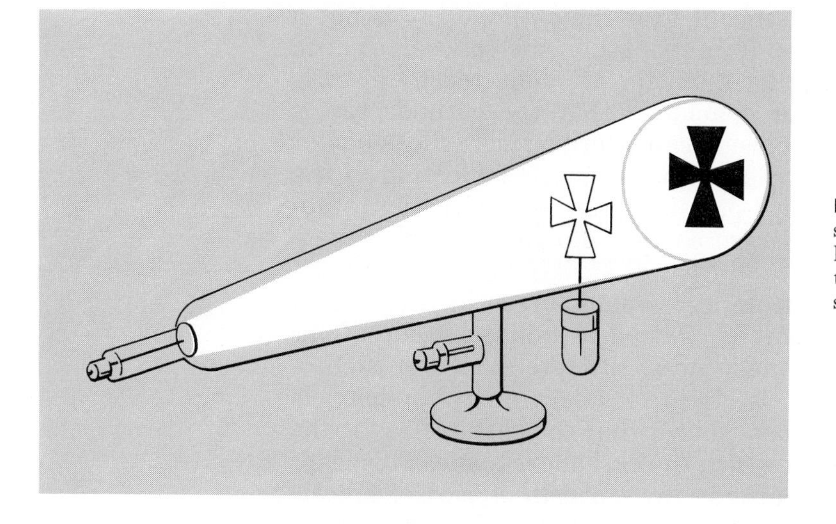

Figure 6–4 Discharge tube to demonstrate that cathode rays travel in straight lines and cast sharp shadows. The cross in the path of the cathode ray casts a sharp shadow.

travel in straight lines and cast sharp shadows. These points are also demonstrated in specially designed tubes as shown in Figure 6–4.

If a positively charged plate is placed above a cathode ray and a negatively charged plate below it, the ray will bend upward toward the

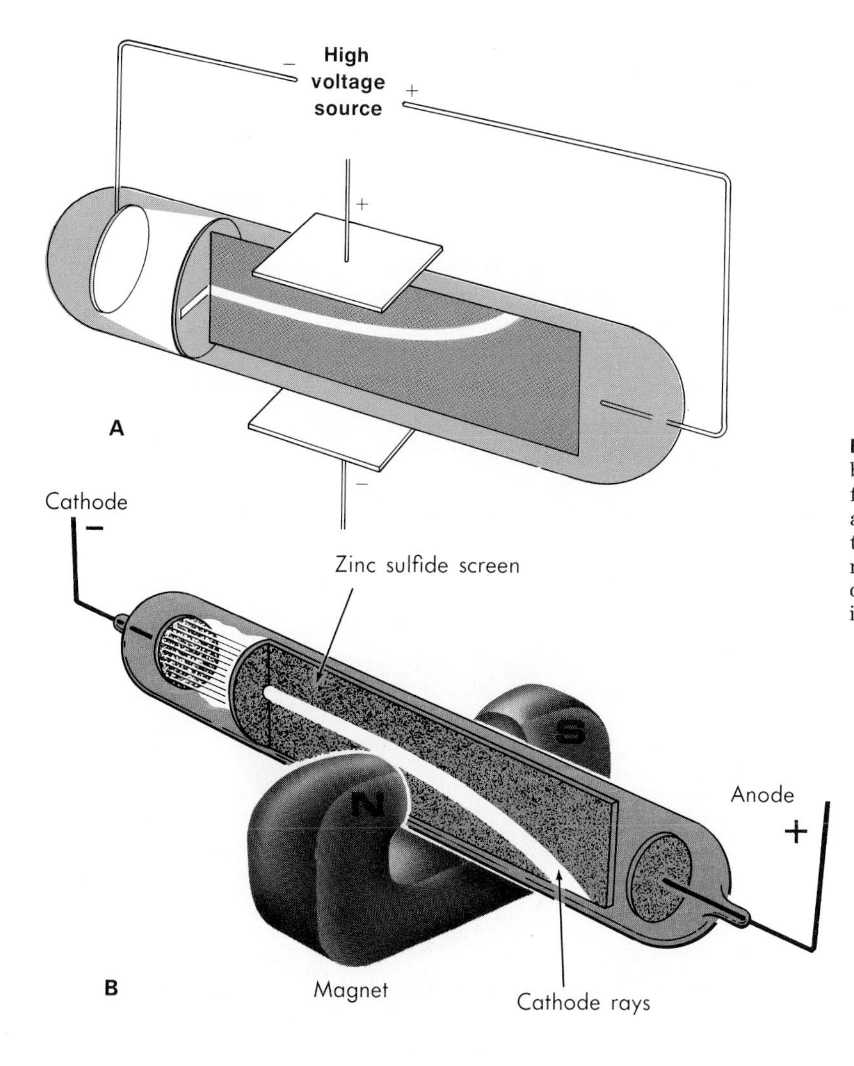

Figure 6–5 Deflection of a cathode ray by an electric field and by a magnetic field. When an external electric field is applied, the cathode ray is deflected toward the positive pole. When a magnetic field is applied, the cathode ray is deflected from its normal straight path into a curved path.

positive charge (Figure 6–5). This means that the cathode ray is not a light ray, since light is not deflected from its path in such a weak electric field. Hence, since charged particles of matter are deflected by electric fields (see Chapter 5), it must be concluded that the cathode ray is negatively charged. This same conclusion can be drawn from the behavior of the cathode ray in a magnetic field. The magnetic deflection of the cathode ray, illustrated in Figure 6–5, is in agreement with the established behavior of a negatively charged particle moving through a magnetic field.

A careful study of the light from a phosphor screen in the path of a cathode ray shows that the light is emitted in minute flashes. This suggests that the cathode ray is made up of discrete particles, each of which (due to its kinetic energy) is able to cause a phosphor particle to glow. Numerous experiments have shown that the cathode ray is the same, regardless of the material of which the cathode is made. This leads one to the conclusion that the cathode ray consists of particles basic to all matter. All the properties of the cathode ray can be explained if one assumes that the ray is composed of negatively charged particles with very high kinetic energy streaming out from the negative electrode. These particles are called ***electrons.***

Cathode rays are streams of the negatively charged particles which we call electrons.

PROPERTIES OF THE ELECTRON

The direct observation of the cathode-ray tube leaves two important questions unanswered: What is the mass of the electron? What is the size of the negative charge that it carries?

In 1897, Sir Joseph John Thomson made use of the basic fact that the path of a charged particle will curve when passing through a magnetic or electric field. Thomson passed a cathode ray through a magnetic field and, as expected, the stream of electrons curved away from a straight-line path, as shown in Figures 6–5 and 6–6. Thomson knew that the path followed depends on the charge, velocity, and mass of a charged particle as well as on the strength of the magnetic field. Both a larger amount of charge on the electron and a stronger magnetic field tend to increase the curvature of the path, whereas an increase in the mass of the particle or a greater velocity tends to reduce the curvature. In addition to the magnetic field, Thomson applied an electric field of such strength that the electric effect exactly opposed the magnetic effect (Figure 6–7).

Thomson discovered the charge-to-mass ratio of the electron.

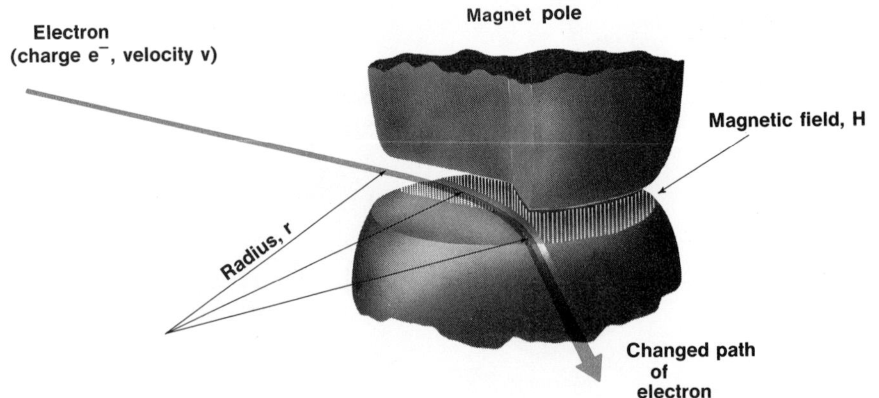

Figure 6–6 Curved path of electron in magnetic field. The radius of curvature measures the curved path and depends on mass of electron, charge of electron, velocity of electron, and strength of magnetic field.

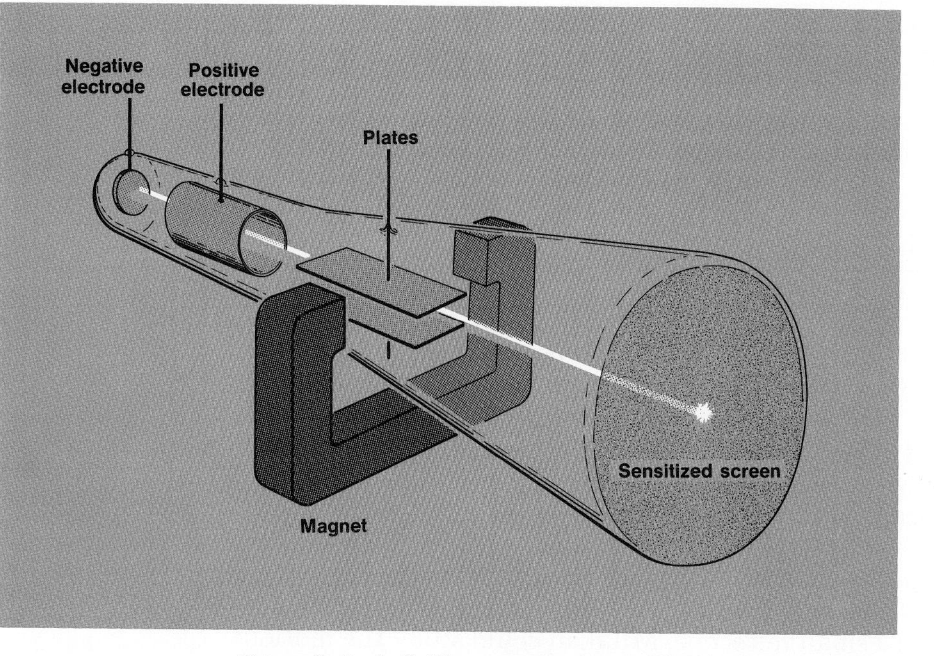

Figure 6–7 J. J. Thomson experiment. Electric field, applied by plates, and magnetic field, applied by magnet, cancel each other to allow cathode ray (electron beam) to travel in straight line.

Hence, the electrons traveled in a straight path as indicated on the zinc sulfide detecting screen. In this balanced state a rather simple equation could be derived from basic laws of magnetism and electricity. The equation is

$$\frac{e}{m} = \frac{E}{H^2 r}$$

where e is the charge on the electron; m is the mass of the electron; E is the strength of the electric field; H is the strength of the magnetic field; and r is the radius of curvature when only the magnetic field is applied. Thomson was able to measure E, H, and r, and as a result was able to calculate the charge-to-mass ratio (e/m) for the electron. Since he still had two unknowns (e and m) and only one equation, he was unable, from this experiment, to determine either. However, he did know the quotient of e divided by m (and if either value could be obtained elsewhere, he could calculate the other). The value of the ratio in this case is 1.758×10^8 coulomb per gram. The ***coulomb*** is a fundamental unit of charge as the gram is a unit of mass.

Thomson and a coworker, J. Townsend, made a notable effort to measure the charge of a single electron by studying clouds composed of negatively charged droplets. The weight of each droplet was estimated by the rate at which it fell through the air. The total weight of the cloud could be obtained by weighing the condensed water from the cloud. As a result, the number of drops in the cloud could be calculated:

$$\text{Total weight of water} \div \frac{\text{weight of water}}{\text{one drop}} = \text{number of drops}$$

Thomson assumed that *each drop* of water contained a negative charge composed of a single electron. Therefore, if he knew the number of drops, he also knew the number of electrons. Now the total charge on all the drops could be obtained by allowing the cloud to condense on an electrometer (a charge-measuring meter). Knowing the total charge and the number of electrons, he could divide to obtain:

$$\frac{\text{Total charge}}{\text{Number of electrons}} = \text{charge per electron}$$

The value obtained was 2.17×10^{-19} coulomb per electron.

It remained for Robert Millikan, with his famous oil-drop experiment in 1909, to establish firmly the charge on a single electron. Millikan's experiment consisted of measuring the number of electric charges carried by tiny drops of oil when suspended in an electric field (Figure 6–8). By means of an atomizer, Millikan distributed the oil droplets in a test chamber. As the droplets settled slowly through the air, high energy x-rays passed through the chamber to charge the droplets negatively (the x-ray caused air molecules to give up electrons to the oil). By means of a beam of light and a small telescope, Millikan could study the motion of single droplets. When the electric field was increased enough to balance the effect of gravity, the droplets could be made to remain motionless. At this point, Millikan could equate the gravitational force with the electrostatic force and, hence, was able to calculate the charge carried by the droplet in order to achieve this balance. Millikan found different sizes of negative charge on different drops, but the charge measured each time was always a whole-number multiple of a very small basic unit of charge. The largest common divisor of all charges measured was 1.602×10^{-19} coulomb. Millikan assumed this to be the charge on the electron.

The charge on a single electron can be determined from the behavior of charged oil droplets in an electric field.

Only a whole number of electrons may be present in a sample of matter.

Millikan's determination of the charge of the electron is analogous to the assignment given to a student who was asked to determine the weight of a marble when given eight sacks, each containing an unknown number of identical marbles. He was not allowed to open the sacks to

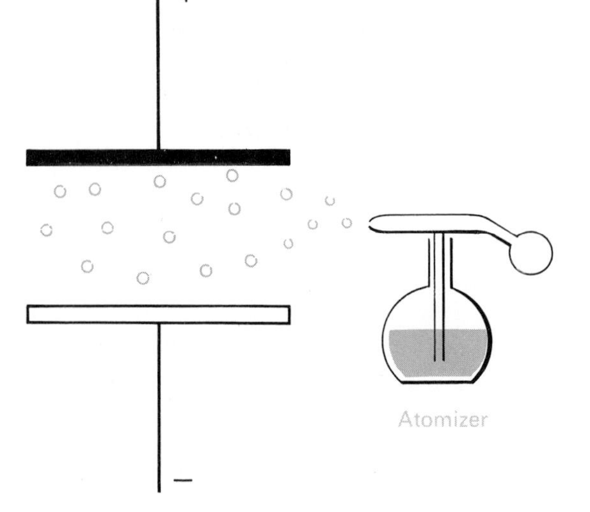

Figure 6–8 Charged oil drops, suspended as a result of opposing gravitational and electrostatic forces, provided Millikan with the means of calculating the charge on the electron.

count the marbles inside. Assuming the sacks to weigh a negligible amount, he weighed each sack of marbles and found them to weigh 8, 14, 18, 20, 24, 28, 36, and 40 grams. Does each marble weigh 1 gram? This is not likely, since chances are that one of the sacks would contain an odd number of marbles and weigh an odd number of grams. Weights of 3, 4, 5, 6, 7, and 8 grams per marble are not possible since the weights of the different sacks are not multiples of any of these numbers. However, if each marble weighs 2 grams, the weights of the sacks could be explained in terms of the sacks containing 4, 7, 9, 10, 12, 14, 18, and 20 marbles, respectively. The number 2 is the largest factor in *all* of the weights measured.

Now, with a good estimate of the charge of the electron and the ratio of charge to mass as determined by Thomson, the mass of the electron can be calculated.

$$\frac{e}{m} = 1.758 \times 10^{8} \frac{\text{coulomb}}{\text{gram}}$$

$$m = \frac{e\ (\text{coulombs})}{1.758 \times 10^{8} \dfrac{\text{coulomb}}{\text{gram}}}$$

Since

$$e = 1.602 \times 10^{-19}\ \text{coulomb}$$

$$m = \frac{1.602 \times 10^{-19}\ \cancel{\text{coulomb}}}{1.758 \times 10^{8} \dfrac{\cancel{\text{coulomb}}}{\text{gram}}}$$

Therefore,

$$m = 9.112 \times 10^{-28}\ \text{gram}$$

The hydrogen atom, the lightest atom, has a mass 1837 times larger than the mass of the electron. Since no charge has ever been found smaller than the charge on the electron, since no particle of ordinary matter has ever been found which has less mass, and since electrons can be obtained from any sample of matter, it must be assumed that the electron is a universally distributed, subatomic particle, which must be considered basic in any modern atomic theory.

Electrons are present in all the elements. An electron has an atomic mass of 0.000549 amu on the atomic weight scale.

POSITIVE RAYS AND THE PROTON

As stated above, many of the experiments performed with cathode-ray tubes were made with tubes that contained small amounts of some gas, such as hydrogen, oxygen, or carbon dioxide. Such tubes emit colored light characteristic of the particular gas. While experimenting with such tubes in 1886, Eugene Goldstein made an interesting discovery. While using a tube with a perforated negative electrode (Figure 6–9 shows a single hole for simplicity), he noticed a glow outside the region between the two electrodes. Goldstein suspected that these rays might be made up of positively charged particles, since such particles could be produced from the gas. On rushing toward the negatively charged electrode, some of them would coast through the opening. The positive character of the ray could easily be demonstrated by its deflection toward a negatively charged plate.

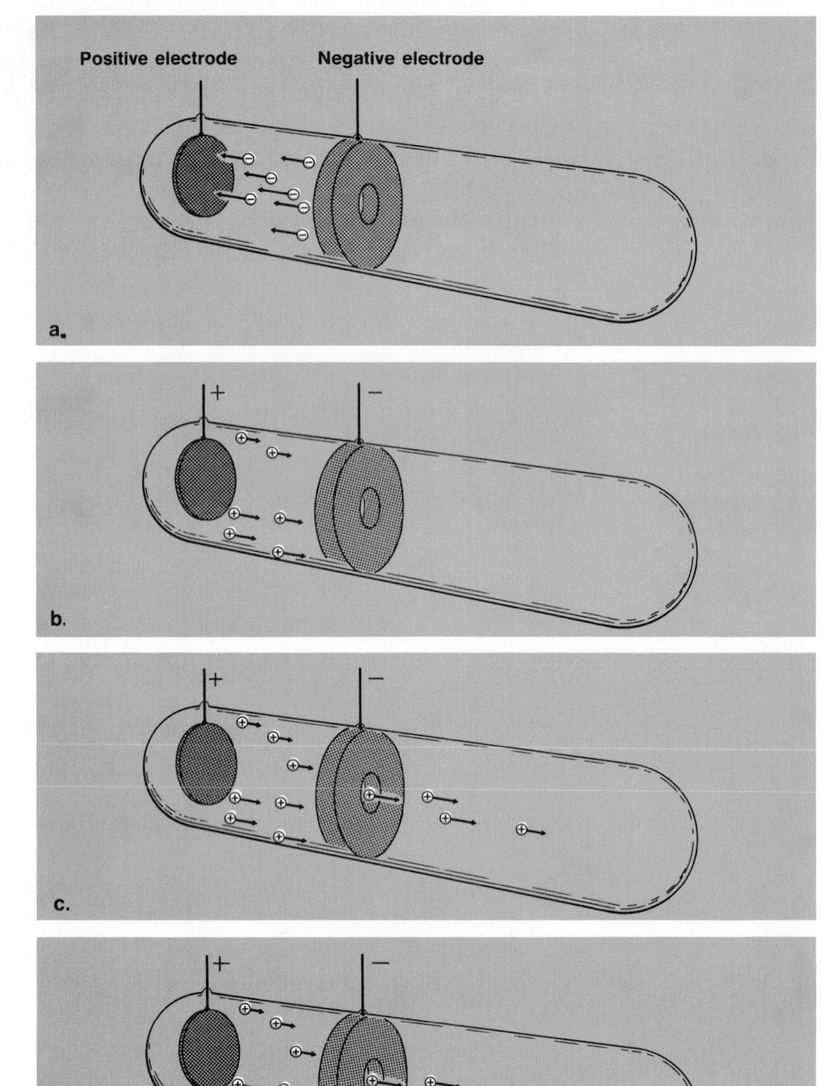

Figure 6–9 *a,* Electrons rush from negative electrode to the positive electrode due to high voltage, *b,* Electrons collide with gas molecules to produce positive ions which are accelerated toward the negative electrode. *c,* Some of the positive ions escape capture by the electrode and rush through opening due to their kinetic energy. *d,* Some of the positive ions in the positive ray collide with the gas molecules to produce a characteristic glow.

By properly constructing the experimental tube, W. Wien was able to obtain a well-defined positive ray (Figure 6–9). As a result, he determined e/m values (charge-to-mass ratios) for the positive ray, just as Thomson had done for the cathode ray. He found the e/m values to be dependent on the gas used in the tube, in contrast to the e/m value for the cathode ray, which remained the same regardless of the materials of construction or the gas that happened to be in the tube. Wien also found that the e/m values for different positive rays were all thousands of times smaller than the e/m value for the cathode ray or the electron. The largest e/m value for a positive ray, however, was found when hydrogen, the lightest atom known, was used as the filler gas.

How could the observations be explained? If the high-energy electron, moving from negative electrode to positive electrode, has enough

An atom or a group of atoms carrying a charge is called an *ion.*

energy to break apart ***(ionize)*** the molecules and atoms composing the filler gas, fragments of the atoms will result. In the case of helium gas, for example, a moving electron may knock another electron out of a helium atom, leaving a positively charged ***ion.*** This reaction can be expressed:

$$He + e^- \xrightarrow{\textit{High voltage}} He^+ + 2e^-$$

Positive rays are streams of positive ions derived from the gases present in the discharge tube.

The resulting helium ions, some of which would move through the opening in the negative electrode, constitute the positive ray in this example.

If, as indicated in the equation above, a helium atom can be ionized (broken) into a He^+ ion and an electron, it is reasonable to conclude that the size of the positive charge on the helium ion is the same as the size of the negative charge on the electron, since the two particles, when combined, form a neutral particle. Now, if the charges on the He^+ ion and the electron are the same, the e/m value for the He^+ ion would have to be smaller, since its m value is larger for the same value of e. If He^+ ions are thousands of times heavier than electrons, then the relative e/m values are satisfactorily explained.

Proton + electron = H atom.

1.007277 amu + 0.000549 amu = 1.007826 amu.

The fact that hydrogen gives rise to a positive ray with the ***largest*** e/m value and ***smallest*** m value suggests that the positive ion from the hydrogen atom, H^+, is a fundamental subatomic particle. This particle is called the ***proton*** and has essentially the same weight (obtained from the e/m ratio) as the hydrogen atom. Relative weights based on the atomic weight scale are: hydrogen atom, 1.007826 amu; hydrogen ion (proton), 1.007277 amu; and electron, 0.000549 amu.

The study of positive rays offers ample evidence that atoms can be broken down and provides a basis for the characterization of another fundamental subatomic particle, the proton. Later experiments in nuclear chemistry show that protons can be obtained from the more massive atoms, establishing the proton as a particle that must be accommodated in a modern atomic theory.

SELF-TEST 6-A

1. Positive rays obtained with different gases are (different/identical) while the cathode rays are (different/identical).

2. Cathode rays are composed of a universal constituent of matter named ______________.

3. When an atom is broken up into an electron and an ion, the charges on them are ______________ in magnitude and ______________ in sign.

4. Electrons can be obtained from (most, some, all) atoms.

5. The mass of the smallest atom is ______________ times greater than the mass of the electron.

6. The two fundamental particles revealed by studies using gas discharge tubes are the __________ and the __________.

7. The charge-to-mass ratio of the electron was discovered by ______________ and the charge on the electron was first accurately measured by ______________.

NATURAL RADIOACTIVITY

In 1896, a French chemist named Henri Becquerel (1852–1908) was experimenting with some compounds of the element uranium. He was interested in the visible fluorescence (light emission) of these materials. In his experiments Becquerel measured the amount of light emitted from a sample by exposing the sample to a photographic film, developing the film, and observing the amount of darkening produced. On one occasion, apparently by accident, he left a sample of uranium ore in contact with an undeveloped film which was still enclosed in its light-tight envelope of black paper. When the film was developed there was evident exposure from the uranium ore. The ore sample was emitting, in addition to the fluorescent light, a radiation that was quite capable of passing through the black paper. Further experimentation proved that it was the element *uranium* that was the source of the mysterious radiation. Later, the element *thorium* was found to exhibit the same property, and Pierre and Marie Curie followed with the famed discovery of the element *radium*. These elements, which were the source of a radiation even more penetrating than visible light, were termed ***radioactive elements.***

Some elements spontaneously give off radiation.

It was soon shown that the radiation from radium was actually composed of three separate rays, which were called **α (*alpha*), β (*beta*),** and **γ (*gamma*) *rays.*** These rays can be demonstrated in the experiment outlined in Figure 6–10. A radioactive sample, such as radium, is placed deep in a block of lead; the lead is necessary since it absorbs stray radiation and hence helps to focus the rays. Slits are used to select a very narrow beam of radiation. The radiation then passes between charged plates and separates, as indicated by a photographic exposure or a phosphor screen. The beta ray is evidently a stream of negative

Radium gives off alpha, beta, and gamma rays.

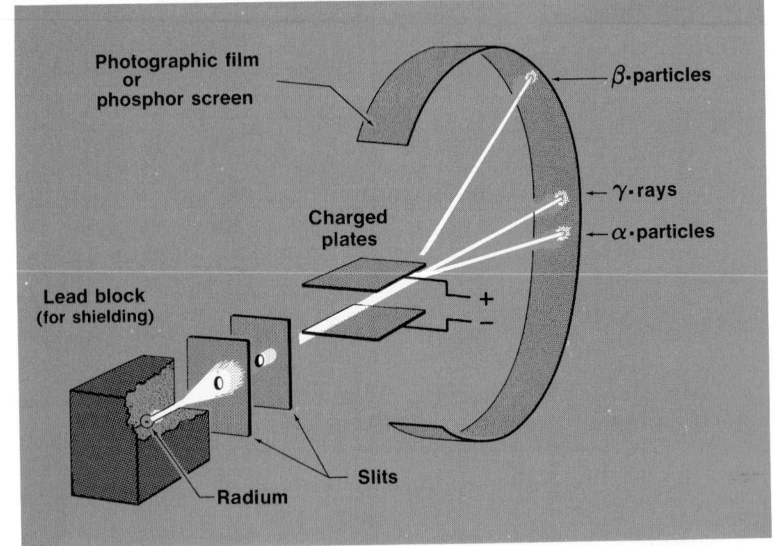

Figure 6–10 Separation of alpha, beta, and gamma rays by electrostatic field.

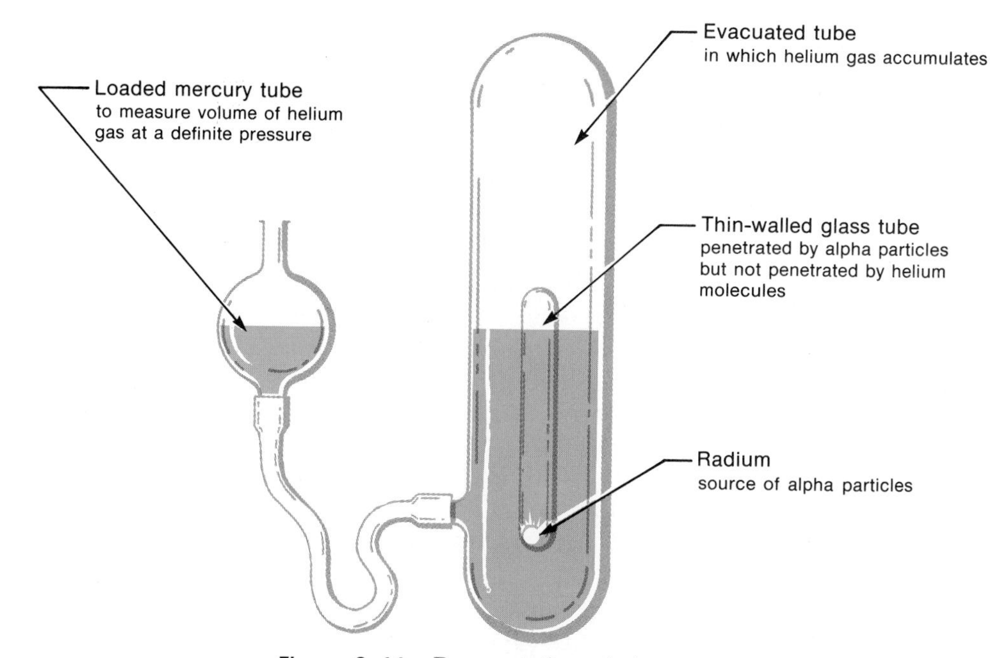

Figure 6–11 Representation of the apparatus designed by Ernest Rutherford and Thomas Royds, used to measure the volume of helium gas obtained from radium in the first direct determination (1910) of Avogadro's number. (From Breschia, F., et al., *Chemistry: A Modern Introduction*, W. B. Saunders Co., Philadelphia, 1974.)

particles, since it is deflected toward a positive plate. Its behavior in a magnetic field leads to the same conclusion. A study of this ray showed that the negative particles were actually electrons, and, except for the different kinetic energies of the particles, the beta ray was the same as the cathode ray.

In 1909, Ernest Rutherford, who had first demonstrated alpha and beta rays 10 years earlier, carried out an experiment that proved beyond all doubt that alpha particles were actually helium ions. Rutherford

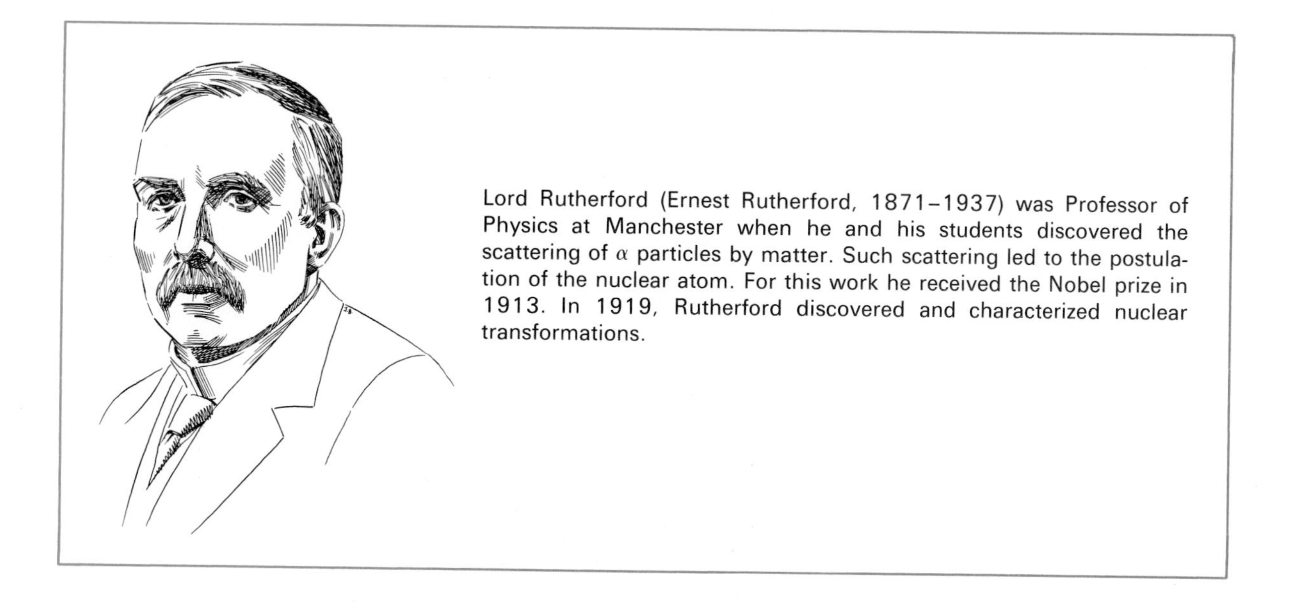

Lord Rutherford (Ernest Rutherford, 1871–1937) was Professor of Physics at Manchester when he and his students discovered the scattering of α particles by matter. Such scattering led to the postulation of the nuclear atom. For this work he received the Nobel prize in 1913. In 1919, Rutherford discovered and characterized nuclear transformations.

placed the radioactive element, radium, which was known to be an alpha emitter, in a closed container and trapped the gas that was produced. The gas had the properties of helium, so Rutherford concluded that the alpha particles were helium ions and that these ions picked up electrons from the environment to form helium atoms.

$$He^{2+} + 2e^- \longrightarrow He$$

The gamma ray, first noted in 1900 by P. Villard, in France, has no detectable mass. It is unaffected by electric or magnetic fields and is extremely penetrating. Whereas alpha radiation can be stopped by aluminum foil and beta rays by sheet aluminum, heavy lead sheets are required to stop gamma rays. Table 6–1 summarizes some of the information about alpha, beta, and gamma rays.

Chapter 24 discusses radioactivity in more detail.

The discovery of natural radioactivity left little doubt that atoms were divisible: some naturally occurring atoms spontaneously decompose, producing fragments which are either atoms of lighter elements or subatomic particles.

Natural radioactivity offers a straightforward approach to the determination of the number of particles in a mole (6.02×10^{23} molecules of helium per 4 grams). If a radioactive element is an alpha emitter, it is possible to set up an experiment whereby the number of alpha particles emitted is counted and, by measuring the volume of helium gas produced, Avogadro's number can be calculated.

How to determine Avogadro's number.

$$\frac{\text{number of atoms}}{\text{mole}} = \frac{\text{number of He atoms counted}}{\text{measured volume of He in liters at STP}} \times \frac{22.4 \text{ liters}}{\text{mole}}$$

Recall that a mole of a gas occupies 22.4 liters at STP.

ISOTOPES AND THE NEUTRON

An extension of the techniques learned in earlier studies of cathode and positive rays led to the development of the ***mass spectrometer*** (Figure 6–12). In this instrument, positive ions are produced from molecules or atoms by subjecting them to an electric discharge or some other source of high energy. The positive ions are accelerated by means of an electric

TABLE 6–1 Characteristics of Alpha, Beta, and Gamma Rays

Ray	Particle	Weight	Charge	Energy
α	Helium ion	4 amu	+2	Low
β	Electron	0.000549 amu	−1	Medium
γ	Radiant energy	none	none	High

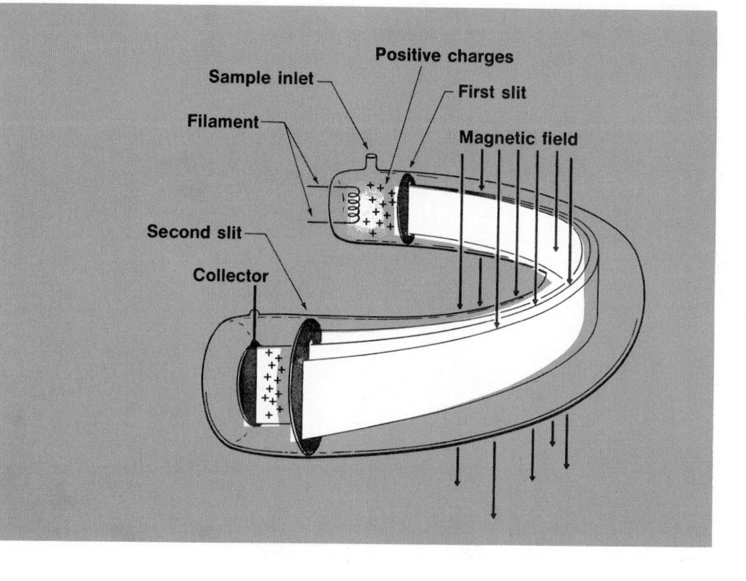

Figure 6–12 Mass spectrometer. Sample to be studied is injected near filament. Electrodes (not shown) subject sample to electron beam which ionizes a part of the sample. Electrodes are arranged to accelerate positive ions toward first slit. The positive ions that pass the first slit are immediately put into a magnetic field perpendicular to their path and follow a curved path determined by charge-to-mass ratio of the ion. A collector plate, behind the second slit, detects charged particles passing through the second slit. The relative magnitudes of the electronic signals are a measure of the numbers of the different kinds of positive ions.

field and then passed through a slit into a magnetic field. The slit selects a beam of ions. Since the chamber is highly evacuated, there will be no air to interfere with the ion paths. As expected, the charged particles follow a curved path in the magnetic field which is determined by the charge-to-mass ratio of the ion. The heavier ions are deflected less. When two ions with the same charge travel through the magnetic field, the one with the greater mass will tend to follow the wider circle. By increasing the strength of the magnetic field, it becomes possible to bring first the less massive ions and then the more massive ions through the second slit, thus producing a signal at the collector plate.

Most elements have atoms with the same chemical properties but different masses (isotopes).

In 1912, Thomson was investigating the positive rays associated with the elemental gas neon. According to the Daltonian concept that all atoms of an element weigh the same, it would have been reasonable for Thomson to have expected separate beams for Ne^{+}, Ne^{++}, and Ne^{+++} ions. This is true because the charge-to-mass ratio would be twice as big for the Ne^{++} ion. These expected results were obtained, but a surprising, additional observation was also made. After the Ne^{+} ions were focused onto the collector, Thomson found that by increasing the magnetic field strength slightly, he picked up another much smaller signal with an e/m value that corresponded to mass 22 on the atomic weight scale. Since neon has an atomic weight of 20, the signal for the mass-22 particle was at first thought to be an impurity in the sample of the neon. Later, Thomson and his student, F. W. Aston, considered the possibility that there were actually two kinds of neon atoms. In an experiment designed to answer this question, Aston allowed neon to diffuse through a porous clay pipe. If some neon atoms were more massive than others, they would tend to move more slowly through the pipe, and, as a result, concentrate in the gas remaining in the pipe. Sure enough, the gas remaining in the tube gave a much increased signal at mass 22. Aston concluded that not all neon atoms weight the same: ***most*** of them (909 out of 1000) have a mass of 20, a ***few*** of them have a mass of 22 and even fewer (3 out of 1000) have a mass of 21 on the atomic weight scale; and he concluded that the average value for all of them is an intermediate value of 20.18. The flood of experimental data that followed proved beyond all doubt that the

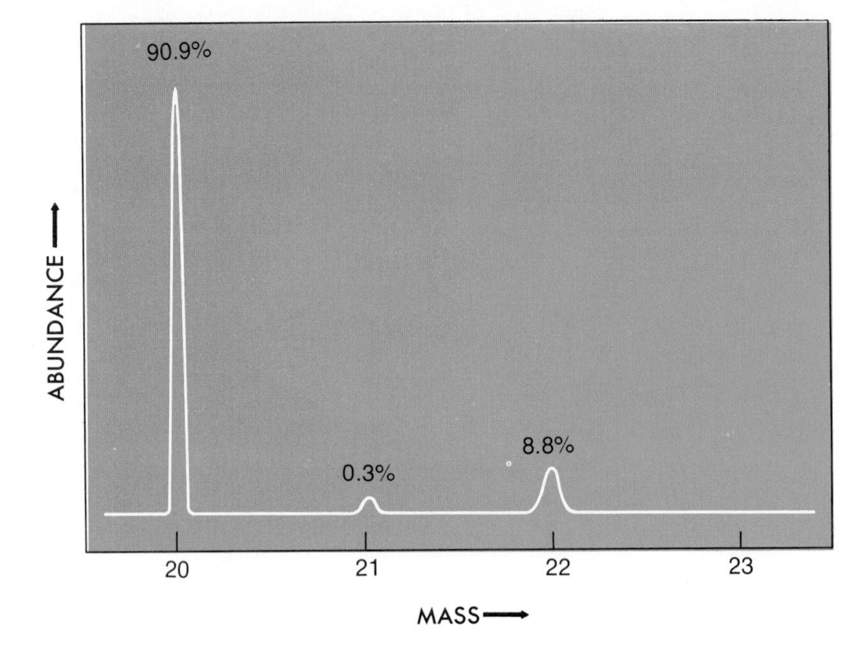

Figure 6–13 Mass spectrum of neon (+1 ions only). The principal peak corresponds to the most abundant isotope, neon-20. (From Masterton, W. L., and Slowinski, E. J., *Chemical Principles,* W. B. Saunders Co., Philadelphia, 1973.)

mass-22 signal was not an impurity and, furthermore, that like neon other elements have more than one kind of atom. Atoms of the same element that differ in mass are called ***isotopes.***

As a result of some of these experiments, Rutherford concerned himself with the problem of how it was possible to have two like-charged positive ions of neon (mass 20 and mass 22) with different masses. If the charges were the same, there would be an equal number of positive particles, protons. The appearance of different masses apparently meant the presence of a different number of some heavy, uncharged particles. In order to explain these mass differences Rutherford utilized a neutral particle called the ***neutron,*** first suggested by Sir J. J. Thomson. Thus, two isotopes of the same mass would contain the same number of protons but a different number of neutrons. The neutron turned out to be a rather elusive particle.

The neutron is needed to explain the existence of isotopes.

It was not until 1932 that a student of Rutherford's, James Chadwick, actually detected the neutron. Chadwick found that when he subjected beryllium (and some other elements as well) to a high-energy ray of alpha particles, massive, uncharged particles were emitted. The particles discovered have essentially the same mass as a proton or a hydrogen atom. These new particles are called ***neutrons*** and have a mass of 1.0087 amu on the atomic weight scale.

It was evident that the solid, indestructible Daltonian atom must give way to structured atoms composed of subatomic particles. The theory must allow for heavy atoms losing or gaining charge to form ions, for two atoms of the same element having different masses, and for subatomic particles. The positively charged proton with a relative mass of 1 amu, the negatively charged electron with a mass of 0.000549 amu, and the neutral neutron with a mass of 1 amu provide the subatomic particles for a revised atomic theory. The question remains: In what numbers and arrangements are these particles arrayed to form atoms of the elements?

RUTHERFORD'S GOLD FOIL EXPERIMENT—WHY BELIEVE IN THE NUCLEUS?

During the characterization of the alpha particle by Rutherford and his students in 1909, considerable interest was expressed in the study of interactions between alpha rays and various kinds of matter. It was found that an alpha ray could pass through a very thin piece of gold foil (4×10^{-5} cm thick) almost as if the gold foil were not present. The alpha particles were detected with a phosphor (scintillation) screen (Figure 6–14).

A close examination of the intensity of the light from the scintillation screen led Rutherford to the conclusion that some of the alpha particles were being scattered or deflected from their original path. When the gold foil was removed from the path of the beam, the light image was relatively sharp and clear in contrast to the slightly larger, and fuzzier, image obtained when the foil was in place. It seemed reasonable to believe that some of the alpha particles were being deflected because they were passing through certain regions of the atoms that caused this effect, whereas other regions of the atom lacked the ability to produce the effect. It was found to Rutherford's astonishment that some of the alpha particles were deflected at rather large angles; some of the particles were even "bounced" back toward their source. Rutherford expressed his surprise by stating he would have been no more surprised if someone had fired a 15-inch artillery shell into tissue paper and then found it in flight back toward the cannon.

Alpha particles are scattered by the atomic nucleus.

Rutherford reasoned that a massive particle with a large positive charge must be somewhere within the atom if alpha particle scattering was to be explained. As a result, the concept of the ***nuclear atom*** was born. If the atom had an extremely small nucleus which contained all the positive charge and most of the atomic mass, and if the negative charges of the electrons were distributed in the relatively vast space around the nucleus, most alpha particles would tend to pass through the gold atoms without ever coming close to a nucleus (Figure 6–15).

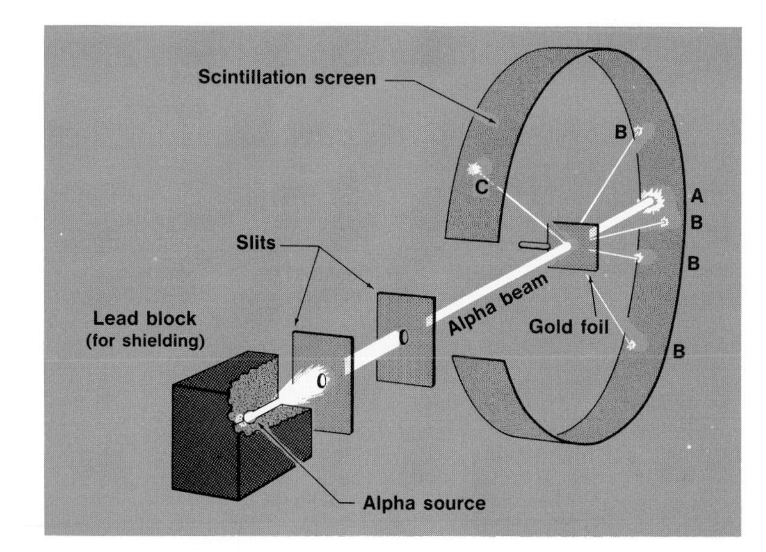

Figure 6–14 Rutherford's gold foil experiment. A circular, scintillation screen is shown for simplicity; actually, a movable screen was employed. Most of the alpha particles pass straight through the foil to strike the screen at point A. Some alpha particles are deflected to Points B, and some are even "bounced" backwards to points such as C.

If the alpha particle passed close to a gold nucleus, which is about 50 times more massive and carries a much higher positive charge than the alpha particle, the repulsion forces between the positive particles would cause the lighter alpha particle to be deflected.

How could an alpha particle be returned toward its source? An alpha particle with a very large amount of kinetic energy will slow down if approaching a positively charged nucleus almost directly, and the kinetic energy will be stored as potential energy between the two positive particles. When all of the kinetic energy is gone, the two particles would be very close together, repelling each other with tremendous force, and with nothing restraining them from flying apart. Hence, the much lighter alpha particle will rush backwards and the stored potential energy will be converted again to kinetic energy.

Knowing the mass of the alpha particle, a good estimate of its kinetic energy, and the magnitude of the charge on both the alpha particles and the gold nucleus, Rutherford was able to calculate just how close the two positive particles would have to approach each other in order to store all the kinetic energy as potential energy. The distance was 10^{-12} cm (0.000000000001 cm). This distance was compared to the size of the entire atom. It was a simple matter to measure the volume of one mole of gold (197 g), and, knowing the number of atoms in a mole, find the volume of a single atom. From the volume of the space occupied by a single atom, the radius was calculated to be approximately 10^{-8} cm. This means that the radius of the atom is 10,000 times

$$\left(\frac{10^{-8}\text{ cm}}{10^{-12}\text{ cm}} = 10^4 = 10{,}000\right)$$

greater than the radius of the nucleus. In terms of objects the sizes of

Alpha particle scattering can be explained if the nucleus occupies a very small volume of the atom.

Is matter really mostly empty space? It depends on the radius of the electron. The Bohr theory calculates an electron radius of 2.8×10^{-13} cm; the modern theory does not provide a way to calculate an electron radius.

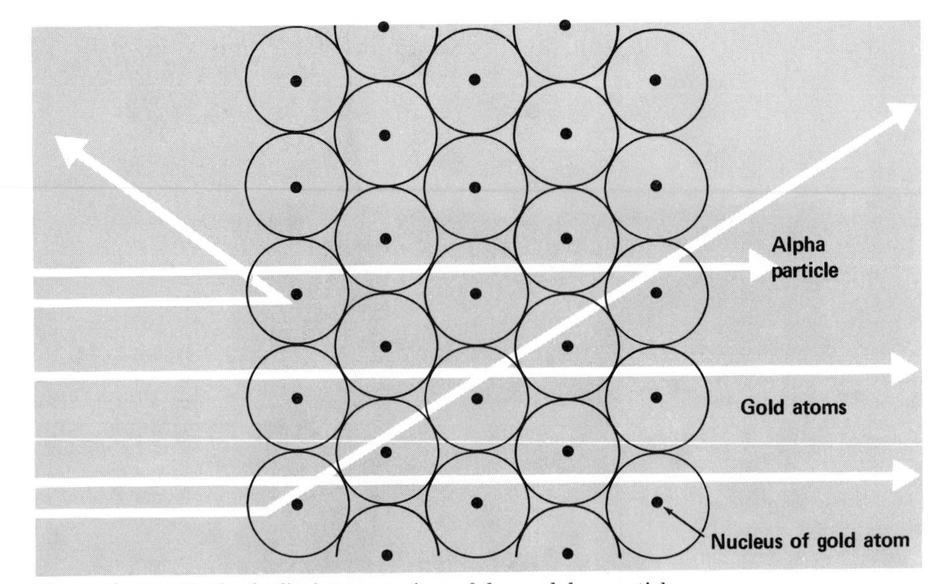

Figure 6–15 Rutherford's interpretation of how alpha particles interact with atoms in a thin gold foil. Actually, the gold foil was about 1000 atoms thick. For illustration purposes, points are used to represent the gold nuclei and the pathwidths of the alpha particles are drawn larger than scale.

which are easily understood, if an average atomic nucleus were the size of a basketball, the center of the nearest electron would be about *two miles* from the nucleus. Rutherford concluded that the volume occupied by an atom is mostly empty space; all the positive charge and practically all the mass of the atom are contained in a nucleus at the center of the atom. The electrons, equal in number to the positive charges (protons) in the nucleus, are distributed in the relatively vast reaches of space outside the nucleus.

THE PERIODIC LAW AND THE ATOMIC NUMBER

The periodic table arranges elements in groups (or families); elements in a given family have similar chemical properties.

Almost simultaneously with Mendeleev, a German chemist named Lothar Meyer showed independently that many of the physical properties of the elements exhibit periodicity and can be correlated with the periodic table. Mendeleev's table was more complete and provided a means for predicting that other elements were yet to be discovered.

At this point, as can be seen in retrospect, there occurred one of those great intellectual syntheses which brought together a great mass of previously unrelated information. Further development of atomic theory produced a model of the atom that began to explain both the physical behavior of atoms and their chemical properties from a unified viewpoint. Here we must note that Rutherford and his co-workers were fully aware of the relationships among the elements which are summarized in the periodic table or chart of the elements. This periodic table was developed over a period of fifty years (1820–1870) by a number of chemists, and ultimately was produced in a near final form by the brilliant Russian chemist, Dmitri Mendeleev. Recall from Chapter 2 that the basic notions of the periodic table are that the chemical properties of the elements are similar for all elements in the same family and that the properties are periodic or recurring if the elements are taken in order and placed into the periodic table.

But what is the order?

Basically Mendeleev used the order of atomic weights. Beginning with the lightest element, hydrogen, the periodic table was put together by adding the next element in order of atomic weight and continuing with the next heaviest atomic weight and so on until all elements were

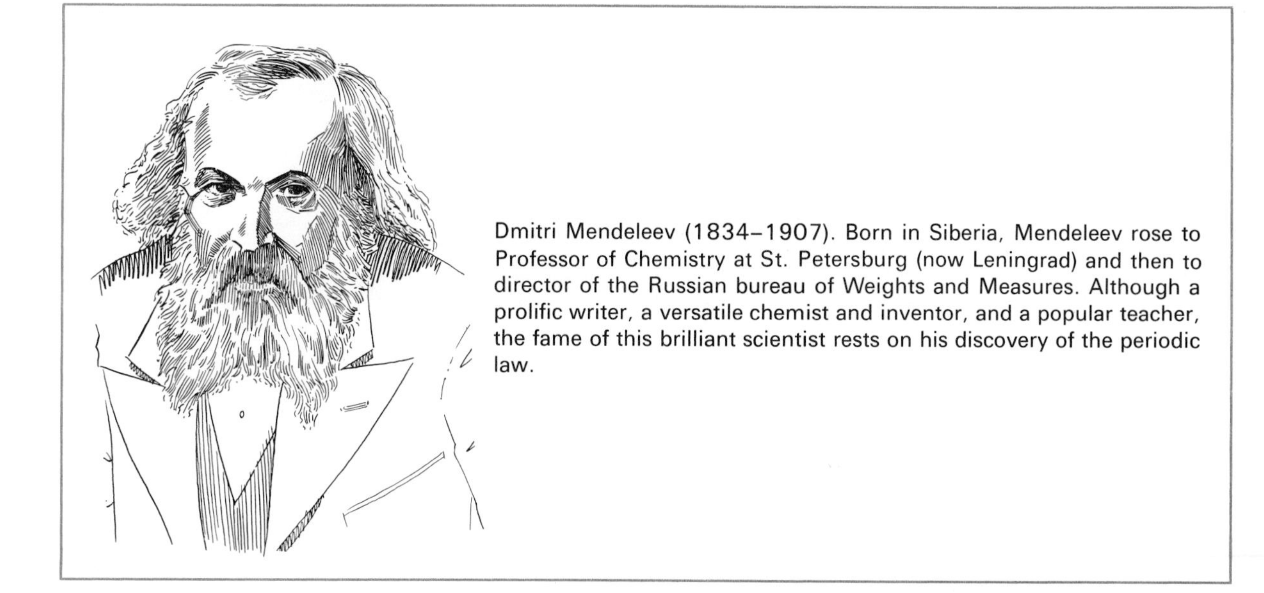

Dmitri Mendeleev (1834–1907). Born in Siberia, Mendeleev rose to Professor of Chemistry at St. Petersburg (now Leningrad) and then to director of the Russian bureau of Weights and Measures. Although a prolific writer, a versatile chemist and inventor, and a popular teacher, the fame of this brilliant scientist rests on his discovery of the periodic law.

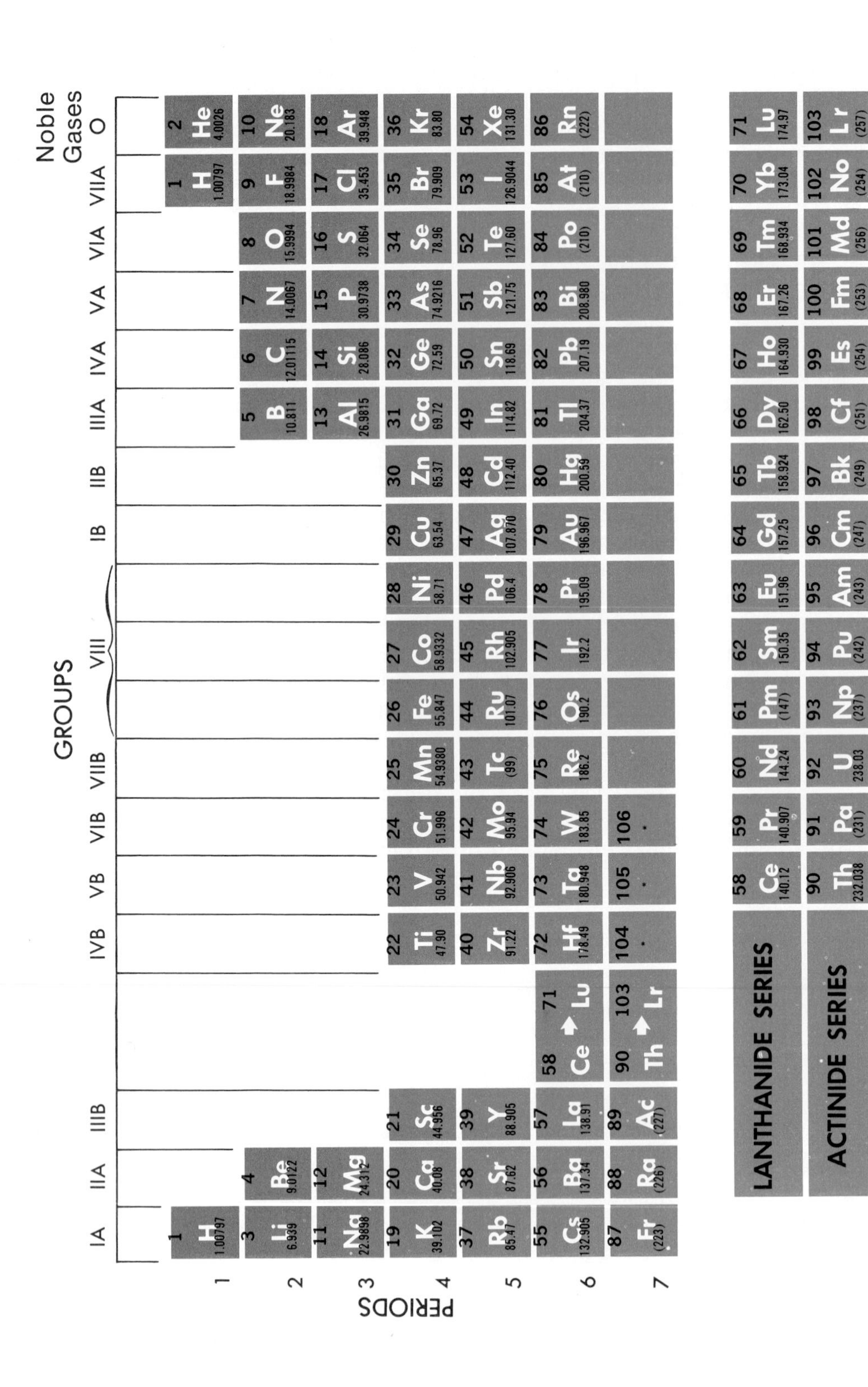

GROUPS

PERIODS	IA	IIA	IIIB		IVB	VB	VIB	VIIB	VIII	VIII	VIII	IB	IIB	IIIA	IVA	VA	VIA	VIIA	O Noble Gases
1	1 H 1.00797																	1 H 1.00797	2 He 4.0026
2	3 Li 6.939	4 Be 9.0122												5 B 10.811	6 C 12.01115	7 N 14.0067	8 O 15.9994	9 F 18.9984	10 Ne 20.183
3	11 Na 22.9898	12 Mg 24.312												13 Al 26.9815	14 Si 28.086	15 P 30.9738	16 S 32.064	17 Cl 35.453	18 Ar 39.948
4	19 K 39.102	20 Ca 40.08	21 Sc 44.956		22 Ti 47.90	23 V 50.942	24 Cr 51.996	25 Mn 54.9380	26 Fe 55.847	27 Co 58.9332	28 Ni 58.71	29 Cu 63.54	30 Zn 65.37	31 Ga 69.72	32 Ge 72.59	33 As 74.9216	34 Se 78.96	35 Br 79.909	36 Kr 83.80
5	37 Rb 85.47	38 Sr 87.62	39 Y 88.905		40 Zr 91.22	41 Nb 92.906	42 Mo 95.94	43 Tc (99)	44 Ru 101.07	45 Rh 102.905	46 Pd 106.4	47 Ag 107.870	48 Cd 112.40	49 In 114.82	50 Sn 118.69	51 Sb 121.75	52 Te 127.60	53 I 126.9044	54 Xe 131.30
6	55 Cs 132.905	56 Ba 137.34	57 La 138.91	58 Ce → Lu 71	72 Hf 178.49	73 Ta 180.948	74 W 183.85	75 Re 186.2	76 Os 190.2	77 Ir 192.2	78 Pt 195.09	79 Au 196.967	80 Hg 200.59	81 Tl 204.37	82 Pb 207.19	83 Bi 208.980	84 Po (210)	85 At (210)	86 Rn (222)
7	87 Fr (223)	88 Ra (226)	89 Ac (227)	90 Th → Lr 103	104 •	105 •	106 •												

LANTHANIDE SERIES	58 Ce 140.12	59 Pr 140.907	60 Nd 144.24	61 Pm (147)	62 Sm 150.35	63 Eu 151.96	64 Gd 157.25	65 Tb 158.924	66 Dy 162.50	67 Ho 164.930	68 Er 167.26	69 Tm 168.934	70 Yb 173.04	71 Lu 174.97
ACTINIDE SERIES	90 Th 232.038	91 Pa (231)	92 U 238.03	93 Np (237)	94 Pu (242)	95 Am (243)	96 Cm (247)	97 Bk (249)	98 Cf (251)	99 Es (254)	100 Fm (253)	101 Md (256)	102 No (254)	103 Lr (257)

Figure 6–16 Periodic table of the elements.

placed in the table. A new period was begun when an element had chemical properties similar to lithium.

The ordering of elements according to atomic weights presented some problems. Mendeleev realized that for some pairs of elements, the element with the higher atomic weight must precede an element with a lower atomic weight in order to match elements with similar chemical properties. In the modern version of the periodic table (Figure 6–16), note that K comes after Ar although the atomic weight of K is less than that of Ar. A similar reversal occurs for Ni-Co and I-Te. In such cases, Mendeleev always relied on the chemical properties in making the final decision on the exact position of such elements in the periodic table. His reliance on similar chemical properties to align vertical groups in the periodic table caused Mendeleev to leave some gaps in his table. He maintained that these blank spaces were elements yet to be discovered. For a number of these he predicted detailed chemical properties, which greatly facilitated their discovery. His hypotheses were completely justified by later discoveries of the new elements and the work of Henry Moseley on the atomic number concept, which is discussed at length in the next chapter.

The chemical properties of the elements are a periodic function of their *atomic number*.

As we shall see in Chapter 7, the ***atomic number*** turns out to be the same as the charge on the nucleus as well as the number of protons in the nucleus. The synthesis of the periodic table with atomic theory was a giant step in understanding matter. The story is begun in this chapter and is completed in Chapter 8.

SELF-TEST 6-B

1. The three types of radiation from a radioactive element such as radium are __________, __________, and __________, of which __________ pass through an electrostatic field without being deflected.
2. Isotopes of an element are atoms that have the same numbers of __________ but different numbers of __________.
3. The nucleus of an atom occupies a relatively (large/small) fraction of the volume of the atom.
4. The positive charges in an atom are concentrated in its __________.
5. The chemical properties of the elements are a periodic function of their ________ ________.
6. In a neutral atom there are equal numbers of __________ and __________.
7. The mass of the proton is __________ times that of the electron.
8. Three elements which were found to be radioactive in early studies in radioactivity are __________, __________, and __________.

MATCHING SET

______ 1.	uranium	a.	cathode ray particle
______ 2.	He^{2+}	b.	number of protons in a nucleus
______ 3.	electron	c.	similar chemical properties
______ 4.	^{22}Ne and ^{20}Ne	d.	proton plus electron
______ 5.	atomic number	e.	uncharged elementary particle
______ 6.	group of elements	f.	alpha particle
______ 7.	gamma ray	g.	scattering experiment
______ 8.	H atom constituents	h.	radioactive element
______ 9.	neutron	i.	radiant energy
______10.	measure nuclear size	j.	isotopes

QUESTIONS

1. What experimental evidence indicates that
   a. cathode rays have considerable energy?
   b. cathode rays have mass?
   c. cathode rays have charge?
   d. cathode rays are a fundamental part of all matter?
   e. two isotopes of neon exist?
   f. atoms are destructible?

2. Describe in detail Rutherford's gold-foil experiment under the following headings:
   a. experimental set-up.
   b. observations.
   c. interpretations.

3. Why was Thomson's charge-to-mass ratio determination for electrons very significant although he did not determine either the charge or the mass of the electron?

4. What part do ions play in explaining positive rays?

5. How do the following discoveries indicate that Daltonian atoms are inadequate?
   a. cathode rays.
   b. positive rays.
   c. nucleus.
   d. natural radioactivity.
   e. isotopes.

6. Characterize the three types of emissions from naturally radioactive substances as to charge, relative mass, and the relative penetrating power.

7. When a charged particle is sent through a magnetic field, which member of the pair would have a more curved path?
   a. an ion of charge +1 or an ion of charge +2 (both have same mass)?
   b. an ion of charge +1 or an uncharged particle?
   c. an ion of 2 amu or an ion of 1 amu (both have same charge)?

8. Use the information on the periodic chart to answer the following:
   a. the nuclear charge on cadmium, Cd.
   b. the atomic number of arsenic, As.
   c. the atomic weight of an isotope of bromine, Br, having 46 neutrons.
   d. the number of electrons in an atom of barium, Ba.

e. the number protons in an isotope of zinc, Zn.
f. the number of protons and neutrons in an isotope of strontium, Sr, atomic weight of 88 amu.
g. the number of electrons in a magnesium (Mg) ion of charge +2.
h. the number of electrons in an oxygen (O) ion of charge −2.
i. an element forming similar compounds to gallium, Ga.

9. Explain what the following terms mean:

   a. isotopes of an element.
   b. atomic number.
   c. an alpha emitter.
   d. chemical group of elements.

10. If electrons are a part of all matter, why are we not electrically shocked continually by the abundance of electrons about and in us?

11. Sodium metal reacts violently with water and forms hydrogen gas in the process. Magnesium metal will react with water only when the water is very hot. Copper metal does not react with water. Suppose you find a bottle containing a lump of metal in a liquid and a label, "Cesium (Cs) Metal." Based on your knowledge of the periodic table, what danger is there, if any, of disposing of the metal by throwing it into a barrel of water?

12. There are more than 1000 atoms with different weights, yet there are only 106 elements. How does one explain this in terms of fundamental particles?

13. What is a practical application of cathode ray tubes?

PROBLEMS

1. A common isotope of lithium (Li) has a mass of 7. The atomic number of lithium is 3. What are the constituent particles in its nucleus?

Ans. 3 protons and 4 neutrons

2. An element has 12 protons in its nucleus. How many extranuclear electrons will the atoms of this element possess?

Ans. 12

3. An isotope of mass number 60 has 33 neutrons in its nucleus. What is its atomic number and what is the element?

Ans. At. no. = 27; element is cobalt

4. An isotope of cerium has 88 neutrons in its nucleus. How many protons plus neutrons does this nucleus contain?

Ans. 146

5. The element iodine occurs as a single isotope of mass 127; its atomic number is 53. How many protons and how many neutrons does it have in its nucleus?

Ans. 53 protons and 74 neutrons

6. An atom with an atomic number of 8 is found to have three isotopes of mass 16, 17, and 18. How many protons and neutrons are present in each nucleus?

Ans. Mass 16; 8 protons and 8 neutrons
Mass 17; 8 protons and 9 neutrons
Mass 18; 8 protons and 10 neutrons

7. Suppose Millikan had determined the following charges on his oil drops:

   1.33×10^{-19} coulomb
   2.66×10^{-19} coulomb
   3.33×10^{-19} coulomb
   4.66×10^{-19} coulomb
   7.92×10^{-19} coulomb

a. What would be the fundamental charge on an electron, assuming these are representative data?
b. Suppose, some time later, someone had determined a charge of 4.29×10^{-19} coulomb. What effect would this have on the fundamental charge?

Ans. a. 0.67×10^{-19} coulomb
b. 0.33×10^{-19} coulomb

8. Suppose an isotope of aluminum has an atomic weight of 27.0 amu. How many protons, neutrons, and electrons are in an atom of this isotope? What is the charge on the nucleus?

Ans. 13 protons, 14 neutrons, 13 electrons in an atom. Charge on nucleus is +13.

9. If the nucleus of a gold atom is 10^{-12} cm in diameter and the diameter of the atom itself is 3×10^{-8} cm, what percentage of the volume of the atom is occupied by the nucleus? The volume of a sphere is $\frac{4}{3}\pi r^3$.

Ans. $3.7 \times 10^{-12}\%$

SUGGESTIONS FOR FURTHER READING

"A Naked Electron," *Chemistry,* Vol. 42, No. 6, p. 25 (1969).
Allen, W. M., "The Diagonal Periodic Relationship," *Chemistry,* Vol. 43, No. 4, p. 22 (1970).
Birks, J. B. (ed.), "Rutherford at Manchester," W. A. Benjamin, Inc., New York, 1963. (A memorial account of Rutherford's work with articles written by Bohr, Marsden, and others. Also contains reprints of Rutherford's original papers.)
Clark, H. M., "The Origin of Nuclear Science," *Chemistry,* Vol. 40, No. 7, p. 8 (1968).
Duggan, J. L., and Yeggl, J. F., "A Rutherford Elastic Scattering Experiment," *Journal of Chemical Education,* Vol. 45, p. 85 (1968).
"Electron Microscope," *Chemistry,* Vol. 43, No. 2, p. 27 (1970).
Garrett, A. B., "The Flash of Genius: The Neutron Identified: Sir James Chadwick," *Journal of Chemical Education,* Vol. 39, p. 638 (1962).
Greenaway, F., "John Dalton and the Atom," Cornell University Press, Ithaca, New York, 1966.
Keller, O. L., "Predicted Properties Of Elements 113 And 114," *Chemistry,* Vol. 43, No. 10, p. 8 (1970).
Millikan, Robert, "Electrons—What They Are And What They Do," *Chemistry,* Vol. 40, No. 4, p. 13 (1967).
Morris, D. L., "Music Of New Spheres," *Chemistry,* Vol. 42, No. 11, p. 10 (1969).
Rich, R., "Periodic Correlations," W. A. Benjamin, Inc., New York, 1965.
Romer, A., "The Restless Atom," Doubleday and Co., Inc., Garden City, N. Y., 1960.
Seaborg, G. T., "Some Recollections of Early Nuclear Age Chemistry," *Journal of Chemical Education,* Vol. 45, p. 278 (1968).
"Spiral Form Of The Periodic Table," *Chemistry,* Vol. 43, No. 1, p. 27 (1970).
"The Periodic Table, 1869-1969," *Chemistry,* Vol. 42, No. 5, p. 26 (1969).
Wallace, H. G., "The Atomic Theory—A Conceptual Model," *Chemistry,* Vol. 40, No. 10, p. 8 (1967).
Young, J. A., and Malik, J. G., "Chemical Queries," *Journal of Chemical Education,* Vol. 45, p. 254 (1968). (The derivations of all the equations for the Millikan oil drop experiments are given.)
Zimmerman, J., "Mendeleev—His Own Man," *Chemistry,* No. 42, No. 11, p. 32 (1969).

CHAPTER 7

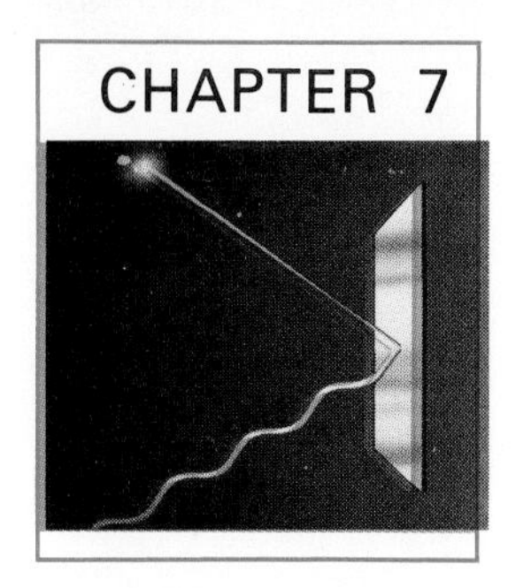

THE BOHR ATOM

LIGHT AND MATTER

A source of light is a source of energy, since light is radiant energy in motion. Light requires no medium through which to move; it can travel through space occupied by no matter at all (a vacuum) at the very high speed of 186,000 miles per second, or 3×10^{10} centimeters per second. When light travels through forms of matter, it does so at lesser speeds, depending on the kind of matter, and some of it is absorbed. This point, coupled with the fact that light is only produced in and by matter, suggests that an understanding of how the absorption and emission are accomplished will reveal a great deal of information on the structure of matter.

Light has a dual nature.

The exact nature of light still remains somewhat of a mystery; no single theory of light has been developed to explain all of its observable properties. It appears to have a dual character. Some of the properties of light, such as color and reflection, can be explained by considering light to be a series of waves, like waves on the surface of water. This wave theory considers the valleys and peaks of the waves to be determined by vibrations in electrical and magnetic fields arranged at right angles to each other (hence the name, electromagnetic energy). Other properties of light can be explained only when light is considered to be

Niels Bohr (1885–1962) received his doctor's degree in the same year that Rutherford announced his discovery of the atomic nucleus, 1911. After studying with Thomson and Rutherford in England, Bohr formulated his model of the atom. Bohr returned to the University of Copenhagen and, as Professor of Theoretical Physics, directed a program that produced a number of brilliant theoretical physicists. He received the Nobel Prize for physics in 1922.

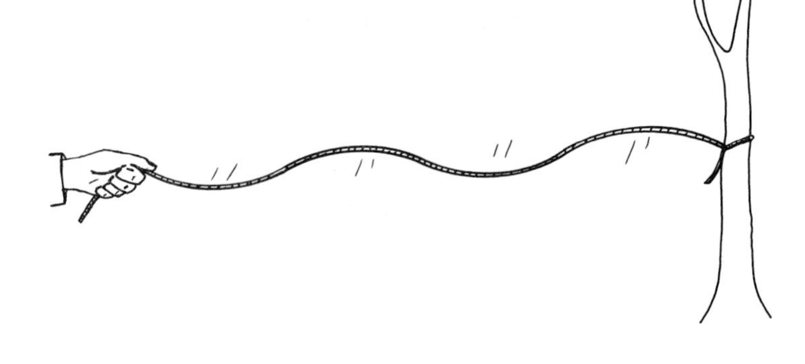

Figure 7–1 The dualistic nature of light: wave nature is similar to waves in a rope; particle nature is similar to individual packages of energy from a machine gun.

a stream of particles. The discrete particles or units of energy are called ***photons*** or ***quanta*** (Figure 7-1). More will be said concerning this theory later in this chapter.

White light, passed through a prism, is dispersed into the various colors of the rainbow (Figure 7-2). The resulting panorama of color is called the ***continuous visible spectrum*** (see Plate II). The triangular prism, which can be made of any transparent material, bends the path of the violet light the most and that of red light the least.

According to the wave theory of light, the different colors have different wavelengths. In traveling a given distance, one of the wavelengths in the blue region, having a shorter wavelength than red light, will trace out more waves than red light (Figure 7-3). Since all colors of light travel at the same speed, the number of waves per second, called ***frequency,*** will be greater for blue light. Since the frequency, ν (the Greek letter nu), refers to the number of waves per second, the product of the frequency times the ***length of one wave,*** λ (lambda), is the distance

Colors were once thought to be qualities of light. Sir Isaac Newton proved the colors to be light.

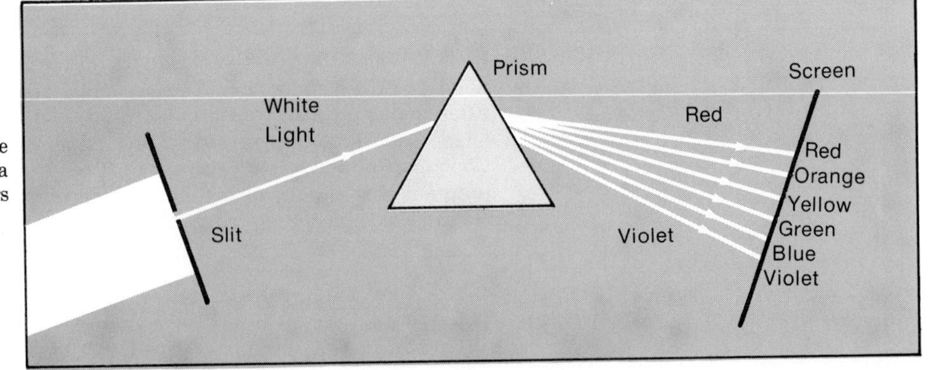

Figure 7–2 Spectrum from white light produced by refraction in a glass prism. The different colors blend into one another smoothly.

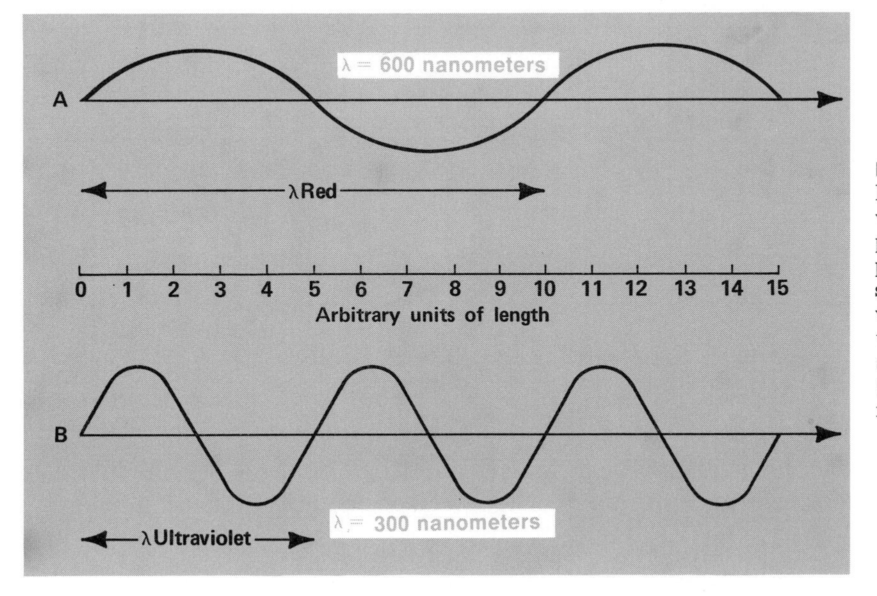

Figure 7–3 The wave theory of light considers light to be waves vibrating at right angles to their path of motion. Red light, *A*, completes one vibration or wave in the same distance and time that involve two complete waves of ultraviolet light. *B*, Ultraviolet light is of shorter wavelength than violet light (see Figure 7–4). A nanometer is 10^{-9} meter.

traveled per second or the speed of light, c. Using λ in cm and ν in cycles per second, this relationship is given by the equation

$$c\left(\frac{\text{cm}}{\text{sec}}\right) = \lambda\ (\text{cm})\ \nu\left(\frac{1}{\text{sec}}\right)$$

A visible wavelength results in a color, but some colors are mixtures of wavelengths.

There are many different shades of each color, each with its unique frequency and wavelength. In the spectrum of white light, each color blends smoothly into the next (Figure 7–2).

The visible portion of the spectrum is only a very small part of the electromagnetic spectrum of radiant energy. Ultraviolet light has a higher frequency than visible light. X-rays and γ-rays (mentioned earlier as an emission in radioactive decay of unstable atoms) have even higher frequencies. Infrared light, microwaves, and radio waves have progressively lower frequencies than visible light. The wavelength of some X-rays are short (about the same length as the distance between atoms in mole-

Some insects see in the ultraviolet part of the spectrum.

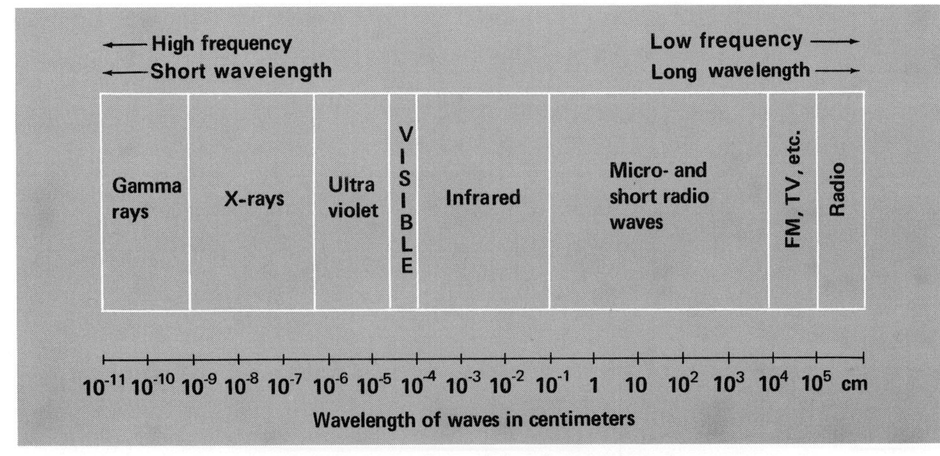

Figure 7–4 Regions of the electromagnetic spectrum.

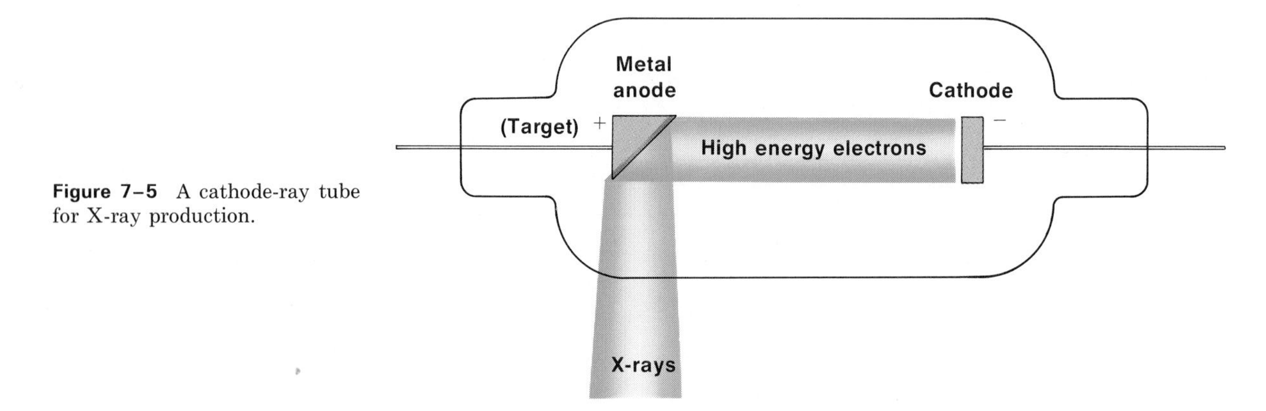

Figure 7–5 A cathode-ray tube for X-ray production.

cules), whereas some radio wavelengths may be as long as a football field. Figure 7–4 summarizes such information about the electromagnetic spectrum.

MOSELEY'S DETERMINATION OF ATOMIC NUMBERS

X-rays can be produced by placing a metal target in the path of a high-energy stream of electrons (Figure 7–5). In addition to the broad background of radiation produced, each element produces certain wavelengths of X-rays in considerable abundance (called lines in the X-ray spectrum), and these wavelengths are characteristic of the element used as the target in the X-ray tube. In 1913, an English scientist, Henry Moseley, measured these characteristic frequencies in the X-ray spectra of the various elements used as targets and discovered that an order existed among the elements (Figure 7–6). Moseley found that if he plotted the square root of the frequency of the X-ray, $\sqrt{\nu}$, against the order of the elements in the periodic table, a straight line resulted, as shown in

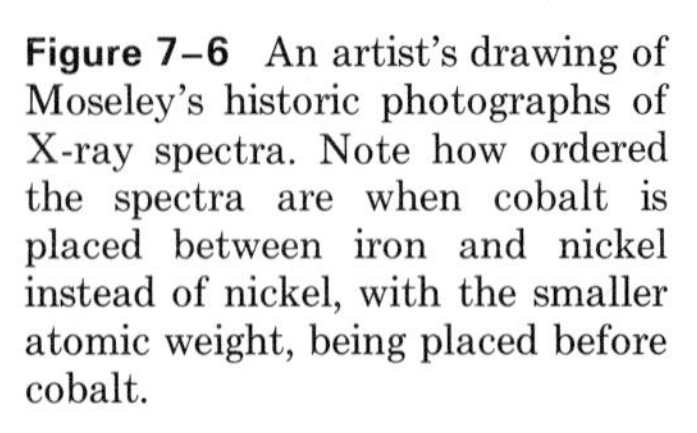
Figure 7–6 An artist's drawing of Moseley's historic photographs of X-ray spectra. Note how ordered the spectra are when cobalt is placed between iron and nickel instead of nickel, with the smaller atomic weight, being placed before cobalt.

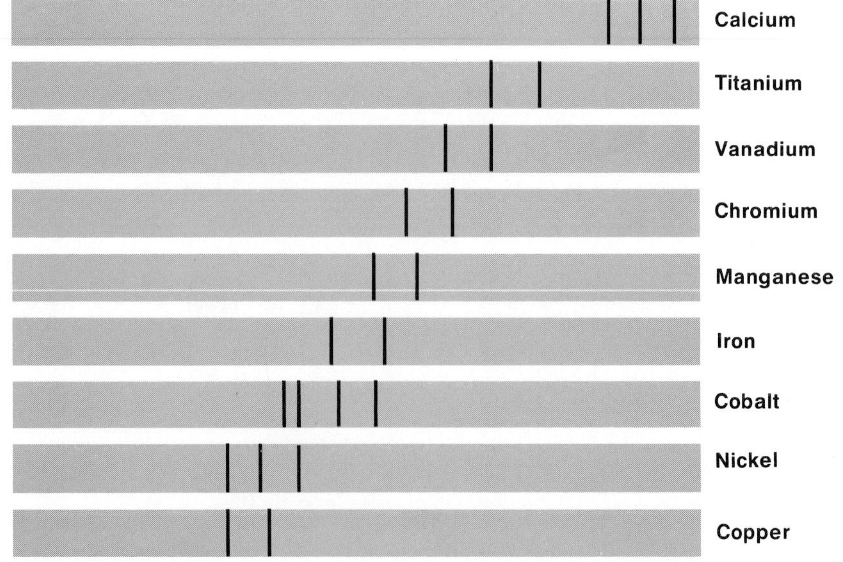

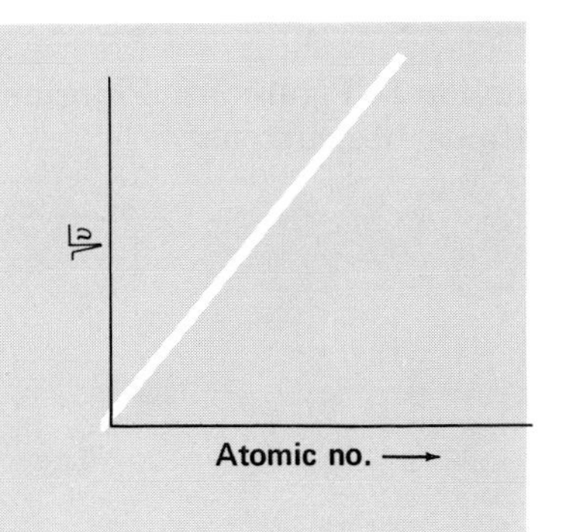

Figure 7–7 A plot of the square root of the frequency of certain X-rays characteristic of the anode material versus the atomic number of the element composing the anode.

Figure 7–7. It is evident that Moseley found a "probe" (X-rays) that revealed a characteristic in the structure of atoms which truly distinguishes one element from another. The ordering of the elements in the periodic table is not trivial but is based on some fundamental property of atoms. His data also verified that the ordering should put cobalt ahead of nickel, although atomic weights indicated otherwise. This characteristic is called the ***atomic number of the element*** and is identified with the positive charge on the nucleus.

With the determination of atomic numbers, it becomes possible to describe the content of the atoms in terms of three subatomic particles: the electron, the proton, and the neutron. For example, the atomic number of 8, for oxygen, indicates that there are eight protons in the nucleus, each of which has a weight of 1 amu. From the atomic weight of oxygen, 16, it can be concluded that oxygen has eight neutrons in the nucleus, each with a weight of 1 amu (16 minus 8). In order for the atom to be electrically neutral there must be eight electrons, each with a very small mass, somewhere in the volume of the atom outside of the nucleus. This information may be symbolized as shown in Figure 7–8.

8e
8p
8n

Figure 7–8 Representation of an oxygen atom.

Consider the case of chlorine, atomic number 17 and atomic weight 35.45. This element consists of two isotopes of masses 35 and 37 (Figure 7–9). The fact that atoms of Cl-35 are more abundant in nature explains why the average atomic weight for this element is closer to 35 than to 37.

Isotopes are sometimes described by a subscript that gives the charge on the nucleus (this is redundant because the symbol of the element also gives this information) and a superscript to give the mass number:

$$^{35}_{17}\text{Cl and } ^{37}_{17}\text{Cl}$$

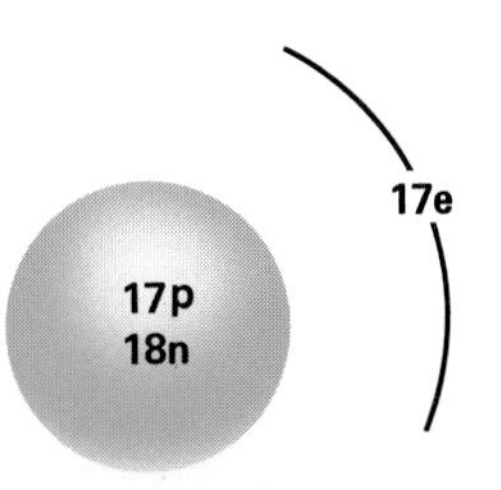

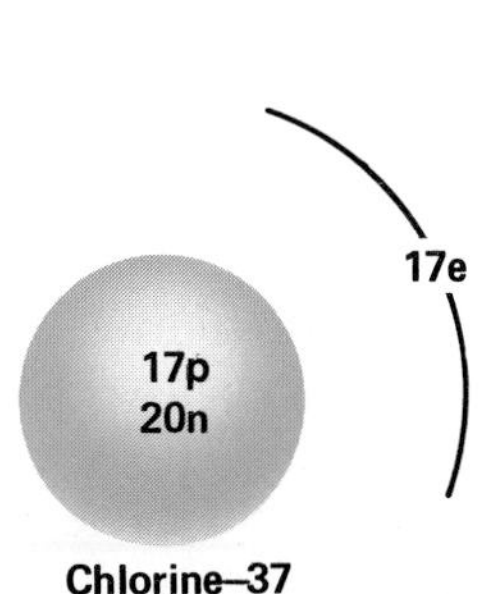

Figure 7–9 Representation of chlorine isotopes.

SELF-TEST 7-A

1. Draw a representation for an atom of fluorine similar to Figure 7–8. Fluorine has an atomic number of 9 and the atom contains 10 neutrons.

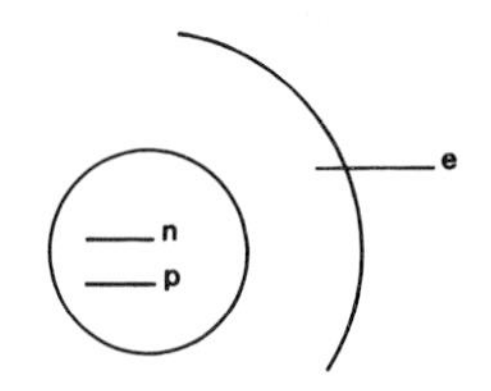

2. Draw a representation for an atom of tin which weighs 118 amu.

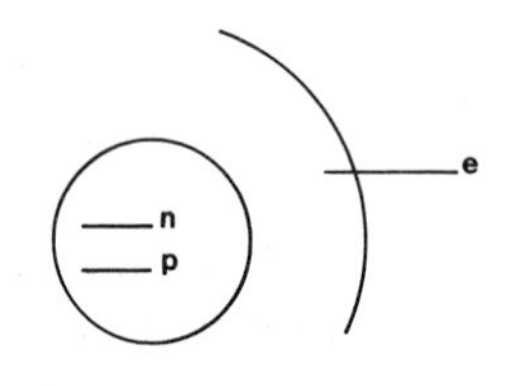

3. Under some conditions, light has properties of ______________ and under other conditions exhibits the properties of ______________.

4. When light is broken up into the different colors composing the light, a ______________ is produced.

5. a. Of the following colors of light, which has the greatest frequency: red, blue, orange, green? ____________

   b. Which has the longest wavelength? ____________

6. All of the following are electromagnetic energy except one. Which is the exception? X-rays, radio waves, gamma rays, light, alpha particles, TV waves. ____________

7. Moseley found an ordering among elements by comparing their ________.

PROBLEMS WITH THE RUTHERFORD ATOM

With the Rutherford model in mind (a small heavy nucleus containing the positive charge and extranuclear electrons at relatively great distances), a question immediately arises concerning the stability of the atom. Why do the negative electrons not collapse into the positive nucleus, since oppositely charged particles attract each other? To avoid this

A radio station operating at 650 kilocycles will have electric charges oscillating 650,000 cycles each second in its sending antenna. As a result, electromagnetic waves are emitted by the antenna with a frequency of 650 kilocycles, which in SI units is 650 kilohertz. The unit, hertz is identical to cycle/sec, and is named in honor of Heinrich Rudolf Hertz, who developed the first receiver for radio waves.

difficulty, Rutherford pictured the atom as a very small solar system; that is, the electrons orbit about the nucleus in much the same way the planets orbit about the sun. However, this is not a completely satisfying explanation because the orbiting charge presents another problem. In 1865, an Englishman, James Clerk Maxwell, showed by some very elegant mathematical reasoning that a moving charged body, when changing its direction in space, will radiate electromagnetic energy. A German physicist, H. Hertz, showed in 1879 that Maxwell's theory was correct. He caused charges to move back and forth (oscillate) in a wire and thereby produced the first radio waves controlled by man. According to this established theory, the orbiting (oscillating) electron within the atom should radiate energy and run down, spiralling into the nucleus. Obviously, this is not the case.

One of the ways out of the dilemma is to postulate that Maxwell's theory and Hertz's experiments apply only to macroscopic units of charge and not to atoms. We are then forced to conclude that the laws of macroscopic matter are not completely adequate to explain the theoretical world of atoms and molecules.

EMISSION SPECTRUM OF HYDROGEN

The beautiful colors of fireworks are produced by mixing various elements (usually in the form of one of their compounds) into the gunpowder charge. Copper in a flame produces green light; sodium, yellow; calcium, brick red; strontium, scarlet red; and so forth. This behavior indicates that if the atoms of various elements are capable of giving off only certain colors or wavelengths of light, this fact must somehow be related to the different structures of the atoms of these elements.

In order to study the characteristic light emitted by an element,

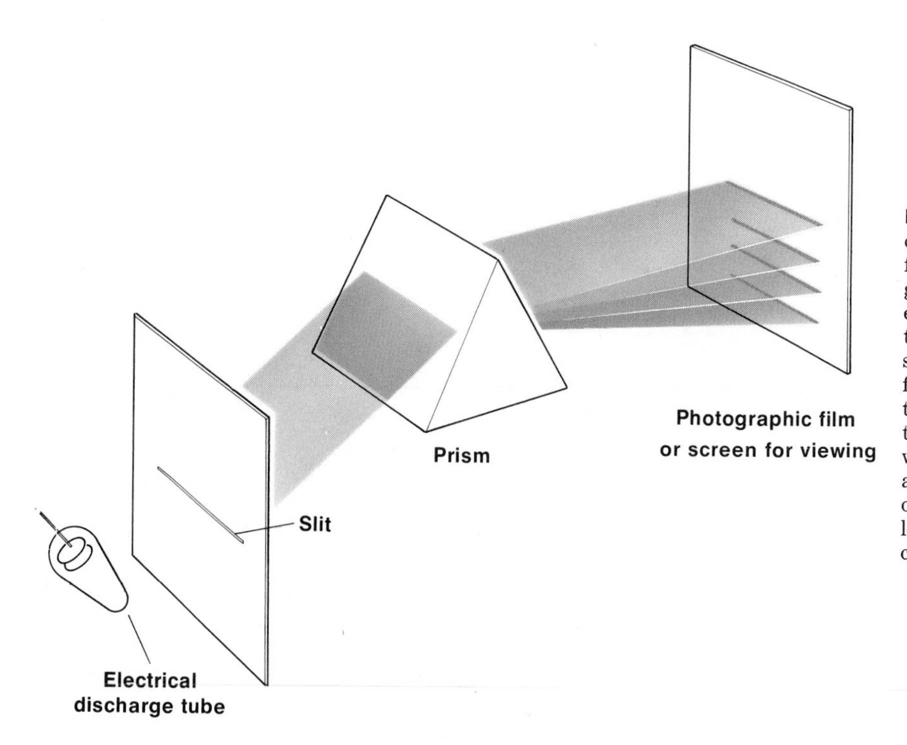

Figure 7–10 Basic components of spectrograph (or spectroscope) for viewing atomic spectra. The gaseous element is energized by electrical discharge. A beam of the emitted light passes through a slit. Each wavelength of light follows a different path through the prism, resulting in an image of the slit on the screen for each wavelength of light. These images are called spectral lines. From the optics of the prism, the wavelength of each line of light can be calculated.

a prism can be used to disperse the light so that the actual wavelengths of the light emitted can be measured. An electrical discharge tube can be used to provide energy to the atoms so that they in turn can emit light. An instrument that separates the wavelengths of light into a visible pattern (Figure 7-10) is called a ***prism spectroscope*** if the different wavelengths (colors) of light are observed with the eye, or a ***spectrograph*** if the spectrum is recorded in some way, such as on photographic film.

Photographic paper can "see" where the eye fails.

Of all of the elemental emission spectra, the hydrogen spectrum is the simplest and, since hydrogen is the simplest of all the atoms, this is another clue that there is a fundamental relationship between atomic structure and atomic spectra. Notice that in the hydrogen visible spectrum (see Color Plate III), the lines are grouped closer together toward the blue end of the spectrum. If the photograph is extended beyond the visible region into the ultraviolet, this series of lines is observed to approach a limiting value (see line A in Figure 7-11). This series of lines is called the ***Balmer series*** in honor of Johann Balmer who, in 1885, discovered a simple mathematical relationship involving the wavelengths of this series of lines in the hydrogen emission spectrum and certain integers. The Balmer equation is

1 nanometer = 10^{-9} m or 10^{-7} cm.

$$\lambda \text{ (in nanometers)} = 364.56 \left(\frac{n^2}{n^2 - 4}\right) \qquad (1)$$

where λ is the wavelength of the spectral line, and n is an integer in a series beginning at 3. If the value of n = 3 is substituted into Balmer's

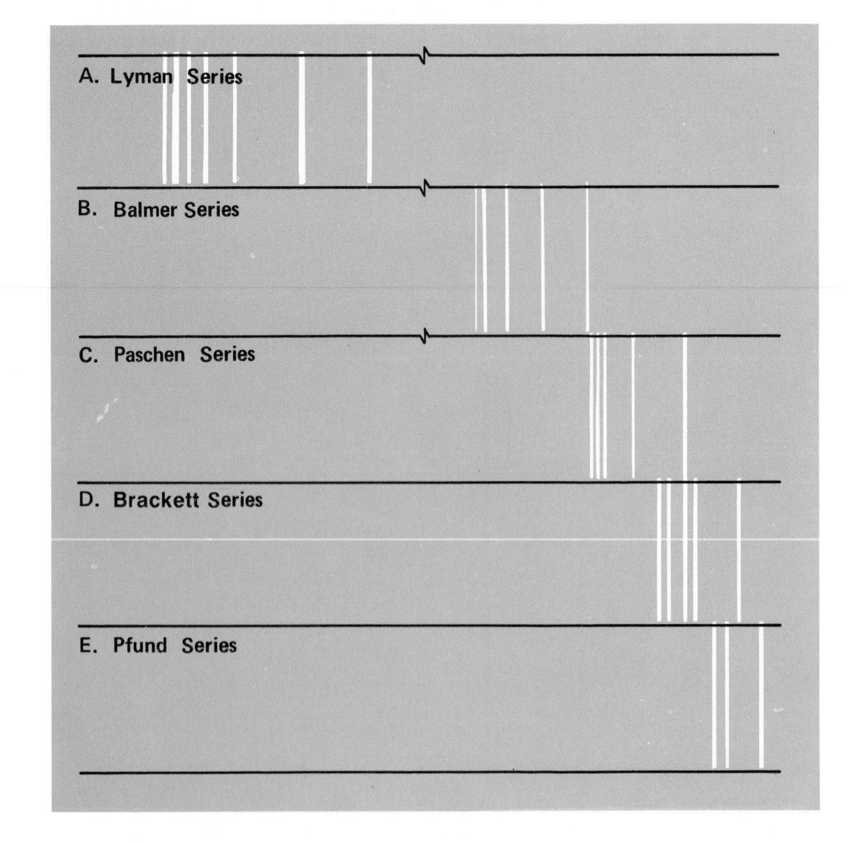

Figure 7-11 The hydrogen spectrum. Frequency increases to the left of the figure. The symbol ⁓ indicates that the Lyman series, in the ultraviolet region, is considerably removed from the Balmer series in the visible region.

equation, the calculated wavelength is 656.21 nanometers. This value differs from the observed wavelength of the spectral line of longest wavelength in this series of spectral lines by only 0.07 nanometer. As the value of n is increased, wavelengths corresponding precisely to other observed lines are obtained. This is truly a remarkable achievement, especially to those who believe there is an inherently simple relationship between the structure of atoms and their properties. There is some property of the hydrogen atom which allows it to emit only certain wavelengths of light, and these wavelengths are related by a simple algebraic equation. The fact that n can have only integral values suggests that there must be a simple orderly change occurring in the atom.

Other spectroscopists—Lyman, Paschen, Brackett, and Pfund—extended the study of the hydrogen spectrum into the infrared and ultraviolet regions of the spectrum. They discovered four more series of lines similar to the Balmer series. In each series a similar, simple mathematical relationship holds; in each series integers are involved just as with the n in the Balmer equation.

PLANCK'S QUANTUM HYPOTHESIS

The observation that hot objects give off visible light is as old as man's knowledge of fire. Another observation is that the color of the visible light emitted by hot objects varies with the temperature. For example, as it is heated, a piece of iron will glow first dull red, then bright red, then orange, and finally white as its temperature is increased.

The relative intensities of the wavelengths of energy emitted from a black object heated to a temperature of about 1500 K are graphed in Figure 7-12. The greatest intensity is at a wavelength of about 2,000 nanometers.

For an object like this, the theory predicted that the intensity of radiation should become greater and greater at shorter and shorter wavelengths. This is shown as the dotted line in Figure 7-12. Obviously, the experiment is in direct contradiction with the theory.

Max Planck (1858–1947), a German physicist, explained how energy radiates from a hot object. In 1900 he announced his quantum hypothesis which was later used to explain other natural phenomena by Einstein (photoelectric effect) and Bohr (hydrogen spectral lines). He received the Nobel Prize in 1918.

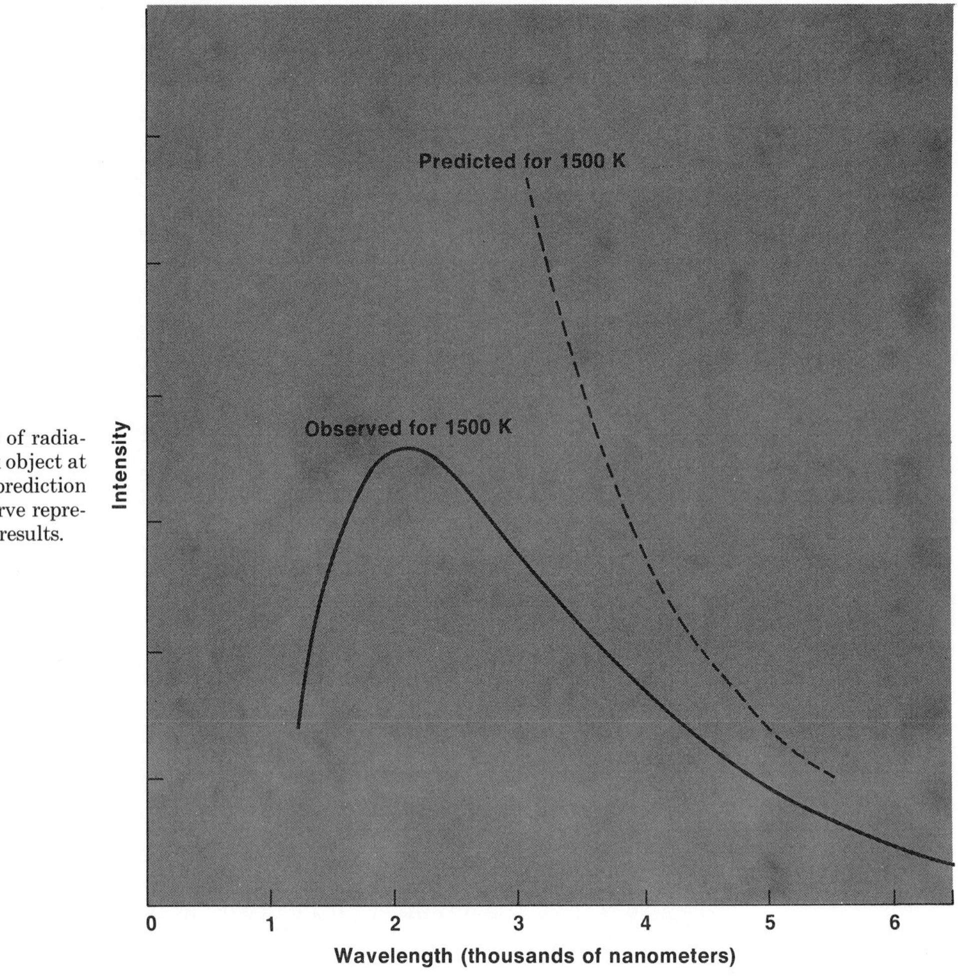

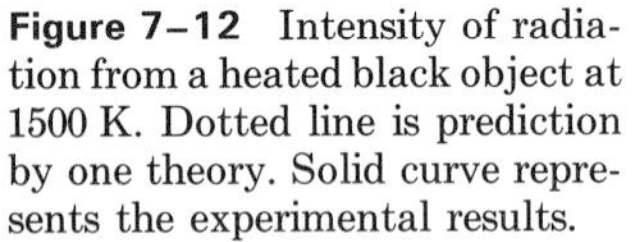
Figure 7–12 Intensity of radiation from a heated black object at 1500 K. Dotted line is prediction by one theory. Solid curve represents the experimental results.

In 1900, Max Planck made the assumption that the light being emitted from the hot object was in the form of packets of energy, which he called ***photons*** or ***quanta.*** The energy of a photon is given by

$$E = h\nu \tag{2a}$$

where h is a constant (later named Planck's constant), and ν is the frequency of the radiation. The energy of n quanta, each having the same energy, is represented by

$$E = nh\nu \tag{2b}$$

where n is any integer.

When Planck's assumption is combined with other mathematical equations, the resulting equation agrees with the experimental curve (heavy line) in Figure 7-12. While this agreement of theory with experiment indicated validity of the theory, Planck was never completely satisfied with this aspect of his theory, and he assumed that after being emitted as quanta the light then behaved in a wavelike manner as expected. In 1905, Albert Einstein made the assumption that if the light quanta were emitted from matter, then they could be absorbed in the

Packets of light energy are called photons or quanta.

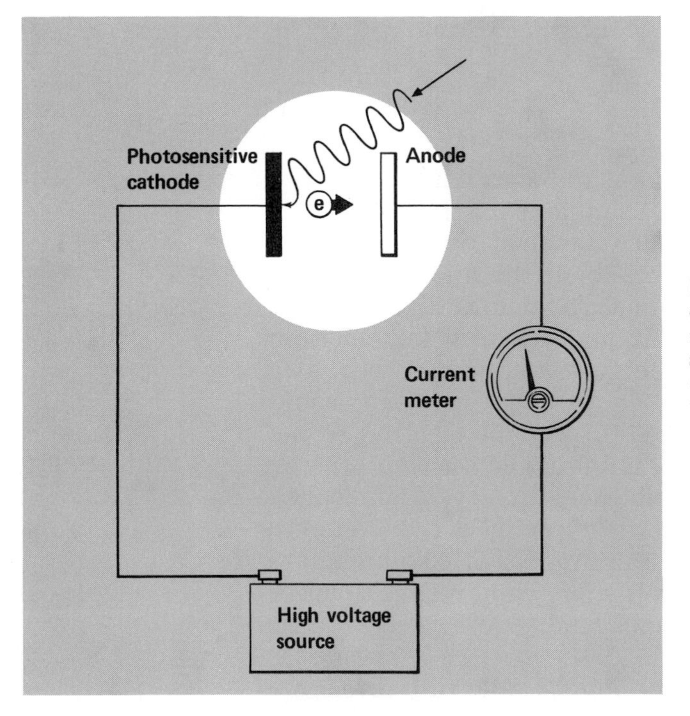

Figure 7–13 Photoelectric effect. Light, with sufficient energy per quantum, strikes a photosensitive cathode and causes electrons to be emitted. A high voltage source provides voltage just under the value for discharge due to voltage alone, and a current meter is used to measure current (electron flow).

same way. His quantum explanation of the ***photoelectric effect,*** which is used today to actuate automatic doors and night-sensing street lights, furnished additional evidence for the existence of quanta of light.

The photoelectric effect can be tested in an apparatus such as the one shown schematically in Figure 7–13. If the light striking the photosensitive metal cathode has a high enough frequency (large enough quanta), electrons are emitted from the metal and a current will flow. Different metals require different frequencies (and thus different-sized quanta) to initiate the flow of electrons. If the frequency is too low (quanta have too little energy), no electrons are emitted, no matter how intense is the beam. This kind of explanation of the photoelectric effect by using quanta helped immensely to validate ***Planck's quantum hypothesis.***

THE BOHR THEORY

Bohr assumed that an atom can exist only in certain energy states.

In 1913, a Danish physicist, Niels Bohr, made a dramatic assumption that seemed, almost overnight, to answer the searching questions that had bothered spectroscopists for many years. That is, why do excited atoms of an element give off only certain wavelengths of light and why are the lines in the hydrogen visible spectrum so neatly ordered by the Balmer equation? Borrowing from Planck, Bohr assumed that an atom can exist in specified energy states only. The energy states possible for a given atom are separated by increments or quanta of energy. If a photon of light having the correct amount of energy ($E = h\nu$) is taken up by the atom, the atom is raised to the next highest energy level. It cannot have a stable existence between the energy levels. If the atom returns to the lower energy level, a quantum of light is emitted. The wavelength of the light radiated depends on the amount of energy involved (ΔE) in the jump from the higher energy level (E_{II}) to the lower one (E_I), according to the equation:

$$\Delta E = E_{II} - E_I = h\nu \quad (3)$$

The explanation advanced by Bohr turned out to be brilliantly accurate in relating the spectrum of hydrogen and the quantum theory.

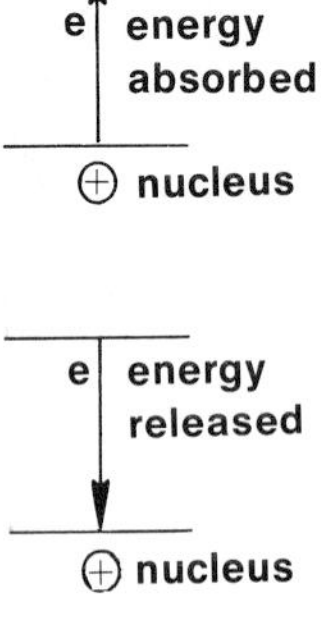

Making use of Rutherford's ideas, Bohr assumed that in a stable atom an electron can travel in a given orbit about the nucleus without any loss or gain in energy. For an atom to move to the next higher energy level means that the electron must jump to another orbit farther from the nucleus. In so doing, a quantum of energy would be absorbed. The size of the quantum and, thereby, the frequency of light involved, would be determined by the size of the jump. A return of the electron to the lower orbit would result in the emission of a photon of the same energy and same wavelength. The various possible energy levels are called ***quantum levels.*** Bohr assigned ***quantum numbers*** (n = 1, 2, 3, etc.) to these levels. Now, if the hydrogen atom has a number of possible quantum levels, and if there is a tendency to emit energy and return to lower levels, a number of different energy jumps and, consequently, light frequencies are possible.

If the hydrogen atom were limited to only six quantum levels, we would have a hypothetical atom with 15 possible downward quantum jumps (Figure 7–14). If we had many millions of excited atoms in an electrical discharge, we would expect all possible jumps to occur, and this would result in a spectrum containing 15 lines. Furthermore, we would expect the lines to be arranged in five series, based on whether the downward transitions were to the lowest, second, third, and so on, energy level, as in Figure 7–14. Actually, the hydrogen atom is not limited to n = 6. The quantum number n may have any integral (whole-number) value.

Using the Bohr structure of the hydrogen atom and basic equations

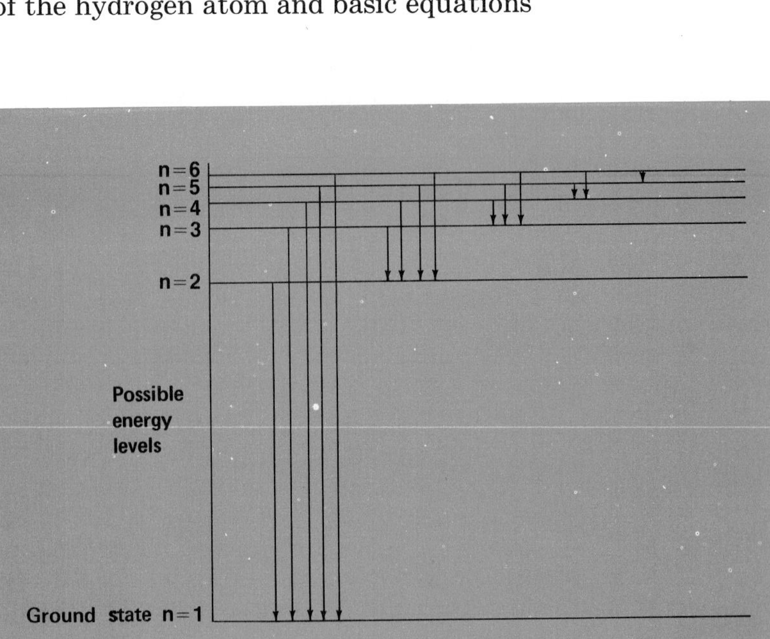

Figure 7–14 Possible downward energy jumps (15) for a hypothetical atom with one electron and five excited quantum levels possible in addition to the ground (unexcited) state.

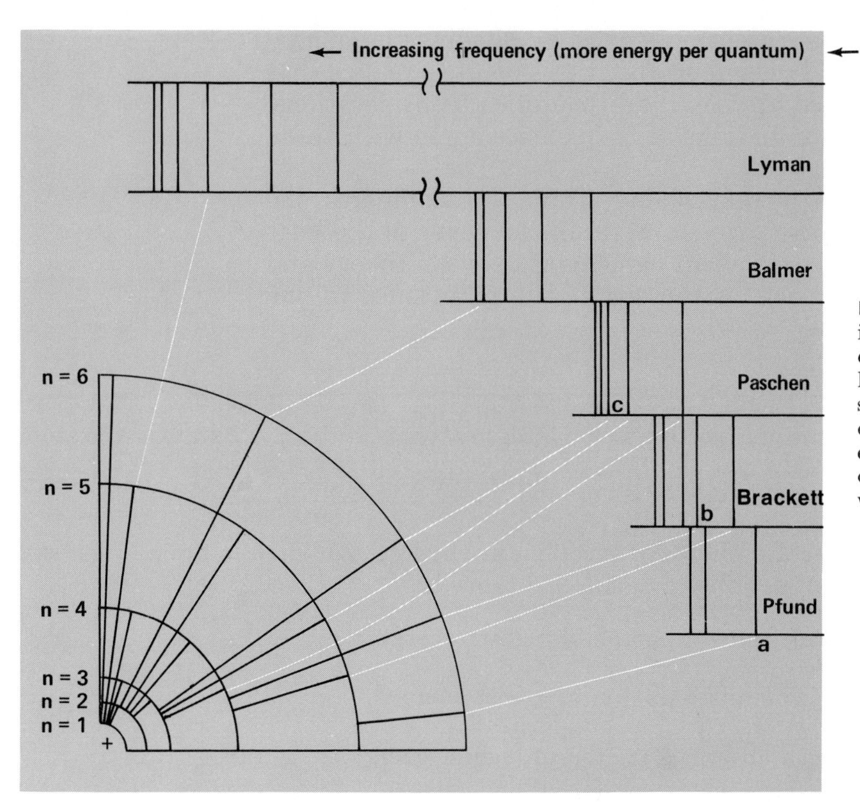

Figure 7–15 Theoretical transitions in Bohr hydrogen atom related to experimentally determined spectral lines. Only six quantum levels are shown for the atom, and only some of the spectral lines are shown in each series. For simplicity, only some of the transitions are connected with white lines to the spectral lines.

of physics, it is possible to calculate the energy for the hydrogen electron in the various possible energy levels. The differences between these energy levels, ΔE in Equation 3, can be converted into light frequencies by rearranging Equation 3 into the form:

$$\nu = \frac{\Delta E}{h} \tag{4}$$

TABLE 7–1 Agreement Between Bohr's Theory and the Hydrogen Spectrum

Quantum Jumps (in Quantum Numbers)	Series	Frequency Predicted by Bohr's Theory $\left(\frac{Cycles}{Sec}\right)$	Frequency Determined from Laboratory Measurement $\left(\frac{Cycles}{Sec}\right)$	Spectral Region
$2 \longrightarrow 1$	Lyman	2.467×10^{15}	2.4660×10^{15}	Ultraviolet
$3 \longrightarrow 1$	Lyman	2.925×10^{15}	2.9228×10^{15}	Ultraviolet
$4 \longrightarrow 1$	Lyman	3.084×10^{15}	3.0827×10^{15}	Ultraviolet
$3 \longrightarrow 2$	Balmer	0.4569×10^{15}	0.45681×10^{15}	Visible red
$4 \longrightarrow 2$	Balmer	0.6167×10^{15}	0.61669×10^{15}	Visible blue-green
$5 \longrightarrow 2$	Balmer	0.6908×10^{15}	0.69069×10^{15}	Visible blue
$4 \longrightarrow 3$	Paschen	0.1599×10^{15}	0.15988×10^{15}	Infrared

NOTE: These lines are typical; other lines could be cited as well with equally good agreement between theory and experiment.

It is evident that we can obtain the frequencies by dividing the energy changes by Planck's constant. It is important at this point to realize that the frequencies thus calculated are based on theoretical considerations. Obviously, if laboratory determinations of frequencies agree with those from theory, the theory seems valid.

Actually, the agreement is remarkable. Table 7–1 compares the theoretical predictions and the experimental results for seven of the lines in the hydrogen spectrum. This excellent agreement between theory and experimental results is truly one of the significant milestones in the development of atomic theory.

SELF-TEST 7-B

1. When the one electron in a hydrogen atom is in the n = 2 state, is the atom in an excited state or its ground state? ______________

2. Consider the following transitions in a hydrogen atom: n = 1 to n = 2, n = 2 to n = 1, n = 2 to n = 3, n = 3 to n = 2.

   a. Which emits a photon of highest energy? ______________

   b. Which emits a photon of lowest energy? ______________

   c. Which absorbs a photon of highest energy? ______________

3. The emission spectrum of hydrogen is a (bright-line/continuous) spectrum. ______________

4. According to Planck's quantum theory, light and other forms of energy are packaged in small quantities of energy called ______________.

5. Planck's quantum theory gained acceptance by its ability to explain consistently which of the following? (There is more than one correct selection.)
   a. photoelectric effect
   b. intensities of various wavelengths of radiation emitted by hot black objects
   c. the number of protons per atom
   d. bright-line spectrum of hydrogen
   e. the shape of an orbit

6. A nanometer is the same length as ________ m and ________ cm.

7. According to Bohr's theory, light is produced as an electron passes from an energy level (closer to/farther from) the nucleus to an energy level (closer to/farther from) the nucleus.

MATCHING SET

________ 1. energy of a photon		a. $\lambda = \frac{hc}{\Delta E}$
________ 2. speed (or velocity) of light		b. $\lambda = 364.56\left(\frac{n^2}{n^2 - 4}\right)$
________ 3. wavelength of light corresponding to a transition in energy in a hydrogen atom		c. $E = h\nu$

________4. wavelength of light of a Balmer line in hydrogen spectrum   d. $c = \lambda\nu$

QUESTIONS

1. What is a quantum? What is a photon?
2. Discuss, in quantum terms, how a ladder works.
3. How do a black object's radiation and the photoelectric effect indicate that electromagnetic energy is composed of quanta?
4. Distinguish between atomic number and atomic weight.
5. Which has more energy per photon, red light or blue light? How is it possible for a beam of red light to contain more energy than a beam of blue light?
6. What part do the following play in formulating the Bohr hydrogen atom?
   a. electrons
   b. Rutherford's nuclear concept
   c. Planck's quantum theory
   d. the hydrogen spectrum
   e. Maxwell's theory
7. Distinguish between a continuous spectrum and a bright-line spectrum under the two headings:
   a. general appearance
   b. source
8. How does the Bohr theory explain the many lines in the spectrum of hydrogen although the hydrogen atom contains only one electron?
9. In the hydrogen atom which electron would be easier to remove from the atom, an electron in the ground state or in energy level number 5? Why?
10. Would it be reasonable to argue that macroscopic energy changes are quantized, but the quantum jumps are too small to be observed?

SUGGESTIONS FOR FURTHER READING

Gamow, G., "The Atom and its Nucleus," Prentice-Hall, Inc., Englewood Cliffs, N.J., 1961, p. 50.

Hochstrasser, R. M., "Behavior of Electrons in Atoms," W. A. Benjamin, Inc., New York, 1964.

Ihde, A. J., "The Development of Modern Chemistry," Harper and Row, New York, 1964, p. 493–507.

Lehman, T. A., "The Planck Radiation Law and the Efficiency of a Light Bulb," "Journal of Chemical Education," Vol. 49, No. 12, p. 833 (1972).

Moore, R., "Niels Bohr: The Man, His Science and the World They Changed," Alfred A. Knopf, Inc., New York, 1966.

Smeaton, W. A., "Moseley and the Numbering of the Elements," *Chemistry in Britain,* Vol. 1, p. 353 (1965).

MODERN ATOMIC THEORY

CHAPTER 8

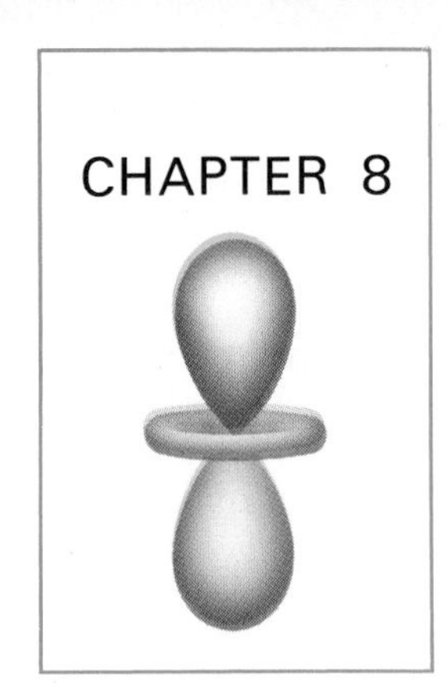

By the early 1920's, it was apparent that the Bohr theory of the atom had several serious shortcomings. Although it provided a very attractive picture of the hydrogen atom, it was not in accord with the spectra of atoms containing two or more electrons, and it proved incapable of explaining how atoms could form molecules. Part of its inadequacy is related to the fact that the Bohr model is simply too exact. It specifies exact orbits, velocities, and energies of the electron in the atom.

WAVE NATURE OF THE ELECTRON

The Bohr theory of the atom was very clearly based upon physical laws which describe the motion of macroscopic pieces of matter. It became apparent by 1927 that electrons and other submicroscopic particles, when in motion, had a wave nature in addition to the properties usually associated with particles. Though not light, electrons, like light, became recognized as having a ***dual nature—particulate and wave.*** This recognition led to a new, more modern theory of the atom.

Electrons are described by both particle and wave theories.

The dual nature of particles was first postulated in 1923 by the French physicist, Louis de Broglie, who suggested that particles should have a characteristic wavelength determined by their momentum. He proposed the relationship:

$$\lambda = \frac{h}{mv}$$

where m is the mass of the particle, v is its velocity, the product mv is its momentum, h is Planck's constant, and λ is the wavelength characteristic of the particle.

In 1927, the wave nature of moving electrons was experimentally verified independently by C. J. Davisson and L. H. Germer in America and by G. P. Thomson in Britain. In both cases, moving electrons were shown to give patterns similar to those produced when light is sent through or reflected from a diffraction grating, or when X-rays are sent through crystals (Figure 8–1). Davisson and Germer reflected a beam of

Sir George Paget Thomson was the son of Sir J. J. Thomson.

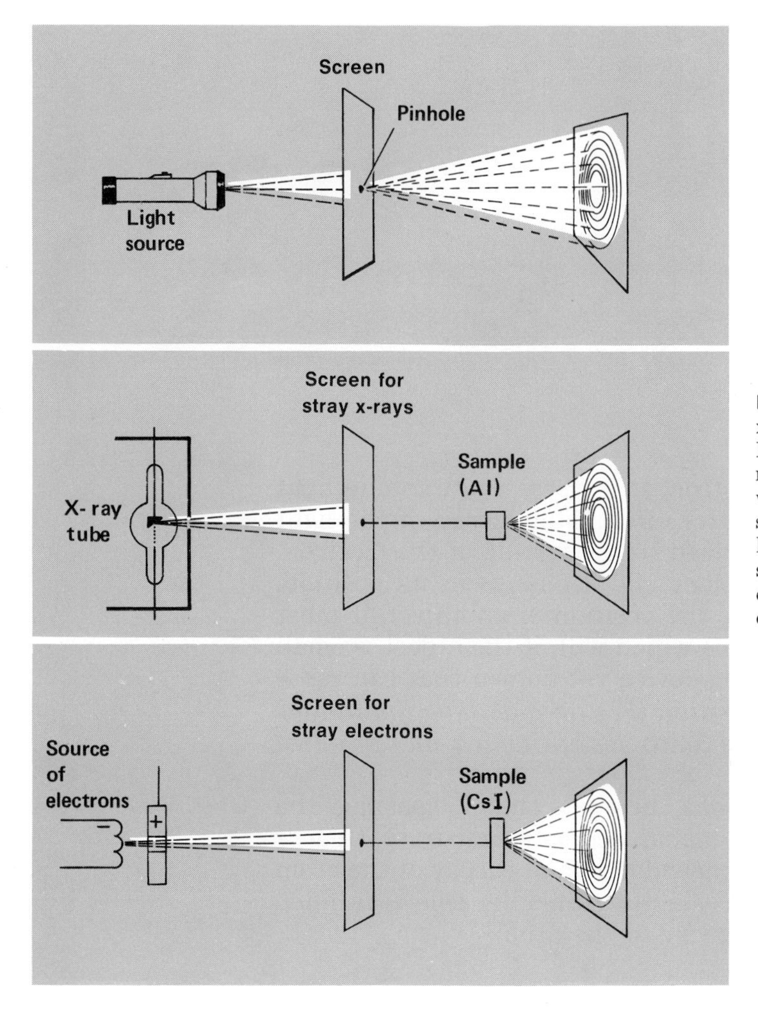

Figure 8–1 Similar patterns are shown by light, x-rays, and electrons as each is diffracted. Diffraction is the bending and spreading of wave motion around edges. The effect is prominent when the wavelength is large compared to the size of the obstacle and small when the wavelength is short compared to the size of the obstacle. Similar effects from light, X-rays, and electrons indicate a property common to all: each has a wave nature.

electrons off a nickel crystal; Thomson shot a stream of high-speed electrons through a very thin crystal of gold. From these studies it was evident that electrons can be diffracted exactly in the same manner as light. Since the diffraction of light or X-rays is very convincing evidence for their wave nature, the diffraction of a stream of electrons indicates at least a partial wave character for electrons. Consequently, electrons have a characteristic wavelength and frequency for a given kinetic energy.

HEISENBERG UNCERTAINTY PRINCIPLE

The accuracy of measurements cannot be increased beyond certain limits inherent in nature.

Another important development in the replacement of the Bohr theory was the formulation of the Heisenberg uncertainty principle by Werner Heisenberg in 1926. This very important principle sets the limits of accuracy which may be attained in simultaneous measurements of the position and velocity of submicroscopic particles such as electrons and atoms, or in simultaneous measurements of the energy and the time for such systems. It can best be appreciated from a consideration of the problems which come up when an attempt is made to measure the position and the velocity of an electron with great accuracy. In order

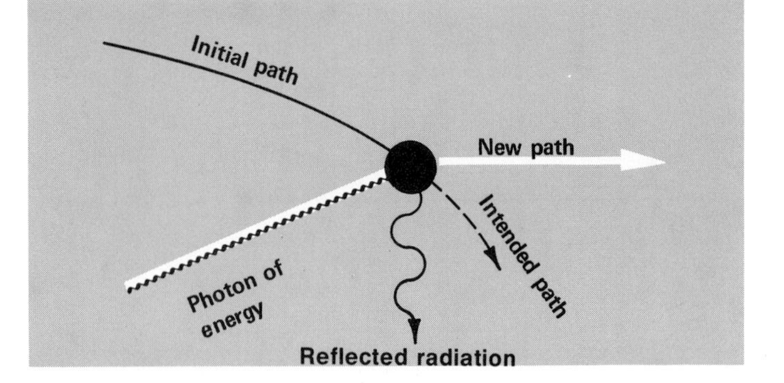

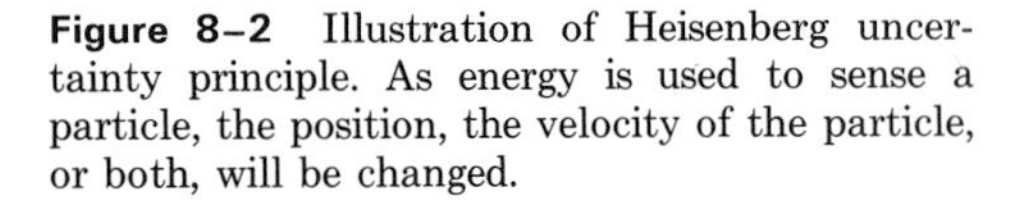
Figure 8–2 Illustration of Heisenberg uncertainty principle. As energy is used to sense a particle, the position, the velocity of the particle, or both, will be changed.

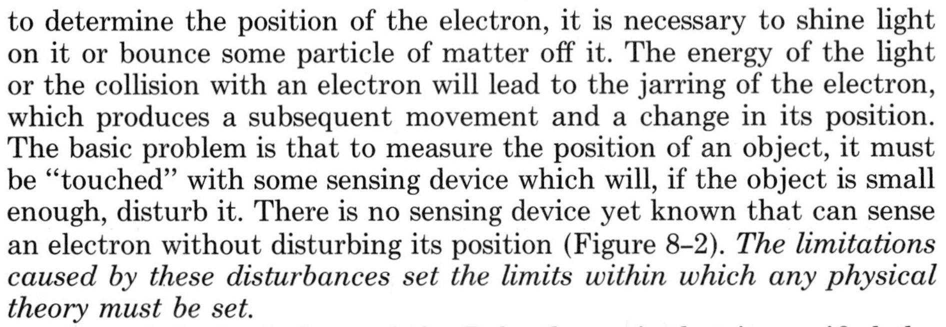
to determine the position of the electron, it is necessary to shine light on it or bounce some particle of matter off it. The energy of the light or the collision with an electron will lead to the jarring of the electron, which produces a subsequent movement and a change in its position. The basic problem is that to measure the position of an object, it must be "touched" with some sensing device which will, if the object is small enough, disturb it. There is no sensing device yet known that can sense an electron without disturbing its position (Figure 8–2). *The limitations caused by these disturbances set the limits within which any physical theory must be set.*

One of the basic flaws of the Bohr theory is that it specified the behavior of atomic systems to a far higher degree of accuracy than is physically attainable. A basic requirement for a valid theory of the atom is that it be consistent with the uncertainty principle; this is fulfilled in the development of the modern theory of the atom.

The Bohr Theory was not consistent with the uncertainty principle.

THE SCHRÖDINGER THEORY AND STANDING WAVES

In 1926, Erwin Schrödinger developed an equation consistent with the wave properties of the electron. His basic equation was similar to an equation that is used to describe standing waves in general. An understanding of standing waves can be appreciated from the behavior of a specific kind of wave, such as a sound wave. When a string on a guitar is plucked, it doesn't produce a variety of tones, but rather, a single tone determined by the length, thickness, tension, and elasticity of the string. This tone is related to the standing wavelengths which are possible when each node of the wave is at each end of the string (Figure 8–3). Both ends of the string must be fixed for standing waves to be produced. These waves are called "standing waves" because the string appears to be permanently shaped into nodes and antinodes—that is, the waves appear to be standing still (Figure 8–3). The Schrödinger equation, then, considers the electron in the atom as a standing wave of energy. In contrast to the guitar string the electron standing wave will be in three dimensions. The resulting geometries of the electron waves should give rise to the geometries involved in chemical bonding and ultimately in macroscopic crystalline form. As we shall see later, the theory is remarkably powerful but not perfectly adequate.

In the absence of friction, standing waves persist.

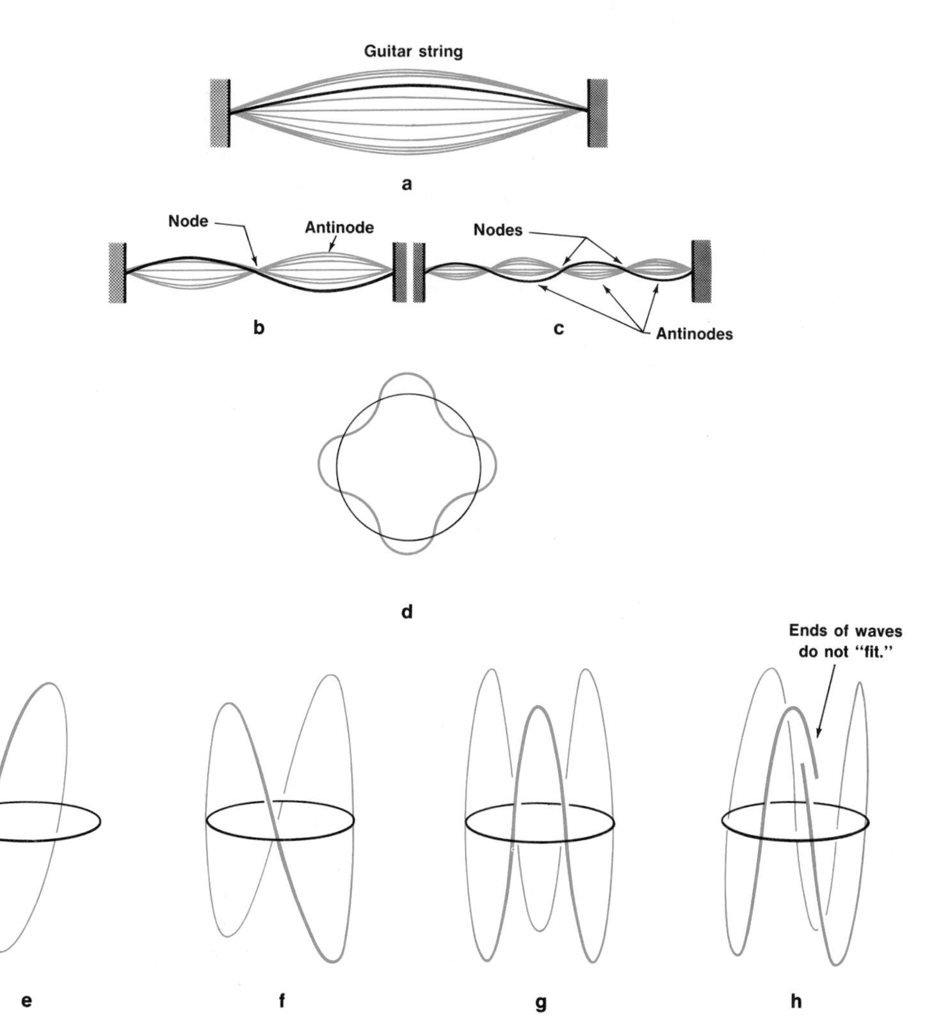

Figure 8–3 Standing waves, nodes, and antinodes. *a* and *e* portray fundamentals; *b, c, d, f,* and *g* are examples of overtones. *h* is not a standing wave since destructive interference would destroy the antinodes.

SELF-TEST 8-A

1. According to the modern theory of the atom, an electron has properties of a ______________ wave.

2. Which of the following led to the modern theory of the atom and were not included in the Bohr theory?

   a. concept of nucleus
   b. quantum theory
   c. particle nature of the electron
   d. wave nature of electron
   e. Heisenberg uncertainty principle

3. According to de Broglie, every moving particle also has a characteristic ______________________.

QUANTUM NUMBERS AND ENERGY LEVELS

When the Heisenberg uncertainty principle, the standing wave nature of the electron, and the de Broglie relationship are combined with the older nuclear concept and quantum theory, the result is the modern theory of the atom. Its usual expression, in the form of the Schrödinger wave equation, is beyond the scope of this text since calculus is required.

Solving the Schrödinger equation yields energy levels and wave functions for electrons in particular orbitals in an atom, specifically in the hydrogen atom. Unlike the Bohr orbit, the ***orbital*** is a region within the atom occupied by one or two electrons. A visualization of the orbitals is developed later (Figure 8–9). Each orbital can contain a maximum of two electrons, and each orbital is a different stable energy level which these electrons can have within the atom. It is important to realize that specifying the orbital in which an electron is located tells us nothing about the path of an electron; if an electron is in a particular orbital, this simply means that the electron has a particular energy and a probable location around the nucleus.

Orbitals have geometries defined by electron standing waves.

Probabilities of locating electrons within an atom and ***quantum numbers*** are two properties which emerge from the solutions of the Schrödinger equation. Quantum numbers are used to describe the energy and the assignment of an electron to possible orbitals within the atom. Probabilities give the chance of finding an electron at a given location within an atom. Much of the next two chapters is devoted to showing how helpful these aspects of the wave theory are in explaining the chemical properties of matter, the periodic table, the structures of molecules, and other chemical phenomena. In the rest of this chapter, we will try to gain a clear concept of quantum numbers, orbitals, probabilities, and the consistency of the theory with the arrangement of the periodic table.

Since the actual solution of the Schrödinger equation involves rather sophisticated calculus, we shall only look at several interesting results that come from its solution. In order to solve the equation for the various energies of an electron in a hydrogen atom, three integers must be substituted into the equation. Each set of three integers determines an orbital, and represents an energy level of an electron in the atom. The three integers, n, ℓ, and m, are called ***quantum numbers.*** The value of n specifies a main energy level such as in the Bohr theory (Figure 8–4).

Quantum numbers specify orbitals and electrons within them.

In its most stable state (called the ***ground state***), the electron in a hydrogen atom will occupy the $n = 1$ level and will be strongly attracted by the nucleus. If the atom picks up enough energy the electron is excited to the $n = 2$ level. If it picks up still larger quanta of energy, it can be excited to a higher energy level or even removed completely from the atom, thus ionizing the atom and producing an ion plus an electron.

$$\underset{\text{ATOM}}{H} \xrightarrow{h\nu} \underset{\text{ION}}{H^+} + \underset{\text{ELECTRON}}{e^-}$$

The value of ℓ can be any integer from 0 to $n-1$, and m can have any integral value from $-\ell$ to $+\ell$. The limitations on ℓ and m required for proper solutions of the Schrödinger equation are summarized in Table 8–1.

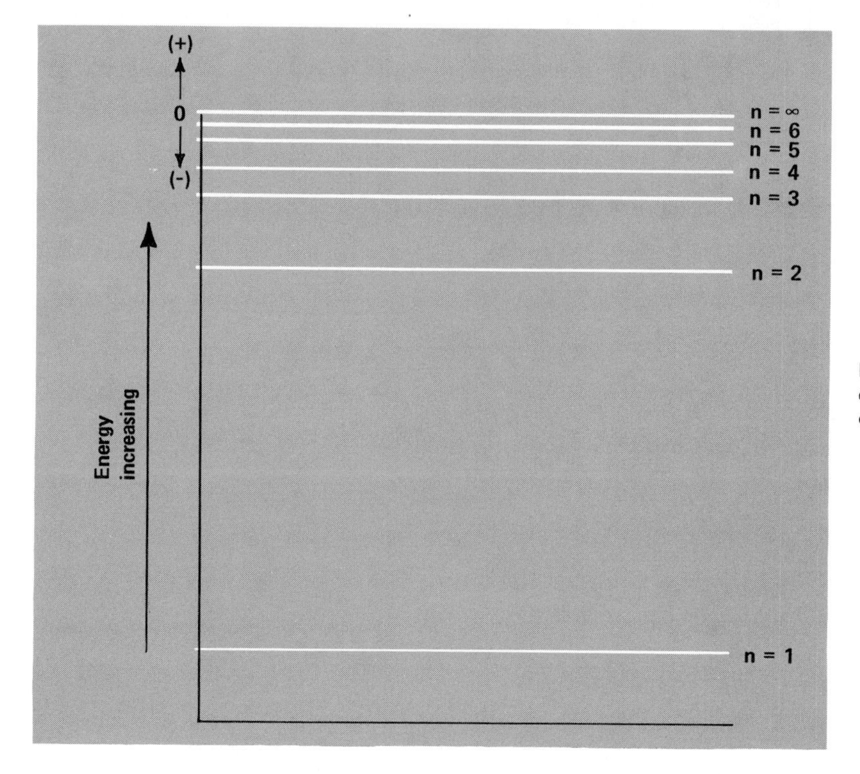

Figure 8–4 The energy levels of an electron for various values of the quantum number, n.

A consistent set of values for n, ℓ, and m designates a particular orbital of an electron. For example, orbitals could be designated by n, ℓ, and m values of 1,0,0, or 2,0,0, or 2,1, − 1, or 2,1,0, or 2,1,1 but not by the combinations 2,2,0 or 2,1,2. The latter two sets would not give meaningful energies upon solution of the Schrödinger equation.

The symbol, $n\ell_m{}^x$, which combines n, ℓ, m and the number of electrons (x = 0, 1, or 2) in the orbital, is sometimes used to designate an orbital. Oftentimes the m is omitted and the symbol becomes $n\ell^x$. In both symbols, ℓ is written as a letter instead of a number, as has been used up to this point. In the symbol without m, the number of electrons is the number in all orbitals having the stated values of n and ℓ. There are

TABLE 8–1 Some Consistent Values for n, ℓ, and m Quantum Numbers

Value of n	Value of ℓ	Value of m
1	0	0
2	0	0
	1	+1,0,−1
3	0	0
	1	+1,0,−1
	2	+2,+1,0,−1,−2
4	0	0
	1	+1,0,−1
	2	+2,+1,0,−1,−2
	3	+3,+2,+1,0,−1,−2,−3

$2\ell + 1$ orbitals in a set of orbitals having given n and ℓ values. The letters to be used for ℓ and the maximum number of electrons for each value of ℓ are as follows:

There are four types of orbitals: s, p, d, and f.

Value of ℓ	Corresponding letter designation for ℓ	Number of orbitals	Maximum number of electrons
0	s	1	2
1	p	3	6
2	d	5	10
3	f	7	14

The letters s, p, d, f come from terms which are descriptive of spectral lines: sharp, principal, diffuse, and fundamental. Orbitals designated by comparable symbolism are shown in the following examples.

n	ℓ	m	$n\ell^x$	
1	0	0	$1s^1$	(for 1 electron)
1	0	0	$1s^2$	(for 2 electrons)
2	1	−1	$2p^2$	
2	1	0	$2p^2$	$2p^6$
2	1	1	$2p^2$	

A fourth quantum number, s, arises when the Schrödinger equation is solved by relativistic mechanics. In this case s is defined as the relative phase angle of an electron wave (that is, $\Delta s = 1$ means a phase angle lag of 180 degrees between waves of the same n, ℓ, and m). Historically, however, the concept of s arose because an electron in an atom gives effects similar to a charged particle which spins on its axis, producing a magnetic field about it. There are only two possible spin quantum numbers, $+\frac{1}{2}$ and $-\frac{1}{2}$. A given orbital can contain only two electrons, and in order to do so, the two electrons are thought to be spinning in opposite directions, that is, one has $s = +\frac{1}{2}$ and the other $s = -\frac{1}{2}$.

$s = +\frac{1}{2}$

$s = -\frac{1}{2}$

Now that we have a way to name the orbitals emerging from solving the Schrödinger equation, it is in order to ask, "What are the relative energies of the various orbitals?" For isolated hydrogen atoms, the energy of an electron in a main energy level, n, increases as the number of the level increases but is independent of the value of ℓ. This is in complete

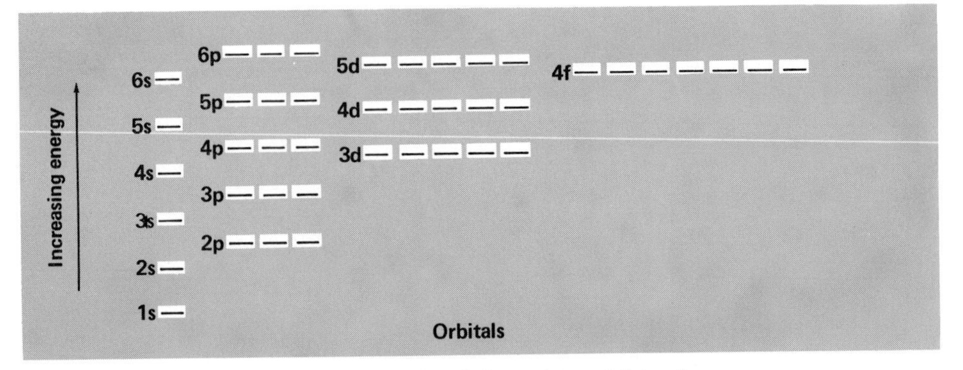

Figure 8–5 Diagram of the relative energies of the various orbitals of energy in an atom. Dashes represent the orbitals within a given set.

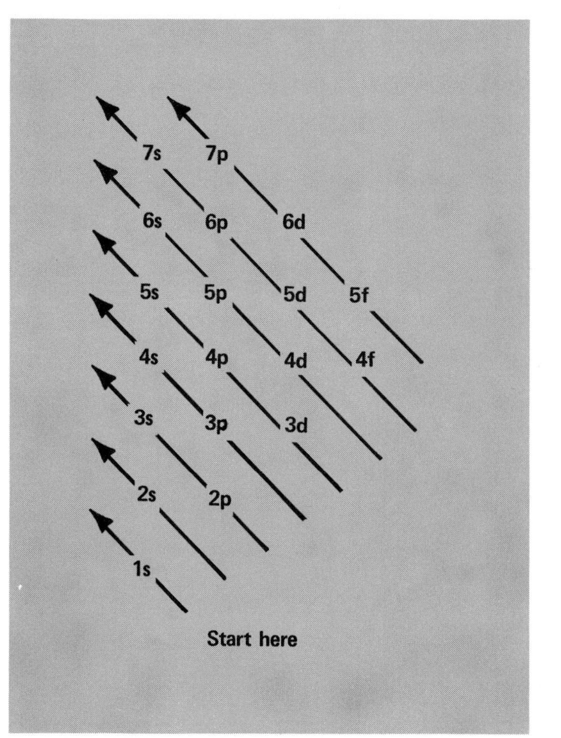

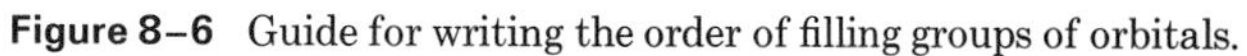

Figure 8-6 Guide for writing the order of filling groups of orbitals.

agreement with the Bohr theory, in which electron transitions between main energy levels are sufficient to explain completely the lines in the emission spectrum of hydrogen. However, for atoms containing several electrons, the orbital divisions within a main energy level are not of the same energy. For isolated gaseous atoms, the approximate order of increasing energy of the orbitals is 1s, 2s, 2p, 3s, 3p, 4s, 3d, 4p, 5s, 4d, 5p, 6s, 4f, 5d, 6p, 7s, 5f, 6d (Figure 8-5). The higher energy levels are very close together, and when occurring in atoms of high atomic number, these energy levels are often in different relative positions from the order listed. Nevertheless, this order is a good indication of relative energies of electron orbitals for the ground state of many atoms. The order of filling the orbitals beginning with the orbital of lowest energy can be written down readily by using the guide shown in Figure 8-6.

To deduce the order in which the various orbitals are filled, write the types of orbitals as shown in the figure and draw lines diagonally across, from lower right to upper left. Then, to derive the orbital electron structure of a particular atom, simply fill up each type of orbital in order starting with the orbital of lowest energy, 1s. For example, arsenic, As (atomic number 33), has 33 electrons, and these electrons would be placed into the following orbitals.

$$1s^2 2s^2 2p^6 3s^2 3p^6 4s^2 3d^{10} 4p^3$$

ELECTRONIC CONFIGURATION OF ARSENIC

Notice the 4p set of orbitals has only the 3 electrons required to give a total of 33 electrons. This means the last set of orbitals need not be filled to its maximum.

A slightly different way of showing the order of filling the orbitals may be given schematically, as in Figure 8-5. Both systems of writing the electronic configuration of an atom can be used to construct its energy level diagram. For example, a nitrogen atom ($1s^2 2s^2 2p^3$, by the previous notation) has the following diagram, where an arrow represents an electron:

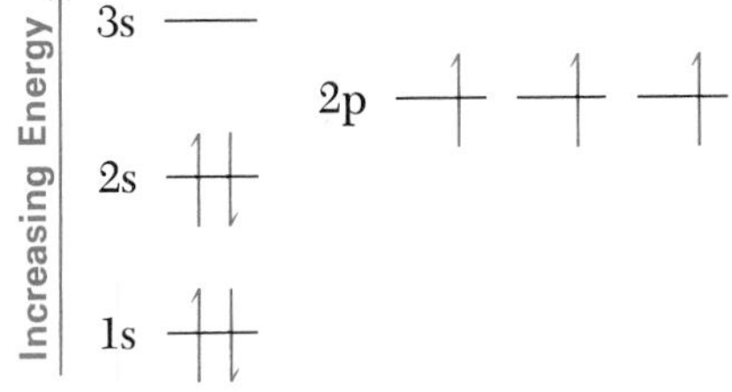

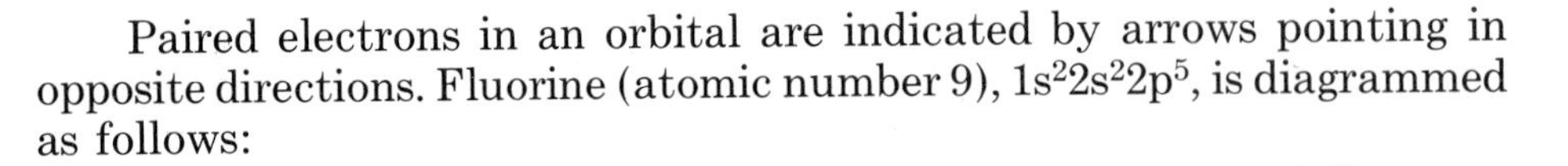
Paired electrons in an orbital are indicated by arrows pointing in opposite directions. Fluorine (atomic number 9), $1s^2 2s^2 2p^5$, is diagrammed as follows:

When one considers atoms in subsequent periods of the periodic table, the same schemes are used. For example, silicon (atomic number 14) has the electronic configuration $1s^2 2s^2 2p^6 3s^2 3p^2$ and its energy levels are occupied as follows:

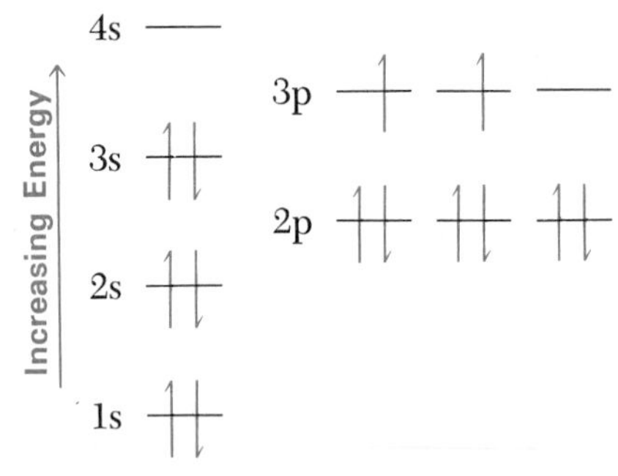

Notice that the 2p electrons of nitrogen and the 3p electrons of silicon are not paired. A state of somewhat lower energy will result if electrons are unpaired in a set of orbitals because energy is required to pair electrons (that is, to give them opposite spins). Thus, electrons will not pair in orbitals at the same energy level (same n and ℓ values) as long as there is an empty orbital in that set.

↿ unpaired electron
⇂↾ paired electrons

SELF-TEST 8-B

1. When electrons are in their lowest energy state, they are said to be in their ____________ state.

2. The maximum number of electrons per orbital is ____________.

3. In the symbol $2p^3$,

   a. what is the main energy level? ____________

b. in what set of orbitals are the electrons?__________

c. how many electrons are represented?__________

4. Fill in the electrons in the ground state for germanium, Ge, atomic number 32.

1s 2s 2p 3s 3p 4s 3d 4p

5. An electron is designated by its main energy level (n), set of orbitals (ℓ), orbital (m), and spin (s). This designation can be compared to a geographical location of a person based on the size (inclusiveness) of the category. Match the similar categories.

Electron designation	Geographical designation
n	city
ℓ	state
m	house number
s	street

6. If a set of p orbitals contains four electrons (p^4), how many of these electrons are unpaired?__________

THE ORDER OF FILLING ORBITALS PREDICTS PERIODS IN THE PERIODIC TABLE

Periodic trends occur in the properties of the elements.

The ordering of the energy of orbitals is in agreement with the divisions of the periodic table remarkably well. It has been pointed out that the periodic chart was originally arranged on the basis that chemical and physical properties show periodicity provided that the elements are taken in the order of their atomic numbers. If the orbitals of the wave theory are filled in the order stated previously (Figures 8–5 and 8–6), periodic changes in electronic arrangement coincide with the periodic changes in chemical properties.

Consider the memory guide (Figure 8–5 or 8–6). The first main energy level (n = 1) contains only one orbital, 1s, and can contain only two electrons; hence, two elements are possible. Period 1 in the chart contains but two elements, hydrogen and helium. The second energy level (n = 2) contains two orbital sets, 2s and 2p, and can contain eight electrons; hence, eight elements are possible. Period 2 in the chart contains eight elements, lithium through neon. Note that both periods 1 and 2 begin with elements formed by adding to an s orbital; period 2 ends with an element, neon, formed by completing the filling of p orbitals.

Trends in electronic arrangement parallel trends in elemental properties.

Period 3 also contains eight elements, sodium through argon; the next orbitals in line to be filled are the 3s and 3p orbitals which can hold the eight electrons for forming these elements. Again, the period begins by adding to an s orbital and ends by filling a set of p orbitals. Period 4 contains 18 elements, potassium through krypton; the orbitals in line to be filled are the 4s (capacity = $2e^-$), 3d (capacity = $10e^-$) and 4p (capacity = $6e^-$) which can contain the 18 electrons for forming these elements. Period 5 contains 18 elements, rubidium through xenon; the orbitals in line to be filled are the 5s, 4d, and 5p, which can contain exactly the 18 electrons necessary to form these elements. Period 6 contains 32 elements; the next groups of orbitals are the 6s, 4f (capacity = $14e^-$), 5d, and 6p, holding exactly the 32 electrons necessary to form these elements. Period 7, thus far, has only 20 elements, which fill up the 7s and 5f orbitals and put 4 electrons in the 6d orbitals.

Each of these periods begins with an element formed by adding an electron to an s orbital. Each period ends with an element formed by completing the filling of a set of p orbitals. Is it any wonder, then, that the periods of elements follow other periods of elements in which similarly positioned members of the various periods have similar chemical properties? The similarly positioned members in the periods have similarly structured atoms. The elements at the first of each period have their highest energy electrons in an s orbital. The elements near the end of each period have their highest energy electrons in a p orbital. Thus, there is a trend of electronic arrangements across the periodic table as there is a trend of chemical properties (Figure 8–7).

Two elements in the same chemical family will have similar electronic arrangements.

We see that the modern theory of the atom agrees with the arrangements of elements into periods of the same sort that were developed when similarities in chemical properties were considered. It is a striking confirmation of the effectiveness of the theory that it agrees so well with experimental observations.

PROBABLE POSITIONS OF ELECTRONS IN ATOMS

In addition to quantum numbers and energies, the solution of the Schrödinger equation gives information about the probable position of the electron within an atom. Because we must accept the Heisenberg uncertainty principle and the limitations that it places upon the kind of information that we can validly possess, the picture of the atom must reflect this lack of precision. Rather than knowing precisely where the electron is located, we can calculate only the probability of finding the electron in a given volume of space around the nucleus. As a consequence, the representations of the atom portray the different ways in which probabilities of finding the electron can be presented. The mathematics involved allows the calculation of the relative probability of finding the electron at a given distance from the nucleus of the atom, as shown in Figure 8–8.

In order to obtain a pictorial representation it is usual to plot the surface that will bound the volume in which the electron will be expected to be found most of the time (e.g., 90 percent of the time). These sketches are important because they tell the preferred locations of the electrons and the way in which these electrons may be expected to influence chemical behavior. The shapes and spatial orientations of the orbitals are designated by the letters s, p, d, and f and, where necessary, with a subscript which relates to the spatial orientations (Figure 8–9). An important aspect of orbital geometry is that an electron is not as likely to be found in some directions out from the nucleus as in other directions. In the case of the s orbitals, which are spherically symmetrical, there is equal probability of finding the electrons, regardless of the direction taken from the nucleus. However, a 2s orbital is farther from the nucleus

PERIODS	GROUPS I A	II A	III B	IV B	V B	VI B	VII B	VIII B			I B	II B	III A	IV A	GROUPS V A	VI A	VII A	Noble Gases VIII A	PERIODS
1	1 H $1s^1$																	2 He $1s^2$	1
2	3 Li $2s^1$	4 Be $2s^2$											5 B $2s^2 2p^1$	6 C $2s^2 2p^2$	7 N $2s^2 2p^3$	8 O $2s^2 2p^4$	9 F $2s^2 2p^5$	10 Ne $2s^2 2p^6$	2
3	11 Na $3s^1$	12 Mg $3s^2$											13 Al $3s^2 3p^1$	14 Si $3s^2 3p^2$	15 P $3s^2 3p^3$	16 S $3s^2 3p^4$	17 Cl $3s^2 3p^5$	18 Ar $3s^2 3p^6$	3
4	19 K $4s^1$	20 Ca $4s^2$	21 Sc $4s^2 3d^1$	22 Ti $4s^2 3d^2$	23 V $4s^2 3d^3$	24 Cr $4s^1 3d^5$	25 Mn $4s^2 3d^5$	26 Fe $4s^2 3d^6$	27 Co $4s^2 3d^7$	28 Ni $4s^2 3d^8$	29 Cu $4s^1 3d^{10}$	30 Zn $4s^2 3d^{10}$	31 Ga $4s^2 4p^1$	32 Ge $4s^2 4p^2$	33 As $4s^2 4p^3$	34 Se $4s^2 4p^4$	35 Br $4s^2 4p^5$	36 Kr $4s^2 4p^6$	4
5	37 Rb $5s^1$	38 Sr $5s^2$	39 Y $5s^2 4d^1$	40 Zr $5s^2 4d^2$	41 Nb $5s^1 4d^4$	42 Mo $5s^1 4d^5$	43 Te $5s^2 4d^5$	44 Ru $5s^1 4d^7$	45 Rh $5s^1 4d^8$	46 Pd $4d^{10}$	47 Ag $5s^1 4d^{10}$	48 Cd $5s^2 4d^{10}$	49 In $5s^2 5p^1$	50 Sn $5s^2 5p^2$	51 Sb $5s^2 5p^3$	52 Te $5s^2 5p^4$	53 I $5s^2 5p^5$	54 Xe $5s^2 5p^6$	5
6	55 Cs $6s^1$	56 Ba $6s^2$	57 La $6s^2 5d^1$	⊗ 72 Hf $6s^2 5d^2$	73 Ta $6s^2 5d^3$	74 W $6s^2 5d^4$	75 Re $6s^2 5d^5$	76 Os $6s^2 5d^6$	77 Ir $6s^2 5d^7$	78 Pt $6s^1 5d^9$	79 Au $6s^1 5d^{10}$	80 Hg $6s^2 5d^{10}$	81 Tl $6s^2 6p^1$	82 Pb $6s^2 6p^2$	83 Bi $6s^2 6p^3$	84 Po $6s^2 6p^4$	85 At $6s^2 6p^5$	86 Hn $6s^2 6p^6$	6
7	87 Fr $7s^1$	88 Ra $7s^2$	89 Ac $7s^2 6d^1$	⊘ 104 *	105 *	106 *													

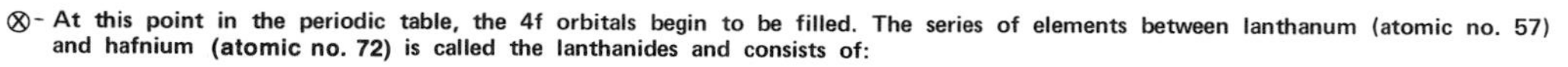

⊗ - At this point in the periodic table, the 4f orbitals begin to be filled. The series of elements between lanthanum (atomic no. 57) and hafnium (atomic no. 72) is called the lanthanides and consists of:

58	59	60	61	62	63	64	65	66	67	68	69	70	71
Ce	Pr	Nd	Pm	Sm	Eu	Gd	Tb	Dy	Ho	Er	Tm	Yb	Lu
$6s^2 5d^1 4f^1$	$6s^2 4f^3$	$6s^2 4f^4$	$6s^2 4f^5$	$6s^2 4f^6$	$6s^2 4f^7$	$6s^2 5d^1 4f^7$	$6s^2 4f^9$	$6s^2 4f^{10}$	$6s^2 4f^{11}$	$6s^2 4f^{12}$	$6s^2 4f^{13}$	$6s^2 4f^{14}$	$6s^2 5d^1 4f^{14}$

⊘ - At about this point in the periodic table the 5f orbitals begin to be filled. The series of elements involved in this process are called the actinides and consists of:

90	91	92	93	94	95	96	97	98	99	100	101	102	103
Th	Pa	U	Np	Pu	Am	Cm	Bk	Cf	Es	Fm	Md	No	Lw
$7s^2 6d^2$	$7s^2 6d^1 5f^2$	$7s^2 6d^1 5f^3$	$7s^2 6d^1 5f^4$	$7s^2 5f^6$	$7s^2 5f^7$	$7s^2 6d^1 5f^7$	$7s^2 5f^9$	$7s^2 5f^{10}$	$7s^2 5f^{11}$	$7s^2 5f^{12}$	$7s^2 5f^{13}$	$7s^2 5f^{14}$	$7s^2 6d^1 sf^{14}$

Figure 8–7 The periodic table including the atomic numbers and the outermost electrons of each element. Group VIIIA is made up of the noble gases.

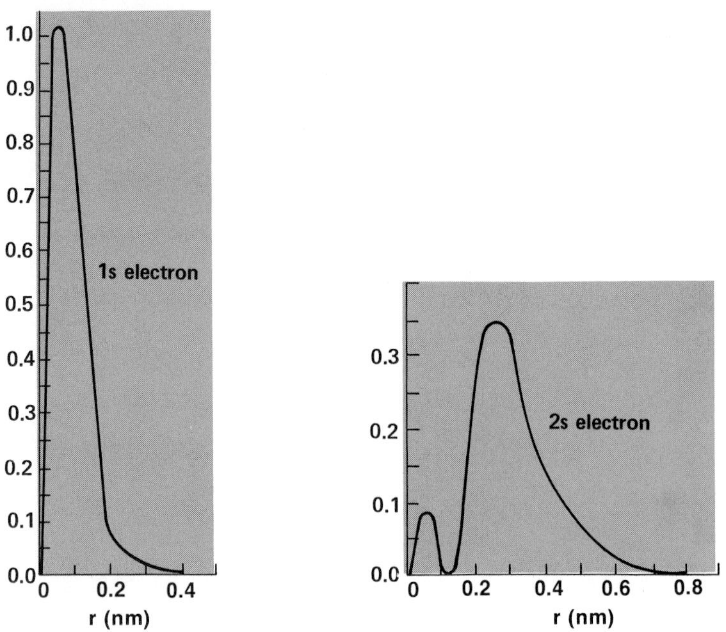

Figure 8–8 Relative probability of finding a 1s and a 2s electron at various distances from the nucleus in a hydrogen atom. The distance, r, from the nucleus is given in nanometers (10^{-9} m).

than a 1s orbital. The three different p orbitals in a set differ only in their orientations about the nucleus, being mutually perpendicular to each other. The five d orbitals have orientations and shapes as shown in Figure 8–9. In each p and d orbital, it is more probable that one might

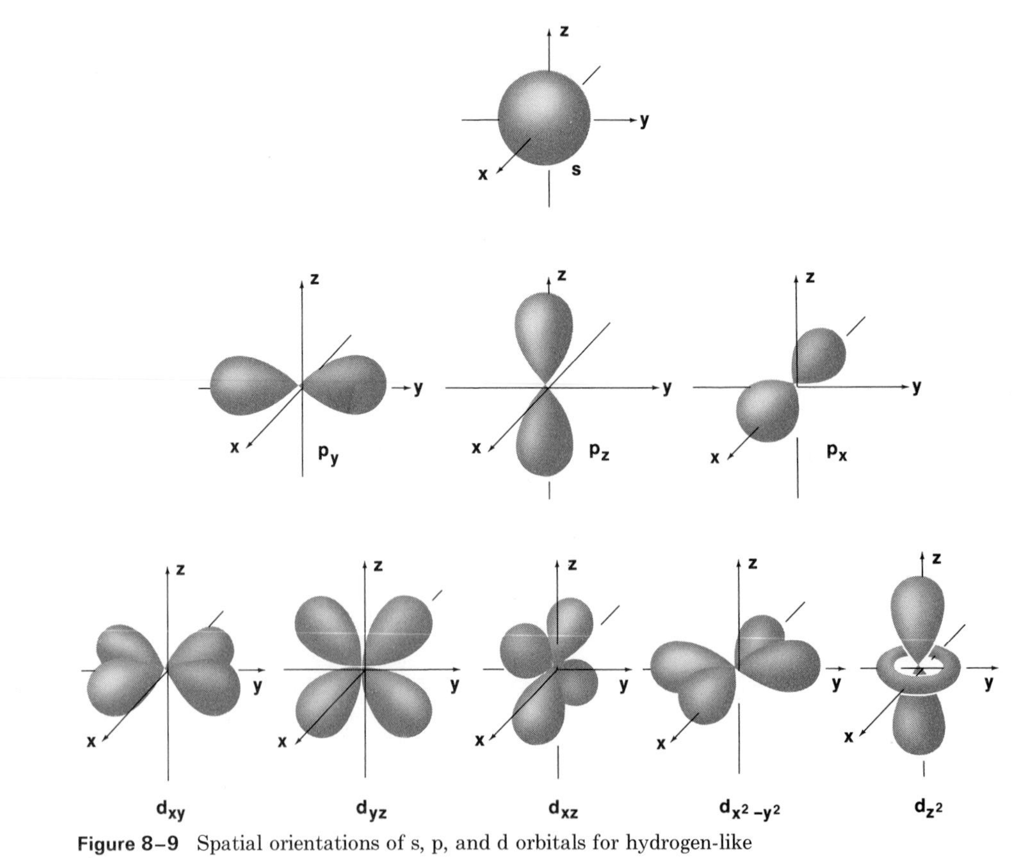

Figure 8–9 Spatial orientations of s, p, and d orbitals for hydrogen-like atoms.

find an electron in particular directions away from the nucleus than in other directions. The validity of these theoretical shapes and spatial orientations will be partially established if they can be used to explain how atoms unite to form molecules and, in turn, if they agree with the shapes of these molecules. We shall look for this in Chapter 9.

SELF-TEST 8-C

1. In which group of the periodic table would elements with the following electron arrangements be found?

   a. $1s^22s^1$ ________ b. $1s^22s^22p^3$ ________ c. $1s^22s^22p^63s^1$ ________

2. A period in the periodic table of the elements begins with an element having its highest-energy electron in a(n) ________ orbital. A period ends with an element that has a full set of ________ orbitals containing its highest-energy electrons.

3. Consider the meaning of the representations of the orbitals shown in Figure 8-9.

   a. Are the representations the paths of electrons? ________

   b. Are the representations the containers of electrons? ________

   c. Do the representations show where an electron is more likely to be found? ________

SUMMARY OF THE MODERN PICTURE OF THE ATOM

The working model of the atom, according to modern atomic theory, involves the following points:

a. An acceptable theoretical treatment of the atom consistent with both the wave nature and the particle nature of the electron is obtained.

b. Sufficient energy levels for an adequate number of transitions to explain the lines in the spectrum of an element are provided.

c. It is no longer possible to speak of the exact position of the electrons in an atom, but only of the relative probability of finding an electron in a given volume. In the same manner, there are limits on the accuracy with which the velocity of the electron, its energy, and other properties of the atom can be specified.

d. According to the Heisenberg uncertainty principle, if one of the two quantities to be measured simultaneously (for example, energy and time, or position and velocity) is specified rather precisely, a large error appears in the other quantity. If a large error in time is allowed, energies can be specified more exactly. This usually allows the specification of the nearly exact energy of the electrons in the various orbitals.

e. The chief factors of chemical interest about the electronic orbitals are their energies, shapes, and positions relative to each other. The shapes of the orbitals are important in that they tell the directions in which electronic density is concentrated.

f. The rules which dictate the manner in which orbitals are to be filled are basic in explaining the periodic table as these rules determine the number of elements in each period and the recurrence of similar electronic structures for elements having similar chemical activity.

g. Since the modern atomic model is the most consistent theory of the atom yet developed, any modern theory of chemical binding between atoms should be built on this theory. That is, the electronic orbitals which are available should be used as a starting point in getting a picture of any molecule. Any changes which are made on the atomic orbitals to build up the molecular picture must be consistent with the known energy changes occurring in chemical reactions.

MATCHING SET

_______ 1. de Broglie	a.	demonstrated wave nature of electron
_______ 2. orbital	b.	quantum numbers: 2,0,0
_______ 3. $2s^1$	c.	changing direction of light or path of particles around a sharp edge
_______ 4. G. P. Thomson, C. J. Davisson, L. H. Germer	d.	predicted wave nature of electron
_______ 5. standing wave	e.	domicile for electrons in an atom
_______ 6. diffraction	f.	no excited electrons
_______ 7. ground state	g.	"permanent" nodes and antinodes
_______ 8. $1s^2 2s^2 2p^1$	h.	electrons in boron atom

QUESTIONS

1. Give two shortcomings of the Bohr theory.
2. What two new discoveries or concepts made it necessary to originate a new atomic theory to replace the Bohr theory?
3. Give experimental evidence for the wave nature of the electron.
4. Which exhibits standing waves: the ringing of a bell or the breaking of a glass?
5. a. Draw the general shape of an s orbital and a p orbital, separately, and then draw a set of three p orbitals in proper relationship to each other.
   b. What do the shapes of these orbitals signify?
6. Using the notations, $1s^2$, and so forth, write the ground state electronic arrangement of lithium (Li), oxygen (O), sodium (Na), and chlorine (Cl).
7. Distinguish between a Bohr orbit and a wave mechanical orbital.

8. a. Neon, argon, krypton, xenon and radon form a group of elements which are similar in that they form very, very few compounds. From their atomic structures, suggest a reason for this similarity in relative inactivity.
   b. How is the atomic structure of helium similar to the structures of the other members of this group?

9. In which orbital does the electron have the greater energy?

   a. 1s or 2s c. 5s or 5f
   b. 2s or 2p d. 3s or 3d

10. The Heisenberg uncertainty principle can be written as

$$\Delta p \Delta v = \frac{h}{2\pi m}$$

where Δp is the uncertainty in the position of a particle, and Δv is the uncertainty in the measurement of its velocity, m is the mass of the particle, and h is Planck's constant. What happens to the magnitude of the uncertainties as the body becomes more massive?

11. A student is arrested for speeding. He claims that the Heisenberg uncertainty principle makes it impossible for the arresting officer to determine his speed and his position at the same time. The judge calls you as an expert witness. What would be your opinion?

12. How many electrons (maximum number) are there in each of the following?

   a. an s orbital
   b. a d orbital
   c. a p orbital
   d. a set of p orbitals
   e. $n = 2$ main energy level
   f. $n = 4$ main energy level

13. What is the significance of this equation?

$$\lambda = \frac{h}{mv}$$

14. What is wrong with the following attempts to write electronic configurations of atoms?

   a. $1s^2 2s^3$
   b. $1s^2 1p^6 2s^2 2p^6$
   c. $1s^2 2s^2 3s^2 3p^6 4s^2 4p^6$
   d. $1s^2 2s^2 2p^8 3s^2 3p^8$

15. Identify the atoms having these electronic configurations:

   a. $1s^2 2s^1$ b. $1s^2 2s^2 2p^6 3s^2 3p^6 4s^2$ c. $1s^2 2s^2 2p^6 3s^2 3p^6 4s^2 3d^{10} 4p^4$

SUGGESTIONS FOR FURTHER READING

Hoffman, B., "The Strange Story of the Quantum," Dover Publications, Inc., New York, 1959.
Kaufman, E. D., "Advanced Concepts in Physical Chemistry," McGraw-Hill Book Company, New York, 1966, pp. 31–49.
Lambert, F. L., "Atomic Orbitals from Wave Patterns," *Chemistry,* Vol. 41, No. 2, p. 10 (1968).

CHEMICAL BONDING

CHAPTER 9

We have already seen how the modern theory of the atom can predict the electronic configurations of multielectron atoms and can explain some of their properties, such as atomic spectra. One very important aspect of atomic theory is its ability to explain the forces that hold atoms together in molecules, molecules together in molecular substances, and ions together in ionic structures. A fixed link between submicroscopic particles, resulting from attractive forces between them, is called a ***chemical bond*** (Figure 9-1). Our knowledge of chemical bonding has emerged through the interaction of two parallel lines of development: (1) the investigation of the properties of substances, and (2) the outgrowth of theories of chemical bonding based on concepts of atomic structure.

Chemical bonds serve to hold atoms, molecules, and ions together.

Studies of the properties of substances, such as chemical reactivity, volatility (ability to pass into gaseous state), melting point, electrical conductivity, and color can often give some indication as to how submicroscopic particles are bonded together. For example, since melting involves a situation in which atoms or molecules are less firmly bound to their neighbors, a high melting point implies that a solid is held together by very stable chemical bonds. As we shall see shortly, compounds composed of a network of tightly bound ions or atoms tend to

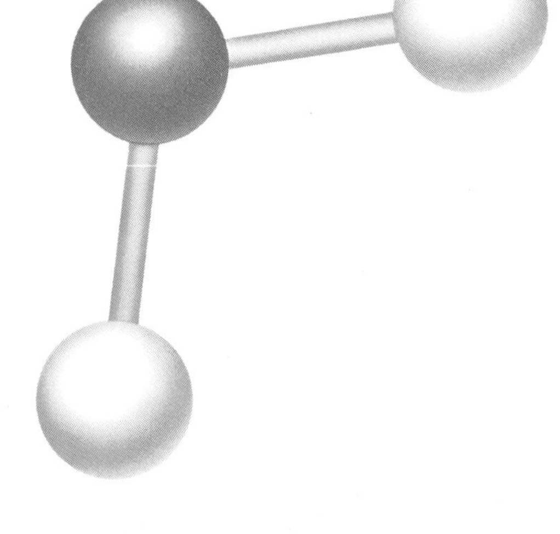

Figure 9–1 The two chemical bonds in a water molecule are often represented by a ball-and-stick model. This model tells which atoms are bonded together and the angles involved, but gives no information as to why the atoms are bonded in a particular pattern, the relative sizes of the atoms, or the distances between them.

have relatively high melting points. The volatility of a substance also indicates how strongly molecules are attracted to each other. For example, in the case of carbon dioxide, CO_2, we must assume that the bonding between molecules (intermolecular bonding) is slight, since it takes relatively little energy to break up the solid CO_2.

CO_2 sublimes at $-78°C$.

There has been a continuing effort to develop theories which can explain chemical bonding. These theories have been advanced in the firm belief that there is a fundamental relationship between the submicroscopic structure of matter and its properties. Theories have been used to establish detailed structure-property relationships for known substances and to predict properties for new substances.

Bonding theories must explain the observed behavior of chemicals.

TYPES OF CHEMICAL BONDING

Chemical bonds are of five major types, each having its own characteristics. These types of bonds, along with common materials in which they occur, are:

1. Ionic bonding	Salts
2. Covalent bonding	Molecular compounds, such as water, and in polymers, such as polyethylene
3. Hydrogen bonding	Water, ammonia, and large molecules in living organisms
4. Van der Waals' attractions	Liquid helium and solid CO_2
5. Metallic bonding	Metals and alloys

We shall now look at these types of chemical bonds in more detail, with a major emphasis on describing various types of compounds and accounting for some of their properties in terms of the bonding that holds them together. We shall see that bonding is explained in terms of the outer electrons in the atoms. Such electrons are referred to as ***bonding*** or ***valence*** electrons.

IONIC BONDING

Ions and Ion Formation

Charged atomic or molecular-sized particles are called ***ions.*** They may be positively charged, as Na^+ (sodium ion), or negatively charged, as Cl^- (chloride ion); multicharged ions, such as Ca^{2+} (calcium ion) and O^{2-} (oxide ion), are also encountered.

Ions may bear one or more positive or negative charges.

Since electrons are in the outermost parts of the atom, it is reasonable to assume that only electrons are transferred between particles to form ions. There are then two ways for ions to be formed: electron loss to form positive ions and electron gain to form negative ions. Let us examine the formation of positive ions first.

Metals generally have electronic configurations in which there are three or fewer electrons in the outermost principal energy level. These electrons have a high probability of being relatively far from the nucleus, as shown for sodium in Figure 9–2. These outermost electrons are s and

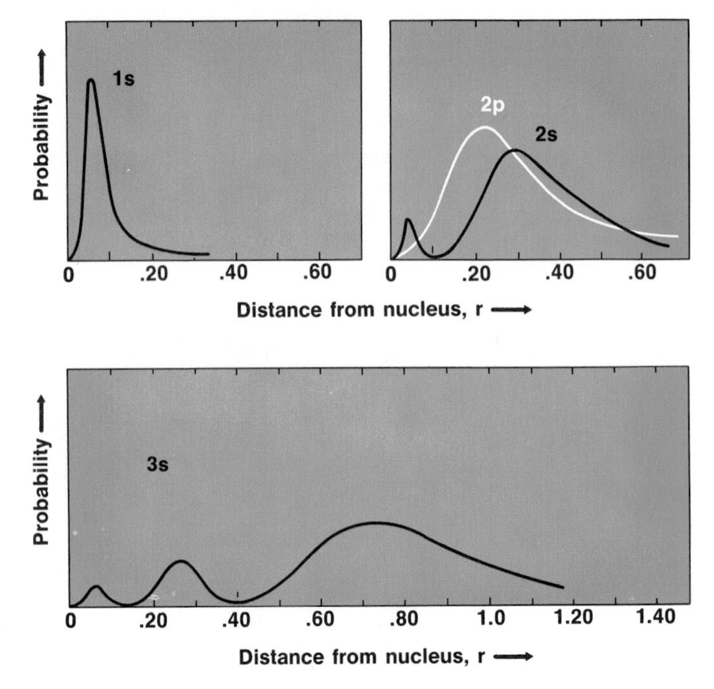

Figure 9–2 Probable distances from the nucleus for the 1s, 2s, 2p, and 3s electrons of a sodium atom. The electronic configuration for sodium is $1s^2\ 2s^2\ 2p^6\ 3s^1$. Distances are in nanometers (nm).

p electrons for the IA, IIA, and IIIA Group metals. Since they are far from the influence of the positively charged nucleus, they have the highest energy of any electrons in the atom. Remember that the potential energy of an electron-nucleus system increases as the distance between the nucleus and the electrons is increased. Thus, it would take a minimum amount of additional energy to remove these electrons completely from the atom. Since metal atoms are neutral (uncharged), the loss of a negative charge results in a positive charge on the ion.

Metals may lose electrons to nonmetals.

$$\text{Na} + \text{Ionization Energy} \longrightarrow \underset{\text{SODIUM ION}}{\text{Na}^+} + 1\text{e}^-$$

The energy required to remove each electron completely from an atom is called its ***ionization energy*** (for the first electron lost, the first ionization energy; for the second electron lost, the second ionization energy; and so on). Metals have low ionization energies; nonmetallic elements generally have higher ionization energies.

Ionization energies are measured experimentally and correspond to the reaction: Atom $\longrightarrow$ Positive ion + electron

As more and more electrons are removed from a single atom, it becomes more difficult to remove the next one. The net positive charge builds up with the loss of each electron and this charge helps to hold the remaining electrons more securely. The difficulty of removing successive electrons from the same atom is emphasized in the ionization energies in Table 9–1.

The table of ionization energies also indicates how many electrons are relatively easy to remove from an atom. Note that Li, Na, and K have relatively low ionization energies for removal of the first electron, but require a big jump in energy for removal of the second. For Be, Mg, and Ca, the big jump is between the second and third ionization energies. This means that two electrons are relatively easy to remove, but the third is in a set of lower energy and requires considerably more energy

Table 9-1 shows the amounts of energy needed to remove one or more electrons from various atoms.

TABLE 9-1 Ionization Energies of Selected Gaseous Atoms

An electron volt (ev)* is the energy acquired by an electron when accelerated by a potential difference of 1 volt. For each element, electrons must be removed to the heavy vertical line in order to attain a noble gas electronic configuration.

Atomic Number	Atom	Ionization Energies (ev)							
		1st	*2nd*	*3rd*	*4th*	*5th*	*6th*	*7th*	*8th*
1	H	13.6							
2	He	24.6	54.4						
3	Li	5.4	75.6	122.4					
4	Be	9.3	18.2	153.9	217.7				
5	B	8.3	25.1	37.9	259.3	340.1			
6	C	11.3	24.4	47.9	64.5	392.0	489.8		
7	N	14.5	29.6	47.4	77.5	97.9	551.9	666.8	
8	O	13.6	35.1	54.9	77.4	113.9	138.1	739.1	871.1
9	F	17.4	35.0	62.6	87.2	114.2	157.1	185.1	953.6
10	Ne	21.6	41.1	64	97.2	126.4	157.9		
11	Na	5.1	47.3	71.7	98.9	138.6	172.4	208.4	264.2
12	Mg	7.6	15.0	80.1	109.3	141.2	186.9	225.3	266.0
13	Al	6.0	18.8	28.4	120.0	153.8	190.4	241.9	285.1
14	Si	8.1	16.3	33.5	45.1	166.7	205.1	246.4	303.9
15	P	10.6	19.7	30.2	51.4	65.0	220.4	263.3	309.3
16	S	10.4	23.4	35.0	47.3	72.5	88.0	281.0	328.8
17	Cl	13.0	23.8	39.9	53.5	67.8	96.7	114.3	348.3
18	Ar	15.8	27.6	40.9	59.8	75.0	91.3	124.0	421
19	K	4.3	31.8	46	60.9	—	99.7	118	155
20	Ca	6.1	11.9	51.2	67	84.4	—	128	147

*One electron volt = 3.83×10^{-20} calories.

to remove it. It can be seen that this pattern of ionization energies is periodic.

Of the elements in a given period, the noble gases have the highest ionization energies.

This periodic behavior of ionization energy, shown in Figure 9-3, is directly related to the electronic configuration of the elements. For example, the elements known as the ***noble gases*** (helium, neon, argon, krypton, xenon, and radon) are at the end of each upward trend. These elements have the highest ionization energies in each period of elements and do not readily enter into chemical change. Indeed, they are called noble gases because of this indifference to chemical combination. It is reasonable to assume that both the high ionization energies and the relative chemical inertness are related to the ns^2np^6 electronic configuration common to this group of elements. (Note that helium has no p electrons; its configuration is $1s^2$.) We must also conclude that the elec-

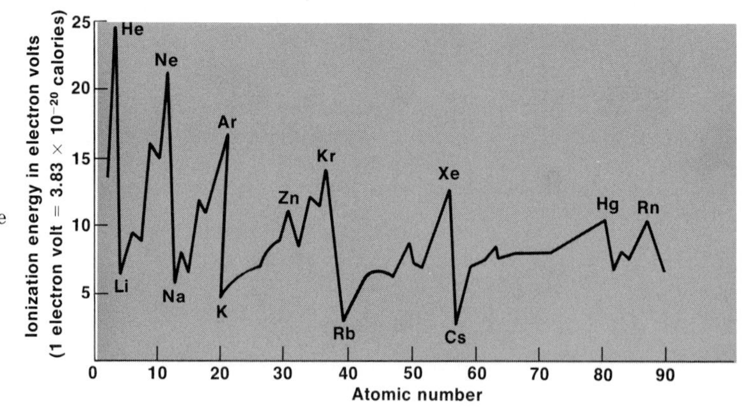

Figure 9–3 First ionization energies of the elements.

tronic configurations of both metals and nonmetals represent more chemically reactive structures than the ns^2np^6 structure of the noble gas.

In the formation of positive ions from metal atoms, the number of electrons a given metal atom will lose in an ordinary chemical process is determined by the electronic configuration of the metal atom. If the metal atom has only one outermost s electron, it could lose that electron, form a singly positive ion, and at the same time achieve the configuration of the noble gas preceding it in the periodic chart. Consider lithium (Li, atomic number 3), with a configuration of $1s^22s^1$. Loss of the 2s electron gives a lithium ion, Li^+, with a $1s^2$ configuration, which is the electronic configuration of the noble gas, helium.

Metal atoms lose electrons to form positive ions.

$$\underset{1s^22s^1}{Li} \longrightarrow \underset{1s^2}{Li^+} + 1e^-$$

(HELIUM CONFIGURATION)

Atoms of metals often lose electrons until they achieve the electronic configuration of a noble gas.

Likewise, sodium (Na, atomic number 11), by the loss of its $3s^1$ electron, gives a sodium ion, Na^+, with the configuration of neon.

$$\underset{1s^22s^22p^63s^1}{Na} \longrightarrow \underset{1s^22s^22p^6}{Na^+} + 1e^-$$

(NEON CONFIGURATION)

A study of the electronic configurations of each of the Group I elements, the alkali metal family, reveals that each of them has the possibility of losing an ns^1 electron to form a singly charged positive ion (Table 9–2).

It appears from the electronic structures of the alkali metals that ions tend to be formed by attaining a closed-shell configuration (the noble gas configuration) by the loss of electrons. This principle applies to some other metals as well. Calcium (Ca, atomic number 20), for example, does this by the loss of two electrons. The stable Ca^{2+} ion has the configuration of the noble gas argon (Ar, atomic number 18, $1s^22s^22p^63s^23p^6$).

$$\underset{1s^22s^22p^63s^23p^64s^2}{Ca} \longrightarrow \underset{1s^22s^22p^63s^23p^6}{Ca^{2+}} + 2e^-$$

(ARGON CONFIGURATION)

TABLE 9–2 Electronic Configurations of the Alkali Metals

Element	Atomic Number	Configuration
Li	3	$1s^22s^1$
Na	11	$1s^22s^22p^63s^1$
K	19	$1s^22s^22p^63s^23p^64s^1$
Rb	37	$1s^22s^22p^63s^23p^64s^23d^{10}4p^65s^1$
Cs	55	$1s^22s^22p^63s^23p^64s^23d^{10}4p^65s^24d^{10}5p^66s^1$

All nonmetals have relatively high first ionization energies. The energies available in ordinary chemical change are simply not great enough to form positive ions from nonmetallic atoms. In order for nonmetals to achieve the noble gas configurations it is reasonable to assume that they would tend to gain electrons. Thus, they could fill their partly empty outer orbitals* and, in so doing, attain the closed-shell configuration of a noble gas.

Consider an atom of fluorine (F, atomic number 9). Its electronic configuration is $1s^22s^22p^5$ with one p orbital being half filled ($2p^5 = 2p^22p^22p^1$). If the fluorine atom gains one electron, a fluoride (F^-) ion would be formed with the closed-shell configuration of neon ($1s^22s^22p^6$).

$$\underset{1s^22s^22p^5}{F} + 1e^- \longrightarrow \underset{1s^22s^22p^6}{F^-}$$

(NEON CONFIGURATION)

Nonmetal atoms gain electrons to form negative ions.

Oxygen (O, atomic number 8) atoms have an electronic configuration $1s^22s^22p^4$, with two half-filled 2p orbitals ($2p^4 = 2p^22p^12p^1$). This means that by the gain of two electrons, an oxygen atom could form an oxide ion (O^{2-}) with the neon configuration.

Stable ions often have "filled" electron subshells.

$$\underset{1s^22s^22p^4}{O} + 2e^- \longrightarrow \underset{1s^22s^22p^6}{O^{2-}}$$

(NEON CONFIGURATION)

On the basis of atomic structures, ionization energies, and the inert character of the noble gases, we have thus far reasoned that metals should form positive ions by losing electrons and nonmetals should form negative ions by gaining electrons. More specifically, we reasoned that Na, Ca, F, and O tend to form Na^+, Ca^{2+}, F^-, and O^{2-} ions, respectively. There is ample experimental evidence that this is actually the case. For example, consider the formulas of the compounds that can be formed from these elements and the equations for the reactions involved:

Chemical formulas of ionic compounds result from the number of electrons lost or gained by the reacting atoms.

$$\begin{array}{ll} NaF: & 2Na + F_2 \longrightarrow 2Na^+ + 2F^- \\ CaF_2: & Ca + F_2 \longrightarrow Ca^{2+} + 2F^- \\ Na_2O: & 4Na + O_2 \longrightarrow 4Na^+ + 2O^{2-} \\ CaO: & 2Ca + O_2 \longrightarrow 2Ca^{2+} + 2O^{2-} \end{array}$$

*The orbitals in the outermost energy level (highest value of n) are called valence orbitals and the electrons therein are termed valence electrons. The word *valence* has historically been used to denote chemical bonding.

Note that in each case the formula is compatible with the ideas developed previously, namely:

a. a sodium atom loses 1 electron to form Na^+,
b. a calcium atom loses 2 electrons to form Ca^{2+},
c. a fluorine atom gains 1 electron to form F^-, and
d. an oxygen atom gains 2 electrons to form O^{2-}.

X-ray studies, as well as chemical properties, indicate that these four compounds are ionic solids. When molten, all four compounds conduct electricity, which is characteristic of mobile ions (see Chapter 10). Mass spectographic studies leave little room for doubt that these ions have a real and stable existence and that they actually have the charges indicated.

There is extensive evidence for the existence of ions in solids and many solutions.

Further considerations of ionic structures along these lines lead to four important conclusions:

1. There is a tendency for atoms with ns^1, ns^2, or ns^2np^1 configurations (fewer than four electrons in outer energy shell) to lose electrons and form positive ions with an electronic configuration of a noble gas;
2. There is a tendency for atoms whose p orbitals have 3, 4, or 5 electrons (5, 6, or 7 electrons in outer energy shell) to gain electrons and form negative ions with an electronic configuration of a noble gas;
3. When metal atoms react with nonmetal atoms (except for noble gases) there is a tendency for the metal atoms to lose electrons to the nonmetal atoms, forming ions;
4. Atoms with an ns^2np^2 configuration (4 electrons in outer energy shell) have no pronounced tendency to lose or gain electrons in the formation of ions.

These generalizations should not be thought of as chemical laws, but merely as memory aids for the cases most frequently encountered.

Electron Dot Formulas

To keep track of the electrons involved in electron-transfer reactions such as those discussed previously, we use a type of notation known as the ***electron dot formula.*** To write the electron dot formula of an atom or ion, the electrons in the outermost principal energy level are counted, and for each electron a dot is placed around the symbol for the element. The dots are paired so that they are consistent with the pairing of electrons in orbitals. For example, fluorine ($1s^22s^22p^5$) is $\cdot\underset{\cdot\cdot}{\overset{\cdot\cdot}{\mathrm{F}}}\colon$ (seven electrons in the outermost 2s and 2p orbitals).

The reaction between sodium and fluorine is written as follows:

$$\mathrm{Na}\cdot + \colon\underset{\cdot\cdot}{\overset{\cdot\cdot}{\mathrm{F}}}\cdot \longrightarrow \mathrm{Na}^+ + \colon\underset{\cdot\cdot}{\overset{\cdot\cdot}{\mathrm{F}}}\colon^-$$

Note that Na^+ has no dots around it, since there are no electrons in the valence (outermost) shell of the sodium atom (n = 3). The reaction between calcium and fluorine can be written using electron dot formulas as follows:

$$\text{Ca}\colon + 2\,\colon\!\ddot{\underset{\cdot\cdot}{\text{F}}}\cdot \longrightarrow \colon\!\ddot{\underset{\cdot\cdot}{\text{F}}}\colon^{-} + \text{Ca}^{2+} + \colon\!\ddot{\underset{\cdot\cdot}{\text{F}}}\colon^{-}$$

Since the calcium ion has two electrons, it must react with two fluorine atoms, each of which takes one electron.

Properties of Ions

Ion Sizes

The sizes (radii) of atoms of practically all the elements can be measured by several experimental methods. Let us consider briefly the sizes of the ions that are produced by either electron gain or electron loss. The sizes of the ions are important because the strength of the forces holding ions together in ionic compounds depends on the sizes of the ions involved.

One nanometer is 10^{-9} meter.

A sodium atom with its single 3s outermost electron has a radius of 0.186 nm. One would expect that when this electron is removed (forming the Na^+ ion) the resulting ion would be smaller. This decrease in size results because there are now only 10 electrons attracted to a charge of +11 on the nucleus, and these electrons are pulled closer to the nucleus by this charge imbalance. This same type of phenomenon is observed for all metal ions. Figure 9–4 compares atomic and ionic sizes for seven elements.

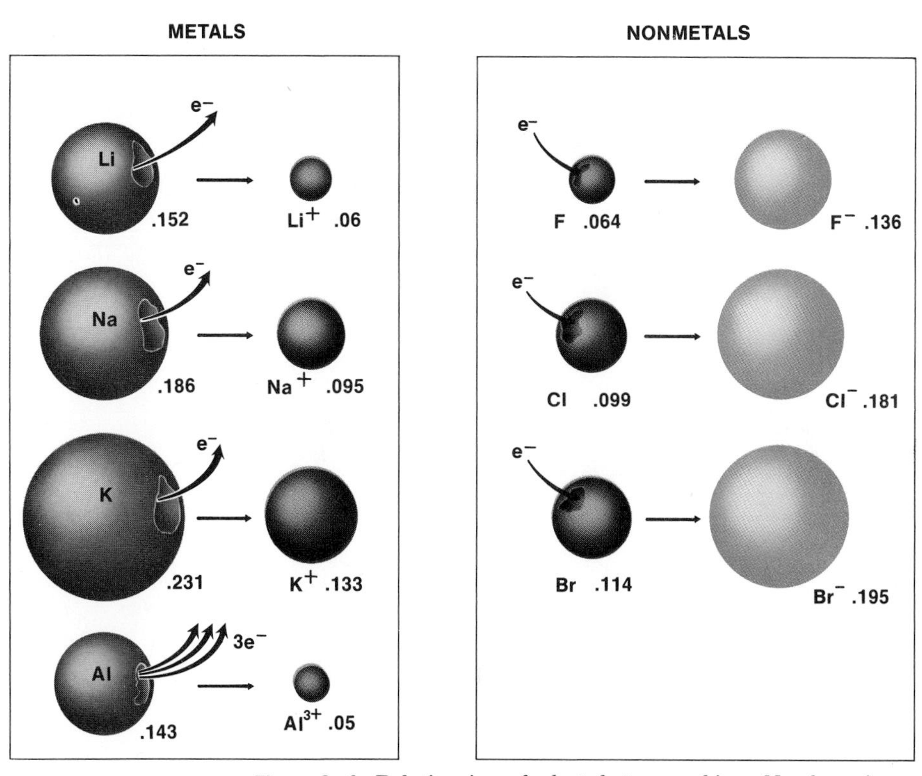

Figure 9–4 Relative sizes of selected atoms and ions. Numbers given are atomic or ionic radii in nanometers.

The metal ions with multiple charges (Al^{3+}, Fe^{2+}, and so on) are much smaller than the corresponding metal atom.

Nonmetals that gain electrons to achieve closed-shell configurations increase in size during this process so that the negative ion is larger than the corresponding atom. Consider the information in Figure 9–4. This behavior is due to the addition of electrons to the orbitals of an atom without increasing the charge on the nucleus. The repulsion of the electrons causes the expansion.

Electrostatic Forces—The Ionic Bond

Ionic bonding is due to electrostatic interaction between ions of opposite charge. The geometrical array of ions is termed an ***ionic lattice,*** and the substance is an ionic solid or salt (Figure 9–5).

Ionic bonds are the attractions between ions of opposite charge.

Recall from Chapter 5 that Coulomb's law tells us that oppositely

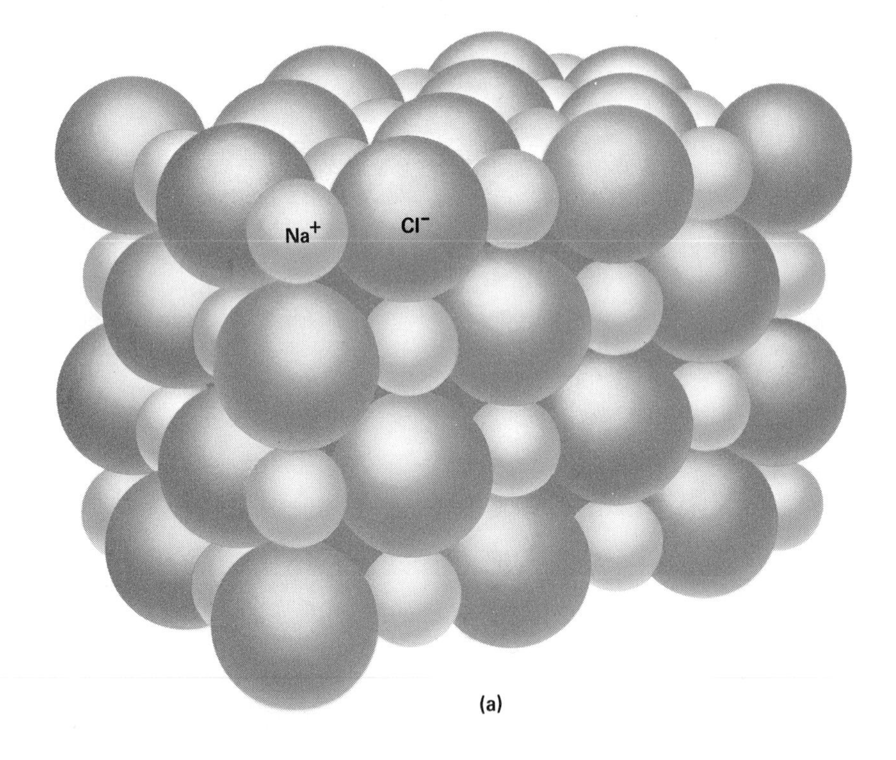

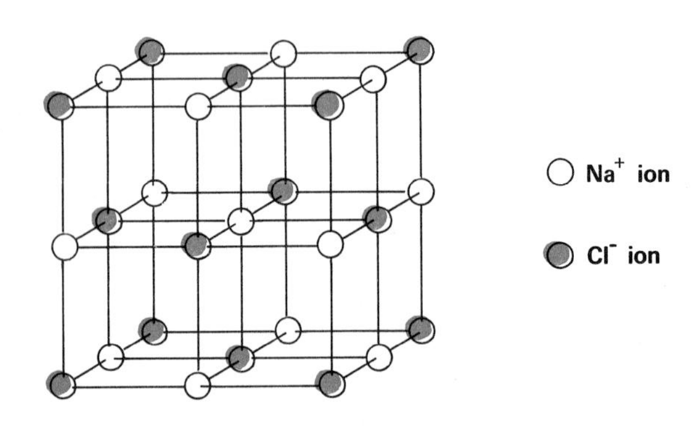

Figure 9–5 Structure of sodium chloride crystal. (a) Model showing relative sizes of the ions; (b) ball-and-stick model showing cubic geometry.

charged ions will attract each other with a force which depends on the charges and the distance between them. We can express this coulombic attraction as:

$$\text{Force} \left\{ \begin{array}{c} \text{is} \\ \text{proportional} \\ \text{to} \end{array} \right\} \frac{(\text{charge on positive ion}) \times (\text{charge on negative ion})}{(\text{distance of separation})^2}$$

Ionic attraction is determined by Coulomb's law.

This expression leads us to expect that strong ionic bonds are formed when the ions possess high charges and are separated by small distances (which will occur for the smaller ions). The effects of ionic charge and size can be seen rather easily if we compare the melting points of several salts, all with similar crystal geometries. Let us compare the melting points of NaCl and KCl (the K^+ ion is larger than the Na^+ ion) and also those of CaO and BaO (the Ba^{2+} ion is larger than the Ca^{2+} ion). All four salts have a sodium chloride (rock salt) crystalline structure (see Figure 9-5). Table 9-3 shows that in both cases the salts with smaller positive ions have the higher melting points. Those with doubly charged ions have dramatically increased melting points.

Ionic Lattices

There are numerous kinds of ionic lattices; a common type is the sodium chloride lattice (Figure 9-5). Here, each Na^+ ion is surrounded by six Cl^- ions, and each Cl^- ion is in turn surrounded by six Na^+ ions. Each ion is attracted to all of the oppositely charged ions and repelled by all of the like charged ions in the lattice. Thus, there are no sodium chloride diatomic molecules in this solid.

An ionic lattice is a regular geometrical array of ions.

One question arises at this point. If the oppositely charged ions attract one another, what keeps them from getting closer and closer to each other? This is prevented by the repulsion which arises when orbitals filled with electrons are moved very close together. There is a very considerable repulsion which prevents significant penetration of filled orbitals from different atoms into the filled orbitals of each other's space. As a consequence, each ion has an ionic radius which is reasonably constant and which represents the closest distance of approach of another ion's filled orbitals.

TABLE 9-3 The Effects of Ionic Charge and Size on Some Salts with Identical Structures

		M^+, Radius, nm	Melting Point
Salts with singly charged ions	NaCl	0.095	801°C
	KCl	0.133	776°C
		M^{2+}, Radius, nm	Melting Point
Salts with doubly charged ions	CaO	0.099	2580°C
	BaO	0.135	1923°C

Other Properties of Ionic Compounds

The strong bonding between the ions in ionic compounds causes them to be rather hard and brittle crystalline solids with relatively high melting points. Since the electrons involved in bonding are localized on the individual ions and cannot move about in the lattice, solid ionic compounds are rather poor electrical conductors as compared to molten ionic compounds. Molten ionic compounds contain free ions since the melting process partially breaks down the structure that had existed in the lattice. These free ions can move in an electrical field and thus transport charge or an electric current from one place to another. Hence, molten salts are generally excellent conductors of electricity.

SELF-TEST 9-A

1. Charged atoms are called ______________.
2. An array of attracting ions make up an ______________ ______________.
3. A sodium atom loses ______________ electrons(s) in achieving a noble gas configuration.
4. Look at Figure 9–3. Which element requires the most energy for ionization? ______________ The least? ______________
5. What is the correct formula for calcium iodide? ______________ The electronic configuration of calcium (Ca) is (Ar electronic structure)$4s^2$ while that of iodine (I) is (Kr electronic structure)$4d^{10}5s^25p^5$. Hint: Find these two elements on the periodic chart.
6. Which ion gained an electron in its formation: Na^+ or Cl^- ______________?
7. Electrons in the shell of highest energy may be called ______________ electrons.
8. Positive ions are formed from neutral atoms by (losing/gaining) electrons.
9. Negative ions are formed from neutral atoms by (losing/gaining) electrons.
10. Predict the number of electrons lost or gained by the following atoms in forming ions. Indicate whether the electrons are gained or lost. Rb ________ ________; S ________ ________; Be ________ ________;

    Mg ________ ________; Al ________ ________; Br ________ ________.

COVALENT BONDING

A second type of chemical bonding, which is found in molecular compounds, nonmetallic elements, and polymers, is covalent bonding. This type of bonding is different from ionic bonding in that electrons are shared between atoms rather than transferred from one atom to another.

The atoms that tend to bond covalently are those with relatively high ionization energies. These atoms do not readily lose electrons. Hence, the atoms must share electrons in their compounds if the atoms

A covalent bond results from sharing an electron pair.

are to acquire noble gas configuration. Such substances, for example, as carbon dioxide (CO_2), carbon tetrachloride (CCl_4), ammonia (NH_3), hydrogen (H_2), chlorine (Cl_2), and water (H_2O) are all composed of molecules which are held together by covalent bonding.

Electron Sharing

Covalent bonding arises from the sharing of electrons between atoms.

The simplest example of covalent bonding is found in diatomic (two atoms) gaseous molecules such as H_2 (hydrogen), F_2 (fluorine), and Cl_2 (chlorine). A hydrogen atom has an electronic configuration of $1s^1$. Comparing this with the electronic configuration of the noble gas nearest to hydrogen in the periodic table, helium (He, $1s^2$), we see that, if the hydrogen atom could somehow obtain one more electron, a very stable configuration could be reached (recall that the Li^+ ion with its $1s^2$ configuration is quite stable). For hydrogen, this configuration can be achieved by two hydrogen atoms sharing their single electrons. The electron dot formula for the resulting H_2 molecule is:

$$2H\cdot \longrightarrow H{:}H$$

Ionic bonding is not a reasonable possibility for the H—H bond since both hydrogen atoms have the same attraction for the electron pair, neither having the ability to remove the electron completely from the other.

For other atoms we may use the tendency to attain a closed-shell configuration (eight electrons) as a guide in writing electron dot formulas. Two fluorine atoms bound together at normal temperatures form a F_2 molecule. This bonding is due to each fluorine atom sharing its unpaired 2p electron.

$$2\ {:}\ddot{\underset{..}{F}}\cdot \longrightarrow {:}\ddot{\underset{..}{F}}{:}\ddot{\underset{..}{F}}{:}$$

$1s^22s^22p^5$ FLUORINE MOLECULE

Orbital Overlap

The theoretical model of sharing electrons involves overlap of atomic orbitals.

Since hydrogen gas is molecular, and since the interaction between the two atoms must be other than ionic, the question arises as to how one hydrogen atom can be bonded to another. We have already concluded, in the modern model of the atom, that two electrons are compatible in an orbital of an atom if their spins are paired. It is reasonable to assume that the close approach of two unpaired electrons on different atoms will result in the formation of a covalent bond where both orbitals are made available to both of the paired electrons. Thus, the bond in the hydrogen molecule can be pictured as resulting from the overlap of two 1s orbitals, one from each atom, and the spin-pairing of the electrons

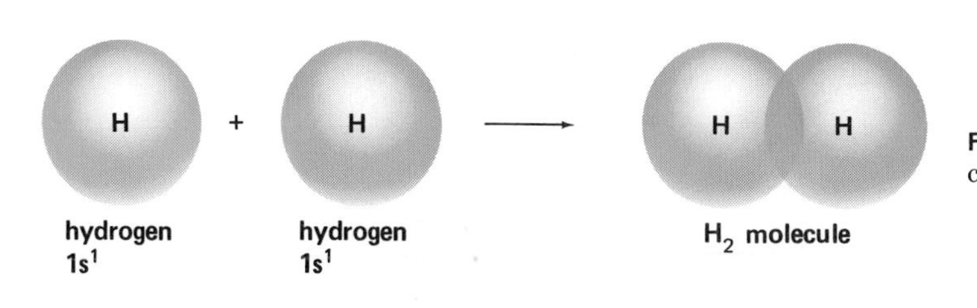

Figure 9–6 Overlap of 1s orbitals in H_2 molecule.

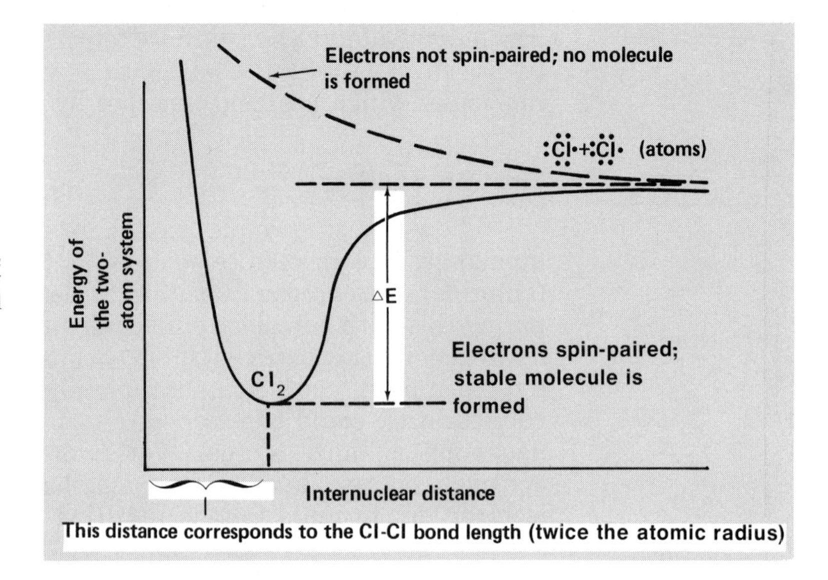

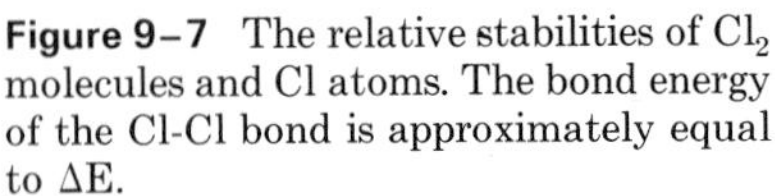

Figure 9–7 The relative stabilities of Cl_2 molecules and Cl atoms. The bond energy of the Cl-Cl bond is approximately equal to ΔE.

involved (Figure 9–6), and the greater the overlap, the more stable the bond will be.

It is assumed that stability is achieved when two atoms attract the same pair of electrons in the formation of a covalent bond. The opposite spin on the two electrons, as well as the attraction to the two positive nuclei add reason to this assumption. Review electric charges in Chapter 5. Figure 9–7 illustrates this for Cl atoms ($3s^2 3p^5$). The amount of energy released, ΔE, is approximately the ***bond energy,*** and is the same as the amount of energy necessary to break the Cl—Cl bond and give gaseous atoms. In general, energy is released in the formation of covalent bonds, and energy is required to break such bonds. Notice in Figure 9–7 that any attempt to force the two atoms closer to each other than their equilibrium bond length, will require a large amount of energy, since there is a repulsion caused by the filled atomic orbitals of the two different atoms.

To break a bond requires energy; when bonds are formed, energy is released.

Single Bonds

A covalent bond formed by the sharing of two electrons (a single pair of electrons) between two atoms is called a ***single bond.*** Many molecules, both simple and complicated, contain one or more single covalent bonds.

A single bond consists of one pair of electrons shared by two atoms.

Hydrogen fluoride (HF) contains a single covalent bond between the two atoms. Fluorine ($2s^2 2p^5$) needs one electron to attain a closed-shell configuration as does hydrogen ($1s^1$). This could be accomplished by an overlap of the 1s hydrogen orbital with the half-filled 2p fluorine orbital. The result is shown in Figure 9–8.

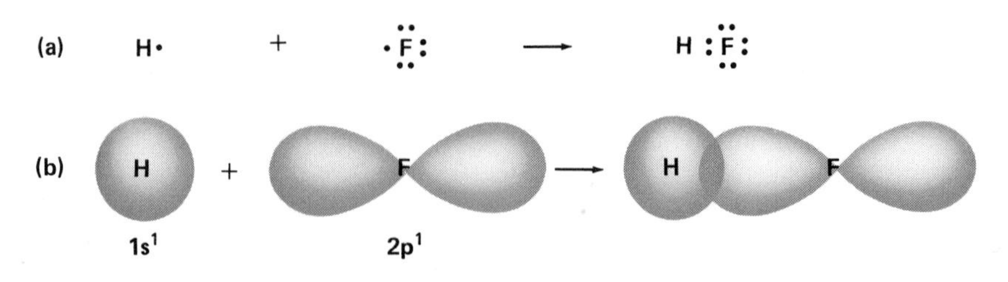

Figure 9–8 Single bond formation in HF. (a) The electron dot representation; (b) the orbital overlap representation.

An oxygen atom has a configuration $2s^2 2p^4$ and, therefore, needs two electrons to attain a closed-shell configuration. (We have already seen how oxygen atoms may gain two electrons and form O^{2-} ions). If two hydrogen atoms are present, there can be sharing of electrons between the atoms. The electron dot formula for the molecule that results, the water molecule, is given below. Oxygen is written as $:\ddot{O}\cdot$ because there are two unpaired electrons in the atom ($2p^4 = 2p^2 2p^1 2p^1$).

Consider next the nitrogen atom. Modern atomic theory predicts a configuration of $2s^2 2p^3$ ($2p^3 = 2p^1 2p^1 2p^1$) with three unpaired electrons. The atom needs three electrons to attain a closed-shell configuration. One way this can be done is by forming three single bonds with some other atoms having one unpaired electron each. With hydrogen atoms, the result is ammonia (NH_3). The nitrogen atom is written as $\cdot\dot{N}\cdot$ to show the three unpaired electrons.

H
|
The choice of H—O over H—O—H will be explained later in this chapter.

$$:\ddot{O}\cdot + 2H\cdot \longrightarrow :\ddot{O}:H$$
$$\text{H}$$

WATER

$$\cdot\ddot{N}\cdot + 3H\cdot \longrightarrow H:\ddot{N}:H$$
$$\text{H}$$

AMMONIA

Note that fluorine, oxygen, and nitrogen all end up with an octet (8) of electrons in HF, H_2O, and NH_3. This leads to the generalization known as the ***octet rule,*** which predicts that second period elements on the periodic chart form compounds in such a way as to complete their octet of electrons. Even though this rule can often be applied usefully to elements with heavier atoms, it does not always apply since some of these atoms can have more than eight electrons in their outer shells. Though generally true for second period elements, there are even a few exceptions among them.

The octet rule: eight electrons in the valence shell represent a stable state.

In the examples just given we can see that for the atoms undergoing covalent bonding, the number of single bonds possible is the same as the number of the unpaired electrons in the atom when it is ready to undergo bonding. This is generally true, as shown in the following examples:

H, hydrogen, 1 unpaired electron ⟶ 1 single bond
F, fluorine, 1 unpaired electron ⟶ 1 single bond
O, oxygen, 2 unpaired electrons ⟶ 2 single bonds
N, nitrogen, 3 unpaired electrons ⟶ 3 single bonds

Multiple Bonding

When an atom has fewer than seven electrons in its outermost shell (ns^2np^4, for example), it can complete its octet in two ways. In the first way, it simply shares a single electron with each of several other atoms which can share a single electron. This leads to ***single*** covalent bonds. But the atom can also share two (or three) pairs of electrons with a single other atom. In this case there will be two (or three) bonds between these two atoms.

One pair of electrons = one covalent bond.

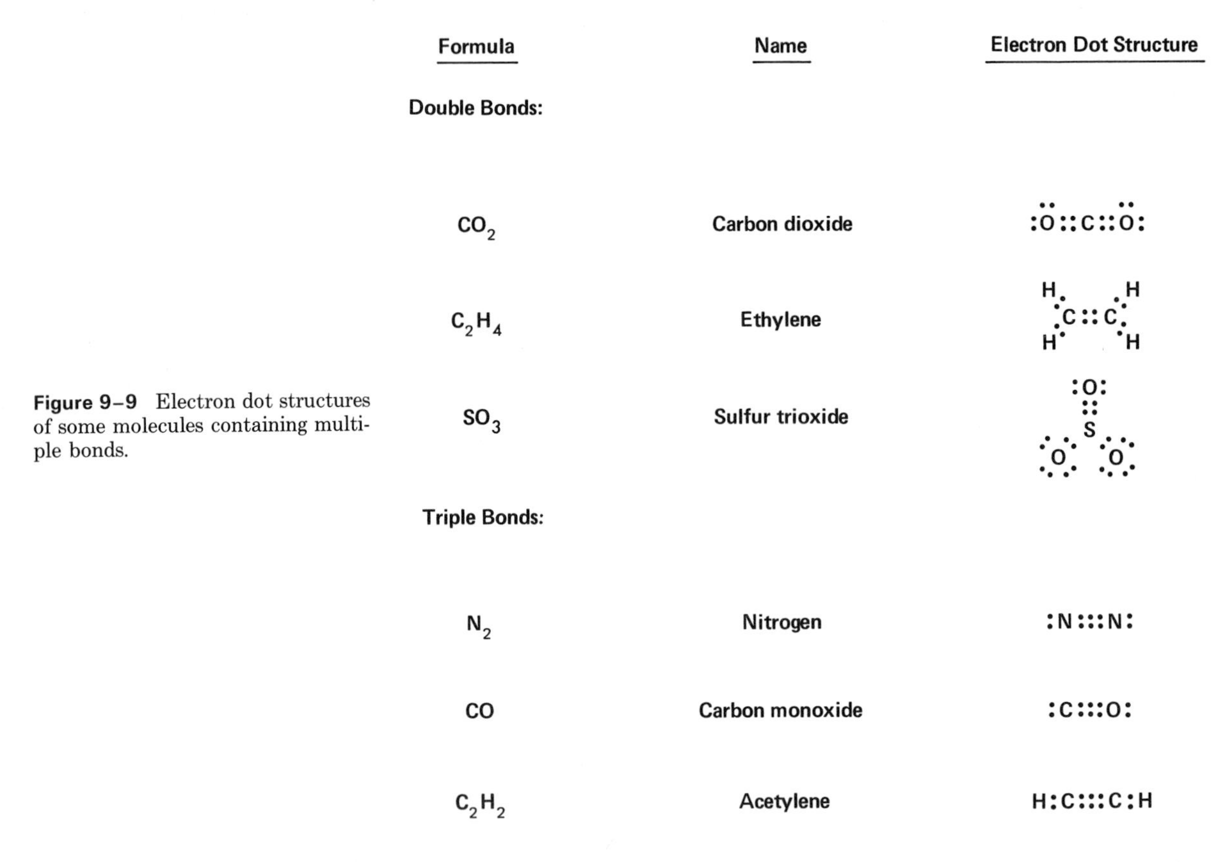

Formula
Name
Electron Dot Structure
Double Bonds:
CO2
Carbon dioxide
C2H4
Ethylene
SO3
Sulfur trioxide
Triple Bonds:
N2
Nitrogen
:N:::N:
CO
Carbon monoxide
:C:::O:
C2H2
Acetylene
H:C:::C:H

Figure 9–9 Electron dot structures of some molecules containing multiple bonds.

When two shared pairs of electrons join together the same two atoms, we speak of a ***double bond,*** and when three shared pairs are involved, the bond is called a ***triple bond.*** Examples of these bonds are found in many compounds of $O(2s^22p^4)$, $N(2s^22p^3)$, and $C(2s^22p^2)$ such as those shown in Figure 9–9.

A double bond consists of two electron pairs shared between two atoms.

As we can see from these structures, molecules may contain several types of bonds. Thus, ethylene (Figure 9–9) contains a double bond between the carbon atoms and single bonds between the hydrogen atoms and the carbon atoms. For convenience, an electron pair bond is often indicated by a dash as follows:

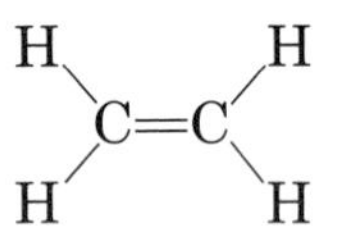

A single bond will be shown as H—H, a double bond as $H_2C{=}CH_2$, and a triple bond as N≡N. Note that in each of these cases the octet rule is obeyed.

A line between two atoms, as in H—H, represents a bonding pair of electrons.

Single, double, and triple bonds differ in length and energy. Triple bonds are shorter than double bonds, which in turn are shorter than single bonds. Bond energies normally increase with decreasing bond length due to greater orbital overlaps. Bond lengths and energies in some typical cases are shown in Table 9–4.

TABLE 9–4 Some Bond Lengths and Bond Energies

Bond	C—C	C=C	C≡C	N—N	N=N	N≡N
Length, nm	0.154	0.134	0.120	0.140	0.124	0.109
Bond energy (kcal/mole)	83	146	200	40	100	225

Kcal/mole = thousands of calories necessary to break 6.02×10^{23} bonds.

Polar Bonds

In a molecule like H_2 or F_2, where both atoms are alike, there is equal sharing of the electron pair. Where two unlike atoms are bonded together, however, the sharing of the electron pair is likely to be unequal, with the result that there is a negative and a positive portion of the bond. Such a bond is termed a ***polar bond.*** The negative pole is in the region of high electron density, and the positive pole is in the region of low electron density. Thus, in the molecule HF, the bonding pair of electrons is more under the control of the fluorine atom than of the hydrogen atom (Figure 9–10). This can also be expressed in terms of the centers of positive and negative charge. In the nonpolar hydrogen molecule, the center of negative charge is in the geometrical center of the molecule since there is a symmetrical distribution of negative charge about that center. The center of positive charge is located there also. In contrast to the nonpolar hydrogen molecule, the polar HF molecule has two distinct centers of positive and negative charge (Figure 9–10).

In a polar bond there is an unequal sharing of the bonding electrons.

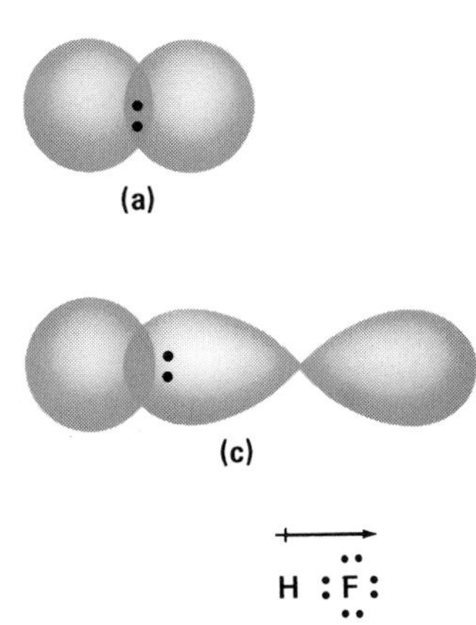

Figure 9–10 Polar bonds in HF and HCl. (a) Symmetrical distribution of electrons in H_2 results in the center of negative charge being identical with the center of positive charge. This is symbolized by the electron dots placed in the overlap area. (b) Overlap of p orbitals in F_2 also results in symmetrical distribution of charge. (c) In HF, the electron pair is displaced toward the fluorine nucleus since fluorine is more electronegative than hydrogen. Note the electron dots to the right of the overlap area which conveys the idea of polarity (separation of charge). (d) δ^+ (delta plus meaning fractional plus charge) and δ^- (delta negative meaning fractional negative charge) are used to indicate poles of charge. In (e) and (f) an arrow is used to indicate electron shift, the arrow having a plus tail to indicate partial positive charge on the hydrogen atom. Note that the longer arrow in HF structure indicates a greater degree of polarity than in HCl. This should not be confused with the greater bond length in HCl.

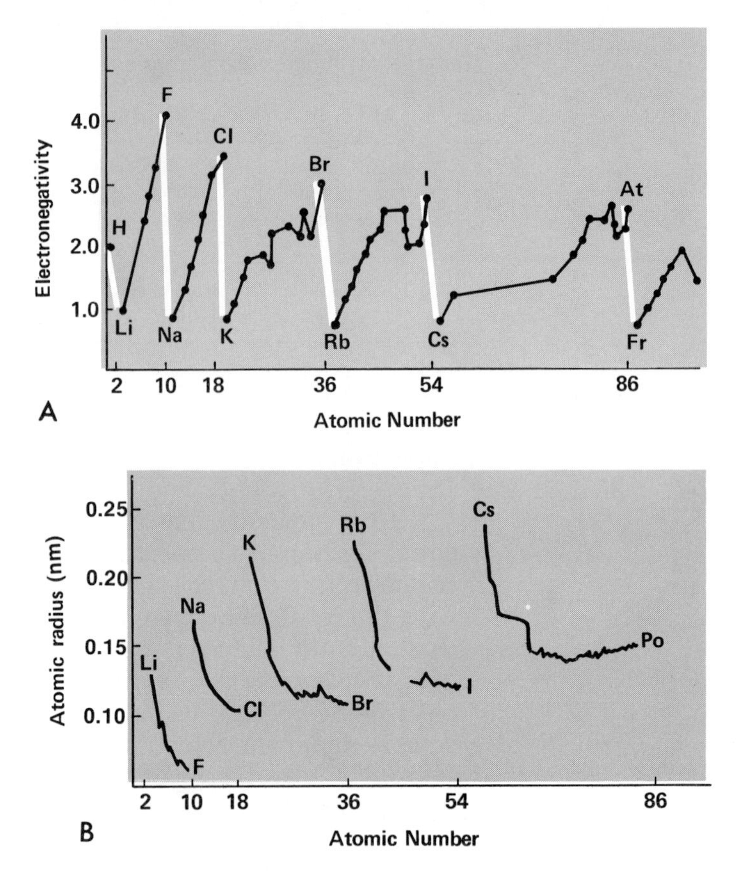

Figure 9–11 A, Electronegativities of the elements. B, Atomic sizes of the elements.

A measure of the net ability of an atom to attract electrons to it in a chemical bond is called its ***electronegativity.*** Since the fluorine atom attracts the shared electrons in the H—F bond more than does the hydrogen atom, we say that the electronegativity of the fluorine atom is greater. Electronegativity is a function of nuclear charge and atomic size. Small atoms with large nuclear charges are the most electronegative. The trend of electronegativities in the periodic table is rather regular and follows an inverse trend in atomic size. Figure 9-11 shows these relationships.

Unequal sharing of electrons results from differences in electron attraction by the bonding atoms.

The electronegativity of an atom is a measure of its ability to attract electrons to itself in a compound.

Covalent bonds joining different atoms are generally polar in that one of the atoms has distorted the electron distribution toward itself due to its greater electronegativity. Thus, polar covalent bonds occur in practically every molecule that has different kinds of atoms covalently bonded together. Additional examples of molecules with polar covalent bonds include NO, SO_2, CO_2, CCl_4, and H_2S (Figures 9-12 and 9-13).

The most electronegative atom is fluorine.

A very common bond that we shall discuss frequently in the remainder of this text is the C—H bond. Since carbon has an electronegativity of 2.5 and hydrogen 2.1, the C—H bond is only slightly polar. The arrangement of C—H bonds around a carbon atom generally make $—CH_2—$ and $—CH_3$ groups nonpolar.

HYDROGEN BONDING

For a series of molecular substances with similar structures, the boiling points ordinarily increase as the molecular weights increase. For

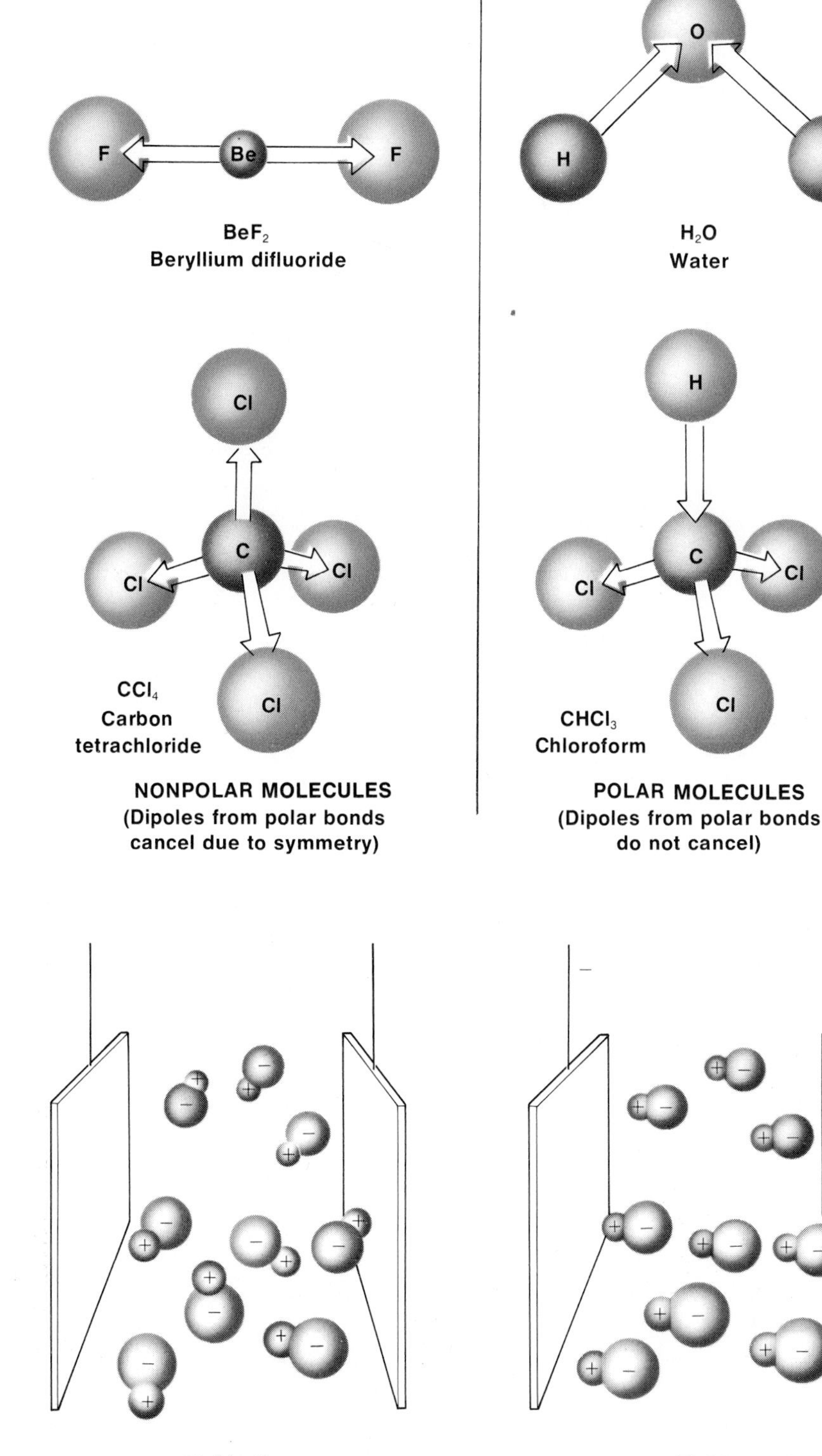

Figure 9–12 Polar bonds may or may not result in polar molecules. The polar bonds in beryllium difluoride and carbon tetrachloride are arranged about the center atom in such a way as to cancel out the polar effect. In contrast, the polar bonds in water and chloroform molecules do not cancel as a result of the molecular shape but combine to give a polar molecule.

Figure 9–13 Physical evidence for both the existence of polar molecules and the degree of polarity is provided by a simple electrical capacitor. The capacitor is composed of two electrically conducting plates with nonconducting material (an electrical insulator) between. The storage of charge by the capacitor is increased in sequence when a vacuum, then nonpolar, and polar materials are placed between the plates. To study individual molecules, the material is placed between the plates in the gas phase.

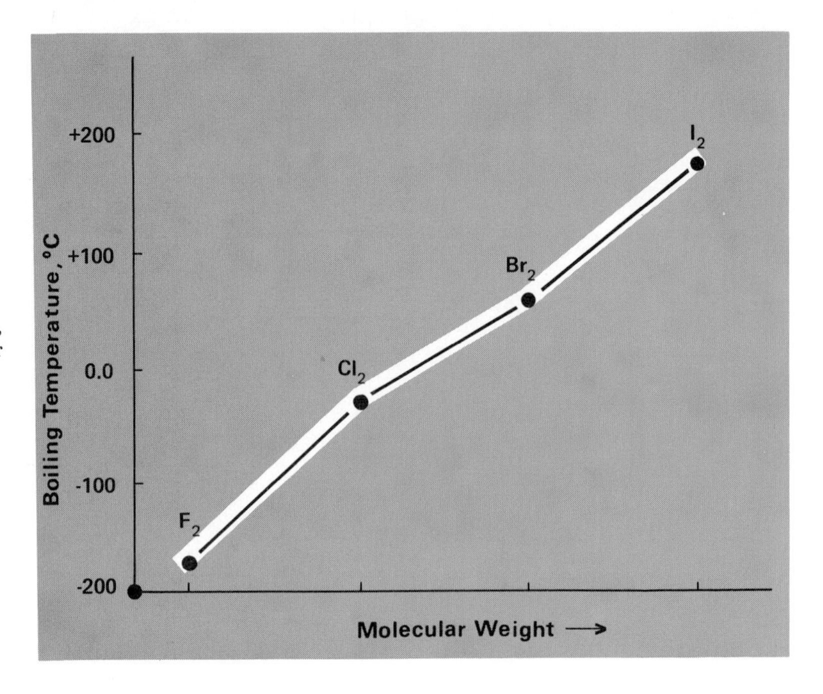

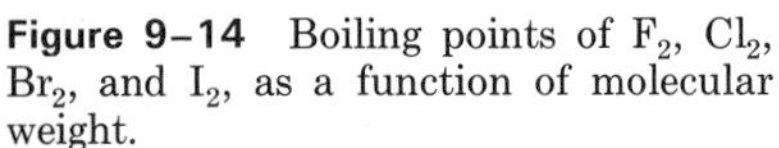
Figure 9–14 Boiling points of F_2, Cl_2, Br_2, and I_2, as a function of molecular weight.

example, the boiling points of fluorine (F_2), chlorine (Cl_2), bromine (Br_2), and iodine (I_2) increase with increasing molecular weight (Figure 9–14).

The general relationship between boiling points and molecular weights also holds for hydrogen chloride (HCl), hydrogen bromide (HBr), and hydrogen iodide (HI). However, the boiling point of hydrogen fluoride (HF), the lightest member of this series of compounds, is abnormally high (Figure 9–15). Irregularities similar to this are also found in other compounds in which hydrogen is bonded to fluorine, oxygen, and nitrogen. The increased interaction between molecules containing H—F, H—O, or H—N bonds is termed ***hydrogen bonding.***

Strongly electronegative atoms compete for a single hydrogen atom to give hydrogen bonds.

The explanation for hydrogen bonding is to be found in the extremely large electronegativity differences of the H—F, H—O, and H—N bonds, the polarity being due to the extreme electronegativity of these

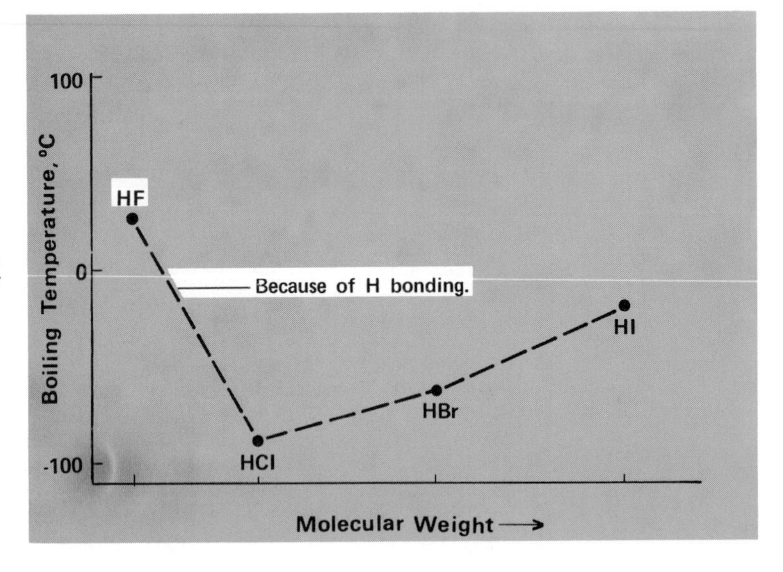

Figure 9–15 Boiling points of HF, HCl, HBr, and HI, plotted against molecular weights.

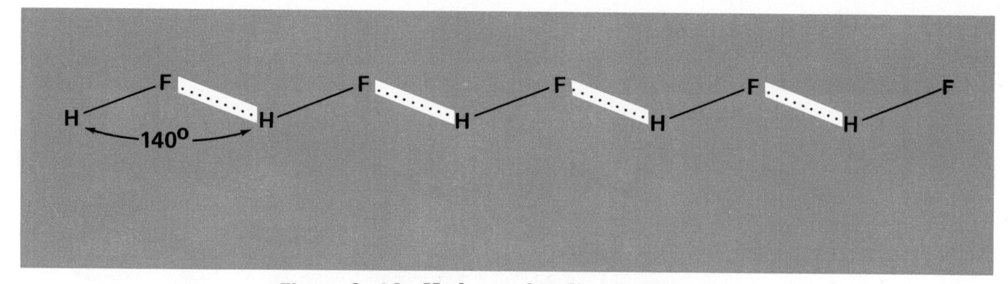

Figure 9–16 Hydrogen bonding in HF.

For a substance to boil, its molecules must gain enough energy to break loose from each other.

three elements. Consider the bonding in liquid HF. Since unlike-charged ends of these molecules (Figure 9–10) should attract each other, we expect HF molecules to be associated with one another. This association is illustrated in Figure 9–16. The increased association between HF molecules compared to that found in HCl offers a ready explanation for the unusually high boiling point of HF.

The structure of water provides another good example of hydrogen bonding. Water molecules are not linear but rather are angular with two nonbonding (unshared) electron pairs located toward one end of the molecule and the partially positive hydrogen atoms located toward the opposite end. The two polar bonds in this geometry result in a distinctly polar molecule.

δ^-

O

H   H

δ^+

In liquid and solid water where the molecules are close enough to interact, the hydrogen atom on one of the water molecules is attracted to the nonbonding electrons on the oxygen atom of an adjacent water molecule (Figure 9–17). This is possible because of the small size of the hydrogen atom. The result of this association is called *hydrogen bonding* because the hydrogen atom acts as a sort of a bridge to hold two

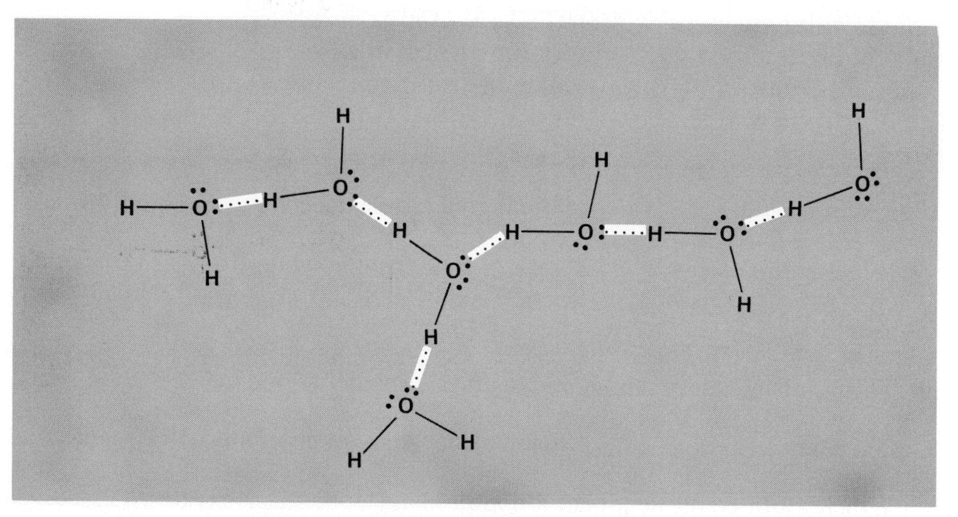

Figure 9–17 Hydrogen bonding in water.

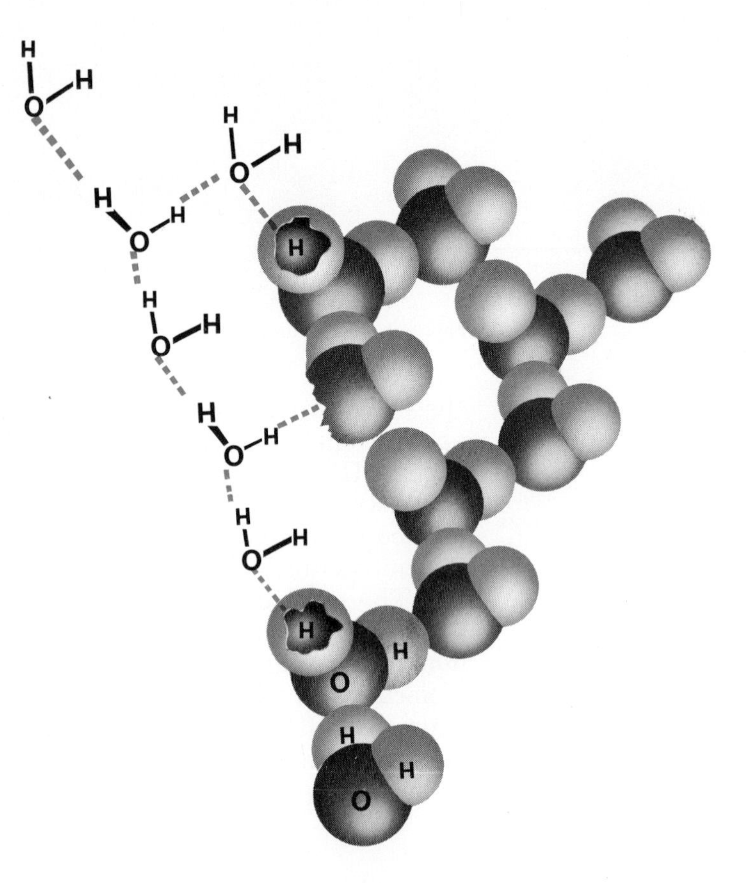

Figure 9–18 Hydrogen bonding in the structure of ice. The hydrogen bonds are indicated by the dashed lines. In liquid water the hydrogen bonding is not as extensive as it is in ice.

molecules together much like electron pairs hold atoms together in molecules. Hydrogen bonds are much weaker than ordinary covalent bonds.

The extent of hydrogen bonding in liquid water varies with temperature. In ice, hydrogen bonding is very extensive and almost all hydrogen atoms are involved in this sort of bonding. This results in a very open structure for ice (Figure 9–18). Consequently, at ordinary pressures ice is *less* dense than water. As ice is melted and the molecules gain more energy and move about, this bonding begins to break down. All the hydrogen bonds are not broken, however, and large aggregates of water molecules exist in liquid water even near 100°C. As water is heated, thermal agitation tends to disrupt the hydrogen bonding until, in gaseous water, there is only a small fraction of the number of hydrogen bonds that are found in liquid or solid water.

Water has an abnormally high boiling point because of hydrogen bonding between the molecules.

Freezing water expands due to hydrogen bonding.

SELF-TEST 9-B

1. (a) An example of covalent bonding where the electrons are equally shared between the atoms is __________; (b) one where they are unequally shared is __________.
2. The number of electrons shared in a triple covalent bond is __________.
3. There are __________ covalent bonds in an ammonia (NH_3) molecule.
4. Which atom could not form a double bond, sodium or oxygen? __________.

5. (a) How many electrons are thought to be in the bonding sphere of covalently bound atoms involving Period 2 and 3 elements? ____________. (b) This is known as the ____________ rule. (c) Is it true most of the time or all of the time? ____________

6. Which is the most electronegative of all of the elements? ____________

7. Hydrogen bonding very probably occurs when hydrogen is bound to atoms of ____________, ____________, or ____________.

VAN DER WAALS ATTRACTIONS

Since every substance known can exist in the solid state (even helium solidifies at $-272°C$ under 26 atmospheres pressure), we must assume that another type of attraction exists between atoms and molecules other than ionic and covalent bonding. In helium, the two 2s electrons are in a closed shell (nonbonding), and yet there is a slight attraction between two helium atoms. There are a large number of other cases in which weak attractions exist between molecules. These interactions, which are much weaker than ionic or covalent bonds, can arise from a variety of causes. For example, as two atoms or molecules approach each other, intermolecular interactions will cause a temporary shifting of electron clouds. An uneven electron distribution in an atom gives rise to a dipole in the atom. These instantaneous dipoles can interact with each other, resulting in the attractive force (Figure 9–19).

Van der Waals forces explain the existence of solid helium.

This type of dipole-dipole interaction is called van der Waals attraction, after the Dutch physicist who suggested its existence. Many molecular crystals, such as solid oxygen and solid nitrogen, evidence van der Waals forces.

METALLIC BONDING

Metals have some properties totally unlike those of other substances. For example, most metals are very good electrical conductors; they are shiny solids and have relatively high melting points (with a few notable exceptions such as mercury). Any theory of the bonding of metal atoms must be consistent with these properties. Structural investigations of

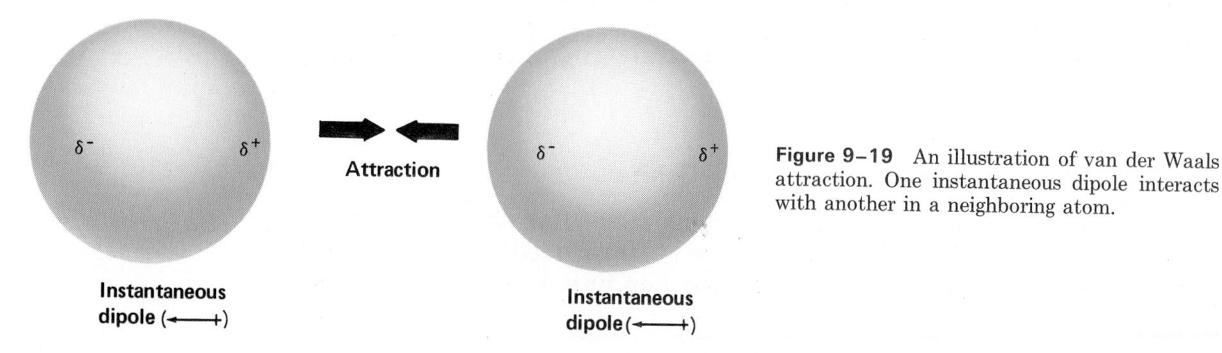

Figure 9–19 An illustration of van der Waals attraction. One instantaneous dipole interacts with another in a neighboring atom.

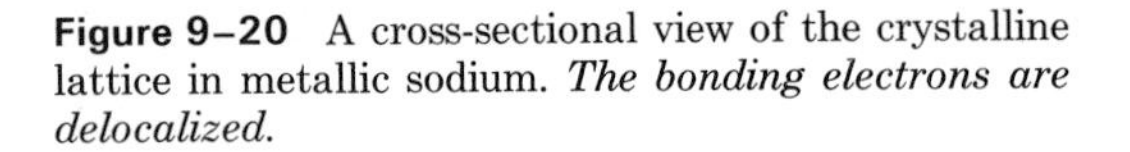

Figure 9–20 A cross-sectional view of the crystalline lattice in metallic sodium. *The bonding electrons are delocalized.*

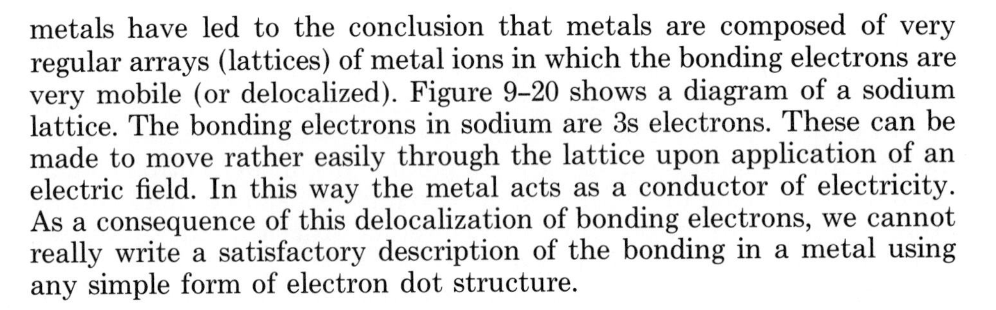

metals have led to the conclusion that metals are composed of very regular arrays (lattices) of metal ions in which the bonding electrons are very mobile (or delocalized). Figure 9–20 shows a diagram of a sodium lattice. The bonding electrons in sodium are 3s electrons. These can be made to move rather easily through the lattice upon application of an electric field. In this way the metal acts as a conductor of electricity. As a consequence of this delocalization of bonding electrons, we cannot really write a satisfactory description of the bonding in a metal using any simple form of electron dot structure.

Delocalized electrons are found in metals. They can move freely and are not confined to the area between any particular pair of atoms.

THE STUDY OF MOLECULAR SHAPES

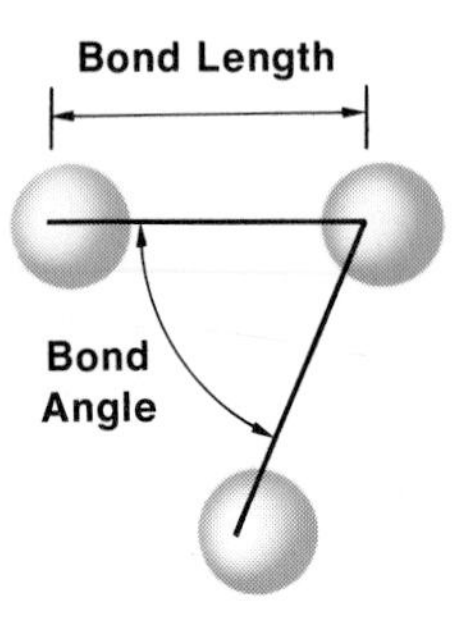

The modern theory of the atom is very effective in explaining bonding and periodicity of the elements; it is also capable of explaining related chemical phenomena, such as molecular shapes and associated properties. The shape of a molecule is determined by the arrangement of the atoms in the molecule with respect to each other, that is, by the angles between and the lengths of its covalent bonds. The angle formed by two intersecting lines drawn from the two nuclei of the attached atoms through the nucleus of the central atom is called the ***bond angle.*** Molecular shapes are established by many different experimental techniques. Here we shall see what molecular shapes are predicted by atomic theory and what modifications are needed for the theoretical and the experimental findings to agree.

Molecular Shapes Predicted by Atomic Orbitals

The starting point for a discussion of molecular shapes is the set of atomic orbitals, since these determine the preferred regions in space for electrons which will be involved in the bonding. Expected molecular geometries based on orbital overlap when s and p atomic orbitals are used to form bonds are shown in Figure 9–21.

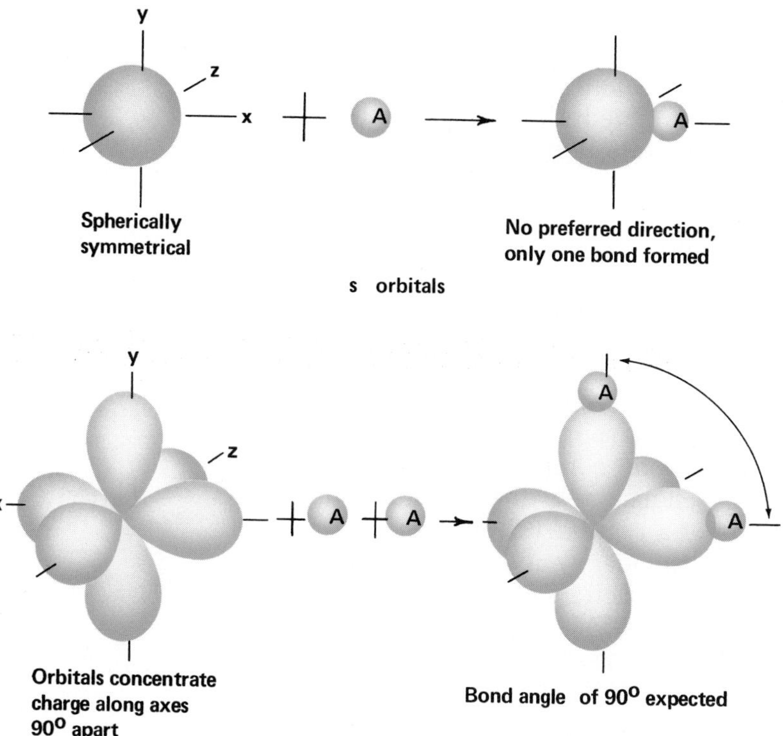

Figure 9–21 Predicted molecular shapes when s and p orbitals are used to form bonds. A represents any atom to be joined to the central atom.

Since an s orbital is spherically symmetrical, there is no preferred direction for the bonding with the other atom. Of course, with only one s orbital per valence shell, only one bond can be formed from this kind of orbital. Thus, lithium, with the electronic configuration $1s^22s^1$, can form only one bond.

There are three p orbitals per valence shell. The electrons in a set of p orbitals concentrate charge along the x, y, and z axes, which are at 90° angles to each other. Any two atoms bonded to a third atom by overlapping two p orbitals would be expected to form bonds enclosing an angle of 90°.

Is this predicted bond angle actually found in molecules? Some molecules such as H_2S and H_2Se come close. However, instead of the expected bond angle of 90°, when s orbitals of two hydrogens overlap two of the p orbitals of S or Se, the angles are 92° for H_2S and 91° for H_2Se. It is rare for the bond angles of molecules to agree this closely with the angles predicted by combinations of simple atomic orbitals. In H_2O, for example, the bond angle is 104.5°.

Before discarding the theory, we should look a little further. With a slight adjustment, the modern atomic theory can be brought into agreement with experimentally derived molecular shapes of relatively simple molecules. The theory is also available for use with complex molecules which contain large numbers of atoms.

Hybridization combines atomic orbitals to give new orbitals with different properties.

The method of adjusting the pure atomic orbitals is called ***hybridization.*** This process allows bond angles to conform with experiment. Although hybridization is a mathematical process, we can gain an insight into its methods and its usefulness through a qualitative examination of its features. An alternative approach to this same problem is the ***valence-shell electron-pair repulsion theory.***

Both of these theories will be examined along with some of their applications and some of the experimental observations used as a guide in their development.

Hybridization

The tetrahedral structure of species such as CH_4 and NH_4^+ requires an adjustment of atomic theory. The isolated carbon atom in its ground state has the electron configuration $1s^2 2s^2 2p_x^1 2p_y^1$ (the 2p electrons are unpaired). If the unpaired electrons in the 2p orbitals are used for bonding, this leads us to expect that carbon would form two bonds at an angle of 90°. Actually carbon commonly forms four single bonds, and each bond angle is 109.5°. This angle directs each bond from carbon toward each of the corners of a regular tetrahedron.

To explain these four equivalent bonds, we assume that prior to or during the formation of the bonds, four orbitals of the carbon atom are hybridized (or mixed) to form a new set of four equivalent orbitals. The carbon atom promotes one of the electrons from the 2s orbital to the empty $2p_z$ orbital to give the electronic arrangement $2s^1 2p_x^1 2p_y^1 2p_z^1$ for the valence shell electrons. These four orbitals are rearranged mathematically according to specific rules to obtain four new equivalent hybrid orbitals. These are designated sp^3 orbitals since they are made from one s and three p orbitals. The new orbitals are directed toward the corners of a tetrahedron. The four valence electrons are distributed equally among the hybrid orbitals (one e^- to each orbital). The hybrid orbitals are shown graphically in Figure 9-22. The energy required for this transition is less than that which is provided by the formation of additional, stronger bonds with the hybrid orbitals. Thus, we have a more favorable energy change than would result from bonding with p orbitals alone.

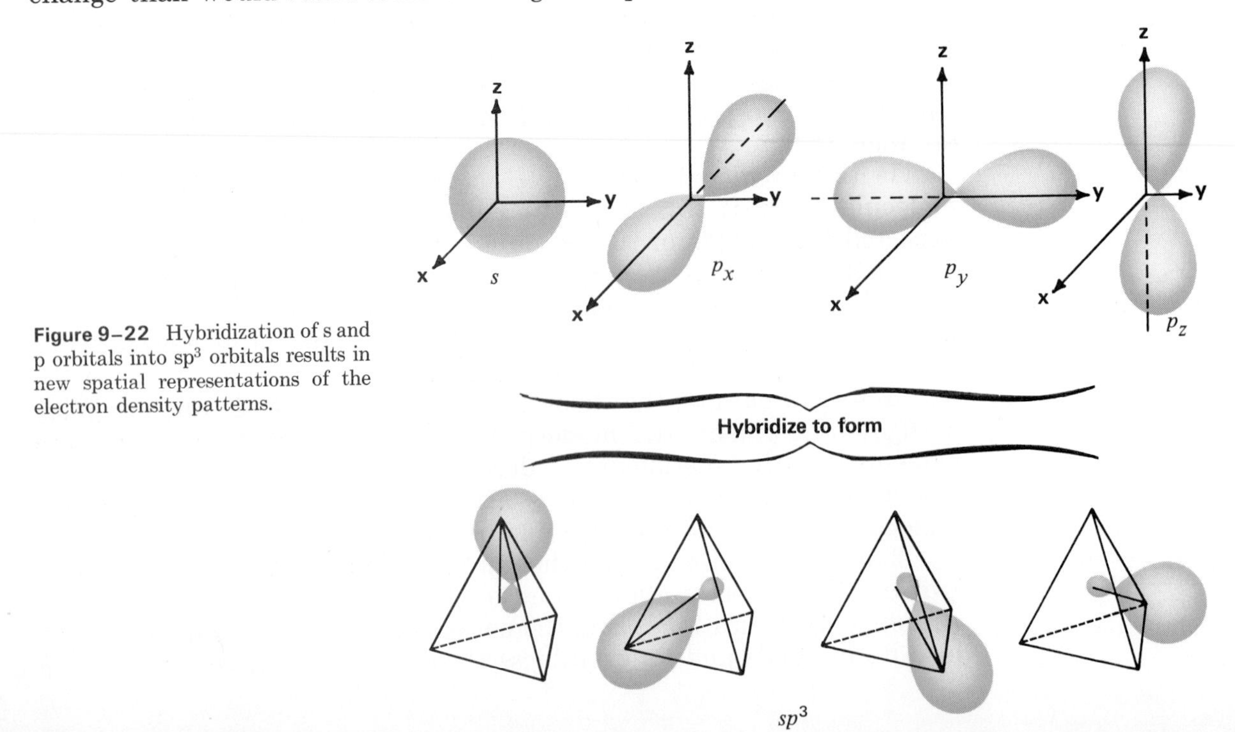

Figure 9-22 Hybridization of s and p orbitals into sp^3 orbitals results in new spatial representations of the electron density patterns.

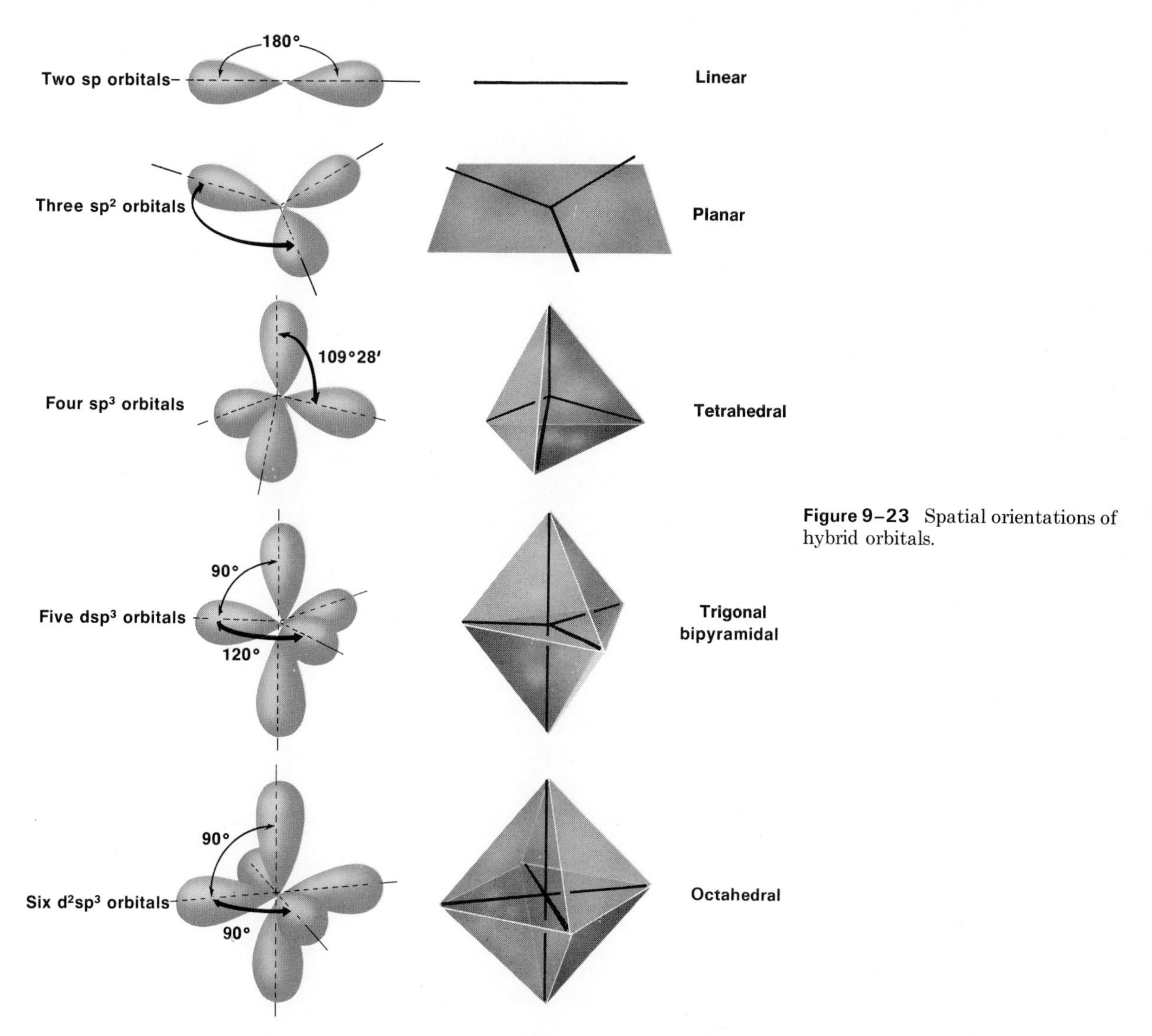

Figure 9–23 Spatial orientations of hybrid orbitals.

Other sets of bonding orbitals with geometries very closely in accord with established molecular geometries can be derived in a similar fashion. Some of these are listed in Figure 9-23. The hybridization analogy in molecules, plants, and animals is illustrated in Figure 9-24.

As examples of the use of hybrid orbitals, consider tetrahedral molecules such as CH_4, $SiCl_4$, and CCl_4. In all of these, the atomic orbitals of the central atom can be described as sp^3 hybrids. For molecules such as H_2O and NH_3, sp^3 hybrid orbitals can also be postulated if it is assumed that the ***nonbonding pairs*** of electrons (those on the O or N which are not involved in bonds to H) are also placed in hybridized orbitals. The experimentally determined bond angles in these molecules are not quite the same as normal tetrahedral bond angles (as shown in Figure 9-25); however, the agreement is relatively close.

NONBONDING PAIRS

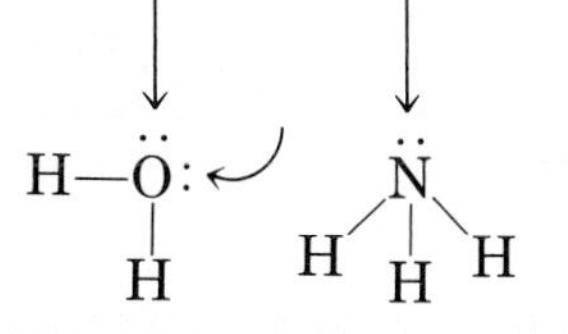

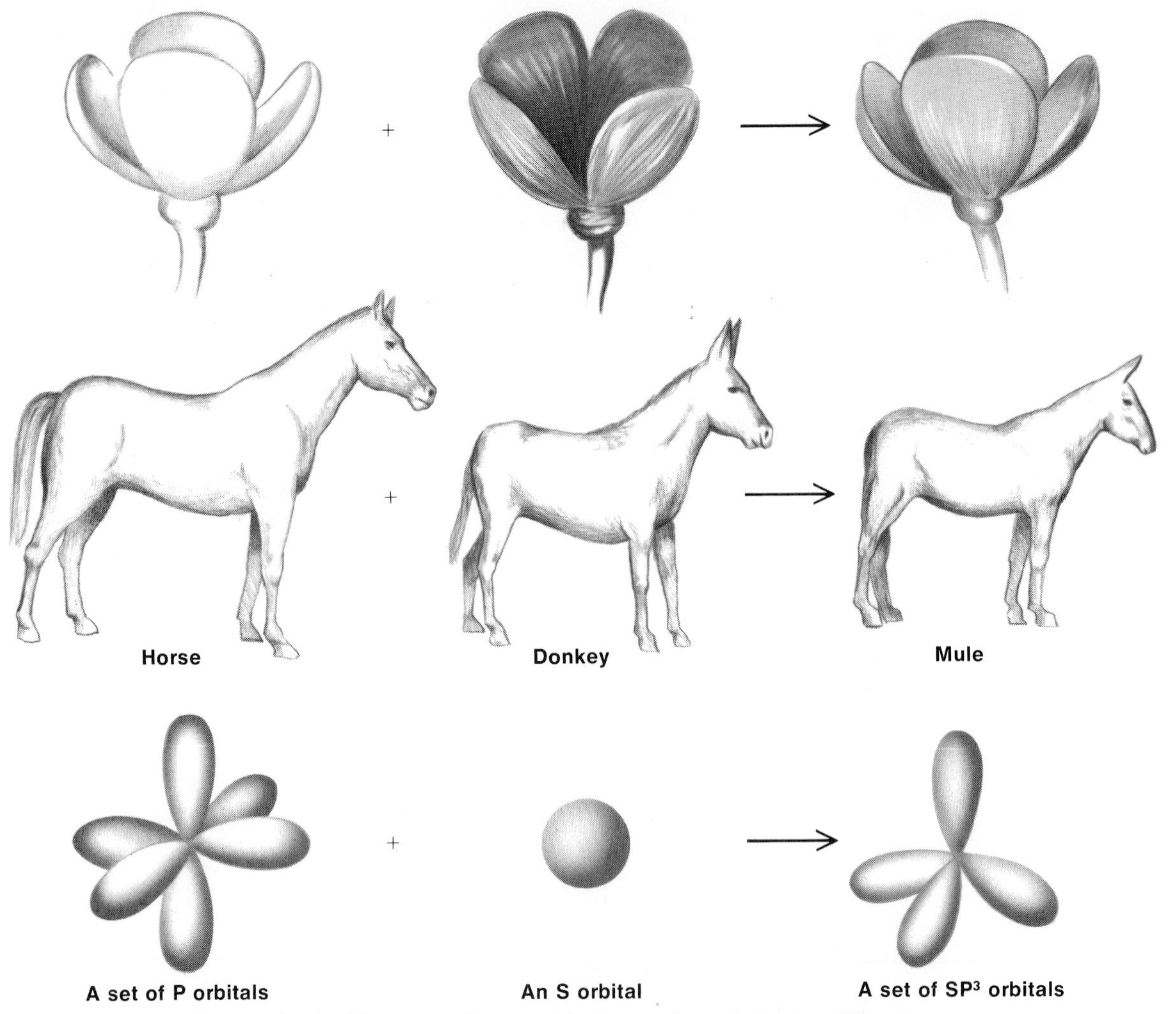

Figure 9–24 Hybridization is a familiar concept in nature. A mixture of two similar but different entities produces an entity with intermediate characteristics. Biologic hybrids are analogous to hybrid orbitals in the sense of being the result of mixtures, yet hybrid orbitals are obtained much more exactly by mathematical calculations.

VALENCE-SHELL ELECTRON-PAIR REPULSION THEORY

There is another, even simpler, notion which can be used to estimate bond angles in covalent molecules, based on the idea that electron pairs repel each other. In fact, this theory assumes that electron pairs in the valence shell of a central atom behave like a group of electrically charged balloons which are connected to a central point by strings. If similarly charged, the balloons would tend to be as far apart from each other as possible. In a similar manner, electron-pair repulsion leads to the arrangements shown in Figure 9–25.

These ideas allow us to predict the shapes of molecules if the underlying electronic structure is known. It is particularly simple when every pair of electrons in the valence shell is a bonding pair. Examples can be seen in BCl_3, $SiCl_4$, and SF_6.

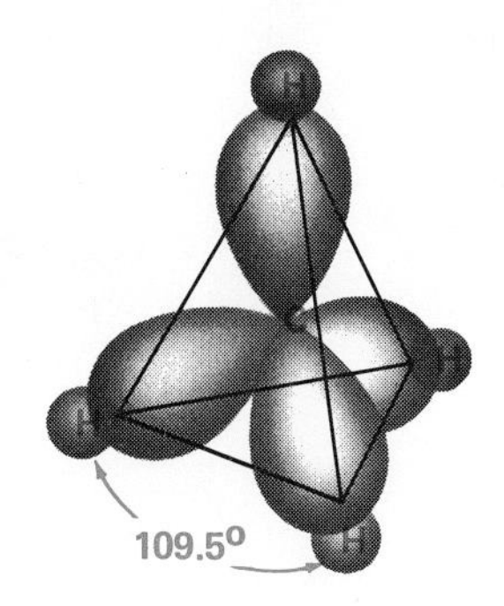

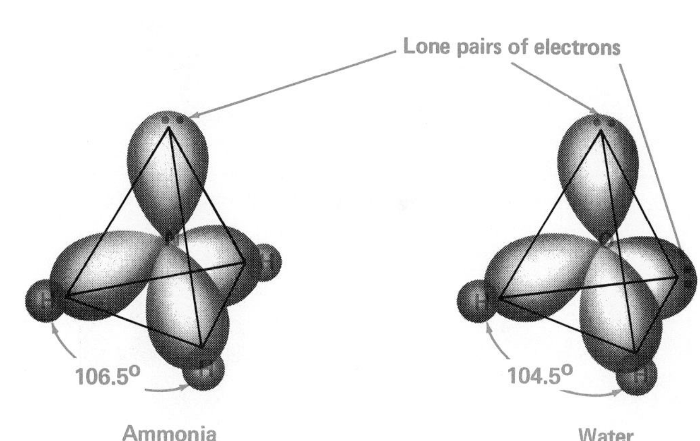

Figure 9–25 Molecules containing sp^3 orbitals. Methane has only bonding pairs of valence electrons; water and ammonia have bonding and nonbonding pairs of valence electrons.

BCl_3

Atomic boron has $1s^22s^22p^1$ as its ground state electronic structure. One of the 2s electrons can be promoted to yield the configuration, $2s^12p_x{}^12p_y{}^1$. When these are paired up with the unpaired electrons of three chlorine atoms, :C̤l·, three electron-pair bonds are formed. This leads to the structure, shown below, which actually is the structure found experimentally.

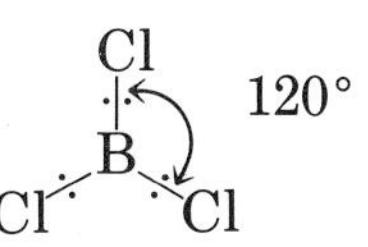

The use of one s and two p orbitals to form sp^2 hybrid orbitals also leads to the same structure.

$SiCl_4$

Atomic silicon has $1s^22s^22p^63s^23p_x{}^13p_y{}^1$ as its electronic structure. If one of the 3s electrons is promoted to the $3p_z$ orbital, four unpaired electrons will be available. When these are paired with the unpaired electrons of four chlorine atoms, :C̤l·, four electron-pair bonds are formed. These four pairs of electrons in the valence shell will repel each other to form the predicted tetrahedral structure.

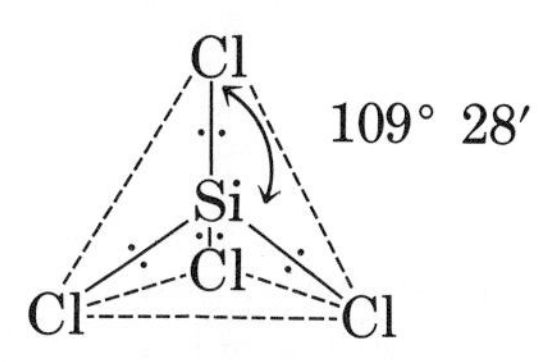

The combination of one s and three p orbitals to form sp^3 hybrids leads to the same structure. The actual structure is, indeed, tetrahedral.

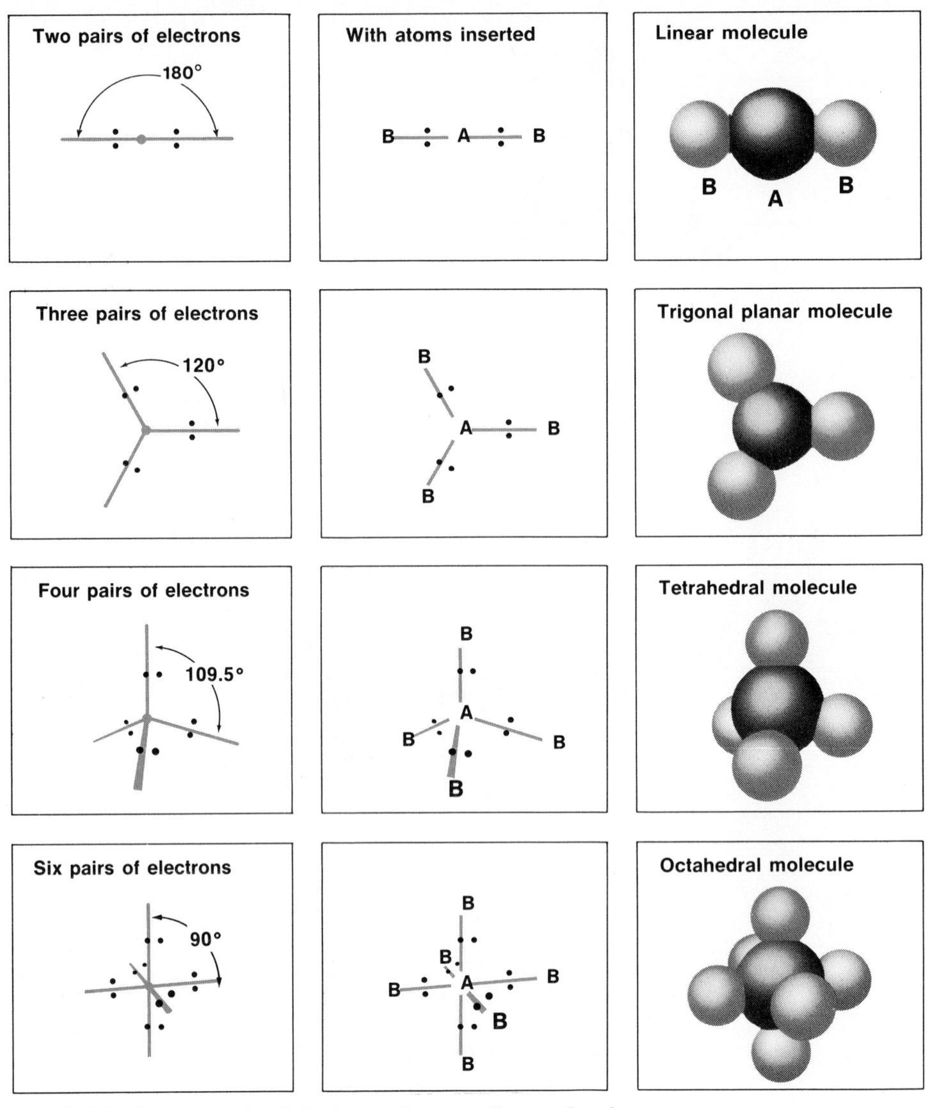

Figure 9–26 Arrangements of electron pairs according to the electron-pair repulsion theory.

SF_6

The sulfur atom has the electronic configuration $1s^22s^22p^63s^23p^4$ prior to formation of the molecule, SF_6. Because it has six fluorines bonded to it, the sulfur atom must use the $3s^2$ and $3p^4$ electrons to form bonding pairs. *Six* electron pairs in the valence shell lead to the prediction of an octahedral structure for the molecule, which is the structure found.

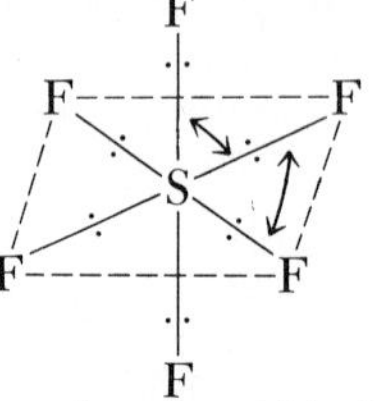

Since the sulfur forms six bonds, six orbitals must be involved. The 3s and 3p orbitals furnish four of these, and the sulfur must also use two of its empty 3d orbitals to obtain a set of sp^3d^2 hybrid orbitals to use in bonding. This also leads to an octahedral structure.

At this point you might ask how the presence of nonbonding pairs of electrons (in the outermost shell of the central atom) affects the disposition of the other electron pairs. In brief, a pair of electrons occupies somewhat more volume when it is not involved in bonding.

Nonbonding electron pairs help to explain the structures of NH_3 and H_2O molecules discussed earlier and shown in Figure 9–25. It can be postulated, for example, that the NH_3 and H_2O molecules have bond angles (106.5° in NH_3 and 104.5° in H_2O) slightly less than the tetrahedral angle of 109.5° for pure sp^3 hybrids because the lone, nonbonding pairs of electrons exert more repulsion than do the bonding pairs. This spreads the nonbonding pairs further apart and squeezes the bonding pairs closer together.

From the typical examples given here, it is obvious that both the electron-pair repulsion theory and the hybridization theory are consistent with experimental structures for molecules. The methods differ, however, in their approach. Hybridization utilizes the mathematical combination of s, p, and d atomic orbitals to form hybrid orbitals that are properly oriented to give the established bond angles. On the other hand, electron-pair repulsion theory assumes that electron pairs simply move as far away from each other as is possible while maintaining the proper bond length. The nonbonding pairs of electrons exert more repulsion than bonding pairs do.

Shapes of Molecules with Double Bonds and Triple Bonds

The electron dot structure for ethylene is:

```
H      H
 :C::C:
H      H
```

Four electrons form a ***double bond*** between the two carbon atoms. Each carbon atom furnishes two unpaired electrons for the double bond. This means two orbitals from each carbon atom are involved in the double bond.

A double bond is one sigma bond and one pi bond.

Sigma (σ) bond: end-to-end overlap.
Pi (π) bond: side overlap.

A pi bond is weaker than a sigma bond.

Experimentally, it is observed that ethylene is a planar molecule (all six atoms are in the same plane) and that all bond angles are close to 120°. This kind of structure around a carbon atom can exist if the 2s and *two* of the 2p orbitals of carbon undergo hybridization to form three sp^2 hybrid orbitals. The hybridization can come about in the following way. Each carbon atom can be excited to the electronic arrangement, $1s^2 2s^1 2p_x{}^1 2p_y{}^1 2p_z{}^1$. When the 2s and two of the p orbitals are mixed, three equivalent orbitals are produced. The three sp^2 orbitals are all in the same plane and form angles of 120° with each other. As shown in Figure 9–27, a sigma bond is formed between the two carbon atoms by overlapping the end of an sp^2 orbital of one carbon atom with the end of an sp^2 orbital of the other carbon atom. The second bond of the double bond between the two carbon atoms is formed by a lateral (or side) overlap of the two p orbitals not involved in the sp^2 hybridizations. The lateral overlap is called a pi (π) bond and prevents free rotation about the bond axis. The two CH_2 groups of ethylene cannot rotate freely with

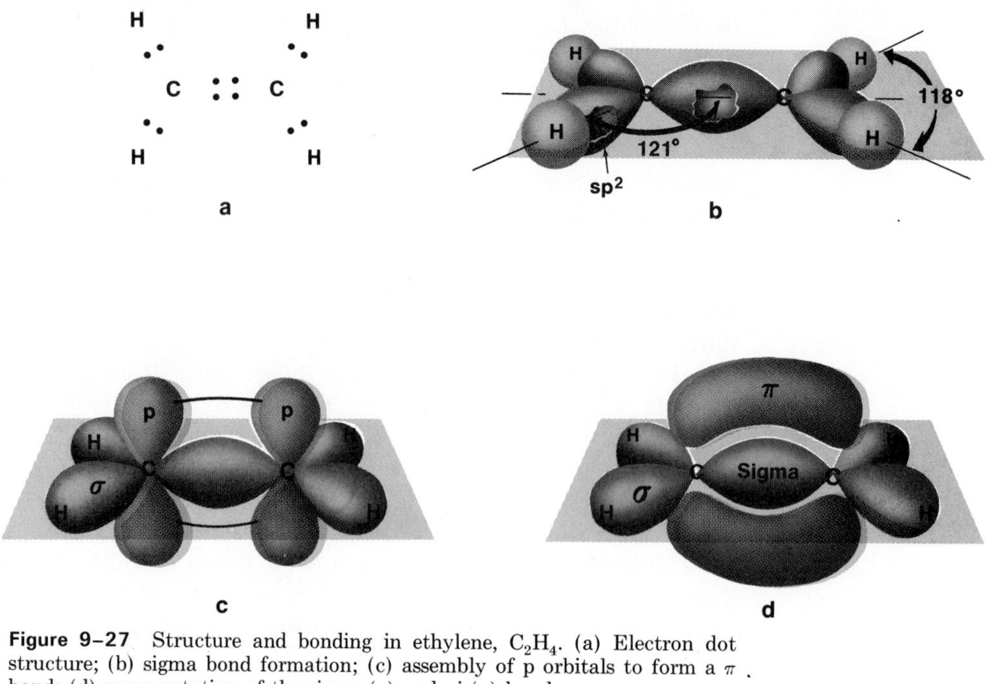

Figure 9–27 Structure and bonding in ethylene, C_2H_4. (a) Electron dot structure; (b) sigma bond formation; (c) assembly of p orbitals to form a π bond; (d) representation of the sigma (σ) and pi (π) bonds.

From two *s* orbitals

s *s*

a sigma bond

From two *p* orbitals (end-to-end overlap)

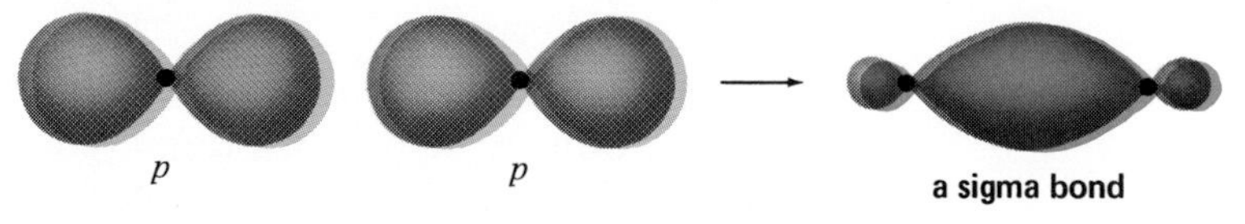

From an *s* and a *p* orbital

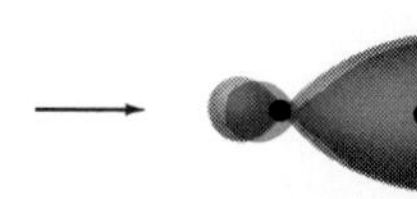

From two *p* orbitals (lateral overlap)

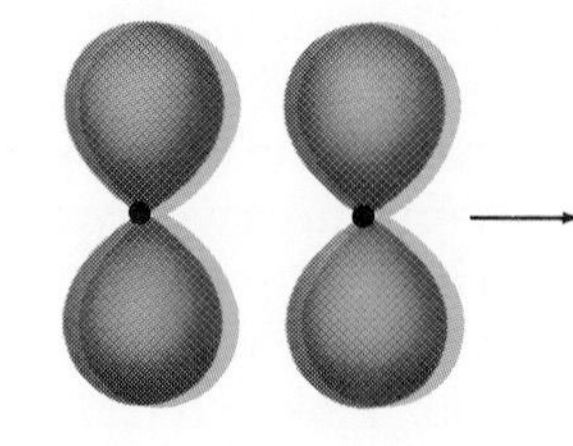

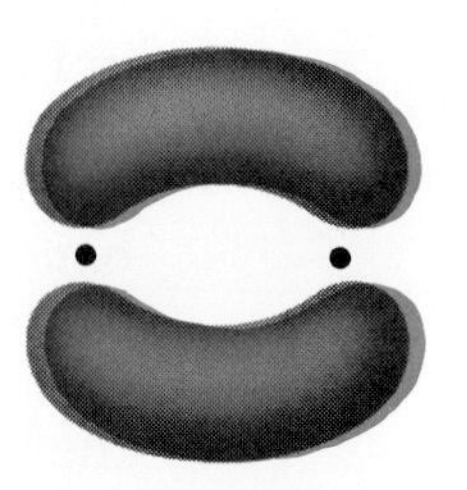

Figure 9–28 Ways of overlapping s and p orbitals to form sigma and pi bonds.

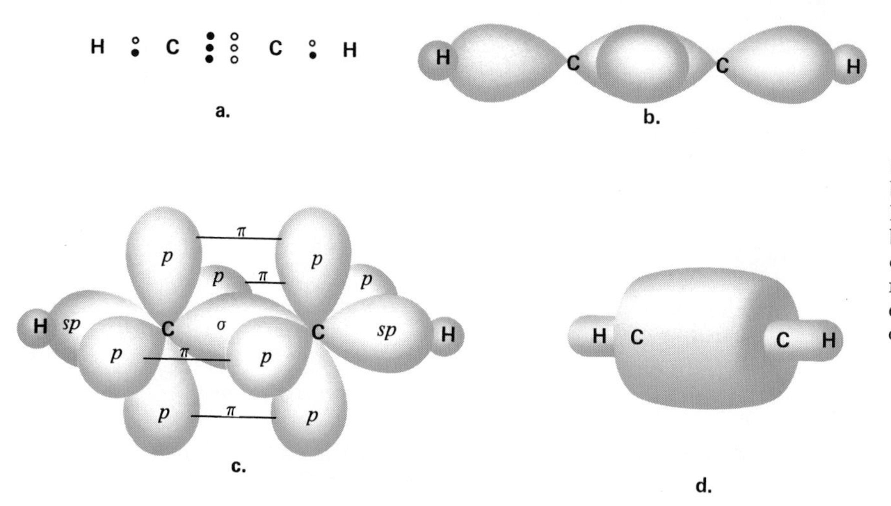

Figure 9–29 Structure and bonding of acetylene, C_2H_2. (a) Electron dot structure; (b) sigma bond formation; (c) assembly of p orbitals to form two pi bonds; (d) representation of the electron density pattern which is cylindrically symmetric in the molecule.

respect to each other because the rotation would reduce or eliminate the orbital overlap of the pi bond. As we shall see later, this rotational limitation about double bonds is quite important in the chemistry of certain compounds.

SELF-TEST 9-C

1. A double bond is composed of one ______________ bond and one ______________ bond.

2. A triple bond is composed of one ______________ bond and two ______________ bonds.

3. An sp^3 hybrid orbital is one of a set of orbitals that can be formed by mixing one ______________ orbital and three ______________ orbitals.

4. (Bonding/nonbonding) pairs of electrons repel more than (bonding/nonbonding) pairs of electrons.

5. Around which kind of bond (single, double, triple) can atoms rotate on an axis drawn through the nuclei of the two atoms bonded?

6. When three orbitals are combined in hybridization ______________ (How many?) new orbitals are produced.

7. Give the molecular shapes expected for the hypothetical molecules: AX_2 ______________, BZ_3 ______________, DQ_6 ______________. There are no nonbonding valence electrons in these structures, and all of the bonds are single bonds.

MATCHING SET I

______	1. double covalent bond	a.	explanation for covalent bonding
______	2. ionic bonding	b.	sharing electrons
______	3. ionization energy	c.	requires O, F, or N
______	4. metallic bonding	d.	ions attracted to each other
______	5. van der Waals attractions	e.	ionic compound
______	6. noble gas	f.	electrons free to move
______	7. covalent bonding	g.	element with eight valence-shell electrons
______	8. NaCl	h.	covalent compound
______	9. metal ion	i.	attraction between nonpolar neutral particles
______	10. hydrogen bonding	j.	measures gaseous atom's hold on electron
______	11. NH_3	k.	smaller than parent atom
______	12. orbital overlap	l.	four electrons shared
______	13. single covalent bond	m.	two electrons shared

MATCHING SET II

______	1. pi bond	a.	an electrically neutral arrangement of covalently bonded atoms
______	2. sigma bond	b.	well-balanced arrangement of parts
______	3. sp^3 orbital	c.	four-sided figure
______	4. hybridization	d.	lateral overlap of p orbitals
______	5. molecule	e.	one pi bond and one sigma bond
______	6. double bond	f.	mixing of orbitals
______	7. symmetry	g.	end-to-end overlap of orbitals
______	8. tetrahedral	h.	hybrid orbital

MATCHING SET III

______	1. $BeCl_2$ (no non-bonding valence electrons)	a.	linear
______	2. CI_4	b.	trigonal planar
______	3. $SiCl_4$	c.	bent
______	4. SF_6	d.	tetrahedral
______	5. BCl_3	e.	octahedral

QUESTIONS

1. Diamond (a form of carbon) has a melting point of 3500°C, whereas carbon monoxide (CO) has a melting point of −207°C. What does this tell about the bonding in these two substances?

2. Write the electronic configuration for the element potassium (atomic number 19). What will be the electronic configuration when a K^+ ion is formed?

3. Is Ca^{3+} a possible ion under normal chemical conditions? Why?

4. Write the symbols for the six elements with the highest ionization potentials, selecting one from each period of the periodic table.

5. Match the electronic configurations which would be expected to lead to similar chemical behavior.

   a. $1s^22s^2$
   b. $1s^22s^22p^3$
   c. $1s^22s^22p^63s^23p^64s^1$
   d. $1s^22s^22p^63s^2$
   e. $1s^22s^1$
   f. $1s^22s^22p^63s^23p^3$

6. Fluorine (atomic number 9) has an electronic configuration of $1s^22s^22p^5$. How many electrons are in the outermost (valence) shell of a fluorine atom? How many electrons will be involved in chemical bonding?

7. Which salt would be expected to have the higher melting point, KCl or RbCl; MgO or BaO? Give a reason for your answer.

8. Write the electronic configuration for iodine (I, atomic number 53). How many covalent bonds should an iodine atom form?

9. Write the electron dot structures for the fluoride ion, F^-, the chloride ion, Cl^-, and the bromide ion, Br^-.

10. Draw the electron dot structure for water. Based on bonding theory, why is water's formula not H_3O?

11. Define the term *bond energy.*

12. Draw electron dot structures for the following molecules:

    a. NF_3
    b. CCl_4
    c. C_2Cl_2
    d. OF_2
    e. H_2S
    f. CO
    g. N_2H_4
    h. CH_3OH

13. The members of the nitrogen family, N, P, As, and Sb, form compounds with hydrogen, NH_3, PH_3, AsH_3, and SbH_3. The boiling points of these compounds are given:

SbH_3	−17°C
AsH_3	−55°C
PH_3	−87.4°C
NH_3	−33.4°C

    Comment on why NH_3 doesn't follow the downward trend of boiling points.

14. Match the substances listed below with the type of bonding responsible for holding units in the solid together.

solid krypton (Kr)	ionic
ice	covalent
diamond	metallic
CaF_2	hydrogen bonding
iron	van der Waals

15. Predict the general kind of chemical behavior (that is, loss, gain, or sharing of

electrons) you would expect from atoms with the following electronic configurations:

a. $1s^22s^22p^63s^1$
b. $1s^22s^22p^5$
c. $1s^22s^22p^2$

16. Show how two fluorine atoms can form a bond with the overlap of their half-filled p orbitals.

17. Select the *polar* molecules from the following list and explain why they are polar: N_2, HCl, CO, and NO

18. How many bonds join the two atoms in the following? CN^-, Cl_2, S_2

19. Boron trichloride has the electron dot formula:

:C̈l:
:C̈l:B:C̈l:

What does this tell you about the rule of eight even for Period 2 elements?

20. Chromium (element no. 24) forms a stable ion in which the chromium has lost three electrons per atom. What does this tell you about a half-filled set of d orbitals?

21. Use your chemical intuition and suggest a reaction that might occur between boron trichloride (problem 19 above) and ammonia,

H:N̈:H
H

22. Sketch separately the spatial distribution of electronic charge associated with an s, a p, and a d atomic orbital.

23. What is the meaning of the notation sp^3?

24. What is the geometry of bonding around a central atom having the following hybridization?

a. sp
b. sp^2
c. sp^3
d. d^2sp^3

25. What is the meaning of hybridization?

26. Account for the fact that carbon in methane, CH_4, forms four equivalent bonds, although the ground state electronic arrangement of an isolated carbon atom is $1s^22s^22p^2$.

27. What is a sigma bond; a pi bond?

28. Give the basic points of the valence-shell electron-pair repulsion theory.

29. Explain the 106.5° H—N—H bond angles of NH_3, the ammonia molecule.

30. How many atomic positions must be specified in order to define a bond angle?

31. $BeCl_2$ is a molecule known to contain polar Be—Cl bonds, yet the $BeCl_2$ molecule is not polar. Explain.

32. What happens to the third p orbital in sp^2 hybridization?

33. In the *Suggestions for Further Reading,* the article "Structures of Noble Gas Compounds" discusses the structure of XeF_6. How many electron pairs would Xe have in its valence shell when bonded to six fluorines in the compound XeF_6? According to electron-pair repulsion theory, would this number of electron pairs predict an octahedral structure for the XeF_6 molecule?

34. There is an article in the *Suggestions for Further Reading* on nuclear magnetic resonance spectroscopy. Use this article, or another, as a basis for a report on how this powerful tool unravels molecular structure.

SUGGESTIONS FOR FURTHER READING

Benfey, T., "Geometry and Chemical Bonding," *Chemistry,* Vol. 40, No. 5, p. 21 (1967).
Bragg, Sir Lawrence, "The Start of X-Ray Analysis," *Chemistry,* Vol. 40, No. 11, p. 8 (1967).
Brey, W., "Physical Methods for Determining Molecular Geometry," Reinhold Publishing Co., New York, 1965.
Companion, A. L., *Chemical Bonding,* McGraw-Hill Book Company, New York, 1964.
Derjaguin, B. V., "The Force Between Molecules," *Scientific American,* Vol. 203, p. 47 (1960).
Eyring, Henry, and MuShik Jhon, "Significant Structure Theory of Water," *Chemistry,* Vol. 39, No. 9, p. 8 (1966).
Garfield, Eugene, Gabrielle S. Reavesz, and Nathin Rubin, "Fixed Valence and Molecular Formula Verification," *Chemistry,* Vol. 43, No. 9, p. 13 (1970).
Gillespie, R. J., "The Valence Shell Electron Pair Repulsion Theory of Directed Valency," *Journal of Chemical Education,* Vol. 40, p. 295 (1963).
Greenwood, N. N., "Chemical Bonds," *Education in Chemistry,* Vol. 4, p. 164 (1967).
Griswold, E., "Chemical Bonding and Structure, Raytheon Education Co. (D. C. Heath), Lexington, Mass., 1968.
Hochstrasser, R. M., *Behavior of Electrons in Atoms,* W. A. Benjamin, Inc., New York, 1964.
House, J. E., Jr., "Ionic Bonding in Solids," *Chemistry,* Vol. 43, No. 2, p. 18 (1970).
Jones, P. R., "Infrared Spectroscopy and Molecular Architecture," *Chemistry,* Vol. 38, No. 2, p. 5 (1965).
Kondratyev, V. N., *The Structure of Atoms and Molecules,* Dover Publications, Inc., New York, 1965.
Lagowski, J. J., *The Chemical Bond,* Houghton Mifflin Co., Boston, 1966.
Luder, W. F., "Electron Repulsion Theory," *Chemistry,* Vol. 42, No. 6, p. 16 (1969).
Mellor, D. P., "The Noble Gases and Their Compounds," *Chemistry,* Vol. 41, No. 10, p. 16 (1968).
Muller, Erwin, W., "Atoms Visualized," *Scientific American,* Vol. 196, p. 113 (1957).
Pauling, L., *The Nature of the Chemical Bond,* Cornell University Press, Ithaca, New York, 1960.
Ryschkewitsch, G. E., *Chemical Bonding and the Geometry of Molecules,* Reinhold Publishing Corp., New York, 1963.
Sanderson, R. T., "Principles of Chemical Bonding," *Journal of Chemical Education,* Vol. 38, p. 382 (1961).
Sheppard, W. J., "The Construction of Solid Tetrahedral and Octahedral Models," *Journal of Chemical Education,* Vol. 44, p. 683 (1967).
"Structures of Noble Gas Compounds," *Chemistry,* Vol. 39, No. 4, p. 17 (1966).
Wagner, J. J., "Nuclear Magnetic Resonance Spectroscopy—An Outline," *Chemistry,* Vol. 43, No. 3, p. 13 (1970).
Wahl, Arnold C., "Chemistry by Computer," *Scientific American,* Vol. 222, p. 54 (1970).
Webb, Valerie J., "Hydrogen Bond 'Special Agent'," *Chemistry,* Vol. 41, No. 6, p. 16 (1968).

HYDROGEN ION (ACID-BASE) AND ELECTRON TRANSFER (OXIDATION-REDUCTION) REACTIONS

CHAPTER 10

Two major classes of chemical reactions which have far-reaching applications in our lives are those of hydrogen ion transfer and electron transfer between molecular or ionic species. ***Acids*** such as vinegar or lemon juice, and ***bases*** such as lime or baking soda are commonly encountered. Reactions of acids with bases can be defined in terms of the transfer of hydrogen ions from an acid to a base. Chemical oxidation and reduction are equally common. The rusting of iron, the burning of wood, and the bleaching of hair or fabrics are examples of oxidation, while the transformation of iron ore in a blast furnace to iron metal is a classic example of chemical reduction. Such ***redox*** (short for ***oxidation-reduction***) reactions can be defined in terms of the transfer of electrons from one particle to another. It is evident that if we can understand how hydrogen ions and electrons are transferred we will be better able to control many processes of importance.

Redox is pronounced REE-DOX.

Acid-base and redox reactions occur in the solid, liquid, or gaseous phase. However, the presentation will be made easier at this point if we begin our study with the liquid (solution) phase. This will require an understanding of some characteristics of the major types of solutions.

FORMATION OF SOLUTIONS

Solution: homogeneous mixture of atoms, ions, or molecules.

The process of ***solution*** is a familiar one. We can see sugar and salt dissolve in water, oil paints dissolve in turpentine, and grease dissolve in gasoline. Though they are less commonly observed, solid solutions such as metal alloys, and gaseous solutions like polluted air are equally possible. In a solution the substance present in greater amount is defined as the ***solvent*** and the one in smaller amount the ***solute.*** For example, in a glass of tea, water is the solvent and sugar, lemon juice, and the tea itself are solutes.

Aqueous solutions are water solutions.

In this presentation of acid-base and redox reactions, most of the chemistry studied will be in water or aqueous solutions, where water is the solvent. Generally, one of the species exchanging hydrogen ions or electrons is the solute. Complicating the study somewhat is the fact that solvent particles themselves can enter into both acid-base and redox reactions. Water molecules can and do exchange hydrogen ions and electrons with solute particles under suitable conditions. In theoretical terms a solution of sugar in water can be thought of in terms of a collection of sugar molecules dispersed among the water molecules (Figure 10–1).

IONIC SOLUTIONS (ELECTROLYTES) AND MOLECULAR SOLUTIONS (NONELECTROLYTES)

Solute particles may be ions or molecules.

When aqueous solutions are examined to see if they conduct electricity, we find that solutions fall into one of two categories: ***electrolytic*** solutions, which conduct electricity, and ***nonelectrolytic*** solutions. A simple apparatus such as that shown in Figure 10–2 can be used to determine into which classification a given solution falls.

The conductance of electrolytic solutions is readily explained, since the solute particles in such solutions are ions rather than molecules. Recall that sodium chloride crystals are composed of sodium ions, which are positively charged, and chloride ions, which are negatively charged. When sodium chloride dissolves in water, ***ionic dissociation*** occurs. The resulting solution (Figure 10–3a) contains positive sodium ions and nega-

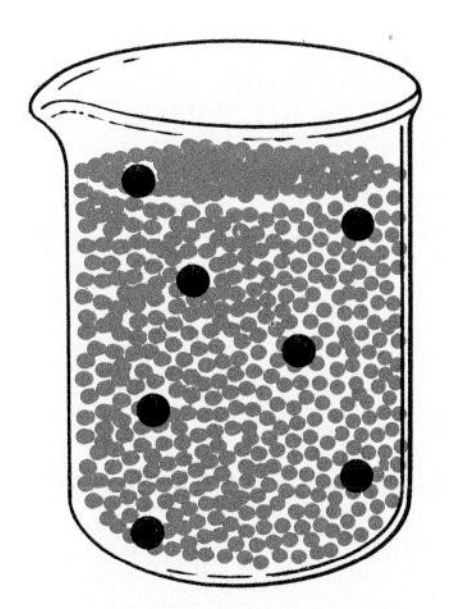

Figure 10–1 A schematic illustration at the molecular level of a sugar solution in water. Large circles represent the sugar molecules and the small circles water. Note that the size of the container and the size of the particles are not to scale.

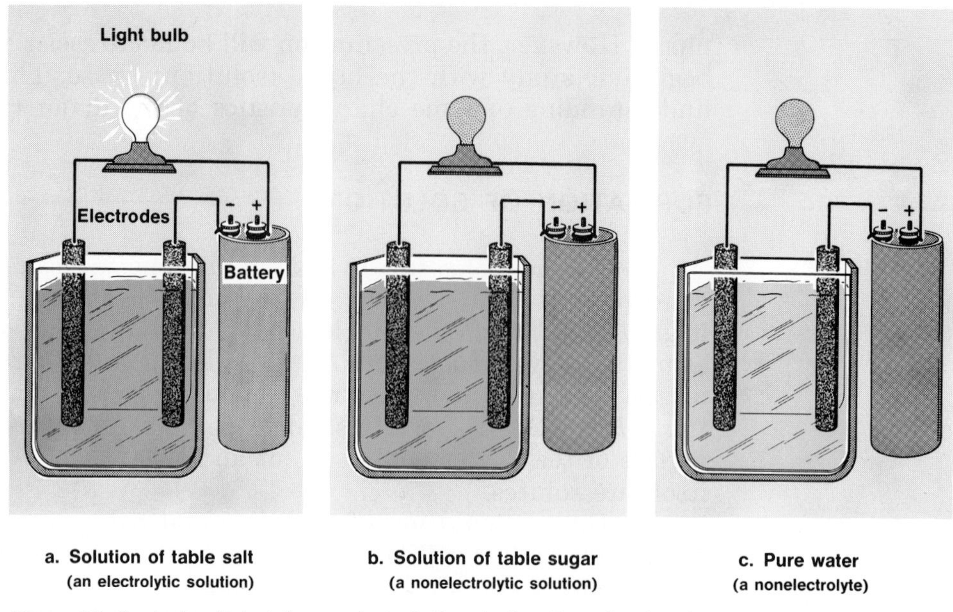

a. Solution of table salt
(an electrolytic solution)

b. Solution of table sugar
(a nonelectrolytic solution)

c. Pure water
(a nonelectrolyte)

Figure 10–2 A simple test for an electrolytic solution. In order for the light bulb to burn (*a*), electricity must flow from one part of the battery and return to the battery via the other part. To complete the circuit, the solution must conduct electricity. A solution of table salt, sodium chloride, results in a glowing light bulb. Hence, sodium chloride is an electrolyte. In (*b*), the light bulb does not glow. Hence, table sugar is a nonelectrolyte. In (*c*), it is evident that the solvent, water, does not qualify as an electrolyte since it does not conduct electricity in this test.

tive chloride ions dispersed in water. Of course, the solution as a whole is neutral since the total number of positive charges is equal to the total number of negative charges.

$$\underset{\text{(SOLID)}}{Na^+Cl^-} \xrightarrow{\text{Water}} \underset{\substack{\text{(AQUEOUS)}\\ \text{SODIUM}\\ \text{ION}}}{Na^+} + \underset{\substack{\text{(AQUEOUS)}\\ \text{CHLORIDE}\\ \text{ION}}}{Cl^-}$$

The random motions of the sodium and chloride ions are not completely independent. The charges on the particles prevent all of the sodium ions from going spontaneously to one side of the container while all of the chloride ions are going to the other side. However, a net motion of ions occurs when charged electrodes are placed in an electrolyte solution (Figure 10–3b). If the negative ions give up electrons to one electrode while the positive ions receive electrons from the other electrode, a flow of electrons or electricity is produced.

Ions migrate toward oppositely charged electrodes in an electric field.

Nonelectrolyte solutions composed of solute molecules dispersed throughout the solvent are insensitive to negatively and positively charged electrodes unless the voltage is so great that it breaks the molecules into ions.

Sometimes ionic solutions arise when a molecular substance dissolves in water. For example, hydrogen chloride, HCl, is a gas composed of diatomic molecules, each having one hydrogen and one chlorine atom. When hydrogen chloride dissolves in water an ***ionization*** reaction occurs, producing ions from molecules. The resulting solution is composed of hydrogen ions and chloride ions dispersed among the water molecules;

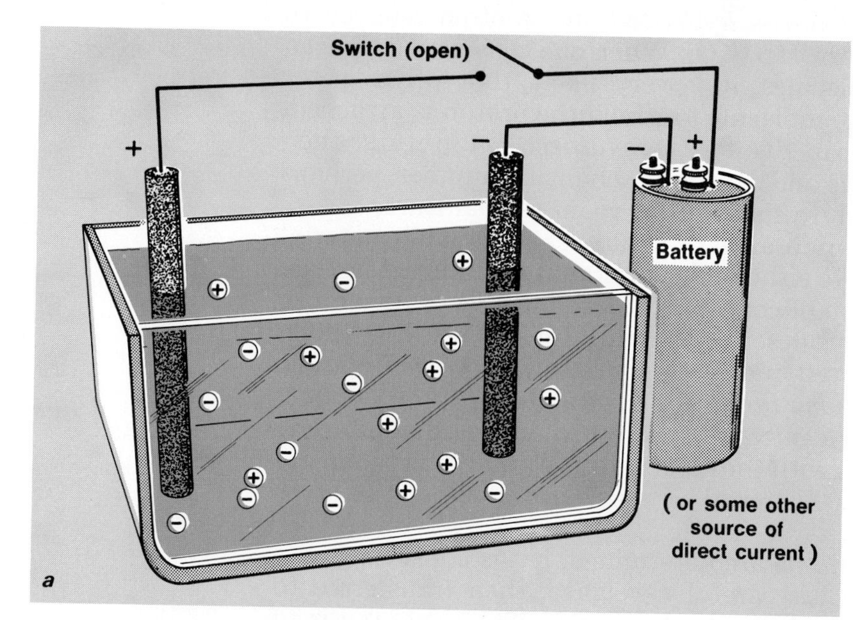

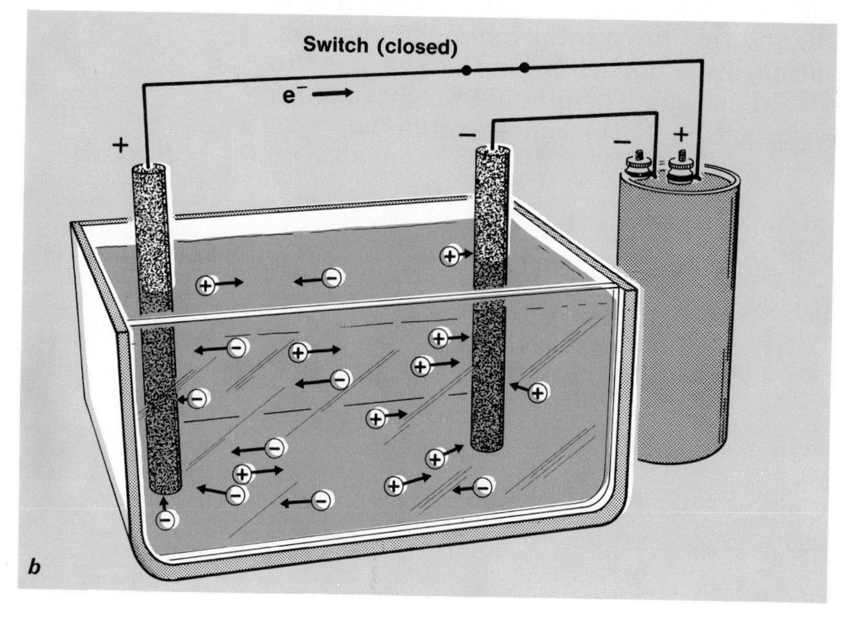

Figure 10–3 Conductance of electricity by ionic solution. (*a*) The hydrated ions are randomly distributed throughout the salt solution; the net charge is zero. (*b*) Negative electrode attracts positive ions (*cations*); positive electrode attracts negative ions (*anions*). If electrons are transferred from negative electrode to cation and from anion to positive electrode, the circuit is complete, and electricity will flow through the circuit.

consequently it is a conducting solution, and hydrogen chloride in water is properly termed an electrolyte.

$$\underset{\text{MOLECULES}}{HCl} \xrightarrow{\text{Water}} \underset{\text{IONS}}{H^+_{(aq)} + Cl^-_{(aq)}}$$

The hydrogen ion in aqueous systems is not free and unattached. Recall that water molecules are polar. A free proton (isolated hydrogen ion) could not exist in such a medium; it becomes attached to the negative end of one of the water dipoles. In fact the attraction of water dipoles for the polar HCl molecule probably brings about its ionization in the first place.

$$H^+ + H:\underset{\textstyle H}{\overset{\cdot\cdot}{\underset{\cdot\cdot}{O}}}: \longrightarrow \underset{\text{HYDRONIUM ION}}{\left[H:\underset{\textstyle H}{\overset{\cdot\cdot}{\underset{\cdot\cdot}{O}}}:H\right]^+}$$

Thus, the hydrogen ion in water is ***hydrated*** and is often referred to as the ***hydronium*** ion, H_3O^+ or $H^+(H_2O)$. When one considers hydrogen bonding between water molecules, it is very likely that other water molecules are attached to the molecule to which the proton is attached. The best representation we can give for the hydrogen ion in water then is $H^+(H_2O)_n$, where n is a large and constantly changing number, perhaps averaging about 4 or 5 in dilute solutions at room temperature.

H_3O^+ is the hydronium ion.

When sugar, sodium chloride, alcohol, or any readily soluble material dissolves in water, we can have either a concentrated or a dilute solution. Such a qualitative description of concentration is much less satisfactory and useful than a quantitative description which tells us just how much of a given substance is dissolved in a specified volume. *In chemical work we often express concentrations in moles of solute per liter of solution.*

Solutions of known concentration are prepared using volumetric flasks. These are glass vessels with the stems precisely marked to indicate specific volumes, such as 1.000 liter. The procedure involves the steps shown in Figure 10–4.

To show how concentrations are determined, let us consider a case where a 25 g sample of NaCl is carefully weighed, then transferred to a one-liter volumetric flask and dissolved in water. The next step is to add water to the flask until the solution has a total volume of one liter. In order to determine the concentration of such a solution, we need to know the number of moles of NaCl present. The formula weight of NaCl is 23.0 + 35.5 or 58.5. We have 25 grams, so the concentration is

$$\frac{25 \ \cancel{\text{grams}}}{58.5 \ \cancel{\text{grams}}/\text{mole}} = 0.43 \text{ mole of NaCl per liter of solution.}$$

Molar concentration: Number of moles of a substance per liter of solution.

We usually indicate this as 0.43 M NaCl, where M stands for moles of solute per liter of solution.

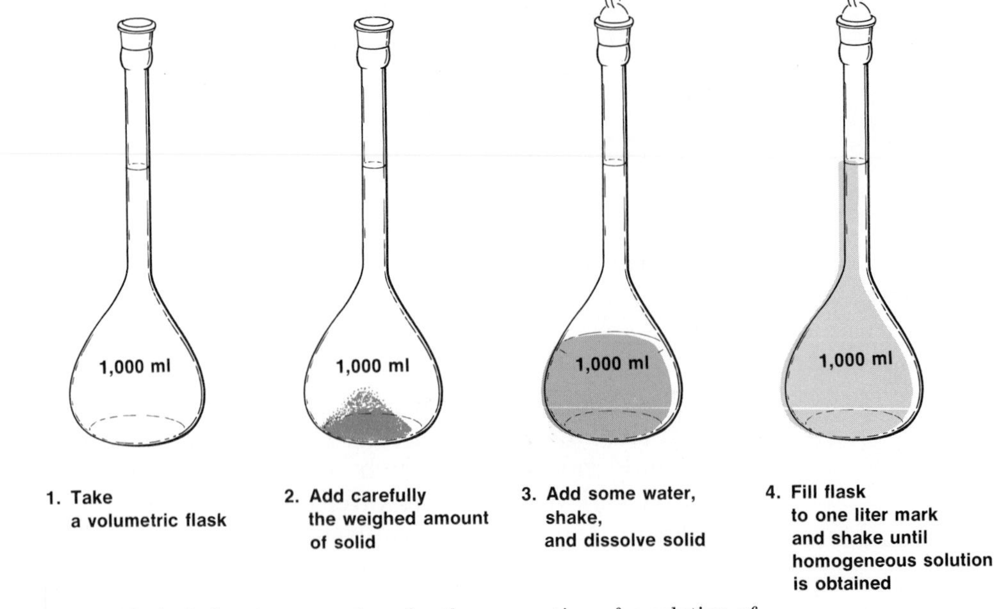
1,000 ml
1,000 ml
1,000 ml
1,000 ml
1. Take a volumetric flask
2. Add carefully the weighed amount of solid
3. Add some water, shake, and dissolve solid
4. Fill flask to one liter mark and shake until homogeneous solution is obtained

Figure 10–4 Laboratory procedure for the preparation of a solution of known concentration.

Suppose we have 86 grams of sucrose (table sugar; $C_{12}H_{22}O_{11}$, molecular weight 342) dissolved in a volume of 500 ml. What is the concentration of sugar in this solution? Since we have ($^{86}/_{342}$) mole of sugar or 0.25 mole of sugar, dissolved in $^{500}/_{1000}$ of a liter, the concentration of the sugar solution is

Molarity = M = $\frac{\text{moles of solute}}{\text{liters of solution}}$

$$\frac{\dfrac{86\ \text{g}}{342\ \text{g/mole}}}{\dfrac{500\ \text{ml}}{1000\ \text{ml/liter}}} = 0.50\ \text{M sucrose.}$$

With this knowledge about solutions and their concentrations in mind, let us look at acid-base reactions occurring in solution.

ACIDS AND BASES

Acids taste sour; bases taste bitter.

The terms "acid" and "base" have been used by chemists for several hundred years. These names were originally given to substances which showed certain properties. For example, acids have long been characterized as substances which are sour tasting, corrosive, and able to react with substances called bases. Acids turn blue litmus red. Bases, on the other hand, have a bitter taste, make the fingers or skin feel slippery on contact, and react with acids. Bases turn red litmus blue. As more and more information was collected on the properties of acids and bases, these simple definitions had to be refined. This has been carried to the point where current definitions of acids and bases are theoretical in nature.

Litmus is an acid-base sensitive dye obtained from lichens.

Proton Transfer and Neutralization

Perhaps the most used definitions of acids and bases are those first given by J. N. Brønsted and T. M. Lowry, in 1923.

Brønsted acid: A chemical species which can *donate* hydrogen ions (also called protons or H^+ ions) is an acid.
Brønsted base: A chemical species which can *accept* hydrogen ions is a base.

To illustrate these definitions we again consider the reaction between gaseous hydrogen chloride (HCl) and water:

$$\underset{\text{ACID}}{\text{HCl(gas)}} + \underset{\text{BASE}}{H_2O} \longrightarrow \underbrace{\underset{\substack{\text{HYDRONIUM ION}\\ \text{(AN ACID)}}}{H_3O^+} + \underset{\substack{\text{CHLORIDE ION}\\ \text{(A WEAK BASE)}}}{Cl^-}}_{\text{HYDROCHLORIC ACID}}$$

Examination of the above reaction shows that the HCl molecule has donated a proton (H^+ ion) to the water molecule. This behavior is understandable in terms of the electronic structures of the reacting molecules.

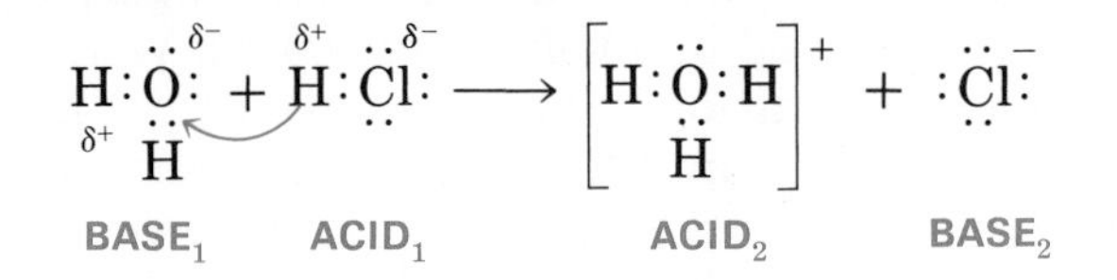

This reaction is reversible, but when equilibrium is established, essentially all the HCl has been converted to H_3O^+ and Cl^-. A concentrated (~12 M) solution of hydrogen chloride in water is mostly a solution of hydronium (H_3O^+) ions and chloride (Cl^-) ions.

If the ionic solid sodium oxide, Na_2O, is dissolved in water, a vigorous reaction occurs producing a solution containing sodium ions (Na^+) and hydroxide ions (OH^-). In this process the oxide ion (O^{2-}) reacts with water to form the hydroxide ion. In this, as in other such aqueous reactions, it is understood that the ions are hydrated (i.e., water molecules are bonded to them on a transitory basis).

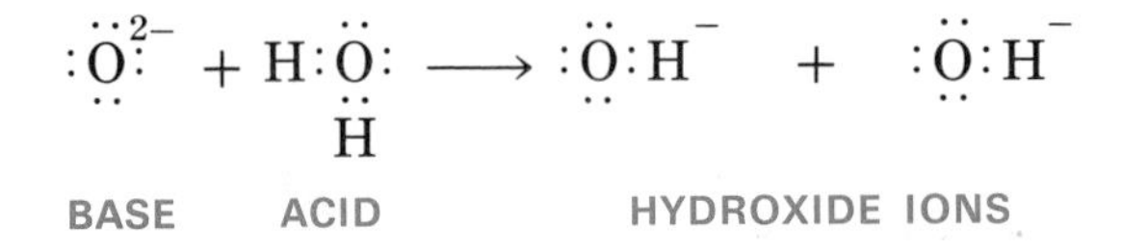

There are many other bases that take a proton from a water molecule in this way. For example:

$$\underset{\text{AMMONIA BASE}}{NH_3} + \underset{\text{ACID}}{H_2O} \rightleftharpoons \underset{\text{ACID}}{NH_4^+} + \underset{\text{BASE}}{OH^-}$$

$$\underset{\text{CYANIDE BASE}}{CN^-} + \underset{\text{ACID}}{H_2O} \rightleftharpoons \underset{\text{ACID}}{HCN} + \underset{\text{BASE}}{OH^-}$$

One view of acid-base: bases competing for protons.

According to the Brønsted-Lowry definition, water acts as an acid in these reactions and donates a proton to the other molecule or ion which acts as a base. A species such as water that can either donate or accept protons is called ***amphiprotic.*** The existence of amphiprotic species implies that acid-base reactions possess a reciprocal nature; an acid and a base react to form another acid (to which a proton has just been added) and another base (from which a proton has just been removed). Because water is the most commonly used solvent, it is also the most usual reference compound for acid-base reactions.

Also, one water molecule can transfer a proton to another water molecule.

$$2H_2O \rightleftharpoons H_3O^+ + OH^-$$

This reaction takes place to only a very small extent, as is indicated by arrows of unequal length. Since H_3O^+ and OH^- are produced in equal amounts when only water is present, pure water is neither acidic nor basic, but is described as ***neutral.***

A chemical species in water solution is commonly spoken of as an acid if it donates protons to water and increases the concentration of H_3O^+ or $H^+_{(aq)}$. Similarly, a base in water solution is commonly described as a compound whose addition to water will increase the concentration of OH^-. Since water is not the only possible solvent, these concepts are

Acids form H^+ ions in water; bases form OH^- ions in water.

When an acid neutralizes a base, acid and base properties are suppressed.

too narrow for general scientific use; they have been extended by the definitions given above, which focus on the essential feature of such acid-base behavior—that is, the donation or acceptance of a proton (H^+) in a reaction.

It has long been known that when acids react with bases the properties of both species disappear. The process involved is called ***neutralization.*** To get a more precise picture of acid-base neutralization reactions, we will consider what happens when a solution of hydrochloric acid is mixed with a solution of sodium hydroxide. The hydrochloric acid contains H_3O^+ and Cl^- ions; the sodium hydroxide solution contains Na^+ and OH^- ions. When these two solutions are mixed, a reaction occurs between H_3O^+ and OH^-.

$$Na^+ + Cl^- + \underset{\text{ACID}}{H_3O^+} + \underset{\text{BASE}}{OH^-} \longrightarrow H_2O + H_2O + Na^+ + Cl^-$$

If we have an equal number of H_3O^+ and OH^- ions, they will react to produce a neutral solution, with the hydronium ions (H_3O^+) donating their protons to the hydroxide ions (OH^-), forming molecules of water. Such reactions are called neutralization reactions because the acids and bases neutralize each other's properties. If we have more H_3O^+ ions than OH^- ions, the extra H_3O^+ will make the resulting solution ***acidic.*** If we have more OH^- ions than H_3O^+ ions, only a fraction of the OH^- ions will be neutralized, and the extra OH^- ions will make the resulting solution ***basic.***

When an acid ionizes, it produces a hydronium ion plus a species which is called the ***conjugate base*** of that acid. For example:

Conjugate acids and bases differ by one hydrogen ion. The conjugate acid has one more hydrogen ion than its conjugate base.

$$\underset{\text{NITRIC ACID}}{HNO_3} + \underset{\text{WATER}}{H_2O} \longrightarrow \underset{\text{HYDRONIUM ION}}{H_3O^+} + \underset{\text{NITRATE ION, THE CONJUGATE BASE OF THE ACID } HNO_3.}{NO_3^-}$$

In the same manner we speak of nitric acid, HNO_3, as being the conjugate acid of the base NO_3^-.

SELF-TEST 10-A

1. When ammonia dissolves in water, the resulting solution conducts electricity. Ammonia in water is therefore a(n) ______________.

2. What is the molar concentration of a solution containing 1.5 g of KCl in 500 ml of solution? The formula weight of KCl is 74.6. ______________

3. A compound HA is found to undergo a reaction forming a product H_2A^+. Therefore HA is a(n) (acid/base). If compound HA reacted to form A^-, then HA would be a(n) (acid/base).

4. If compound HA mentioned above undergoes both reactions described, then HA is termed ______________.

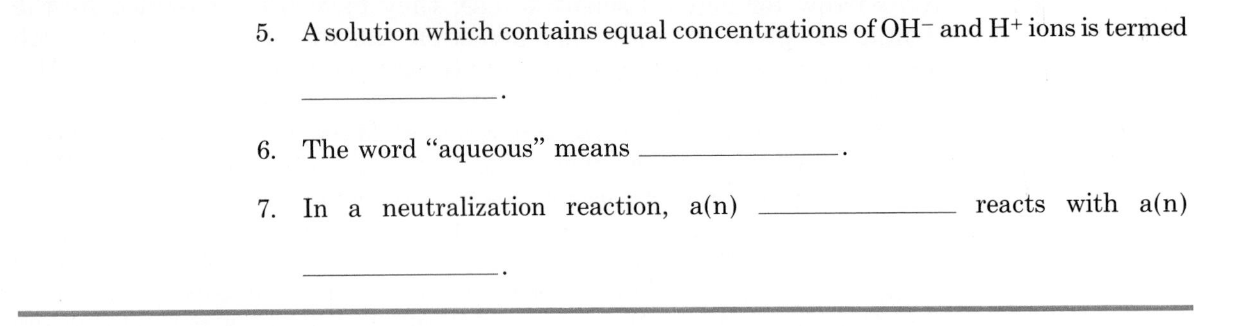

5. A solution which contains equal concentrations of OH^- and H^+ ions is termed ________.

6. The word "aqueous" means ________.

7. In a neutralization reaction, a(n) ________ reacts with a(n) ________.

The Strengths of Acids and Bases

Because the water molecule is itself a weak base, the strongest acid (that is, the best proton donor) that can exist in water is the hydronium ion (H_3O^+). If strong acids such as perchloric acid ($HClO_4$) or nitric acid (HNO_3) are added to water they donate all of their protons to water, and the resultant solutions contain H_3O^+ and ClO_4^-, or H_3O^+ and NO_3^-. All of these reactions are reversible, at least in principle, but when strong acids are dissolved in water the equilibria that result favor the reaction products to a very great extent.

Reactants			Products	
HNO_3 +	H_2O	$\longrightarrow$	H_3O^+ +	NO_3^-
H_2SO_4 +	H_2O	$\longrightarrow$	H_3O^+ +	HSO_4^-
ACID	BASE		ACID	BASE

Since there are a large number of ions present, the solution conducts electricity well; HNO_3 and H_2SO_4 are termed ***strong electrolytes.***

Strong electrolytes dissociate completely in water.

Not all acids lose protons as readily to water as do nitric acid and sulfuric acid. Some anions are capable of competing with water for the proton being exchanged. The result of this competition is the establishment of an equilibrium between neutral acid molecules and hydronium ions in water solution.

Acetic acid, the acid found in vinegar, is one of these weak acids. The molecular structure for acetic acid ($HC_2H_3O_2$) is as follows:

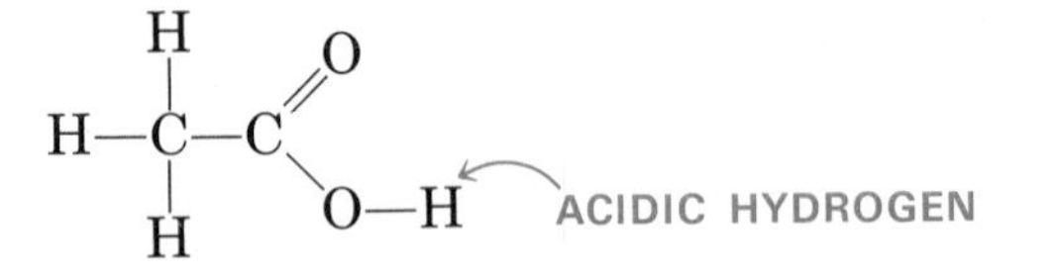

The hydrogen atom bonded to the oxygen in the molecule is the only one that is donated to a base in water solution; for that reason it is designated an acidic hydrogen. When acetic acid is dissolved in water, some ions are produced. However, most of the acetic acid molecules do not donate protons to water molecules. The result is an equilibrium mixture containing a few H_3O^+ and $C_2H_3O_2^-$ ions; acetic acid molecules are more abundant than acetate ions in the solution. The reaction is:

$$HC_2H_3O_2 + H_2O \rightleftharpoons H_3O^+ + C_2H_3O_2^-$$

ACETIC ACID — WATER — HYDRONIUM ION — ACETATE ION

The relatively few ions in an acetic acid solution do not conduct electricity very effectively; consequently, acetic acid is a ***weak electrolyte.***

The strongest bases have the weakest conjugate acids.

Another way of looking at this reaction is to realize that in the equilibrium mixture there are two bases: the water molecule H_2O, and the acetate ion $C_2H_3O_2^-$. Since the reverse reaction is more important in determining the position of equilibrium, the acetate ion must be a stronger base than the water molecule.

The same kind of considerations can be made for bases. Ammonia, dissolved in water, is a weak base. The resulting reaction produces relatively few ions; at equilibrium the ammonia is mostly in the molecular form:

$$\underset{\text{AMMONIA}}{NH_3} + \underset{\text{WATER}}{H_2O} \rightleftharpoons \underset{\text{AMMONIUM ION}}{NH_4^+} + \underset{\text{HYDROXIDE ION}}{OH^-}$$

Consequently, the relatively few ions present do not conduct electricity well and ammonia may thus be called a weak electrolyte.

Table 10–1 gives some common acids and bases ranked according to their relative strengths.

The pH Scale

Because water solutions of dilute acids and bases are used so extensively, it is convenient to have a simple way of designating the acidity or basicity of these solutions. The pH scale was devised for this purpose; it furnishes a number which describes the acidity of a solution. The ***pH*** is defined as pH = $-\log[H_3O^+]$. The brackets mean moles per liter of hydronium ions. The pH scale, for practical purposes, runs from 0 to 14. A pH of 7 indicates a neutral solution (such as pure water), a pH below 7 indicates an acidic solution, and a pH above 7 indicates a basic solution.

At pH = 7; $[H_3O^+] = [OH^-]$

The pH number is related to the concentration of the hydrogen ions in an aqueous solution expressed as the negative power of 10 (logarithm). A hydrogen ion concentration of 0.00001 or 1×10^{-5} M corresponds to a pH of 5. To have a pH of 7 for pure water, the hydrogen ion concentration would have to be 1×10^{-7} M.

TABLE 10–1 Relative Strengths of Some Acids and Bases

Acid (Increasing Acid Strength ↑)						Conjugate Base (Increasing Base Strength ↓)	
perchloric acid	$HClO_4$	+	H_2O	$\rightleftharpoons H_3O^+$	+	ClO_4^-	perchlorate ion
hydrochloric acid	HCl	+	H_2O	$\rightleftharpoons H_3O^+$	+	Cl^-	chloride ion
nitric acid	HNO_3	+	H_2O	$\rightleftharpoons H_3O^+$	+	NO_3^-	nitrate ion
hydronium ion	H_3O^+	+	H_2O	$\rightleftharpoons H_3O^+$	+	H_2O	water
hydrofluoric acid	HF	+	H_2O	$\rightleftharpoons H_3O^+$	+	F^-	fluoride ion
acetic acid	$HC_2H_3O_2$	+	H_2O	$\rightleftharpoons H_3O^+$	+	$C_2H_3O_2^-$	acetate ion
water	H_2O	+	H_2O	$\rightleftharpoons H_3O^+$	+	OH^-	hydroxide ion
hydroxide ion	OH^-	+	H_2O	$\rightleftharpoons H_3O^+$	+	O^{2-}	oxide ion

To understand this we will need to reexamine the very slight acid-base reaction between water molecules:

$$H_2O + H_2O \rightleftharpoons H_3O^+ + OH^-$$

The reaction is an ionization reaction producing ions which should cause water to conduct electricity. But, we have said that water is a nonelectrolyte. Actually, with very sensitive electrical equipment, pure water can be shown to conduct electricity *very slightly* due to the very small amount of ionization. Laboratory measurements reveal that 0.0000001 or 1×10^{-7} mole per liter of water ionizes at 25°C. Pure water contains 55.5 moles of water per liter. Consequently, we can see that the actual amount of the water that ionizes is very small compared to the total amount of water present. Pure water, then, which is defined as being neutral, contains 1×10^{-7} mole per liter of hydrogen ions (hydronium ions), and its pH is 7. In pure water the concentrations of hydronium ions (H_3O^+) and hydroxide ions (OH^-) must be equal since each time a hydronium ion (H_3O^+) is produced in the ionization reaction a hydroxide ion (OH^-) is also produced. Therefore, in pure water the concentration of the hydroxide ion (OH^-) is also 1×10^{-7} mole per liter.

One liter of water weighs 1000 grams and is 55.5 molar: $\dfrac{1000 \not{g}}{\dfrac{18 \not{g}}{\text{mole}}} = 55.5$ moles

Figure 10–5 graphically displays the relationship between the pH number and the concentration of the hydrogen ion. It also gives the approximate pH values for common solutions.

A close examination of Figure 10–5 reveals that basic solutions such

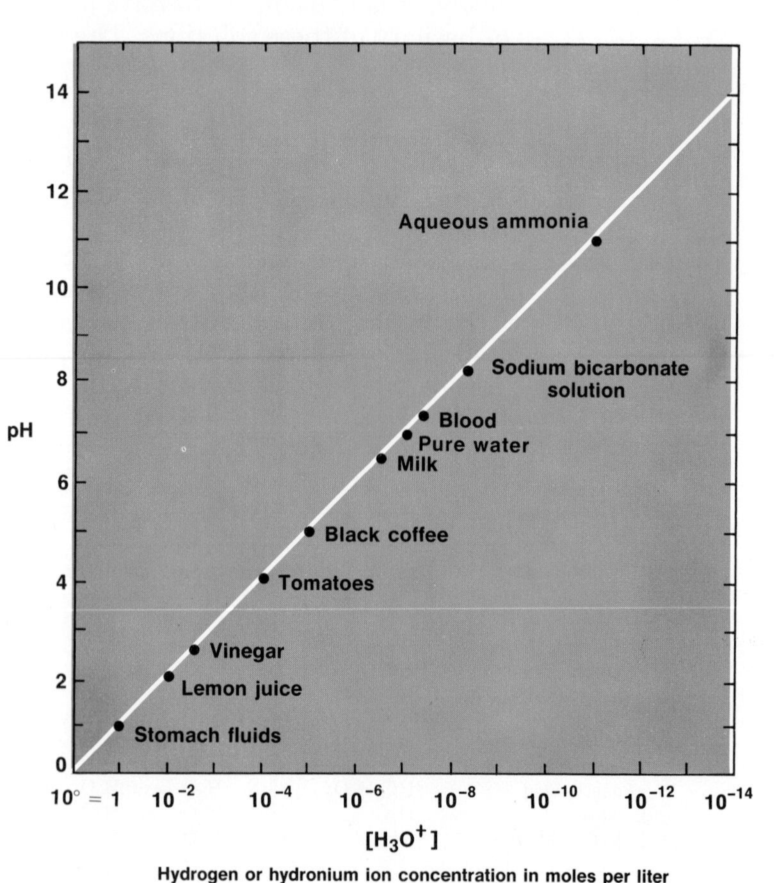

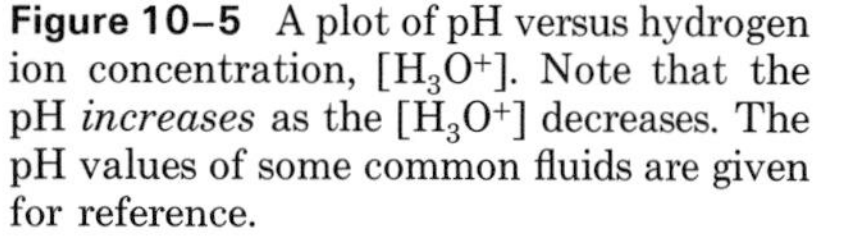
Figure 10–5 A plot of pH versus hydrogen ion concentration, $[H_3O^+]$. Note that the pH *increases* as the $[H_3O^+]$ decreases. The pH values of some common fluids are given for reference.

as ammonia water have hydrogen ion concentrations less than that of pure water. Note that the pH of a typical sample of ammonia water is 11. This is equivalent to a hydrogen ion concentration of 1×10^{-11} or 0.00000000001 mole per liter. You may wonder how it is possible for ammonia water to have fewer hydrogen ions than an equal volume of water. The answer to this puzzle lies in the fact that this ionization reaction, like all chemical reactions, will reach a state of chemical equilibrium. In the following paragraphs an explanation is given, but a facility with exponential notation and logs is required of the reader.

As stated, the quantitative definition of pH is:

$$pH = -\log[H_3O^+]$$

where $[H_3O^+]$ is the hydrogen ion concentration of the solution in moles per liter. A 0.1 M solution of HCl has $[H_3O^+] = 0.1$ M, because HCl ionizes completely in water:

$$HCl + H_2O \xrightarrow[100\%]{} H_3O^+ + Cl^-$$

Short review of logarithms (base = 10): $\log 1 = 0$; $\log 10 = 1$; $\log 10^{-1} = -1$; $-\log 10^{-1} = -(-1) = 1$

For this solution, the pH is given by the expression:

$$pH = -\log[H_3O^+] = -\log(0.1)$$

The log of 0.1 is -1, so

$$pH = -(-1) = 1$$

For neutral water we find $[H_3O^+] = [OH^-] = 10^{-7}$ mole per liter, so its pH is:

$$pH = -\log[H_3O^+] = -\log(10^{-7}) = -(-7) = 7$$

As you can see, solutions with pH values *less* than 7 are acidic. What about basic solutions? What happens to the hydrogen ion concentration in a water solution which is also 0.1 M in NaOH? First we have to recall the equilibrium which occurs in *all* aqueous solutions:

$$2H_2O \rightleftarrows H_3O^+ + OH^-$$

It has been shown experimentally that the product of the hydrogen ion concentration and hydroxide ion concentration is the same constant value for all aqueous solutions at the same temperature.

$$[H_3O^+][OH^-] = \text{a constant}$$

Solutions with a pH less than 7 are acidic; those with a pH greater than 7 are basic.

This constant is called the ***ion product*** of water, is given the symbol K_W, and has the experimental value of 10^{-14} at 25°C.

$$K_W = [H_3O^+][OH^-] = 10^{-14}$$

Since the product is always 10^{-14} at 25°C, an acid with a high $[H_3O^+]$ will have a low $[OH^-]$, and a basic solution with a high $[OH^-]$ will have a low $[H_3O^+]$. The adjustment of the concentrations that ensures the constant product comes through shifting the water equilibrium. For example, if sufficient acid is added to water to provide a hydrogen ion concentration of 0.1 M, enough of the hydrogen ions react with the hydroxide ions present in water to lower the $[OH^-]$ from its value of 10^{-7} M in neutral water to the much smaller concentration of 10^{-13} M.

Returning to our 0.1 M NaOH solution, we know that this compound produces Na^+ and OH^- ions in solution and that it is completely dissociated. From this we can say:

$$NaOH \xrightarrow[100\%]{H_2O} Na^+_{(aq)} + OH^-_{(aq)}$$

$$[OH^-] = 0.1\ M$$

$[OH^-]$ means the concentration of hydroxide ions in moles per liter.

The $[H_3O^+]$ value can now be obtained because we know that

$$[H_3O^+][OH^-] = 10^{-14}, \text{ so}$$
$$[H_3O^+](0.1) = 10^{-14}, \text{ and}$$
$$[H_3O^+] = 10^{-13}\ M$$

The pH of this solution is then:

$$pH = -\log[H_3O^+] = -\log(10^{-13}) = -(-13) = 13$$

A 0.1 M NaOH solution is more basic than any of the entries in Figure 10-5.

SALTS

Preparation of Salts

The chemical compounds known as ***salts*** play a vital role in nature, in animal growth and life, and in the manufacture of various chemicals for human use. They are formed as the products of acid-base neutralizations, as in the following example:

$$(K^+ + OH^-) \quad + \quad (H_3O^+ + Cl^-) \rightleftharpoons \underbrace{K^+ + Cl^-}\ + 2H_2O$$

POTASSIUM HYDROXIDE IN WATER BASE

HYDROCHLORIC ACID IN WATER ACID

CRYSTALLIZE OUT THE SALT BY REMOVAL OF SOLVENT, WATER

KCl (SOLID) SALT

Salts are the result of acid-base neutralization.

Salts contain species held together by ***ionic bonding*** (see Chapter 9). Solid potassium chloride, for example, is composed of an equal number of K^+ ions and Cl^- ions arranged in definite positions with respect to one another in a lattice (see Figure 9-5). Since the salt crystal must be electrically neutral, it can have neither an excess nor a deficiency of positive or negative charge.

Let us imagine that we have at our disposal the ions listed below, and let us see what salts could result.

Ions		Possible Salts	Salt Name
Na^+	sodium	$NaCl$	sodium chloride
Ca^{2+}	calcium	$NaNO_3$	sodium nitrate
Cl^-	chloride	Na_2SO_4	sodium sulfate
NO_3^-	nitrate	$NaC_2H_3O_2$	sodium acetate
SO_4^{2-}	sulfate	$CaCl_2$	calcium chloride
$C_2H_3O_2^-$	acetate	$Ca(NO_3)_2$	calcium nitrate
		$Ca(C_2H_3O_2)_2$	calcium acetate
		$CaSO_4$	calcium sulfate

In the formula of a salt, the positive and negative charges are equal.

In the examples just given, notice that in order to attain an electrically neutral lattice, it is necessary to balance the charges of the ions. A sodium (Na^+) ion requires just one chloride ion (Cl^-), and the NaCl lattice contains an equal number of Na^+ and Cl^- ions. A sulfate ion (SO_4^{2-}) with two negative charges must have its negative charge balanced by two positive charges. This may be done by using two Na^+ ions:

$$2Na^+ + SO_4^{2-} \longrightarrow Na_2SO_4$$

or one Ca^{2+} ion:

$$Ca^{2+} + SO_4^{2-} \longrightarrow CaSO_4$$

It is possible to form many solid salts by mixing water solutions of different soluble salts with each other. For example, both lead acetate and sodium chloride are soluble in water. If we prepare solutions of these salts and then mix the solutions, we find the insoluble salt, lead (II) chloride, precipitates from the mixture.

$$\underbrace{2Na^+ + 2Cl^-}_{\text{SODIUM CHLORIDE IN SOLUTION}} + \underbrace{Pb^{2+} + 2C_2H_3O_2^-}_{\text{LEAD (II) ACETATE IN SOLUTION}} \longrightarrow \underbrace{PbCl_2}_{\text{SOLID LEAD (II) CHLORIDE}} + \underbrace{2Na^+ + 2C_2H_3O_2^-}_{\text{SODIUM ACETATE IN SOLUTION}}$$

Sodium acetate may be removed from the solution later by evaporating off the water.

This reaction illustrates an important principle based upon differences in solubility. If the different ions of a compound of low solubility are brought together in solution by mixing, and if their concentrations are great enough, that compound containing those ions will precipitate from solution.

Salts in Solution

While some salts are very soluble in water, others are quite insoluble. Salts are found with a wide range of water solubilities.

An important property of salts is their ability to dissolve in suitable solvents to yield solutions. A salt solution is a homogeneous mixture of solvent molecules and ions (frequently bonded to solvent molecules). The amount of a salt that will dissolve in a given quantity of solvent tells us the salt's solubility in that solvent. The preparation of lead (II) chloride shown above is an example of a reaction made possible by the differences in solubilities of different salts in the same solvent. It is because of these differences in solubilities that we can make roads out of calcium carbonate (limestone), which is insoluble in water, but not calcium chloride, which is water soluble.

Because of the amount of water on the earth and the importance of water as a major component of living matter, the study of aqueous solutions of salts is important in understanding ourselves and the changing earth about us. Consider what happens at the ionic level when a sodium chloride crystal is placed in contact with water. We know that sodium chloride is soluble in water. This means that most, if not all, of the potential energy in the crystal lattice is somehow overcome in the solution process.

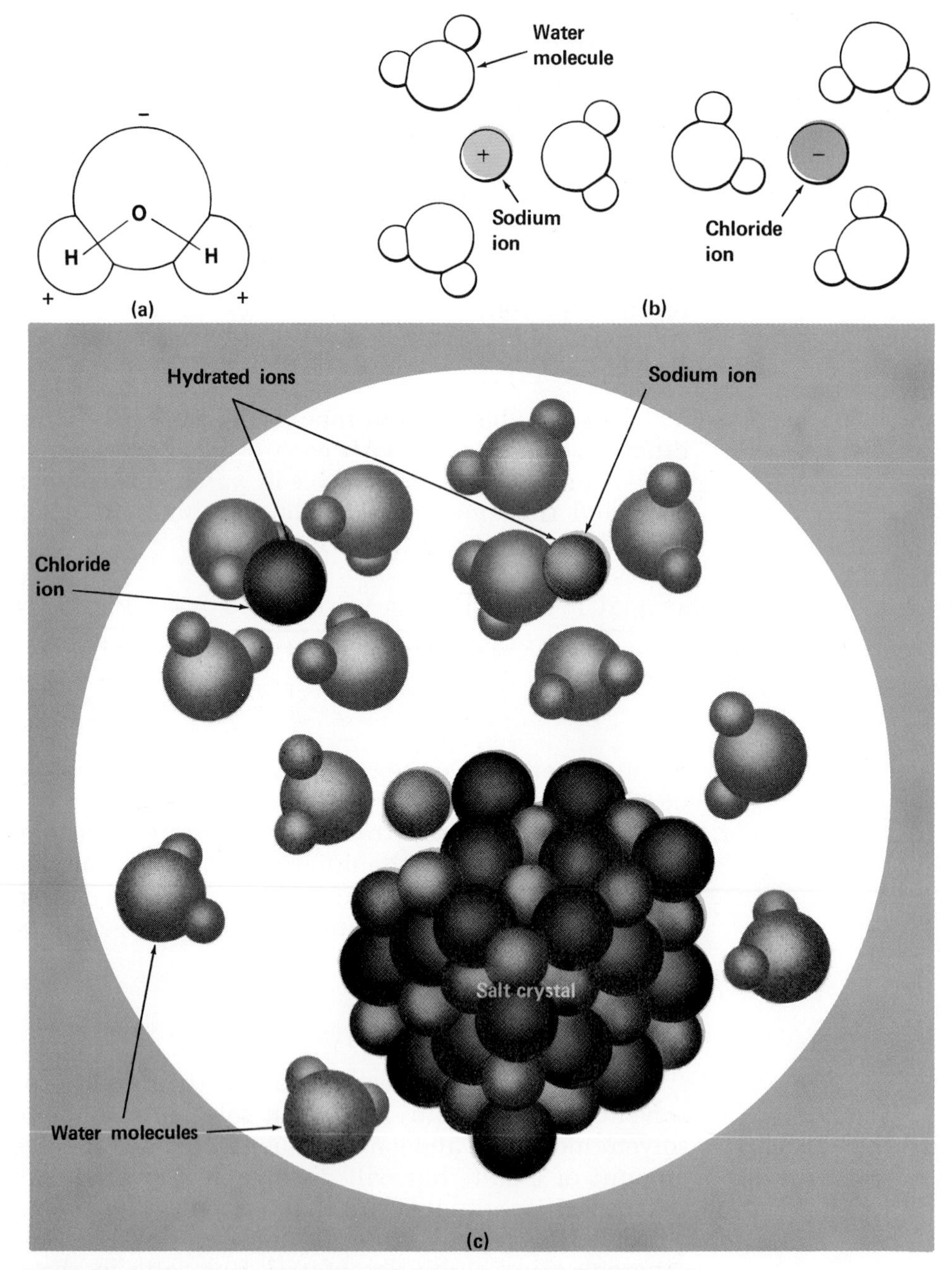

Figure 10–6 Dissolution of sodium chloride in water. (*a*) Geometry of the polar water molecule. (*b*) Solvation of sodium and chloride ions due to interaction (bonding) between these ions and water molecules. (*c*) Dissolution occurs as collisions between water molecules and crystal ions result in the removal of the crystal ion. In the process the ion becomes completely solvated.

PLATE III

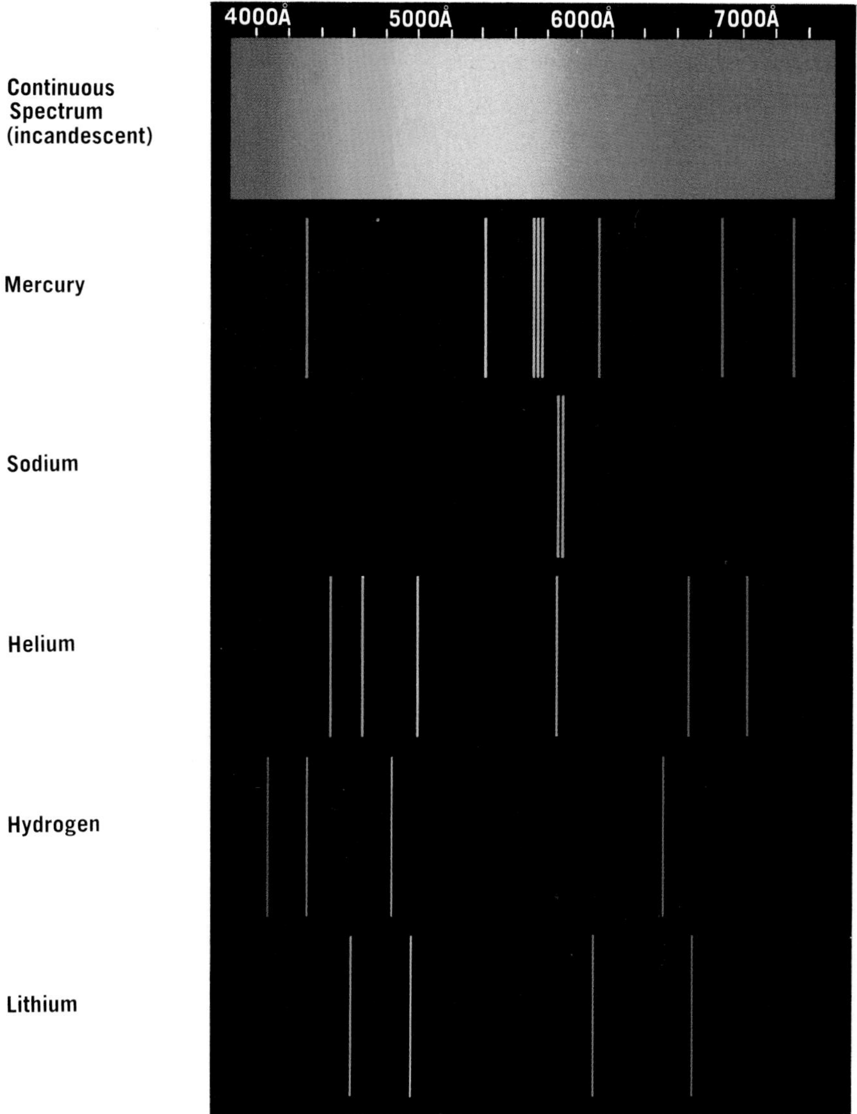

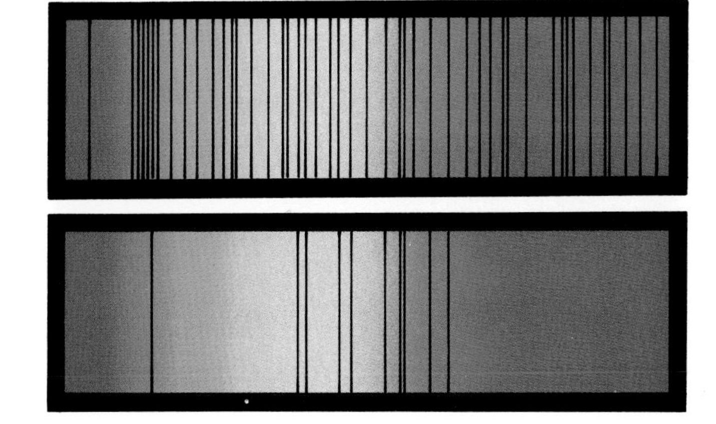

In the continuous spectrum, note that the line emission spectra from various elements differ from each other and that they consist of well-defined colors.
The hydrogen spectrum contains only a few well-defined colors in the visible portion.
The solar spectrum is almost a continuous spectrum, but it has many absorption lines in it.
In the spectrum of iron, note the correspondence with the solar spectrum above it, indicating the presence of iron in the sun's atmosphere.
(From Faughn, J. S., and Kuhn, K. F.: *Physics for People Who Think They Don't Like Physics*. Philadelphia, W. B. Saunders Co., 1976.)

PLATE IV

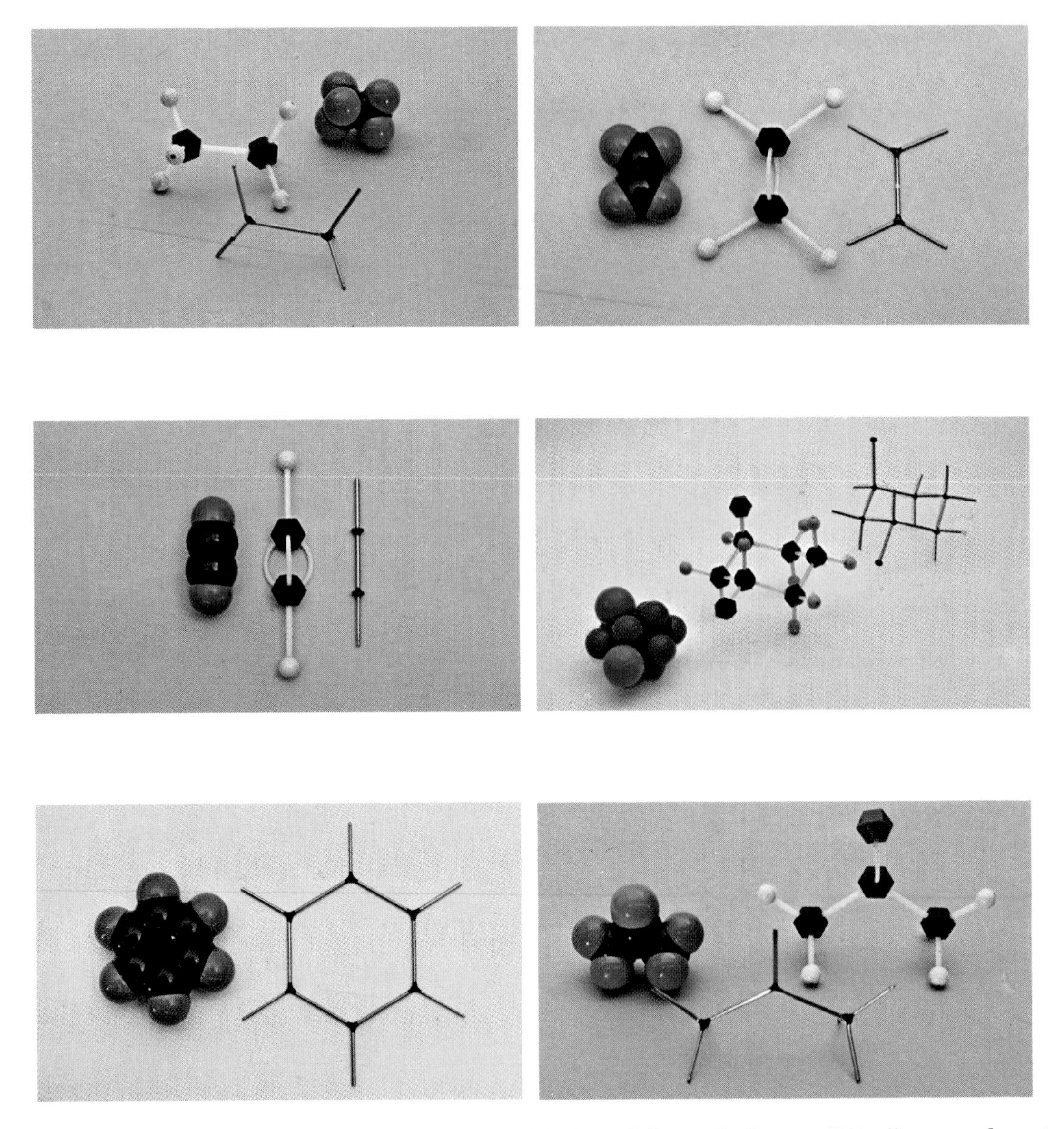

Three types of molecular models. The orange and black models are the "space-filling" type and most closely approximate the overall molecular shape. Wire models on the right in each photo show bonds only, and are useful in visualizing spatial relationships. (From Moore, J. A.: *Elementary Organic Chemistry*. Philadelphia, W. B. Saunders Co., 1974.)

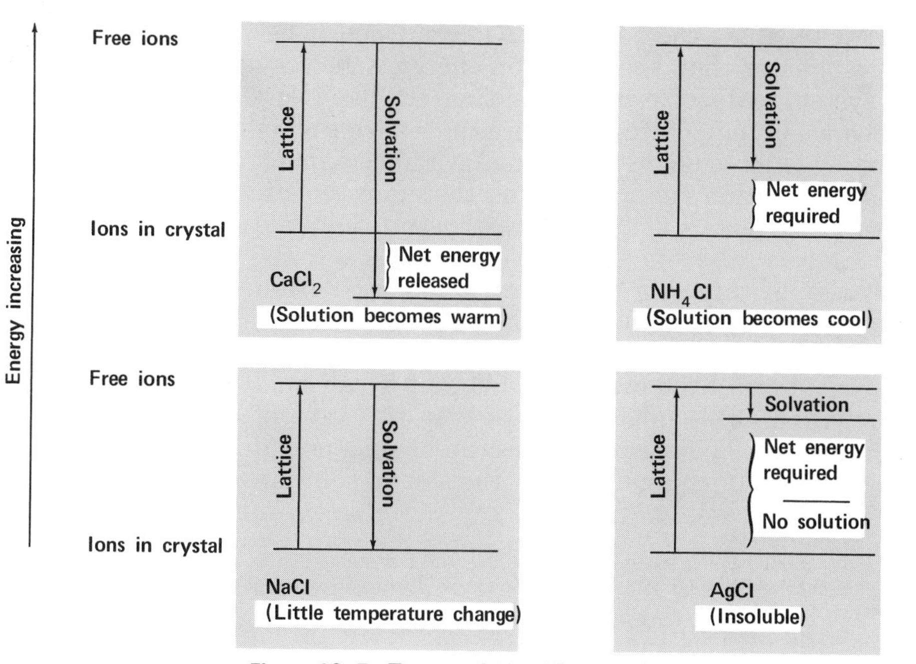

Figure 10–7 Energy relationships associated with the solution of salts. Free ions are the ions of the salt separated from each other.

The surface of the salt crystal appears calm when the crystal is placed in water, but on the ionic level there is a great deal of agitation. Water molecules are trying to pull the ions of the salt away from the lattice and into solution. Water molecules have sufficient polarity to interact strongly with the ions and bond with them. Once this takes place the ion is removed from the crystal and the crystal lattice now has a gap in it where the ion was removed. As stated earlier, ions that are bonded by solvent molecules are termed solvated ions and the process of ion-solvent interaction is called ***solvation.*** But just how are these ions solvated? What causes this solvation?

Because of its polar nature, the water molecule is ideally suited to interact with ions.

The answers lie in the structure of the water molecule and its ability to interact with ions. As we saw in Chapter 9, water is a *polar* molecule. The negative end of the molecule is attracted to positive ions, and the positive end is attracted to negative ions. As a result, several water molecules will interact with each ion. Figure 10–6 shows several water molecules solvating a Na^+ ion. The positive ends of the water molecule will tend to interact with a negative ion; this is shown for the Cl^- ion in Figure 10–6.

We have learned earlier that energy is required to break a chemical bond, or conversely, energy is released when a bond is formed. The solvation process we have just described involves the breaking of ionic bonds and the formation of ion-solvent molecule bonds. The energy required to separate all of the ions of a salt is a measure of the ***lattice energy.*** Lattice energies are relatively large, but they are partially compensated for by the energy liberated in the solvation of the ions called the ***solvation energy,*** Figure 10–7.

When the dissolving of a salt such as calcium chloride ($CaCl_2$) causes the solution to become warmer, the solvation energy must be greater than the lattice energy. On the other hand, the dissolving of a salt like ammonium chloride (NH_4Cl) causes the solution to become cooler because the lattice energy is greater than the solvation energy, so energy must be taken from the system to dissolve the salt. Some salts like NaCl dissolve without much change in temperature because the lattice energy approximately equals the solvation energy. Salts like AgCl have a much greater lattice energy than solvation energy and are, therefore, practically insoluble in water. These energy changes are illustrated in the graphs of Figure 10–7.

Salts which absorb heat when they dissolve, such as ammonium chloride, are used in cold packs for the emergency medical treatment of athletic injuries.

Every salt has what may be termed a "solubility limit" for a given solvent: at a given temperature, a certain number of grams of salt, and *no more,** will dissolve in a certain quantity (frequently 100 grams) of solvent. A solution that contains all the dissolved salt that it can hold is termed a "saturated solution." One might ask just why this type of solubility limit is found in nature. The reason for this can be understood if we remember that the oppositely charged ions of the salt in solution

*Supersaturated solutions can be formed under special conditions, but are not stable in the presence of the solid solute. The presence of the solid solute causes the excess dissolved solute to crystallize and a saturated solution is formed.

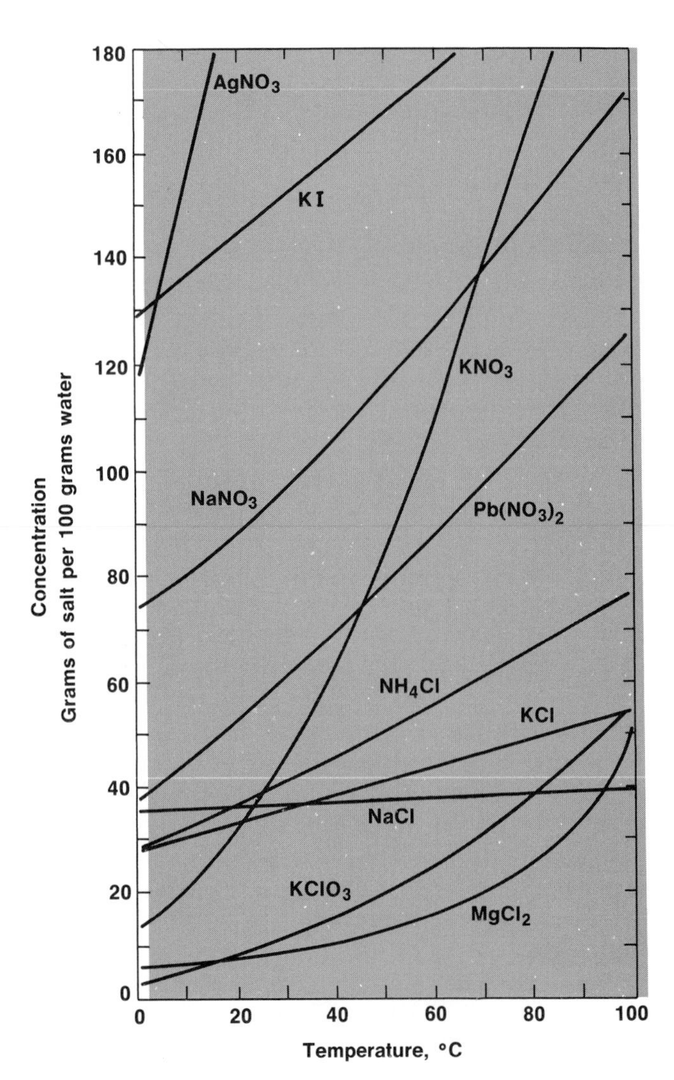

Figure 10–8 The effect of temperature on the solubility of some common salts in water.

actually attract one another. If one crowds the solution with solvated ions to too great an extent and "ties up too many solvent molecules," the ions will begin re-forming the crystal lattice. This is ***crystallization,*** or solvation in reverse. When undissolved salt is in contact with a saturated solution of that salt, there is a dynamic equilibrium established with the salt crystal being broken down at one point while being formed in another. The effect of temperature on the solubility of some common salts is shown in Figure 10–8.

Uses of Salts

Salts are used for so many different applications that it would be impossible to list them all. We can, however, look at the uses of several salts and see just how versatile they are.

Sodium Chloride (Table Salt)

Many body fluids contain potassium ions (K^+) in greater concentration than Na^+ ions.

This is the best-known salt and perhaps the most important to man. Physiologically, mammals need this salt for an ionic balance in the bloodstream and for the production of hydrochloric acid in the stomach to aid in the digestion of food. Very early in man's history, sodium chloride was used in barter, and wars have actually been fought for control of salt deposits. Commonly, sodium chloride is used for medicinal purposes, such as in water solutions, to wash wounds and to treat burns.

Nitrate Salts

Salts containing the nitrate ion (NO_3^-), such as sodium nitrate ($NaNO_3$), potassium nitrate (KNO_3), and ammonium nitrate (NH_4NO_3), find special uses in fertilizers and in the manufacture of explosives. Sodium nitrate and potassium nitrate occur in natural deposits while ammonium nitrate can be prepared by a reaction involving ammonia (NH_3) and nitric acid.

$$\underset{\text{BASE}}{NH_3} + \underset{\text{ACID}}{HNO_3} \rightleftarrows \underset{\substack{\text{SALT} \\ \text{(AMMONIUM NITRATE)}}}{NH_4NO_3}$$

Gunpowder contains potassium nitrate (KNO_3), sulfur, and charcoal.

Explosions are possible when nitrate salts and combustible materials are mixed. Gunpowder contains KNO_3. Ammonium nitrate will explode by itself under certain conditions, and a number of disasters have occurred when large amounts of this salt have exploded in warehouses and in the holds of ships in harbors. Under more controlled conditions, NH_4NO_3 decomposes to yield dinitrogen oxide (N_2O) and water. Dinitrogen oxide is called laughing gas and is used as an anesthetic in certain forms of surgery.

$$NH_4NO_3 \xrightarrow{\text{Heat}} \underset{\substack{\text{DINITROGEN} \\ \text{OXIDE} \\ \text{(LAUGHING GAS)}}}{N_2O} + 2H_2O$$

This gas also has a limited solubility in milk and cream and finds a use in canned whipped cream. When the valve is opened, the cream and gas escape together, and upon reaching the atmosphere the gas expands, thus "whipping" the cream.

Sodium Hydrogen Carbonate

Sodium hydrogen carbonate, also called sodium bicarbonate ($NaHCO_3$), has the common name, baking soda. The negative ion of the salt, the hydrogen carbonate ion, is a base and is capable of reacting with water in an acid-base reaction.

$$\underset{\text{BASE}}{HCO_3^-} + \underset{\text{ACID}}{H_2O} \rightleftharpoons \underset{\text{ACID}}{H_2CO_3} + \underset{\text{BASE}}{OH^-}$$

When this salt is placed in a mixture containing water, such as biscuit batter, this reaction takes place to a slight extent. The product of the reaction, carbonic acid (H_2CO_3), is quite unstable and decomposes under the influence of heat, yielding the products carbon dioxide (CO_2) and water. The decomposition of the H_2CO_3 causes the reaction of bicarbonate with water to take place to a greater extent. In effect, the equilibrium is shifted to the products side, and the overall reaction is essentially a complete conversion of the bicarbonate ion to carbon dioxide and water. The reaction is:

$$HCO_3^- + H_2O \longrightarrow H_2CO_3 + OH^-$$

$$H_2CO_3 \xrightarrow{\text{Heat}} H_2O + \underset{\text{(GAS)}}{CO_2}$$

The CO_2 gas which is produced in the reaction causes the biscuit batter to rise. To hasten these reactions, pioneer women often added sour milk or buttermilk, which contain lactic acid, a weak acid. Baking powder contains baking soda along with a weak acid which shifts the equilibrium to the right. Self-rising flour contains a small amount of baking soda and a weak acid to produce a controlled amount of CO_2 gas as the mixture is heated in the oven.

The HCO_3^- ion can also act as a weak acid in the presence of a strong base:

$$\underset{\text{ACID}}{HCO_3^-} + \underset{\text{BASE}}{OH^-} \rightleftharpoons \underset{\text{ACID}}{HOH} + \underset{\text{BASE}}{CO_3^{2-}}$$

This explains its use as a neutralizer of spilled acids and bases in the laboratory.

Barium Sulfate

This salt may be prepared by taking advantage of its insolubility in water. If a solution of barium chloride ($BaCl_2$) is mixed with sodium sulfate solution (Na_2SO_4), insoluble $BaSO_4$ precipitates:

$$(Ba^{2+} + 2Cl^-) + (2Na^+ + SO_4^{2-}) \longrightarrow BaSO_4\ \text{(Solid)} + (2Na^+ + 2Cl^-)$$

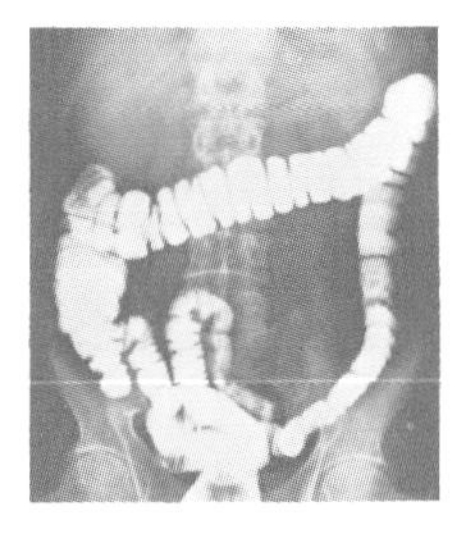

$BaSO_4$ is used in intestinal x-ray studies to minimize the danger of barium poisoning, since the very insoluble salt passes through the body unaltered.

In the diagnosis of stomach ulcers and intestinal disorders x-rays are often used. Since these organs are normally transparent to x-rays, a suitable material must be placed in the organ to render it partially opaque. Barium is a good element for this since it is essentially opaque toward the type of x-rays used in medicine. Barium ions, however, are quite poisonous and cannot be taken internally. The solution to this problem lies in the extreme insolubility of $BaSO_4$. A syrup containing $BaSO_4$ is swallowed to introduce the barium into the digestive system.

SELF-TEST 10-B

1. Complete the table, filling in the blanks.

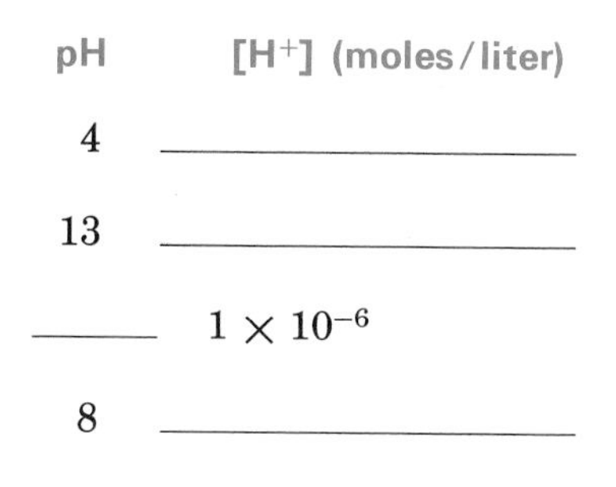

pH	$[H^+]$ (moles/liter)
4	________
13	________
________	1×10^{-6}
8	________

2. If the solubility of KI is 140 grams per 100 grams of water at 20°C, how much KI will dissolve in 1000 grams of water? ________

3. Write a chemical reaction showing water to be an acid. Use ammonia (NH_3) as the base.

$$H_2O + NH_3 \longrightarrow \_\_\_\_\_ + \_\_\_\_\_$$

4. Write a chemical reaction showing water to be a base. Use HCl as the acid.

$$H_2O + HCl \longrightarrow \_\_\_\_\_ + \_\_\_\_\_$$

5. Which is more acidic, a pH of 6 or a pH of 2? ________

6. Is a pH of 8 more basic than water? ________

ELECTRON TRANSFER (OXIDATION-REDUCTION)

The loss of electrons is called *oxidation.*

A very common type of reaction is one in which the control of valence-shell electrons passes from one species to another. These reactions are called oxidation-reduction reactions. Examples include:

$$2Na + Cl_2 \longrightarrow 2NaCl\ (Na^+ + Cl^-)$$
$$2H_2 + O_2 \longrightarrow 2H_2O$$
$$2Fe + 3Br_2 \longrightarrow 2FeBr_3$$

If the atom has either lost electrons or lost some control over the electrons, we say it has been ***oxidized.*** If it has gained electrons or gained a greater degree of control over the valence shell electrons, we say it has been ***reduced.*** In the reactions above, the sodium, hydrogen, and iron atoms have all lost electrons, so these elements have been oxidized. The species which have caused this change (Cl_2, O_2, and Br_2) are ***oxidizing agents.*** Since these molecules have picked up electrons, they have been reduced, and the species which caused this change (Na, H_2, and Fe) are ***reducing agents.***

The gain of electrons is called *reduction.*

The term oxidation has its historical background in the fact that the first oxidizing agent whose chemical behavior was thoroughly studied was oxygen. The phenomenon was then named after the element. As the understanding of chemical reactions deepened, it became apparent that the reaction of a metal with oxygen was very similar to its reaction with fluorine, chlorine, or bromine. Thus, in each of the following reactions:

Oxidizing agents gain electrons at the expense of reducing agents.

$$4Fe + 3O_2 \longrightarrow 2Fe_2O_3$$
$$2Fe + 3F_2 \longrightarrow 2FeF_3$$
$$2Fe + 3Cl_2 \longrightarrow 2FeCl_3$$

the iron loses valence electrons to the other reactant. After this similarity was noted, the concept of oxidation was generalized to cover all situations in which an atom, ion, or molecule lost electrons. In the same manner the concept of reduction is now used to cover all situations in which an atom, ion, or molecule gains electrons.

The formation of a salt by the direct combination of the elements is an example of an oxidation-reduction reaction. When sodium reacts with chlorine to form sodium chloride, the following reactions take place.

$$Na \longrightarrow Na^+ + e^- \quad \text{(OXIDATION OF SODIUM)}$$
$$Cl_2 + 2e^- \longrightarrow 2Cl^- \quad \text{(REDUCTION OF CHLORINE)}$$

Here, chlorine is the oxidizing agent and gets reduced; sodium is the reducing agent and gets oxidized.

Since two electrons are gained per chlorine molecule, two sodium atoms must be oxidized for every Cl_2 molecule reduced; the overall chemical reaction is

$$2Na + Cl_2 \longrightarrow 2Na^+ + 2Cl^-$$

Reduction of Metals

One of the most practical applications of the oxidation-reduction principle is the winning of metals from their ores. The majority of metals are found in nature in compounds; that is, the metals are in an oxidized state. In order to obtain the metal, it must be reduced from the positive oxidation state to the neutral elemental state. This requires a gain of electrons. In addition to electricity there are a variety of chemicals which can supply the electrons for the reduction process. Some of the ways that metals can be reduced from their ores are given in Table 10–2.

The reduction of metals for making tools is as old as recorded history.

Electrolysis

Several metals are either won from their ores or purified by electrolysis as noted in Table 10–2. ***Electrolysis*** is the term used to designate a chemical reaction carried out by supplying electrical energy.

Figure 10–9 outlines the principal parts of an electrolysis apparatus. Electrical contact between the external circuit and the solution is ob-

TABLE 10–2 Some Ways Used to Win Metals from Their Ores by Reduction

Metal	Occurrence	A Reduction Process	Uses of Metal
Cu	Cu_2S, chalcocite	Air blown through melted ore $Cu_2S + O_2 \longrightarrow 2\,Cu + SO_2\uparrow$ $(Cu^+ + e^- \longrightarrow Cu)$	Electrical wiring, boilers, pipes, brass (Cu 85%, Zn), bronze (Cu 90%, Sn, Zn)
Na	NaCl, rock salt	Electrolysis of fused chloride $2NaCl \longrightarrow 2\,Na + Cl_2$ $(Na^+ + e^- \longrightarrow Na)$	Coolant in nuclear reactors, orange street lights
Mg	Mg^{2+}, seawater	Electrolysis of fused chloride $MgCl_2 \longrightarrow Mg + Cl_2$ $(Mg^{2+} + 2\,e^- \longrightarrow Mg)$	Light alloys such as duralumin (0.5% Mg, Al), Dowmetal H (90.7% Mg, Al, Zn, Mn), flares, some flash bulbs
Ca	$CaCO_3$, chalk, limestone, marble	Thermal reduction for very pure Ca $CaCO_3 + 2\,HCl \longrightarrow CaCl_2 + H_2O + CO_2$ $3\,CaCl_2 + 2\,Al \xrightarrow{heat} 3\,Ca + 2\,AlCl_3$ $(Ca^{2+} + 2\,e^- \longrightarrow Ca)$	Bearing metal alloys (0.7% Ca, Pb, Na, Li)
Al	$Al_2O_3 \cdot H_2O$, bauxite	Electrolysis in fused cryolite, Na_3AlF_6, at 800–900°C. $2\,Al_2O_3 \longrightarrow 4\,Al + 3\,O_2$ $(Al^{3+} + 3\,e^- \longrightarrow Al)$	Packaging, airplane parts, alloys, roofing, siding
Ag	Ag_2S, argentite	Reduction by a more active metal $Ag_2S + 4\,CN^- \rightleftharpoons 2\,Ag(CN)_2^- + S^{2-}$ $2\,Ag(CN)_2^- + Zn \longrightarrow Zn(CN)_4^{2-} + 2\,Ag\downarrow$ $(Ag^+ + e^- \longrightarrow Ag)$	Jewelry, electrical wiring
Zn	ZnS, zinc blende	Roasting in air with carbon $ZnS + 2\,O_2 + C \longrightarrow Zn + SO_2\uparrow + CO_2\uparrow$ $(Zn^{2+} + 2\,e^- \longrightarrow Zn)$	Galvanizing iron sheet
Hg	HgS, cinnabar	Roasting in air $HgS + O_2 \xrightarrow{heat} Hg\uparrow + SO_2\uparrow$ $(Hg^{2+} + 2\,e^- \longrightarrow Hg)$	Thermometers, electric switches, blue-green street lights, amalgams in dentistry, fluorescent lighting
Fe	Fe_2O_3, hematite	Reduction by carbon monoxide $Fe_2O_3 + 3\,CO \longrightarrow 3\,CO_2 + 2\,Fe$ $(Fe^{3+} + 3\,e^- \longrightarrow Fe)$	Alloyed with C (0.1 to 1.5%) to make steels: (stainless steel has 8% or more Cr)

tained by means of ***electrodes,*** which are often made of graphite or metal. The electrode at which electrons enter an electrolysis cell is termed the ***cathode,*** and this is the electrode at which reduction takes place. The electrode at which the electrons leave the cell is the ***anode.*** At the anode oxidation takes place. The battery produces a direct current of electrons which flow toward one electrode (the cathode) and away from the other electrode (the anode). A salt solution can easily be observed to conduct electricity if a light bulb is part of the circuit. When the switch is closed, the positive ions in solution migrate toward the cathode. Soon, evidence of a chemical reaction can be seen at the electrodes. Depending on the substances present in the solution, gases may be evolved, metals deposited, or ionic species changed at the electrodes. The ions that migrate to the electrodes are not necessarily the species undergoing reaction at the electrodes because sometimes the solvent undergoes reaction more easily. Whatever happens, the chemical reactions that are taking place at

-lysis **is a suffix meaning splitting or decomposition; electrolysis—decomposition by electricity.**

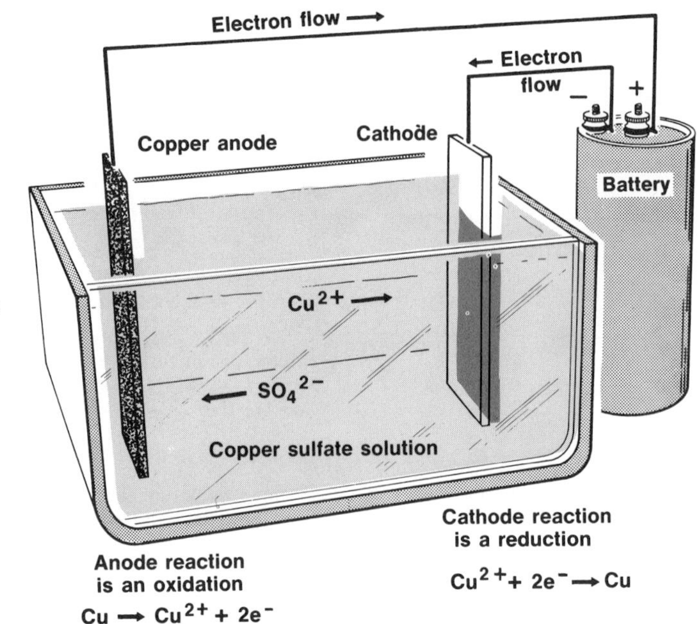

Figure 10–9 Electroplating from a copper sulfate solution.

the cathode and anode are due to electrons going into and coming out of the solution. The chemical reaction at the cathode is one that furnishes electrons to solution species (reduction). At the anode, electrons are taken from species in solution, so the chemical reaction at the anode is one that gives up electrons (oxidation).

Figure 10–9 illustrates the electroplating of copper. Such an electrolysis can be used either to plate an object with a layer of pure copper or to purify an impure sample of copper metal; copper metal is transferred from the positive electrode to the negative electrode. If the positive electrode is impure copper to be purified, electrolysis deposits the copper as very pure copper on the negative electrode.

Now let us examine how the electrolysis transfers the copper from the positive electrode to the negative electrode. Electrons flow out of the negative terminal of the battery, through the wire, and into the negative electrode. Somehow this negative charge must be used up at the surface of the electrode.

Consider what happens when the electrons build up on the negative electrode. Since the positive copper ions are nearby, they will be attracted to the surface and take electrons from it. Thus, the Cu^{2+} ions are reduced:

$$Cu^{2+} + 2e^- \longrightarrow Cu \qquad \text{(CATHODE REACTION)}$$

In a similar way the negative sulfate ions migrate to the positive electrode (anode). However, it turns out that it is easier to get electrons from the copper metal of the electrode than it is from the sulfate ions. As each copper atom gives up two electrons, it passes into solution:

$$Cu \longrightarrow Cu^{2+} + 2e^- \qquad \text{(ANODE REACTION)}$$

In effect, then, the copper of the positive electrode (anode) passes into

In any electrochemical cell, oxidation occurs at the anode; reduction occurs at the cathode.

Copper can be plated onto an object by making that object the electrode at which electrons enter a cell containing dissolved copper salts.

solution; the copper ions in solution migrate to the negative electrode (cathode) and plate out as copper metal. Large amounts of copper are purified in this way each year. Silver and gold purification can be done similarly.

If one desires to plate an object with copper, he has only to render the surface conducting and make it the negative electrode; it will become coated with copper, with the copper coating growing thicker as the electrolysis is continued. If the object is a metal, it will conduct electricity by itself. If it is a nonmetal, its surface can be lightly dusted with graphite powder to render it conducting.

A potentially very important electrolysis reaction is the electrolysis of water. When electricity is passed into graphite electrodes immersed into dilute salt solution, water is reduced to hydrogen gas and hydroxide ions at the cathode:

$$2H_2O + 2e^- \longrightarrow H_2\,(g) + 2OH^- \quad \text{(CATHODE REACTION)}$$

At the anode, water is oxidized to oxygen and hydrogen ions:

$$2H_2O \longrightarrow O_2 + 4H^+ + 4e^- \quad \text{(ANODE REACTION)}$$

The overall, or net cell reaction is:

$$2H_2O \xrightarrow{\text{Electricity}} 2H_2\,(g) + O_2\,(g)$$

The hydrogen gas produced by the reduction of water can be stored and used as a fuel—for example, to power rockets into space. Someday, if

Stamping masters for making high quality phonograph records are made by electroplating nickel onto a plastic record finely coated with silver. (Courtesy of GRT Record Pressing, Nashville.)

electricity becomes low enough in cost (see Chapter 25), water may be electrolyzed to produce hydrogen which can then be piped to the point of use, just like natural gas is today.

Relative Strengths of Oxidizing and Reducing Agents

When a piece of metallic zinc is placed in a solution containing blue copper ions (Cu^{2+}), an oxidation-reduction reaction occurs:

$$Zn + Cu^{2+} \longrightarrow Zn^{2+} + Cu$$

Evidence for this reaction is the deposit of copper on the zinc and the gradual decrease in the intensity of the blue color of the solution indicating that the Cu^{2+} ions are being removed.

The oxidation of zinc by copper ions can be thought of as a competition between zinc ions (Zn^{2+}) and copper ions (Cu^{2+}) for the two electrons. Since the reaction proceeds almost to completion, the Cu^{2+} ions obviously win out in the competition.

The ***activity*** of a metal is a measure of its tendency to lose electrons. Zinc is a more active metal than copper on the basis of this experiment. This means that given an equal opportunity, the first reaction will take place to a greater extent:

A copper atom has a greater attraction for the valence electrons than does the zinc atom.

$$Zn \longrightarrow Zn^{2+} + 2e^-$$
$$Cu \longrightarrow Cu^{2+} + 2e^-$$

Experiments of this type with various pairs of metals and other reducing agents yield an ***activity series*** of the elements which ranks each oxidizing and reducing agent according to its *strength* or *tendency* for the electron transfer to take place. An iron nail will be partly dissolved in a solution of a copper salt containing Cu^{2+} ions, with copper being deposited on the nail that remains. From this, it is determined that iron, like zinc, is more active than copper. The reaction that takes place is:

The active metals lose electrons more easily; hence, these free metals are not found in nature.

$$Fe + Cu^{2+} \longrightarrow Fe^{2+} + Cu$$

Now, which is more active, zinc or iron? This question can be answered by placing an iron nail in a solution containing Zn^{2+} ions and, in a separate container, a strip of zinc in a solution containing Fe^{2+} ions. It is found that the zinc strip is eaten away in the solution containing Fe^{2+} ions. The reaction is

$$Zn + Fe^{2+} \longrightarrow Fe + Zn^{2+}$$

Nothing happens to the iron nail in the solution of Zn^{2+} ions. We deduce that Zn loses electrons more readily than Fe.

Such an activity series can be extended to include other metals and even nonmetals as well. The concentrations of the ions in solution and other factors often must be considered for accurate work, but for our purposes these will be ignored. Table 10–3 gives an activity series of some oxidizing and reducing agents.

Activity: $Zn > Fe > Cu$

An application of the activity series can be seen in the use of

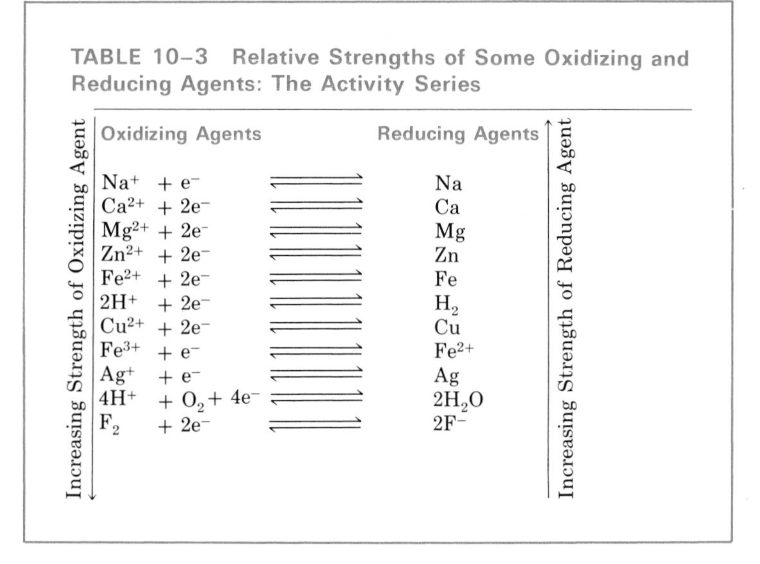

TABLE 10–3 Relative Strengths of Some Oxidizing and Reducing Agents: The Activity Series

Oxidizing Agents			Reducing Agents
Na^+	$+ e^-$	$\rightleftharpoons$	Na
Ca^{2+}	$+ 2e^-$	$\rightleftharpoons$	Ca
Mg^{2+}	$+ 2e^-$	$\rightleftharpoons$	Mg
Zn^{2+}	$+ 2e^-$	$\rightleftharpoons$	Zn
Fe^{2+}	$+ 2e^-$	$\rightleftharpoons$	Fe
$2H^+$	$+ 2e^-$	$\rightleftharpoons$	H_2
Cu^{2+}	$+ 2e^-$	$\rightleftharpoons$	Cu
Fe^{3+}	$+ e^-$	$\rightleftharpoons$	Fe^{2+}
Ag^+	$+ e^-$	$\rightleftharpoons$	Ag
$4H^+$	$+ O_2 + 4e^-$	$\rightleftharpoons$	$2H_2O$
F_2	$+ 2e^-$	$\rightleftharpoons$	$2F^-$

Increasing Strength of Oxidizing Agent ↓

Increasing Strength of Reducing Agent ↑

cathodic protection to reduce corrosion. For this, the metal to be protected, say an iron pipeline, is connected to a more active metal, such as magnesium. In moist earth, an electrolytic cell is set up in which the magnesium is oxidized and transfers electrons to the iron. This keeps the iron from being oxidized by the air or any other reagent which might otherwise remove electrons from it.

Corrosion is discussed in detail in Chapter 12.

The value of the activity series is that it allows us to predict the feasibility of reactions involving the species in it. Thus, a reducing agent in the table (right-hand column) is able to reduce the oxidized form of any species below it. For example, magnesium can reduce Cu^{2+} to Cu:

$$Mg + Cu^{2+} \longrightarrow Cu + Mg^{2+}$$

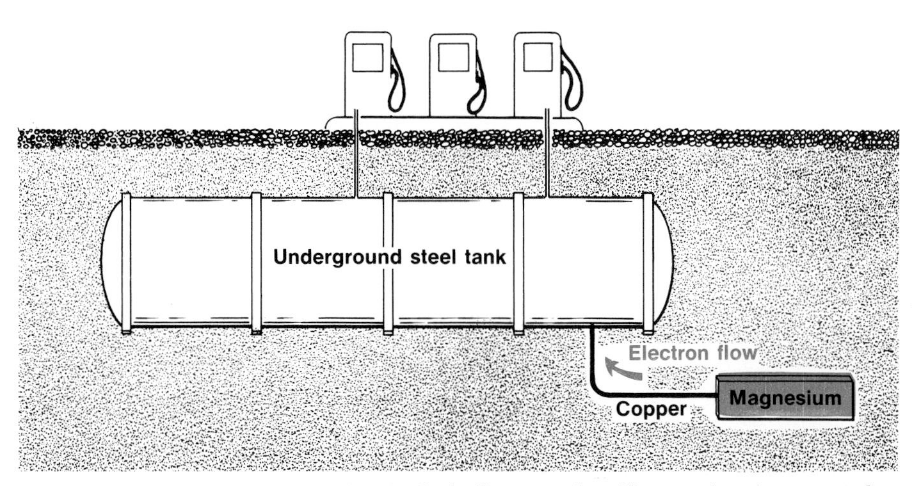

Figure 10–10 Cathodic protection. If magnesium is connected to the steel tank to be protected, the magnesium is more easily oxidized than the iron or copper connecting wire. The magnesium serves as the anode. Hence, the cathode is protected with no points of oxidation occurring on its surface. The anode is the electrode where oxidation occurs; reduction occurs at the cathode. When the magnesium is used up, it is replaced by another block. The replacement is much easier and cheaper than replacing the tank.

Magnesium can also reduce Ag^+ to Ag:

$$Mg + 2Ag^+ \longrightarrow Mg^{2+} + 2Ag$$

Zinc can also reduce silver ions:

$$Zn + 2Ag^+ \longrightarrow 2Ag + Zn^{2+}$$

Zinc cannot reduce calcium ions, since calcium is above zinc in the series:

$$Zn + Ca^{2+} \longrightarrow \textit{No reaction}$$

The series arranges oxidizing agents in the order of their effectiveness also. Fluorine, F_2, can oxidize water, silver, ferrous ion, or any species above it in the series; cupric ion, Cu^{2+}, can oxidize H_2, Fe, Zn, and the metals above Cu in the table, because it has a greater tendency to take on electrons than the ions that are formed. Thus:

The activity series is useful in predicting many chemical reactions.

$$Cu^{2+} + Fe \longrightarrow Fe^{2+} + Cu$$
$$Cu^{2+} + Mg \longrightarrow Mg^{2+} + Cu$$
$$Cu^{2+} + Ag \longrightarrow \textit{No reaction}$$

SELF-TEST 10-C

1. In a redox reaction, the oxidizing agent (gains/loses) electrons and gets (reduced/oxidized); the reducing agent (gains/loses) electrons and gets (reduced/oxidized).

2. Consider the reaction:

$$HgS + O_2 \longrightarrow Hg + SO_2$$

Oxygen is the (oxidizing/reducing) agent. Mercuric sulfide (HgS) is the (oxidizing/reducing) agent.

3. When silver-plated dinnerware is made, silver is deposited by the reaction $[Ag(CN)_2]^- + e^- \longrightarrow Ag + 2CN^-$. The silver is (oxidized/reduced). This reaction takes place at the (cathode/anode) of an electrolysis cell.

4. Considering the activity series in Table 10–3, write yes or no beside the following reactions, depending on whether or not they will proceed as written.

$2Ag + Cu^{2+} \longrightarrow Cu + 2Ag^+$ ______

$F_2 + Zn \longrightarrow Zn^{2+} + 2F^-$ ______

$Fe^{2+} + Ag^+ \longrightarrow Fe^{3+} + Ag$ ______

$Mg^{2+} + 2Fe^{2+} \longrightarrow 2Fe^{3+} + Mg$ ______

5. When electrons are written on the left-hand side of an equation such as in $Cu^{2+} + 2e^- \longrightarrow Cu$, the reaction is a(n) (oxidation/reduction).

6. When electrons are written on the right-hand side of an equation such as in $Zn \longrightarrow Zn^{2+} + 2e^-$, the reaction is a(n) (oxidation/reduction).

Batteries

One of the most useful applications of oxidation-reduction reactions is in the production of electrical energy. A device that produces an electron flow (current) is called an ***electrochemical cell.*** Although a series of cells is a ***battery,*** we shall use the term battery, exclusively. Consider the reaction between zinc and copper ions that was previously discussed. If zinc is placed in a solution containing Cu^{2+} ions, the electron transfer takes place between the zinc metal and the copper ions, and the energy liberated simply causes a slight heating of the solution and the zinc strip.

If the zinc can be separated from the copper solution, and the two connected in such a way to allow current flow, the reaction proceeds, but now the electrons are transferred through the connecting wires. Figure 10–11 shows a battery that can be constructed to make use of the energy evolved in the reaction of Zn with Cu^{2+}.

The anode reaction is the oxidation of zinc to Zn^{2+} ions.

$$Zn \longrightarrow Zn^{2+} + 2e^-$$

The electrons flow from the Zn electrode through the connecting wire, light the lamp in the circuit, and then flow into the copper cathode where reduction of Cu^{2+} ions takes place:

$$Cu^{2+} + 2e^- \longrightarrow Cu$$

Historic battery used in the early telegraph.

The copper is deposited on the copper cathode.

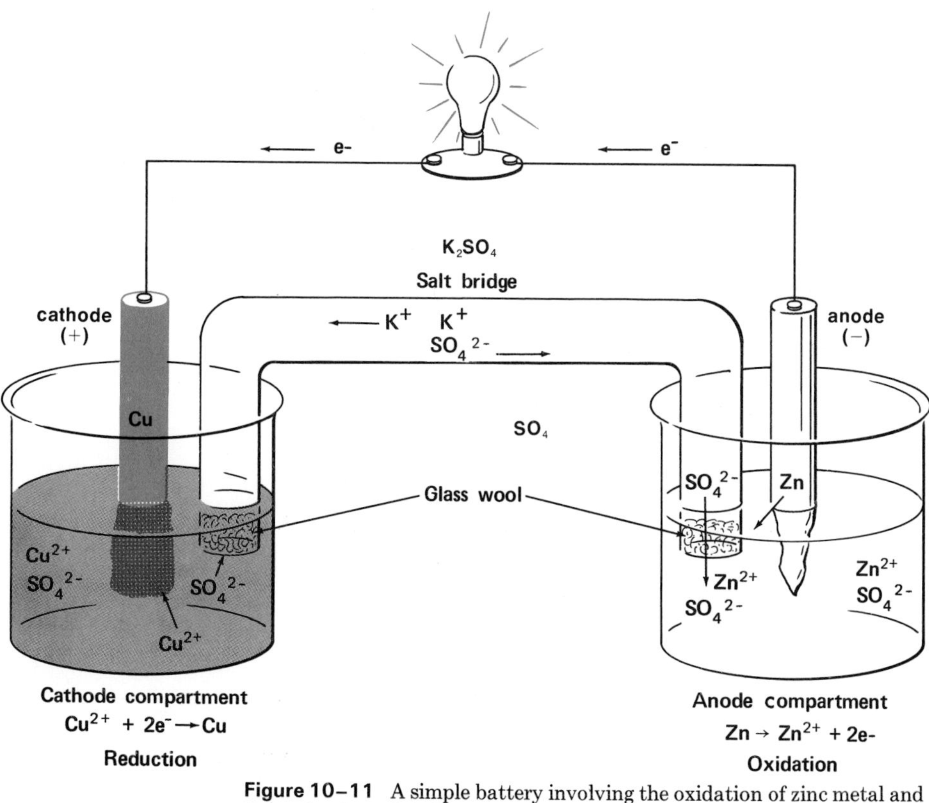

Figure 10–11 A simple battery involving the oxidation of zinc metal and the reduction of Cu^{2+} ions.

TABLE 10–4 Characteristics of Some Batteries

System	Anode (Oxidation)	Cathode (Reduction)	Electrolyte	Typical Operating Voltage Per Cell
Dry cell	Zn	MnO_2	NH_4Cl-$ZnCl_2$	0.9–1.4 volts
Edison storage	Fe	Ni oxides	KOH	1.2–1.4
Nickel-Cadmium—NiCad	Cd	Ni oxides	KOH	1.1–1.3
Silver cell	Cd	Ag_2O	KOH	1.0–1.1
Lead storage	Pb	PbO_2	H_2SO_4	1.95–2.05
Mercury cell	Zn(Hg)	HgO	KOH—ZnO	1.30
Alkaline cell	Zn(Hg)	MnO_2	KOH	0.9–1.2

This flow of electrons (negative charge) from the anode to the cathode compartment in the battery must be neutralized electrically. A "salt bridge" is provided which connects the two compartments. Negative ions (SO_4^{2-}) flow through this bridge readily since it is constructed of a salt solution such as K_2SO_4 in a glass tube with porous plugs at either end. During the operation of the battery, a number of negative charges just equal to the number of electrons used at the cathode will pass through the bridge to the Zn^{2+} solution. Likewise, the copper solution is kept neutral by negative SO_4^{2-} ions flowing through the salt bridge to the right and positive K^+ ions flowing into the left solution. The reaction of zinc with Cu^{2+} continues until the battery runs down; that is, until equilibrium is attained.

In commercial batteries, the salt bridge is often replaced by a porous membrane.

Many different oxidation-reduction combinations are used in a suitable battery to produce a flow of electrons. A few of the more popular ones are listed in Table 10–4.

MATCHING SET

______	1. chemicals in automobile battery	a. 7
______	2. most active metal in Table 10–3	b. gain of electrons
______	3. ore of mercury	c. proton donor
______	4. pH of pure water	d. Na
______	5. $[H^+] = 1$ mole/liter	e. proton acceptor
______	6. strong acid	f. pH = 0
______	7. oxidation	g. $HC_2H_3O_2$
______	8. reduction	h. Pb, PbO_2, H_2SO_4
______	9. acid	i. causes solution to conduct
______	10. base	j. loss of electrons
______	11. weak acid	k. cinnabar
______	12. electrolyte	l. H_2SO_4

QUESTIONS

1. Define acid-base reactions in terms of protons, and oxidation-reduction reactions in terms of electrons.

2. Indicate the solute and solvent in (a) coffee, (b) a 5 percent alcohol in water solution, (c) a 5 percent water in alcohol solution, and (d) a solution of 50 percent alcohol and 50 percent water.

3. What is the one test that all aqueous electrolytes must pass?

4. Give an example of ionic dissociation. Give an example of ionization. What is the difference between the two?

5. What is the difference between the hydrogen ion and the hydronium ion?

6. Write the equation for a chemical reaction in which water acts as a Brønsted acid; as a Brønsted base.

7. A solution of hydrochloric acid is electrolyzed. The products are hydrogen at the cathode and chlorine at the anode. Write the reactions occurring at each electrode and tell which ions move toward the cathode and which toward the anode.

8. Classify each of the following as acids or bases, using the Brønsted-Lowry definitions: H_2SO_4, CO_3^{2-}, Cl^-, HCO_3^-, O^{2-}, H_2O.

9. Predict the formulas of salts formed between the following pairs of ions:

$$Na^+ \text{ and } SO_4^{2-}$$
$$Ca^{2+} \text{ and } I^-$$
$$Mg^{2+} \text{ and } NO_3^-$$
$$Ca^{2+} \text{ and } PO_4^{3-}$$
$$K^+ \text{ and } Br^-$$

10. Moist baking soda is often put on acid burns. Why? Write an equation for the reaction assuming the acid to be hydrochloric (HCl).

11. Using the table of relative strengths of acids and bases, predict whether the products or reactants would be favored in the following reactions:

$$HClO_4 + C_2H_3O_2^- \rightleftharpoons HC_2H_3O_2 + ClO_4^-$$
$$HF + OH^- \rightleftharpoons H_2O + F^-$$
$$O^{2-} + HNO_3 \rightleftharpoons OH^- + NO_3^-$$

12. Using the table giving relative oxidizing and reducing strengths, predict whether or not the following reactions would be expected to proceed to the right.

$$Ag + Na^+ \rightleftharpoons Ag^+ + Na$$
$$H_2 + Cu^{2+} \rightleftharpoons Cu + 2H^+$$
$$F_2 + 2Fe^{2+} \rightleftharpoons 2Fe^{3+} + 2F^-$$
$$Mg + 2Ag^+ \rightleftharpoons 2Ag + Mg^{2+}$$

13. A certain salt, MX, is found to raise the temperature when dissolved in water. Which is the larger quantity, the solvation energy or the lattice energy? Explain.

14. Describe what happens when an ionic solid dissolves in water.

15. What ions are present in water solutions of the following salts: Na_2SO_4, $CaBr_2$, $Mg(NO_3)_2$?

PROBLEMS

1. NaCl has a formula weight of 58.5. If 58.5 g of this salt is dissolved in enough water to make a liter of solution, what is the molarity of the solution? What would be the molarity if 2 liters of solution were made with the same amount of salt? 0.5 liter?

   Ans. 1 molar, 0.5 molar, 2 molar

2. What is the molar concentration of a solution prepared by dissolving 90 grams of acetic acid, CH_3COOH, in a liter of solution?

   Ans. 1.5 molar

3. What is the pH of a 0.001 M solution of HCl?

   Ans. Approx. 3

4. What is the hydrogen ion concentration in a solution with a pH of 8?

   Ans. Approx. 10^{-8}M

5. What is the hydroxide ion concentration in a solution with a pH of 5?

   Ans. Approx. 10^{-5}M

6. Analysis of 20 ml of a solution shows that it contains 1.0 gram of NaCl. What is the molarity of this solution?

   Ans. 1.25 M

7. A solution is prepared by mixing 500 ml of 2 M NaOH with 500 ml of 2 M HCl. What is its pH?

   Ans. Approx. 7

8. Show by calculation that pure water is 55.6 M.

   Ans. $\frac{1000\text{ g}}{\text{liter}} \bigg/ \frac{18\text{ g}}{\text{mole}} = \frac{55.6\text{ mole}}{\text{liter}}$

9. How many grams of salt will be required to make each of the following solutions?

   a. 20 ml of 0.1 M KCl
   b. 250 ml of 0.3 M $AgNO_3$
   c. 2 l of 4.0 M Na_2CO_3
   d. 400 ml of 0.23 M $Fe(NO_3)_3 \cdot 9H_2O$

   Ans. (a) 0.149 g KCl; (b) 12.74 g $AgNO_3$; (c) 848 g Na_2CO_3; (d) 32.72 g $Fe(NO_3)_3 \cdot 9H_2O$

10. A saturated solution of KCl, at 20°C, contains 347 grams of KCl per liter. What is the molarity of a solution which is prepared when 100 ml of this concentrated solution is taken and diluted up to one liter with distilled water?

    Ans. 0.046 M in KCl

11. What is the molar concentration of water in a solution which contains 48 grams of water in sufficient ethyl alcohol to bring the total volume up to one liter?

    Ans. 2.67 M in H_2O

12. When water trickles through limestone ($CaCO_3$), the water dissolves some of it. Much drinking water is obtained from such sources. If you drink 4 liters of water a day, all of which is saturated with calcium carbonate (0.00153 gram of $CaCO_3$ per 100 ml at 25°C), how many moles of calcium carbonate will you ingest from this source in the course of a week? If the recommended daily adult intake of calcium is 800 mg per day (as Ca^{2+}), is it feasible to satisfy your calcium requirement by drinking such water?

    Ans. 0.0043 mole $CaCO_3$ per week. This is only 24 mg of Ca^{2+} per week. No.

13. A 0.01 M solution of sodium hydroxide is combined with an equal volume of 0.005 M hydrochloric acid. What is the pH of the resultant solution?

    Ans. If 1 liter of each solution mixed, the pH would be 11.4.

SUGGESTIONS FOR FURTHER READING

Eyring, H., and MuShik, J., "Significant Structure Theory of Water," *Chemistry,* Vol. 39, No. 9, p. 8 (1966).
Haggin, J., "New Method of Iron Production," *Chemistry,* Vol. 39, No. 5, p. 24 (1966).
Manufacturing Chemists' Association, "Ammonia, Industry Profile," *Chemistry,* Vol. 39, No. 6, p. 14 (1966).
Morris, D. L., "Brønsted-Lowry Acid-Base Theory—A Brief Survey," *Chemistry,* Vol. 43, No. 3, p. 18 (1970).
Newman, D. S., "Fused Salts," *Chemistry,* Vol. 39, No. 4, p. 7 (1966).
Schmuckler, J. S., and Mogue, P. H.: "Dilute Solutions of Strong Acids: The Effect of Water on pH," *Chemistry,* Vol. 42, No. 9, p. 14 (1969).
Steele, D., "The Chemistry of the Metallic Elements," Pergamon Press, New York, 1966.

CHAPTER 11
SCIENCE AND TECHNOLOGY AS A NEW PHILOSOPHY

A PHILOSOPHICAL AND HISTORICAL BACKGROUND

The various factors which led to the systematic development of science and its associated technologies in the Western World are both numerous and complex. Rather than attempt to develop an overall "complete" view, we will discuss only a few of the empirical results. One of the key points seems to be the development of *printing techniques* which enormously increased the availability of the accumulated knowledge of mankind (see Figure 11–1). Another seems to be the continuous *refusal to accept authority as the ultimate judge of truth* in certain areas; this indeed still seems to be one of the major differences between mathematics and the sciences and other areas of human knowledge. No one ever states, as proof of the validity of a piece of scientific information, that so-and-so said this and therefore it must be true. This skeptical attitude is to be found throughout all nations at all times; however, it was only in Western Europe that such a frame of mind came to be cultivated systematically and ultimately formed the basis of certain intellectual organizations whose primary purpose was the search for and propagation of a kind of objective knowledge.

Authority is not the ultimate judge of truth in the sciences.

The most important aspects of this type of knowledge are the ways in which it changes. With the passage of time, it becomes *more extensive, more accurate, more broadly disseminated, and,* most important of all, *more concisely summarized* in terms of generalizations or their equivalent: mathematical equations giving the quantitative relationships connecting a set of properties.

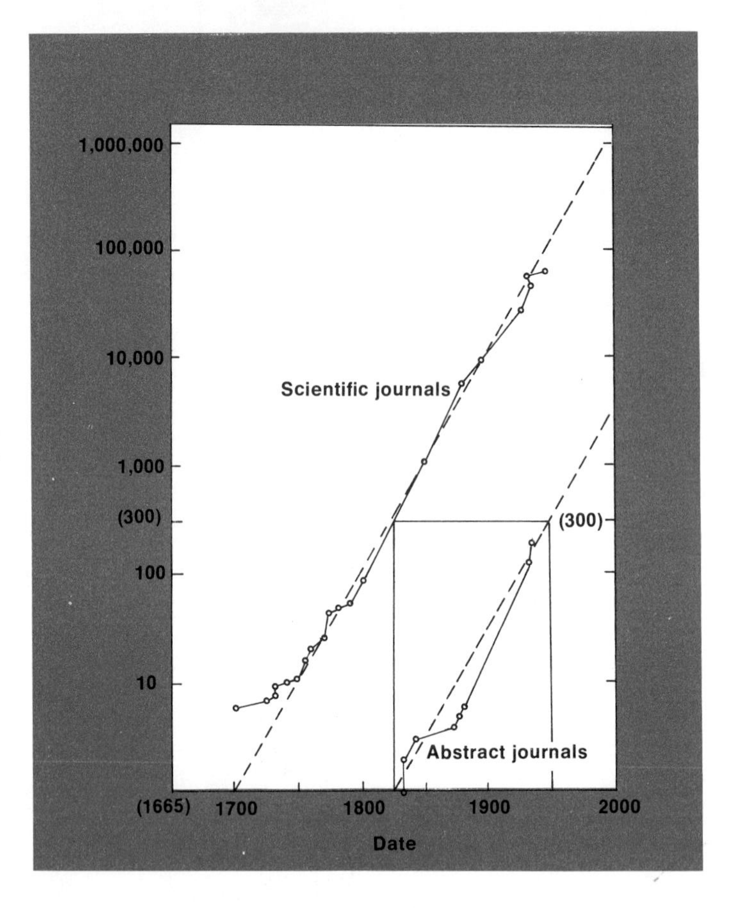

Figure 11–1 Rate of increase in number of scientific journals since 1665. (Reproduced with permission of Yale University Press.)

Because the physical sciences deal with our environment, this growth of knowledge leads to an ever increasing understanding of the physical world and the ways it can be manipulated to obtain desired ends. This manipulation of our environment can attain a high degree of sophistication. Presently, modern ***technology*** has two types of roots. The one derives from the practical knowledge of craftsmen and frequently has been transmitted orally or by apprenticeships. The other is in the sciences and has been transmitted by publications of various sorts (Figure 11–1) and formal education. In practice the dual roots tend to fuse and reinforce each other. The older method of oral transmission had its ups and downs, as is especially obvious in certain areas where knowledge of techniques has died out (e.g., the early Middle Ages in Europe as contrasted with the Roman Empire). The newer method of transmission by publication tends to make such knowledge available to all who can read and who have the required training for understanding (see Figure 11–2).

Technology: the manipulation of our environment.

As time has passed, the effectiveness of this new method of collecting and extending knowledge has become more apparent to the ordinary citizen. It has also become accepted by scholars in more disciplines and has led to the extension of this method to new areas of knowledge which previously had not been the subject of systematic study.

Mankind is in the midst of a knowledge explosion.

At this point one might well ask for a definition of ***science*** or the scientific method. While it is difficult to give such definitions which are acceptable to all, a close approximation may be achieved. First of all, "science" consists of that which has been collected by the use of the ***scientific method.*** The term "scientific method" refers more to a mental

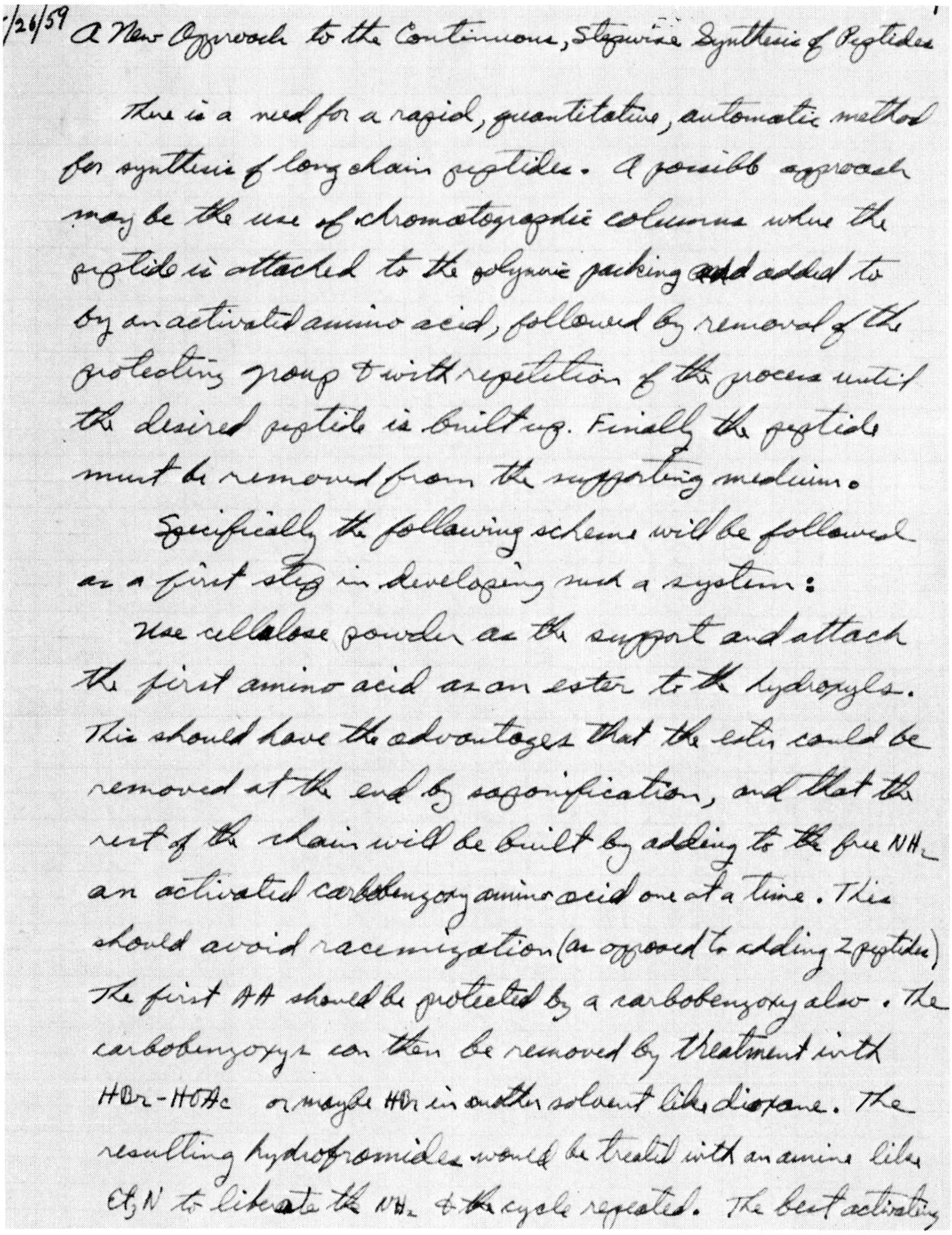

/26/59 A New Approach to the Continuous, Stepwise Synthesis of Peptides

There is a need for a rapid, quantitative, automatic method for synthesis of long chain peptides. A possible approach may be the use of chromatographic columns where the peptide is attached to the polymeric packing and added to by an activated amino acid, followed by removal of the protecting group & with repetition of the process until the desired peptide is built up. Finally the peptide must be removed from the supporting medium.

Specifically the following scheme will be followed as a first step in developing such a system:

Use cellulose powder as the support and attach the first amino acid as an ester to the hydroxyls. This should have the advantages that the ester could be removed at the end by saponification, and that the rest of the chain will be built by adding to the free NH_2 an activated carbobenzoxy amino acid one at a time. This should avoid racemization (as opposed to adding 2 peptides) The first AA should be protected by a carbobenzoxy also. The carbobenzoxys can then be removed by treatment with HBr-HOAc or maybe HBr in another solvent like dioxane. The resulting hydrobromides would be treated with an amine like Et_3N to liberate the NH_2 & the cycle repeated. The best activating

Figure 11–2 A page from the notes of Dr. Bruce Merrifield (Professor of Biochemistry at Rockefeller University), the designer of a protein-making machine. Progress in science is made through disciplined human thought, taking advantage of previous knowledge and useful theories. See Figure 16–27. (From Chemical and Engineering News, August 2, 1971.)

attitude than to a procedure. The attitude is one which requires the strictest intellectual honesty in the collection of data and in the arrangement of this data in a pattern that reveals the underlying basis of the *observed* behavior. The data normally must be collected under conditions which can be reproduced anywhere in the world, so that new data can be obtained to confirm or to refute the correctness of the

Figure 11–3 Name some discoveries that led to this change in transportation. How many of them are chemical in nature?

suggested pattern (geologists and astronomers are excused from these rigid requirements). The results obtained are thus independent of differences in language, culture, religion, or economic status in the various observers and represent a unique type of truth.

Over the last 200 years this accumulated knowledge has been put to use on a very extensive scale in Europe and in those areas of the world which have had the means to follow the example set by England, which underwent this "industrial revolution" first (see Figure 11–3). The result has been the development of a society largely dependent upon and supported by a technology which is itself undergoing constant change. The first consequence of this technology has been to increase the *rate* at which things can be produced. This in turn has continually changed the occupational patterns of millions of human beings, and has brought forcefully to mind the persistence of *change* in our pattern of life.

Technology increases the rate at which things can be produced.

These changes have influenced profoundly the way in which people think about their material wants and the ways in which they can be satisfied. For example, there seems to be little argument with the statement, "If the number of human beings on the earth could be stabilized, a much higher standard of living could prevail over most of the earth." A statement such as this would have been greeted with widespread derision 500 years ago. While depending on technology, people today are beginning to doubt its ability to solve both personal and social problems on a long-range basis. It is obvious that confusion exists on this point since the cries about the curses of technology come from people who are highly dependent on it and who are constantly asking for even more help from technology.

TECHNOLOGY: ITS TRIUMPHS AND PROBLEMS

Almost as soon as the industrial revolution began in England, the public realized that technological progress brought with it a series of problems. The first to be noticed was the necessity for progress to be accompanied by changing patterns of employment.

It is obvious that if a machine makes as much thread as 100 men can make, the men are released to do other work. The 100 men, however, do not look on this as an advantage, especially if they are settled in their

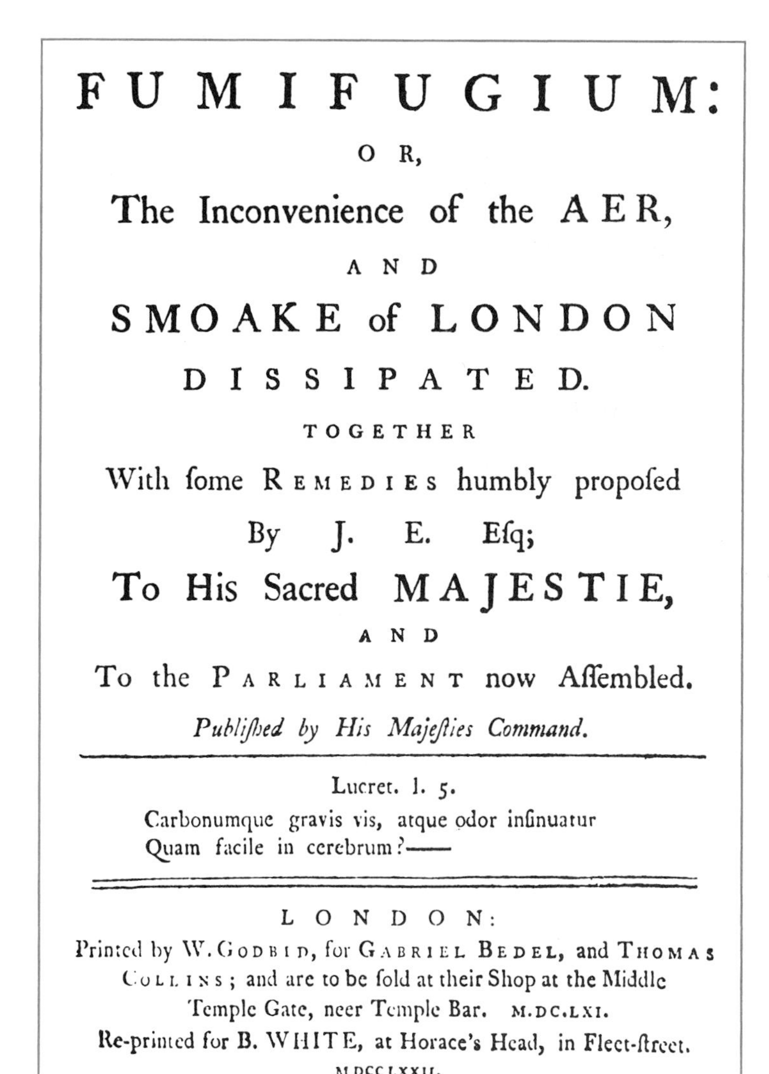
FUMIFUGIUM:

OR,

The Inconvenience of the AER,

AND

SMOAKE of LONDON

DISSIPATED.

TOGETHER

With ſome REMEDIES humbly propoſed

By J. E. Eſq;

To His Sacred MAJESTIE,

AND

To the PARLIAMENT now Aſſembled.

Publiſhed by His Majeſties Command.

Lucret. l. 5.

Carbonumque gravis vis, atque odor inſinuatur
Quam facile in cerebrum?——

LONDON:

Printed by W. GODBID, for GABRIEL BEDEL, and THOMAS COLLINS; and are to be ſold at their Shop at the Middle Temple Gate, neer Temple Bar. M.DC.LXI.

Re-printed for B. WHITE, at Horace's Head, in Fleet-ſtreet. MDCCLXXII.

Figure 11–4 Title page from J. Evelyn, F.R.S., *The Smoake of London.* The Latin quotation is from the Roman poet Lucretius (97–53 B.C.). It may be translated, "How easily the heavy potency of carbons and odors sneaks into the brain!" (Courtesy A. E. Gunther and the University Press, Oxford.)

Technological unemployment results when men are displaced from their jobs by machines.

place of employment with their families. The new opportunities that result from such a machine are rarely of benefit directly to the men displaced. The wealth of their country is increased since there are now 100 men able to do other work. However, the initial reaction of the men in 18th century England was to riot and break up the machinery.

The increased use of fuels of all sorts, especially the introduction of coal and coke into metallurgical plants, led to widespread problems with air pollution which were recognized and discussed over 200 years ago.

The most important point of these results is the realization that technological progress is always obtained at some cost, and the cost may not be obvious at the outset.

A very important technological development was recognized as necessary in 1890 by Sir William Crooks, who addressed the British Association for the Advancement of Science on the problem of the fixed nitrogen supply (that is, nitrogen in a chemical form which plants are capable of using in growth). At that time, scientists recognized that nitrogen compounds were necessary in fertilizers and that the future food supply of mankind would be determined by the amount of nitrogen compounds which could be made available for this purpose. The source of these nitrogen supplies was then limited to rapidly depleting supplies of guano (bird droppings) in Peru and to sodium nitrate in Chile. It was realized that, when these were exhausted, widespread famine would result unless an alternative supply could be developed. This problem was recognized first by English scientists as a potentially acute one because by the 1890's England had become very dependent upon imported food supplies.

The amount of food which can be grown is related to the nitrogen content of the soil.

Widespread interest in this problem led to research on a number of chemical reactions by which nitrogen can be obtained from the relatively inexhaustible supply present in the air. Air is 21 percent oxygen (O_2) and 79 percent nitrogen (N_2). The nitrogen in the air is present as the rather unreactive molecule N_2, and in this form it can be used as a source of other nitrogen compounds only by a few kinds of bacteria. Some is also transformed into NO by lightning, and when this is washed into the soil by rain it can be utilized by plants. Needless to say, the amount of nitrogen transformed into chemical compounds useful to plants by these processes is quite limited and cannot be increased easily.

The N_2 molecule is rather unreactive because of its strong triple bond ($N{\equiv}N$).

Several chemical reactions were developed to form useful compounds from atmospheric nitrogen, but the best known and most widely used one has an ironic history. While England was interested in nitrogen for fertilizers, Germany was interested in nitrogen for explosives. The German General Staff realized that the British Navy could blockade German ports and cut them off from the sources of nitrogen compounds in South America. As a consequence, when a German chemist named Fritz Haber showed the potential of an industrial process in which nitrogen reacts with hydrogen in the presence of a suitable catalyst to form ammonia, the German General Staff was quite interested and furnished support through the German chemical industry for the study of the reaction and the development of industrial plants based on it. The first such plant was in operation by 1911, and by 1914 such plants were being built very rapidly by Germany.

The Haber process transforms the nitrogen of the air into ammonia.

$N_2(g) + 3H_2(g) \rightleftharpoons 2NH_3(g)$

When the First World War broke out in August, 1914, many people

Figure 11–5 An early German ammonia plant. (From "The Realm of Chemistry," Econ-Verlag Gmb H, Düsseldorf, 1965.)

Figure 11-6 Mankind usually has the capability of making very large quantities of materials that possess useful properties. Sometimes mass production precedes a complete understanding of the consequences. In this industrial process, ribbons of soap are produced and are subsequently converted into flakes or bars. (Courtesy of the Proctor & Gamble Company.)

thought that a shortage of explosives based on nitrogen compounds would force the war to end within a year. Unfortunately, by this time the nitrogen-fixing industry in Germany was capable of supplying the needed compounds in large amounts. This process thus prolonged the war considerably and resulted in an enormous increase in mortality. Subsequently, the ammonia process has been used on a huge scale to prepare fertilizers and now is largely responsible for the fact that the earth can support a population of four billion (see Figure 11-7). Ammonia production by this process exceeds 40,000 tons per day in the United States alone.

Understanding nature does not necessarily lead to a better world.

The same type of problem seems to arise from the development of many technological processes. The control of nuclear energy brings with it the ability to make nuclear explosives. The development of rapid and convenient means of transportation such as the automobile and the airplane also brings forth new weapons of war and air pollution problems. Man, however, must learn to control his technology in such a manner as to maximize its benefits and minimize its disadvantages. These problems arise with *all* technological developments, even the most primitive. The discovery of the techniques necessary to the manufacture of iron led first to the development of new weapons (swords) by their discoverers, the Hittites, who then proceeded to conquer their neighbors and lead the first successful invasion of Egypt (ca. 1550 B.C.).

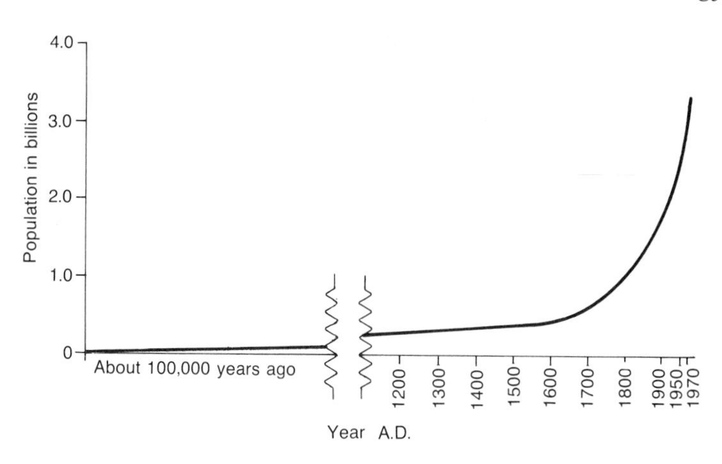

Figure 11-7 World population size. (From Turk, A., Turk, J., and Wittes, J. T.: *Ecology, Pollution and Environment.* Philadelphia, W. B. Saunders Co., 1972.)

TECHNOLOGY AND THE HUMAN ENVIRONMENT

The growth in large scale technology has a very large number of effects, both direct and indirect, on the human environment. The examination of a few of these shows just how complex these consequences can be.

An obvious case is the development of atomic energy. When the incredibly large amounts of energy released by nuclear reactions were first recognized, the development of such energy sources was placed on a top priority basis. After some nuclear reactors had been built and actually placed in operation, it was evident that their operation was accompanied by some serious potential risks to their human users. The first was the danger of some potential disaster such as the explosion of a boiler, with the consequent dispersal of radioactive material. The second danger was in the generation of radioactive products as the nuclear reaction proceeded. The uranium used in such reactors was transformed into a wide variety of fission products which made the operation of the pile more difficult as time went on. The repurification of the uranium could be accomplished chemically, but what was to be done with the radioactive wastes generated in the process? Obviously

Technology usually drastically alters the environment.

Figure 11–8 Land burial trench at the Oak Ridge National Laboratory reservation. Each day's accumulation of waste containers is buried by 3 or more feet of earth. (From *Radioactive Wastes.* Courtesy· of the U.S. Atomic Energy Commission, Washington, D.C.)

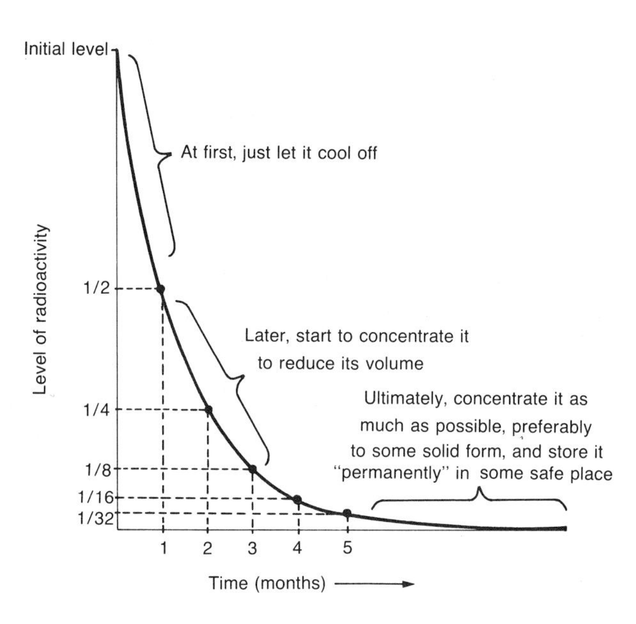

Figure 11–9 Disposal of radioactive wastes, for a hypothetical waste product with a one-month half-life. (From Turk, A., Turk, J., and Wittes, J.: *Ecology, Pollution and Environment.* Philadelphia, W. B. Saunders Co., 1972.)

Nuclear wastes constitute an unsolved problem. See Chapters 24 and 25.

they could not be dumped into a sewer since many of them have relatively long half-lives. Present practice calls for solid radioactive wastes to be placed in vaults and buried at carefully chosen sites (Figure 11-8). Liquid wastes are placed in underground storage tanks. The problem which faces us here is obtaining the benefits of the technology with minimum disruption of our own environment. The level of radioactivity of such wastes as a function of time is shown in Figure 11-9.

The same type of problem arises whenever we introduce a specific chemical compound into our environment to accomplish one single thing. The compound is often capable of a variety of actions and can lead to consequences undreamed of by those who introduce it. Many such examples can be found in the fields of drugs and insecticides. Thalidomide was designed to be an effective tranquilizer, and for this purpose it is an unqualified success. Unfortunately, when taken by pregnant women, it often prevents the normal development of the children they bear.

Technological procedures have both obvious and not so obvious consequences.

Man must always combat insects in his struggle for a food supply (Table 11–1), and the most effective aid he now receives in this struggle

TABLE 11–1 Estimated Annual Crop Loss Due to Pests and Disease (Percent of Total U.S. Crop Destroyed)*

Crop	Insects	Weeds	Disease
Corn	12	10	12
Rice	4	17	7
Wheat	6	12	14
Potatoes	14	3	19
Cotton	19	8	12

*From Scientific Aspects of Pest Control, National Academy of Science Publication No. 1402.

is from insecticides. Unfortunately, many insecticides are rather unspecific poisons.

Arsenic compounds were used on a very large scale as insecticides, but a realization of their great toxicity for man has led to their replacement by other compounds equally effective but usually less dangerous to human beings. However, some of these compounds which are most effective in protecting crops are nevertheless toxic to humans and are even capable of causing death when improperly handled or ingested. Even those insecticides which are not obviously harmful to man often have side effects which make them unattractive. DDT, for example, is very effective for the control of a wide variety of insects and, as far as can be ascertained, is nontoxic to humans. Because of its widespread use and the fact that it is accumulated in fish and animal fats, it is found to concentrate enormously in birds of prey (eagles, pelicans, and ospreys), whose diet is principally fish, and to disrupt their reproductive cycles (Figure 11–10). A side effect such as this has led to extensive agitation to replace DDT with other compounds which are equally effective as insecticides yet free from this particular side effect. It is possible, however, that the compounds used to replace DDT will have other side effects which will not become obvious until after they have been used for some time. It would seem that the use of any insecticide will have a considerable effect upon the bird population of an area, if only through the changes it introduces into the food supply of the birds.

DDT is far more harmful to many species of wild animals than it is to man.

Most technical problems require the efforts of several disciplines for a solution.

Chemistry, however, is not the only science which can furnish tech-

Figure 11–10 The concentration of DDT in birds of prey, such as hawks, results in poor egg shell formation along with other abnormalities in the reproductive cycle.

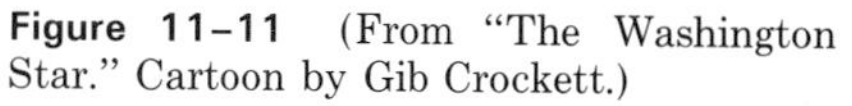

Figure 11–11 (From "The Washington Star." Cartoon by Gib Crockett.)

nological processes for the control of insects. There is a wide variety of biological processes which can be used for the same purposes. One is to furnish assistance to the biological species which destroy insects, such as other parasitical insects. Another is the introduction of large numbers of sterile insects. When these mate with normal insects, no offspring are produced; the number of a type of undesirable insect can be reduced considerably by this process, at least temporarily. These examples are given to emphasize the fact that there are usually several very different kinds of processes which can be used to solve any given practical problem. In the future these will be assessed, at least in part, on their ability to leave the environment undamaged as well as on their monetary cost.

TECHNOLOGICAL DEVELOPMENT AND ITS ENVIRONMENTAL CONSEQUENCES

A very good example of a highly desirable technological development which has consequences that are obviously not so desirable is seen in the development of fertilizers. Most of us agree that an abundant food supply is good. Most would admit, too, that every crop harvested from a field removes essential nutrients from that field: nitrogen, phosphorus, potassium, and so forth. By replacing these lost elements with fertilizer, we can restore or enhance the amount of food we obtain from the field. The increased yields obtained with increased application of fertilizers have been well established and commonly held to be desirable. It is also well established that increased fertilization of a field *increases* the con-

Figure 11–12 Human thought versus chemicals. Will man ever completely control chemicals—or will they control him? (Courtesy of Varian, Palo Alto, California.)

centration of these essential nutrients in the rainwater which runs off such land.

The change in the mineral concentration in the runoff water is capable of causing drastic changes in the rivers and lakes into which it drains (Figure 11–13). By increasing the amount of minerals available, we have greatly stimulated the growth of surface algae in rivers and lakes, and as these grow they choke out other forms of life. The growth of such slimes makes the water much less capable of supporting its normal population of fish, and ultimately they die off. The algae also make the water more difficult to purify for drinking purposes. After a time the increased food supply is paid for by a general disruption of the biology of the waterways which drain an area. This disruption has occurred in many areas in the United States where the movement of water through lakes and rivers is slow.

Fertilizers increase the growth of both desirable and undesirable organisms.

Other examples of this same kind of process can be seen in the development of energy sources, the use of antibiotics, weed killers, and, in fact, any kind of process which releases chemical compounds into the environment in sufficient quantities to cause an appreciable percentage increase in its composition in the environment. This holds for the hydrocarbons released from automobile engines and for the carbon dioxide and sulfur dioxide released by electric power plants. The change which these increased concentrations of various compounds cause cannot always be

What are the questions that should be asked before a new chemical is widely used?

A

B

Figure 11–13 This pond was fertilized with chemicals to produce an excessive growth of algae. Part *A* shows the water covered with the plant growth: Part *B* shows the decay that follows when the concentrated form of life cannot be sustained. (Courtesy of Dr. D. L. Brockway, Federal Water Quality Administration.)

estimated on the basis of experiments covering only a short period of time since some of the effects may be very long range ones which build up slowly.

IS TECHNOLOGY ESCAPING CONTROL?

It is quite easy to imagine a situation in which a technological process can introduce into our environment drastic and irreversible changes which set a chain of events into action before we can stop them. This is especially easy to visualize in the case of changes which might be provided by the continuous increase in the carbon dioxide content of our atmosphere (Figure 11–14). This is caused by the increased use of "fossil" fuels, such as coal and petroleum, to furnish power for energy generation and transportation. What will be the long range effects, if any, from this steadily increasing amount of carbon dioxide in our atmosphere? At the present time, no one really knows.

It is not presently clear that man will effectively control technology; he may destroy himself with it.

There are other situations where the consequences of technology seem more obviously to be moving beyond the control of man. The ability to build nuclear weapons is now spreading quite rapidly and soon may be well within the power of all but the smallest nations. How are these to be controlled? What mechanisms can be developed to prevent mankind from destroying itself in an atomic holocaust triggered by some insignificant local dispute? This is obviously a very urgent problem.

The application of technology to the problems of warfare has already produced some frightening developments, including chemical and biological warfare agents whose discovery has closely followed the extension of human knowledge in these sciences. The same kind of knowledge that allows more effective drugs and insecticides to be synthesized facilitates the synthesis of more effective agents for gas warfare. The understanding of cellular behavior that allows us to produce new varieties of high yielding grains also can be used to develop new techniques of biological warfare.

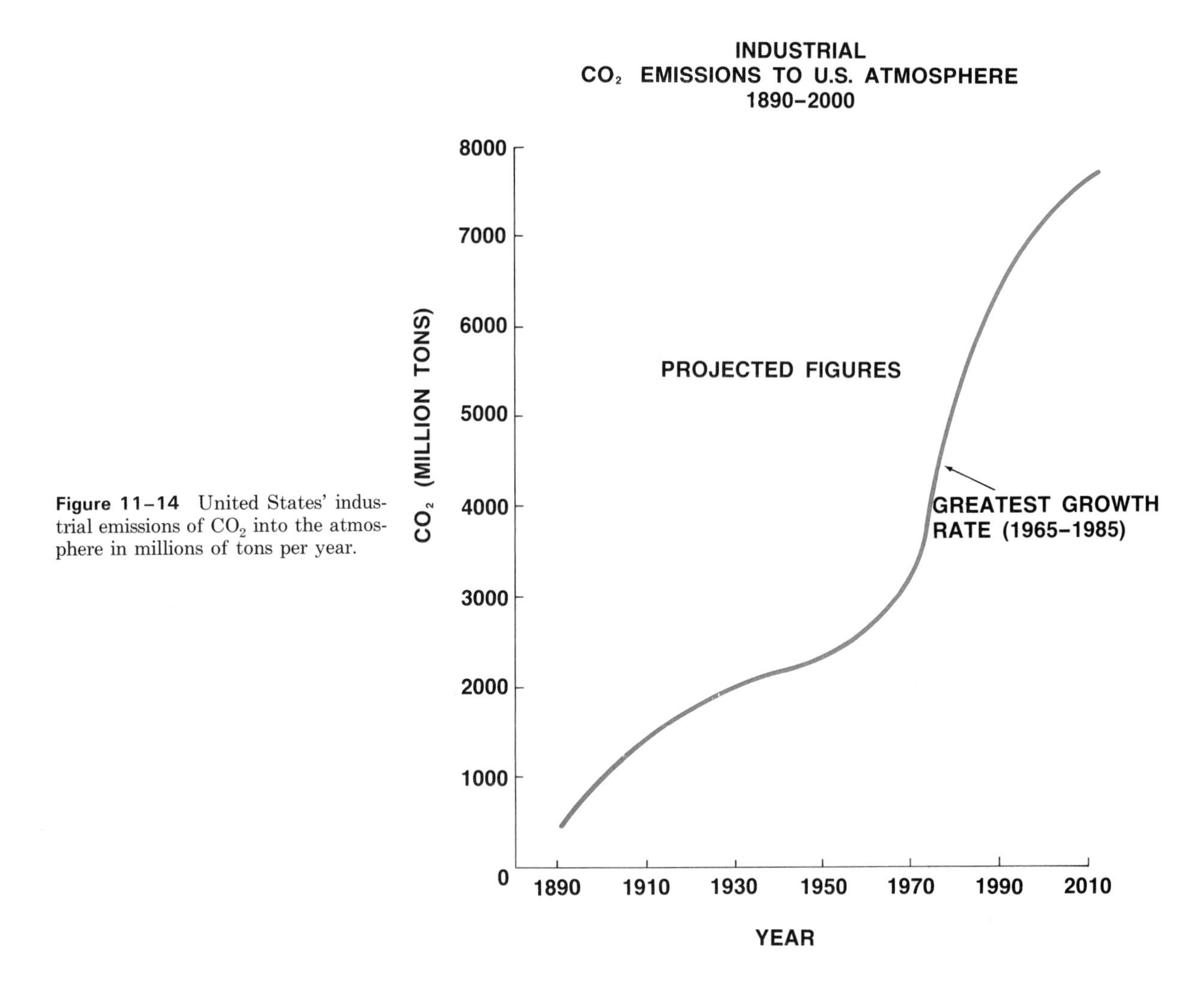

Figure 11–14 United States' industrial emissions of CO_2 into the atmosphere in millions of tons per year.

Obviously, a distinction must be made between scientific knowledge which can be used for good or evil purposes and the types of technology that are developed specifically for destructive purposes. The kind of emotional thinking which puts all technological developments under some kind of moral ban can only lead back to a new Dark Age. Most of the materials of the world can be used for a variety of purposes, which are ethical to varying degrees. In mankind's struggles he has always used his intelligence as well as his emotions in order to succeed. Selection must be made between good developments, the indifferent ones, and the bad ones. This has *always* been a problem for mankind and probably always will be. In such a situation, ignorance can be catastrophic; only by a study of such processes and an evaluation of their probable consequences can rational selections be made. In this case it makes more sense to ask if man is being given control over things beyond his understanding than to ask, "Is technology escaping control?" Technology is always *initiated* by man and is under some sort of actual or potential control by him.

Technology is always initiated by man and is therefore under his control in principle.

There is still another manner in which fears arise over the developing course of technology and man's ability to control it. The vast majority of mankind has no knowledge of the scientific principles upon which technology is based and accordingly must accept the opinions of others on technological matters. This dependence upon experts, in turn, arouses fears of these experts and the damage they can do either by error or by evil intent. This fear, in some people, is an unreasoning, blind, driving force which leads them to condemn all technology. The only antidote to such fear is an understanding of technology based on study. Mankind's perennial enemy has been ignorance and the prejudice it generates. Human progress has always been the result of activities of that small percentage of people who accept neither the ignorance nor the popular prejudices of their fellow human beings.

In all fairness, it must be noted that the benefits of technology far outweigh its disadvantages, and also that its transformation of the human condition is still in its infancy. There is, even now, an enormous number of practical problems facing mankind which can be solved only by new scientific discoveries and the technological advances which they will make possible. It is the responsibility of all of us to learn about the sciences and to develop an understanding of the possible ways in which this knowledge may be developed to the advantage of mankind. In these learning processes, we will discover an avenue to responsible participation in decision making, an understanding of technological advances, and a thorough enjoyment of our investigations.

No person should hesitate to probe into technology to find how it affects his life. This is necessary if he is to make good decisions on many basic questions.

SELF-TEST 11-A

1. The ultimate test of a scientific theory is the agreement with ______________.
2. Different workers, in different countries, who carry out a particular laboratory experiment in exactly the same way should get the ______________ result.
3. Technology allows us to produce things at a greater ______________.
4. We now expect that over the period of our own lifetime our way of life will ______________.
5. Nuclear power plants produce competitively priced energy but dangerously ______________ by-products.
6. The principal justification for the use of chemical fertilizers and pesticides is that they allow us to produce more ______________.

SUBJECTIVE TEST

Rate the following on a scale: GOOD, BAD, INDIFFERENT. Be prepared to state your reasons.

1. DDT _______
2. Nuclear energy _______
3. Coal as a fuel _______
4. Petroleum as a fuel _______

5. Fertilizers ______
6. Birth control pills ______
7. Plastic containers ______
8. Synthetic foods ______
9. Solar energy ______
10. Genetic manipulation ______

MATCHING SET

______	1. transmission of scientific knowledge	a. appeal to authority
______	2. allowed World War I to be prolonged	b. DDT
______	3. source of the element N_2	c. technology
______	4. alternative to insecticides	d. nitrogen compounds
______	5. unsatisfactory way to establish scientific truth	e. used to study nature
______	6. insecticide harmful to wildlife	f. printed journals
______	7. increases rate of change in life	g. Haber process
______	8. used in explosives	h. air
______	9. needed to grow crops	i. observation
______	10. scientific methods	j. biological controls
______	11. basis of scientific truth	

QUESTIONS

Note: Specific answers to these questions are not found in this chapter; rather, they are based on the concepts presented.

1. Select a law or generalization from a book on chemistry and track down the supporting evidence. Do the same from a book on economics or sociology.
2. Cite three technological advances, give their direct benefits, and list a problem arising from each.
3. List 10 technological advances since 1940 which affect your life.
4. Name two specific changes that have been made in the past few years to decrease the amount of air pollution caused by automobiles.

5. Try to think of a reason not to dump liquid chemical wastes in abandoned mine shafts.

6. Distinguish between science and technology.

7. Discuss three of the most important practical problems facing mankind, in terms of how science and technology have helped to bring them about and how they are being solved.

8. Look up the number of deaths in the U.S. due to malaria for the past century. Did a "break" occur? When? See: "Statistical Abstract of the United States: National Data Book and Guide to Sources," U.S. Dept. of Commerce, Bureau of the Census, Washington, D.C.

9. Is there any human evidence of technological advances resulting in the decline of civilization? Consider an article in the *Journal of Occupational Medicine,* Vol. 7, pp. 53–60 (1965).

10. Over coffee one morning, a friend states, "The results of chemical technology have all been harmful to mankind!" He goes on to list smog, water pollution, DDT, chemical and biological warfare agents, and so forth. Could you balance the argument by listing some advances in this area which have been for the general good of man?

SUGGESTIONS FOR FURTHER READING

Commoner, B. (Ed.), "Science and Survival," Viking Press, New York, 1966.

Fischer, R. B., "Science, Man and Society," W. B. Saunders Co., Philadelphia, 1975.

Helfrich, H. W., Jr. (Ed.), "The Environmental Crisis," Yale University Press, New Haven, 1970.

Kranzberg, M., and Pursell, C. W., Jr. (Eds.), "Technology in Western Civilization," Oxford University Press, New York, 1967. (In 2 volumes; a comprehensive collection of well-illustrated essays.)

National Science Foundation, "Chemicals and Health," Washington, D.C., 1973.

Novich, S., "A New Pollution Problem," *Environment,* Vol. 11, No. 4, p. 2 (1969).

Turk, A., Turk, J., and Wittes, J. T., "Ecology, Pollution and Environment," W. B. Saunders Co., Philadelphia, 1972.

U.S. Atomic Energy Commission, "Radioactive Wastes," 1969.

Vavoulis, A., and Colver, A. (Eds.), "Science and Society—Collected Essays," Holden-Day, Inc., San Francisco, 1966.

Wade, N., "Sahelean Drought: No Victory for Western Aid," Science, Vol. 85, p. 234 (1974).

USEFUL MATERIALS FROM THE EARTH, SEA, AND AIR

CHAPTER 12

A large number of the things we use in everyday life result from chemical reactions which are carried out on an industrial scale. They include iron and steel products, fabricated items of aluminum, magnesium, copper, glass, ceramics, and cement. Also included are fertilizers, detergents, chemicals used for purifying water, and numerous chemicals used to alter our surroundings. The purpose of this chapter is to discuss some of the chemistry involved in preparing these products from natural resources found in the earth, sea, and air.

Industrial chemistry should be and probably will be limited by two relatively new demands. Not only must it produce a useful arrangement of atoms, but it must also provide for the return of the unused atoms to nature in a desirable form—that is, the eventual recycling of the atoms. A reasonable economic package must place enough demand on the desirable arrangement of atoms to provide for negating the undesirable arrangements.

Recycling decreases our demands on many natural resources.

METALS AND THEIR PREPARATION

Metals occur in the crust of the earth mostly as compounds, though some of the less active ones such as copper, silver, and gold can be found also as free elements. Because of extensive and continuing geological processes which have formed the crust of the earth over the last billion years, the distribution of elements is anything but uniform. In order to obtain a metal from the crust of the earth, one must first find an ***ore,*** which is defined simply as a mineral from which the element can be extracted economically. Color Plate II pictures some natural ***minerals.*** Table 12–1 lists some common ones.

A continual search is under way for new ore deposits.

TABLE 12-1 Some Common Ores

Metal	Chemical Formula of Compound of the Element	Name of Ore
Aluminum	$Al_2O_3 \cdot xH_2O$	bauxite
Copper	Cu_2S	chalcocite
Zinc	ZnS	sphalerite
	$ZnCO_3$	smithsonite
Iron	Fe_2O_3	hematite
	Fe_3O_4	magnetite
Manganese	MnO_2	pyrolusite
Chromium	$FeO \cdot Cr_2O_3$	chromite
Calcium	$CaCO_3$	limestone
Lead	PbS	galena

Some elements which are not particularly abundant in the earth's crust are nevertheless very familiar to us because they tend to occur in very concentrated, localized deposits from which they can easily be extracted. Examples of these are lead, copper, and tin, none of which is among the more abundant elements in the crust of the earth (Figure 12-1). Other elements that actually form a much larger percentage of the crust of the earth are almost unknown to us because concentrated deposits of their ores are less commonly found. An example is titanium, the tenth most abundant element in the crust of the earth.

Reduction of magnesium: $Mg^{2+} + 2e^- \rightarrow Mg$

The preparation of metals from their ores involves chemical reduction (Chapter 10). Indeed, the concept of oxidation and reduction developed from metallurgical operations. As presented earlier, iron in iron oxide is in the form of Fe^{3+}. If we *reduce* Fe^{3+} ions to Fe atoms, we must find a source of electrons. (Recall that oxidation is the loss of electrons and that reduction is the gain of electrons.) Sometimes the desired metal is in solution (e.g., magnesium in the sea), where it exists in the oxidized form (Mg^{2+} ions). To obtain the free metal magnesium we must add electrons to these ions (reduction) to produce neutral atoms.

IRON

Iron ores are iron compounds. To get iron from the ores, the compounds must be reduced.

The sources of most of the world's iron are large deposits of the iron oxides in Minnesota, Sweden, France, Venezuela, Russia, Australia, and England. In nature these oxides frequently are mixed with impurities, so the production of iron usually incorporates steps to remove such impurities. Iron ores are reduced to the metal by using carbon, in the form of coke, as the reducing agent.

The reduction of the iron ore is carried out in a blast furnace (Figure 12-2). The solid material fed into the top of the blast furnace consists of a mixture of an oxide of iron (Fe_2O_3), coke (C), and limestone ($CaCO_3$). A blast of heated air is forced into the furnace near the bottom. The reactions which occur within the blast furnace are:

$$\underset{\text{CARBON}}{2C} + \underset{\text{OXYGEN}}{O_2} \longrightarrow \underset{\text{CARBON MONOXIDE}}{2CO} + \text{heat}$$

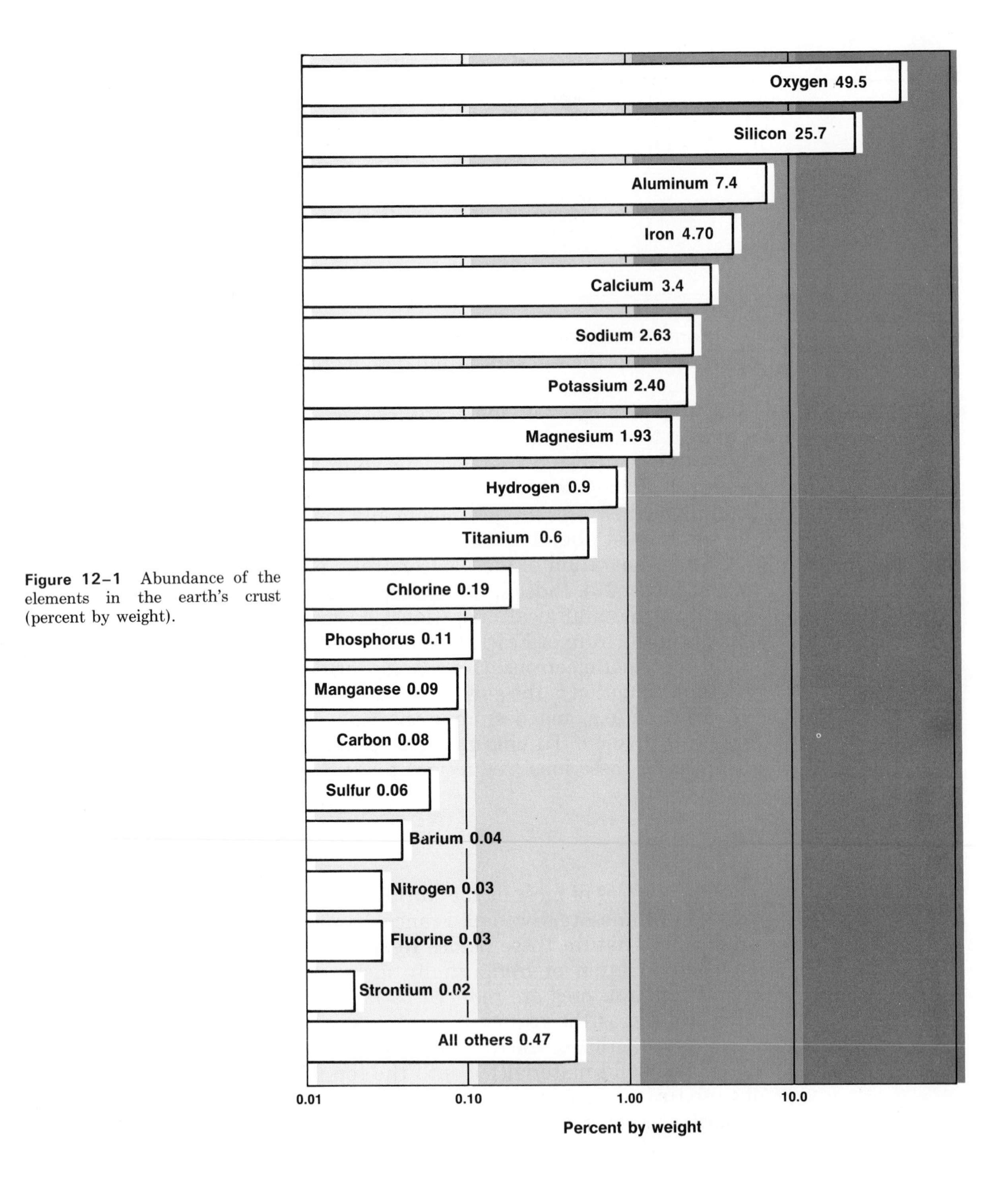

Figure 12–1 Abundance of the elements in the earth's crust (percent by weight).

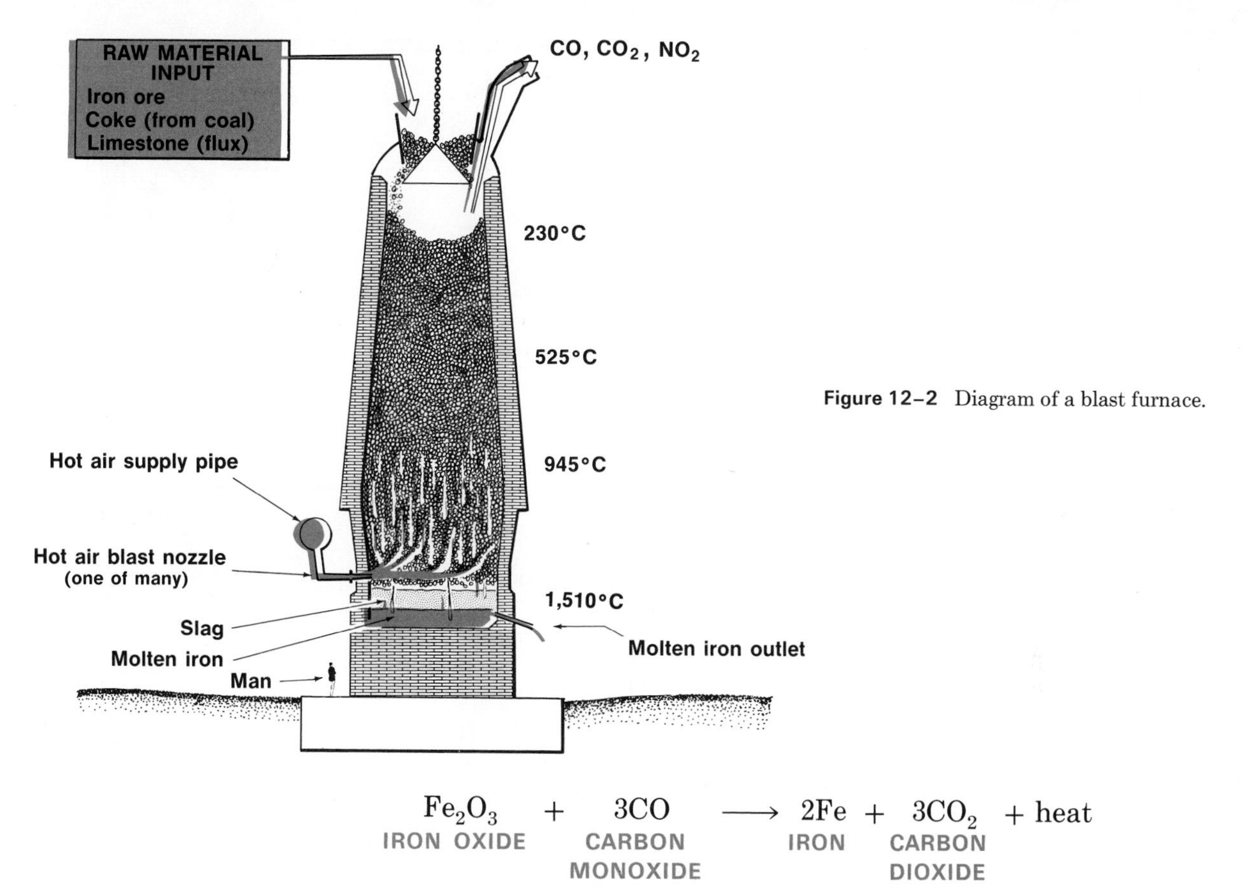

Figure 12–2 Diagram of a blast furnace.

$$\underset{\text{IRON OXIDE}}{Fe_2O_3} + \underset{\text{CARBON MONOXIDE}}{3CO} \longrightarrow \underset{\text{IRON}}{2Fe} + \underset{\text{CARBON DIOXIDE}}{3CO_2} + \text{heat}$$

Limestone is added because iron ores contain silica (SiO_2) as an impurity. The limestone reacts as follows:

$$\underset{\text{CALCIUM CARBONATE}}{CaCO_3} \xrightarrow{heat} \underset{\text{CALCIUM OXIDE}}{CaO} + \underset{\text{CARBON DIOXIDE}}{CO_2}$$

$$\underset{\text{CALCIUM OXIDE}}{CaO} + \underset{\text{SILICON DIOXIDE}}{SiO_2} \longrightarrow \underset{\text{CALCIUM SILICATE}}{CaSiO_3}$$

The calcium silicate, or ***slag,*** has a melting point low enough to allow it to exist as a liquid in the furnace. Consequently, as the blast furnace operates, two molten layers collect in the bottom. The lower, denser layer is mostly liquid iron which contains a fair amount of dissolved carbon and often smaller amounts of other impurities. The upper, lighter layer is primarily molten calcium silicate with some impurities. From time to time the furnace is tapped at the bottom and the molten iron is drawn off. Another outlet somewhat higher in the blast furnace base can be opened to remove the liquid slag.

The iron that is obtained from the blast furnace contains too much carbon for most uses. If some of the carbon is removed, the mixture becomes structurally stronger and is known as steel. ***Steel*** is an alloy of iron with a relatively small amount of carbon (less than 1.5 percent); it may also contain other metals. In order to convert iron into steel, the excess carbon is burned out with oxygen.

Alloy: a metal consisting of two or more elements.

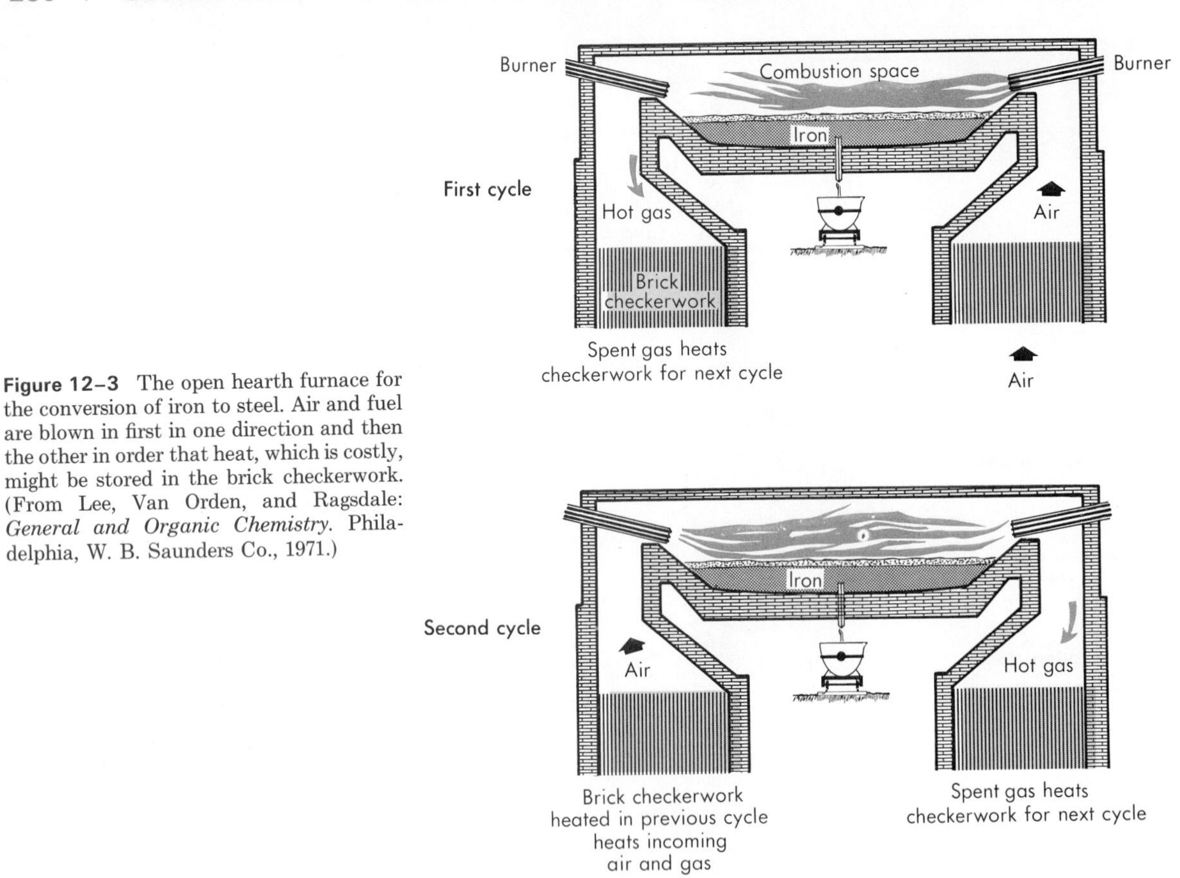

Figure 12–3 The open hearth furnace for the conversion of iron to steel. Air and fuel are blown in first in one direction and then the other in order that heat, which is costly, might be stored in the brick checkerwork. (From Lee, Van Orden, and Ragsdale: *General and Organic Chemistry*. Philadelphia, W. B. Saunders Co., 1971.)

There are several techniques for burning the excess carbon (see Figures 12-3 and 12-4). A recent development that has been very widely adopted is the basic oxygen process (Figure 12-4). In this process pure oxygen is blown into molten iron through a refractory tube (oxygen gun) which is pushed below the surface of the iron. At elevated temperatures,

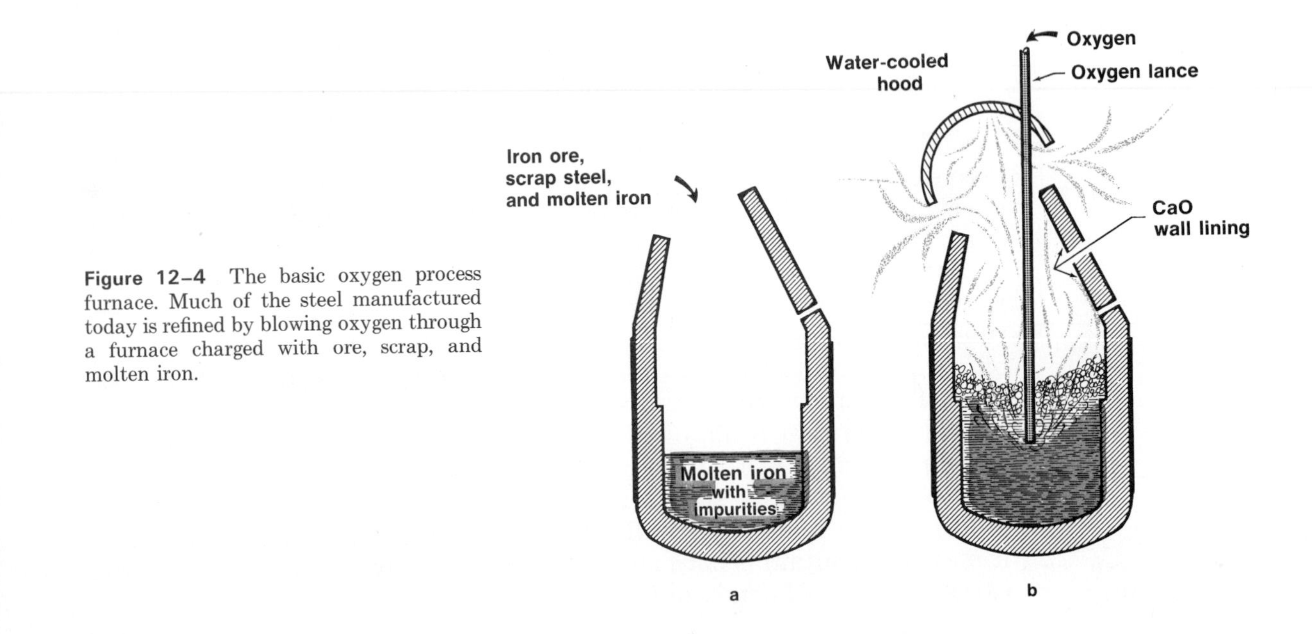

Figure 12–4 The basic oxygen process furnace. Much of the steel manufactured today is refined by blowing oxygen through a furnace charged with ore, scrap, and molten iron.

the dissolved carbon reacts very rapidly with the oxygen to give gaseous carbon monoxide or dioxide, which then escapes.

After the carbon content has been reduced to a suitable level, the molten steel is formed into desired shapes. During the processing it is subjected to carefully controlled heat treatment to insure that the steel has the desired mechanical properties. This is necessary in order to control the crystalline form of the steel, which determines its pliability, toughness, and other properties.

CORROSION, RUSTING, AND STAINLESS STEEL

Billions of dollars worth of valuable materials are destroyed each year by corrosion. Several techniques are now being used to try to reduce this destruction.

Many metals undergo corrosion when exposed to moist air over a long period of time. Typically, corrosion is a reaction with the oxygen and water of the air which transforms a metal into its oxide or hydroxide. Corrosion of iron is called rusting and leads to the transformation of iron into rust ($Fe_2O_3 \cdot xH_2O$). The initial reaction is the oxidation of iron to iron (II) hydroxide:

$$\underset{\text{MOIST AIR}}{2Fe + O_2 + 2H_2O} \longrightarrow \underset{\text{IRON (II) HYDROXIDE}}{2Fe(OH)_2}$$

Iron (II) hydroxide is itself subject to further oxidation in moist air to give iron (III) hydroxide:

$$4Fe(OH)_2 + \underset{\text{MOIST AIR}}{O_2 + 2H_2O} \longrightarrow \underset{\text{RUST}}{4Fe(OH)_3 \text{ (or } 2Fe_2O_3 \cdot 3H_2O)}$$

The iron (III) hydroxide loses water readily to form iron (III) oxides with variable amounts of water. The rusting process transforms iron metal back into a compound very similar in composition to its ore and thus undoes all the effort expended in obtaining the metal.

Rusting also occurs when we make an iron object into an electrochemical cell or battery, often without realizing it. This happens when part of the iron surface becomes wet in contact with air (Figure 12–5). The oxygen in the air will dissolve in the water and remove electrons from the iron via the reaction:

$$O_2 + 2H_2O + 4e^- \longrightarrow 4OH^-$$

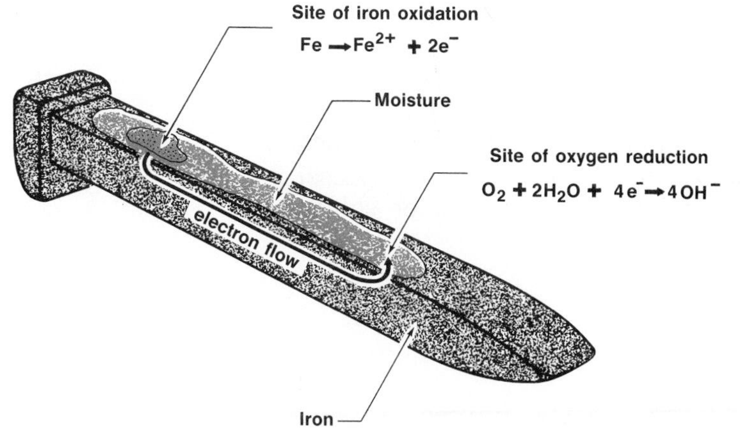

Figure 12–5 The site of iron oxidation may be different from the point of oxygen reduction owing to the ability of the electrons to flow through the iron. The point of oxygen reduction can be located with an acid-base indicator because of the OH^- ion produced.

Corrosion and rusting are obviously very serious problems which have occupied the attention of scientists for years. There are many ways of preventing or reducing corrosion of a metal object. Three of these are: (1) protective coatings; (2) cathodic protection; and (3) alloying (stainless steel is a corrosion-resistant alloy of iron).

Cathodic protection is discussed on page 223.

Protective coatings are applied to prevent the access of atmospheric oxygen to the iron surface. Such coatings may be of paint, enamel, grease, or another more resistant metal, such as chromium. This method is successful as long as cracks or holes do not develop in the coating. Galvanized iron contains a surface coating of the more active metal zinc, which forms an oxide that tends to be hard and impervious to the air.

Stainless steels are alloys of iron which contain metals such as nickel, chromium, or cobalt. They are made by melting iron and the alloying elements together in an electric furnace. The resulting alloys are resistant to corrosion. In the presence of oxygen they form a very thin, tough, and impervious adherent layer of metal oxide on their surfaces. The layer is *so* thin it is essentially transparent (the luster of the metal is retained). This protects the underlying metal from further contact with the oxygen of the air and renders the objects "stainless," or very resistant to corrosion under normal circumstances.

A stainless steel alloy which is widely used is the so-called 18-8: 18 percent chromium, 8 percent nickel, and the rest iron.

Stainless steels are used in household items; larger amounts are used in the construction of industrial plants which handle great quantities of hydrochloric acid and other highly corrosive materials.

ALUMINUM

Seven and four tenths percent of the crust of the earth is aluminum, in the form of Al^{3+} ions. However, because of the difficulty of reducing Al^{3+} to Al, only recently has man learned to isolate and use this abundant element. Aluminum metal is soft and has a low density. Many of its alloys, however, are quite strong. Hence, it is an excellent choice when a light weight, strong metal is required. In structural aluminum, the high chemical reactivity of the element is offset by the fact that a transparent, hard film of aluminum oxide, Al_2O_3, forms over the surface, protecting it from further oxidation:

When aluminum was first made, it was very expensive and rare. A bar of aluminum was displayed next to the Crown Jewels at the Paris Exposition in 1855.

$$4Al + 3O_2 \longrightarrow 2Al_2O_3$$

The principal ore of aluminum is ***bauxite,*** a hydrated aluminum oxide, $Al_2O_3 \cdot xH_2O$. Because impurities in the ore have undesirable effects on the properties of aluminum, these must be removed, generally by the purification of the ore. This is accomplished with the Bayer process, which is based upon the fact that aluminum oxide and aluminum hydroxide are weak acids. The principal impurities are iron oxides, which are not soluble in sodium hydroxide solution. The Bayer process separates these by treating the mixture with a sodium hydroxide solution, which dissolves the aluminum and leaves the iron:

$$\underset{\text{(SOLID)}}{Al_2O_3 \cdot xH_2O} + \underset{\text{(SOLID)}}{Fe_2O_3} \xrightarrow[\text{solution}]{\text{NaOH}} \underset{\text{(SOLUTION)}}{Al(OH)_4^- + Na^+} + \underset{\text{(SOLID)}}{Fe_2O_3}$$

The mixture is filtered; the $Al(OH)_3$ is then carefully precipitated out of the clear solution by adding carbon dioxide (an acid), thus making the solution less basic:

$$CO_2 + Al(OH)_4^- \longrightarrow Al(OH)_3\downarrow + HCO_3^-$$

The aluminum hydroxide is then heated to transform it into pure anhydrous aluminum oxide:

$$2Al(OH)_3 \xrightarrow{\text{Heat}} Al_2O_3 + 3H_2O$$

Aluminum metal is obtained from the purified oxide by electrolysis in molten cryolite (Figure 12-6). Cryolite, Na_3AlF_6, has a melting point of 1006°C; the molten compound dissolves considerable amounts of aluminum oxide, which in turn lowers its melting point. This mixture of cryolite and aluminum oxide is electrolyzed in a cell with carbon anodes and a carbon cell lining that serves as the cathode on which aluminum is deposited. As the operation of the cell proceeds, the molten aluminum sinks to the bottom of the cell. From time to time the cell is tapped and the molten aluminum is run off into molds.

The top of the Washington monument is a casting of aluminum made in 1884.

Aluminum is used both as a structural metal and as an electrical conductor in high voltage transmission lines. It competes with copper as an electrical conductor because of its lower cost. The lower cost allows larger diameter aluminum wires to be used to offset the fact that an aluminum wire has a lower electrical conductivity than a copper wire of the same diameter.

COPPER

Although copper metal occurs in the free state in some parts of the world, the supply available from such sources is quite insufficient for the world's need. The majority of the copper obtained today is from various copper sulfide ores, most of which must be concentrated prior to the

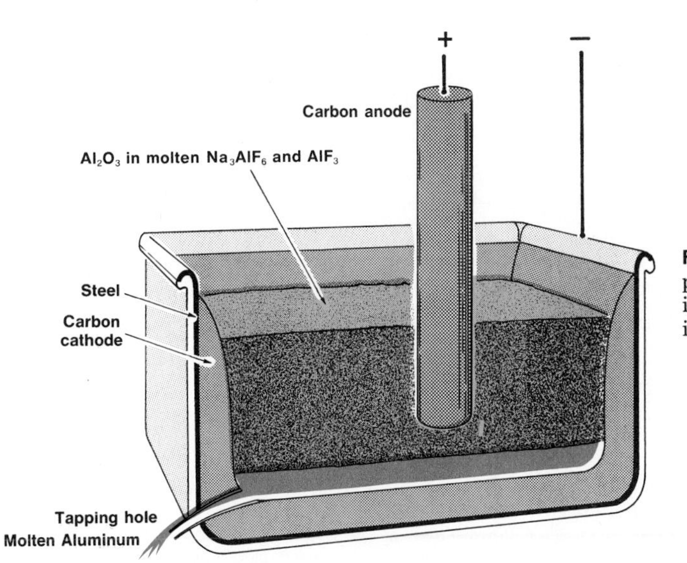

Figure 12-6 Schematic drawing of a furnace for producing aluminum by electrolysis of a melt of Al_2O_3 in Na_3AlF_6 and AlF_3. The molten aluminum collects in the carbon cathode container.

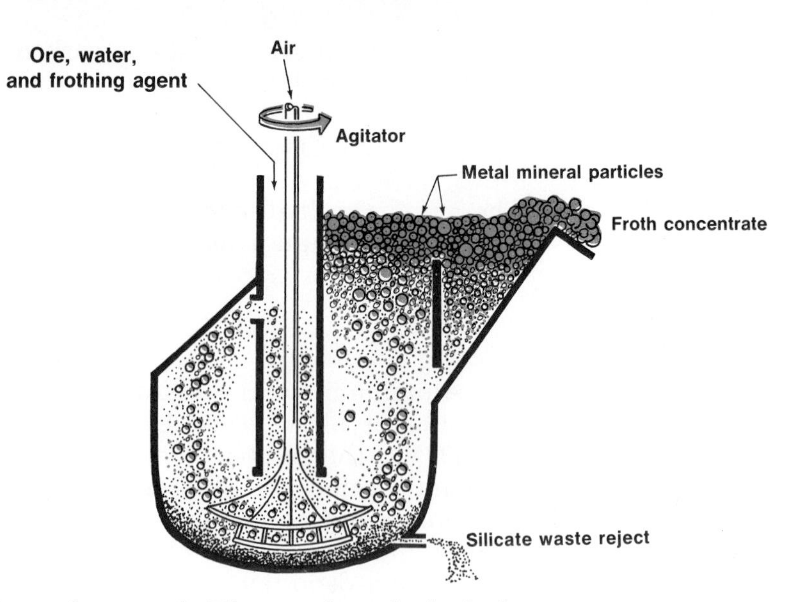

Figure 12–7 Apparatus for flotation concentration.

chemical processes which produce the metal. These minerals include $CuFeS_2$ (chalcopyrite), Cu_2S (chalcocite), and CuS (covellite). Because the copper content of these ores is around 1 to 2 percent, the powdered ore is first concentrated by a ***flotation process*** (Figure 12–7).

In the flotation process, the powdered ore is mixed with water and a frothing agent such as pine oil. A stream of air is blown through to produce froth. The ***gangue*** in the ore, which is composed of sand, rock, and clay, is easily wet by the water and sinks to the bottom of the container. In contrast, a copper sulfide particle is hydrophobic—it is not wet by the water. The copper sulfide particle becomes coated with oil and is carried to the top of the container in the froth. The froth is removed continuously, and the copper sulfide minerals are recovered from it.

Gangue: commercially worthless material associated with valuable mineral in an ore.

The preparation of copper metal from a copper sulfide ore involves roasting in air to oxidize some of the copper sulfide and any iron sulfide present:

$$2Cu_2S + 3O_2 \longrightarrow 2Cu_2O + 2SO_2\uparrow$$
$$2FeS + 3O_2 \longrightarrow 2FeO + 2SO_2\uparrow$$

Subsequently the mixture is heated to a higher temperature, and some copper is produced by the reaction:

$$Cu_2S + 2Cu_2O \longrightarrow 6Cu + SO_2\uparrow$$

The iron oxides then form a slag. The product of this operation is a ***matte,*** a mixture of copper metal and sulfides of copper, iron, other ore constituents, and slag. The molten matte is heated in a converter with silica materials. When air is blown through the molten material in the converter, two reaction sequences occur. First, the iron is converted into a slag:

$$2FeS + 3O_2 \longrightarrow 2FeO + 2SO_2$$
$$FeO + SiO_2 \longrightarrow \underset{\text{(MOLTEN SLAG)}}{FeSiO_3}$$

The remaining copper sulfide is converted to copper metal:

$$2Cu_2S + 3O_2 \longrightarrow 2Cu_2O + 2SO_2$$
$$Cu_2S + 2Cu_2O \longrightarrow 6Cu + SO_2$$

The blister copper contains metals such as nickel, gold, and platinum!

The copper produced in this manner is crude or "blister" copper and is purified electrolytically.

In the electrolytic purification of copper, the crude copper is first cast into anodes; these are placed in a water solution of copper sulfate and sulfuric acid. The cathodes are made of pure copper. As electrolysis proceeds, copper is oxidized at the anode, moves through the solution as Cu^{2+} ions and is deposited on the cathode (Chapter 10). The voltage of the cell is regulated so that more active impurities (such as iron) are left in the solution and less active ones are not oxidized at all. These less active impurities include gold and silver, and they collect as "anode sludge," an insoluble residue. The anode sludge is subsequently worked up to recover these rarer metals.

The copper produced by the electrolytic cell is 99.95 percent pure and is suitable for use as an electrical conductor. Copper for this purpose must be pure because very small amounts of impurities, such as arsenic, considerably reduce the electrical conductivity of copper.

MAGNESIUM

Magnesium is active chemically; hence it is not often used structurally in the pure state.

There are about 331 million cubic miles of sea water.

Magnesium, with a density of 1.74 grams per cubic centimeter, is the lightest structural metal in common use. For this reason it is most often used in alloys designed for light weight and great strength. It is a relatively active metal chemically because it loses electrons easily. Magnesium "ores" include sea water, which has a magnesium concentration of 0.13 percent, and dolomite, a mineral with the composition $CaCO_3 \cdot MgCO_3$. Because there are six million tons of magnesium present as Mg^{2+} salts in every cubic mile of sea water, the sea can furnish an almost limitless amount of this element.

The recovery of magnesium from sea water (Figure 12–8) begins with the precipitation of magnesium hydroxide by the addition of lime to sea water:

$$CaO + H_2O \longrightarrow Ca^{2+} + 2OH^-$$
$$Mg^{2+} + 2OH^- \longrightarrow \underline{Mg(OH)_2}$$

The magnesium hydroxide is removed by filtration and then neutralized with hydrochloric acid to form the chloride:

$$\underline{Mg(OH)_2} + 2H^+ + 2Cl^- \rightleftharpoons Mg^{2+} + 2Cl^- + 2H_2O$$

The water is evaporated; this is followed by the electrolysis of molten magnesium chloride in a huge steel pot that serves as the cathode (Figure 12–9). Graphite bars serve as the anodes.

ELECTROLYSIS

$$\underset{\text{(MELTED)}}{Mg^{2+} + 2Cl^-} \xrightarrow{\textit{Electricity}} Mg + Cl_2\uparrow$$

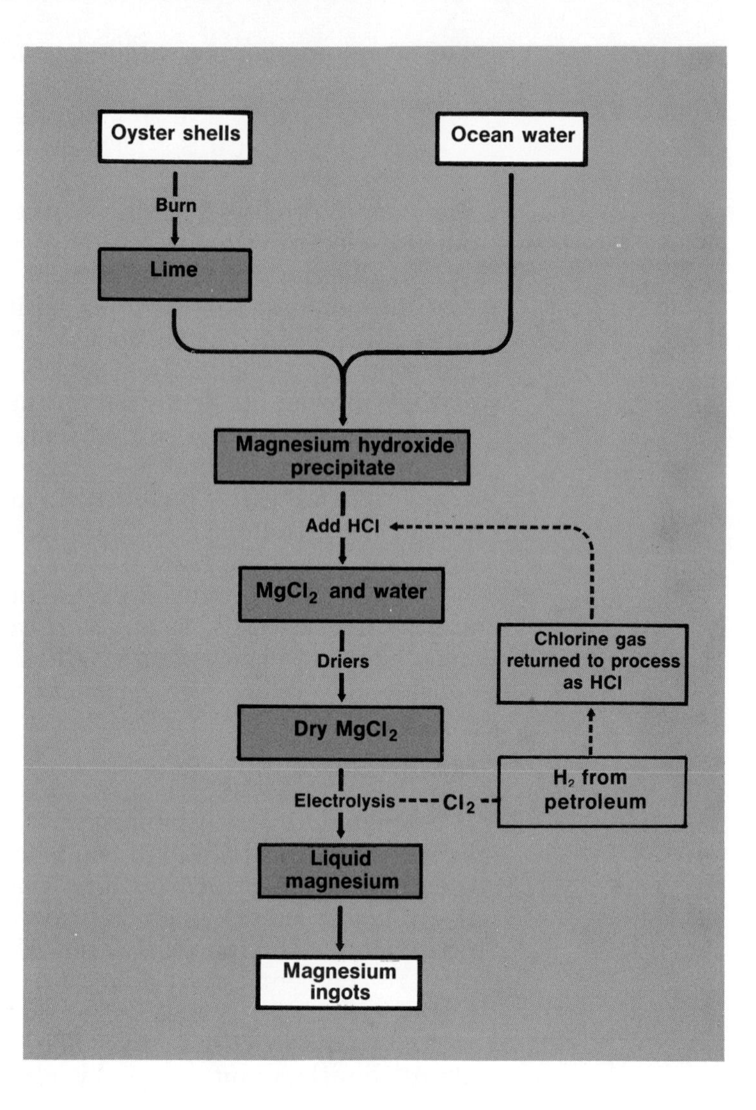

Figure 12–8 Flow diagram showing how magnesium metal is produced from sea water.

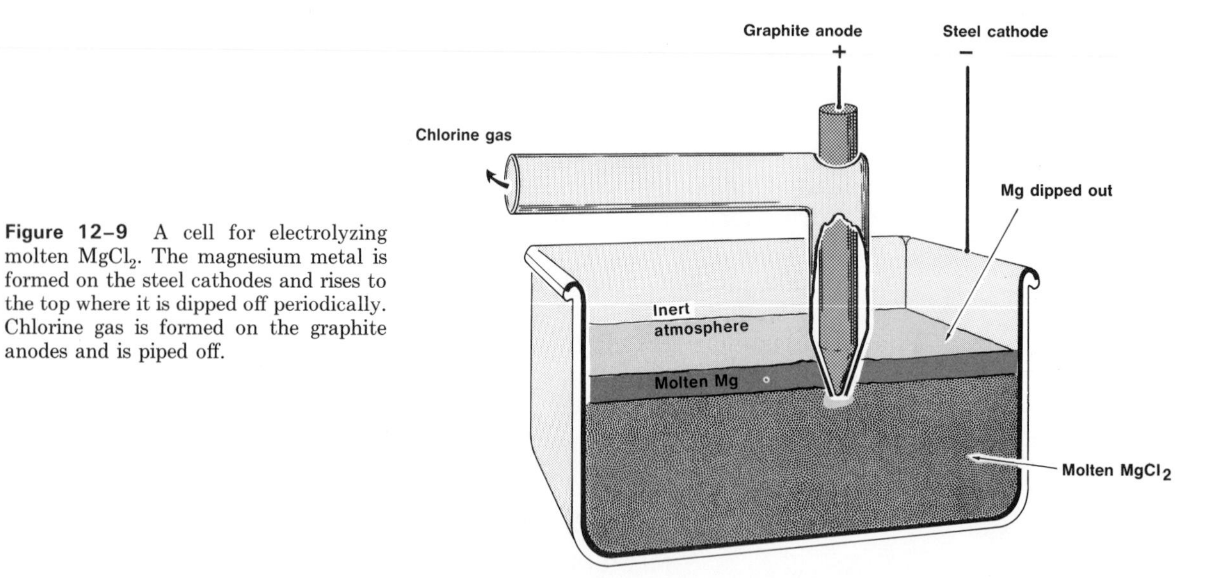

Figure 12–9 A cell for electrolyzing molten $MgCl_2$. The magnesium metal is formed on the steel cathodes and rises to the top where it is dipped off periodically. Chlorine gas is formed on the graphite anodes and is piped off.

AT THE CATHODE:

$$Mg^{2+} + 2e^- \longrightarrow Mg$$

AT THE ANODE:

$$2Cl^- \longrightarrow Cl_2\uparrow + 2e^-$$

To make magnesium one can use sea water, lime from oyster shells, methane from natural gas, and electricity.

As the melted magnesium forms, it floats to the surface and is periodically removed. The chlorine is recovered and reacted with air and natural gas (methane, CH_4) to form hydrochloric acid for use in dissolving the magnesium hydroxide:

$$4Cl_2 + \underset{\text{METHANE}}{2CH_4} + O_2 \longrightarrow 2CO + 8HCl$$

The lime used to precipitate the magnesium as the hydroxide is obtained by heating limestone or oyster shells:

$$CaCO_3 \xrightarrow{Heat} \underset{\text{LIME}}{CaO} + CO_2$$

The total world production of magnesium is only about 250,000 tons per year, although it is potentially available on a larger scale.

SELF-TEST 12-A

1. The most abundant element in the earth's crust is ______________.
2. The most abundant metal in the earth's crust is ______________.
3. In the United States, the largest iron deposits are found in the state of ______________.
4. A natural material which is almost pure calcium carbonate is ______________.
5. Which of the following metals occurs in the free or metallic state in mineral deposits? Iron, copper, aluminum, magnesium. ______________.
6. Which of the metals listed are either produced or purified using electricity? Iron, copper, aluminum, magnesium. ______________.
7. Another name for calcium silicate as it applies to production of iron is ______________.
8. In order for most metals to be prepared from their ores, they must be (oxidized/reduced).
9. In an electrical refining process for metals, the purest metal will always be found at the (anode/cathode).
10. Which metal is sufficiently concentrated in the oceans to be extracted commercially? Magnesium, aluminum, copper, iron. ______________

THE KINGDOM OF SILICON: SILICATES, GLASS, CERAMICS, AND CEMENT

While the carbon-carbon bonds of organic compounds serve as the backbone for the enormous number of organic compounds, a different kind of linkage serves the same purpose in a large number of important ***inorganic*** structures. This is the —Si—O— linkage which is found in silica itself (sand) and silicates, which are basic constituents of glasses, ceramics, and concrete.

Carbon compounds are discussed in Chapter 13.

Silica

Silicon dioxide, SiO_2, occurs naturally in large amounts as sand or more rarely in much larger crystals (quartz) (Figure 12–10). It has a melting point of 1710°C. If the melted material is cooled, a ***glass*** is usually obtained. Crystalline quartz consists of an extended structure in which each silicon atom is bonded tetrahedrally to four oxygen atoms (Figure 12–11a), and each oxygen atom is bonded to two silicon atoms. The bonding thus extends throughout the crystal (Figure 12–11b). When silica is melted, some of the bonds are broken and the units move with respect to each other. When the liquid is cooled, the re-formation of the original solid entails a reorganization which is hard to achieve because of the difficulty the groups experience in moving. The very viscous liquid structure is thus partially preserved on cooling to give the characteristic feature of a ***glass,*** which is an apparently solid material (pseudo-solid) with some of the randomness in structure characteristic of a liquid. This unique structure accounts for one of the typical properties of a glass: it breaks irregularly rather than splitting along a plane like a crystal.

Glass will flow slowly like a liquid. Many colonial homes contain glass window panes thicker at the bottom than at the top.

Glass

The difference between a crystalline solid and a glass can be more easily appreciated by use of a *two*-dimensional analogy. If we consider a two-dimensional model, G_2O_3, in which each G atom* is bonded to three

*A hypothetical atom representing Si in a two-dimensional model.

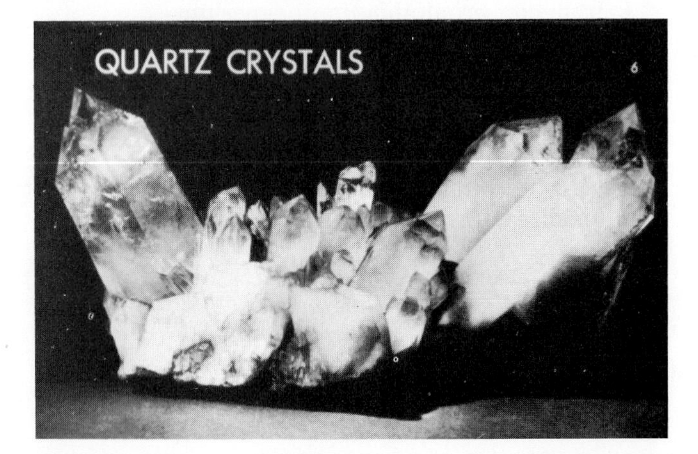

Figure 12–10 Quartz. (Courtesy of McGraw-Hill.)

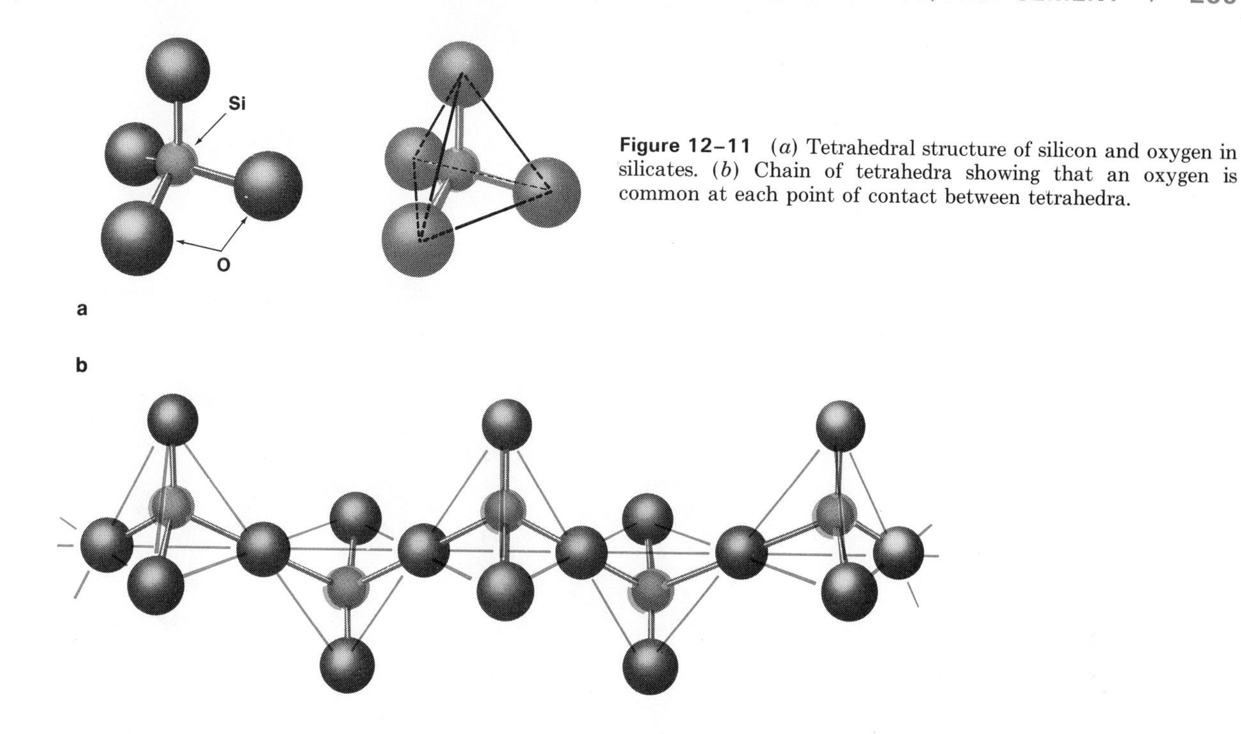

Figure 12–11 (*a*) Tetrahedral structure of silicon and oxygen in silicates. (*b*) Chain of tetrahedra showing that an oxygen is common at each point of contact between tetrahedra.

oxygen atoms and each oxygen to two G atoms, we have a regular crystalline structure with the arrangement shown in Figure 12-12.

If the substance having this structure is melted, some of the G—O bonds are broken and new ones are formed. In the liquid we do not have the long-range order that we have in the crystalline solid. When the substance is cooled to form a glass, some irregular atomic arrangements

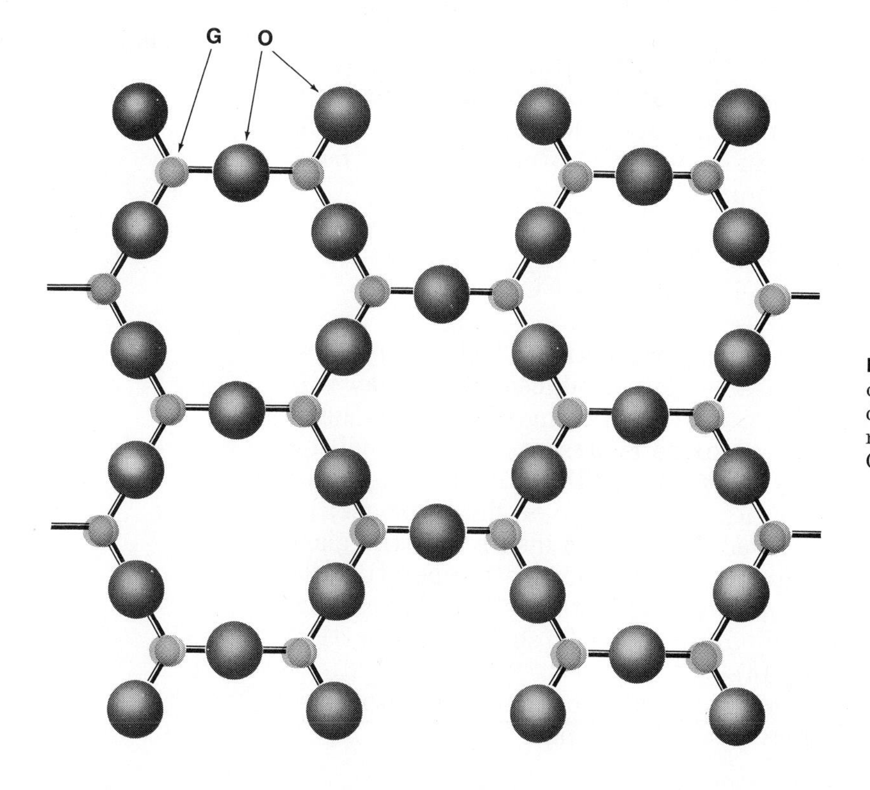

Figure 12–12 Hypothetical regular crystalline structure for SiO_2 in two dimensions. In two dimensions the ratios of atoms would be 2:3; hence, G_2O_3.

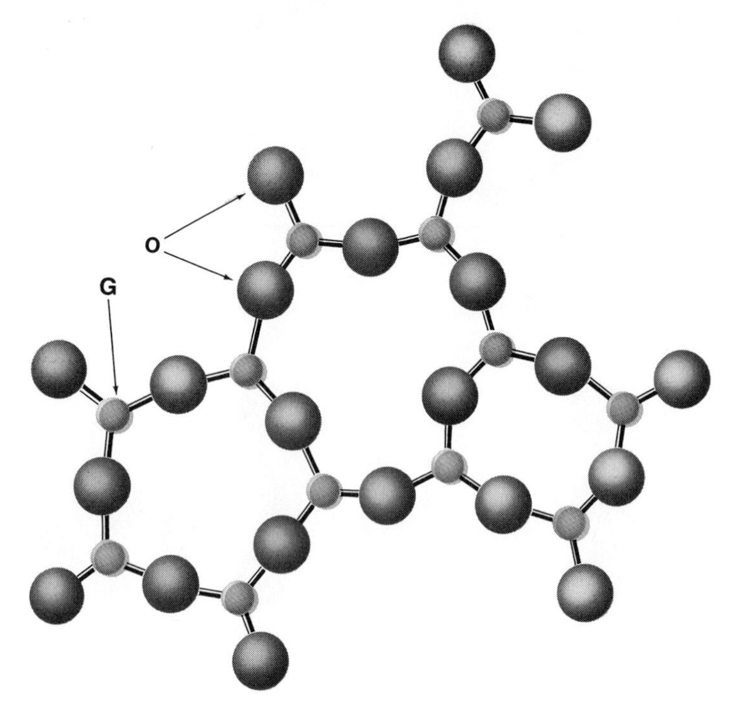

Figure 12–13 Hypothetical irregular crystalline structure for SiO_2 in two dimensions. In the two-dimensional analogy the formula would be G_2O_3.

are formed. Each G atom is still bonded to three oxygens and each oxygen bonded to two G atoms; but the structure is irregular as shown in Figure 12–13.

Glass has an analogous irregular structure in *three* dimensions. This structure can be modified by the addition of molten metal oxides because the metal ions form ionic (nondirectional) bonds with oxygen atoms that had previously been bonded rigidly to Si atoms. The resultant mixture has a lower melting point and viscosity than pure SiO_2 (melting point = 1710°C).

Most glass consists of a mixture of the oxides of silicon, sodium, and calcium, which are fused. Colored glass is produced by the addition of other metal oxides.

By the addition of metal oxides the melting temperature of glass is reduced to about 700°C. The oxides used to effect this change are sodium oxide (added as Na_2CO_3) and calcium oxide (added as $CaCO_3$). Because SiO_2 is an acidic oxide, it can displace CO_2 from sodium carbonate and calcium carbonate. For example,

$$CaCO_3 + SiO_2 \longrightarrow CaSiO_3 + CO_2$$
$$Na_2CO_3 + SiO_2 \longrightarrow Na_2SiO_3 + CO_2$$

The ions of sodium and calcium occupy spaces in the lattice, breaking some of the Si—O bonds. The soda-lime glass produced in this way will be clear and colorless only if the purity of the ingredients has been carefully controlled. If too much iron oxide is present, the glass will be green; other metal oxides produce other colors (Table 12–2). The materials for the glass are melted together in a gas- or oil-fired furnace; as they react, bubbles of CO_2 are evolved. In order to insure that the glass is homogeneous, it is heated to about 1500°C; at this temperature its viscosity is low, and the bubbles of gas trapped in it are allowed to escape. The mixture is cooled somewhat and drawn off to be shaped into bottles, sheets, or other forms (Figure 12–14).

It is possible to incorporate a wide variety of materials into glass for special purposes. Table 12–3 gives some examples.

TABLE 12–2 Substances Used in Colored Glasses

Substance	Color
Copper(I) oxide	red, green, or blue
Tin(IV) oxide	opaque
Calcium fluoride	milky white
Manganese dioxide	violet
Cobalt(II) oxide	blue
Finely divided gold	red, purple, or blue
Uranium compounds	yellow, green
Iron(II) compounds	green
Iron(III) compounds	yellow

Ceramics

Ceramic materials have been made since well before the dawn of recorded history. They are generally materials fashioned at room temperature from clay or other natural earths and then permanently hardened by heat. Clays with a wide variety of properties are found in a considerable range of ceramic materials, from bricks to table china. The

Figure 12–14 Craftsman working with molten glass. (Courtesy of the Corning Glass Company.)

TABLE 12–3 Special Glasses

	Special Addition or Composition	Desired Property
1.	Large amounts of PbO with silica and Na_2CO_3	crystal or flint glass
2.	SiO_2, B_2O_3, and small amounts of Al_2O_3 (borosilicate glass)	small thermal expansion "Pyrex," "Kimax," etc.
3.	One part SiO_2 and four parts PbO	stops (absorbs) large amounts of x-rays and γ-rays—"lead glass"
4.	Large concentration of CdO	neutron-absorbing glass
5.	High concentration of pure As_2O_3	transparent to infrared radiation

techniques developed with natural clay have been applied to a wide range of other inorganic materials in recent years. The result has been a considerable increase in the kinds of ceramic materials available. One can now obtain ceramic magnets as well as ceramics suitable for rocket nozzles—both were developed from mixtures of inorganic oxides by the use of ceramic technology, which includes as an indispensable process the heating of the materials to make them hard and resistant to wear.

The three basic ingredients of common pottery are silicate minerals: clay, sand, and feldspar. ***Clays*** contain a wide range of chemicals, which are produced from the weathering of granite and other rocks. ***Feldspars*** are aluminosilicates containing aluminum, potassium, sodium and other ions in addition to silicon and oxygen. If we write feldspar as a combination of oxides ($K_2O \cdot Al_2O_3 \cdot 6SiO_2$), we may then approximate one of the many chemical reactions involved in the weathering process; the reaction is that of the mineral with water containing dissolved carbon dioxide to form clay:

$$K_2O \cdot Al_2O_3 \cdot 6SiO_2 + 2H_2O + CO_2 \longrightarrow \underset{\text{A CLAY MINERAL}}{Al_2O_3 \cdot 2SiO_2 \cdot 2H_2O} + 4SiO_2 + 2K^+ + CO_3^{2-}$$

Clays result from the weathering of rocks.

The essential feature of the clay mineral is that it occurs in the form of extremely minute platelets which, when wet, are plastic and can easily be shaped. When dry they are rigid; if heated to an elevated temperature they become permanently rigid and are no longer subject to easy dispersion in water. When these clays are mixed with feldspars and silica, heating them produces a mixture of crystals held together by a matrix of glass-like material. The clays can be used by themselves to make bricks, flowerpots, and clay pipe, but finer quality ceramic materials contain purified clays and other ingredients in carefully controlled proportions. Clay is used to make the mixture pliable; the silica decreases the amount of shrinkage which occurs after drying and firing in a kiln (Figure 12–15); the feldspar lowers the temperature needed for adequate firing, and acts as a glass-like material to hold the grains of clay and silica together. It is usually necessary to dry clay before firing it; otherwise the rapid loss of water from the surface and the slower loss from the interior of the object will cause cracks in the clay.

Figure 12–15 Fired pottery. Different colors and surface textures can be achieved by fusing the coating material with the clay itself. The coating ceases to be a "coat" and becomes a part of the whole, the entire structure being held together by covalent bonds. (Courtesy of the Robinson-Ransbottom Pottery Company.)

Natural clays are generally extremely complex mixtures. If these are used in ceramics without treatment, the finished materials have a color and physical properties characteristic of the impurities present. The first pieces of fine oriental chinaware arrived in Europe during the late Middle Ages, and European potters envied and admired the obviously superior product. This led to the beginning of systematic studies on the effect of composition on the nature of the ceramic produced and to a keen appreciation of the role of the purity of the clay in determining the color and potential decorative development of the piece.

Alchemists made notable achievements in this area. One of them, Johann Friedrich Bottger, worked from about 1705 to 1719 for King Augustus of Saxony, who kept him almost as a prisoner. The king hoped to gain power from the alchemist's discoveries. Bottger succeeded in developing several novel ceramic materials, of which the most important was the first white glazed porcelain made in Europe (in 1709). Bottger devoted the rest of his life to the perfection of the manufacture and decoration of this material, in which he enjoyed considerable success.

The china was made in Meissen and was both ***glazed*** and ***vitrified***. The glazing was accomplished by coating the pieces with a material which melted and produced an impermeable layer on the surface. Vitrification was produced by firing the clay at a temperature sufficient to melt a portion of the material and, in effect, produce an impermeable glass which held the remaining particles together.

Glazing and vitrification are melting processes which produce glass-like materials.

In the past few decades new ceramic materials have been developed and used on an increasingly wide scale. Nearly pure alumina (Al_2O_3) and zirconia (ZrO_2) are now used as bases for ceramic materials which are excellent electrical or thermal insulators. Magnetic ceramics, or ***ferrites,*** containing iron compounds are used as memory elements in computers.

In recent years a new class of materials, the ***glass ceramics,*** has been discovered; these have unusual but very valuable properties. Normally glass breaks because once a crack starts, there is nothing to stop it from spreading. It was discovered that if glass is treated by heating until a very large number of tiny crystals has developed in it, the resulting material, when cooled, is much more resistant to breaking than normal glass. The process has to be controlled carefully to obtain the desired

Figure 12–16 Cookware made of Pyroceram. (Courtesy of Corning Glass Company.)

Many new ceramics were developed for nose cones for space vehicles. Now they are being used in the kitchen.

properties. The materials produced in this way are generally opaque and are used for cooking utensils and kitchenware. They include materials marketed under the name "Pyroceram" (Figure 12–16). The initial manufacturing process is similar to that of other glass objects, but once they have been formed into their final shapes, they are heat-treated to develop their special properties.

Cements

A colloid is a classification of matter based on particle size (1-10,000 nm in diameter).

A cement is a material used to bind other materials together. Portland cement contains calcium, iron, aluminum, silicon, and oxygen in varying proportions. It has a structure somewhat similar to that described earlier for glass, except that in cement some of the silicon atoms have been replaced by aluminum atoms. Cement reacts in the presence of water to form a hydrated colloid of large surface area which subsequently undergoes recrystallization and reaction to bond to itself and to bricks or stone.

Cement is made by roasting a powdered mixture of calcium carbonate (limestone or chalk), silica (sand), an aluminosilicate mineral (kaolin, clay, or shale), and iron oxide. The roasting is carried out at a temperature of up to 870°C in a rotating kiln (Figure 12–17). As the materials pass through the kiln, they lose water and carbon dioxide and ultimately form a "clinker," in which the materials are partially fused together. The "clinker" is then ground to a very fine powder after the addition of a small amount of calcium sulfate (gypsum).

The composition of Portland cement is 60 to 67 percent CaO, 17 to 25 percent SiO_2, 3 to 8 percent Al_2O_3, up to 6 percent Fe_2O_3, and small amounts of magnesium oxide, magnesium sulfate, and potassium and sodium oxides.

Setting cement requires water and carbon dioxide.

The reactions which occur during the setting of cement are quite complex and involve the reaction of the various constituents with water and, subsequently at the surface, with the carbon dioxide in air. The initial reaction of cement with water gives a sticky gel which results from the hydrolysis of the calcium silicates. This sticks to itself and to the other particles (sand, crushed stone, or gravel). The gel has a very large surface area and is responsible for the strength of concrete. The setting process also involves the formation of small densely interlocked crystals after the initial solidification of the wet mass. This continues for years after the initial setting and increases the compressive strength of the cement. Water is required since the setting reactions involve hydration. For this reason freshly poured concrete is kept moist for several days. Over 400,000,000 tons of cement are manufactured each year, most of which is used to make concrete. Concrete, like many other materials containing Si—O bonds, is highly noncompressible but lacks tensile strength. If concrete is to be used where it is subject to tension, it must be reinforced with steel.

CHEMICAL FERTILIZERS—KEYS TO WORLD FOOD PROBLEMS

Regardless of the method used to predict future world populations, everyone agrees that food production for that population will be a major

A

B

Figure 12–17 *A*, A cement kiln. Note the rollers on the supports which allow the giant cylinder to rotate. As the kiln turns, the powder moves down the cylinder because it is at a slight angle. Intense heat is produced by the combustion of gaseous fuels. The powder loses volatile materials as it moves along and the finished product is discharged from the lower end. *B*, The first kiln used for making cement in the United States was constructed by David Saylor over a century ago at Coplay, Pennsylvania. The kiln still stands as pictured here. (Courtesy of the Portland Cement Northwestern States Company, Mason City, Iowa.)

TABLE 12-4 Nutrients Needed by Plants

Element	Compound
	Macronutrients
C	CO_2, carbon dioxide
H	H_2O, water
O	H_2O, water
N	NH_3, ammonia; NH_4NO_3, ammonium nitrate; H_2NCONH_2, urea
P	$Ca(H_2PO_4)_2$, calcium dihydrogen phosphate
K	KCl, potassium chloride
Ca	$Ca(OH)_2$, slaked lime
Mg	$MgCO_3$, magnesium carbonate
S	Elemental sulfur
	Micronutrients
B	$Na_2B_4O_7 \cdot 10H_2O$, borax
Cu	$CuSO_4 \cdot 5H_2O$, copper(II) sulfate pentahydrate
Fe	$FeSO_4$, iron(II) sulfate
Mn	$MnSO_4$, manganese sulfate
Zn	$ZnSO_4$, zinc sulfate
Mo	$(NH_4)_2MoO_4$, ammonium molybdate
Cl	KCl, potassium chloride

Soil nutrients must be replaced every one or two growing cycles.

problem. Indeed, adequate food production is a problem even today as starvation is a reality for as many as a billion of the four billion persons on earth. Food production is complicated, involving the fields of management, economics, mechanical and civil engineering, genetics, and chemistry. In this section we will deal with the chemical nutrients of plants as they relate to food production.

Man fertilized his crops with manure, dead fish, or straw for centuries before it was understood that growing plants require 16 different nutrients (Table 12-4). Nine of these, the macronutrients, are needed in large amounts, while seven, the micronutrients, are needed in smaller amounts. Three of the macronutrients, carbon, hydrogen, and oxygen, come from the atmosphere and are available to all plants, while the remainder of all nutrients must come from the soil. Growing plants deplete the available nutrients rather drastically (Table 12-5). Within the space of one or two growing cycles the soil nutrients must be replaced.

With today's modern requirements on yields, the use of chemical fertilizers is the only* way to continue the high productivity of the world's farmland (Figure 12-18).

There is a little less than one acre under cultivation for every individual on the earth.

There are about 3.5 billion acres presently under cultivation worldwide. It has been calculated that the application of $30 worth of fertilizer per acre would increase crop production by 50 percent, equivalent to 1.7 billion more acres under cultivation. Of course this would be expensive, requiring about $35 per capita worldwide, and might be impossible due to energy limitations since present fertilizer manufacture uses a

*Organic materials such as feedlot animal wastes, slaughter-house wastes, and plant wastes (stalks, leaves, etc.) could satisfy much of the soil nutrient demand, but social and economic restrictions have retarded their extensive application in this country.

TABLE 12–5 Approximate Amounts of Nutrients Required to Produce 150 Bushels of Corn

Nutrient	Approximate Pounds per Acre	Source
Oxygen	10,200	air
Carbon	7,800	air
Water	3,225 to 4,175 tons	29 to 36 inches of rain
Nitrogen	310	1,200 lbs. of high-grade fertilizer
Phosphorus	120 (as phosphate)	1,200 lbs. of high-grade fertilizer
Potassium	245 (K_2O)	1,200 lbs. of high-grade fertilizer
Calcium	58	150 lbs. of agricultural limestone
Magnesium	50	275 lbs. of magnesium sulfate (epsom salt)
Sulfur	33	33 lbs. of powdered sulfur
Iron	3	15 lbs. of iron sulfate
Manganese	.45	1.3 lbs. of manganese sulfate
Boron	.05	1 lb. of borax
Zinc	trace	small amount of zinc sulfate
Copper	trace	small amount of copper sulfate
Molybdenum	trace	trace of ammonium molybdate

significant portion of our fossil fuels. As one result of this, poorer countries such as India have almost been priced out of the world fertilizer markets due to the recent steep increase in the prices of petroleum products.

The large amount of energy needed to make fertilizer causes fertilizer prices to rise with the price of petroleum.

Nitrogen Chemistry

Since the atmosphere is 78 percent nitrogen by volume, it might appear that this nutrient would be quite readily available to plants.

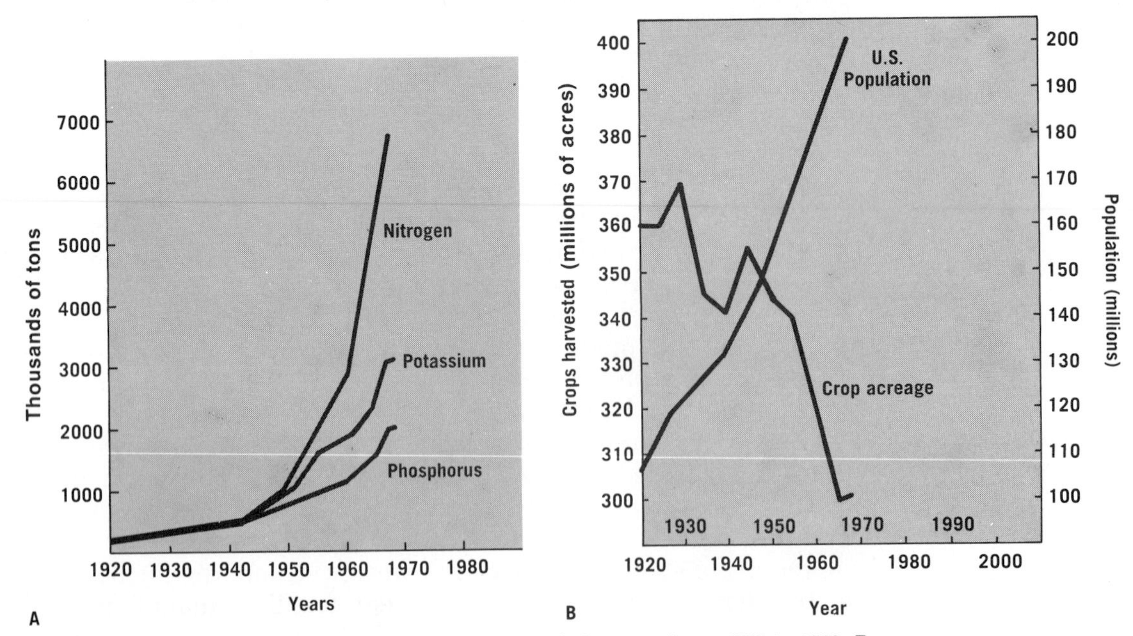

Figure 12–18 Fertilizers in U.S. agriculture. *A*, Use of plant nutrients, 1920 to 1968. *B*, Acreage of 59 principal crops harvested plus acreages in fruits, tree nuts, and farm gardens. Total U.S. population. (From Turk, A., Turk, J., and Wittes, J.: *Ecology, Pollution, and Environment.* Philadelphia, W. B. Saunders Co., 1972.)

Fixed nitrogen refers to nitrogen present in chemical compounds other than N_2.

However, due to the nitrogen-nitrogen triple bond in the N_2 molecule, nitrogen is rather unreactive and as such is not directly available to plants as a nutrient. To be useful, the nitrogen must be ***fixed***—that is, oxidized or reduced to some species containing nitrogen such as nitrate ion, NO_3^- (oxidized form), or ammonia, NH_3 (reduced form), either of which is chemically more reactive than the nitrogen molecule. Only a few organisms, such as some bacteria and blue-green algae, are able to fix atmospheric nitrogen. Bacteria of the genus *Rhizobium* can fix nitrogen when they grow in nodules in the roots of legumes such as peas, beans, and alfalfa. Generally more nitrogen is fixed than the plants need; thus, the soil nitrogen is increased. In some parts of the world, crop rotation between legumes and nonlegumes works quite nicely for soil nitrogen requirements. Two of the most important cereal crops, wheat and corn, are not favored by nitrogen-fixing bacteria and presently chemists and biologists are working on a newly discovered bacteria which grows in the root structures of certain South American grasses in hopes of establishing a new ***synergistic*** relationship between nitrogen-fixing bacteria and plants.

Synergism is a mutually cooperative action producing an effect greater than either action by itself.

Commercial fertilizers contain nitrogen fixed by the Haber process, the direct reaction of nitrogen with hydrogen to produce ammonia:

$$N_2 + 3H_2 \rightleftharpoons \underset{\text{AMMONIA}}{2NH_3}$$

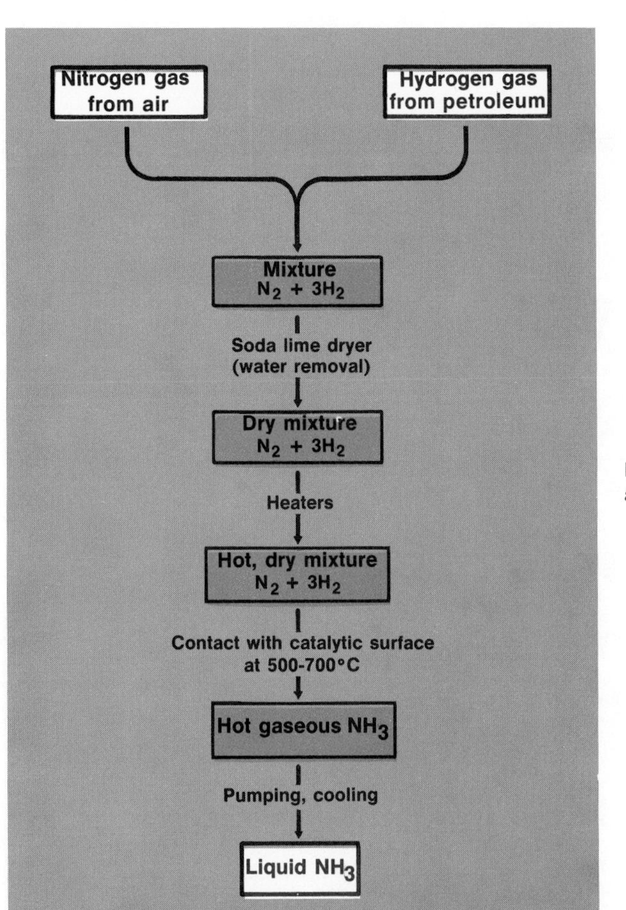

Figure 12–19 The Haber process for synthesizing ammonia.

Pure nitrogen is obtained by distilling oxygen and other gases from liquid air. Hydrogen is more difficult to obtain. At present, petroleum products such as propane, $CH_3CH_2CH_3$, are made to react with steam in the presence of catalysts to produce hydrogen:

$$\underset{\text{STEAM}}{CH_3CH_2CH_3 + 6H_2O} \xrightarrow{\text{Catalysts}} 3CO_2 + 10H_2$$

This is one of the principal reasons why ammonia fertilizer costs are so closely tied to petroleum prices. Hydrogen can also be prepared by the electrolysis of water,

$$2H_2O \xrightarrow[\text{KOH}]{\text{Electricity}} 2H_2 + O_2$$

and by several other methods, but all of these require great quantities of energy. So, as energy costs continue to rise, food costs will necessarily rise due to the added costs of fertilizers.

Nitrogen-containing fertilizers are all based upon ammonia. In the U.S., much of the ammonia is injected directly into the ground. Other useful fertilizers are prepared from ammonia. Ammonium nitrate,

NH_4NO_3, contains 35 per cent nitrogen and is the salt of the reaction between ammonia and nitric acid:

$$\underset{\text{AMMONIA}}{NH_3} + \underset{\text{NITRIC ACID}}{HNO_3} \longrightarrow \underset{\text{AMMONIUM NITRATE}}{NH_4NO_3}$$

Ammonium nitrate is a solid fertilizer. It is often applied to crops as a solution.

Ammonium nitrate, being a solid, has an advantage over ammonia in ease of handling, although it is sensitive to shock and can explode violently. Nitric acid used to prepare ammonium nitrate is itself prepared from ammonia by oxidation to nitrogen dioxide followed by reaction with water:

$$4NH_3 + 5O_2 \xrightarrow[\text{Pt catalyst}]{700°C} 4NO + 6H_2O$$

$$2NO + O_2 \longrightarrow 2NO_2$$

$$\underset{\text{NITROGEN DIOXIDE}}{3NO_2} + H_2O \longrightarrow \underset{\text{NITRIC ACID}}{2HNO_3} + \underset{\text{NITROGEN OXIDE (RECYCLED)}}{NO}$$

Urea (NH_2CONH_2) is probably one of the world's most important chemicals because of its wide use as a fertilizer and as a feed supplement for cattle. Ammonia and carbon dioxide react under high pressure near 200°C to produce first ammonium carbamate, which then decomposes into urea and water:

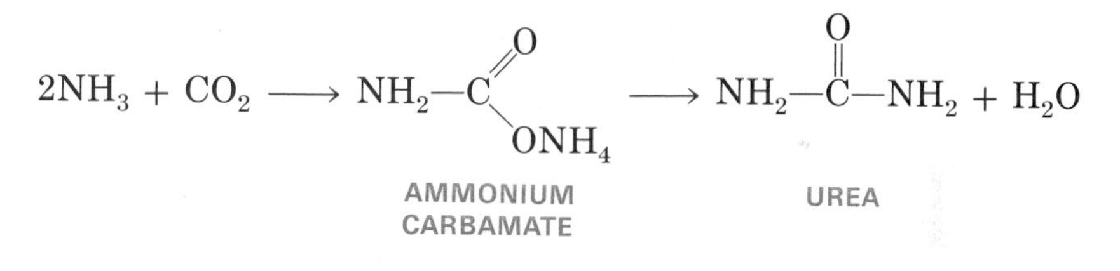

Urea synthesis also produces the compound biuret as a side product. Biuret must be removed from fertilizer-grade urea because it retards seed germination.

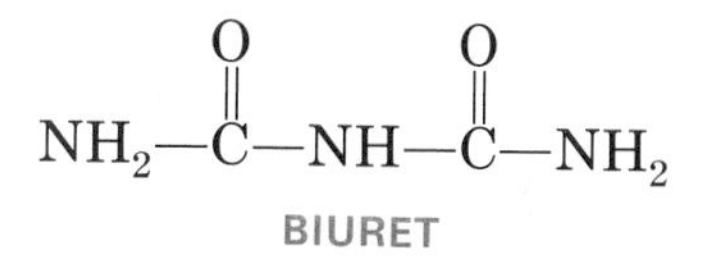

A slurry of urea and ammonium nitrate is often applied to crops under the name of "liquid nitrogen." Such a solution can contain up to 30 per-cent nitrogen and is very easy to store and apply.

Phosphate and Potash

These two mineral fertilizers can be mined, pulverized, and dusted directly onto the deficient soil. Often they are specially treated to produce desirable mixing properties. Phosphorus, for example, is found throughout the world in deposits of ***phosphate rock,*** which when treated with sulfuric acid produces a product of greater phosphorus availability called "superphosphate."

$$\underset{\text{PHOSPHATE ROCK}}{Ca_3(PO_4)_2} + 2H_2SO_4 + H_2O \longrightarrow \underbrace{Ca(H_2PO_4)_2 + 2CaSO_4}_{\text{"SUPERPHOSPHATE"}}$$

The world has limited deposits of the phosphate rock essential to the manufacture of fertilizers.

Phosphate rock itself cannot be applied due to its very low solubility. Recently, it has become evident that phosphate rock demand will eventually exceed supply unless large new deposits are discovered.

Potassium in the form of ***potash,*** K_2CO_3, exists in enormous quantities throughout the world. A soluble form of potassium is its chloride (KCl), also called potash, but because it often occurs with sodium chloride, which is toxic to plants, the potash ores must be treated by some process such as recrystallization to separate the two compounds.

The Trace Nutrient Iron

Certain of the micronutrients listed in Table 12–4 are quite interesting when their importance to plants is considered. For example, iron is an essential component of the catalyst involved in the formation of chlorophyll, the green plant pigment. When the soil is iron deficient, or when soil conditions exist such as too much "lime," $Ca(OH)_2$*, being present, iron availability will decrease. This condition is usually present when plant leaves lighten in color or even turn yellow. Often a gardener or lawn worker will apply lime, for calcium or to adjust soil acidity, and phosphate only to see his green plants turn yellow. What is happening is that both phosphate and the hydroxide from the lime tie up the iron,

The addition of lime, a basic substance, raises the pH of the soil.

*Actually, lime is calcium oxide, CaO, while hydrated lime, called slaked lime, is $Ca(OH)_2$. Slaked lime is sold as "lime."

Figure 12–20 Phosphate from agricultural uses is a principal cause of water pollution.

making it unavailable to the plants. The complex formation reactions are:

$$Fe^{3+} + \underset{\text{PHOSPHATE}}{2PO_4^{3-}} \longrightarrow \underset{\text{TIGHTLY BOUND COMPLEX}}{Fe(PO_4)_2^{3-}}$$

$$Fe^{3+} + 3OH^- \longrightarrow \underset{\text{INSOLUBLE HYDROXIDE}}{Fe(OH)_3}$$

These problems can be overcome since iron can be ***chelated*** with certain organic molecules which hold the iron quite tightly but will release the iron at the root structure. The word chelate comes from the Greek, *chela,* meaning claw.

Certain large negative ions, such as ethylenediaminetetraacetate (EDTA) (shown below in color), will react with an iron ion and chelate it, holding the iron tightly.

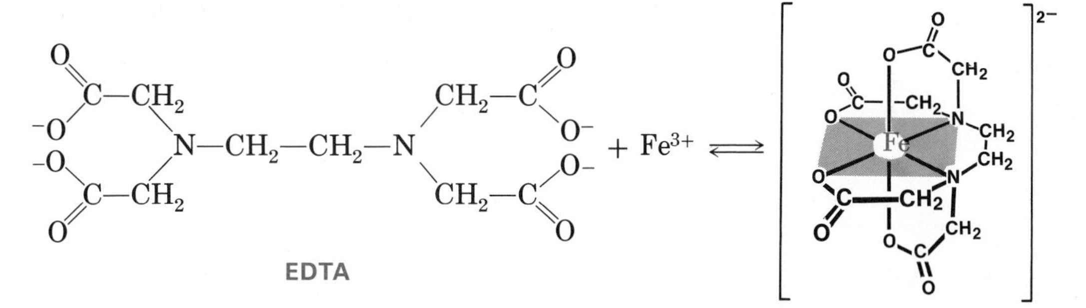

Since the resulting chelate is an ion, the iron can move through the soil to the root structures.

Chelate structures are important in many aspects of chemistry. Chlorophyll (see Chapter 17) is a chelate of magnesium, while hemoglobin (see Chapter 16) is an iron chelate. In Chapter 18, chelating agents such as EDTA will be discussed in relation to metal poisoning.

SELF-TEST 12-B

1. The principal element in glass is ______________.
2. Flint or crystal glass contains a large amount of the element ______________.
3. The fertilizer produced by the Haber process is ______________.
4. Three macronutrients which must be replaced in the soil and are the main ingredients in chemical fertilizers are ______________, ______________, and ______________.
5. The principal source of hydrogen used to make ammonia is (water/petroleum hydrocarbons).
6. Superphosphate is manufactured from phosphate rock and (nitric/sulfuric) acid.
7. Potassium is usually added to the soil in the form of (K_2O/KCl/KNO_3).

MATCHING SET

________ 1.	copper	a.	a nutrient necessary for the manufacture of chlorophyll
________ 2.	aluminum	b.	calcium fluoride added to glass
________ 3.	milk glass	c.	a nutrient in short supply
________ 4.	magnesium	d.	source of carbon for plants
________ 5.	oyster shells	e.	made from nitrogen and hydrogen
________ 6.	atmosphere	f.	a limitless supply in sea water
________ 7.	ammonia	g.	supply calcium hydroxide for magnesium production
________ 8.	iron	h.	purified electrolytically
________ 9.	phosphate	i.	the most abundant metal in the earth's crust

QUESTIONS

1. Name three metals that you would expect to find free in nature. Name three that you would not.
2. What is the primary reducing agent in the production of iron from its ore?
3. Why is CaO necessary for the production of iron in a blast furnace?
4. What is the chemical difference between iron and steel?
5. Both iron and magnesium will oxidize in the air. Why is the oxidation of iron a much greater problem than the oxidation of magnesium?
6. How is it possible that oxidation and reduction can occur at different points on a piece of iron?
7. Give examples of three ways in which metals can be protected from corrosion.
8. Describe the solution used in the commercial cell for the electrolytic reduction of aluminum.
9. Why is it so important to purify electrolytically industrial quantities of copper to a level above 99.9 percent pure?
10. What chemical is obtained from oyster shells in the production of magnesium from sea water? What is the role of this chemical in the process?
11. What is the difference between quartz and sand?
12. Explain how the structures of glass and a liquid are similar.
13. Give reactions involved in the preparation of:

    $Ca(OH)_2$ from $CaCO_3$
    NH_3 from N_2 and H_2
    HNO_3 from NH_3

14. Explain what is meant by the term "flotation." Why is this an important process?

15. In the production of copper from its sulfide ores, large amounts of SO_2 gas are formed as a by-product. What useful product can be prepared from this material? What are the consequences of allowing the gas to escape into the atmosphere?

16. Name four items in everyday use which contain (a) silicon, (b) aluminum, (c) calcium.

17. A typical lime-soda glass has a composition reported as 70 percent SiO_2, 15 percent Na_2O, and 10 percent CaO. What weights of sand (SiO_2), sodium carbonate (Na_2CO_3), and calcium carbonate ($CaCO_3$) must be melted together to make this glass? The carbonates are decomposed by heat to evolve carbon dioxide gas.

18. What is the maximum weight (in pounds) of magnesium that can be obtained from 1000 pounds of sea water?

19. Explain why fertilizer costs are directly tied to petroleum costs.

20. Of the three major plant macronutrients, nitrogen, potassium, and phosphorus, which is in the most danger of running out? Explain.

SUGGESTIONS FOR FURTHER READING

Andrade, J., "Materials Science and Engineering—A Modern Multidiscipline," *Chemistry,* Vol. 43, No. 4, p. 13 (1970).
Brown, G. H., "Liquid Crystals," *Chemistry,* Vol. 40, No. 9, p. 10 (1967).
Chandler, M., "Ceramics in the Modern World," Doubleday & Co., Inc., Garden City, New York, 1968.
"Encyclopedia of Chemical Technology," Second Edition, Interscience Publishers, New York, 1963.
Farber, E., "Oxygen—The Element With Two Faces," *Chemistry,* Vol. 39, No. 5, p. 17 (1966).
"Industry Profile," *Chemistry,* Vol. 39, No. 6, pp. 17–18 (June, 1966).
"McGraw-Hill Encyclopedia of Science and Technology," McGraw-Hill Book Co., New York, 1960.
Maloney, F. J. T., "Glass in the Modern World," Doubleday & Co., Inc., Garden City, New York, 1968.
Safrany, David, "Nitrogen Fixation," *Scientific American,* Vol. 224, No. 10, p. 64 (1974).
Sanchelli, V., "Trace Elements in Agriculture," Van Nostrand Reinhold Co., New York, 1969.
Schaar, B. E., "Dental Amalgams," *Chemistry,* Vol. 40, No. 8, p. 21 (1967).
Stone, J. K., "Oxygen in Steel Making," *Scientific American,* Vol. 218, No. 4, p. 24 (1968).
Turk, Amos, et al., "Environmental Science," W. B. Saunders Co., Philadelphia, 1974, p. 230.

THE UBIQUITOUS CARBON ATOM—AN INTRODUCTION TO ORGANIC CHEMISTRY

CHAPTER 13

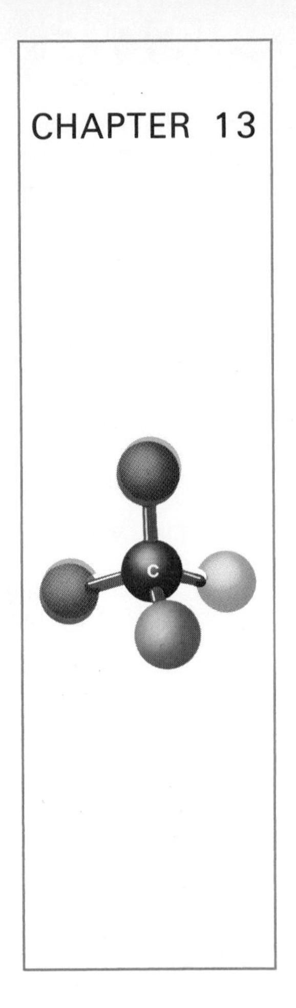

The importance of carbon compounds to life on earth cannot be overestimated. Consider what the world would be like if all the carbon and carbon compounds were suddenly removed. The result would be somewhat like the barren surface of the moon! Many of the little everyday things often taken for granted would be quite impossible without this versatile element. In an ordinary pencil, for example, the "lead" in the pencil (made from graphite, an elementary form of carbon), the wood, the rubber in the eraser, and the paint on the surface are all carbon or carbon compounds. The paper in this book, the cloth in its cover, and the glue holding it together are also made of carbon compounds. All of the clothes one wears, including the leather in shoes, would not exist. If carbon compounds were removed from the human body, there would be nothing left except water and a small residue of minerals, and the same is true of all forms of living matter. Fossil fuels, foods, and most drugs are essentially made of carbon compounds. In addition, many carbon compounds such as plastics and detergents, which are not directly connected with the life processes, play a vital role in one's life.

Carbon and its compounds are vital to life on this planet.

There are about two million different carbon compounds that have been studied and described in the chemical literature, and thousands of new ones are reported every year. Although there are 88 other naturally occurring elements, the number of known carbon compounds is many times greater than that of the known compounds which contain no carbon. The very large and important branch of chemistry devoted to the study of carbon compounds is called ***organic chemistry.*** The name "organic" is actually a relic of the past, when chemical compounds produced from once-living matter were called "organic" and all other compounds were called "inorganic."

Organic chemistry is the study of the compounds of carbon.

THE BASIS FOR THE LARGE NUMBER OF ORGANIC COMPOUNDS

The enormous number of organic compounds has intrigued chemists for over a hundred years. The atomic theory, as developed earlier for all atoms, describes a structure for the carbon atom which explains this multiplicity of carbon compounds. The peculiar structure of this atom allows it ***to form covalent bonds with other carbon atoms in a seemingly endless array of possible combinations.*** A simple organic molecule may contain a single carbon-carbon bond, whereas a complex one may contain literally thousands of such bonds. A few other elements are capable of forming stable bonds between like atoms. These include such elements as nitrogen, N_2, oxygen, O_2, and sulfur, S_8, to name a few. But only S, Sn, Si, and P can form long chain molecules, and none of these can approach carbon in the ability to do this.

The large number of carbon compounds is due to:
1. stability of chains of carbon atoms;
2. occurrence of isomers;
3. importance and reactivity of functional groups.

An additional factor in the large number of carbon compounds lies in the stability of carbon chains. The carbon chains are not normally subject to attack by water or, at ordinary temperatures, by oxygen. Chains formed by atoms of other elements undergo reaction with either water or oxygen, or both, much more easily than do carbon chains.

A further reason for the large number of organic compounds is the ability of a given number of atoms to combine in more than one molecular pattern and, hence, produce more than one compound. Such compounds, each of which has molecules containing the same number and kinds of atoms, but arranged differently relative to each other, are called ***isomers.*** For example, the molecular structure represented by A—B—C is different from the molecular structure A—C—B, as is C—A—B; these three species are isomers. If one considers the number of possible ways the digits one through nine can be ordered to make nine-digit numbers, he can begin to imagine how a single group of atoms could possibly form hundreds of different molecules. Carbon, with its ability to bond to other carbon atoms, is especially well suited to form isomers.

Isomers are two or more different compounds with the same number of each kind of atom per molecule.

A final factor explaining the large number of organic compounds is the ability of the carbon atom to form strong covalent bonds with atoms of numerous other elements, such as nitrogen, oxygen, sulfur, chlorine, fluorine, bromine, and iodine. As a result there are large classes of organic compounds, each of which has a ***functional group,*** a particular combination of atoms, which appears in each member of that class. For example, organic acids have a carboxyl group attached to another carbon atom.

$$\left(-C\begin{matrix}\nearrow O \\ \searrow OH\end{matrix}\right)$$

CARBOXYL GROUP, A FUNCTIONAL GROUP

The remainder of this chapter will be devoted to a closer examination of these three basic reasons for the great number of organic compounds. Some interesting properties of a few of these compounds will be discussed.

A dash in a formula represents a single bond; two electrons are shared.

CHAINS OF CARBON ATOMS—THE HYDROCARBONS

Carbon has an intermediate electronegativity value of 2.5. This intermediate value is not large enough to enable a carbon atom to remove electrons completely from metals, but it is sufficient to keep even the most electronegative atom, fluorine, from removing an electron from carbon. As a result, carbon atoms tend not to form ionic bonds but rather to share electrons in the formation of covalent bonds. In addition to the tendency for carbon to form covalent bonds with many other atoms, there is also a remarkable inclination for carbon atoms to form relatively strong covalent bonds with each other.

The electronegativity of an element is a measure of its ability to attract electrons to itself in a covalent bond.

In Chapter 9 the structure around a carbon atom bonded to either three or four other atoms was discussed. If three atoms are bonded to a carbon atom (one by a double bond such as in $H_2C{=}CH_2$), the arrangement is trigonal planar. If four atoms are bonded to a carbon atom (such as CH_4), the arrangement is tetrahedral. These structures are consistent with experimental data, hybridization theory, and valence-shell electron-pair repulsion concepts.

A trigonal planar structure

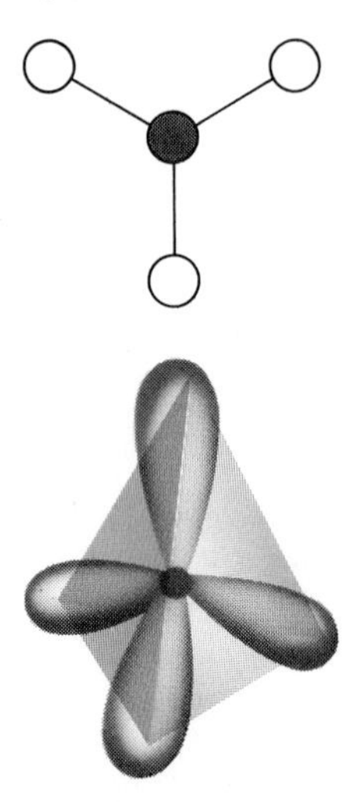

A tetrahedron around a carbon atom

Also in Chapter 9, the formation of sigma and pi bonds was discussed. A sigma bond is formed either by the overlap of an s orbital with any other orbital or by the end-to-end overlap of the *ends* of any two orbitals such as p-p, sp^3-sp^3, and p-sp^3. A pi bond is formed by the overlap of the *sides* of two orbitals, generally two p orbitals.

Only two elements, hydrogen and carbon, are required to explain the existence of literally thousands of compounds known as ***hydrocarbons.*** The simplest hydrocarbon is methane, CH_4. Methane is tetrahedral and has four C—H sigma bonds, as shown in Figure 13–1.

A hydrocarbon with two carbon atoms is ethane, C_2H_6. Ethane has six C—H sigma bonds formed by sp^3-s overlap as in methane and one C—C sigma bond. The C—C single bond is formed by sp^3-sp^3 overlap. The structure of ethane is illustrated by the following oversimplified formulas:

$$\begin{array}{c}\mathrm{H}\ \mathrm{H}\\ \mathrm{H}:\ddot{\underset{..}{\mathrm{C}}}:\ddot{\underset{..}{\mathrm{C}}}:\mathrm{H}\\ \mathrm{H}\ \mathrm{H}\end{array} \quad \text{or} \quad \begin{array}{ccccc} & \mathrm{H} & & \mathrm{H} & \\ & | & & | & \\ \mathrm{H}- & \mathrm{C} & - & \mathrm{C} & -\mathrm{H}\\ & | & & | & \\ & \mathrm{H} & & \mathrm{H} & \end{array}$$

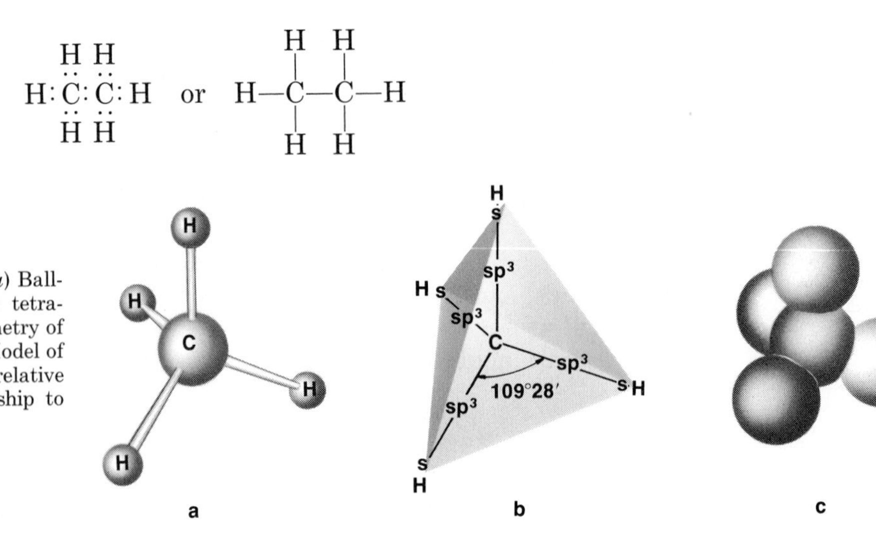

Figure 13–1 Methane. (*a*) Ball-and-stick model showing tetrahedral structure. (*b*) Geometry of regular tetrahedron. (*c*) Model of methane, CH_4, showing relative size of atoms in relationship to interatomic distances.

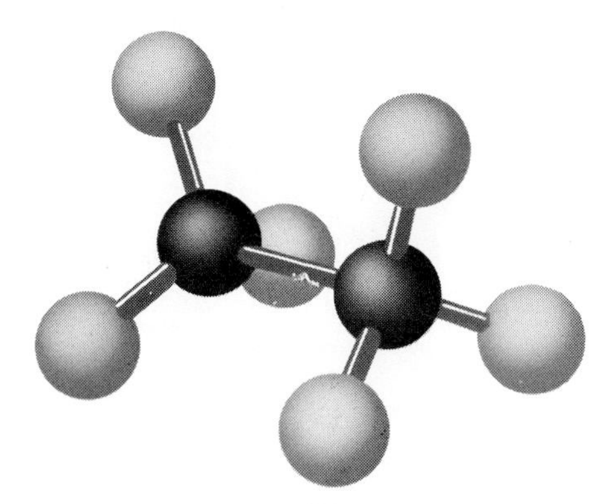

Figure 13–2 Ball-and-stick model of ethane.

The three-dimensional shape of an ethane molecule is indicated in Figure 13–2.

Ethane is composed of two CH_3 groups (methyl groups, from methane) joined by a single carbon-carbon sigma bond. Since this sigma bond is symmetrical about the bond axis (the line drawn between the carbon atoms), there is only a little interference when one methyl group spins around the bond axis relative to the other methyl group. In Figure 13–3 it is evident that the hydrogen atoms can be lined up, in a staggered relationship, or in any intermediate position. The eclipsed (or lined-up) form is actually a slightly higher energy form due to atomic repulsions; the staggered form, in which repulsions are less important, is a slightly lower energy form. This rotation about a carbon-carbon single bond is a common feature in organic compounds.

By applying what we have learned, it is a simple matter to extend the concept of carbon-carbon bonding to a three-carbon molecule such as propane (C_3H_8).

$$\begin{array}{c} \mathrm{H\ H\ H} \\ \mathrm{H{:}\ddot{\underset{..}{C}}{:}\ddot{\underset{..}{C}}{:}\ddot{\underset{..}{C}}{:}H} \\ \mathrm{H\ H\ H} \end{array} \quad \text{or} \quad \begin{array}{c} \mathrm{\ \ H\ \ \ H\ \ \ H} \\ \mathrm{\ \ |\ \ \ \ |\ \ \ \ |} \\ \mathrm{H{-}C{-}C{-}C{-}H} \\ \mathrm{\ \ |\ \ \ \ |\ \ \ \ |} \\ \mathrm{\ \ H\ \ \ H\ \ \ H} \end{array}$$

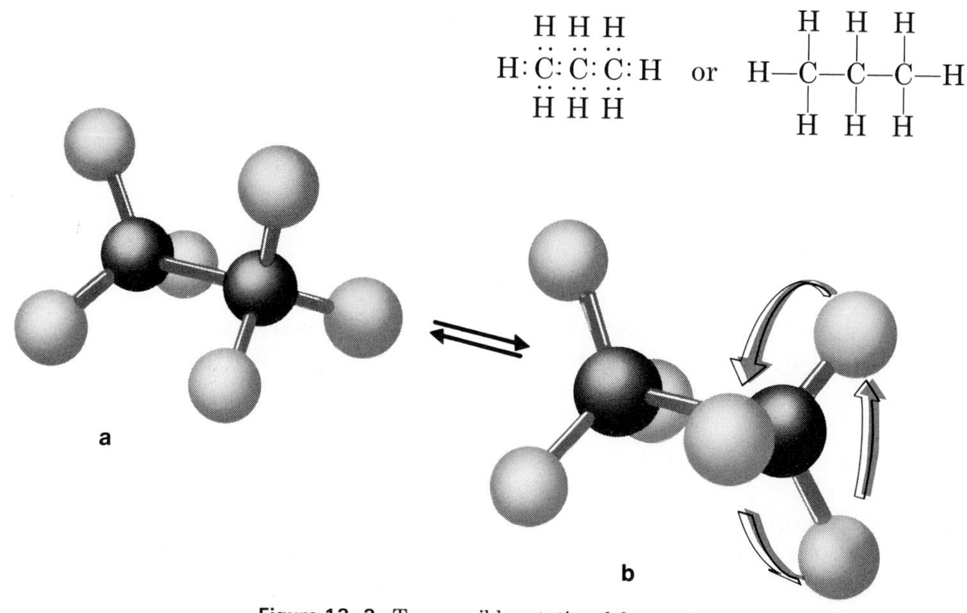

Figure 13–3 Two possible rotational forms of the ethane molecule. The hydrogen atoms in the methyl groups may be in an eclipsed position (a), staggered (b), or in any intermediate position. In ethane the two methyl groups can rotate easily about the carbon-carbon bond.

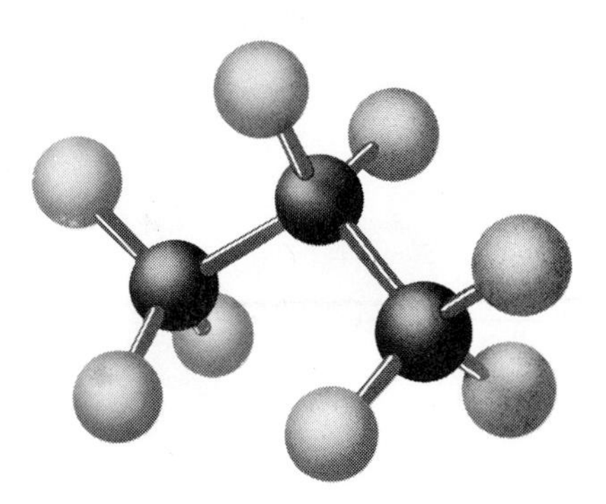

Figure 13–4 Ball-and-stick model of propane.

Propane

In Figure 13–4, note that the three carbon atoms in propane do not lie in a straight line because of the tetrahedral bonding about each carbon atom. Also, because of the rotation about the two C—C sigma bonds, the molecule is "flexible."

It is apparent that these bonding concepts can be extended to a four-carbon molecule and to a limitless number of even larger hydrocarbon molecules. Actually, many such compounds are known; some, such as natural rubber, are known to contain over a thousand carbon atoms in a chain.

STRUCTURAL ISOMERS

Structural isomers have the same molecular formulas but a different pattern of bonds.

When given the task of writing the structures for butane, C_4H_{10}, the student will soon discover that two structures are possible.

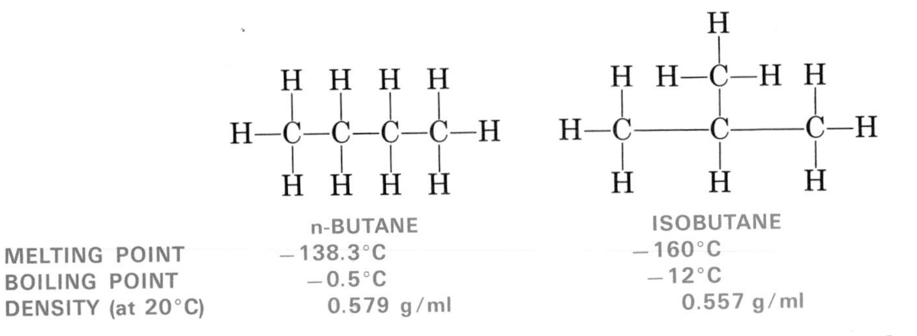

n-BUTANE
ISOBUTANE
MELTING POINT
BOILING POINT
DENSITY (at 20°C)
−138.3°C
−0.5°C
0.579 g/ml
−160°C
−12°C
0.557 g/ml

The two formulas represent two distinctly different compounds. Both are well known, each with its own particular set of properties. It must be concluded, then, that molecular formulas such as C_4H_{10} are sometimes ambiguous, and that structural formulas are necessary. If no carbon atom is attached to more than two other carbon atoms, the carbon chain is said to be a ***straight chain*** structure. Actually, as shown in Figure 13–5, the carbon chain is bent (109° 28′) at each carbon atom, but it is called a straight chain because the carbon atoms are bonded together in succession one after the other. The student might note that many molecular shapes are possible for n-butane because of the possible rotational motion about the single bonds. These arrangements (called ***conformations***) do not constitute different molecules; rather, there tends to be a mixture of all possible shapes at a given temperature due to the ease of bond rotation.

Rotational forms resulting from the twisting around C—C single bonds are called conformations.

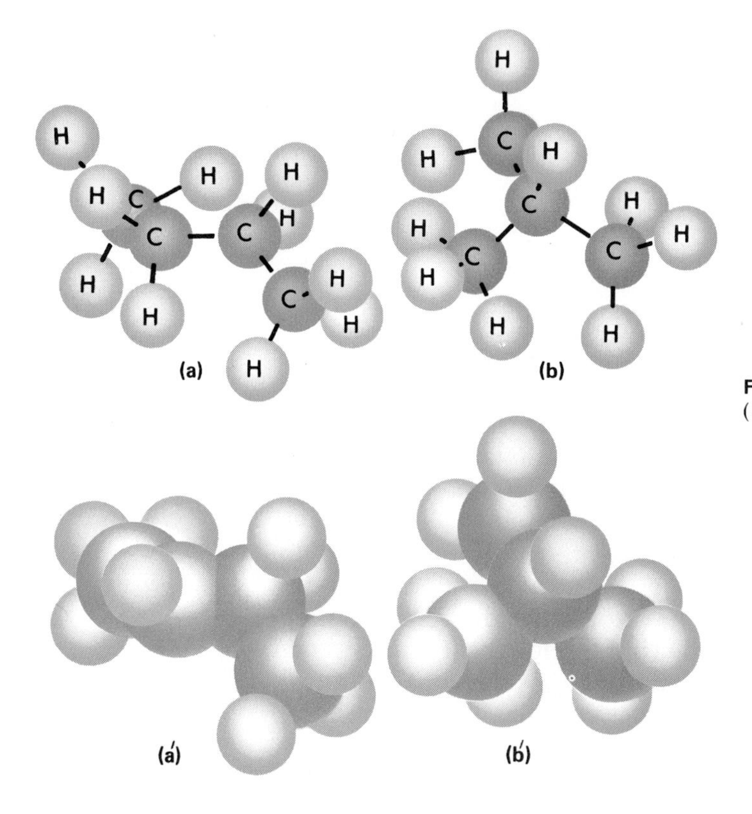

Figure 13–5 (*a*) and (*a′*) Normal butane. (*b*) and (*b′*) Isobutane.

In all branched chain structures, at least one carbon atom is bonded to three or four other carbon atoms.

If one carbon atom is bonded to either three or four other carbon atoms in a molecule, the molecule is said to have a ***branched chain.*** Isobutane is an example of a branched-chain hydrocarbon (Figure 13–5). Isobutane and n-butane are called ***structural isomers*** because both molecules contain exactly the same number and kinds of atoms, C_4H_{10}, but the molecules are put together differently. Structural isomerism is somewhat like a child building many different structures with the same collection of building blocks, and using all of the blocks in each structure.

The two butanes (and all hydrocarbon molecules) are essentially nonpolar since the C—C bonds are nonpolar, and the slightly polar C—H bonds are symmetrically arranged to cancel each other out. The forces holding these molecules together in the liquid, therefore, are van der Waals attractions, which depend upon the surface area of a molecule and the closeness of approach of the molecules to each other. In general, a branched-chain isomer has a lower boiling point than a straight-chain isomer since the branched-chain isomer does not permit intermolecular distances as short as those of a straight chain and has less surface area. Both of these mean less intermolecular attraction. Melting points of isomers generally do not follow the same pattern since they also depend upon the ease with which the molecules fit into a crystalline array.

It is important to distinguish between different bond rotational arrangements (conformations) and structural isomers (configurations). To change from one rotational arrangement to another, only motion about a bond is required. However, to change from one structural isomer to another (for example, from isobutane to n-butane), it is necessary to break bonds and to form new ones. Structural isomers are "permanent" arrangements; conformations are transient.

Consider the isomeric pentanes with the formula, C_5H_{12}. There are three of these:

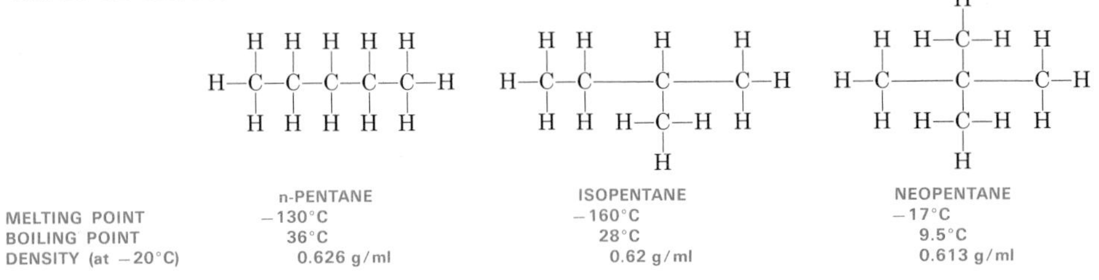

	n-PENTANE	ISOPENTANE	NEOPENTANE
MELTING POINT	−130°C	−160°C	−17°C
BOILING POINT	36°C	28°C	9.5°C
DENSITY (at −20°C)	0.626 g/ml	0.62 g/ml	0.613 g/ml

Note that the three isomers are predicted by bonding theory. Furthermore, the theory predicts no other possible isomers for C_5H_{12} since there are no other ways to unite the 17 atoms and satisfy the octet rule for Period II. These three isomers of pentane are well known, and no others have ever been found.

According to the octet rule, each atom (except H) shares or controls eight valence shell electrons. Hydrogen shares only two electrons.

Table 13–1 gives the number of isomers predicted for some larger molecular formulas, starting with C_6H_{14}. Every predicted isomer, *and no more,* has been isolated and identified for the C_6, C_7, and C_8 groups. However, not all of the C_{15}'s and C_{20}'s have been produced, but there is sufficient belief in the theory to presume that if enough time and effort were spent, all of the isomers could eventually be produced. Structural isomerism certainly helps to explain the vast number of carbon compounds.

Structural isomers can also exist in molecules with double and triple carbon-carbon bonds. For example, two of the six isomers of C_4H_8, 1-butene and *trans*-2-butene, have the following structures:

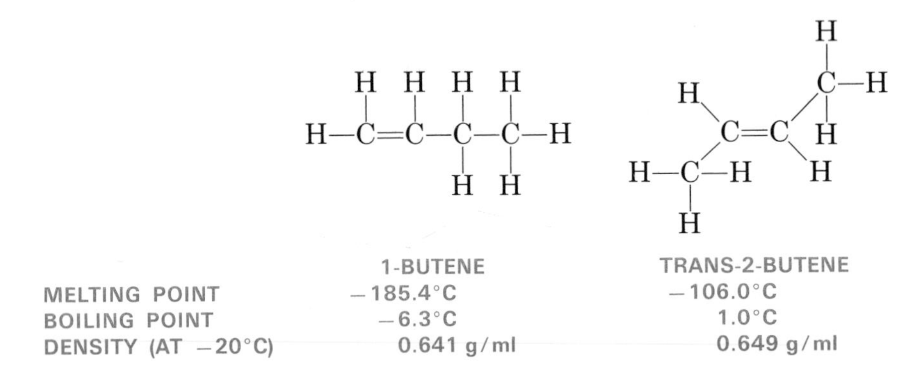

	1-BUTENE	TRANS-2-BUTENE
MELTING POINT	−185.4°C	−106.0°C
BOILING POINT	−6.3°C	1.0°C
DENSITY (AT −20°C)	0.641 g/ml	0.649 g/ml

The number placed before the name butene indicates the position number of the double bond. *Trans*-2-butene is an example of a geometrical

TABLE 13–1 Structural Isomers of Some Hydrocarbons

Formula	Isomers Predicted	Found
C_6H_{14}	5	5
C_7H_{16}	9	9
C_8H_{18}	18	18
$C_{15}H_{32}$	4347	—
$C_{20}H_{42}$	366,319	—
$C_{30}H_{62}$	4,111,846,763	—

isomer which results from double bonds within a molecule. Geometric isomers will be considered later. Note that the properties of 1-butene and 2-butene definitely indicate two *different* compounds. 1-Butyne and 2-butyne illustrate structural isomerism due to the positioning of triple bonds.

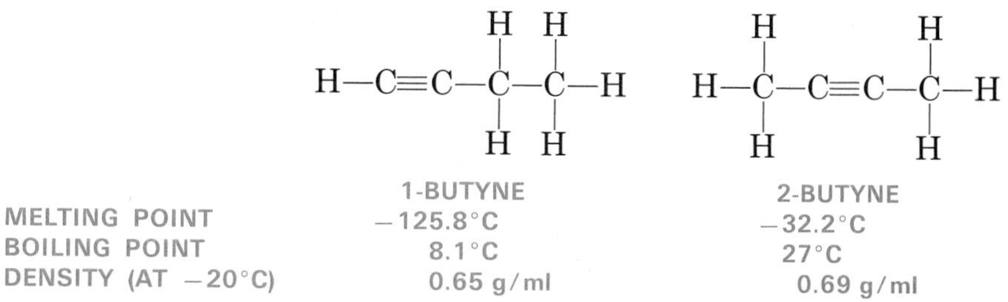

	1-BUTYNE	2-BUTYNE
MELTING POINT	−125.8°C	−32.2°C
BOILING POINT	8.1°C	27°C
DENSITY (AT −20°C)	0.65 g/ml	0.69 g/ml

There are large numbers of single bonded ring structures which contain three or more carbon atoms in the rings.

Hydrocarbons containing a double bond have isomers with single bonded ring-type structures. For example, in addition to the isomers given before for C_4H_8, another isomer exists which has a cyclic structure with single bonds.

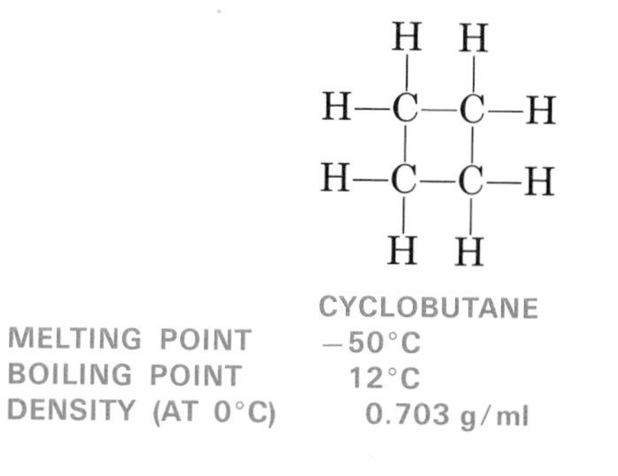

	CYCLOBUTANE
MELTING POINT	−50°C
BOILING POINT	12°C
DENSITY (AT 0°C)	0.703 g/ml

The compounds with double and triple bonds are much more reactive than those with only single bonds. For example, such compounds readily undergo ***addition reactions*** with halogens and compounds such as HI:

$$CH_2{=}CH_2 + Br_2 \longrightarrow CH_2Br{-}CH_2Br$$
$$CH_2{=}CH_2 + HI \longrightarrow CH_3CH_2I$$
$$HC{\equiv}CH + 2Br_2 \longrightarrow CHBr_2{-}CHBr_2$$

Unsaturated compounds contain double or triple bonds; saturated compounds do not contain such bonds.

The greater reactivity of alkenes and alkynes is one of the reasons why processes to transform the alkanes of petroleum into alkenes and alkynes are so important. Because of their ability to add on other molecules, alkenes and alkynes are often called ***unsaturated*** compounds while alkanes are referred to as ***saturated*** compounds. Like the alkanes, the alkenes and alkynes undergo combustion reactions with oxygen and hence burn in air:

$$CH_3{=}CH_2 + 3O_2 \longrightarrow 2CO_2 + 2H_2O + \text{heat}$$
$$2HC{\equiv}CH + 5O_2 \longrightarrow 4CO_2 + 2H_2O + \text{heat}$$

NOMENCLATURE

With so many organic compounds, a system of common names quickly fails due to the shortage of unique names. It is evident that a system of nomenclature is needed that makes use of numbers as well as names. Much attention has been given to this problem, and several international conventions have been held to work out a satisfactory system that can be used throughout the world. The International Union of Pure and Applied Chemistry has given its approval to a very elaborate nomenclature system (**IUPAC System**), and this system is now in general use.

A few of these IUPAC names will be needed, and an appreciation for the basic simplicity of the approach in naming organic compounds is desirable. Inscrutable names, such as some of those encountered on medicine bottles, are converted to systematic names that are descriptive of the molecules involved. The names of a few of the hydrocarbons are presented in Table 13–2, and additional points in nomenclature will be presented in later sections.

Alkanes: hydrocarbons with single bonds only; name ending: -ane.

Alkenes: hydrocarbons with one or more double bonds; name ending -ene.

Alkynes: hydrocarbons with one or more triple bonds; name ending -yne.

When writing and interpreting structural formulas for organic compounds remember that there are many equivalent ways of writing the same structural formula. For example,

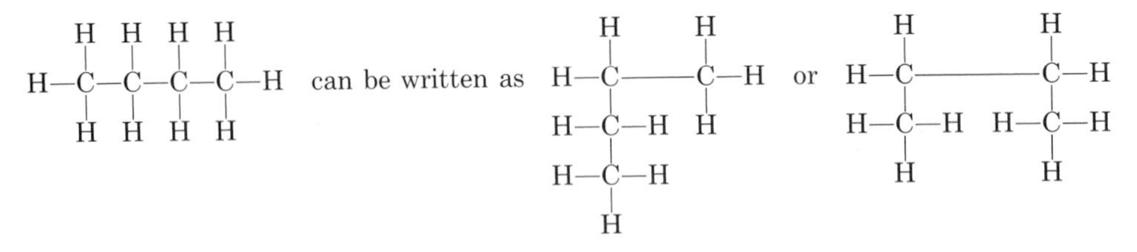

For branched-chain hydrocarbons it becomes necessary to name submolecular groups. The $—CH_3$ group is called the methyl group; this name is derived from methane by dropping the -ane and adding -yl. Any of the nine other hydrocarbons listed in Table 13–2 could give rise to a similar group. For example, the propyl group would be $—C_3H_7$. As an illustration of the use of the group names, consider this formula:

The -yl ending indicates an attached group such as $—CH_3$, methyl.

An aliphatic compound is any hydrocarbon that has an open chain of single-bonded carbon atoms.

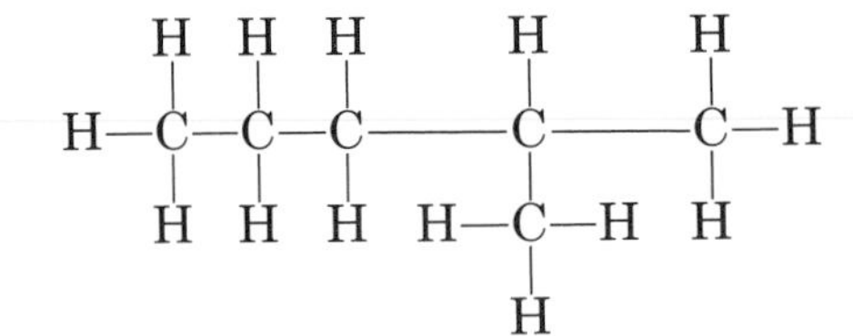

The longest carbon chain in the molecule is five carbon atoms long; hence, the root name is pentane. Furthermore, it is a methylpentane (written as one word) because a methyl group is attached to the pentane structure. In addition, a number is needed because the methyl group could be bonded to either the second or third carbon atom.

position of group on chain

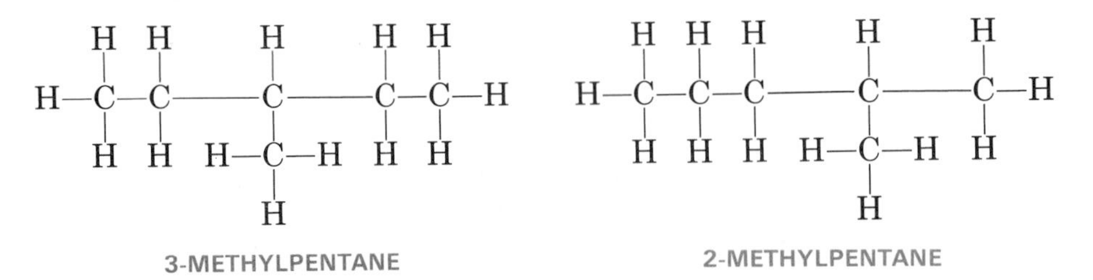

3-METHYLPENTANE

2-METHYLPENTANE

TABLE 13–2 The First Ten Straight-Chain Saturated Hydrocarbons

Name	Formula	Structural Formula
Methane	CH_4	$\mathrm{H-\overset{H}{\underset{H}{C}}-H}$
Ethane	C_2H_6	$\mathrm{H-\overset{H}{\underset{H}{C}}-\overset{H}{\underset{H}{C}}-H}$
Propane	C_3H_8	$\mathrm{H-\overset{H}{\underset{H}{C}}-\overset{H}{\underset{H}{C}}-\overset{H}{\underset{H}{C}}-H}$
n-Butane	C_4H_{10}	$\mathrm{H-\overset{H}{\underset{H}{C}}-\overset{H}{\underset{H}{C}}-\overset{H}{\underset{H}{C}}-\overset{H}{\underset{H}{C}}-H}$
n-Pentane	C_5H_{12}	$\mathrm{H-\overset{H}{\underset{H}{C}}-\overset{H}{\underset{H}{C}}-\overset{H}{\underset{H}{C}}-\overset{H}{\underset{H}{C}}-\overset{H}{\underset{H}{C}}-H}$
n-Hexane	C_6H_{14}	$\mathrm{H-\overset{H}{\underset{H}{C}}-\overset{H}{\underset{H}{C}}-\overset{H}{\underset{H}{C}}-\overset{H}{\underset{H}{C}}-\overset{H}{\underset{H}{C}}-\overset{H}{\underset{H}{C}}-H}$
n-Heptane	C_7H_{16}	$\mathrm{H-\overset{H}{\underset{H}{C}}-\overset{H}{\underset{H}{C}}-\overset{H}{\underset{H}{C}}-\overset{H}{\underset{H}{C}}-\overset{H}{\underset{H}{C}}-\overset{H}{\underset{H}{C}}-\overset{H}{\underset{H}{C}}-H}$
n-Octane	C_8H_{18}	$\mathrm{H-\overset{H}{\underset{H}{C}}-\overset{H}{\underset{H}{C}}-\overset{H}{\underset{H}{C}}-\overset{H}{\underset{H}{C}}-\overset{H}{\underset{H}{C}}-\overset{H}{\underset{H}{C}}-\overset{H}{\underset{H}{C}}-\overset{H}{\underset{H}{C}}-H}$
n-Nonane	C_9H_{20}	$\mathrm{H-\overset{H}{\underset{H}{C}}-\overset{H}{\underset{H}{C}}-\overset{H}{\underset{H}{C}}-\overset{H}{\underset{H}{C}}-\overset{H}{\underset{H}{C}}-\overset{H}{\underset{H}{C}}-\overset{H}{\underset{H}{C}}-\overset{H}{\underset{H}{C}}-\overset{H}{\underset{H}{C}}-H}$
n-Decane	$C_{10}H_{22}$	$\mathrm{H-\overset{H}{\underset{H}{C}}-\overset{H}{\underset{H}{C}}-\overset{H}{\underset{H}{C}}-\overset{H}{\underset{H}{C}}-\overset{H}{\underset{H}{C}}-\overset{H}{\underset{H}{C}}-\overset{H}{\underset{H}{C}}-\overset{H}{\underset{H}{C}}-\overset{H}{\underset{H}{C}}-\overset{H}{\underset{H}{C}}-H}$

Positional numbers of groups should be the smallest set possible.

Note that 2-methylpentane is the same as 4-methylpentane since the latter would be the same molecule turned around; the accepted rule requires numbering from the end of the carbon chain that will result in the smallest numbers. Therefore, 2-methylpentane is the correct name.

Any number of substituted groups can be handled in this same fashion. Consider the name and the formula for 4,4-diethyl-3,5,6,6-tetramethyloctane.

$$\mathrm{CH_3CH_2-\overset{CH_3}{\underset{CH_3}{C}}-\overset{H}{\underset{CH_3}{C}}-\overset{C_2H_5}{\underset{C_2H_5}{C}}-\overset{H}{\underset{CH_3}{C}}-CH_2CH_3}$$

If a double bond appears in a hydrocarbon, the root name, which indicates the number of carbon atoms, must be modified to reflect the double bond structure and its position. Changing -ane to -ene indicates the presence of the double bond, and a number is used to indicate its position. For example:

No number is necessary for ethene or propene because there is only one possible position for the double bond.

$CH_2{=}CH_2$ ETHENE (COMMON NAME: ETHYLENE)

$CH_2{=}CHCH_3$ PROPENE

$CH_2{=}CHCH_2CH_3$ 1-BUTENE

$CH_3CH{=}CHCH_3$ 2-BUTENE

$$\begin{array}{l} \quad\;\; CH_3 \\ \quad\;\;\; | \\ CH_3{-}C{=}CHCH_2CH_3 \end{array}$$ 2-METHYL-2-PENTENE

If a triple bond is present the -ane is changed to -yne. Examples:

$H{-}C{\equiv}C{-}H$ ETHYNE (COMMON NAME: ACETYLENE)

$CH_3CH_2C{\equiv}CH$ 1-BUTYNE

A point of difficulty often arises because the use of common names persists. Consider, for example:

$$\begin{array}{c} CH_3 \\ | \\ CH_3{-}C{-}CH_3 \\ | \\ H \end{array}$$

The correct name is methylpropane. However, this compound is more often referred to by its common name, isobutane. Only under special circumstances does the multiplicity of names for a single compound create confusion.

Some compounds are known by two or more different names.

When other groups are present in an organic compound the names are developed on the same basis as those of the hydrocarbons, as can be seen from the following examples:

$CH_3CH_2CH_2Br$ 1-BROMOPROPANE

$CH_3CHBrCH_3$ 2-BROMOPROPANE

$BrC{\equiv}CH$ BROMOETHYNE

$$\begin{array}{c} CH_3 \\ | \\ CH_3{-}C{-}H \\ | \\ CH_2Br \end{array}$$ 1-BROMO-2-METHYLPROPANE

$CH_3CHBrCH_2Br$ 1,2-DIBROMOPROPANE

SELF-TEST 13-A

1. The branch of chemistry that deals with compounds of carbon is known as ______________ chemistry.

2. The structure of the CH_4 molecule is described as ______________.

3. When s and p orbitals are mathematically mixed to form sp^2 or sp^3 orbitals, the process is called ______________.

4. True or False: A straight-chain hydrocarbon, such as pentane, actually has all of its carbon atoms in a straight line.

5. How many different isomers of C_5H_{12} are shown below?

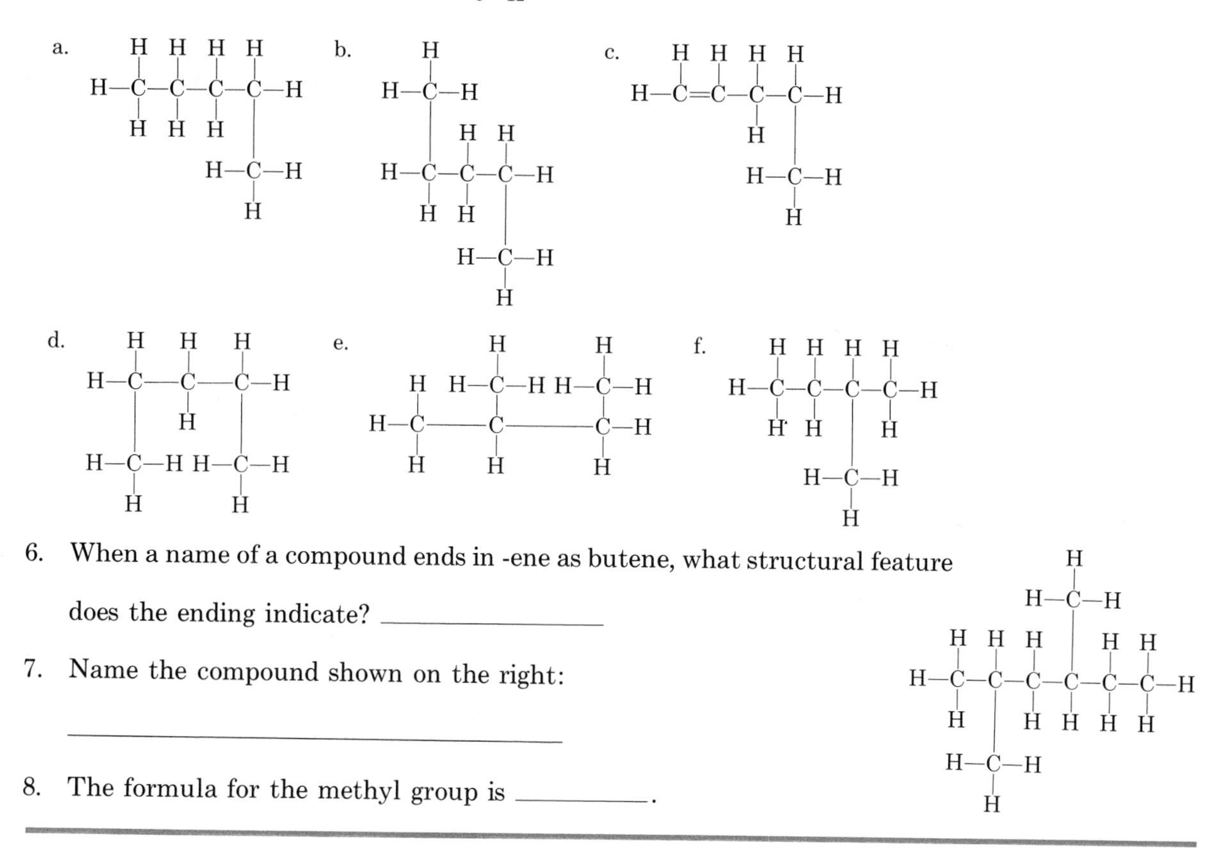

6. When a name of a compound ends in -ene as butene, what structural feature does the ending indicate? ______________

7. Name the compound shown on the right:

____

8. The formula for the methyl group is __________.

OPTICAL ISOMERS

In the preceding section, it was pointed out that structural isomers differ because there is more than one way in which a given set of atoms can be held together. However, it is possible for some sets of atoms to form two isomeric molecules, both of which have the same set of bonds as well as the same set of atoms. This special case of structural isomerism is called ***optical isomerism.***

Stereoisomers have the same atoms and the same bonds but the atoms are arranged in space differently.

Optical isomerism is possible when a molecular structure is asymmetric (without symmetry). One common example of an asymmetric molecule is one containing a tetrahedral carbon atom bonded to four

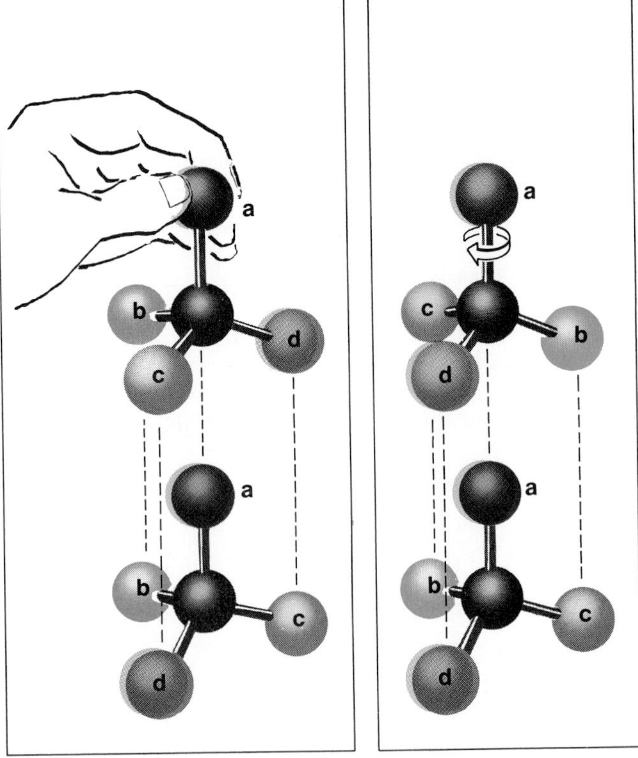

Figure 13–6 Optical isomers. Four different atoms, or groups of atoms, are bonded to tetrahedral center atoms so that the upper isomeric form cannot be turned in any way and exactly match the lower structure. These are nonsuperimposable mirror images.

different atoms or groups of atoms. Such a carbon atom is called an ***asymmetric*** carbon atom; an example is the carbon atom in the molecule CBrClIH.

Figure 13–6 shows the two ways to arrange the four different atoms in the tetrahedral positions about the central carbon atom. These result in two nonsuperimposable, mirror-image molecules which are optical isomers.

There are many examples of nonsuperimposable mirror images in the macroscopic world. Consider right- and left-hand gloves for instance.

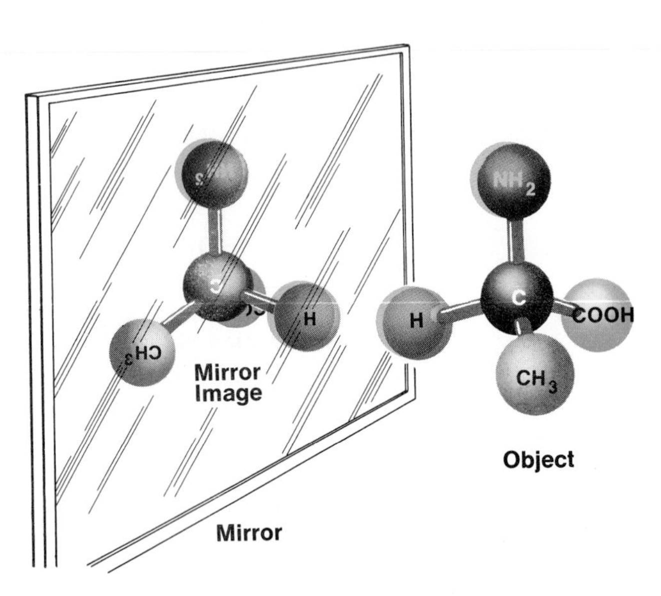

Figure 13–7 Optical isomers of the amino acid, alanine.

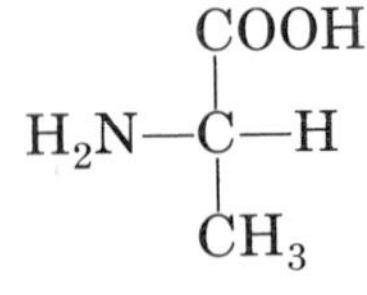

The D-form is the mirror image of the L-form.

They are mirror images of one another and are nonsuperimposable. In Figure 13–7, this mirror image relationship is shown for isomeric forms of alanine, the molecules of which contain a tetrahedral carbon atom surrounded by an amino group (—NH_2), a methyl group (—CH_3), an acid group (—COOH), and a hydrogen atom. Note in Figure 13–7 that

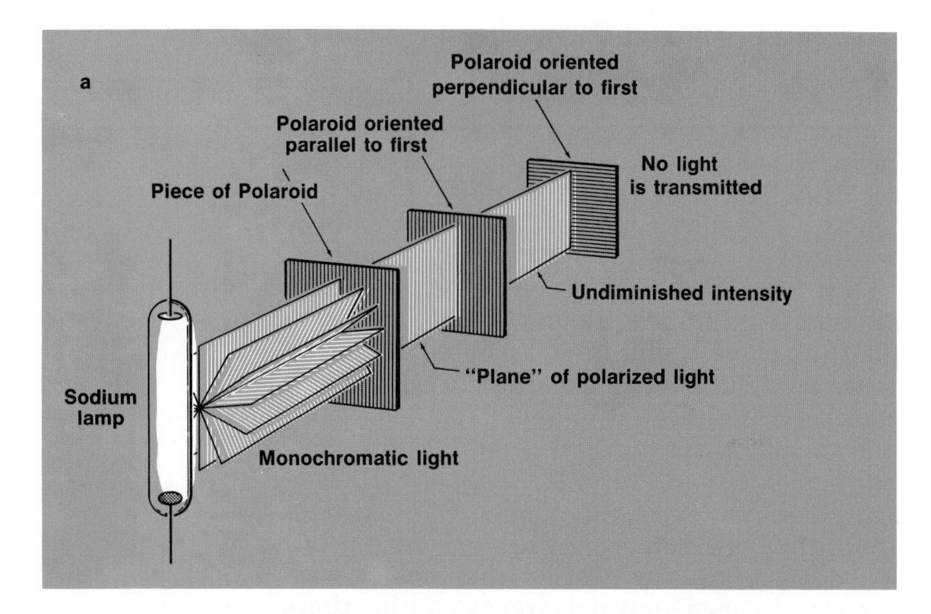

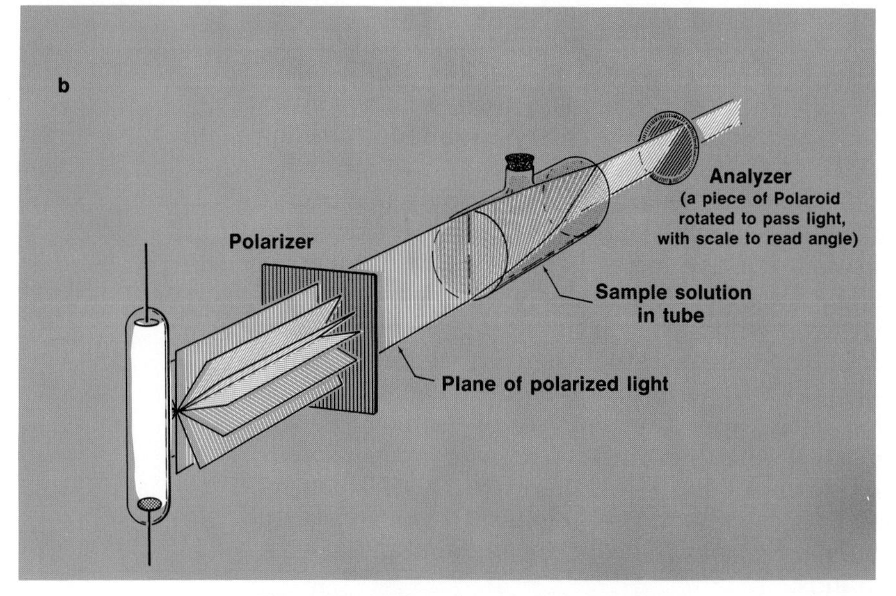

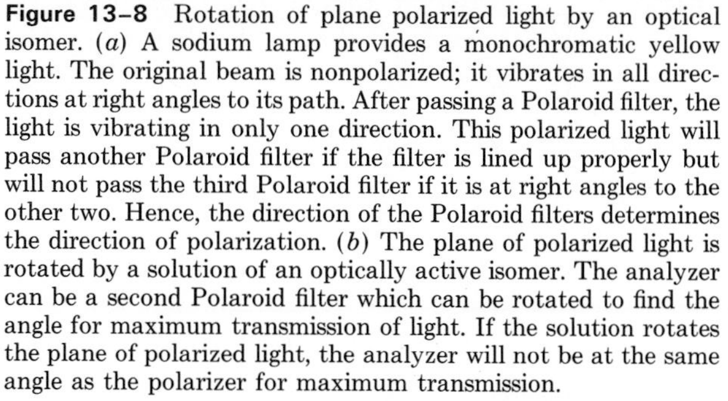

Figure 13–8 Rotation of plane polarized light by an optical isomer. (*a*) A sodium lamp provides a monochromatic yellow light. The original beam is nonpolarized; it vibrates in all directions at right angles to its path. After passing a Polaroid filter, the light is vibrating in only one direction. This polarized light will pass another Polaroid filter if the filter is lined up properly but will not pass the third Polaroid filter if it is at right angles to the other two. Hence, the direction of the Polaroid filters determines the direction of polarization. (*b*) The plane of polarized light is rotated by a solution of an optically active isomer. The analyzer can be a second Polaroid filter which can be rotated to find the angle for maximum transmission of light. If the solution rotates the plane of polarized light, the analyzer will not be at the same angle as the polarizer for maximum transmission.

the carbon atoms in the methyl and acid groups are not asymmetric since these atoms are not bonded to four different groups.

The properties of optical isomers are almost identical. Different compounds whose molecules are mirror images of one another have the same melting point, the same boiling point, the same density, and will be the same in many other physical and chemical properties. However, they always differ in one physical property; they rotate the plane of ***polarized*** light in opposite directions. According to the wave theory of light, a light wave traveling through space vibrates at right angles to its path (Figure 13–8). A group of such rays traveling together vibrate in random directions, all of which are at right angles to the path of travel. If such a group of waves is passed through a polarizing crystal, such as Iceland spar (a form of $CaCO_3$), the light is split into two rays and the emerging waves along the incoming axis will vibrate in only one plane, perpendicular to the light path. Such light is said to be ***plane polarized.*** When plane polarized light is passed through a solution of D-lactic acid, the light is still polarized but the plane of vibration is rotated somewhat in one direction. If the other lactic acid isomer is substituted (L-lactic acid), just the opposite rotation of the light is obtained.

Optical isomers are a type of stereoisomer (or spatial isomer). The same atoms are bonded together with the same bonds but their mirror images cannot be superimposed.

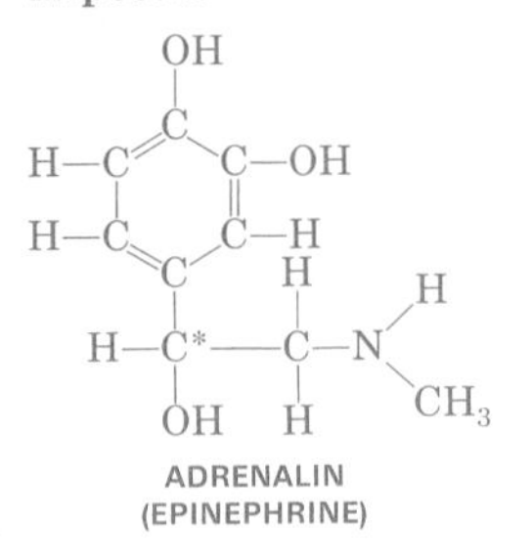

ADRENALIN (EPINEPHRINE)

D- and L- simply indicate that two structures are possible around an asymmetric C atom. The D- and L- notations do not indicate which way the substance will rotate the plane polarized light.

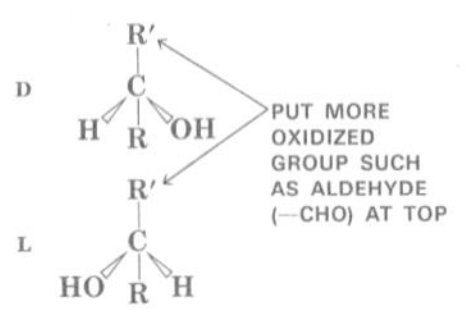

Optical isomers can also differ in biological properties. An example is the hormone adrenalin (or epinephrine). Adrenalin is one of a pair of optical isomers and its structure is shown in the margin. C* designates the asymmetric carbon atom. Only the isomer that rotates plane polarized light to the left is effective in starting a heart that has stopped beating momentarily or in giving a person unusual strength during times of great emotional stress.

It is also interesting to note that during the contraction of muscles the body produces only the L-form of lactic acid and not the D-form. The concentration of this lactic acid in the blood is associated with the feeling of tiredness, and a period of rest is necessary to reduce the concentration of this chemical.

Large organic molecules may have many asymmetric carbon atoms within the same molecule. At each such carbon atom there exists the possibility of *two* arrangements of the molecule. The total number of possible molecules, then, increases exponentially with the number of asymmetric centers. With two asymmetric carbon atoms there are 2^2 or four possible structures; for three, there are 2^3 or eight possible structures. It should be emphasized that each of the eight isomers could be made from the *same* set of atoms with the *same* set of chemical bonds. Glucose, a simple blood sugar also known as dextrose, contains four asymmetric carbon atoms per molecule. Thus, there are 2^4 or 16 isomers in the family of compounds to which glucose belongs. Obviously, then, the concept of optical isomerism is significant in explaining the vast number of carbon compounds.

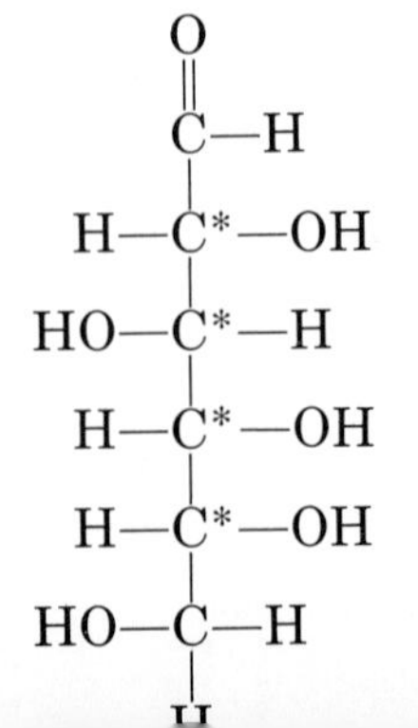

The symbols + (or D, dextrorotatory) and − (or L, levorotatory) are used to indicate rotations of plane polarized light to the right and to the left, respectively.

GEOMETRIC ISOMERS

Where carbon-carbon double bonds exist in a molecule, geometric isomerism is possible. The lack of rotation about a double bond provides a structural basis for this type of isomerism. Recall from the discussion in Chapter 9 that a carbon-carbon double bond is composed of a sigma bond and a pi bond. The side overlap of the two p orbitals forming the pi bond prevents free rotation about the bond axis.

Consider the compound ethylene, C_2H_4. Its six atoms lie in the same plane, with bond angles of approximately 120°:

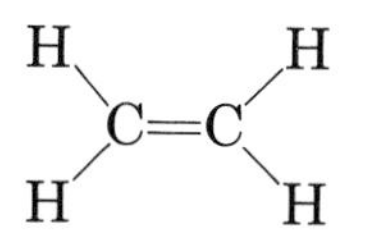

If two chlorine atoms replace two hydrogen atoms, one on each carbon atom of ethylene ($H_2C{=}CH_2$), the result is CHCl=CHCl. Experimental evidence confirms that there are two compounds with this general arrangement. If the two chlorine atoms are close together, this is one isomer (the ***cis*** isomer), and if they are far apart, this forms another isomer (the ***trans*** isomer). Both compounds are called 1,2-dichloroethene (the 1 and 2 indicate that the two chlorine atoms are attached to different carbon atoms). They are distinguished by the designations *cis* and *trans*. Note that the two isomeric compounds have significant differences in their properties.

Geometric isomers are another type of stereoisomer.

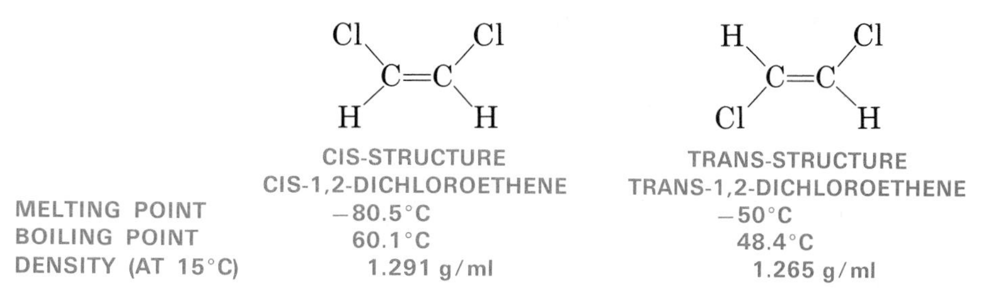

	CIS-STRUCTURE CIS-1,2-DICHLOROETHENE	TRANS-STRUCTURE TRANS-1,2-DICHLOROETHENE
MELTING POINT	−80.5°C	−50°C
BOILING POINT	60.1°C	48.4°C
DENSITY (AT 15°C)	1.291 g/ml	1.265 g/ml

As a general rule, *trans* isomers are higher melting than *cis* isomers due to the greater ease with which the *trans* molecules can fit into the lattice and form strong intermolecular bonds.

Often, when there is a carbon-carbon double bond in an organic molecule, the possibility exists for *cis* and *trans* isomers. Sometimes a number of such bonds can be found in the same molecule, giving rise to numerous isomeric compounds.

FUNCTIONAL GROUPS

As pointed out earlier, carbon also forms covalent bonds with a number of other elements. As a result, certain groups of atoms called

TABLE 13–3 Alcohols Derived From Pentane (C_5H_{12})

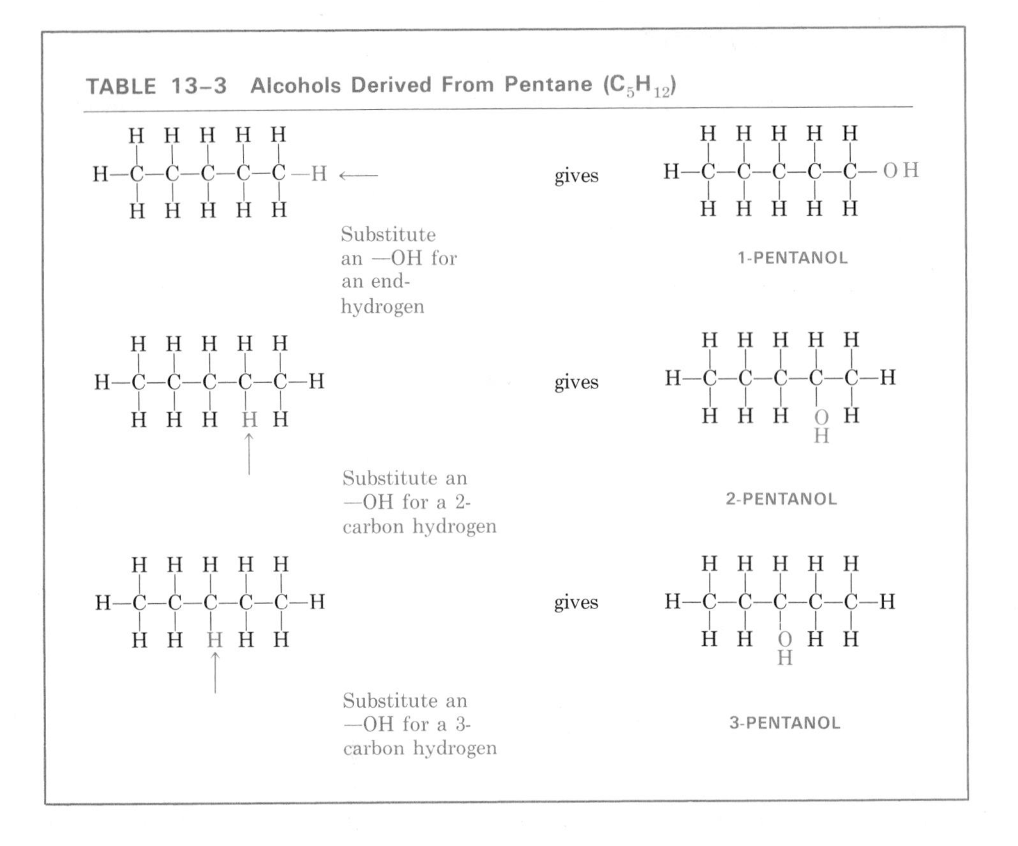

Alcohols contain the —OH functional group.

functional groups appear over and over again in different organic compounds. Consider, for example, the —OH group, called the hydroxyl group. A large number of molecules containing an —OH group have properties which are characteristic of a class of organic compounds called ***alcohols.***

The —OH group has the same combining power as an atom of hydrogen. This means that, in any of the hydrocarbons considered thus far, any of the hydrogen atoms could be replaced by an —OH group to form an alcohol. A single hydrocarbon molecule can give rise to a number of alcohols if there are different isomeric positions for the —OH group. Three different alcohols result when a hydrogen atom is replaced by an —OH group in n-pentane, depending on which hydrogen atom is replaced (Table 13–3). When one or more functional groups appear in a molecule, the IUPAC name reveals the functional group name and position. For example, the name of an alcohol will use the root of the name of the analogous hydrocarbon to indicate the number of carbon atoms and the suffix -ol to denote an alcohol. As before, a number is used to indicate the position of the alcohol group.

Some major functional groups that we will consider in following chapters are shown in Table 13–4. The symbol —R is a hydrocarbon group such as methyl (—CH_3) or ethyl (—C_2H_5).

AROMATIC COMPOUNDS

All of the hydrocarbons we have discussed up to this point have localized electronic structures; that is, the bonding electrons are essen-

TABLE 13–4 Classes of Organic Compounds Based on Functional Groups*

General Formulas of Class Members	Class Name	Typical Compound	Compound Name	Common Use of Sample Compound
R—OH	Alcohol (Hydroxyl)	H_3C—OH	Methanol (Wood alcohol)	Solvent
R—C(=O)—H	Aldehyde	H—C(=O)—H	Methanal (Formaldehyde)	Preservative
R—C(=O)—OH	Carboxylic acid	H_3C—C(=O)—OH	Ethanoic Acid (Acetic acid)	Vinegar
R—C(=O)—R	Ketone	H_3C—C(=O)—CH_3	Propanone (Acetone)	Solvent
R—O—R	Ether	C_2H_5—O—C_2H_5	Diethyl ether (Ethyl ether)	Anesthetic
R—O—C(=O)—R	Ester	CH_3—CH_2—O—C(=O)—CH_3	Ethyl ethanoate (Ethyl acetate)	Solvent in fingernail polish
R—NH_2	Amine	HO—C(=O)—CH_2—NH_2	2-Aminoethanoic acid (Glycine)	An amino acid found in proteins

*R = H or a hydrocarbon group such as CH_3—, C_2H_5—, etc.

Delocalized electrons can occupy orbitals on several nuclei.

tially fixed between two atomic centers as in C—C or C=C bonds. For a large group of organic compounds known as ***aromatic*** compounds, this type of complete electron localization is not found. Rather, these compounds have some delocalized electrons which are spread over the entire molecule, a behavior which leads to some interesting chemical properties.

The simplest aromatic compound is *benzene* (C_6H_6). The molecular structure of benzene is a ring of carbon atoms in a plane. The bonds between these carbon atoms are shorter than single bonds, but longer than double bonds. Since the measured bond angles are 120°, this implies sp^2 hybridization in the carbon atoms.

One way to account for the bonding and structure of the benzene molecule is to use two of the three sp^2 hybrid orbitals of each carbon to form ***sigma*** bonds between the carbon atoms and to use the six remaining sp^2 orbitals to form six sigma bonds with six hydrogens. The six p orbitals not involved in the formation of sp^2 orbitals can overlap laterally to form ***pi*** bonds. These pi bonds are not exactly the same as those discussed earlier for double and triple bonds because there is a ring system of p orbitals in an aromatic molecule. Each p orbital overlaps with the p orbitals on both neighboring carbon atoms. This causes the

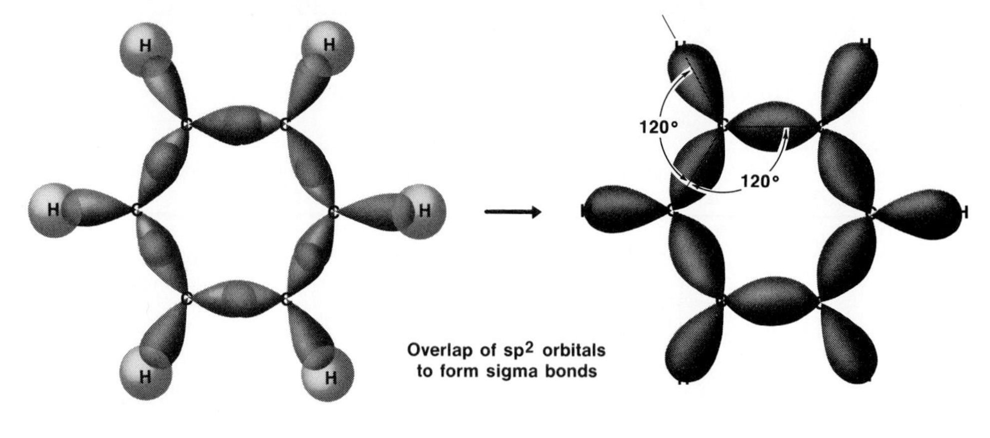

Figure 13–9 Bonding in an aromatic compound, benzene, C_6H_6.

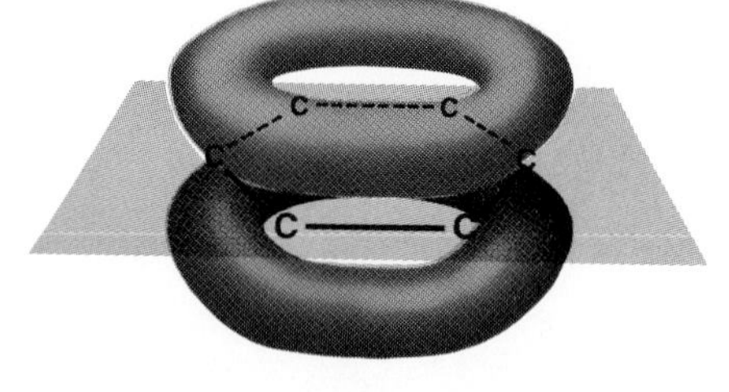

pi bonds to be less well fixed (or localized) between alternating pairs of p orbitals. Figure 13–9 illustrates this type of pi bonding. The pi bonds are averaged evenly around the ring; we say they are ***delocalized.***

Aromatic compounds have electrons delocalized over the orbitals of carbon atoms which are arranged in a ring.

Various symbols have been used to represent the benzene ring, but the symbol at the right below is perhaps the most used. The hydrogen atoms are not shown.

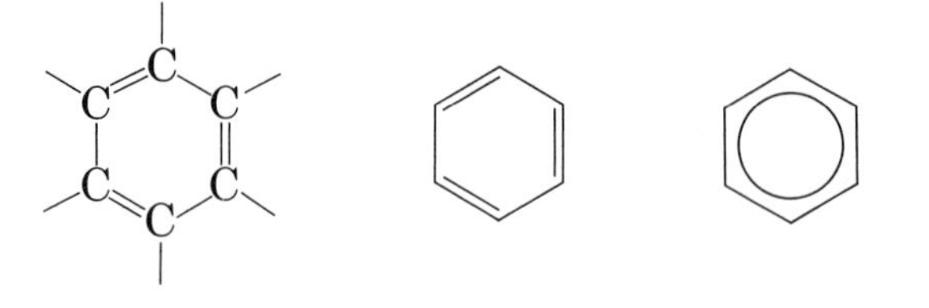

There is a large number of aromatic compounds that can be derived from benzene by replacing one or more of the hydrogen atoms about the ring. These molecules exhibit a wide variety of properties and differ greatly in their chemical reactivity. Some interesting examples of isomerism are also possible. Consider the three different compounds with the formula C_8H_{10} found in some coal tars. There are several names given to these isomers:

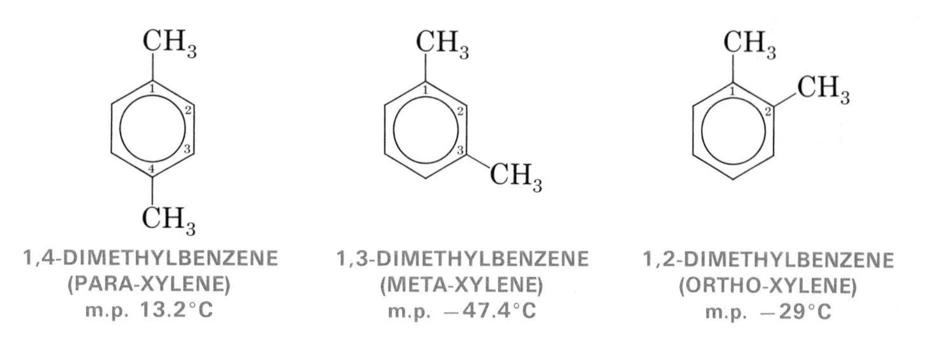

Nomenclature in aromatic compounds:

ortho—groups on adjacent carbon atoms (1,2 positions)

meta—groups have one carbon atom between them (1,3 positions)

para—groups have two carbon atoms between them (1,4 positions)

Each of these isomers has two methyl groups substituted for hydrogen atoms on the ring. The prefixes *para-*, *meta-*, and *ortho-* are used if there are only two groups on the benzene ring.

If more than two groups occur, a number system is most useful. Consider the following compounds:

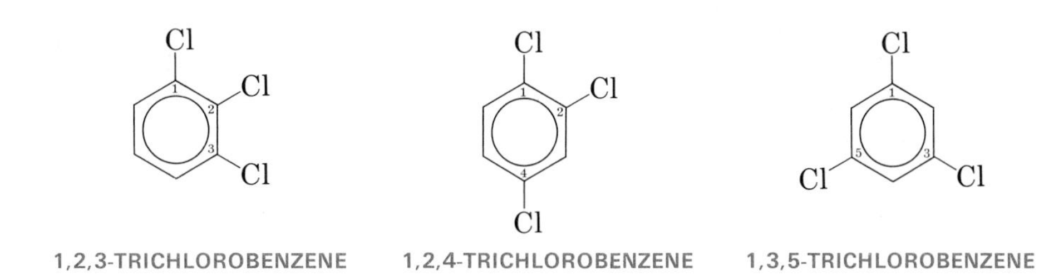

There are no other ways of drawing these three isomers and only three trichlorobenzenes have been isolated in the laboratory. We shall look at some of the reactions of aromatic compounds in this and the next chapter.

No wonder carbon is ubiquitous; there are about two million recorded compounds containing carbon. These millions of carbon compounds are due to the following reasons:

The large number of carbon compounds is due to some specific reasons.

1. The ability of carbon to form covalent bonds to other carbon atoms almost without limit.
2. The ability of a given number of carbon atoms to combine in more than one molecular pattern.
3. The ability of carbon to form stable covalent bonds to a large number of other atoms.
4. The stability of carbon chains in the presence of reagents such as oxygen and water which usually destroy chains of other atoms.

Structural differences explain how two compounds can have the same chemical composition by weight, and yet have different physical and chemical properties. Some order can be brought out of chaos with an understanding of isomerism and a systematic approach to the possible molecular structures.

SOURCES OF ORGANIC COMPOUNDS

Carbon compounds come from numerous sources. They are separated by chemical and physical methods, such as distillation and extraction, from natural materials which were once alive or which resulted from living organisms; alternatively they may be prepared in the laboratory from inorganic chemicals such as carbon monoxide and water or from other organic compounds. The largest sources of naturally occurring organic compounds are the fossil fuels, coal and petroleum, and vegetable matter such as wood, plant stalks, and so forth. A series of distillations (fractional distillation) separates petroleum ether, gasoline, kerosene, oils, and asphalt from petroleum (Table 13–5); each of these materials can be further separated into a multitude of individual compounds.

Petroleum is rich in nonaromatic hydrocarbons.

By heating coal or wood in the absence of air (destructive distillation), coal tar or wood tar can be separated from the coke or charcoal (carbon). A ton of a typical soft coal produces about 140 lbs of coal tar, which is about one-half pitch, the rest being composed of chemicals such as naphthalene, benzene, phenol, cresols, toluene, and xylenes.

Coal is one natural source of aromatic compounds.

The enormous increase in man's use of organic compounds in the last century forces us to recognize that the sources of these materials are limited and subject to eventual exhaustion. Thus, petroleum sources, from which we draw so much of our energy and which serve as starting materials for the synthesis of industrial organic compounds, cannot be expected to last beyond 50 to 100 years at the most. The transformation of many of the natural sources of organic compounds into disposable convenience items which soon end up in a garbage dump represents an extremely short-sighted use of irreplaceable natural resources. Most of us realize that we must begin ultimately to recycle a much larger percentage of the material we use in our daily lives, but few seem ready to begin effective efforts in this direction now. It seems very probable that the next decade will force us to change many of our casual attitudes toward the utilization of our natural sources of organic compounds.

Our supplies of coal, and especially petroleum, are limited.

SYNTHETIC ORGANIC COMPOUNDS

The naturally occurring organic compounds are not only important in and of themselves, but they also serve as starting materials for making

TABLE 13–5 Petroleum Fractions

Fraction	Composition	Distillation Range °C
Natural gas	C_1—C_4	Less than 20°
Liquefied petroleum	C_5—C_6	20–60°
Gasoline	C_5—C_{10}	40–200°
Kerosene	C_{10}—C_{16}	175–275°
Fuel oil, diesel oil	C_{15}—C_{20}	250–400°
Lubricating oils	C_{18}—C_{22}	Above 300°
Asphalt or petroleum coke	C_{20} and above (complicated structures)	Nonvolatile

numerous organic compounds that do not occur in nature. Modification of existing molecules can take the form of rearranging atoms within molecules, making giant molecules from small ones ***(polymerization)***, making small molecules from large ones ***(cracking)***, substituting one functional group for another, and similar changes. Plastics, synthetic fibers, and drugs are important organic compounds that result from these kinds of reactions.

The synthesis of new compounds is seldom done by trial and error; the synthetic chemist often has as an aid in planning a synthesis some idea of the step by step structural changes the reactants undergo in becoming products. A description of the step by step structural changes is the ***mechanism*** of a reaction. Regardless of the reaction, all mechanisms have some things in common. For example, every mechanism must have the reactant molecules colliding with proper orientation (turned correctly in relationship to each other) and with sufficient energy to react. In addition, most organic reaction mechanisms obey the ***principle of minimum structural change,*** which means that most of the structure of the reacting species goes through the reaction unchanged. Only a few specific bonds are broken or made. A mechanism usually involves breaking the weakest bond first unless the attacking species is specifically reactive toward a group containing stronger bonds. A specific mechanism may contain some unique features, but all mechanisms adhere to these general principles.

Let us briefly look at some of the more important types of organic reactions and their mechanisms and see how they can lead the organic chemist to create new compounds.

Free Radical Substitution of Simple Hydrocarbons

Under the proper conditions it is possible to substitute a halogen (most often Cl or Br) atom for a hydrogen atom in simple hydrocarbon molecules.

$$-\overset{|}{\underset{|}{C}}-H + Cl_2 \longrightarrow -\overset{|}{\underset{|}{C}}-Cl + HCl$$

Photons of ultraviolet light contain sufficient energy to initiate many chemical reactions.

This is an important reaction since it is useful in preparing molecules that are much more reactive (and hence, more valuable in other synthetic processes) than the hydrocarbons from which they came. A typical reaction might involve the chlorination of ethane at 25°C in the presence of light:

$$H-\overset{H}{\underset{H}{\overset{|}{\underset{|}{C}}}}-\overset{H}{\underset{H}{\overset{|}{\underset{|}{C}}}}-H + Cl_2 \xrightarrow[25^\circ]{\text{Light}} H-\overset{H}{\underset{H}{\overset{|}{\underset{|}{C}}}}-\overset{H}{\underset{H}{\overset{|}{\underset{|}{C}}}}-Cl + HCl$$

CHLOROETHANE

The mechanism of this type of reaction is an interesting one. It involves a series of reaction steps and chemical species called ***free radicals*** that have a single unpaired valence electron.

Free radical reactions involve an atom or a molecule with an unpaired electron such as Cl•, CH_3•, and H•. Such species are called free radicals.

Step 1.

The chlorine molecule is broken apart into two chlorine atoms (free radicals) by a photon of light:

$$Cl_2 \xrightarrow{\text{Light}} 2Cl\cdot$$

(Other outer shell electrons omitted)

Step 2.

The chlorine free radicals, being very reactive species, attack the ethane molecules creating new (ethyl) free radicals:

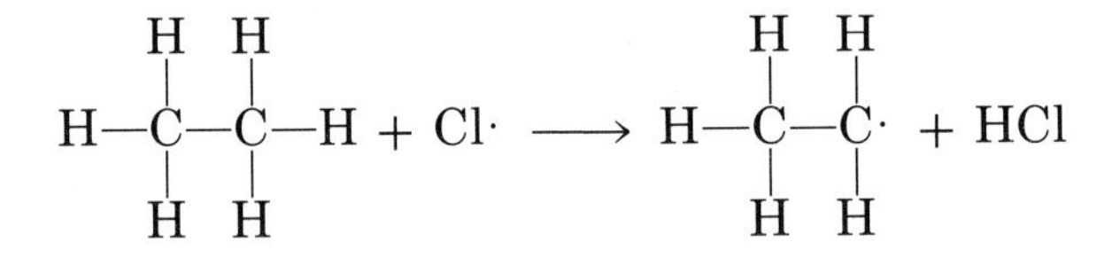

Step 3.

The ethyl radicals in turn can react with chlorine molecules to produce product:

$$\begin{array}{c}\text{H}\;\;\text{H}\\ |\;\;\;|\\ \text{H—C—C}\cdot\\ |\;\;\;|\\ \text{H}\;\;\text{H}\end{array} + Cl_2 \longrightarrow \begin{array}{c}\text{H}\;\;\text{H}\\ |\;\;\;|\\ \text{H—C—C—Cl}\\ |\;\;\;|\\ \text{H}\;\;\text{H}\end{array} + Cl\cdot$$

(Main product)

$$Cl\cdot + Cl\cdot \longrightarrow Cl_2$$

Reactions (2 and 3) continue until all the molecules have reacted and the process is terminated. Steps 2 and 3 constitute what is called a ***chain reaction.***

Step 4.

Termination of the reaction can take place in a number of ways other than using up all the reactants:

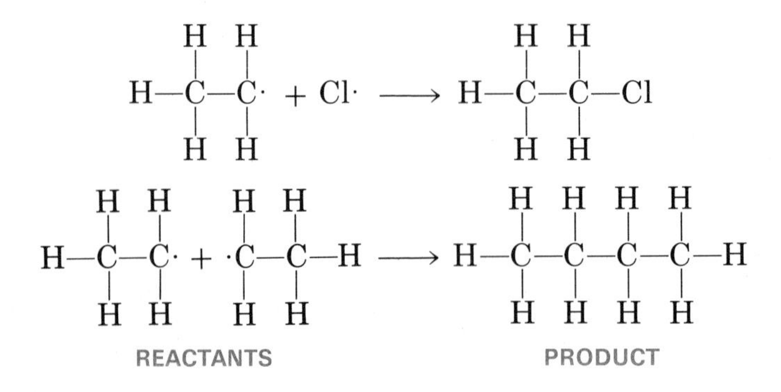

This free radical mechanism for halogenation of hydrocarbons is also a prototype for the polymerization reactions which we shall study in Chapter 15. There we shall see that many useful products result from reactions of this type.

Addition Reactions

As mentioned previously, halogens react with organic molecules containing double or triple carbon-carbon bonds in a special manner. For example, chlorine reacts with ethylene to produce 1,2-dichloroethane.

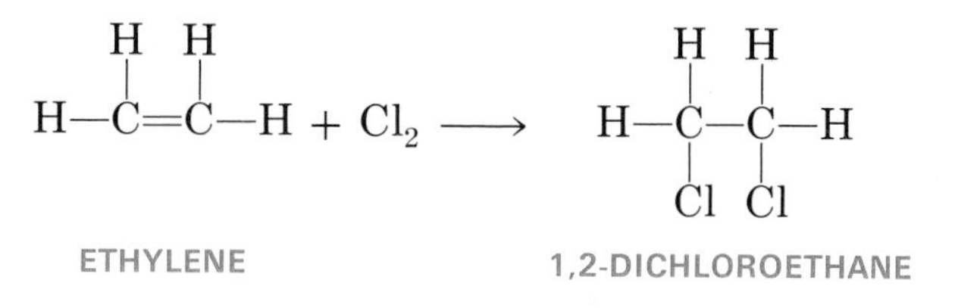

In effect, this is an ***addition*** reaction since two chlorine atoms have been added to the molecule. The mechanism for this reaction makes use of a special property of the C=C bond. Since there are four electrons localized between the two carbon atoms, they can interact with a chlorine molecule and cause its electron distribution to be distorted. This electron distortion is called ***polarization*** and can result in the breaking of the Cl—Cl bond and forming a bond between a carbon and a chlorine atom.

Carbon-carbon double bonds react more readily than carbon-carbon single bonds.

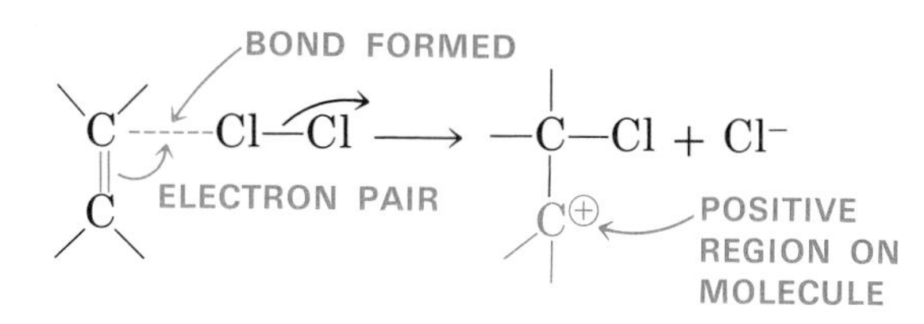

δ^+ denotes a partial positive charge.

The resulting positive region on the molecule then attracts the negative chloride ion:

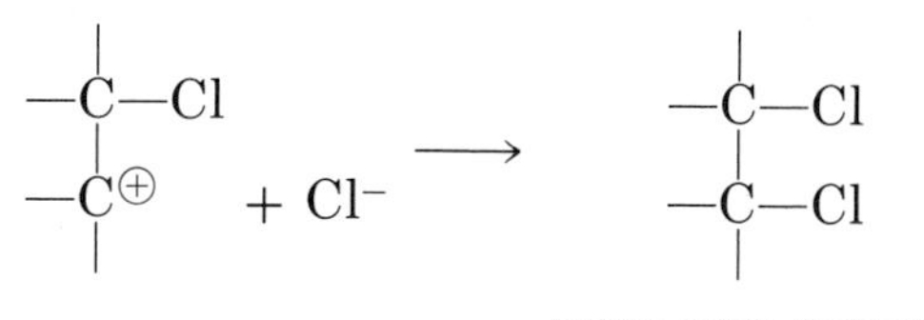

Aromatic compounds typically undergo substitution reactions in which a hydrogen atom is replaced by another atom or group.

Substitution Reactions of Aromatic Compounds

When benzene reacts with bromine in the presence of iron, the product which results is bromobenzene, a substitution product in which a hydrogen atom has been replaced by a bromine atom.

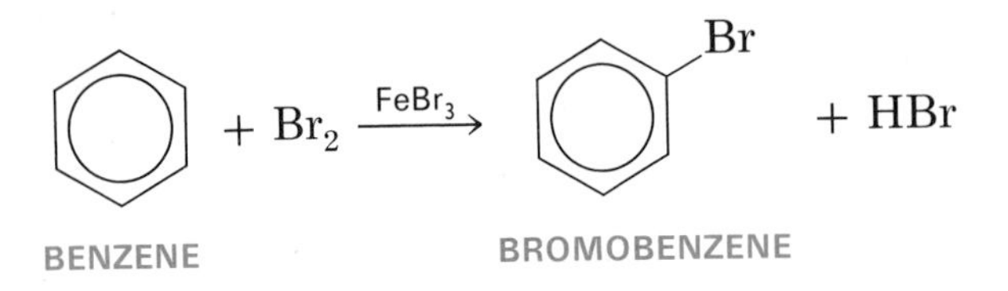

On the surface, this reaction appears to be like the substitution reactions discussed for simple hydrocarbons. A difference exists, however, because the conditions are not those that give free radicals. This reaction is actually more like the addition reactions just described.

The typical reactions of aromatic compounds are substitution processes.

Some of the $FeBr_3$ reacts with bromine to give $FeBr_4^-$ and Br^+ ions:

$$Br_2 + FeBr_3 \rightleftharpoons FeBr_4^- + Br^+$$

The Br^+ species then attacks the electron-rich pi electron orbitals on the benzene ring:

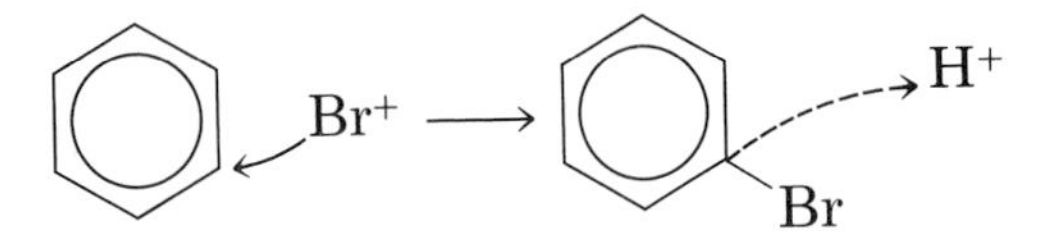

Hydrogen leaves the ring as H^+ ion and reacts with the $FeBr_4^-$ giving HBr:

$$H^+ + FeBr_4^- \longrightarrow HBr + FeBr_3$$

SELF-TEST 13-B

1. In order to have optical isomers in carbon compounds, a carbon atom must have ________ different groups attached.
2. In what physical property do optical isomers that are mirror images differ?

   ____
3. How many optical isomers would be possible if five asymmetric carbon atoms are contained in a structure? ________
4. The benzene ring has both localized electrons (sigma bonds) and ________ ________ electrons (pi bonds).
5. How many atoms does the symbol, ⬡, represent? ________
6. Name the following compound. ____________

   CH_3 — CH_3 — CH_3

7. A natural source of aromatic compounds is ________ while a natural source of aliphatic compounds is ________________.

8. The principle of minimum structural change states that most of the molecule remains intact during a chemical change; only a few bonds are broken or made. (True/False)

9. Bromine can be put onto a benzene ring by a (substitution/addition) reaction.

MATCHING SET

______ 1.	organic chemistry	a.	$—\overset{\overset{\displaystyle O}{\|\|}}{C}—O—H$
______ 2.	isomers	b.	occurs when pi bonds overlap over three or more atomic centers
______ 3.	functional group	c.	compound such as C_3H_8
______ 4.	hydrocarbon	d.	lateral overlap of p orbitals
______ 5.	methyl group	e.	step-by-step structural changes that occur during a chemical reaction
______ 6.	asymmetric carbon atom	f.	chemistry of carbon compounds
______ 7.	delocalized electron	g.	$—CH_3$
______ 8.	sigma bond	h.	same number and kinds of atoms differently arranged
______ 9.	pi bond	i.	end-to-end overlap of pi orbitals
______ 10.	synthesis	j.	species with an unpaired valence electron
______ 11.	mechanism	k.	has four different groups attached
______ 12.	free radical	l.	the process of making or putting together

---

QUESTIONS

1. *Saturated hydrocarbons* are so named because they have the maximum amount of hydrogen present for a given amount of carbon. The saturated hydrocarbons have the general formula C_nH_{2n+2} where n is a whole number. What are the names and formulas of the first four members of this series of compounds?

2. Using diagrams and bonding theory, explain why nearly free rotation is allowed around a carbon-carbon single bond but not around a carbon-carbon double bond.

3. Using the periodic chart and electron-dot formulas, illustrate a sigma bond in a compound other than those given in this chapter.

4. Draw the structural formula for each of the five isomeric hexanes, C_6H_{14}.

5. Write the structural formulas for:

   a. 2-methylbutane
   b. ethylpentane
   c. 4,4-dimethyl-5-ethyloctane
   d. methylbutane
   e. 2-methyl-2-hexene
   f. 3-methyl-3-hexanol

6. Give the names for:

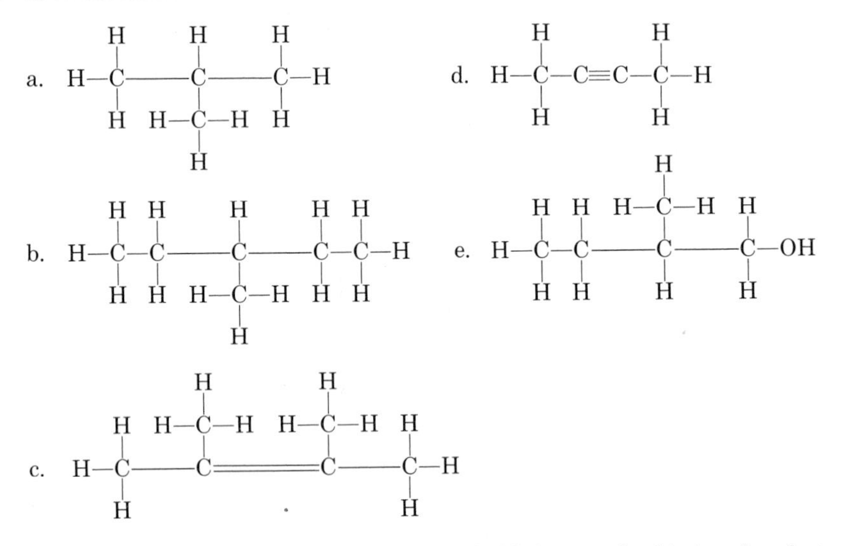

7. If cyclohexane has the formula C_6H_{12}, and if it has *no* double bonds, what would you suggest as its structural formula?

8. How can optical isomers be distinguished from each other experimentally?

9. a. Which arrangement has a mirror image which is nonsuperimposable?
   b. Explain the term: asymmetric carbon atom.

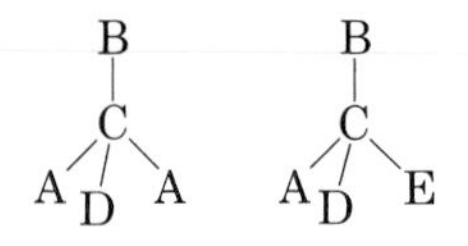

10. If a solution of the D optical isomer of a pair of optical isomers rotates plane polarized light clockwise, and the L isomer does the opposite, predict what an equimolar mixture of the two would do?

11. How many optical isomers can there be for a molecular structure containing eight asymmetric carbon atoms?

12. Explain the statement: The formation of a pi bond between two carbon atoms does not release as much energy as the formation of a sigma bond, yet more energy is released in the formation of a carbon-carbon double bond than in the formation of a carbon-carbon single bond.

13. Which do you think would be better to use for medicinal purposes, pure adrenalin obtained from natural products (found in nature) or the pure compound as synthesized in the laboratory? Why?

14. Ethyl alcohol, $H{-}\overset{H}{\underset{H}{C}}{-}\overset{H}{\underset{H}{C}}{-}OH$, can be oxidized to acetic acid, $H{-}\overset{H}{\underset{H}{C}}{-}C{\overset{\large O}{\underset{OH}{\lessgtr}}}$. How does this illustrate the *principle of minimum structural change?*

15. Methane reacts with chlorine at room temperature in the presence of light to form chloromethane. Write the mechanistic steps for this reaction. Name the species involved.

16. What structural feature characterizes aromatic compounds?

17. Indicate the functional groups present in the following molecules:

a. $CH_3CH_2CH_2COOH$
b. $CH_3CH_2NH_2$
c. $CH_3\underset{NH_2}{\underset{|}{C}}HCH_2CH_2COOH$
d. $CH_3\underset{OH}{\underset{|}{C}}HCH_2COOH$
e. $CH_3\underset{O}{\underset{\|}{C}}CH_2CH_2COOH$
f. $CH_3\underset{NH_2}{\underset{|}{C}}HCH_2OH$

18. Name the following compounds:
$CH_3OCH_2CH_3$, $CH_3CH_2CH_2COOH$, $CH_3CH_2CH_2C{\equiv}CH$, $CH_3CH_2CH_2NH_2$.

19. Write structural formulas for butanoic acid, aminomethane, 2-butanol, and 3-aminopentane.

20. Give an example of:

a. an alkane
b. an amine
c. a carboxylic acid
d. an ether
e. an ester
f. an alkene
g. an alkyne
h. an alcohol
i. a ketone

21. Write two structural formulas for compounds which can have each of the molecular formulas listed:

a. $C_5H_{12}O$
b. C_3H_6O
c. $C_5H_{10}O_2$

22. Draw the *cis* and *trans* isomers for:

a. 1,2-dibromoethene
b. 1-bromo-2-chloroethene

PROBLEMS

1. Write the structure for a compound that contains 81.8% C and 18.2% H, and has a molecular weight of 44.

Ans. $CH_3CH_2CH_3$

2. Two isomeric compounds, A and B, are found to have molecular weights of 46 and percent composition as follows: 52.2% C, 13.0% H, and 34.8% O. Compound A is found to give every evidence of hydrogen bonding to itself in the liquid state while compound B does not. Suggest structural formulas for A and B on the basis of these data.

Ans. A: CH_3CH_2OH
B: $CH_3{-}O{-}CH_3$

SUGGESTIONS FOR FURTHER READING

Dodge, B. S., "Louis Pasteur's Looking Glass World," *Chemistry,* Vol. 41, No. 2, p. 16 (1968).
Mills, G. A., "Ubiquitous Hydrocarbons," *Chemistry,* Vol. 44, No. 2, p. 8; No. 2, p. 12 (1971).
Moore, J. A., "Elementary Organic Chemistry," W. B. Saunders Co., Philadelphia, 1974.
Orchin, M., "Determining the Number of Isomers From A Structural Formula," *Chemistry,* Vol. 42, No. 5, p. 8 (1969).
Roberts, J. D., "Organic Chemical Reactions," *Scientific American,* Vol. 197, p. 117 (1957).
van Tamelen, E. E., "Benzene, The Story of Its Formula, 1865–1965," *Chemistry,* Vol. 38, No. 1, p. 6 (1965).
Westheimer, F. H., "The Structural Theory of Organic Chemistry," *Chemistry,* Vol. 38, No. 6, p. 13 (1965), and Vol. 38, No. 7, p. 10 (1965).

CHAPTER 14

SOME APPLICATIONS OF ORGANIC CHEMISTRY

Chemical synthesis is the preparation of compounds.

It is difficult to overestimate the importance of carbon chemistry to man. Every month hundreds of new carbon compounds are prepared. A few of these new compounds become important as medicines, plastics, textiles, solvents, food additives, cosmetics, or some other consumer product. A precious few may provide an important clue to the mechanism of a fundamental chemical reaction in the human body. Most, however, become laboratory curiosities and for the present, at least, have no practical application.

Organic compounds can be made from inorganic compounds.

The preparation of new and different organic compounds through chemical reactions is called ***organic synthesis.*** A million or so organic compounds now known and characterized have been synthesized in the laboratories of the world during the past 150 years. Prior to 1828, it was widely believed that chemical compounds synthesized by living matter could not be made without living matter—a "vital force" was necessary for the synthesis. In 1828, a young German chemist, Friedrich Wöhler, destroyed the vital force myth and opened the door to the organic syntheses that we now know. Wöhler heated a solution of silver cyanate and ammonium chloride, neither of which had been derived from any living substance. From these he prepared urea, a major waste product found in urine.

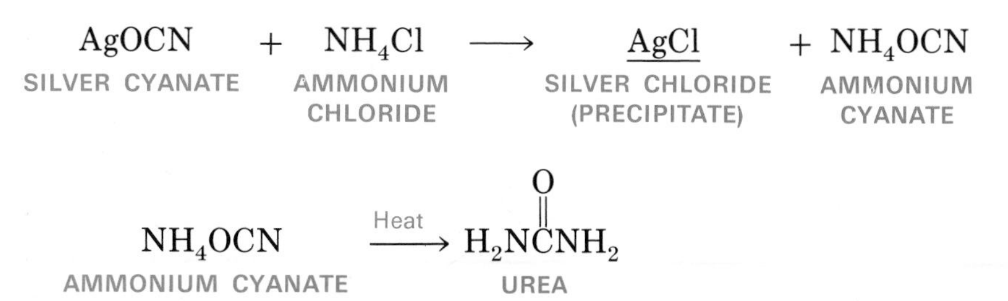

$$\underset{\text{SILVER CYANATE}}{AgOCN} + \underset{\text{AMMONIUM CHLORIDE}}{NH_4Cl} \longrightarrow \underset{\text{SILVER CHLORIDE (PRECIPITATE)}}{\underline{AgCl}} + \underset{\text{AMMONIUM CYANATE}}{NH_4OCN}$$

$$\underset{\text{AMMONIUM CYANATE}}{NH_4OCN} \xrightarrow{\text{Heat}} \underset{\text{UREA}}{H_2N\overset{\overset{\displaystyle O}{\|}}{C}NH_2}$$

Friedrich Wöhler (1800–1882) was Professor of Chemistry at the University of Berlin and then at Göttingen. His preparation of the organic compound urea from ammonium cyanate, an inorganic source, did much to overturn the theory that organic compounds must be prepared in living organisms. One of the first to study the properties of aluminum, he discovered the element beryllium and is known for many other outstanding contributions to chemistry.

The notion of a mysterious vital force declined as other chemists began to synthesize more and more organic chemicals without the aid of a living system. Soon it was shown that chemists could do more than imitate the products of living tissue; they could form unique materials of their own design.

Advances in understanding the structure of organic compounds gave organic synthesis a tremendous boost. Knowing the structure of compounds, the organic chemist could predict by analogy with simpler molecules what reactions might take place when organic reagents were used. Very elegant and reliable schemes of synthesis could now be constructed.

In this chapter emphasis will be given to some important organic compounds that not only are useful in themselves but are necessary for the synthesis of many other organic compounds. The dependence of synthesis upon a knowledge of structure will be pointed out from time to time.

SOME USES OF HYDROCARBONS

Common sources of the hydrocarbons are the "fossil fuels" (coal, petroleum, and natural gas).

Complex mixtures of hydrocarbons, compounds containing only carbon and hydrogen, occur in enormous quantities in nature as petroleum and natural gas. These materials were formed from organisms that lived millions of years ago. After their death, they became covered with layers of sediment and ultimately were subjected to high temperatures and pressures in the depths of the earth's crust. In the absence of free oxygen, these conditions converted once living tissue into petroleum and coal. After petroleum and natural gas are brought to the surface of the earth by man, they can be separated into various fractions with different boiling points by the use of fractional distillation (Figure 14–1). Thus, pure hydrocarbons become available for synthesis of many other organic compounds. From the simplest hydrocarbons come such diverse consumer products as plastic dishes, acrylic fibers, vinyl paints, and neoprene rubber, and such industrial products as Teflon and cattle feed.

Natural gas consists primarily of low molecular weight hydrocarbons. It is predominantly methane (CH_4), but ethane (C_2H_6), propane (C_3H_8), and butane (C_4H_{10}) are also present. This mixture is conveyed

CHAPTER 14

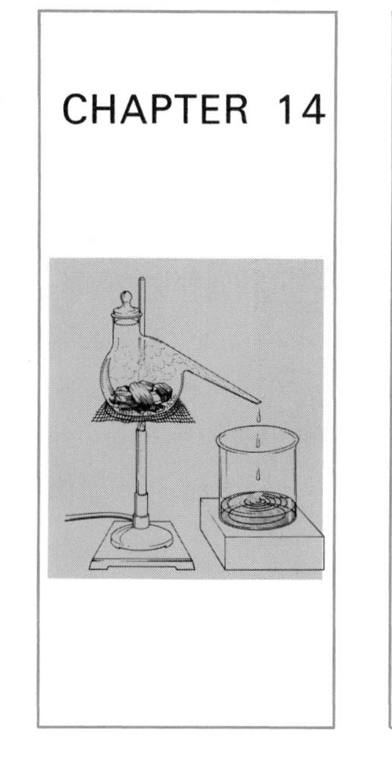

SOME APPLICATIONS OF ORGANIC CHEMISTRY

It is difficult to overestimate the importance of carbon chemistry to man. Every month hundreds of new carbon compounds are prepared. A few of these new compounds become important as medicines, plastics, textiles, solvents, food additives, cosmetics, or some other consumer product. A precious few may provide an important clue to the mechanism of a fundamental chemical reaction in the human body. Most, however, become laboratory curiosities and for the present, at least, have no practical application.

Chemical synthesis is the preparation of compounds.

The preparation of new and different organic compounds through chemical reactions is called ***organic synthesis.*** A million or so organic compounds now known and characterized have been synthesized in the laboratories of the world during the past 150 years. Prior to 1828, it was widely believed that chemical compounds synthesized by living matter could not be made without living matter—a "vital force" was necessary for the synthesis. In 1828, a young German chemist, Friedrich Wöhler, destroyed the vital force myth and opened the door to the organic syntheses that we now know. Wöhler heated a solution of silver cyanate and ammonium chloride, neither of which had been derived from any living substance. From these he prepared urea, a major waste product found in urine.

Organic compounds can be made from inorganic compounds.

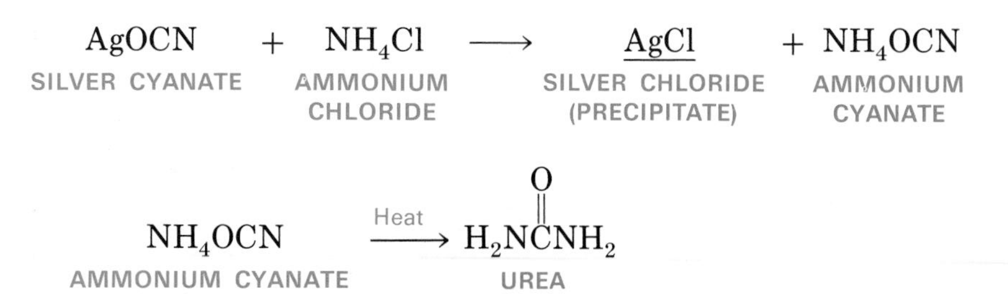

$$\underset{\text{SILVER CYANATE}}{AgOCN} + \underset{\text{AMMONIUM CHLORIDE}}{NH_4Cl} \longrightarrow \underset{\text{SILVER CHLORIDE (PRECIPITATE)}}{\underline{AgCl}} + \underset{\text{AMMONIUM CYANATE}}{NH_4OCN}$$

$$\underset{\text{AMMONIUM CYANATE}}{NH_4OCN} \xrightarrow{\text{Heat}} \underset{\text{UREA}}{H_2N\overset{\overset{\displaystyle O}{\|}}{C}NH_2}$$

Friedrich Wöhler (1800–1882) was Professor of Chemistry at the University of Berlin and then at Göttingen. His preparation of the organic compound urea from ammonium cyanate, an inorganic source, did much to overturn the theory that organic compounds must be prepared in living organisms. One of the first to study the properties of aluminum, he discovered the element beryllium and is known for many other outstanding contributions to chemistry.

The notion of a mysterious vital force declined as other chemists began to synthesize more and more organic chemicals without the aid of a living system. Soon it was shown that chemists could do more than imitate the products of living tissue; they could form unique materials of their own design.

Advances in understanding the structure of organic compounds gave organic synthesis a tremendous boost. Knowing the structure of compounds, the organic chemist could predict by analogy with simpler molecules what reactions might take place when organic reagents were used. Very elegant and reliable schemes of synthesis could now be constructed.

In this chapter emphasis will be given to some important organic compounds that not only are useful in themselves but are necessary for the synthesis of many other organic compounds. The dependence of synthesis upon a knowledge of structure will be pointed out from time to time.

SOME USES OF HYDROCARBONS

Common sources of the hydrocarbons are the "fossil fuels" (coal, petroleum, and natural gas).

Complex mixtures of hydrocarbons, compounds containing only carbon and hydrogen, occur in enormous quantities in nature as petroleum and natural gas. These materials were formed from organisms that lived millions of years ago. After their death, they became covered with layers of sediment and ultimately were subjected to high temperatures and pressures in the depths of the earth's crust. In the absence of free oxygen, these conditions converted once living tissue into petroleum and coal. After petroleum and natural gas are brought to the surface of the earth by man, they can be separated into various fractions with different boiling points by the use of fractional distillation (Figure 14–1). Thus, pure hydrocarbons become available for synthesis of many other organic compounds. From the simplest hydrocarbons come such diverse consumer products as plastic dishes, acrylic fibers, vinyl paints, and neoprene rubber, and such industrial products as Teflon and cattle feed.

Natural gas consists primarily of low molecular weight hydrocarbons. It is predominantly methane (CH_4), but ethane (C_2H_6), propane (C_3H_8), and butane (C_4H_{10}) are also present. This mixture is conveyed

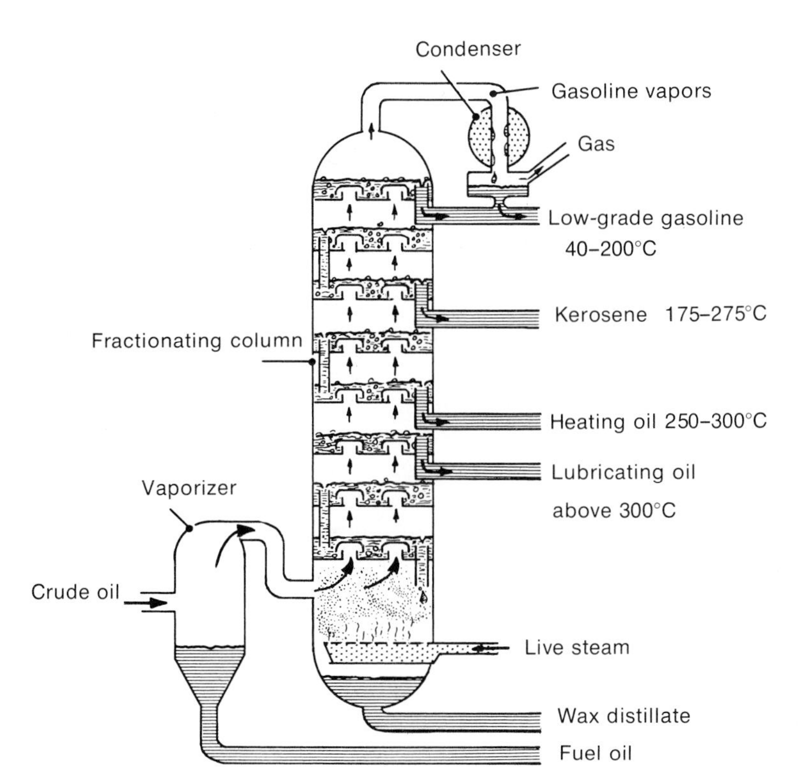

Figure 14–1 A diagram of a fractionating column for distilling petroleum. Notice that the higher boiling substances condense out at the lower levels and the lower boiling substances do not condense until the higher, cooler levels. (From Routh, J. I., Eyman, D. P., and Burton, D. J.: *A Brief Introduction to General, Organic and Biochemistry,* Philadelphia, W. B. Saunders Co., 1971.)

in long pipelines from the areas in which it occurs to cities where it is used as a fuel.

Energy is obtained by burning the constituents of natural gas in air:

$$CH_4 + 2O_2 \longrightarrow CO_2 + 2H_2O + 213\ \text{kcal}$$

$$C_2H_6 + 3\tfrac{1}{2}O_2 \longrightarrow 2CO_2 + 3H_2O + 372.8\ \text{kcal}$$

"Bottled gas" is actually stored in the liquid state under pressure. In order for bottled gas to be burned in stoves or furnaces, it must be in the vapor state.

The energy released here can be used to heat homes, run electric power plants, or power special internal combustion engines.

Propane and butane, principal components of bottled gas, can be liquefied by the use of moderate pressure. Because propane is more volatile than butane, bottled gas contains more propane in colder climates and more butane in warmer ones.

Cracking breaks down larger molecules into smaller ones.

The lower molecular weight hydrocarbons can also be prepared from higher molecular weight hydrocarbons by a process known as ***cracking.*** In this process, the gaseous hydrocarbon is passed over a catalyst at elevated temperatures and pressures and is broken down into smaller molecular weight fragments. The cracking of n-butane is a simple example:

$$CH_3CH_2CH_2CH_3 \xrightarrow[600^\circ]{\text{Catalyst}} CH_2{=}CH_2 + CH_2{=}CH_2 + H_2$$

These small molecules are most useful in the chemical synthesis of other products. If large hydrocarbon molecules in heavy petroleum oils are cracked, molecules in the C_5 to C_{10} range are produced. Molecules in this range are useful as fuels.

GASOLINE

From the time petroleum was first discovered in the United States in 1859 until 1900 when the automobile became popular, most oil was refined to yield kerosene, a mixture of hydrocarbons in the C_8 to C_{13} range. This liquid was used principally as a fuel for lamps.

Hydrocarbons in the C_6 to C_{10} range are known as *gasolines.*

The internal combustion engines used in early automobiles were designed to burn a more volatile mixture of hydrocarbons, the C_6 to C_{10} fraction, which became known as ***gasoline.*** These lower molecular weight hydrocarbons evaporate easily, so that they mix readily with air in simple carburetors and burn fairly completely.

With the increasing popularity of the automobile, petroleum refiners had to shift the output of a barrel of crude oil from a reasonably large fraction of kerosene to almost no kerosene and a much greater fraction of gasoline (Table 14–1). This dramatic increase in the amount of gasoline from a barrel of crude oil was accomplished by the discovery of chemical processes which convert nongasoline molecules into ones which burn well in an automobile engine.

The first conversion discovered was ***thermal cracking,*** a process in which long chain hydrocarbon molecules were heated under pressure. This heating strains the bonds of the molecule so that it may break or "crack," thus forming two smaller molecules, some of which will be in the gasoline range.

$$\underset{\text{AN ALKANE}}{C_{16}H_{34}} \xrightarrow[\text{Heat}]{\text{Pressure}} \underset{\text{AN ALKANE}}{C_8H_{18}} + \underset{\text{AN ALKENE}}{C_8H_{16}}$$

IN THE GASOLINE RANGE

Later, cracking was carried out in the presence of certain catalysts which allowed the processes to proceed at lower pressures and to produce even higher yields of gasoline. Today, these catalysts include aluminum oxide and platinum, as well as certain acids and specially processed clays. Each refiner of petroleum has his own special methods which offer different advantages in cost and type of crude oil handled. The hydrocarbon molecules produced are much the same regardless of the methods used.

A barrel of crude oil is 42 gallons.

Cracking of petroleum fractions brought about not only an increase in the quantity of gasoline available from a barrel of crude oil, but also an increase in *quality*. That is, gasoline from a cracking process can be

TABLE 14–1 Division of a Barrel of Crude Oil

	1920 %	1967 %
Gasoline	26.1	44.8
Kerosene	12.7	2.8
Jet Fuel	—	7.6
Heavy Distillates	48.6	22.2
Other	12.6	22.6
Total	100.0	100.0

used at higher efficiency (in a high-compression engine) than can "straight run" gasoline, because the molecular structures of the hydrocarbons in the cracked gasoline allow them to oxidize more smoothly at high pressure. When the burning of the gasoline-air mixture is too rapid or irregular, preignition occurs in the combustion chamber, resulting in a small detonation which is heard as a "knock" in the engine. This knocking can spell trouble to the owner of an automobile, because it will eventually lead to the breakdown of the internal parts of the automobile's engine.

Knocking in gasoline engines is a sign of improper combustion.

An arbitrary scale for rating the relative knocking properties of gasolines has been developed. Normal heptane, typical of straight run gasoline, knocks considerably and is assigned an octane rating of 0,

$$CH_3CH_2CH_2CH_2CH_2CH_2CH_3 \text{ (octane rating = 0)}$$

n-HEPTANE

while isooctane (2,2,4-trimethylpentane) is far superior in this respect and is assigned an octane rating of 100.

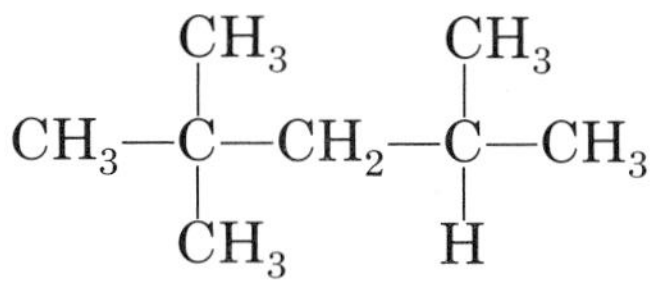

$$CH_3-\underset{CH_3}{\overset{CH_3}{\underset{|}{\overset{|}{C}}}}-CH_2-\underset{H}{\overset{CH_3}{\underset{|}{\overset{|}{C}}}}-CH_3 \text{ (octane rating = 100)}$$

2,2,4-TRIMETHYLPENTANE

The octane scale measures the ability of a mixture to burn without knocking in a gasoline engine.

To determine the octane rating of a gasoline, it is used in a standard engine and its knocking properties are recorded. This is compared to the behavior of mixtures of n-heptane and isooctane, and the percentage of isooctane in the mixture with identical knocking properties is called the octane rating of the gasoline. Thus, if a gasoline has the same knocking characteristics as a mixture of 9 percent n-heptane and 91 percent isooctane, it is assigned an octane rating of 91. This corresponds to a regular grade of gasoline. Since the octane rating scale was established, fuels have been developed which are superior to isooctane, so the scale has been extended well above 100. A primary source of the present high-octane gasolines is a chemical process known as ***catalytic reforming.*** Under the influence of certain catalysts, such as finely divided platinum, straight-chain hydrocarbons with low octane numbers can be reformed into their branched-chain isomers, which have higher octane numbers.

A catalyst increases the speed of a reaction without being consumed in the reaction.

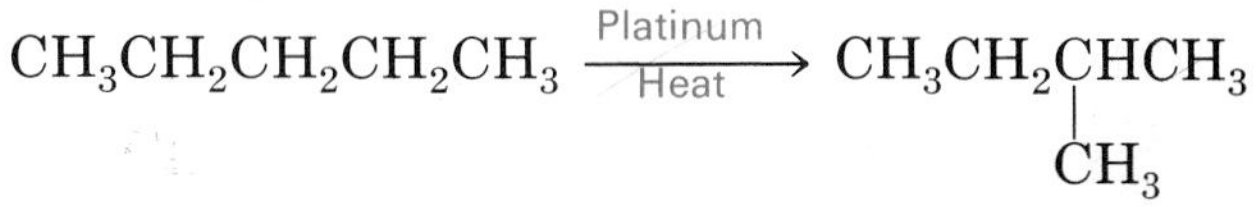

$$\underset{\text{n-PENTANE}}{CH_3CH_2CH_2CH_2CH_3} \xrightarrow[\text{Heat}]{\text{Platinum}} \underset{\text{2-METHYLBUTANE}}{CH_3CH_2\underset{CH_3}{\underset{|}{C}}HCH_3}$$

A great deal of research has been directed toward reducing the knocking caused by gasoline, and it has been discovered that this can be avoided by two procedures. The first is to use only hydrocarbons with structures known to burn satisfactorily in a high-compression engine. Branched-chain structures are particularly desirable in this respect, and gasolines with appreciable amounts of aromatic hydrocarbons (20 to 40 percent) also are less prone to knock. The second is to add "antiknock" agents to a more motley mixture of hydrocarbons to achieve the same effect. The best-known antiknock additive is tetraethyllead, $(C_2H_5)_4Pb$. Tetraethyllead is extremely toxic (see later section) and gasolines containing it ordinarily contain a dye to warn of its presence. Gasoline containing this compound is known as "ethyl" or "high octane" gasoline.

Tetraethyl lead, $Pb(C_2H_5)_4$, is the most common gasoline additive to suppress knocking. Its use will gradually decrease, as new cars, since 1975, are designed to run on lead-free gasoline.

Antiknock additives in general act to slow down free radical reactions that take place in the combustion chamber of an engine. By slowing these reactions down, the gasoline-air mixture can burn more smoothly, producing more power with a reduced likelihood of damaging the engine. When tetraethyllead is used in gasolines, it also is necessary to add a halogen compound such as 1,2-dibromoethane, $BrCH_2—CH_2Br$, to assist in the removal of the lead from the engine. The burning of a mixture of these compounds produces lead bromide, $PbBr_2$, which is eliminated as a vapor in the engine exhaust and released into the atmosphere.

A typical antiknock mixture for gasolines consists of about 62 percent tetraethyl lead, 18 percent 1,2-dibromoethane, 18 percent 1,2-dichloroethane, and 2 percent other ingredients, such as dye, petroleum solvent, and stability improver. Ethyl gasoline normally contains only a few tenths of a percent of such a mixture, corresponding to about 2 to 4 g of lead per gallon. Nevertheless, the total consumption of lead (and discharge into the atmosphere) amounts to millions of pounds per year. The chemistry of automobile fuels is discussed in greater detail in Chapter 21.

SOME IMPORTANT COMPOUNDS OF CARBON, HYDROGEN, AND OXYGEN

Carbon, hydrogen, and oxygen can be combined to form an enormous number of compounds. As we have seen, considerable order is introduced into the study of these compounds when they are divided into classes on the basis of the functional groups they contain. The ***alcohols,*** the ***organic acids*** and their derivatives (compounds that can be made from them), and the ***esters*** and ***soaps*** are very important compounds of carbon, hydrogen, and oxygen since they find such wide application in our everyday lives.

Alcohols

When a hydroxyl (—OH) group is attached to a nonaromatic carbon skeleton (an R-group), the resulting R—OH molecule has properties common to a class of compounds called ***alcohols.*** Formulas and names for some important molecules of this type are given in Table 14–2.

Alcohols are compounds with the structure R—O—H, where R is a hydrocarbon group.

Methanol (Methyl Alcohol)

Methanol was originally called wood alcohol since it was obtained by the destructive distillation of wood. It is the simplest of all alcohols and has the formula CH_3OH. In the older method for the production of wood alcohol, hardwoods such as beech, hickory, maple, or birch are heated in the absence of air in a retort (Figure 14–2). Methanol, which

TABLE 14–2 Some Important Alcohols

Formula	IUPAC Name	Common Name
H H—C—OH H	Methanol	Methyl alcohol (wood alcohol)
H H H—C—C—OH H H	Ethanol	Ethyl alcohol (grain alcohol)
H H H H—C—C—C—OH H H H	1-Propanol	n-Propyl alcohol
H H H H—C—C—C—H H O H H	2-Propanol	Isopropyl alcohol (rubbing alcohol)
H H—C—OH H—C—OH H	1,2-Ethanediol	Ethylene glycol (permanent antifreeze)
H H—C—OH H—C—OH H—C—OH H	1,2,3-Propanetriol	Glycerol (glycerin)

is 92 to 95 percent pure, can be obtained by fractional distillation of the liquid which results.

In 1923, the price of wood alcohol in the United States was 88¢ per gallon. In that year German chemists discovered how to produce this

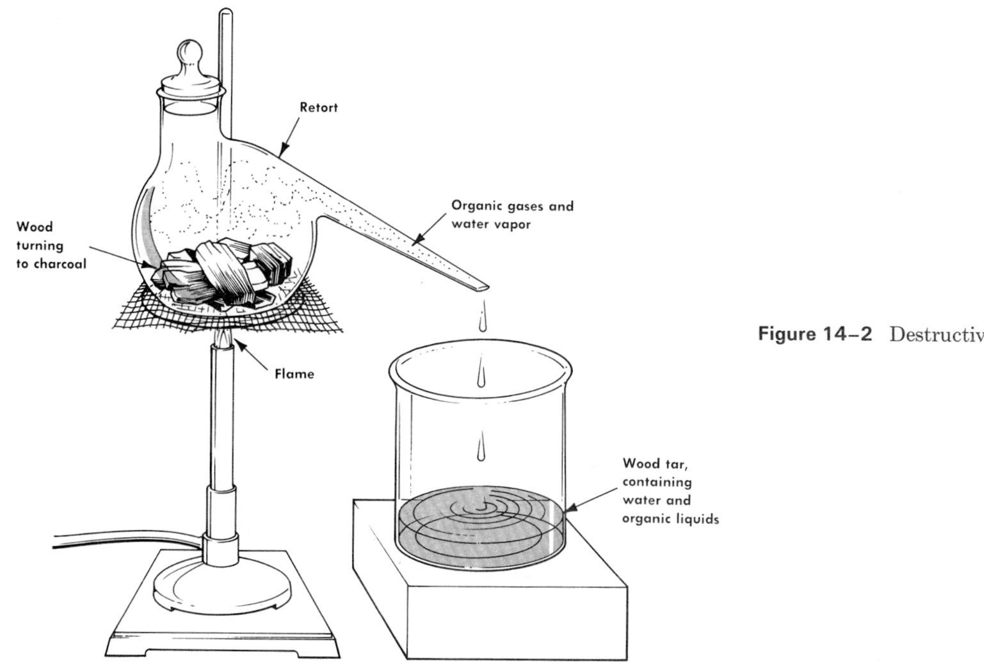

Figure 14–2 Destructive distillation of wood.

useful compound synthetically. Methanol is formed when carbon monoxide, CO, and hydrogen are heated at a pressure of 200 to 300 atmospheres over a catalyst of mixed oxides (90% ZnO–10% Cr_2O_3).

$$CO + 2H_2 \xrightarrow[\text{300°C; Pressure}]{\text{ZnO–Cr}_2\text{O}_3} CH_3OH$$

As a result of this synthetic process, German industrialists were able to sell pure methanol at 20¢ per gallon. Even a high tariff was not able to save the wood distillers in their outdated operations. The synthetic product soon dominated the market in the United States.

Industrial chemists are continually searching for less costly ways to prepare important chemicals such as methanol.

The production of synthetic methanol in the United States is over 4.0 billion pounds per year. About one half of this is used in the production of formaldehyde (used in plastics), 30 percent in the production of other chemicals, and smaller amounts for jet fuels, antifreeze mixtures, solvents, and as a denaturant (a poison added to ethanol to make it unfit for beverages). Methanol is a *deadly poison;* it causes blindness in less than lethal doses. Many deaths and injuries have resulted when this alcohol was mistakenly substituted for ethanol in beverages.

Ethanol

Ethanol (ethyl alcohol) is called grain alcohol because it can be fractionally distilled from the fermented mash made from corn, rice, barley, or other grains which are sources of carbohydrates. Fermentation is a breakdown of complex organic molecules, such as carbohydrates, brought about by means of enzymes. Enzymes are catalysts that are complex organic molecules produced by living cells. If the enzyme, diastase, is mixed with ground grain and water, and the mixture is allowed to stand at 40°C for a period of time, the starch in the grain will be changed into the sugar, maltose. (The more detailed chemistry of carbohydrates is discussed in Chapter 16.)

The subscript n in the formula for starch indicates that starch is made up of many $C_6H_{10}O_5$ units.

$$\underset{\text{STARCH (A CARBOHYDRATE)}}{2(C_6H_{10}O_5)_n} + nH_2O \xrightarrow{\text{diastase}} \underset{\text{MALTOSE (A SUGAR)}}{nC_{12}H_{22}O_{11}}$$

Brewers call the resulting mixture of maltose and water the wort. The wort is diluted and mixed with yeast and held at a temperature of 30°C for 40 to 60 hours. The living yeast cells secrete enzymes, maltase and zymase. The maltase causes the sugar, maltose, to decompose into a simple sugar, glucose:

$$\underset{\text{MALTOSE}}{C_{12}H_{22}O_{11}} + H_2O \xrightarrow{\text{maltase}} \underset{\text{GLUCOSE}}{2C_6H_{12}O_6}$$

The glucose, in turn, is converted by zymase to ethanol and carbon dioxide:

$$\underset{\text{GLUCOSE}}{C_6H_{12}O_6} \xrightarrow{\text{zymase}} 2CO_2 + \underset{\text{ETHANOL}}{2C_2H_5OH}$$

A solution of 95 percent ethanol and 5 percent water can be recovered from the mash by fractional distillation.

Synthetic ethanol is produced on a large scale for industrial use. The direct addition of water to ethylene accounts for more than 80 percent of all ethanol production. Under high pressure in the presence of a large excess of water, ethylene produces ethanol:

About 2 billion pounds of ethanol are produced synthetically in the U.S. each year.

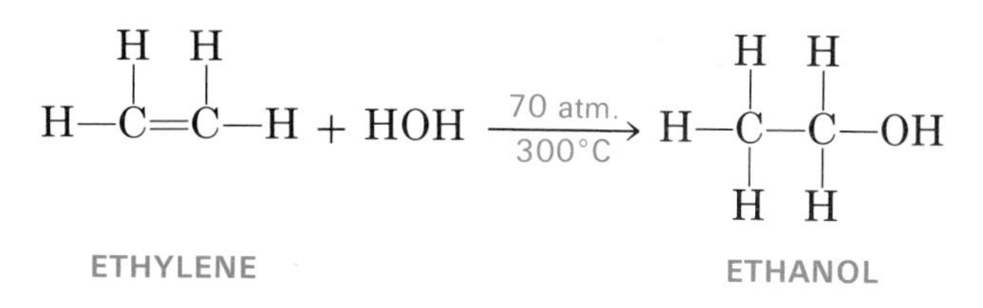

The "proof" of an alcoholic solution is twice the percentage of alcohol by volume.

Some cough syrups contain as much as 20% ethanol.

Pure ethanol is 200 proof (exactly twice the percentage of alcohol). Apart from the alcoholic beverage industry, ethyl alcohol is used widely in solvents and in the preparation of chloroform, ether, and many other organic compounds.

Some of the commonly encountered alcoholic beverages and their characteristics are presented in Table 14–3.

TABLE 14–3 Common Alcoholic Beverages

Name	Source of Fermented Carbohydrate	Amount of Ethyl Alcohol
Beer	Barley	3.2–9%
Wine	Grapes or other fruit	12% maximum, unless fortified*
Brandy	Distilled wine	40–45%
Whiskey	Barley, rye, corn, etc.	45–55%
Rum	Molasses	~45%

*The growth of yeast is inhibited at alcohol concentrations of over 12% and fermentation comes to a stop. Beverages with a higher concentration are prepared either by distillation or by fortification with alcohol which has been obtained by the distillation of another fermentation product.

Propanols (Propyl Alcohols)

When one considers the possible structures for propyl alcohol, it is apparent that two isomers are possible.

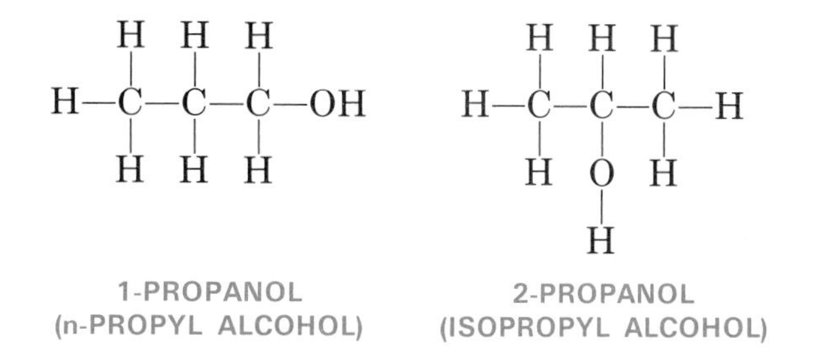

1-PROPANOL
(n-PROPYL ALCOHOL)

2-PROPANOL
(ISOPROPYL ALCOHOL)

Of the two propanols, 1-propanol is the more expensive; it is prepared by the oxidation of simple hydrocarbons. It finds uses as a solvent and as a raw material in the manufacture of other organic compounds.

The hydration of propylene yields 2-propanol (isopropyl alcohol) which is sold as rubbing alcohol. It has a greater germicidal activity than the other simple alcohols and is used as an antiseptic.

Ethylene Glycol and Glycerol (Glycerin)

More than one alcohol group (—OH) can be present in a single molecule. Ethylene glycol, the base of permanent antifreeze, and glycerol are examples of such compounds.

Permanent antifreeze is ethylene glycol:

$$\underset{\text{OH}}{CH_2}-\underset{\text{OH}}{CH_2}$$

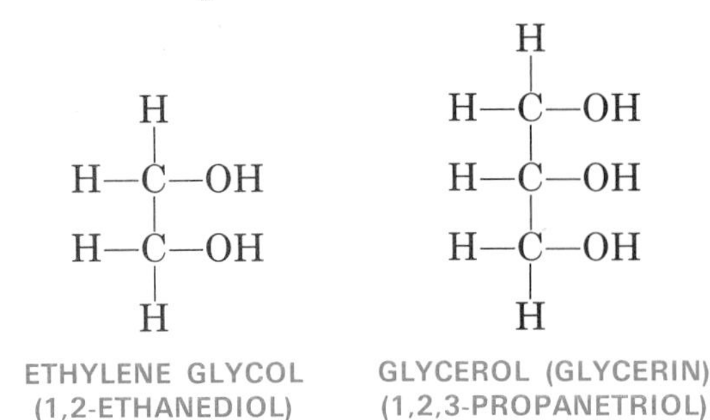

ETHYLENE GLYCOL (1,2-ETHANEDIOL) GLYCEROL (GLYCERIN) (1,2,3-PROPANETRIOL)

Glycerol has many uses in the manufacture of drugs and cosmetics, in the production of nitroglycerin and numerous other chemicals. Perhaps the most important compounds of glycerol are its natural esters (fats and oils), which we shall discuss later in this chapter.

Hydrogen Bonding in Alcohols

The physical properties of water, methanol, ethanol, the propanols, ethylene glycol, and glycerol offer another interesting example of the effects of ***hydrogen bonding*** between molecules in liquids. For a more complete discussion of hydrogen bonding, see Chapter 9. In Table 14-4 the boiling points for these compounds are listed.

Since boiling involves overcoming the attractions between liquid molecules as they pass into the gas phase, a higher boiling point indicates stronger intermolecular forces holding the molecules together. Another factor is also present: as the molecules become larger, higher boiling points result, since more energy is required to change the longer-chain molecules from the liquid to the gaseous phase, owing in part to the larger van der Waals attraction. A graph showing the boiling points of the normal alcohols (straight carbon chains with the —OH group on an end carbon) as a function of chain length is given in Figure 14-3.

Hydrogen bonding is responsible for the fact that an alcohol has a higher boiling point than its parent hydrocarbon.

Methanol, like water, has an —OH group, and some hydrogen bonding is to be expected, as shown in Figure 14-4. Hydrogen bonding explains why methanol (molecular weight = 32) is a liquid, whereas propane (C_3H_8, molecular weight = 44), an even heavier molecule, is a gas at room

TABLE 14-4 Boiling Points for Some —OH Compounds

Water	HOH	100°C
Methanol	CH_3OH	64.6°
Ethanol	CH_3CH_2OH	78.5°
1-Propanol	$CH_3CH_2CH_2OH$	97.2°
2-Propanol	$CH_3CHOHCH_3$	82.3°
Ethylene glycol	CH_2OHCH_2OH	198°
Glycerol	$CH_2OHCHOHCH_2OH$	290°

Note: The parent hydrocarbon of methanol is methane (bp −182.5°C); of ethanol and ethylene glycol, ethane (bp −88.6°C); and of the propanols and glycerol, propane (bp −44.5°C).

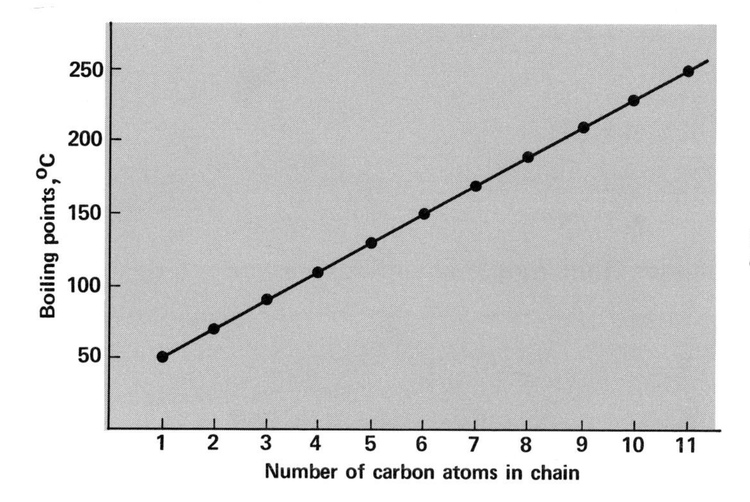

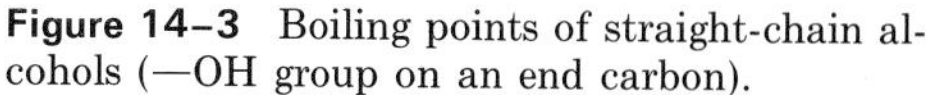
Figure 14–3 Boiling points of straight-chain alcohols (—OH group on an end carbon).

temperature. Methanol has only one hydrogen through which it can hydrogen bond, while water can hydrogen bond from either of its two hydrogen atoms. Thus, water, with more extensive intermolecular bonding, has the higher boiling point even though it is made up of lighter molecules.

Both methyl and ethyl alcohol can be used as antifreeze, but they tend to distill out of the coolant at the temperatures of a hot gasoline engine. Protection against freezing is then lost over a period of time. Ethylene glycol is equally effective (molecule for molecule) in lowering the freezing point of water, and its high boiling point (198°C) makes it a permanent type antifreeze. This property makes ethylene glycol more desirable even though it takes almost twice as much ethylene glycol by weight as methanol for the same amount of protection of a car's cooling system. Ethylene glycol with suitable additives to protect the radiator system is sold under a number of brand names. The higher boiling point of ethylene glycol is readily explained in terms of the two —OH groups per molecule and the enhanced possibility for hydrogen bonding. Glycerol, with three —OH groups per molecule, has an even higher boiling point, as well as a very high viscosity (resistance to flow).

Alcohols can be used as the starting material for the synthesis of other organic compounds.

Alcohols can serve as the starting material for the synthesis of many other types of organic compounds. For example, the oxidation of ethanol can be used to make acetaldehyde and acetic acid:

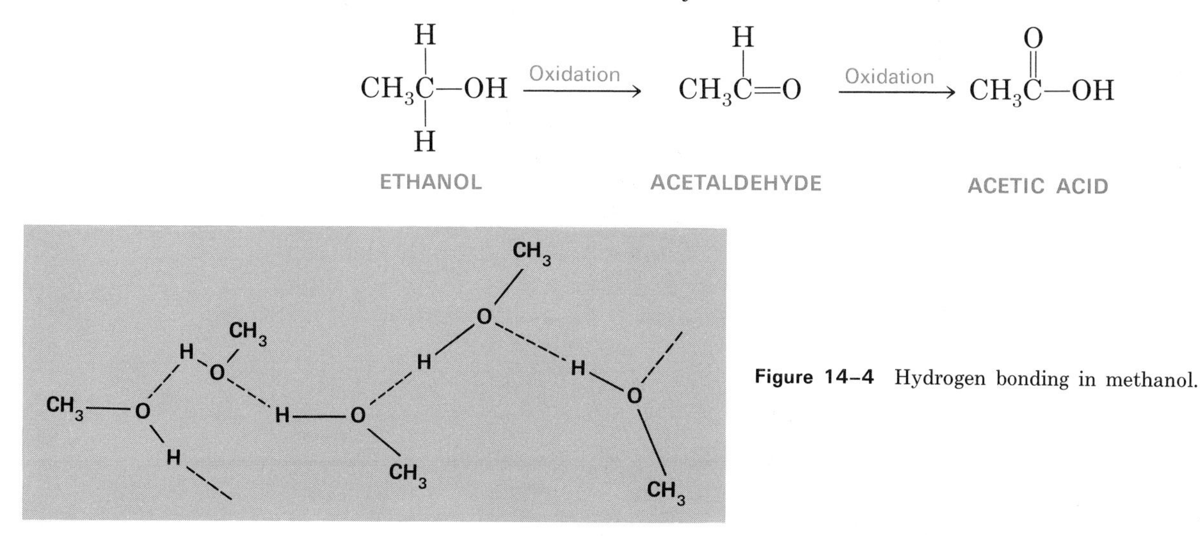

Figure 14–4 Hydrogen bonding in methanol.

The oxidation of 2-propanol provides the ketone, acetone:

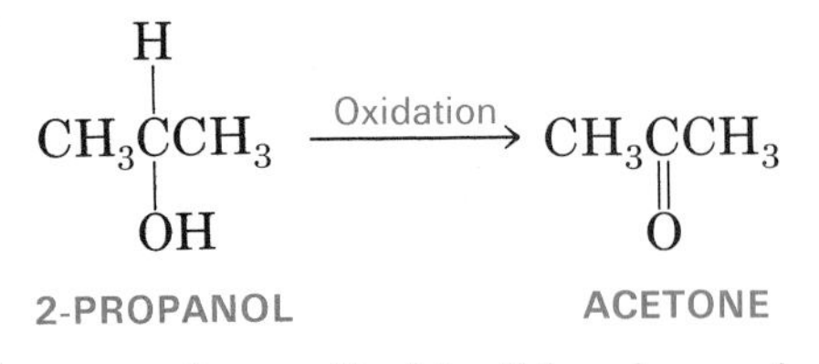

Substitution reactions produce alkyl halides from simple alcohols:

$$CH_3CH(OH)CH_3 + HBr \longrightarrow CH_3CHBrCH_3 + H_2O$$

2-PROPANOL 2-BROMOPROPANE

These represent just a very few of the reactions which can be used to transform alcohols into related organic compounds containing other functional groups. The reactions found are related to the type of alcohol present. Thus, primary alcohols like ethanol, which bear their —OH group at the end of the chain, are oxidized to aldehydes and then acids; secondary alcohols, which bear their —OH group on a carbon atom bonded to two other carbon atoms, are oxidized to ketones. Tertiary alcohols, which have their —OH group on a carbon atom bonded to three other carbon atoms, exhibit other reactions. These structural types are illustrated by the compounds:

$$CH_3CH_2CH_2CH_2OH \qquad CH_3CH_2CH(OH)CH_3 \qquad CH_3C(CH_3)(OH)CH_3$$

1-BUTANOL (A PRIMARY ALCOHOL) — 2-BUTANOL (A SECONDARY ALCOHOL) — 2-METHYL-2-PROPANOL (A TERTIARY ALCOHOL)

Organic Acids

Earlier an acid was defined as a species that has a tendency to donate hydrogen ions, or protons. We will now consider the carboxylic acids, which contain the carboxyl group, $-C(=O)OH$. The electronegative character of the C=O group tends to drain electron density away from the region between the oxygen and hydrogen atoms. The partial positive charge assumed by the hydrogen makes it possible for polar water molecules to remove the hydrogen ions from some of the carboxyl groups. The strength of an organic acid depends on the group which is attached to the carboxyl group. If the attached group has a tendency to pull electrons away from the carboxyl group, the acid is a stronger acid. For example, trifluoroacetic acid is a much stronger acid than acetic acid:

Organic acids are compounds of the type $R-C(=O)-OH$. They are generally weak acids.

$$F_3C-C(=O)-O-H \qquad H_3C-C(=O)-O-H$$

TRIFLUOROACETIC ACID — ACETIC ACID

The electronegative fluorine atoms withdraw electron density from the region of the carboxyl group and facilitate the loss of the hydrogen ion. However, trifluoroacetic acid is still weaker than a strong mineral acid such as sulfuric acid.

The ionization of carboxylic acids in water is as follows:

$$R{-}\overset{\overset{\displaystyle O}{\|}}{C}{-}OH + H_2O \rightleftharpoons R{-}\overset{\overset{\displaystyle O}{\|}}{C}{-}O^- + H_3O^+$$

They are neutralized by bases to form salts:

$$R{-}\overset{\overset{\displaystyle O}{\|}}{C}{-}OH + Na^+ + OH^- \longrightarrow \underset{\text{SALT}}{R{-}\overset{\overset{\displaystyle O}{\|}}{C}{-}O^-} + Na^+ + H_2O$$

Formic Acid (Methanoic Acid)

The simplest organic acid is formic acid, in which the carboxyl group is attached directly to a hydrogen atom.

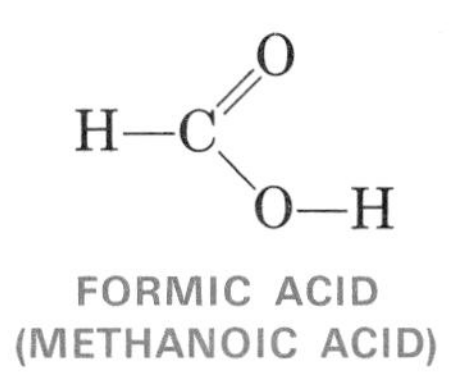

FORMIC ACID
(METHANOIC ACID)

This acid is found in ants and other insects and is part of the irritant that produces itching and swelling after a bite. The sodium salt of formic acid is readily prepared by heating carbon monoxide (CO) with sodium hydroxide (NaOH):

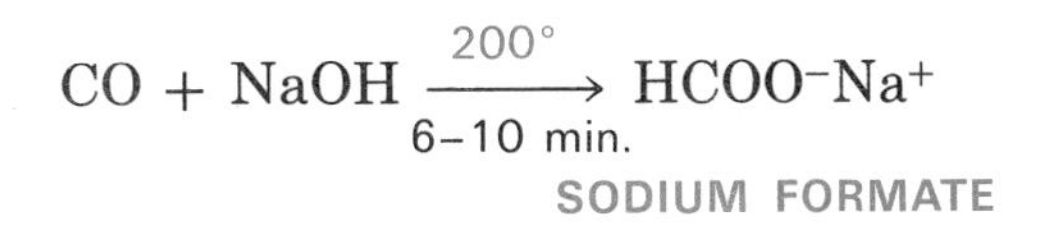

$$CO + NaOH \xrightarrow[\text{6–10 min.}]{200^\circ} \underset{\text{SODIUM FORMATE}}{HCOO^-Na^+}$$

If the resulting salt is mixed with a mineral acid, formic acid can be distilled from the mixture:

$$\underset{\text{SODIUM FORMATE}}{HCOO^-Na^+} + \underset{\text{HYDROCHLORIC ACID}}{H_3O^+ + Cl^-} \longrightarrow \underset{\text{FORMIC ACID}}{HCOOH} + \underset{\text{SODIUM CHLORIDE}}{Na^+Cl^-} + H_2O$$

Acetic Acid (Ethanoic Acid)

Acetic acid can be made by oxidizing ethanol.

Acetic acid is the most widely used of the organic acids. It is found in vinegar, an aqueous solution containing 4 to 5 percent acetic acid. Flavor and colors are imparted to vinegars by the constituents of the alcoholic solutions from which they are made. Ethanol in the presence of certain bacteria and air is oxidized to acetic acid:

$$\underset{\text{ETHANOL}}{CH_3CH_2OH} + \underset{\text{OXYGEN}}{O_2} \xrightarrow{\text{Bacteria}} \underset{\substack{\text{ACETIC ACID} \\ \text{(ETHANOIC ACID)}}}{CH_3COOH} + \underset{\text{WATER}}{H_2O}$$

The bacteria, called mother of vinegar, form a slimy growth in a vinegar solution. The growth of bacteria can sometimes be observed in a bottle of commercially prepared vinegar after it has been opened to the air.

Acetic acid is an important starting material for making other chemicals and is a convenient acidic material when a cheap organic acid is needed.

Fatty Acids

A fatty acid contains a carboxyl group attached to a long hydrocarbon chain. The chains often contain only single carbon-carbon bonds, but may contain carbon-carbon double bonds as well. Examples are stearic acid, palmitic acid, and oleic acid.

Fatty acids may be obtained from animal and vegetable fats. The hydrocarbon chain in fatty acids is generally 8 to 18 carbon atoms in length.

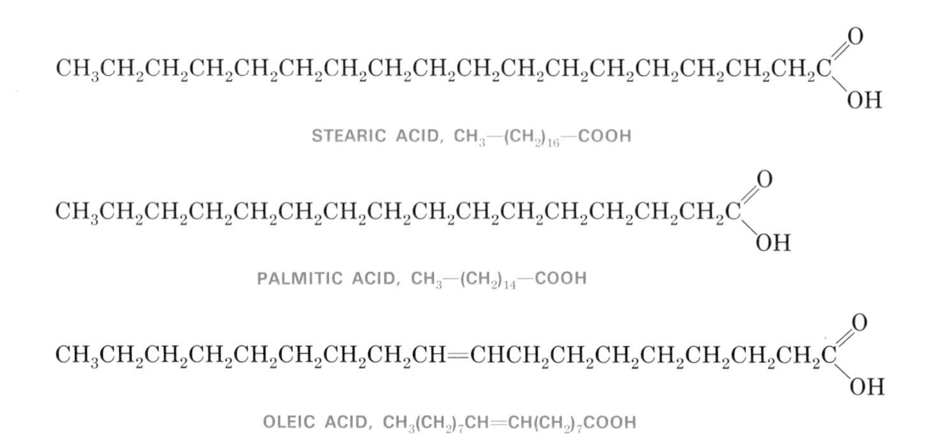

STEARIC ACID, $CH_3—(CH_2)_{16}—COOH$

PALMITIC ACID, $CH_3—(CH_2)_{14}—COOH$

OLEIC ACID, $CH_3(CH_2)_7CH{=}CH(CH_2)_7COOH$

Stearic acid is obtained by the hydrolysis of animal fat, palmitic acid results from the hydrolysis of palm oil, and oleic acid is obtained from olive oil. These reactions are given in the next two sections. Stearic and palmitic acids are especially important in the manufacture of soaps.

SELF-TEST 14-A

1. Which of the following compounds would be expected to have the highest octane rating?

a. $CH_3CH_2CH_2CH_2CH_2CH_2CH_3$

b. $CH_3—CH_2—CH(CH_3)—CH_2CH_2CH_3$

c. $CH_3—C(CH_3)_2—CH(CH_3)—CH_3$

2. Name the following compounds:

   a. CH_3OH ________________

   b. CH_3CH_2OH ________________

   c. $HCOOH$ ________________

   d. $CH_3CH_2\underset{|}{C}HCH_3$ (with OH on the third carbon) ________________
      OH

   e. CH_3COOH ________________

   f. $CH_2{-}CH_2$ (each with OH) ________________
      OH OH

   g. CH_3CHO ________________

   h. CH_3CHCH_3 (Br on the central carbon) ________________
      Br

3. Indicate products in the reactions:

   a. $CH_3CHO \xrightarrow{\text{Oxidation}}$ ________________

   b. $CH_3COOH + Na^+ + OH^- \longrightarrow$ ________________

   c. $CH_3CHCH_3 + HBr \longrightarrow$ (OH on the central carbon) ________________
      OH

4. Two methods by which ethanol is made on a large scale are:

   a. ________________

   b. ________________

5. An example of a carboxylic acid that contains a long hydrocarbon chain is

   ________________.

6. Identify the functional groups present in each of the following molecules:

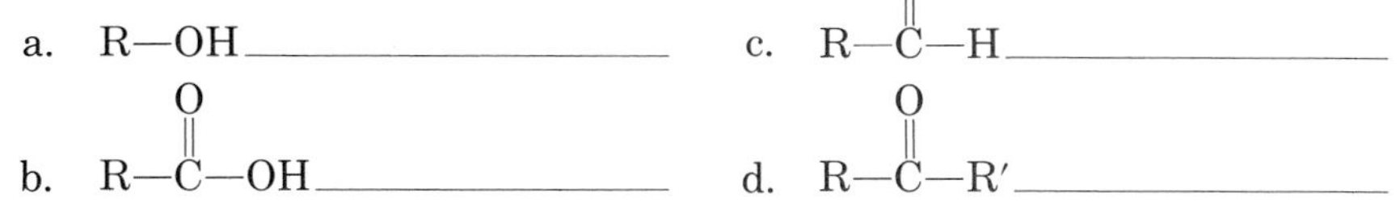

Esters

Esters are compounds of the type R—O—C(=O)—R′ formed by the reaction of organic acids and alcohols.

In the presence of certain strong mineral acids, organic acids react with alcohols to form compounds called ***esters.*** For example, when ethyl alcohol is mixed with acetic acid in the presence of sulfuric acid, ethyl acetate is formed. This reaction is a dehydration in which sulfuric acid acts as a catalyst.

$$CH_3CH_2O\text{-}H + HO\text{-}\underset{\underset{O}{\|}}{C}CH_3 \xrightleftharpoons{H_2SO_4} CH_3CH_2O\underset{\underset{O}{\|}}{C}CH_3 + H_2O$$

Ethyl acetate is a common solvent and is often used as fingernail polish remover.

Some of the odors of common fruits are due to the presence of mixtures of volatile esters (Table 14–5). In contrast, higher molecular weight esters often have a distinctly unpleasant odor.

TABLE 14–5 Some Alcohols, Acids, and Their Esters

Alcohol	Acid	Ester	Odor of the Ester
$CH_3CH(CH_3)CH_2CH_2OH$ ISOPENTYL ALCOHOL	CH_3COOH ACETIC ACID	$CH_3CH(CH_3)CH_2CH_2{-}O{-}C({=}O)CH_3$ ISOPENTYL ACETATE	Banana
$CH_3CH(CH_3)CH_2CH_2OH$ ISOPENTYL ALCOHOL	$CH_3CH_2CH_2CH_2COOH$ n-VALERIC ACID	$CH_3CH(CH_3)CH_2CH_2{-}O{-}C({=}O){-}CH_2CH_2CH_2CH_3$ ISOPENTYL n-VALERATE	Apple
$CH_3CH_2CH_2CH_2OH$ n-BUTYL ALCOHOL	$CH_3CH_2CH_2COOH$ n-BUTYRIC ACID	$CH_3CH_2CH_2CH_2{-}O{-}C({=}O){-}CH_2CH_2CH_3$ BUTYL n-BUTYRATE	Pineapple
$CH_3CH(CH_3)CH_2OH$ ISOBUTYL ALCOHOL	CH_3CH_2COOH PROPIONIC ACID	$CH_3CH(CH_3)CH_2{-}O{-}C({=}O){-}CH_2CH_3$ ISOBUTYL PROPIONATE	Rum

Fats, Oils, and Soaps

Fats and oils are esters of glycerol (glycerin) and a fatty acid. R, R′, and R″ stand for the hydrocarbon chains in the following equation.

Lipids are soluble in fats and oils.

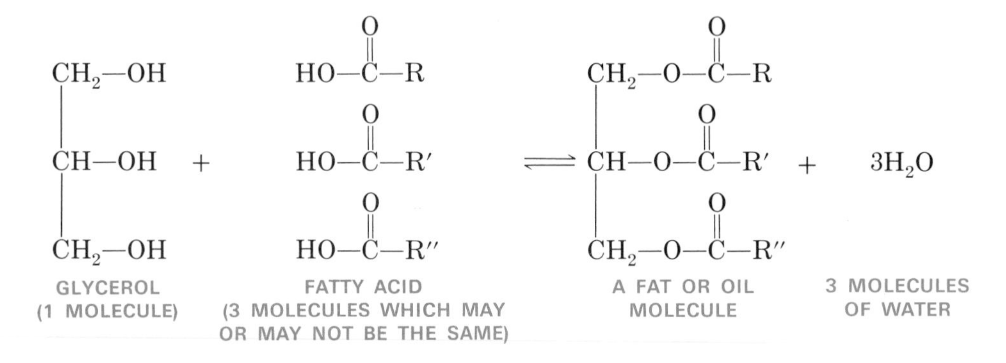

The term ***fat*** is usually reserved for solids (butter, lard, tallow) and ***oil*** for liquids (castor, olive, linseed, tung, and so forth). The term ***lipid*** includes fats, oils, and other fat-soluble compounds.

Saturation (all single bonds with maximum hydrogen content) in the carbon chain of the fatty acids is usually found in solid or semi-solid fats, whereas unsaturated fatty acids (containing one or more double bonds) are usually found in oils. Hydrogen can be catalytically added to the double bonds of an oil to convert it into a semisolid fat. For example, liquid soybean and other vegetable oils are hydrogenated to produce cooking fats and margarine.

Fats and oils are esters of fatty acids and glycerol.

Consumers in Europe and North America have historically valued butter as a source of fat. As the population increased, the advantages of a substitute for butter became apparent, and efforts to prepare such a product began about a hundred years ago. One problem which arose was the fact that common fats are almost all *animal* products with very pronounced tastes of their own. Analogous compounds from vegetable oils, which have mixed flavors, were generally *unsaturated* and consequently *oils*. A solid fat could be made from the much cheaper vegetable oils if an inexpensive way could be discovered to add hydrogen across their double bonds. After extensive experiments, many catalysts were found, of which finely divided nickel is among the most effective. The nature of the process can be illustrated by the reaction:

Catalytic hydrogenation can convert a liquid oil into a solid fat.

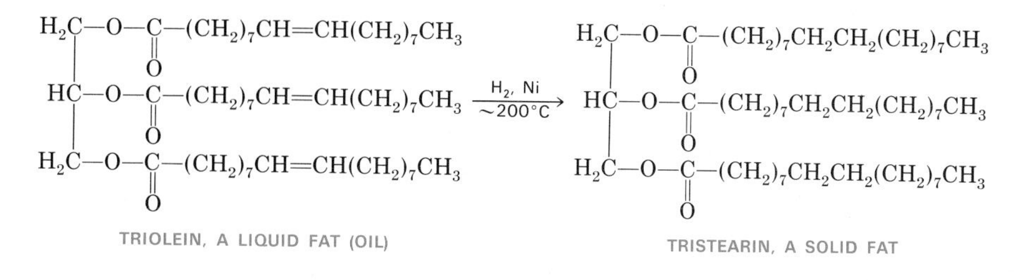

TRIOLEIN, A LIQUID FAT (OIL)

TRISTEARIN, A SOLID FAT

Oils commonly subjected to this process include those from cottonseed, peanut, corn germ, soybean, coconut, and safflower seeds. In recent years, as it became apparent that hydrogenation destroys essential fatty acids, soft margarines and cooking oils, which still contain some of the unhydrogenated essential fatty acid, have been placed on the market.

There is considerable interest in the relationship between saturated fats in the diet and heart disease. It is known that an increase in solid fats increases the concentration of the complicated biochemical cholesterol in the blood stream. Since it is thought by some physicians that a rise in cholesterol is associated with hardening of the arteries, there has been much interest in replacing solid fats with liquid oils in the diet.

Soaps can be made by treating fats or oils with sodium hydroxide.

Naturally occurring fats and oils can be hydrolyzed in strongly basic solutions to form glycerol and salts of the fatty acids. Such hydrolysis reactions are called ***saponification*** reactions; the sodium or potassium salts of the fatty acids formed are ***soaps.*** Pioneers prepared their soap by boiling animal fat with an alkaline solution obtained from the ashes of hard wood. The resulting soap could be "salted out" by adding sodium chloride, making use of the fact that soap is less soluble in a salt solution than in water.

$$\begin{array}{l} CH_3(CH_2)_{16}COO{-}CH_2 \\ CH_3(CH_2)_{16}COO{-}CH \\ CH_3(CH_2)_{16}COO{-}CH_2 \end{array} + 3NaOH \longrightarrow 3CH_3(CH_2)_{16}COO^-Na^+ + \begin{array}{l} HO{-}CH_2 \\ HO{-}CH \\ HO{-}CH_2 \end{array}$$

STEARIC ACID ESTER (GLYCERYL TRISTEARATE) — SODIUM STEARATE (A SOAP) — GLYCEROL

The cleansing action of soap can be explained in terms of its molecular structure. Material that is water soluble can be readily removed from the skin or a surface by simply washing with an excess of water. To remove a sticky sugar syrup from one's hands, the sugar is dissolved in water and rinsed away. Many times the material to be removed is oily and water will merely run over the surface of the oil. Since the skin has natural oils, even substances, such as ordinary dirt, which are not oily themselves, can cover the skin in a greasy layer. The cohesive forces (forces between like molecules tending to hold them together) within the water layer are too large to allow the oil and water to intermingle (Figure

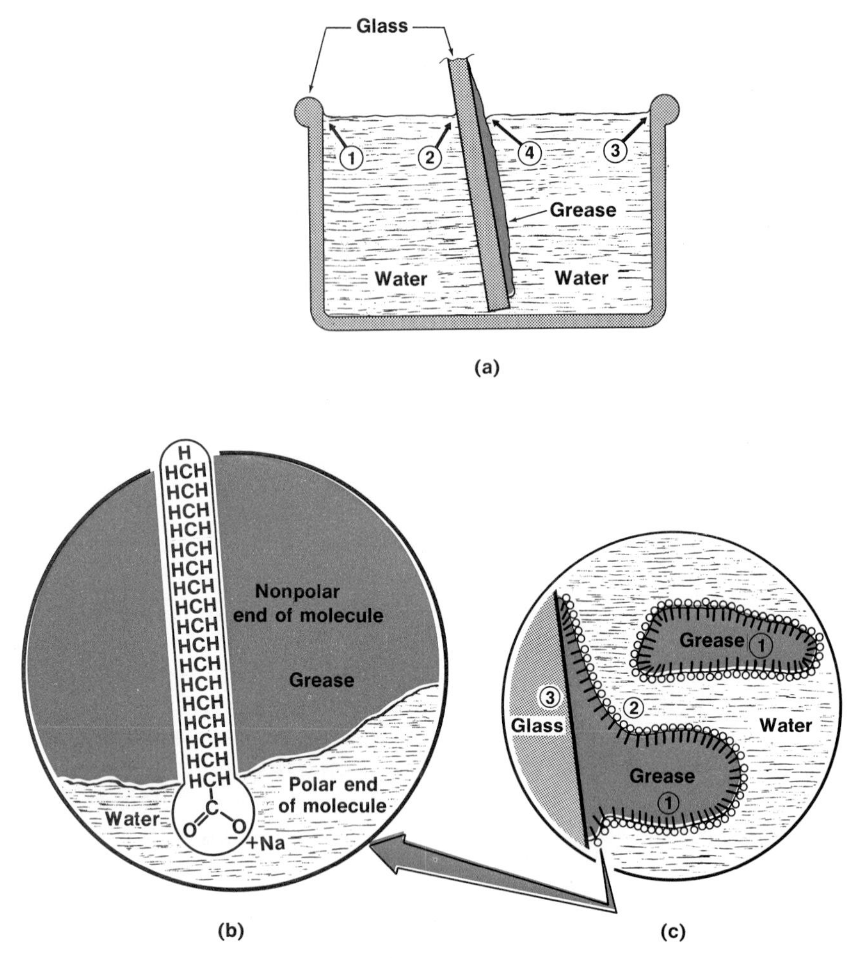

Figure 14–5 The cleansing action of soap. (*a*) A piece of glass coated with grease inserted in water gives evidence for the strong adhesion between water and glass at 1, 2 and 3. The water curves up against the pull of gravity to wet the glass. The relatively weak adhesion between oil and water is indicated at 4 by the curvature of the water away from the grease against the force tending to level the water. (*b*) A soap molecule, having oil-soluble and water-soluble ends, will orient at an oil-water interface such that the hydrocarbon chain is in the oil (with molecules that are electrically similar, nonpolar) and the COO^-Na^+ group is in the water (highly charged polar groups interacting electrically). (*c*) In an idealized molecular view, a grease particle, 1, is surrounded by soap molecules which in turn are strongly attracted to the water. At 2 another droplet is about to break away. At 3 the grease and clean glass interact before the water moves between.

14–5). When present in an oil-water system, soap molecules such as sodium stearate,

$$CH_3CH_2CH_2CH_2CH_2CH_2CH_2CH_2CH_2CH_2CH_2CH_2CH_2CH_2CH_2CH_2CH_2C\begin{matrix}\diagup\!\!O \\ \diagdown O^-Na^+\end{matrix}$$

will move to the interface between the two liquids. The carbon chain, which is a nonpolar structure, will readily mix with the nonpolar grease molecules, whereas the highly polar $—COO^-Na^+$ group enters the water layer (Figure 14–5*b*). The soap molecule will then tend to lie across the oil-water interface. The grease is broken up into small droplets, each surrounded by hydrated soap molecules (Figure 14–5*c*). The surrounded oil droplets cannot come together again since the exterior of each is covered with $—COO^-Na^+$ groups which strongly interact with the surrounding water. If enough soap and water are available, the oil will be swept away forming a clean and water-wet surface.

SYNTHETIC DETERGENTS (SYNDETS)

Ordinary soaps give precipitates with the hard water ions such as Ca^{2+}, Mg^{2+} and Fe^{3+}.

If Ca^{2+} or Mg^{2+} ions are present in water, an ordinary soap will precipitate as an insoluble salt and the water is said to be ***hard.*** The ring on the bathtub and the scum in the washer are visible signs of this precipitate. Only after all these interfering ions have been precipitated can the added soap cause the water to form suds. Water softeners remove the hard water ions, replacing them with ions such as Na^+ which do not interfere with the ordinary soap's action, but this is often an expensive process. Synthetic detergents (syndets) are similar to soaps in that they are composed of molecules having water-soluble and oil-soluble ends, but syndets have an advantage over soaps in hard water in that they tend not to form precipitates with Ca^{2+} or Mg^{2+}. The synthesis of a typical syndet molecule with a water-soluble sodium sulfonate ($—SO_3^-Na^+$) group and an oil-soluble hydrocarbon group is shown below.

$$\underset{\text{LAURYL ALCOHOL}}{CH_3(CH_2)_{11}OH} \xrightarrow{H_2SO_4} \underset{\text{LAURYL HYDROGEN SULFATE}}{CH_3(CH_2)_{11}OSO_3H} \xrightarrow{NaOH} \underset{\text{SODIUM LAURYL SULFATE}}{CH_3(CH_2)_{11}OSO_3^-Na^+}$$

The most widely used syndets are sodium salts of substituted benzene-sulfonic acids. Normally, a long hydrocarbon chain, designated —R, is attached to the benzene ring by a substitution reaction (Chapter 13).

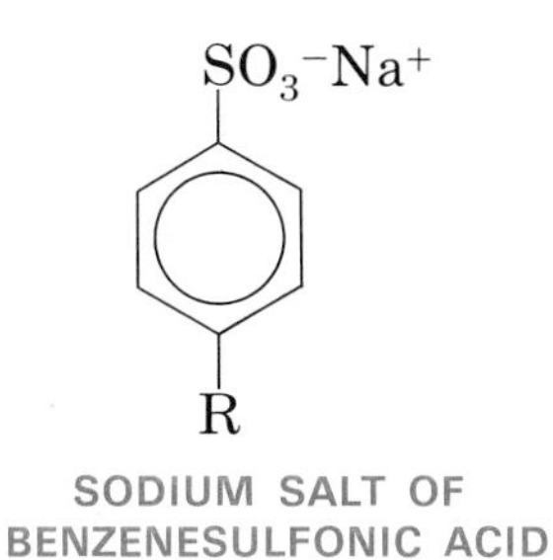

SODIUM SALT OF BENZENESULFONIC ACID

Syndets are effective cleansing agents, but they pose problems in water treatment and purification. By careful synthesis of these molecules, chemists have been able to produce syndets which, once "used," can be broken down into simpler molecules by the action of bacteria. Since about 1965, these biodegradable syndets have been used almost exclusively in the United States.

USEFUL PRODUCTS FROM AROMATIC ORGANIC REACTIONS

Many chemists are engaged in the synthesis of organic compounds. In educational and industrial laboratories throughout the world they prepare new and different compounds on a small scale, expand the scale to pilot plant operation, or set up and manage large-scale industrial processes (Figure 14–6).

The thousands of chemical changes required to synthesize the many organic compounds have a few characteristics in common.

Organic syntheses are usually carried out so that only one part of the molecule changes in each reaction step.

1. Several chemical changes are usually required to synthesize a single organic compound. Each chemical change is called a step, and each step produces an intermediate compound that is used in the next step. The final step produces the desired end product.
2. Normally only one functional group undergoes change in each step. The rest of the molecule remains intact and unchanged.

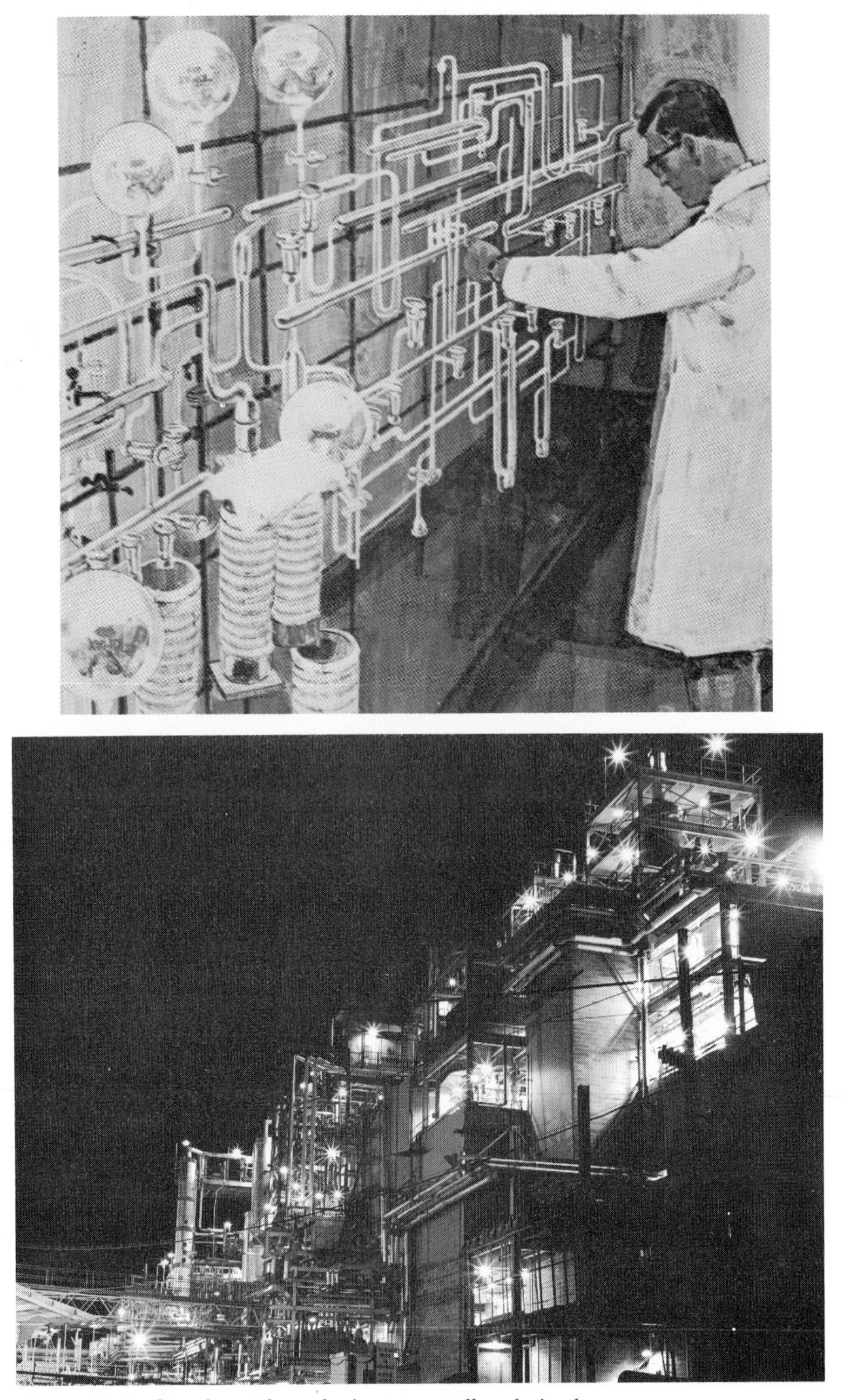

Figure 14–6 Organic syntheses begin on a small scale in the laboratory (*top*), but graduate to a very large scale if they become commerically important. The top view shows a typical laboratory bench. The bottom view shows an automatic, modern chemical plant. In both cases the chemical reactions are the same; the scope is different. (*bottom,* courtesy of the Du Pont Company, Textile Fiber Department.)

This is known as the principle of minimum structural change.

3. From one principal starting material can come many diverse products. The kind of product depends upon the reactants and the conditions imposed.
4. The more steps in the synthesis, the less the percent yield of final product. The starting material and each intermediate are only partially converted to the next intermediate because of equilibrium considerations, side reactions, or both, which convert some of the starting material into undesirable products. If an intermediate product must be removed and purified before proceeding to the next step, some additional material is lost. The principal purification methods are recrystallization and extraction for solids and distillation for liquids.

These principles will be illustrated by the preparation of chlorobenzene from a natural product, benzene, and then the use of chlorobenzene to prepare such diverse and useful products as aspirin, oil of wintergreen, sulfa drugs, and an organic dye.

About 500 million pounds of chlorobenzene were produced in the U.S. in 1970.

In the previous chapter we saw how bromobenzene could be prepared from benzene in the presence of a suitable catalyst, such as iron(III) bromide, $FeBr_3$. Benzene undergoes the same kind of a reaction with chlorine to produce chlorobenzene.

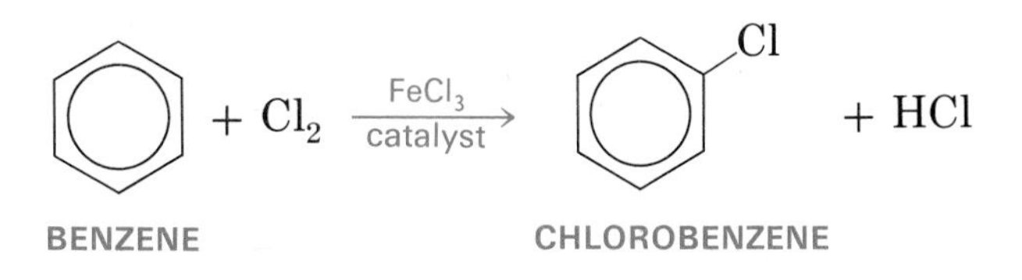

Chlorobenzene is manufactured on an enormous scale and is widely used in industrial processes to prepare other organic compounds.

Aspirin, Oil of Wintergreen, and Salicylamide

The series of steps required to synthesize aspirin, oil of wintergreen, and salicylamide from chlorobenzene is outlined in Figure 14–7. Each structure represents the beginning or end of a step in the synthesis. Aspirin and oil of wintergreen require five steps; salicylamide requires six. Only the principal organic substance is given as the product for each step. Other products such as NaCl (a coproduct with phenol) and water (a coproduct with sodium phenolate) are sometimes important in the synthesis because they have to be removed to avoid interference with subsequent steps. However, in giving a broad outline of the synthetic process, the coproducts are generally omitted; only those products made in a previous step and then required for subsequent steps are included. In Figure 14–7 the step-by-step structural changes can be followed by noting groups in color. Conditions and additional reactants for each step are written with the arrow. These conventions will be used to summarize the organic syntheses presented in this chapter.

Some intermediates are useful compounds in their own right. Phenol, commonly called carbolic acid, is used to prepare plastics such as Bakelite, drugs, dyes, and other compounds. Phenol also has medicinal application as a topical anesthetic for some types of lesions and in the treat-

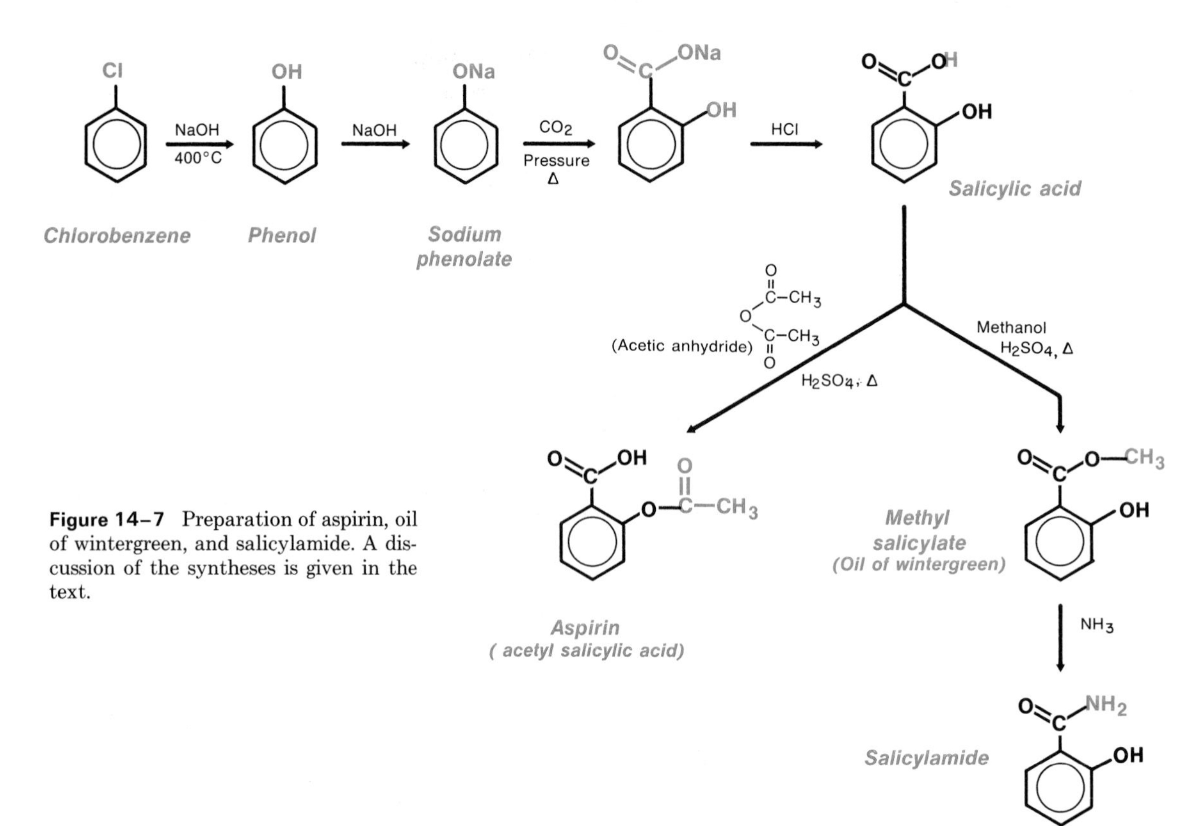

Figure 14–7 Preparation of aspirin, oil of wintergreen, and salicylamide. A discussion of the syntheses is given in the text.

ment of mange and colic in animals. Methyl salicylate, or oil of wintergreen, is used as a flavoring agent in addition to being an intermediate in the synthesis of salicylamide, a compound with many of the properties of aspirin but reportedly less toxic.

Any given organic compound can usually serve as the starting material for the synthesis of many other organic compounds.

The conversion of phenol into sodium phenolate is an acid-base reaction. Phenol, with an acidic hydrogen in the hydroxyl group, reacts with a base, sodium hydroxide, to give a salt, sodium phenolate, and water. The reaction of salicylic acid to form oil of wintergreen is an *esterification*. The organic acid reacts with an alcohol in the presence of a strong mineral acid to produce an ester, methyl salicylate, and water.

Sulfa Drugs

An important group of useful products from aromatic organic reactions is the sulfa drugs. Sulfa drugs represent a group of compounds discovered in a conscious search for materials which could control bacterial infections in living animals. In 1904, the German chemist, Paul Ehrlich, realized that infectious diseases could be conquered if toxic chemicals could be found that attacked parasitic organisms rather than host cells. Ehrlich achieved some success toward his goal—the use of dyes against African sleeping sickness and arsenic compounds against syphilis. The mass outbreak of typhus during World War I, the loss of many wounded due to secondary bacterial infection, and the great influenza epidemic of 1917–1918 emphasized the need for such compounds that might eradicate infectious diseases.

Ehrlich's search for selectively toxic chemicals to kill infectious organisms inside of man was a brilliant, novel idea which has been extensively exploited.

After experimenting with many drugs, Gerhard Domagk, a pathologist in the I. G. Farbenindustrie Laboratories in Germany, found, in 1935,

TABLE 14–6 Production of Sulfa Drugs and Antibiotics per Year in Thousands of Pounds

	1942	1943	1946	1952	1956	1966	1969
Sulfa drugs	5,435	10,006	5,104	5,786	3,817	5,450	4,916
Antibiotics	0	0	38	1,487	1,967	9,652	13,199

that prontosil, a coloring matter or dye, was active against bacterial infection in mice. Prontosil as such is not effective in killing bacteria, but it hydrolyzes into sulfanilamide, which is effective.

The medicinal action of sulfanilamides is described in Chapter 21.

This discovery led to the synthesis and testing of a large number of related compounds in the search for drugs which were more effective and less toxic to the infected animal. By 1964, more than 5000 sulfa drugs had been prepared and tested (Table 14-6). Some of the more effective ones are listed in Table 14-7. Sulfanilamide was first prepared by Schroeter in 1906; Fritz Mietzsch, Joseph Klarer, and Gerhard Domagk are credited with the synthesis of several other sulfonamides.

Many sulfa drugs are derivatives of sulfanilamide, which is prepared

The sulfonamide grouping is the essential part of all of the various sulfa drugs.

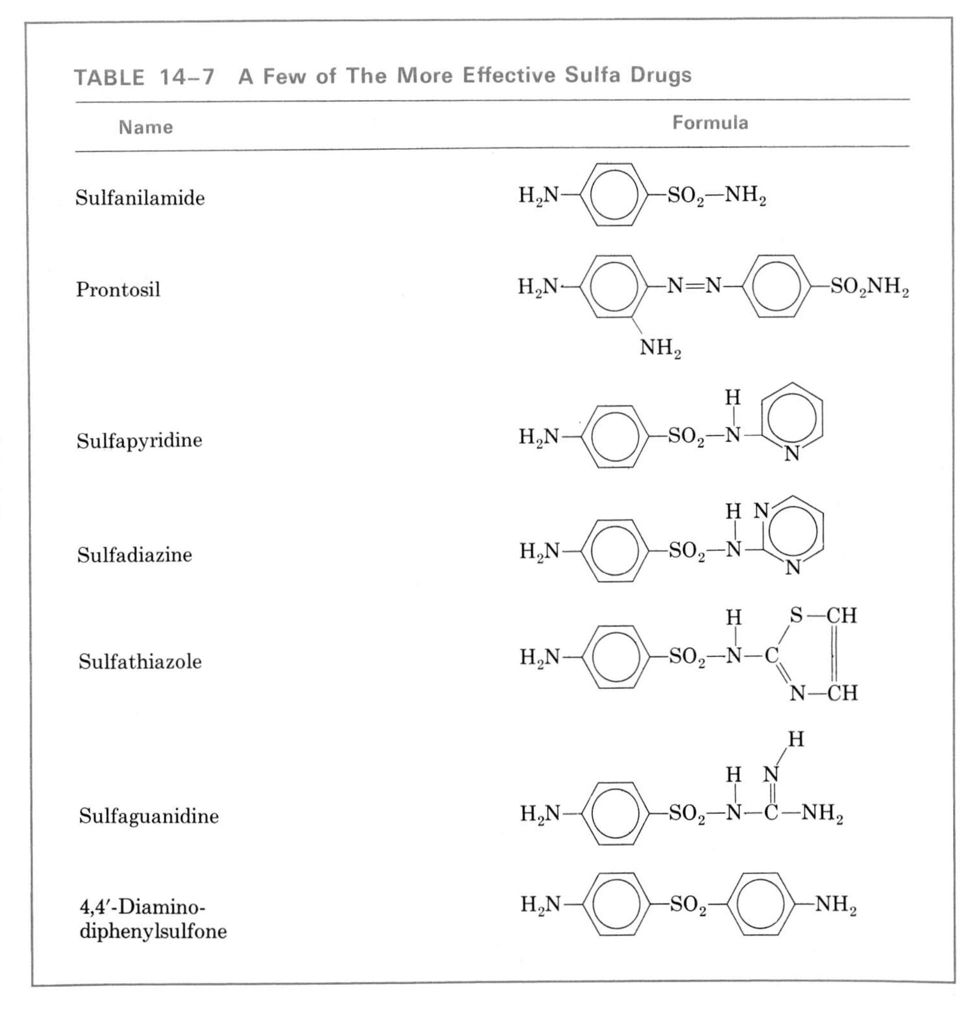

TABLE 14–7 A Few of The More Effective Sulfa Drugs

Name	Formula
Sulfanilamide	H_2N–(ring)–SO_2–NH_2
Prontosil	H_2N–(ring, NH_2)–N=N–(ring)–SO_2NH_2
Sulfapyridine	H_2N–(ring)–SO_2–N(H)–(pyridyl, N)
Sulfadiazine	H_2N–(ring)–SO_2–N(H)–(pyrimidinyl, N, N)
Sulfathiazole	H_2N–(ring)–SO_2–N(H)–C(S–CH=CH–N)
Sulfaguanidine	H_2N–(ring)–SO_2–N(H)–C(=NH)–NH_2
4,4′-Diamino-diphenylsulfone	H_2N–(ring)–SO_2–(ring)–NH_2

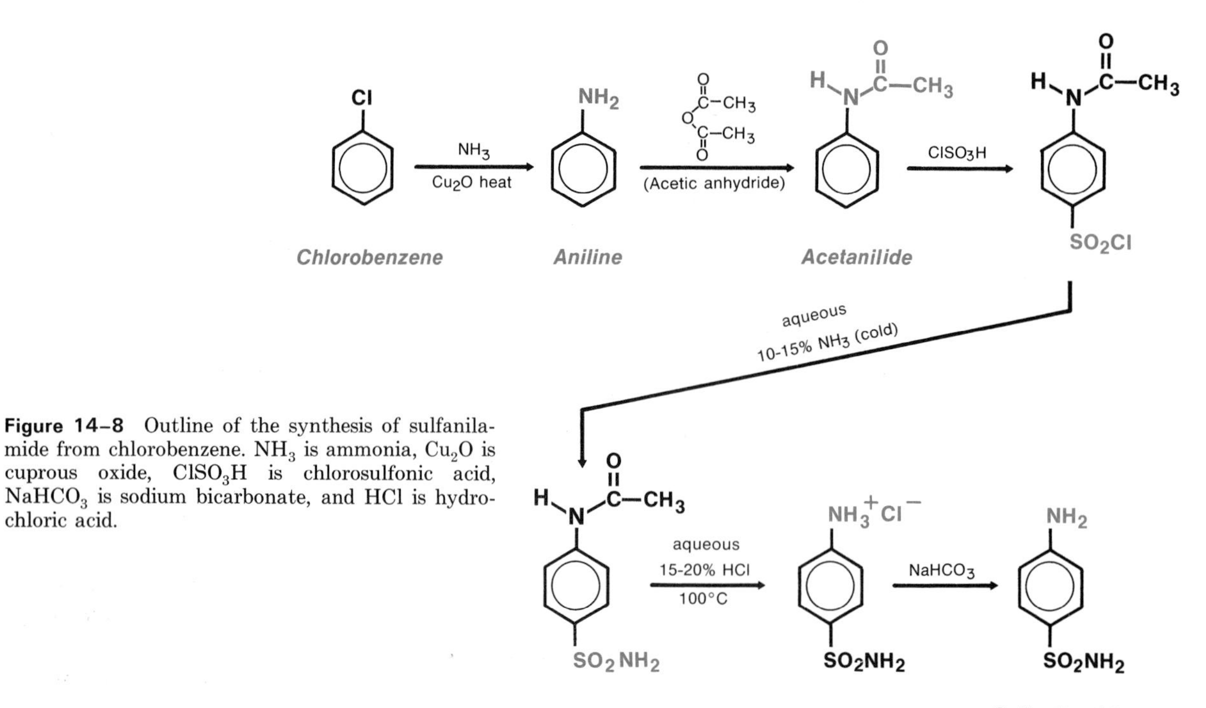

Figure 14–8 Outline of the synthesis of sulfanilamide from chlorobenzene. NH_3 is ammonia, Cu_2O is cuprous oxide, $ClSO_3H$ is chlorosulfonic acid, $NaHCO_3$ is sodium bicarbonate, and HCl is hydrochloric acid.

from chlorobenzene as summarized in Figure 14–8. This synthesis, beginning with acetanilide, is commonly repeated in school laboratories. Great care must be exercised in handling the chlorosulfonic acid, $ClSO_3H$, since it is very corrosive and can cause severe burns. The synthesis employs an often-used technique in organic synthesis. The amine group in aniline is first "blocked" with an acetyl group prior to adding the chlorosulfonic acid. Blocking of the —NH_2 group is necessary to prevent its reaction with chlorosulfonic acid. With the —NH_2 blocked, the acid reacts at the para position of the ring, substituting a —SO_2Cl group for a hydrogen; the hydrogen reacts with an OH of the chlorosulfonic acid to form a molecule of water as a coproduct. The technique of blocking one group while another is attacked is necessary in many organic syntheses.

Organic Dyes

Natural dyes have been used since before the dawn of recorded history.

Until the middle of the last century all of the dyes available to man came from natural sources. Most of these were vegetable extracts of one sort or another; a few were animal products. The range of colors was limited as was the utility of the dyes. If a natural dye did not have the chemical groupings necessary to react with the chemical groupings of a particular fabric, the fabric could not be dyed with that material.

The history of natural dyes is very interesting. Egyptian mummies have been found wrapped in cloth dyed with juice from the madder plant. Alexander the Great is supposed to have deceived the Persians into thinking that his army was badly wounded by splashing his soldiers with a red dye, probably madder juice. The dye in madder juice is the compound, alizarin. Dark blue indigo dye has been known for over 4000 years. When the Romans invaded England, they found it inhabited by the Picts, who both tattooed and painted themselves with indigo. The word *Briton* is apparently the Latin form of a Celtic word meaning painted men. A legend recorded on coins of Tyre attests that Hercules discovered Tyrian purple (royal purple) when his dog bit a snail which stained the dog's jaws purple. Mark Anthony's flight from the crucial naval battle of Actium was especially conspicuous because he fled in Cleopatra's barge, which had sails of royal purple. The value of dyes in the ancient world was indicated by the fact that Cortez extorted cochineal, a crimson dyestuff, as well as gold from the Aztecs.

One of the earliest dyes synthesized in the laboratory was the purple dye, mauve.

The first useful synthetic dye was prepared in 1856 by William Perkin, a young English chemist. Only 18 years old at the time, Perkin prepared mauve in his small home laboratory by the oxidation of impure aniline. Charmed by the purple color, he tested it as a dye and found that it resisted both sunshine and soap. Perkin called the dye aniline purple, but it became known popularly as mauve. Perkin's discovery clearly showed the possibilities of synthetic dyes, and within a few years a large number of new dyes were developed. Well over 100,000 dyes of various sorts have been prepared, and quite often new ones must be developed before new polymeric fibers can be dyed. Of those developed and tested, perhaps 2000 or so are in production at various chemical plants throughout the world.

The basic problem in the synthesis of a dye is to prepare a molecule which will absorb some wavelengths from white light and reflect back a desired color. Certain atomic groupings must be present in aromatic dyes to absorb light and produce colored compounds. These groups are called ***chromophores*** (Greek: *chroma* = color + *phoros* = bearer); some are listed in Table 14–8.

Chromophores are groupings of atoms that absorb particular wavelengths of visible light.

The absorption of certain wavelengths of light by the double bonds in chromophores causes the coloration in these compounds. The electrons in a double bond can absorb certain small quanta of energy and be promoted to a higher energy level. The particular wavelengths absorbed are extracted from the incoming light, and the reflected light is the color of the combined unabsorbed wavelengths (Figure 14–9). If only one double bond is present in a molecule, the electrons are held too tightly, and high-energy ultraviolet light is necessary to excite them. If, however, a chromophore is bonded to an aromatic system or to another conjugated system (alternating double and single bonds which lead to a pi system

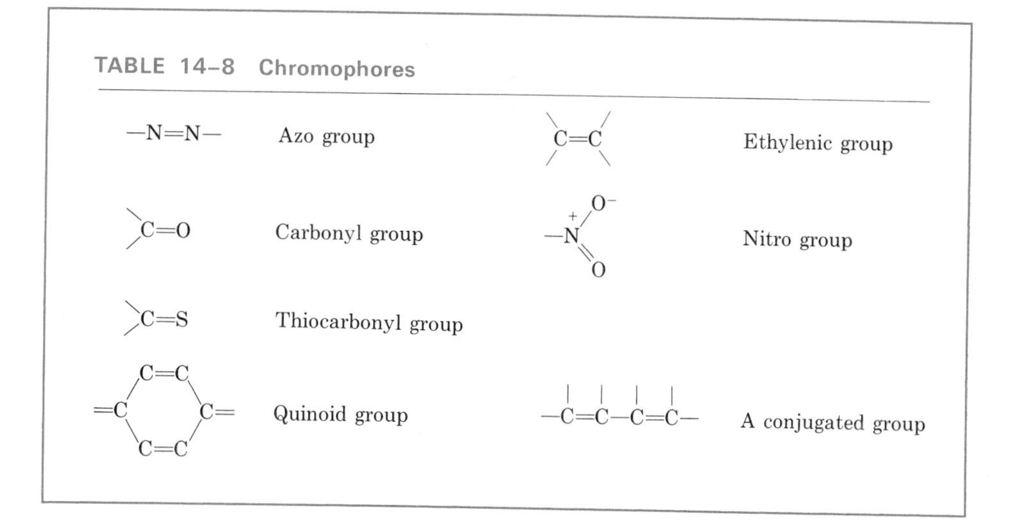

TABLE 14–8 Chromophores

Structure	Name	Structure	Name
—N=N—	Azo group	>C=C<	Ethylenic group
>C=O	Carbonyl group	$-\overset{+}{N}(=O)O^-$	Nitro group
>C=S	Thiocarbonyl group		
=C(C=C)(C=C)C=	Quinoid group	—C=C—C=C—	A conjugated group

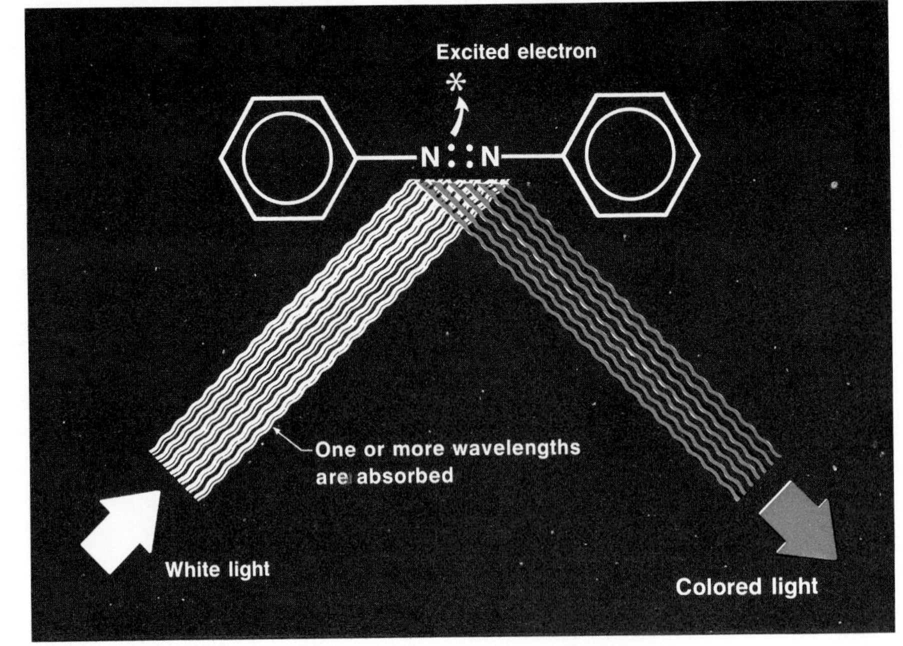

Figure 14–9 An explanation for the color of aromatic dyes. Electrons in the conjugated system absorb certain wavelengths of light and reflect the rest. The assortment of wavelengths reflected determines the color of the dye. A wavelength is absorbed only if that amount of energy will take an electron to an existing higher energy level.

of electrons), the electrons move more freely, are excited more easily, and absorb in the lower energy, visible light range of the spectrum. For example, $CH_3—N=N—CH_3$ has only a double bond, is not conjugated, and absorbs ultraviolet light at 340 nanometers. This compound is colorless in visible light. When the azo group is attached to two benzene molecules, two "racetracks" of electrons join to give one longer one, and the energy required to excite an electron to a higher-energy orbital is decreased; bluish light is absorbed. Azobenzene, $C_6H_5—N=N—C_6H_5$, is orange and absorbs light in the visible region at 445 nanometers.

Auxochromes are groupings in dyes which intensify the colors due to the chromophores.

In addition to chromophores, dyes must also contain reactive groups that allow them to form bonds with the fiber molecule. These groups, when attached to the conjugated system, often increase the wavelength at which the dye absorbs light. These anchoring and color-intensifying groups are called ***auxochromes*** (Latin: *auxilium* = aid); some are listed in Table 14-9.

Dyes are synthesized by incorporating appropriate combinations of chromophores and auxochromes into a molecule. An example of dye synthesis is the preparation of a red dye, para red, from chlorobenzene (Figure 14–10). Note in the structure of para red that the azo and nitro chromophoric groups and the phenolic hydroxyl auxochromic group are bonded to the conjugated ring systems, the structural pattern necessary to produce an aromatic dye.

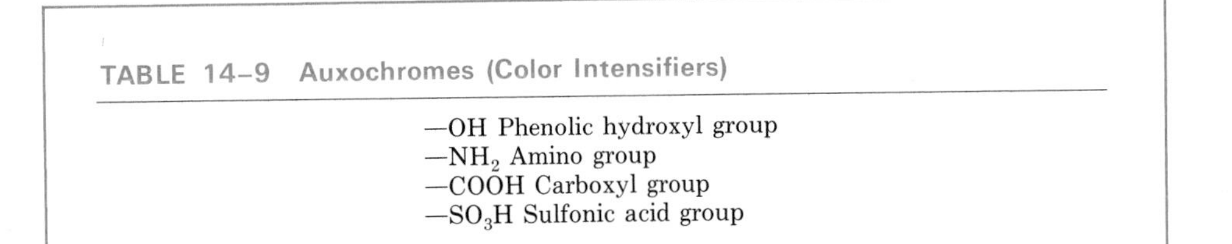

TABLE 14–9 Auxochromes (Color Intensifiers)

Group	Name
—OH	Phenolic hydroxyl group
$—NH_2$	Amino group
—COOH	Carboxyl group
$—SO_3H$	Sulfonic acid group

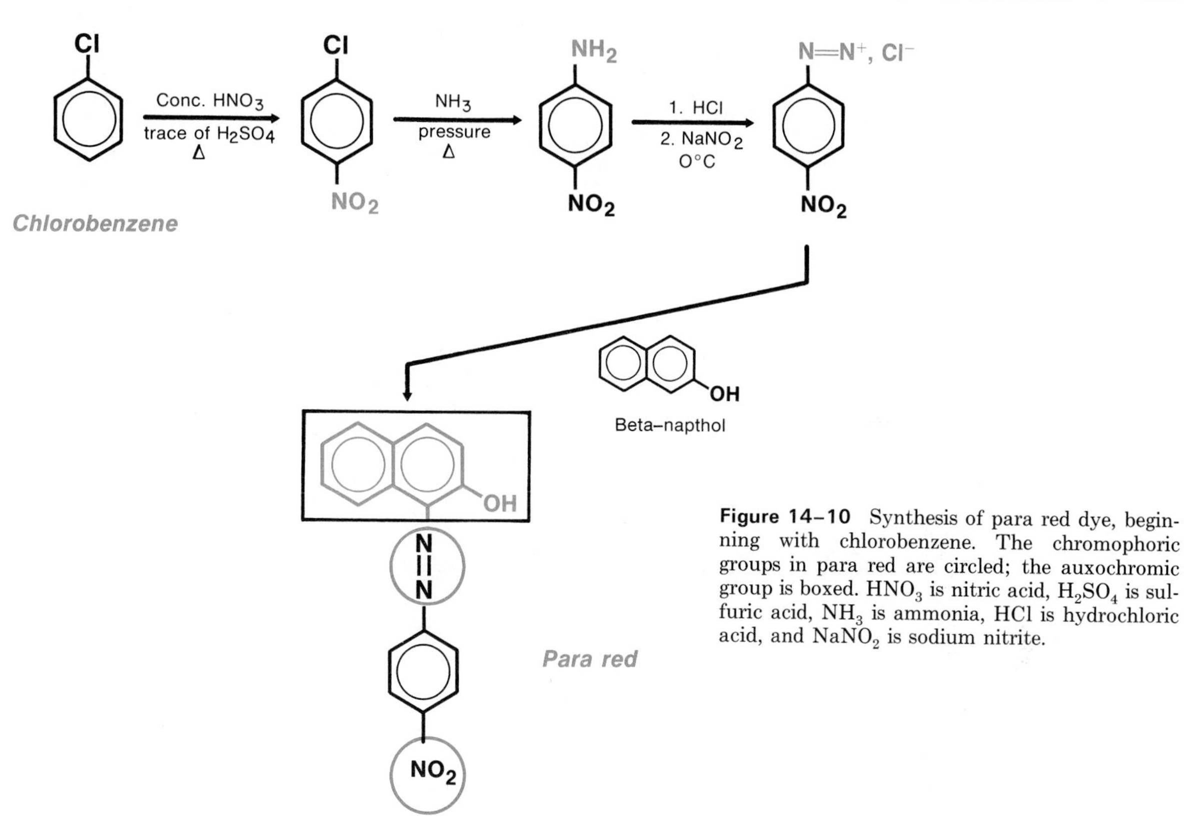

Figure 14–10 Synthesis of para red dye, beginning with chlorobenzene. The chromophoric groups in para red are circled; the auxochromic group is boxed. HNO_3 is nitric acid, H_2SO_4 is sulfuric acid, NH_3 is ammonia, HCl is hydrochloric acid, and $NaNO_2$ is sodium nitrite.

In dyeing, the basic process is the attachment of a colored dye to a colorless substrate.

Para red is used to dye cotton by a technique known as ***ingrain dyeing.*** In this technique the last step of the synthesis is carried out directly on the fiber rather than by synthesizing the dye first and then trying to apply it to the fiber. This is an effective dyeing technique because smaller molecules can penetrate crevices in the fiber which might be very difficult for a large molecule to enter. After the synthesis, the large molecules are trapped where they are formed, like a ship in a bottle. Besides the physical entrapment of the dye molecule, cotton holds onto para red by means of hydrogen bonds. In addition to some other methods of dyeing, this technique is illustrated in Figure 14–11.

Direct dyeing is a chemical reaction in which a reactive part of the dye molecule, usually an acidic or basic group, reacts directly with some group on the fiber itself. Wool and silk are particularly easy to dye in this way since they are composed of proteins with available amino (—NH_2) and carboxyl (—COOH) groups. The amino groups, because they are basic, will react with and bind to the fiber any dye that has an acidic group. Likewise the carboxyl groups, because they are acidic, will react with and bind any dye that has a basic group. Picric acid will dye wool and silk because it has an acidic phenolic hydrogen which reacts with the amino groups of wool and silk, as shown in Figure 14–11.

When a fiber is relatively unreactive to dyes, it can often be dyed by ***vat dyeing.*** In this process the cloth is immersed in a vat of the soluble dye, and the dye penetrates into the inner crevices of the fiber. While imbedded in the fibers, the dye is rendered insoluble to prevent it from being easily washed from the cloth. This method is particularly useful for dyeing cotton; the chemistry of this technique is illustrated with indigo in Figure 14–11.

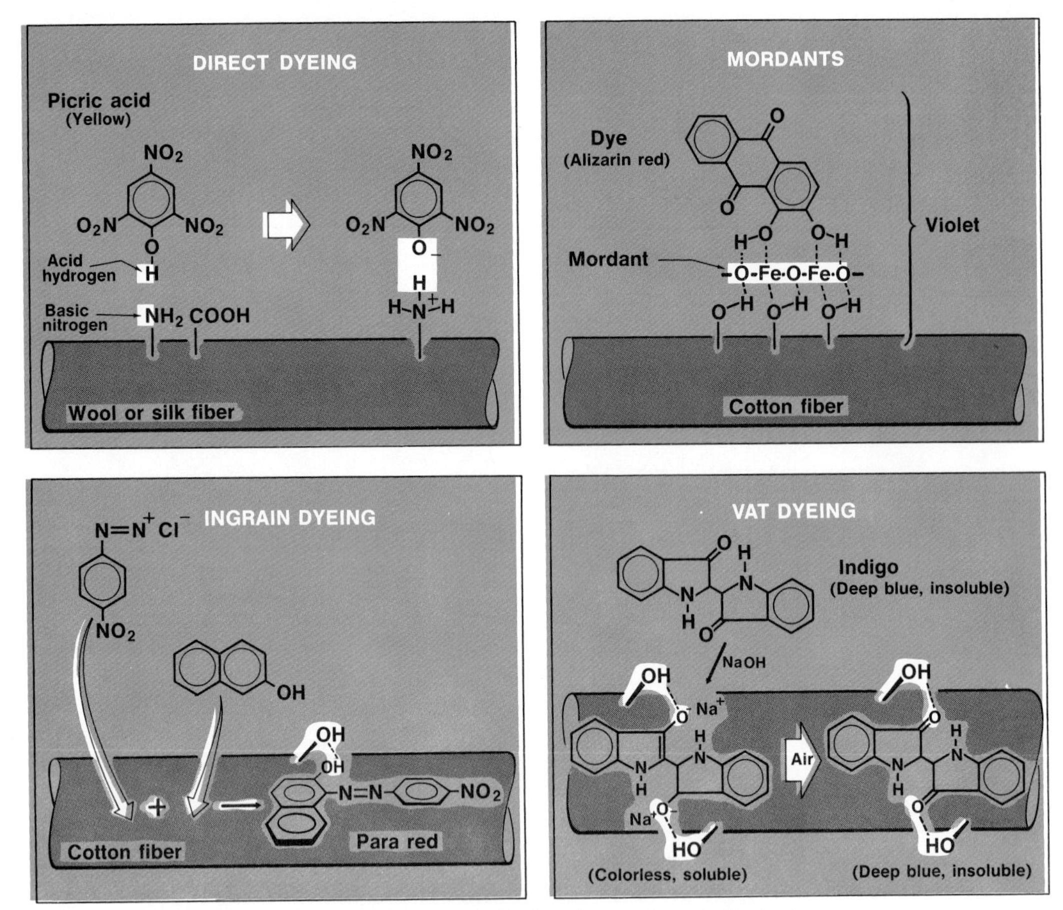

Figure 14–11 Methods of dyeing cloth. Details are given in the text.

Some dyes do not adhere very well to a fiber unless it is first treated with a substance known as a ***mordant*** (Latin: *mordere* = to bite). Mordants interact with both the fiber and the dye, forming a link between them. Metal oxides are often used as mordants, and anyone who has tried to remove an iron rust stain from cotton can testify to the strong affinity between the two. Reactive centers in a dye, suitable for causing a reaction with mordants, are the acidic carboxyl and phenolic hydroxyl groups. Alizarin with phenolic hydroxyl groups will dye cotton that has previously been treated with a metallic oxide, as shown in Figure 14-11. The mordant frequently changes the color of the dye since it reacts chemically with it. For example, alizarin is blood-red, but it gives violet colors with iron mordants and brownish red colors with chromium mordants.

SELF-TEST 14-B

1. Complete the following equations:

   a. $CH_3CH_2CH_2OH + CH_3CH_2COOH \xrightarrow{H_2SO_4}$

   b. (benzene) $+ Cl_2 \xrightarrow{FeCl_3}$

c. 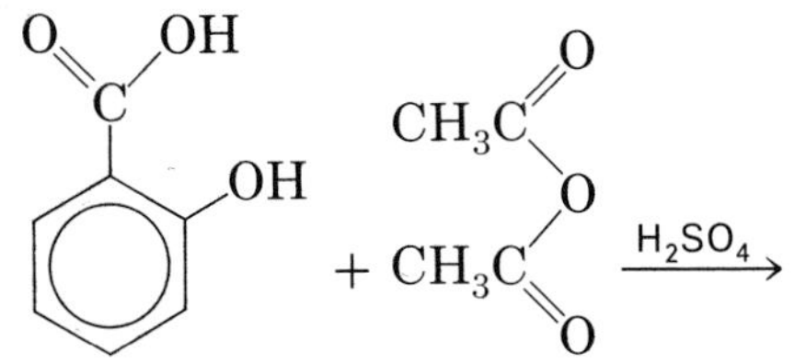 (salicylic acid: benzene ring bearing $\mathrm{C(=O)OH}$ and OH) + $CH_3C(=O)-O-C(=O)CH_3$ $\xrightarrow{H_2SO_4}$

2. Define the terms:
   a. chromophore ____________________
   b. sulfa drug ____________________
   c. dye ____________________
   d. soap ____________________

3. a. What is the difference between a fat and an oil when referring to edible lipids? ____________________

   b. How can the melting points of most edible oils be increased?

   ____

MATCHING SET

____	1. synthesized from NH_4OCN by Wöhler	a.	measures knocking behavior in engine
____	2. RCOOH	b.	breaks hydrocarbons into smaller molecules
____	3. R—OH	c.	antiknocking agent
____	4. $\mathrm{R{-}O{-}\overset{\overset{\large O}{\|\|}}{C}{-}R'}$	d.	aldehyde
____	5. octane rating	e.	carboxylic acid
____	6. RCHO	f.	intensifies color of dye
____	7. cracking	g.	soap
____	8. auxochrome	h.	urea
____	9. $Pb(C_2H_5)_4$	i.	alcohol
____	10. sodium stearate	j.	ester

QUESTIONS

1. Would you expect $CH_3CH_2CH_2CH_2CH_2CH_2CH_2CH_2CH_3$ to be an important useful constituent of bottled gas? Explain your answer. Would it be useful in gasoline? Explain.

2. Indicate what products would be formed in the reaction of the following:
   a. methanol and acetic acid
   b. 1-propanol and stearic acid
   c. ethylene glycol and acetic acid

3. Explain how hydrogen bonding could play a significant role in fixing the boiling point of acetic acid.

4. Write structural formulas for the four alcohols with the composition C_4H_9OH.

5. Draw a structural formula for each of the following:
   a. an alcohol
   b. an organic acid
   c. an ester
   d. glycerol

6. What is meant by:
   a. proof rating of an alcohol
   b. octane rating of a gasoline
   c. denatured alcohol

7. Beginning with petroleum, outline the steps and write out the chemical equations for the production of ethyl acetate.

8. How does a soap cleanse?

9. Discuss the fundamental characteristics of the following:
   a. synthetic detergents
   b. soaps

10. Which would you expect to boil at a higher temperature, $CH_3CH_2CH_2CH_3$ or CH_3CH_2OH? Why?

11. Why will the addition of strong alkali aid in unstopping a greasy sink drain?

12. Draw structural fomulas for:
    a. aspirin
    b. oil of wintergreen
    c. phenol
    d. aniline

13. What chemical reactions can be used to distinguish between:
    a. C_2H_5OH and CH_3COOH
    b. CH_3NH_2 and CH_3OH
    c. $CH_3COOC_2H_5$ and CH_3COOH

14. Tell how you could prepare the following compounds:
    a. chlorobenzene
    b. urea
    c. acetanilide

15. Describe the types of binding between dye and wool fibers when dyed by the direct dyeing method.

16. Describe three ways a dye is held on a fiber.

PROBLEMS

1. Suppose in the synthesis of aspirin from chlorobenzene (Figure 14-7) each step produces a 50 percent yield. What fraction of an original amount of starting material would actually remain as a final product?

   Ans. 0.031 of the original.

2. A gaseous compound which is 85.6 percent C and 14.4 percent H by weight reacts with water to form a new compound which has the composition of 52.1 percent C, 34.8 percent O, and 13.1 percent H. What would be the properties of this new compound?

   Ans. New compound would contain an OH group.

Ans. 62.07 g ethylene glycol = 1 mole of molecules; 32.04 g methanol = 1 mole of molecules.

3. Antifreezes work on a principle that the freezing point of a solution is lowered owing to the presence of dissolved molecules. It is the number of dissolved molecules relative to the number of molecules of solvent which is of primary importance. Why does it take 62.07 grams of ethylene glycol and only 32.04 grams of methanol to lower the freezing point of a sample of water by a given amount?

SUGGESTIONS FOR FURTHER READING

Campaigne, E., "Wöhler and the Overthrow of Vitalism," *Journal of Chemical Education,* Vol. 32, No. 8, p. 403 (1955).

Collier, H. O., "Aspirin," *Scientific American,* Vol. 209, No. 5, p. 97 (1963).

Ferguson, L. N., "Hydrogen Bonding and the Physical Properties of Substances," *Journal of Chemical Education,* Vol. 33, No. 6, p. 267 (1956).

Juster, N. J., "Color and Chemical Constitution," *Journal of Chemical Education,* Vol. 39, No. 11, p. 596 (1962).

Rossini, F. D., "Hydrocarbons in Petroleum," *Journal of Chemical Education,* Vol. 37, No. 11, p. 554 (1960).

Schaar, B. E., "Aniline Dyes," *Chemistry,* Vol. 39, No. 1, p. 12 (1966).

MAN-MADE GIANT MOLECULES—THE SYNTHETIC POLYMERS

CHAPTER 15

It is impossible for most Americans to get through a day without using a dozen or more materials based on synthetic ***polymers.*** Many of these materials are ***plastics*** of one sort or another. Examples of these include plastic dishes and cups, combs, automobile steering wheels and seat covers, telephones, pens, plastic bags for food and wastes, plastic pipes and fittings, plastic water-dispersed paints, false eyelashes and wigs, a wide range of synthetic fibers for suits and stockings, synthetic glues, and flooring materials. In fact, these materials are so widely used they are usually taken for granted. All these materials are composed of *giant molecules.* The purpose of this chapter is to examine some of the chemistry and technology that lie behind this flood of "indispensable" plastic objects.

A plastic is a substance capable of being molded into various shapes. All plastics are polymers, but not all polymers are plastic.

WHAT ARE GIANT MOLECULES?

Most giant molecules are formed by the bonding together of a large number of smaller molecules. The smaller molecules are bonded to each other by strong covalent bonds. A giant molecule in a plastic is analogous to a long train of individual railroad cars hooked together.

Many of the properties of the giant molecules are caused by their high molecular weights. By the late nineteenth century, techniques had been developed that enabled researchers to determine the approximate molecular weights of rubber, cellulose, proteins, and some other natural

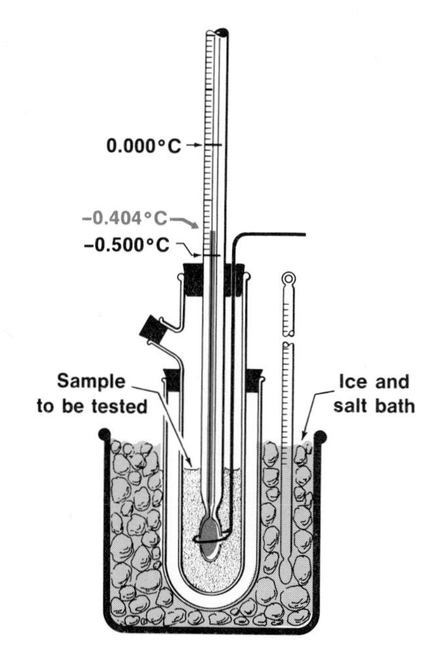

Figure 15–1 Molecular weights of solutes can be determined by the lowering of the freezing point below that of the pure solvent. In this example, water, which freezes at 0°C, dissolves a solute, and the resulting solution freezes at −0.404°C. An ice and salt bath is used in this instance to cool the sample.

materials of very high molecular weights. One of these methods, involving the change in the freezing point of water, is illustrated in Figure 15–1. Although this method is not suitable for very high molecular-weight polymers, the basic idea is quite simple: the freezing point of water is depressed if substances are dissolved in it, and the depression of the freezing point is directly proportional to the number of molecules (or moles) of solute added, as long as their total concentration is not too high. One mole of any nonionized solute dissolved in 1000 grams of water depresses the freezing point by 1.86°C. The results of studies using this approach, usually with organic solvents, and other methods for the determination of molecular weight, have led to values of about 12,000 for natural rubber; materials such as starch have molecular weights of 40,000 or higher, and the molecular weights of proteins range from a thousand up to a million.

A macromolecule is a molecule with a very high molecular weight.

Many chemists were reluctant to accept the concept of giant molecules, but in the 1920's a persistent German chemist, Hermann Staudinger (1881–1965; Nobel Prize, 1953), championed the idea and introduced a new term, ***macromolecule,*** for these giant molecules. Staudinger devised experiments that yielded accurate molecular weights, and, in addition, he synthesized "model compounds" to test his theory. One of his first model compounds was prepared from styrene.

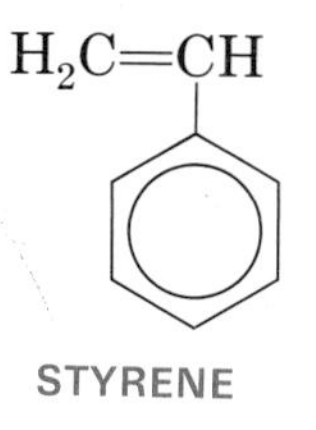

STYRENE

A polymer is a molecule composed of a large number of similar units.

Under the proper conditions, styrene molecules use the "extra" electrons of the double bond to undergo a ***polymerization*** reaction to yield polystyrene, a giant molecule. The word ***polymer*** means "many parts" (Greek: *poly* = many, *meros* = parts). The molecules of styrene are the ***monomers*** (Greek: *mono* = one); they provide the recurring units in the giant molecule.

The macromolecule polystyrene is represented as a long chain of monomer units bonded to each other. The polymer chain is not an endless one; some polystyrenes made by Staudinger were found to have molecular weights of about 600,000, corresponding to a chain of about 5,700 styrene units. The polymer chain can be indicated as

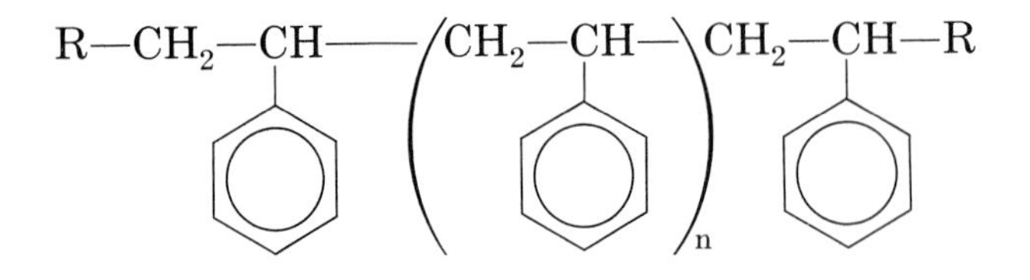

where R represents some terminal group, often an impurity, and n is a large number.

Synthetic polymers are commonly called *plastics* when in a solid form.

Polystyrene is a clear, hard, colorless solid at room temperature. Since it can be molded easily at 250°C, the term ***plastic*** has become associated with it and similar materials. Polystyrene has so many useful properties that its commercial production, which began in Germany in 1929, today exceeds 2.5 billion pounds per year. It is used to make combs, bowls, toys, and many other items.

Thermoplastic polymers can be repeatedly softened by merely heating.

Thermosetting polymers form cross-linking bonds when heated and then become rigid.

There are two broad categories of plastics. One, when heated repeatedly, will soften and flow; when it is cooled, it hardens. Materials which undergo such reversible changes when heated and cooled are called ***thermoplastics;*** polystyrene is one example. The other type is plastic when first heated, but when heated further it forms a set of interlocking bonds. When reheated, it cannot be softened and reformed without extensive degradation. These materials are called ***thermosetting plastics*** and include rigid-foamed polyurethane, a polymer which is finding many new uses as a construction material.

In order to gain a better understanding of polymers it is necessary to look at representative examples of the different types of polymerization processes.

ADDITION POLYMERS

In the previous section it was noted that some polymers, such as polystyrene, are made by adding monomer to monomer to form a polymer chain of great length. Perhaps the easiest addition reactions to understand chemically are those involving monomers containing double bonds. The simplest monomer of this group is ethylene, C_2H_4. When ethylene is heated under pressure in the presence of oxygen, polymers with molecular weights of about 30,000 are formed. In order to enter into reaction, the double bond of ethylene must be broken. This forms ***reactive sites*** composed of unpaired electrons at either end of the molecule.

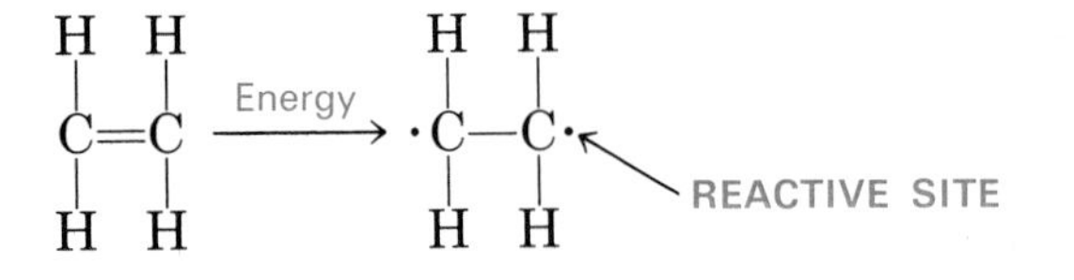

The partial breaking of the double bond can be accomplished by physical means such as heat, ultraviolet light, x-rays, and high-energy electrons. The *initiation* of the polymerization reaction can also be accomplished with chemicals such as organic peroxides. These initiators, which are very unstable, break apart into pieces with unpaired electrons. These fragments (called ***free radicals***) are ravenous in trying to find a "buddy" for their unpaired electrons. They react readily with molecules containing carbon-carbon double bonds. Benzoyl peroxide is one commonly used free radical initiator:

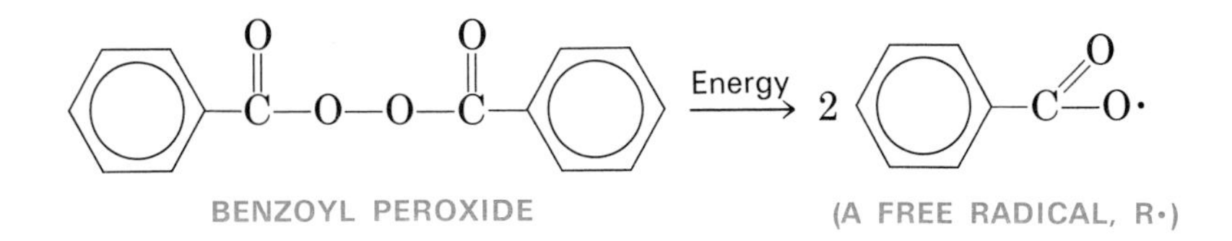

BENZOYL PEROXIDE (A FREE RADICAL, R·)

An addition polymerization can be started by a free radical.

The odd electron in the free radical pairs with an electron in ethylene and, in turn, forms another free radical.

$$\text{Peroxide} \xrightarrow{\text{Energy}} R\cdot$$

$$R\cdot + CH_2{=}CH_2 \longrightarrow R{-}CH_2{-}CH_2\cdot$$

ETHYLENE (ANOTHER FREE RADICAL)

The polymer grows as the resulting free radical reacts with other ethylene molecules to form a long hydrocarbon chain:

$$R{-}CH_2{-}CH_2\cdot + CH_2{=}CH_2 \longrightarrow R{-}CH_2{-}CH_2{-}CH_2{-}CH_2\cdot$$

$$R{-}CH_2{-}CH_2{-}CH_2{-}CH_2\cdot + CH_2{=}CH_2 \longrightarrow R{-}CH_2{-}CH_2{-}CH_2{-}CH_2{-}CH_2{-}CH_2\cdot,$$

etc.

Some time after the chain has begun to form, *termination* of the polymerization process occurs. Occasionally, two long chains may meet and link up their reactive sites.

$$R{-}CH_2{-}CH_2{-}(CH_2{-}CH_2)_n{-}CH_2{-}CH_2\cdot + \cdot CH_2{-}CH_2{-}(CH_2{-}CH_2)_m{-}R \longrightarrow$$

$$R{-}CH_2{-}CH_2{-}(CH_2{-}CH_2)_n{-}CH_2{-}CH_2{-}CH_2{-}CH_2{-}(CH_2{-}CH_2)_m{-}R$$

An addition polymer is formed when many small molecules containing a double bond add to each other.

(In this example n and m are large, and probably different, numbers.)

In addition to this process, the initiator free radicals not used in the initiation process begin to terminate the build-up of some of the polymer chains.

$$R{-}(CH_2{-}CH_2)_n{-}CH_2{-}CH_2\cdot + \cdot R \longrightarrow R{-}(CH_2{-}CH_2)_n{-}CH_2{-}CH_2{-}R$$

Polyethylenes formed under various pressures and catalytic conditions have different molecular structures and hence different physical properties. For example, chromium oxide as a catalyst yields almost exclusively the linear polyethylene shown previously. If ethylene is heated to 230°C at a pressure of 200 atm, irregular branches result. Under these conditions, free radicals undoubtedly attack the chain at random positions, thus causing the irregular branching.

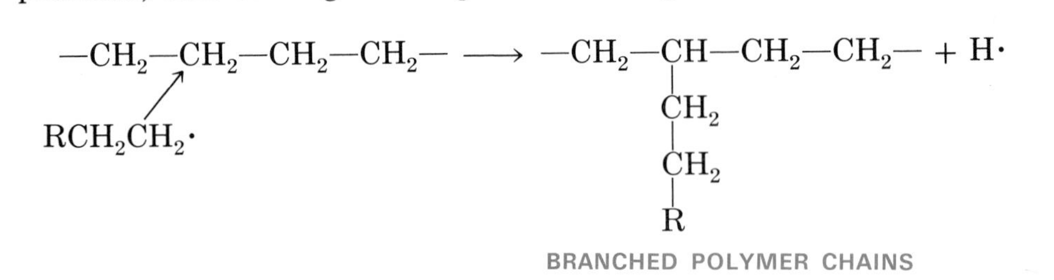

BRANCHED POLYMER CHAINS

The molecules in linear polyethylene can line up with one another very easily, yielding a tough, high-density compound which is useful in making toys, bottles, and so forth. The polyethylene with irregular branches is less dense, more flexible, and not nearly as tough as the linear polymer, since the molecules are generally farther apart and their arrangement is not as precisely ordered. This material is used for squeeze bottles and other similar applications.

TABLE 15–1 Ethylene Derivatives Which Undergo Addition Polymerization

Formula	Monomer Name	Polymer Name	Uses
$CH_2{=}CH_2$	Ethylene	Polyethylene	coats, milk cartons, wire insulation, bread boxes
$HC{=}CH_2$ (attached to benzene ring, C_6H_5)	Styrene	Polystyrene	combs, toys, bowls
$CH_2{=}CHCl$	Vinyl chloride	Polyvinylchloride (PVC)	as a vinyl acetate copolymer in phonograph records, credit cards, rain wear
$CH_2{=}CH-O-C({=}O)-CH_3$	Vinyl acetate	Polyvinylacetate	latex paint
$CH_2{=}CH-CN$	Acrylonitrile	Polyacrylonitrile (PAN)	rug fibers
$CH_2{=}CH-CH{=}CH_2$	Divinyl (1,3-Butadiene)	BUNA rubbers	tires and hoses
$CH_2{=}C(CH_3)-C({=}O)-O-CH_3$	Methyl methacrylate	Polymethyl methacrylate (Plexiglas, Lucite)	transparent objects, lightweight "pipes"
$CF_2{=}CF_2$	Tetrafluoroethylene	Polytetrafluoroethylene (TFE) (Teflon)	Insulation, bearings, nonstick fry pan surfaces

There is a large group of derivatives of ethylene which undergo addition polymerization, usually via a free radical mechanism. Table 15–1 summarizes some pertinent information on these materials.

TAILOR-MADE MOLECULES

Stereochemical control regulates the structural and molecular properties of the polymer.

When the structure of polypropylene is drawn out to illustrate the three-dimensionality of the carbon-carbon bonds, three unique structures appear (Figure 15–2). In the first structure, called ***isotactic,*** all the methyl (—CH_3) groups are in identical positions along the polymer chain. When the methyl groups extend in alternate directions away from the chain, a ***syndiotactic*** arrangement results; finally, there is the possibility of a random, or ***atactic,*** arrangement. Polypropylenes of these three types were actually prepared and named by Professor Giulio Natta (who shared the Nobel prize with Karl Ziegler in 1963), in Italy in 1955. Using novel catalysts such as aluminum or iron attached to organic molecules, Natta was the first to control the growth phase of a polymerization reaction. The types of catalysts used by Natta were first used by Professor Karl Ziegler in Germany to increase the yields of straight-chain, high-density polyethylene. This type of control is called ***stereochemical control.*** As a result of this discovery, chemists were able to choose the extent of polymerization (molecular weight) and also the fine, structural features of the polymer chain itself.

Natta's three polypropylenes are different in ways that can be related to their structures. The isotactic material (named by Natta's wife) actually has a helical chain structure, owing to the bulkiness of the methyl groups interacting with each other (Figure 15–3); this allows the chains to approach each other closely. Isotactic polypropylene melts

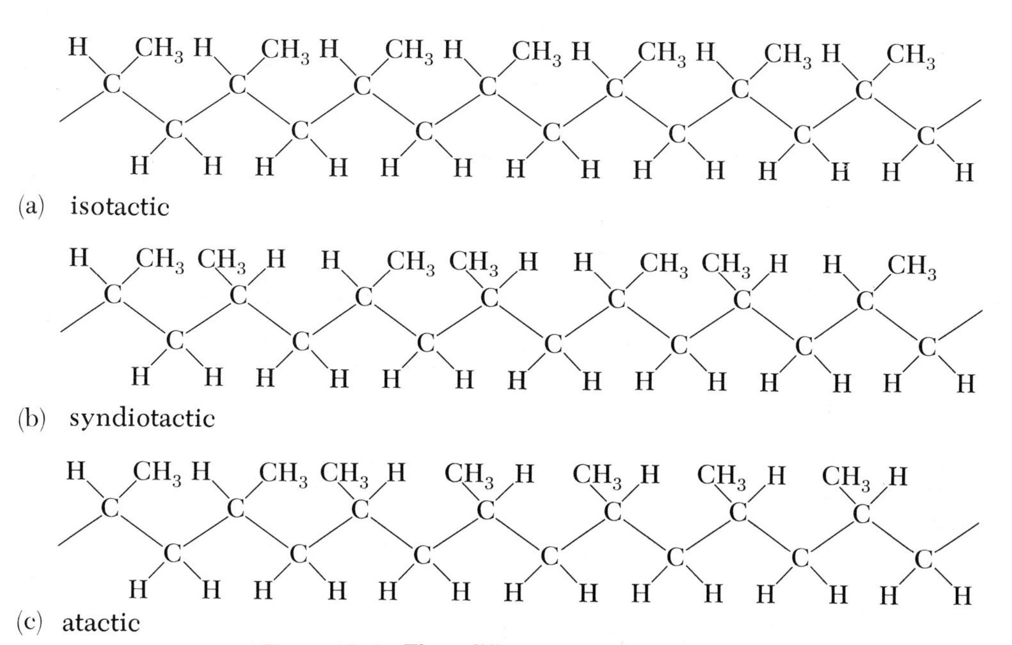

Figure 15–2 Three different structures of polypropylene. Each structure imparts different properties to the plastic. See text for discussion.

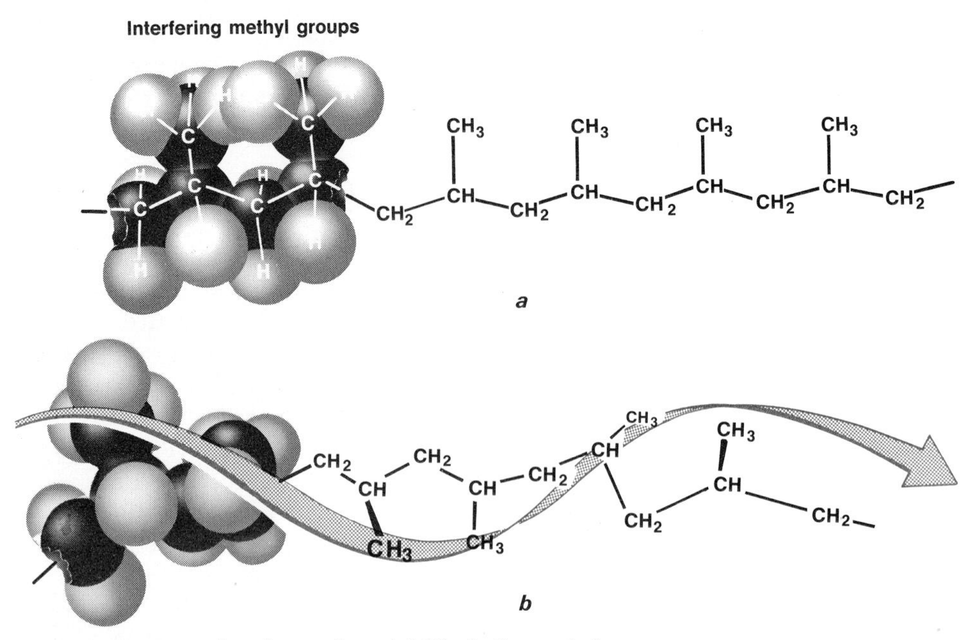

Figure 15–3 Isotactic polypropylene. (*a*) The bulky methyl groups are close to each other in this arrangement. (*b*) To eliminate crowding of the methyl groups on one side of the polymer chain, the chain flexes to produce a helical arrangement.

at 170°C and can be easily formed into fibers. The atactic polypropylene chains cannot approach as closely because of the randomness of the methyl groups; this irregular structure renders the material rubbery and less dense.

Synthetic "Natural" Rubber

A very interesting application of stereochemical control over polymerization is the manufacture of synthetic rubber. Natural rubber has the composition $(C_5H_8)_n$, and when decomposed in the absence of oxygen, yields isoprene:

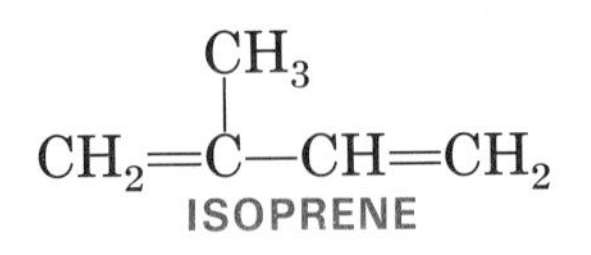

Natural rubber occurs as latex (a suspension of rubber particles in water) that oozes from rubber trees when they are cut. When the rubber particles are precipitated from the latex, a gummy mass is obtained that is not only elastic and water-repellent but also very sticky, especially when warm. In 1839, after 10 years' work on this material, Charles Goodyear (1800–1860) discovered that heating rubber latex with sulfur produced a material that was no longer sticky, but still elastic, water-repellent, and resilient.

Rubber is vulcanized by heating it with sulfur, which forms links between the polymer chains.

Vulcanized rubber, as Goodyear called his product, contains short chains of sulfur atoms which bond together the polymer chains of the natural rubber. The sulfur chains help to align the polymer chains, so

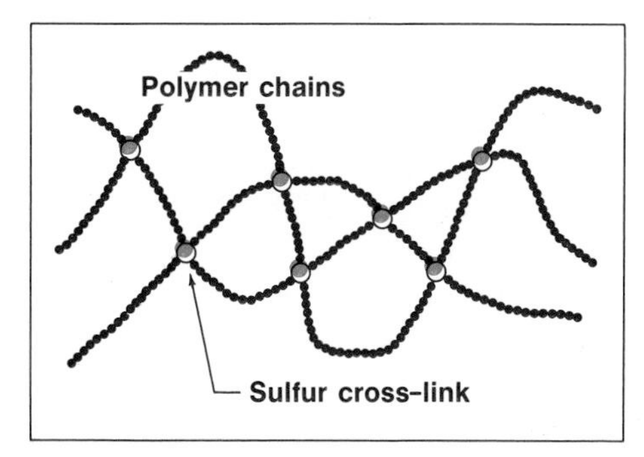

a. Before stretching

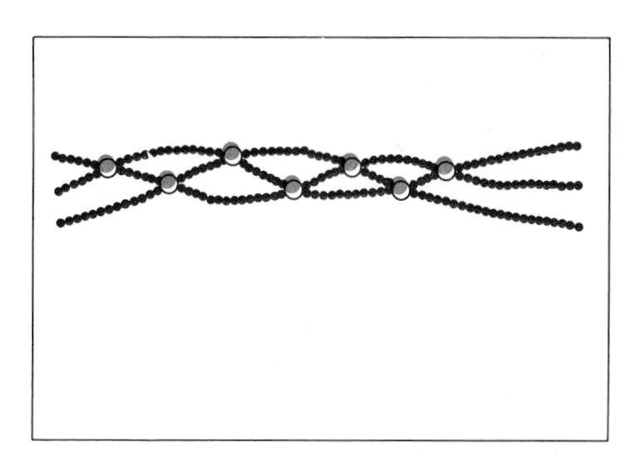

b. Stretched

Figure 15-4 Stretched vulcanized rubber retains its elasticity.

the material does not undergo a permanent change when stretched, but springs back to its original shape and size when the stress is removed (Figure 15-4).

In later years chemists searched for ways to make a synthetic rubber so we would not be completely dependent on natural rubber during emergencies, as during the first years of World War II. In the mid-1920's, German chemists polymerized butadiene (structurally similar to isoprene and obtained from petroleum) to produce Buna rubber, so named because it was made from butadiene (Bu—) catalyzed by sodium (—Na).

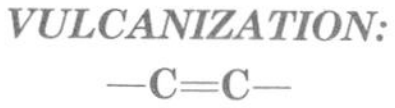

VULCANIZATION:

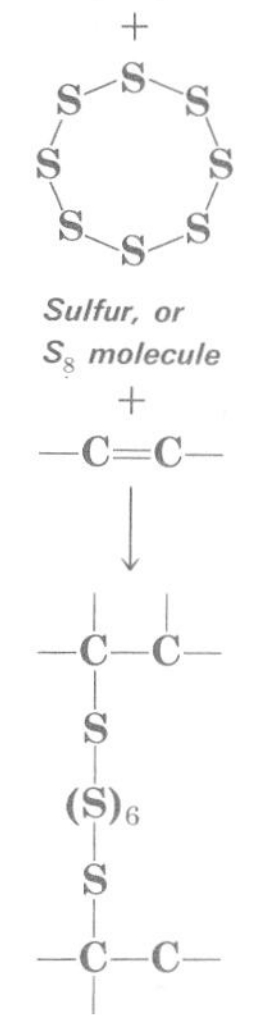

The behavior of natural rubber (polyisoprene), it was learned later, is due to the specific arrangement within the polymer chain. We can write the formula for polyisoprene with the CH_2 groups on opposite sides of the double bond (the *trans* arrangement),

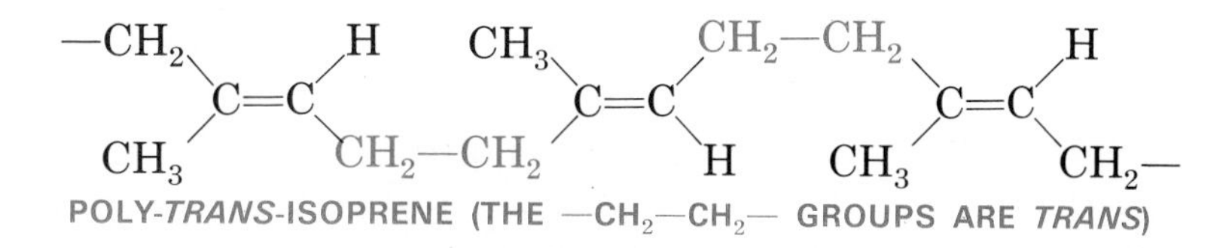

or with the CH_2 groups on the same side of the double bond (the *cis* from Latin meaning on this side).

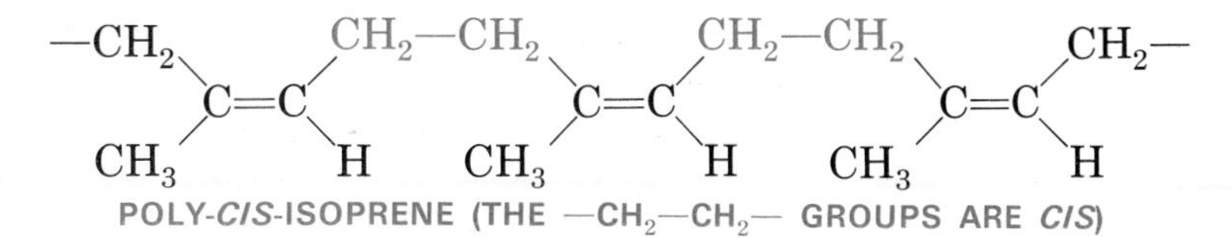

POLY-*CIS*-ISOPRENE (THE $—CH_2—CH_2—$ GROUPS ARE *CIS*)

Natural rubber is poly-*cis*-isoprene. However, the *trans* material also occurs in nature, in the leaves and bark of the sapotacea tree, and is known as *gutta-percha*. It is used as a thermoplastic for golf ball covers, electrical insulation and other such applications. Without an appropriate catalyst, polymerization of isoprene yields a solid that is like neither rubber nor gutta-percha. Neither the *trans* polymer nor the randomly arranged material is as good as natural rubber (*cis*) for making automobile tires.

Natural rubber is poly-*cis*-isoprene.

In 1955, chemists at Goodyear and Firestone discovered, almost simultaneously, how to use stereoregulating catalysts similar to those developed by Ziegler to prepare synthetic poly-*cis*-isoprene. This material is, therefore, structurally identical to natural rubber. Today, synthetic poly-*cis*-isoprene can be manufactured cheaply and is used almost equally well (there is still an increased cost) when natural rubber is in short supply. More than 2.4 million tons of synthetic rubber are produced in the United States yearly.

Copolymers

After examining Table 15–1, one might well wonder what would happen if a mixture of two monomers was polymerized. This type of reaction has been studied in detail and the products are called ***copolymers.*** If we polymerize pure monomer A, we get a ***homopolymer,*** poly A:

—AAAAAAAAAA—

Likewise, if pure monomer B is polymerized, we get a homopolymer, poly B:

—BBBBBBBBBB—

In contrast, if the monomers A and B are mixed and then polymerized, we get copolymers such as the following:

A copolymer is made by polymerizing two or more different monomers together.

—AABABAAABB—

—AABABABABB—

—BABABBAABA—

In such polymers the order of the units is often completely random, in which case the properties of the copolymer will be determined by the ratio of the amount of A to the amount of B.

A copolymer can have useful properties that are different from and often superior to those of the polymers of its pure constituents. As an example, let's go back to our discussion of rubber and pick up with synthetic rubbers. During World War II it was apparent to our military planners that we would be hard-pressed if our rubber supplies were cut off by Japan. A crash program was begun to develop synthetic rubber which would be as good as natural rubber. The Germans had earlier polymerized styrene, but this is a hard thermoplastic with little elasticity.

They had also polymerized butadiene to make the first synthetic rubber (Buna rubber), although it was not very serviceable. American chemists found, however, that a 1 to 3 copolymer of styrene and butadiene possessed properties closer to those of natural rubber.

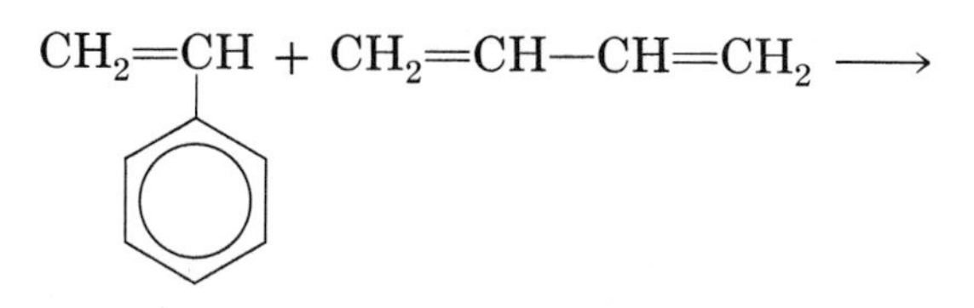

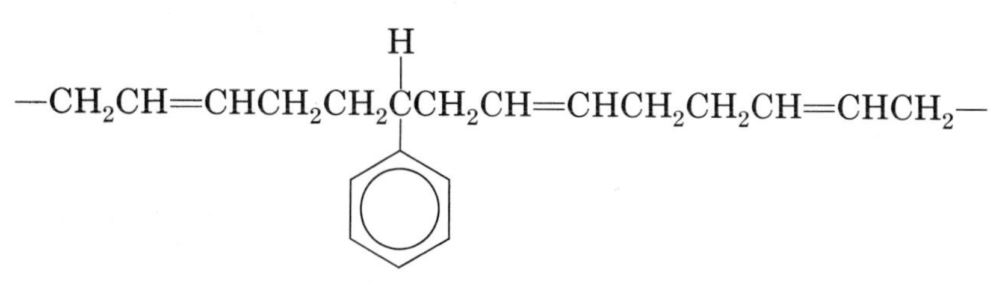

SBR COPOLYMER (STYRENE-BUTADIENE RUBBER)

Synthetic rubber is now made on a large scale.

The double bonds remaining in the polymer chain allow them to undergo vulcanization like natural rubber polymer chains (Figures 15–4 and 15–5).

SBR rubber is today manufactured on a large scale. About two million tons are used each year in manufacturing automobile and truck tires. A pure form of this polymer has even found its way into the marketplace as the replacement for the latex in chewing gum.

CONDENSATION POLYMERS

Polyesters

A chemical reaction in which two molecules react by splitting out or eliminating a small molecule is called a ***condensation reaction.*** For example, acetic acid and ethyl alcohol will react, splitting out a water molecule, to form ethyl acetate, an *ester.*

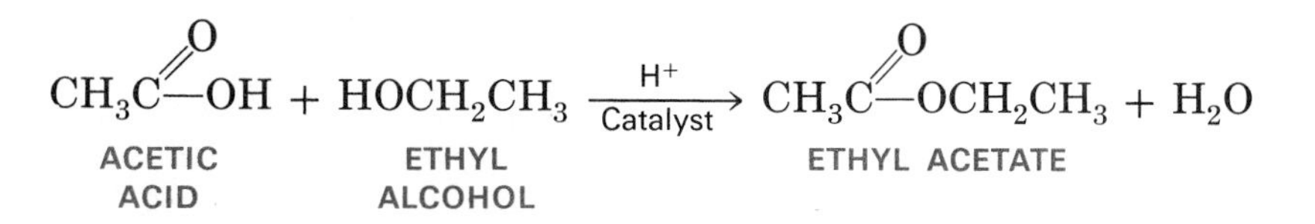

In a condensation polymerization, molecules are linked when they react to split out a small molecule such as water.

This important type of chemical reaction does not depend upon the presence of a double bond in the reacting molecules. Rather, it requires the presence of two kinds of functional groups on two different molecules. If each reacting molecule has two functional groups, both of which can react, it is then possible for condensation reactions to lead to a polymer. If we take a molecule with two carboxyl groups, such as terephthalic acid, and another molecule with two alcohol groups, such as ethylene glycol, each molecule can react at both ends. The reaction of one acid group of terephthalic acid with one alcohol group of ethylene glycol

Figure 15–5 Ready to take the cure, these "green" truck tires at Goodyear's Danville, Virginia, plant will assume their familiar "doughnut" shape when they are cured, or vulcanized. This involves placing the tires in the huge molds, just behind the technician, and subjecting them to heat and pressure. Note the familiar cured product in the open molds. Goodyear estimated that the industry shipped 28 million truck tires to motor vehicle manufacturers and operators during 1971. (Photo through the courtesy of the Goodyear Tire and Rubber Company, Akron, Ohio.)

initially produces an ester molecule with an acid group left over on one end and an alcohol group left over on the other:

The esterification of a dialcohol and a diacid involves two positions on each molecule.

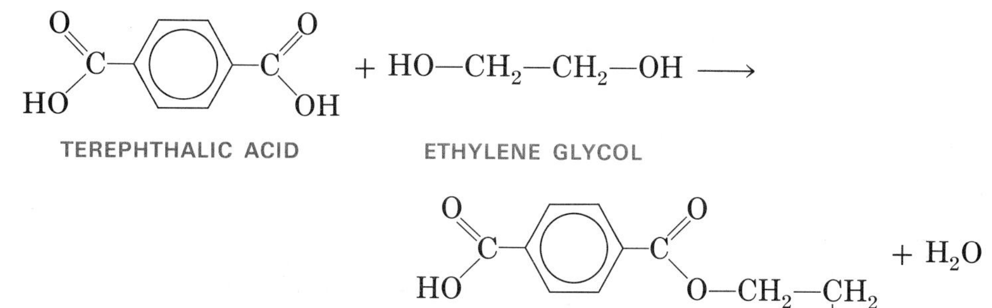

Subsequently, the remaining acid group can react with another alcohol group, and the alcohol group can react with another acid molecule. The process continues until an extremely large polymer molecule, known as a ***polyester,*** is produced with a molecular weight in the range of 10,000–20,000.

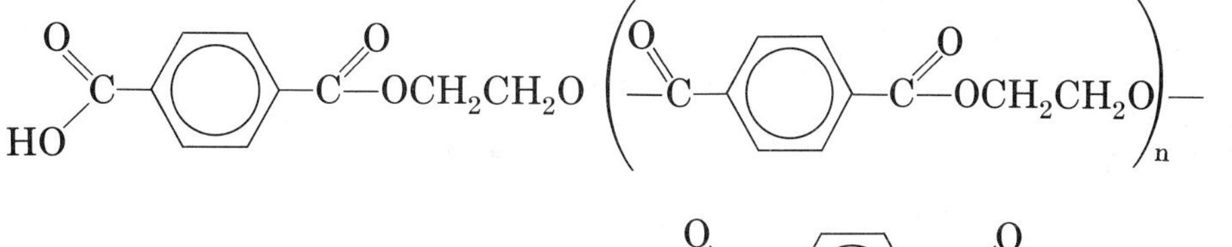

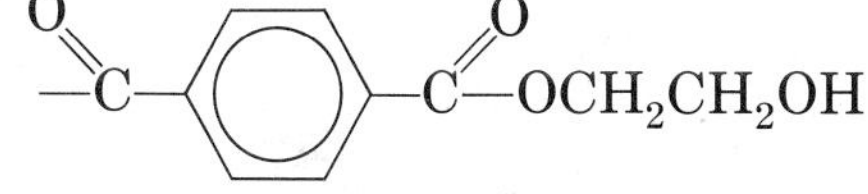

POLY(ETHYLENE GLYCOL TEREPHTHALATE)

A typical polyester is produced from a dialcohol and a diacid.

Poly(ethylene glycol terephthalate) is used in making textile fibers marketed under such names as "Dacron" and "Terylene," and films such as "Mylar." The film material has unusual strength and can be rolled into sheets one-thirteenth the thickness of a human hair. In film form this polyester is often used as a base for magnetic recording tape and for frozen food packaging.

Polyamides (Nylons)

Another useful condensation reaction is that occurring between an acid and an amine to split out a water molecule and form an ***amide.*** Reactions of this type yield a group of polymers which perhaps have had a greater impact upon society than any other type. These are the ***polyamides,*** or nylons.

In 1928, the Du Pont Company embarked upon a program of basic research headed by Dr. Wallace Carothers (1896–1937) who had been hired from the Harvard University faculty. His research interests were high molecular weight compounds, such as rubber, proteins, resins, and the reaction mechanisms that produced these compounds. In February, 1935, his research produced a product known as nylon 66, prepared from adipic acid and hexamethylenediamine.

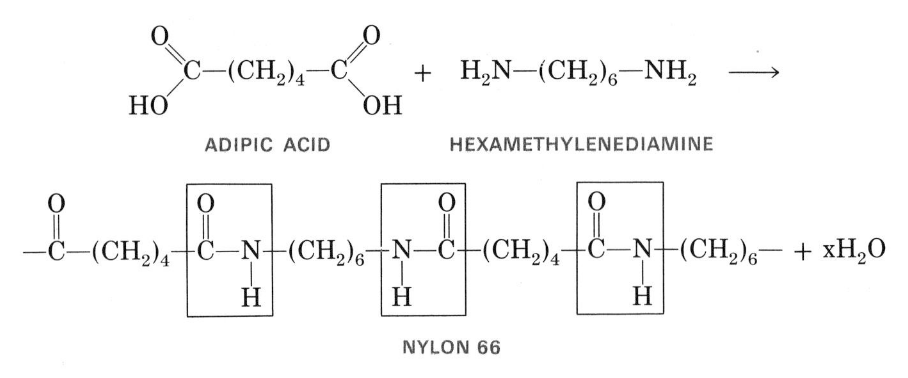

(The amide groups are outlined for emphasis.)

This material could easily be extended into fibers that were stronger than natural fibers and more chemically inert. The discovery of nylon jolted the American textile industry at almost precisely the right time. Natural fibers were not meeting the needs of twentieth-century Americans. Silk was not durable and was very expensive, wool was scratchy, linen crushed easily, and cotton did not lend itself to high fashion. As women's hemlines rose in the mid-1930's silk stockings were in great demand, but they were very expensive. Nylon changed all that almost overnight. It could be woven into the sheer hosiery women wanted, and it was much more durable than silk. The first public sale of nylon hose took place in Wilmington, Delaware (the hometown of Du Pont's main office), on October 24, 1939. They were so popular they had to be rationed. World War II caused all commercial use of nylon to be abandoned until 1945. Not until 1952 was the nylon industry able to meet the demands of the hosiery industry and to release nylon for other uses as a fiber and as a thermoplastic.

Common nylon can be made by the reaction of adipic acid and hexamethylenediamine.

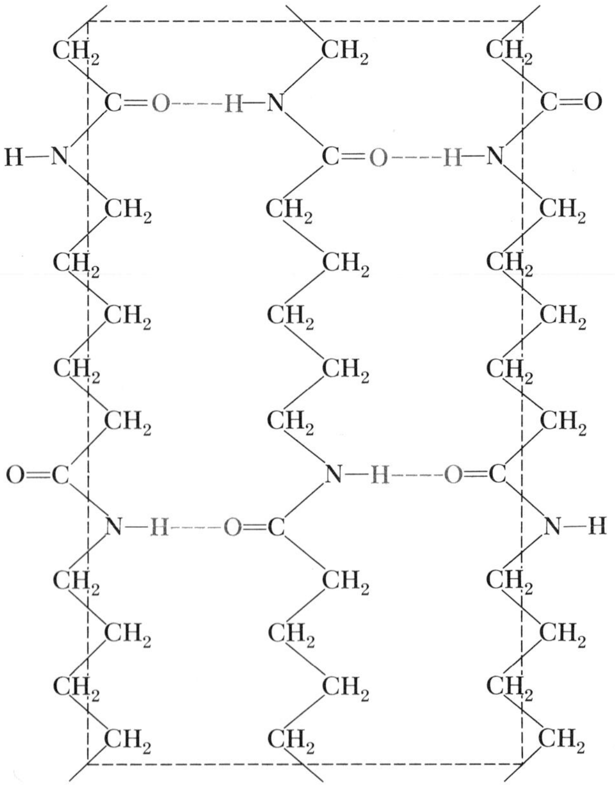

Figure 15–6 Structure and hydrogen bonding in nylon 6.

Many nylons have been prepared and tried on the consumer market, but two, nylon 66 and nylon 6, have been most successful. Nylon 6 is prepared from caprolactam, which comes from aminocaproic acid. Notice how aminocaproic acid contains an amine group on one end of the molecule and an acid group on the other end.

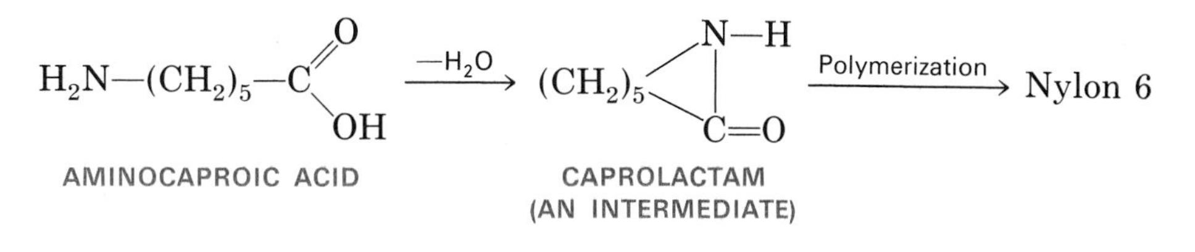

Hydrogen bonding is important in determining the properties of nylon fibers.

Figure 15–6 illustrates another facet of the structure of nylon—***hydrogen bonding.*** This type of bonding explains why the nylons make such good fibers. In order to have good tensile strength, the chains of atoms in a polymer should be able to attract one another, but not so strongly that the plastic cannot be initially extended to form the fibers. Ordinary covalent chemical bonds linking the chains together would be too strong. Hydrogen bonds, with a strength about one tenth that of an ordinary covalent bond, link the chains in the desired manner. We will see later that this type of bonding is also of great importance in protein structures.

SELF-TEST 15-A

1. Write the formulas of the monomers used to prepare the polymers listed below. For example, $CH_2{=}CH_2$ is used to prepare polyethylene.

   a. polypropylene

   ____

   b. polystyrene

   ____

   c. Teflon

   ____

2. Polypropylene may exist in three different structural forms which are designated as ______________, ______________, and ______________.

3. An addition polymerization is started by an initiator compound which forms a ________ ________.

4. Natural rubber is a polymer of ______________.

5. When styrene and butadiene are polymerized together the product is called a ______________.

6. Nylon is an example of a ______________ polymer.

7. Polyamides are formed when water is split out from the reaction of

   ______________________ and ______________________.

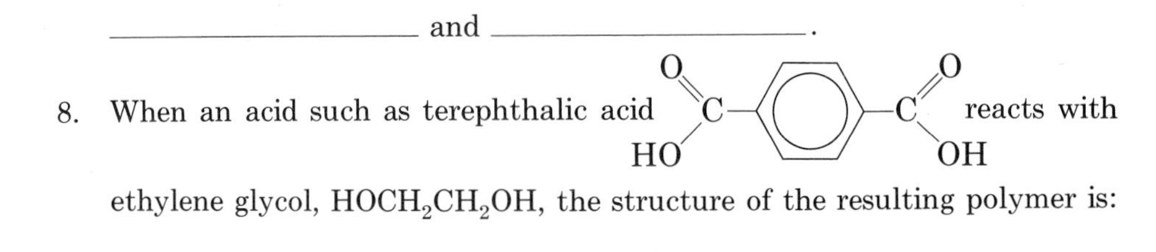

8. When an acid such as terephthalic acid reacts with ethylene glycol, $HOCH_2CH_2OH$, the structure of the resulting polymer is:

   ____

SILICONES

The element silicon, in the same chemical family as carbon, also forms many compounds with numerous Si—Si and Si—H bonds analogous to C—C and C—H bonds. However, the Si—Si bonds and the Si—H bonds are reactive toward both oxygen and water; hence, there are no useful silicon counterparts to most hydrocarbons. Silicon does form stable bonds with carbon, and especially oxygen, and this fact gives rise to an interesting group of condensation polymers containing silicon, oxygen, carbon, and hydrogen (bonded to carbon).

In 1945, E. G. Rochow, at the General Electric Research Laboratory, discovered that a silicon-copper alloy will react with organic chlorides to produce a whole new class of compounds, the ***organo-silanes.***

Silane, SiH_4, is structurally like methane, CH_4, in that both are tetrahedral.

$$2CH_3Cl + Si(Cu) \longrightarrow (CH_3)_2SiCl_2 + Cu$$

METHYL CHLORIDE — SILICON-COPPER ALLOY — DIMETHYLDICHLOROSILANE

These compounds readily react with water to replace the chlorine atoms with hydroxyl (—OH) groups. The resulting molecule is like a dialcohol.

$$(CH_3)_2SiCl_2 + 2H_2O \longrightarrow (CH_3)_2Si(OH)_2 + 2HCl$$

Two dihydroxysilane molecules undergo a condensation reaction in which a water molecule is split out. The resulting Si—O—Si linkage is very strong; the same linkage holds together all the natural silicate rocks and minerals. Continuation of this condensation process results in polymer molecules with molecular weights in the millions:

Silicones are polymers held together by a series of covalent Si—O bonds.

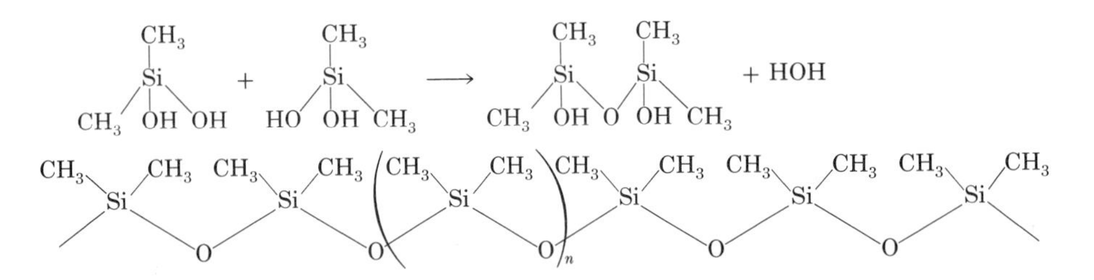

By using different starting silanes, polymers with very different properties result. For example, methyl groups on the silicon atoms result in ***silicone oils*** which are more stable at high temperatures than hydrocarbon oils and also have less tendency to thicken at low temperatures.

Silicone rubbers are composed of very high molecular weight chains bridged together. Room-temperature-vulcanizing (RTV) silicone rubbers are commercially available. These contain readily hydrolyzable groups which cross-link in the presence of atmospheric moisture:

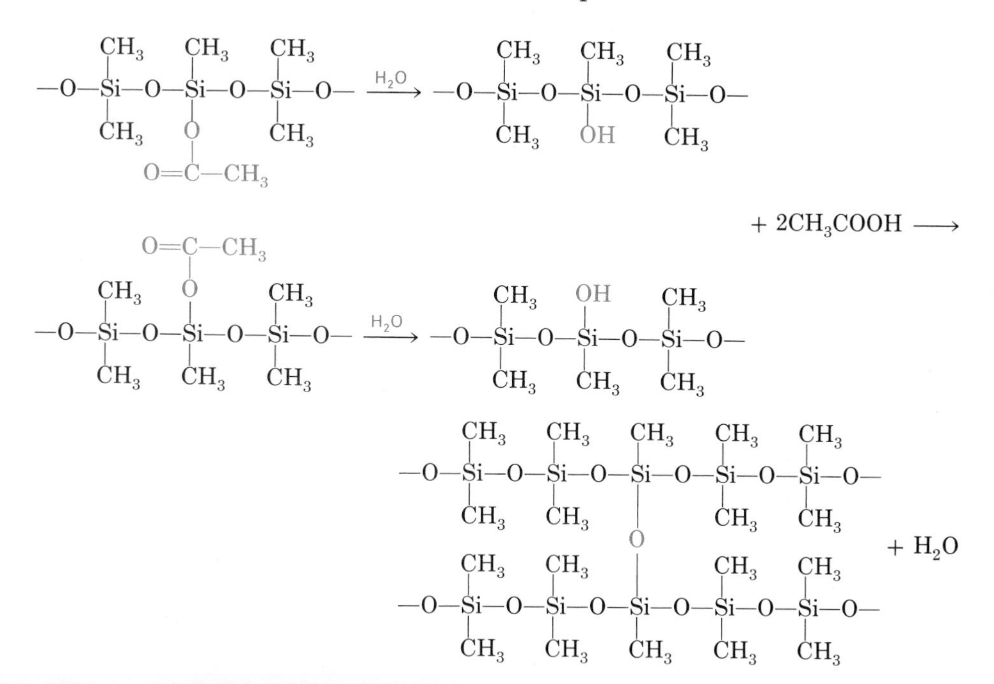

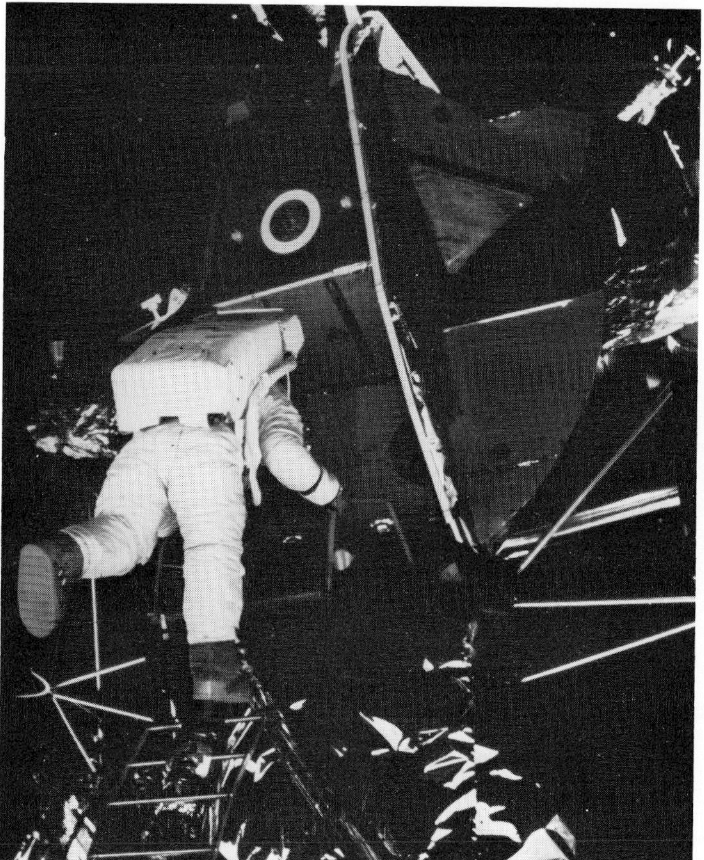

Figure 15–7 Examples of the use of silicone in the space program. Soles of lunar boots worn by the Apollo astronauts are made of high-strength silicone rubber. A silicone compound is also used for the air-tight seal of the lunar module hatch from which Astronaut Edwin E. Aldrin, Jr., has just emerged in this photo of the first manned landing on the moon on July 20, 1969.

The —OH groups which are produced are then made to condense, resulting in a cross-linking "cure" which is similar to the vulcanization of organic rubbers.

Silicone rubbers and oils find many medical uses.

Over 3,000,000 pounds of silicone rubber are produced each year in the United States. The uses include window gaskets, o-rings, insulation, sealants for buildings, space ships, and jet planes and even some wearing apparel. The first footprints on the moon were made with silicone rubber boots which readily withstood the extreme surface temperatures.

"Silly Putty," a silicone widely distributed as a toy, is intermediate between silicone oils and silicone rubber. It is an interesting material with elastic properties on sudden deformation, but its elasticity is quickly overcome by its ability to flow like a liquid when allowed to stand.

REARRANGEMENT POLYMERS

Some molecules polymerize by rearrangement reactions to yield very useful products. Molecules containing the isocyanate group (—NCO), for example, will react with almost any other molecule containing an active hydrogen atom (such as in an —OH or —NH_2 group) in a rearrangement process. An example is the reaction of hexamethylene diisocyanate and butanediol. The urethane linkage $\left(\begin{array}{c}-\underset{\displaystyle H}{\underset{|}{N}}-\overset{}{\underset{\displaystyle O}{\underset{\|}{C}}}-O-\end{array}\right)$ is a rearrangement of the same atoms in the isocyanate and alcohol groups and is similar to, but not the same as, the amide bond in nylons.

$$OCN(CH_2)_6NCO + HO(CH_2)_4OH \longrightarrow$$

HEXAMETHYLENE DIISOCYANATE — 1,4-BUTANEDIOL

$$OCN(CH_2)_6-\overset{\displaystyle H}{\overset{|}{N}}-\overset{\displaystyle O}{\overset{\|}{C}}-O-(CH_2)_4OH$$

PRODUCT MOLECULE (A URETHANE)

The continued reaction of the other groups gives rise to a polymer chain—a polyurethane.

Polyurethanes are structurally similar to many polyamides.

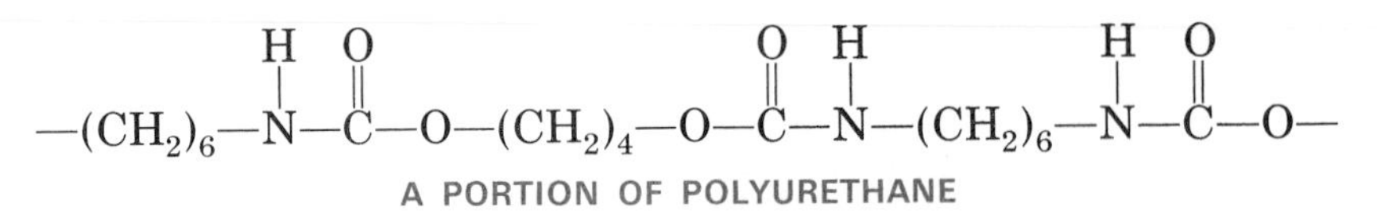

$$-(CH_2)_6-\overset{\displaystyle H}{\overset{|}{N}}-\overset{\displaystyle O}{\overset{\|}{C}}-O-(CH_2)_4-O-\overset{\displaystyle O}{\overset{\|}{C}}-\overset{\displaystyle H}{\overset{|}{N}}-(CH_2)_6-\overset{\displaystyle H}{\overset{|}{N}}-\overset{\displaystyle O}{\overset{\|}{C}}-O-$$

A PORTION OF POLYURETHANE

A polyurethane is structurally similar to a polyamide (nylon), and in Europe it is used similarly to nylon in this country. Polyurethanes have viscosities and melting points that make them useful for foam applications. Foamed polyurethanes are known as "foam rubber" and "foamed plastics."

POLYMER ADDITIVES—TOWARD AN END USE

Few plastics produced today find end uses without some kind of modification. Polyurethanes are a good example. In order for polyurethane to be used as insulation in refrigerators, refrigerated trucks and railroad cars, and as construction insulation, it is foamed.

Foaming Agents

Two methods of producing foamed plastics are commonly used. Hydrocarbons such as pentane or fluorocarbons like trichlorofluoromethane are ***physical foaming agents*** because they can foam a plastic simply by boiling to produce bubbles in the plastic. Polyurethane can be foamed by dissolving pentane in the liquid under pressure. When this mixture is extruded from an outlet into the atmosphere, the pentane volatilizes, leaving the polyurethane full of small holes. The polyurethane is quickly solidified by cooling. Planks, boards, or logs of foamed polystyrene (Styrofoam) can also be made in this way.

Foaming agents are used to make low-density plastics which are full of gas bubbles.

Chemical foaming agents produce a gas by chemical reaction, as their name implies. For example, when polyurethanes are formed, it is sometimes possible to make use of gaseous carbon dioxide, generated when an isocyanate group reacts with a water molecule.

$$\underset{\text{AN ISOCYANATE}}{R\text{—}NCO} + H_2O \longrightarrow \underset{\text{AN AMINE}}{R\text{—}NH\text{—}H} + CO_2$$

Plasticizers

Many times a plastic such as polyethylene or polypropylene turns out to be too stiff for its intended application. Chemists have found that the addition of certain compounds, called ***plasticizers,*** can render a polymer more flexible. The structure of a polymer largely determines its flexibility. The polymer chains are intertwined and somewhat aligned in the solid state. This is enough to prevent adequate flexing of the solid in many instances. Certain compounds such as di-(2-ethylhexyl) phthalate (DOP) and di-(2-ethylhexyl) adipate (DOA),

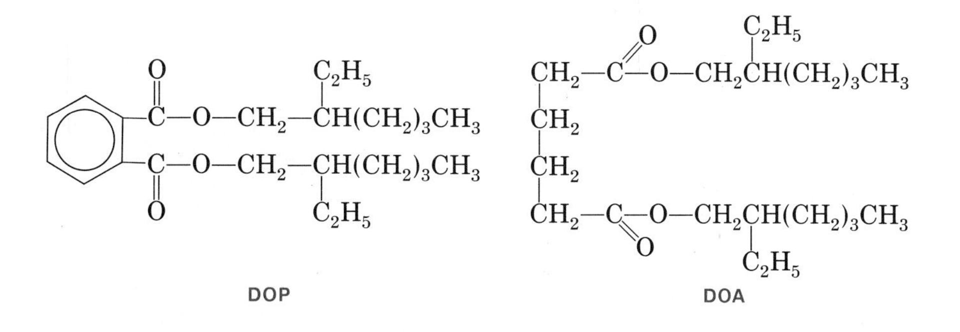

Plasticizers are added to plastics to make them more flexible.

exert a partial dissolving effect by fitting in between the polymer chains and weakening the attractions between chains, thereby increasing the flexibility (Figure 15-8).

In 1970, 500 million pounds of plasticizers were used in plastic formulation in the United States. The extent of plastic food wraps, for example, would be much more limited without DOA, which has Food and Drug Administration approval. DOP and similar compounds are found in many areas of our environment. Recently, DOP and similar plasticizers were discovered in the heart muscle of cattle, dogs, rabbits, and rats and in the bloodstreams of a group of laboratory workers at one laboratory.

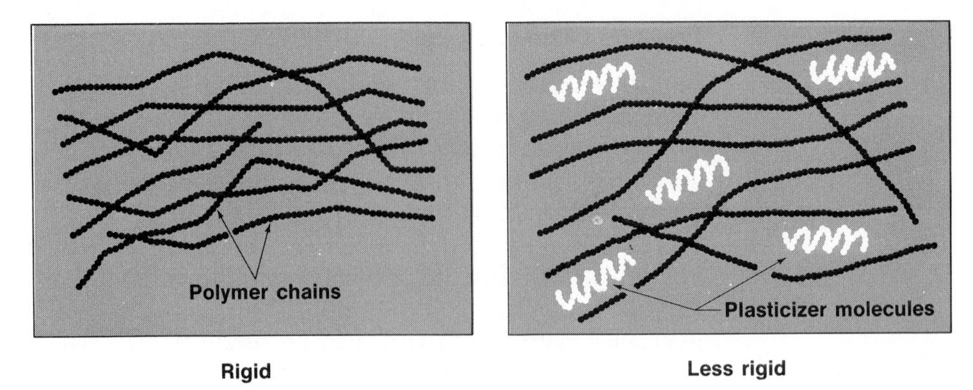

Figure 15–8 Schematic representation of how a plasticizer works.

Possible adverse side effects of their presence are currently being studied. One form of the synthetic styrene-butadiene rubber (SBR, discussed earlier in this chapter) requires only ordinary hydrocarbon oil as a plasticizer.

Stabilizers

Most plastics used in outside locations, such as signs, tarpaulins, indoor-outdoor carpeting, auto seat covers, and toys, must be protected

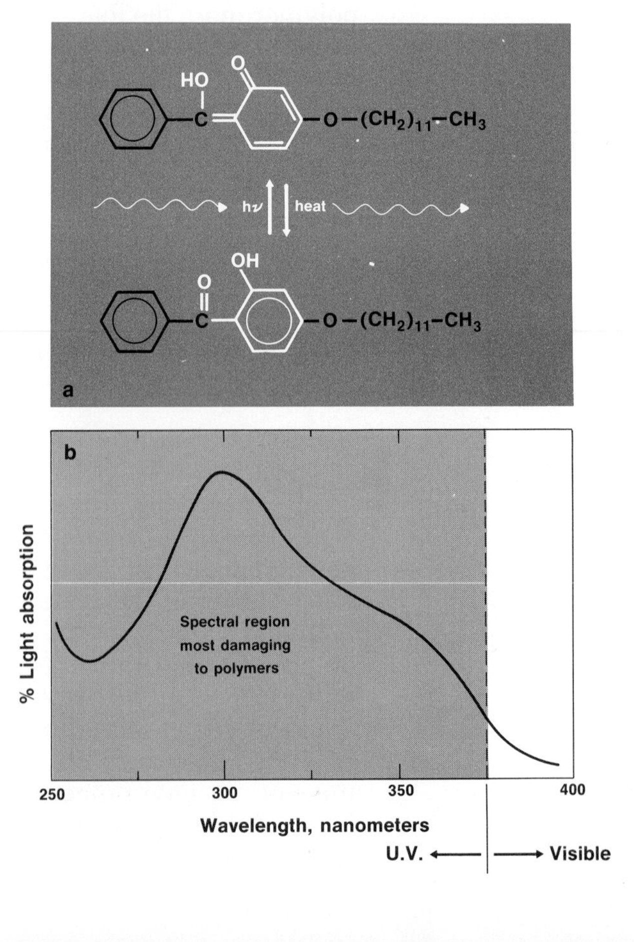

Figure 15–9 (a) A possible mechanism by which a molecule can absorb ultraviolet radiation and thus protect a plastic. This particular molecule is almost 100 percent efficient in the process and finds wide use in plastics. (b) Typical ultraviolet absorption by a 2-hydroxybenzophenone-type compound used to protect plastics from sunlight.

Figure 15–10 Sunlight (mostly ultraviolet) damages most plastics such as polypropylene webbing. The results are shorter product life and higher costs to the consumer. Plastics containing ultraviolet absorbing chemicals have a longer outdoor life.

from sunlight, as anyone who has repeatedly parked his plastic-upholstered convertible in the outdoors knows.

Stabilizers are added to plastics to reduce the rate of damage from exposure to sunlight.

Photons of the 290 to 400 nanometer spectral region have sufficient energy to break some chemical bonds found in organic molecules. However, there are other molecules that absorb ultraviolet light and liberate the energy as heat, thus preventing broken bonds in the polymer structure. These may be added directly to the plastic (about 0.1 percent by weight) that is to be subjected to sunlight. Unfortunately, many plastic formulations intended for outdoor use do not contain sufficient ultraviolet stabilizer to prevent decay.

PROCESSING METHODS

In order to make full use of a polymer's useful properties, the material must generally be processed into some desired form. Foaming, which results in many useful plastic articles, has already been mentioned. A number of other forming processes have been developed and are applicable depending on the qualities of the plastic used and the form to be achieved. Some techniques are illustrated in Figure 15-12.

THE FUTURE OF POLYMERS

As we have seen in this chapter, the development and use of synthetic polymers is quite recent. Polyethylene, for example, was not discovered until 1933, yet by 1974, its production in the United States amounted to billions of pounds. Chemists are constantly synthesizing new polymers and finding applications for them. The space age has brought with it the need for new polymers, especially in electronics and as special coatings which can withstand high temperatures without breaking down. Among the newcomers are the polyimides, prepared from the polycondensation of a diacid anhydride and a diamine. Some of these polymers have very high service temperatures (Figure 15-11).

TABLE 15–2 Composition of Some Polymers*

Trade Name	Composition
Acrilan	85% Acrylonitrile plus vinyl acetate or vinyl pyridine
Acrylic	At least 85% acrylonitrile
Arnel	Cellulose triacetate
Bakelite	Phenol plus formaldehyde
Caprolan	Nylon 6
Cellophane	Regenerated cellulose
Celluloid	Cellulose nitrate
Creslan	Copolymer of acrylonitrile and acrylamide
Dacron	Ethylene glycol plus terephthalic acid
Delrin	Polyacetal
Carvan	Vinylidene dinitrile plus vinyl acetate
Dynel	60% Vinyl chloride plus 40% acrylonitrile
Epoxy	Phenol plus acetone plus epichlorohydrin
Formica	Phenol plus formaldehyde
Fortrel	Polyester similar to Dacron
Herculon	Polypropylene
Kodel	Polyester; terephthalic acid plus 1,4-cyclohexane dimethanol
Lucite	Methyl methacrylate
Melinex	Polyethylene terephthalate
Melmac	Melamine plus formaldehyde
Meraklon	Polypropylene
Mylar	Polyethylene terephthalate
Neoprene	2-Chlorobutadiene
Nylon 501	Nylon 66
Nytril	At least 85% vinylidene dinitrile
Orlon	Originally pure acrylonitrile; now up to 14% of another monomer
Plexiglas	Methyl methacrylate
Polythene	Polyethylene
Polyzote	Polystyrene
Rayon	Regenerated cellulose
Saran	Vinylidene chloride plus vinyl chloride
Spandelle	Polyurethane; ethylene glycol plus diisocyanate
Spandex	Polyurethane; ethylene glycol plus diisocyanate
Teflon	Polytetrafluoroethylene
Terylene	Polyester similar to Dacron
Vectra	Polypropylene
Vectra nylon	Nylon 6
Velon	Vinylidene chloride plus vinyl chloride; see Saran
Vinyon	Vinyl chloride or copolymer with vinyl acetate
Zantrel	Rayon fiber

$—CH_2C(OC_2H_5)_2—$

Acetal

$—CH_2CH(CONH_2)—$

Acrylamide

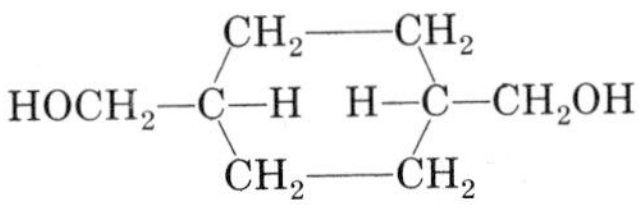

$HOCH_2—CH<(CH_2—CH_2)_2>CH—CH_2OH$

1,4-Cyclohexane Dimethanol

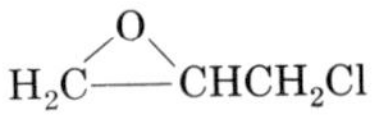

$H_2C—CHCH_2Cl$ (epoxide ring: O bridging the two ring carbons)

Epichlorohydrin

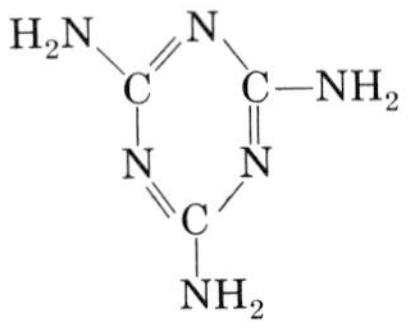

$C_3N_3(NH_2)_3$ (triazine ring with three NH_2 groups)

Melamine

$—CH_2CH(OCOCH_3)—$ (O above the carbonyl C: $OC(=O)CH_3$)

Vinyl Acetate

$—CH_2C(CN)_2—$

Vinylidene Dinitrile

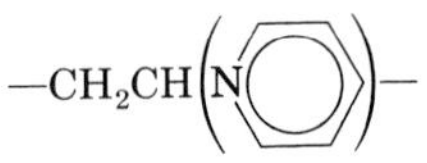

$—CH_2CH(C_5H_4N)—$ (pyridine ring)

Vinyl Pyridine

*Structures are given for those monomers not found elsewhere in the text. Consult the index for formulas not given.

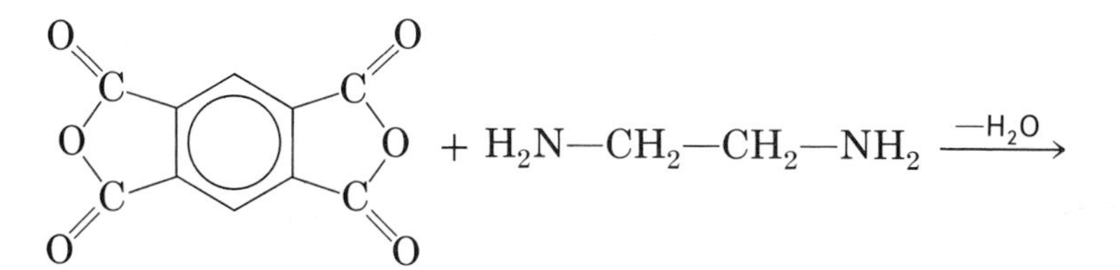

PHTHALIC ANHYDRIDE

1,2-DIAMINOETHANE

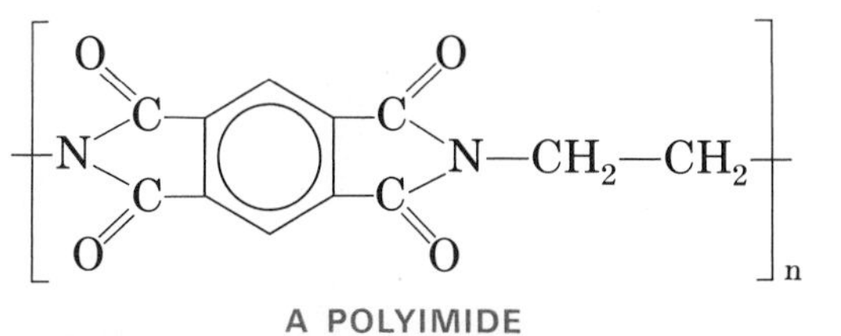

A POLYIMIDE

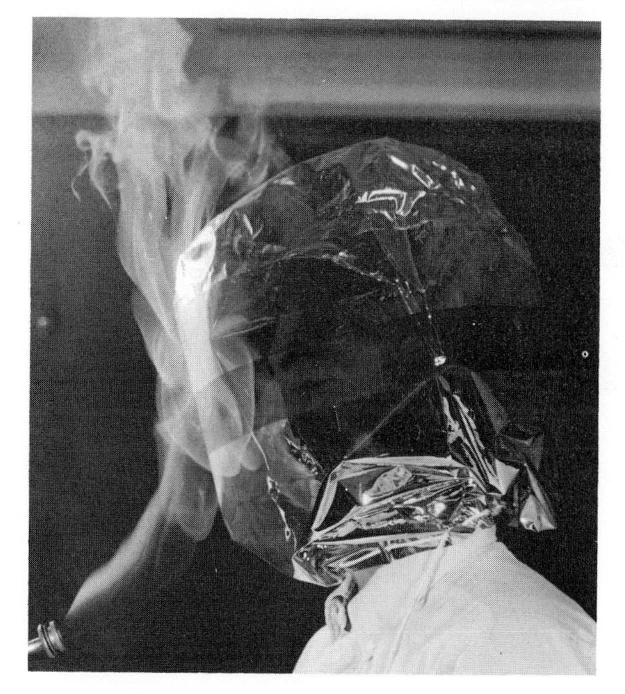

Figure 15–11 Preparation of a polyimide (*above*) and an example of how a polyimide film can withstand, for a short period, the flame of a blowtorch.

Because polymers are used so extensively in the world today, the problem of waste disposal is inevitable. Engineers have envisioned plants in which solid wastes from cities would undergo first a magnetic separation to remove iron and steel objects, then a ballistic separation based on density, since glass and aluminum objects are more dense than plastics. The plastics thus separated would be treated in two ways. If suitable separation methods could be developed, thermoplastics could be reprocessed into new items (e.g., if all the nylon could be separated from polystyrene). Thermosetting plastics could not be treated this way, however, because breaking the cross-linking would cause complete molecular degradation. If separation and reuse were not feasible, combustion units built near cities could actually use plastics as fuels since they are mostly carbon and hydrogen. There is a danger, though, in that some plastics contain elements which could create massive pollution if released into the atmosphere. An example is polyvinylchloride, which on burning

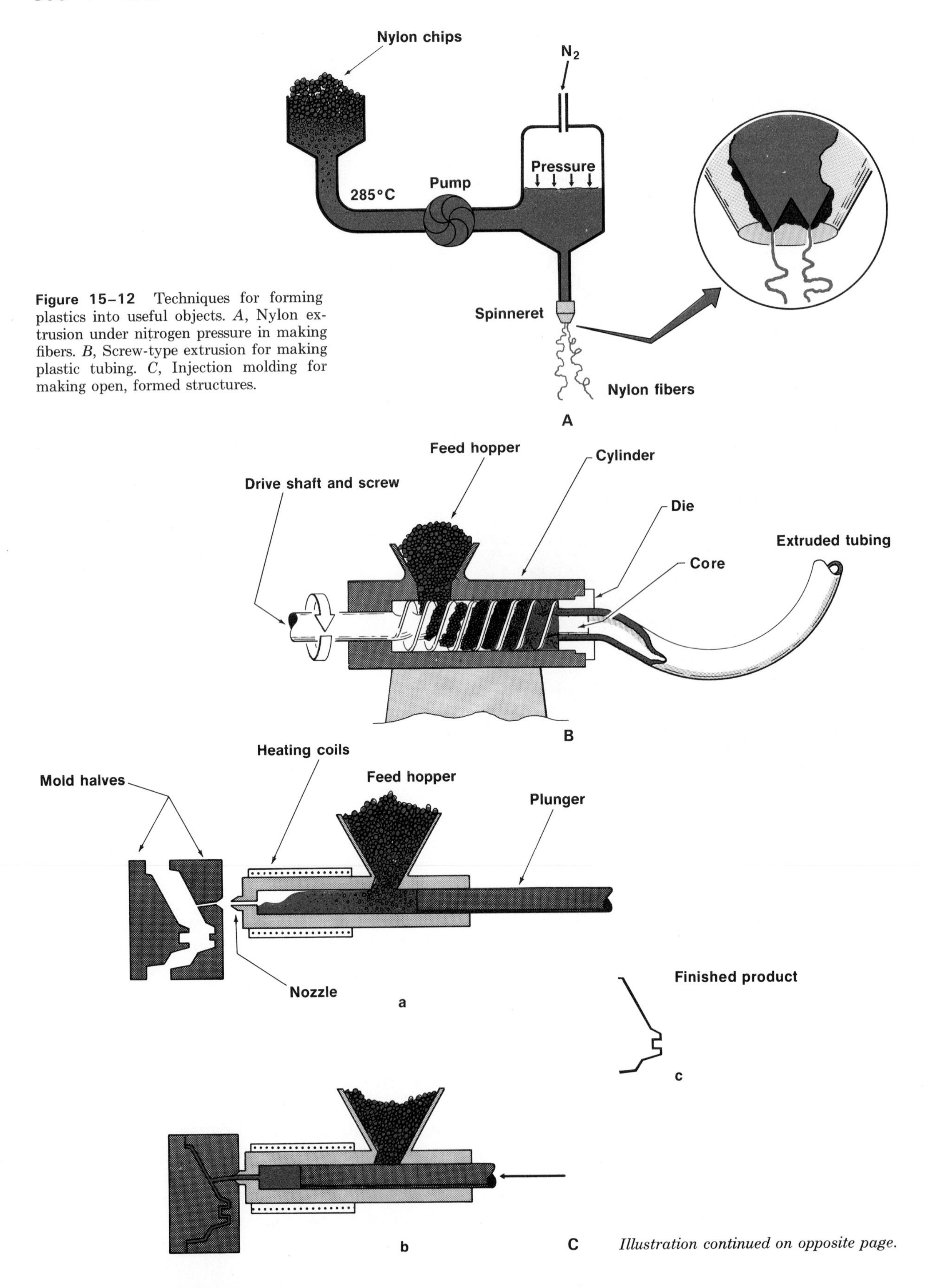

Figure 15–12 Techniques for forming plastics into useful objects. *A*, Nylon extrusion under nitrogen pressure in making fibers. *B*, Screw-type extrusion for making plastic tubing. *C*, Injection molding for making open, formed structures.

Illustration continued on opposite page.

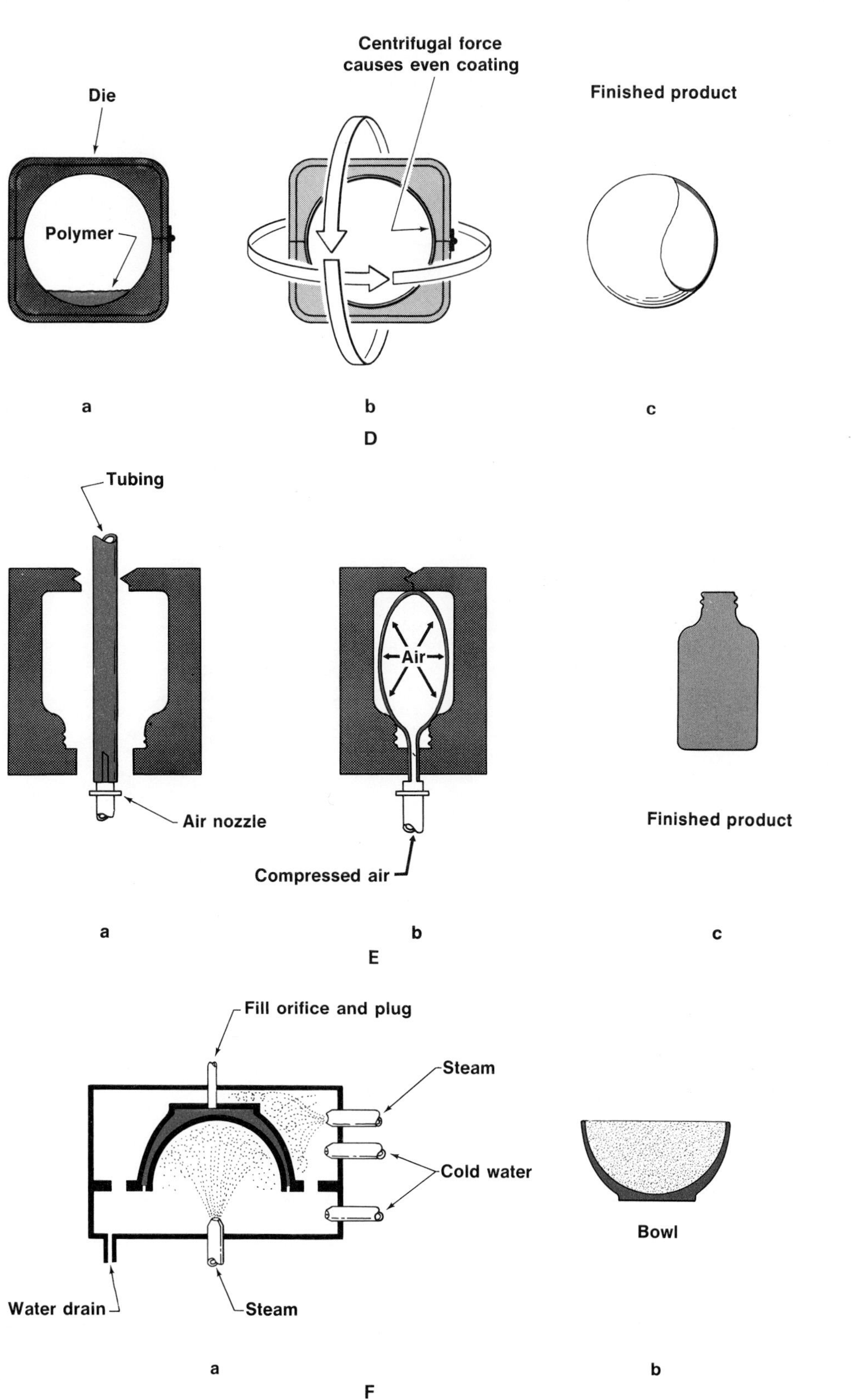

Figure 15-12 ***Continued*** *D*, Rotational molding for making closed plastic structures. *E*, Blow molding of a hollow plastic article for making plastic bottles. *F*, Beaded plastic particles can be formed and "cooked."

yields hydrogen chloride, a very corrosive gas. However, the combustion products could be recycled as raw materials for other chemical syntheses.

Hopefully, these and similar problems will be solved as man begins to understand more fully how to use what he has on this planet, and how to live in greater harmony with nature.

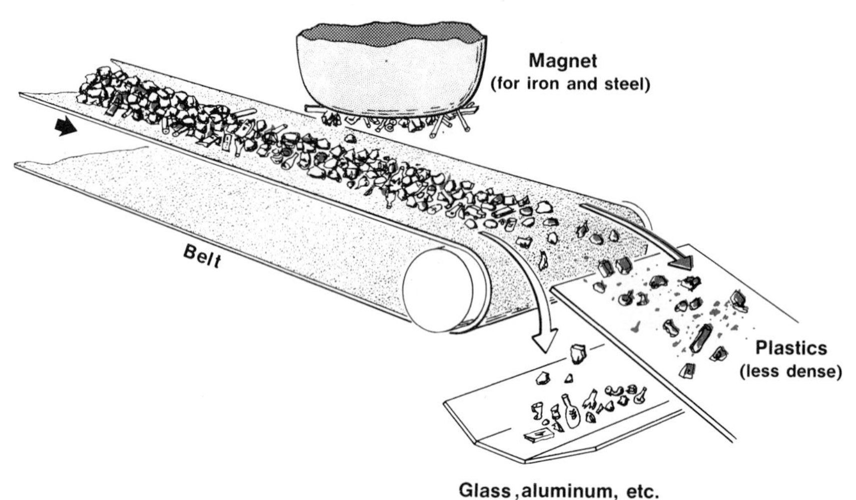

Figure 15–13 A method of separation of plastics from other wastes prior to recycling the plastics for reuse.

SELF-TEST 15-B

1. When $(CH_3)_2SiCl_2$ reacts with water, a representative portion of the structure of the polymer obtained is

2. Stabilizers protect plastics against the action of ______________________.

3. A plastic which is too stiff can be rendered more flexible by the addition of a ______________________.

4. A silicone polymer contains Si— ________ bonds.

MATCHING SET

______	1. nylon	a.	plastic which forms interlocking bonds when heated
______	2. monomer	b.	has an unpaired electron
______	3. thermoplastic	c.	hardening via reaction with sulfur
______	4. thermosetting plastic	d.	form polymers of desired structure
______	5. free radical	e.	natural rubber
______	6. vulcanize	f.	a synthetic rubber
______	7. stereochemical control	g.	building unit for a polymer
______	8. styrene-butadiene copolymer	h.	plastic softened by heat
______	9. poly-*cis*-isoprene	i.	a polyamide
______	10. freezing point depression	j.	formed from a dialcohol and a diacid
______	11. polyester	k.	used to determine molecular weights

QUESTIONS

1. Explain how polymers could be prepared from each of the following compounds. (Other substances may be used.)

   a. $CH_3—CH{=}CH—CH_3$ (each double-bonded C bearing H)

   b. $HO—C(=O)—CH_2—CH_2—C(=O)—OH$

   c. $CH_2(OH)—CH(OH)—CH_2(OH)$

   d. $H_2N—CH_2—C_6H_4—CH_2—NH_2$ (para-substituted benzene ring)

2. Beginning with petroleum, outline the steps involved in the preparation of polyethylene.

3. Draw the monomer used to prepare:

   a. polyethylene
   b. neoprene
   c. polystyrene
   d. Teflon

4. What are the monomers used to prepare the following polymers?

   a. $-CH_2CH_2CH_2CH_2CH_2CH_2CH_2CH_2CH_2-$

   b.
   $$\begin{array}{l}\;\;CH_3\quad\;\; CH_3\quad\;\; CH_3\\ \;\;\;|\qquad\quad\;\; |\qquad\quad\;\; |\\ -CHCH_2CHCH_2CHCH_2-\end{array}$$

   c.
   $$\begin{array}{l}\qquad\;\; H\qquad\;\; H\qquad\;\; H\qquad\; H\\ \qquad\;\; |\qquad\quad |\qquad\quad |\qquad\;\; |\\ -CH_2CCH_2-CCH_2-CCH_2-C-\\ \qquad\;\; |\qquad\quad |\qquad\quad |\qquad\;\; |\\ \qquad\; C_6H_5\quad C_6H_5\quad C_6H_5\quad C_6H_5\end{array}$$

5. Write equations showing the formation of polymers by the reaction of the following pairs of molecules:

   a. $HOOC-C_6H_4-COOH$ (COOH groups at opposite ring positions) and $HOCH_2CH_2OH$

   b. $HOOCCH_2CH_2COOH$ and $H_2NCH_2CH_2NH_2$

   c.

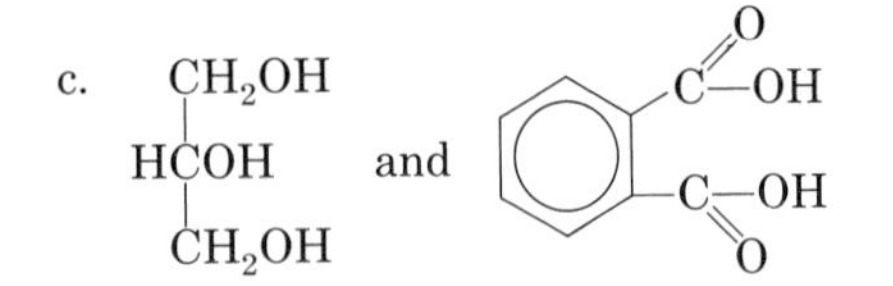

6. You are given two specimens of plastic, A and B, to identify. One is known to be nylon, and the other poly(methylmethacrylate). Analysis of A shows it to contain C, H, and O, while B contains C, H, O, and N. What are A and B?

7. What structural features are necessary in a molecule for it to undergo addition polymerization?

8. What is meant by the term macromolecule?

9. Orlon has a polymeric chain structure of
   What is the monomer from which this can be made?

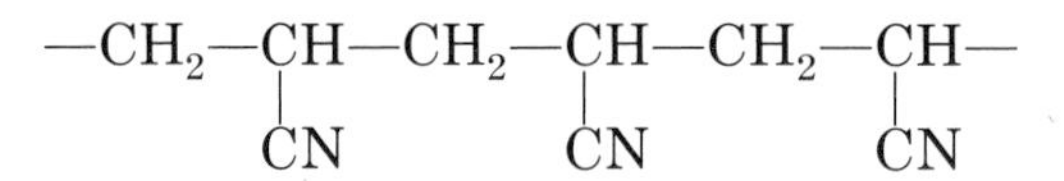

10. Illustrate how the presence of too much peroxide initiator could cause an addition polymerization to stop prematurely.

11. What is a copolymer? Give an example of one.

12. What feature do all condensation polymerization reactions have in common?

13. Give an example of the possibilities that exist if a trifunctional acid reacts with a difunctional alcohol.

14. What are the starting materials for nylon 66?

15. Suggest a major difference in the bonding of thermosetting and thermoplastic polymers. Which is more likely to have an interlacing of covalent bonds throughout the structure? Which is more likely to have weak bonds between large molecules?

16. Explain how a plasticizer can render a polymer more flexible.

17. Write the reaction for the formation of a silicone from $(C_2H_5)_2SiCl_2$.

PROBLEMS

1. How many moles of ethylene gas are required to manufacture a 100-gram polyethylene bottle?
   Ans. 3.6 moles

2. A specimen of polystyrene is found to consist of molecules with an average molecular weight of 12,000 amu. What is the average number of styrene monomers in each unit?
   Ans. 115

3. How many molecules of ethylene are contained in a polyethylene molecule with a molecular weight of 28,000?
   Ans. 1000

4. Which requires the larger number of molecules of monomer for its preparation: 100 g of polystyrene or 100 g of polypropylene?
   Ans. 100 g of polypropylene

5. What is the maximum molecular weight polymer molecule that can be prepared from 0.01 g of ethylene?
   Ans. 6.0×10^{21}

SUGGESTIONS FOR FURTHER READING

Billmeyer, F. W., Jr., "Measuring the Weight of Giant Molecules," *Chemistry,* Vol. 39, No. 3, p. 8 (1966).
Factor, A., "The Chemistry of Polymer Burning and Flame Retardance," *Journal of Chemical Education,* Vol. 51, No. 7, p. 453 (1974).
Franer, A. H., "High Temperature Plastics," *Scientific American,* Vol. 221, No. 1, p. 96 (1969).
Kaufman, M., "Giant Molecules," Doubleday, New York, 1968.
Mark, H. F., "Giant Molecules," Time Incorporated, New York, 1966.
Mark, H. F., "The Nature of Polymeric Materials," *Scientific American,* Vol. 217, No. 3, p. 98 (1967).
Natta, G., "How Giant Molecules are Made," *Scientific American,* Vol. 197, No. 3, p. 98 (1957).
"Nylon: The First 25 Years," The Du Pont Company, Wilmington, 1963.

CHAPTER 16

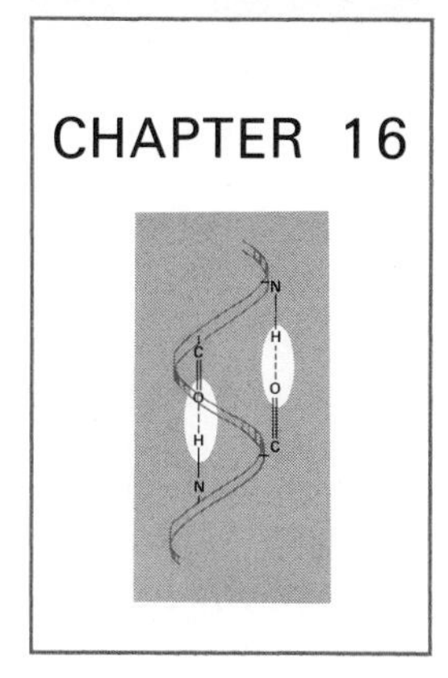

BIOCHEMISTRY—BASIC STRUCTURES

The goal of biochemistry is to develop a chemically based understanding of living cells of all types. This includes the determination of the kinds of atoms present, the investigation of how they are joined together to form the larger structural units present in cells, and the study of the chemical reactions by which living cells obtain the energy required for the life processes of growth, movement, and reproduction.

In this chapter emphasis will be placed on fundamental biochemical structures and their preparation. These structures are (1) ***carbohydrates;*** (2) ***fats;*** (3) ***amino acids*** and ***proteins;*** and (4) ***nucleic acids.***

CARBOHYDRATES

Carbohydrates contain the elements carbon, hydrogen, and oxygen, with hydrogen atoms and oxygen atoms generally in the ratio of 2 to 1.

Carbohydrates are composed of the three elements carbon, hydrogen, and oxygen. Three structural groups are prevalent in carbohydrates: alcohol (—OH), aldehyde $\left(\text{—}\overset{\displaystyle O}{\overset{\|}{C}}\text{H}\right)$, and ketone $\left(\text{—}\overset{\displaystyle O}{\overset{\|}{C}}\text{—}\right)$. The carbohydrates can be classified into three main groups: ***monosaccharides*** (Latin, *saccharum,* sugar), ***oligosaccharides,*** and ***polysaccharides.*** Monosaccharides are simple sugars that cannot be broken down into smaller units by mild acid hydrolysis. Hydrolysis of a molecule of an oligosaccharide yields two to six molecules of a simple sugar; complete hydrolysis of a polysaccharide produces many monosaccharide units.

Carbohydrates are synthesized by plants from CO_2 and H_2O, using energy from the sun.

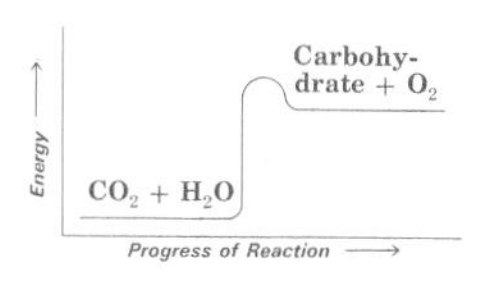

Carbohydrates are synthesized by plants from water and atmospheric carbon dioxide. The process is called ***photosynthesis*** since it is a synthetic reaction that occurs when energized by photons of light energy. It is an endothermic process; consequently, carbohydrates are energy-rich compounds. These compounds serve as important sources of energy for the metabolic processes of plants and animals. Glucose, $C_6H_{12}O_6$, along with some of the other simple sugars, are quick energy sources for the cell. Polysaccharides, such as starch, store large amounts of energy that can be used by the cell only when the complex unit is broken down into monosaccharides.

Some complex carbohydrates are also used by cells for structural purposes. Cellulose, for example, partially accounts for the structural properties of wood.

Monosaccharides

Approximately 70 monosaccharides are known; 20 of these simple sugars occur naturally. Unlike many organic compounds these sugars are very soluble in water owing to the numerous —OH groups present which can form hydrogen bonds with water.

The most common simple sugar is ***D-glucose.*** This monosaccharide requires three structures for its adequate representation (Figure 16–1). Structure (*a*), in which the carbon atoms are numbered for later reference, depicts the "straight-chain" structure with the aldehyde group (—CHO) in position 1. The properties of a water solution of D-glucose cannot be explained by this structure alone. At any given time, most of the molecules exist in the ring form, structures (*b*) and (*c*), which results from a molecular rearrangement in which carbon 1 bonds to carbon 5 through an oxygen atom. Both ring structures are possible since the OH group on carbon 1 may form in such a way to point either along the plane of the molecule or out of the plane. It should be emphasized that a solution of D-glucose contains a mixture of three forms in a dynamic state of equilibrium. However, equilibrium is shifted towards the ring forms; very limited amounts of the straight-chain form are present.

D-glucose, the most important monosaccharide is found in fruits, blood, and in living cells.

Because it is sweet, D-glucose is used in the manufacture of candy and in commercial baking. This simple sugar, also called *dextrose, grape sugar,* and *blood sugar,* is prevalent in fruits, vegetables, blood, and tissue fluids. A solution of D-glucose is fed intravenously when a readily availa-

Glucose has a relative sweetness of 74.3, compared to sucrose having an assigned value of 100.0. The value for fructose is 173.3.

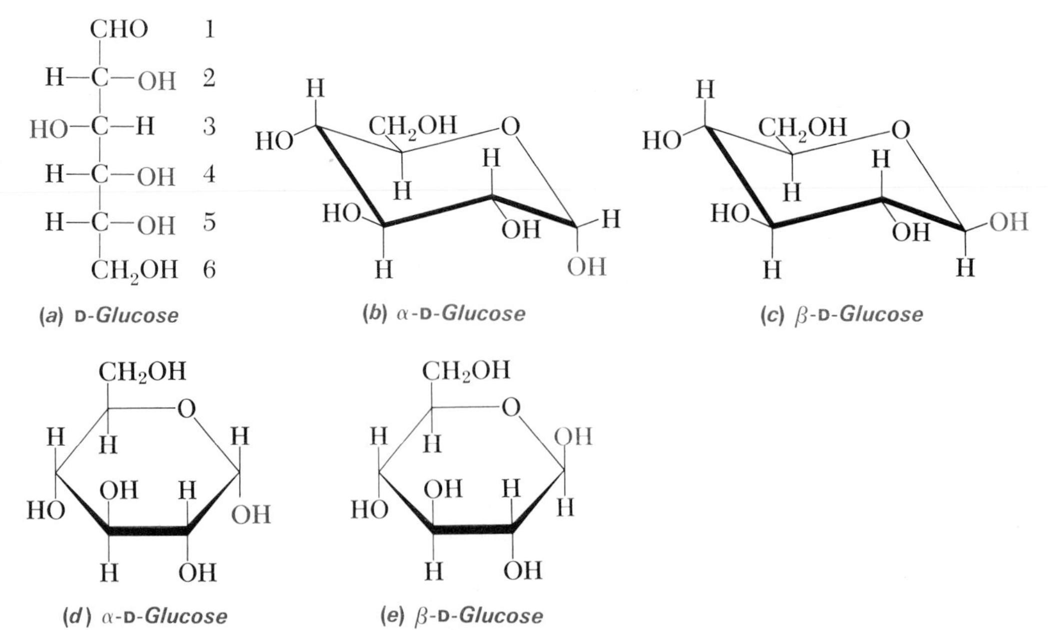

Figure 16–1 The structures of D-glucose; *d* and *e* are two-dimensional representations of *b* and *c*, respectively. Note the difference in the position of the —OH group (color) in the α and β forms of glucose.

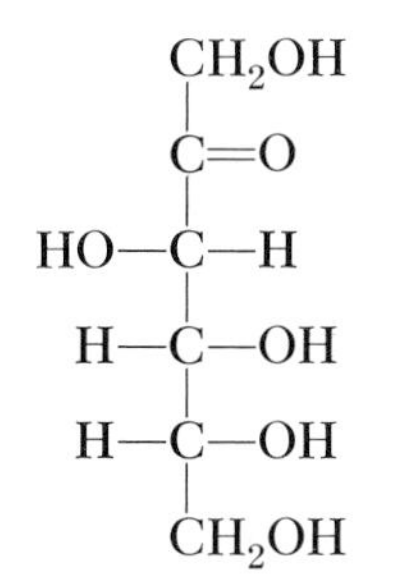

(a) *Ketone structure*

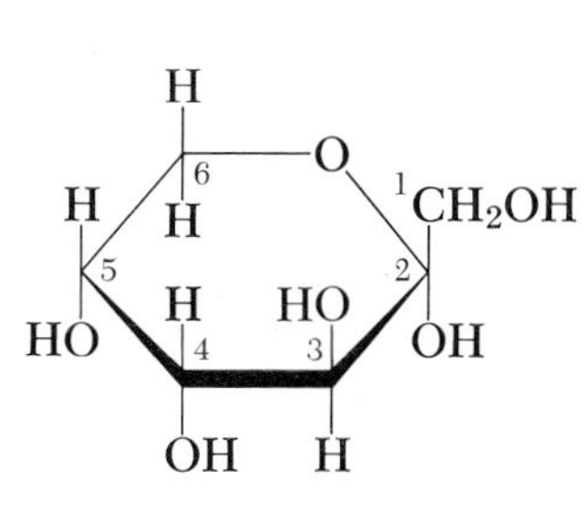

(b) *α-Ring structure*

Figure 16–2 The structures of D-fructose. The β-ring structure (not shown) differs from the α-ring structure in that the CH_2OH and OH groups are in reversed positions on carbon 2.

ble source of energy is needed to preserve life. As will be discussed later, many polysaccharides, including starch, are composed of glucose units and serve as a source of this important chemical upon hydrolysis of the complex structures.

Another very important monosaccharide is ***D-fructose.*** Its structure is given in Figure 16–2.

Oligosaccharides

The most important oligosaccharides are the disaccharides (two simple sugar units per molecule). Examples include the following widely used sugars:

sucrose (from sugar cane or sugar beets), which consists of a glucose unit and a fructose unit,
maltose (from starch), which consists of two glucose units, and
lactose (from milk), which consists of a glucose unit and a galactose (an isomer of glucose) unit.

Disaccharides are molecules containing two simple sugars bound together, such as in sucrose, which contains a glucose and a fructose unit in each molecule. A water molecule is eliminated when the bond forms.

The formula for these disaccharides, $C_{12}H_{22}O_{11}$, is not simply the sum of two monosaccharides, $C_6H_{12}O_6 + C_6H_{12}O_6$. A water molecule must be added (hydrolysis) to obtain the monosaccharides. The structures for sucrose, maltose, and lactose, along with their hydrolysis reactions, are given in Figure 16–3.

The disaccharides are important as foods; sucrose is produced in a very high state of purity on an enormous scale. The annual production of this food amounts to over 80 million tons per year. Originally produced in India and Persia, sucrose is now used universally as a sweetener. About 40 percent of the world sucrose production comes from sugar beets and 60 percent from sugar cane. Sugar provides a high caloric value (1794 kcal per pound); it is also used as a preservative in jams, jellies, and other foods.

Polysaccharides

There is an almost limitless number of possible structures in which monosaccharide units (monosaccharide molecules minus one water molecule at each bond between units) can be combined. Molecular weights are known to go above 1,000,000 amu. Apparently nature has been very selective in that only a few of the many possible monosaccharide units are found in polysaccharides.

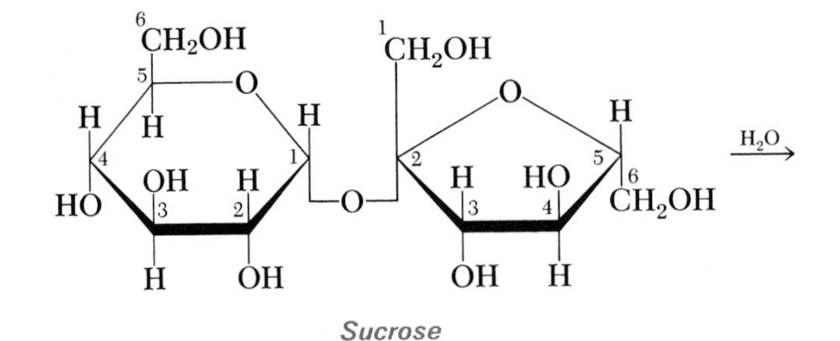

Sucrose

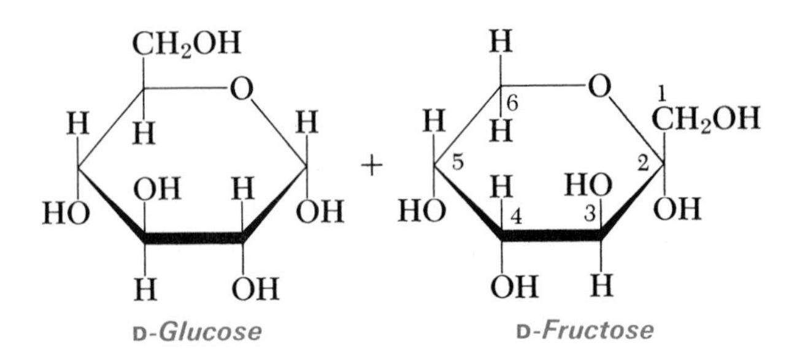

D-*Glucose* D-*Fructose*

(Note: D-Fructose forms a six-membered ring when isolated and a five-membered ring in the sucrose structure.)

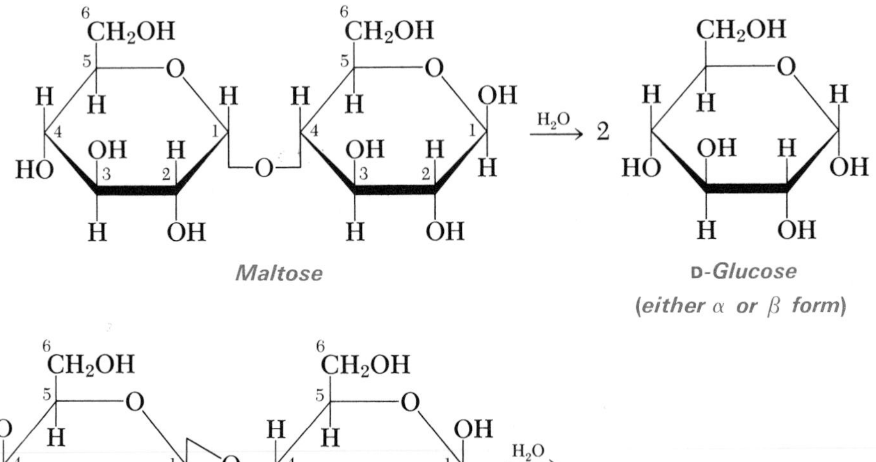

Maltose

D-*Glucose* (*either* α *or* β *form*)

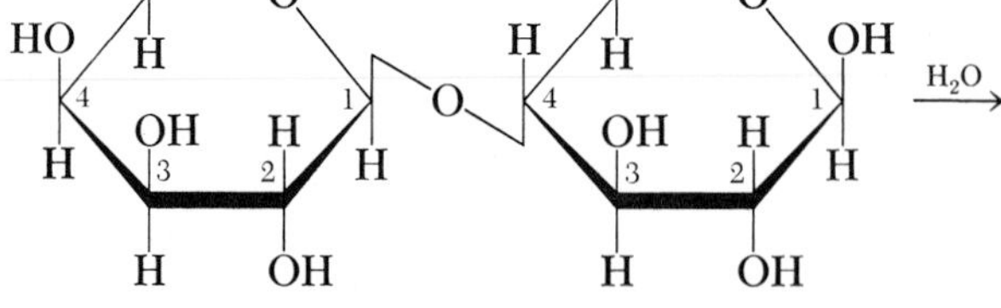

Lactose

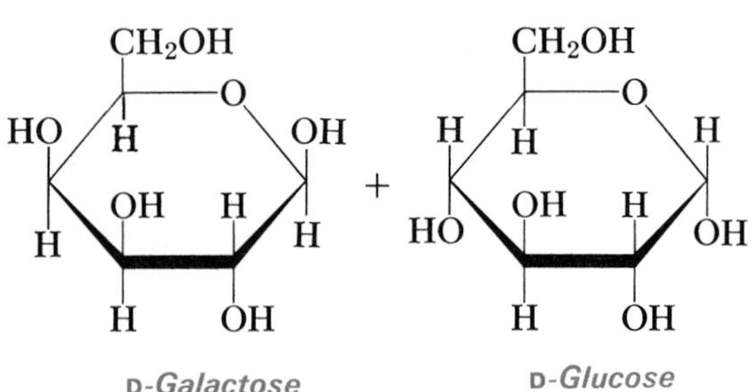

D-*Galactose* D-*Glucose*

Figure 16–3 Hydrolysis of disaccharides (sucrose, maltose, and lactose).

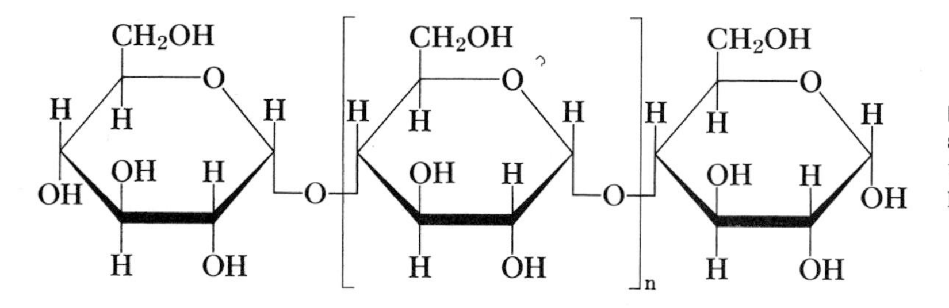

Figure 16–4 Amylose structure. n represents a large number of α-D-glucose units; the value of n determines the type of polysaccharide.

Starches and Glycogen

Starch molecules consist of many glucose units bonded together.

Starch is found in plants in protein-covered granules. These granules are disrupted by heat, and part of the starch content is soluble in hot water. Soluble starch is ***amylose*** and constitutes 22 to 26 per cent of most natural starches; the remainder is ***amylopectin.*** Amylose gives the familiar blue-black starch test with iodine solutions; amylopectin turns red on contact with iodine.

Structurally amylose is a straight chain of α-D-glucose units, each one bonded to the next, just as the two units are bonded in maltose (Figure 16–3). Molecular weight studies on amylose indicate the average chain contains about 200 units. The structure of amylose is illustrated in Figure 16–4.

Amylopectin is made up of branched chains of α-D-glucose units (Figure 16–5). Its molecular weight generally corresponds to about 1000 glucose units. Partial hydrolysis of amylopectin yields mixtures called ***dextrins.*** Complete hydrolysis, of course, yields D-glucose. Dextrins are used as food additives, mucilages, paste, and in finishes for paper and fabrics.

Glycogen serves as an energy reservoir in animals as does starch in plants. Glycogen has essentially the same structure as amylopectin

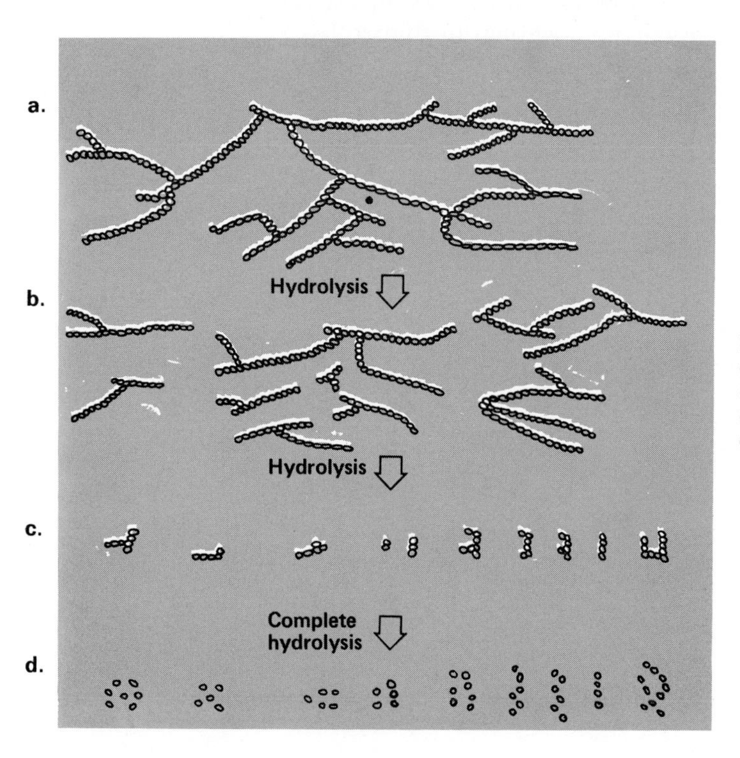

Figure 16–5 (*a*) Partial amylopectin structure. (*b*) Dextrins from incomplete hydrolysis of *a*. (*c*) Oligosaccharides from hydrolysis of dextrins. (*d*) Final hydrolysis product: D-glucose. Each circle represents a glucose unit.

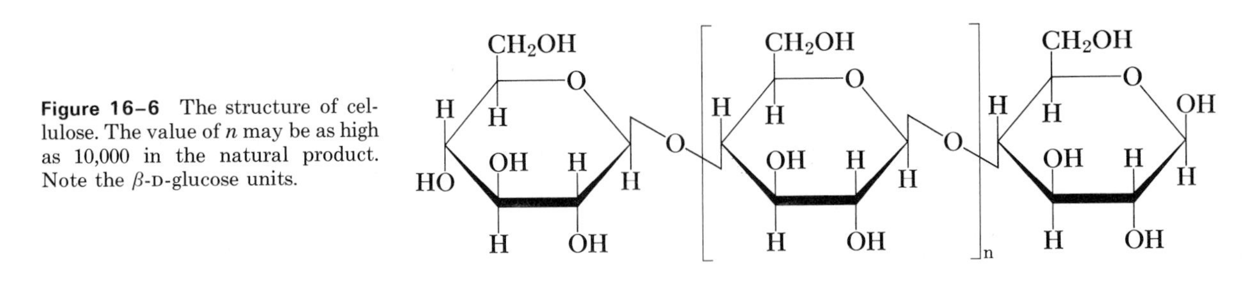

Figure 16–6 The structure of cellulose. The value of n may be as high as 10,000 in the natural product. Note the β-D-glucose units.

(branched chains of glucose units) except that glycogen is even more highly branched.

Cellulose

Cellulose is the most abundant polysaccharide in nature. Like amylose, it is composed of D-glucose units. The difference between the structure of cellulose and that of amylose lies in the bonding between the D-glucose units. In cellulose, all of the glucose units are in the β-ring form in contrast to the α-ring form in amylose. Review the ring forms in Figure 16–1 and compare the structures in Figures 16–4 and 16–6.

Although cellulose consists of a large number of glucose units bonded together, humans do not possess an enzyme capable of splitting it.

The different structures of starch and cellulose account for their difference in digestibility. Human beings and carnivorous animals do not have the necessary enzymes (biochemical catalysts) to break down the cellulose structure as do numerous microorganisms. Cellulose is readily hydrolyzed to D-glucose in the laboratory by heating a suspension of the polysaccharide in the presence of a strong acid. Unfortunately, at present this is not an economically feasible solution to man's growing need for an adequate food supply.

Paper, rayon, cellophane, and cotton are essentially cellulose. A representative portion of the structure of cotton is shown in Figure 16–7. Note the hydrogen bonding between cellulose chains.

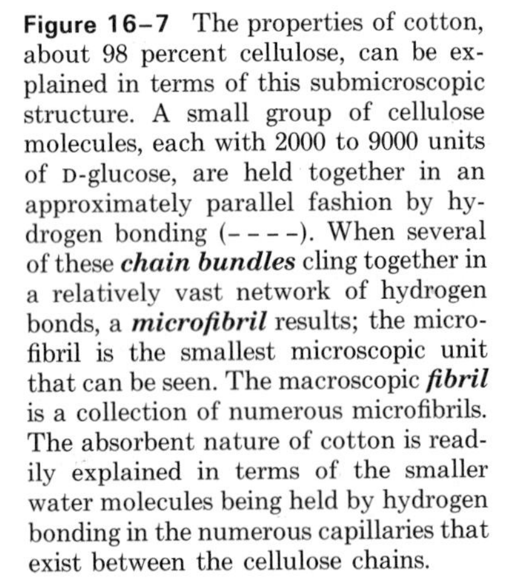

Figure 16–7 The properties of cotton, about 98 percent cellulose, can be explained in terms of this submicroscopic structure. A small group of cellulose molecules, each with 2000 to 9000 units of D-glucose, are held together in an approximately parallel fashion by hydrogen bonding (- - - -). When several of these ***chain bundles*** cling together in a relatively vast network of hydrogen bonds, a ***microfibril*** results; the microfibril is the smallest microscopic unit that can be seen. The macroscopic ***fibril*** is a collection of numerous microfibrils. The absorbent nature of cotton is readily explained in terms of the smaller water molecules being held by hydrogen bonding in the numerous capillaries that exist between the cellulose chains.

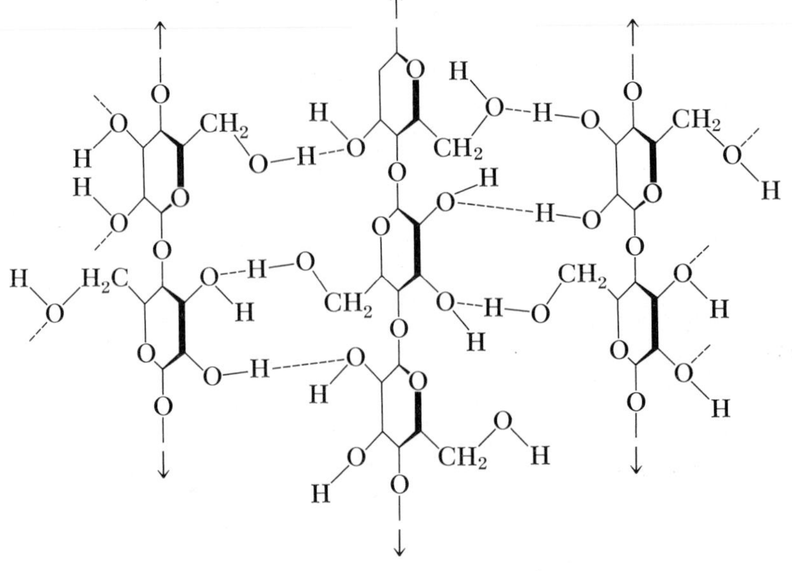

DIETARY FATS AND ESSENTIAL FATTY ACIDS

Saturated fats contain the maximum amount of hydrogen (no C=C double bonds) and are usually solids at room temperature.

Most diets in the United States furnish 40 to 50% of their calories in the form of fats or oils. This is rather high when compared to diets in most other parts of the world. Natural fats and oils are generally mixtures of various esters of glycerol with more than one kind of fatty acid. In our diets most of the fatty acids are ***saturated*** fatty acids. Such fatty acids can be (1) used as a source of energy if the body burns them to CO_2 and H_2O, (2) stored for possible future use in fat cells, or (3) used as starting materials for the synthesis of other compounds needed by the body. Fats are the most concentrated source of food energy in our diets, as they furnish about 9000 calories/g when burned for energy. The human body can make some fats from carbohydrates and carries out such processes to store the excess energy furnished in the diet.

A high intake of dietary fat has been implicated as one of the factors which can give rise to ***atherosclerosis,*** a complex process in which the walls of the arteries suffer damage and ultimately develop scar tissue and fatty deposits. Atherosclerosis is generally considered as a precursor to certain types of heart disease and strokes. Atherosclerosis is also related to the amount of cholesterol in the diet, but the relationship of both dietary fat and cholesterol intake to atherosclerosis is not a simple one.

A high intake of saturated fats has been correlated with atherosclerosis, which leads to heart disease.

It has been known for about 50 years that the human body has a small requirement for certain types of fatty acids (called ***essential fatty acids***) and in recent years the basis for the need for these essential fatty acids has been determined.

The essential fatty acids are ***linoleic, linolenic,*** and ***arachidonic*** acids.

$$\underset{18}{CH_3}\underset{17}{CH_2}\underset{16}{CH_2}\underset{15}{CH_2}\underset{14}{CH_2}\underset{13}{CH}{=}\underset{12}{CH}\underset{11}{CH_2}\underset{10}{CH}{=}\underset{9}{CH}\underset{8}{CH_2}\underset{7}{CH_2}\underset{6}{CH_2}\underset{5}{CH_2}\underset{4}{CH_2}\underset{3}{CH_2}\underset{2}{CH_2}\underset{1}{C}(=O)OH$$

LINOLEIC ACID ($C_{18}\Delta_{9,12}$)

($C_{18}\Delta_{9,12}$ means that there is a chain of 18 carbon atoms with double bonds at carbons 9 and 12.)

$$\underset{18}{CH_3}\underset{17}{CH_2}\underset{16}{CH}{=}\underset{15}{CH}\underset{14}{CH_2}\underset{13}{CH}{=}\underset{12}{CH}\underset{11}{CH_2}\underset{10}{CH}{=}\underset{9}{CH}\underset{8}{CH_2}\underset{7}{CH_2}\underset{6}{CH_2}\underset{5}{CH_2}\underset{4}{CH_2}\underset{3}{CH_2}\underset{2}{CH_2}\underset{1}{C}(=O)OH$$

LINOLENIC ACID ($C_{18}\Delta_{9,12,15}$)

$$\underset{20}{CH_3}\underset{19}{CH_2}\underset{18}{CH_2}\underset{17}{CH_2}\underset{16}{CH_2}\underset{15}{CH}{=}\underset{14}{CH}\underset{13}{CH_2}\underset{12}{CH}{=}\underset{11}{CH}\underset{10}{CH_2}\underset{9}{CH}{=}\underset{8}{CH}\underset{7}{CH_2}\underset{6}{CH}{=}\underset{5}{CH}\underset{4}{CH_2}\underset{3}{CH_2}\underset{2}{CH_2}\underset{1}{C}(=O)OH$$

ARACHIDONIC ACID ($C_{20}\Delta_{5,8,11,14}$)

The presence of any one of these in the diet permits the body to synthesize a very essential group of compounds, the prostaglandins. The key compound here is linoleic acid, which the body cannot make from more saturated fatty acids. If linoleic acid is available, the body can make arachidonic acid and linolenic acid.

Prostaglandins are synthesized from the essential fatty acids. Even in very small amounts, prostaglandins have very powerful effects on the human body.

Prostaglandins are a group of more than a dozen related compounds with very potent effects on physiologic activity such as blood pressure, relaxation and contraction of smooth muscle, gastric acid secretion, body temperature, food intake, and blood platelet aggregation.

Their potential use as drugs is currently under widespread investigation. Two of the prostaglandins which have been characterized are Prostaglandin E_1 (used to induce labor to terminate pregnancy) and Prostaglandin E_2.

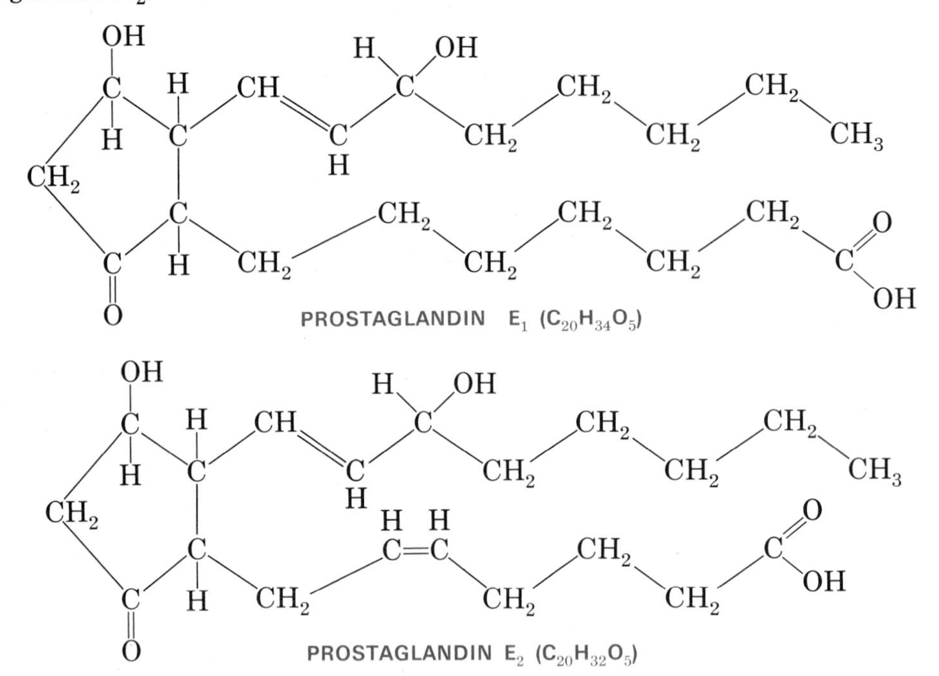

Note that both of these prostaglandins contain exactly the same number of carbon atoms as arachidonic acid.

SELF-TEST 16-A

1. Carbohydrates contain the elements ________, ________, and ________.
2. The complete hydrolysis of a polysaccharide yields a ________ ________.
3. When a molecule of sucrose is hydrolyzed one obtains one molecule of each of the monosaccharides ________ and ________.
4. The sugar referred to as blood sugar, grape sugar, or dextrose is actually the compound ________.
5. Starch is a polymer built up out of ________ units.
6. Essential fatty acids are needed by the body to synthesize ________.

PROTEINS, AMINO ACIDS, AND THE PEPTIDE LINKAGE

Proteins occur in all the major regions of living cells. These compounds serve a wide variety of functions including motion of the organism, defense mechanism against foreign chemicals, metabolic regulation of cellular processes, and cell structure. The close relationship between protein structures and life was first noted by the German chemist, G. T. Mulder, in 1835. He named these compounds proteins from the Greek *proteios,* meaning first, indicating this to be the starting point in the chemical understanding of life.

Proteins are high molecular weight compounds made up of amino acid units.

Proteins are macromolecules with molecular weights ranging from 5000 to several million amu. Like the polysaccharides, these macrostructures are composed of recurring submicroscopic units. The fundamental units in the case of proteins are ***amino acids.*** Proteins and amino acids are made primarily from four elements: carbon, oxygen, hydrogen, and nitrogen. Other elements occur in trace amounts; the one most often encountered is sulfur.

Amino Acids

Amino acids are compounds which generally have the structure

$$\mathrm{R{-}CH(NH_2){-}C({=}O){-}OH}$$

Essential amino acids are amino acids which the body needs but cannot make from other amino acids.

The complete hydrolysis of a typical protein yields a mixture of about 20 different amino acids. Some proteins lack one or more of these acids, others have small amounts of other amino acids characteristic of a given protein, but the 20 given in Table 16–1 are predominant. In a few instances, one amino acid will constitute a major fraction of a protein (the protein in silk, for example, is 44 per cent glycine), but this is not common.

As the name suggests, amino acids contain an amino group (—NH_2) and an acid (carboxyl) group (—COOH). In all of the amino acids listed in Table 16–1, the amine group and the acid group are bonded to the same carbon atom. Of these acids, 18 have the general formula

$$\mathrm{R{-}CH(NH_2){-}C({=}O){-}OH}$$

There are about 20 common amino acids.

where R is a characteristic group for each amino acid. The simplest amino acid is ***glycine,*** in which R is a hydrogen atom:

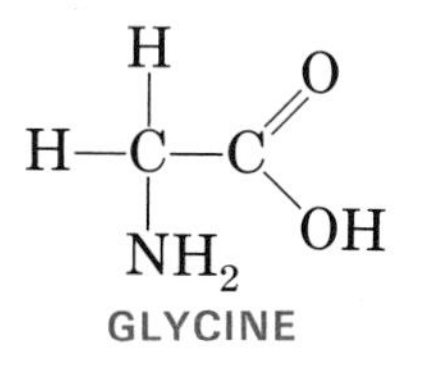

GLYCINE

For good nutrition we require *all* of the essential amino acids in our daily diet, but the amount required does not exceed 1.5 g for any of them.

The human body is capable of synthesizing some amino acids needed for protein structures, but it is unable to provide others necessary for normal growth and development. The latter are designated ***essential amino acids*** and must be ingested in the food supply. The ***nonessential amino acids*** are just as necessary as the essential amino acids for life but can be made from other molecules by the body. The essential amino acids are indicated in Table 16–1.

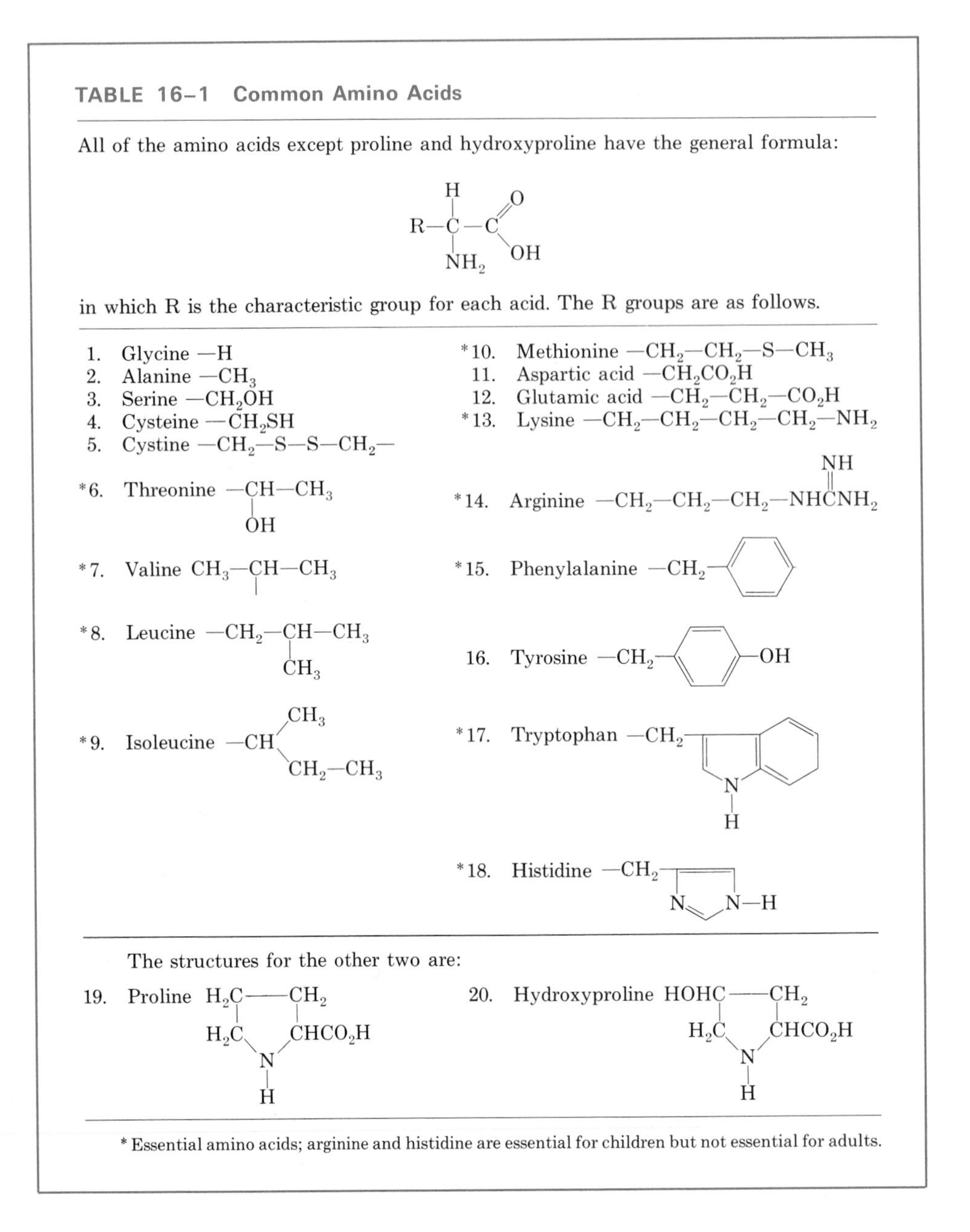

TABLE 16–1 Common Amino Acids

All of the amino acids except proline and hydroxyproline have the general formula:

$$R-\underset{NH_2}{\overset{H}{C}}-C\begin{smallmatrix} \nearrow O \\ \searrow OH \end{smallmatrix}$$

in which R is the characteristic group for each acid. The R groups are as follows.

1. Glycine $-H$
2. Alanine $-CH_3$
3. Serine $-CH_2OH$
4. Cysteine $-CH_2SH$
5. Cystine $-CH_2-S-S-CH_2-$

*6. Threonine $-\underset{OH}{CH}-CH_3$

*7. Valine $CH_3-\underset{|}{CH}-CH_3$

*8. Leucine $-CH_2-\underset{CH_3}{CH}-CH_3$

*9. Isoleucine $-CH\begin{smallmatrix} CH_3 \\ CH_2-CH_3 \end{smallmatrix}$

*10. Methionine $-CH_2-CH_2-S-CH_3$

11. Aspartic acid $-CH_2CO_2H$

12. Glutamic acid $-CH_2-CH_2-CO_2H$

*13. Lysine $-CH_2-CH_2-CH_2-CH_2-NH_2$

*14. Arginine $-CH_2-CH_2-CH_2-NH\overset{NH}{\overset{\|}{C}}NH_2$

*15. Phenylalanine $-CH_2-C_6H_5$ (benzene ring)

16. Tyrosine $-CH_2-C_6H_4-OH$ (benzene ring)

*17. Tryptophan $-CH_2-$ (indole ring, N–H)

*18. Histidine $-CH_2-$ (imidazole ring, N, N–H)

The structures for the other two are:

19. Proline (ring: H_2C-CH_2, H_2C, $CHCO_2H$, N–H)

20. Hydroxyproline (ring: $HOHC-CH_2$, H_2C, $CHCO_2H$, N–H)

* Essential amino acids; arginine and histidine are essential for children but not essential for adults.

The Peptide Linkage

The peptide linkage, $-\overset{O}{\overset{\|}{C}}-\underset{H}{\underset{|}{N}}-$, serves to bind amino acid units together in proteins.

Amino acid units are linked together in protein structures by the peptide linkage. This same linkage was illustrated in nylon 66, in which a carboxylic acid and an amine were condensed to form the polymer and the peptide bond. As it applies to proteins, this type of chemical bond can be understood in terms of the reaction between two glycine molecules.

If the acid group of one glycine molecule reacts with the basic amine group of another, the two are joined through the peptide linkage, and one molecule of water is eliminated for each link formed.

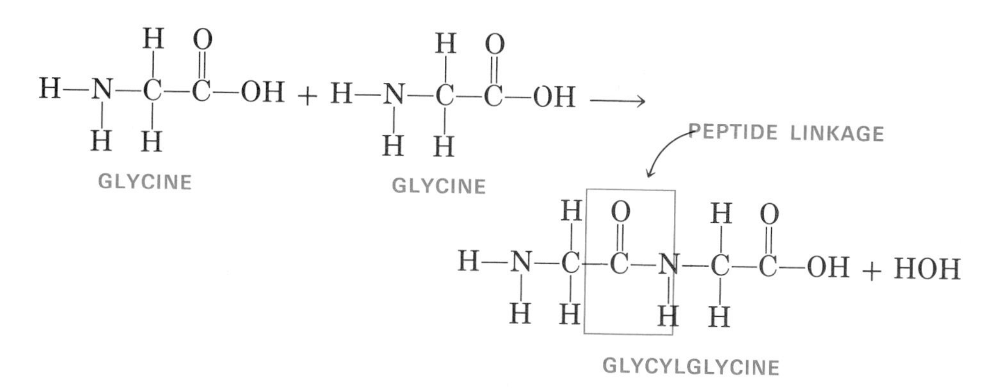

If this hypothetical reaction is carried out with two different amino acids, glycine and alanine, two different ***dipeptides*** are possible.

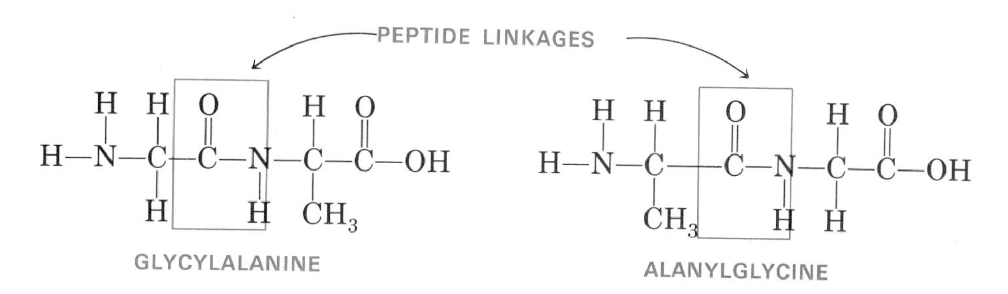

Twenty-four ***tetra***peptides are possible if four amino acids (for example, glycine, Gly; alanine, Ala; serine, Ser; and cystine, Cy) are linked in all possible combinations. They are:

Gly-Ala-Ser-Cy	Ala-Gly-Ser-Cy	Ser-Ala-Gly-Cy	Cy-Ala-Gly-Ser
Gly-Ala-Cy-Ser	Ala-Gly-Cy-Ser	Ser-Ala-Cy-Gly	Cy-Ala-Ser-Gly
Gly-Ser-Ala-Cy	Ala-Ser-Gly-Cy	Ser-Gly-Ala-Cy	Cy-Gly-Ala-Ser
Gly-Ser-Cy-Ala	Ala-Ser-Cy-Gly	Ser-Gly-Cy-Ala	Cy-Gly-Ser-Ala
Gly-Cy-Ser-Ala	Ala-Cy-Gly-Ser	Ser-Cy-Ala-Gly	Cy-Ser-Ala-Gly
Gly-Cy-Ala-Ser	Ala-Cy-Ser-Gly	Ser-Cy-Gly-Ala	Cy-Ser-Gly-Ala

A very large number of different proteins can be prepared from a small number of different amino acids.

If 17 different amino acids are used, the sequences alone would make 3.56×10^{14} uniquely different 17-unit molecules.* Although there are numerous protein structures in nature, these represent an extremely small fraction of the possible structures. Of all the many different proteins which could possibly be made from a set of amino acids, a living cell will make only a relatively small, select number.

Protein Structures

The ***primary structure*** of a protein indicates only the sequence of amino acid units in the polypeptide chain. Since the single bonds in the chain allow free rotation around the bond, it is reasonable to assume that there is an almost infinite number of possible conformations. Because of interactions, such as hydrogen bonding, between atoms in the same chain, certain conformations, called ***secondary structures,*** are favored. Linus

*If the amino acids are all different, the number of arrangements is n! (read n factorial). For five different amino acids, the number of different arrangements is 5! (or $5 \times 4 \times 3 \times 2 \times 1 = 120$).

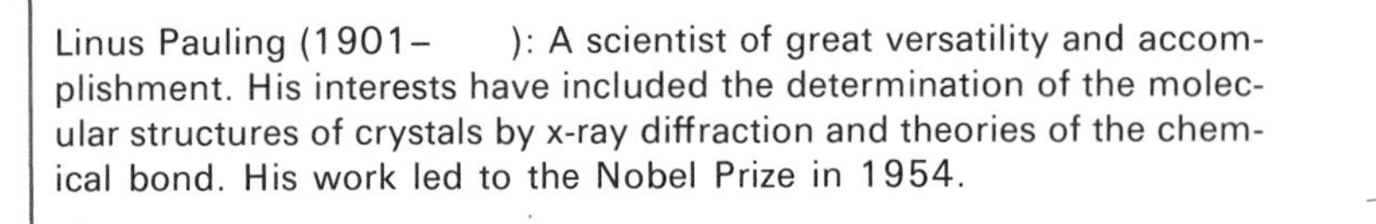

Linus Pauling (1901–): A scientist of great versatility and accomplishment. His interests have included the determination of the molecular structures of crystals by x-ray diffraction and theories of the chemical bond. His work led to the Nobel Prize in 1954.

Pauling, along with R. B. Corey, suggested the two secondary structures for polypeptides discussed below, the sheet structure and the helical structure.

Polyglycine is the synthetic protein made entirely of the amino acid

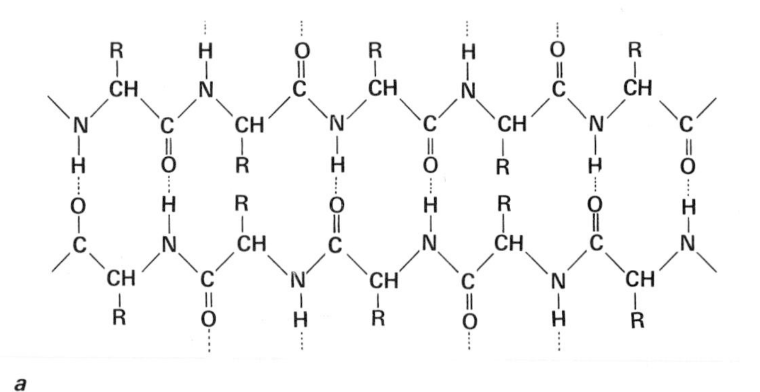

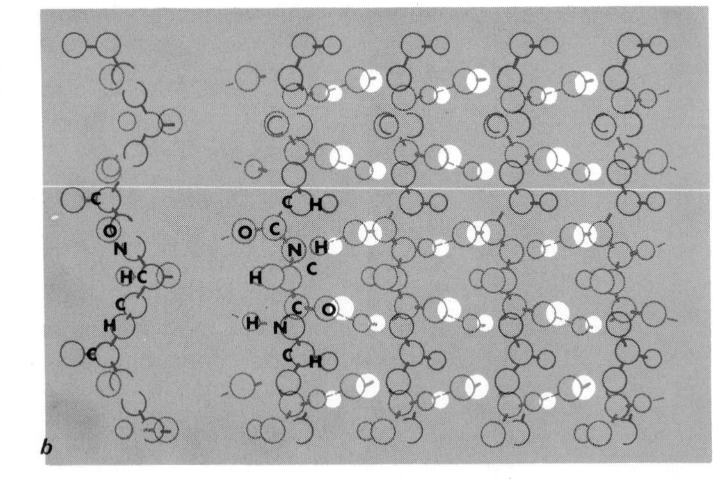

Figure 16–8 Sheet structure for polypeptide. In (*a*) the two-dimensional drawing emphasizes that all of the oxygen and nitrogen atoms are involved in hydrogen bonding for the most stable structure. (*b*) Illustrates the bonding in perspective showing that the sheet is not flat; rather, it is sometimes called a pleated sheet structure.

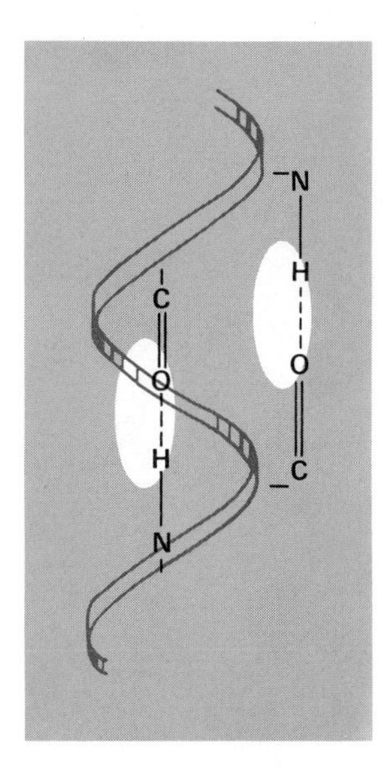

Figure 16–9 Helix structure for a polypeptide in which each oxygen atom can be hydrogen bonded to a nitrogen atom in the third amino acid unit down the chain.

glycine. In polyglycine the hydrogen attached to the nitrogen atom and the oxygen bonded to the carbon are both well suited to engage in hydrogen bonding. In the two stable conformations of polyglycine, maximum advantage is taken of the hydrogen bonding available. In the sheet structure, the hydrogen bonding is between adjacent chains of the polypeptide; in the helical structure, hydrogen bonding occurs between atoms within the same chain.

The amino acids in a protein chain interact with each other via hydrogen bonding.

Figure 16–8 illustrates a sheetlike structure in which several chains of the polypeptide are joined by hydrogen bonding. Note that all the oxygen and nitrogen atoms are involved in hydrogen bonding. Most of the properties of silk can be explained in terms of this type of structure for fibroin, the protein of silk.

Hydrogen bonding is possible within a single polypeptide chain if the secondary structure is helical (Figure 16–9). Bond angles and bond lengths are such that the nitrogen atom forms hydrogen bonds with the oxygen atom in the third amino acid unit down the chain (Figure 16–10).

A coiled spring is helical in structure.

Collagen is the principal fibrous protein in mammalian tissue. It has remarkable tensile strength which makes it useful in structuring bones, tendons, teeth, and cartilage. Three polypeptide chains, each of which is twisted into a left-handed helix, are twisted into a right-handed super helix to form an extremely strong fibril, as shown in Figure 16–11. A bundle of such fibrils forms the macroscopic protein.

The structure of collagen illustrates a third level of protein structure, ***tertiary structure.*** The primary structure is the sequence of amino acids in the protein, the secondary structure is the helical form of the protein chain, and the tertiary structure is the twisted or folded form of the helix. Another tertiary structure is found in globular proteins. In these structures, the helix chain (secondary structure) is folded and twisted into a definite geometric pattern. This pattern may be held in place by one or more of several different kinds of chemical bonds, such as —S—S— bonds, depending on the particular functional groups in the amino acids involved (Table 16–1). Figure 16–12 illustrates the folded structure of a typical globular protein. Abnormal hemoglobin structures are unable to transfer oxygen in the blood if the wrong amino acid is in a given position

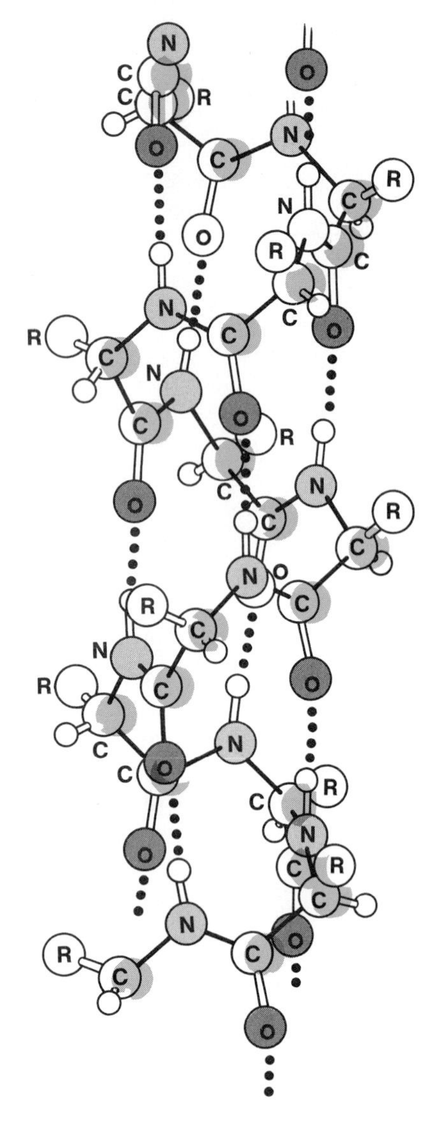

Figure 16–10 α-Helix structure of proteins. The sketch represents the actual position of the atoms and shows where intra-chain hydrogen bonding occurs.

in the polypeptide structure (Figure 16–13). Such genetic mistakes disrupt the secondary and tertiary structures of proteins. An important class of globular proteins is the ***enzymes,*** molecules which function as catalysts for reactions in living cells. These are discussed in detail in the next chapter.

Enzymes are protein molecules that speed up chemical reactions.

Figure 16–11 The structure of collagen.

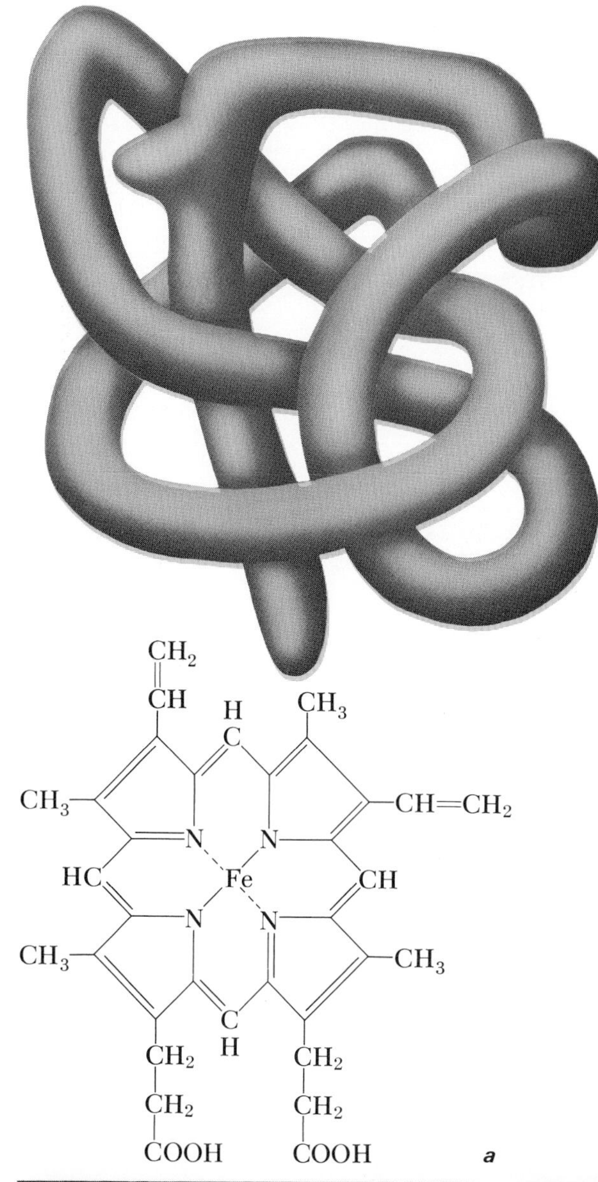

Figure 16–12 Typical folded structure of the helix in a globular protein.

Figure 16–13 (*a*) The structure of heme; (*b*) a model, from two views, of the hemoglobin structure. The heme structures are indicated by disks. (Courtesy of M. F. Perutz and *Science,* 140:863, 1963.)

SELF-TEST 16-B.

1. The fundamental building units in proteins are the ______________________.

2. Amino acids which the body cannot synthesize from other molecules are called ______________________.

3. The peptide linkage which bonds amino acids together in protein chains has the structure

4. The basic structure present in almost all of the amino acids can be represented as

5. Polypeptides which are naturally occurring catalysts are called __________.

6. The formula for glycylglycine is:

7. a. The primary structure of a protein refers to its ______________________;

   b. the secondary structure refers to its ______________________;

   c. while its tertiary structure refers to ______________________.

8. a. If we have three different amino acids, we can make a total of ______ different tripeptides from them if we can use an amino acid up to three times in any given tripeptide.

   b. If we can use each amino acid only once, there are still ______ possible different tripeptides.

NUCLEIC ACIDS

Like the polysaccharides and the polypeptides, the ***nucleic acids*** are high molecular weight substances, with molecular weights up to several million. Nucleic acids are found in all living cells, with the exception of the red blood cells of mammals. The structures of these compounds are believed to be directly related not only to the characteristics of the individual cell but of the gross organism itself. The almost infinite variety of possible structures for nucleic acids allows information in coded form to be recorded in molecular complexes in a somewhat similar fashion to the way a few language symbols can be used to convey the many ideas in this book. Such stored information controls the inherited characteristics of the next generation as well as many of the ongoing life processes of the organism.

Nucleic acids contain the coded information that tells cells which molecules to synthesize.

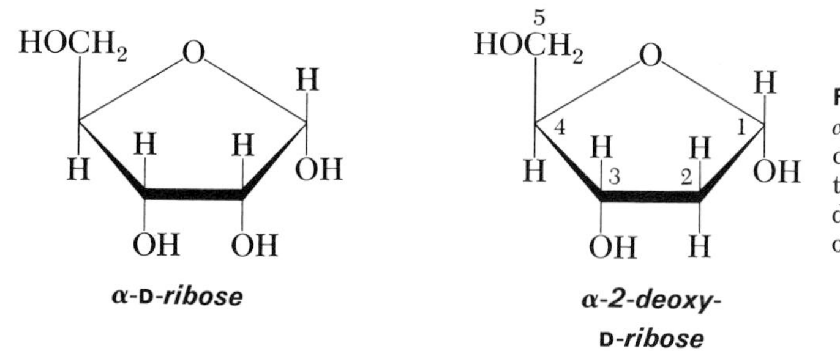

Figure 16–14 The structure of α-D-ribose and α-2-deoxy-D-ribose. In the IUPAC names given, α indicates the one of two ring-forms possible; D distinguishes the isomers that rotate plane polarized light in opposite directions, and the 2 indicates the carbon to which no oxygen is attached in the second sugar.

Hydrolysis of nucleic acids yields one of two simple sugars, phosphoric acid (H_3PO_4), and a group of nitrogen compounds that have basic (alkaline) properties. Depending upon these hydrolysis products, the nucleic acids can be classified as either those that contain the sugar ***α-2-deoxy-D-ribose,*** or those that contain ***α-D-ribose.*** The former are called ***deoxyribonucleic acids*** (DNA) and the latter ***ribonucleic acids*** (RNA). DNA is found primarily in the nucleus of the cell, whereas RNA is found mainly in the cytoplasm outside of the nucleus.

Ribose and Deoxyribose

The structures for the two sugars in nucleic acids are shown in Figure 16–14. The names and formulas for the basic nitrogen compounds in nucleic acids are given in Figure 16–15.

One nucleotide is joined to another by an ester-forming reaction.

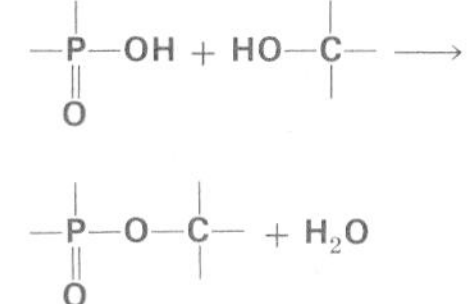

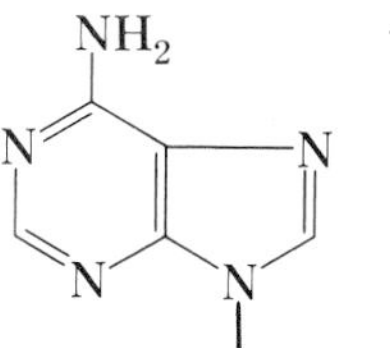

Nucleotides

Incomplete hydrolysis of DNA or RNA yields ***nucleotides.*** These substances contain a simple sugar unit, one of the nitrogenous base units, and one or two units of phosphoric acid. An example of the nucleotide structure is illustrated by inosinic acid, Figure 16–16.

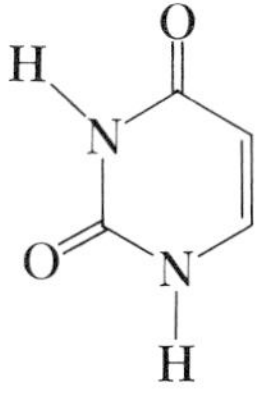
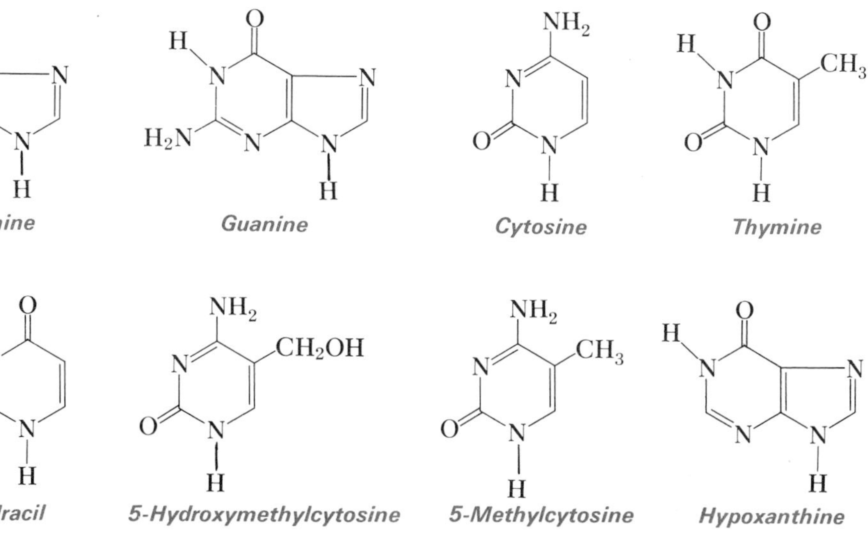

Figure 16–15 Nitrogenous bases obtained from the hydrolysis of nucleic acids.

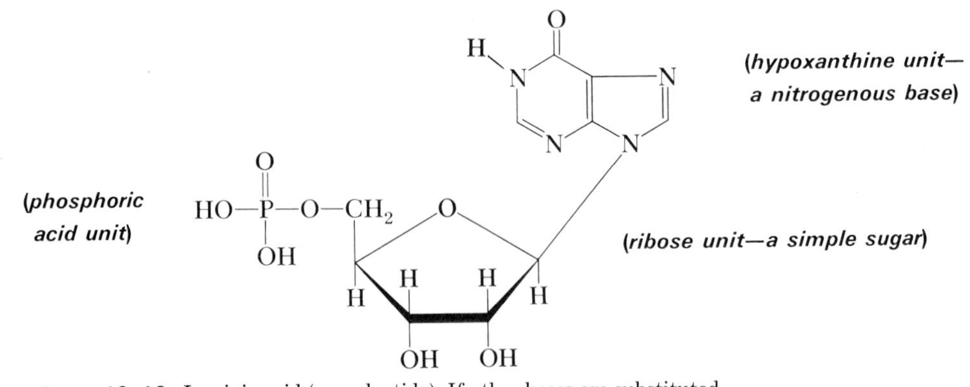

Figure 16–16 Inosinic acid (a nucleotide). If other bases are substituted for hypoxanthine, a number of nucleotides are possible for each of the two sugars. There is ample evidence that the nucleotides found in both DNA and RNA have the general structure indicated for inosinic acid.

Polynucleotides

In addition to the mononucleotides, partial hydrolysis of DNA or RNA yields oligonucleotides which have a few nucleotide units in their molecular structure. The structure for a trinucleotide is illustrated in Figure 16–17. Obviously, a large number of oligonucleotides is possible

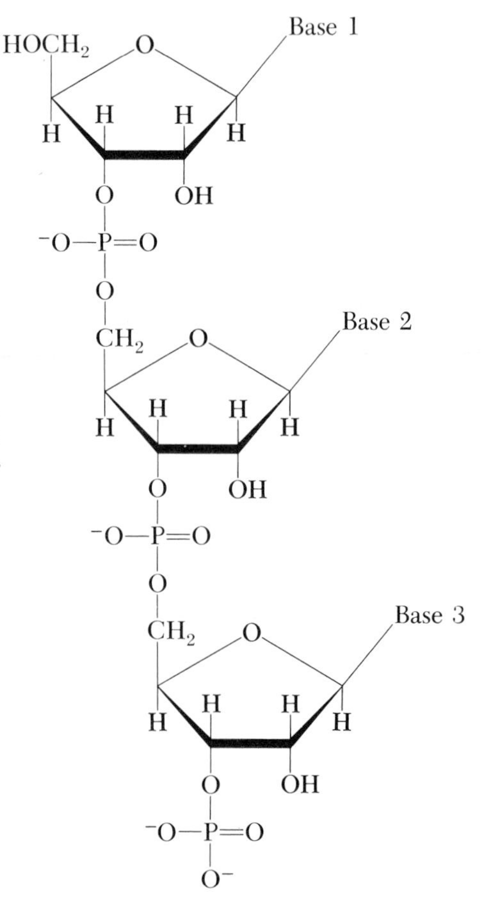

Figure 16–17 Bonding structure of a trinucleotide. Bases 1, 2, and 3 represent any of the nitrogenous bases obtained in the hydrolysis of DNA and RNA. The primary structure of both DNA and RNA is an extension of this structure to produce molecular weights as high as a few million amu.

when one considers the choice of base structures and the different sequence possibilities for the chain of nucleotides.

DNA and RNA are polynucleotides. The number of possible structures for these molecules, which have molecular weights as high as a few million amu, appears to be almost limitless. Since DNA is a major part of the chromosome material in the nucleus of a cell, it seems reasonable to assume that the organism's characteristics are coded in the DNA structure. It has been estimated that there are over two million different species of organisms. Even if each individual of each species requires a different DNA structure, there are ample combinations of nucleotides for each individual to be unique. It is now believed that some kinds of RNA transfer the information coded in the DNA structure to the cytoplasmic region of the cell, where they control the thousands of reactions that occur.

The heredity of an organism is contained in DNA molecules.

Three major types of RNA have been identified. They are messenger RNA (mRNA), transfer RNA (tRNA), and ribosomal RNA (rRNA). Each has a characteristic molecular weight and base composition. Messenger RNA's are generally the largest, with molecular weights between 25,000 and one million amu. They contain from 75 to 3000 mononucleotide units. Transfer RNA's have molecular weights in the range of 23,000 to 30,000 amu and contain 75 to 90 mononucleotide units. Ribosomal RNA's have molecular weights between those of mRNA's and tRNA's and make up as much as 80 percent of the total cell RNA.

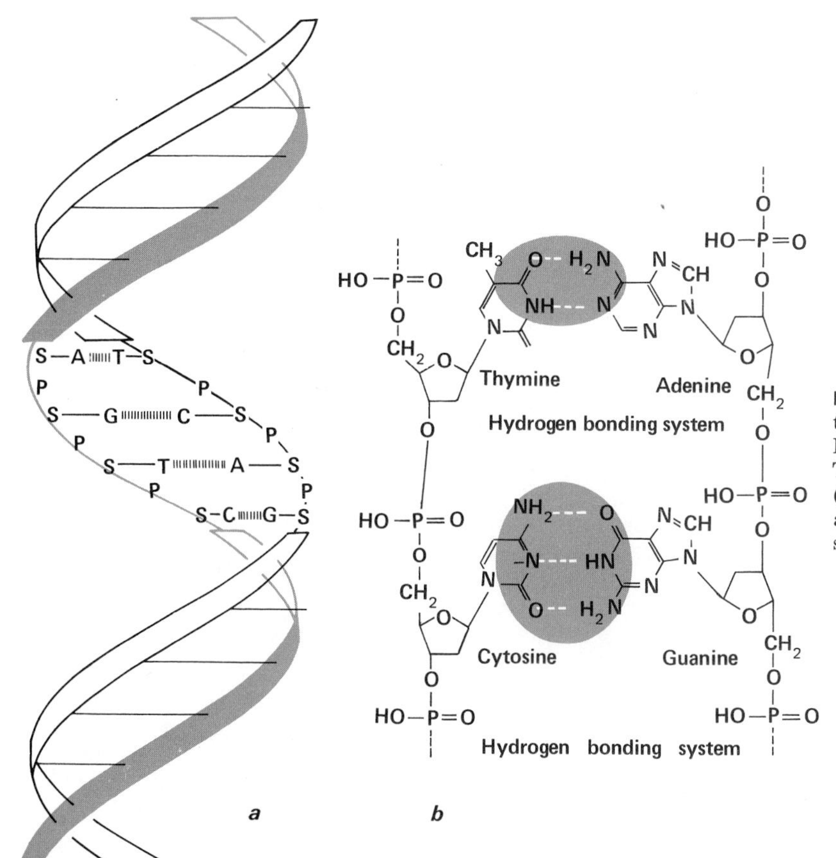

Figure 16–18 (*a*) Double helix structure proposed by Watson and Crick for DNA. S-sugar, P-phosphate, A-adenine, T-thymine, G-guanine, C-cytosine. (*b*) Hydrogen bonds in the thymine—adenine and cytosine—guanine pairs stabilize the double helix.

Most RNA is found in the cytoplasm and ribosomes of the cell (Figure 16–22), but in liver cells as much as 11 percent (largely mRNA) of the total cell RNA is found in the nucleus. Besides having different molecular weights, the three types of RNA appear to differ in function. One difference in function is described later in this chapter in the discussion of natural protein synthesis.

Secondary Structure of DNA and RNA

In 1953, J. D. Watson and F. H. C. Crick (Figure 16–20) proposed a secondary structure for DNA that has since gained wide acceptance. Figure 16–18 illustrates the proposed structure in which two polynucleotides are arranged in a double helix stabilized by hydrogen bonding between the base groups opposite to each other in the two chains. RNA is generally a single strand of helical polynucleotide.

Virus Structure

A virus is a parasitic chemical complex that can reproduce only when it has invaded a host cell. It has the ability to disrupt the life processes of the host cell and order the cell contents therein to reproduce the virus structure. The isolated virus unit has neither the enzymes nor the smaller molecules necessary to reproduce itself alone.

A virus is a polynucleotide surrounded by a layer of protein. One virus that has been studied in detail is the tobacco mosaic virus, illustrated in Figure 16–19.

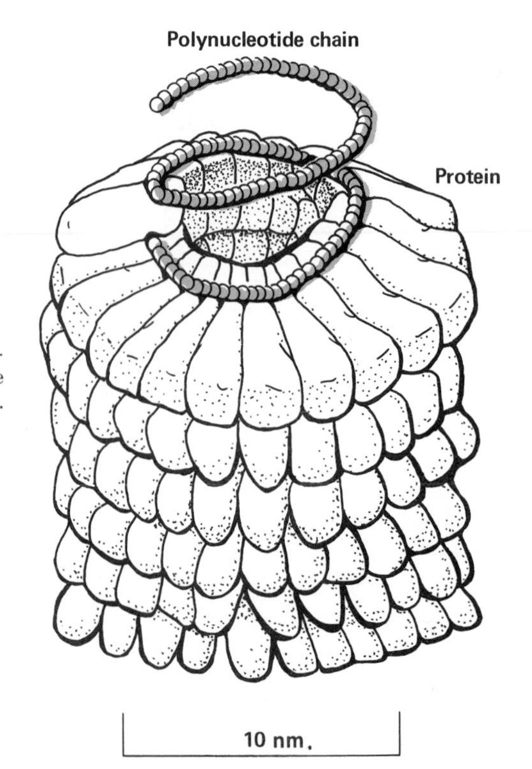

Figure 16–19 The structure of the tobacco mosaic virus. The polynucleotide chain is coiled, and there is one protein unit for each three nucleotide units. Part of the polynucleotide chain is exposed for clarity. This structure is based on x-ray studies.

SYNTHESIS OF LIVING SYSTEMS

Our current knowledge is far from sufficient to allow the synthesis of a living cell.

In the quest for a molecular understanding of living systems, theories must finally be put to the ultimate test of synthesis. If some of the complex molecules described above can be synthesized and can successfully participate in the life processes, we can be reassured that we are on the right track. It should be emphasized that the biochemist is presently working at the molecular level, and as yet only a relatively few of the giant molecules have been characterized. The syntheses in a living cell are far too complex for our present methods to duplicate. However, in spite of the enormity of this undertaking, remarkable strides have been made recently, and interest in current research in this area is intense.

Replication of DNA

The DNA molecule is capable of causing the synthesis of its duplicate.

To make a replicate is to make a complement (something that fits) of the original.

Replicates:
object—cast of object
screw—hole for screw
bolt—bolt hole
adenine—thymine

Almost all of the cells of one organism contain the same chromosome structure in their nuclei. This structure remains constant regardless of whether the cell is starving or has an ample supply of food materials. Each organism begins life as a single cell with this same chromosome structure; in sexual reproduction one-half of this structure comes from each parent. These well known biological facts, along with recent discoveries concerning polynucleotide structures, lead to the conclusion that the DNA structure is faithfully copied during normal cell division (mitosis—both strands) and is only partly copied in cell division producing reproductive cells (meiosis—only one strand).

A prominent theory of DNA replication, based on verifiable experimental facts, suggests that the double helix of the DNA structure unwinds and each half of the structure serves as a template or pattern to reproduce the other half from the molecules in the cell environment (Figure 16–21). The location of the DNA in a typical cell is shown in Figure 16–22.

Figure 16–20 F. H. C. Crick (1916–) (*right*) and J. D. Watson (1928–) (*left*), working in the Cavendish Laboratory at Cambridge, built scale models of the double helical structure of DNA based on the x-ray data of M. H. F. Wilkins. Knowing distances and angles between atoms, they compared the task to the working of a three-dimensional jigsaw puzzle. Watson, Crick, and Wilkins received the Nobel Prize in 1962 for their work relating to the structure of DNA.

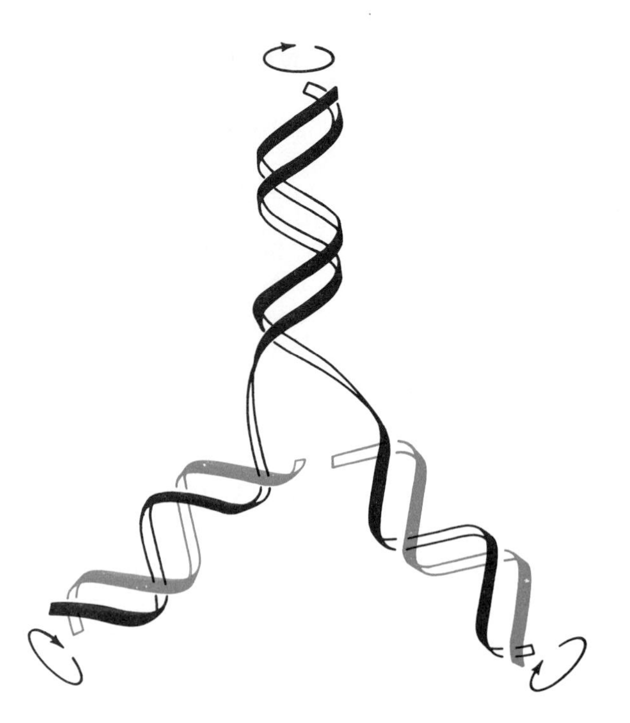

Figure 16–21 Replication of DNA structure. When the double helix of DNA (black) unwinds, each half serves as a template on which to assemble subunits (color) from the cell environment.

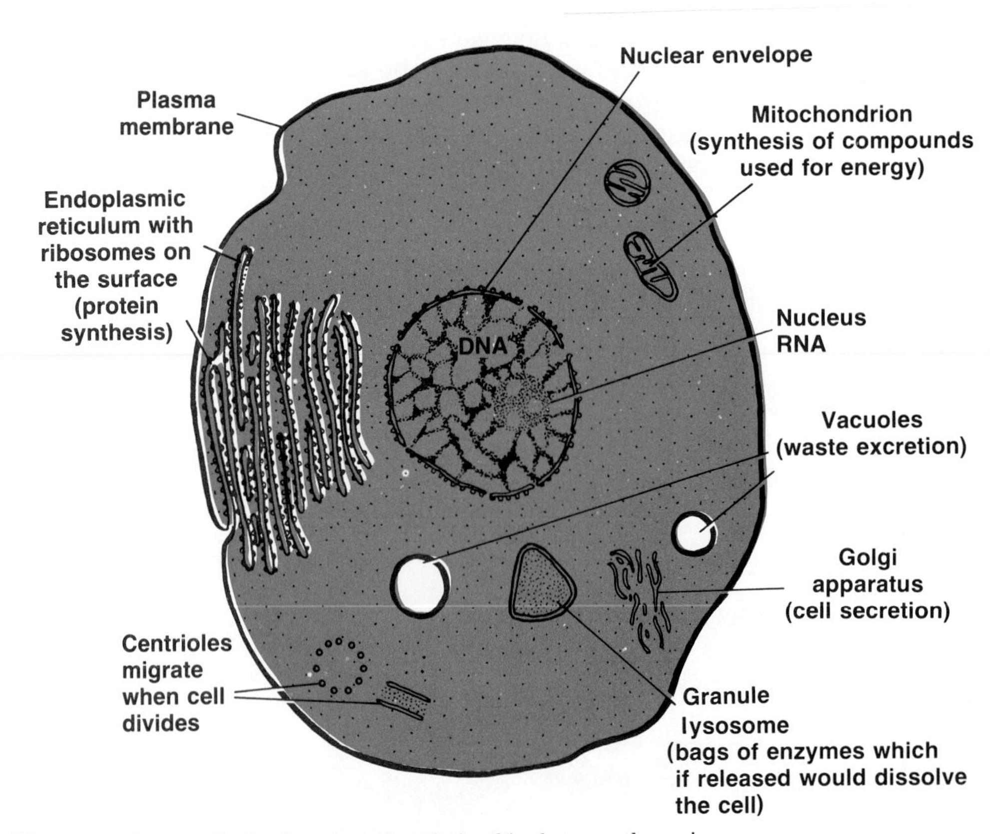

Figure 16–22 Diagrammatic generalized cell to show the relationships between the various components of the cell.

Natural Protein Synthesis

The proteins of the body are being replaced and resynthesized continuously from the amino acids available to the body. The amino acids and proteins in the body can be considered as constituents of a "nitrogen pool"; additions to and losses from the pool are shown in Figure 16–23.

The proteins in the human body are being continuously replaced.

The use of isotopically labeled amino acids has made possible studies on the average lifetimes of amino acids as constituents in proteins—that is, the time it takes the body to replace a protein in a tissue. For a process that must be extremely complex, it is very rapid. Only minutes after radioactive amino acids are injected into animals, radioactive protein can be found. Although all the proteins in the body are continually being replaced, the rates of replacement have been found to vary. Half of the proteins in the liver and plasma are replaced in *six days*. The time is longer for muscle proteins, about 180 days, and replacement of protein in other tissues, such as bone collagen, takes even longer.

Recall that each organism has its own kinds of proteins. The number of possible arrangements of 20 amino acid units is 2.43×10^{18}, yet proteins characteristic of a given organism can be synthesized in a matter of a few minutes. It should come as no surprise, then, that a vast amount of research has been devoted to this problem in recent years. Although it is thought that the general scheme of protein synthesis is now understood, you should realize that many of the details are still unknown.

The DNA molecule tells the cell what kind of protein to synthesize.

The DNA in the cell nucleus holds the code for protein synthesis. Messenger RNA, like all forms of RNA, is synthesized in the cell nucleus. The sequence of bases in one strand of the chromosomal DNA serves as the template for monoribonucleotides to order themselves into a single strand of mRNA (Figure 16–24). The bases of the mRNA strand complement those of the DNA strand. A pair of complementary bases is so structured that each one fits the other and forms one or more hydrogen bonds. Messenger RNA contains only the four bases adenine (A), guanine (G), cytosine (C), and uracil (U). DNA contains principally the four bases adenine (A), guanine (G), cytosine (C), and thymine (T). The base pairs are as follows:

DNA	mRNA
A	U
G	C
C	G
T	A

Amino Acids from Foods
↓
Products of Amino Acid Metabolism ⟵ Nitrogen Pool ⟶ Urinary Nitrogen
↑↓
Tissue Protein

Figure 16–23 The nitrogen pool.

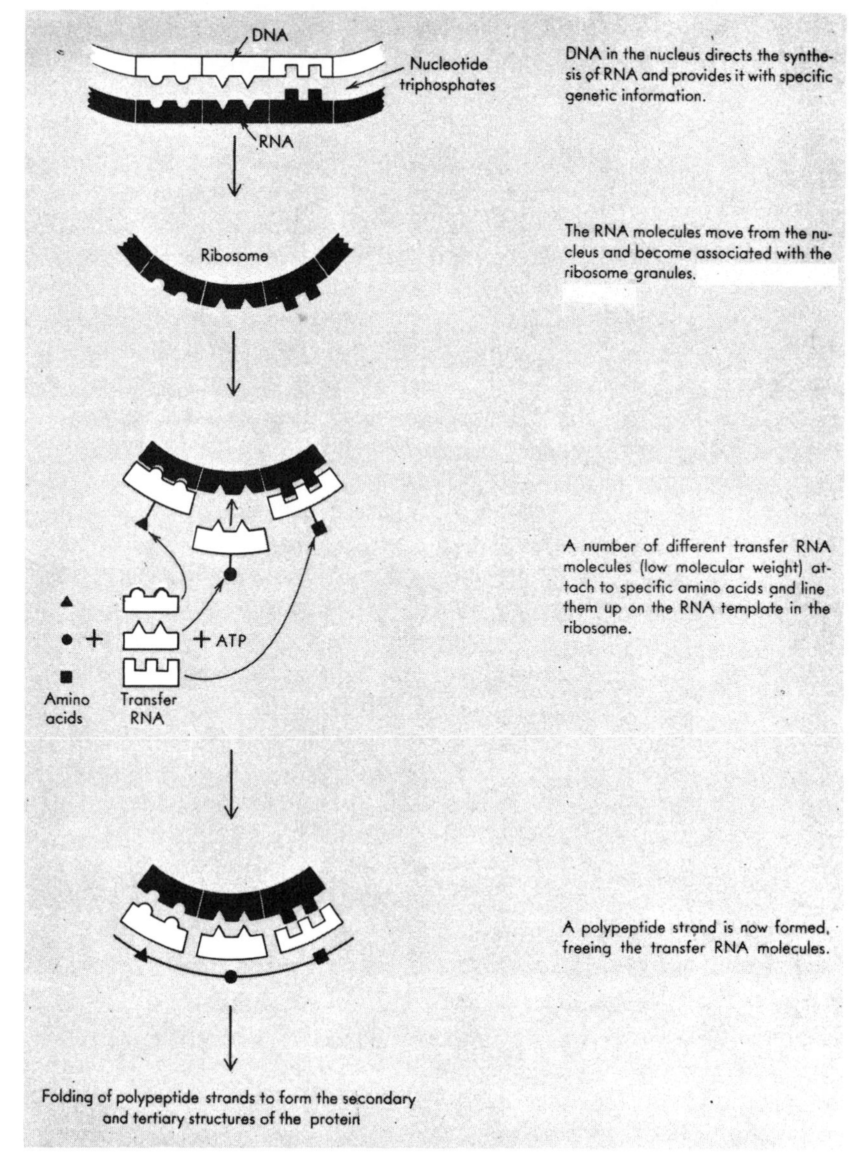
DNA
Nucleotide triphosphates
RNA
DNA in the nucleus directs the synthesis of RNA and provides it with specific genetic information.
Ribosome
The RNA molecules move from the nucleus and become associated with the ribosome granules.
+ ATP
Amino acids
Transfer RNA
A number of different transfer RNA molecules (low molecular weight) attach to specific amino acids and line them up on the RNA template in the ribosome.
A polypeptide strand is now formed, freeing the transfer RNA molecules.
Folding of polypeptide strands to form the secondary and tertiary structures of the protein

Figure 16–24 A schematic illustration of the role of DNA and RNA in protein synthesis.

This means that every place a DNA has an adenine base (A), the mRNA will transcribe a uracil base (U), and so on, provided the necessary enzymes are present.

After transcription, mRNA passes from the nucleus of the cell to a ribosome, where it serves as the template for the sequential ordering of amino acids during protein synthesis. As its name implies, messenger RNA contains the sequence message for ordering amino acids into proteins. Each of the thousands of different proteins synthesized by cells is coded by a specific mRNA or segment of a mRNA molecule.

Transfer RNA carries specific amino acids to the messenger RNA.

Transfer RNA's carry the specific amino acids to the messenger RNA. Each of the 20 amino acids found in proteins has at least one corresponding tRNA, and some have multiple tRNA's. For example, there are five distinctly different tRNA molecules specifically for the transfer of the amino acid leucine in cells of the bacterium *Escherichia coli.* At one end of a tRNA molecule is a trinucleotide base sequence that fits a trinucleotide base sequence on mRNA. At the other end of a tRNA molecule is a specific base sequence of three terminal nucleotides—CCA—with a hydroxyl group exposed on the terminal adenine nucleotide group. This hydroxyl group reacts with a specific amino acid by an esterification reaction with the aid of enzymes.

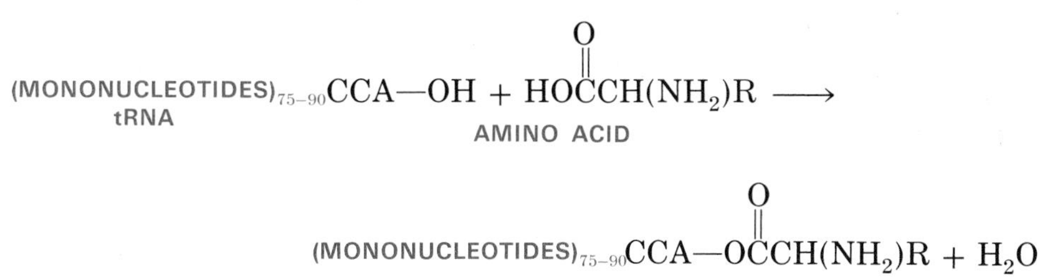

$$\underset{\text{tRNA}}{(\text{MONONUCLEOTIDES})_{75-90}\text{CCA—OH}} + \underset{\text{AMINO ACID}}{\text{HO}\overset{\text{O}}{\overset{\|}{\text{C}}}\text{CH(NH}_2\text{)R}} \longrightarrow$$

$$\underset{\text{tRNA-AMINO ACID}}{(\text{MONONUCLEOTIDES})_{75-90}\text{CCA—O}\overset{\text{O}}{\overset{\|}{\text{C}}}\text{CH(NH}_2\text{)R}} + H_2O$$

As described in more detail in the next chapter, the P—O bonds of adenosine triphosphate (ATP) furnish energy for carrying out an enormous number of reactions in living systems. ATP provides the energy for the reaction between an amino acid and its transfer RNA. A molecule of ATP first activates an amino acid.

$$\text{ATP} + \text{Amino acid} \longrightarrow \text{ATP—amino acid-activated species}$$

The activated complex then reacts with a specific transfer RNA and forms the products shown in Figure 16–25.

The ribosome is the part of the cell in which protein synthesis takes place.

The transfer RNA and its amino acid migrate to the ribosome where the amino acid is given up in the formation of a polypeptide. The transfer RNA is then free to migrate back and repeat the process.

Messenger RNA is used only once, or at most a few times, before it is depolymerized. While this may seem to be a terrible waste, it allows the cell to produce different proteins on very short notice. As conditions change, a different type of messenger RNA comes from the nucleus, a different protein is made, and the cell adequately responds to a changing environment.

Synthetic Protein

The proteins in living things are not a random sequence of amino acids.

A mixture of amino acids can react to form polypeptide structures outside of a living cell. Only relatively simple catalytic agents are required, and the complex enzymatic environment of cytoplasm is not

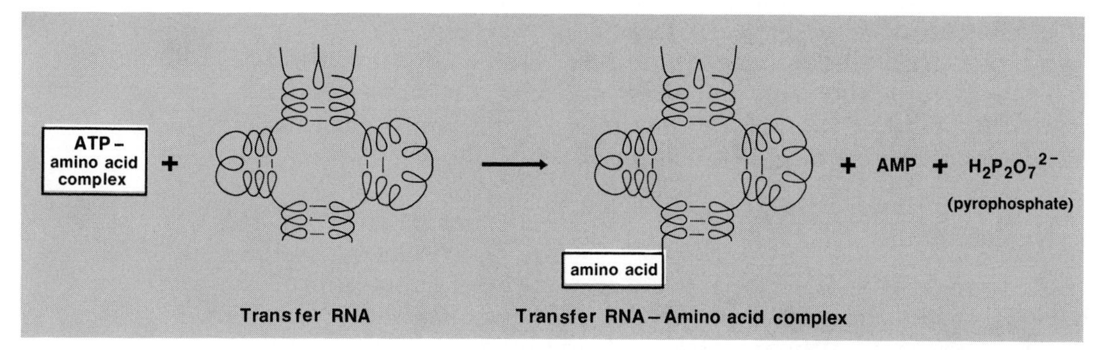

Figure 16–25 Bonding of activated amino acid to transfer RNA. AMP is adenosine monophosphate (see Chapter 17).

necessary. However, a mixture of only a few amino acids will result in a multitude of different protein structures. The synthesis of a particular sequence of amino acids presents many problems.

Methods too complex to be detailed here have been devised to construct protein structures with a desired sequence of amino acids. In order to obtain the desired bond between an amino acid and the peptide already constructed, it is necessary to block all the functional groups except those which are to undergo the peptide reaction. With the chemical blocking groups in place, the particular amino acid is added and the peptide bond is made. This is followed by removal of the blocking groups. To build a polypeptide with 20 amino acid units, this complicated process would have to be repeated 19 times. As the peptide chain grows, it becomes increasingly difficult to carry out the chemical operations without disturbing the bonds previously formed.

In spite of the difficulties, the customized synthesis of a prescribed protein has progressed steadily. In 1953, Vincent du Vigneaud synthesized a hormone from 9 amino acids. For his work he received a Nobel Prize in 1955. Other, more lengthy molecules have been synthesized similarly by starting with the individual amino acids and adding them in proper sequence to make the desired protein. Notable hormone syntheses were β-corticotropin with 39 amino acid units, and insulin, with 51 units. In 1969, two teams of researchers synthesized the first enzyme, ribonuclease (Figure 16–26), to be assembled outside the living cell from individual amino acids. Ribonuclease contains 124 amino acid units. One team was led by Rockefeller University's Robert B. Merrifield and Bernd Gutte, and the other team at Merck Sharp & Dohme research laboratories was led by Robert G. Denkewalter and Ralph F. Hirschmann.

Many large protein molecules have now been made in the laboratory.

Merrifield's method has had wider application. In this automated technique, an insoluble solid support, polystyrene, acts as an anchor for the peptide chain during the synthesis (Figure 16–27). The first amino acid is firmly bonded to a small polystyrene bead, and each of the other amino acids is then added one at a time in a stepwise manner. The synthesis of ribonuclease required 369 chemical reactions and 11,931 steps in a continuous operation on a machine developed for this purpose.

In 1970, a group led by Choah Hao Li at the University of California used Merrifield's method to synthesize human growth hormone (HGH) (Figure 16–28). HGH has 188 amino acid units and a molecular weight of about 21,500 amu. This hormone is produced naturally by the front lobe of the pituitary gland, a pea-sized body located at the base of the brain. In humans, it has a dual function—it controls milk formation and it regulates many aspects of growth. In childhood, excess secretions of HGH can cause giantism; too little HGH causes dwarfism.

These synthetic proteins are identical in every respect to the natural products. They perform the same functions and give the same results on analysis. The limits of the present methods are unknown; in time even longer molecules are expected to be duplicated.

Synthetic Nucleic Acids

Progress in the synthesis of polynucleotides has been difficult, principally because of the difficulties involved in determining the proper blocking groups. Progress, although slow, is being made.

In 1959, Arthur Kornberg synthesized a DNA-type polynucleotide,

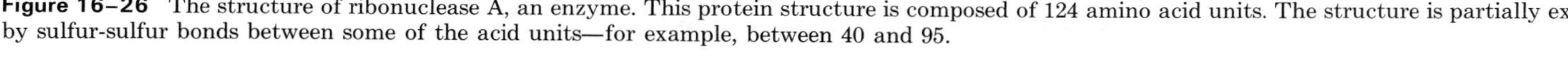

Figure 16–26 The structure of ribonuclease A, an enzyme. This protein structure is composed of 124 amino acid units. The structure is partially explained by sulfur-sulfur bonds between some of the acid units—for example, between 40 and 95.

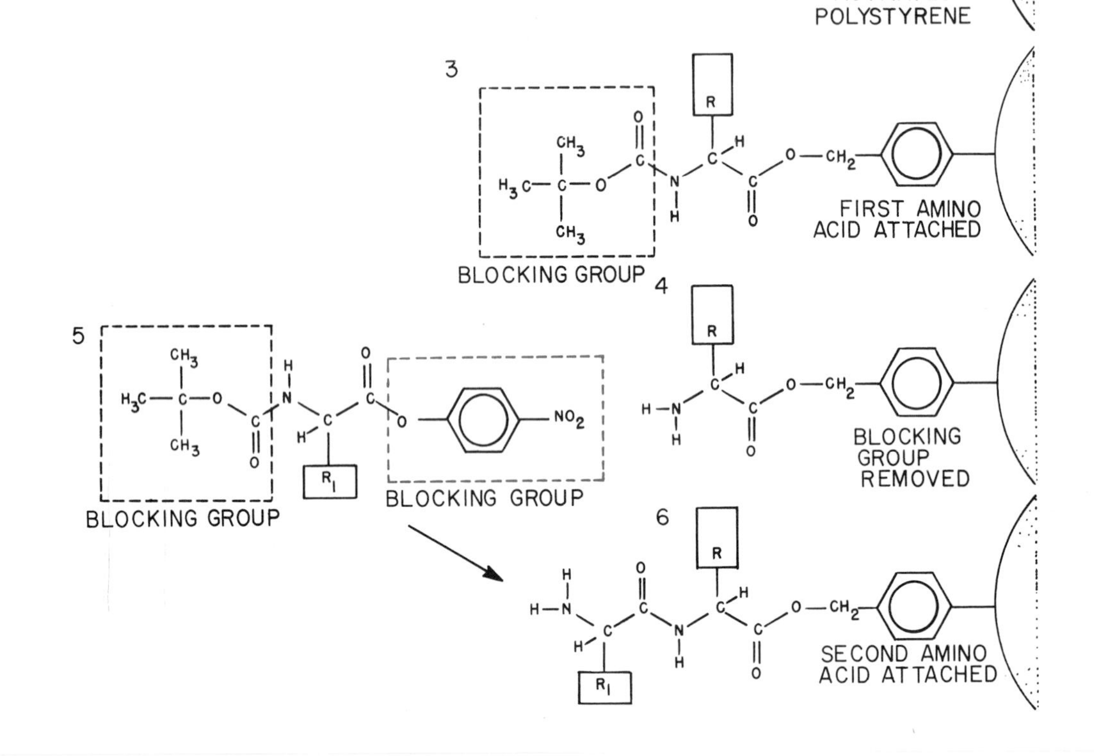

Figure 16–27 Solid phase method of synthesizing a protein is carried out stepwise from the carboxyl end toward the amino end of the peptide. An aromatic ring of the polystyrene (1) is activated by attaching a chloromethyl group (2). The first amino acid (black), protected by a butyloxycarbonyl (Boc) group (black box), is coupled to the site (3) by a benzyl ester bond and is then deprotected (4). Subsequent amino acid units are supplied in one of two activated forms; a second unit is shown in one of these forms, the nitrophenyl ester of the amino acid (5). The ester (colored box) is eliminated as the second unit couples to the first. Then the second unit is deprotected, leaving a dipeptide (6). These processes are repeated to lengthen the peptide chain.

for which he received a Nobel Prize. He used natural enzymes as templates to arrange the nucleotides in the order of a desired polynucleotide. His product was not biologically active. In 1965, Sol Spiegelman synthesized the polynucleotide portion of an RNA virus. This polynucleotide was biologically active and reproduced itself readily when introduced into living cells. In 1967, Mehran Goulian and Kornberg synthesized a fully infectious virus of the more complicated DNA type.

DNA-type polynucleotides have been synthesized in the laboratory.

In 1970, Gobind Khorana synthesized a complete, double-stranded, 77-nucleotide gene. He, too, used natural enzymes to join previously synthesized, short, single-stranded polynucleotides into the double-stranded gene.

If scientists can construct DNA, can they then control the genetic code? Genes are the submicroscopic, theoretical bodies proposed by early geneticists to explain the transmission of characteristics from parents to progeny. It was thought that genes composed the chromosomes, which are large enough to be observed through the microscope as the central

Figure 16–28 Dr. Li (*center*) checks experiment with Dr. Richard Noble (*left*) and Dr. Donald Yanashiro (*right*). This group was the first to synthesize the HGH polypeptide. (From *Chemical and Engineering News,* January 11, 1971.)

figures in cell division. It is now generally believed that DNA structures carry the message of the genes; hence, DNA contains the ***genetic code.*** If scientists can construct DNA, they could very well alter its structure and thereby control the genetic code.

A mutation results when there has been an alteration of the genetic code contained within the DNA molecule.

A ***mutation*** has occurred when an individual characteristic appears that has not been inherited but has been passed along as an inherited factor to the next generation. A mutation can readily be accounted for in terms of an alteration in the DNA genetic code; that is, some force alters the nucleotide structure in a reproductive cell. Some sources of energy, such as gamma radiation, are known to produce mutations. This is entirely reasonable because certain kinds of energy can disrupt some bonds, which can re-form in another sequence.

If scientists can control the genetic code, can they control hereditary diseases such as sickle cell anemia, gout, some forms of diabetes, or mental retardation? If the understanding of detailed DNA structure and the enzymatic activity in building these structures continues to grow, it is reasonable to believe that some detailed relationships between structure and gross properties will emerge. If this happens, it may be possible to build compounds which, when introduced into living cells, can combat or block inherited characteristics.

SELF-TEST 16-C

1. a. The sugar in RNA is ______________,

   b. while the one in DNA is ____________________.

2. A nucleotide contains ______________, ______________, and ______________.

3. The secondary structure of DNA consists of a __________ __________.

4. A virus is a chemical compound that can reproduce itself. (True/False)

5. a. The basic code for the synthesis of protein is contained in the ________ molecule.

   b. The synthesis of the protein is carried out when ________ molecules bring up the required amino acids.

6. When DNA duplicates itself, each nitrogenous base in the chain is matched to another one via __________ bonds.

MATCHING SET

______ 1. D-glucose	a.	made from essential fatty acids
______ 2. linoleic acid	b.	polymer consisting of α-D-glucose units
______ 3. methionine	c.	proteins
______ 4. enzymes	d.	sugar present in blood
______ 5. carbohydrate storage in animals	e.	an essential amino acid
______ 6. starch	f.	an unsaturated fatty acid
______ 7. polypeptides	g.	a polynucleotide
______ 8. prostaglandins	h.	biochemical catalysts
______ 9. DNA	i.	glycogen
______ 10. fibrous protein	j.	collagen
______ 11. cellulose	k.	polymer consisting of β-D-glucose units

---

QUESTIONS

1. Show the structure of the product which would be obtained if two alanine molecules (Table 16–1) react to form a dipeptide.

2. What is an essential amino acid?

3. The ketone structure of D-fructose has three asymmetric carbon atoms per molecule. How many isomers result from these asymmetric centers?

4. Name a polysaccharide that yields only D-glucose upon complete hydrolysis. Name a disaccharide that yields the same hydrolysis product.

5. What is the difference between the starch, amylopectin, and the "animal starch," glycogen?

6. What is the chief function of glycogen in animal tissue?

7. Explain the basic difference between starch, amylose, and cellulose.

8. What functional groups are always present in each molecule of an amino acid?

9. Give the name and formula for the simplest amino acid. What natural product has a high percentage of this amino acid?

10. If six different amino acids form all the possible different tripeptides, how many would there be?

11. What is the meaning of the terms: primary, secondary, and tertiary structures of proteins?

12. What three molecular units are the parts of nucleotides?

13. Based on the structures in Figure 16–14, explain the meaning of the prefix *deoxy-* in deoxyribonucleic acid.

14. How many trinucleotides with the structure indicated in Figure 16–17 could be made with the nitrogenous bases listed in Figure 16–15?

15. What are the basic differences between DNA and RNA structures?

16. What stabilizing forces hold the double helix together in the secondary DNA structure proposed by Watson and Crick?

17. Consult a medical dictionary and determine the difference between atherosclerosis and arteriosclerosis.

18. What two molecular structures are present in viruses?

19. Does a strand of DNA actually duplicate itself base for base in the formation of a strand of messenger RNA? Explain.

20. Distinguish between the three types of RNA as to molecular weight, number of mononucleotide units per molecule, and function.

21. a. Describe the general method of synthesizing DNA ***in vitro*** (in a test tube) at present.
    b. Why would the synthesis of a polynucleotide from the individual phosphoric acid, sugar, and nitrogenous bases be a breakthrough in controlling the genetic code?

22. Discuss the feasibility of solving sociological and psychological problems via religion, DNA alteration, chemical suppressants, political pressures, and/or persuasive dialogue.

23. Check the recent issues of *Science* or other scientific news publications to update the work done on synthesis of proteins and polynucleotides.

SUGGESTIONS FOR FURTHER READING

"A Step Toward Synthetic Life," *Chemistry,* Vol. 41, No. 2, p. 27 (1968).
"Anatomy of a Virus," *Chemistry,* Vol. 39, No. 4, p. 20 (1966).
Battista, O. A., "Sugar—The Chemical with a Thousand Uses," *Chemistry,* Vol. 38, No. 4, p. 12 (1965).
"Bonding Habits of DNA Bases," *Chemistry,* Vol. 41, No. 8, p. 34 (1968).
Crick, F., "The Genetic Code: III," *Scientific American,* p. 55, October (1966).
Davies, D. R., "X-ray Diffraction and Nucleic Acids," *Chemistry,* Vol. 40, No. 2, p. 8 (1967).
Fraenkel-Conrat, H., "The Genetic Code of a Virus," *Scientific American,* p. 46, October (1964).
Hofmann, K., Khorana, H., and Spiegelman, S., "The Synthesis of Living Systems," *Chemical and Engineering News,* Vol. 45, p. 144, August (1967).
"Interlocking Rings of DNA," *Chemistry,* Vol. 41, No. 2, p. 26 (1968).
"Is DNA the Master Molecule?" *Chemistry,* Vol. 41, No. 6, p. 23 (1968).
Mazur, A., and Harrow, B., *Biochemistry: A Brief Course,* W. B. Saunders Company, Philadelphia, 1968.
Merrifield, R. B., "The Automated Synthesis of Proteins," *Scientific American,* Vol. 218, No. 3, p. 56 (1968).
Neckers, D. C., "Solid Phase Synthesis," *Journal of Chemical Education,* Vol. 52, No. 11, p. 695 (1975).
Pauling, L., Corey, R., and Hayward, R., "The Structure of Protein Molecules," *Scientific American,* p. 51, July (1954).
"Portrait of a Gene," *Chemistry,* Vol. 42, No. 8, p. 20 (1969).
Raw, I., "Enzymes, How They Operate," *Chemistry,* Vol. 40, No. 6, p. 8 (1967).
"Trouble on the DNA Front," *Chemistry,* Vol. 43, No. 9, p. 24 (1970).
"Viruses as Invaders of Living Cells," *Chemistry,* Vol. 41, No. 5, p. 29 (1968).
Watson, J. D., *The Double Helix,* Atheneum, New York, 1968.
Zimmerman, J., "First Synthesis of an Enzyme, Ribonuclease," *Chemistry,* Vol. 42, No. 4, p. 21 (1969).

CHAPTER 17

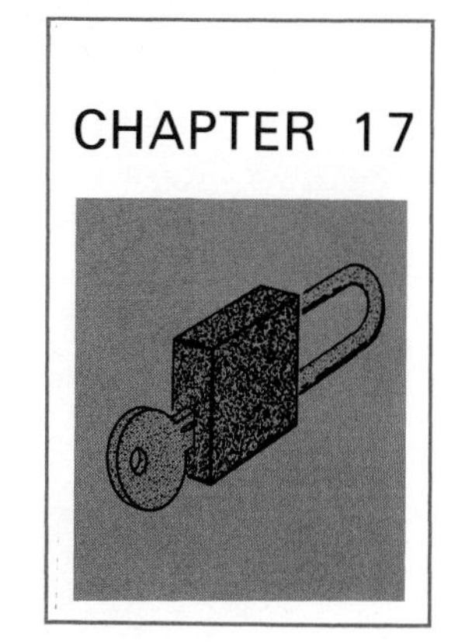

BIO-CHEMICAL PROCESSES

Similar biochemical reactions occur in a wide variety of living cells.

In Chapter 16 we examined the structures of some molecules that are important in biochemical systems. We will now look at some of the reactions by which such molecules are constructed, serve their function, and are subsequently destroyed. There are three aspects of biochemical systems which should be kept in mind in this study. First, the contents of a living cell are in a dynamic state; the molecules are constantly being synthesized and degraded. However, the healthy living cell is characterized by a "steady state" condition in which the rates of buildup and breakdown are nearly the same at any time. Second, biochemical processes are very general in that the same basic chemistry is employed by a wide variety of cells. For example, the same types of reactions utilized to obtain energy from a compound in man also occur in simple unicellular organisms. Third, biochemical systems are composed of literally thousands of different kinds of molecules. Out of such an apparent chaos comes an ordered array of reactions that supports all life forms. In view of this complexity, it is evident that an elementary treatment of what is presently known about biochemical reactions will have to deal only with selected highlights.

ENZYMES

An ***enzyme*** is a biochemical catalyst. Like other catalysts, a given enzyme increases the rate of a reaction without requiring an increase in temperature. As an example of a simple type of catalysis, consider the oxidation of glucose, a sugar which burns in air with some difficulty and is hard to light with a match. If cigarette ashes, or other catalysts, are placed on its surface, combustion can be initiated easily with a match. When the glucose burns, it liberates a large amount of energy, 688 kilocalories, or 688,000 calories per mole.

$$\underset{\text{GLUCOSE}}{C_6H_{12}O_6} + \underset{\text{OXYGEN}}{6O_2} \longrightarrow \underset{\text{CARBON DIOXIDE}}{6CO_2} + \underset{\text{WATER}}{6H_2O} + 688\text{ kcal}$$

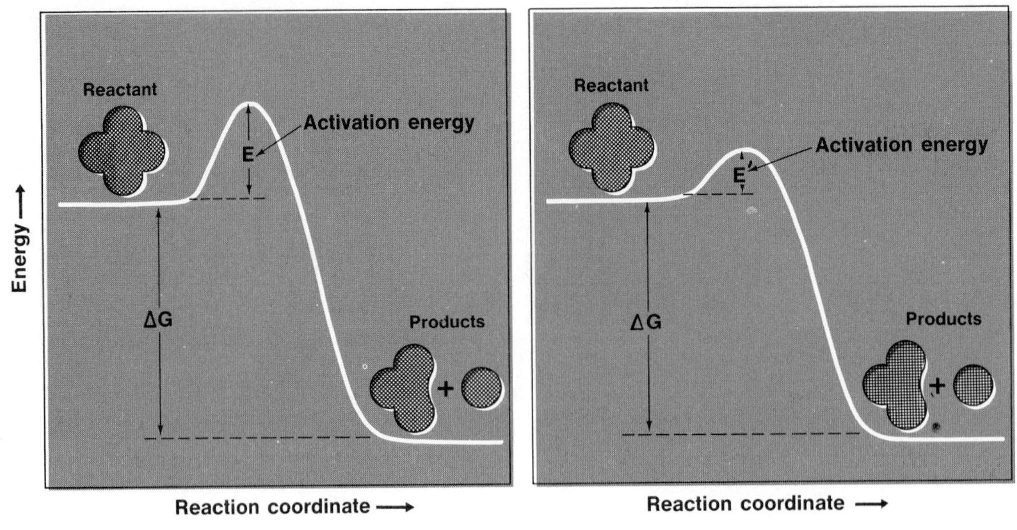

Figure 17–1 Enzyme effect on activation energy. The vertical coordinate represents increasing energy and the horizontal one the course of the reaction in going from reactants to products. For energy-producing reactions the reactant molecules are at a higher energy than the product molecules, as illustrated in *a*. The difference between these energies is the net free energy change of the reaction (ΔG). However, it is necessary for the reactant molecules to "get over" the energy barrier (acquire the activation energy, E) in going from reactants to products. Note that the activation energy is given back along with the free energy. The enzyme lowers the activation energy, E′, as illustrated in *b*, while the free energy change remains the same. The net effect is to obtain the free energy of the reaction with a smaller expenditure of activation energy. Free energy is discussed later in this chapter.

The energy required to get the reaction started is the activation energy; catalysts, in general, work by lowering the activation energy. If an enzyme can lower the activation energy to a point where the average kinetic energy of the molecules in a living cell (or in a laboratory system) is sufficient for reaction, then the reaction can proceed rapidly. Glucose is oxidized rapidly and efficiently at ordinary temperatures in the presence of the proper enzymes. To be sure, the oxidation of glucose in a living cell requires many enzymes and many steps, but enzymatic catalysis produces the same final result as combustion at elevated temperature, namely carbon dioxide, water, and 688 kilocalories of usable energy per mole of sugar. Figure 17-1 graphically illustrates the concepts of activation energy, the energy available from an energy-producing reaction, and the reduction of the activation energy by an enzyme.

Enzymes are very specific catalytic molecules with a specific catalytic task.

Enzymes are remarkable catalysts in that they are highly specific for a given reaction. Maltase is an enzyme that catalyzes the hydrolysis of maltose into two molecules of D-glucose. This is the only known function of maltase, and no other enzyme can substitute for it. The explanation for the specific activity of enzymes can be found in their molecular geometry.

Enzymes are globular proteins.

Enzymes are globular proteins with very definite tertiary structures. The highly specific action of maltase can be explained if its globular structure accurately accommodates a maltose molecule at the point where the reaction occurs, the reactive site. When the two units come

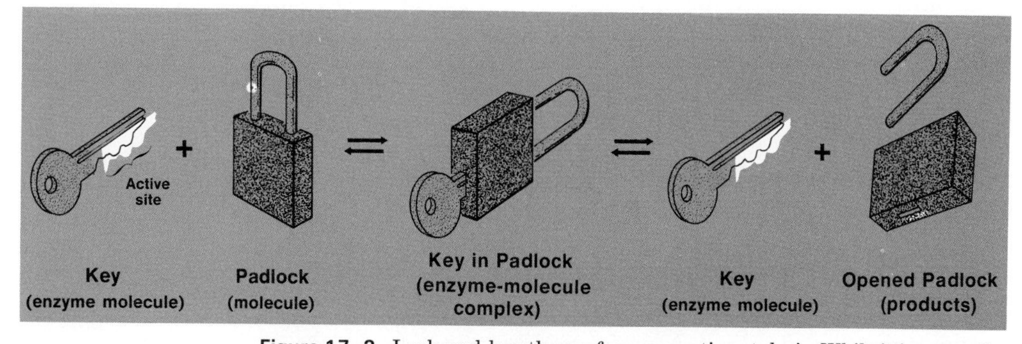

Figure 17–2 Lock-and-key theory for enzymatic catalysis. While it is generally agreed that this analogy is an oversimplification, it does make one very important point: the enzyme makes a difficult job easy by reducing the energy required to get the job started. It also suggests that the enzyme has a particular structure at an active site which will allow it to work only for certain molecules, similar to a key that fits the shape of a particular keyhole.

together, strain is placed on the bonds holding the two simple sugar units together. As a result, water is allowed to enter and hydrolysis occurs. Sucrose cannot be hydrolyzed by maltase because of the different geometry involved. Another enzyme, sucrase, hydrolyzes sucrose effectively. Some enzymes, however, are less specific. The digestive enzyme, trypsin, for example, acts predominantly on peptide bonds in proteins, but it will also catalyze the hydrolysis of some esters because of somewhat similar geometry and polarity at the active site.

How does an enzyme work? How can it lower the activation energy and be so specific for a given reaction? While a definitive answer cannot be given at this time (the matter is presently under intensive research), a "lock-and-key" analogy has been a fruitful approach to the problem. Just as a key can separate a padlock into two parts and subsequently remain unchanged, ready to unlock other identical locks, so the enzyme makes possible a molecular change (Figure 17–2). With enough energy the lock could be separated without the key, and with enough energy the molecular alteration could occur without the enzyme. An enzyme cannot make a nonspontaneous chemical reaction occur.

The tremendous speed of enzyme-catalyzed reactions requires more than just random collisions to fit "the key in the lock." For example, a molecule of β-amylase catalyzes the breaking of four thousand bonds in amylose per second. Speed like this requires something to attract the key into the lock, such as electrically polar regions, partially charged groupings, or ionic sections on the enzyme and the ***substrate*** (the reactant molecule). These regions attract as well as guide the substrate to the proper position on the enzyme and thereby speed up the reaction. The electrically charged portions of the enzyme are believed to be the chemically ***active sites*** in the enzyme.

Enzyme structure is the key to specific catalytic activity.

Sometimes an enzyme is more than a globular protein alone; in such cases, the protein is not a catalyst by itself. In addition to the protein part of the enzyme there is also another chemical species called a ***coenzyme.*** The coenzyme, required for catalytic activity, may be an ion (e.g., Co^{3+}, Fe^{3+}, Mg^{2+}, or another of the essential minerals) or may be in part derived from a vitamin (as we shall see later in this section). The protein part of such an enzyme is called the ***apoenzyme.*** The coenzyme alone does not have enzymatic activity; neither does the apoenzyme.

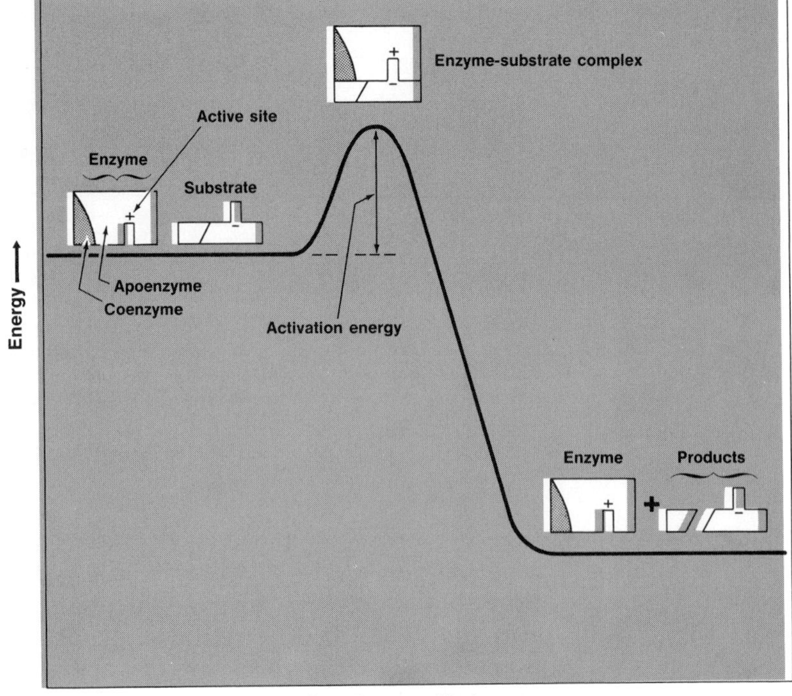

Figure 17–3 A reaction utilizing an enzyme which requires a coenzyme. Here the essential feature is that the combination of enzyme plus coenzyme allows the reaction to proceed with a lower activation energy. Apoenzyme is the name given to the enzyme structure that combines with the coenzyme.

Before the enzyme becomes active, the apoenzyme and the coenzyme must combine like the two keys required to open a bank lock-box. Neither your key nor the bank's key will open the box, when used alone, but both together will (Figure 17–3).

Let's use a simple chemical example to demonstrate the relationship between coenzyme, substrate, and apoenzyme. The peptide glycylglycine is hydrolyzed very slowly by water to give two molecules of glycine.

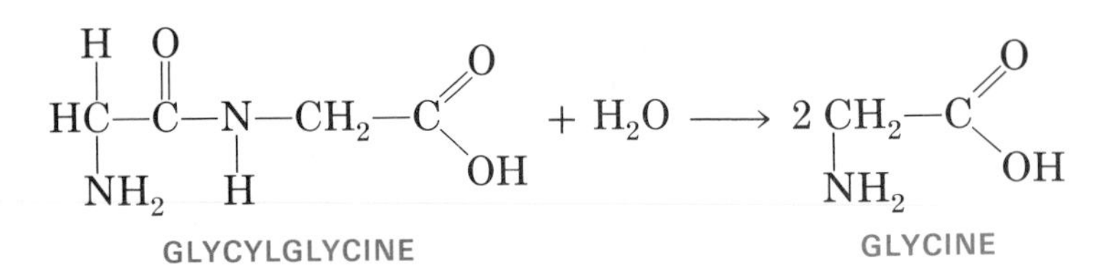

The glycylglycine substrate forms an intermediate compound with an enzyme, and as a result the substrate is activated for further reaction. Activation can result from extensive hydrogen bonding, interaction with a metal ion in the enzyme, or a number of other processes. In this case the coordination of the glycylglycine to the positive charge of a cobalt ion (Co^{2+}), the coenzyme, makes it more susceptible to attack by the negative end of a water molecule (Figure 17–4).

The names of most enzymes end in *ase.*

Most of the names of enzymes end in ***ase.*** There are a few exceptions, such as pepsin and trypsin, which are both digestive enzymes. Hydrolases promote the breakdown of foodstuffs and other substances by hydrolysis. Carbohydrases (such as maltase, lactase, sucrase, ptyalin and amylase) help to effect hydrolysis of carbohydrates. Proteases (such as pepsin, trypsin, and chymotrypsin) hydrolyze large protein molecules into smaller groups of proteins. Lipases hydrolyze esters such as fats and

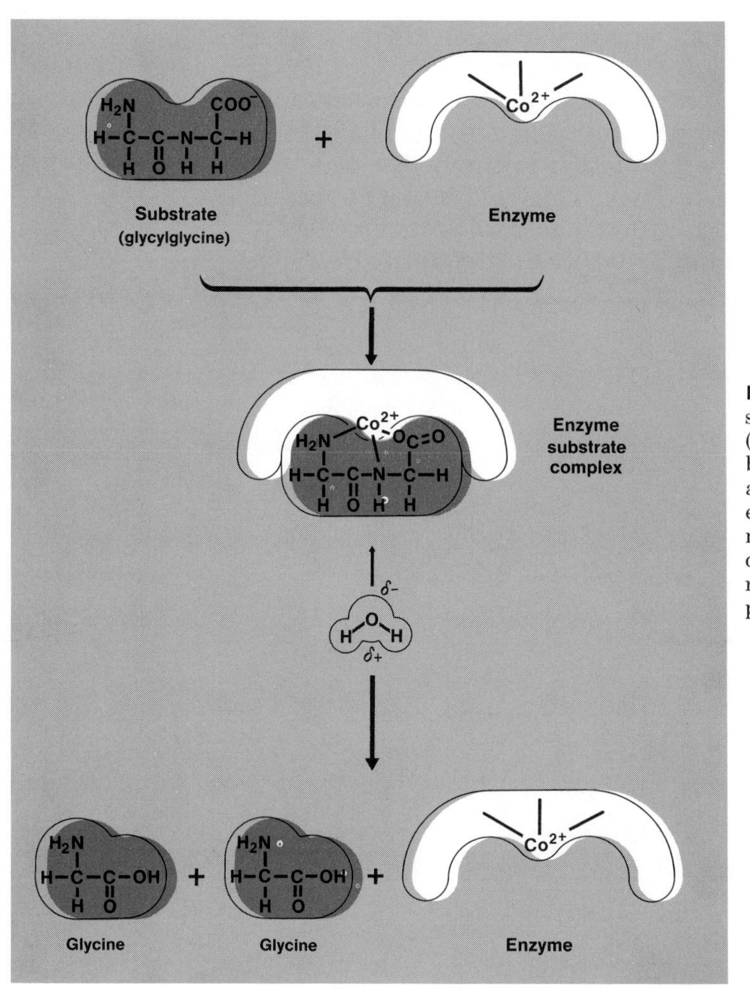

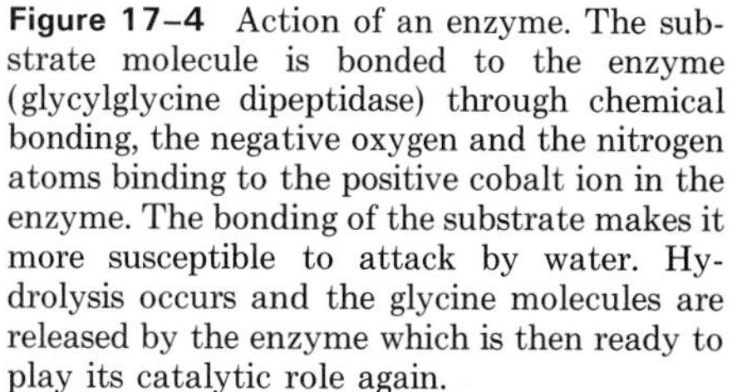

Figure 17–4 Action of an enzyme. The substrate molecule is bonded to the enzyme (glycylglycine dipeptidase) through chemical bonding, the negative oxygen and the nitrogen atoms binding to the positive cobalt ion in the enzyme. The bonding of the substrate makes it more susceptible to attack by water. Hydrolysis occurs and the glycine molecules are released by the enzyme which is then ready to play its catalytic role again.

oils. Examples of the oxidizing enzymes, called oxidases, are catalase, which speeds up the conversion of hydrogen peroxide to water and oxygen, and dehydrogenases, which assist in the removal of hydrogen from molecules. There are other categories of enzymes, but these illustrate the wide variety of biochemical catalysts.

Besides being biochemical middlemen, speeding up and directing all the chemical reactions that go into the continuous breakdown and buildup of our cells (three million red blood cells are renewed in the human body every second), enzymes may be the answer to future food problems. Scientists have already developed a way to produce sugar (on a limited experimental basis) by bubbling carbon dioxide into water containing enzymes. Trash fish can be converted into palatable animal feed by using enzymes. Work is underway to convert oil spills into edible products for sea organisms. A little later in this text we will encounter the use of enzymes as meat tenderizers.

Enzymes are also important in commercial products.

Many vitamins function as coenzymes. They not only trigger specific enzymes, they pitch in and help out. The B vitamins are found in every cell as coenzymes in various oxidative processes. For example, niacin (vitamin B_5) becomes part of an enzyme which prevents pellagra, at one time a common vitamin-deficiency disease in the United States. Doctors and biochemists now know that the body suffers from pellagra when it lacks sufficient niacin. Some foods such as yeast, liver, meats, fish, eggs, whole wheat, brown rice, and peanuts contain niacin. The body needs only one two-thousandth of an ounce of niacin each day in order to prevent pellagra. This isn't much but it is vital. The coenzyme of which niacin is a part is vital to the energy production in the body. If energy is not provided, the whole process of renewing cells and building needed compounds slows down and eventually stops. The coenzyme involved is quite large, as is its name, ***nicotinamide adenine dinucleotide*** (NAD^+).

Niacin (Nicotinic acid)

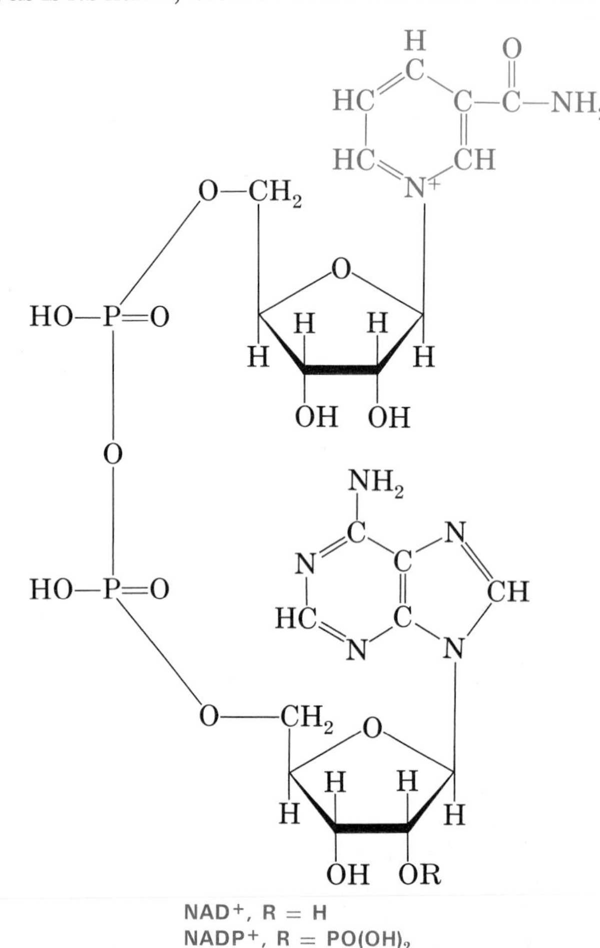

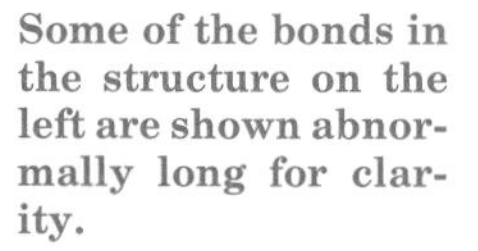
Some of the bonds in the structure on the left are shown abnormally long for clarity.

NAD^+, R = H
$NADP^+$, R = $PO(OH)_2$

Niacin is also an integral part of another oxidative enzyme, ***nicotinamide adenine dinucleotide phosphate*** ($NADP^+$). In NAD^+, R = H; in $NADP^+$, R = H_2PO_3. The part of the structure printed in color above comes from niacin and is the active part in oxidation-reduction. It accepts hydrogen during biological oxidation processes.

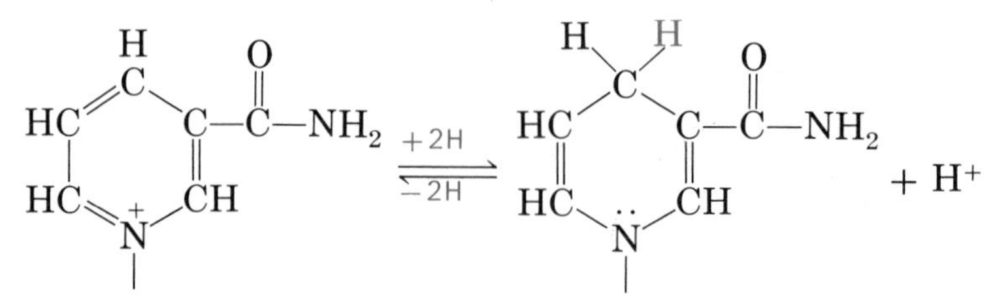

Riboflavin (vitamin B_2) is a necessary part of another important hydrogen acceptor coenzyme, ***flavin adenine dinucleotide*** (FAD).

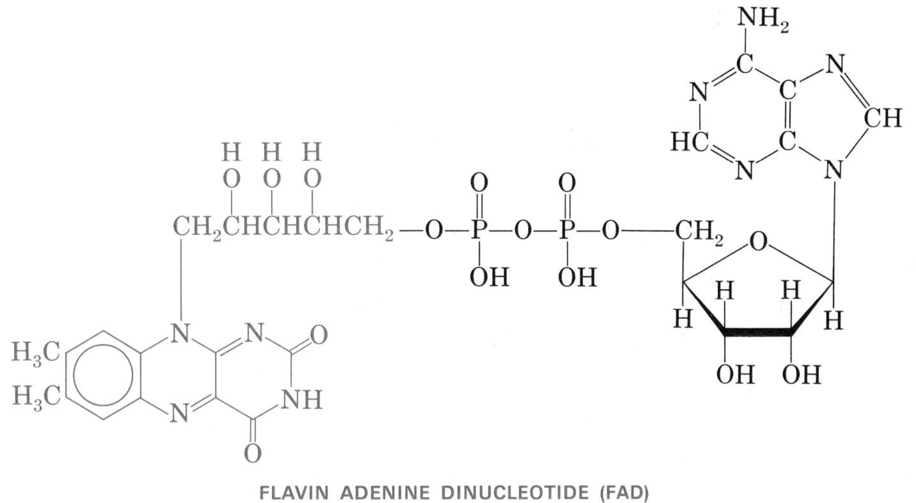

FLAVIN ADENINE DINUCLEOTIDE (FAD)
(The portion in color is riboflavin.)

FAD accepts hydrogens in the following manner:

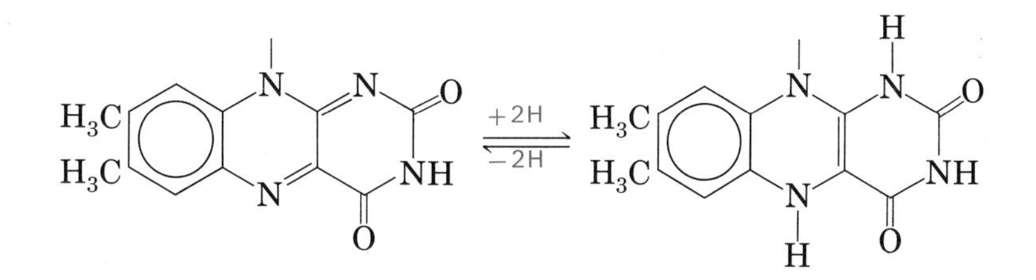

Glucose and glycogen are the principal sources of energy in the body. How NAD^+, $NADP^+$, and FAD fit into the oxidation of glucose and glycogen will be considered later.

SELF-TEST 17-A

1. The best term to describe the general function of enzymes is (catalyst/intermediate/oxidant).
2. In the lock-and-key analogy of enzyme activity, the enzyme functions as the __________ while the substrate molecule serves as the __________.
3. Pellagra can be prevented by intake of the vitamin named __________.
4. The activation energy of many biological reactions is decreased if a(n) __________ is present.
5. Apoenzyme + coenzyme ⟶ __________.
6. Riboflavin is a vitamin which is needed because it is part of an essential __________.

7. The coenzyme nicotinamide adenine dinucleotide (NAD^+) cannot be made by the human body unless it has a supply of ________________.

8. That portion of the enzyme at which the reaction is catalyzed is called the ______________ ______________.

BIOCHEMICAL ENERGY AND ATP

In biochemical processes energy is required to drive the process to yield the desired products. This energy comes from the ultimate energy source in our environment, the sun. How is the energy of sunlight stored in various compounds, and in turn supplied to feed the life processes? This is accomplished primarily by adenosine triphosphate, often written ATP (Figure 17–5). The energy obtained by the oxidation of foods is mostly used to synthesize ATP. ATP is the immediate source of energy in muscular contraction, and the energy released by it allows many energy-requiring biochemical reactions to occur.

ATP molecules each contain two so-called high-energy phosphate bonds. These are marked by wiggle bonds (~) in Figure 17–5. In the presence of a suitable catalyst, ATP will undergo a three-step hydrolysis. The hydrolysis of ATP to adenosine diphosphate (ADP) and phosphoric acid releases about 12 kilocalories per mole. The second hydrolysis of ADP to adenosine monophosphate (AMP) also produces about 12 kilocalories of energy per mole. Finally, the hydrolysis of AMP to adenosine, which involves a low energy bond, releases only about 2.5 kilocalories per mole.

ATP is an energy-rich molecule that furnishes the energy required by many biochemical processes.

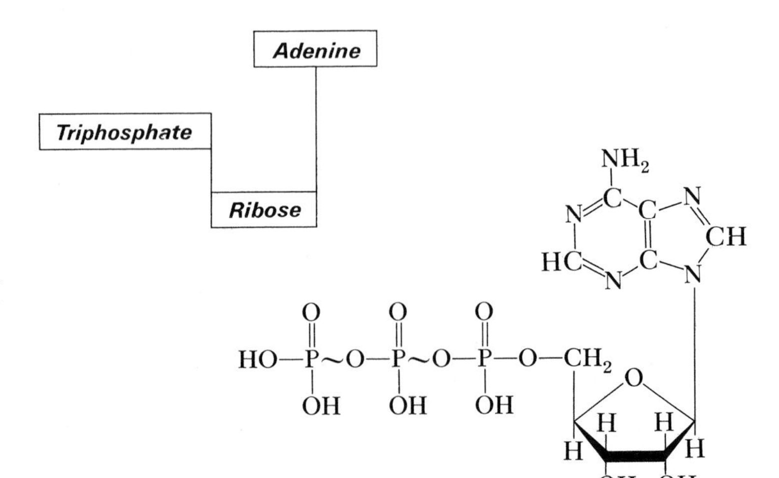

Figure 17–5 Molecular structure of adenosine triphosphate (ATP). Note the similarity between ATP and the fundamental unit of a nucleic acid, the nucleotide (see Chapter 16).

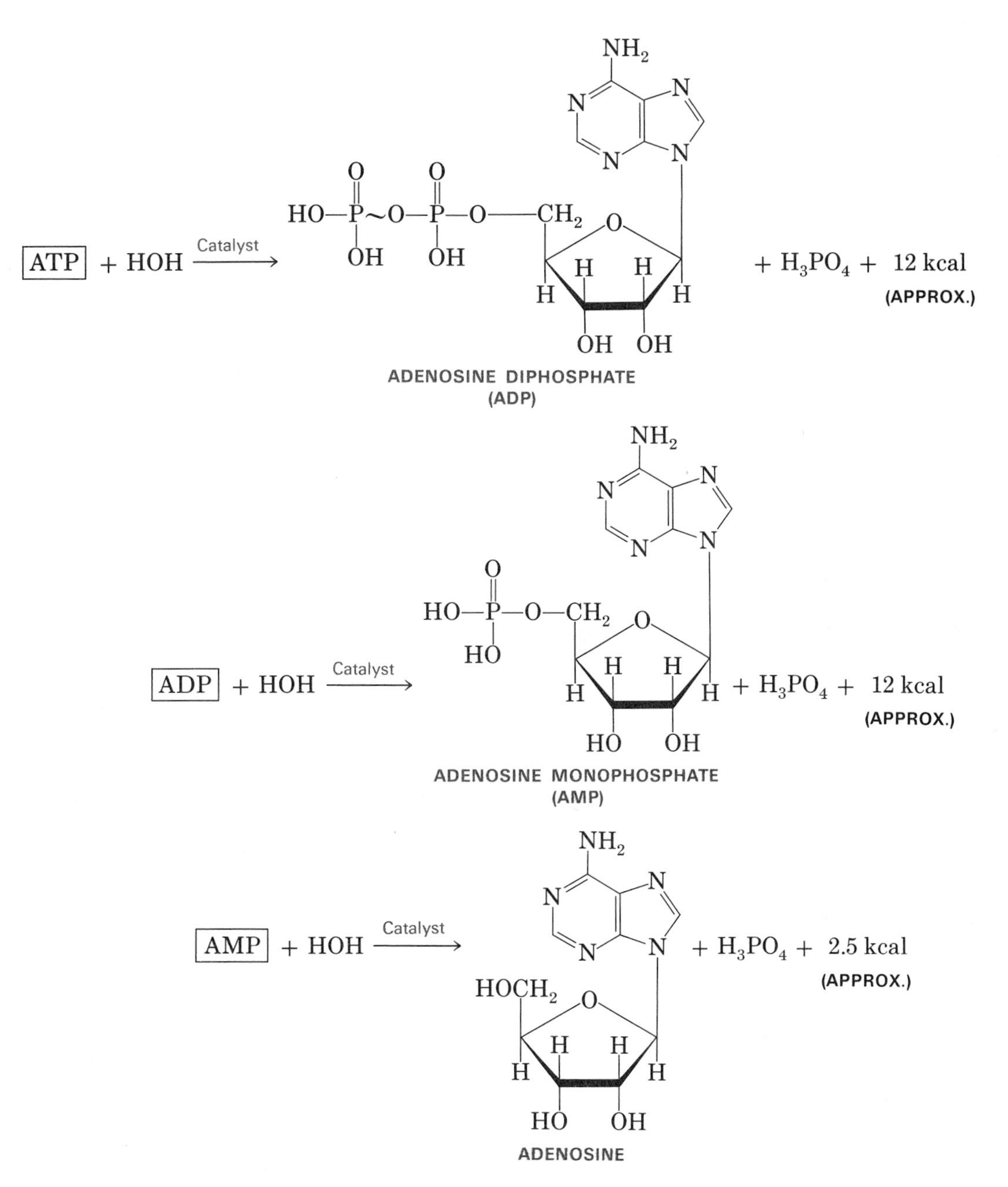

Studies of the energetics of chemical reactions show that not all the energy released is available to initiate other chemical change. The energies for these three reactions are given only to show the relatively large amount of energy associated with the hydrolysis of the first two phosphate units in the ATP structure. A more meaningful value is the energy change that is available to initiate other chemical change; this is termed the change in ***free energy.*** A change in free energy is symbolized by ΔG. For the three hydrolytic reactions of ATP the approximate free energy changes are:

The change in free energy, ΔG, is the energy available to do work from a reaction.

ATP hydrolysis: $\Delta G = -7.4$ kcal per mole
ADP hydrolysis: $\Delta G = -6.8$ kcal per mole
AMP hydrolysis: $\Delta G = -2.2$ kcal per mole

(Note: The negative sign means that energy is evolved.)

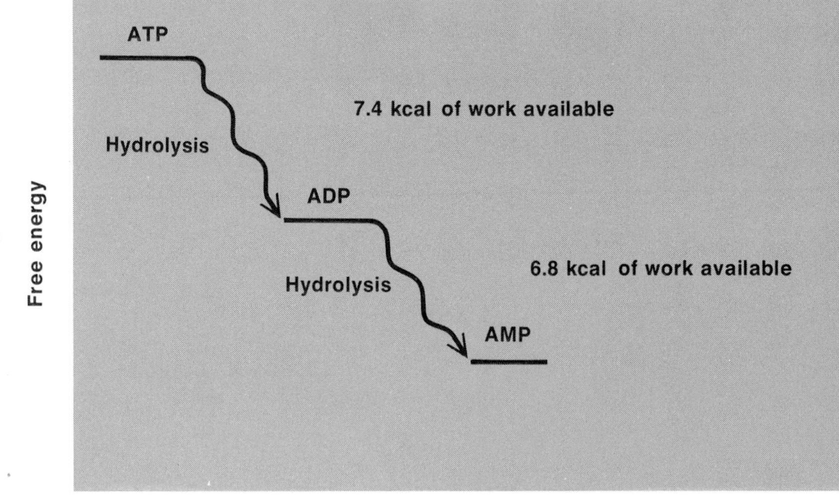

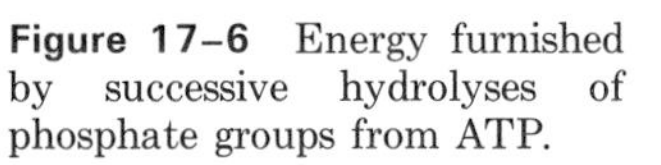
Figure 17-6 Energy furnished by successive hydrolyses of phosphate groups from ATP.

These free energy values indicate the amount of chemical potential energy in the phosphate bonds in the ATP molecule. Such energy-producing reactions provide the energy necessary for energy-requiring biochemical reactions to occur, and thus supply the energy necessary for the basic life processes.

PHOTOSYNTHESIS

Photosynthesis is a very complex process which belies the relatively simple overall reaction in which carbon dioxide and water are converted into energy-rich carbohydrates by solar energy.

$$\underset{\text{CARBON DIOXIDE}}{6CO_2} + \underset{\text{WATER}}{6H_2O} + \underset{\text{ENERGY (SUNLIGHT)}}{688\text{ kcal}} \longrightarrow \underset{\text{GLUCOSE}}{C_6H_{12}O_6} + \underset{\text{OXYGEN}}{6O_2}$$

In photosynthesis carbon dioxide is ***reduced*** to sugar

$$6CO_2 + 24H^+ + 24e^- \longrightarrow C_6H_{12}O_6 + 6H_2O$$

Reduction: gain of electrons or hydrogen.
Oxidation: loss of electrons or hydrogen.

and water is ***oxidized***

$$12H_2O \longrightarrow 6O_2 + 24H^+ + 24e^-$$

Note that these two half reactions, the first reduction and the second oxidation, give the overall reaction when added together.

All the details of photosynthesis are not fully understood. However, some aspects are presented here because it is the beginning of the energy flow through biochemical systems.

Photosynthesis involves a number of different steps and is a very complex process.

Photosynthesis is generally considered in terms of the ***light reaction,*** which can occur only in the presence of light energy, and the ***dark reaction,*** which can occur in the dark, but feeds on the high energy structures produced in the light reaction. Actually, both the light and dark reactions are a series of reactions, all occurring simultaneously in the green plant cell. The light reaction is unique to green plants while the dark reaction is characteristic of both plant and some animal cells.

The Light Reaction

Photosynthesis is initiated by a quantum of light energy. The green plant contains certain pigments that readily absorb light in the visible region of the spectrum. The most important of these are the chlorophylls, ***chlorophyll a*** and ***chlorophyll b.*** Note that both chlorophylls are com-

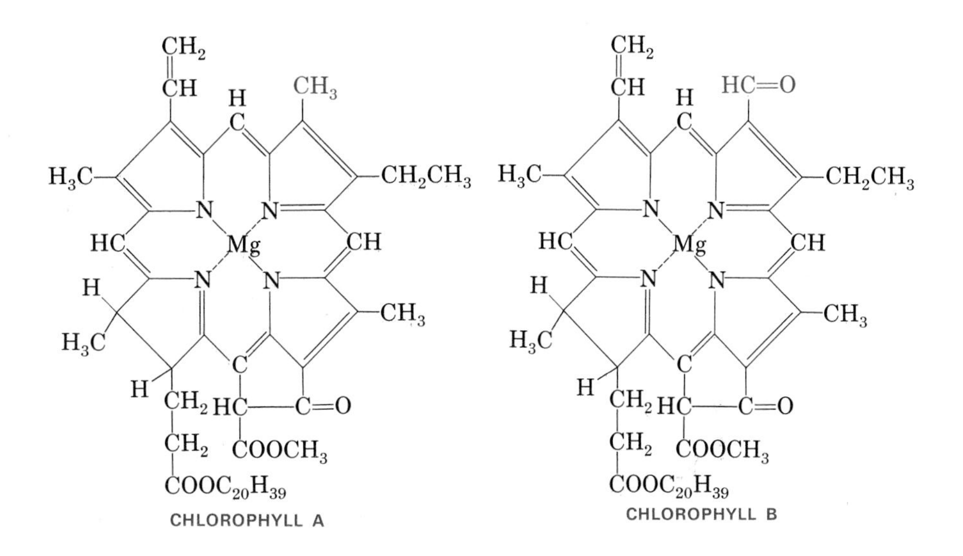

pounds of magnesium and both have complex ring systems. Such ring systems usually absorb light in the visible region of the spectrum; consequently, they are colored. For example, chlorophyll is green because it absorbs light in the violet region (about 400 nanometers) and the red region (about 650 nanometers) and allows the green light in between those wavelengths to be reflected or transmitted.

The energy of a photon is captured by chlorophyll by raising an electron to a higher energy state.

When chlorophyll absorbs photons of light, electrons are raised to higher energy levels. As these electrons move back down to the ground state, very efficient subcellular components of the plant cell known as chloroplasts grab this energy and, through a series of steps which are not all completely known, store the energy as chemical potential energy. As shown in Figure 17–7, one of the chemicals used to store this energy is nicotinamide adenine dinucleotide phosphate ($NADP^+$).

Energy is absorbed by $NADP^+$ in the process of being reduced by a hydrogen donor to $NADPH + H^+$.

$$NADP^+ + 2H + \text{energy} \longrightarrow NADPH + H^+$$

NADPH eventually transfers its precious energy to the universal storehouse of biochemical energy, adenosine triphosphate, ATP.

Thus far little mention has been made of the electrochemical charge balance and the oxygen produced by photosynthesis, both of which are

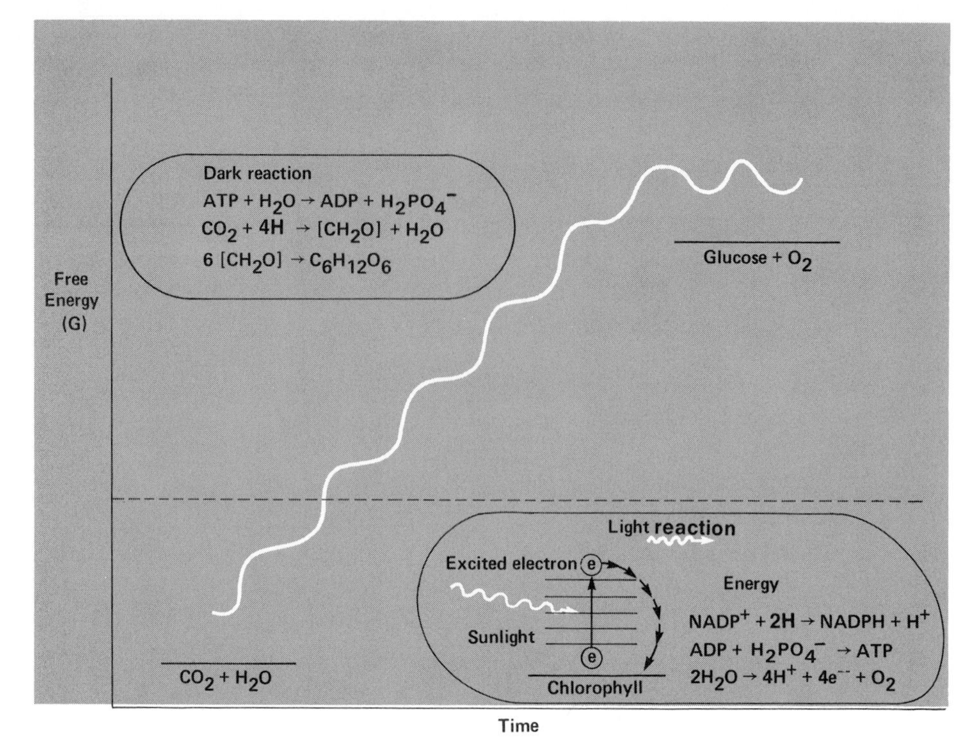

Figure 17–7 Free energy within the chemical system is increased as carbon dioxide and water are converted into glucose and oxygen by photosynthesis. This results in stored, useful energy in glucose. The very simplified mechanism for photosynthesis is discussed in the text.

important parts of the light reaction. Water, in the presence of chloroplasts and light, is decomposed into oxygen, hydrogen ions, and electrons:

Green plants produce oxygen by oxidizing water.

$$2H_2O \longrightarrow 4H^+ + 4e^- + O_2$$

The hydrogen ions and electrons are available for maintaining the balance of charge, and oxygen is liberated from the plant cell.

At this point we no longer need the energy of the sun. We have the energy necessary to run biochemical systems stored in the ATP structure. If the plant cell is given the minerals along with carbon dioxide and water, it, or subsequent living cells, can use the energy in ATP in the dark to provide energy for the complex biochemical reactions that take place.

The Dark Reaction

The dark reaction is responsible for the ultimate conversion of gaseous carbon dioxide to glucose. It was discovered by Melvin Calvin (1911– ; Nobel Prize in 1961). Calvin studied the uptake of radioactive carbon in carbon dioxide by plant cell chloroplasts. He illuminated the plants for definite short periods of time and then analyzed the plant cells to determine which compounds contained the most radioactive carbon. As the time periods were reduced, more of the radioactive carbon was found in those compounds into which it had been initially incorporated and less in compounds which had been formed in subsequent reactions.

For example, after only five seconds' illumination, radioactive carbon is found in the compound 3-phosphoglyceric acid:

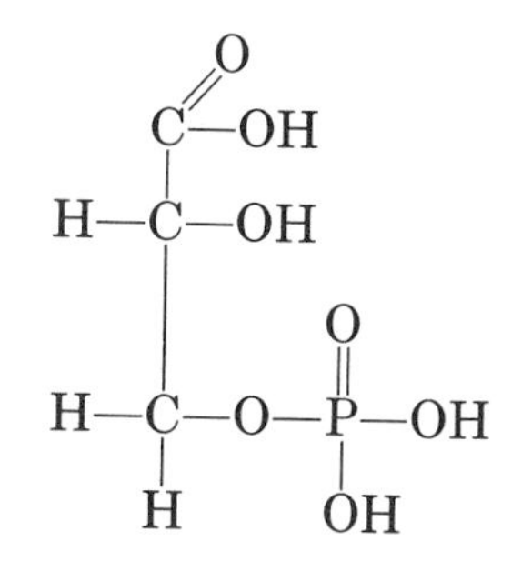

3-PHOSPHOGLYCERIC ACID

The mechanisms of many reactions can be studied by the use of radioactive atoms known as *tracers.*

This compound is apparently formed in the initial reaction in which CO_2 from the air reacts with some molecule present in the plant. Calvin discovered that the key was the reaction of atmospheric CO_2 with ribulose 1,5-diphosphate to give two molecules of 3-phosphoglycerate:

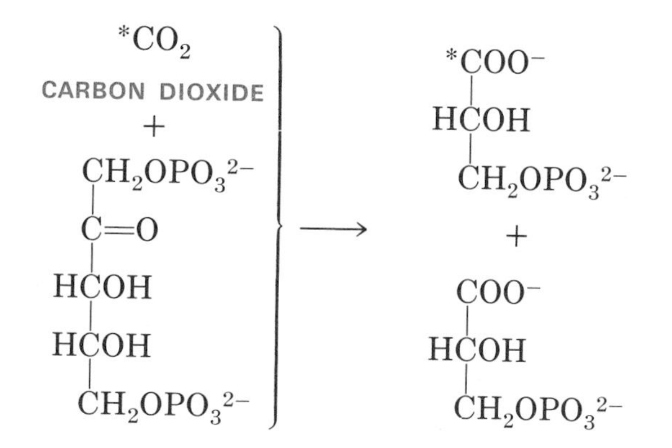

RIBULOSE 1,5-DIPHOSPHATE 3-PHOSPHOGLYCERATE

*C indicates the fate of radioactive carbon as determined by Calvin's experiments.

The 3-phosphoglycerate is then transformed into other carbohydrates in reactions which produce glucose and regenerate ribulose 1,5-diphosphate (for further uptake of more atmospheric CO_2). The energy needed to carry out these reactions is furnished by the NADPH and ATP generated from the light reaction.

Figure 17–8 presents the cyclic character of the dark reaction of photosynthesis. Note that carbon dioxide enters the cycle at the upper left and that sugars are removed at the lower right. Since just one carbon atom enters the cycle per revolution, only one fructose 6-phosphate molecule out of seven is converted to glucose.

When the light and dark reactions are considered, one can readily see that the net equation for photosynthesis, given at the beginning of this section, is a tremendous simplification of the actual process. The light reaction may be summarized as follows:

Oxygen and ATP are the products of the light reaction.

$$(12 - n)H_2O + 12NADP^+ + nADP + nH_3PO_4 \longrightarrow nATP + 12H^+ + 12NADPH + 6O_2$$

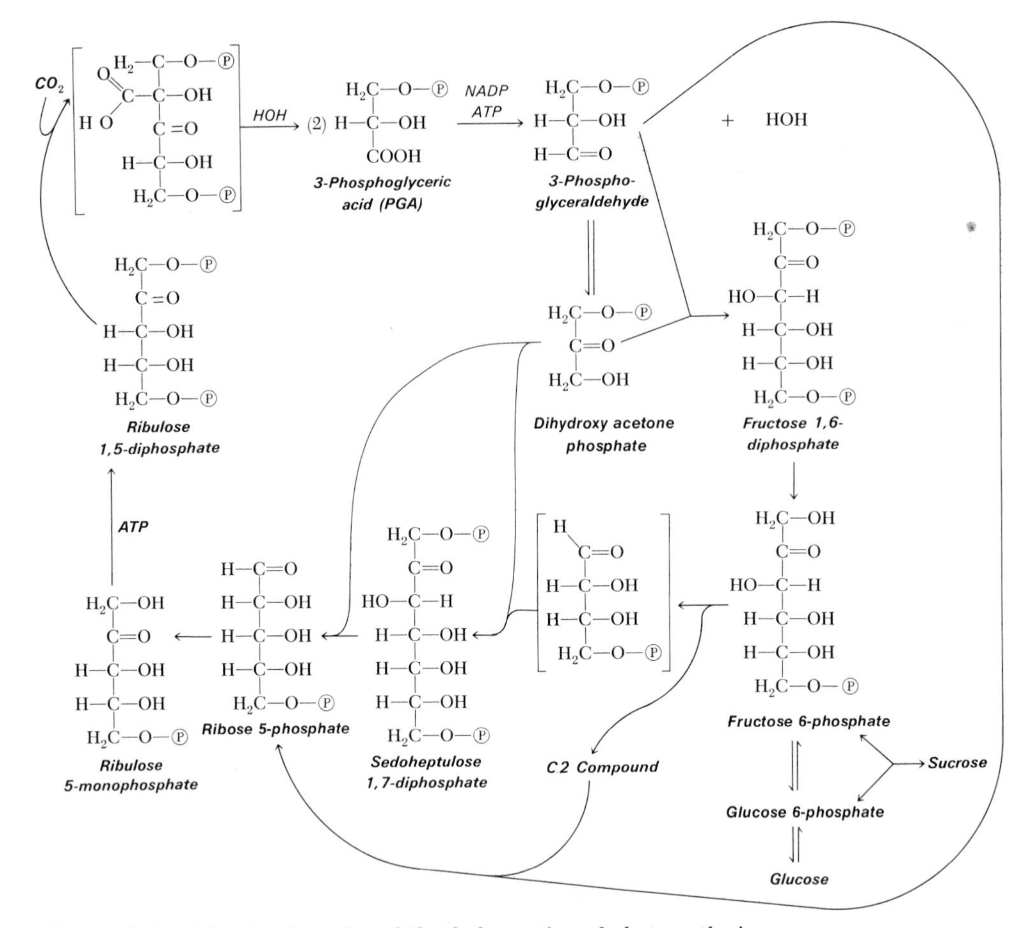

Figure 17–8 Abbreviated version of the dark reaction of photosynthesis. The energy needed to carry out the dark reactions is furnished from the high energy compounds produced in the light reaction. Ⓟ indicates a phosphate group.

The dark reaction (carbon-fixation) may be represented as follows:

$$6CO_2 + 18ATP + 12NADPH + 12H^+ + 12H_2O \longrightarrow$$

$$C_6H_{12}O_6 + 18ADP + 12NADP^+ + 18H_3PO_4$$

GLUCOSE

Glucose is a product of the dark reaction.

SELF-TEST 17-B

1. The source of energy for photosynthesis is ________________.
2. Most of the energy obtained by food oxidation is used immediately to synthesize the molecule __________.
3. The hydrolysis of ATP results in the molecules __________ and ________________. The other "product" is ________________.
4. Energy available to do work is called ______________ energy.

5. The reactants in the photosynthesis process are ________ and ________; ____________ must also be supplied.

6. Energy absorbed by chloroplasts in the green cells of a plant is transferred by means of the molecule ____________ to ATP.

7. The first chemical product of the dark reaction, as determined by Calvin, is ________________________.

8. The final product of the dark reaction is the substance ______________.

DIGESTION

Digestion is the hydrolysis of carbohydrates, fats, and proteins to provide small molecules which can be absorbed.

From a chemical point of view, digestion is the breakdown of ingested foods through hydrolysis. The products of these hydrolytic reactions are relatively small molecules that can be absorbed through the intestinal walls into the body fluids where they are used for metabolic processes. The hydrolytic reactions of digestion are catalyzed by enzymes, there being a specific enzyme for each hydrolysis. The hydrolysis of carbohydrates ultimately yields simple sugars, proteins yield amino acids, and fats yield fatty acids. The activities of some of the more important enzymes are described in Table 17–1. Note that the pH of the stomach fluid is much lower than the pH of the intestines.

Carbohydrate Digestion and Absorption

The principal forms of carbohydrates in our food are (1) high molecular weight polymers such as starch and glycogen, (2) disaccharides such as sucrose and lactose, and (3) simple sugars such as glucose and fructose. The first enzyme capable of facilitating the splitting of polysaccharides and of acting on ingested food is furnished by the saliva and is named salivary amylase, or ptyalin. Its action on starch or glycogen produces limited amounts of the disaccharide maltose. Ptyalin is inactivated by the high acidity in the stomach, so its activity is stopped there. The stomach furnishes no enzymes which can catalyze the splitting of carbohydrate polymers.

When food passes from the stomach, the acidity is neutralized by a secretion of the pancreas, which also contains enzymes which can facilitate the splitting of some of the polysaccharides to maltose, the splitting of maltose to glucose, and the catalysis of other hydrolytic reactions. The final result is a mixture of simple sugars such as glucose, fructose, and galactose. These simple sugars are then absorbed into the blood stream where the control of blood sugar is regulated by the hormone insulin. If the sugar level is too high, the polysaccharide glycogen is produced in the liver; if the blood sugar is low, the stored glycogen is hydrolyzed.

Insulin is a protein.

TABLE 17–1 Principal Digestive Enzymes

Enzyme	Source	Substrate	Products	Optimal pH
Ptyalin	Salivary glands	Starch	Smaller carbohydrate polymers (minor physiologic role)	6–7
Pepsin	Chief cells of stomach	Protein	Polypeptides	1.6–2.4
Gastric lipase	Stomach	Fat	Glycerides, fatty acids (minor physiologic role)	—
Enterokinase	Duodenal mucosa	Trypsinogen	Trypsin	—
Trypsin	Exocrine pancreas	Denatured proteins and polypeptides	Small polypeptides (also activates chymotrypsinogen to chymotrypsin)	8.0
Chymotrypsin	Exocrine pancreas	Proteins and polypeptides	Small polypeptides	8.0
Nucleases	Exocrine pancreas	Nucleic acids	Nucleotides	—
Carboxypeptidases	Exocrine pancreas	Polypeptides	Smaller polypeptides*	—
Pancreatic lipase	Exocrine pancreas	Fat	Glycerides, fatty acids, glycerol	8.0
Pancreatic amylase	Exocrine pancreas	Starch	Maltose units	6.7–7.0
Aminopeptidases	Intestinal glands	Polypeptides	Smaller polypeptides†	8.0
Dipeptidase	Intestinal glands	Dipeptide	Amino acids	—
Maltase	Intestinal glands	Maltose	Hexoses	5.0–7.0
Lactase	Intestinal glands	Lactose	(glucose, galactose	5.8–6.2
Sucrase	Intestinal glands	Sucrose	and fructose)	5.0–7.0
Nucleotidase	Intestinal glands	Nucleotides	Nucleosides, phosphoric acid	—
Nucleosidase	Intestinal glands	Nucleosides	Purine or pyrimidine base, pentose	—
Intestinal lipase	Intestinal glands	Fat	Glycerides, fatty acids and glycerol	8.0

*Removal of C-terminal amino acid.
†Removal of N-terminal amino acid.

Lipid Digestion and Absorption

The term ***lipid*** denotes a group of compounds which include fats and oils and other substances whose solubility characteristics are similar to those of fats and oils. These compounds are not all structurally related, and we will consider only the triglycerides (triesters of fatty acids and glycerol) in this chapter. A typical triglyceride is palmitooleostearin. Its structure and its hydrolytic products are shown in the following reaction:

Acidic solutions: pH below 7
Basic solutions: pH above 7

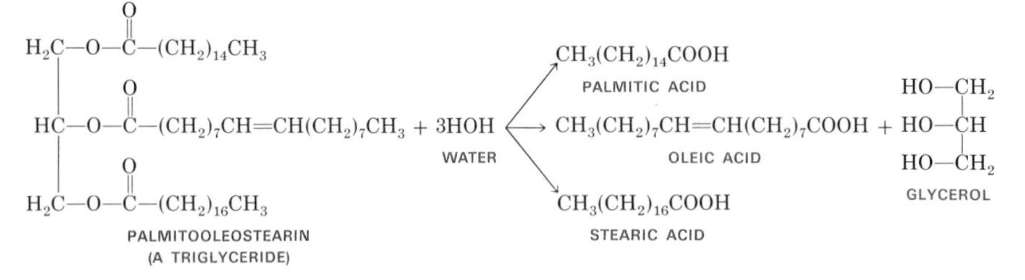

Lipid digestion occurs primarily in the intestinal tract where bile salts secreted by the liver aid in the process. The enzyme which aids in the hydrolysis of the fatty acid esters is water-soluble while the fats and oils are insoluble in water. Bile salts emulsify the oil, that is, they break up the oil into very tiny drops and prevent the drops from recombining readily. The tiny drops provide more surface area for the enzyme to attack so digestion can occur. The bile salts form an interface between the nonpolar oil and the polar water and make it possible for the oil to "dissolve" in water. For a molecule to be an emulsifier between polar and nonpolar molecules, the emulsifier must have characteristics of both. One of our principal bile salts is derived from glycocholic acid.

Bile salts act much like detergent molecules.

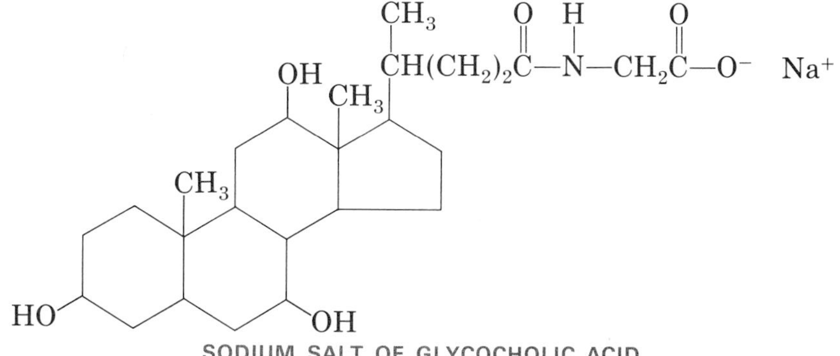

SODIUM SALT OF GLYCOCHOLIC ACID

Notice that the bulky hydrocarbon groups of this bile acid are compatible with oil or fat and that the —OH and ionic groups anchor to water molecules. The bile salts emulsify oil in a manner similar to the action of a soap or detergent during the cleaning process (Chapter 14).

Protein Digestion and Absorption

The hydrolysis of proteins begins in the stomach and continues in the small intestine. Several different types of enzymes are known to be involved. These enzymatic systems must be controlled very carefully, for they have the potential of digesting the walls of the stomach and intestines. A number of these enzymes are secreted in an inactive form. For example, pepsin, which is secreted in the stomach, is first present in a form called pepsinogen. The molecular weight of pepsinogen is 42,600 amu. In the presence of the acid of the stomach, pepsinogen is broken by still another enzyme to pepsin. The molecular weight of pepsin is 34,500 amu. It is reasonable to believe that pepsinogen, the enzyme that breaks it down, and the stomach acid normally have no effect on the stomach wall. However, pepsin would have considerable action on the stomach protein if it were formed under the mucous lining, a lining which is constantly sloughing off like the outer skin.

The stomach is protected from protein-splitting enzymes by a mucous lining.

Pepsin facilitates the breakdown of only about 10 percent of the bonds in a typical protein, leaving polypeptides with molecular weights from 600 up to 3000 amu. In the small intestine, hydrolysis is completed to amino acids which are absorbed through the intestinal wall.

Some protein enzymes are sold commercially. Meat tenderizers are proteases, materials that speed up partial digestion of meat. Enzymes are used as stain removers in detergents, although they may irritate the skin of some individuals. Related enzymes are also used to free the lens of the eye prior to cataract surgery.

Enzymes and Heredity

Genetic effects are often observed in the patterns of enzymes produced by individuals or races. An example of this is found in "lactose intolerance," common in certain peoples of Asia (e.g., Chinese and Japanese) and Africa (many black tribes) whose diets have traditionally contained little milk after the age of weaning. While infants such people manufacture the enzyme ***lactase*** which is necessary to digest lactose, a sugar occurring in all mammals' milk. As they grow older their bodies stop producing this enzyme because their diets normally contain no milk, and the ingestion of milk products containing lactose can lead to considerable discomfort in the form of stomach aches and diarrhea. People whose ancestral adult diets contained substantial amounts of milk or milk products (African tribes such as the Masai, Mongols, Caucasians, etc.) continue to produce lactase as adults and can eat such foods and digest the lactose they contain. It is quite possible that this is only one of several similar cases in which a traditional tribal diet has altered the pattern of production of digestive enzymes.

GLUCOSE METABOLISM

The biochemical process by which living cells obtain energy from glucose and similar compounds involves many chemical reactions. The initial sequence of reactions can follow two courses; one of these does not use oxygen (***anaerobic***) while the other one does (***aerobic***).

Aerobic: use oxygen. Anaerobic: use no oxygen.

When a muscle is used, glucose is anaerobically converted by a series of steps known as the ***Embden-Meyerhof pathway*** to lactic acid (Figure 17–9). We are concerned here only with the starting material, the energy flow, and the final products. The overall reaction can be represented by the equation:

$$\underset{\text{GLUCOSE}}{C_6H_{12}O_6} + 2ADP + 2H_3PO_4 \longrightarrow \underset{\text{LACTIC ACID}}{2CH_3-\overset{\overset{\displaystyle H}{|}}{\underset{\underset{\displaystyle OH}{|}}{C}}-\overset{\overset{\displaystyle O}{\|}}{C}-OH} + 2ATP + 2H_2O$$

If the muscle is used for a sufficiently long period of time, the lactic acid buildup will produce tiredness and a painful sensation. Oxygen is needed to convert the lactic acid to carbon dioxide and water, which are excreted. This is accomplished by the slower, aerobic (with air) process known as the ***Krebs cycle.*** As you look at Figure 17–10 imagine how a molecule of citric acid reacts with the enzyme aconitase. The citric acid has —OH and —H exchange sites which exchange and produce isocitric acid. The isocitric acid dissociates from aconitase and goes to the next enzyme, isocitric dehydrogenase, where two hydrogen atoms are removed by $NADP^+$, and so on around the cyclic pathway. Hydrogen atoms are removed, carboxyl groups are destroyed as carbon dioxide is formed, and hydrolysis occurs. Since two carbon atoms are fed into the cycle at oxaloacetic acid by acetyl coenzyme A, two carbons in the form of two molecules of carbon dioxide (CO_2) must be eliminated before the cycle returns to oxaloacetic acid. This is exactly what happens. See whether you can find the two molecules of CO_2 formed in the Krebs cycle. Figure 17–11 will help.

Muscular activity converts glucose to lactic acid which produces fatigue in muscles as it accumulates.

If the Krebs cycle is aerobic, where does the oxygen enter the cycle? Oxygen is required to regenerate the coenzyme $NADP^+$ from NADPH. It is also required to remove the hydrogen atoms from NADH and $FADH_2$ as well. There are about seven known steps involved in the

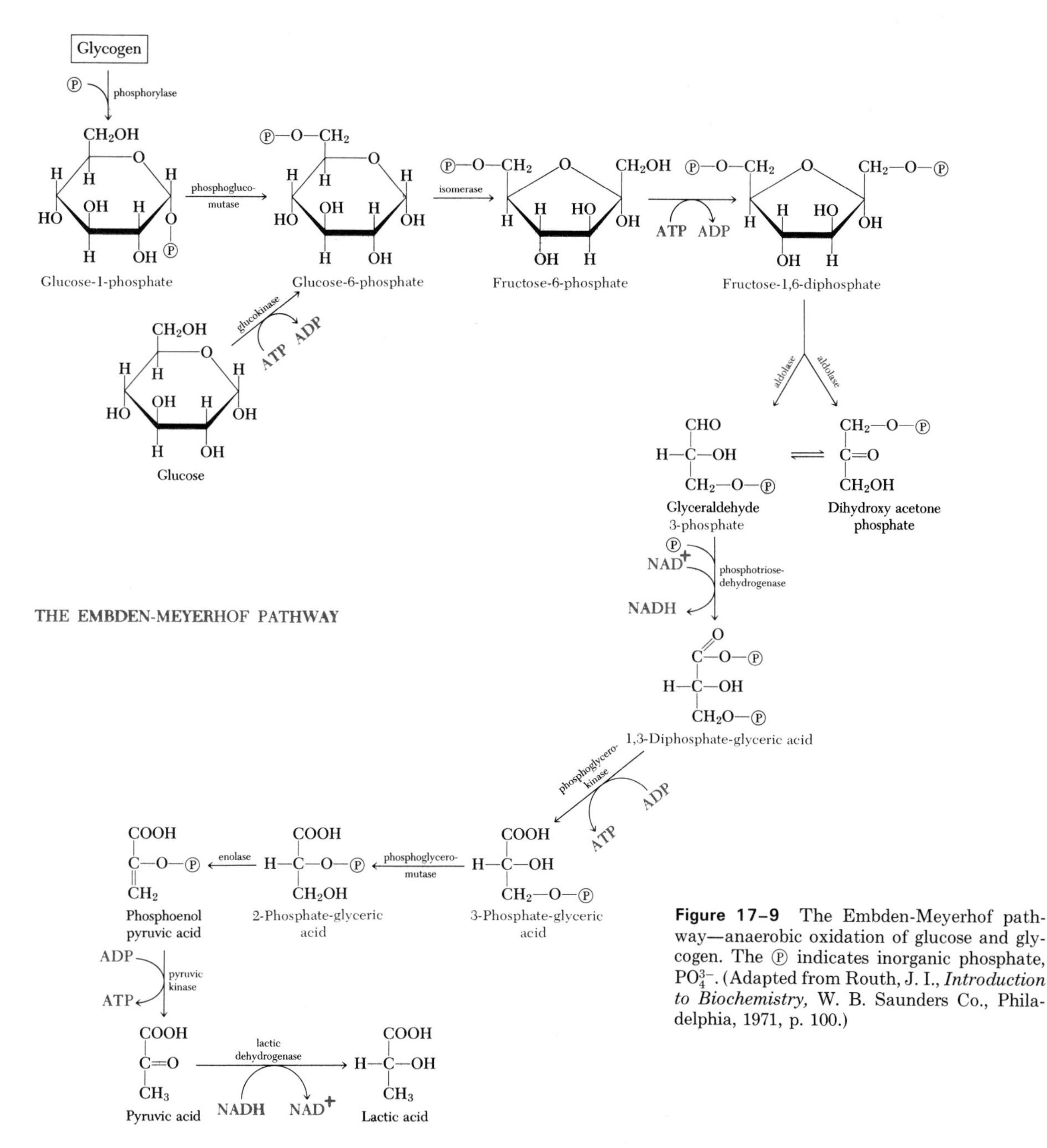

Figure 17–9 The Embden-Meyerhof pathway—anaerobic oxidation of glucose and glycogen. The Ⓟ indicates inorganic phosphate, PO_4^{3-}. (Adapted from Routh, J. I., *Introduction to Biochemistry,* W. B. Saunders Co., Philadelphia, 1971, p. 100.)

removal of hydrogen from NADH and NADPH and six steps for $FADH_2$. The steps can be summarized by the following equations:

$$NADPH + 3ADP + 3H_3PO_4 + H^+ + \tfrac{1}{2}O_2 \longrightarrow NADP^+ + 3ATP + 4H_2O$$

$$NADH + 3ADP + 3H_3PO_4 + H^+ + \tfrac{1}{2}O_2 \longrightarrow NAD^+ + 3ATP + 4H_2O$$

$$FADH_2 + 2ADP + 2H_3PO_4 + \tfrac{1}{2}O_2 \longrightarrow FAD + 2ATP + 3H_2O$$

The $NADP^+$, NAD^+, and FAD are now ready to be fed back into the Krebs cycle, where the process is continued by extracting other hydrogen atoms.

When these equations are combined with the principal sequence of

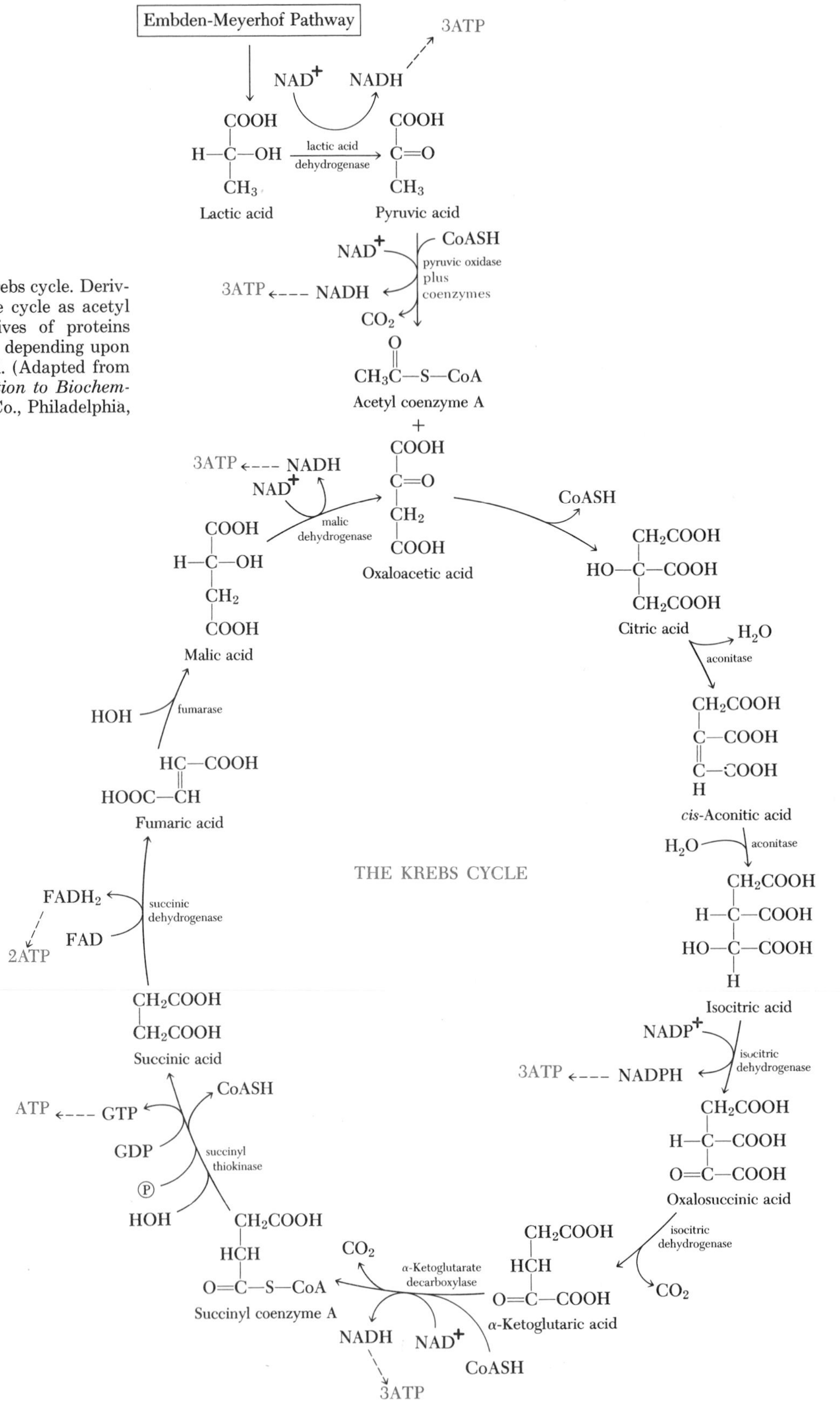

Figure 17–10 The Krebs cycle. Derivatives of fats enter the cycle as acetyl coenzyme A. Derivatives of proteins enter at various points depending upon the specific amino acid. (Adapted from Routh, J. I., *Introduction to Biochemistry,* W. B. Saunders Co., Philadelphia, 1971, p. 102.)

reactions for the Krebs cycle, we have an equation for the conversion of lactic acid into carbon dioxide and water aerobically.

$$\underset{\text{LACTIC ACID}}{C_3H_6O_3} + 18ADP + 18H_3PO_4 + 3O_2 \longrightarrow 3CO_2 + 21H_2O + 18ATP$$

The key products of glucose metabolism are ATP (energy), CO_2, and H_2O.

The key product of this oxidation reaction is ATP. Energy derived from glucose (and lactic acid) is stored and transported in the phosphorus-oxygen bonds of ATP. This is the usable product of the Krebs cycle. Waste products are carbon dioxide and water. Since two lactic acid molecules are formed from one glucose molecule, 36 ATP's are formed per glucose molecule oxidized via the Krebs cycle. Two more ATP's are formed in the Embden-Meyerhof pathway, making a total of 38 ATP molecules formed in the complete oxidation of a glucose molecule. If we burn a mole of glucose to carbon dioxide and water outside the body, 688,000 calories of energy are released. This is the total energy available. Earlier in this chapter, the point was made that breaking a P—O—P bond and forming $H_2PO_4^-$ gives up about 7,400 calories of free energy (ΔG). Thirty-eight ATP molecules would provide 281,200 calories of free energy.

$$38 \,\cancel{\text{moles ATP}} \times 7{,}400 \frac{\text{calories}}{\cancel{\text{mole ATP}}} = 281{,}200 \text{ calories}$$

This is an efficiency of 41 percent for obtaining stored energy from the total usable energy available.

$$\frac{281{,}200 \,\cancel{\text{calories}} \times 100\%}{688{,}000 \,\cancel{\text{calories}}} = 41\%$$

This figure is remarkable when you consider that the efficiency of the automobile engine is only about 20 percent, and real heat engines of any size seldom go above 35 percent.

When ATP converts back to ADP by losing a phosphate group, the energy is used to move muscles, such as the pumping action of the heart, twitching the nose, moving the diaphragm so we can breathe, and hundreds of other movements required for the everyday ongoing of life.

The energy is stored in ATP and released from ATP by means of ***coupled reactions.*** In Figures 17-10 and 17-11 curved arrows denote reactions in which sufficient free energy is provided by one reaction to drive another reaction. These are examples of coupled reactions. In fact, the very first step in the oxidation of glucose is a specific example of a coupled reaction. The substitution of a phosphate group for a hydrogen on glucose is an unfavorable reaction, as far as energy is concerned.

Coupled reactions allow the energy from one reaction to be used to run another reaction.

$$\text{Glucose} + H_3PO_4 \rightleftharpoons \text{Glucose-6-}PO_4 + H_2O; \ \Delta G = \text{about } +3{,}000 \text{ cal}$$

(UNFAVORABLE—REQUIRES ENERGY)

If this reaction is coupled with the favorable hydrolysis of ATP,

$$ATP + H_2O \rightleftharpoons ADP + H_3PO_4; \quad \Delta G = \text{about } -7{,}400 \text{ cal}$$

(FAVORABLE—RELEASES ENERGY)

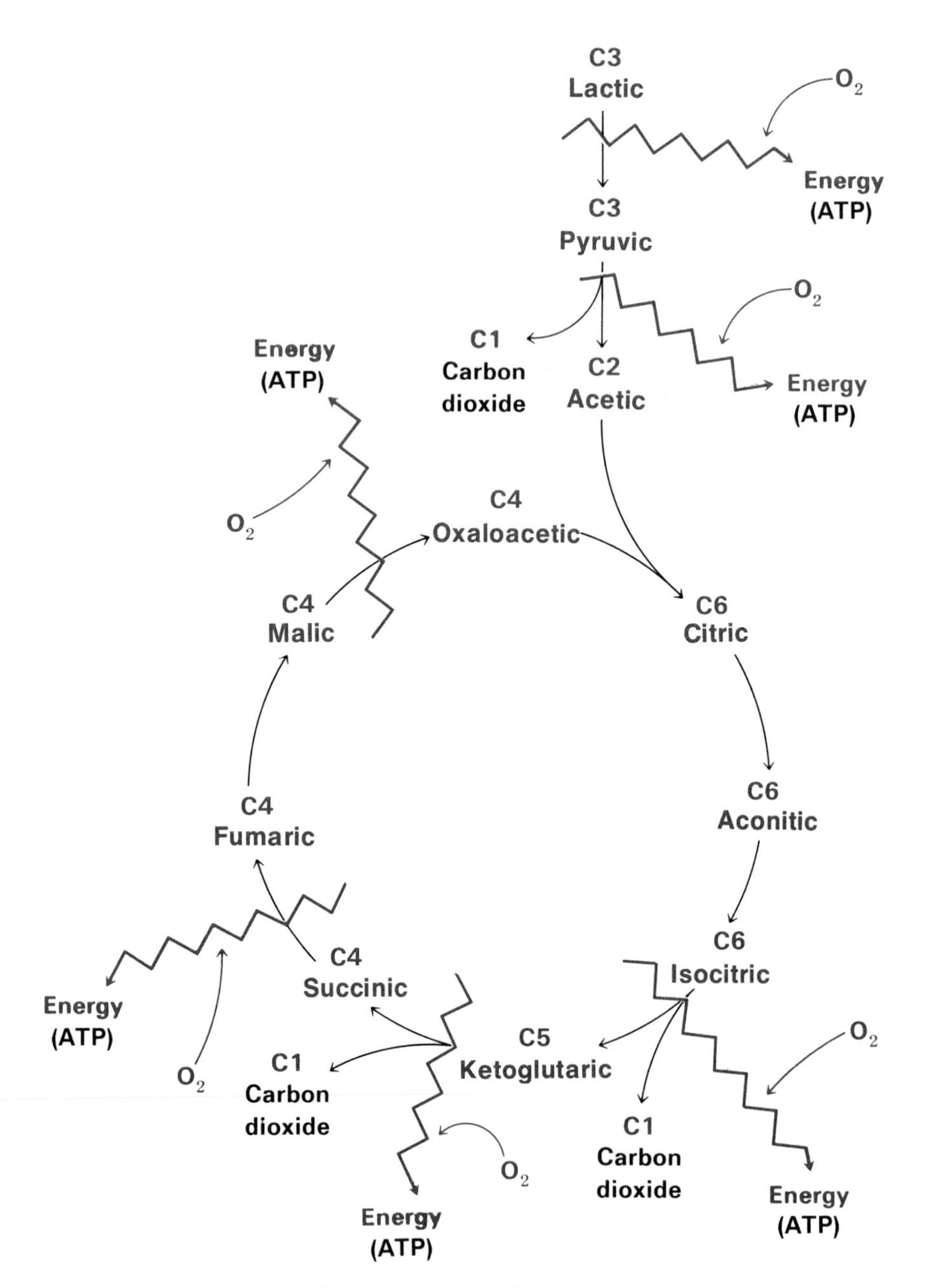

Figure 17–11 A summary of the Krebs cycle showing the number of carbon atoms per molecule for the principal substances in the cycle.

the net reaction is favorable:

$$
\begin{array}{llllll}
\text{Glucose} + \cancel{H_3PO_4} & \rightleftharpoons & \text{Glucose-6-}PO_4 & + \cancel{H_2O} & \Delta G = +3{,}000 \text{ cal} \\
\text{ATP} + \cancel{H_2O} & \rightleftharpoons & \text{ADP} & + \cancel{H_3PO_4} & \Delta G = -7{,}400 \text{ cal} \\
\hline
\text{Glucose} + \text{ATP} & \rightleftharpoons & \text{Glucose-6-}PO_4 & + \text{ADP} & \Delta G = -4{,}400 \text{ cal}
\end{array}
$$

Even with this favorable free energy change, the enzyme glucokinase is required to hasten the process.

Reactions which are favored energetically may still require a catalyst to proceed at a reasonable rate.

Fats and proteins (as amino acids) can enter the Krebs cycle also. Amino acids can enter at several places after first being converted by a series of reactions into one of the compounds in the cycle. Each amino acid has its particular point of entry. For example, aspartic acid is converted into α-ketoglutaric acid and enters at that point in the Krebs cycle.

Fatty acids go through a series of at least five reactions in which the hydrocarbon chain of the fatty acid is decreased by a two-carbon fragment. The fragment completes a molecule of acetyl coenzyme A. Palmitic acid, a 16-carbon fatty acid (see Chapter 14), would do seven turns around the cycle to form eight acetyl CoA molecules.

In conclusion, you now have seen some of the detailed chemistry involved in simply raising your arm, and you are now aware of what happens to some of the sugars and starches that disappear down the hatch. Of course, there is much more known than is presented here and there appears to be no end to what is left to be discovered.

SELF-TEST 17-C

1. Digestion is the breakdown of foodstuffs by ________________.
2. Substances whose solubility characteristics are similar to those of fats and oils are termed ________________.
3. Bile salts act as (catalysts/emulsifying agents/enzymes).
4. The two products of the Embden-Meyerhof pathway are __________ and ________________.
5. The end products of the Krebs energy cycle are the molecules __________ and __________.
6. Energy is stored and released from ATP by chemical reactions called ________________ reactions.

MATCHING SET

______	1. energy "cash" in the living cell	a. ptyalin
______	2. enzyme which splits maltose into two molecules of glucose	b. ADP + energy
______	3. enzyme which splits polysaccharides in the mouth	c. Krebs cycle
______	4. compound which prevents pellagra	d. chlorophylls
______	5. first oxidation product of glucose	e. coupled reaction

_______	6. occurs under aerobic (with air) conditions	f.	maltase
_______	7. reaction utilizing ATP	g.	lactic acid
_______	8. product of ATP hydrolysis	h.	niacin
_______	9. metal ion in an enzymatic system	i.	ATP
_______	10. molecules which absorb light energy	j.	coenzyme
_______	11. due to inadequate supply of lactase	k.	lactose intolerance

QUESTIONS

1. Write a basic equation for the digestion of:
   a. starch to a disaccharide
   b. a disaccharide to a simple sugar
   c. a protein to amino acids
   d. a triglyceride to fatty acids

2. What is the metal in chlorophyll? Recall a similar metal complex from Chapter 16.

3. If you were to "feed" radioactive carbon dioxide to a green plant, what would be the first radioactive carbon compound formed? Who made this discovery?

4. Give the structure of ATP and point out the region of the molecule that contains bonds which are hydrolyzed in coupled reactions.

5. What is meant by coupled reactions? Give an example. Why do the energetics of biochemical systems make coupled reactions necessary?

6. Using the energy diagrams, explain the concept of activation energy for a chemical reaction and show the effect of an enzyme on the activation energy.

7. Point out three important similarities of the lock-and-key analogy to enzymatic activity.

8. What is a coenzyme and why are they sometimes necessary?

9. What type of compound first absorbs light energy in photosynthesis? Give an example.

10. In photosynthesis why is it partially correct to say that light is an oxidizing agent?

11. What are the two major divisions in photosynthesis? Express in words what is accomplished in each.

12. What is the source of oxygen in photosynthesis?

13. What part of photosynthesis could even take place in an animal cell?

14. Since chlorophyll loses electrons because of light, it must subsequently gain electrons from somewhere. Where do they come from?

15. What is the basic nature of the digestion processes for large molecules?

16. The chemical changes in the Krebs cycle can be classified as dehydrogenations (removal of hydrogens, a type of oxidation), dehydration (removal of water), hydrolysis (reaction with water in which water loses its molecular identity), decarboxylation (removal of COOH group and formation of CO_2), and phosphorylations (adding a phosphate group, such as $H_2PO_4^-$). Beginning with citric acid and

progressing around the cycle to oxaloacetic acid, determine the total number of each kind of chemical change.

17. What compound produces soreness after a period of exercise?

18. Which compounds in the CO_2 fixation scheme in photosynthesis could enter directly the reactions of the Embden-Meyerhof pathway or the Krebs cycle?

19. If protein digestion is facilitated by enzymes, and these enzymes are produced in body organs made of proteins, explain why the enzymes do not cause rapid digestion of the organs themselves.

20. What is pyruvic acid? Why is it so important in getting energy from sugars?

21. What happens if an amino acid is needed for protein synthesis and the amino acid can neither be made by the body nor obtained from the diet? Does the modern theory of protein synthesis include an explanation of the role of essential amino acids? (These are amino acids which cannot be manufactured by the human body; they must be obtained in the diet.)

PROBLEMS

1. If a certain muscle requires 10 calories for contraction and obtains this ultimately by the hydrolysis of ATP to ADP, what is the minimum number of molecules of ATP that are needed to furnish the energy for such a contraction process?

Ans. 8.1×10^{20} moles

2. What volume of O_2 gas (measured under standard conditions of temperature and pressure) is released to the atmosphere for each gram of CO_2 transformed to glucose by photosynthesis?

Ans. 0.5 liter

SUGGESTIONS FOR FURTHER READING

Baldwin, E., "The Nature of Biochemistry," Second Edition, Cambridge University Press, Cambridge, England, 1967.
"Bio-Organic Chemistry," Readings From *Scientific American,* W. H. Freeman and Company, San Francisco, 1968.
Govindjee, A., and Govindjee, R., "The Primary Events of Photosynthesis," *Scientific American,* Vol. 231, No. 6, p. 68 (1974).
Harrison, K., "Guidebook to Biochemistry," Second Edition, Cambridge University Press, Cambridge, England, 1965.
Horecker, R. L., "Pathways of Carbohydrate Metabolism and Their Biological Significance," *J. Chem. Ed.* Vol. 42, p. 244 (1965).
Lehninger, A. L., "Bioenergetics," W. A. Benjamin Inc., N.Y., 1965.
Levine, "The Mechanism of Photosynthesis," *Scientific American,* Vol. 221, No. 6, p. 58 (1969).
Maxwell, K. E., "Chemicals and Life," Dickerson Publishing Co., Inc., Belmont, Calif., 1970.
Mazur, A., and Harrow, B., "Biochemistry: A Brief Course," W. B. Saunders Co., Philadelphia, 1968.
Oesper, P., "Error and Trial: The Story of the Oxidative Reaction of Glycolysis," *J. Chem. Ed.* Vol. 45, p. 607 (1968).
Routh, J. I., "Introduction to Biochemistry," W. B. Saunders Co., Philadelphia, 1971.

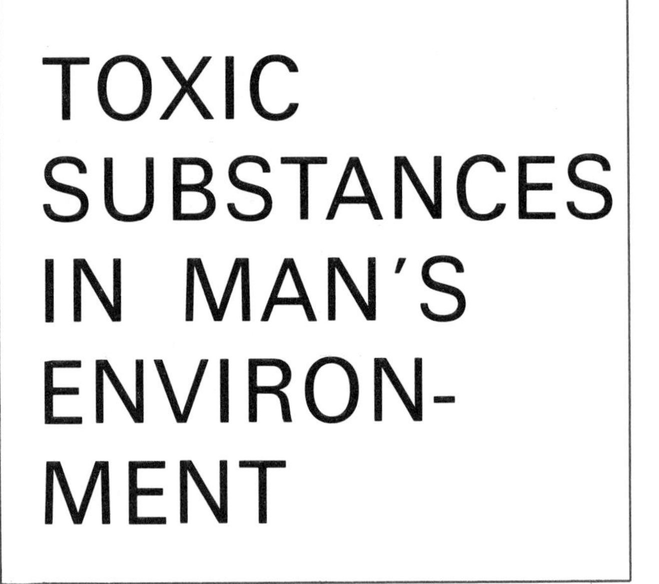

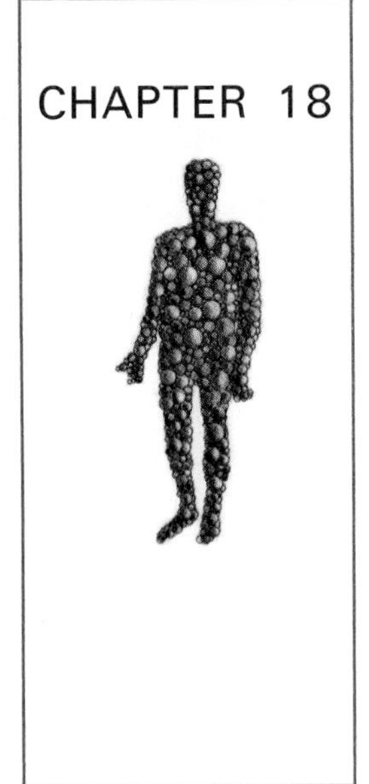

TOXIC SUBSTANCES IN MAN'S ENVIRONMENT

Toxic substances are materials which upset the incredibly complex system of chemical reactions occurring in the human body. Sometimes toxic substances cause mere discomfort; sometimes they cause illness, or even death. Toxic symptoms can be caused by very small amounts of toxic materials (an example is sodium cyanide), and the term ***toxic substances*** usually is limited to these materials. However, as most of us know, ill effects can be caused by excessive intake of substances normally considered harmless (eating too much candy, for example). Fortunately, in most cases the human body is capable of recognizing "foreign" chemicals and ridding itself of them. In this chapter, we shall focus on the chemical mechanisms by which toxic substances act and the chemical ways by which the body protects itself against toxic substances.

A large enough dose of any compound can result in poisoning.

DOSE

Lethal doses of toxic substances are customarily expressed in milligrams (mg) of substance per kilogram (kg) of body weight of the subject. For example, the cyanide ion (CN^-) is generally fatal to human beings in a dose of 1 mg of CN^- per kg of body weight. For a 200-pound (90.7 kg) person, about one tenth of a gram of cyanide is a lethal dose. Examples of somewhat less toxic substances and the range of lethal doses for human beings are the following:

"Dosis sola facit veneum"—the dose makes the poison.

Morphine	1–50 mg per kg
Aspirin	50–500 mg per kg
Methyl alcohol	0.5–5 g per kg
Ethyl alcohol	5–15 g per kg

A quantitative measure of toxicity is obtained by introducing various dosages of substances to be tested into laboratory animals (such as rats). That dosage which would be lethal in 50 percent of a large number of the animals under controlled conditions is called the LD_{50} (lethal dosage—50%) and is reported in milligrams of poison per kilogram of body weight. Thus, if a statistical analysis of data on a large population of rats showed that a dosage of 1 mg per kg was lethal to 50 percent of the population tested, the LD_{50} for this poison would be 1 mg per kg. Obviously, metabolic variations and other differences between species will produce different LD_{50} values for a given poison in different kinds of animals. For this reason such data cannot be extrapolated to humans with any assurance, but it is safe to assume that a substance with a low LD_{50} value for several animal species will also be quite toxic to humans.

Metabolism (from the Greek, *metaballein,* to change or alter) is the sum of all the physical and chemical changes by which living organisms are produced and maintained.

Toxic substances can be classified according to the way in which they disrupt the chemistry of the body. Some of the modes of action of toxic substances can be described as ***corrosive, metabolic, neurotoxic, mutagenic,*** and ***carcinogenic,*** and these will serve as the bases of our discussion.

CORROSIVE POISONS

Toxic substances that actually destroy tissues are corrosive poisons. Examples include strong acids and alkalies, and many oxidants such as those found in laundry products, which can destroy tissues. Sulfuric acid (found in auto batteries) and hydrochloric acid (also called muriatic acid; used for cleaning purposes) are very dangerous corrosive poisons. Death has resulted from the swallowing of 1 ounce of concentrated (98 percent) sulfuric acid, and much smaller amounts can cause extensive damage and severe pain.

Concentrated mineral acids such as sulfuric acid act by first dehydrating cellular structures. After the cell dies, its protein structures are destroyed by the acid-catalyzed hydrolysis of the peptide bonds.

$$R-\overset{\overset{\Large O}{\|}}{C}-\overset{\overset{\Large H}{|}}{N}-R' + H_2O \xrightarrow[\text{from acid}]{H^+} R-\overset{\overset{\Large O}{\|}}{C}-OH + H-\overset{\overset{\Large H}{|}}{N}-R$$

PEPTIDE LINK (IN PROTEIN) — CARBOXYL END OF SMALLER PEPTIDE OR AMINO ACID — AMINE END OF SMALLER PEPTIDE OR AMINO ACID

Strong acids and bases destroy cell protoplasm.

In the early stages of this process there will be a large proportion of larger fragments present. Subsequently as more bonds are broken, smaller and smaller fragments result, leading to the ultimate disintegration of the tissue.

Some poisons act by undergoing chemical reaction to produce corrosive poisons. Phosgene, the deadly gas used during World War I, is an example. When inhaled, it is hydrolyzed in the lungs to hydrochloric acid, which causes pulmonary edema (a collection of fluid in the lungs) due to the dehydrating effect of the strong acid on tissues. The victim dies of suffocation because oxygen cannot be absorbed effectively by the damaged tissues.

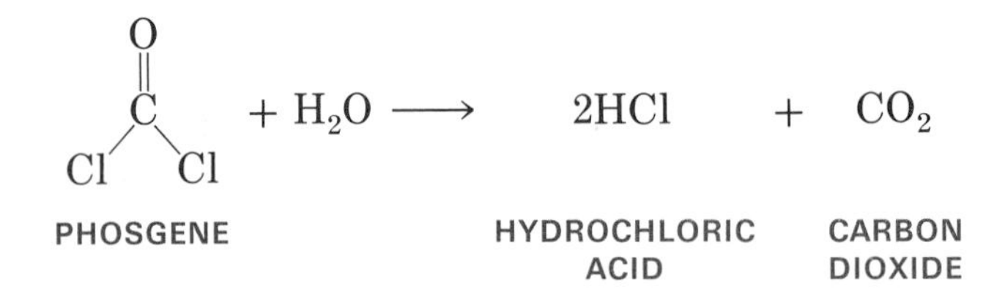

Chemical "warfare gases," such as phosgene, were outlawed by an international conference in 1925.

Sodium hydroxide, NaOH (caustic soda—a component of drain cleaners), is a very strongly alkaline, or basic, substance that can be just as corrosive to tissue as strong acids. The hydroxide ion also catalyzes the splitting of peptide linkages:

$$\mathrm{R{-}\overset{\overset{\displaystyle O}{\|}}{C}{-}\overset{\overset{\displaystyle H}{|}}{N}{-}R} + H_2O \xrightarrow[\text{base}]{OH^-} \mathrm{R{-}\overset{\overset{\displaystyle O}{\|}}{C}{-}OH} + \mathrm{H{-}\overset{\overset{\displaystyle H}{|}}{N}{-}R}$$

Both acids and bases, as well as other types of corrosive poisons, continue their action until they are consumed in chemical reactions.

Some corrosive poisons destroy tissue by oxidizing it. This is characteristic of substances such as ozone, nitrogen dioxide, and possibly iodine, which destroy enzymes by oxidizing their functional groups. Specific groups, such as the —SH and —S—S— groups in the enzyme, are believed to be converted by oxidation to nonfunctioning groups; alternatively, the oxidizing agents may break chemical bonds in the enzyme, leading to its inactivation.

A summary of some common corrosive poisons is presented in Table 18-1.

There are many common uses for corrosive poisons.

TABLE 18–1 Some Corrosive Poisons

Substance	Formula	Toxic Action	Possible Contact
Hydrochloric acid	HCl	Acid hydrolysis	Tile and concrete floor cleaner
Sulfuric acid	H_2SO_4	Acid hydrolysis, dehydrates tissue—oxidizes tissue	Auto batteries
Phosgene	ClCOCl	Acid hydrolysis	Combustion of chlorine-containing plastics (PVC or Saran)
Sodium hydroxide	NaOH	Base hydrolysis	Caustic soda, drain cleaners
Trisodium phosphate	Na_3PO_4	Base hydrolysis	Detergents, household cleaners
Sodium perborate	$NaBO_3 \cdot 4H_2O$	Base hydrolysis—oxidizing agent	Laundry detergents, denture cleaners
Ozone	O_3	Oxidizing agent	Air, electric motors
Nitrogen dioxide	NO_2	Oxidizing agent	Polluted air
Iodine	I_2	Oxidizing agent	Antiseptic
Hypochlorite ion	OCl^-	Oxidizing agent	Bleach
Peroxide ion	O_2^{2-}	Oxidizing agent	Bleach
Oxalic acid	$H_2C_2O_4$	Reducing agent, precipitates Ca^{2+}	Bleach, ink eradicator, leather tanning, rhubarb, spinach, tea
Sulfite ion	SO_3^{2-}	Reducing agent	Bleach

METABOLIC POISONS

Metabolic poisons are more subtle than the tissue-destroying corrosive poisons. In fact, many of them do their work without actually indicating their presence until it is too late. Metabolic poisons can cause illness or death by interfering with a vital biochemical mechanism to such an extent that it ceases to function or is prevented from functioning efficiently.

Carbon Monoxide

The interference of carbon monoxide with extracellular oxygen transport is one of the best understood processes of metabolic poisoning. As early as 1895, it was noted that carbon monoxide deprives body cells of oxygen (asphyxiation), but it was much later before it was known that carbon monoxide, like oxygen, combines with hemoglobin:

$$O_2 + \text{hemoglobin} \longrightarrow \text{oxyhemoglobin}$$

$$CO + \text{hemoglobin} \longrightarrow \text{carboxyhemoglobin}$$

Laboratory tests show that carbon monoxide reacts with hemoglobin to give a compound (carboxyhemoglobin) which is 140 times more stable than the compound of hemoglobin and oxygen (oxyhemoglobin). (See Figure 18–1.)

An organic material which undergoes incomplete combustion will always liberate carbon monoxide. Sources include auto exhausts, smoldering leaves, lighted cigars or cigarettes, and charcoal burners. In the United States alone, combustion sources of all types dump about two hundred million tons of carbon monoxide per year into the atmosphere.

ppm—parts per million—a measure expressing concentration. 50 ppm CO means 50 g CO for every million grams of air.

While the best estimates of the maximum global background level of carbon monoxide are of the order of 0.1 ppm, the background concentration in cities is higher. In heavy traffic, sustained levels of 50 or more

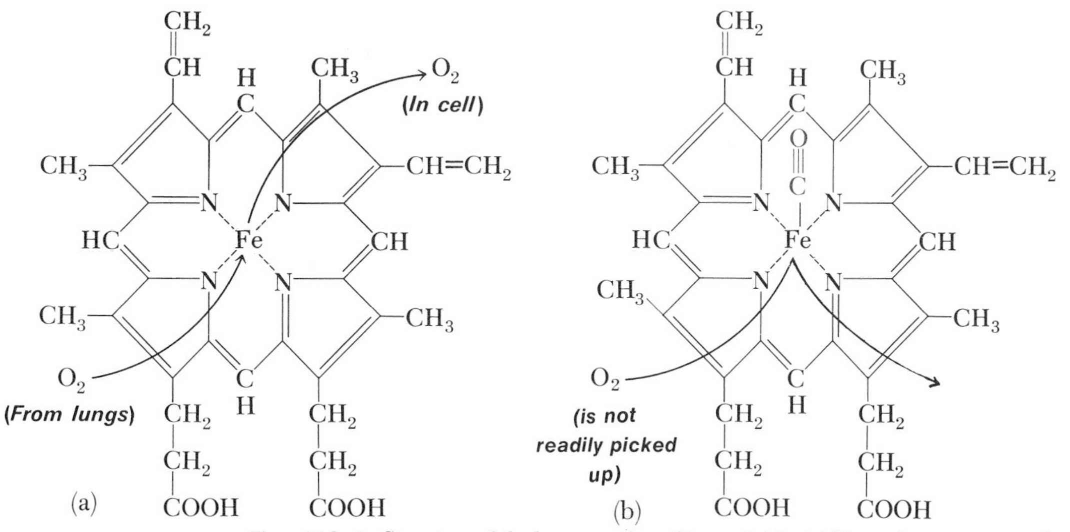

Figure 18–1 Structure of the heme portion of hemoglobin. (a) Normal acceptance and release of oxygen. (b) Oxygen blocked by carbon monoxide.

TABLE 18–2 Concentration of CO in Atmosphere versus Percentage of Hemoglobin (Hb) Saturated*

CO concentration in air	0.01% (100 ppm)	0.02% (200 ppm)	0.10% (1,000 ppm)	1.0% (10,000 ppm)
Percentage of hemoglobin molecules saturated with CO†	17	20	60	90

*A few hours of breathing time is assumed.
†Normal human blood contains up to 5% of the hemoglobin as carboxyhemoglobin (HbCO).

ppm are common; for offstreet sites an average of about 7 ppm is typical for large cities. A concentration of 30 ppm for eight hours is sufficient to cause headache and nausea. Breathing an atmosphere which is 0.1 percent (1,000 ppm) carbon monoxide for four hours converts approximately 60 percent of the hemoglobin of an average adult to carboxyhemoglobin (Table 18–2), and death is likely.

To convert ppm to per cent, divide by 10,000.

Since both the carbon monoxide and oxygen reactions with hemoglobin involve equilibria, concentrations, as well as relative strengths of bonds, affect the positions of the equilibria. In air that contains 0.1 percent CO, oxygen molecules outnumber CO molecules 200 to 1. The larger concentration of oxygen helps to counteract the greater combining power of CO with hemoglobin. Consequently, if a carbon monoxide victim is exposed to fresh air or, still better, pure oxygen (provided he is still breathing), the carboxyhemoglobin (HbCO) is gradually decomposed, due to the greater concentration of oxygen:

Air is 21% O_2 by volume so, in 1,000,000 "air molecules" there would be 210,000 O_2 molecules.

$$HbCO + O_2 \rightleftharpoons HbO_2 + CO$$

EQUILIBRIUM SHIFTED TO RIGHT BECAUSE OF GREATER CONCENTRATION OF OXYGEN

Although carbon monoxide is not a cumulative poison, permanent damage can occur if certain vital cells (e.g., brain cells) are deprived of oxygen for more than a few minutes.

Individuals differ in their tolerance of carbon monoxide, but generally those with anemia or an otherwise low reserve of hemoglobin (e.g., children) are more susceptible. No one is helped by carbon monoxide, and smokers suffer chronically from its effects. It is a subtle poison, since it is odorless and tasteless.

Cyanide

The cyanide ion, CN^-, is the toxic agent in cyanide salts such as sodium cyanide used in electroplating. Since the cyanide is a relatively strong base, it reacts easily with many acids (weak and strong) to form the volatile hydrogen cyanide gas, HCN:

$$\underset{\text{ACETIC ACID}}{CH_3COOH} + NaCN \rightleftharpoons \underset{\text{HYDROGEN CYANIDE}}{HCN} + CH_3COO^- + Na^+$$

Since HCN boils at a relatively low temperature (26°C), it is a gas at temperatures slightly above room temperature. It is often used as a fumigant in storage bins and holds of ships because it is toxic to most forms of life and, in gaseous form, can penetrate into tiny openings, even into insect eggs.

Natural sources of cyanide ions include the seeds of the cherry, plum, peach, apple, and apricot fruits. Hydrocyanic acid is produced by hydrolysis of certain compounds, such as amygdalin, contained in the seeds:

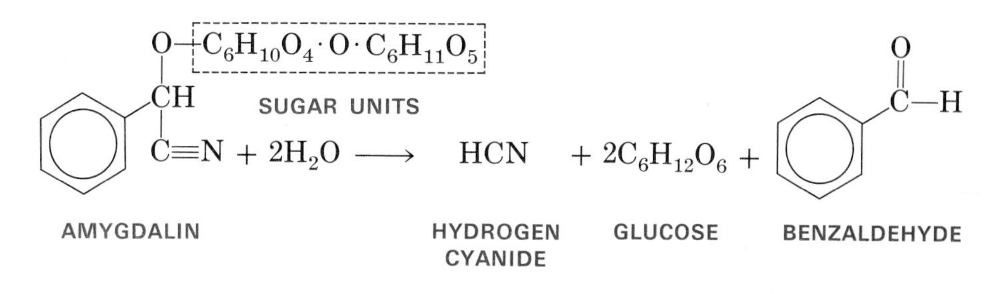

The cyanide is not toxic as long as it is tied up in the amygdalin, but presumably if enough apple or peach seeds were hydrolyzed in warm acid, sufficient HCN would result to cause considerable danger. There are a few recorded instances of humans poisoned by eating large numbers of apple seeds. Amygdalin is not confined to the seeds; amounts as high as 66 mg per 100 g have been reported in peach leaves.

The cyanide ion is one of the most rapidly working poisons. Lethal doses taken orally act in minutes. Cyanide poisons by asphyxiation, as does carbon monoxide, but the mechanism of cyanide poisoning is different (Figure 18–2). Instead of preventing the cells from getting oxygen, cyanide interferes with oxidative enzymes, such as cytochrome oxidase. Oxidases are enzymes containing a metal, usually iron or copper. They catalyze the oxidation of substances such as glucose:

A metabolite is any substance in a metabolic process.

$$\text{Metabolite } (H)_2 + \tfrac{1}{2}\, O_2 \xrightarrow{\text{Oxidase}} \text{Oxidized metabolite} + H_2O + \text{Energy}$$

The iron atom in cytochrome oxidase is oxidized from Fe^{2+} to Fe^{3+} to provide electrons for the reduction of O_2. The iron regains electrons from other steps in the process. The cyanide ion forms stable cyanide complexes with the metal ion of the oxidase and renders the enzyme incapable of reducing oxygen or oxidizing the metabolite.

$Fe^{2+} \longrightarrow Fe^{3+} + e^-$ Oxidation

$$\text{Cytochrome oxidase (Fe)} + CN^- \longrightarrow \text{cytochrome oxidase } \underbrace{(\text{Fe})\cdots CN^-}_{\text{complex}}$$

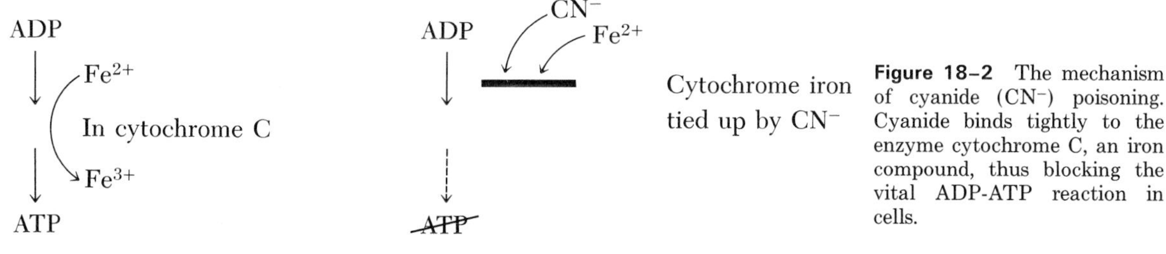

Figure 18–2 The mechanism of cyanide (CN^-) poisoning. Cyanide binds tightly to the enzyme cytochrome C, an iron compound, thus blocking the vital ADP-ATP reaction in cells.

In essence the electrons of iron are "frozen"—they cannot participate in the oxidation-reduction processes. Plenty of oxygen gets to the cells, but the mechanism by which the oxygen is used in the support of life is stopped. Hence the cell dies, and if this occurs fast enough in the vital centers, the victim dies.

The body has a mechanism for ridding itself slowly of cyanide ions. The cyanide-oxidative enzyme reaction is reversible, and other enzymes such as rhodanase, found in almost all cells, can convert cyanide ions to relatively harmless thiocyanate ions. For example,

The body can rid itself of a large number of toxic substances if the dose is small enough and sufficient time is allowed.

$$CN^- + \underset{\text{THIOSULFATE}}{S_2O_3^{2-}} \xrightarrow{\text{Rhodanase}} \underset{\text{THIOCYANATE}}{SCN^-} + SO_3^{2-}$$

This mechanism is not as effective in protecting a cyanide-poison victim as it might appear, since there is only a limited amount of thiosulfate available in the body at a given time.

Fluoroacetic Acid

Nature has used the synthesis of fluoroacetic acid as part of a defense mechanism for certain plants. Native to South Africa, the *gilfbaar* plant contains lethal quantities of fluoroacetic acid. Cattle that eat these leaves usually sicken and die.

$$\begin{array}{c} \text{F} \quad \text{O} \\ | \quad \| \\ \text{H—C—C—O—H} \\ | \quad\quad\quad \\ \text{H} \quad\quad\quad \end{array}$$

FLUOROACETIC ACID

Sodium fluoroacetate, the sodium salt of this acid (Compound 1080), is a potent rodenticide. Because it is odorless and tasteless it is especially dangerous, and its sale in this country is strictly regulated by law.

Fluoroacetate is toxic because it enters the Krebs cycle, producing fluorocitric acid, which in turn blocks the Krebs cycle (Figure 18–3). The C—F linkage apparently ties up the enzyme aconitase, thus preventing it from converting citrate to isocitrate.

In this instance, the poison is similar enough to the normal substrate to compete effectively for the active sites on the enzyme. If a poison has sufficient affinity for the active site on the enzyme, it blocks the normal function of the enzyme. The blocking of the Krebs cycle by fluorocitrate is a typical example of this affinity. If fluoroacetate is not present in excessive amounts, its action can be reversed simply by increasing the concentration of available citrate.

Some poisons can act by mimicking other compounds.

Heavy Metals

Heavy metals are perhaps the most common of all the metabolic poisons. These include such common elements as lead and mercury, as well as many less common ones such as cadmium, chromium, and thallium. In this group we should also include the infamous poison, *arsenic,*

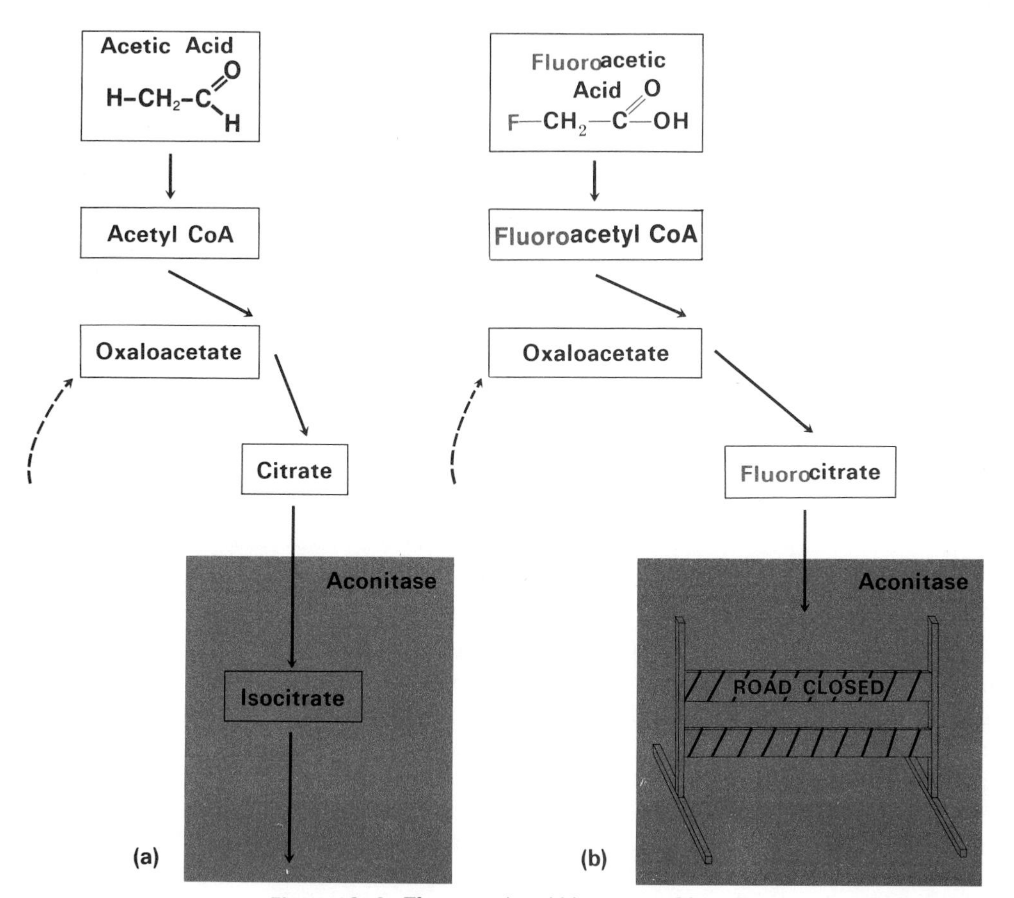

Figure 18–3 Fluoroacetic acid is converted into fluorocitrate, which then forms a stable bond with the enzyme aconitase. This blocks the normal Krebs cycle, a portion of which is shown.

which is really not a metal but is metal-like in many of its properties, including its toxic action.

Arsenic, a classic homicidal poison, occurs in small amounts in many foods, since arsenic compounds are still used occasionally to spray fruits and vegetables. The Federal Food and Drug Administration (FDA) has a limit of 0.15 mg of arsenic per pound of food, and this amount apparently causes no harm. Several drugs, such as arsphenamine, which has found some use in treating syphilis, contain covalently bonded arsenic. In its ionic forms, arsenic is much more toxic.

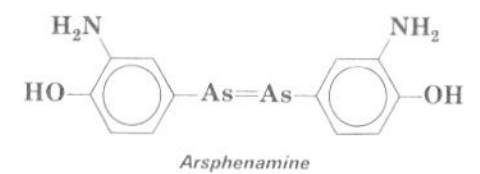

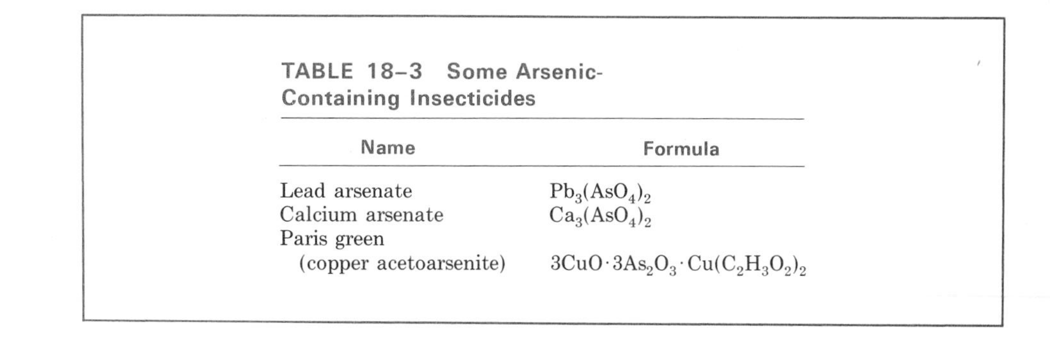

TABLE 18–3 Some Arsenic-Containing Insecticides

Name	Formula
Lead arsenate	$Pb_3(AsO_4)_2$
Calcium arsenate	$Ca_3(AsO_4)_2$
Paris green (copper acetoarsenite)	$3CuO \cdot 3As_2O_3 \cdot Cu(C_2H_3O_2)_2$

$$2 \text{ Glutathione} + \text{Metal ion } (M^{2+}) \longrightarrow M(\text{Glutathione})_2 + 2H^+$$

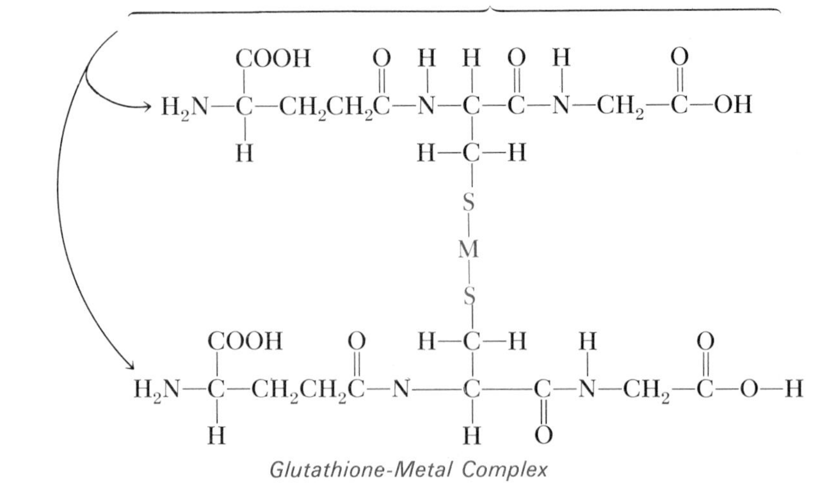

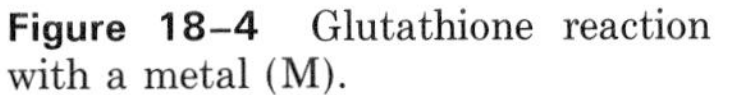
Figure 18–4 Glutathione reaction with a metal (M).

Arsenic and heavy metals owe their toxicity primarily to their ability to react with and inhibit sulfhydryl (—SH) enzyme systems, such as those involved in the production of cellular energy. For example, glutathione (a tripeptide of glutamic acid, cysteine, and glycine) occurs in most tissues; its behavior with metals illustrates the interaction of a metal with sulfhydryl groups. The metal replaces the hydrogen on two sulfhydryl groups on adjacent molecules (Figure 18–4), and the strong bond which results effectively eliminates the two glutathione molecules from further reaction.

The typical forms of toxic arsenic compounds are inorganic ions such as arsenate (AsO_4^{3-}) and arsenite (AsO_3^{3-}). The reaction of an arsenite ion with sulfhydryl groups results in a complex in which the arsenic unites with two sulfhydryl groups, which may be on two different molecules of protein or on the same molecule:

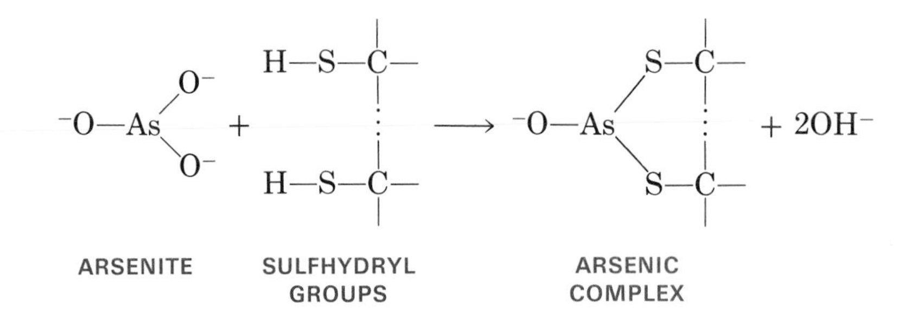

The problem of developing a compound to counteract *Lewisite,* an arsenic-containing poison gas used in World War I, led to an understanding of how arsenic acts as a poison and subsequently to the development of an antidote. Once it was understood that Lewisite poisoned people by the reaction of arsenic with protein sulfhydryl groups, British scientists set out to find a suitable compound that contained highly reactive sulfhydryl groups which could compete with sulfhydryl groups in the poisoned molecule for the arsenic, and thus render the poison ineffective. Out of this research came a compound now known as British Anti-Lewisite (BAL).

CH_2—CH—CH_2 (OH, SH, SH)

BAL
British Anti-Lewisite

A chelating agent encases an atom or ion like a crab or an octopus surrounds a bit of food.

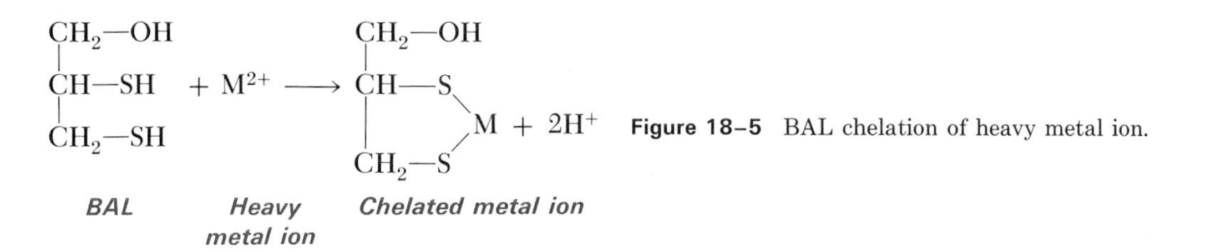

Figure 18–5 BAL chelation of heavy metal ion.

The BAL, which bonds to the metal at several sites, is called a ***chelating agent*** (Greek, *chela,* claw), a term applied to a reacting agent that envelops a species such as a metal ion. BAL is one of a large number of compounds which can act as chelating agents for metals (Figure 18–5).

With the arsenic or heavy metal ion tied up, the sulfhydryl groups in vital enzymes are freed and can resume their normal functions. BAL is a standard therapeutic item in a hospital's poison emergency center and is used routinely to treat heavy metal poisoning.

A vivid description of the psychic changes produced in an individual by mercury poisoning can be found in the Mad Hatter, a character in Lewis Carroll's *Alice in Wonderland.* An old practice in the fur felt industry involved the use of mercury (II) nitrate, $Hg(NO_3)_2$, to stiffen the felt. This not only accounted for the Mad Hatter's odd behavior, but also gave the workers in the hat factories chronic mercury poisoning, with symptoms known as "hatter's shakes."

Mercury deserves some special attention because it has a rather peculiar fascination for some people, especially children, who love to touch it. It is poisonous and, to make matters worse, mercury and its salts accumulate in the body. This means the body has no quick means of ridding itself of this element and there tends to be a build-up of the toxic effects leading to ***chronic*** poisoning.

While mercury is rather unreactive compared to other metals, it is quite volatile and easily absorbed through the skin. In the body the metal atoms are oxidized to Hg_2^{2+} [mercury (I) ion] and Hg^{2+} [mercury (II) ion]. Compounds of both Hg_2^{2+} and Hg^{2+} are known to be toxic.

Today mercury poisoning is a potential hazard to those working with or near this metal or its salts, such as dentists (who use it in making amalgams for fillings), various medical and scientific laboratory personnel (who routinely use mercury compounds or mercury pressure gauges), and some agricultural workers (who employ mercury salts as fungicides).

Figure 18–6 The first step to coating a coin with mercury. Children love to coat coins with metallic mercury—a very dangerous practice since mercury is easily inhaled and also passes through the skin.

Mercury can also be a hazard when it is present in food. It is generally believed that mercury enters the food chain through small organisms that feed at the bottom of bodies of water. These in turn are food for bottom-feeding fish. Game fish in turn eat these fish and accumulate the largest concentration of mercury, the accumulation of poison building up as the food chain progresses.

Amalgam: any mixture or alloy of metals in which mercury is a constituent.

Lead is another widely encountered heavy-metal poison. The body's method of handling lead provides an interesting example of a "metal equilibrium" (Figure 18–7). Lead often occurs in foods (100 to 300 μg per kg), beverages (20 to 30 μg per liter), public water supplies (100 μg per liter, from lead-sealed pipes), and even air (2.5 μg per cubic meter, from lead compounds in auto exhausts). With this many sources and contacts per day, it is obvious that the body must be able to rid itself of this poison; otherwise everyone would have died long ago of lead poisoning! The average person can excrete about two milligrams of lead a day through the kidneys and intestinal tract; the daily intake is normally less than this. However, if intake exceeds this amount, accumulation and storage results. In the body lead not only resides in soft tissues but is also deposited in bone. In the bones lead acts on the bone marrow, while in tissues it behaves like other heavy-metal poisons, such as mercury and arsenic. Lead, like mercury and arsenic, can also affect the central nervous system.

1 μg (microgram) = 10^{-6} gram

1 mg (milligram) = 10^{-3} gram = 1,000 μg

Lead salts, unless they are very insoluble, are always toxic, and their toxicity is directly related to the salt's solubility. One common covalent lead compound, tetraethyl lead, $Pb(C_2H_5)_4$, a component of most anti-knock gasolines, is different from most other metal compounds in that

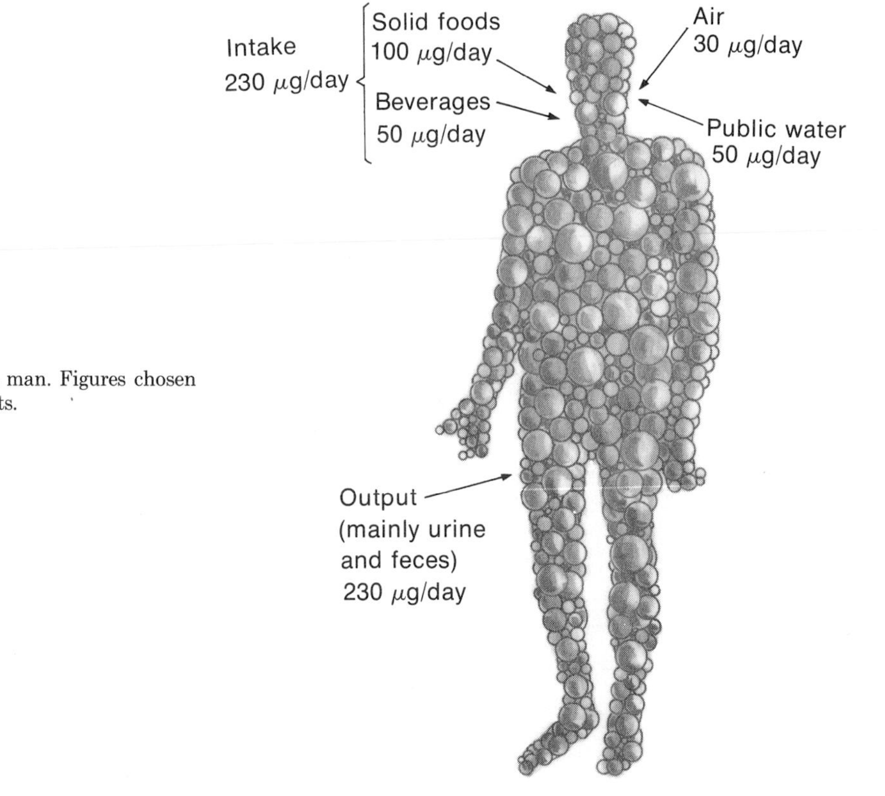

Figure 18–7 Lead equilibrium in man. Figures chosen for intake are probable upper limits.

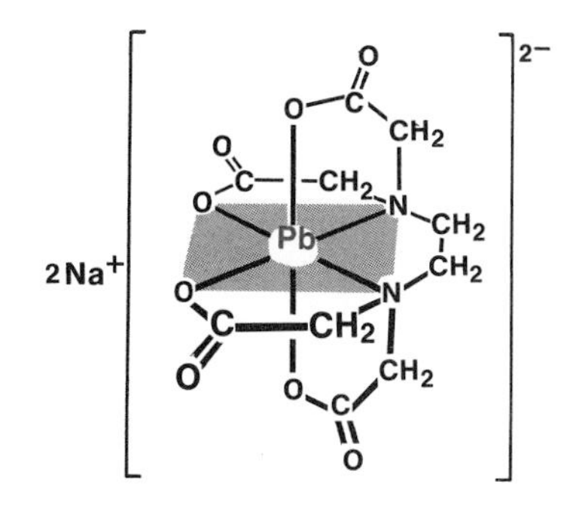

Figure 18–8 The structure of the chelate formed when the anion of EDTA envelops a lead (II) ion.

it is readily absorbed through the skin. Even metallic lead can be absorbed through the skin; cases of lead poisoning have resulted from repeated handling of lead foil, bullets, and other lead objects.

One of the truly tragic aspects of lead poisoning is the fact that even though lead-pigmented paints have not been used in this country for interior painting during the past 30 years, children are still poisoned by lead from old paint. Health experts estimate that up to 225,000 children become ill from lead poisoning each year, with many experiencing mental retardation or other neurological problems. The reason for this is twofold. Lead-based paints still cover the walls of many older dwellings. Coupled with this is the fact that many children in poverty-stricken areas are ill-fed and anemic. These children develop a peculiar appetite trait called ***pica,*** and among the items that satisfy their cravings are pieces of flaking paint, which may contain lead. In 1969, about 200 children in the United States alone died of lead poisoning and untold thousands suffered permanent damage.

In order to remove lead from the human body toxicologists have discovered a very effective chelating agent—ethylenediaminetetraacetic acid, also called EDTA (Figure 18–8).

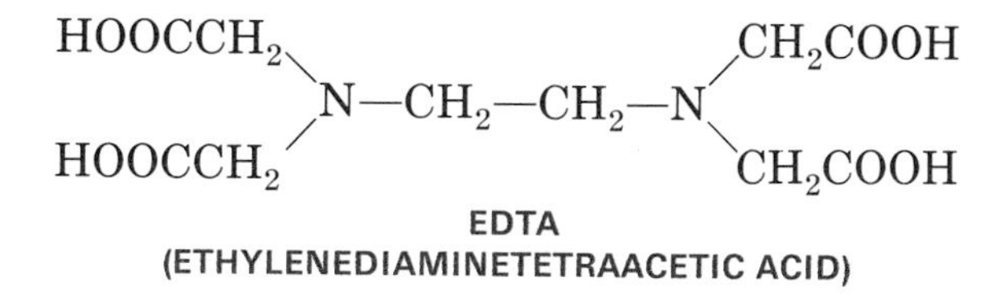

The calcium disodium salt of EDTA is used in the treatment of lead poisoning because EDTA by itself would remove too much of the blood serum's calcium. In solution EDTA has a greater tendency to complex with lead (Pb^{2+}) than with calcium (Ca^{2+}). As a result, the calcium is released and the lead is tied up in the complex:

$$[CaEDTA]^{2-} + Pb^{2+} \longrightarrow [PbEDTA]^{2-} + Ca^{2+}$$

The lead chelate is then excreted in the urine.

SELF-TEST 18-A

1. Corrosive poisons such as sulfuric acid destroy tissue by ________________ followed by ________________ of proteins.

2. Corrosive poisons, such as ozone, nitrogen dioxide, and iodine, destroy tissue by ________________ it.

3. Carbon monoxide poisons by forming a very strong bond with iron in ________________ and thus prevents the transport of ________________ from the lungs to the cells throughout the body.

4. CO is a cumulative poison. True/False.

5. The cyanide ion has the formula, ________. It poisons by complexing with iron in the enzyme, ________________ ________________, and thus prevents the use of ________________ in the oxidative processes in the cells.

6. Give an example of a metabolic poison which is toxic because its structure is so similar to a useful substance that it can mimic the useful substance.

   ____

7. BAL is an antidote for ________________________________. It is effective because its sulfhydryl (—SH) groups ________________ arsenic and heavy metals and render them ineffective toward enzymes.

8. Mercury is a cumulative poison. True/False.

NEUROTOXINS

Some metabolic poisons are known to limit their action to the nervous system. These include poisons such as strychnine and curare (a South American Indian dart poison), as well as the dreaded nerve gases developed for chemical warfare. The exact modes of action of most neurotoxins are not known for certain, but investigations have discovered the action of a few.

Investigations of the actions of neurotoxins have provided insight into how the nervous system works.

A nerve impulse or stimulus is transmitted along a nerve fiber by electrical impulses. The nerve fiber connects with either another nerve fiber or with some other cell (such as a gland, or cardiac, smooth, or skeletal muscle) capable of being stimulated by the nerve impulse (Figure 18-9). Neurotoxins often act at the point where two nerve fibers come together, called a ***synapse.*** When the impulse reaches the end of certain nerves, a small quantity of acetylcholine is liberated. This activates a receptor on an adjacent nerve or organ. The acetylcholine is thought to activate a nerve ending by changing the permeability of the nerve cell membrane. The method of increasing membrane permeability is not clear, but it may be related to an ability to dissociate fat-protein complexes or to penetrate the surface films of fats. Such effects can be brought about by as little as 10^{-6} mole of acetylcholine which could alter the permeability of a cell so ions can cross the cell membrane more freely.

$$CH_3\overset{O}{\overset{\|}{C}}OCH_2CH_2\overset{CH_3}{\overset{|}{\underset{CH_3}{\underset{|}{N^+}}}}CH_3,\ OH^-$$

Acetylcholine

Permeability: the ability of a membrane to let chemicals pass through it.

10^{-6} of a mole of acetylcholine is 6×10^{17} molecules.

To enable the receptor to receive further electrical impulses, the

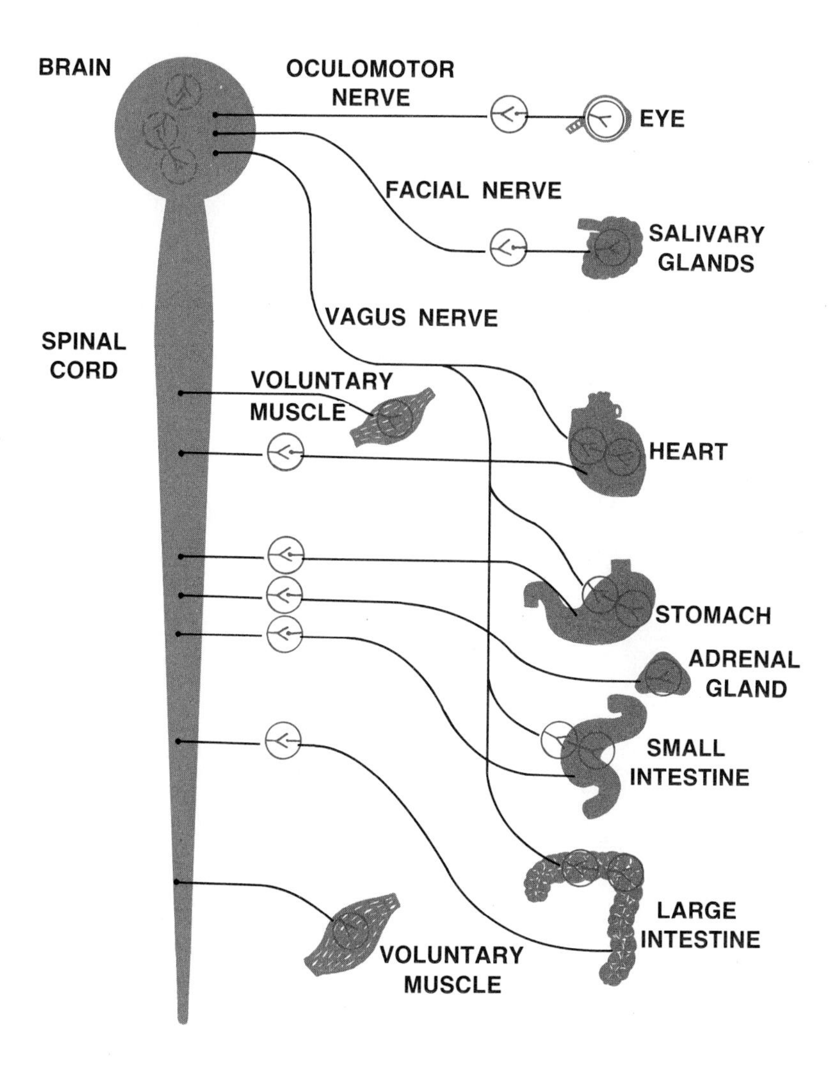

Figure 18–9 "Cholinergic" nerves, which transmit impulses by means of acetylcholine, include nerves controlling both voluntary and involuntary activities. Exceptions are parts of the "sympathetic" nervous system that utilize norepinephrine instead of acetylcholine. Sites of acetylcholine secretion are circled in color; poisons that disrupt the acetylcholine cycle can interrupt the body's communications at any of these points. The role of acetylcholine in the brain is uncertain, as is indicated by the broken circles.

enzyme ***cholinesterase*** breaks down ***acetylcholine*** into acetic acid and choline (Figure 18–10):

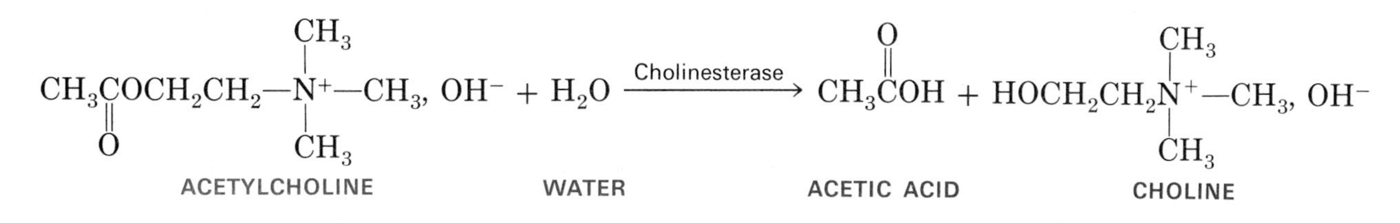

$$CH_3C(=O)OCH_2CH_2{-}N^+(CH_3)_2{-}CH_3,\ OH^- + H_2O \xrightarrow{\text{Cholinesterase}} CH_3C(=O)OH + HOCH_2CH_2N^+(CH_3)_2{-}CH_3,\ OH^-$$

In the presence of potassium and magnesium ions, other enzymes, such as acetylase, resynthesize new acetylcholine from the acetic acid and the choline within the incoming nerve ending:

$$\text{Acetic acid} + \text{Choline} \xrightarrow{\textit{Acetylase}} \text{Acetylcholine} + H_2O$$

The new acetylcholine is available for transmitting another impulse across the gap.

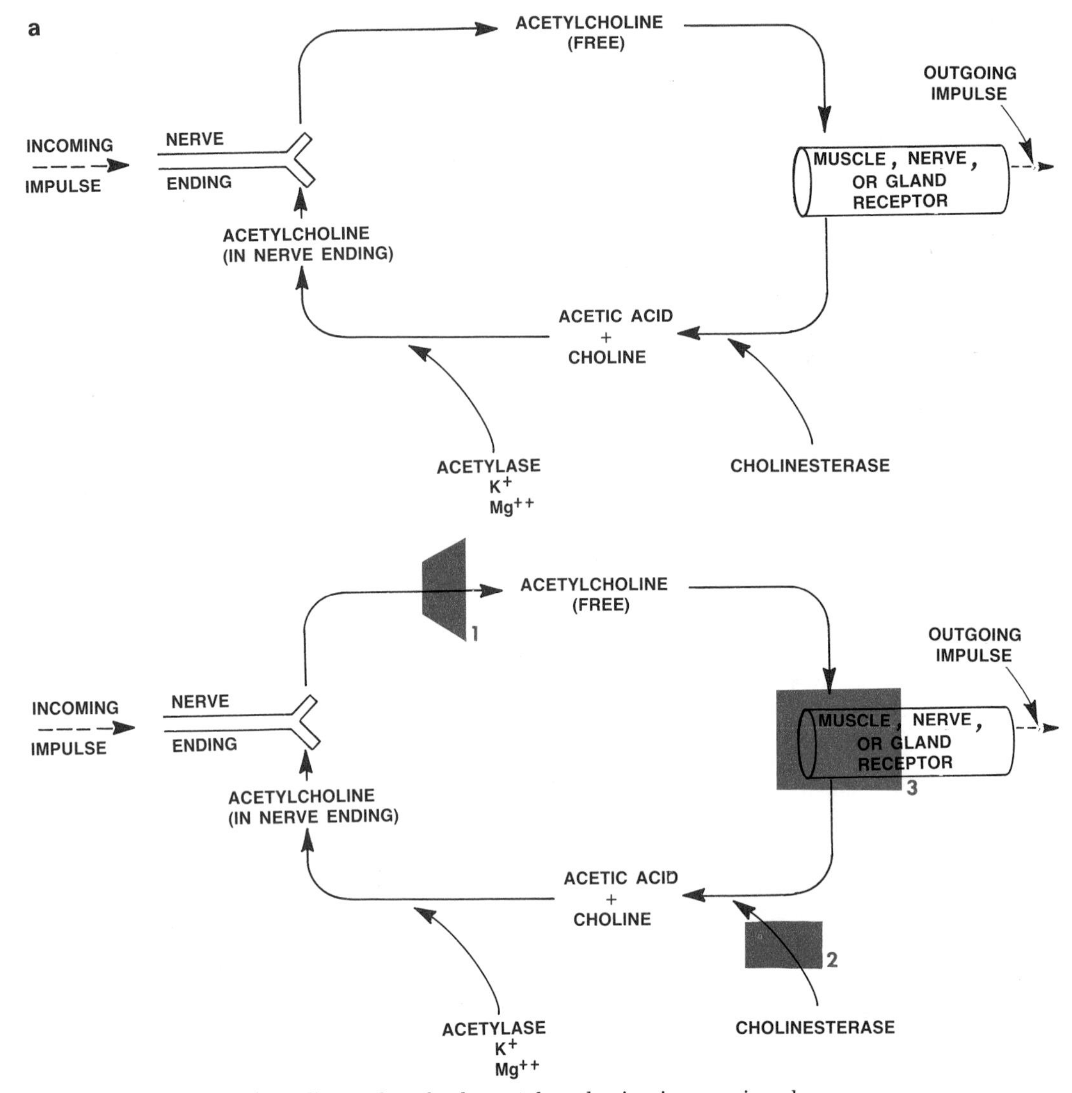

Figure 18–10 The acetylcholine cycle, a fundamental mechanism in nerve impulse transmission, is affected by many poisons. An impulse reaching a nerve ending in the normal cycle (*a*) liberates acetylcholine, which then stimulates a receptor. To enable the receptor to receive further impulses, the enzyme *cholinesterase* breaks down acetylcholine into acetic acid and choline; other enzymes resynthesize these into more acetylcholine. (*b*) Botulinus and dinoflagellate toxins inhibit the synthesis, or the release, of acetylcholine (*1*). The "anticholinesterase" poisons inactivate cholinesterase, and therefore prevent the breakdown of acetylcholine (*2*). Curare and atropine desensitize the receptor to the chemical stimulus (*3*).

Neurotoxins can affect the transmission of nerve impulses at nerve endings in a variety of ways. The ***anticholinesterase poisons*** prevent the breakdown of acetylcholine by deactivating cholinesterase. These poisons are usually structurally analogous to acetylcholine, so they bond to the enzyme cholinesterase and deactivate it (Figure 18–11). The cholinesterase molecules bound by the poison are held so effectively that the restoration of proper nerve function must await the manufacture of new cholinesterase. In the meantime, the excess acetylcholine overstimulates nerves, glands, and muscles, producing irregular heart rhythms, convulsions, and death. Many of the organic phosphates which are widely

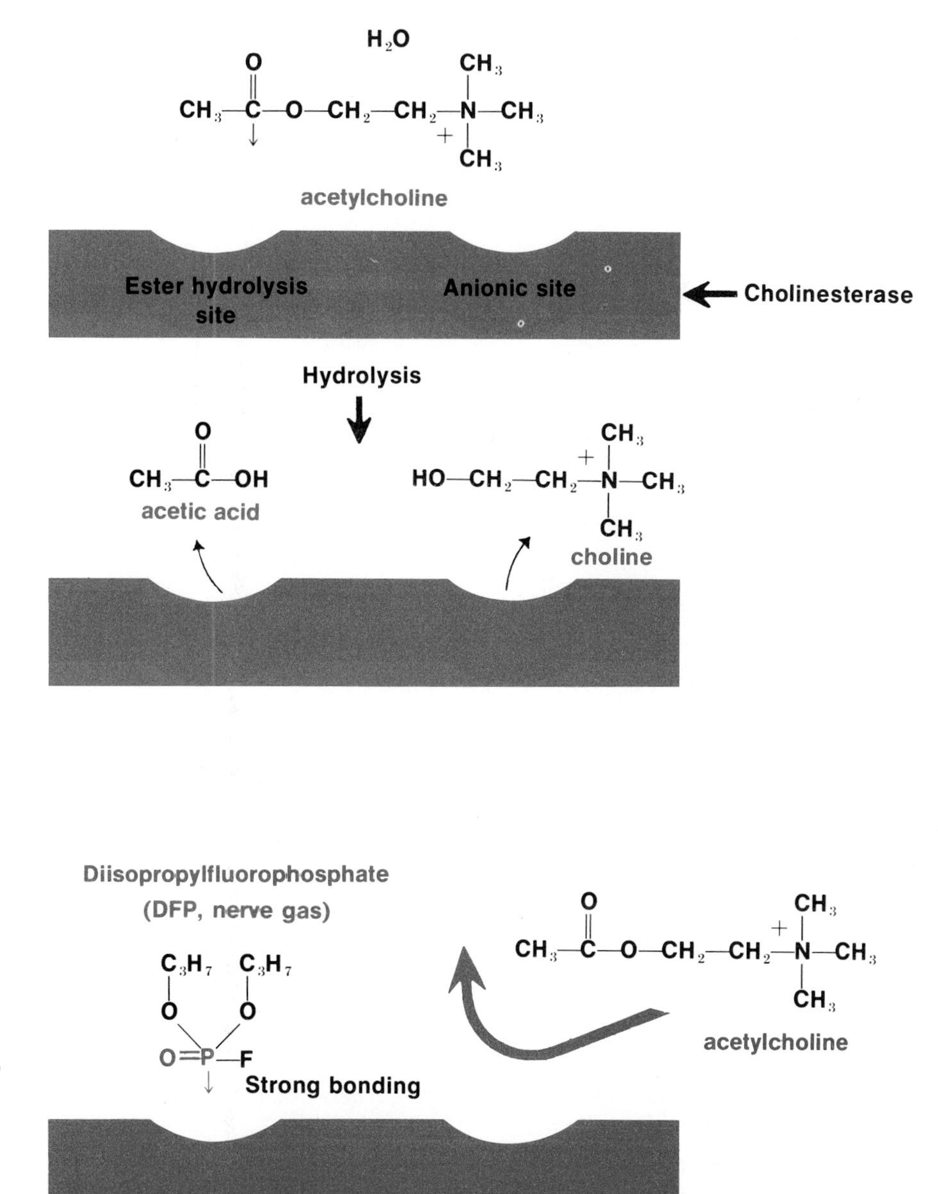

Figure 18–11 (a) The mechanism of cholinesterase breakdown of acetylcholine. (b) The tie-up of cholinesterase by an anticholinesterase poison like the nerve gas DFP blocks the normal hydrolysis of acetylcholine since the acetylcholine cannot bind to the enzyme.

used as insecticides are metabolized in the body to produce anticholinesterase poisons. For this reason, they should be treated with extreme care. Some poisonous mushrooms also contain an anticholinesterase poison. Figure 18–12 contains the structures of some anticholinesterase poisons.

Curare, used by South American Indians in poison darts, was brought to Europe by Sir Walter Raleigh in 1595. It was purified in 1865, and its structure was determined in 1935.

Neurotoxins, such as ***atropine*** and ***curare,*** are able to occupy the receptor sites on nerve endings of organs which are normally occupied by the impulse-carrying acetylcholine. When atropine or curare occupies the receptor site, no stimulus is transmitted to the organ. Acetylcholine in

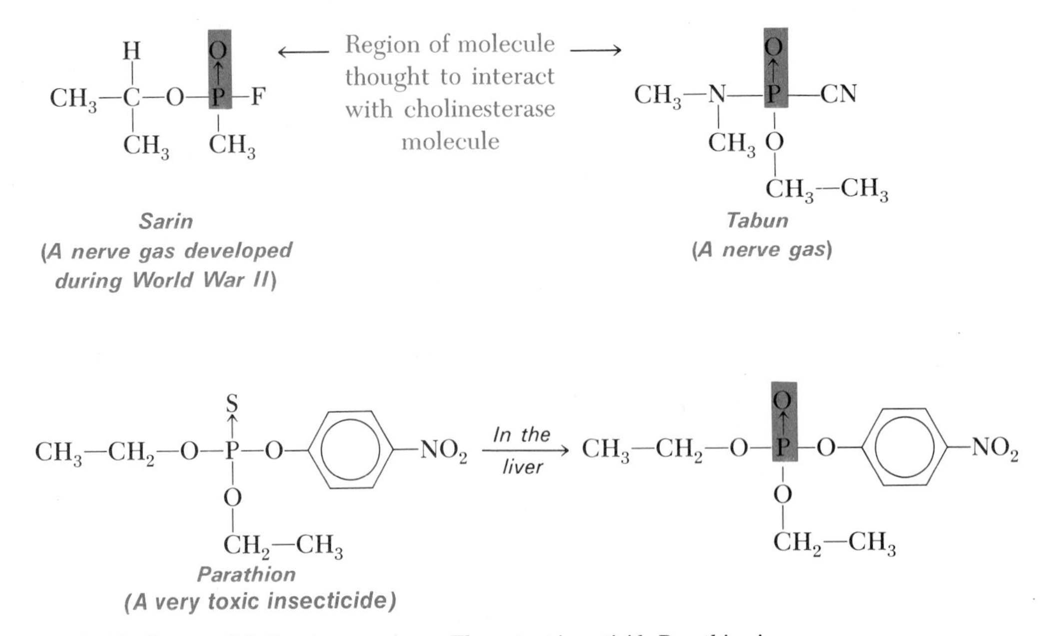

Figure 18–12 Some anticholinesterase poisons. The potent insecticide Parathion is converted in the liver to a molecule much like the nerve gases (colored area).

excess causes a slowing of the heart beat, a decrease in blood pressure, and excessive saliva, whereas atropine and curare produce excessive thirst and dryness of the mouth and throat, a rapid heart beat, and an increase in blood pressure. The normal responses to acetylcholine activation are absent, and the opposite responses occur when there is sufficient atropine present to block the receptor sites.

Neurotoxins of this kind can be extremely useful in medicine. For example, atropine is used to dilate the pupil of the eye to facilitate examination of its interior. Applied to the skin, atropine sulfate, or other atropine salts, relieves pain by deactivating sensory nerve endings on the skin. Atropine is also used as an antidote for anticholinesterase poisons. Curare has long been used as a muscle relaxant.

A well-known, natural organic compound that blocks receptor sites in a manner similar to that of curare and atropine is ***nicotine.*** This powerful poison causes stimulation and then depression of the central nervous system. The probable lethal dose for a 70-kilogram man is less than 0.3 gram. It is interesting to note that pure nicotine was first extracted from tobacco and its toxic action observed *after* tobacco was established as an acceptable habit.

Morphine is the most effective pain killer known.

Natural or synthetic morphine is the most effective pain reliever known. It is widely used for short-term acute pain resulting from surgery, fractures, burns, etc., as well as to reduce suffering in the later stages of terminal illnesses such as cancer. The manufacture and distribution of these narcotic drugs are stringently controlled by the federal government through laws designed to keep these products available for legitimate medical use. Under federal law, some preparations containing small amounts of narcotic drugs may be sold without a prescription (for example, cough mixtures containing codeine), but not many.

In spite of stringent controls, drugs like ***morphine, heroin, Meperidine,*** and ***Methadone*** are abused and illicitly used. As of June 30, 1973,

TABLE 18–4 Alkaloid Neurotoxins That Compete with Acetylcholine for the Receptor Site*

Name	Normal Contact	Lethal Dose (for a 70 kg man)	Formula
Atropine	Dilation of pupil of the eye	0.1 g	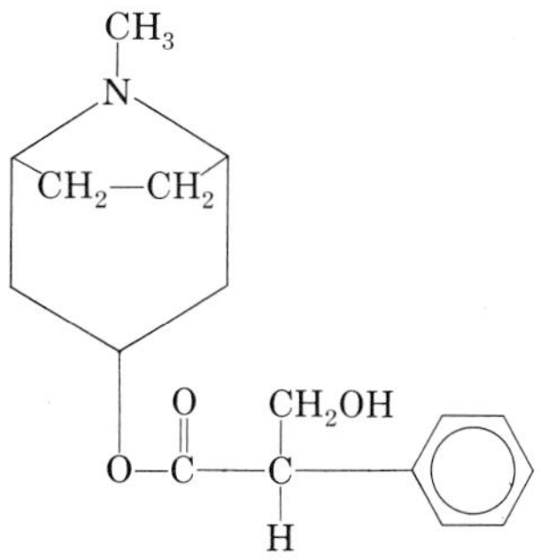
Curare	Muscle relaxant	20 mg	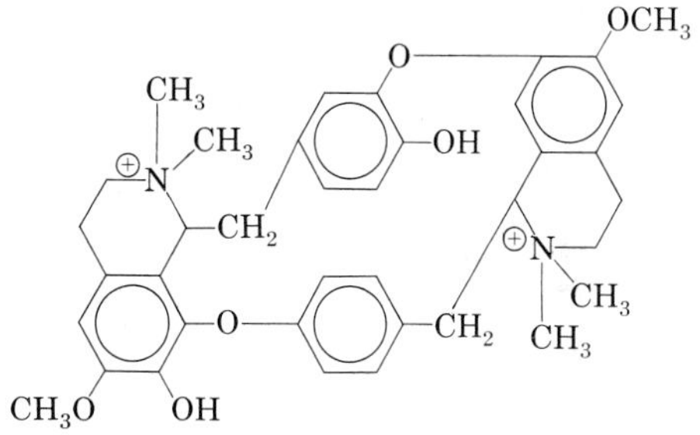
Nicotine	Tobacco, insecticide	75 mg	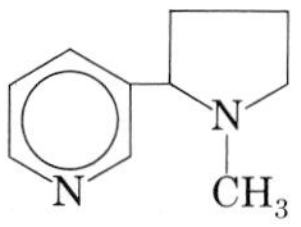
Caffeine	Coffee, tea, cola drinks		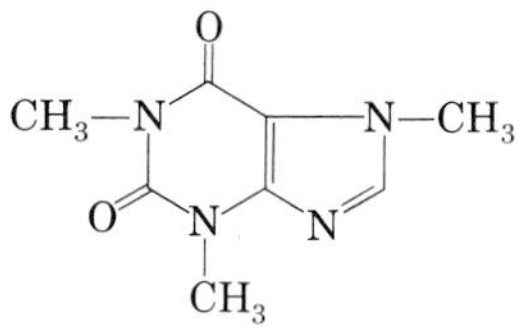
Morphine	Opium—pain killer	100 mg	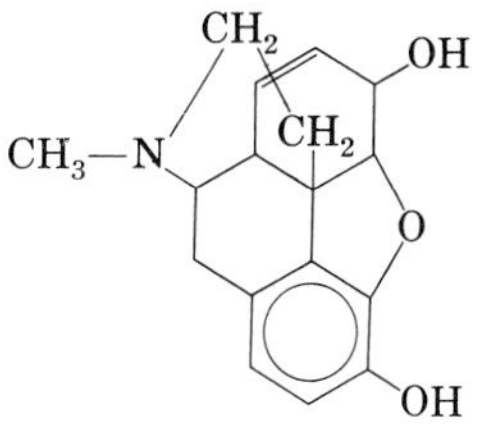
Codeine	Opium—pain killer	0.3 g	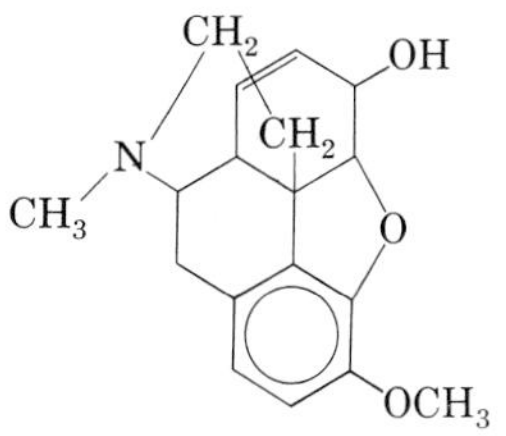
Cocaine	Leaves of *Erythroxylon coca* in South America	1 g	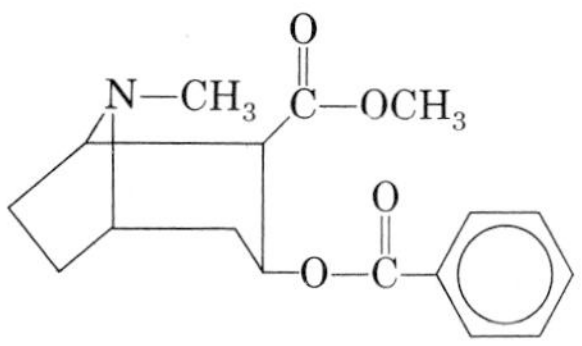

****Alkaloid*** is broadly defined as a physiologically active compound found in plants and containing amino nitrogen atoms (consequently it has basic properties). The nitrogen atom, or atoms, are frequently found as part of rings.

there were 95,897 recorded drug addicts in the U. S. These figures are not intended to represent all addicts. Estimates by the Drug Enforcement Administration show that a more realistic figure for June 30, 1973 might be 612,478. The average addict requires 50 mg of heroin per day to sustain the habit, at a cost of about $58 per day, or $21,170 per year. Heroin is prepared from morphine, which is derived from sap in the opium poppy. It takes about 10 pounds of opium to prepare one pound of morphine. Morphine reacts with acetic anhydride in a one-to-one reaction to form heroin. Street-grade heroin is only 9 to 10% pure.

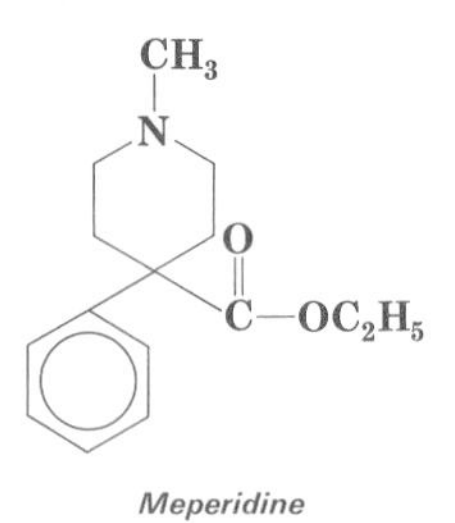

Meperidine

Meperidine and Methadone are products of chemical laboratories rather than of poppy fields. Meperidine was claimed to be nonaddictive when first produced. Experience, however, proved otherwise (as it did with morphine and heroin). A major difference between Methadone and morphine and heroin is that when Methadone is taken orally, under medical supervision, it prevents withdrawal symptoms for approximately 24 hours.

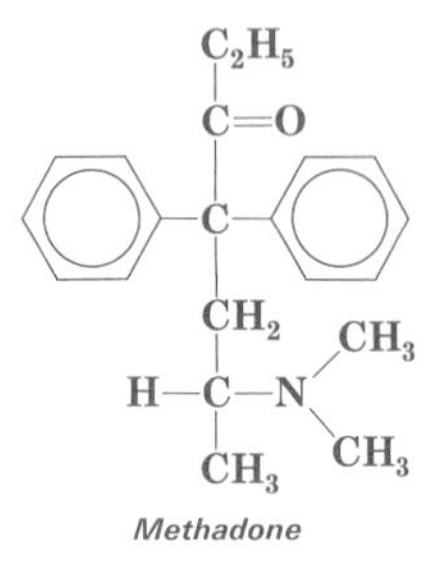

Methadone

Some other natural products which affect the central nervous system and can be neurotoxic in comparatively small amounts are listed in Table 18–4.

MUTAGENS

Mutagens are chemicals capable of altering the genes and chromosomes sufficiently to cause abnormalities in offspring. Chemically mutagens alter the structures of DNA and RNA, which compose the genes (and, in turn, the chromosomes) which transmit the traits of parent to offspring. Mature sex, or germinal, cells of humans normally have 23 chromosomes; body, or somatic, cells have 23 *pairs* of chromosomes.

Although many chemicals are under suspicion because of their mutagenic effects on laboratory animals, it should be emphasized that no one has yet shown conclusively that any chemical induces mutations in the germinal cells of man. Part of the difficulty of determining the effects of mutagenic chemicals in man is the extreme rarity of mutation. A specific genetic disorder may occur as infrequently as only once in 10,000 to 100,000 births. Therefore, to obtain meaningful, statistical data, a carefully controlled study of the entire population of the United States would be required. In addition, the very long time between generations presents great difficulties, and there is also the problem of tracing a medical disorder to a single, specific chemical out of the tens of thousands of chemicals with which man comes in contact.

A mutagen is a chemical which can change the hereditary pattern of a cell.

If there is no direct evidence for specific chemical mutagenic effects in man, why, then, the interest in the subject? The possibility of a deranged, deformed human race is frightening; the chance for an improved human body is hopeful; and the evidence for chemical mutation in plants and lower animals is established. A wide variety of chemicals is known to alter chromosomes and to produce mutations in rats, worms, bacteria, fruit flies, and other plants and animals. Some of these are listed in Table 18–5.

The horrible Thalidomide disaster in 1961 focused worldwide attention on chemically induced birth defects. Thalidomide, a tranquilizer and sleeping pill, caused gross deformities (flipper-like arms, shortened arms, no arms or legs, and other defects) in children whose mothers used this

TABLE 18–5 Mutagenic Substances as Indicated by Experimental Studies on Plants and Animals

Substance	Experimental Results
Aflatoxin (from mold, *Aspergillus flacus*)	Mutations in bacteria, viruses, fungi, parasitic wasps, human cell cultures, mice
Benzo(α)pyrene (from cigarette and coal smoke)	Mutations in mice
Caffeine	Chromosome changes in bacteria, fungi, onion root tips, fruit flies, human tissue cultures
Captan (a fungicide)	Mutagenic in bacteria and molds; chromosome breaks in rats and human tissue cultures
Dimethyl sulfate (used extensively in chemical industry to methylate amines, phenols, and other compounds)	Methylates DNA base guanine; potent mutagen in bacteria, viruses, fungi, higher plants, fruit flies
LSD (lysergic acid diethylamide)	Chromosome breaks in somatic cells of rats, mice, hamsters, white blood cells of humans and monkeys
Maleic hydrazide (plant growth inhibitor; Trade names Slo-Gro, MH-30)	Chromosome breaks in many plants and in cultured mouse cells
Mustard gas (dichlorodiethyl sulfide)	Mutations in fruit flies
Nitrous acid (HNO_2)	Mutations in bacteria, viruses, fungi
Ozone (O_3)	Chromosome breaks in root cells of broadleaf plants
Solvents in glue (glue sniffing) (toluene, acetone, hexane, cyclohexane, ethyl acetate)	4% more human white blood cells showed breaks and abnormalities (6% versus 2% normal)
TEM (triethylenemelamine) (anticancer drug, insect chemosterilants)	Mutagenic in fruit flies, mice

drug during the first two months of pregnancy. The use of this drug resulted in more than 4000 surviving malformed babies in West Germany, more than 1000 in Great Britain, and about 20 in the United States. With shattering impact, this incident demonstrated that a compound can appear to be remarkably safe on the basis of animal studies (so safe, in fact, that Thalidomide was sold in West Germany without prescription) and yet cause catastrophic effects in humans. While the tragedy focused attention on chemical mutagens, Thalidomide presumably does not cause genetic damage in the germinal cells, and is really not mutagenic. Rather, Thalidomide, when taken by a woman during early pregnancy, causes direct injury to the developing embryo. Even though Thalidomide is generally believed not to be mutagenic, the disaster spotlighted the serious potential hazards of supposedly nontoxic chemicals in our environment.

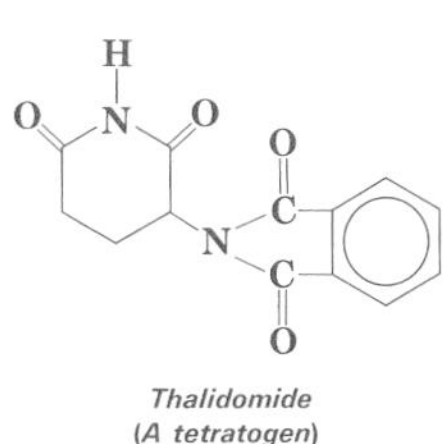

Thalidomide
(A tetratogen)

Experimental work on the chemical basis of the mutagenic effects of nitrous acid (HNO_2) has been very revealing. Repeated studies have shown that nitrous acid is a potent mutagen in bacteria, viruses, molds, and other organisms. In 1953, at Columbia University, Dr. Stephen Zamenhof demonstrated experimentally that nitrous acid attacks DNA. Specifically, nitrous acid reacts with the adenine, guanine, and cytosine bases of DNA by removing the amino group of each of these compounds.

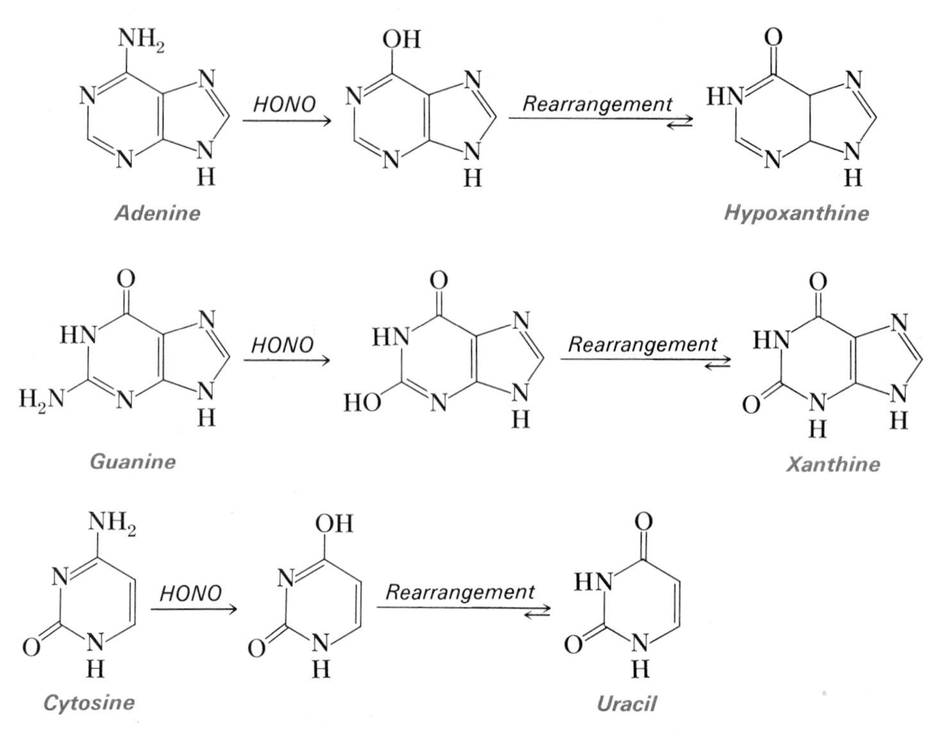

Figure 18–13 Reaction of nitrous acid with nitrogenous bases of DNA. Nitrogen gas (N_2) and water are also products of each reaction.

The eliminated group is replaced by an oxygen atom (Figure 18–13). The changing of the bases may garble a part of DNA's genetic message, and in the next replication of DNA the new base may not form a base pair with the proper nucleotide base.

For example, adenine (A) typically forms a base pair with thymine (T) (Figure 18–14). However, when adenine is changed to hypoxanthine, the new compound forms a base pair with cytosine (C). In the second replication, the cytosine forms its usual base pair with guanine (G). Thus, where an adenine-thymine (A-T) base pair existed originally, a guanine-

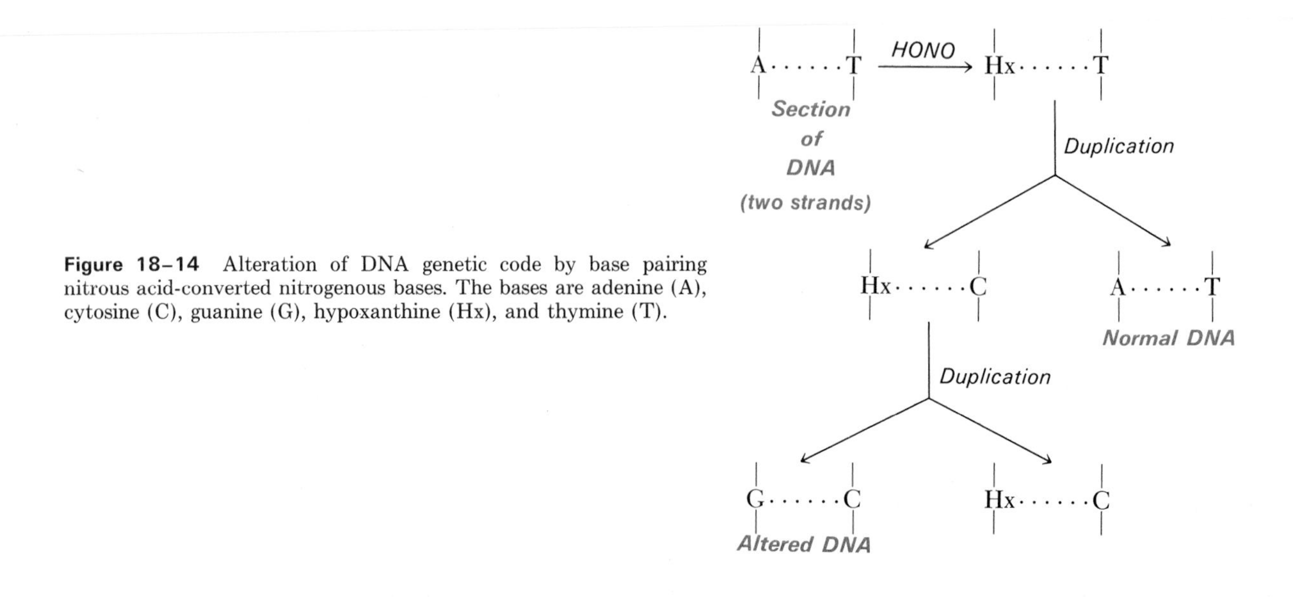

Figure 18–14 Alteration of DNA genetic code by base pairing nitrous acid-converted nitrogenous bases. The bases are adenine (A), cytosine (C), guanine (G), hypoxanthine (Hx), and thymine (T).

cytosine (G-C) pair now exists. The result is an alteration in the DNA's genetic coding, so that a different protein is formed later.

Do all of these findings mean that nitrous acid is mutagenic in humans? Not necessarily. We do know that sodium nitrite has been widely used as a preservative, color enhancer, or color fixative in meat and fish products for at least the past 30 years. It is currently used in such foods as frankfurters, bacon, smoked ham, deviled ham, bologna, Vienna sausage, smoked salmon, and smoked shad. The sodium nitrite is converted to nitrous acid by hydrochloric acid in the human stomach:

$$NaNO_2 + HCl \longrightarrow HNO_2 + NaCl$$

Some scientists theorize that the mutagenic effects of nitrous acid in lower organisms are sufficiently ominous to suggest strongly that the use of sodium nitrite in foods be severely curtailed, if not completely banned. A number of European countries already restrict the use of sodium nitrite in foods. The concern is that this compound, after being converted in the body to nitrous acid, may cause mutation in somatic cells (and possibly in germinal cells) and thus could possibly produce cancer in the human stomach. Other scientists doubt that nitrous acid is present in germinal cells and, therefore, seriously question whether this compound could be a cause of genetically produced birth defects in man. The uncertainty of extrapolating results obtained in animal studies to human beings hovers over the mutagenic substances.

Thus far research has concentrated on the action of chemicals in causing mutations in bacterial viruses, molds, fruit flies, mice, rats, human white blood cells, and so on. Perhaps in the next 10 to 20 years it will be demonstrated that these chemicals can produce transmissible alteration of chromosomes in human germinal cells. Meanwhile, many scientists are pressing for a more vigorous research effort to expand our knowledge of chemically induced mutations and of their potentially harmful effects in man. One intriguing theory that will surely invoke experimental examination is the belief that some compounds cause cancer because they are first and foremost mutagenic. The supporting evidence at present is still extremely inconclusive.

CARCINOGENS

Cancer of the epithelial tissue —*carcinoma.*

Cancer of the connective tissue—*sarcoma.*

Carcinogens are chemicals that cause cancer. ***Cancer*** is an abnormal growth condition in an organism that manifests itself in at least three ways. The rate of cell growth (that is, the rate of cellular multiplication) in cancerous tissue differs from the rate in normal tissue. Cancerous cells may divide more rapidly or more slowly than normal cells. Cancerous cells spread to other tissues; they know no bounds. Normal liver cells divide and remain a part of the liver. Cancerous liver cells may leave the liver and be found, for example, in the lung. Most cancer cells show partial or complete loss of specialized functions. Although located in the liver, cancer cells no longer perform the functions of the liver.

Attempts to determine the cause of cancer have evolved from early studies in which the disease was linked to a person's occupation. It was first noticed in 1775 that persons employed as chimney sweeps in England had a higher rate of skin cancer than the general population. It was not until 1933 that benzo(α)pyrene, $C_{20}H_{12}$, (a 5-ringed aromatic hydrocarbon) was isolated from coal dust and shown to be carcinogenic. In 1895, the German physician Rehn noted three cases of bladder cancer, not in a random population, but in employees of a factory which manufactured dye intermediates in the Rhine Valley. Rehn attributed these cancers to his patients' occupation. These and other cases confirmed that at times as many as 30 years passed between the time of the initial employment and the occurrence of bladder cancer. The principal product of these factories was aniline. While aniline was first thought to be the carcinogenic agent, it was later shown to be noncarcinogenic. It was not until 1937 that continuous, long-term treatment with 2-naphthylamine, one of the suspected dye intermediates, in dosages of up to 0.5 g per day produced bladder cancer in dogs. Since then other dye intermediates have been shown to be carcinogenic.

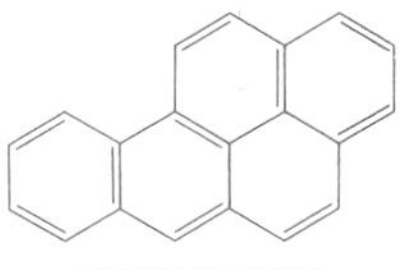

BENZO(α)PYRENE

NH_2 NH_2

ANILINE 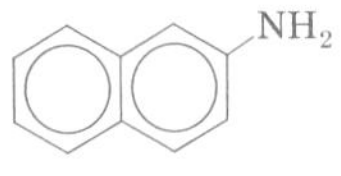 2-NAPHTHYLAMINE

A vast amount of research effort has verified the carcinogenic behavior of a large number of diverse chemicals. Some of these are listed in Table 18-6. This research has led to the formulation of a few generalizations concerning the relationship between chemicals and cancer. For example, carcinogenic effects on lower animals are commonly extrapolated to man. The mouse has come to be the classic animal for studies of carcinogenicity. Strains of inbred mice and rats have been developed which are uniform and show a standard response.

Some carcinogens are relatively nontoxic in a single, large dose, but may be quite toxic, often increasingly so, when administered continuously. Thus, much patience, time, and money must be expended in carcinogenic studies. The development of a sarcoma in humans, from the activation of the first cell to the clinical manifestation of the cancer, takes from 20 to 30 years. With life expectancy of an average person in the United States now set at about 70 years, it is not surprising that the number of deaths due to cancer is increasing.

Cancer spreads from one tissue to another via *metastases*.

An abnormal growth is classified as cancerous or *malignant* when examination shows it is invading neighboring tissue.

Cancer does not occur with the same frequency in all parts of the world. Breast cancer occurs with lower frequency in Japan than in the United States or Europe. Cancer of the stomach, especially in males, is more common in Japan than in the United States. Cancer of the liver is not widespread in the Western Hemisphere but accounts for a high proportion of the cancers in the Bantu in Africa and in certain populations in the Far East. The widely publicized incidence of lung cancer is higher in the industrialized world and is increasing at an appreciable rate.

Some compounds cause cancer at the point of contact: benzo(α)pyrene, for example, which has already been mentioned. Other compounds cause cancer in an area remote from the point of contact. The liver, the common site of detoxification, is particularly susceptible to such compounds. Since the original compound does not cause cancer on contact, some other compound made from it must be the cause of cancer. For example, it appears that the substitution of an NOH group for an N—H group in an aromatic amine derivative produces at least one of the active intermediates for carcinogenic amines. If R denotes a two- or three-ring aromatic system, then the process can be represented as follows:

A growth is *benign* if it is localized at its original site.

$$\underset{\text{INACTIVE ON CONTACT}}{\overset{\text{H}}{\text{RNCOCH}_3}} \longrightarrow \underset{\text{ACTIVE ON CONTACT}}{\overset{\text{OH}}{\text{RNCOCH}_3}} \longrightarrow \underbrace{\text{RX?} \longrightarrow \text{RY?}}_{\text{OTHER UNKNOWN INTERMEDIATES}} \xrightarrow{\text{Tissue}} \text{Tumor cell}$$

TABLE 18–6 A Sample of Carcinogenic Compounds*

Compound	Use or Source	Formula of a Representative Compound
Carbon tetrachloride	Dry cleaning agent, solvent	CCl_4
Dioxane	Solvent in chemical industry, cosmetics, glues, deodorants	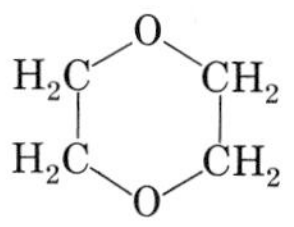 H_2C, CH_2, O, H_2C, CH_2, O (ring)
Aromatic amines		
2-Naphthylamine	Optical bleaching agent	 $-NH_2$
Benzidine	Polymer formation, manufacture of dyes, stain for tissues, rubber compounding	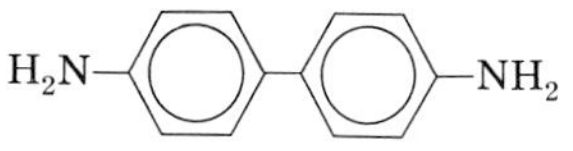 H_2N- $-NH_2$
N-2-Fluorenylacetamide	Herbicide	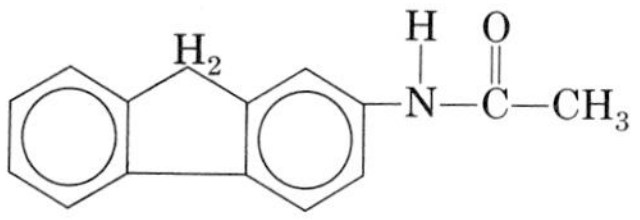 H_2; H O; $-N-C-CH_3$
Azo dyes		
p-Dimethylamino-azobenzene	Yellow dye (once used as a food color)	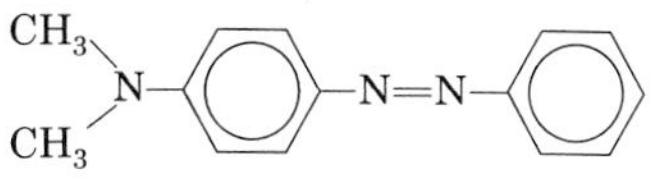 CH_3, N, CH_3; $-N{=}N-$
Nitrosoamines		
N-Ethyl-N-nitroso-n-butylamine	Insecticide, gasoline and lubricant additive	$C_2H_5-N(N{=}O)-CH_2-(CH_2)_2-CH_3$
Polynuclear compounds		
Benzo(α)pyrene	Cigarette and coal smoke	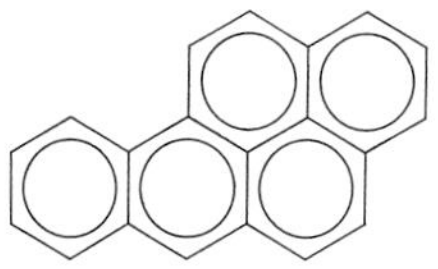
Aflatoxin	Mold on peanuts and other plants (*Aspergillus flavus*)	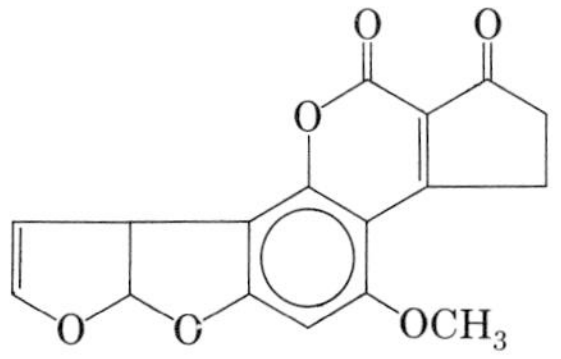 O O; O; O O; OCH_3

*Most of the evidence is based on studies with mice.

As indicated by the variety of chemicals in Table 18-6 many molecular structures produce cancer while closely related ones do not. The 2-naphthylamine mentioned earlier is carcinogenic, but repeated testing gives negative results for 1-naphthylamine.

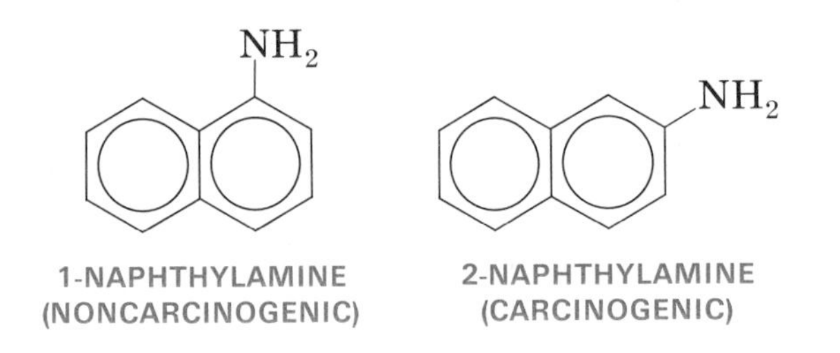

For some types of cancer there are discrete and distinct stages which ultimately result in cancer. These may be identified as the *initiation period,* the *development* or *promotion period,* and the *progression period.* A single, minute dose of a carcinogenic polynuclear aromatic hydrocarbon, such as benzo(α)pyrene, applied to the skin of mice produces the permanent change of a normal cell to a tumor cell. This is the initiation step. No noticeable reaction occurs unless further treatment is made. If the area is painted repeatedly with noncarcinogenic croton oil, even up to one year later, carcinomas appear (Figure 18-15). This is the development period. Additional fundamental alterations in the nature of the cells occur during the progression period. If there is no initiator, there are no tumors. If there is initiator but no promoter, there are no tumors. If the initiator is followed by repeated doses of promoter, tumors appear. This seems to indicate that cancer cannot be contracted from chemicals unless repeated doses are administered or applied.

Almost all chemical carcinogens have an induction period.

Just how do these toxic substances work? Cancer might be caused if the carcinogen combines with growth control proteins, rendering them inactive. During the normal growth process the cells divide and the

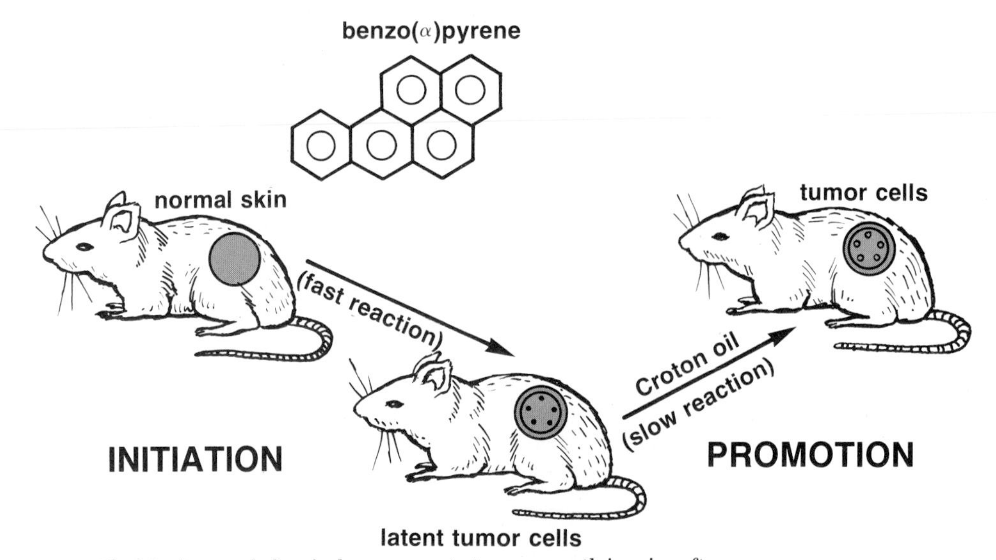

Figure 18-15 A second chemical can promote tumor growth in mice after an initiation period. Treatment of mice with croton oil produces no tumor nor does treatment with small quantities of benzo (α) pyrene alone. Croton oil is an irritant oil similar to castor oil. Both are derived from plants.

organism grows to a point and stops. Cancer is abnormal in that cells continue to divide and portions of the organism continue to grow. One or more proteins are thought to be present in each cell with the specific duty of preventing replication of DNA and cell division. Virtually all of the carcinogens bind firmly to proteins, but so do some similar compounds which are noncarcinogenic. The specific growth proteins involved are not yet known for any of the carcinogens, despite considerable efforts to find them.

Another theory suggests that carcinogens react with and alter nucleic acids so the proteins ultimately formed on the messenger-RNA are sufficiently different to alter the cell's function and growth rate. The carcinogen may be included in the DNA or RNA strands by covalent bonding or it may be entangled in the helix and held by weak van der Waals attractions. The carcinogenic compounds nitrosodimethylamine and mustard gas have been shown to react with nucleic acids.

While researchers collect data in their laboratories and speculate on the theoretical structural causes of cancer, we can studiously avoid compounds known to cause cancer in man. It has been proposed that as much as 80% of all human cancer has its origin in carcinogenic chemicals.

HALLUCINOGENS

Hallucinogens can produce temporary changes in perception, thought, and mood. These substances include *mescaline, lysergic acid*

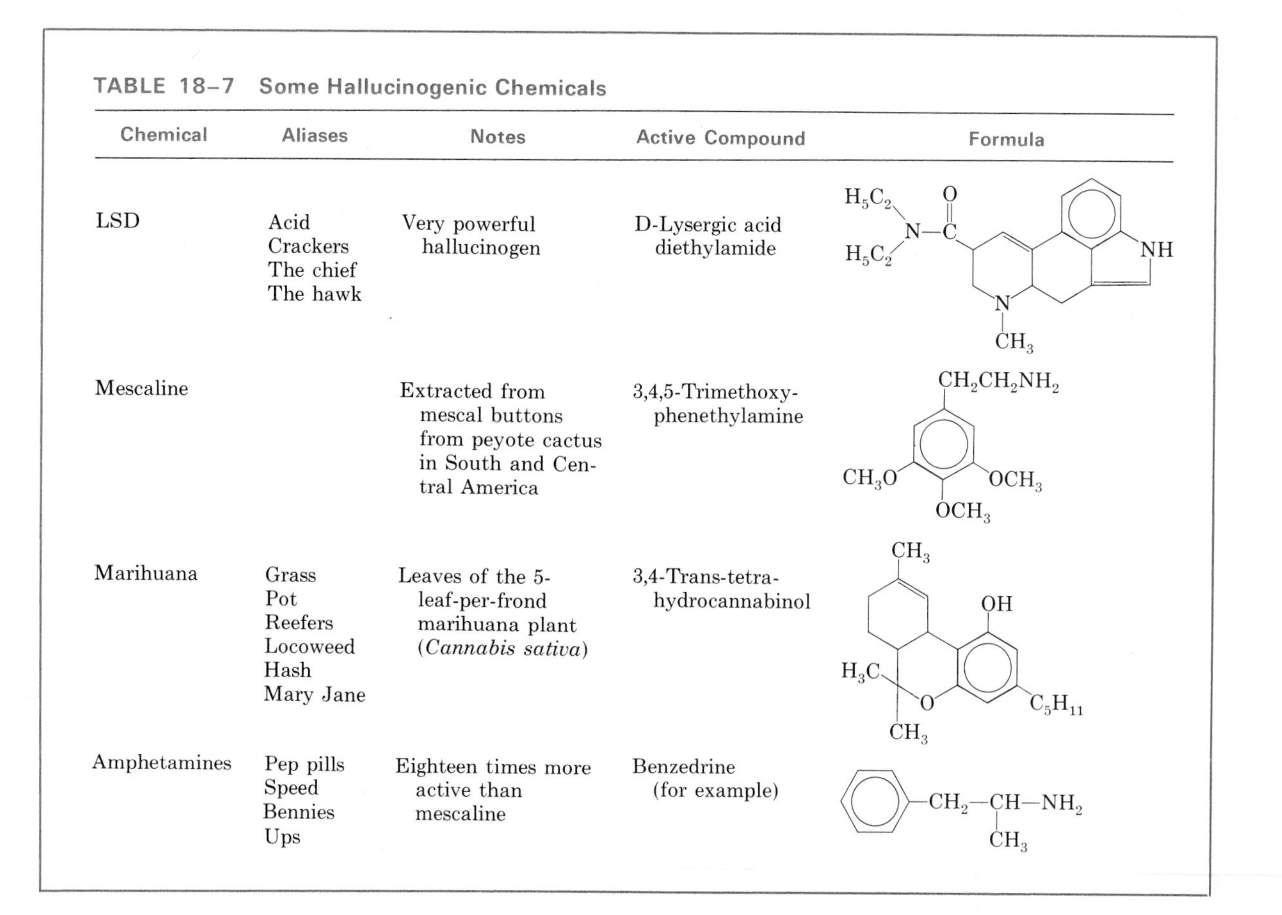

TABLE 18–7 Some Hallucinogenic Chemicals

Chemical	Aliases	Notes	Active Compound	Formula
LSD	Acid Crackers The chief The hawk	Very powerful hallucinogen	D-Lysergic acid diethylamide	
Mescaline		Extracted from mescal buttons from peyote cactus in South and Central America	3,4,5-Trimethoxy-phenethylamine	
Marihuana	Grass Pot Reefers Locoweed Hash Mary Jane	Leaves of the 5-leaf-per-frond marihuana plant (*Cannabis sativa*)	3,4-Trans-tetra-hydrocannabinol	
Amphetamines	Pep pills Speed Bennies Ups	Eighteen times more active than mescaline	Benzedrine (for example)	

diethylamide (*LSD*), *tetrahydrocannabinol* (the active component of marihuana), and a broadening field of more than 50 other substances. They are included in this chapter as a separate section because they can be toxic in comparatively small doses.

All the hallucinogenic drugs are toxic.

Several characteristics of some of the more famous hallucinogens are given in Table 18–7. Each of these substances is capable of disturbing the mind and producing bizarre and even colored interpretations of visual and other external stimuli. Mescaline is one of the oldest known hallucinogens, having been isolated from the peyote plant in 1896 by Heffter. Indeed, as early as 1560 the Mexican Indians who ate or drank the peyote were described by Spanish explorers as experiencing "terrible or ludicrous visions; the inebriation lasting for two or three days and then disappearing."

Although known to man for nearly 5,000 years, marihuana is one of the least understood of all natural drugs. Very early in China's history it was used as a medicine, and in the United States it was early used as an analgesic and a poultice for corns. However, marihuana no longer has any acceptable medical use in this country.

The body does not become dependent on the continuing use of marihuana like it does with heroin or other narcotics. However, reliable scientific data are not available with regard to chronic toxicity resulting from long-term use of the drug.

Today there are many substances that produce hallucinogenic experiences; the most powerful one known is LSD. Our brief discussion will be restricted to this compound.

LSD

Literally hundreds of scientists are doing research on the effects of hallucinogens, including LSD. They are investigating the influence of these drugs on nerve and brain function, the possibility of chromosome alteration, tissue damage, psychological changes, and a host of other physiological and psychological effects. Scientists are not near a consensus on the toxic effects that LSD can cause. Collecting valid data is very difficult; many users of LSD overestimate the quantity of the drug they have used, the purity is often in question, and some take other drugs in addition. Even if the purity of the LSD is known, the results are difficult to interpret. For example, a study in which mice were given 0.05 to 1.0 microgram of LSD on the seventh day of pregnancy showed a 5 percent incidence of badly deformed mouse embryos. Another study, probably equally valid and meticulous, showed no unusual fetal damage to rat embryos when 1.5 to 300 micrograms of LSD was administered during the fourth or fifth day of pregnancy. Babies with depressed skulls were aborted from two mothers who had taken LSD during the early weeks of pregnancy. Two other women, also LSD users, delivered full term, apparently normal babies.

LSD has been linked with birth defects.

There are, however, some dangers that are well documented. These drugs destroy one's sense of judgment. Such things as height, heat, or even a moving truck may seem to hold no danger for the person under the influence of a hallucinogen. A dose of 50 to 200 micrograms will take the user on a "trip" for approximately eight to sixteen hours. Excessive or prolonged use of LSD can cause a person to "freak out" and possibly to sustain permanent brain damage. After one "trip," a user can experience

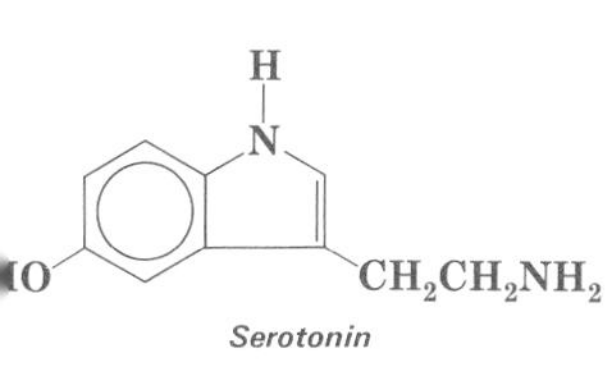

Serotonin

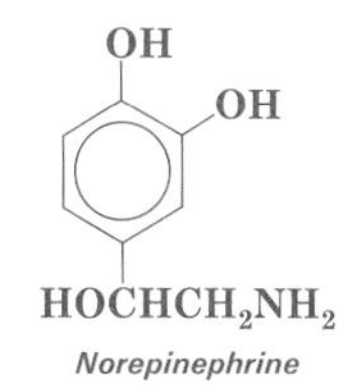

Norepinephrine

another "trip" some time later, unexpectedly, without taking any more of the drug. The debate over LSD's dangers is continuing and is not likely to be resolved soon.

Out of the darkness of LSD-induced suicides and brain damage has come some new understanding of how the brain works. It is an interesting coincidence that soon after LSD was found localized in areas of the brain responsible for eliciting man's deep-seated emotional reactions, the compounds that are probably responsible for transmitting the impulse across synapses in the brain were discovered. Serotonin and norepinephrine are thought to act in a manner similar to acetylcholine, which was discussed earlier. These compounds carry the message from the end of one neuron to the end of another across the synapse. Somewhat later Dilworth Wooley discovered that LSD blocks serotonin action (Figure 18-16). This has led to an interesting theory to explain the LSD trip.

The theory of the hallucinogenic mechanism is, roughly, as follows. The LSD releases in some chemical way those emotional experiences that are generally hidden away, or chemically stored in the lower midbrain and the brain stem. Serotonin or norepinephrine, or both, inhibit the escape of these experiences into consciousness, as they turn off certain excitations. LSD interacts with serotonin or norepinephrine at the synapse and nullifies their blocking effect. Thus, these stored, previously rejected thoughts are allowed to enter the conscious part of the brain where the imaginary trip occurs.

This theory is built on the experimental fact that LSD blocks the action of serotonin, but there are other theories as well as some unexplained phenomena involved. Psychiatry is fraught with theories of why LSD causes an uplifting experience in some people and a frightening one in others. There are theories about dosage, purity of the compound, psychiatric state of the tripper, and conditions of administration. If there is anything close to a consensus, it is that the more controlled and relaxed the surroundings, the "better" the trip will be; even this, however, is debatable. Many trips, even under medical supervision, have not been satisfactory.

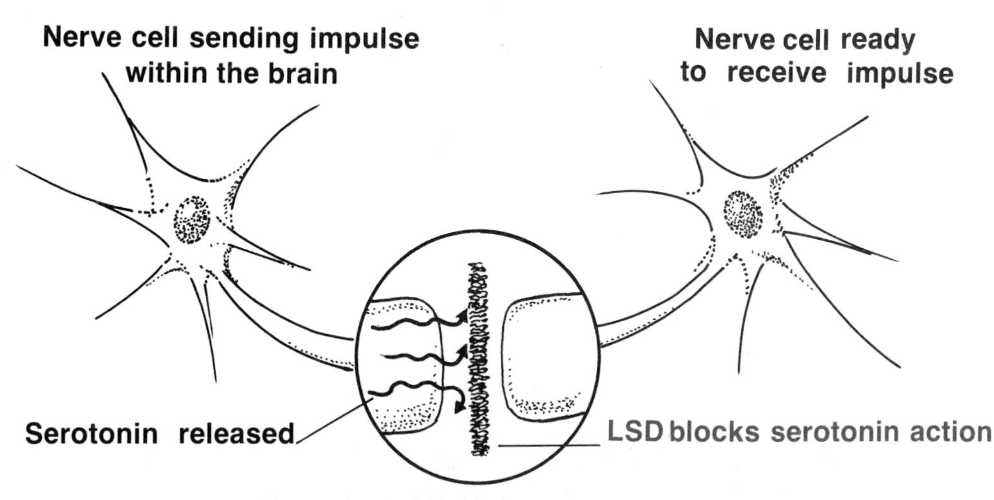

Figure 18-16 LSD blocks the flow of serotonin from one brain nerve cell to another.

ALCOHOLS

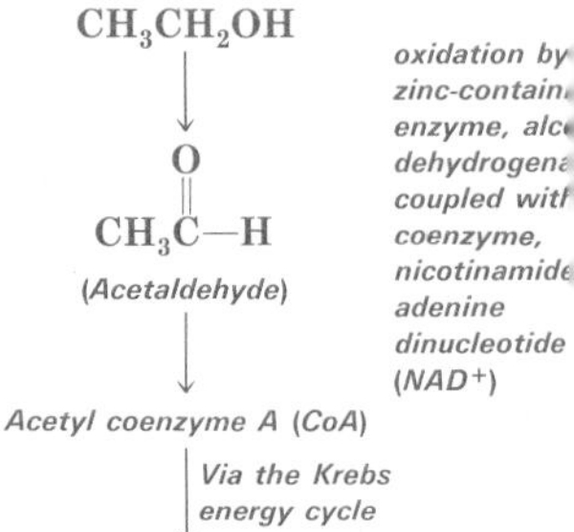

Some well known toxic effects are found with the alcohols, but a complete chemical explanation has eluded scientists so far.

Methyl alcohol (methanol or wood alcohol) is highly poisonous, and unlike the other simple alcohols, it is, in effect, a cumulative poison in humans. It has a specific toxic effect on the optic nerve, causing blindness with large doses. After its rapid uptake by the body, oxidation occurs in which the alcohol is first converted into formaldehyde and then to formic acid, which is eliminated in the urine.

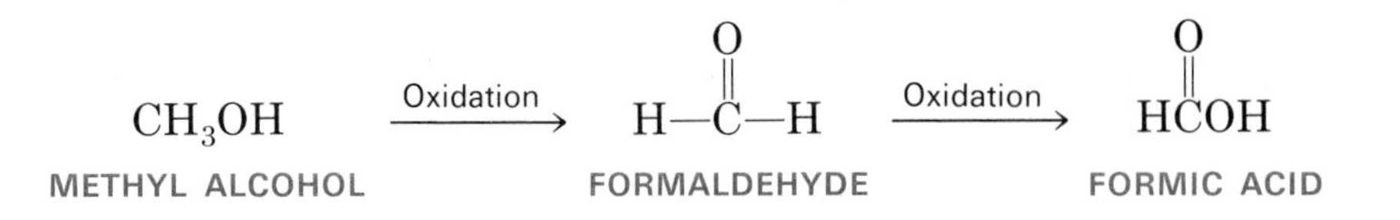

This is a slow process and, for this reason, daily exposure to methyl alcohol can cause an extremely dangerous buildup of the alcohol in the body. The toxic effect on the optic nerve is thought to be caused by the oxidative products.

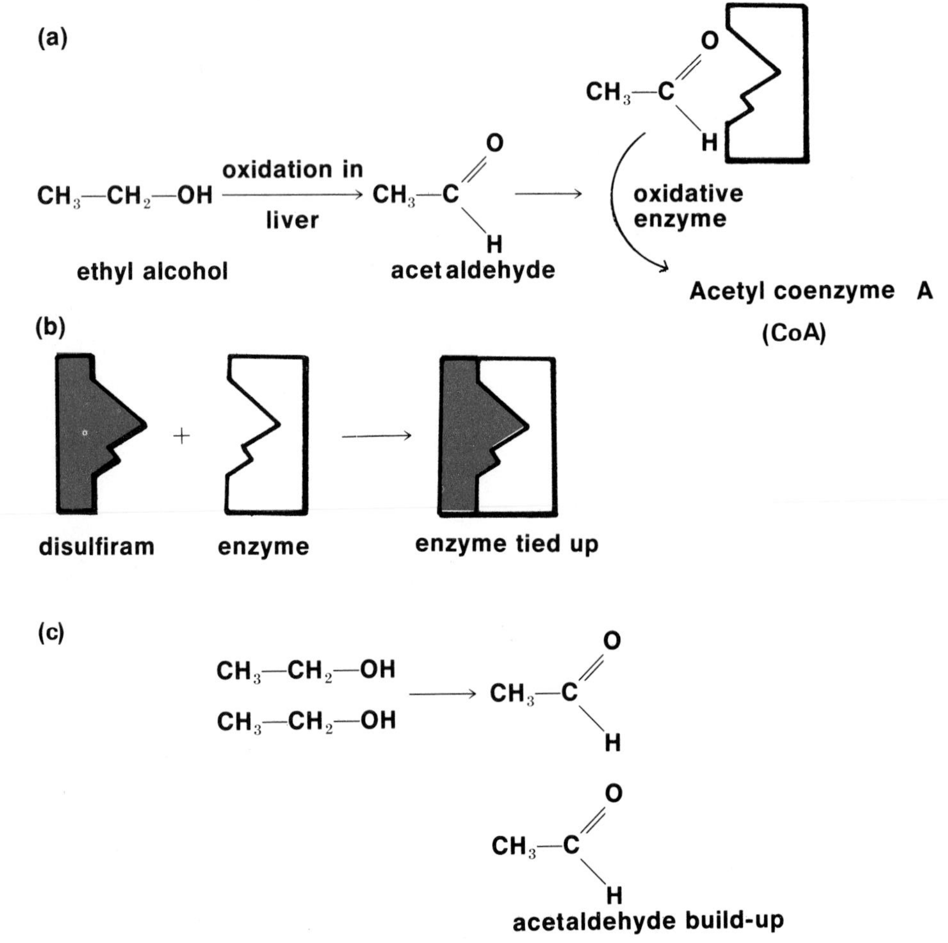

Figure 18–17 The blocking of ethyl alcohol oxidation by disulfiram (Antabuse). When the normal oxidative process (*a*) is stopped by blockage of the oxidative enzyme (*b*), acetylaldehyde builds up in the body (*c*).

Ethyl alcohol (ethanol, grain alcohol) is found in alcoholic beverages, yet it is toxic like the other simple alcohols and is quantitatively absorbed by the gastrointestinal tract. About 58 percent of a dose is absorbed in 30 minutes, 88 percent in 1 hour, and 93 percent in 90 minutes. Over 90 percent of the ethyl alcohol is then slowly oxidized to carbon dioxide and water, mainly in the liver.

$$(CH_3CH_2)_2N{-}\overset{S}{\overset{\|}{C}}{-}S{-}S{-}\overset{S}{\overset{\|}{C}}{-}N(CH_2CH_3)_2$$

Disulfiram

The intoxicated person's staggering gait, stupor, and nausea are caused by the presence of acetaldehyde, but the chemical reactions involved are not fully understood. The compound disulfiram (Antabuse) is sometimes given as a treatment for chronic alcoholism because it blocks the oxidative steps beyond acetaldehyde. The accumulation of acetaldehyde causes nausea, vomiting, blurred vision, and confusion. This is supposed to encourage the partaker to avoid this severe sickness by avoiding alcohol. Interestingly, this drug was discovered by two researchers who took a dose for another purpose and got violently ill that evening at a cocktail party (Figure 18–17).

IMMUNOCHEMISTRY

The large group of foreign protein or protein-polysaccharide substances frequently found in the bloodstream of human beings is collectively referred to as ***antigens.*** These substances generally have high molecular weights and come from such sources as some bacteria (such as typhoid and tuberculosis), pollen, and insect bites. The body's counterattack against antigens is the synthesis of ***antibodies.*** The principal antibodies are ***immunoglobulins,*** which are high molecular weight, globular proteins.

If we are immune to a certain type of antigen, such as the typhoid bacteria, we have in our bloodstream antibodies which react with and destroy the typhoid bacteria. We can be immunized to an extraordinarily large number of different antigens, even to various synthetically prepared proteins which we would normally never encounter in our environment. Even with the large number of antigens that the body is able to combat, there is a different, specific antibody for each specific antigen.

Antigens are toxic substances.

The immunoglobulins, or antibodies, are presently classified according to molecular weight. As more is learned about their specific structures and individual functions, more specific classifications will no doubt emerge. Before 1966, three immunoglobulins were known. These were designated IgG, IgA, and IgM (Ig for immunoglobulin), and they have molecular weights of about 150,000, 180,000, and 950,000 amu, respectively. In 1966, IgE (molecular weight of 196,000 amu) was isolated. This globular protein comprises only about 0.001 to 0.01 percent of the globular proteins in blood serum, and it escaped discovery for a long time by hiding as an impurity in IgA antibodies. Another immunoglobulin, IgD, has been recently discovered.

The IgG immunoglobulins are the most abundant and best understood. IgG molecules have the general structure shown in Figure 18–18. Each molecule has two identical halves. Each half is composed of two polypeptide chains. The heavy chain has a molecular weight of about 50,000 amu (about 400 amino acids) and the light chain has a molecular weight of about 25,000 amu (about 214 amino acids). The four chains are bonded to each other by disulfide bonds. The thousands of different kinds

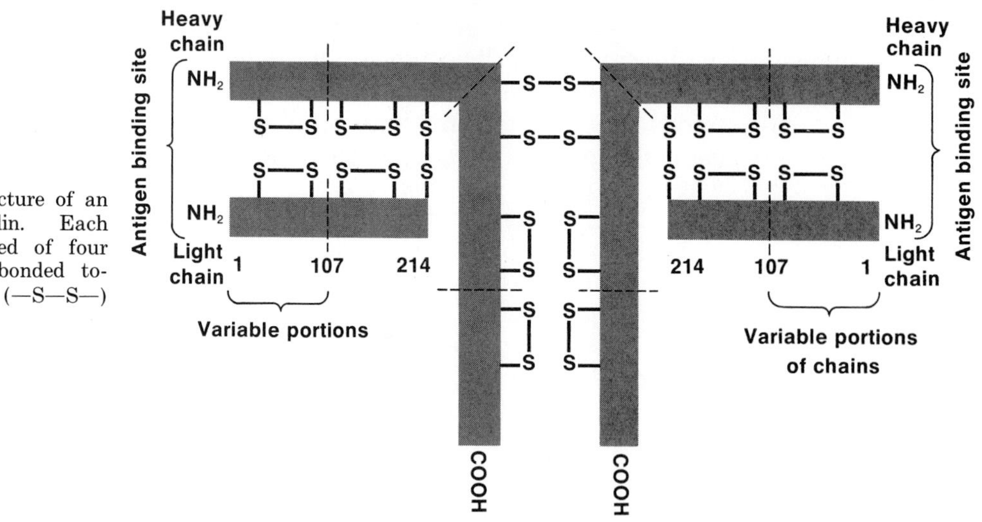

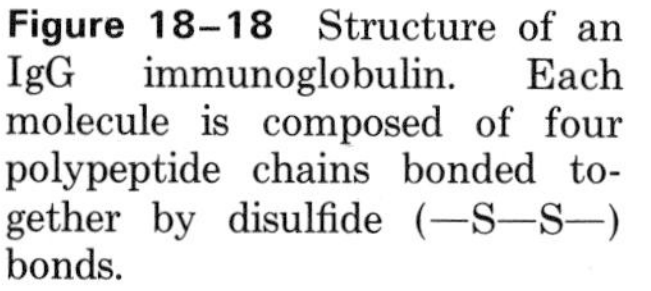

Figure 18–18 Structure of an IgG immunoglobulin. Each molecule is composed of four polypeptide chains bonded together by disulfide (—S—S—) bonds.

of IgG molecules come from the different primary sequences and kinds of amino acids in the last 107 amino acids on the end of each light chain. While the 107 amino acids on the ends of the light chain are known to change from molecule to molecule, the rest of the amino acids in IgG molecules are thought to be in the same primary sequence from molecule to molecule.

The sites where the antibody reacts with the antigens are at the ends of the light and heavy chains. The two identical sites allow each antibody to react with two antigens. Information available at present indicates the sites are formed from portions of both a heavy chain and a light chain.

Lock-and-Key Picture of How Antibodies Work

Paul Ehrlich was the first to use chemicals to combat bacterial infection.

Since each immunoglobulin is constructed to be specific for combining with one particular kind of antigen, the lock-and-key analogy as proposed for enzymes and substrates by Emil Fischer in 1894 seems appropriate for antibodies and antigens. The application of the lock-and-key concept to antibodies and antigens was first made by Paul Ehrlich in 1906 and emphasizes the great specificity of antibodies for antigens. A key fits a particular lock because the two are structurally complementary. So it is with antibodies and antigens; they react because they fit—they are structurally complementary.

To be structurally complementary, two conditions must be met (Figure 18–19). First, the reacting parts of the two molecules must come close together; that is, they must be sterically complementary. Second, the reacting groups must be compatible. Positive groups (such as $—NH_3^+$) must be opposite negative groups (such as $—\overset{\displaystyle O}{\overset{\|}{C}}—O^-$). Positive ends of polar bonds must be opposite negative ends of polar bonds. Hydrogen atoms bonded to electronegative atoms (N or O) must be opposite the unshared pair of electrons on another electronegative atom (hydrogen bonds), or atoms with large electron clouds which can be distorted must be opposite atoms capable of distorting the clouds (by van der Waals forces). For an antibody to react with an antigen and thus nullify its

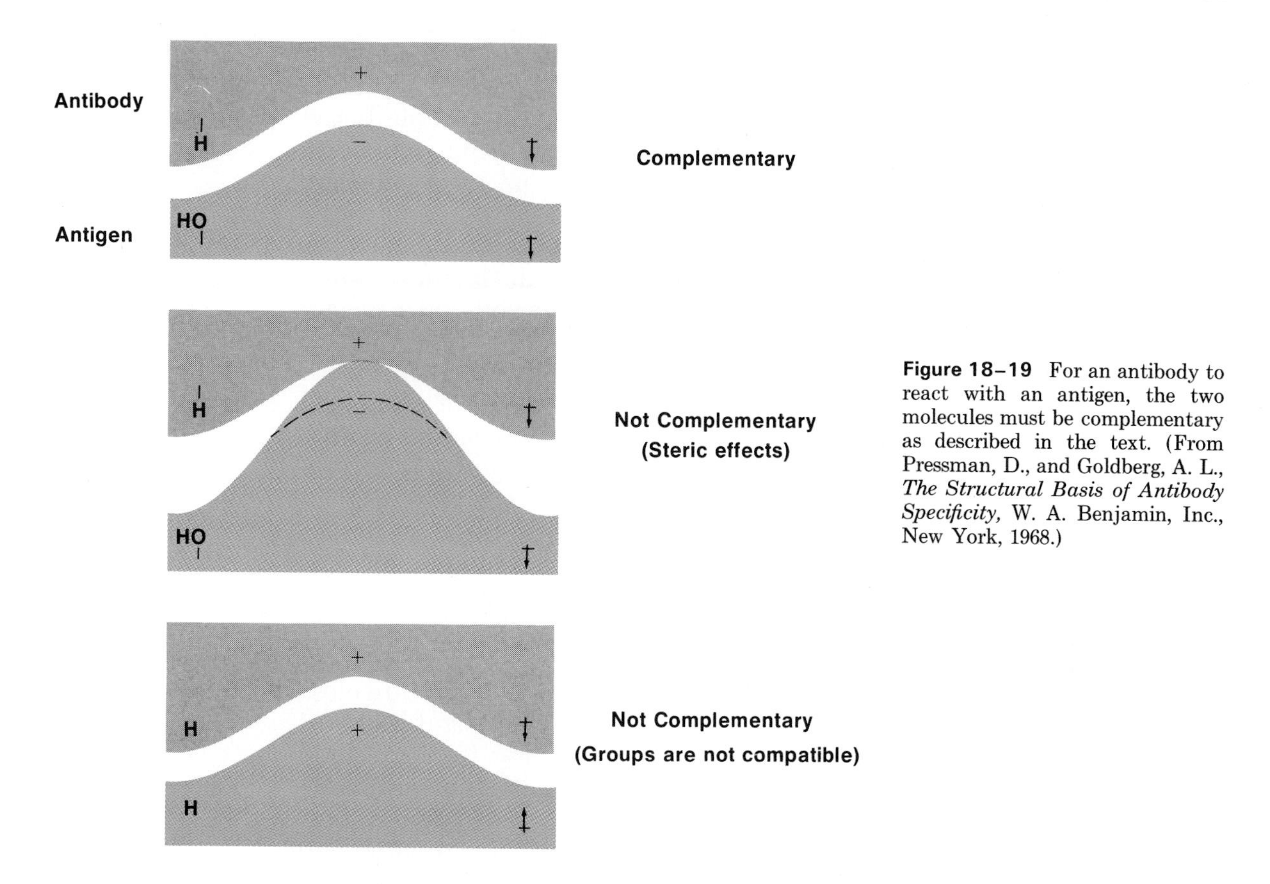

Figure 18–19 For an antibody to react with an antigen, the two molecules must be complementary as described in the text. (From Pressman, D., and Goldberg, A. L., *The Structural Basis of Antibody Specificity,* W. A. Benjamin, Inc., New York, 1968.)

toxicity, both steric effects and compatibility of reacting groups must be satisfied.

The study of how a specific antibody reacts with a specific antigen is complicated by the fact that antigens generally have many reactive groups, and it is difficult to know exactly which groups are participating in the bonds between antibody and antigen. Furthermore, it is difficult to isolate a particular antibody for individual study since only a very small amount of a particular antibody exists in the blood serum at any given time and is mixed with thousands of other antibodies of similar size and molecular weight.

In spite of the difficulties, knowledge has progressed in antibody-antigen chemistry by having antibodies react with proteins which have groups of known structure chemically attached. These groups are called ***haptens*** (Greek, stuck on). By this method it can be determined which

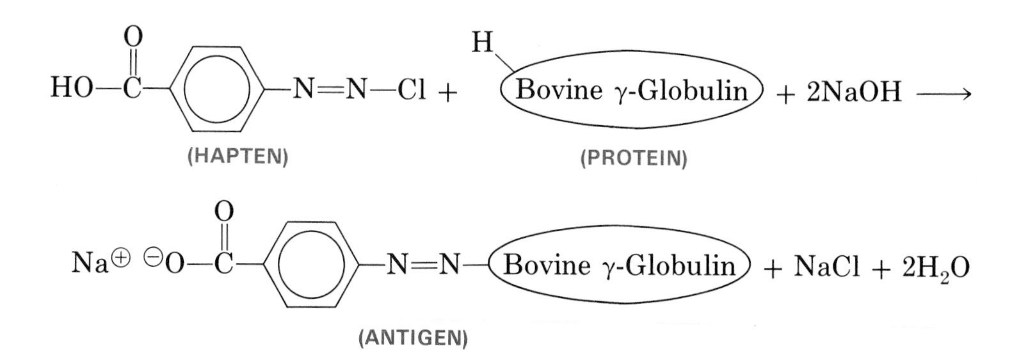

kinds of groups must be present on the antibody in order to react with a hapten of a particular structure. It works this way. First, a hapten is bonded covalently to a protein of known structure. An example is the attachment of a *p*-azobenzoate group to the protein, bovine γ-globulin. The *p*-azobenzoate group reacts with a hydrogen atom on a tyrosine, histidine, lysine, or $—NH_2$ group in the protein.

The antigen is then injected into the bloodstream of a rabbit (or other animal). The rabbit's body chemistry forms antibodies, which combine with the azobenzoate group on the antigen. Various experiments can now be performed on the antigen-antibody precipitate and with the blood serum of the rabbit to ascertain if the antibody actually bonds with the hapten and, if so, how it does. For example, if some of the blood serum is removed from the animal, and if some of the original protein alone (without the azobenzoate group) is added to the serum, a relatively small amount of precipitate forms. If a completely different protein (such as ovalbumin) is added to the serum, no precipitate is formed. If, on the other hand, the azobenzoate group is combined with the ovalbumin, the product gives a precipitate with the serum. These experiments show that the action of the antibody is definitely with the azobenzoate group. Structural studies can be done on the antibody-antigen precipitate in order to establish which groups of the antibody and the antigen are close enough to each other to react. These kinds of studies give insight into how the antibody attaches to the hapten on the antigen.

Hundreds of different haptens have been examined in antibody-antigen interaction experiments. Much has been learned about how antibodies react with antigens. General shapes of antibody sites and locations of structurally complementary groups have been elucidated for many haptens. An example is shown schematically in Figure 18–20 for the *p*-azobenzoate hapten attached to a tyrosine amino acid, which is part of a protein.

The manner in which the body provides a different antibody for each kind of antigen is a mystery at this time. Does the body hold enough different antibodies in storage to react with the many antigens to which the body is immune? There are not enough immunoglobulins in the blood to satisfy this requirement. Then how does the body provide the antibodies? One prevalent theory is the clone theory. ***Clones*** are antibody-forming cells, each of which possesses the capacity to make only a single

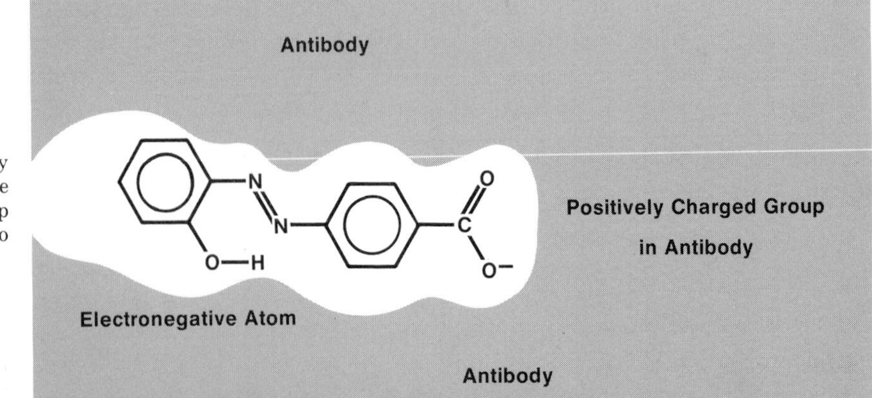

Figure 18–20 Possible antibody fit and interaction around the *p*-azobenzoate antigenic group attached to a tyrosine amino acid.

type of antibody with a characteristic amino acid sequence. Clones are special cells with part of their DNA unique for producing RNA, which in turn produces the particular antibody characteristic of that particular clone. This is the one cell-one antibody hypothesis.

The details of how a specific antibody, such as the typhoid antibody, actually attaches itself to the typhoid antigen are not yet known. Neither are the details known as to how any antibody attaches to any *natural* antigen. But progress is being made in this area, not the least of which are some important applications in other areas. For example, we now have a means of detecting specific antigens. Since antibodies are generally specific and can be labeled with a color-producing compound (such as fluorescein) or with a radioactive isotope, a particular antigen can be located and identified at the microscopic level. Also, antigen-antibody research has led to the realization that complementary structures are the basis of cell-cell interaction in the development of an organ or other structures in the body, cell-cell interaction in fertilization, adsorption of viruses onto susceptible cells, and the aforementioned enzyme-substrate interactions.

SELF-TEST 18-B

1. Substrates which poison the nervous system are called ____________.
2. Most neurotoxins affect chemical reactions which occur in the opening between two nerve cells. These openings are called ________________.
3. The electrical impulse is carried across a synapse by the chemical, ____________________.
4. Mutagens alter the structures of ____________ or ____________.
5. If a substance is mutagenic in test animals, particularly dogs, it must necessarily be mutagenic in human beings. True/False.
6. The first occupation definitely linked to cancer was ____________ ____________.
7. An active component of marihuana is ____________________.
8. The nausea and stupor of drunkenness from ethyl alcohol is caused by ____________________ and not by the alcohol itself.
9. How many chains does an IgG antibody have? ______ How many antigen molecules can an IgG molecule combine with? ______
10. Antibodies are mostly (protein, carbohydrate, fat).
11. The distinctive difference between two IgG antibodies is the composition of the 107 amino acids on the ends of the ____________________ chains.

MATCHING SET

_______ 1. metabolic poison	a.	lysergic acid diethylamide
_______ 2. metabolism	b.	cyanide ion
_______ 3. corrosive poison	c.	cancer in connective tissue
_______ 4. neurotoxin	d.	immunoglobulin
_______ 5. mutagen	e.	benzo(α)pyrene
_______ 6. carcinogen	f.	cancerous growths of lung tissue located in liver
_______ 7. carcinoma	g.	use of chemicals in the body
_______ 8. metastases	h.	cancer of skin
_______ 9. sarcoma	i.	sodium hydroxide (caustic soda)
_______ 10. antigen	j.	atropine
_______ 11. chelating agent	k.	good fit and groups are compatible
_______ 12. hallucinogen	l.	typhoid bacteria
_______ 13. antibody	m.	alters DNA
_______ 14. structurally complementary	n.	EDTA

QUESTIONS

1. Give an example of a toxic substance that is toxic as a result of

   a. binding to an oxygen-carrying molecule.
   b. disguising itself as another compound.
   c. attack on an enzyme.
   d. hydrolysis.

2. True/False. Explain each answer concisely.

   a. Lead is a corrosive poison.
   b. Carbon monoxide and cyanide poison in the same way.
   c. There are no known chemical compounds that cannot be toxic under some circumstances.

3. The application of a single, minute dose of a fused ring hydrocarbon such as benzo(α)pyrene fails to produce a tumor in mice. Does this mean this compound is definitely noncarcinogenic? Give a reason for your answer.

4. Describe the chemical mechanism by which the following substances show their toxic effects.

   a. fluoroacetic acid
   b. phosgene
   c. curare

5. Should any laws and regulations be placed on the use of any of the following? Justify your answers.

   a. LSD

b. marihuana
c. cyclamates
d. ethanol

6. Discuss some of the pros and cons of testing toxic substances on animals.

7. Give chemical reactions in words for

a. action of NaOH on tissue.
b. action of carbon monoxide in blood.
c. reaction of EDTA and lead ion.

8. What questions do you think need to be answered before the action of ethyl alcohol is understood?

9. Assume a normal diet has the quantity of lead in a given quantity of food as stated in the text. What would a person's total food intake of lead be per day?

10. What is the meaning of the symbolism LD_{50}?

11. Write chemical equations for:

a. the hydrolysis of acetylcholine.
b. acid hydrolysis of a protein having a glycine-glycine primary structure.

PROBLEMS

The following problems are included because of the importance of dose when discussing poisons.

1. If as little as 10^{-6} mole of acetylcholine can activate a nerve ending, what weight in grams of acetylcholine can activate a nerve ending? *Ans.* 1.63×10^{-4} g

2. If air contains 50 ppm CO, what is the percentage of CO in the air? *Ans.* 0.005%

3. If a soft drink contains 20 μg of Pb per liter of solution, how many moles of Pb are contained in 12 ounces of the soft drink? (33.8 liquid ounces per liter) *Ans.* 3.4×10^{-8} moles Pb

4. Assuming the lethal dose of morphine is 5 mg per kg of body weight, what is the lethal dose for a 200-pound person in moles of morphine. The molecular weight of morphine is 303.35 amu? *Ans.* 0.0015 mole morphine

5. Calculate the weight of $PbCl_2$ in grams exhausted from an automobile when 10 gallons of gasoline are burned. Assume that the gasoline contains 5.00 grams of tetraethyl lead [$Pb(C_2H_5)_4$] per gallon of gasoline: *Ans.* 43 grams

$$Pb(C_2H_5)_4 + C_2H_4Cl_2 + 16O_2 \longrightarrow PbCl_2 + 10CO_2 + 12H_2O$$

SUGGESTIONS FOR FURTHER READING

Adams, E., "Poisons," *Scientific American,* Vol. 201, No. 5, p. 76 (1959).
Brooks, V. J., and Jacobs, M. B., "Poisons," 2nd ed., D. Van Nostrand Co., Inc. Princeton, 1958.
Chisolm, J. J., Jr., "Lead Poisoning," *Scientific American,* Vol. 224, No. 2, p. 15 (1971).
Christensen, H. E., and Luginbyhl, T. T., editors, "The Toxic Substances List, 1974 Edition," U. S. Department of Health, Education, and Welfare, Center for Disease Control, Rockville, Maryland (1974); publication number (NIOSH) 74-134.
"Drug Enforcement Administration Fact Sheets," U.S. Government Printing Office, Washington, D.C., 1973, 0-507-933.
"Environmental Lead and Public Health," U.S. Environmental Protection Agency, Air Pollution Control Office, Research Triangle Park, N. C., 1971.
Goldwater, L. J., "Mercury in the Environment," *Scientific American,* Vol. 224, No. 5, p. 15 (1971).
Hoffer, A., and Osmond, H., "The Hallucinogens," Academic Press, New York, 1967.
Labianca, D. A., "Acetaldehyde Syndrome and Alcoholism," *Chemistry,* Vol. 47, No. 9, p. 21 (1974).
Loomis, T. A., "Essentials of Toxicology," Lea and Febiger, Philadelphia, 1968.
Maxwell, K. E., "Chemicals and Life," Dickenson Publishing Co., Inc., Belmont, Cal., 1970.
McGrady, P., "The Savage Cell; A Report on Cancer and Cancer Research," Basic Books, New York, 1964.

Prescott, D. M., "Cancer, The Misguided Cell," The Bobbs-Merrill Co., Indianapolis, 1973.
Pressman, D., and Grossberg, A. L., "The Structural Basis of Antibody Specificity," W. A. Benjamin, Inc., New York, 1968.
Robinson, T., "Alkaloids," *Scientific American,* Vol. 201, No. 1, p. 113 (1959).
Sanders, H. J., "Chemical Mutagens," *Chemical and Engineering News,* Vol. 47, No. 21, p. 50 (1969); and Vol. 47, No. 23, p. 54 (1969).
Schroeder, H. A., "The Poisons Around Us," Indiana University Press, Bloomington, 1974.
Weisburger, J. H., and Weisburger, E. K., "Chemicals as Causes of Cancer," *Chemical and Engineering News,* Vol. 44, No. 7, p. 124 (1966).

CHAPTER 19

WATER: ITS USE AND MISUSE

Water is certainly one of the most important of all chemical compounds. Its intimate participation in life processes marks it as indispensable to all forms of life known to man. The rapidly expanding scope of human life, with its associated industrial development, is increasing the demand for water supplies so rapidly that future needs may easily be underestimated. The time has passed when man can take an abundant water supply for granted.

Pollution causes something that is normally clean to be unfit for direct use.

Over the centuries the world water supply has remained essentially constant. However, the availability of relatively pure water is being diminished by pollution. When the Pilgrims landed at Plymouth Rock, drinkable water could be obtained directly from most of the streams in the Colonies without endangering health because of polluted water. Now there are only a few streams with drinkable water, and some bodies of water, especially those near large cities, are not even fit for swimming. Some of the most polluted are the Hudson River, the lower Mississippi River, and Lake Erie; all are near major metropolitan areas.

Water—the universal solvent.

Water pollution arises from two main sources. First, water, being an unusually good solvent, readily dissolves materials from the soil. Second, and by far the most serious, are the activities of man; mercury and raw sewage effluents are but two examples.

It is important to realize at the outset that the technical know-how is available to solve water pollution problems. Much of this know-how is patterned after natural purification processes.

WATER REUSE

It is estimated that an average of 4350 billion gallons of rain (and snow) fall on the contiguous United States each day. Of this amount, 3100 billion gallons return to the atmosphere by evaporation and transpiration. The discharge to the sea and underground reserves amounts to 800 billion gallons daily, leaving 450 billion gallons of surface water each day for domestic or commercial use. The 48 contiguous states withdrew from natural sources 40 billion gallons per day in 1900, 325 billion gallons per day in 1960, and it is estimated that the demand will be at least 900 billion gallons per day by the year 2000. It is evident that the reuse of surface water is the only way that human needs for water can be met in the future.

The increasing reuse of surface water by cities contributes to the pollution problem. The fact that rivers serve as a source of water for many cities as well as receivers for their sewage and industrial waste effluents means that a series of cities on the banks of a river will use some of the same water over and over again. If we choose not to purify our water after use or, in some cases, do not give nature enough time to do the job, the river becomes progressively more polluted downstream. The polluted water may contain excessive heat, dissolved chemicals, and an undesirable population of microorganisms, along with silt. However, it should be noted that what is pollution to one user may not be to the next. The industrial user who needs water for cooling may primarily be concerned with heat content. The director of the municipal water plant is likely to be concerned with microorganisms, chemicals, and then heat, in that order.

Since the reuse of water is necessary, and since the purification of used water before reuse is often necessary, the question arises as to where the purification should occur. Purification immediately prior to reuse actually may lead to greater chemical problems because of the complex mixtures that result. Purification immediately after use is likely to involve only one type of problem characteristic of that use.

Nature recycles water—why can't man?

Pure water supplies could be maintained more easily if users would purify the used waters before returning them to rivers, lakes, and even the oceans. Society will have a rich bonus if it can expand its ability to use purified water several times before returning it to nature. An increasing number of industrial users are learning to use treated sewage for their purposes prior to future purification and eventual return to the river. Much chemistry is involved in the purification and reuse of water. Obviously, the abatement of water pollution calls for some understanding of the chemical problems involved.

NATURAL WATER PURIFICATION

Rainwater in clean air is very pure.

Volatile: goes easily into the gaseous state.

Water is a natural resource which, within limitations, is continuously renewed. The familiar hydrologic or water cycle (Figure 19–1) offers a number of opportunities for nature to purify its water. The world-wide ***distillation*** process results in rain water containing only traces of nonvolatile impurities, along with gases dissolved from the air. ***Crystallization*** of ice from ocean salt water results in relatively pure water in the form of icebergs. ***Aeration*** of ground water as it trickles over rock surfaces, as in a rapidly running brook, allows volatile impurities, previously dissolved from mineral deposits, to be released to the air. ***Sedimentation*** of solid particles occurs in slow-moving streams and lakes. ***Filtration*** of water through sand rids the water of suspended matter such as silt. Next, and of very great importance, are the ***oxidation processes.*** Essentially all naturally occurring organic materials—plant and animal tissue, as well as their waste materials—are changed through a complicated series of oxidation steps in surface waters to simple molecules common to the environment. Finally, another process that has been

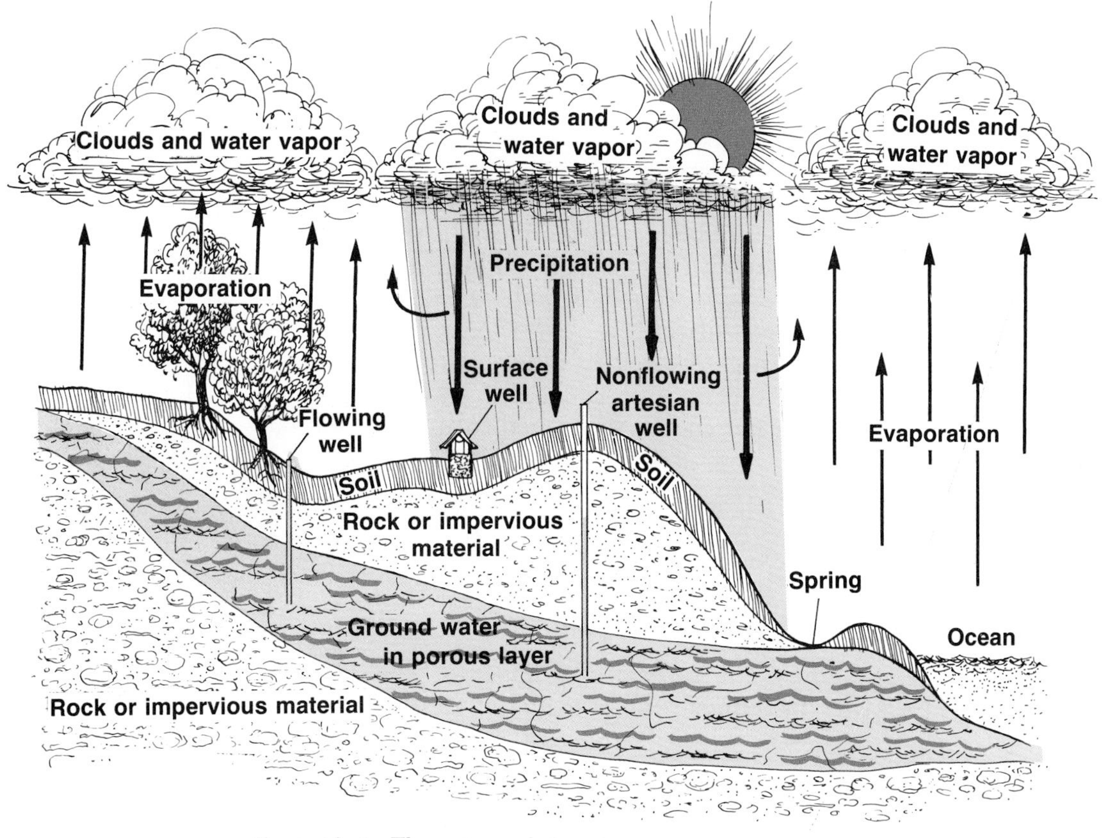

Figure 19–1 The water cycle in nature.

used by nature is ***dilution.*** Most, if not all, pollutants found in nature are rendered harmless if reduced below certain levels of concentration by dilution with pure water.

Pure water:
Chemist: "Pure H_2O —no other substance."
Housewife: "Nothing harmful to human beings."
Game and Fish Commission: "Nothing harmful to animals."
Sunday boater: "Pleasing to the eye and nose, no debris."
Ecologist: "Natural mixture."

Before the advent of the exploding human population and the industrial revolution, natural purification processes were quite adequate to provide ample water of very high purity in all but the arid regions. Nature's purification processes can be thought of as massive but somewhat delicate. In many instances the activities of man push the natural purification processes beyond their limit, and polluted water accumulates.

A simple example of nature's inability to handle man-made pollution is in dragging gravel from stream beds. This excavation leaves large amounts of suspended matter in the water. For miles downstream from a source of this pollutant, aquatic life is essentially destroyed. Eventually, the solid matter settles, and normal life can be found again in the stream.

A more complex example, one for which there is not nearly so much reason to hope for the eventual solution by natural purification, is the degradability of organic materials. A ***biodegradable*** substance is composed of molecules which are broken down to simpler ones in the natural environment. For example, cellulose suspended in water will eventually be converted to carbon dioxide and water.

$$(C_6H_{10}O_5)_x \xrightarrow[\text{Biochemical catalysts}]{\text{Oxidizing agents}} CO_2 + H_2O$$

CELLULOSE — CARBON DIOXIDE — WATER

The overall oxidation process is a complicated one, involving microorganisms and their enzymatic secretions. Many reaction steps are required; some are not yet understood. Some organic molecules, notably some of those synthetically produced, are not easily biodegradable; these molecules simply stay in the natural waters or are absorbed by life forms and remain intact for long periods of time.

Some man-made molecules are not biodegradable and therefore are very persistent in natural waters.

Detergents offer an interesting contrast in biodegradability. Recall that detergents are composed of long organic molecules, one end of which is "soluble" in water and the other in oil. Branched-chained detergent molecules, such as

$$CH_3-\underset{\displaystyle CH_3}{\underset{|}{CH}}-CH_2-\underset{\displaystyle CH_3}{\underset{|}{CH}}-CH_2-\underset{\displaystyle CH_3}{\underset{|}{CH}}-CH_2-\underset{\displaystyle CH_3}{\underset{|}{CH}}-C_6H_4-SO_3^{\ominus}\ Na^{\oplus}$$

are not readily biodegradable. The enzymes of most of the bacteria present do not catalyze this decomposition. When such detergents were

Figure 19–2 Foam in natural waters. A high concentration of organic molecules dissolved in natural water causes foam because of the lowered surface tension of the water. If the water is badly polluted, the foam may exist miles from the pollution source. (From Singer, S. F., "Federal interest in estuarine zones builds," *Environmental Science and Technology,* Vol. 3, p. 2, 1969.)

widely used, foaming problems occurred in sewers, streams, and, in extreme cases, in the purified fresh water supply (Figure 19–2). With the subsequent advent of straight-chain detergents, such as

$$CH_3CH_2CH_2CH_2CH_2CH_2CH_2CH_2CH_2CH_2CH_2CH_2OSO_3^{\ominus}Na^{\oplus}$$

Bacteria are capable of breaking detergent molecules down quite rapidly if the detergents have only straight chains of carbon atoms present.

biodegradability was readily achieved. Even so, a major river can still be overloaded and caused to foam from detergent for a mile or more downstream from a raw sewage exit.

Some chemically stable materials, (DDT) and radioisotopes (strontium-90) are made in such large quantities that they cannot be dealt with effectively by dilution. If such materials are judged to be harmful to the natural environment, we must choose not to make or use them or we must assist nature in dealing with these species. No one really knows what catastrophic effects would be caused by the sinking of a ship loaded with DDT or the release of strontium-90 that would be produced in a nuclear war.

Even nature's pure rainwater is in jeopardy in isolated instances. In areas in which heavy concentrations of automobile fumes collect, poisonous lead compounds have been found in rainwater in concentrations many times higher than the 0.01 ppm generally allowed in drinking water. The concentration of the lead can be correlated with the concentration of exhaust fumes in the air. Fortunately, lead does not long remain in water, since it generally forms insoluble compounds.

Pollution occurs when the natural purification processes are overrun. For those areas of the world where population is concentrated, it appears that we must either apply supplementary purification techniques to a greater degree than we are presently doing or learn to live with the pollution.

When natural purification processes cannot cope with materials added to water, pollution results.

THE SCOPE OF WATER POLLUTANTS

There was a time when polluted water could be thought of in terms of dissolved minerals, natural silt, and contaminants associated with the

TABLE 19–1 Classes of Water Pollutants, with Some Examples

Class	Examples
1. Oxygen-demanding wastes	Plant and animal material
2. Infectious agents	Bacteria and viruses
3. Plant nutrients	Fertilizers, such as nitrates and phosphates
4. Organic chemicals	Pesticides, such as DDT, detergent molecules
5. Other minerals and chemicals	Acids from coal mine drainage, inorganic chemicals such as iron from steel plants
6. Sediment from land erosion	Clay silt on stream bed may reduce or even destroy life forms living at the solid-liquid interface
7. Radioactive substances	Waste products from mining and processing of radioactive material, radioactive isotopes after use
8. Heat from industry	Cooling water used in steam generation of electricity

natural wastes of animals and humans. As our use of water has increased, the pollution has become more diversified. The U.S. Public Health Service has classified water pollutants into the eight broad categories listed in Table 19–1.

Biochemical Oxygen Demand

The way in which organic materials are oxidized in the natural purification of water deserves special attention. The process opposes ***eutrophication.*** Even in the natural state, living organisms found in natural waters are constantly discharging organic debris into the water. To change this organic material into simple inorganic molecules (such as CO_2 and H_2O) requires oxygen. The amount of oxygen required to oxidize a given amount of organic material is called the ***biochemical oxygen demand*** **(*BOD*).** The oxygen is required by microorganisms, such as many forms of bacteria, to utilize organic matter as food in their metabolic reactions. Ultimately, given near normal conditions and enough time, the microorganisms will convert huge quantities of organic matter into the following end-products:

Eutrophication is the enrichment of a body of water with nutrients such that growth of organisms in the water makes it unfit for human purposes.

$$
\begin{aligned}
&\text{Organic carbon} && \longrightarrow CO_2 \\
&\text{Organic hydrogen} && \longrightarrow H_2O \\
&\text{Organic oxygen} && \longrightarrow H_2O \\
&\text{Organic nitrogen} && \longrightarrow NO_3^-
\end{aligned}
$$

One way to determine the amount of organic pollution is to determine how much oxygen a given sample of polluted water will require. For example, a known volume of the sewage is diluted with a known volume of standardized sodium chloride solution of known oxygen content. This mixture is then held at 20°C for five days in a closed bottle. At the end of this time the amount of oxygen which has been consumed is taken to be the biochemical oxygen demand.

A standardized solution is one of known concentration.

The waste waters of modern society often have very high concentrations of organic material, which have a very large biochemical oxygen demand (Figure 19–3). In extreme cases, more oxygen is required than

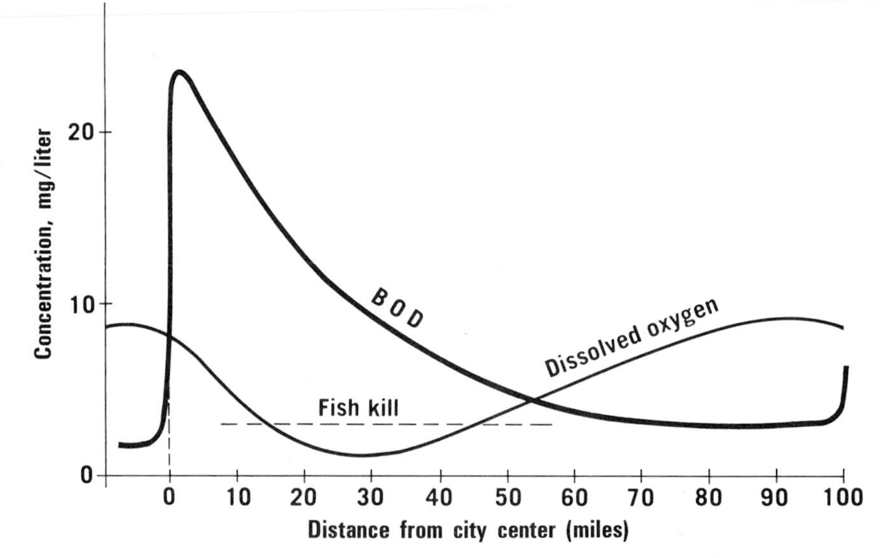

Figure 19–3 Graph showing oxygen content and dissolved nutrients (BOD) as a result of sewage introduced by a city. The results are approximated on the basis of a river flow of 750 gallons per second. Note that it takes 70 miles for the stream to recover from a BOD of .023 gram oxygen per liter. (From Turk, A., et al.: *Environmental Science.* Philadelphia, W. B. Saunders Co., 1974.)

A quantitative relationship exists between oxygen needs and organic pollutants to be destroyed. This is BOD.

Fish cannot live in water that has less than 0.004 gram O_2 per liter (4 ppm).

Temperature °C	Solubility g_{O_2}/liter H_2O
0	0.0141
10	0.0109
20	0.0092
25	0.0083
30	0.0077
35	0.0070
40	0.0065

These data are for water in contact with air.

is available from the environment, and putrefaction results. Fish and other freshwater aquatic life can no longer survive. The aerobic bacteria (those that require oxygen for the decomposition process) die. As a result of the death of these organisms, even more lifeless organic matter results and the BOD soars. Nature, however, has a back-up system for such conditions. A whole new set of microorganisms (anaerobic bacteria) takes over; these organisms take oxygen from oxygen-containing compounds to convert organic matter to CO_2 and water. Organic nitrogen is converted to elemental nitrogen by these bacteria. Given enough time, enough oxygen may become available, and aerobic oxidation will then return.

A stream containing 10 parts per million (ppm) by weight (just 0.001 percent) of an organic material, the formula of which can be represented by $C_6H_{10}O_5$, will contain 0.01 g of this material per liter. The calculation used to obtain this is:

$$?\text{g} = 1 \text{ liter of water}$$

$$?\text{g} = 1\,\cancel{\text{liter}} \times \frac{1000\,\cancel{\text{ml}}}{1\,\cancel{\text{liter}}} \times \frac{1\text{ g}}{\cancel{\text{ml}}} = 1000\text{ g}$$

0.001 percent of this is the pollutant:
0.001 percent of 1000 g = (0.00001)(1000 g) = 0.010 g

Figure 19–4 Fish kills can be caused by the lack of a substance necessary for life, such as oxygen, or the presence of toxic materials that interfere with the life processes. A heavy concentration of organic matter in a stream may depress the oxygen concentration below that required to support fish life. (From Tidwell, P.: Anti-pollution: A Fish Management Priority. The Tennessee Conservationist, *37*:4, 1971.)

To transform this pollutant to CO_2 and H_2O, the bacteria present use oxygen as described by the equation:

$$\underset{\substack{\text{RELATIVE}\\\text{WEIGHT}\\162}}{C_6H_{10}O_5} + \underset{\substack{\text{RELATIVE}\\\text{WEIGHT}\\192}}{6O_2} \longrightarrow 6CO_2 + 5H_2O$$

The 0.01 g of pollutant requires 0.012 g of dissolved oxygen.

$$\frac{?\text{g oxygen}}{\text{liter}} = \frac{0.010\ \cancel{\text{g pollutant}}}{\text{liter}} \times \frac{192\text{ g oxygen}}{162\ \cancel{\text{g pollutant}}} = \frac{0.012\text{ g oxygen}}{\text{liter}}$$

At 68°F the solubility of oxygen in water under normal atmospheric conditions is 0.0092 g of oxygen per liter.

Since the BOD (0.012 g per liter) is greater than the equilibrium concentration of dissolved oxygen (0.009 g per liter), as the bacteria utilize the dissolved oxygen in this stream, the oxygen concentration of the water will soon drop too low to sustain any form of fish life (Figure 19–4). Life forms can survive in water where the BOD exceeds the dissolved oxygen if the water is flowing very vigorously in a shallow stream (this facilitates the absorption of more oxygen from the air via aeration).

Characteristic Levels	*BOD* g_{O_2}/liter
Untreated municipal sewage	0.1 to 0.4
Run-off from barnyards and feed lots	0.1 to 10
Food processing wastes	0.1 to 10

High concentration of organic pollutants
↓
Low oxygen concentration
↓
Dead organisms
↓
Higher concentration of organic pollutants
↓
Lower oxygen concentration
↓
Anaerobic conditions

Thermal Pollution

Thermal pollution results when water is used for cooling purposes and in the process has its own temperature raised. Water has a very high ***heat capacity*** (the heat required to raise a unit weight of water one degree) and a high ***heat of vaporization*** (the heat required to change a unit weight of liquid water to gaseous water); this combination makes water an ideal cooling fluid for thermal power stations, nuclear energy generators, and industrial plants. The extent to which thermal pollution can occur is illustrated by the Thames River, the center of an industrial complex in England. The yearly average temperature in this river, corrected for changes in climatic conditions, rose from 53°F in 1930 to 60°F in 1950. The heat released into this river in 1930 was at the average rate of 132,000 kilocalories per second. In 1950 the rate became 888,000 kilocalories per second. The river, even with various means of cooling (evaporation, conduction, and discharge to the sea), simply could not dissipate

Water will absorb large amounts of heat before it is vaporized.

Vast amounts of heat are released to nature by industrial processes.

TABLE 19–2 Heat Contributors to Increased Heat Content of Thames River in 1950

	Percent
Fossil fuel power stations	75
Sewage effluents	9
Industrial effluents	6
Fresh water discharge	6
Biochemical activity	4

Figure 19–5 Thermal pollution. Plant effluent causes the Calumet River to "steam." Such heat discharges can change the temperature of a natural body of water enough to vastly alter the ecological balances. (From "Technology," *Chemical and Engineering News,* Feb. 24, 1969.)

the heat as fast as it was fed into the river. Table 19-2 lists the types of contributors to this heat inflow.

The most obvious result of thermal pollution is to make the water less efficient for further cooling applications. Far more important, however, are the biological and biochemical implications.

The solution of oxygen in water is facilitated by:
1. exposed surface area of water;
2. cool temperature;
3. low concentration of oxygen in the water.

The oxygen content of water in contact with air is dependent on the temperature of the water. More oxygen can dissolve in a quantity of cold water than in the same quantity of warm water, as shown by the data in the margin of p. 465. Also, the *rate* at which water dissolves oxygen is directly proportional to the difference between the actual concentration of oxygen present and the equilibrium value. This is extremely fortunate, since it means the rate of solution of oxygen increases sharply as the oxygen is consumed. Since a larger surface area will allow quicker absorption of oxygen, the rate of absorption of oxygen from the air is more greatly facilitated in a shallow, cold mountain stream than in a deep lake behind a dam in a warm river.

Aquatic life is very sensitive to temperature. Lethal temperatures for various species of fish in Wisconsin and Minnesota:

trout	77°F
white sucker	84–85°F
walleye	86°F
yellow perch	84–88°F
fathead minnow	93°F

The increased temperature of the Thames River decreased the oxygen content by four percent over what it otherwise would have been. However, the biochemical results of this factor alone could not be determined, since the river was anaerobic as a result of other pollutants in 1950. Even though more is to be learned about thermal pollution, two conclusions seem obvious: (1) thermal pollution aggravates the problem of oxygen supply; and (2) a significant rise in the temperature of a stream can drastically change or even destroy entire biological populations.

Solutions to thermal pollution involve cooling the water in evaporation towers or storage lakes before returning it to the natural body of water. Cycling the water for reuse after cooling has obvious advantages. Table 19–3 shows a quantitative forecast of the thermal pollution problem in the United States.

TABLE 19–3 A Projection of the Thermal Pollution Problem in the United States*

Year	Per Capita Electrical Energy Consumption in U.S. (48 States, Actual or Projected)	Cooling Water Needs (Billion Gallons per Day)
1950	2,000 Kwh	~30
1968	6,500 Kwh	~100
1980	11,500 Kwh	200†
2000	24,000 Kwh	450†

*Notes: a. Annual surface runoff is 1250 billion gallons per day.
b. Generation of electrical energy is doubling every 10 years.
c. 500 new power stations will be needed by the year 2000, at the present growth rate.
d. Kwh is kilowatt-hour, an energy unit equivalent to 860 kilocalories.

†Projected.

Fertilizers

Fertilizers are made from chemicals such as sodium nitrate, $NaNO_3$, and "superphosphate," $CaH_4(PO_4)_2 \cdot CaSO_4$. These substances are soluble enough to be carried away by the surface run-off during and after a hard rain. On the other hand, natural fertilizers are not as water-soluble, but release nitrate, phosphate, and metal ions at a rate commensurate with the rate of uptake by the plants. New fertilizer products are now on the market that imitate nature in releasing their soil nutrients at a slower rate.

Contamination of water by fertilizers leads to very undesirable effects. In a large measure, this contamination results from the phosphate (PO_4^{3-}) and nitrate (NO_3^-) present in fertilizers. It is generally believed that phosphate and nitrate encourage the growth of large amounts of algae. It is certain that nitrate in sufficient concentration is toxic to most higher organisms.

Limnology is the study of physical, biochemical, and biological aspects of freshwater systems. One of the conclusions of limnology is that massive but relatively delicate balances exist in such systems. The growth of a limited kind and quantity of algae is good for a lake; such growth is a source of needed oxygen via photosynthesis:

Limnology—the study of freshwater systems.

$$\underset{\text{CARBON DIOXIDE}}{6CO_2} + \underset{\text{WATER}}{6H_2O} \xrightarrow[\text{Catalysts}]{\text{Sunlight}} \underset{\text{SUGAR}}{C_6H_{12}O_6} + 6O_2$$

However, in waters that contain excessive amounts of phosphate and nitrate leached from fertilized farm land, algae growths are sometimes so massive that they tend to choke out desirable life forms. Such "blooms" of algae may lead to eutrophication. In this state, the dead oxygen-producing algae actually give rise to an oxygen-deficient environment, especially at lower depths. This is because the decaying algae quickly consume the available oxygen produced by the algae; the BOD rises sharply.

Is there a danger of too much phosphate (PO_4^{3-}) and nitrate (NO_3^-)?

As yet it is not clear that the nitrate and phosphate are the only or even the primary causes of the algae blooms. Some have suggested the primary cause could be an excessive organic pollution. Bodies of water in which the phosphate concentrations were kept abnormally low—less than 10 parts per billion—produced algae blooms. It was reasoned that the higher-than-normal levels of carbon dioxide resulting from the bacterial oxidation of organic materials caused the excessive algae growth. It appears at this stage of studying algae blooms that there can be causes other than excess nitrate and phosphate. The confusion exists because the limnological problem is so complex and our present knowledge is so limited.

Although the evidence is inconclusive on whether excess phosphates must be removed from water, the W. R. Grace Company, at the end of a ten-year project, has developed a process that would reduce the great bulk of phosphate in municipal sewage.

The process (Figure 19–6) can be accomplished in most cases without any addition of chemicals. The Grace process involves heating the sewage to 70°C. The solution becomes alkaline as a result of the decomposition of ammonium bicarbonate and other chemicals.

$$\underset{\text{AMMONIUM BICARBONATE}}{NH_4HCO_3} \longrightarrow \underset{\text{AMMONIA}}{NH_3} + \underset{\text{WATER}}{H_2O} + \underset{\text{CARBON DIOXIDE}}{CO_2}$$

$$\underset{\text{AMMONIA}}{NH_3} + \underset{\text{WATER}}{H_2O} \longrightarrow \underset{\text{AMMONIUM ION}}{NH_4^+} + \underset{\text{HYDROXIDE ION}}{OH^-}$$

The ammonia causes the solution to be basic (alkaline) and the carbon dioxide is removed with a vacuum. Phosphate is then precipitated as magnesium ammonium phosphate.

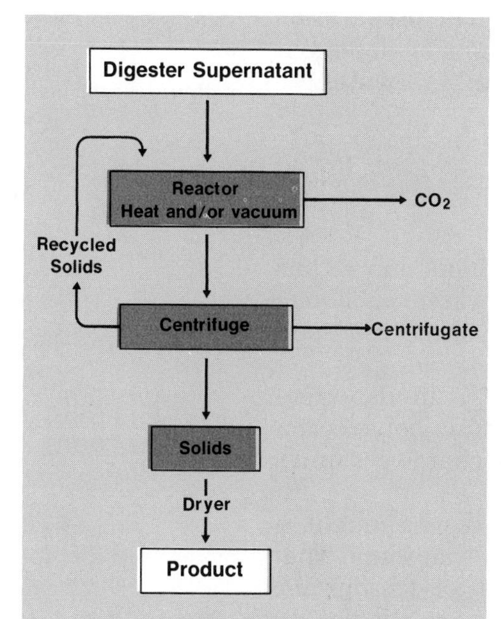

Figure 19–6 W. R. Grace process for removing phosphates from sewage.

$$Mg^{2+} + NH_4^+ + PO_4^{3-} \longrightarrow MgNH_4PO_4$$

MAGNESIUM ION · AMMONIUM ION · PHOSPHATE ION · MAGNESIUM AMMONIUM PHOSPHATE

The solid product is centrifuged and dried; it is a marketable product. The process also reduces nitrogen, BOD, and COD (chemical oxygen demand) levels.

Normal procedures used in purifying water for municipal supplies do not remove nitrate, and this ion can be more dangerous to human life than phosphate. When the concentration of nitrate is over a few parts per million (ppm), it is toxic because it destroys the ability of human hemoglobin to carry oxygen, as a result of the reduction of nitrate (NO_3^-) to nitrite (NO_2^-). The nitrite combines with iron in hemoglobin and prevents the combination of iron with oxygen. This leads to ***cyanosis*** (a condition in which the surface of the body becomes blue because of a lack of oxygen in the blood). Nitrate poisoning is especially dangerous to infants.

Detergent Builders

Detergents contain ***detergent builders,*** which are additives that make the surfactant (detergent) molecules more efficient and effective. In some detergents, as much as 40 percent of the total weight is in the form of detergent builders. A common detergent builder is sodium tripolyphosphate.

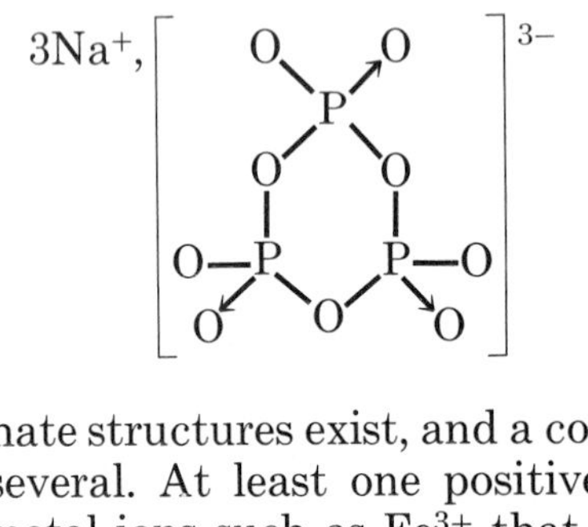

Detergents and detergent builders both may present water pollution problems.

Many polyphosphate structures exist, and a commercial preparation is likely to contain several. At least one positive action of detergent builders is to tie up metal ions such as Fe^{3+} that cause hard water.

$$Fe^{3+} + 2PO_4^{3-} \longrightarrow Fe(PO_4)_2^{3-}$$

IRON-PHOSPHATE COMPLEX

Because of the use of phosphorus in detergent builders, more than 500 million pounds of this element end up in our waste waters each year in the form of phosphate. A typical detergent contains about nine percent phosphorus by weight.

Replacements have been suggested for phosphates in detergent builders. These include nitrilotriacetic acid (NTA), organic polyelectrolytes (organic structures with a relatively large negative charge per unit), and sodium citrate.

$$N(CH_2COOH)_3$$
NTA

Since NTA is readily biodegradable, detergent makers thought it would be a suitable substitute for phosphates until it was found that NTA solubilizes heavy metals, producing toxic effects for various life forms. The degradability of the organic polyelectrolytes is not yet established, and sodium citrate, though readily degradable, is not as good a detergent builder as are the phosphates. A multimillion dollar industry awaits the development of a suitable chemical.

SELF-TEST 19-A

1. What liquid is sometimes called the universal solvent? ______________

2. Which process does nature not use in purifying natural water: distillation, crystallization, sedimentation, chlorination, or aeration?

   ____

3. Which is more likely to be biodegradable, a soap or a detergent?

   ____

4. BOD stands for ______________.

5. BOD is the amount (grams) of ______________ per ______________ required to oxidize a given amount of organic material.

6. ______________ is the study of physical, biochemical, and biological aspects of freshwater systems.

7. Detergents contain detergent builders, such as ______________, in addition to the detergent chemical itself.

Pesticides

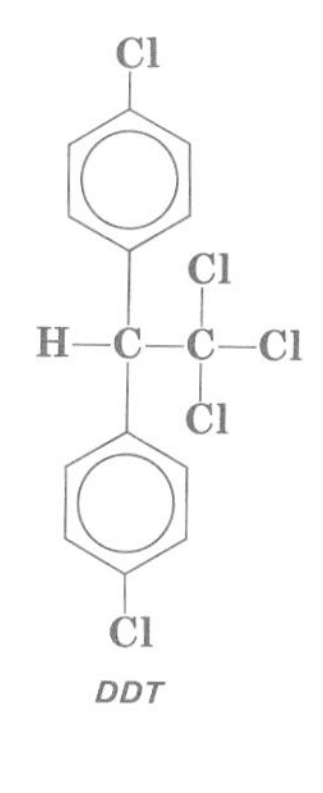

DDT

The use of synthetic insecticides increased enormously on a worldwide basis after World War II. As a result, insecticides such as DDT have found their way into lakes and rivers. There is a great variety of pesticides, and their use frequently leads to severe damage to other forms of animal life, such as fish and birds. The widespread use of pesticides was originally justified by the need for more food for a growing human population. However, the toxic reactions and peculiar biological side effects of many of the pesticides were not thoroughly studied or understood prior to their widespread use.

A good case in point is DDT. This insecticide, which had not been shown to be toxic to man in doses up to those received by factory workers involved in its manufacture (400 times the average dosage over years of exposure), does have peculiar biological consequences. The structure of DDT is such that it is *not* metabolized (broken down) very rapidly by animals; it is deposited and stored in the fatty tissues. The biological half-life of DDT is about eight years; that is, it takes about eight years for an animal to metabolize one-half of an amount it assimilates. The enzymes are just not present to catalyze the breakup of this molecule. If ingestion continues at a steady rate, it is evident that DDT will build up within the animal over a period of time. For many animals this is not a problem, but for some predators, such as eagles and ospreys, which feed on other animals and fish, the consequences are disastrous. The DDT in the fish eaten by such birds is concentrated in the bird's body,

Half-life is the time required for half of the substance to disappear (see Chapter 24).

Do we really have to decide between pests and eagles?

which attempts to metabolize the large amounts by an alteration in its normal metabolic pattern. This alteration involves the use of molecules which normally regulate the calcium metabolism of the bird and are vital to its ability to lay eggs with thick shells. When these molecules are diverted to their new use, they are chemically modified and are no longer available for the egg-making process. As a consequence, the eggs the bird does lay are easily damaged, and the survival rate decreases drastically. This process has led to the nearly complete extinction of eagles and ospreys in some parts of the United States where formerly they were numerous.

City water treatment generally does not remove DDT.

Not only does DDT have a long biological half-life, but it also is not readily biodegradable, and there is a resultant buildup of this molecule in natural waters. DDT and other insecticides such as ***dieldrin*** and ***heptachlor*** are referred to as ***persistent pesticides.*** Substitutions of other substances with biodegradable structures are now made where possible. The compound ***chlordan*** is an example of just such a substitution. It is interesting to note that the structural differences between heptachlor (persistent) and chlordan (short-lived) are relatively slight (look at the chlorine atom on the lower five-membered ring).

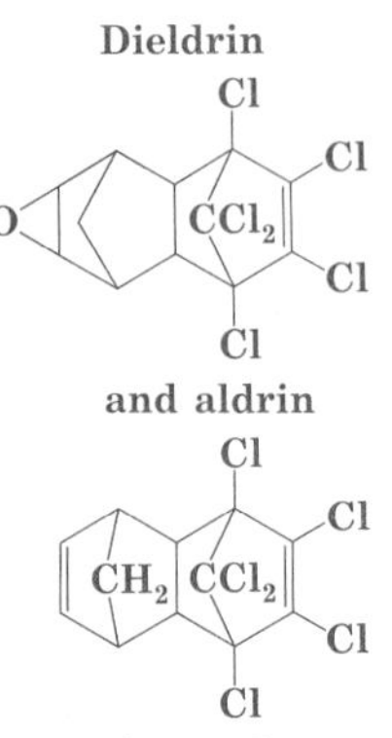

were banned by the courts from their major uses (such as on corn) in 1974.

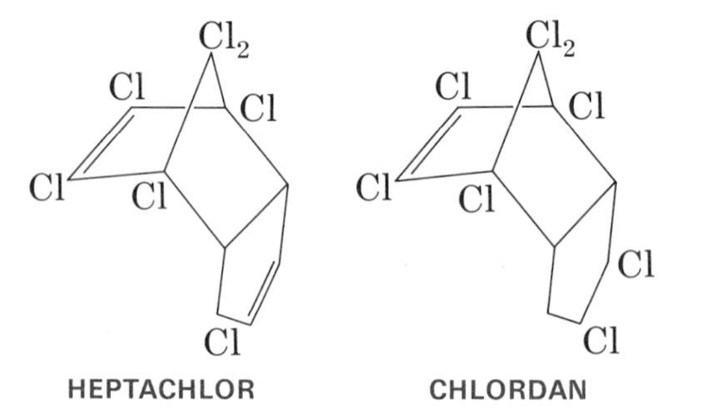

HEPTACHLOR CHLORDAN

In a study made under the sponsorship of the U.S. Department of Health, Education and Welfare from 1964 through 1967, 10 municipal water supplies were examined for pesticides. These waters were taken from the Mississippi and Missouri rivers. Over 40 percent of the samples contained dieldrin, over 30 percent contained endrin, and 20 percent contained chlordan. The amounts present were barely detectable and in the parts per billion (ppb) concentration range. Conclusion: these pesticides are present in the waters but not at levels considered to be dangerous.

Heptachlor and chlordan were banned for most garden and home use in December, 1975.

Endrin is an optical isomer of dieldrin (see Chapter 13).

There are many other insecticides which are actually much more toxic to man than is DDT. These include inorganic materials based on arsenic compounds, as well as a wide variety of phosphorus derivatives based on structures of the type

$(C_2H_5O)_2P(=S)-O-C_6H_4-NO_2$

Parathion

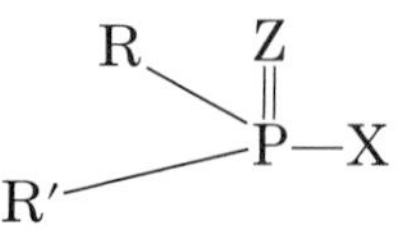

where Z = O or S, R and R′ are alkyl, alkoxy, alkylthio or amide groups, and X is a group which can be easily split from the phosphorus. Insecticides of this type include parathion, which is effective against a very large number of insects but is also *very* poisonous to human beings. These compounds, as we saw in Chapter 18, are anticholinesterase poisons. One of their most important properties, however, is that they are readily hydrolyzed to less toxic substances which are not residual poisons.

While it might occur to us to forbid completely the use of such insecticides, we would soon find that there are many areas of the United States in which agricultural production would drop precipitously were this to be done. An enlarged human population requires that crops be protected against the ever-present threat of destruction by insects. Furthermore, the use of insecticides is essential if we are to control malaria, plague, and other diseases. It must be recognized that a single insect species, the tsetse fly, alone has retarded the development of over 4,000,000 square miles of Africa by means of the diseases which it transmits to people and cattle.

The goal of the insecticide quest: a selectively toxic chemical that is quickly biodegradable.

The choice of solutions to our problems with insects is not an easy one. The use of insecticides introduces them into our environment and our water supplies. A refusal to use insecticides means that man must tolerate malaria, plague, sleeping sickness, and consumption of a large part of his food supply by insects. It is obvious that continuing research is needed on new methods and materials for the control of insect populations. A few of the new replacements for insecticides are discussed in Chapter 11.

Industrial Wastes

Industrial wastes can be an especially vexing sort of pollution problem because they often are not removed or are removed very slowly by naturally occurring purification processes and are generally not removed at all by a typical municipal water treatment plant. Table 19–4 lists some of the large variety of industrial pollutants that have been put into our natural waters on a large-scale basis.

Industrial wastes—only a cost problem; the solution is clear.

The technology necessary to remove industrial wastes from the water before it is returned to natural waters is available now. The limiting factor is the cost. We will simply have to decide if we are willing to spend a greater fraction of our resources in order to have a less polluted environ-

TABLE 19–4 A Partial List of Industrial Wastes that Have Been Dumped into Natural Waters

Acids
Bases
Salts
Various metal solutions
Oils
Bacteria
Emulsifying agents
Grease
Plant and animal wastes
Dyes
Waste solvents
Poisons such as cyanides and mercury
Numerous chemicals from washing operations

ment. The control of this pollution is made easier by the use of national and even international standards for all parties to observe. For example, steel mills have been and, to a lesser extent, still are notorious polluters of water. Some companies would find it very difficult to compete if they had to clean up all of their wastes while other companies did not. Concerted action, whether voluntary or enforced, is necessary. While the problem has not been completely solved in fact as it has been in theory, it only requires the will and commitment of the people.

Chemical methods capable of eliminating industrial water pollution are known for most pollutants.

An example of an industrial pollution problem that can be solved by a classical as well as by a recently developed method comes from metal processing and finishing operations, which produce large quantities of ions in solution. Chromium plating, for example, produces a waste solution containing chromium in the form of chromate (CrO_4^{2-}), cyanide (CN^-), and other less objectionable ions. Such wastes formerly were emptied into the natural waters. Chemical treatment has been understood for many years and has been used when efforts were made to clean up these wastes. The cyanide can be oxidized by an alkaline chlorine solution:

$$\underset{\text{HYDROXIDE (BASE)}}{10OH^-} + \underset{\text{CYANIDE}}{2CN^-} + \underset{\text{CHLORINE}}{5Cl_2} \xrightarrow[H_2O]{\text{Base}} \underset{\text{NITROGEN}}{N_2} + \underset{\text{BICARBONATE}}{2HCO_3^-} + \underset{\text{CHLORIDE}}{10Cl^-} + 4H_2O$$

The chromium-containing ion (CrO_4^{2-} [chromate]) is reduced by sulfur dioxide (SO_2):

$$4H^+ + \underset{\text{CHROMATE}}{2CrO_4^{2-}} + \underset{\text{SULFUR DIOXIDE}}{3SO_2} \longrightarrow \underset{\text{CHROMIUM(III) SALTS}}{2Cr^{3+}} + \underset{\text{SULFATE IONS}}{3SO_4^{2-}} + 2H_2O$$

The products of these reactions either are insoluble (Cr(III) salts) or are not objectionable (SO_4^{2-} ions), being common to the natural environment. However, the waste is not chemically pure, since it is a solution of NaCl containing a high concentration of CO_2.

Electrical methods of water purification are often quite costly.

It has been known for some time that an electrolytic reduction would produce chemically pure water. This method, however, has not been used because water is such a poor conductor of electricity. Excessive amounts of time and power would be required to complete the electrolysis. Recently, Resource Control, Inc., of West Haven, Conn., developed an electrolytic cell packed with an insoluble carbonaceous material (Figure 19-7). In this cell, the carbonaceous material provides good electrical conductivity between the electrodes; carbon is a good conductor of electricity. The electrical resistance is not a function of the salt concentration, and the salt concentration can be reduced essentially to zero in a short period of time with minimal power requirements. While the theory is as yet not well understood, there appear to be many microcells on the surface of the carbon. The positive ions (cations) are reduced at each cathode and the negative ions (anions) are oxidized at each anode.

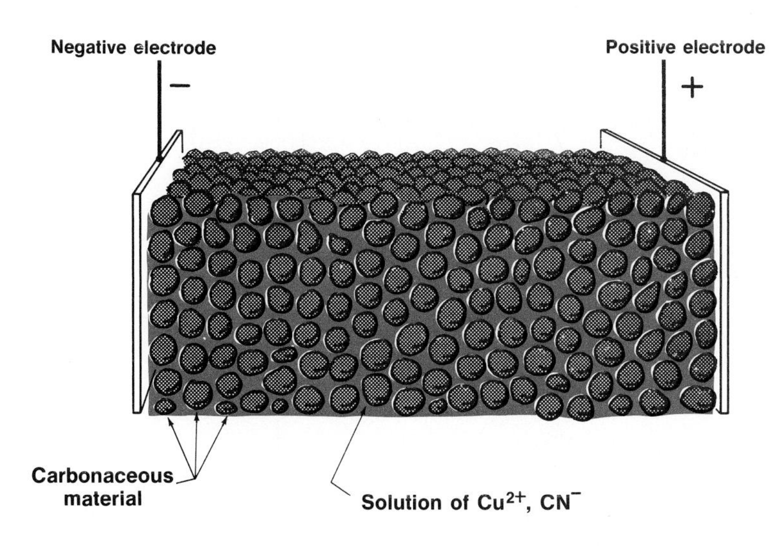

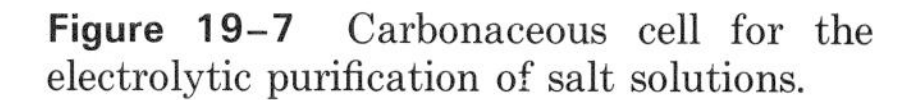
Figure 19–7 Carbonaceous cell for the electrolytic purification of salt solutions.

Table 19–5 gives before-and-after results for a copper-plating waste solution.

Another pollution problem, summarized in Table 19–6, concerns pickle liquor, a solution of HCl used to clean steel.

Heavy metals such as lead, mercury, cadmium, chromium, manga-

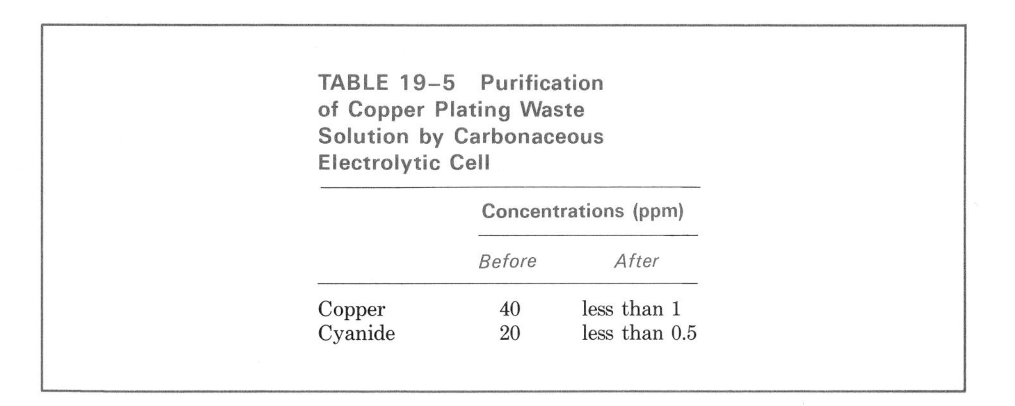
TABLE 19–5 Purification of Copper Plating Waste Solution by Carbonaceous Electrolytic Cell

	Concentrations (ppm)	
	Before	*After*
Copper	40	less than 1
Cyanide	20	less than 0.5

TABLE 19–6 The Fate of Spent Pickle Liquor

Content of pickle liquor:	From 0.5 up to 10% acid, 12% iron
Pollution problem:	Lowers pH of stream and produces solid slime
Scope of problem:	8 to 15 gallons produced per ton of steel; 50 million tons of steel produced in U.S. per year.
Method of disposal used:	1. Discharge into natural waterway. 2. Discharge into deep well. 3. Neutralization of acid before release to waterway. Sludge is produced and iron is released. 4. Hauling away liquor—a dilution approach. 5. Recovery. Cool liquid from 180°F. Iron crystallizes as $FeSO_4 \cdot 7H_2O$. Acid is concentrated and recycled. 6. Regeneration: One step beyond step 5. Commercial HCl and Fe_2O_3 recovered.
Immediate costs:	$6 > 5 > 4 > 3 > 2 > 1$
Long-range costs:	$1 > 2 > 3 > 4 > 5 > 6$

TABLE 19–7 Mercury Consumption in the United States Tops Two Thousand Tons*

1972 Consumption (tons)	
Electrical apparatus (batteries)	591
Electrolytic chlorine	438
Paint	311
Instruments	248
Dental preparations	114
Agriculture	70
Pharmaceuticals	22
All other uses	216
Total	2010

* Mineral Yearbook, 1972.

nese, copper, zinc, etc., are generally toxic to life forms in greater than trace amounts. The mechanism of their toxic action was described in Chapter 18. Here we will discuss the water pollution of these metals using mercury (Table 19–7) as an example.

About 10,000 tons of mercury are mined annually worldwide, and about 5,000 tons are somehow lost. Some of the loss is discharged as waste effluent from manufacturing plants and some of the rest is incorporated as traces into products in which it does not belong. For example, one chlorine plant is known to have been discharging as much as 200 pounds of mercury a day. In the production of chlorine, mercury is used as an electrode in the electrolysis of a solution of sodium chloride (brine) as described in Chapter 12. When the salt solution becomes too dilute, it is discarded and this discharge carries some mercury along with it. The hydrogen gas discharged to the atmosphere also carries some mercury vapor with it, and the sodium hydroxide produced is contaminated with mercury as well.

About 5000 tons of mercury are lost into the environment each year.

The upper allowable limit for mercury in public water is 0.5 ppm, set by both the United States and Canada. In polluted regions, much less mercury is actually in solution, the figure of 0.03 ppm or less being given. The mercury must collect elsewhere.

Mercury in solution is often in the form of methyl mercury, $(CH_3)_2Hg$. This form is especially hazardous to man. Research indicates that metallic and inorganic mercury compounds can be converted to methyl mercury by anaerobic bacteria in the mud of lake bottoms, as well as by fish and mammals.

Fish and other aquatic organisms concentrate the mercury which is present in the organisms they eat.

One way to remove mercury from industrial wastes before they are discharged into public waters is by the process developed by the FMC Corporation. The effluent is run into a large concrete basin where sodium sulfide is added. Insoluble mercury sulfide is precipitated and collected by filtration:

$$\underset{\text{MERCURY(II), SOLUBLE}}{Hg^{2+}} + \underset{\text{SULFIDE}}{S^{2-}} \longrightarrow \underset{\text{MERCURY(II) SULFIDE, INSOLUBLE}}{HgS\downarrow}$$

The filtrate, which contains less than 3 parts per billion (ppb) mercury, is discharged into public waters. Less than 0.1 pound mercury is discharged per day.

The next step is to solubilize the mercury sulfide. The mud is stirred in a solution of sodium hydroxide and chlorine. The sulfide ion is oxidized by the hypochlorite ion (formed by the sodium hydroxide and chlorine), and mercury ions go into solution.

This reaction is profit-making and provides economic incentives to pollution control.

$$\underset{\text{INSOLUBLE MERCURY (II) SULFIDE}}{HgS} + \underset{\text{HYPOCHLORITE}}{OCl^-} + \underset{\text{WATER}}{H_2O} \longrightarrow \underset{\text{MERCURY (II)}}{Hg^{2+}} + \underset{\text{CHLORIDE}}{Cl^-} + \underset{\text{SULFUR}}{S} + \underset{\text{HYDROXIDE}}{2OH^-}$$

The mercury ions are then reduced to elemental metallic mercury and returned to the mercury cathode to be used again.

Silt

Silt, which is finely divided solid material, is picked up and carried along by flowing water until the velocity of the water is reduced as it enters a lake, dam area, or the sea. The silt is stabilized by two factors: (1) the kinetic energy of the moving water keeps at least some of the silt moving against gravity, and (2) charges build up on the silt particles, causing them to repel each other (Figures 19–8 and 19–9). In the quiet water of a lake, silt tends to settle out. The salt water of the sea is very effective in removing silt; hence, the great river deltas at the river mouths. The ions in salt water neutralize the electrostatic charge on the particles and then collisions between them result in coagulation. The larger particles settle readily.

Silt is undesirable in municipal water because it is esthetically unappealing and also because the silt particles can be carriers of chemicals or microorganisms, or both, that are harmful to man. Silt is readily removed in holding tanks by sedimentation and by filtration beds. Sedi-

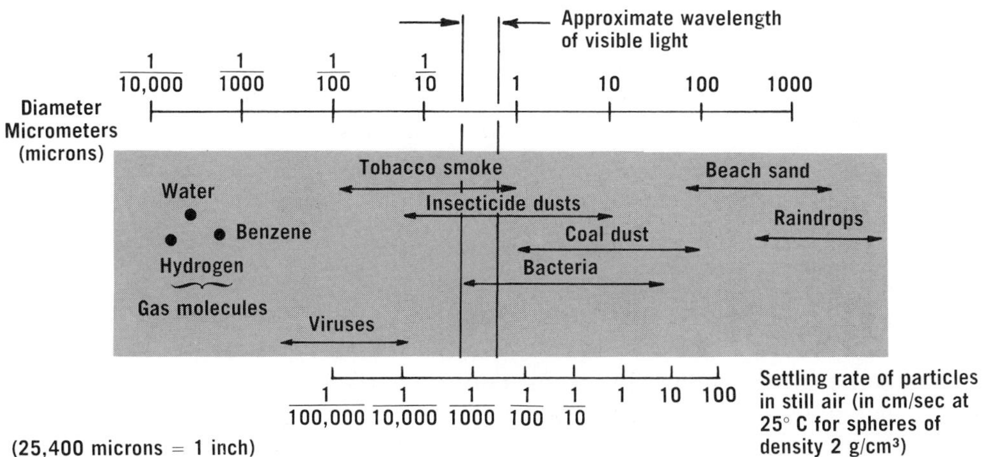

Figure 19–8 The settling rates in water of particles as a function of their size. Note that very long times are needed for the smaller particles. For practical purposes, colloidal particles will not settle. Of course, true solution species never settle. (From Turk, A., et al.: *Environmental Science*. Philadelphia, W. B. Saunders Co., 1974.)

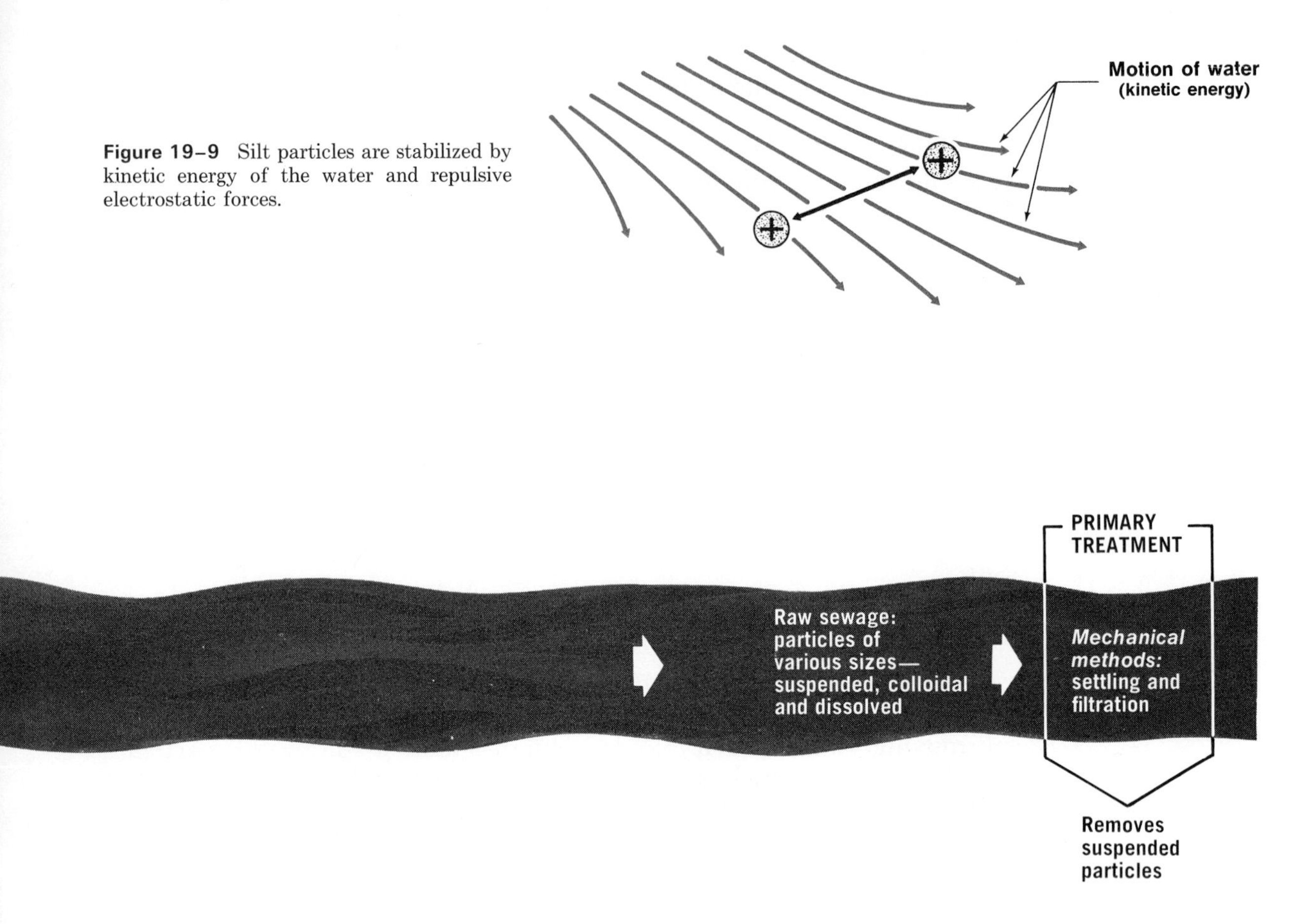

Figure 19–9 Silt particles are stabilized by kinetic energy of the water and repulsive electrostatic forces.

mentation is greatly enhanced if aluminum sulfate and calcium hydroxide, (slaked lime) are added to the settling (holding) tank; this results in the formation of aluminum hydroxide, which is a very sticky precipitate. The salt solution tends to neutralize static charges, and the sticky $Al(OH)_3$ collects the silt particles by physical attraction. Most of it settles and filtration removes the rest.

A sticky precipitate will settle silt.

WATER PURIFICATION: CLASSICAL AND MODERN PROCESSES

The outhouses of some rural dwellers had their counterparts in city cesspools. The terrible job of cleaning led to the development of cesspools that could be flush-cleaned with water, followed by a connecting series of such pools that could be flushed from time to time. City sewer systems with no holding of the wastes were the next step.

Cesspools were an early and crude form of the modern activated sludge process.

Since there were not enough pure wells and springs to serve the growing population, water purification techniques were developed. The classical, and what is now termed ***primary water treatment*** involved settling and filtration (Figures 19–10 and 19–11).

Sewage is still 99.9 percent water!

If the intake water is polluted enough with biological wastes, the primary treatment, even with chlorination, cannot render the water safe. To be sure, enough chlorine or other oxidizing agents could be added to kill all life forms but the result would be water loaded with a wide variety of noxious chemicals, especially chlorinated organics, many of which are suspected carcinogens. Some way had to be found to coagulate and separate out the organic material that passed through the primary filters.

Secondary water treatment revives the old cesspool idea under a more controlled set of conditions and only acts on the material that will not settle or cannot be filtered (Figure 19–12). It operates in an oxygen-rich environment (aerobic) whereas the cesspool operates in an oxygen-

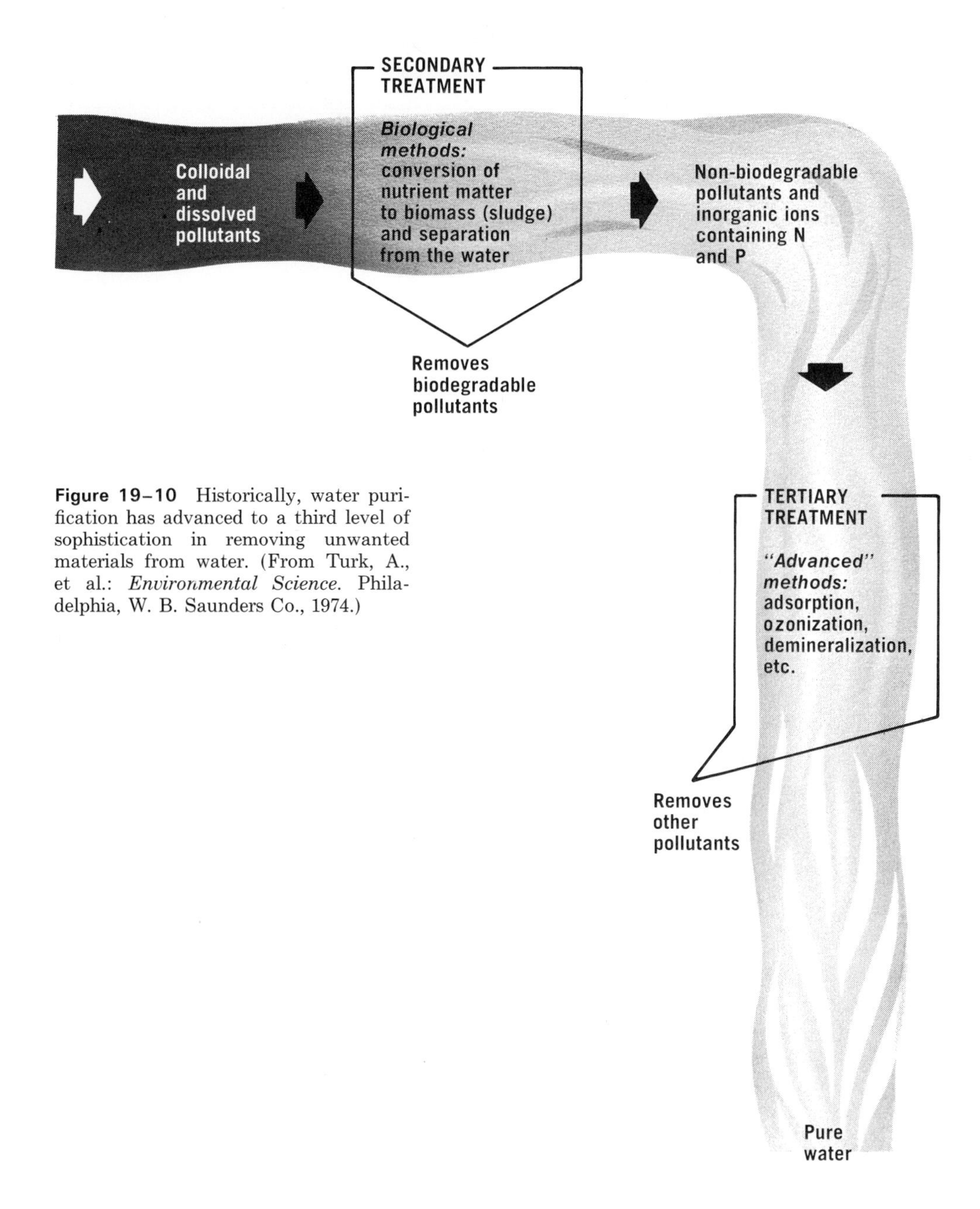

Figure 19–10 Historically, water purification has advanced to a third level of sophistication in removing unwanted materials from water. (From Turk, A., et al.: *Environmental Science*. Philadelphia, W. B. Saunders Co., 1974.)

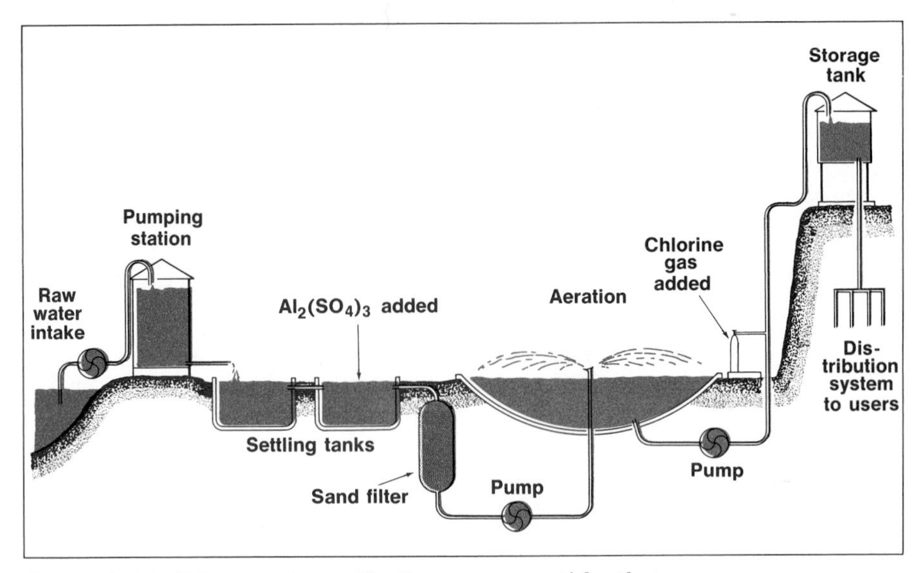

Figure 19–11 Primary water purification removes particles that will settle or can be filtered, usually by sandbed filters. Aeration adds oxygen and gets rid of foul gases, and chlorination kills microbes. These are more recent additions. The system outlined here is almost as common as communities in the civilized world.

poor environment (anaerobic). The results are the same: the organic molecules that will not settle are consumed by organisms; the resulting sludge will settle. Bacteria and even protozoa are introduced into the oxygen-rich environment for this purpose. Two techniques, the trickle filter (Figure 19–13) and the activated sludge method (Figure 19–14), have been widely used in secondary water treatment.

Primary and secondary water treatment systems will not remove dissolved inorganic materials such as poisonous metal ions or even residual amounts of organic materials. These materials are removed by a variety of ***tertiary water treatments.***

Primary, secondary, and tertiary water treatment methods can be used in both the purification of water to be consumed and the preparation of sewage to be sent back into a stream.

Filtration through carbon pellets absorbs most organic and odor-causing compounds. The method practically eliminates carcinogens from drinking water.

Since oceans are an "infinite" water supply, we shall examine some of

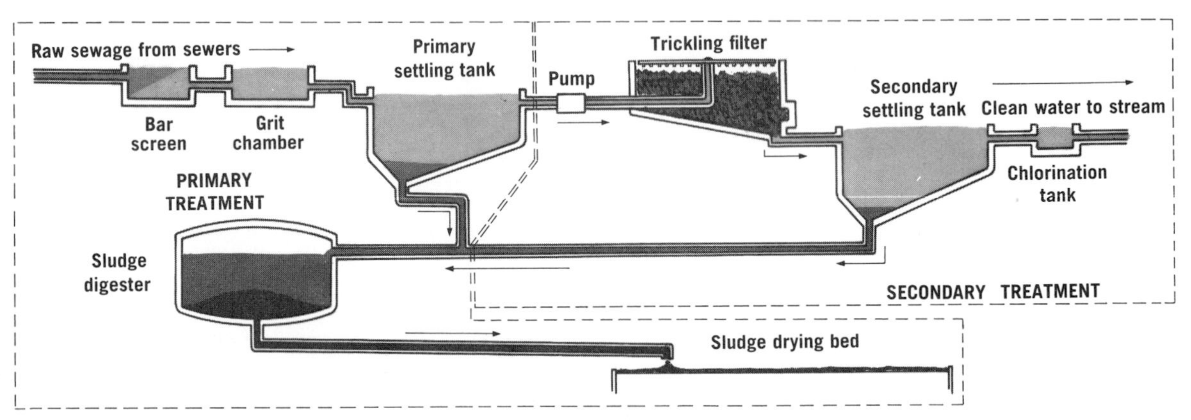

Figure 19–12 Sewage plant schematic, showing facilities for primary and secondary treatment. (From *The Living Waters.* U.S. Public Health Service Publication No. 382.)

Figure 19–13 Picture of a trickle filter with a section removed to show construction details. Rotating pipes discharge the water over a bed of stones. As a result, the organic molecules are "eaten" by microorganisms. (From Warren: *Biology and Water Pollution Control.* Philadelphia, W. B. Saunders Co., 1971. Photo courtesy of Link-Belt/FMC.)

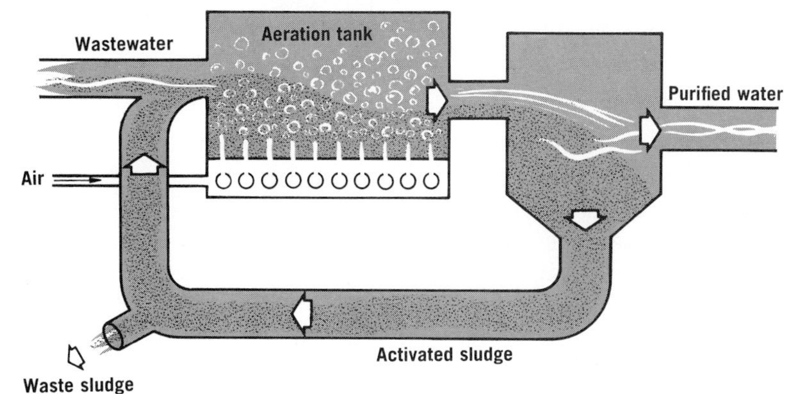

Figure 19–14 The activated sludge process provides a closed-system environment in which organic molecules can be consumed by organisms that will readily settle. (From Turk, A., et al.: *Environmental Science.* Philadelphia, W. B. Saunders Co., 1974.)

the tertiary water purification treatments of sea water. Of course, these same techniques serve as the tertiary stage in reclaiming sewage or any other polluted water supply.

Fresh Water from the Sea

As we look for new supplies of water, attention is inevitably drawn to the sea, as well as to the vast supplies of brackish water frequently found by drilling in arid lands. Sea water contains 3.5 percent salts: brackish waters have smaller salt concentrations. If we can convert such water to potable (drinkable) water inexpensively, we have a potentially valuable process. Thus far, the cost of purification of salt water is high in comparison with the cost of water supplied by a typical European or American municipal water system. However, one can expect the cost of fresh water purification to increase if pollution continues to increase, and the cost of purifying salt water to decrease as technology is improved.

The sea is a largely untapped water source.

The purification of salt water is at present approached in two general ways. Water may be separated from the salt by evaporation or by freezing. A second approach is to use the influence of electrical charge, chemical attraction, or selective membranes to cause the ions in the salt water to move out of the stream of water flow. If this process continues long enough, pure water results. Changing the state of water is expensive because of the large amount of energy that must move either into or out of the water. Recall that 540 cal are required to vaporize 1 gram of water and 80 cal must be removed to freeze each gram of water. The removal of ions from water is expensive because it is relatively slow for large flows of water and involves new techniques that are still under development.

Because water needs a high heat for vaporization, distillation requires a great deal of energy and is expensive.

Distillation

Distillation processes for the purpose of obtaining fresh water from sea water have been developed to a considerable state of refinement. Distillation in its simplest sense is illustrated in Figure 1–2 (Chapter 1). In view of the large amounts of heat required, various methods have been devised to heat the water with "waste heat" prior to entering the still. Since the steam has to lose 540 cal per gram in order to liquefy, this energy can be used to heat up the raw water to near the boiling point. Solar and nuclear energy have also been used to heat the water. Distillation plants are now the most common method in use to refine sea water and can produce water at a cost of less than $1 per 1000 gallons. This cost is competitive with the cost of older methods in many areas of the world.

Freezing

When cold sea water is sprayed into a vacuum chamber, the evaporation of some of the water cools the remainder, and ice crystals form in the brine. When the crystals of ice form, they tend to exclude the salt ions. Any solid separating from a liquid will tend to take only the molecules or ions that fit into the particular solid pattern. Recall that this generalization is the basis for purification by recrystallization. Even though the separation of salt and water is not complete in one step, the ice has less salt than the same weight of liquid solution. The ice crystals can be collected on a filter, washed with a small amount of fresh water, and then melted to obtain "pure" water. The process is repeated until the degree of purity desired is achieved. Plants have been built and successfully operated which produce as much as 250,000 gallons of pure water per day by this purification technique.

The ice in icebergs can be melted and drunk by humans.

Hydrate Formation

Propane (C_3H_8), a simple hydrocarbon, offers an interesting approach to the purification of water. In comparison to water, propane molecules have relatively little attraction for each other. Recall that water molecules interact through hydrogen bonding, while weaker van der Waals forces are the only attractive forces available between propane molecules. Consequently, theoretical considerations predict the heat changes in the freezing and melting of propane to be considerably less than for water. Experiment shows this to be the case. Experiment also shows that when propane is solidified in the presence of water, a hydrate is formed which is a crystalline solid containing water molecules trapped in the "holes" between the molecules of the solid propane. The salt ions which are attached to a hydrated sphere of water would be too large to fit into the "holes." If the solid hydrate is initially separated from the salt water and then heated, it first melts; the propane then vaporizes, leaving the "pure" water behind. This purification method for salt water has not been used widely because of the higher costs; these costs result from the relatively complex equipment required and the lack of experience in this new technology.

Ion Exchange

In the process called ion exchange, brackish water or sea water is first passed through a cation exchange resin to replace the cations with H^+ and then through an anion exchange resin to replace the anions with OH^-; these two ions then neutralize each other.

Modern ion-exchange resins are high molecular weight polymers which contain firmly bonded functional groups which can exchange one ion for another as an ionic solution is passed over the giant molecules. ***Cation exchangers,*** or positive-ion exchangers, swap their hydrogen ions for the positive ions present in solution. They are usually organic derivatives of sulfuric acid, and their action can be depicted as:

The chloride ion is shown on both sides of this reaction to indicate that it is not removed.

$$\text{Polymer-}SO_3^-H^+ + \underbrace{Na^+ + Cl^-}_{\text{BRACKISH WATER}} \longrightarrow \underbrace{\text{Polymer-}SO_3^-\,Na^+}_{\text{FROM CATION EXCHANGER}} + H^+ + Cl^-$$

When the polymer is saturated with cations, it can be regenerated by treatment with strong acid, which reverses the above reaction. Because they generate weakly acidic solutions, cation exchange resins themselves do not do a complete job. When they are followed by ***anion exchange resins,*** they provide a good route to very pure water (Figure 19–15). An

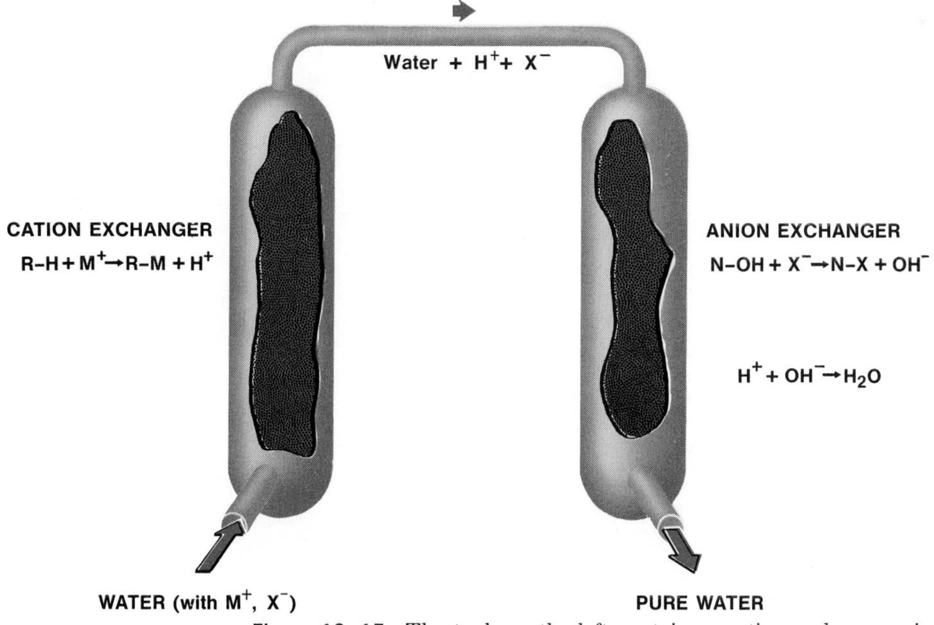

Figure 19–15 The tank on the left contains a cation exchange resin which replaces metal ions in solution with hydrogen ions. The tank on the right contains an anion exchanger resin which replaces nonmetal ions in solution with hydroxide ions. One might well ask why this kind of process cannot be used to turn sea water into pure water. The answer is that the process can be used, but its cost is *greater* than that of distillation. It is best suited to the removal of small amounts of salts from water which is already quite pure.

anion exchange resin can replace the anions in solution by hydroxide ions. These resins are again high molecular weight polymers, but now the polymer contains a nitrogen atom bonded to the polymer and three other groups. They function as shown:

$$\text{Polymer—}\overset{|}{\underset{|}{\text{N}^+}}\text{—},\ \text{OH}^- + \underbrace{\text{H}^+ + \text{Cl}^-}_{\text{FROM CATION EXCHANGER}} \longrightarrow \text{Polymer—}\overset{|}{\underset{|}{\text{N}^+}}\text{—},\ \text{Cl}^- + \underbrace{\text{H}_2\text{O}}_{\text{PURE WATER FROM CATION AND ANION EXCHANGER}}$$

When the anion exchange resin has exchanged all its hydroxide, it can be regenerated by treatment with a strong solution of sodium hydroxide:

Ion exchange resins can be regenerated.

$$\text{Polymer—}\overset{|}{\underset{|}{\text{N}^+}}\text{—},\ \text{Cl}^- + \text{Na}^+ + \text{OH}^- \longrightarrow \text{Polymer—}\overset{|}{\underset{|}{\text{N}^+}}\text{—},\ \text{OH}^- + \text{Na}^+ + \text{Cl}^-$$

Unfortunately, the amount of sea water which can be purified by a given amount of ion exchange resin is quite small, and the resulting water is relatively costly.

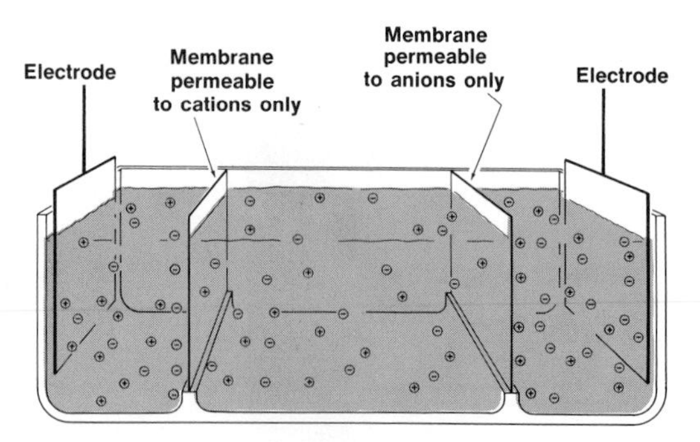

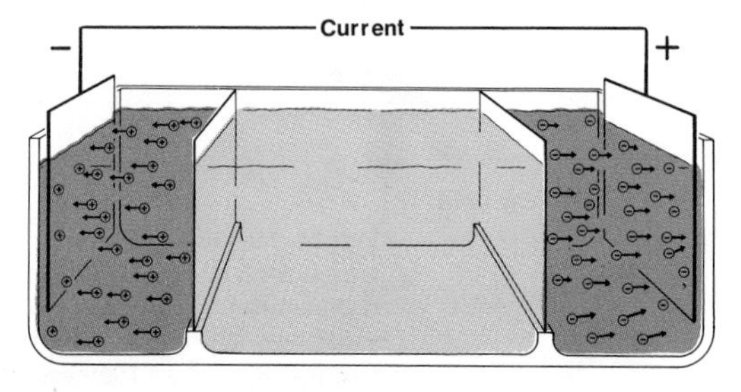

Figure 19–16 The essential features of the electrodialysis process. Each compartment of the cell contains brackish water. Application of an electrical potential across the cell causes the cations to move from the center into the left compartment and the anions to move into the right compartment. The salt content of the water in the center compartment is thus reduced.

Electrodialysis

In electrodialysis, ions are attracted to oppositely charged electrodes on the other side of a membrane.

We have learned that in the electrolysis of salt water, positive ions migrate toward the negative electrode and negative ions move toward the positive electrode. If an electrolysis cell is divided into three compartments by semipermeable membranes, one permeable to positive ions and the other permeable to negative ions (Figure 19–16), the process is called ***electrodialysis.*** Dialysis is the passage of selected species in solution through membranes while other species are excluded. Electrodialysis is the special case in which the passage of ions is influenced by an electrical field.

In Figure 19–16, note that positive ions (cations) can move to the left out of the center compartment but cannot move through the negative ion (anion) membrane from the right compartment to the center one. In a similar way, anions move out of the center compartment to the right. Thus, the ionic concentration of the water in the central compartment is reduced. If the process is continued long enough, the water in the central compartment loses most of its salt content.

Reverse Osmosis

Osmosis is the process whereby water will move through a semipermeable membrane from a region of relatively pure water into a region containing a concentrated solution. For example, water will move through a cell membrane into the cell's protoplasm, causing the cell to become turgid. The resulting pressure inside the cell is often very large. If enough mechanical pressure is brought to bear on the solution inside the

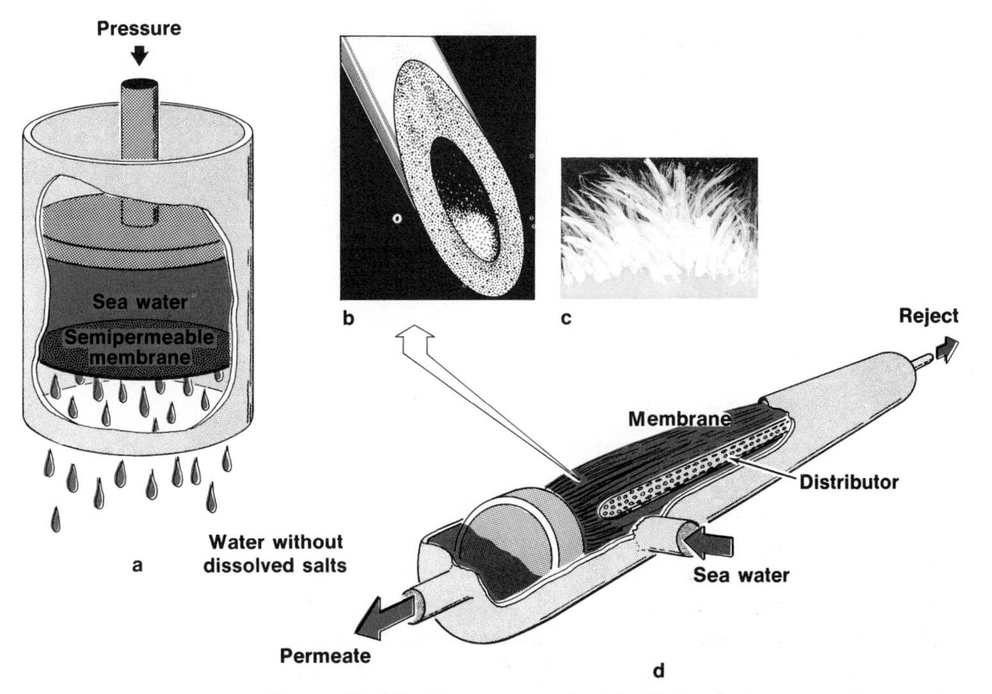

Figure 19–17 Reverse osmosis. (*a*) Mechanical pressure forces water against osmotic pressure to region of pure water. (*b*) Enlargement of individual membrane. (*c*) Mass of many membranes. (*d*) Industrial unit; feed water (salt) that passes through membranes collects at left end (permeate). The more concentrated salt solution flows out to the right as the reject.

membrane (in other words, if a pressure greater than the osmotic pressure is applied in reverse), the water can actually be made to flow from the concentrated solution inside to the region outside (Figure 19–17).

Reverse osmosis is a very promising process. Costs are down to 25¢ per 1000 gal for a plant producing 30,000,000 gal of pure water per day. This compares to 7.5¢ for the typical municipal treatment and 11¢ for the activated sludge treatments. Membranes used thus far are mostly cellulose acetate and work very well for salt water. Organic materials in the water tend to foul these films when used for municipal sewage. However, much effort is being made in developing new films for a wide variety of applications of reverse osmosis.

Softening of Hard Water

The presence of Ca^{2+}, Mg^{2+}, Fe^{2+}, or Mn^{2+} will impart "hardness" to waters. Hardness in water is objectionable because (1) it causes precipitates (scale) to form in boilers and hot water systems, (2) it causes soaps to form insoluble curds (this reaction does not occur with most synthetic detergents), and (3) it can impart a disagreeable taste to the water.

Hard water contains metal ions that react with soaps and give precipitates.

Hardness consisting of calcium or magnesium, present as their bicarbonates, is produced when water containing dissolved carbon dioxide trickles through limestone or dolomite:

$$\underset{\text{LIMESTONE}}{CaCO_3} + CO_2 + H_2O \longrightarrow Ca^{2+} + 2HCO_3^-$$

$$\underset{\text{DOLOMITE}}{CaCO_3 \cdot MgCO_3} + 2CO_2 + 2H_2O \longrightarrow Ca^{2+} + Mg^{2+} + 4HCO_3^-$$

Such "hard water" can be softened by removing these compounds. The principal processes for achieving this are (1) the lime-soda process and (2) ion-exchange processes.

Water softeners which act like ion exchange resins are used to make soft water. They remove the hard water ions, Ca^{2+}, Mg^{2+}, and Fe^{3+} and put Na^+ ions in the water in exchange.

The lime-soda process is based on the fact that calcium carbonate ($CaCO_3$) is much less soluble than calcium bicarbonate [$Ca(HCO_3)_2$] and that magnesium hydroxide is much less soluble than magnesium bicarbonate. The raw materials added to the water in this process are hydrated lime [$Ca(OH)_2$] and soda (Na_2CO_3). In the system, several reactions take place, which can be summarized as follows:

$$HCO_3^- + OH^- \longrightarrow CO_3^{2-} + H_2O$$

$$Ca^{2+} + CO_3^{2-} \longrightarrow CaCO_3\downarrow$$

$$Mg^{2+} + 2OH^- \longrightarrow Mg(OH)_2\downarrow$$

The overall result of the lime-soda process is to precipitate all the calcium and magnesium ions and to leave sodium ions as replacements.

Iron present as Fe^{2+} and manganese present as Mn^{2+} can be removed from water by oxidizing them with air (aeration) to higher oxidation states. If the pH of the water is 7 or above (either naturally or by adding lime), the insoluble compounds $Fe(OH)_3$ and $MnO_2(H_2O)_x$ are produced.

Chlorination

Chlorine is introduced as the gaseous free element (Cl_2), and it acts as a powerful oxidizing agent for the purpose of killing bacteria that remain in water after preliminary purification. The great majority of the bacteria in a water supply are removed physically by aluminum hydroxide precipitation. When the filtered water is allowed to stand, the bacterial population often decreases still more since its food supply is inadequate.

Chlorine—an oxidizing agent—destroys microbes in water. As an oxidizing agent, chlorine is reduced:
$Cl_2 + 2e^- \longrightarrow 2Cl^-$

When water is heavily contaminated by sewage, the possibility exists that preliminary treatment will not be sufficient to remove *all* of the bacteria capable of causing disease. As cities grew during the late nineteenth century, their water supplies came more and more from waters that had already been used by other cities or from wells which were very close to cesspools or sewers. This meant that the possibility increased for an epidemic caused by a water-borne disease. The principal diseases which are spread in this fashion include cholera, typhoid, paratyphoid, and dysentery.

The fact that cholera is caused by drinking polluted water was proved before the discovery of bacteria. An outbreak of cholera in London in 1854 was examined in great detail by Dr. John Snow. This was a relatively "small" epidemic, in which about 700 died. Dr. Snow made a very careful study of the last few weeks in the lives of about 600 of those killed in the epidemic. Where information was available, he was able to show that in every case all the victims of the plague had drunk water from a particular well. An examination of the well itself showed a crack in its wall through which sewage from persons sick with cholera had drained before the outbreak of the epidemic. This study showed that polluted water sources are a source of disease.

Figure 19–18 This apparatus adds chlorine in sufficient amounts to meet health standards (1 ppm residual) for a 60-million gallon per day water treatment plant. (Courtesy of the Robert L. Lawrence, Jr., Filtration Plant.)

Subsequent identification of bacteria and proof that they can cause disease led to the chemical treatment of drinking water to kill bacterial populations. The chemical first used for this purpose was chloride of lime [a crude form of calcium hypochlorite, $Ca(OCl)_2$]. This chemical was first used to treat the water supply of Louisville, Kentucky, in 1896. Within a decade and a half, the use of hypochlorites of various sorts spread and became quite common. In 1910, Major C. R. Darnall of the U.S. Army Medical Corps demonstrated that the addition of small amounts of chlorine to partially purified water brought about almost complete elimination of bacteria. Because of the convenience of using elemental chlorine and its ability to remove bacteria even from waters heavily polluted with sewage, chlorine is now widely used in the purification of municipal water supplies. Both bromine and iodine can also be used for this purpose, but these are usually less convenient and more expensive.

Chlorine was used in military water supplies in World War I and thus became more acceptable for domestic use.

Chlorine acts as an oxidizing agent, not only on bacteria and other living microscopic inhabitants of water, but also on dead organic matter. For this reason, chlorine is added to a water supply in sufficient quantity to kill the bacteria, oxidize the remains, oxidize any other organic material the water may contain, and leave a small residual amount of free chlorine. The presence of this small residual amount is a guarantee that the bacterial population has been destroyed. Although there is no direct evidence on this point, it is possible that chlorine is also capable of destroying many, if not all, potentially disease-causing viruses that may be present in water drawn from a polluted source.

Like many chemicals, chlorine has its good and its bad side. In moderate amounts, it kills harmful bacteria (as well as the harmless). In excess, it is a deadly poison. At Ypres, France (now Belgium), April 22, 1915, chlorine was the first chemical to be used as a poisonous gas in warfare.

Chlorination of industrial waste waters offers a danger that is only beginning to be understood. Since chlorine is a powerful oxidizing agent it will react with many organic and inorganic chemicals to produce potential hazards that did not exist in the untreated water. This has prompted some to say that the tap water in cities like New Orleans is more dangerous than the raw Mississippi River water. However, the evidence in the form of statistical occurrences of cancer and trace detection of many chemical species is not yet clear.

Fluoridation

Fluorine is added to water in the form of solid sodium fluoride or sodium fluorosilicate to introduce the fluoride ion, F^-, or its derivatives, such as SiF_6^{2-} (fluorosilicate), for the purpose of reducing tooth decay.

Fluoride, too, is a poison in large concentrations, but a valuable additive in trace quantities.

Widespread fluoridation of water followed the discovery that certain cities in the western United States, where the water supplies contained abnormally high concentrations of natural fluoride, had populations with a less than average incidence of tooth decay. A thorough investigation showed that the ingestion of such fluorides by young children was especially effective in protecting their teeth against cavities. As a consequence, the addition of small amounts of fluoride to drinking water to bring the fluoride concentration up to about 1 part per million was instituted on an experimental basis in some cities. The results bore out the original observation, that small amounts of fluoride in the drinking water could be effective in reducing cavities to about 40 percent of their normal incidence. Excessive amounts of fluoride, however, can cause damage to teeth.

Sodium fluorosilicate is Na_2SiF_6.

The mineral content, or the hard part, of bones and teeth consists of two compounds of calcium. Calcium carbonate ($CaCO_3$) is present in

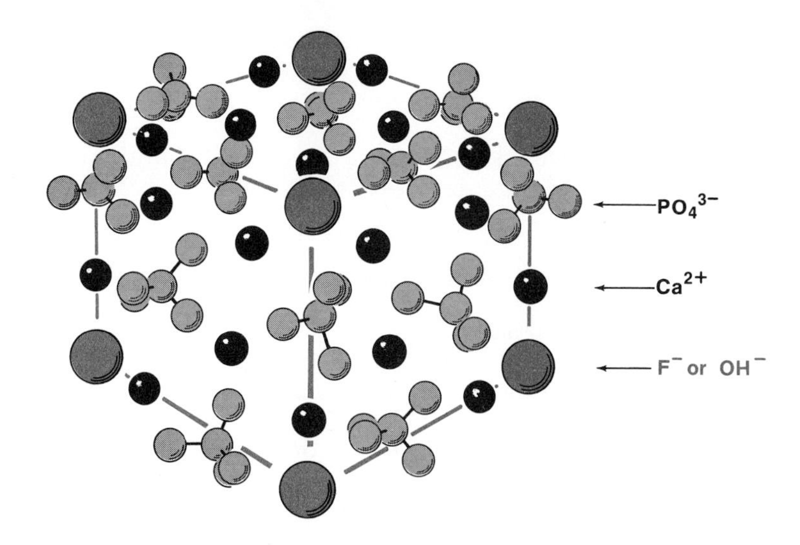

Figure 19–19 Structure of hydroxyapatite and apatite. The dark circles denote Ca^{2+} ions, the groups of five circles tied together by lines represent the PO_4^{3-} groups, and the largest circles represent OH^- groups in hydroxyapatite and F^- ions in apatite.

bones and teeth in the crystalline form, known to mineralogists as aragonite. The second calcium compound found in teeth is calcium hydroxyphosphate, [$Ca_5(OH)(PO_4)_3$], or hydroxyapatite, and it is this crystalline structure that is modified and made more inert in the presence of fluoride ions (F^-). Both aragonite and hydroxyapatite are susceptible to acid attack from food decay, but hydroxyapatite is the more so because of its basic characteristics associated with the OH^- ion. If fluoride ions are present in sufficient concentrations during the formative period of the teeth, fluoroapatite [$Ca_5F(PO_4)_3$] is laid down instead of hydroxyapatite. The two apatites have the same general structure, so the tooth structure is not physically changed (Figure 19–19). Since apatite does not have the base properties of hydroxyapatite and is not readily attacked by mouth acids, the teeth are less likely to develop cavities.

SELF-TEST 19-B

1. Pick the pesticide with the shortest half-life in nature: DDT, dieldrin, heptachlor, or chlordan. ______________
2. Heavy metal ions in more than trace concentrations are usually ______________ to life forms.
3. Silt is often stabilized in suspension by ______________ forces.
4. Select the ions that may cause water to be hard: sodium, calcium, magnesium, potassium. ______________
5. The element ______________ is added to water to kill microbes in water, and a compound of ______________ is added to protect the teeth against cavities.
6. Electrodialysis involves ion migration in a(n) ______________ field.

7. Ion-exchange systems for removing salts from water involve ________ (how many) types of resins.

8. Ice formed from impure water will likely be (more pure, equally pure, or less pure) than the original water? ________________

9. Primary water treatment involves ________________ and ________________ of particles.

10. In one secondary water treatment, ________________ in a(an) ____________-rich environment, remove a portion of material (principally organic) that does not settle and cannot be filtered.

11. Tertiary water treatment removes ____________ ions and trace amounts of ________________.

MATCHING SET

______	1. sedimentation	a.	tertiary purification process
______	2. biodegradable	b.	a measure of organic material in water
______	3. detergent	c.	widely used as a detergent builder
______	4. BOD	d.	results mostly from water used to cool a process
______	5. thermal pollution	e.	caused by metal ions such as Ca^{2+} and Mg^{2+} in solution
______	6. phosphate	f.	primary purification process
______	7. NTA	g.	persistent pesticide
______	8. mercury	h.	secondary purification process
______	9. activated sludge process	i.	naturally reduced to simpler compounds
______	10. reverse osmosis	j.	soap substitute
______	11. hard water	k.	metal poison
______	12. DDT	l.	nitrilotriacetic acid

QUESTIONS

1. What are some processes that *decrease* the amount of dissolved oxygen in a stream? What are some processes that *increase* the amount of dissolved oxygen in a stream? Which ones are most readily subject to man's control?

2. Explain why each of the following introduces a pollution problem when its wastes are emptied into a stream:

   a. a chlorine-producing plant.
   b. a steel mill.
   c. an electricity-generating plant burning oil or coal.
   d. an agricultural area which is intensively cultivated.

3. An old rule of thumb is, "Water purifies itself by running two miles from the source of incoming waste." What processes are active in purifying the water? Is this adage foolproof? Explain.

4. What is a detergent builder? What property should it have to increase the cleansing power of the detergent? What property must it have to avoid environmental problems?

5. What are some consequences of requiring by law that all water put into our national waters be clinically and chemically pure H_2O? Do you think this is the proper solution to water pollution? If not, what is the proper compromise?

6. What are some ecological consequences of thermal pollution?

7. From your experience, add one additional example for each of the classes of pollutants listed in Table 19-1.

8. What pertinent facts would you try to gather if it were your responsibility to vote on a bill to regulate water pollution?

9. At what point should pollutants be removed from used water? Who should be responsible for this removal? Would you distinguish between industrial wastes and household wastes?

10. The most abundant elements in organic compounds are carbon, hydrogen, oxygen, and nitrogen. What are the oxidation products for these elements in the natural decomposition that occurs in nature?

11. Relate molecular structure to biodegradability for detergent molecules.

12. Classify water pollutants into as few major groups as you can. Describe some effects of each group and a removal process.

13. From your study of biochemistry, Chapters 16 and 17, explain why there is no question concerning the biodegradability of the proposed detergent builder, sodium citrate.

14. In your judgment, what are the most serious water pollution problems? Be ready to defend your points in class discussion.

15. What is natural osmosis? Explain the significance of the word "reverse" in reverse osmosis.

PROBLEMS

1. If electrical energy costs 0.0023¢ per kilocalorie, what will be the cost to distill 4 liters (approximately 1 gal) of water from sea water if no attempt is made to utilize the heat of condensation of the water vapor? (The heat of vaporization of water is 540 calories per gram.)

   Ans. \$0.05

Ans. 0.003 g O_2/liter 2. What will be the BOD of a stream that contains the equivalent of 0.0001 percent C as oxidizable organic matter?

Ans. 40 years 3. If the biological half-life of DDT is 8 years, how long will it take to reduce the amount of DDT in a human from 100 mg/kg to 3.12 mg/kg?

SUGGESTIONS FOR FURTHER READING

Behrman, A. S., *Water is Everybody's Business,* Doubleday & Co., Garden City, N.Y., 1968.
Chisholm, J. J., Jr., "Lead Poisoning," *Scientific American,* Vol. 224, p. 15 (Feb., 1971).
Clark, J. R., "Thermal Pollution and Aquatic Life," *Scientific American,* Vol. 220, p. 18 (March, 1969).
Dowry, B., Carlisle, D., Laseter, J., and Storer, J., "Halogenated Hydrocarbons in New Orleans Drinking Water," *Science,* Vol. 187, p. 75 (1975).
Encyclopedia of Chemical Technology, Vol. 14, John Wiley and Sons, Inc., New York, 1970, p. 926.
"Eutrophication: Causes, Consequences, Correctives," Proceedings of a Symposium, National Academy of Sciences, Washington, D.C., 1969.
Goldwater, L., "Mercury in the Environment," *Scientific American,* Vol. 224, p. 15 (May, 1971).
Hills, E. S. (ed.), *Arid Lands: A Geographical Appraisal,* Methuen & Co., New York, 1969.
Howells, G. P., and Kneipe, T. J., "Water Quality in Industrial Areas: Profile of a River," *Environmental Science and Technology,* Vol. 4, p. 26 (1970).
Keller, E., "Fish Kills," *Chemistry,* Vol. 41, No. 9, p. 8 (1968).
Keller, E., "Nuclear-Powered Desalting in the Middle East," *Chemistry,* Vol. 42, No. 2, p. 7 (1969).
Keller, E., "The DDT story," *Chemistry,* Vol. 43, No. 2, p. 8 (1970).
Kirk, R. E., and Othmer, D. F., "Insecticides," *Encyclopedia of Chemical Technology,* Vol. 11, p. 677 (1970).
Leopold, L. B., and Langheim, W. B., "A Primer on Water," U.S. Government Printing Office, Washington, D.C., 1960.
Newman, Frank, "A Water Pollution Study," *Chemistry,* Vol. 42, No. 1, p. 28 (1969).
Overman, M., *Water,* Doubleday & Co., Garden City, N.Y., 1968.
Shaheen, E. I., *Environmental Pollution: Awareness and Control,* Engineering Technology, Inc., Mahomet, Ill., 1974.
Slabaugh, W. H., "Clay Colloids," *Chemistry,* Vol. 43, No. 4, p. 8 (1970).
Williams, C. M., "Third-Generation Pesticides," *Scientific American,* Vol. 217, p. 13 (July, 1967).

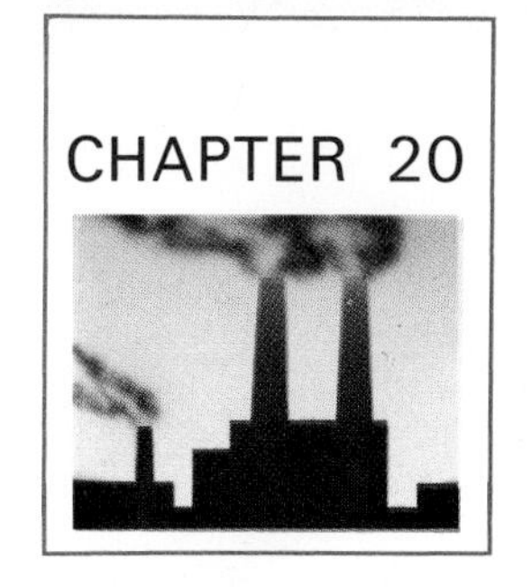

CHAPTER 20

AIR POLLUTION

Prior to 1960 there was little national or international concern about air pollution. Most smoke, carbon monoxide, sulfur dioxide, nitrogen oxides, and organic vapors were emitted into the air with little apparent thought of these pollutants being harmful as long as they were scattered into the atmosphere. Few controls were used to curtail the increasing

Figure 20–1 Air pollution covers the base of the St. Louis arch. (World Wide Photos, Inc.)

Figure 20–2 Region of Donora, Pennsylvania. In late October, 1948, Donora had a five-day siege of extreme air pollution. Before rain cleansed the air, over 800 domestic animals died and 43 percent of the population, or 5910 people, became ill. Eighteen died; the normal rate was two deaths every five days. (From Turk, A., et al.: *Environmental Science*. Philadelphia, W. B. Saunders Co., 1974.)

Figure 20–3 Coppertown Basin (Ducktown), Tennessee, as photographed in 1943. Copper ore (principally copper sulfide, Cu_2S) had been mined and smelted in this area since 1847. In the early years, large quantities of sulfur dioxide, a by-product, were discharged directly into the atmosphere and killed all vegetation for miles around the smelter. Today the sulfur is reclaimed in the exhaust stacks to make sulfuric acid, but the denuded soil remains a monument to the misuse of the atmosphere.

TABLE 20–1 Some Air Pollution Disasters*

Location and Date	Pollutants: Type	Pollutants: Range	Meteorology	Health Effects: Diseases and Symptoms	Health Effects: Excess Deaths
Meuse Valley, Belgium Dec., 1930	H_2SO_4 HF NO_2 CO CO_2	Concentrations unknown	Fog Wind speed—0.62 mph Ceiling—246 ft Temp. inversion	Cardiovascular hypotension Alkalosis Sore throat Cough Nausea Vomiting	60 to 80
St. Louis, Mo. Nov. 28, 1939 ("Black Tuesday")	Smoke	(Nighttime darkness through the day)	Temp. inversion Windless period		
Donora, Pa. Oct. 27–31, 1948	SO_2 (last day) H_2SO_4 mist Other sulfur compounds O_3 NO_x Organic compounds Smoke	Concentrations unknown	Temp. inversion Fog Stagnant anticyclone	Cough Respiratory irritation Sore throat Chest constriction Dyspnea Eye irritation Vomiting Nausea Heart diseases Bronchitis (43% of population became ill)	18
London Nov. 26–Dec. 1, 1948	Black suspended matter SO_2	200 to 2800 mg/m³ 0.09 to 0.75 ppm	Fog Wind speed—0 to 4.6 mph Visibility—27 to 440 yd		700 to 800
London Dec. 5–9, 1952	SO_2 Black suspended matter	0.09 to 1.34 ppm 400 to 4500 mg/m³	Wind vel.—9 to 5.8 mph Visibility—22 to 240 yd Fog	Bronchitis Emphysema Cardiovascular (ischemic heart disease) Acute wheezy chests Dyspnea Fever Yellow-black sputum	4000
London Jan. 3–6, 1956	SO_2 Black suspended matter	0.19 to 0.55 ppm 700 to 2400 mg/m³	Wind speed—0 to 8.1 mph Visibility—5 to 12,000 yd		1000
Rotterdam Dec. 2–7, 1962	SO_2	5 times normal			Increase in mortality noted
London Dec. 5–10, 1962	Smog SO_2	1.98 ppm, highest hourly concentration		Wheezy chests Bronchitis Dyspnea Fever Yellow-black sputum	700
Osaka, Japan Dec. 7–10, 1962	Pollution levels high				60
New York Jan. 29–Feb. 12, 1963	SO_2 Smoke shade	0.2 to 0.5 ppm 1.5 to 7 Coh units†	Few winds < 5 mph Few inversions Stagnating anticyclones were absent	Influenza Pneumonia Vascular lesions Cardiac arrests	200 to 400
New York Feb. 27–Mar. 10, 1964	SO_2 Smoke shade	0.0 to 7 ppm 2 to 5 Coh units†	A few inversions Few winds < 5 mph		168

*From "Guide for Pollution Episode Avoidance," Office of Air Programs, U.S. Environmental Protection Agency, 1971. Publication No. AP-76.

†Coh is the abbreviation for coefficient of haze, a unit of measurement of visibility interference.

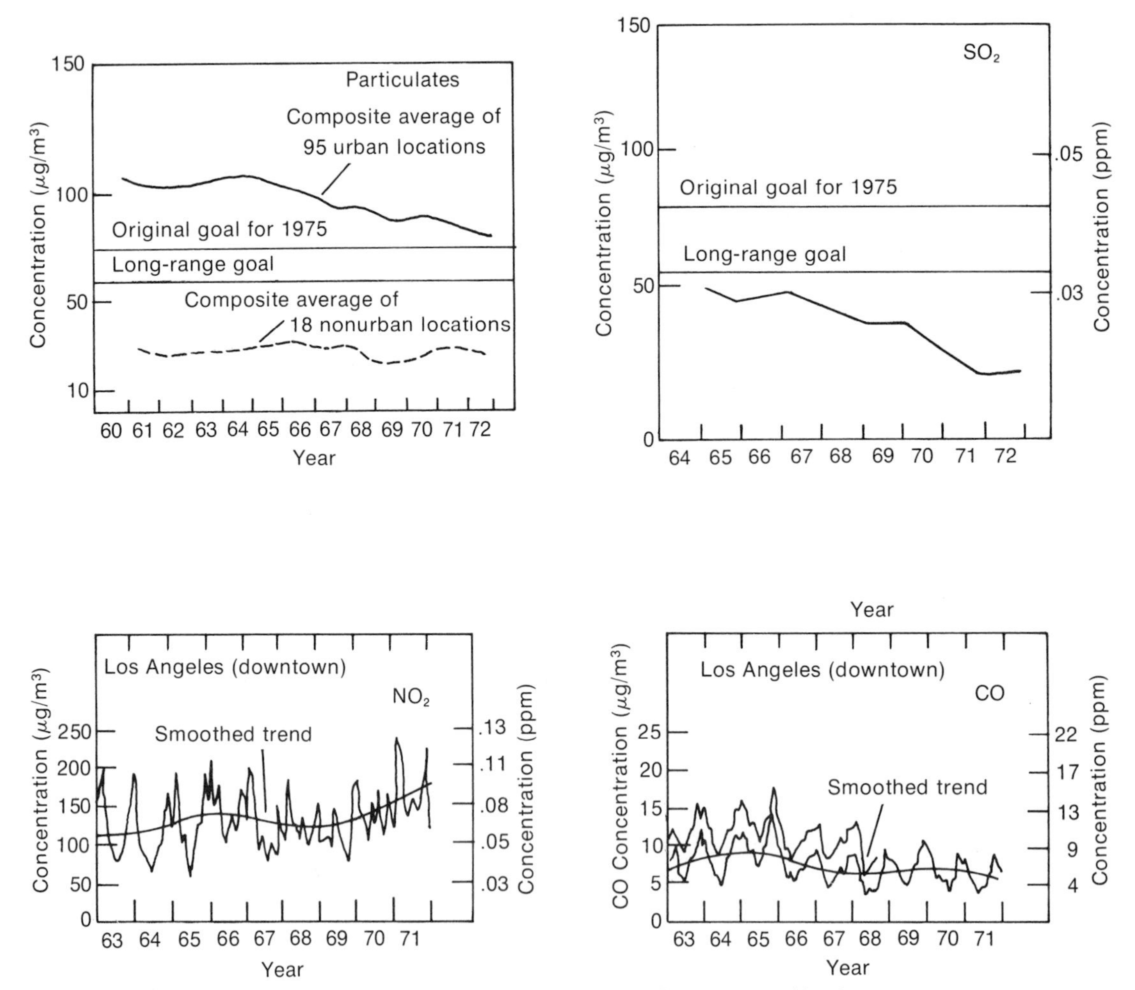

Figure 20–4 Trends in air content of particulates, SO_2, CO_2, and NO_2 as measured by the National Aerometric Surveillance Network (NASN). Since NASN stations are located primarily in downtown areas, these data do not necessarily reflect the "worst" air quality to be found in heavily industrialized portions of the cities. ($\mu g/m^3$ is micrograms (10^{-6} grams) per cubic meter; mg/m^3 is milligrams per cubic meter.)

pollution of the air brought on by increased urbanization, industrialization, and transportation via the automobile. Yet the worst air pollution disasters of all time occurred prior to 1960 (Table 20–1).

During the 1960's more public and governmental concern about air pollution led to controls being placed on emissions from automobiles, industries, and power companies. These efforts caused a general downward trend in most air pollutants (Figure 20–4).

The energy crisis of the 1970's has caused a relaxation of air quality standards (Table 20–2). Obviously a choice had to be made—do we want cheap energy and the attendant air pollution, or are we willing to pay for higher priced energy in order to obtain abatement of air pollution? A serious study of the material in this chapter, combined with a study of the chapter on energy (Chapter 25), provides a sound basis for the decisions society must make concerning energy and pollution, and its willingness to pay for them.

Shakespeare's impression of 17th century London air pollution: ". . . this most excellent canopy, the air, look you, this excellent o'erhanging firmament, this majestical roof fretted with golden fire, why, it appears no other thing to me but a foul and pestilent congregation of vapors."
—*Hamlet,* Act II, Scene II

The long-range effects of air pollution on materials and the health of

TABLE 20–2 Federal Air Quality Standards (As of April 30, 1971)

	Concentration
Sulfur Dioxide	
Arithmetic mean (annual)	0.03 ppm
24-hour concentration not to be exceeded more than once per year	0.14 ppm
Suspended Particulates	
Geometric mean (annual)	75 $\mu g/m^3$
24-hour concentration not to be exceeded more than once per year	260 $\mu g/m^3$
Carbon Monoxide	
8-hour concentration not to be exceeded more than once per year	9 ppm
1-hour concentration not to be exceeded more than once per year	35 ppm
Photochemical Oxidants	
1-hour concentration not to be exceeded more than once per year	0.08 ppm
Hydrocarbons	
3-hour concentration not to be exceeded more than once per year	0.24 ppm
Nitrogen Dioxide	
Arithmetic mean (annual)	0.05 ppm

Air pollutants tend to attack the site where they first enter the body, i.e., the lungs.

plants, animals, and human beings are beginning to emerge. The lung cancer rate in large metropolitan areas is twice as great as the rate in rural areas, even after full allowance is made for differences in cigarette smoking habits. The incidence of the serious pulmonary disease, emphysema, shot up eightfold during the decade of the sixties. However, only after we comprehend the short- and long-range effects of air pollution can we evaluate wisely its relative importance. A logical beginning is to understand some basic facts about our atmosphere.

THE ATMOSPHERE

The atmosphere is a mixture of an estimated 5500 trillion tons of gases, mostly nitrogen (78 percent) and oxygen (21 percent) with small quantities of water vapor and carbon dioxide and still smaller quantities of other materials. About 99 percent of the total mass is below an altitude of 19 miles, and there is sufficient oxygen to sustain life only 4 miles above sea level (Figure 20–5). The region that contains most of the oxygen and supplies most of our weather is the troposphere, which extends to an average altitude of 7 miles.

To change from percent to ppm multiply by 10,000.

A representative composition of dry air near sea level is given in Table 20–3. Near industrial complexes and communities some of the substances, normally present in very small amounts (less than 1 part per million), often increase to abnormal amounts of 50 or more ppm.

ppm = number of molecules of pollutant per 1,000,000 molecules of air.

Pollutants litter the troposphere, move both vertically and horizontally, often reacting chemically with themselves or with materials on the surface of the earth, and in due time generally return to land or water. The troposphere—the rug in the sky under which we try to hide our wastes—receives more than 140,000 tons of combustion products per day from the United States alone. How pollutants behave chemically while in the atmosphere and what their ultimate fate is are interesting chemical

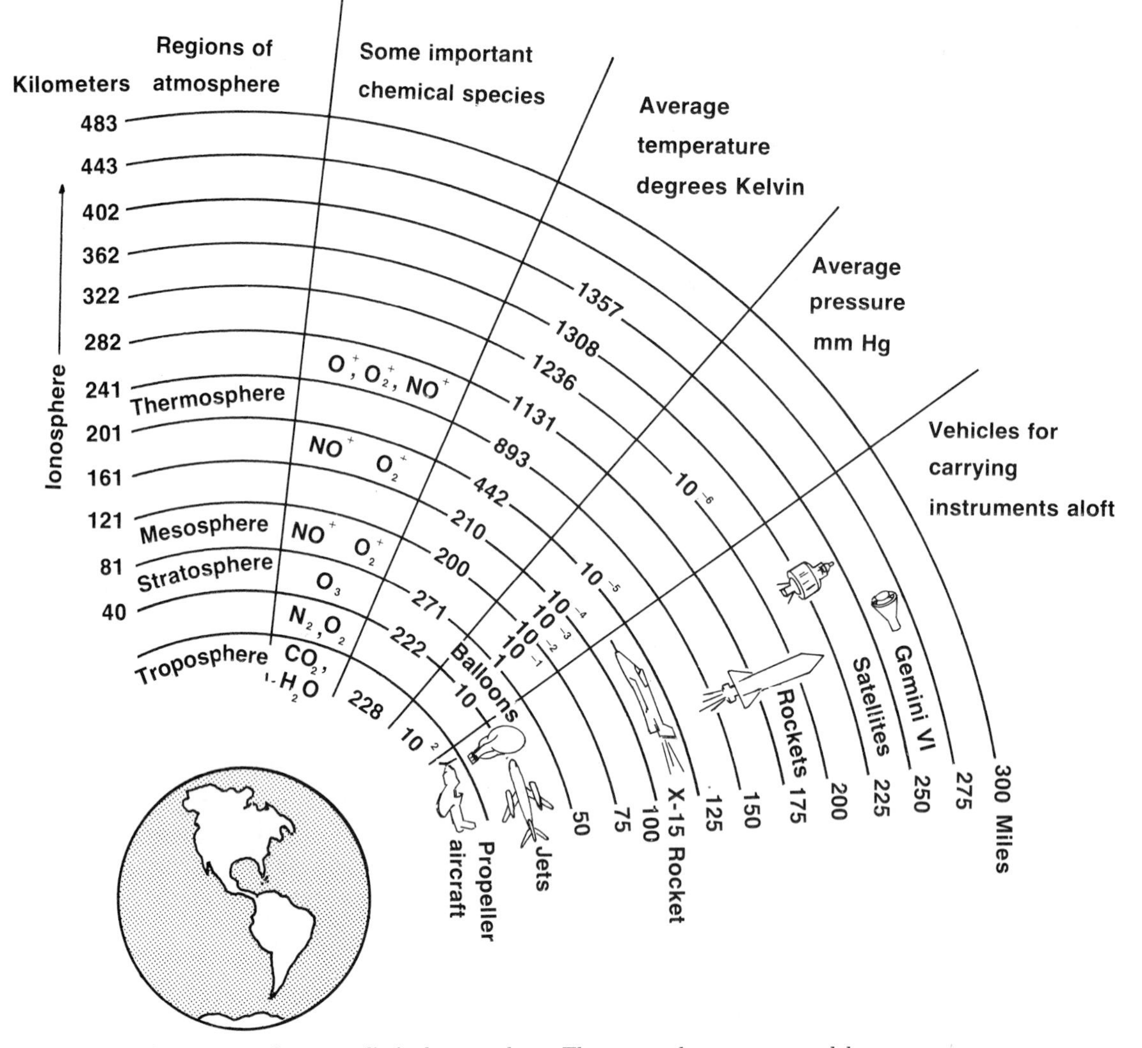

Figure 20–5 Some facts about our limited atmosphere. The troposphere was named by British meteorologist Sir Napier Shaw from the Greek word *tropos,* meaning turning. The stratosphere was discovered by the French meteorologist Leon Philippe Teisserenc de Bort, who believed that this region consisted of an orderly arrangement of layers with no turbulence or mixing. The word *stratosphere* comes from the Latin word *stratum,* meaning layer.

TABLE 20–3 Composition of Clean, Dry Air Near Sea Level (ppm)

Component	Content	Component	Content
Nitrogen	780,900	Nitrous oxide	0.25
Oxygen	209,400	Carbon monoxide	0.10
Argon	9,300	Xenon	0.08
Carbon dioxide	318	Ozone	0.02
Neon	18	Ammonia	0.01
Helium	5.2	Acetone	0.001
Methane	1.5	Nitrogen dioxide	0.001
Krypton	1.0	Sulfur dioxide	0.0002
Hydrogen	0.5	Lead	0.00000013

TABLE 20–4 Composition of Inhaled and Exhaled Air

	Inhaled Air (%)	Exhaled Air (%)
Oxygen	20.96	15.8
Carbon dioxide	.04	4.0
Nitrogen and other gases	79.00	80.2

questions, with which we shall deal later. Except for the freons, pollutants do not seem to go beyond the heights of the troposphere.

One of the most important parts of the atmosphere is the part we actually breathe. With every breath, the average adult inhales (and exhales) about one half liter of air. This quantity contains about 1.3×10^{22} (13 sextillion) molecules. Most of these molecules leave the lungs just as they entered. Only those that linger can influence our health.

A quadrillion is 10^{15}.

Although pollution occurs in relatively small amounts compared to oxygen, nitrogen, and carbon dioxide, about 200 quadrillion ***pollutant*** molecules are inhaled per breath on a ***clear*** day in Los Angeles where the average breath would contain the following:

Carbon monoxide	175,000,000,000,000,000 molecules
Hydrocarbons	10,000,000,000,000,000
Peroxides	5,000,000,000,000,000
Nitrogen oxides	4,000,000,000,000,000
Lower aldehydes	3,500,000,000,000,000
Ozone	3,000,000,000,000,000
Sulfur dioxide	2,500,000,000,000,000

On a smoggy day, the numbers increase by a factor of five or more. The breath you are now inhaling could contain pollutant molecules in comparable amounts, give or take a few quadrillion.

TABLE 20–5 Air Pollutant Emissions in the United States in 1970 (millions of tons per year)*

	Totals	% of Totals	Carbon Monoxide	Sulfur Oxides	Hydrocarbons	Nitrogen Oxides	Particulates
Transportation	144.0	54	111.0	1.0	19.5	11.7	0.8
Fuel Combustion In Stationary Sources	44.5	17	0.8	26.4	0.6	10.0	6.7
Industry	36.8	14	11.4	6.4	5.5	0.2	13.3
Solid Waste Disposal	11.1	4	7.2	0.1	2.0	0.4	1.4
Miscellaneous	30.3	11	18.3	0.2	7.3	0.5	4.0
Totals:	266.7	100	148.7	34.1	34.9	22.8	26.2

*From "Air Quality and Emissions Trends Annual Report," Volume I, U. S. Environmental Protection Agency, August, 1973. Publication No. 450/1-73-001a.

SOURCES OF POLLUTANTS IN THE AIR

Automobiles, industry, and electric power plants are the main sources of air pollutants from man-controlled processes. Volcanic action, forest fires, and dust storms are natural sources of air pollutants, but these contribute very little compared to the man-made sources. A summary of the principal sources of emissions in the United States in 1970 is shown in Table 20–5.

POLLUTANT PARTICLE SIZE

Pollutant particles may be grouped in three sizes: fundamental (such as single molecules, ions, or atoms), aerosol (ranging from a thousand to about a trillion atoms, ions, or small molecules per particle), and particulate (suspended particles composed of more than a trillion atoms, ions, or molecules). Aerosols range in size from 1 to about 10,000 nanometers (nm) in diameter (1 nanometer = 10^{-9} meter); molecules are smaller; particulates are larger. Aerosols and particulates serve as collectors of the chemically active sulfur oxides, nitrogen oxides, ozone, hydrocarbons, and other pollutant molecules.

Aerosols

Aerosols can be either liquid or solid particles, and are small enough to remain suspended in the atmosphere for long periods of time. The particles composing aerosols cannot be seen individually by even the most powerful optical microscopes. Smoke, dust, clouds, fog, mist, and sprays are typical aerosols. Aerosols have much greater surface area than a single lump of the material. Consider a cube of coal 1 centimeter on an edge. Such a cube would have a surface area of 6 square centimeters. If this cube is subdivided into a sextillion (1,000,000,000,000,000,000,000) smaller cubes, each cube would be 1 nanometer on an edge, about the size of an aerosol particle. The surface area is now 60,000,000 square centi-

Aerosol particles are intermediate in size between small molecules and easily visible small particles.

Adsorption **is the attachment of particles to a surface.**

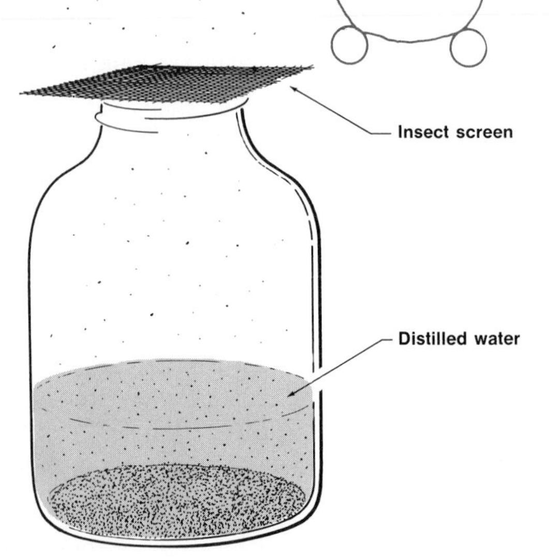

Figure 20–6 A dust sample collector for use in the backyard. The open jar containing water is exposed to the air for a known interval. Fifteen days is sufficient exposure time. The water is then evaporated and the residue weighed. Balances sensitive to 0.001 g should be used; the area (cm^2) of the opening must be known.

meters—about 1.5 acres. Because of this large surface area, aerosols have enormous capacities to adsorb and concentrate gases on their surfaces. Many times aerosols adsorb toxic gases, provide the water medium for a reaction to occur, and cause devastating results when breathed. Specific examples and the important role that aerosols play in smog formation will be cited later.

Absorption is pulling particles inside.

Particulates

Particulates are generally large enough to be seen. They range in size from 1 to 10 microns in diameter. Over 26 million tons of soot, dust, and smoke particulates were deposited into the atmosphere of the United States in 1970. Average suspended particulate concentrations in the United States range from 0.00001 gram per cubic meter of air (g per m^3) in remote rural areas to about six times that value in urban locations. In heavily polluted areas concentrations up to 0.002 g per m^3, or 200 times the usual value, have been measured.

1 micron = 10^{-6} meters.

Particulates may cause physical damage to certain materials. Particles whipped by the wind grind exposed materials by abrasive action. Particles settling in electronic equipment can cause short circuits and foul contacts and switches. Dust settling in paint detracts from its beauty and hastens its deterioration by allowing water to reach the surface underneath. Particles may interfere physically with one or more of the clearance mechanisms in the respiratory tract of man and animals (inhibiting the ciliary transport of mucus, for example). People who have asthma or emphysema know that heavy concentrations of particles in the air increase discomfort. In extremely polluted regions, these diseases often lead to death.

Figure 20–7 The General Electric CF6-6D engines that power this McDonnell Douglas DC-10 were designed to eliminate exhaust smoke and reduce by one half the noise level at takeoff. American Airlines advertises that it spent in a ten year period an amount equal to 43 percent of its profits on noise and air pollution control systems. (Courtesy of American Airlines.)

Particulates may also injure humans or animals because they are intrinsically toxic. The effects of lead and arsenic were described in Chapter 18. Lead compounds are emitted in automobile exhaust. Arsenic compounds are used as insecticides to dust plants. Particles containing fluorides, commonly emitted from aluminum-producing and fertilizer factories have caused weakening of bones and loss of mobility in animals which have eaten plants covered with the dust. Leaves of bean plants dusted with cement kiln particles (0.00047 g per cm^2 per day) wilt significantly.

Particulate effects depend heavily on the chemical nature of the particle.

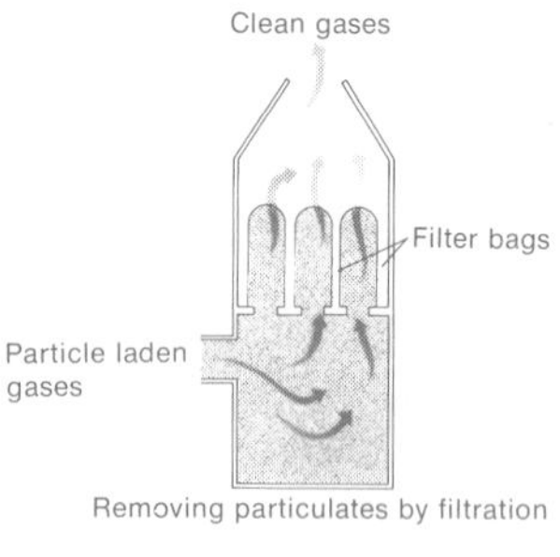

Removing particulates by filtration

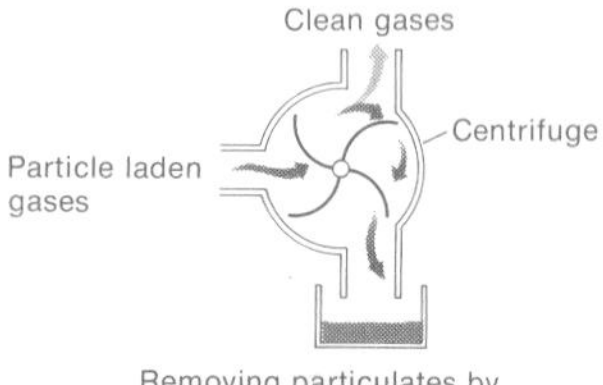

Removing particulates by centrifugal separation

Like aerosol particles, larger particulates may cause damage because they adsorb toxic substances. Sulfur dioxide, nitrogen oxides, hydrocarbons, and carbon monoxide do their greatest damage when concentrated on the surface of particulates.

Particles are removed naturally from the atmosphere by gravitational settling and by rain and snow. They can be prevented from entering the atmosphere by treating industrial emissions by one or more of a variety of physical methods such as filtration, centrifugal separation, spraying, ultrasonic vibration, and electrostatic precipitation.

Glass fiber or silicone treated textile bags filter hot exhaust gases

Figure 20–8 Old post office building being cleaned in St. Louis, Mo., 1963. (Photo by H. Neff Jenkins. From Stern, A. C.: *Air Pollution*. New York, Academic Press, 1968.)

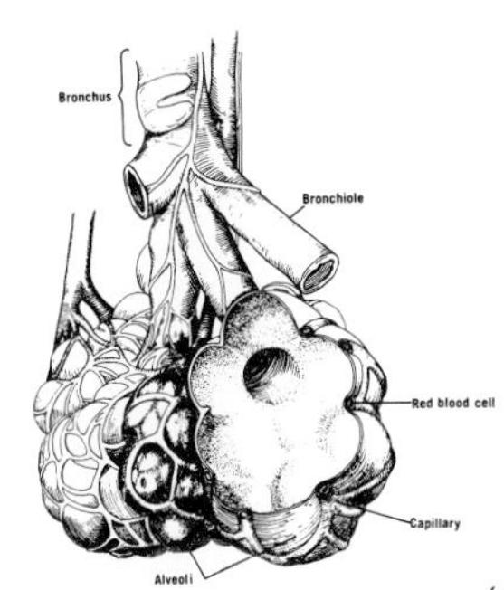

Figure 20–9 Diagram of a small section of the lung. The walls of the airways are lined with cells, some of which secrete mucus. The walls are also lined with hairlike projections called cilia, which wave in rhythm and serve to propel mucus and impurities of the air out of the lungs. (From Turk, A., et al.: *Environmental Science*. Philadelphia, W. B. Saunders Co., 1974.)

(290°C maximum); cotton, nylon, acrylics, wool, and felts are used to filter out cold aerosols.

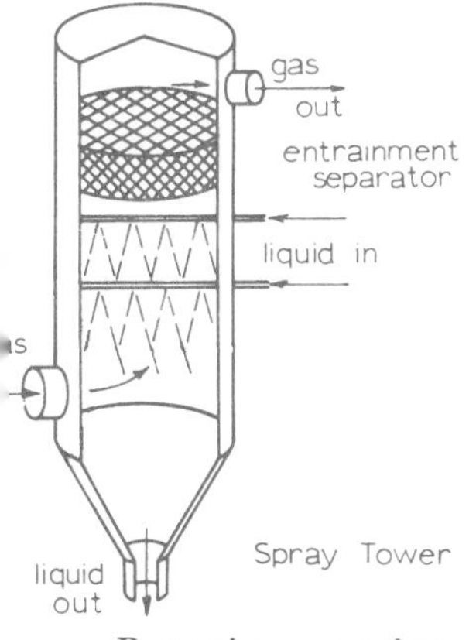

Removing particulates and aerosols by scrubbing. Schematic drawing of a spray collector, or scrubber.

Centrifugal separators whirl the gases and sling the particles against the walls where they collect. The same result is obtained when fast-moving exhaust is made to change direction. The dust particles bump against the wall where they collect.

Spray treatment (scrubbing) is utilized by the petroleum industry, among others. The smoke from a waste gas burner is collected in the water from several small spray nozzles set a few inches above the tip of the burner.

Ultrasonic vibration operates on the principle that increased collisions of the particles and aerosols cause adherence and condensation of the suspended particles. The ultrahigh frequency vibrations cause the increased collisions.

Electrostatic precipitators can remove aerosols and dust particles smaller than 1 micron from plant exhaust gases. A diagram of a Cottrell electrostatic precipitator is shown in Figure 20–10. The central wire is

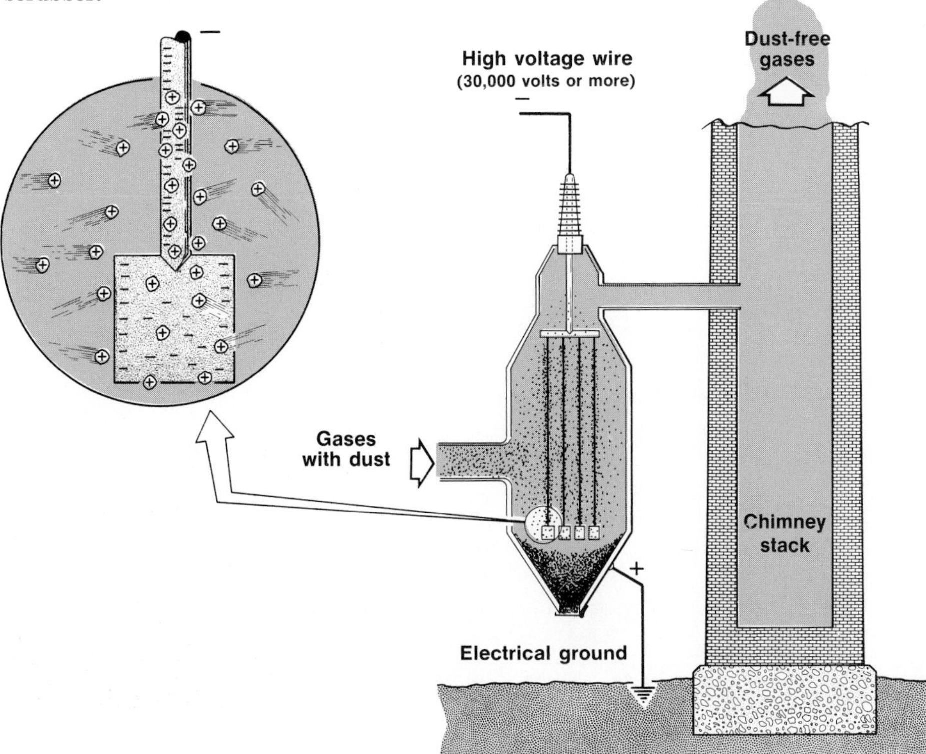

Figure 20–10 The Cottrell electrostatic precipitator.

connected to a source of direct current and high voltage (about 50,000 volts). As dust or aerosols pass through the strong electrical field, the particles attract ions which have been formed in the field, become strongly charged, and are attracted to the walls. The precipitated solid falls to the bottom where it is collected. Electrostatic precipitators are better than 98% effective in removing most particulates and aerosols from exhaust gases.

SMOG—INFAMOUS AIR POLLUTION

The poisonous mixture of smoke, fog, air, and other chemicals was first called ***smog*** in 1911 by Dr. Harold de Voeux in his report on a London air pollution disaster that caused the deaths of 1150 people. Through the years, smog has been a technological plague in many communities and industrial regions.

Two general kinds of smog have been identified. One is the chemically reducing type that is derived largely from the combustion of coal and oil, and contains sulfur dioxide mixed with soot, fly ash, smoke, and partially oxidized organic compounds. This is the ***London type,*** which is diminishing in intensity and frequency as less coal is burned and more controls are installed. A second type of smog is the chemically oxidizing type, typical of Los Angeles; it is called ***photochemical*** smog because light—in this instance sunlight—is important in initiating the photochemical process. This smog is practically free of sulfur dioxide but contains substantial amounts of nitrogen oxides, ozone, ozonated olefins, and organic peroxide compounds, together with hydrocarbons of varying complexities.

Industrial or London-type smog: fog + SO_2
Photochemical smog: fog + NO_x + hydrocarbons

"Olefin" is another name for an unsaturated hydrocarbon.

Figure 20–11 Smog over New York City. A heavy haze hangs over Manhattan Island, viewed from the roof of the RCA Building. The Empire State Building is barely visible in the background. (Wide World Photos, Inc.)

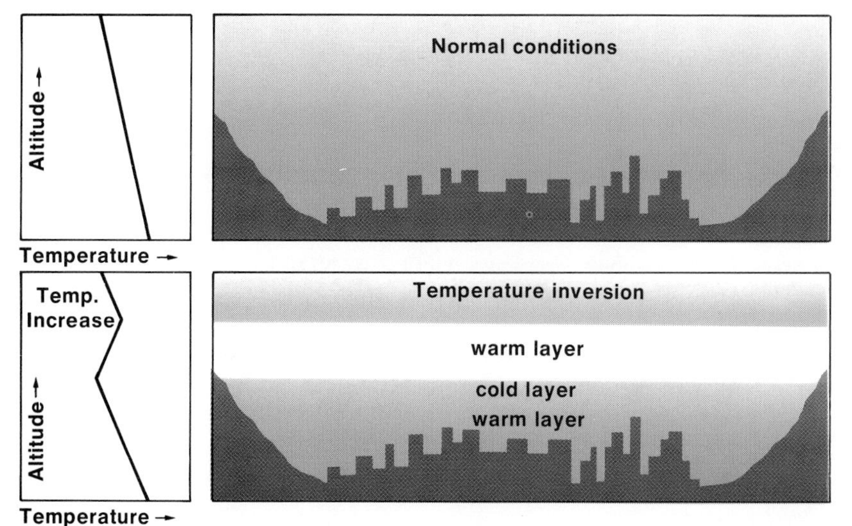

Figure 20–12 A diagram of a temperature inversion over a city. Warm air over a polluted air mass effectively acts as a lid, holding the polluted air over the city until the atmospheric conditions change. The line on the left of the diagram indicates the relative-air temperature.

What general conditions are necessary to produce smog? Although the chemical ingredients of smogs often vary, depending upon the unique sources of the pollutants, certain geographical and meteorological conditions exist in nearly every instance of smog.

Thermal inversion: mass of warmer air over a mass of cooler air.

There must be a period of windlessness so that pollutants can collect without being dispersed vertically or horizontally. This lack of movement in the ground air can occur when a layer of warm air rests on top of a layer of cooler air. This sets the conditions for a ***thermal inversion,*** which is an abnormal temperature arrangement for air masses. If the warmer air is on the bottom nearer the warm earth, which is usual, the warmer, less dense air rises and transports most of the pollutants to the upper troposphere where they are dispersed. When the warmer air is on top, as in thermal inversion, the cooler, more dense air retains its position nearer the earth and vertical movement is stagnated. If the land is bowl-shaped (surrounded by mountains, cliffs, or the like), horizontal movement of the air mass is also hindered.

When these natural conditions exist, man supplies the pollutants by combustion and evaporation in automobiles, electrical power plants, space heating, and industrial plants. The chief pollutants are sulfur dioxide (from burning coal and some oils), and nitrogen oxides, carbon monoxide, and hydrocarbons (chiefly from the automobile). Add to these ingredients the radiation from the sun, and a massive smog is in the offing.

Photochemical Smog

Primary pollutants: pollutants emitted into the air.

Secondary pollutants: pollutants formed in the air by chemical reaction.

A city's atmosphere is an enormous mixing bowl of frenzied chemical reactions. Ferreting out the exact chemical reactions that produce smog has been a tedious job, but in 1951, insight into the formation process was gained when smog was first duplicated in the laboratory. Detailed studies have subsequently revealed that the chemical reactions involved in the smog-making process are photochemical and that aerosols serve as breeders, participants, and products in the formation of the secondary pollutants, which are formed by chemical reaction in the atmosphere. Light provides the energy of activation for the series of photochemical

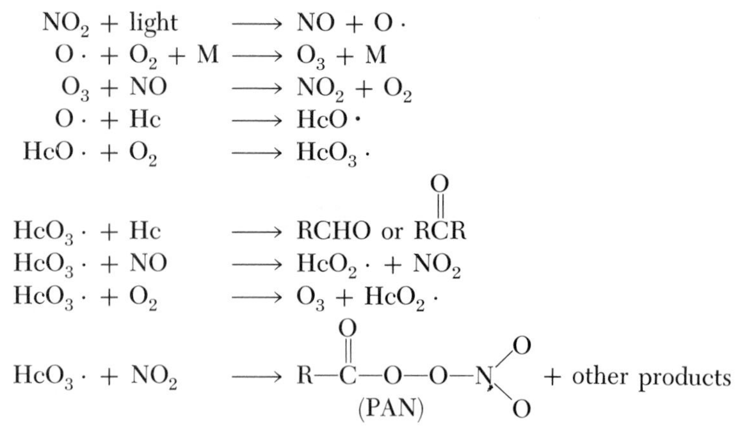

Figure 20–13 Simplified reaction scheme for photochemical smog. Hc is hydrocarbon (olefin or aromatic); M is a third body to absorb the energy released from forming the ozone; among many possibilities, M could be a N_2 molecule, O_2 molecule, or solid particle. A species with a dot, as HcO ·, is a chemical radical—a very reactive chemical species; R is a hydrocarbon group; PAN is peroxy acyl nitrate.

reactions, and ultraviolet radiation from the sun is the energy source for the formation of ***photochemical smog.***

The exact reaction scheme by which primary pollutants are converted into the secondary pollutants found in smog is still not completely understood, but the reactions shown in Figure 20–13 account for the major secondary pollutants in photochemical smog. The process is thought to begin with the absorption of a quantum of light by nitrogen dioxide, which causes its breakdown into nitrogen oxide and atomic oxygen, a free radical. The very reactive atomic oxygen reacts with molecular oxygen to form ozone (O_3), which is then consumed by reacting with nitrogen oxide to form the original reactants—nitrogen dioxide and molecular oxygen. Atomic oxygen, however, also reacts with reactive hydrocarbons—olefins and aromatics—to form free radicals (species with unpaired electrons). Chemical radicals, in turn, react to form other radicals and secondary pollutants such as aldehydes (e.g., formaldehyde). While it is known that about 0.2 ppm of nitrogen oxides and 1 ppm of reactive hydrocarbons suffice to initiate the process, the sources of all of the hydrocarbons that enter the smog-forming process cannot be accounted for. Furthermore, the exact nature and behavior of the radicals are not well understood.

Photochemical smog

A secondary pollutant:

H—C(=O)—H
Formaldehyde

On the other hand, it is known that aldehydes, ketones, and PAN, at least, are found in the air but are not emitted by any identifiable sources. Measurements made in several United States cities reveal that the

RCOON(O)O
Peroxy Acyl Nitrate

TABLE 20–6 Approximate Concentrations of Pollutants in Los Angeles on a Clear Day (Visibility 7 Miles) and on a Smoggy Day (Visibility 1 mile)

	Concentration (ppm)		
Pollutant	*Clear Day*	*Smoggy Day*	*Increase*
Carbon monoxide	3.5	23.0	×6.5
Hydrocarbons	0.2	1.1	×5.5
Peroxides	0.1	0.5	×5.0
Oxides of nitrogen	0.08	0.4	×5.0
Lower aldehydes	0.07	0.4	×6.0
Ozone	0.06	0.3	×5.0
Sulfur dioxide	0.05	0.3	×6.0

amount of airborne aerosols increased tenfold during the 1960's. Despite the more strict controls effected during the last decade, Los Angeles did not show a downward trend in the occurrence and amount of smog, which is present about 300 days a year. A comparison of the levels of pollutants on a clear day and on a smoggy day is presented in Table 20–6.

Industrial Smog

The type of smog formed in London and around some industrial and power plants is thought to be caused by sulfur dioxide. Laboratory experiments have shown that sulfur dioxide increases aerosol formation, particularly in the presence of irradiated mixtures of hydrocarbons, nitrogen oxides, and air. For example, irradiated mixtures of 3 ppm olefin, 1 ppm NO_2, and 0.5 ppm SO_2 at 50 percent relative humidity forms aerosols which have sulfuric acid as a major product. Even with 10 to 20 percent relative humidity, sulfuric acid is a major product. Sulfuric acid, which is formed in this kind of smog, is very harmful to people suffering from bronchial diseases such as asthma or emphysema. At a concentration of 5 ppm for one hour, SO_2 can cause constriction of bronchial tubes. A level of 10 ppm for one hour can cause severe distress. In the 1962 London smog, readings as high as 1.98 ppm of SO_2 were recorded. The sulfur dioxide and sulfuric acid are thought to be the primary causes of deaths in the London smogs.

Figure 20–14 Photochemical smog (a brown haze) enveloping the city of Los Angeles. (Los Angeles Times photo.)

We shall now examine some of the principal components of air pollution and their chemistry. Emphasis will be placed on how the compounds are produced and how they can be eliminated from our atmosphere.

SELF-TEST 20-A

1. ______ ______ ______ is abbreviated ppm. To change from percent to ppm, multiply by ______. Thus 0.092 percent is ______ ppm.
2. According to the graphs of Figure 20–4, which pollutant shows an increase during the last two or three years? ______.
3. Three substances normally considered as primary pollutants are ______, ______, and ______.
4. In what region of the atmosphere does most air pollution exist? ______
5. Particles that can remain suspended in air for long periods of time and are intermediate in size between individual molecules and particulates are called ______.
6. Much of the effect of aerosols is due to their large ______ ______.
7. A device that removes charges from dust and aerosol particles so they will form larger particles heavy enough to fall out of the air is known as a(n) ______ ______.
8. The initial step in the formation of photochemical smog occurs when ultraviolet light decomposes a molecule of ______ ______.
9. During a thermal inversion, a (warm/cool) mass of air is above a mass of (warm/cool) air.
10. The ingredient that distinguishes industrial (or London-type) smog from other kinds is the substance ______.

SULFUR DIOXIDE

Sulfur dioxide is produced when sulfur or sulfur-containing substances are burned in air:

$$S + O_2 \longrightarrow SO_2(g)$$

CH_3—SH
Methyl mercaptan

Most of the sulfur dioxide in the atmosphere comes from sulfur-containing fuels burned in power plants, from smelting plants treating sulfide ores, and from sulfuric acid plants. The average sulfur content of all coal shipped from mines in the United States is about 2.0%, although most coal has either less than 1% or more than 3% sulfur. Coal contains sulfur in several forms: as elemental sulfur (S), as iron pyrite (FeS_2), and as organic compounds, such as mercaptans (compounds containing —SH groups). According to the National Air Pollution Control Administration (NAPCA), the burning of fuels by the nation's power plants accounted for 70 percent of the SO_2 emitted into the atmosphere in 1974. One study found a concentration of 2200 ppm SO_2 in the stack of a coal-burning power plant. Concentrations as high as 2.9 ppm have been recorded up to half a mile from power plants. If coal and petroleum containing up to 5 percent sulfur are burned, a 1000-megawatt electric power plant could emit 600 tons of SO_2 per day. Note how the SO_2 concentration in the air has decreased in Bayonne, New Jersey, as the sulfur content of the fuel has been reduced (Figure 20–15). Much of the SO_2 pollution in this country is isolated in seven industrialized states (New York, Pennsylvania, Michigan, Illinois, Indiana, Ohio, and Kentucky), which account for almost 50 percent of the nation's total SO_2 output.

A large modern power station (e.g., a station of 2,000 megawatts capacity) annually produces about the same amount of SO_2 as an industrial city of a million inhabitants.

More than 34 million tons of SO_2 were put into the air in 1970.

In spite of the large output of SO_2 in the United States per year, its concentration rarely exceeds a few parts per million. This is because SO_2 has a relatively short atmospheric life. In the presence of oxygen, sun-

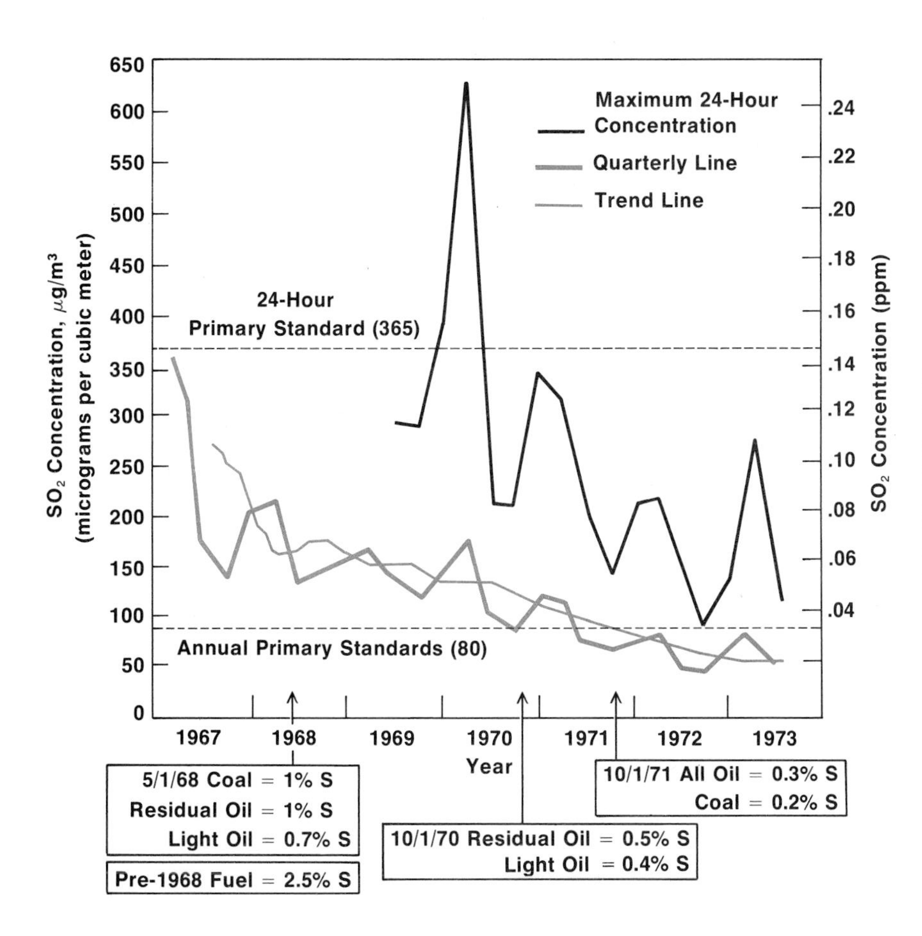

Figure 20–15 Comparison of SO_2 trends at Bayonne, New Jersey, with regulations governing percent sulfur content in fuel. (From "Monitoring and Air Quality Trends Report," U.S. Environmental Protection Agency, 1972. Publication No. 450 1-73-004, pp. 4–5.)

light, and water vapor, SO_2 is oxidized and converted to sulfuric acid (H_2SO_4):

$$2SO_2 + O_2 \longrightarrow 2SO_3$$
$$SO_3 + H_2O \longrightarrow H_2SO_4$$

The SO_2 and sulfuric acid are readily dissolved in rivers, lakes, and streams and can increase the acidity considerably.

Sulfurous acid, H_2SO_3, is a weak acid; sulfuric acid, H_2SO_4, is a strong acid. Both can make the water in streams and lakes too acidic for fish.

$$H_2O + SO_2 \rightleftharpoons \underset{\text{WEAK ACID}}{H_2SO_3} \rightleftharpoons H^+ + HSO_3^-$$

and

$$\underset{\text{STRONG ACID}}{H_2SO_4} \longrightarrow H^+ + HSO_4^-$$

Often streams contain ions that help to offset the acidity increase. Sulfite is one such ion.

$$\underset{\text{SULFITE}}{SO_3^{2-}} + H^+ \longrightarrow \underset{\text{BISULFITE}}{HSO_3^-}$$

If the pH of streams varies too much, aquatic life suffers. Salmon, for example, cannot survive if the pH is as low as 5.5. The lower limit of tolerance for most organisms is a pH of 4.0. In the late 1950's and early 1960's, certain sections of the Netherlands had precipitation with a pH less than 4.

TABLE 20–7 Physiologic and Corrosive effects of SO_2*

SO_2 Exposure, ppm	Duration	Effect	Comment
0.03 to 0.12	Annual average	Corrosion	Moist temperate climate with particulate pollution
0.3	8 hr	Vegetation damage (alfalfa mostly, but many other species are similarly sensitive)	Laboratory experiment; other environmental factors optimal. Field studies are consistent but dose is difficult to estimate
0.47 ppm. 50% *of* subjects detect	< 1 hr	Odor threshold	May be higher for many persons or when other methods are used
0.2	Daily average	Respiratory symptoms	Community exposure exceeding 0.2 ppm more than 3% of the time
> 0.05	Long-term average	Respiratory symptoms	With particulates $> 100\ \mu g/m^3$
0.2	Daily average	Respiratory symptoms	With particulates
0.9	Hourly average	Respiratory symptoms	With particulates
> 0.05	Monthly average	Respiratory symptoms, plus impairment of lung function in children	With particulates

* From "Guide for Control of Air Pollution," Office of Air Programs, U.S. Environmental Protection Agency, 1971. Publication No. AP-78, p. 47.

The corrosion of steel is promoted by sulfur dioxide. In a study involving 100-gram panels of steel exposed at several sites in Chicago in 1964, the amount of weight loss by corrosion was very definitely related to the SO_2 concentration in the air. Corrosive action is also promoted by particulate matter, high humidity, and high temperatures.

Sulfur dioxide in the air is harmful to people, animals, plants, and buildings.

The effects of SO_2 on vegetation are manifested as bleached spots, suppression of growth, leaf drop, and reduction in yield. Plant damage has been noticed 52 miles downwind from a smelting operation which emitted large quantities of SO_2. The SO_2, and SO_3 formed from it, enter the stomata of the leaves, H_2SO_4 is produced, and leaf tissue is hydrolyzed and dehydrated, causing yellowing and leaf drop (Figure 20–16).

There are many ways to control SO_2 pollution of the air. For example, tall stacks (Figure 20–17) emit the SO_2 into the higher atmosphere and away from the ground. The results of a study in Great Britain showed that tall stacks effectively reduce ground level SO_2 concentrations. During a 10-year period, the SO_2 emissions from power plants *increased* by 35 percent, but, as a result of the construction of tall stacks, the ground level concentrations *decreased* by as much as 30 percent.

Chemical means are more efficient than high stacks in removing SO_2. One method involves heating limestone to produce lime (calcium oxide) and allowing this to react with the SO_2 to produce calcium sulfite, which can be removed from the stack by an electrostatic precipitator.

$$\underset{\text{LIMESTONE}}{CaCO_3} \xrightarrow{\text{Heat}} \underset{\text{LIME}}{CaO} + CO_2$$

$$CaO + SO_2 \longrightarrow \underset{\text{CALCIUM SULFITE}}{CaSO_3 \text{ (solid)}}$$

Another efficient method involves passing the combustion gases through molten Na_2CO_3 (sodium carbonate), in which solid sodium sulfite is formed.

Figure 20–16 Rose leaves in Independence, Missouri, show marginal and interveinal necrotic injury from sulfur dioxide. (From National Air Pollution Control Administration, HEW Public Health Service, 1970. Publication No. AP-71.)

Figure 20–17 World's largest chimney (as of 1972) standing 1,250 feet high (as tall as the Empire State Building) was built at a cost of $5.5 million for the Copper Cliff smelter in the Sudbury District of Ontario, Canada. (From Turk, A., et al.: *Environmental Science*. Philadelphia, W. B. Saunders Co., 1974.)

$$SO_2 + Na_2CO_3 \xrightarrow{800°C} Na_2SO_3 + CO_2$$

The use of sulfur-free fuels is one obvious method of eliminating SO_2 emissions (Figure 20–15). However, most low-sulfur coals are far from major metropolitan areas, and removal of the sulfur from coal can cost up to one dollar per ton.

Most of the sulfur in coal is a part of organic matter or pyrite (FeS_2, fool's gold). Pulverized coal, with a consistency of talcum powder, can be cleansed partially of pyrite by magnetic separation, electrostatic separation, froth flotation, and dry centrifugation. But none of these methods completely eliminates the sulfur dioxide emissions.

Much of the petroleum used in the Eastern United States, where SO_2 pollution is a problem, is Caribbean residual fuel oil, which has an average sulfur content of 2.6 percent. While technology has developed refining techniques for reducing the sulfur content to 0.5 percent, this procedure would increase the cost of the oil by about 35 percent. The process involves the formation of hydrogen sulfide by bubbling hydrogen through the oil in the presence of metallic catalysts (platinum-palladium, for example). Residual fuel oil produces less than 10 percent of the total utility power. In 1968, New York's Consolidated Edison switched to low-sulfur oil from Liberia and Nigeria at a cost of $3 million to change over from coal and an increase of $7.5 million in a $63 million fuel bill.

NITROGEN OXIDES

There are eight known oxides of nitrogen, two of which are recognized as important components of the atmosphere: dinitrogen oxide (N_2O) and nitrogen dioxide (NO_2).

Fixed nitrogen is nitrogen chemically bonded to another element.

Fixed nitrogen (nitrogen oxides are one type) is necessary to perpetuate nature's nitrogen cycle (Figure 20–18). In this respect, nitrogen oxides are useful. However, too large a quantity of nitrogen oxides in the air can lead to photochemical smog and bronchial problems. In these respects and many others, nitrogen oxides are harmful.

Most of the nitrogen oxides emitted are in the form of NO, a colorless reactive gas. In a combustion process involving air as the oxidant, some of the atmospheric nitrogen reacts with oxygen to produce NO:

$$N_2 + O_2 + \text{Heat} \longrightarrow 2NO$$

Nitrogen oxide is formed in this manner during electrical storms. Since the formation of nitrogen oxide is endothermic (21,600 cal per mole of NO), it follows that a higher combustion temperature would produce relatively more NO. This will be an important point to consider later in our discussion of the automobile and its nitrogen oxide emissions, since one way to achieve greater burning efficiency of fuels in automobile engines is to operate at higher temperatures.

In the atmosphere NO reacts rapidly with atmospheric oxygen to produce NO_2:

$$2NO + O_2 \longrightarrow \underset{\text{NITROGEN DIOXIDE}}{2NO_2}$$

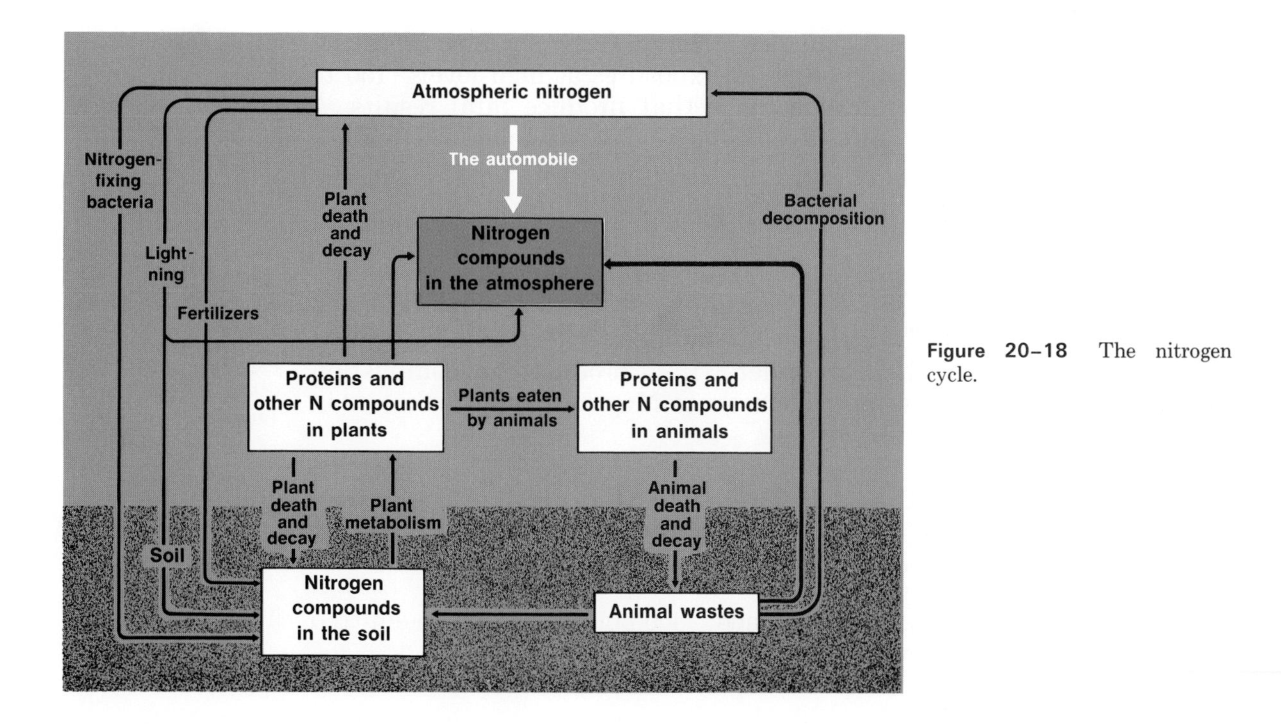

Figure 20–18 The nitrogen cycle.

Nitrogen dioxide, a brown, choking gas, is a necessary component of photochemical smog (Figure 20–13). Normally the atmospheric concentration of NO_2 is a few parts per billion (ppb) or less.

If NO_2 does not react photochemically, it can react with water vapor in the air to form nitric and nitrous acids:

$$2NO_2 + H_2O \longrightarrow \underset{\text{NITRIC ACID}}{HNO_3} + \underset{\text{NITROUS ACID}}{HNO_2}$$

In addition, nitrogen dioxide and oxygen yield nitric acid:

$$4NO_2 + 2H_2O + O_2 \longrightarrow 4HNO_3$$

These acids in turn can react with ammonia or metallic particles in the atmosphere to produce nitrate or nitrite salts. For example,

$$\underset{\text{AMMONIA}}{NH_3} + HNO_3 \longrightarrow NH_4NO_3$$

The acids or the salts, or both, ultimately form aerosols, which eventually settle from the air or dissolve in raindrops. Nitrogen dioxide, then, is a primary cause of haze in urban or industrial atmospheres because of its participation in the process of aerosol formation. Normally nitrogen dioxide has a lifetime of about three days in the atmosphere.

Nitrates are important components of fertilizers.

At present, the emission of nitrogen oxides by human beings and our machines is only a minor part of that emitted by natural processes (Figure 20–18).

In laboratory studies, nitrogen dioxide in concentrations of 25 to 250 ppm inhibits plant growth and causes defoliation. The growth of tomato and bean seedlings is inhibited by 0.3 to 0.5 ppm NO_2 applied continuously for 10 to 20 days.

In a concentration of 3 ppm for 1 hour, nitrogen dioxide causes bronchioconstriction in man, and short exposures at high levels (150 to 220 ppm) produce changes in the lungs that produce fatal results. A seemingly harmless and painless exposure one day can cause death a few days later.

CARBON MONOXIDE

Carbon monoxide (CO) is the most abundant and widely distributed air pollutant found in the atmosphere. It is produced in combustion processes when carbon or some carbon-containing compound is burned in an insufficient amount of oxygen:

The toxic effects of carbon monoxide are discussed in Chapter 18.

$$2C + \underset{\text{LIMITED SUPPLY}}{O_2} \longrightarrow 2CO$$

For every 1000 gallons of gasoline burned, 2300 pounds of CO are emitted if no emission controls are placed on the engine.

The background concentration in cities is higher (Figure 20–4); in

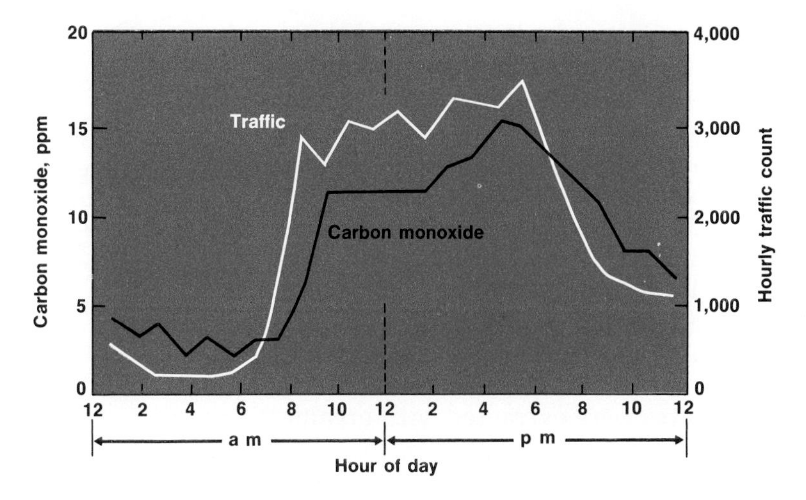

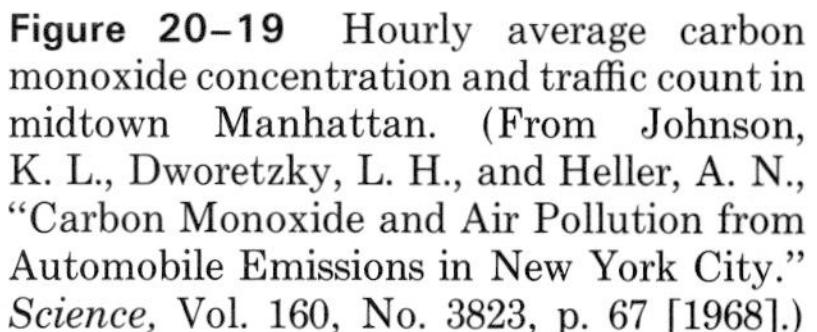
Figure 20–19 Hourly average carbon monoxide concentration and traffic count in midtown Manhattan. (From Johnson, K. L., Dworetzky, L. H., and Heller, A. N., "Carbon Monoxide and Air Pollution from Automobile Emissions in New York City." *Science,* Vol. 160, No. 3823, p. 67 [1968].)

heavy traffic sustained levels of 50 or more ppm are common, and instantaneous concentrations of 150 ppm and higher have been found. The variation of CO levels over a 24-hour period in an urban atmosphere is shown in Figure 20–19. However, in spite of increased emission of CO (enough to raise the concentration 0.03 ppm yearly), the global level is not increasing at all. This is because the natural cycle dominates the pollutant emissions.

HYDROCARBONS

As we have seen in Chapter 13, hydrocarbons come in all shapes and sizes, beginning with methane, CH_4, and continuing to molecules containing many carbon atoms. Some have all single bonds, some have double bonds, and a few have triple bonds. Aromatic hydrocarbons have bonds that can be described as a partial double bond (see Chapter 13). Literally hundreds of these hydrocarbons and their oxygen, sulfur, nitrogen, and halogen derivatives find their way into the atmosphere.

Hydrocarbons enter the atmosphere from natural sources as well as from man's activities.

Trees and plants silently release turpentine, pine oil, and thousands of other fragrances into the air. These are hydrocarbons or hydrocarbon derivatives. Bacterial decomposition of organic matter emits very large amounts of marsh gas, principally methane. Man contributes his share of 15 percent (of the total global emissions; a greater quantity in urban areas) through incomplete incineration, through leakage of industrial solvents, through unburned fuel from the automobile, through incomplete combustion of coal and wood, and through petroleum processing, transfer, and use.

In Chapter 18 we saw that polynuclear aromatic hydrocarbons like benzo(α)pyrene are capable of causing cancer in mice and in humans. In the late 1950's, the U.S. Public Health Service, Division of Air Pollution, surveyed 103 urban and 28 nonurban areas of the United States and found the air in all of the 103 urban areas contained benzo(α)pyrene (BaP). Concentrations ranged from 0.11 to 61 micrograms per 1000 cubic meters of air, with the average concentration being 6.6. In 1967, the estimated annual emission of BaP in the United States was 422 tons from burning coal, oil, and gas, 20 tons from refuse burning, 19 tons from industries (petroleum catalytic cracking, asphalt road mix, and the like), and 21 tons from motor vehicles. British researchers report that lung cancer in nonsmokers closely parallels the 10 times greater amount of BaP in city air than in rural air; there is 9 times more lung cancer in cities than in rural areas. A resident of a large town may inhale 0.20 g BaP a year. If he is a heavy smoker (two packs a day without filters), add another 0.15 g for a total of 0.35 g. This is about 40,000 times the amount of BaP necessary to produce cancer in a mouse. Coal smoke contains about 300 ppm BaP. Every million tons of coal burned in England in 1958 produced smoke laden with 750 tons of BaP. Many authorities attribute England's high lung cancer rate to this enormous production of BaP.

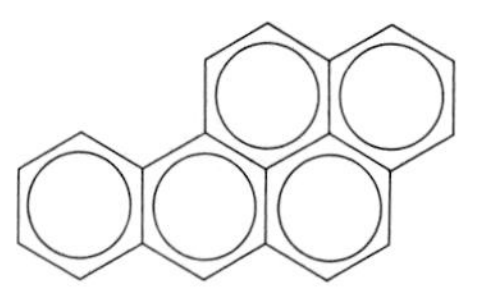

Figure 20–20 Benzo (α) pyrene, a carcinogenic polynuclear aromatic hydrocarbon found in smoke.

Other polynuclear aromatics have also shown carcinogenic activity. Particulates from the atmosphere around Los Angeles, London, Newcastle, Liverpool, and eight other urban sites were extracted with organic solvents, usually benzene. The extracts produced cancer in mice.

Although BaP and its polynuclear counterparts have received considerable publicity, other hydrocarbons and hydrocarbon derivatives play important roles in air pollution. The chlorinated hydrocarbons, widely used as pesticides, at least partially counteract their beneficial aspects by killing birds and fish, and by generally polluting our streams. Some of these, such as DDT, are found in the atmosphere.

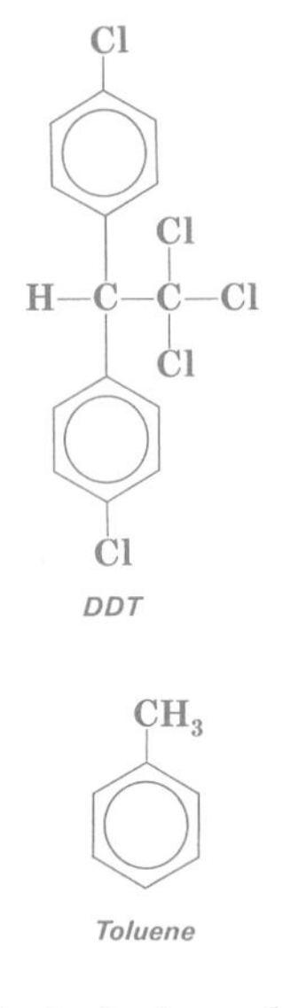

In the section on smog we discussed the role of olefins and aromatic hydrocarbons in photochemical smog formation. In a study made in Los Angeles in 1970, an average of 0.106 ppm (maximum of 0.33 ppm) aromatics was found in that city's atmosphere. (About 38 percent of the total was toluene and 40 percent the more reactive dialkyl- and trialkylbenzenes.) These compounds are about as reactive as propylene and higher molecular weight olefins in causing smog formation. The automobile is responsible for emitting most of these hydrocarbons to the atmosphere. In fact, the automobile emits more than 200 different hydrocarbons and hydrocarbon derivatives.

Alkyl: hydrocarbon group such as ethyl, $—C_2H_5$, and octyl, $—C_8H_{17}$.

SELF-TEST 20-B

1. When sulfur is burned in air, the major product is ______________.
2. The major source of sulfur dioxide is the burning of ____________.
3. How does the production of electrical power rank as a producer of sulfur dioxide? ____________
4. Fixed nitrogen is nitrogen combined with ____________ ____________.
5. The reaction of nitrogen with oxygen is (endothermic/exothermic).
6. In all combustion processes in air, some nitrogen ____________ are formed.
7. Man's emission of nitrogen oxides has greatly disrupted the nitrogen cycle. (True/False)
8. Nitrogen oxides remain in the atmosphere for an indefinite length of time. (True/False)

9. Carbon monoxide is always formed when hydrocarbons are burned in a (limited/plentiful) amount of air.

10. In the human body, carbon monoxide is a formidable competitor with oxygen for ______________.

11. The amount of carbon monoxide is definitely increasing in the atmosphere. (True/False)

12. A polynuclear aromatic hydrocarbon commonly found in the air and in cigarette smoke that is known to cause cancer is ______________.

OZONE

Ozone is the pungent smelling gas often noticed around electric motors. A concentration of only 0.02 to 0.05 ppm is required to detect the unique odor. Sparking (including lightning) and even silent electric discharges convert oxygen into ozone, which is a more reactive form of oxygen:

$$68{,}400\ \text{cal} + 3O_2 \longrightarrow \underset{\text{OZONE}}{2O_3}$$

Ozone attacks mercury and silver, which are not affected by oxygen at room temperature. Even with all of the electric sparks from lightning, electric motors and such, practically no ozone is emitted into the air. It reacts too quickly to leave its source. A typical reaction might be

$$6Ag + O_3 \longrightarrow 3Ag_2O$$

but

$$Ag + O_2 \longrightarrow \text{No reaction}$$

The stratosphere contains ozone. This "ozone layer" can be destroyed by certain man-made pollutants.

Ozone is found in the lower troposphere only as a secondary pollutant; that is, it is formed from other substances, as in photochemical smog (Figure 20–13). When sunlight impinges on automobile exhaust fumes, considerable ozone is produced. The stratosphere contains about 10 ppm ozone in a layer that has the important function of filtering out some of the ultraviolet light. There is considerable concern that freons (fluorocarbon propellants) released from aerosol cans (see Chapter 22) may interact with and destroy the ozone layer in the stratosphere.

Pure oxygen can be breathed for weeks by man and animals without apparent injurious effects. Several studies have shown that concentrations of 0.3 to 1.0 ppm ozone, well within the recorded range of photochemical oxidant levels, after 15 minutes to 2 hours cause marked respiratory irritation accompanied by choking, coughing, and severe fatigue. For these reasons, outdoor recreation classes in Los Angeles public schools are cancelled on days when the ozone level reaches 0.35 ppm. Ozone at these levels for 1 hour depresses the body temperature, perhaps by an impairment of the brain center that regulates body temperature or by opening the pores of the skin. These levels (0.2 to 0.5 ppm) cause a considerable decrease in night vision in addition to other effects on vision.

CARBON DIOXIDE

Carbon dioxide is not usually considered a pollutant because it is a normal component of the air and directly involved in the give-and-take between animal life and plant life (a product of respiration, a reactant in photosynthesis). There is concern, however, because the concentration in the atmosphere is increasing (Figure 20–21). Studies at Scripps Institute of Oceanography have shown that the concentration of CO_2 in the atmosphere increased by 1.3 percent (or 3.7 ppm) from 1958 to 1962. Electric power plants, internal combustion engines, and the manufacture of cement are the principal technological sources of CO_2 emission, but there are numerous other sources—home heating, trash burning, and bacterial oxidation of soil humus, for example. These emissions have been entering the atmosphere faster than oceans and plants can remove them.

CO_2 is a necessary ingredient for photosynthesis.

In some ways an increase in CO_2 can be very beneficial. Experiments at Michigan State University showed an increase of 115 per cent in the dry weight of lettuce and tomatoes when the CO_2 was increased from the normal 325 ppm (0.032 percent by volume) to 800 to 2000 ppm during daylight hours. Fruit size increased and quality was enhanced; reproduction improved; tomatoes had higher vitamin C and sugar content; and plants were more resistant to some fungi, virus diseases, and insects. Plants have even responded favorably to 20,000 to 30,000 ppm CO_2 levels.

A substantial increase of CO_2 in the air is potentially detrimental in two ways: it can increase the temperature of the atmosphere as well as increase the acidity of the oceans.

Carbon dioxide in the atmosphere produces a greenhouse effect.

Carbon dioxide in the atmosphere acts similarly to the glass (or plastic) on a greenhouse. Both let the shorter wavelengths of light through but absorb the longer wavelengths as they are emitted from the surfaces of objects. The carbon dioxide (and the glass) in turn emit the

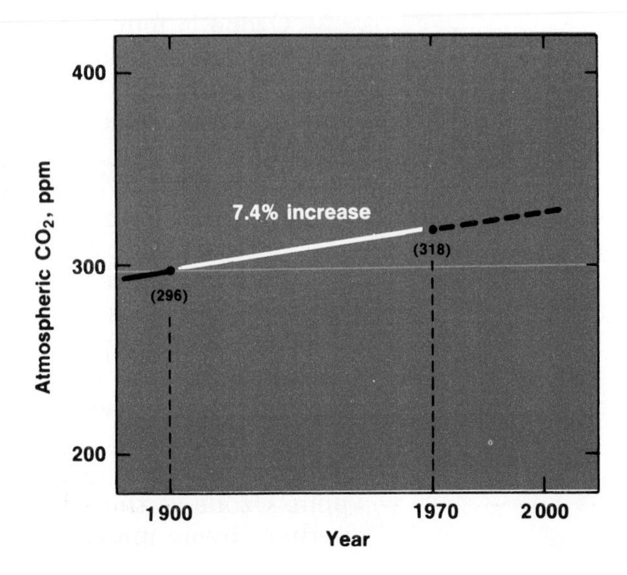

Figure 20–21 Atmospheric carbon dioxide. It is estimated that doubling the concentration would increase the average world temperature by 3.8°C.

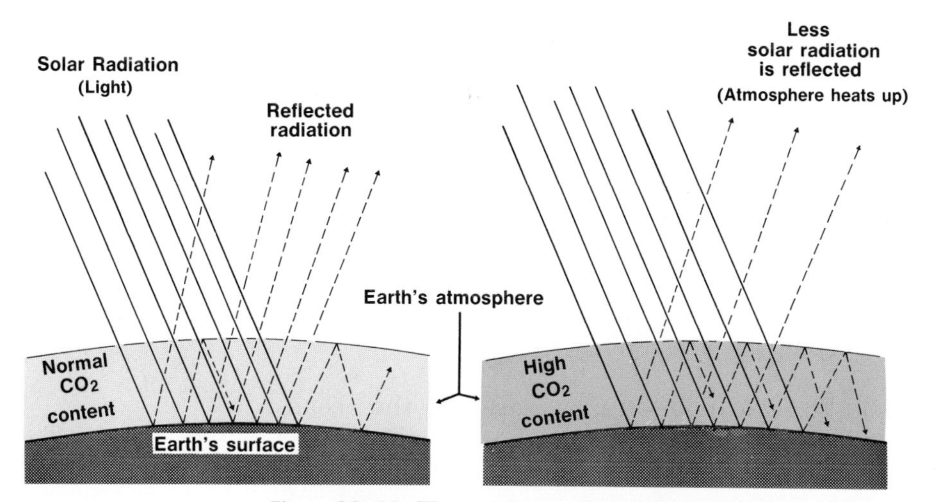

Figure 20–22 The greenhouse effect. Owing to a balance of incoming and outgoing energy in the earth's atmosphere, the mean temperature of the earth is 14.4°C (58°F). Carbon dioxide permits the passage of visible radiation from the sun to the earth but traps some of the heat radiation attempting to leave the earth.

absorbed heat radiation, some of which returns to earth and some of which energizes other molecules via collisions. The net effect is an increase in temperature of the lower atmosphere, a phenomenon known as the ***"greenhouse effect"*** (Figure 20–22).

In contrast, solid and liquid aerosols scatter incoming sunlight of all wavelengths, thus decreasing the amount of solar energy that reaches the earth. The net effect is a cooling of the lower atmosphere. Calculations show that a 25 percent increase in aerosols (turbidity) would counteract a 100 percent increase in carbon dioxide concentration.

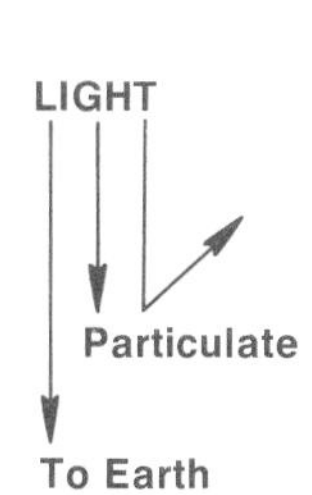

It is generally agreed that the global temperature increased about 0.4°C between 1880 and 1940 and then decreased by nearly 0.2°C by 1967. The CO_2 concentration is believed to have increased by 7.4 percent during a comparable interval of time—from 296 ppm in 1900 to 318 ppm in 1969, a net gain of 7.4 percent. The cooling during the 1950's possibly was due in part to the particulates emitted by gigantic volcanic eruptions of Mt. Spurr in Alaska in 1953 and the Bezymyannaya (Kamchatka, U.S.S.R.) eruption in 1956. While an increasing CO_2 concentration could explain the increase in temperature and the increasing turbidity could explain the decrease in temperature, other factors may also have contributed significantly. For example, it is assumed in this explanation that solar radiation has been constant over this period of time. Actually, it may not have been; we simply do not know. (Carbon dioxide is the only substance whose global background concentration is known to be rising.)

Three mechanisms account for the removal of CO_2 from the atmosphere. These are photosynthesis, dissolving in surface water (principally the oceans), and weathering of silicate rocks. None of these seems adequate since the CO_2 level is on the rise. While a higher level of CO_2 increases plant activity and growth, it is apparent that the photosynthesis process has not picked up enough to maintain the status quo. It is known that pollutants such as SO_2 hinder photosynthesis and that much of our forest area is being cleared for housing and development. Perhaps these factors partially explain why photosynthesis cannot absorb enough

CO_2. At the rates prevailing in 1962, we would need to add 2.7 billion acres of new, growing forest (a little more than the area of the United States) to absorb the CO_2 generated by the combustion of fossil fuel.

Perhaps as much as half of the CO_2 emitted by burning is absorbed by the surface waters. The oceans serve as a huge sink for carbon dioxide, which reacts with water to form carbonic acid. The acid ionizes to raise the acidity and produce bicarbonate ions:

$$H_2O + CO_2 \rightleftharpoons \underset{\text{CARBONIC ACID}}{H_2CO_3} \rightleftharpoons H^+ + \underset{\text{BICARBONATE ION}}{HCO_3^-}$$

Carbonic acid can disrupt the limestone-forming process in the oceans. Limestone is mostly calcium carbonate, $CaCO_3$. The insoluble $CaCO_3$ reacts with carbonic acid to form soluble calcium bicarbonate:

$$\underset{\text{INSOLUBLE}}{CaCO_3} + H_2CO_3 \longrightarrow \underset{\text{SOLUBLE}}{Ca^{2+} + 2HCO_3^-}$$

The overall effect of the solution of CO_2 in the oceans is to reduce its concentration in the air and, consequently, to reduce the atmospheric temperature; but this raises the acidity, dissolves limestone, or both.

Some CO_2 is dissipated by weathering silicate rocks to carbonates:

$$CaAl_2(SiO_3)_4 + 4CO_2 \longrightarrow CaCO_3 + Al_2(CO_3)_3 + 4SiO_2$$

This process is very slow and could not help to relieve a relatively sudden increase in CO_2 emissions.

When the CO_2 level increases, it does so at the expense of the oxygen supply. While this is another disadvantage of an increasing CO_2 concentration, the increase is far more detrimental than the equivalent decrease in oxygen. For example, approximately 60 percent of our oxygen need is being restored by photosynthesis within the United States. The rest comes from bacteria that reduce sulfates in anaerobic environments and from plankton upwellings in the sea. Some upwellings—containing literally millions of tons of diatoms, bacteria, and algae (called plankton)—are found off the coasts of California, Peru, Morocco, and Southwest Africa.

A major consumer of CO_2 (through photosynthesis) is phytoplankton (small ocean plants).

Carbon dioxide is not normally considered detrimental to man's health. Obviously our systems contain it at all times as a product of respiration. Prolonged exposure to about 5000 ppm CO_2 (15 times present levels), however, is considered unsafe primarily because the greater CO_2 concentration diminishes the O_2 concentration in our lungs.

THE AUTOMOBILE—A SPECIAL CASE IN AIR POLLUTION

More than 100 million automobiles jam our roadways, each adding its share of pollutants to the atmosphere. Like it or not, the automobile, as we know it, is *the* major source of air pollution (see Table 20–5).

Automobiles are the major emitters of air pollutants in the United States.

Automobile air pollutants enter the atmosphere in three major ways: from the exhaust, from the crankcase blowby (gases that escape around the piston rings), and by evaporation from the fuel tank and carburetor.

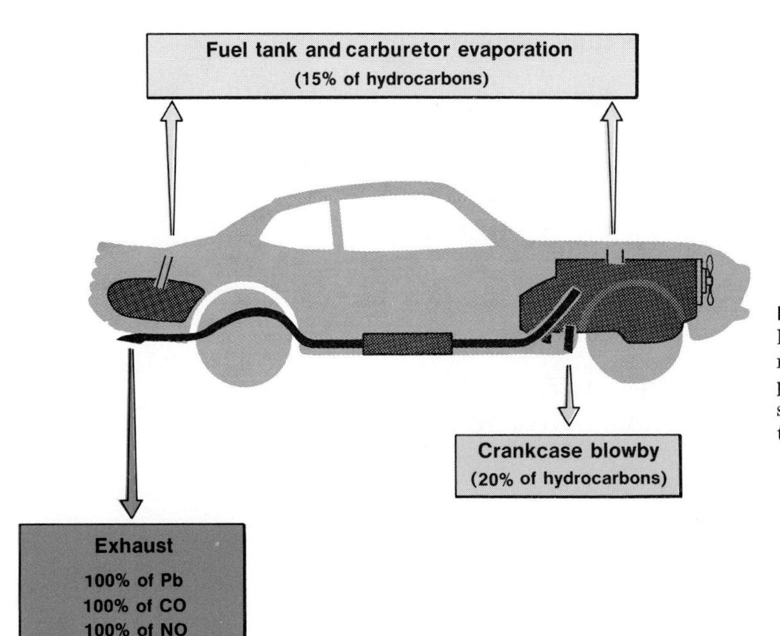

Figure 20–23 Sources of automobile pollutant emissions. The percentages show the relative amounts of the various types of pollutants from the three major automobile sources for automobiles which have no pollution controls.

Gasoline is a mixture of hydrocarbons (isomers of octane, for example), tetraethyl lead, 1,2-dibromoethane, 1,2-dichloroethane (in a molar ratio of 2 to 1 in ethyl gasoline) and various "additives" such as TCP (tricresyl phosphate). When this mixture is compressed with air in an automobile cylinder and ignited (Figure 20–24), many of the hydrocarbon molecules combine with oxygen to form CO, CO_2, and H_2O. Some hydrocarbons react with oxygen, other hydrocarbons, or both, to form a large variety of organic substances. Other hydrocarbons do not react at all.

For smooth driving, gasoline hydrocarbons should be a mixture of C_4 to C_{10}.

Nitrogen combines with oxygen to form a variety of nitrogen oxides, principally NO and a little NO_2. The tetraethyl lead, $Pb(C_2H_5)_4$, forms PbClBr, $PbCl_2$, and $PbBr_2$ by reacting with 1,2-dibromoethane and 1,2-dichloroethane. These combustion products, added to the evaporation

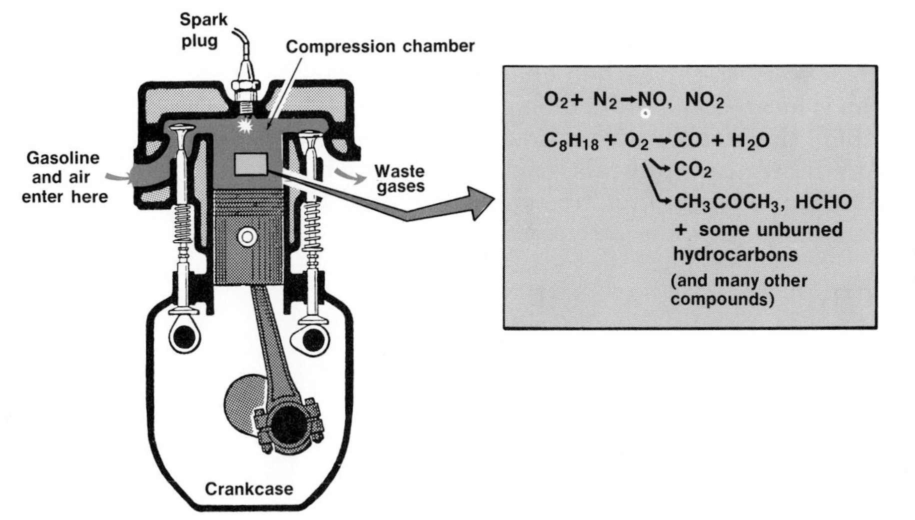

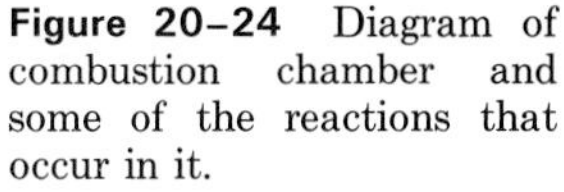
Figure 20–24 Diagram of combustion chamber and some of the reactions that occur in it.

losses of gasoline from the carburetor and gas tank, comprise the noxious waste products of the automobile.

Tetraethyl lead is used as a component of gasoline because it causes the mixture to burn evenly and all at once, thus raising the octane rating. When the burning of the gasoline-air mixture is too rapid or irregular, preignition occurs in the combustion chamber and causes a small explosion which is heard as a "knock" in the engine. Straight-chain hydrocarbons can react with oxygen in an uncontrollably fast reaction to give the products carbon dioxide and water. The first step in this process is the abstraction of an atom of hydrogen:

The octane scale is basedontheknocking produced by a fuel. The standards are: n-heptane, octane rating = 0; and 2,2,4-trimethylpentane, octane rating = 100. $Pb(C_2H_5)_4$ added to gasoline raises its octane rating.

$$CH_3CH_2CH_2CH_2CH_2CH_2CH_3 \longrightarrow CH_3CH_2CH_2CH_2\dot{C}HCH_2CH_3 + H\cdot$$

TYPICAL HYDROCARBON FREE RADICAL

The straight-chain free radical then reacts with a molecule of oxygen to form another free radical:

$$CH_3CH_2CH_2CH_2\dot{C}HCH_2CH_3 + O_2 \longrightarrow CH_3CH_2CH_2CH_2\underset{}{\overset{O-O\cdot}{\overset{|}{C}}}HCH_2CH_3$$

The free radical with oxygen then proceeds to react very rapidly in a series of steps to form carbon dioxide and water. However, if tetraethyl lead is present in the mixture, it decomposes at cylinder temperatures quickly into lead and ethyl free radicals:

$$Pb(CH_2CH_3)_4 \longrightarrow Pb + 4CH_3CH_2\cdot$$

The metallic lead quickly reacts with the straight-chain free radical:

The lead atoms in tetraethyl lead act to slow down some of the very rapid free radical reactions which can occur in burning gasoline-air mixtures.

$$CH_3CH_2CH_2CH_2\overset{OO\cdot}{\overset{|}{C}}HCH_2CH_3 + Pb \longrightarrow CH_3CH_2CH_2CH_2\overset{OOPb}{\overset{|}{C}}HCH_2CH_3$$

This product decomposes at a slower rate. As a result the explosive decomposition of straight-chain hydrocarbons is retarded and the "knock" is diminished. Gasoline containing tetraethyl lead is known as "ethyl" gasoline.

When branched-chain hydrocarbons burn, the first step is the loss of a hydrogen atom at a branch point in the carbon chain. Free radicals in which the unpaired electron is located at a branching carbon in the chain tend to be more stable than the straight-chain free radicals, so they decompose more slowly. As a result these radicals react with oxygen more slowly and smoothly. An example is the first free radical formed from 2,2,4-trimethylpentane:

$$CH_3-\underset{CH_3}{\overset{CH_3}{\overset{|}{\underset{|}{C}}}}-CH_2-\underset{CH_3}{\overset{CH_3}{\overset{|}{\underset{|}{C}}}}\cdot + O_2 \longrightarrow CH_3-\underset{CH_3}{\overset{CH_3}{\overset{|}{\underset{|}{C}}}}-CH_2-\underset{CH_3}{\overset{CH_3}{\overset{|}{\underset{|}{C}}}}OO\cdot$$

The purpose of the 1,2-dibromoethane and 1,2-dichloroethane is to keep lead from reacting with oxygen to form lead (II) oxide (PbO_2), which is a nonvolatile, white solid that fouls spark plugs. Instead lead forms $PbBr_2$, $PbCl_2$, and PbBrCl, which are volatile and pass out of the exhaust system of the automobile and into the atmosphere:

$$Pb + BrCH_2CH_2Br + 3O_2 \longrightarrow 2CO_2 + 2H_2O + PbBr_2$$

Ways to Control Emissions

What can be done to abate the emissions of the automobile? For one thing, governments, both state and federal, have recently passed legislation regulating automobile emissions. Significant data regarding effects of governmental laws and control devices on levels of auto emissions are summarized in Table 20–8 and Figure 20–25.

Two approaches are presently being developed to meet these standards: mechanical changes in automobiles and changes in fuel. Mechanical changes are made at relatively low cost, but fuel modifications offer advantages over mechanical devices. The most obvious advantage is that fuel changes would apply to all cars, old and new.

TABLE 20–8 Levels of Auto Emissions in Grams per Mile†

	Hydrocarbons, HC	Carbon Monoxide, CO	Nitrogen oxides, NO_x
Prior to control	11	80	4.0
California Pure Air Act (AB 357) 1972	1.5	23	3.0
EPA Standards for 1975 Model Cars*	0.41	3.4	3.0
EPA Standards for 1976 Model Cars*	0.41	3.4	0.4

*By decision of the Administrator of the EPA, the 1975 standards were extended one year to 1976 models on April 11, 1973. The decision on extending the 1976 standards to 1977 was made on July 30, 1973. Further relaxations occurred in 1975.

†From Wildeman, T. R., "The Automobile and Air Pollution," *Journal of Chemical Education,* Vol. 51, No. 5, p. 290 (1974).

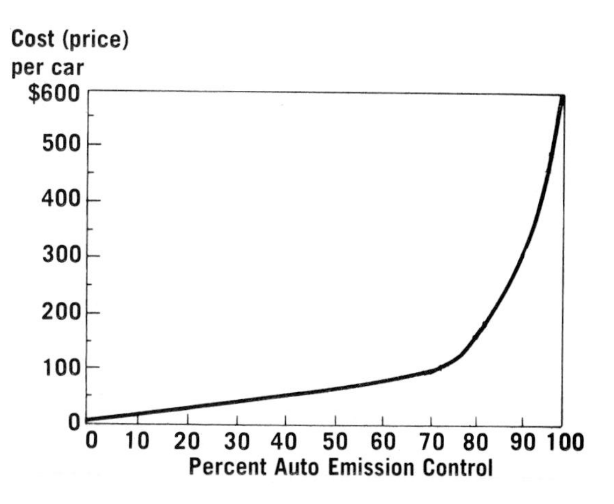

Figure 20–25 Estimated cost of automobile emission control. Based on 1971 costs. (Courtesy of Chrysler Corporation Research and Editorial Services, Detroit, Michigan.)

Beginning with the 1963 models, all automobiles have been equipped with a PCV (positive crankcase ventilation) valve. Fresh air is drawn through the crankcase by the vacuum system, through the PCV valve and connecting hose into the engine. This system reduces crankcase blowby exhaust. Newer models of automobiles have gas tank caps which allow air to come into the gas tank but do not allow gasoline vapors to pass out through the cap.

Methods for reduction of hydrocarbon and carbon monoxide emissions in the exhaust are being developed along three lines: (1) the adjustment of air-fuel ratio, spark timing, and other variables; (2) the injection of air into the hot exhaust gases; and (3) the use of a catalytic converter. The first two methods have physical limits that will prevent complete conversion of hydrocarbons and carbon monoxide to carbon dioxide and water. Ideally, if the stoichiometric amount of oxygen is present, complete conversion to CO_2 is expected. For example, the combustion of a mole of octane requires 12.5 moles of oxygen:

Serious causes of hydrocarbon and CO emissions from automobiles are bad spark plugs and engines not properly timed.

$$C_8H_{18} + 12.5O_2 \longrightarrow 8CO_2 + 9H_2O + \text{heat}$$

But the time for reaction in an automobile engine is so rapid that the octane and oxygen molecules cannot react completely. In the chaotic split second of ignition and reaction, many undesirable side reactions occur. These reactions produce the 200 or so hydrocarbons found in an automobile's exhaust which were not originally in the gasoline.

Adjustment of the air-fuel ratio can reduce CO and hydrocarbon emissions, but this can mean loss of power. For example, it may be seen in Figure 20–26 that maximum power is obtained when the air-fuel ratio is 12.5:1. As shown in Figure 20–27, this ratio does not give the lowest

Note that 13.5 molecules of reactants produce 17 molecules of products (or 27 produce 34). The heat produced and the increased number of molecules cause the expansion that pushes the piston and thereby does the work that drives the car.

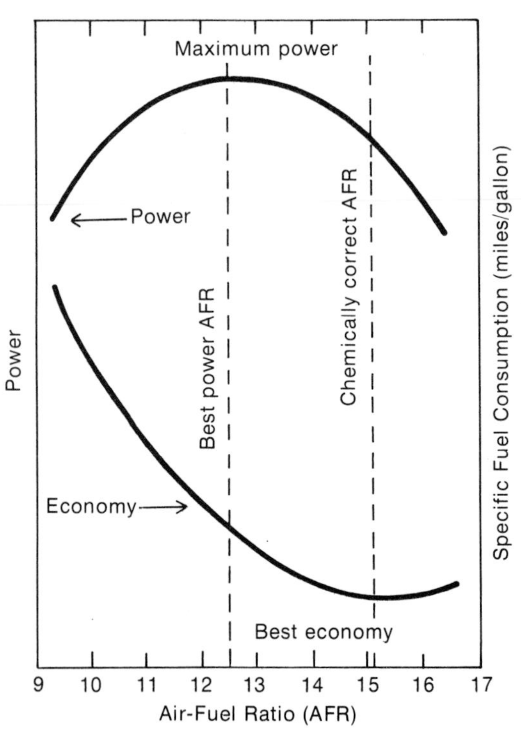

Figure 20–26 The effect of air-fuel ratio on power and economy for a particular engine. (From Wildeman, T. R.: "The Automobile and Air Pollution." *Journal of Chemical Education,* Vol. 51, No. 5, p. 290 1974.)

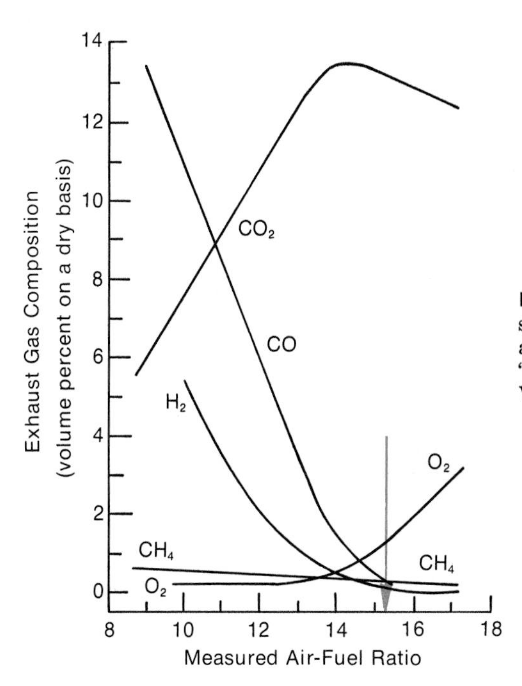

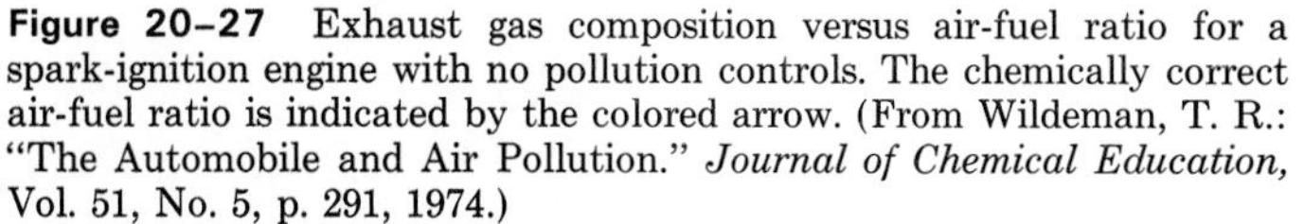
Figure 20–27 Exhaust gas composition versus air-fuel ratio for a spark-ignition engine with no pollution controls. The chemically correct air-fuel ratio is indicated by the colored arrow. (From Wildeman, T. R.: "The Automobile and Air Pollution." *Journal of Chemical Education,* Vol. 51, No. 5, p. 291, 1974.)

emissions of CO and hydrocarbons. A ratio of 15.1:1 would produce about eight times less CO (than an air-fuel ratio of 12.5:1) and slightly less hydrocarbons. Therefore, the 15.1:1 air-fuel ratio would give a more efficient use of the fuel—more complete burning. Why don't the air-fuel ratios for maximum power and most efficient use of fuel match? The answer is well-known to automotive engineers. All of the oxygen (not all of the fuel) must be consumed to produce maximum power. Part of the reason for this is that too much oxygen produces more NO_x, with a resultant loss of power.

Auto carburetors vaporize the fuel and mix it with air.

The nitrogen oxide problem is simple to understand but hard to correct. Reactions of hydrocarbons and CO with oxygen are exothermic. Reactions of nitrogen with oxygen to form nitrogen oxides are endothermic. Therefore, if additional air is provided to burn the unburned hydrocarbons and CO, the heat produced provides the energy for the formation of nitrogen oxides. Thus, in the process of reducing the amounts of CO and hydrocarbons, nitrogen oxides are increased. Reducing the air from a ratio of 15.1:1 to a ratio of 12.5:1 can cause up to a 60% reduction in nitrogen oxide emissions.

Endothermic:
$N_2 + O_2 + 4.32\text{ kcal} \longrightarrow 2NO$

Exothermic:
$C + O_2 \longrightarrow CO_2 + 94.1\text{ kcal}$

The formation of nitrogen oxides deducts heat energy that could be used for expansion. This reduces power.

Less nitrogen is oxidized to oxides of nitrogen if the temperature within the combustion chamber is reduced to the lowest practical limit. One way to do this is to recycle part of the exhaust gas back into the engine. This reduces the peak combustion temperature and the amount of nitric oxide that forms. Prototypes of such recycling systems have reduced nitrogen oxides emissions by 80 percent without causing the exhaust to exceed 275 ppm hydrocarbons and 1.5 percent CO. Coincident with the recirculation of about 15 percent of the exhaust and the reduction of nitrogen oxides by 80 percent is a cut in power output by 16 percent and a decrease in fuel economy by 15 percent. Changes in timing can increase power and economy, but the amount of nitrogen oxides produced also increases. Fuel injected specifically at the spark plug gap will pro-

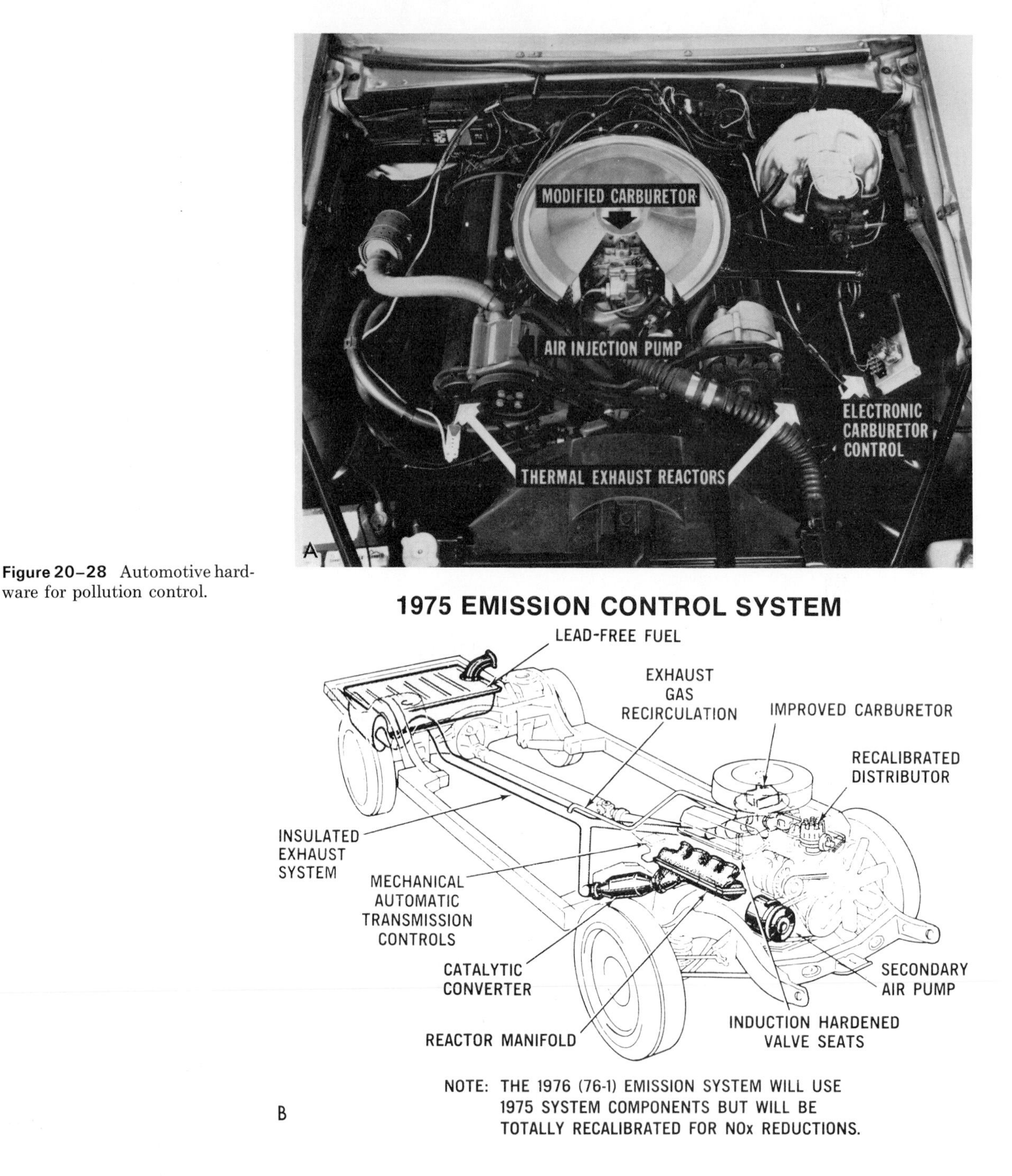

Figure 20–28 Automotive hardware for pollution control.

duce low hydrocarbon, CO, and NO_x emissions, but then particulate emissions become a problem.

The amount of CO and unburned hydrocarbons in the exhaust can be reduced by injecting air into the hot gases just after they leave the combustion chamber—that is, the exhaust valve. This is known as the "afterburner" technique and does not affect the piston and drive train of the automobile. Some of the unburned hydrocarbons and carbon monoxide react with oxygen in the injected air and complete the combustion *outside* the power chamber. Although this reduces hydrocarbon and CO

Figure 20–29 Cutaway view of catalytic exhaust muffler showing catalyst pellets. (From Turk, A., et al.: *Environmental Science.* Philadelphia, W. B. Saunders Co., 1974.)

emissions, heat is evolved that must be dissipated without gaining any additional power to run the automobile. It also increases the nitrogen oxide problem.

Considerable research is being done to develop catalysts for converting CO, hydrocarbons, and NO_x to less harmful products. For example, in a laboratory mixture of 5 percent NO, 5 percent CO, and 90 percent helium run over a barium-promoted copper chromite catalyst at 200°C, 92 percent of the NO and CO was converted to N_2 and CO_2, neither of which is considered a pollutant in the ordinary sense. At 350°C, 98 percent conversion is effected.

$$2CO + 2NO \xrightarrow{Ba\text{-}CuCrO_2} 2CO_2 + N_2$$

All 1975 cars equipped with catalytic mufflers use a platinum-based catalyst which is inactivated by lead. For this reason they must use lead-free gasoline.

Using the exhaust from a 1947 Ford V-8 engine, complete reaction to CO_2 and N_2 can be achieved with a copper chromite catalyst if the temperature is between 370° and 460°C and if the reactants are in the stoichiometric amounts—a most important condition. The perfect balance is difficult to achieve because an engine produces the most nitric oxide under conditions that make very little carbon monoxide (high temperatures, much oxygen). Furthermore, if too much oxygen is present in the exhaust, CO reacts with O_2 in preference to NO.

New Kinds of Cars?

Several thousand automobiles in this country are now burning natural gas instead of gasoline. Natural gas, which is mostly methane, CH_4, burns cleanly to CO_2 and water. A regulating device, positioned over the carburetor, costs about $350. The problem here is that our surplus of natural gas in this country is rapidly diminishing, a potential long-range problem.

At the turn of the century, a Baker Electric Coupe cost $2,600 and a Borland Electric Delux could be bought for $5,500.

One of the very few ways to eliminate combustion processes is to use electricity in some way to power our transportation devices. Early attempts to develop electric cars in the United States were made in Boston in 1888, 28 years after the development of the first storage battery in 1860. By 1912, about 6000 electric passenger and 4000 commercial vehicles were being manufactured annually in this country. The electric car lost out in competition with the internal combustion engine during the 1920's. The problems of short range (about 20 miles), low speeds (20 miles per hour maximum), 8 to 12 hours recharge time for the batteries, and relatively high price combined to eliminate the electric car from the race for leader in transportation.

At present, more than 100,000 rider-type, battery powered, materials-handling vehicles are operating in U.S. plants and warehouses where it is vital to avoid air pollution from internal combustion engines. Despite

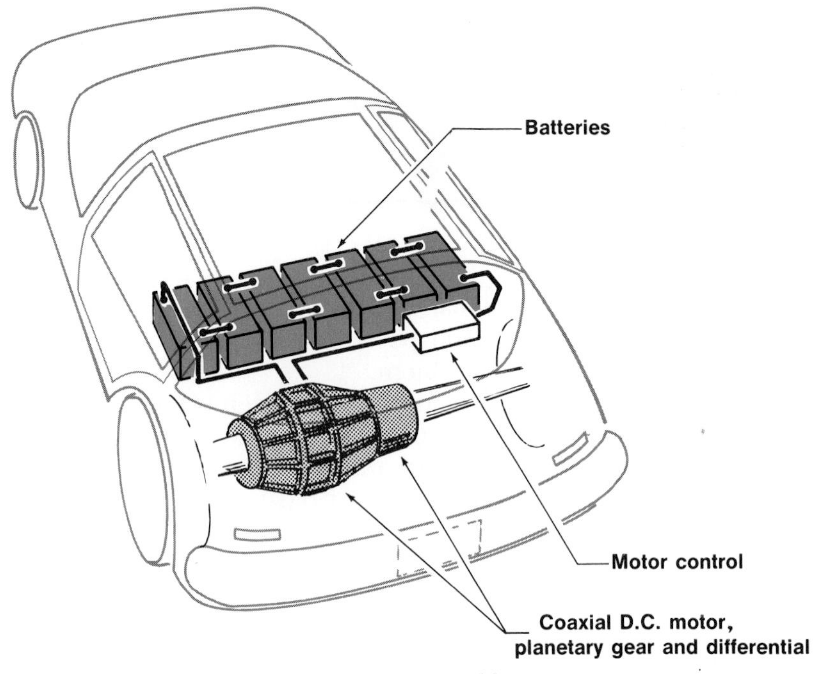

Figure 20–30 Experimental electric car. Powered by conventional lead-acid batteries, it has a range of 50 miles. Lighter, more energetic batteries would increase its range significantly. Present-day batteries account for 30 percent of the total weight of the car.

this positive start, attempts to employ electric vehicles for street and highway use generally have been commercial failures. The energy storage capacities of conventional batteries (lead-acid, nickel-iron, and silver-zinc) are too limited or expensive to provide an acceptable energy source for electric passenger cars (Figure 20–30). New battery designs may prove competitive for short-range urban travel.

Other developments currently either coming on the market or being seriously considered include the stratified charge engine, the Wankel (rotary) and the turbine-generator engine. These relatively new inventions are reported to be improvements in transportation, fuel economy, and pollution control.

SELF-TEST 20-C

1. How many carbon atoms are in the most plentiful hydrocarbon molecule in gasoline?____________
2. The purpose of tetraethyl lead in gasoline is to prevent ____________.
3. $Pb(CH_2CH_3)_4$ performs its function by slowing down free radical reactions of ____________-chain hydrocarbons.

4. Combustion of gasoline occurs in the ________________ of the automobile.

5. Mixing of air and gasoline occurs in the ________________ of the automobile.

6. "Ethyl" gasoline necessarily contains ________________.

7. Gasoline having the same knocking characteristics as a mixture of 7% n-heptane and 93% 2,2,4-trimethylpentane has an octane rating of ______.

8. In the process of adding more air to burn the residual CO and hydrocarbons, are more nitrogen oxides formed? ________

9. If all of the fuel is burned in a regular automobile engine, maximum power is produced. (True/False)

10. Injecting air into the hot exhaust gases reduces the amount of ________ and ________________ but increases the amount of ________________.

11. A metal catalyst that is capable of converting CO and unburned hydrocarbons into CO_2 and H_2O is ________________ (specific metal).

WHAT DOES THE FUTURE HOLD?

At this time, it is uncertain whether the knowledge of the harmful, long-range effects of air pollution will bring people to the point where they are willing to give up the immediate activities that give rise to it. No one seems willing to use less energy or to give up his automobile.

There will undoubtedly be an abatement of air pollution in the future; the sheer pressures of population increase will demand it. But life also will undoubtedly have to be different. Perhaps the first major change will be the disappearance of the automobile from the city, followed by a gradual modification of the power plant of the automobile until it is relatively nonpolluting.

Most pollution exists because we demand the benefits of a technology which, for the most part, has given little consideration to the long-range effects of its products. When industry, automobile manufacturers, or power plant operators add equipment to stop noxious waste products from getting into the air, the costs are added to the already considerable manufacturing expense without adding one cent to the market value of the product being made. The cost to the consumer, however, will go up and will be reflected in the increased price of consumer goods, automobiles, and electrical rates. This is a high price, of course, but what is the value of clean air?

In the final analysis, we all pollute the atmosphere, and much of the pollution is due to the misapplication of chemical techniques. Although we are becoming aware of the problems and although it is within the capabilities of chemical technology to eradicate most forms of air pollution, the process will be very slow.

MATCHING SET

(Use each choice no more than once)

______ 1.	smog	a.	heat absorbing
______ 2.	primary pollutant	b.	prevents lead from fouling spark plugs
______ 3.	secondary pollutant	c.	mixture of fog, SO_2, and (or) hydrocarbons and NO_2
______ 4.	aerosol	d.	parts per million
______ 5.	micron	e.	disease of lungs
______ 6.	ppm	f.	hydrocarbon with double bond
______ 7.	chemical radical	g.	eradication
______ 8.	olefin	h.	CO
______ 9.	photochemical	i.	molecular-sized species with an unpaired valence electron
______ 10.	abatement	j.	n-heptane
______ 11.	emphysema	k.	process of decreasing
______ 12.	brown gas	l.	intermediate in size between individual molecules and particulates
______ 13.	endothermic	m.	benzo(α)pyrene
______ 14.	polynuclear hydrocarbon	n.	PAN, peroxy acyl nitrate
______ 15.	straight-chain hydrocarbon.	o.	chemical reaction energized by light
______ 16.	1,2-dibromoethane	p.	10^{-6} meter
		q.	NO_2

QUESTIONS

1. The formation of photochemical smog involves very reactive free radicals. What structural feature makes free radicals very reactive?
2. Write a balanced chemical equation for the burning of iron pyrite (FeS_2) in coal to sulfur dioxide and Fe_2O_3, iron(III) oxide.
3. What conditions are necessary for thermal inversion?
4. What are the basic chemical differences between industrial smog and Los Angeles-type smog?

5. What are the major sources of the following pollutants?

   a. carbon monoxide
   b. sulfur dioxide
   c. nitrogen oxides
   d. ozone

6. What is a photochemical reaction? Give an example.

7. If air pollutants rise from the earth into the atmosphere, why do they not continue on into space?

8. Why is it so difficult to avoid the formation of either carbon monoxide or nitrogen oxides in the combustion process in an automobile engine?

9. What effects does weather have on local air pollution problems? on regional air pollution problems?

10. Knowing the chemistry of photochemical smog formation, list some ways to prevent its occurrence.

11. What part do aerosols play in the formation of smogs?

12. Describe the antipollution devices on your car.

13. Describe an air pollution problem in your community. How can this problem be solved?

14. Discuss the merits of abatement versus eradication of air pollution.

15. Of the following air pollutants—particulates, sulfur dioxide, carbon monoxide, ozone, and nitrogen dioxide—

   a. which is normally a secondary pollutant?
   b. which is emitted almost exclusively by man-controlled sources?
   c. which can be removed from emissions by centrifugal separators?
   d. which two react with water to form acids?

16. Define:

   a. micron
   b. ppm
   c. PAN

17. What compound gives "ethyl" gasoline its name?

18. What two products are produced by the perfect combustion of hydrocarbons in gasoline?

19. Which is more effective in producing smog, paraffinic (all single bonds) or olefinic (some double bonds) hydrocarbons? Why?

20. Why is natural gas a good substitute for oil and coal as far as air pollution is concerned? What problem is related to its substitution?

21. What is the approximate concentration of air pollutants in the atmosphere during smoggy conditions?

22. Describe an effective way of eliminating carbon monoxide from the exhaust of an automobile.

23. What is the chief source of air pollution in the United States?

24. Account for the peaks and valleys in the graphs of CO in Figure 20–4.

SUGGESTIONS FOR FURTHER READING

Beard, R. R., and Wertheim, G. A., "Behavioral Impairment Associated with Small Doses of Carbon Monoxide," *American Journal of Public Health,* Vol. 57, p. 2012 (1967).

Cook, L. M. (ed.), "Cleaning Our Environment—A Chemical Basis For Action," A. C. S. Committee on Chemistry and Public Affairs, American Chemical Society, Washington, D.C., 1970.

Hall, H. J., and Bartok, W., "NO_x Control From Stationary Sources," *Environmental Science and Technology,* Vol. 5, No. 4, p. 320, (1971).

National Tuberculosis and Respiratory Disease Association, "Air Pollution Primer," New York, 1969.

Newell, R. E., "The Global Circulation of Atmospheric Pollutants," *Scientific American,* Vol. 224, No. 1, p. 32, (1971).

Perkins, H. C., "Air Pollution," McGraw Hill Book Co., New York, 1974.

Turk, A., Turk, J., Wittes, J. T., and Wittes, R., "Environmental Science," W. B. Saunders Co., Philadelphia, 1974.

Wildeman, Thomas R., "The Automobile and Air Pollution," *Journal of Chemical Education,* Vol. 51, No. 5, p. 290, (1974).

Williamson, Samuel J., "Fundamentals of Air Pollution," Addison-Wesley Publishing Co., Reading, Mass., 1973.

Wolf, P. C., "Carbon Monoxide-Measurement and Monitoring in Urban Air," *Environmental Science and Technology,* Vol. 5, No. 3, p. 212, (1971).

Among the many technical pollution publications of the U. S. Environmental Protection Agency are the following:

"Air Quality Criteria for Particulate Matter"
"Air Quality Criteria for Sulfur Oxides"
"Air Quality Criteria for Carbon Monoxide"
"Air Quality Criteria for Photochemical Oxidants"
"Air Quality Criteria for Hydrocarbons"
"Air Quality Criteria for Nitrogen Oxides"
"Control Techniques for Particulate Air Pollutants"
"Control Techniques for Sulfur Oxide Air Pollutants"
"Air Quality And Emissions Trends Annual Reports"
"Compilation of Air Pollutant Emission Factors"
"Guide for Control of Air Pollution Episodes"

CHAPTER 21

CONSUMER CHEMISTRY—FOOD AND MEDICINE

A major portion of the household budget is spent on consumer products. We buy chemical products to feed, cleanse, disinfect, wax, deodorize, paint, remove spots, relieve pain, cure disease, fertilize, preserve, destroy, and do hundreds of other things for us. You already know about the undesirable qualities of certain chemicals as pollutants (Chapters 19 and 20) and as toxic substances (Chapter 18). You are also familiar with many of the chemical products that make our lives more pleasant, healthy, and convenient, provided we use them in the proper amounts and for their intended purposes. You are aware that some, if not all, consumer products ultimately become pollutants, and it is a question of values as to how we use them in our society.

Consideration of consumer products must involve disposal as well as preparation and use.

To reap the full benefit of the products we purchase, we should know the types of raw materials in the products, the ways in which products perform their jobs, and the precautions to be taken when using the products. Many formulations under different brand names are essentially the same and cost about the same to produce. Competitive products vie for sales through the cleverness of the name, the attractiveness of the container, and the effectiveness of advertisements. These costs are generally added to the price without, of course, increasing the true value of the product. When you have some knowledge of the chemistry of the product, the small print on the label becomes important. A comparison of lists of ingredients sometimes uncovers better formulations, and even harmful substances. Chemical knowledge can then protect the consumer.

Most consumer products are mixtures, and their properties are not necessarily blends of individual properties of the chemicals. In some formulations, the individual properties of certain chemicals are reinforced and enhanced by other chemicals in the mixture. This is called ***synergism,*** defined as the cooperative action of discrete agencies such that the total effect is greater than the sum of the effects of each used alone. Citric acid has very little antioxidant effect on foods; butylated hydroxyanisole (BHA) has considerable effect on preventing oxidation of foods. When these two substances are used together in foods, the antioxidative

Synergism — cooperative action so that the total effect is greater than the sum of the parts.

powers of the BHA are increased several-fold. The citric acid is said to have synergistic action on BHA. The reasons for this will be discussed in the section on food chemistry.

A closely related term is ***potentiation.*** Potentiators do not have a particular effect themselves but exaggerate the effect in other chemicals. The 5′-nucleotides, for example, have no taste, but they enhance the flavor of meat or the effectiveness of salt. This phenomenon will also be discussed in the next section.

We will discuss the chemistry of only a few of the hundreds of consumer products. However, the application of the principles involved transfer readily to other products. The discussion is divided into three chapters. In each chapter, emphasis will be placed on the purposes of the specific chemicals in the formulations. Sometimes a chemical is a part of a formulation because it undergoes a desired chemical reaction; sometimes it is simply there to dilute or dissolve the active ingredients. Chemicals are added to improve color, taste, viscosity, texture, hardness, crispness, dryness, and many other properties. As the various formulations are discussed, and the purpose of the individual chemicals are given, molecular theory will be used to explain the effects of the chemicals.

THE CHEMICALS WE EAT

Preservation of Foods

Dry foods tend to be stable.

Foods generally lose their usefulness and appeal a short time after harvest. Bacterial decomposition and oxidation are the prime reasons steps must be taken to lengthen the time a foodstuff remains edible. Any process that prevents the growth of or presence of microorganisms and excludes oxygen or water, or both, is generally an effective preservative process for food. Perhaps the oldest technique is the drying of grains, fruits, fish, and meat. Water is necessary for the growth and metabolism of microorganisms. Without water they dehydrate and die. Water is also important in food oxidation and, in this capacity, will be discussed in the next section. Dryness then thwarts both the oxidation of food and the microorganisms which feed on it.

A hypertonic solution is more concentrated than solutions in its immediate environment.

Osmosis is the flow of water from a more dilute solution through a membrane into a more concentrated solution.

Chemicals may also be added as preservatives. Salted meat, and fruit preserved in a concentrated sugar solution, are protected from microorganisms. The abundance of sodium chloride or sucrose in the immediate environment of the microorganisms forms a ***hypertonic*** condition in which water flows by ***osmosis*** from the microorganism to its environment (Figure 21–1). The salt and sucrose have the same effect on the microorganism as does dryness. Both dehydrate the microorganism and thus destroy its ability to grow, reproduce, and exude distasteful and dangerous products into the food.

The canning process for preserving food was developed by 1810, and involves first the heating of food to kill all bacteria and then sealing in bottles or cans to prevent access of other microorganisms and oxygen. This technique is widely used for fish, meat, and vegetables. Some canned meat has been successfully preserved for over a century.

With the increasing growth of urban population and the need for preserving food for long periods of time, the necessity of developing improved techniques for food preservation became urgent. As a result of

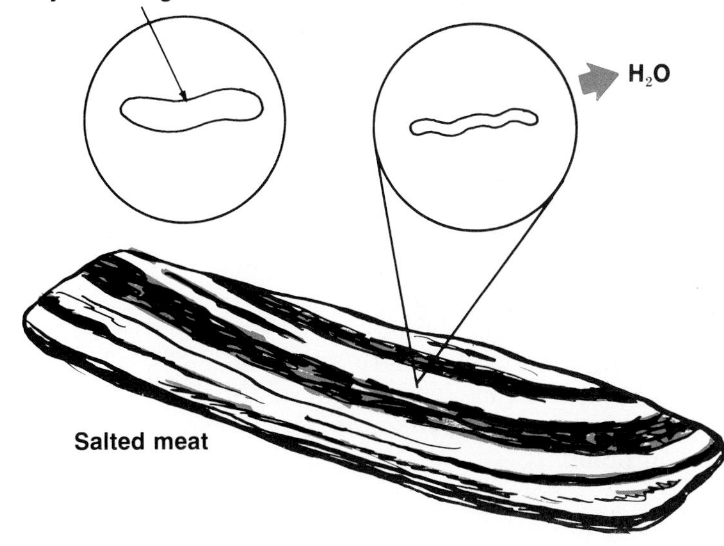

Figure 21–1 A salt solution dehydrates microorganisms by osmosis. Without water, microorganisms cannot carry on the chemical reactions required for growth and reproduction.

this need, newer techniques have been developed, such as dehydration, freezing, pasteurization, cold storage, irradiation, and chemical preservation. Nevertheless, approximately 30 percent of all food harvested is spoiled by bacterial decomposition, fungal attack, or is eaten by animals before it reaches the consumer.

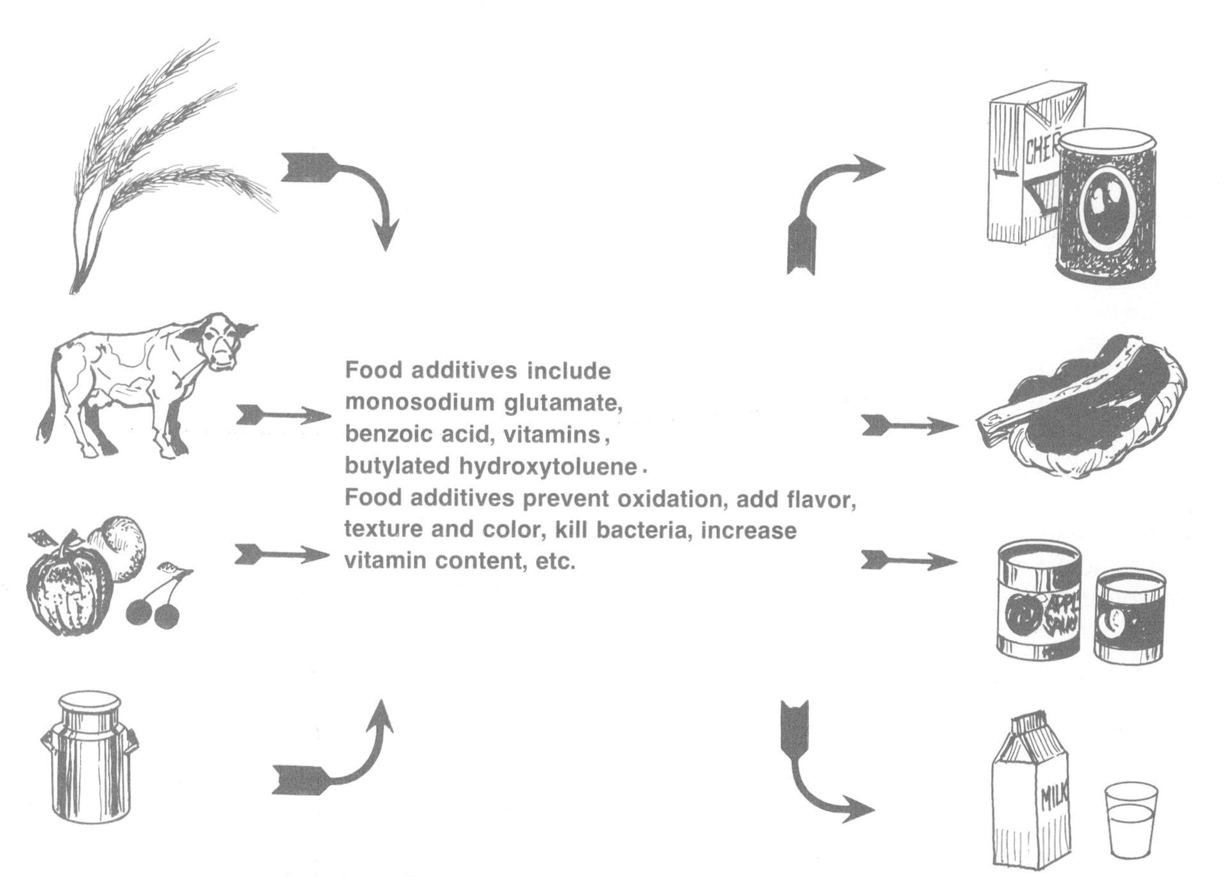

Food additives include
monosodium glutamate,
benzoic acid, vitamins,
butylated hydroxytoluene.
Food additives prevent oxidation, add flavor,
texture and color, kill bacteria, increase
vitamin content, etc.
MILK

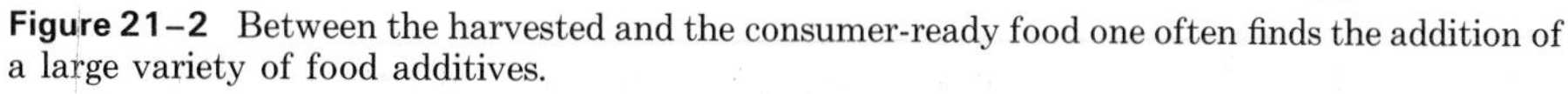
Figure 21–2 Between the harvested and the consumer-ready food one often finds the addition of a large variety of food additives.

Antimicrobial Preservatives

Food spoilage caused by microorganisms is a result of the excretion of toxins. A preservative is effective if it prevents multiplication of the microbes during the shelf-life of the product. Sterilization by heat or radiation or inactivation by freezing is often undesirable since the quality of the food is impaired. Chemical agents seldom achieve sterile conditions but can preserve foods for considerable lengths of time.

Antimicrobial preservatives are widely used in a large variety of foods. For example, in the United States sodium benzoate is permitted in nonalcoholic beverages and in some fruit juices, fountain syrups, margarines, pickles, relishes, olives, salads, pie fillings, jams, jellies, and preserves. Sodium propionate is legal in bread, chocolate products, cheese, pie crust, and fillings. Depending on the food, the weight of the preservative permitted ranges up to a maximum of 0.1 percent for sodium benzoate and 0.3 percent for sodium propionate.

A preservative must interfere with microbes but be harmless to the human system—a delicate balance.

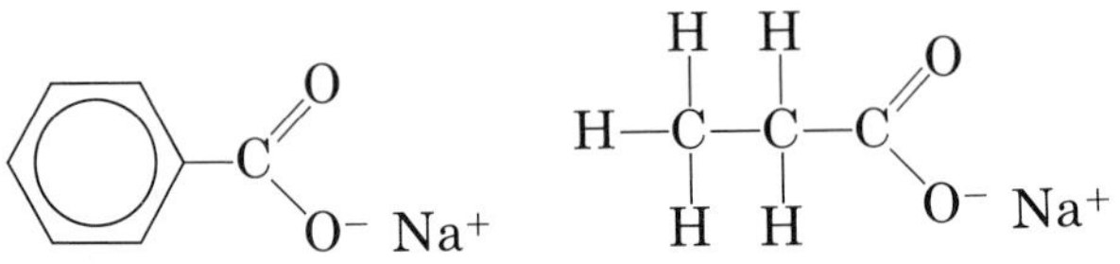

$C_6H_5{-}C(=O){-}O^- \; Na^+$

SODIUM BENZOATE

$H{-}CH_2{-}CH_2{-}C(=O){-}O^- \; Na^+$

SODIUM PROPIONATE

Despite the fact that chemical preservatives of various kinds have been used for several decades, their mode of action remains largely unexplained. Postulated mechanisms may be grouped into three categories: (1) interference with the permeability of cell membranes of the microbes for foodstuffs, so the bacteria die of starvation; (2) interference with genetic mechanisms so the reproduction processes are hindered; and (3) interference with intracellular enzyme activity so that metabolic processes such as the Krebs cycle grind to a halt.

Interference with the cell membrane and cell wall permeability can have vast effects on the flow of cell nutrients into and waste products out of the cell. For example, propionic acid is known to have an antimicrobial action on certain types of bacteria. It is thought that the propionic acid coats the cell surface with a substance that reduces the permeability of the surface. The coat effectively blocks passage of materials into and out of the cell, and thus the cell ceases to function. On the other hand, oxytetracycline, an antibiotic added to meats, may destroy bacteria by creating a leakage in the cell membrane. Some antibiotics, such as penicillin, are known to prevent cell wall synthesis in microbes. Benzoic acid and salicylic acid are known to accumulate on the cell membranes of microorganisms and are thought to render the membrane less permeable.

A permeable cell membrane or wall is easy to penetrate.

There is very little evidence that the generally accepted food preservatives interfere with genetic mechanisms of microbes. However, certain antibiotics such as streptomycin and chloramphenicol are known to exert their effects by chemically combining with the ribosome and inhibiting protein synthesis.

There is some evidence that food preservatives interfere with cellular enzymes in microbes. They may interrupt the metabolism of enzymes, and thus destroy them. For example, sorbic acid is known to interfere with cellular dehydrogenases, enzymes which dehydrogenate fatty acids as the first step in the metabolism of molds that grow on cheese. This

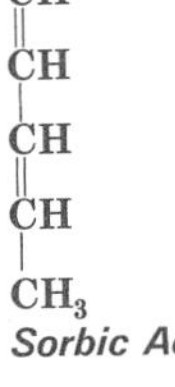

$COOH{-}CH{=}CH{-}CH{=}CH{-}CH_3$

Sorbic Acid

interrupts the use of food by the molds and inhibits growth. Several food preservatives are known to inhibit enzyme systems isolated from the cell, but such observations cannot necessarily be extrapolated to cellular conditions. The preservative either may not penetrate the cell wall or may not attain sufficient concentration at the enzyme site.

"Spoiled" food may not be all bad.

It is interesting to note that, at times, certain social groups, for one reason or another, develop a taste for food that has been partially decomposed by microorganisms. Limburger cheese is considered a gourmet food, while the Eskimos bury some of their fish to allow bacterial decomposition to the point that it is a foul semiliquid. This *titmuck* is considered a delight to the Eskimos, but not to their dogs, which refuse to eat it. Sometimes, when the wrong bacteria grow on the fish, the titmuck kills those who eat it.

Microbial activity results in oxidative decay of food, but it is not the only means of oxidizing food. The direct action of oxygen in the air, ***atmospheric oxidation,*** is the chief factor in destroying fats and fatty portions of foods. Chemically, oxygen reacts with the fat to form a hydroperoxide (R—OOH).

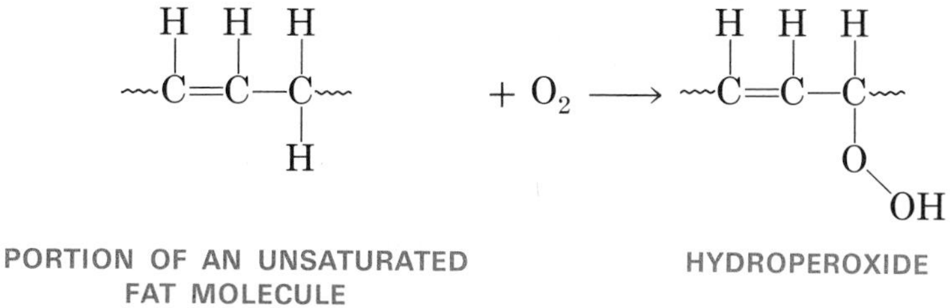

PORTION OF AN UNSATURATED FAT MOLECULE HYDROPEROXIDE

Free radicals usually have short lives; in rare cases they have a stable structure.

The mechanism involves the formation of very reactive free radicals (species with one or more unpaired electrons) in a chain reaction process. For example:

$$RH + O_2 \longrightarrow R\cdot + HO_2\cdot$$

FAT FREE RADICALS

$$R\cdot + O_2 \longrightarrow ROO\cdot$$

$$ROO\cdot + RH \longrightarrow ROOH + R\cdot$$

HYDROPEROXIDE

Foods kept wrapped, cold, and dry are relatively free of oxidation. An antioxidant added to the food can also hinder oxidation. Antioxidants most commonly used in edible products contain various combinations of butylated hydroxyanisole (BHA), butylated hydroxytoluene (BHT), or propyl gallate:

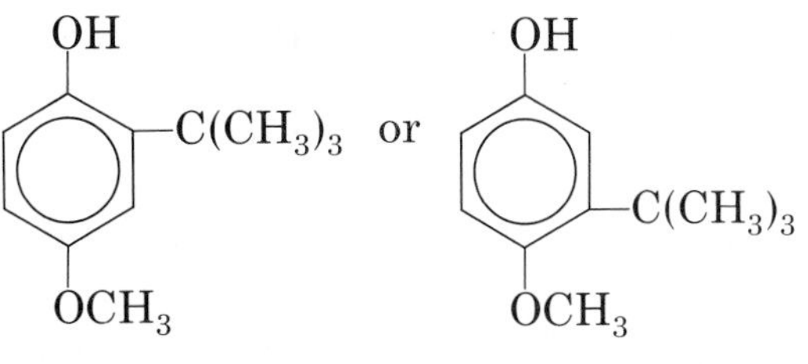

BHA

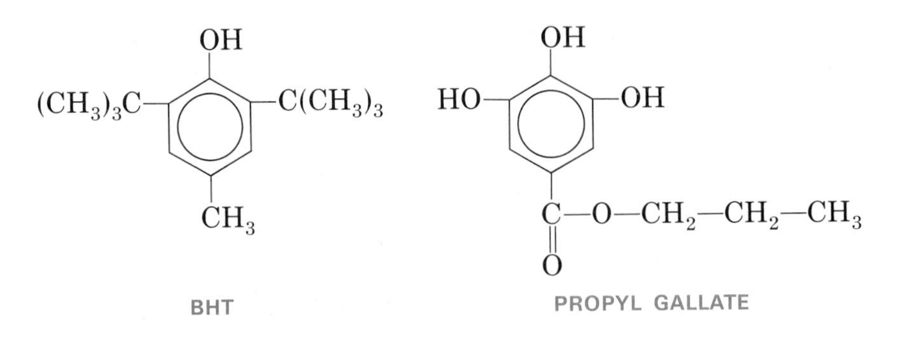

The word ***butylated*** is not widely used in organic chemical nomenclature. It is applied here because the usual names for BHA and BHT include the words *cresol* or *phenol,* and these names are associated with toxic materials. To avoid consumer rejection of these "safe" compounds —added within the prescribed limits (maximum of 0.02 percent or less, depending on the food)—the names were made "safe," too.

Most chemicals added to foods are considered safe.

To prevent the oxidation of fats, the antioxidant can donate the hydrogen atom in the —OH group to the free radicals and stop the chain reactions. The bulky aromatic radicals formed are relatively stable and unreactive; they add the unpaired electrons to their supply of delocalized electrons:

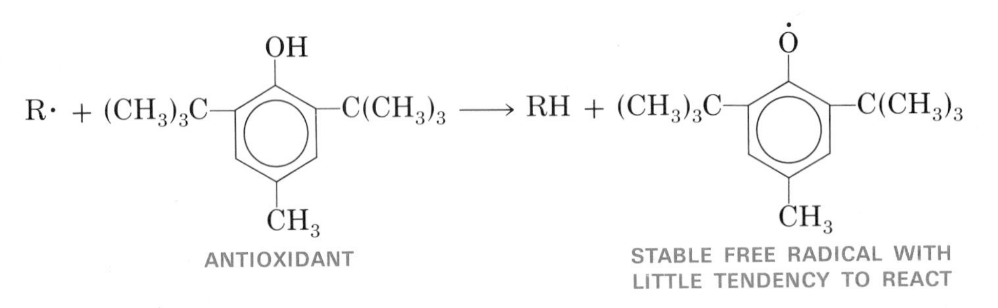

If antioxidants are not present, the hydroperoxy group will attack a double bond. This reaction leads to a complex mixture of volatile aldehydes, ketones, and acids, which cause the odor and taste of rancid fat.

Sequestrants

Metals get into food from the soil and from machinery during harvesting and processing. Copper, iron, and nickel, and their ions, catalyze the oxidation of fats. However, a molecule of citric acid bonds with the metal ion, thereby rendering it ineffective as a catalyst. With the competitor metal ions tied up, antioxidants such as BHA and BHT can accomplish their task much more effectively.

Citric acid belongs to a class of food additives known as ***sequestrants.*** For the most part sequestrants react with trace metals in foods, tying them up in complexes so the metals will not catalyze the decomposition or oxidation of food. Sequestrants like sodium and calcium salts of EDTA (ethylenediaminetetraacetic acid) are permitted in beverages, cooked crab meat, salad dressing, shortening, lard, soup, cheese, vegetable oils, pudding mixes, vinegar, confectioneries, margarine, and other foods. The amount ranges from 0.0025 to 0.15 percent. The structural formula of EDTA bonded to a metal ion is shown in Figure 21–3. Note the five-

To sequester means "to withdraw from public use."

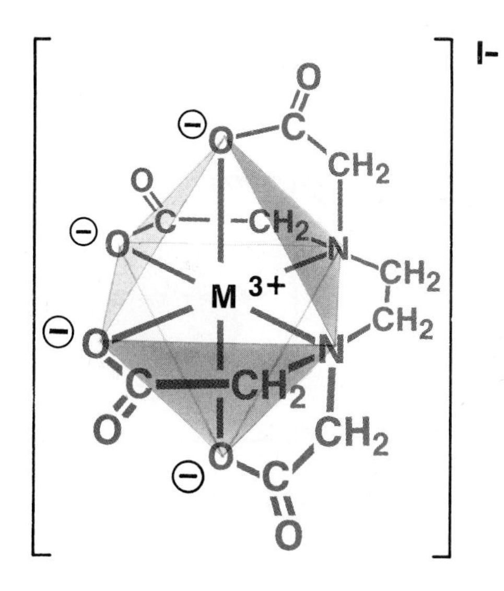

Figure 21–3 The structural formula for the metal chelate of ethylenediaminetetraacetic acid (EDTA).

membered rings (five- or six-membered rings are usually required for stable bonding angles), and the octahedral arrangement of the atoms bonded to the central metal ion, M^{3+}.

Flavor in Foods

Natural Flavors

Flavors result from a complex mixture of volatile chemicals. Since we have only four tastes (sweet, sour, salt, bitter), much of the sensation of food is smell. For example, the flavor of coffee is determined largely by its aroma, and this in turn is due to a very complex mixture of over 100 compounds. The reproduction of a particular flavor is usually attained by mixing coffee from different sources rather than trying to match it chemically. Because roasted coffee loses its volatile compounds in air (and, consequently, its flavor), it must be kept sealed to prevent staleness. Roasted chicory, which is often mixed in with coffee, contains no caffeine, but it does provide color and flavor.

Tea may be made by steeping the dried tea leaves in hot water for three to five minutes. This extracts water-soluble compounds, including

There are more than 100 different chemicals in coffee aroma

Figure 21–4 That just-right coffee aroma is a mixture of many different compounds.

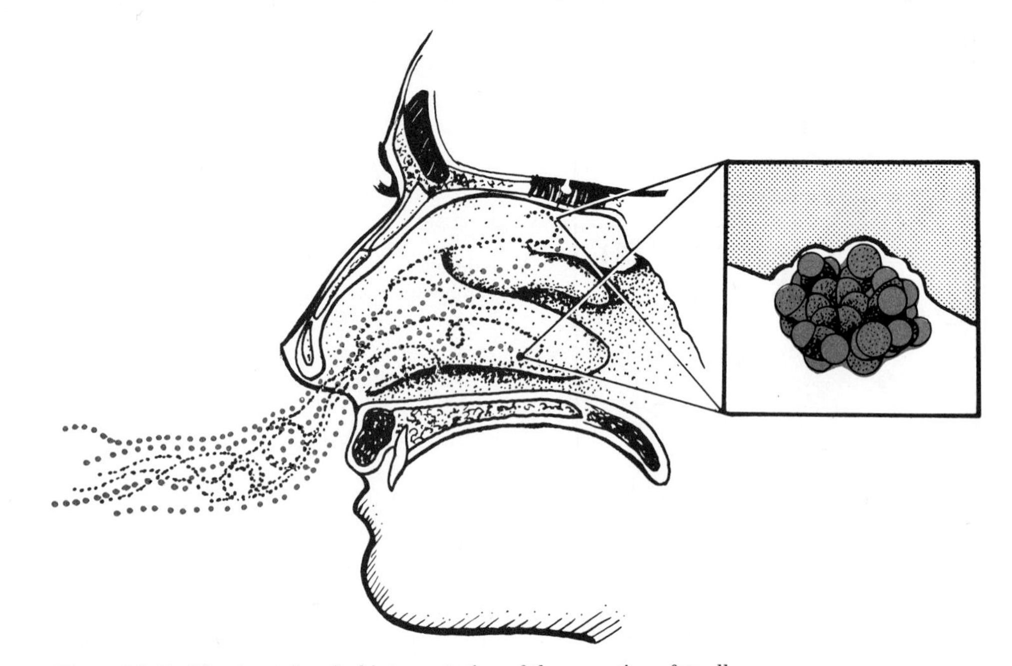

Figure 21-5 The stereochemical interpretation of the sensation of smell. The substance fits a cavity in the back of the oral cavity. If the atoms are properly spaced, they sensitize nerve endings which transmit impulses to the brain. The brain identifies these sensations as a particular smell. Most tastes are largely odors.

caffeine. The caffeine content of tea is considerably higher than that of coffee, but coffee contains a variety of other ingredients and generally produces more side effects than tea. Because of the way tea is made, it ordinarily does not contain any insoluble matter suspended in it, as coffee does. Tea absorbs flavors and odors readily, and this fact can be used to advantage in the preparation of specially flavored teas. It also means that tea should not be stored near onions, garlic, or soaps since it will also absorb these aromas.

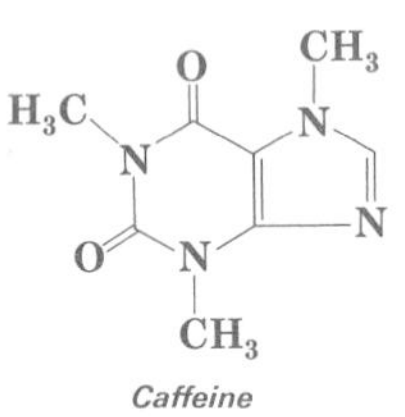

Artificial Flavors

Most flavor additives originally came from plants. The plants were crushed and the compound extracted with various solvents such as ethanol and carbon tetrachloride. Sometimes a single compound was

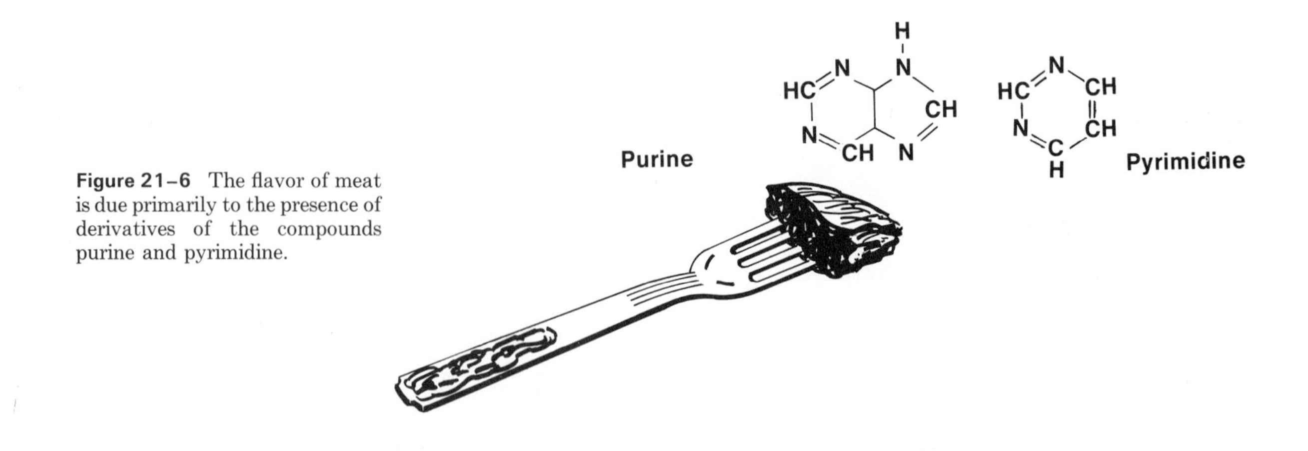

Figure 21-6 The flavor of meat is due primarily to the presence of derivatives of the compounds purine and pyrimidine.

extracted; more often, a mixture of several compounds occurred in the residue. By repeated efforts, relatively pure oils were obtained. Oils of wintergreen, peppermint, orange, lemon, and ginger, among others, are still obtained this way. These oils, alone or in combination, are then added to foods to obtain the desired flavor. Gradually analyses of the oils and flavor components of plants revealed the active compounds responsible for the flavor. Today, the synthetic preparation of flavors actively competes with natural sources.

O—CH_2 O CH_2—CH=CH_2

Safrol

The Food and Drug Administration has recently banned some of the naturally occurring flavoring agents which were formerly used, including safrol, the primary root beer flavor, found in the root of the sassafras tree.

Flavor Enhancers

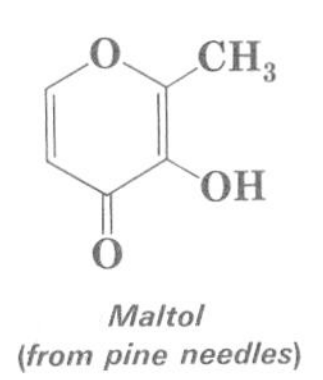

Maltol
(from pine needles)

Flavor enhancers have little or no taste of their own but amplify the flavors of other substances. They exert synergistic and potentiation effects. Potentiators were first used in meats and fish. Now they are also used to intensify the flavor and cover unwanted flavors in vegetables, bread, cakes, fruits, nuts, and beverages. Three common flavor enhancers are *monosodium glutamate* (*MSG*), 5′-*nucleotides* (similar to inosinic acid; Chapter 16), and *maltol.*

MSG is the best known and most widely used flavor enhancer. More than 220 million pounds were sold world-wide in 1966. Glutamic acid was isolated in 1866 by the German chemist Ritthausen.

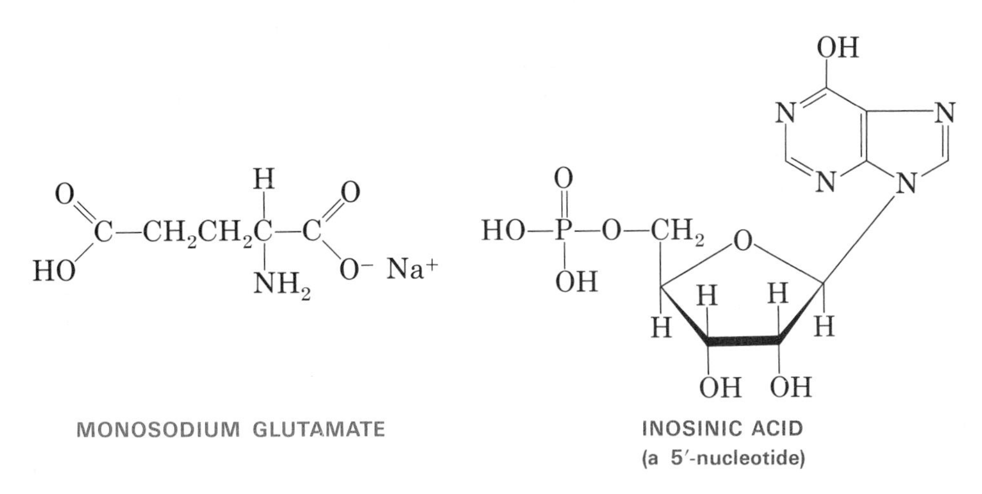

MONOSODIUM GLUTAMATE

INOSINIC ACID
(a 5′-nucleotide)

In 1908, a Japanese chemist at the University of Tokyo, Dr. Kikunae Ikeda, discovered the flavor-enhancing properties of MSG. Japanese cooks had used the seaweed *Laminaria japonica* for centuries to improve the flavor of soups and certain other foods. Dr. Ikeda discovered that the ingredient in the seaweed that made the difference was MSG, and that it had an unusual ability to enhance or intensify the flavor of many high protein foods.

MSG is a natural constituent of many foods, such as tomatoes and mushrooms.

If MSG imparts no flavor of its own in the concentrations allowed in foods, then how does MSG, or any enhancer or potentiator, work? Some chemists say that it acts on certain nerve endings to make the taste buds more sensitive and, therefore, increases the flavor of food. Others claim that it increases salivation and that this leads to increased flavor perception. Although the detailed chemical and biological action of flavor enhancers is only partially understood, they are very effective, and we use them without apparent harm, unless the sodium intake must be limited, as in low-salt diets.

Brain damage is caused when MSG is injected in very high dosage under the skin in 10-day-old mice. When these laboratory results were reported, considerable discussion ensued concerning the merits of MSG. National investigative councils have suggested that it be removed from baby foods since infants do not seem to appreciate enhanced flavor. However, in the absence of hard evidence that MSG is harmful in the amounts used in regular food, no recommendations were made relative to its use.

In some people, MSG causes the so-called "Chinese restaurant syndrome," an unpleasant reaction that includes headaches, sweating, and other symptoms usually occurring after an MSG-rich Chinese meal. Tomatoes and strawberries affect some individuals in the same way.

The flavor-enhancing property of the 5′-nucleotides was discovered in 1913 by Dr. Shintara Kodama of the University of Tokyo, an associate of Dr. Ikeda. The 5′-nucleotides were discovered to be the ingredients responsible for the flavor-enhancing qualities of bonita tuna. These were approved by the FDA in 1962. While MSG is effective in enhancing the flavor of foods, in concentrations of only a few parts per thousand, the 5′-nucleotides significantly enhance the flavor of foods in smaller concentrations than MSG. In liquid foods they create a sense of increased viscosity.

Flavor enhancers were enjoyed long before they were characterized chemically.

SELF-TEST 21-A

1. Butylated hydroxyanisole plus citric acid in food is an example of (synergism/potentiation) ________________.
2. Name two of the oldest means for preserving food. ________________ ________________
3. Antimicrobial preservatives make foods sterile. (True/False)
4. Name a food which depends on microbial action for its taste. ________________
5. Antioxidants are (more/less) easily oxidized than the food into which they are placed. ____________
6. Ethylenediaminetetraacetic acid is an example of a(n) ____________ ____________. Such compounds tie up metals in stable complexes.
7. A flavor in a food can usually be traced to a single compound. (True/False)
8. Monosodium glutamate is a(n) ____________ ____________.
9. Salt is effective in preserving foods because it kills microorganisms by ________________ them.
10. BHT serves as an antioxidant by destroying ____________ ____________.

Sweetness

Sweetness is characteristic of a wide range of compounds, many of which are completely unrelated to sugars. Lead acetate, $Pb(CH_3COO)_2$, is sweet but poisonous. A number of ***artificial sweeteners*** are allowed in foods. These are primarily used for special diets such as those of diabetics. Artificial sweeteners have no known metabolic use in the body and do not require insulin.

Insulin is a hormone which regulates glucose metabolism.

The sweetness factor is determined by panels of tasters who sample different dilutions of solutions of the sweeteners in water. If a 0.01 M artificial sweetener solution provides the same sweetness sensation as a 1 M solution of sucrose, the artificial sweetener is said to be 100 times sweeter than sucrose.

Saccharin

The most common artificial sweetener is saccharin. Its sweet taste was discovered accidentally when the chemist who synthesized it noted that a piece of bread he ate had an obvious sweet taste. He traced the taste back to the saccharin he had prepared. The material was found in the course of studies on the oxidation of orthotoluenesulfonamide, which is the starting material used for its preparation today:

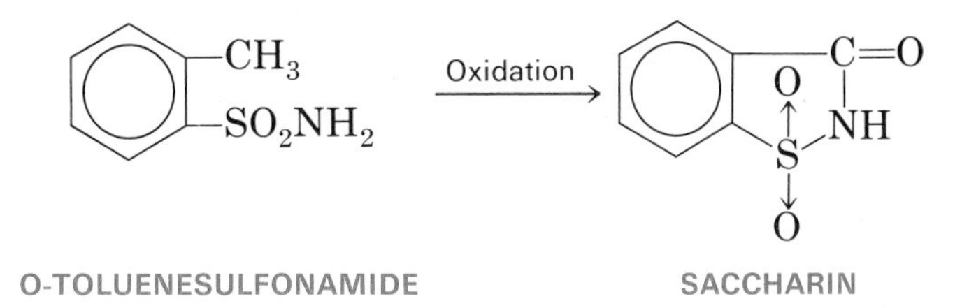

O-TOLUENESULFONAMIDE SACCHARIN

Saccharin is about 300 times sweeter than ordinary sugar (sucrose). When ingested, saccharin passes through the body unchanged. It therefore has no food value other than to render an otherwise bland mixture more tasty. Saccharin has a somewhat bitter aftertaste which renders it unpleasant to some users. Glycine, the simplest amino acid, which is also sweet-tasting, is often added to counteract this bitter taste.

Saccharin is a synthetic chemical.

Cyclamates

The sweetness of cyclamate salts was also discovered accidentally. Sodium cyclamate is about 30 times sweeter than sugar. Because a cyclamate does not have the aftertaste of saccharin, and because it could be used in cooked or baked products, it rapidly replaced saccharin in a wide variety of dietary products.

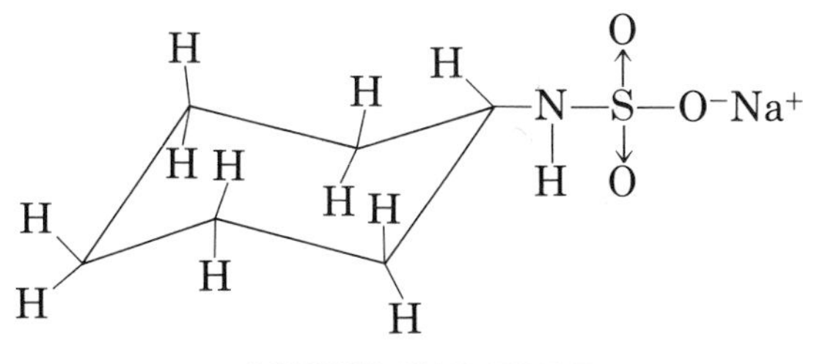

SODIUM CYCLAMATE

In 1969, experimental evidence indicated that cyclamates, under certain conditions, caused cancer of the bladder in laboratory mice. When this evidence was coupled with the fact that 20 percent of all cyclamate users metabolized cyclamate to a metabolite which caused chromosome breakage in rat cells, the Food and Drug Administration banned cyclamates from foods. There is no evidence that it causes cancer in human beings.

Recently the FDA allowed cyclamates back on the market.

Aspartame

With cyclamates off the market and saccharin under scrutiny by the FDA as a possible carcinogen, a new entry into the sweetener market was approved by the FDA in 1974. Aspartame is about 180 times sweeter than sugar.

Sold under the trademark Equa, the sweetener's chemical name is L-aspartyl-L-phenylalanine methyl ester, and its common name is aspartame. The caloric value of aspartame is similar to that of proteins. It is metabolized in the body as a peptide. The caloric intake of consumers using the product is greatly reduced, since much smaller amounts are needed to produce the same sweetening effect. Aspartame does not have the bitter aftertaste associated with other artificial sweeteners.

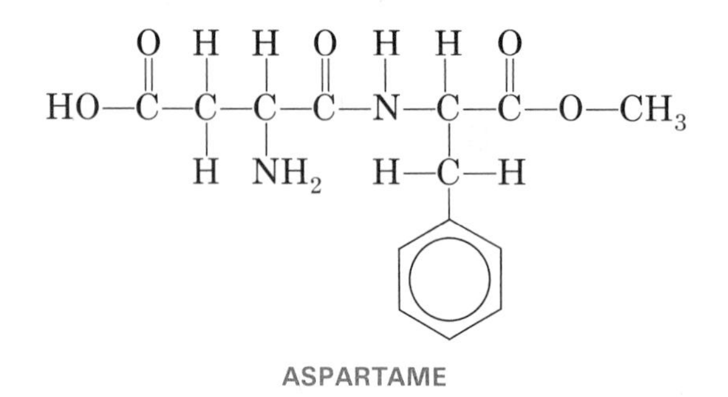

ASPARTAME

Food and Esthetic Appeal

Food Colors

Food colors are generally large organic molecules having several double bonds and aromatic rings. Consistent with the discussion of dyes in Chapter 14 these conjugated structures are chromophores and can absorb only certain wavelengths of light and pass the rest; this behavior gives the substance its characteristic color. β-Carotene is an orange-red substance that occurs in a variety of plants, the carrot in particular, and is commonly used as a food color. It has a conjugated system of delocalized electrons. Most of the other food colors have similar conjugated systems.

Colored organic substances often are conjugated molecules having alternating double and single bonds in the carbon chain or ring.

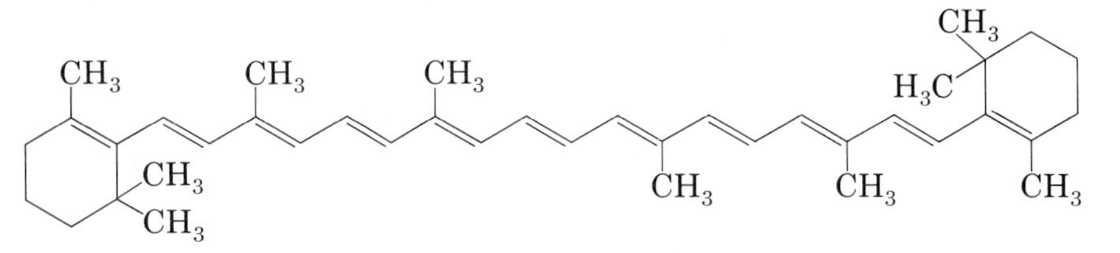

β-CAROTENE

β-Carotene is a *precursor* of vitamin A. (It is changed into vitamin A in the body.)

pH Control

Weak organic acids are added to such foods as cheese, beverages, and dressings to give a mild acidic taste. They often mask undesirable aftertastes. Weak acids and acid salts, such as tartaric acid and potassium acid tartrate, react with bicarbonate to form CO_2 in the baking process.

Some acid additives control the pH of food during various stages of processing as well as of the finished product. In addition to single substances, there are several combinations of substances that will adjust and then maintain a desired pH. These mixtures are called ***buffers.*** An example of one type of buffer is potassium acid tartrate, $KHC_4H_4O_6$, which is used to adjust the pH in candymaking. Most sugar syrups —maple, cane, corn, molasses, and so on—are acidic. Hydrogen ions accelerate the conversion of ordinary sugar (sucrose) to noncrystalline invert sugar (glucose and fructose) (Chapter 16). Since invert sugar inhibits the crystallization of sugar, the candymaker must be sure that the pH of his syrup is such that crystallization will be prevented in the case of creamy fondants and will not be interfered with in the case of hard candies. Potassium acid tartrate achieves a particular pH by means of the following equilibrium:

Buffer solutions resist change in acidity and basicity; pH remains constant.

Invert sugar is made by hydrolyzing sucrose.

$$HC_4H_4O_6^- + H_2O \rightleftharpoons H_3O^+ + C_4H_4O_6^{2-}$$

If the syrup is too acidic, the hydrogen ions of the syrup (as H_3O^+) force the equilibrium to the left, and the acidity is reduced. It is unlikely that the syrup would be too basic but if it is, the equilibrium is shifted to the right to resupply the hydrogen ions used in the neutralization of the basic hydroxide ions, OH^-.

$$H_3O^+ + OH^- \longrightarrow 2H_2O$$

To react with either acid or base, and hence to shift the equilibrium in either direction, the buffer must have an ample supply of both $HC_4H_4O_6^-$ and $C_4H_4O_6^{2-}$ ions. In a 0.003 molar solution of potassium acid tartrate, there are about four $HC_4H_4O_6^-$ ions for every one $C_4H_4O_6^{2-}$, and the pH is between 3 and 4. Under these conditions 0.0006 mole of acid would use up the 0.0006 mole of $C_4H_4O_6^{2-}$ in a liter of the 0.003 M solution. The FDA limit of 0.25 percent potassium acid tartrate is usually sufficient to adjust the acidity of most syrups. If adjustment to another pH range is desired, other buffers must be used. The requirement is that the buffer ionize sufficiently to give the desired hydrogen ion concentration.

Adjustment of the pH of a fruit juice is allowed by the FDA. If the pH of the fruit is too high, it is permissible to add acid (called an ***acidulant***). Citric acid and lactic acid are the most common acidulants used since they are believed to impart good flavor; but phosphoric, tartaric, and malic acids are also used. These acids are often added at the end of the cooking time to prevent extensive hydrolysis of the sugar. In the making of jelly they are sometimes mixed with the hot product immediately after pouring. To raise the pH of a fruit which is unusually acidic, buffer salts such as sodium citrate or sodium potassium tartrate are used.

Small amounts of certain acids are allowed to be added to some foods.

The versatile acidulants also function as preservatives to prevent growth of microorganisms, as synergists and antioxidants to prevent rancidity and browning, as viscosity modifiers in dough, and as melting point modifiers in such food products as cheese spreads and hard candy.

Anticaking Agents

Anticaking agents are added to hygroscopic foods—in amounts of 1 percent or less—to prevent caking in humid weather. Table salt (sodium chloride) is particularly subject to caking unless an anticaking agent is present. The additive (magnesium silicate, for example) incorporates water into its structure as hydrated water and does not appear wet as sodium chloride does when it absorbs water physically on the surface of its crystals. As a result, the anticaking agent keeps the surface of sodium chloride crystals dry and prevents crystal surfaces from co-dissolving, which joins the crystals together.

Hygroscopic substances absorb moisture from the air.

Stabilizers and Thickeners

Stabilizers and thickeners improve the texture and blends of foods. The action of carrageenin (a polymer from edible sea weed) is shown in Figure 21–7. Most of this group of food additives are polysaccharides (Chapter 16) having numerous hydroxyl groups as a part of their structure. The hydroxyl groups form hydrogen bonds with water to prevent the segregation of water from the less polar fats in the food, and to provide a more even blend of the water and oils throughout the food. Stabilizers and thickeners are particularly effective in icings, frozen desserts, salad dressing, whipped cream, confectionery, and cheeses.

Stabilizers and thickeners are types of emulsifying agents.

Surface Active Agents

Surface active agents are similar to stabilizers, thickeners, and detergents in their chemical action. They cause two or more normally incompatible (nonpolar and polar) chemicals to mix. If the chemicals are liquids, the surface active agent is called an emulsifier. If the surface active agent has a sufficient supply of hydroxyl groups, such as cholic acid has, the groups form hydrogen bonds to water. The cholic acid and its associated group of water molecules are distributed throughout dried egg

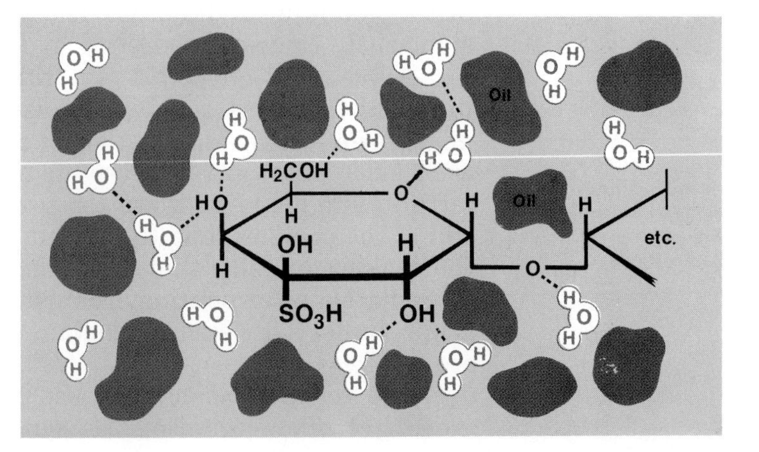

Figure 21–7 The action of carrageenin to stabilize an emulsion of water and oil in salad dressing. An active part of carrageenin is a polysaccharide, a portion of which is shown above. The carrageenin hydrogen-bonds to the water, which keeps it dispersed. The oil, not being very cohesive, disperses throughout the structure of the polysaccharide. Gelatin (a protein) undergoes similar action in absorbing and distributing water to prevent ice crystals in ice cream.

in a manner quite similar to that of carrageenin and water in salad dressing.

Hydrogen bonding plays a major role in stabilizers, thickeners, surface active agents, and humectants.

Some surface active agents have both hydroxyl groups and a relatively long nonpolar hydrocarbon end. Examples are diglycerides of fatty acids, polysorbate 80, and sorbitan monostearate. The hydroxyl groups on one end of the molecule are anchored via hydrogen bonds in the water, and the nonpolar end is held by the nonpolar oils or other substance in the food. This provides tiny islands of water held to oil. These islands are distributed evenly throughout the food.

Polyhydric Alcohols

Polyhydric alcohols are allowed in foods as humectants, sweetness controllers, dietary agents, and softening agents. Their chemical action is based on their multiplicity of hydroxyl groups that hydrogen-bond to water. This holds water in the food, softens it, and keeps it from drying out. Tobacco is also kept moist by the addition of polyhydric alcohols. An added feature of polyhydric alcohols is their sweetness. Two particularly effective alcohols added to sweeten sugarless chewing gum are mannitol and sorbitol. Compare the structures of these alcohols with the structure of glucose presented in Chapter 16. It is not surprising that the very similar structures of these polyhydric alcohols and the isomers of glucose produce a similar taste sensation.

D-Sorbitol	D-Mannitol
CH_2OH	CH_2OH
H—C—OH	HO—C—H
HO—C—H	HO—C—H
H—C—OH	H—C—OH
H—C—OH	H—C—OH
CH_2OH	CH_2OH

Kitchen Chemistry

Leavened Bread

Sometimes cooking causes a chemical reaction that releases carbon dioxide gas. The carbon dioxide causes breads and pastries to rise. Yeast has been used since ancient times to make bread rise, and remains of bread made with yeast have been found in Egyptian tombs and the ruins of Pompeii. The metabolic processes of the yeast furnish gaseous carbon dioxide, which creates bubbles in the bread and makes it rise:

$$\underset{\text{GLUCOSE}}{C_6H_{12}O_6} \xrightarrow[\text{from yeast}]{\text{Zymase}} \underset{\text{(GAS)}}{2CO_2} + \underset{\text{ETHANOL}}{2C_2H_5OH}$$

When the bread is baked, the CO_2 expands even more to produce a light airy loaf.

Carbon dioxide can be generated in cooking by other processes. For example, baking soda (which is simply sodium bicarbonate, $NaHCO_3$, a base) can react with acidic ingredients in a batter to produce CO_2 (Figure 21–8).

Leavened bread is as old as recorded history.

$$NaHCO_3 + H^+ \longrightarrow Na^+ + H_2O + CO_2(\text{gas})$$

Baking powders contain sodium bicarbonate and an added acid salt or a salt which hydrolyzes to produce an acid. Some of the compounds used for this purpose are potassium hydrogen tartrate, $KHC_4H_4O_6$, calcium dihydrogen phosphate monohydrate, $Ca(H_2PO_4)_2 \cdot H_2O$, and sodium acid pyrophosphate, $Na_2H_2P_2O_7$. The reactions of these white, powdery salts

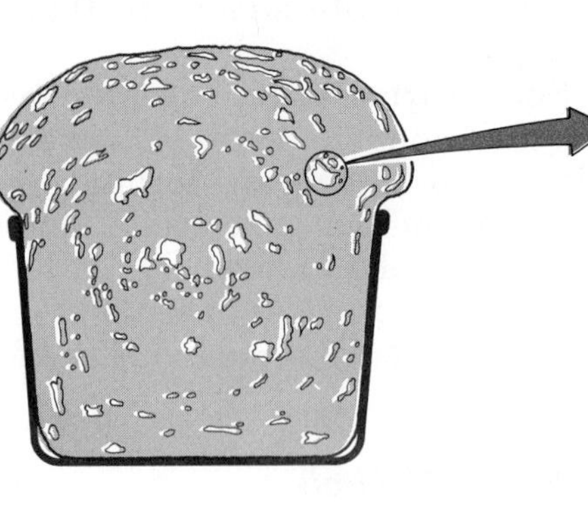

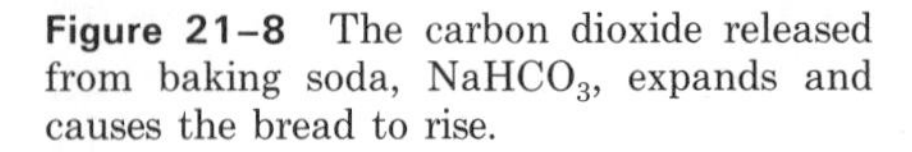

Figure 21–8 The carbon dioxide released from baking soda, $NaHCO_3$, expands and causes the bread to rise.

with sodium bicarbonate are similar, although the compounds all have somewhat different appearances. For example:

$$KHC_4H_4O_6 + NaHCO_3 \xrightarrow{\text{Water}} KNaC_4H_4O_6 + H_2O + CO_2(\text{gas})$$

Cooking and Precooking—"Preliminary Digestion"

Cooking starts the digestive process, although it is rarely needed for this purpose. Since it may destroy nutrients, it is done mostly for pleasure.

The cooking process involves the partial breakdown of proteins or carbohydrates by means of heat and hydrolysis (Figure 21–9). The polymers which must be degraded if cooking is to be effective are the carbohydrate cellular wall materials in vegetables and the collagen or connective tissues in meats. Both types of polymers are subject to hydrolysis in hot water or moist heat. In either case, only partial depolymerization is required. A pan of glucose instead of baked bread or a skillet of amino acids in place of broiled steak is not the goal of cooking. However, the partial hydrolysis of starch releases some glucose, which gives the food a sweeter taste. The partial hydrolysis into smaller fragments also breaks many of the bonds which must be broken before food can be digested. As a consequence, cooked food is easier to digest—that is, to break down into its monosaccharide or amino acid units. These basic units were discussed in Chapter 16. Of course, overcooking can and does destroy some nutrients, such as vitamins.

In recent years several precooking additives have become popular; the ***meat tenderizers*** are a good example. These are simply enzymes which catalyze the breaking of peptide bonds in proteins via hydrolysis at room temperature. As a consequence, the same degree of "cooking" can be obtained in a much shorter heating time. Meat tenderizers are usually

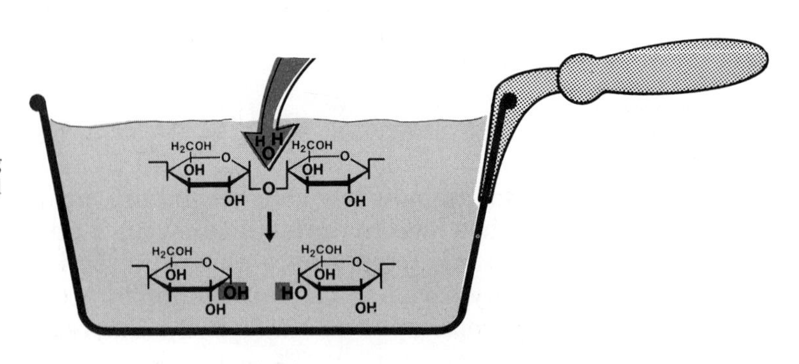

Figure 21–9 The hydrolysis of starch during the cooking of foods such as potatoes and rice.

plant products such as papain, a proteolytic (protein-splitting) enzyme from the unripe fruit of the papaw tree. Papain has considerable effect on connective tissue, mainly collagen and elastin, and shows some action on muscle fiber proteins. On the other hand, microbial protease enzymes (from bacteria, fungi, or both) have considerable action on muscle fibers. A typical formulation for the surface treatment of cuts of beef contains 2 percent commercial papain or 5 percent fungal protease, 15 percent dextrose, 2 percent monosodium glutamate (MSG), and salt.

Vitamins and Minerals—Nutrient Supplements

A food additive that has received considerable attention is the nutrient supplement. The addition of iodide (as KI) to common table salt to prevent goiter is one example. Many food products now contain added vitamins: vitamin D is added to milk, the B vitamins are added to wheat flour, vitamin C is added to certain beverages, and minerals are added to many cereal products. Since these compounds are needed by the body, they fall into a somewhat different category from the other nonessential additives. The purposes of the various vitamins are given in Table 21-1.

Professor Linus Pauling's announcement, in 1971, that vitamin C appears to prevent the common cold has focused much attention on this vitamin. However, since some vitamins (D, for example) can cause adverse effects when taken in excess, *the unlimited use of vitamins is not advised.* Here, as with other food additives, the amount, the substance itself, and its synergistic effects must all be considered before the additive can be labeled safe to use.

Only fat-soluble vitamins A and D cause adverse effects; water-soluble vitamins are excreted in the urine.

TABLE 21-1 The Purposes of Vitamins*

Vitamin	Biochemical Function	Deficiency Effects
A	Regeneration of rhodopsin (visual purple)	Excessive light sensitivity; night blindness; increased susceptibility to infection
B_1 (Thiamine)	Nerve activity; carbohydrate metabolism	Beriberi; serious nervous disorders; muscular atrophy; serious circulatory changes
B_2 (Riboflavin)	Coenzyme; affects sight	Sores on the lips; bloodshot and burning eyes; excessive light sensitivity
B_3 (Nicotinamide)	Metabolism of ATP	Stunted growth; pellagra
B_5 (Pantothenic acid)	Growth factor; component of coenzyme A	Retarded growth
B_6 (Pyridoxine)	Coenzyme; metabolism of fatty acids	Retarded growth; anemia; leukocytosis; insomnia; lesions about eyes, nose, mouth
B_7 (Biotin)	Growth factor; affects scaly and greasy skin; CO_2 fixation; coenzyme	Scaly and greasy skin
B_9 (Folic acid)	Coenzyme; tyrosine metabolism	Anemia
B_{12} (Cobalamin)	Growth factor; involved in synthesis of DNA, hemoglobin	Degeneration of the spinal cord, pernicious anemia
C (Ascorbic acid)	Coenzyme; reducing agent; cholesterol metabolism	Hemorrhages; lesions in the mouth; muscular degeneration; sterility; scurvy
D	Ca and P metabolism	Abnormal development of bones and teeth; rickets
E (α-Tocopherol)	Antioxidant; cofactor between cytochromes b and c	Sterility
K	Synthesis of prothrombin, coenzyme Q	Hemorrhages; slow clotting of blood

*The vitamins participate in biochemical changes concerned with the utilization of foodstuffs.

The GRAS List

To protect the food consumer, the U.S. Government has enacted many pieces of legislation called the Food, Drug, and Cosmetic Acts. The Federal Food and Drug Administration (FDA) administers these laws and, in this sense, is the guardian of our health. For example, the FDA is required by law (Delaney Amendments of 1958) to withdraw sanction of or to ban any food additive which "is found to induce cancer when ingested by man or animal." To recognize a food additive as safe, the substance must not be harmful in the amounts used in the food.

Because of the scope of the tests the most difficult point to check is the synergistic action of food additives. An additive may produce no harmful effects by itself but, in combination with other substances in the foodstuff, effects could be amplified much beyond those of the individual chemicals. Our increasing understanding of molecular structure and interaction facilitates the predictability of synergism.

At this time the FDA lists about 600 chemical substances ***"generally recognized as safe"*** **(*GRAS*)** for their intended use. A small portion of this list is given in Table 21–2. It must be emphasized that an additive on the GRAS list is safe *only if it is used in the amounts and in the foods specified.* The GRAS list was published in several installments in 1959 and 1960. It was compiled by asking experts in nutrition, toxicology, and related fields to give their opinions about the safety of using various materials in foods. Since its publication, few substances have been added to the GRAS list and some, such as the cyclamates, have been removed.

The GRAS list is a noble effort—but at the present time it is not foolproof.

It is evident, in view of the more than 2500 known food additives, that many more chemicals than those that appear on the GRAS list are approved (or at least, not banned) for use as food additives by the FDA. It is quite expensive to introduce a new food additive with the approval of the FDA. Allied Chemical Corporation began research in 1964 on a new synthetic food color, Allura Red AC. It was approved by the FDA and went on the market in 1972. The cost for introducing this product was $500,000 and about half of this amount was spent on safety testing.

Each food additive is included in a consumer foodstuff for a specific purpose. Table 21–2 is organized according to some of the functions of the additives. Some additives have more than one function; only the major function is given in this table. An understanding of some of the chemistry, among other things, can make the morning's reading of the cereal box more enlightening.

SELF-TEST 21-B

1. Which has the sweetest taste when an equal amount of each is tasted: table sugar, a cyclamate, or saccharin?________________
2. Does sweetness have a qualitative or a quantitative base? ________________
3. Which widely used sweetener has been subject to the most criticism as a health hazard?________________

TABLE 21–2 A Partial List of Food Additives Generally Recognized as Safe*

Food Colors

Annatto (yellow)
Carbon black
Carotene (yellow-orange)
Cochineal (red)
Food dyes and colors
Red No. 2‡
Red No. 3
Blue No. 1
Titanium dioxide (white)

Acids, Alkalies, and Buffers

Acetates: Ca, K, Na
Acetic acid
Calcium lactate
Citrates: Ca, K, Na
Citric acid
Fumaric acid
Lactic acid
Phosphates, CaH, Ca_3, Na_2, Na_3, NaAl
Potassium acid tartrate
Sorbic acid
Tartaric acid

Surface Active Agents

Cholic acid
Glycerides: mono- and diglycerides of fatty acids
Polyoxyethylene (20) sorbitan monopalmitate
Sorbitan monostearate

Polyhydric Alcohols

Glycerol
Sorbitol
Mannitol
Propylene glycol

Nonnutritive Sweeteners

Saccharin: NH_4^+, Ca^{2+}, Na^+
Aspartame

Flavor Enhancers

Monosodium glutamate (MSG)
5′-nucleotides
Maltol

Preservatives

Benzoic acid
  Na benzoate
Methylparaben
Oxytetracycline
Propylparaben
Propionic acid
  Ca propionate
  Na propionate
Sorbic acid
  Ca sorbate
  K sorbate
  Na sorbate
Sulfites, Na^+, K^+

Antioxidants

Ascorbic acid
  Ca ascorbate
  Na ascorbate
Butylated hydroxyanisole (BHA)
Butylated hydroxytoluene (BHT)
Lecithin
Propyl gallate
Sulfur dioxide and sulfites
Trishydroxybutyrophenone (THBP)

Sequestrants

Citrate esters:
  isopropyl, stearyl
Citric acid
EDTA, Ca and Na salts
Pyrophosphate, Na^+
Sorbitol
Tartaric acid
  Na tartrate

Stabilizers and Thickeners

Agar-agar
Algins: NH_4^+, Ca^{2+}, K^+, Na^+
Carrageenin
Gum acacia
Gum tragacanth
Sodium carboxymethyl cellulose

Flavorings

Acetanisole (slight haylike)
Allyl caproate (pineapple)
Amyl acetate (banana)
Amyl butyrate (pearlike)
Bornyl acetate (piney, camphor)
Carvone (spearmint)
Cinnamaldehyde (cinnamon)
Citral (lemon)
Ethyl cinnamate (spicy)
Ethyl formate (rum)
Ethyl propionate (fruity)
Ethyl vanillin (vanilla)
Eucalyptus oil (bittersweet)
Eugenol (spice, clove)
Geraniol (rose)
Geranyl acetate (geranium)
Ginger oil (ginger)
Linalool (light floral)
Menthol (peppermint)
Methyl anthranilate (grape)
Methyl salicylate (wintergreen)
Orange oil (orange)
Peppermint oil (peppermint) (menthol)
Pimenta leaf oil (allspice) (eugenol, cineole)
Vanillin (vanilla)
Wintergreen oil (wintergreen) (methyl salicylate)

*For precise and authoritative information on levels of use permitted in specific applications, the regulations of the U.S. Food and Drug Administration and the Meat Inspection Division of the U.S. Department of Agriculture should be consulted.

‡Red No. 2 was removed from the GRAS list in 1976.

If past history is a guide, at least some of these compounds will be taken off the GRAS list in the future.

4. The gas that leavens bread is ______________________.

5. Cooking ______________ some chemical bonds.

6. Meat tenderizers are enzymes that assist in breaking peptide bonds in proteins; hence the one word ______________________ is apt.

7. Is it true or false that excesses of all of the vitamins are equally dangerous? ______________

8. The GRAS list is the FDA list of foods that are ______________________ ______________________ ________ ______________.

9. Hydrogen bonding generally plays a very important role in the action of surface ________________ agents.

10. In order to produce CO_2 in bread, bicarbonate reacts with a(n) ____________________________.

PARADE OF MEDICINES

The average life expectancy in the United States has risen from age 49 in 1900 to age 70 in 1970. It is expected to go beyond the age of 79 by the year 2000. The major contributing factor is the widespread use of a large assortment of new medicinal compounds.

Medicines rise and fall in popularity.

The contents of the medicine cabinet have changed drastically in the past few years. As the parade of new drugs continues, the indispensable drugs of one decade frequently become obsolete in the next. A survey of physicians shortly before World War I revealed the ten most essential drugs (or drug groups) to be ether, opium and its derivatives, digitalis, diphtheria antitoxin, smallpox vaccine, mercury, alcohol, iodine, quinine, and iron. When another survey was made at the end of World War II, at the top of the list were sulfonamides, aspirin, antibiotics, blood plasma and its substitutes, anesthetics and opium derivatives, digitalis, antitoxins and vaccines, hormones, vitamins, and liver extract. Today there is such a wide array of medicinal chemicals that similar surveys have not been significant statistically, but drugs for reducing fever, relieving pain, and fighting infection still head the list in all areas of medical practice (Tables 21–3 and 21–4).

Many new drugs are tested each year, but few ever reach the market place.

Medicines are costly (Table 21–4), as anyone knows when he has to buy them. Still, the research required to put a new drug on the market is becoming increasingly more complex because of recent government regulations, and this alone is some justification for the prices of medicines. In 1970, for example, drug producers in this country tested 3620 new drugs, although only 16 new products came onto the market.

TABLE 21–3 The Most Widely Prescribed Generic Drugs in 1973*

Generic Drug	Medical Use
Tetracycline HCl	Antibiotic
Ampicillin	Antimicrobial (kills bacteria but is not derived from a plant, as is an antibiotic)
Phenobarbital	Sedative, hypnotic, anticonvulsant
Thyroid	Increases rates of metabolism
Prednisone	Antiinflammatory, antiallergic agent similar to cortisone
Digoxin	Decreases the rate of the heartbeat but increases the force of the heartbeat, similar to digitalis
Meprobamate	Tranquilizer
Erythromycin	Antimicrobial
Penicillin G potassium	Antibiotic
Nitroglycerin	Dilates the blood vessels of the heart
Penicillin VK	Antibiotic
Quinidine sulfate	Slows the heartbeat, also used for malaria and hiccups
Paregoric	A tincture (alcoholic solution) of camphorated opium used as an analgesic
Reserpine	Tranquilizer
Nicotinic acid (Niacin)	Dilates blood vessels; also, essential B vitamin with antipellagra activity

*The generic name for a drug is its generally accepted chemical name rather than a specific brand name. For example, the generic drug, tetracycline HCl, is sold under such brand names as Achromycin V, Ambryon, Artomycin, Dumocyclin, Subamycin, Unimycin, and others. Medical doctors can either prescribe the generic name or a brand name. If the generic name is used, the prescription often is cheaper, particularly if the drug is not protected by patents and can be manufactured and marketed competitively by several companies. Only about 12% of the prescriptions written in 1974 used generic names.

The generic name for a drug is its widely accepted chemical name.

TABLE 21–4 Drug Sales in the United States for Several Important Therapeutic Groups

Therapeutic Group	1960*	1970*
Analgesics (pain reducers)	$ 65	$250
Antacids	41	95
Antibiotics	375	660
Cough and cold preparations	65	200
Hormones	141	405
Tranquilizers	140	490
Sulfonamides	44	50

*Manufacturers' sales, in millions of dollars.

Antacids

The contents of the stomach are highly acidic.

The walls of a human stomach contain thousands of cells which secrete hydrochloric acid, the main purposes of which are to suppress growth of bacteria and to aid in the hydrolysis (digestion) of certain foodstuffs. Normally, the stomach's inner lining is not harmed by the presence of this hydrochloric acid, since the mucosa, the inner lining of

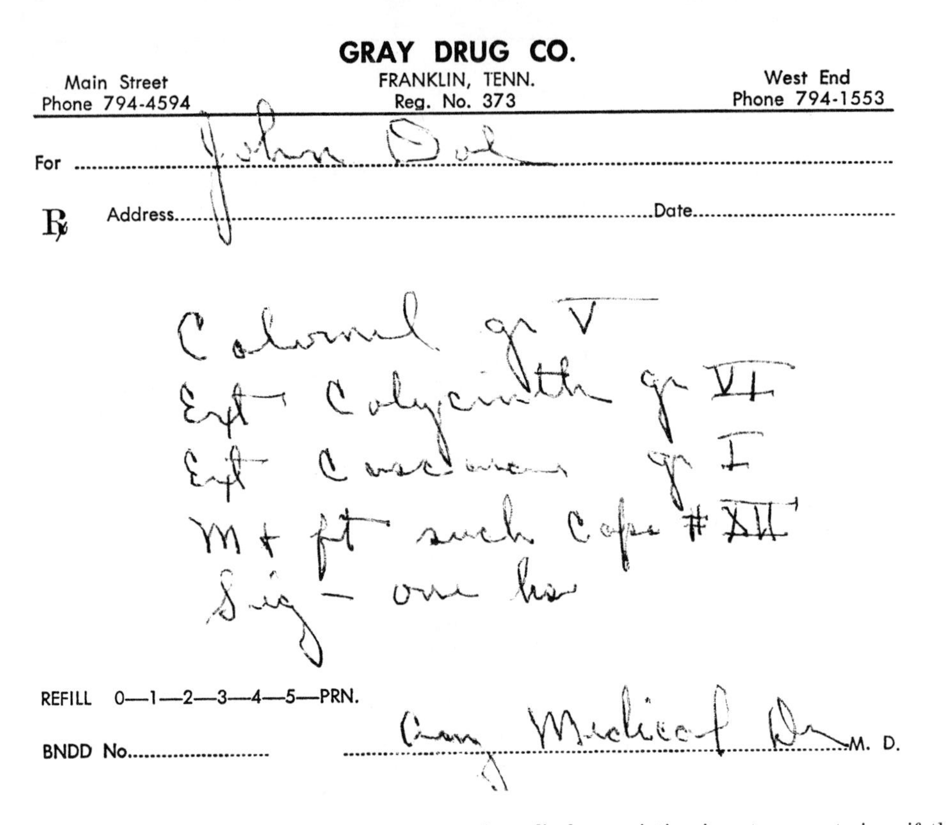

GRAY DRUG CO.

Main Street
Phone 794-4594

FRANKLIN, TENN.
Reg. No. 373

West End
Phone 794-1553

For John Doe

℞ Address.......... Date..........

Calomel gr V
Ext Colycinth gr VI
Ext Cascara gr I
M + ft such Caps # XII
Sig — one hs

REFILL 0—1—2—3—4—5—PRN.

BNDD No.......... Any Medical Dr M. D.

Figure 21–10 A medical prescription is not so mysterious if the handwriting is legible and if the symbols are understood. For example, the R_x (Latin for *recipe*) above calls for 5 grains (or 0.32 gram) of calomel (Hg_2Cl_2), 6 grains (or 0.39 gram) of extract of colocynth, 1 grain (or 0.06 gram) of extract of cascara per each of 12 capsules. The fourth line of the prescription means mix and fix such that each of the 12 capsules contains the formulation given in the first three lines. The label should signify "one at bedtime." The BNDD No. at the bottom is for the doctor's number, as assigned by the Bureau of Narcotics and Dangerous Drugs. His number must be on the prescription if he prescribes any narcotic or other dangerous drug. In this illustration, this relatively new practice is contrasted with the relatively old recipe of the prescription.

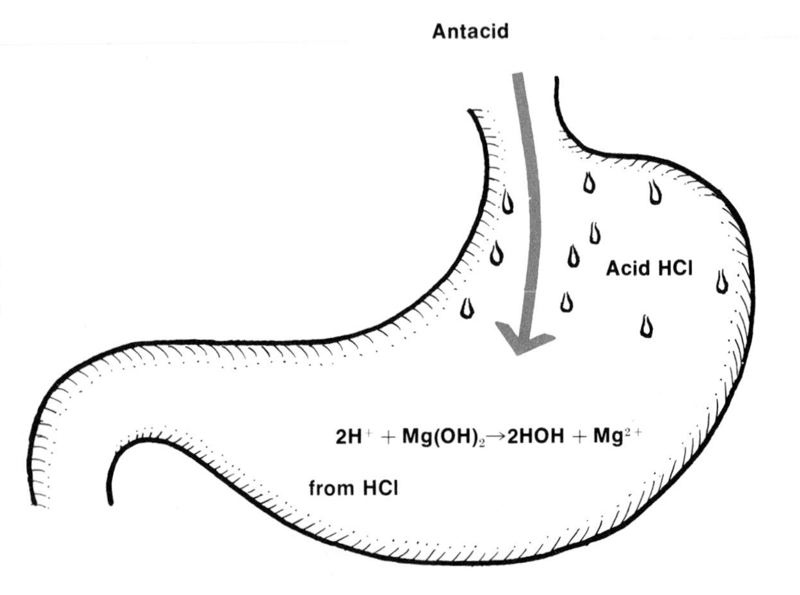

Figure 21–11 The chemical action of milk of magnesia. This antacid, which is magnesium hydroxide, neutralizes acid in the stomach.

TABLE 21–5 The Chemistry of Some Antacids

Compound	Reaction in Stomach	Comments
Magnesium oxide MgO	$MgO + 2H^+ \longrightarrow Mg^{2+} + H_2O$	MgO is white and tasteless
Milk of magnesia $Mg(OH)_2$ in water	$Mg(OH)_2 + 2H^+ \longrightarrow Mg^{2+} + 2H_2O$	The water suspension has an unpleasant chalky consistency
Calcium carbonate $CaCO_3$	$CaCO_3 + 2H^+ \longrightarrow Ca^{2+} + H_2O + CO_2$	Calcium carbonate is purified limestone
Sodium bicarbonate $NaHCO_3$	$NaHCO_3 + H^+ \longrightarrow Na^+ + H_2O + CO_2$	Baking soda, like $CaCO_3$, produces CO_2 gas in the stomach
Aluminum hydroxide $Al(OH)_3$	$Al(OH)_3 + 3H^+ \longrightarrow Al^{3+} + 3H_2O$	$Al(OH)_3$ is a clear gel
Dihydroxyaluminum sodium carbonate $NaAl(OH)_2CO_3$	$NaAl(OH)_2CO_3 + 4H^+ \longrightarrow Na^+ + Al^{3+} + 3H_2O + CO_2$	Sold as Rolaids, will not ordinarily cause pH to go above 5
Sodium citrate $Na_3C_6H_5O_7 \cdot 2H_2O$	$Na_3C_6H_5O_7 \cdot 2H_2O + 3H^+ \longrightarrow 3Na^+ + H_3C_6H_5O_7 + 2H_2O$	Mild

the stomach, is replaced at the rate of about a half million cells per minute. When too much food is eaten the stomach often responds with an outpouring of acid which lowers the pH to a point where discomfort is felt. Fortunately, the mucosa is protected by cells which contain a thick fatty layer.

When presented with this problem of minor stomach upset, some people respond by taking one of the commonly available antacids.

If the reduction of acidity is too great the stomach responds by secreting an excess of acid. This is "acid rebound."

Antacids are compounds used to decrease the amount of hydrochloric acid in the stomach. The normal pH of the stomach ranges from 1.2 to 0.3. Various compounds with basic properties can decrease stomach acidity. Some compounds used for antacid purposes and their modes of action are given in Table 21–5.

Analgesics

Analgesics relieve pain, but they are harmful in large doses.

The most important group of therapeutic agents is the *analgesics,* or *pain killers.* Most people need these compounds at one time or another. When we have a headache we take aspirin. When we have a tooth filled or extracted, the dentist uses novocaine. Intense suffering requires a strong pain killer, such as codeine or morphine. While these compounds are immensely useful, they are nevertheless dangerous if taken or used improperly. Most of them can even become killers if taken in overdose.

Early man may well have used opium. Although not all opium derivatives have therapeutic value, most of them are very efficient pain killers. Their chief disadvantage lies in their addictive properties.

Opium is obtained from the opium poppy by scarring the seed pod with a sharp instrument. From this scar flows a sticky mass which contains about 20 different compounds called ***alkaloids*** (organic nitrogenous bases which contain basic nitrogen atoms like ammonia). About 10 percent of this mass is the alkaloid ***morphine,*** which is primarily responsible for opium's effects.

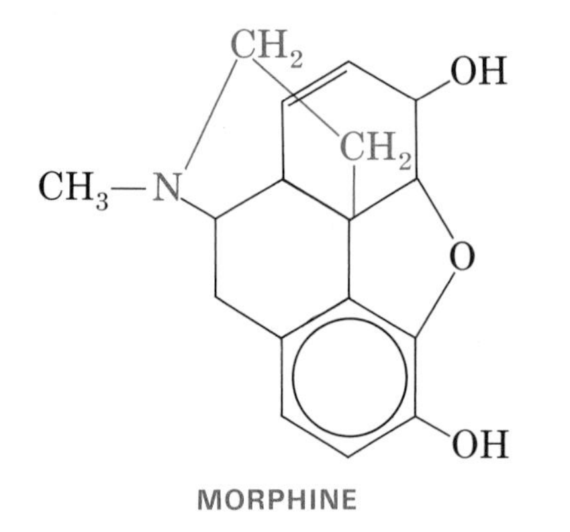

MORPHINE

Morphine and its derivatives are addictive drugs.

Two derivatives of morphine are of interest. One of these is ***codeine,*** a methyl ether of morphine, which is less addictive than morphine and is about as powerful an analgesic. The other compound is ***heroin,*** the diacetate ester of morphine. Heroin is much more addictive than morphine and for that reason finds no medical uses in the United States.

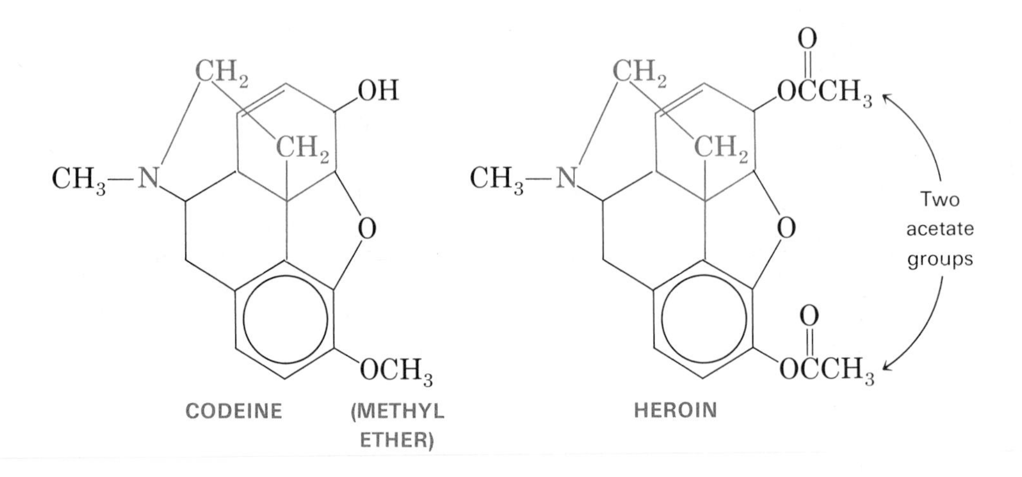

CODEINE (METHYL ETHER) HEROIN

The action of morphine, codeine, and other powerful analgesics is not completely understood. Over the years, numerous chemically similar drugs have been synthesized in attempts to test various theories of the mechanisms of action of these drugs.

Analgesics may or may not be habit-forming.

One of the most effective substitutes for morphine is ***meperidine,*** first reported in 1931, and now sold as Demerol. It is less addictive than morphine. Two other relatively strong pain relievers used today are ***pentazocine*** (Talwin) and ***propoxyphene*** (Darvon). Talwin is slightly addictive while Darvon has not been shown to be. Note that in the structures of these compounds there is a strong resemblance (colored portion) to the morphine structure.

It is not known why some analgesics are addictive.

Many of the ***local analgesics,*** or local anesthetics, are nitrogen compounds, like the alkaloids (Table 21–6). Local analgesics include the naturally occurring ***cocaine,*** derived from the leaves of the coca plant of South America, and the familiar novocaine (procaine). All of these drugs

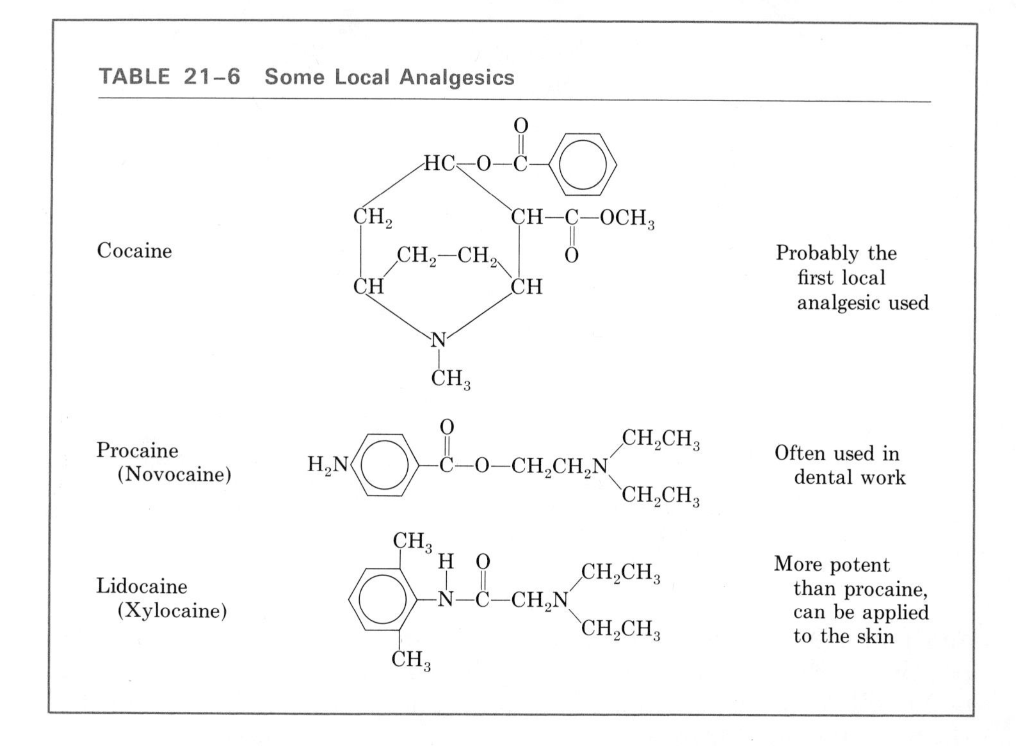

TABLE 21–6 Some Local Analgesics

Compound	Structure	Remarks
Cocaine		Probably the first local analgesic used
Procaine (Novocaine)		Often used in dental work
Lidocaine (Xylocaine)		More potent than procaine, can be applied to the skin

act by some mechanism of blockage of the nerves which transmit pain. Acetylcholine appears to be the "opener of the gate" for sodium (Na^+)

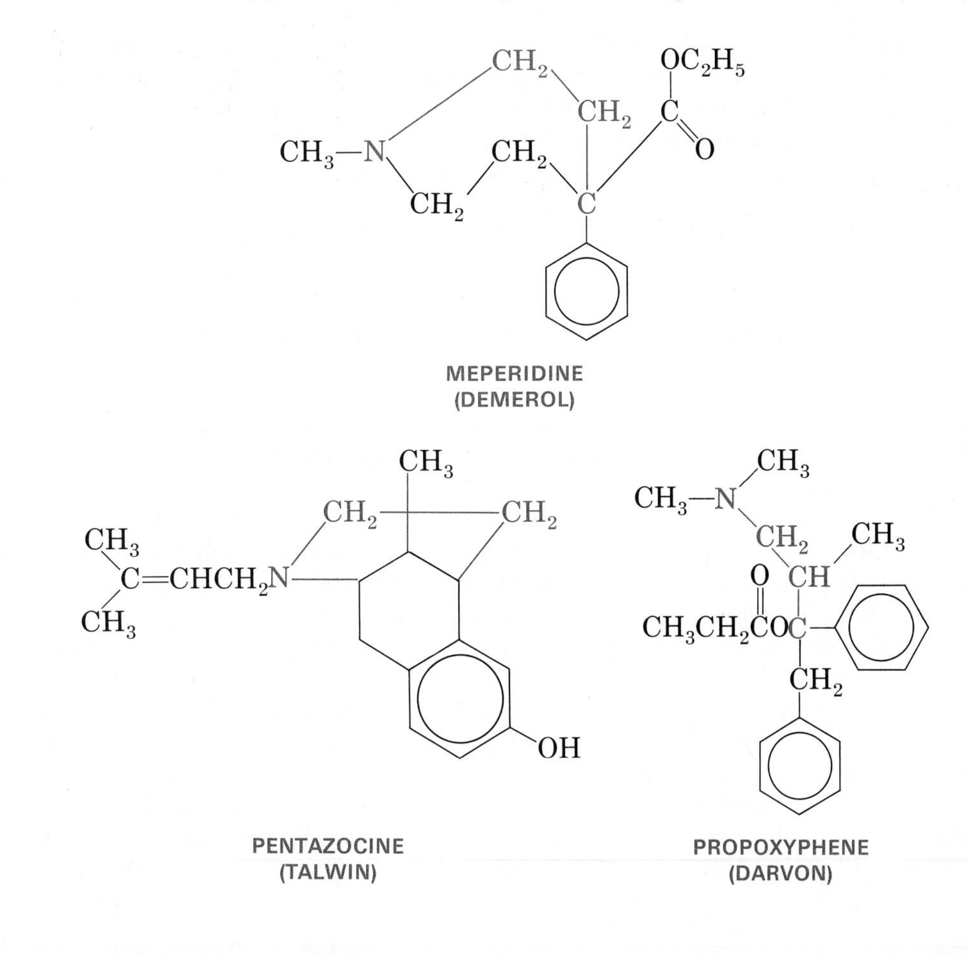

MEPERIDINE
(DEMEROL)

PENTAZOCINE
(TALWIN)

PROPOXYPHENE
(DARVON)

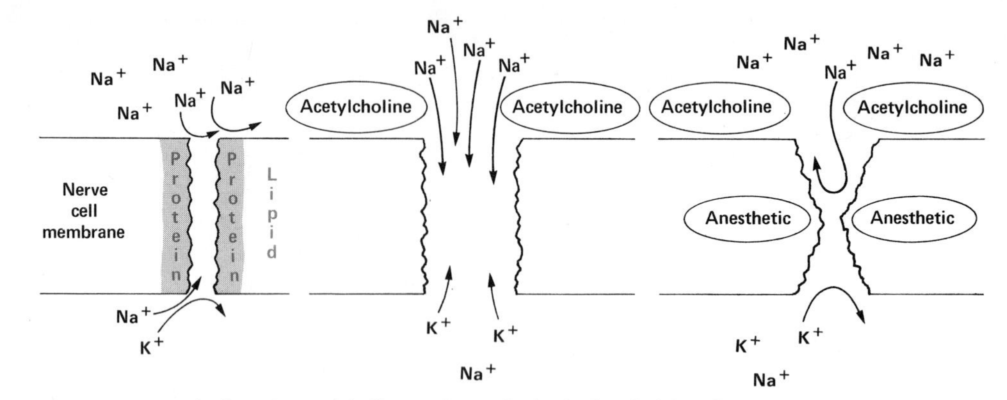

Figure 21–12 Action of acetylcholine and anesthetics in depolarizing the membrane of a nerve cell. Acetylcholine makes it possible for sodium and potassium ions to neutralize the negative charge associated with a nerve impulse so another impulse can be transmitted. Anesthetics block the action of acetylcholine and do not allow repetitive impulses to travel along the nerve.

and potassium (K^+) ions to flow into a nerve cell (Figure 21–12). The positive ions neutralize the charge on the cell membrane and prepare the nerve cell for another impulse. The anesthetics contain a nonpolar part of the molecule that fits into the nonpolar fatty tissue of the cell wall. In the process they constrict the pores, reduce the Na^+ flow, and nullify the effect of acetylcholine. Higher concentrations may also affect the sense of touch, including the ability to estimate temperature.

Antipyretic compounds reduce fever.

There are times when milder, general analgesics are required, and few compounds work as well for as many people as ***aspirin.*** Highlights in the history of aspirin go back to 1763, when Edward Stone noticed that the bark of the willow, when chewed, helped relieve the symptoms of malaria. In 1838 Raffaele Pivia isolated salicylic acid from the active ingredient in the willow. It turned out to be this compound which had the analgesic and antipyretic properties attributed to the willow bark.

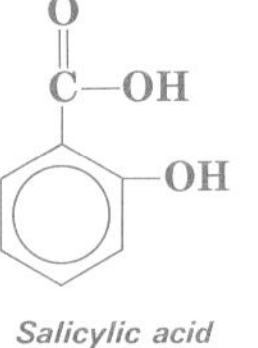

Salicylic acid

Attempts to administer salicylic acid and its salts to patients proved unsuccessful because of their disagreeable tastes. By 1893, Felix Hofmann, a chemist working for the Bayer Company in Germany, discovered a way to attach an acetyl group ($—O\overset{O}{\overset{\|}{C}}CH_3$) in place of the hydroxyl hydrogen to make acetylsalicylic acid. The importance of the acetyl group is that it renders the molecule relatively tasteless and reduces the acidity enough so that it can be taken orally.

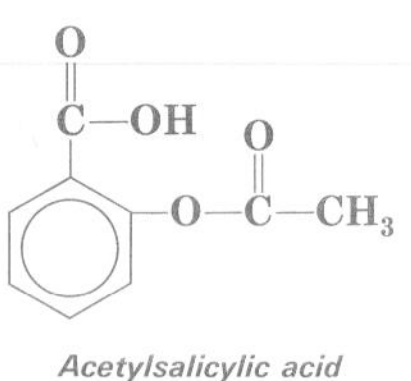

Acetylsalicylic acid (aspirin)

The synthesis of aspirin was outlined in Chapter 14. Each year, about 40 million pounds are manufactured in the United States.

Taking an aspirin tablet appears to be a simple act; yet much research goes into making a tablet which will cause a minimum of bad side effects for the user. The greatest danger presented by aspirin is that of stomach bleeding, caused when an undissolved aspirin tablet lies on the stomach wall. As the aspirin molecules pass through the fatty layer of the mucosa, they appear to injure the cells, causing small hemorrhages. The blood loss for most individuals taking two 5-grain tablets is between 0.5 ml and 2 ml. Some people are more susceptible. Early aspirin tablets were not particularly fast-dissolving, which aggravated this problem

Many aspirin tablets contain starch to hasten their disintegration in the stomach.

greatly. Today, aspirin tablets are formulated to disintegrate and dissolve quickly, although dissolving the tablet in a little water might not be a bad idea.

Aspirin and Headaches

Americans spend more than $250 million a year on headache remedies. More than 200 kinds of tablets and powders are on the market for headache relief. According to the National Health Service, one out of every 12 Americans has severe headaches regularly.

The basic ingredient of most headache formulations is aspirin. Commercially available headache remedies include caffeine, antacids, extra pain killers, antihistamines, vitamins, and tranquilizers.

Headaches (Figure 21–13) are triggered by such diverse factors as emotional problems (tension), heredity (migraine), and, less frequently, by eye strain, acute sinus conditions, inflammation of the lining of the brain, infection of a cranial nerve, carbon monoxide poisoning, and poorly positioned teeth.

In tension headaches, muscles are tightened and strained. An overworked head muscle can ache just as much as an overworked arm or leg muscle. The tight muscle may also squeeze arteries and reduce blood flow through the muscle, adding to the pain.

Experimentation has established that aspirin produces its effect on the central nervous system, and that salicylic acid from the hydrolysis of aspirin is the active chemical. Besides relieving pain, the antipyretic effects of aspirin apparently affect the hypothalamus gland, though hard

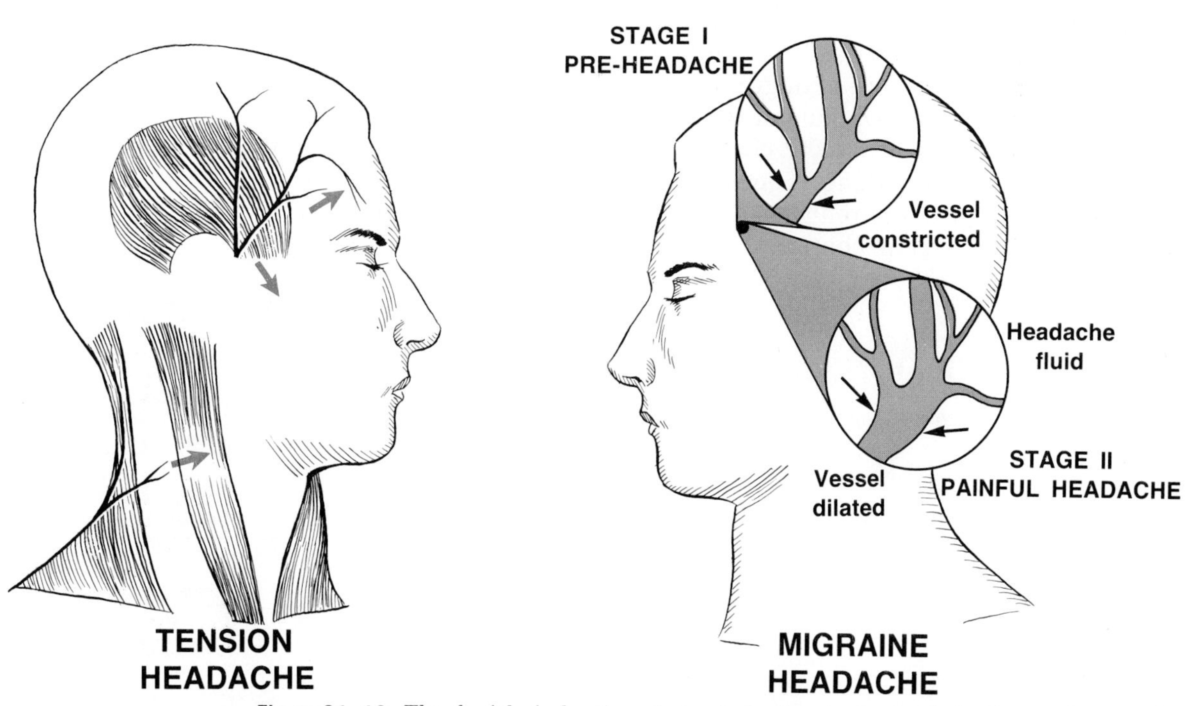

Figure 21–13 The physiological action of a headache. In a tension headache the muscles constrict the arteries. The muscles tire and ache, and this causes pain which spreads throughout the head. In a migraine headache the arteries constrict (this is painless) and then expand (this is painful). (In McGraw-Hill Yearbook of Science—Adapted by permission from The New York Times.)

data are not available to establish this action. The hypothalamus gland, attached to the pituitary gland near the center of the brain, is the thermostat of the body. Fever is thought to be produced by a chemical secreted by white blood cells when they engulf bacteria. The chemical enters lymph vessels, migrates to the brain, and resets the thermostat to raise the body temperature. Aspirin in some way adjusts the body temperature back to normal.

Aspirin appears to be able to regulate the hypothalamus gland.

The pain of a migraine headache appears to be caused by a headache "fluid" and constriction followed by expansion of the arteries in the head. Each time the heart pumps, the arteries expand further and more pain is produced by the pulsations. The headache fluid contains two small proteins, bradykinin and neurokinin, which are believed to make nerves sensitive to pain. When the substances are extracted from the headache fluid and injected elsewhere in the body, the person will sense pain in the new area. The constriction and subsequent expansion of the arteries may be initiated by the release of serotonin, which is known to contract smooth muscle. Normally the serotonin is bound, but its sudden release into the bloodstream could cause the initial contraction of the arteries. An enzyme, monoamine oxidase, metabolizes the serotonin, causing the arteries to relax and expand. When the normal supply of serotonin is depleted, the arteries remain expanded until a new supply is made available. Australian scientists reported in 1967 that the content of serotonin in the blood does fall sharply at the onset of a migraine attack.

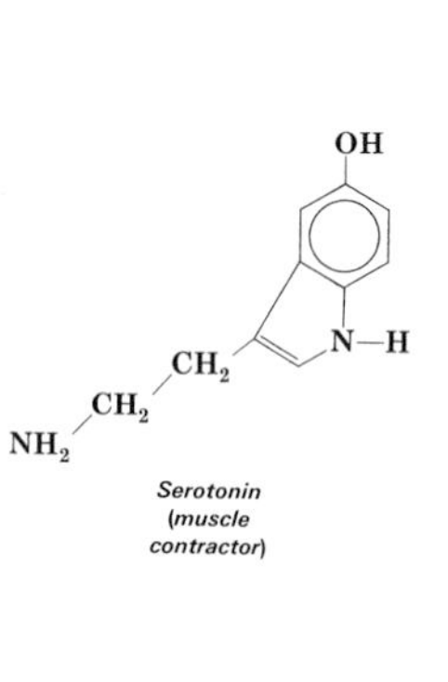

Serotonin (muscle contractor)

The migraine headache can be relieved by ergotamine tartrate if it is taken early enough. It acts by constricting the muscles of the blood vessels, preventing painful stretching. This property of ergotamine tartrate has been known for over 40 years. A newer drug, methysergide, is effective in about 70 percent of patients. It is taken between headaches. Neither drug is always effective; both are sometimes associated with serious side effects when used for long periods.

The hereditary nature of migraine headaches is indicated by a study made by Dr. Adrian M. Ostfield of the University of Illinois. Dr. Ostfield reports that there is a 70 percent chance that the child will have migraine if both parents have migraine. If one parent has migraine, there is a 45 percent chance; if neither parent has migraine but there is a history of it in the family, a 25 percent chance.

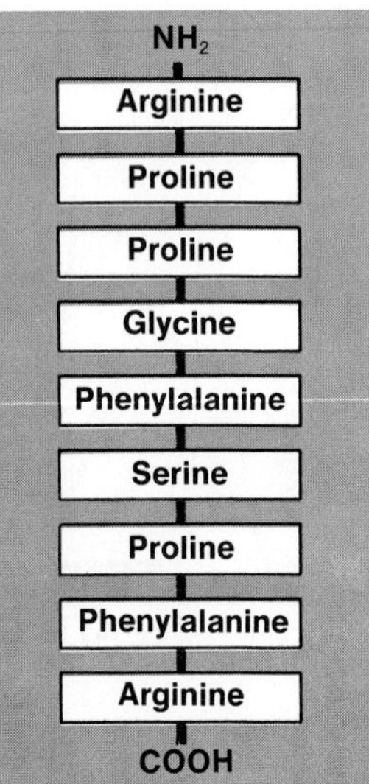

Figure 21–14 The amino acid sequence in bradykinin, a pain-causing protein, containing nine amino acids.

Antiseptics and Disinfectants

Pathogenic bacteria cause many illnesses.

An antiseptic is a compound which prevents the growth of microorganisms. It now has the legal meaning ***"germicide,"*** or a compound which *kills* microorganisms. A disinfectant is a compound which destroys pathogenic bacteria or microorganisms, but usually not bacterial spores. These materials are generally poisonous, and, therefore, suitable only for external use as on the skin or a wound.

$$\mathrm{R{-}\overset{\displaystyle R}{\underset{\displaystyle R}{N^{+}}}{-}R_1Cl}$$

A quaternary ammonium chloride

Some common germicides are listed in Table 21–7. Some of these, such as the halogens, sodium hypochlorite, hydrogen peroxide, and potassium permanganate, are effective because of their oxidizing properties. This is a general property and allows them to oxidize any kind of cell, including human cells. For this reason they are used mostly as disinfectants in destroying the germs on nonliving objects. Phenol (carbolic acid), which is no longer widely used, is readily absorbed by cells and is a general poison. The quaternary ammonium compounds are surface active agents, and their bactericidal ability seems to be related to their ability to weaken the cell wall so the cell contents cannot be contained.

A tincture is an alcoholic solution. Tincture of iodine is a solution of water, iodine, and potassium iodide in ethanol.

One of the newer developments solves the problem of applying antiseptics to children. You may remember the sting of "iodine" when applied to a scratch or a wound. Old-fashioned iodine is a solution of iodine (I_2) in alcohol with a little potassium iodide (KI) to increase the solubility of the iodine. The alcohol causes most of the pain. Now there are polymers, such as polyvinylpyrrolidone, which complex iodine molecules (Figure 21–15); the products (iodophors) are soluble in water. The resultant solution is a very efficient and painless disinfectant. The iodophors are active ingredients in a popular mouthwash and in restaurant glassware disinfectants.

Because most of these compounds are generally toxic to living matter,

TABLE 21–7 Some of the More Common Antiseptics

Iodine	Mercurochrome
Sodium hypochlorite	Metaphen
Potassium permanganate	Merthiolate
Hydrogen peroxide	Pine oil
Iodophors	Soap
Ethanol	Hexylresorcinol
Quaternary ammonium compounds	Hexachlorophene
Chloramine-T	Mercuric chloride
Phenols	

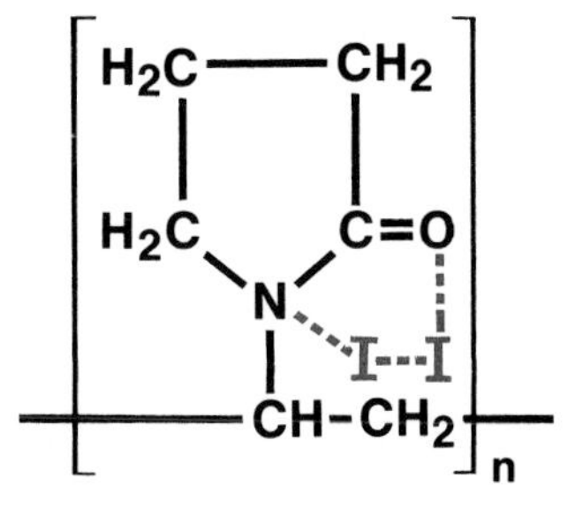

Figure 21–15 An iodophor: complex of polyvinylpyrrolidone (PVP) and iodine.

it is necessary to utilize only dilute solutions and then only on the skin. While they help to prevent the spread of disease, they are practically useless in its treatment because they act nonspecifically against all cells with which they come in contact. They are to be distinguished from antibiotics, which act more selectively against infecting bacteria than against "organic" cells within the human body.

Antimicrobial Medicines

In our time, the quest for drugs to wipe out disease due to microorganisms has virtually been fulfilled by the ***antibiotics*** (Table 21–8). In the original sense, an antibiotic is a substance such as penicillin, produced by a microorganism, that inhibits the growth of another organism. It has become common practice to include synthetic chemicals such as the sulfa drugs in a discussion of antibiotics.

Since the antibiotics are so efficient, they were the first of what came to be called "miracle" drugs. Their job generally is to aid the white blood cells by stopping bacteria from multiplying. When a person is sick or is killed by a disease, it means that the bacteria have multiplied faster than the white blood cells could devour them, and that the bacterial toxins increased more rapidly than the antibodies could neutralize them. The action of the white blood cells and *antibodies* plus an antibiotic is generally enough to repulse an attack of the germs.

An antibody is a specific protein produced to protect the organism from harmful invading molecules (see Chapter 18).

The Sulfa Drugs

Sulfa drugs represent a group of compounds discovered in a conscious search for antibiotics. In 1904, the German chemist Paul Ehrlich (1854–1915; Nobel prize in 1908) realized that infectious diseases could be conquered if toxic chemicals could be found that attacked parasitic organisms within the body to a greater extent than they did host cells. Ehrlich achieved some success toward his goal; he found that certain dyes which were used to stain bacteria for microscopic examination could also kill the bacteria. This led to the use of dyes against African sleeping sickness and arsenic compounds against syphilis. The mass outbreak of typhus during World War I, the loss of many wounded due to secondary bacterial infection, and the great influenza epidemic of 1917–1918 prepared the medical world for discoveries that might eradicate infectious diseases.

TABLE 21–8 Deaths per 100,000 Americans Due to Different Causes

	1900	1970
Infectious diseases	500	50
Influenza and pneumonia	210	30
Diphtheria	40	fewer than 1
Typhoid and paratyphoid	30	fewer than 1
Whooping cough	10	fewer than 1
Gastrointestinal problems	150	fewer than 10

The synthesis of sulfanilamide is outlined in Chapter 14.

Large doses of sulfa drugs are required compared to the doses of "true antibiotics."

After experimenting with several drugs, Gerhard Domagk, a pathologist in the I. G. Farbenindustrie Laboratories in Germany, found, in 1935, that prontosil, a coloring matter or dye, was active against bacterial infection in mice. *Prontosil* as such is not effective in killing bacteria, but it is changed to *sulfanilamide,* which is effective.

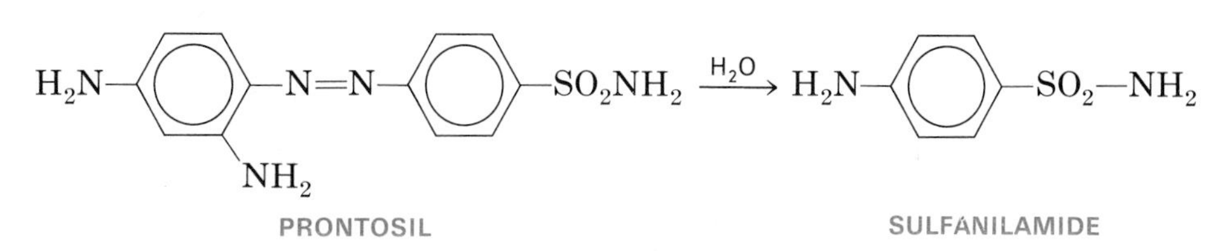

This discovery led to the synthesis and testing of a large number of related compounds in the search for drugs which were more effective or less toxic to the infected animal. By 1964, more than 5000 sulfa drugs had been prepared and tested.

A sulfa drug mimics an essential compound.

Sulfa drugs inhibit bacteria by preventing the synthesis of folic acid, a vitamin essential to their growth. The drugs' ability to do this apparently lies in their structural similarity to a key ingredient in the folic acid synthesis, para-aminobenzoic acid.

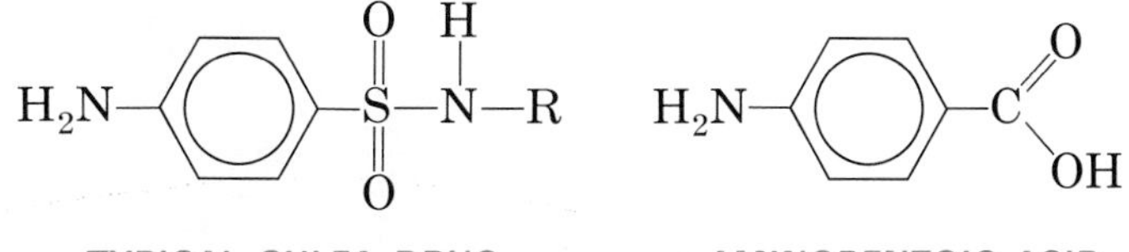

Gram-negative bacteria, e.g., *Escherichia coli,* lose stain or are decolorized by alcohol in Gram's method of staining. Gram-positive bacteria, e.g., *Diplococcus pneumoniae,* retain stain and resist decolorization by alcohol in Gram's method of staining. Gram was a man and is not to be confused with the unit of mass of the same name.

The close structural similarity of sulfanilamide and p-aminobenzoic acid permits sulfanilamide to be incorporated into the enzymatic reaction sequence instead of p-aminobenzoic acid. By bonding tightly, sulfanilamide shuts off the production of the essential folic acid, and the bacteria die of vitamin deficiency. Not all bacteria are susceptible to sulfa drugs. However, the drugs are effective on streptococci, staphylococci, many gram-negative and gram-positive bacteria, and protozoa such as coccidia. In man and the higher animals, p-aminobenzoic acid is not necessary for folic acid synthesis, and so sulfa drugs have no effect on this mechanism. But sulfa drugs must be buffered (e.g., by $NaHCO_3$) to prevent digestive upset, and they attack and pass through several kinds of protein linings. They also are suspected of modifying the base sequence of nucleic acids (by incorporation) and acting as potential mutagens.

The Penicillins

Penicillin was first discovered in 1928 by Alexander Fleming, a bacteriologist at the University of London, who was working with cultures of *Staphylococcus aureus,* a germ that causes boils and some other types of infections. In order to examine the cultures with a microscope, he had to remove the cover of the culture plate for a while. One day as he started work he noticed that the culture was contaminated by a blue-green mold. Many people would have discarded the contaminated culture, but Fleming's trained eye had noticed something else. For some distance around the mold growth, the bacterial colonies were being

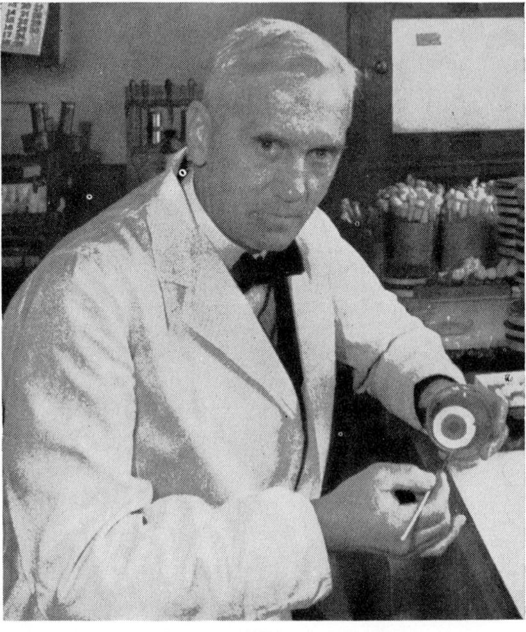

A

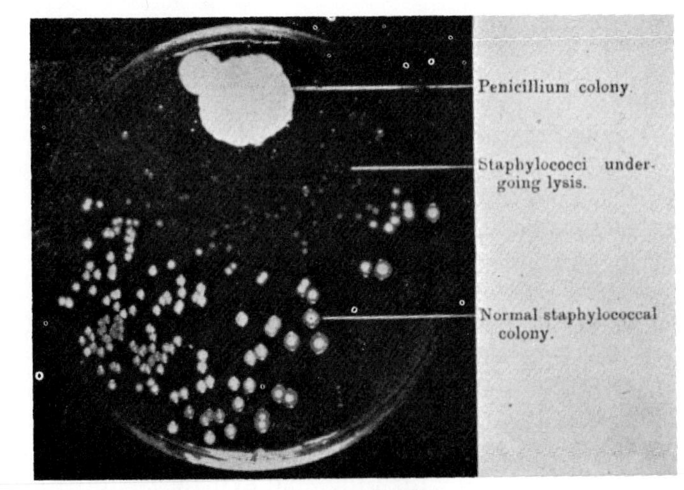

B

Figure 21–16 *A*, Sir Alexander Fleming. *B*, A culture-plate showing the dissolution of staphylococcal colonies in the neighborhood of a colony of penicillium. (From article by Fleming, *British Journal of Experimental Pathology,* June, 1929.)

destroyed. Upon further investigation, Fleming found that the broth in which the mold was grown had an inhibitory or lethal effect against many pathogenic organisms. The mold was later identified as *Penicillium notatum* (the spores sprout and branch out in pencil shapes, hence the name).

Penicillin, the name given to the antibacterial substance produced by the mold, apparently had no toxic effect on animal cells, and its activity was selective. Because Fleming's extracts from the mold were crude, clinical results were discouraging, and his brilliant discovery made little impact on the medical world for almost a decade. Eventually, as a result of the interest and further research of Howard Florey and Ernst Chain, penicillin was developed as a practical drug. In 1941, penicillin was used for the first time on a human being, a London policeman who was hospitalized with a serious case of blood poisoning contracted from

Fleming, Florey, and Chain shared a richly deserved Nobel Prize in medicine and physiology in 1945.

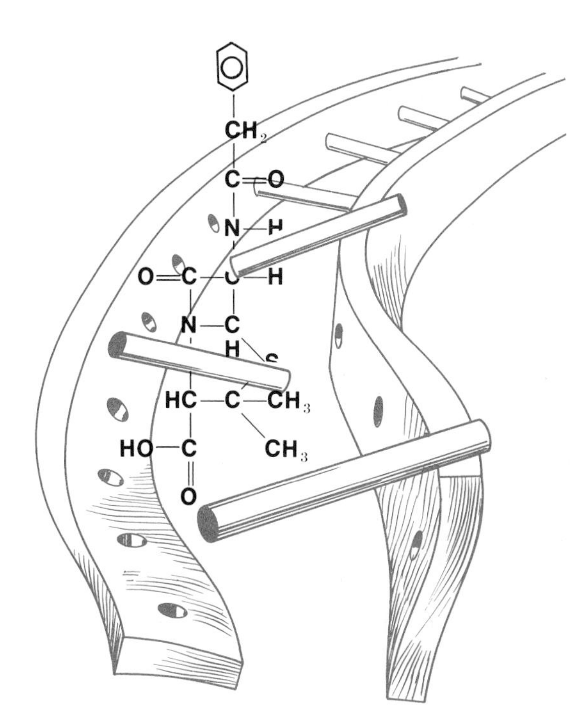

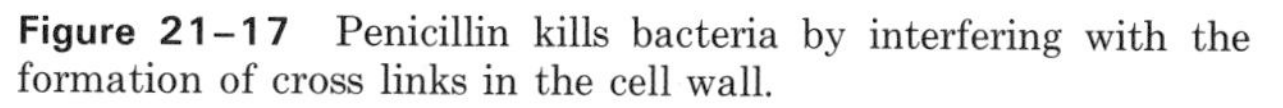
Figure 21–17 Penicillin kills bacteria by interfering with the formation of cross links in the cell wall.

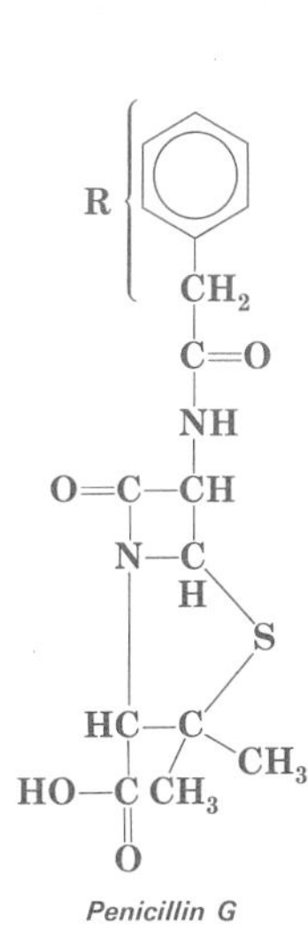

Penicillin G

a shaving cut. Since he could not recover by normal means, the doctors decided to try the new drug. The effect was immediately favorable.

Because of wartime needs, a large supply of penicillin was urgently needed. Through cooperation between American and British firms, the supply was provided and thousands of lives and limbs were saved.

The structure of penicillin (see margin) has now been determined. Many different penicillins exist, differing in the structure of the R group. Penicillin G is the most widely used in medicine.

Several antibiotics, such as penicillin and bacitracin, are known to prevent cell-wall synthesis. The cell walls of some pathogenic bacteria are composed of mucoprotein. Mucoprotein is a combination of proteins and mucopolysaccharides, in which some of the monosaccharide units (usually glucose or galactose) contain a $—NHCOCH_3$ group substituted for a hydroxyl group (—OH). Only bacterial cells have mucoprotein walls.

Penicillin interferes with the synthesis of the mucoprotein cell wall of the bacteria by interfering with the formation of cross-links between layers of the cell wall. The cell wall protects and supports the delicate cell components enclosed within it. The cytoplasmic membrane, immediately inside the cell wall, regulates the flow of nutrients and water in and out of the cell. The layers are reinforced by a series of chemical cross-links connecting one layer to another. When the cross-links are not formed properly, the weakened wall, unable to hold its size and shape, expands as water comes into the cell by osmosis. Eventually the cytoplasmic membrane bursts, causing the cell to die.

Streptomycin and the Tetracyclines

In 1937, following collaboration with René Dubos, Selman Waksman isolated a compound from a soil organism, *Streptomyces griseus,* which came to be known as ***streptomycin*** and was released to physicians in 1947.

This compound was quite successful in controlling certain types of bacteria but later had to be withdrawn, due to its adverse side effects.

In 1945, B. M. Duggar discovered that a gold-colored soil fungus, *Streptomyces aureofaciens,* produced a new type of antibiotic, ***Aureomycin,*** the first of the ***tetracyclines.*** Research then stepped up to a fever pitch. Pfizer laboratories tested 116,000 different soil samples before they discovered the next antibiotic, which they named ***Terramycin.***

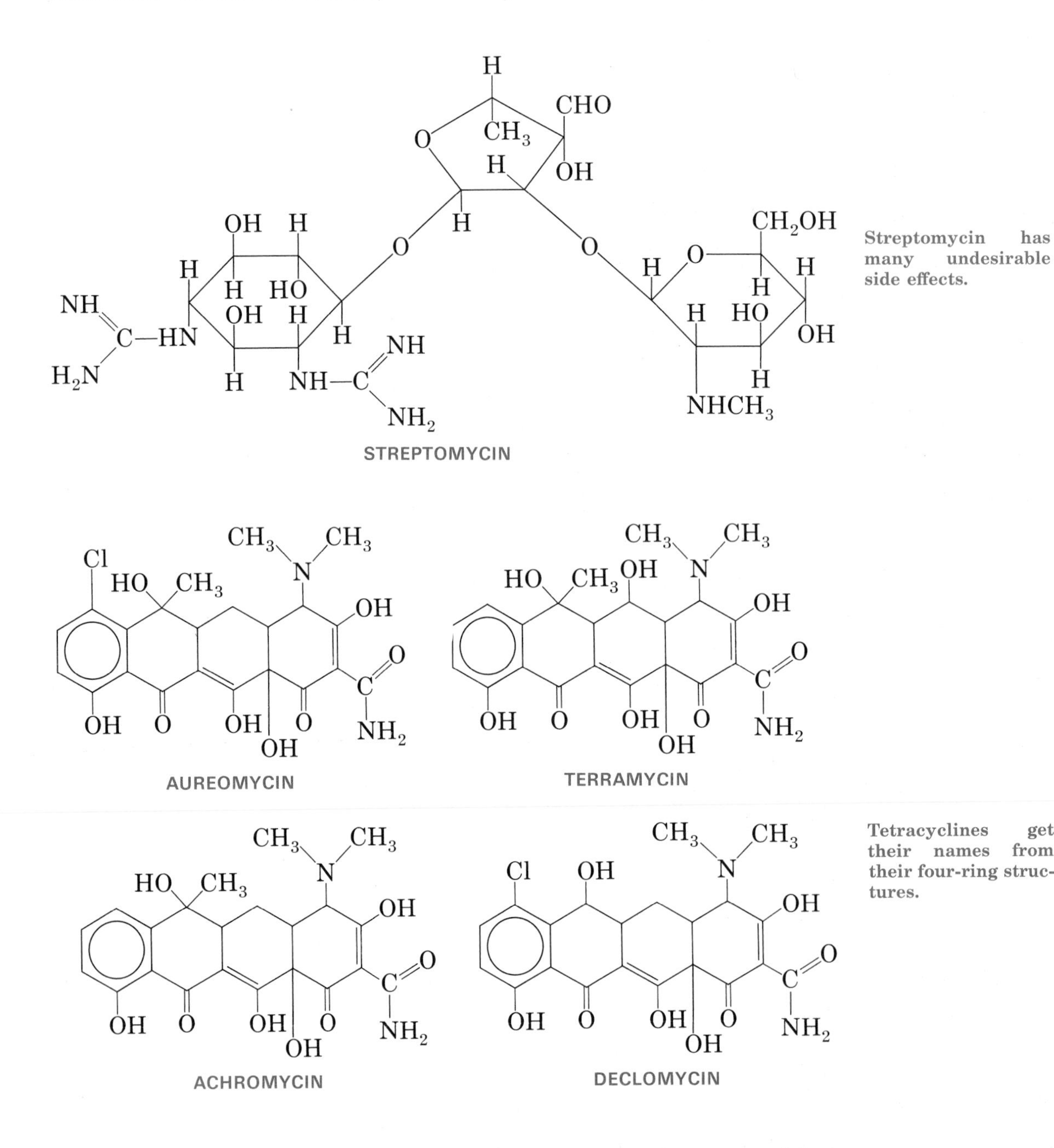

Streptomycin has many undesirable side effects.

Tetracyclines get their names from their four-ring structures.

Compounds of the tetracycline family are so named because of their four-ring structure. One side effect of taking these drugs is the diarrhea caused by the killing of the patient's intestinal flora (the bacteria normally residing in the intestines).

The Steroid Drugs

A large and important class of naturally occurring compounds is derived from the tetracyclic structure given below.

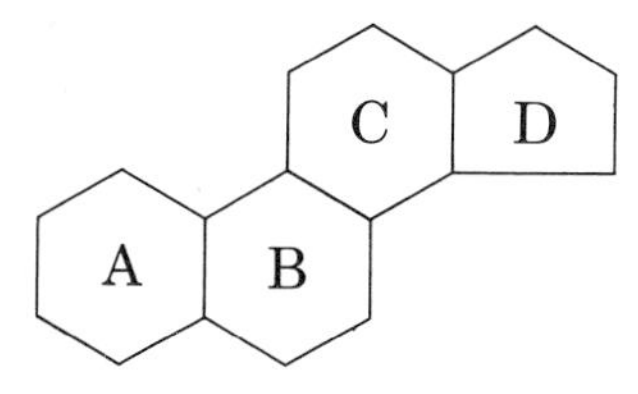

These compounds are known as ***steroids,*** and they occur in all plants and animals. The most abundant animal steroid is cholesterol, $C_{27}H_{46}O$. The human body synthesizes cholesterol and also readily absorbs dietary cholesterol through the intestinal wall. It is associated with gallstones and hardening of the arteries.

Cortisone is a "powerful drug" having a major effect on biological systems.

Biochemical alteration or degradation of cholesterol leads to many steroids of great importance in human biochemistry. When ***cortisone,*** one of the adrenal cortex hormones, is applied topically or injected into a diseased joint, it acts as an antiinflammatory agent and is of great use in treating arthritis.

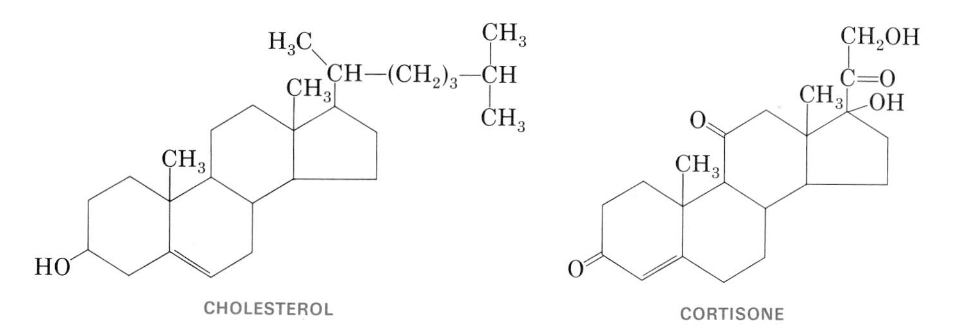

CHOLESTEROL CORTISONE

Structurally related to cholesterol and cortisone are the sex hormones. The female sex hormone, ***progesterone,*** differs only slightly from the male hormone, ***testosterone.***

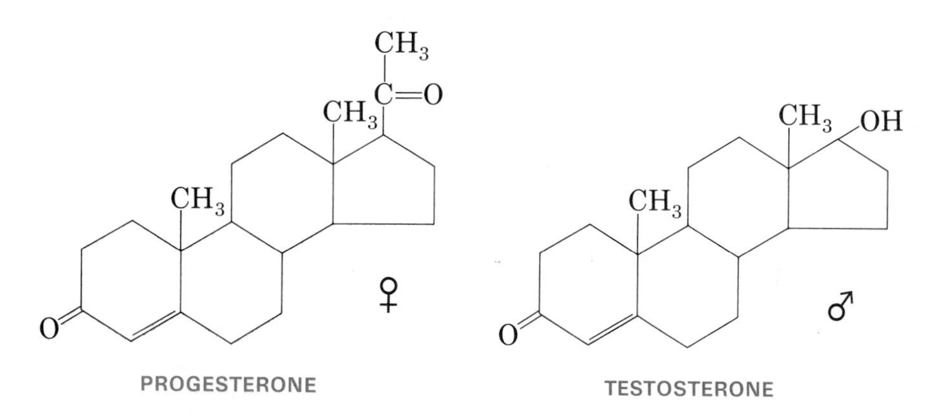

PROGESTERONE TESTOSTERONE

Other female hormones are estradiol and estrone, called ***estrogens.*** The estrogens differ from the other steroids discussed earlier in that they contain an aromatic A ring (in color).

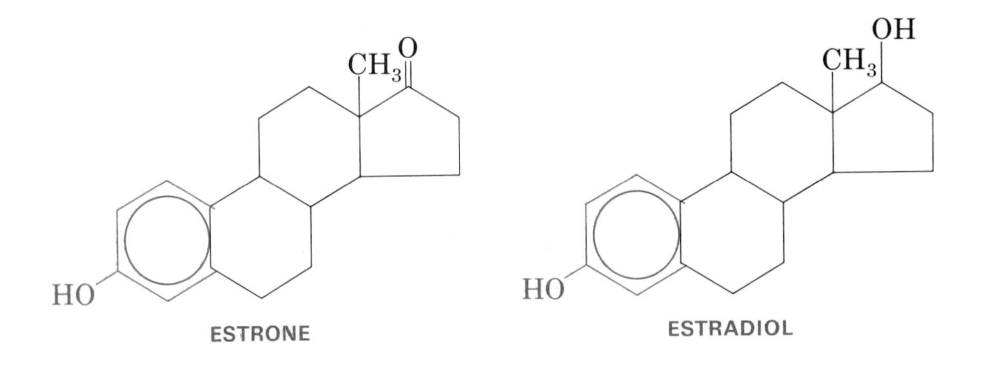

Birth Control Pills

One of the most revolutionary medical developments of the 1960's was the worldwide introduction and use of "The Pill." An estimated 8.5 million women in the United States use birth control pills. The basic feature of oral contraceptives for women is their chemical ability to simulate the hormonal processes resulting from pregnancy and, in so doing, prevent ovulation. Ovulation, the production of eggs by the ovary, ceases at the onset of pregnancy because of hormonal changes (Figure 21–18). This same result can be produced by the administration of a variety of steroids, some of which are effective when taken orally, although the mechanism of their action and their long-term effects are not known in detail.

The active ingredients of the Pill are the hormones progesterone and estrogen, or their derivatives. Enovid, a product of this sort made by G. D. Searle and Co., is a mixture of norethindrone and mestranol. Notice the structural similarities between these compounds and progesterone and estrogen:

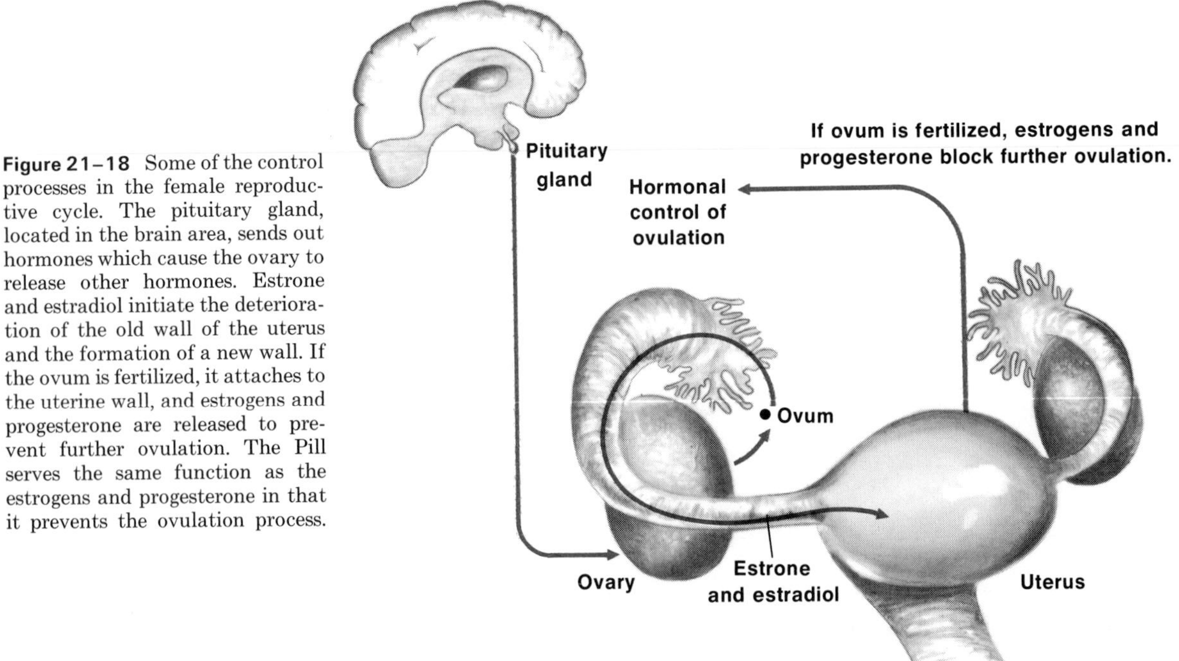

Figure 21–18 Some of the control processes in the female reproductive cycle. The pituitary gland, located in the brain area, sends out hormones which cause the ovary to release other hormones. Estrone and estradiol initiate the deterioration of the old wall of the uterus and the formation of a new wall. If the ovum is fertilized, it attaches to the uterine wall, and estrogens and progesterone are released to prevent further ovulation. The Pill serves the same function as the estrogens and progesterone in that it prevents the ovulation process.

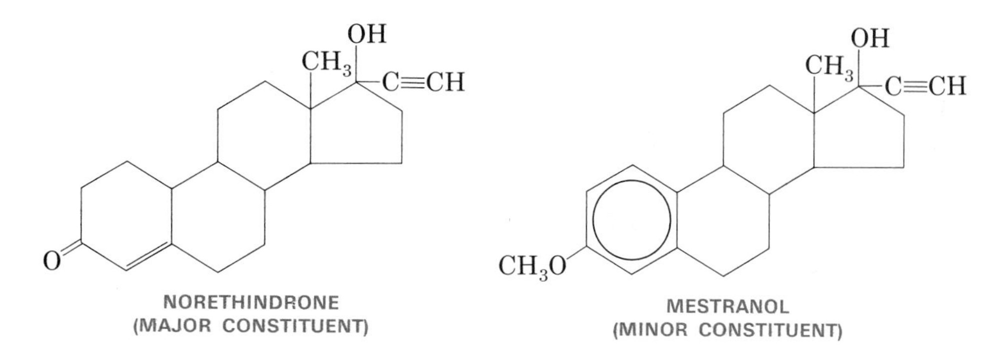

The safety of the hormone-like compounds used in the Pill has been under investigation for many years.

Do these pills cause cancer? After the first 15 years of usage there was no conclusive evidence that oral contraceptives are carcinogenic. Nor was there evidence that they cause, to a significant degree, diabetes, sterility, eye disorders, mental illness, or any of a number of other diseases to which they have been linked by critics. Since the latency period for cancer is thought to range from 10 to 20 years, it seems probable that if oral contraceptives do cause cancer, this will become evident during the late 1970's.

There is, however, one serious disorder that has been linked with the Pill: thromboembolic (blood clotting) disease. Such clots are potentially lethal. If they block a major blood vessel in a limb, amputation may be necessary; if they block a vessel to the lungs or brain, death may result. Studies done in Great Britain and the United States indicate that blood clotting is the cause of about three deaths per 100,000 users each year. Statistics indicate that this risk is considerably less than the risk of thromboembolism that would accompany the number of pregnancies averted. However, because there is a danger, all women who use the Pill should have medical checkups at least once every six months.

Perhaps the major problem facing our world is the population explosion. As the human population doubles every 40 years and associated problems intensify, wider efforts are being made to study procedures for controlling human fertility. Antifertility drugs for males are under active study, and there is every reason to believe that these will be on the market in the near future. Some of the first drugs of this sort to be studied act by temporarily suppressing the formation of sperm cells in the male.

Allergens and Antihistamines

A person may have an unpleasant physiological response to poison ivy, pollen, mold, food, cosmetics, penicillin, and even cold, heat, and ultraviolet light. In the United States about 5000 people die yearly from bronchial asthma, at least 30 from the stings of bees, wasps, hornets, and other insects, and about 300 people die from ordinary doses of penicillin. The reason: ***allergy.*** About one in 10 suffers from some form of allergy; more than 16 million Americans suffer from hay fever. This means that many of you already are familiar with the symptoms of headaches, inflamed eyes, congested sinuses, sneezing, and the raw, endlessly runny nose that accompanies hay fever.

An allergy is a physiological response such as sneezing, runny nose, coughing, dermatitis, etc., to the introduction of a foreign substance. This foreign substance is called an allergen.

An allergy is an adverse response to a foreign substance or to a physical condition that produces no obvious ill effects in *most* other organisms, including man. An ***allergen*** (the substance that initiates the allergic reaction) is, in most cases, a highly complex substance—usually a protein. Some are polysaccharides or compounds formed by combining a protein and a polysaccharide. Usually allergens have a molecular weight of 10,000 amu or more.

Most allergens are high molecular weight substances.

What is really the chemical cause of an allergy such as, say, hay fever? The details are at best sketchy now. But the overall process and a few of the details have been worked out reasonably well.

For a patient with pollen allergy, a pollen grain enters his nose and clings to the mucous membrane. The nasal secretions, acting on the pollen grain, release the grain's allergens and other soluble components, which penetrate the outer layer of the mucous membrane. The principal allergen of ragweed pollen, a major allergy-producer, has been isolated and is named ragweed antigen E. It is a protein with a molecular weight of about 38,000 amu; it represents only about 0.5 percent of the solids in ragweed pollen, but contributes about 90 percent of the pollen's allergenic activity. A mere 1×10^{-12} gram of antigen E injected into an allergic person is enough to induce a response. The reason why antigen E, of all the ragweed pollen proteins, is so unusually reactive is as yet not understood.

The allergens come in contact with special cells in the nose and breathing passages to which a particular type of antibody is attached. This is the IgE antibody (Chapter 18), which has a molecular weight of about 196,000. An allergic person has 6 to 14 times more IgE in his blood serum than a nonallergic person. In the nasal secretion of an allergic person, the concentration of IgE is 100 or more times greater than in the blood serum of the same person. The IgE is formed in the nose, bronchial tubes, and gastrointestinal tract, and binds firmly to specific cells, called mast cells, in these regions.

Antigen E from ragweed reacts with the IgE antibody attached to

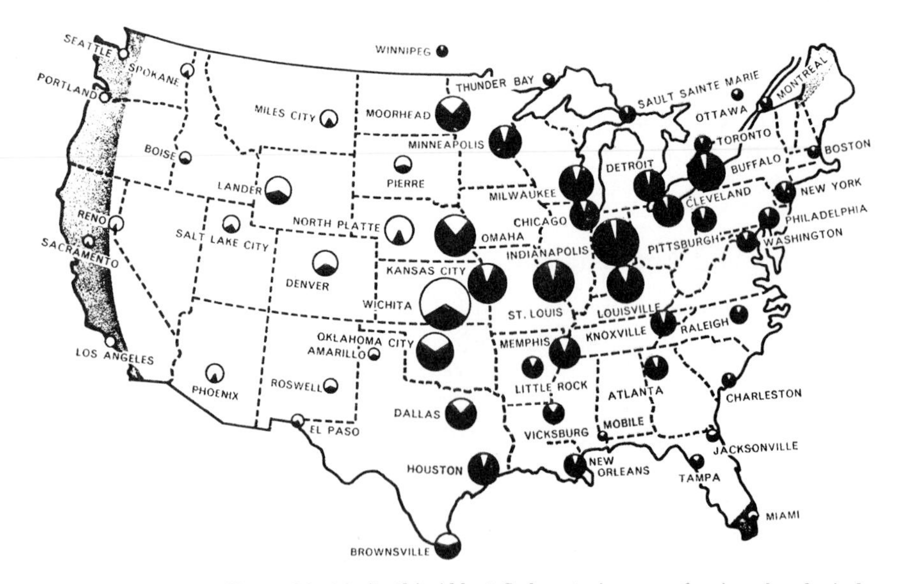

Figure 21–19 In this Abbott Laboratories map, the size of each circle represents the amount of all late-summer and fall pollens found in the air in each city. Dark portions show amount of ragweed pollen. Shaded areas are regions of low pollen count.

the mast cells, forming antigen-antibody complexes. Some think that the ragweed antigen E forms a bridge between two or more IgE molecules attached to a mediator-containing cell. This bridge, in turn, alters the configuration of parts of the attached IgE molecules.

By a series of events that are only crudely understood, the formation of these antigen-antibody complexes leads to the release of so-called "allergy mediators" from special granules in the mast cells. Some believe that the changed configuration of the cell caused by the antigen-antibody complex activates enzymes that ultimately cause the cell to release the allergy mediators. The most potent of these mediators found so far is ***histamine,*** which is formed by the breakdown of the amino acid, histidine. Although it is widely distributed in the body, it is especially concentrated in the 250 to 300 granules of the mast cells. Histamine accounts for

Histamine causes runny noses, red eyes, and other hay fever symptoms.

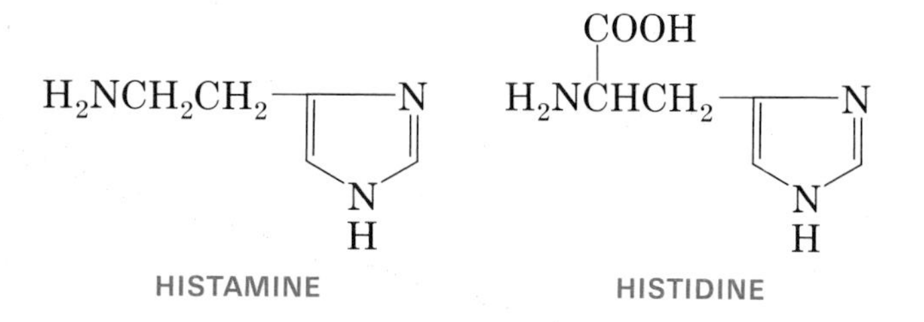

many, if not most, of the symptoms of hay fever, bronchial asthma, and other allergies. This compound causes dilation of blood capillaries. In addition, it makes the capillaries more permeable to blood fluids. Thus, these fluids can readily leak out of the capillaries and cause swelling of the tissues. The compound causes contraction and spasm of smooth muscles (a serious difficulty in the bronchial tubes in asthma). It can produce skin swellings (hives) and stimulate the glands that secrete watery nasal fluids, mucus, tears, saliva, and so on.

The chemical mediators such as histamine must be released from the cell to cause the symptoms of allergy. The release mechanism is an energy-requiring process in which the granules may move to the outer edge of the living cell and, without leaving the cell, discharge their contents of histamine through a temporary gap in the cell membrane. This sends histamine on its way to produce the toxic effects of hay fever.

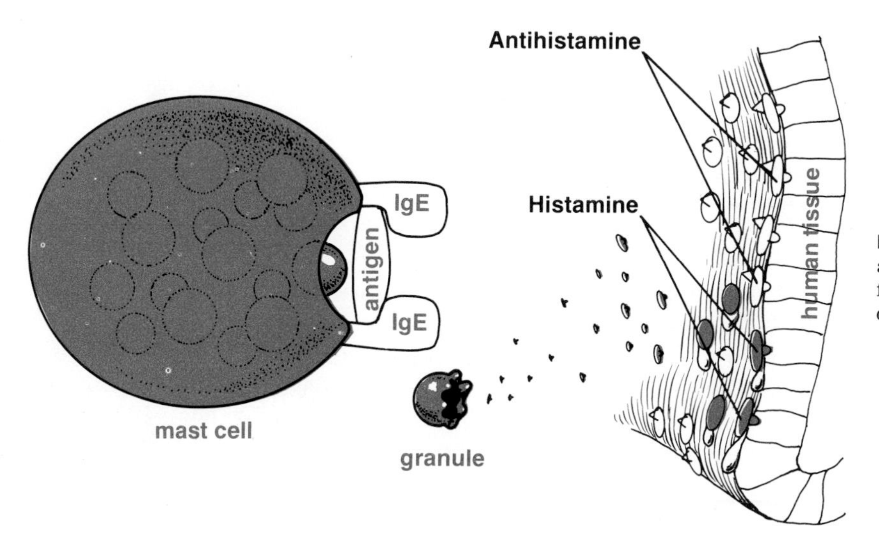

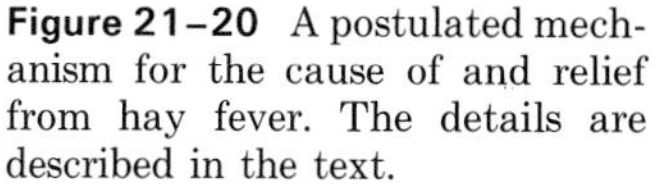
Figure 21–20 A postulated mechanism for the cause of and relief from hay fever. The details are described in the text.

Although histamine, an allergenic compound, has been isolated and some understanding of the mechanism that leads to allergy is now known, many questions are still unanswered. Since IgE is a class of several proteins, what structural relationships cause an IgE to react, for example, with the allergen in ragweed pollen and another IgE to react with the allergen in lobster? Does heredity play a part in allergy? Are people allergic because their membranes are inherently thinner or weaker than those of nonallergics? If antigen E causes 90 percent of the ragweed allergic activity, what causes the other 10 percent? What chemical reactions does histamine undergo in order to produce its symptoms? These are but a few questions, and there are many others. Although hay fever, bronchial asthma, and other allergies have not been conquered, they are better understood and better treated today.

Treatment consists of three procedures: avoidance, desensitization, and drug therapy. The idea of avoidance is simple enough. If you are allergic to strawberries, do not eat strawberries; if you are allergic to pollen, move, say, from Decatur, Illinois (ragweed pollen index: 114), to Seattle, Washington (index: 0.02; see Figure 21–19). Sometimes, however, avoidance is impractical.

Desensitization therapy is costly and inconvenient, since 20 or more injections are required to achieve what usually is a partial cure. One chemical idea of desensitization is to inject a blocking antibody that preferentially reacts with the allergen so that it cannot react with the IgE allergy-sensitizing antibody. This breaks the chain of events leading to the release of histamine or other allergy-producing mediators. Many small injections, spaced in time, are required to build up a sufficient level of the blocking antibody.

Epinephrine (adrenalin), steroids, and antihistamines are effective drugs in treating allergies. The first two are particularly effective in treating bronchial asthma, while the ***antihistamines,*** introduced commercially in the United States in 1945, are the most widely used drugs for treating allergies. More than 50 antihistamines are offered commercially in the United States. Many of these contain, as does histamine, an ethylamine group, ($—CH_2CH_2N<$):

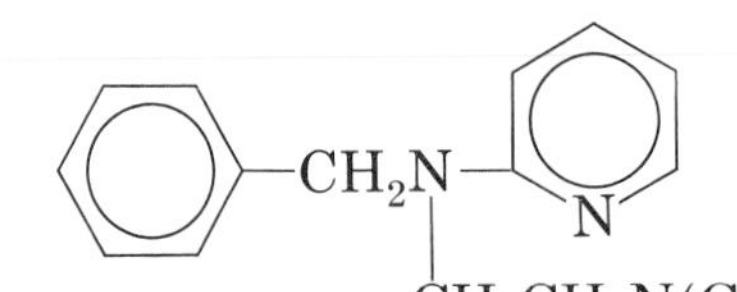

PYRIBENZAMINE
(AN IMPORTANT ANTIHISTAMINE)

These drugs act competitively by occupying the receptor sites normally occupied by histamine on cells. This, in effect, blocks the action of histamine. A new drug, disodium cromoglycate, acts not by blocking the action of histamine (as do the antihistamines), but by blocking the release of histamine and other mediators from the granules. Efforts are being made to find drugs with less troublesome side effects than those often associated with antihistamines. The most troublesome side reactions are drowsiness, mental confusion, dizziness, headache, nervousness, rapid

Antihistamines mimic histamine in their chemical reactions.

Poison ivy

Poison sumac

Figure 21–21 Poison ivy and poison sumac. The toxic nature of poison ivy is usually attributed to four substances, all related chemically to phenol (carbolic acid), of which the principal ones are urushiols.

pulse, nausea, depression, blurred vision, and dryness of the mouth. No proven drug is available now that will relieve the allergy and completely avoid the side effects.

Phenol

Another particularly annoying allergy is the response that some people have to poison ivy. The allergenic characteristics of poison ivy are usually attributed to four substances, all related chemically to phenol (carbolic acid), and known collectively as urushiols, so named because they were first discovered in the sap (the "kiurshi") of the Japanese lac tree. Two of the most potent of the urushiols are the diene and the triene shown below.

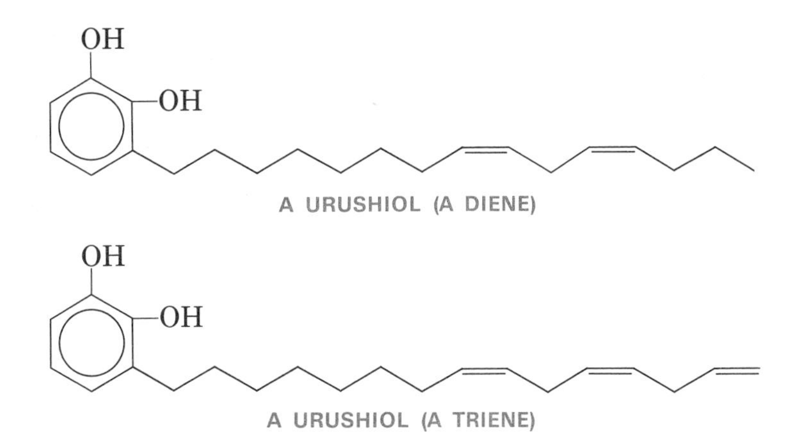

A URUSHIOL (A DIENE)

A URUSHIOL (A TRIENE)

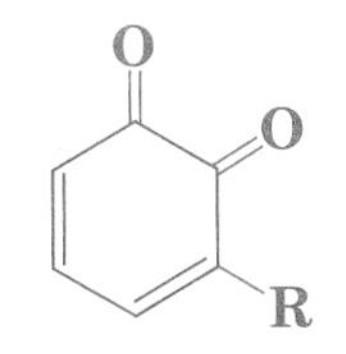

An o-benzoquinone

In 1968, J. S. Byck and C. R. Dawson proposed a mechanism which partially accounts for the allergenic nature of urushiols in poison ivy. They proposed that upon entering the host, the urushiols are enzymatically oxidized to the corresponding *o*-benzoquinones, which react with certain terminal amino (NH_2) or sulfhydryl (SH) groups on a protein. This alteration of what was an acceptable protein to the body now presents a "foreign," unacceptable protein to the chemical guardians of the chemistry of the body. The foreign protein, or antigen, could be responsible for the poison ivy allergy.

There is no known cure for poison ivy. Since it takes about 15 minutes for the poison to penetrate the skin, a quick and thorough washing with water and soap or trisodium phosphate (to neutralize the acidic urushiol) is highly desirable.

Na_3PO_4 is trisodium phosphate.

SELF-TEST 21-C

1. What acid is secreted by the walls of the stomach? ____________

2. Is an analgesic a pain killer or a germ fighter? ____________

3. Select the natural compound from which the other is made: morphine, heroin. ____________

4. A molecule of aspirin has two functional groups. What are they? ____________ and ____________

5. An antipyretic reduces ____________.

6. Penicillin was discovered by ____________ ____________.

7. Allergens are complex proteins with a usual molecular weight of 10,000 amu or more. (True/False)

8. Two groups of medicines that act by mimicking other compounds are ____________, which mimic *p*-aminobenzoic acid, and ____________, which mimic histamine.

Diet Pills

There are many kinds of special diets—diets to lower cholesterol, diets to correct an inborn error of metabolism, diets for ulcers, diabetes, obesity, underweight, hormone deficiencies, high blood pressure, kidney disorders, and many others. The most popular diets in our affluent society are those designed for overweight. Only a very small percentage of obese individuals can attribute their overweight to endocrine gland deficiencies. All other overweight people have only their lack of activity and excess food intake to blame for their obesity.

The desired rate of weight loss for most people ranges from about 1 to 2 pounds per week. This rate may vary depending on such factors as water balance, activity, heat loss, and the presence of disease. For most women, diets ranging between 1200 and 1500 food calories a day will bring about a satisfactory weight loss. For men, the range is between 1500 and 2000 food calories a day. An intake of less than 1200 calories is generally not advised. Young children vary greatly from one individual to the next; no general rule can be applied to them, and calorie levels should be prescribed individually by a dietitian or doctor.

A food calorie is actually a kilocalorie, or 1000 calories.

A variety of special, rather unusual diets have been reported to be

Many reducing diets are not nutritionally adequate.

effective in weight reduction. These diets have been known under such names as the grapefruit diet, the water diet, the 10-day diet, and others. The problem with the vast majority of these diets is that they are often made up of only a few foods, so that they are not nutritionally adequate. Five basic types of nutrients are essential: protein, fat, carbohydrate, vitamins, and certain minerals. They must be included in the diet for normal health. If the diet is lacking in any one category, the body suffers. Therefore, a balanced diet plus water is required, even if the amounts are reduced.

Many individuals have difficulty staying on a diet, and they must be highly motivated in order to maintain a caloric level low enough to bring about a loss of weight. Appetite-depressing drugs, known as ***anorexigenic drugs,*** have been used by some physicians to curb the appetite. More than two billion diet pills are distributed per year. The ingredients include amphetamines (which suppress the appetite), digitalis (which affects the heart), various diuretics (which increase the amount of water excreted), thyroid (which increases rates of metabolism), and prednisone (an antiinflammatory and antiallergic agent). Some of the drugs are potentially addictive; others tax the heart or cause potassium loss in the body.

Figure 21–22 A limited understanding of the biochemistry of smoking, foods, and medicine produces after-the-fact problems. (Editorial cartoon by Hugh Haynie of the Louisville *Courier-Journal.* Copyright, Los Angeles Times Syndicate. Reprinted with permission.)

By themselves anorexigenic drugs will not control obesity. Reliance on such drugs, rather than proper diet, leads to failure in a weight-reduction plan, and their unsupervised use may be harmful.

Diet pills should be used under medical supervision.

In addition to the appetite-depressing drugs, there are many other products that are advertised to help overweight people lose weight. These products include methyl cellulose (remember from Chapter 16 that the body cannot digest cellulose), vitamins, iron and calcium compounds, benzocaine (a local anesthetic), dextrose, potassium p-aminobenzoate, caffeine, flavorings, and a few other substances. By themselves, however, none of these products will cause a reduction in weight.

Drugs in Combination

Drugs, like some food additives, can have enhanced effects when placed in certain chemical environments. Sometimes the effects are harmful, sometimes helpful. Take the case of an aging business executive who took an antidepressant and then ate a meal that included aged cheese and wine. The antidepressant is an inhibitor of monoamine oxidase, an enzyme that helps to control blood pressure. Both the aged cheese he ate and the wine he drank contained pressor amines, which raise blood pressure. Without the controlling effect of the enzyme, these amines skyrocketed his blood pressure and caused a stroke. Neither the amines nor the antidepressant alone would likely cause the stroke, but the combination did.

Pressor amines tend to increase blood pressure.

Likewise, people who take digitalis for heart trouble and for reducing the sodium level should take aspirin only under medical supervision. Aspirin can cause a 50 percent reduction in salt excretion for three or four hours after it is taken.

Alcohol increases the action of many antihistamines, tranquilizers, and drugs such as reserpine (for lowering blood pressure) and scopolamine (contained in many over-the-counter nerve and sleeping preparations), making such combinations extremely dangerous. Staying away from dangerous alcohol-drug combinations is not as easy as it may seem. Many people fail to realize that a large number of over-the-counter preparations —such as liquid cough syrup and tonics—contain appreciable amounts of alcohol.

Not all drug combinations are bad. Doctors have been highly successful in prolonging the lives of leukemia and other cancer victims with combinations of drugs that individually could not do the job. Resistant kidney disease has also responded to drug combinations in cases in which single drugs were ineffective.

Perhaps the best advice is to take medicine only when you are sick, making sure that a physician knows what you are taking.

MATCHING SET

______	1. histamine	a.	interferes with cell wall structure
______	2. cortisone	b.	antiseptic
______	3. tetracycline	c.	pain-causing protein

______ 4.	penicillin	d.	source of morphine
______ 5.	sulfa drug	e.	analgesic
______ 6.	iodine	f.	nutrient supplement in food
______ 7.	bradykinin	g.	antacid
______ 8.	procaine (novocaine)	h.	a steroid
______ 9.	opium poppy	i.	food color
______ 10.	dihydroxyaluminum sodium carbonate	j.	flavor enhancer
______ 11.	mineral	k.	catalyze oxidation of fats
______ 12.	β-carotene	l.	causes symptoms of hay fever
______ 13.	monosodium glutamate	m.	antimicrobial preservative
______ 14.	copper, nickel, and iron	n.	antibiotic containing a four-ring structure
______ 15.	sodium benzoate	o.	exaggerates some chemical effects
______ 16.	potentiator	p.	sulfanilamide

QUESTIONS

1. What weight of MgO is required to neutralize all the acid in a stomach which contains 0.05 mole of hydrochloric acid?
2. Which of the following food additives should be avoided? Give your reasons.

Butter yellow	Glycerol
Propionic acid	Sodium cyclamate

3. Why does it take less time to cook food in a pressure cooker than in an open pot of boiling water?
4. What would happen to your ability to digest protein if you kept the acid in your stomach neutralized all the time? Is acid bad for your stomach?
5. Why is saccharin preferable to chloroform as an artificial sweetener?
6. A label on a brand of breakfast pastries contains the following additives: dextrose, glycerine, citric acid, potassium sorbate, Vitamin C, sodium iron pyrophosphate, and BHA. What is the purpose of each substance?
7. Choose a label from a food item and try to identify the purpose of each additive.
8. Describe some of the chemical changes that occur during the cooking of
   (a) a carbohydrate,
   (b) a protein,
   (c) a fat.
9. What causes fat in foods to become rancid? How can this be avoided?
10. What causes bread to rise?

11. What are the pros and cons of eating "natural" foods as opposed to foods containing chemical additives?

12. Why were cyclamates taken off the market?

13. Do you think it is wise to use animals in safety tests for drugs and food additives? Should mental patients and prisoners be used for this purpose?

14. Many consumer products are almost identical in chemical composition but are sold at widely different prices under different trade names. Do you think the products should be identified by their chemical names or their trade names? Why?

15. See what you can find out about correlation between taste and smell. Are they the same sensation? Are they independent of each other?

16. Which has more caffeine, tea or coffee?

17. Bixin (annatto extract from the seeds of the tropical tree *Bixa orellana*) is added to food to give it a yellow color. What part of the structure is primarily responsible for the color? If white light is incident on the substance, why is the substance yellow?

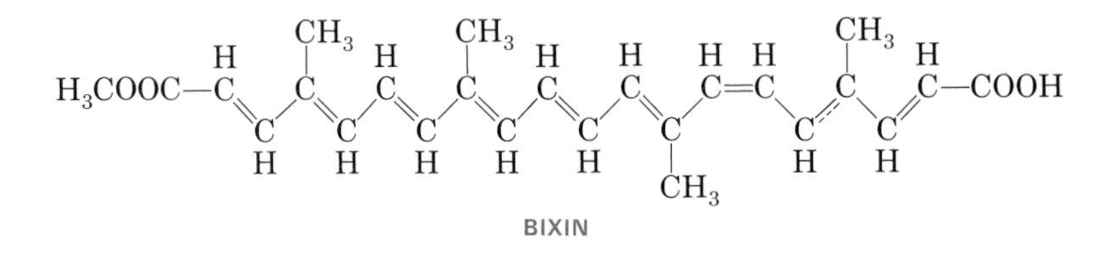

BIXIN

18. Write a chemical equation that will show how the tartrate ion ($C_4H_4O_6^{2-}$) can raise the pH when added to an unusually acidic fruit juice.

19. Suggest a way that citric acid sequesters metals.

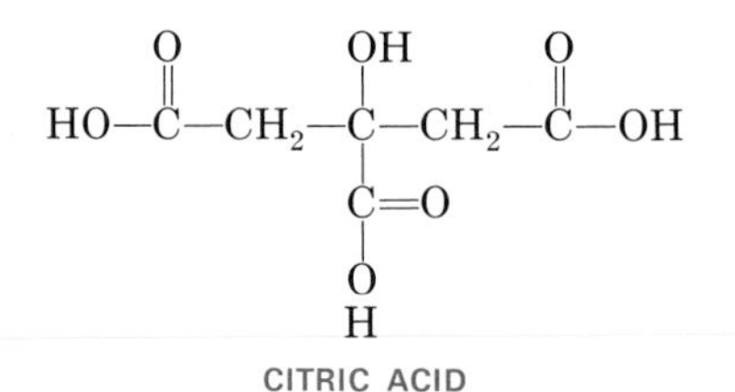

CITRIC ACID

20. Using the structure of a protein given in Chapter 16, show how cooking can affect its structure.

21. How is the action of antibiotics different from the action of antiseptics in killing bacteria?

22. What is a synergist? Give an example.

23. What part, if any, does hydrogen bonding play in the activities of surface active agents, humectant action of polyhydric alcohols, and stabilizing effect of gelatin in ice cream?

24. If a flavoring agent is extracted from a plant by using nonpolar carbon tetrachloride, CCl_4, what does this indicate about the polarity and degree of ionic character of the flavoring agent?

SUGGESTIONS FOR FURTHER READING

Amoore, J. E., Johnston, J. W., Jr., and Rubin, M., "The Stereochemical Theory of Odor," *Scientific American,* Vol. 210, No. 2, p. 42 (1964).

"Artificial Sweetener Gets FDA Approval," *Chemical and Engineering News,* p. 5, August 5, 1974.

"Chinese Restaurant Syndrome," *Chemistry,* Vol. 42, No. 8, p. 4 (1969).

"Flavor of a Potato Chip," *Chemistry,* Vol. 43, No. 7, p. 2 (1970).

Furia, T. E. (ed.), "Handbook of Food Additives," The Chemical Rubber Co., Cleveland, 1968. (The GRAS chemicals are listed and discussed on pages 565 to 751.)

Gates, M., "Analgesic Drugs," *Scientific American,* Vol. 215, No. 6, p. 131 (1969).

Hodge, H. C., and Smith, R. P., "Clinical Toxicology of Commercial Products," 3rd edition, the Williams & Wilkins Co., Baltimore, 1969. (Contains a wealth of information on the toxic aspects of various commercial products.)

"How Penicillin Kills Bacteria," *Chemistry,* Vol. 41, No. 7, p. 44 (1968).

Jacobson, M. F., "Eater's Digest: The Consumer's Factbook of Food Additives," Doubleday and Co., Garden City, N.Y., 1972.

Kirk, R. E., and Othmer, D. F., "Encyclopedia of Chemical Technology," 2nd edition, Interscience Publishers, New York, 1963. (Contains detailed information on various aspects of applied chemistry.)

Majtenyi, J. Z., "Antibiotics—Drugs from the Soil," *Chemistry,* Vol. 48, No. 1, p. 6 (1975).

"New Artificial Sweeteners," *Chemistry,* Vol. 43, No. 6, p. 23 (1970).

Pirie, N. W., "Orthodox and Unorthodox Methods of Meeting World Food Needs," *Scientific American,* Vol. 216, No. 2, p. 27 (1970).

Pyke, M., "Man and Food," McGraw-Hill Book Co., New York, 1970.

Sanders, H. J., "The Tasteless Condiment," *Chemistry,* Vol. 40, No. 1, p. 23 (1967).

Schubert, J., "Chelation In Medicine," *Scientific American,* Vol. 214, No. 5, p. 40 (1968).

Solmssen, U. V., "The Chemist and New Drugs," *Chemistry,* Vol. 40, No. 4, p. 22 (1967).

"The Allergenic Mechanism," *Chemistry,* Vol. 44, No. 5, p. 23 (1971).

Weiss, H. S., "Aspirin—A Dangerous Drug?" *Journal of the American Medical Association,* Vol. 229, p. 1221 (1974).

CONSUMER CHEMISTRY– BEAUTY AIDS AND CLEANSING AGENTS

CHAPTER 22

BEAUTY AIDS

People find many reasons for applying various chemical preparations (***cosmetics***) to their skin and hair. We wish to be clean, beautiful, healthy, and pleasing to others. Often we find that a cosmetic can do something easier than we can do it ourselves, such as take off unwanted hair or hold hair in place on a windy day.

Considerable progress has been made in producing chemical products which color hair and skin, disinfect our body surfaces, and keep down unwanted body odor. In this chapter we are going to look at a few of them in some detail. The purpose is not so that you can make face creams or hair sprays, but to give you a better understanding of the basic chemistry involved. In fact, amateur chemical preparations should be avoided, since toxic reactions often are encountered when impure or otherwise harmful concoctions are used without proper prior testing. Even the professionals have their problems. Scarcely a cosmetic producer is in business today who hasn't received a letter of complaint saying, in effect, "Your shampoo caused my hair to fall out!"

Cosmetic products require careful testing before they are used on humans.

SKIN AND HAIR

The skin, hair, and nails are protein structures. Our skin (Figure 22–1), like the other organs of the body, is not uniform tissue. Rather it is composed of layers, each parallel to the surface. The outermost layer, called the ***stratum corneum,*** or ***corneal layer,*** is where most cosmetic preparations for the skin act. The corneal layer is composed of essentially

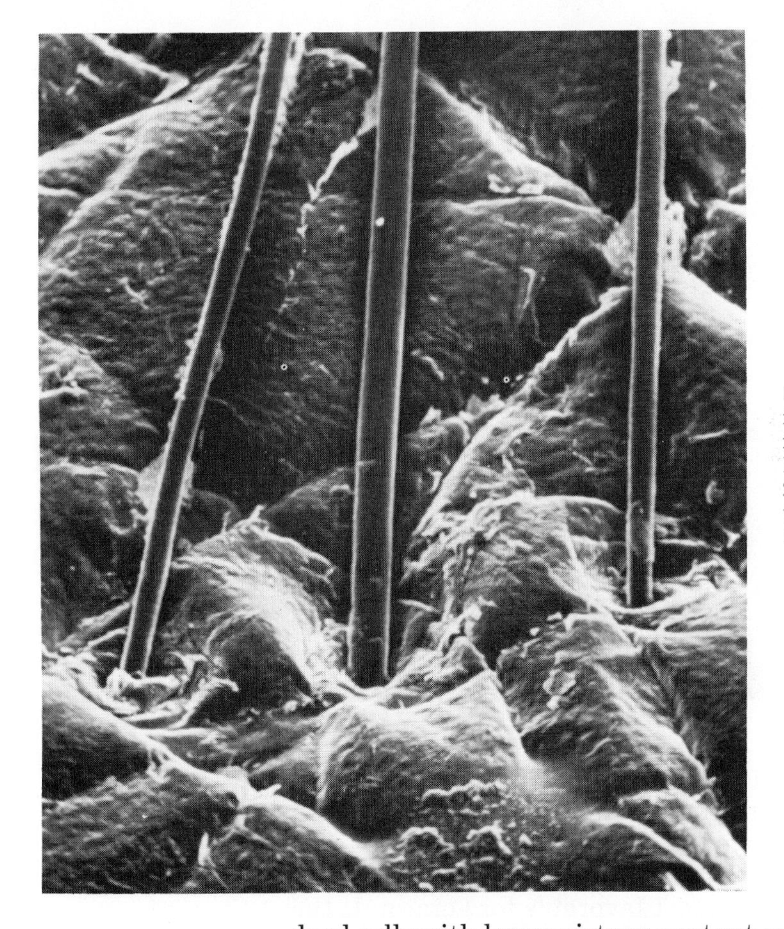

Figure 22–1 Replica of the surface of human forearm skin, showing three hairs emerging from the skin (× 225). (Courtesy of E. Bernstein and C. B. Jones, *Science,* 166: 252–253, 10 October, 1969. Copyright 1969 by the American Association for the Advancement of Science.)

dead cells with low moisture content and a surface pH of about 4, slightly acid. Depending upon its location on the body, the corneal layer is populated with as many as a million microorganisms per square centimeter. The principal protein of the corneal layer is ***keratin.*** Keratin is composed of about 22 different amino acids. Its structure renders it insoluble in, but slightly permeable to, water. This is important since we do not wish to dissolve in rain, but we do need to perspire! In order to control the moisture content of the corneal layer so it does not dry out and slough off too fast, moisturizers are added to the skin.

Keratin is skin protein.

$$\begin{array}{l} \quad\quad\quad NH_2 \\ \quad\quad\quad | \\ HOOC—CH—CH_2—S \\ \quad\quad\quad\quad\quad\quad\quad\quad | \\ HOOC—CH—CH_2—S \\ \quad\quad\quad | \\ \quad\quad\quad NH_2 \end{array}$$

Cystine

Hair (Figure 22-2) is composed principally of keratin. An important difference between hair keratin and other proteins is its high content of the amino acid ***cystine.*** About 16 to 18 percent of hair protein is cystine, while only 2.3 to 3.8 percent of the keratin in corneal cells is cystine. This amino acid plays an important role in the structure of hair.

The toughness of both skin and hair is due to the bridges between different protein chains, such as hydrogen bonds (Chapters 9 and 16) and —S—S— linkages (called ***disulfide*** linkages) between different chains.

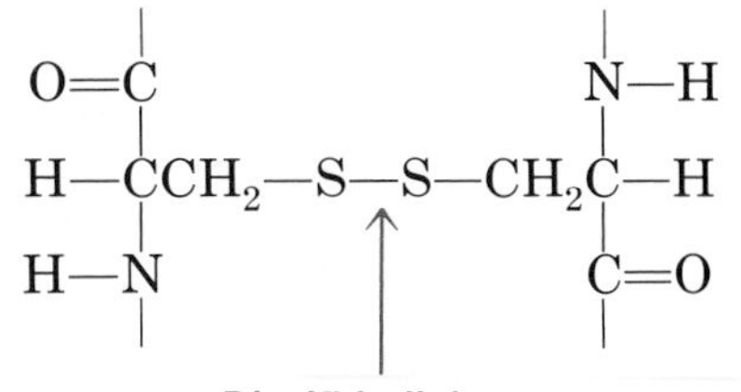

Disulfide linkage

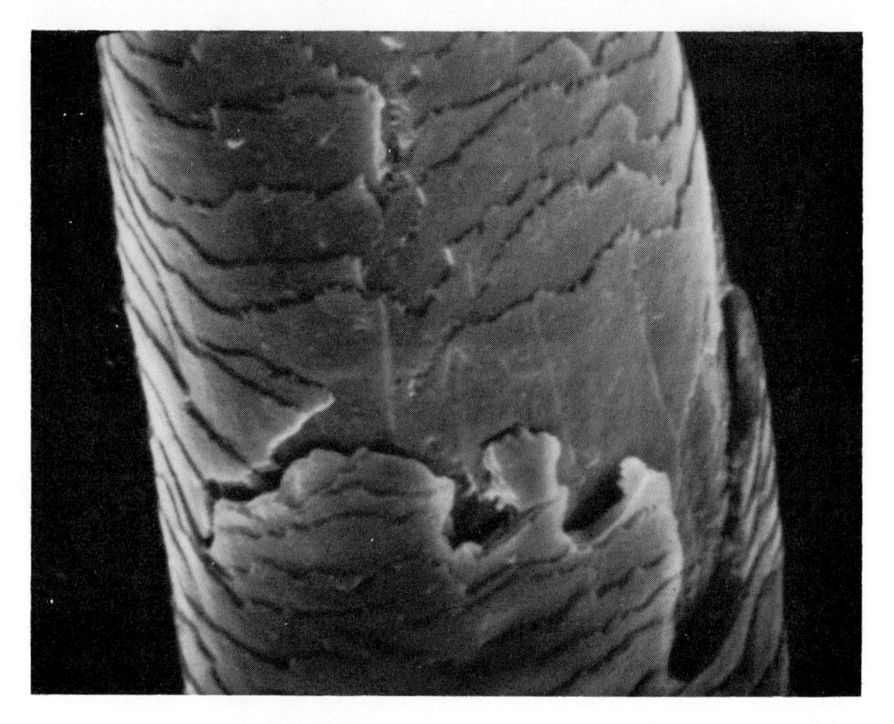

Figure 22–2 Electron micrograph of human hair. Note the layers of keratinized cells.

Another type of bridge between two protein chains which is important in keratin as in all proteins is the ***salt bridge.*** Consider the interaction between a lysine $-NH_2$ group and a carboxylic $-COOH$ group of glutamic acid on a neighboring protein chain. At pH 4.1, the presence of a $-NH_3^+$ group and a $-COO^-$ group is most favorable for keratin. If the two charges approach closely, an ionic bond is formed.

The structures of protein tissues are due in part to disulfide bridges and to salt bridges between "molecules."

$$HCCH_2CH_2CH_2CH_2NH_2 + HOOCCH_2CH_2CH \xrightarrow{\text{At pH 4.1}}$$

LYSINE GLUTAMIC ACID

$$HCCH_2CH_2CH_2CH_2NH_3{}^+ \; {}^-OOCCH_2CH_2CH$$

(IONIC BOND—A "SALT" BRIDGE)

As the pH rises above 4, keratin will swell and become soft as these salt bridges are broken. This is an important aspect of hair chemistry.

Changing the Shape of Hair

From the time of ancient Egypt, Greece, and Rome, there have always been those who considered their hair too straight or too curly. The properties of hair in this respect are determined by the extent and nature of the cross-linking. This is dictated as the messenger-RNA constructs the protein structure from the amino acids available (Chapter 16).

Changing the shape of hair is a story in redox reactions.

When hair is wet it can be stretched to one and a half times its dry length because water (pH = 7) destroys some of the salt bridges and causes swelling of the keratin. Imagine the disulfide cross-links remaining

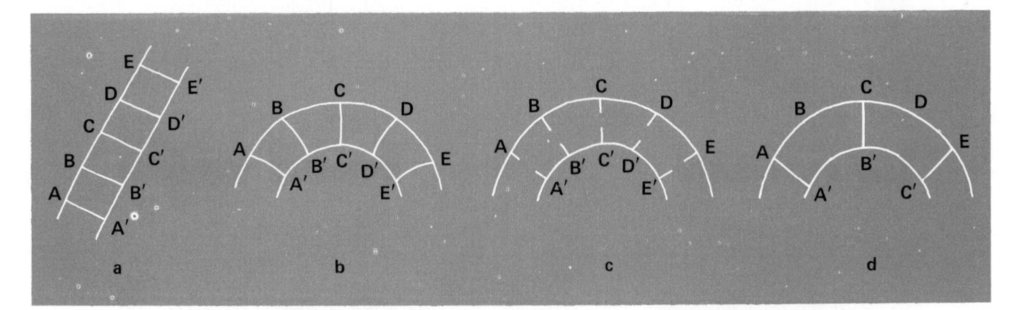

Figure 22–3 A schematic diagram of a permanent wave.

between two protein chains in hair as in Figure 22–3*a*. Winding the hair on rollers causes tension to develop at the cross-links (*b*). In "cold" waving, these cross-links are broken by a reducing agent (*c*), relaxing the tension. Then, an oxidizing agent regenerates the cross-links (*d*) and the hair holds the shape of the roller. The chemical reactions in simplified form are shown in Figure 22–4.

The most commonly used reducing agent is thioglycolic acid. The

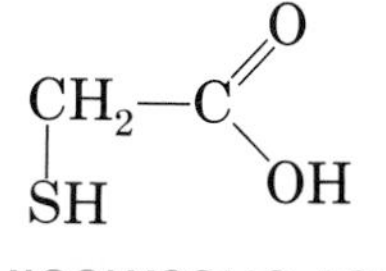

THIOGLYCOLIC ACID

common oxidizing agents used include hydrogen peroxide, perborates ($NaBO_2 \cdot H_2O_2 \cdot 3H_2O$), and sodium or potassium bromate ($KBrO_3$). A typical neutralizer solution contains one or more of the oxidizing agents dissolved in water. The presence of water and strong base in the oxidizing solution also helps to break and re-form hydrogen bonds between adjacent protein molecules.

Hair can be straightened by the same solutions. It is simply "neutralized" (or oxidized) while straight (no rolling up).

Various additives are present in both the oxidizing and the reducing solutions in order to control pH, odor, and color, and for general ease of application. A typical waving lotion contains 5.7 percent thioglycolic acid, 2.0 percent ammonia, and 93.3 percent water.

Hair Coloring and Bleaches

Melanin—black.

Iron pigment—red.

Hair contains two pigments: brown-black melanin and an iron-containing red pigment. The relative amounts of each actually determine the color of the hair. In deep-black hair melanin predominates, while in light-blond, the iron pigment predominates. The depth of the color depends upon the size of the pigment granules.

Permanent dyes are fixed to hair in a manner similar to the ingrain dyeing of cloth.

In recent years hair bleaches and colors have become very popular, largely because of newer formulations which can produce a much more uniform coloration of human hair. The formulations vary from temporary coloring (removable by shampoo), which is usually achieved by means of a water-soluble dye which acts on the surface of the hair, to semi-permanent dyes, which penetrate the hair fibers to a great extent (Figure 22–5). These often consist of cobalt or chromium complexes of dyes dissolved in an organic solvent. Permanent dyes are generally "oxidation" dyes. They penetrate the hair, and then are oxidized to give a colored product which is permanently attached to the hair by chemical bonds or which is much less soluble than the reactant molecule.

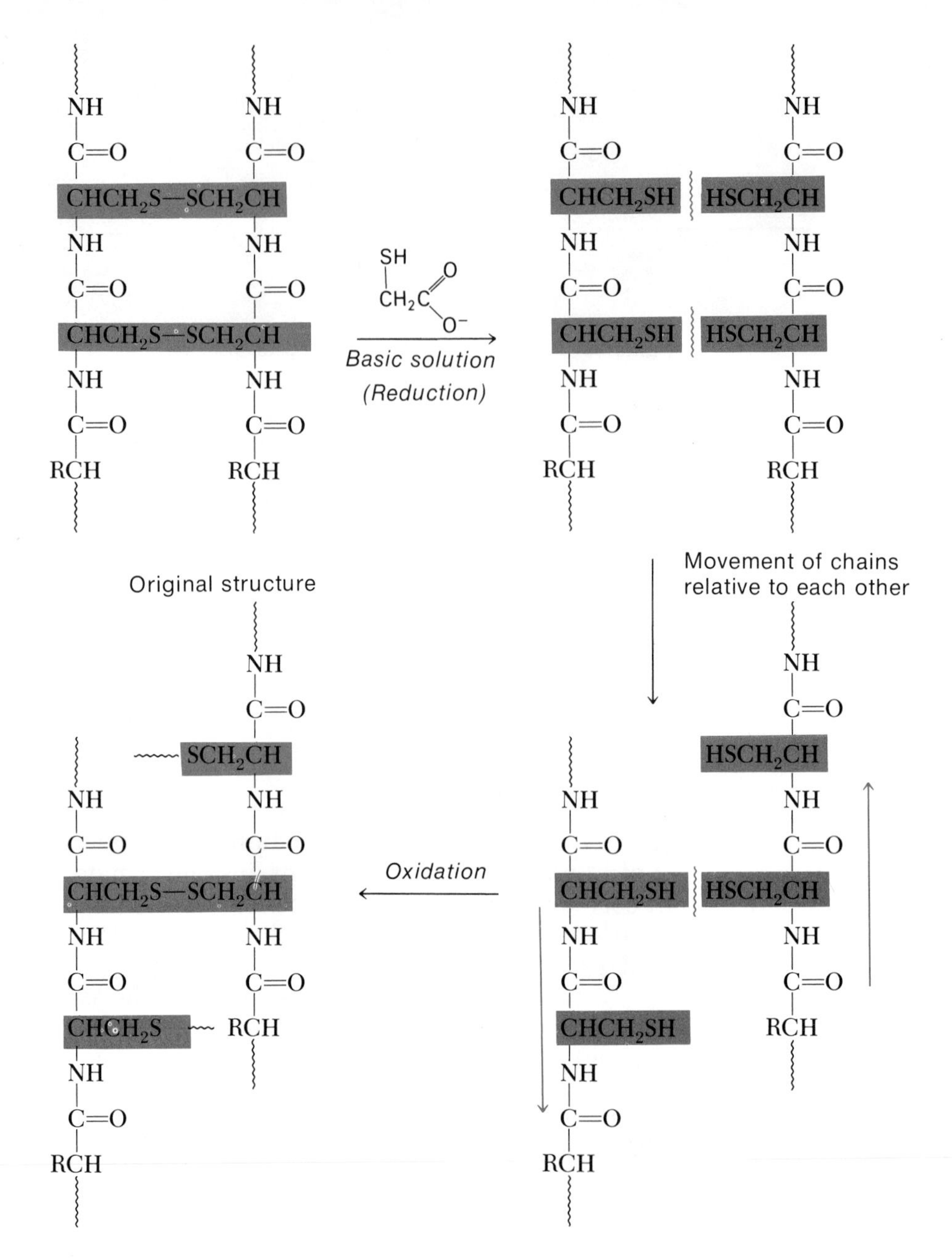

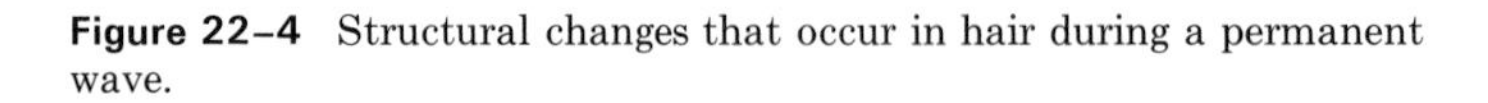
Figure 22–4 Structural changes that occur in hair during a permanent wave.

Permanent hair dyes generally are derivatives of phenylenediamine.

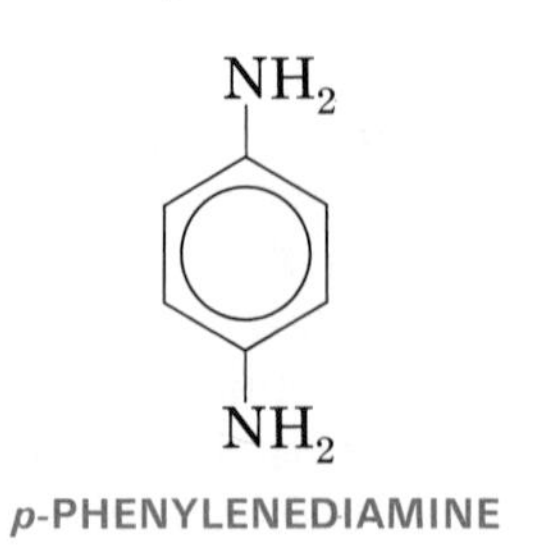

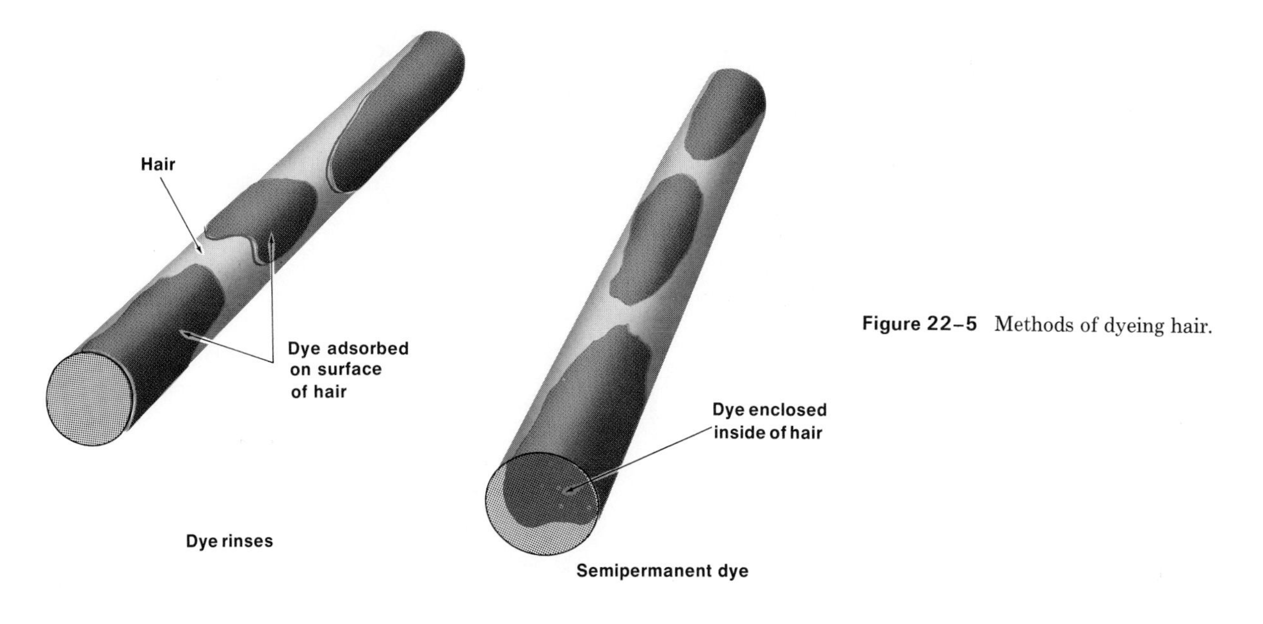

Figure 22–5 Methods of dyeing hair.

Phenylenediamine dyes hair black. A blond dye can be formulated with *p*-aminodiphenylaminesulfonic acid,

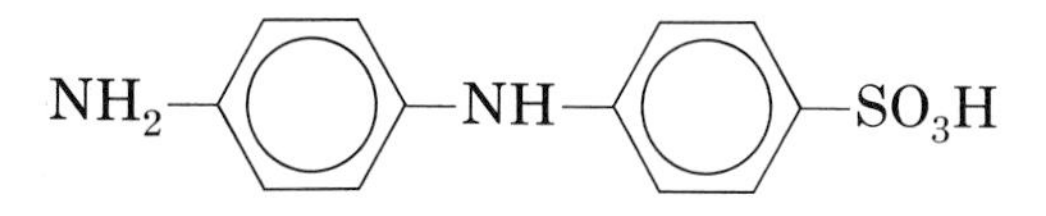

Just about any shade of hair color can be prepared by varying the chromophore groups on certain basic dye structures.

or *p*-phenylenediaminesulfonic acid.

NH_2, SO_3H, NH_2 (structure: benzene ring with NH_2 at position 1, SO_3H at position 2, NH_2 at position 4)

One blond formulation contains *p*-phenylenediamine (0.3 percent), *p*-methylaminophenol (0.5 percent), *p*-aminodiphenylamine (0.15 percent), *o*-aminophenol (0.15 percent), pyrocatechol (0.25 percent), resorcinol (0.25 percent), and inert solvent (98.40 percent). These compounds are applied in an aqueous soap or detergent solution containing ammonia to make the solution basic. The dye material is then oxidized by hydrogen peroxide to develop the desired color. The amines are oxidized to nitro compounds.

$$\underset{\text{AMINE}}{-NH_2} + 3H_2O_2 \xrightarrow{\text{Oxidation}} \underset{\text{NITRO COMPOUND}}{-NO_2} + 4H_2O$$

Hair can be bleached by hydrogen peroxide, which destroys the hair pigments by oxidation. The solutions are made basic with ammonia to enhance the oxidizing power of the peroxide. Parts of the chemical

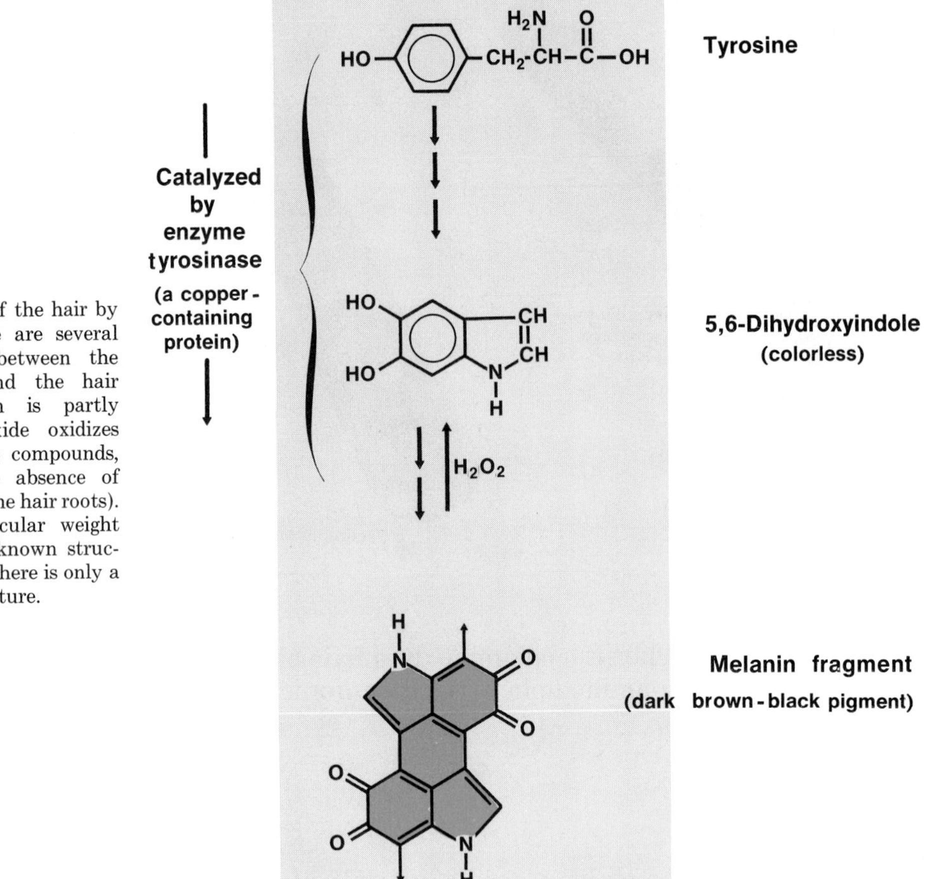

Figure 22–6 Bleaching of the hair by hydrogen peroxide. There are several chemical intermediates between the amino acid—tyrosine—and the hair pigment—melanin, which is partly protein. Hydrogen peroxide oxidizes melanin back to colorless compounds, which are stable in the absence of tyrosinase (found only in the hair roots). Melanin is a high molecular weight polymeric material of unknown structure. The structure shown here is only a segment of the total structure.

process are given in Figure 22–6. This drastic treatment of hair does more than just change the color. It may destroy sufficient structure to render the hair brittle and coarse.

Hair Sprays

Hair sprays are essentially solutions of a resin in a very volatile solvent whose purpose, when sprayed on hair, is to furnish a film with sufficient strength to hold the hair in place after the solvent has evaporated. After early experiments with shellac, the introduction of the aerosol can (Figure 22–7) allowed the use of a wider variety of resins and solvents and provided greater control over the application of the product.

A very common resin in hair sprays is the addition polymer, polyvinylpyrrolidone (PVP).

Hair spray coats the hair with a plastic film.

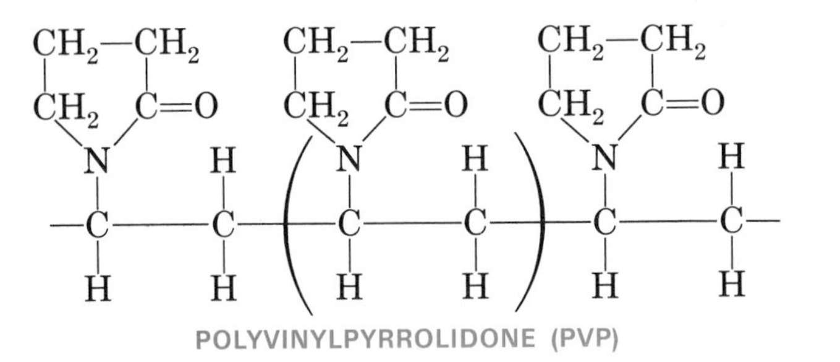

POLYVINYLPYRROLIDONE (PVP)

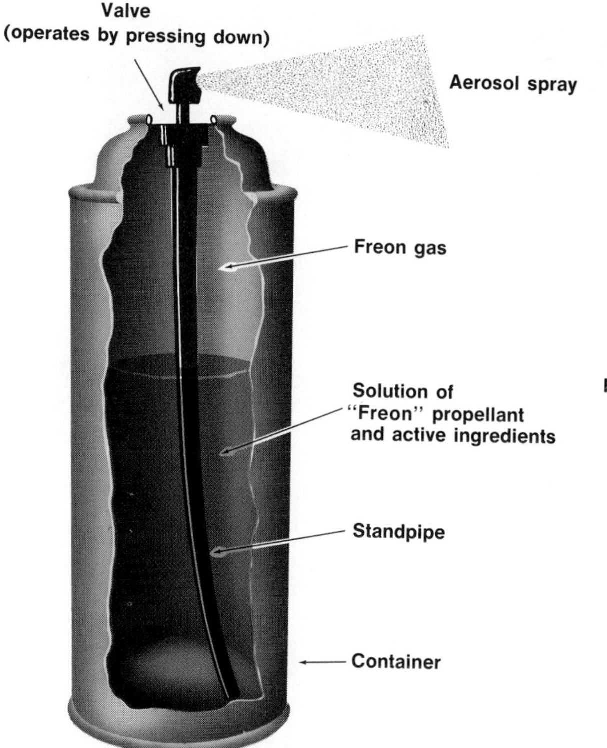

Figure 22–7 Cross-section of a typical aerosol spray can.

The aerosal propellants are suspected of altering the protective ozone layer in the upper atmosphere.

The resin is blended in hair spray formulations with a plasticizer, a water repellent, and a solvent-propellant mixture. The plasticizer makes the plastic more pliable, as described and illustrated in Chapter 15. The solvent-propellant system is a solvent such as anhydrous ethanol mixed with a liquefied propellant, such as Freon 11 or Freon 12.

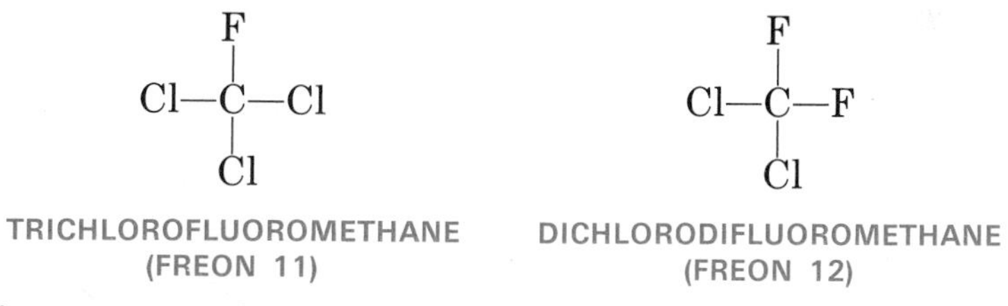

There are several dangers in breathing hair sprays, such as the danger of possible carcinogens acting on delicate lung tissue and the danger of asphyxiation by the plastic coating lining the lungs.

The resin concentration of hair sprays is of the order of 4 percent with a ratio of 30 percent ethanol to 70 percent propellant for the liquid phase.

Other additives are often put into hair sprays to give the hair a sheen (silicone oils).

A typical hair spray formulation contains the following ingredients and amounts:

PVP	(Resin)	4.60 parts by weight
Dimethyl phthalate	(Plasticizer)	0.20
Silicone oil	(Sheen)	0.10
Ethanol	(Solvent)	25.00
Freon 11	(Propellant)	45.00
Freon 12	(Propellant)	25.00
Perfume	(For effect)	0.10
		100.00 parts by weight

Figure 22-8 Film of hair spray. Hair spray was allowed to dry on white surface and was then pulled up to reveal film.

Since PVP tends to pick up moisture, other, less hygroscopic polymers are beginning to replace PVP in hair sprays. For example, significantly better moisture control is obtained with a copolymer made from a 60/40 ratio of vinyl pyrrolidone and vinyl acetate.

Depilatories

The purpose of a depilatory is to remove hair chemically. Since skin is sensitive to the same kind of chemical attack as hair, such preparations should be used with caution and, even then, some attack on the skin is almost unavoidable. Because of this, the interval between applications of a depilatory should be of the order of a week or so. It should never be used on skin which is infected or which has a rash, and should not be followed by application of a deodorant with its *astringent* (contracting) action. If the sweat pores are closed by the deodorant, the caustic chemicals are retained and can do considerable harm. If the sweat pores are open, the body fluids will dilute and wash the caustic chemicals to the outside of the body.

The chemicals used as depilatories include sodium sulfide, calcium sulfide, strontium sulfide (water-soluble sulfides), and calcium thioglycolate [$Ca(HSCH_2COO)_2$], the calcium salt of the compound used to break S-S bonds between protein chains in permanent waving. A typical cream depilatory contains calcium thioglycolate (7.5 percent), calcium carbonate (filler, 20 percent), calcium hydroxide (provides basic solution, 1.5 percent), cetyl alcohol ([$CH_3(CH_2)_{15}OH$], skin conditioner, 6 percent), sodium lauryl sulfate (detergent, 0.5 percent), and water (64.5 percent).

The water-soluble sulfides are all strong bases in water, as indicated by the hydrolysis of the sulfide ion:

$$\underset{\text{SULFIDE}}{S^{2-}} + H_2O \longrightarrow HS^- + \underset{\text{HYDROXIDE}}{OH^-}$$

For example, a 0.1 M solution of Na_2S has a pH of about 13—a strongly basic solution. The compounds act chemically on the hair to disrupt bonds in the protein chains and cause it to "cure," that is, to hydrolyze to soluble amino acids and small peptides, which may be wiped off with a damp cloth.

0.1 M means 0.1 mole of Na_2S (7.81 g) dissolved in sufficient water to make one liter of solution.

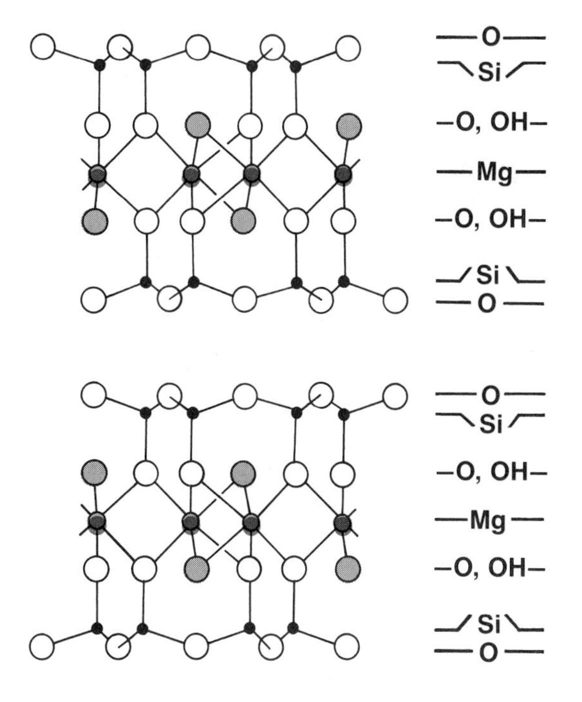

Figure 22–9 A partial structure of talc ($Mg_3Si_4O_{10}(OH)_2$). The open circles are atoms of oxygen; the small black circles are silicon; shaded circles are OH groups; brown-filled circles are magnesium. Only a portion of two sheets is shown. (From Evans, R. C.: Introduction to Crystal Chemistry, 2nd ed., New York, Cambridge University Press, 1964.)

The area on which a depilatory has been used should be washed with soap and water, dried, and then treated with small amounts of talcum powder. Talcum powder is a mixture of finely ground talc [$Mg_3(OH)_2Si_4O_{10}$] and a perfume. Talc has a porous structure and a great capacity for adsorbing liquids. Its adsorbing power results in large part from the tremendous surface area of a finely ground powder. This provides very many opportunities for liquid molecules to "rest" on the surface via physical adsorption. In addition, smaller molecules can be absorbed into the open spaces of the structure of talc (Figure 22–9). Talc is an effective collector of liquids, such as the serum and mucus secreted after the use of depilatories. Talc is relatively inert chemically and, consequently, it does not undergo chemical reactions that might irritate exposed nerve endings. It forms a protective cover that excludes air and other nerve-irritating chemicals.

Killing Germs

Harmful microorganisms (germs) are always with us in large numbers. Fortunately, there are so many nonpathogenic microorganisms among them that there is little room left for the pathogenic kinds.

Alcohols dehydrate microbes.

Things we do to our skin, such as shaving, can upset the balance between harmful and harmless bacteria and promote infection by pathogenic microorganisms. ***Disinfectants*** are chemicals used to kill these organisms before they can overcome the skin's defenses. Most commonly used are the alcohols. These are the only disinfectants used in many after-shave preparations. Maximum effectiveness of ethanol is reached at a concentration of 70 percent, while isopropyl (rubbing) alcohol is most effective at a concentration of 50 percent. These alcohols kill germs apparently by hydrogen bonding with water, which dehydrates the cellular structure of the germ.

One widely used disinfectant is phenol; its aqueous solution is known as carbolic acid. An —OH group attached to the benzene ring is slightly acidic. The disinfectant action of phenol was discovered in 1867 by Sir Joseph Lister, who introduced it into surgery. It appears that phenol kills bacteria by denaturing their cellular proteins. Today, about one third of all toilet soaps sold contain some derivative of phenol. One famous compound in this class is commonly known as ***hexachlorophene.*** This compound is extremely effective against staphylococci and streptococci bacteria which cause most so-called "body odors" and small infections.

PHENOL

To denature a protein is to break down its structure.

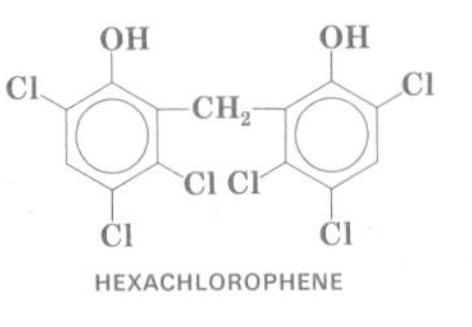

HEXACHLOROPHENE

In 1971 it became known that hexachlorophene in some way damaged the brain cells of baby monkeys, and in December of that year the FDA warned against the use of products containing hexachlorophene. Later, in 1972, the FDA placed a ban on hexachlorophene in all but prescription uses.

There are still other phenol derivatives which find wide use in soaps, deodorants, facial creams, and other cosmetics.

Deodorants

The 2,000,000 sweat glands on the body surface are primarily used to regulate body temperature via the cooling effect produced by the evaporation of the water they secrete. This evaporation of water leaves solid constituents, mostly sodium chloride, as well as smaller amounts of proteins and other organic compounds. Body odor results largely from amines and hydrolysis products of fatty oils (fatty acids, acrolein, etc.) emitted from the body, and from bacterial growth on the body. Sweating is both normal and necessary for the proper functioning of the human body.

Body odor is promoted by bacterial action.

There are three kinds of deodorants: those which directly "dry up" perspiration or act as astringents, those which have an odor to mask the odor of sweat, and those which remove odorous compounds by chemical reaction. Among those which have astringent action are hydrated aluminum sulfate, hydrated aluminum chloride, ($AlCl_3 \cdot 6H_2O$), aluminum chlorohydrate [actually aluminum hydroxychloride, $Al_2(OH)_5Cl \cdot 2H_2O$ or $Al(OH)_2Cl$ or $Al_6(OH)_{15}Cl_3$], and alcohols. Those compounds that act as deodorizing agents include zinc peroxide, essential oils and perfumes, and a variety of mild antiseptics. Zinc peroxide removes odorous compounds by oxidizing the amines and fatty acid compounds. The essential oils and perfumes absorb or otherwise mask the odors, and the antiseptics are generally oxidizing or reducing agents.

The most widely advertised deodorants contain aluminum salts as the active ingredients. Other materials are added to assist the application or to provide a fragrance. Aluminum salts are astringents in that they can reduce or close the openings of the sweat glands by affecting the hydrogen bonds that hold protein molecules together or (less often) by precipitating skin proteins. The ones that precipitate skin proteins cause skin-irritating effects. A typical spray deodorant will have the aluminum salt and minor ingredients dissolved in an alcoholic solution. A typical cream antiperspirant deodorant contains aluminum hydroxychloride (astringent; 20 percent), sorbitan monostearate (hydrogen bonds water and absorbs it; 5 percent), polyoxyethylene sorbitan monostearate (hydrogen bonds water and absorbs it; 5 percent), stearic acid (precipitates amines; 15 percent),

An astringent closes the pores, thus stopping the flow of perspiration.

propylene glycol (precipitates fatty acids; 5 percent), and water (for desired consistency; 50 percent).

Skin Preparations for Health and Beauty

Skin with a low fat content tends to be dry.

Healthy skin is important if we are to ward off bacterial invasion and disease. To remain healthy, the moisture content of skin must stay near 10 percent; if it is higher, microorganisms grow too easily; if lower, the corneal layer breaks down. Washing skin removes fats which help retain the right amount of moisture. If dry skin is treated with a fat after washing, it will be protected until enough natural fats have been regenerated.

Lanolin is grease from wool.

Lanolin is an excellent skin softener (***emollient***) and is a component of many cosmetics. It is a complex mixture of esters from hydrated wool fat. The alcohols in the esters have up to 33 carbon atoms, and the fatty acids have up to 37 carbon atoms. Cholesterol, a common alcohol in lanolin, is found both free and in esters (Chapters 17 and 21). Cholesterol appears to endow fat mixtures with the property of absorbing water. This is one factor that makes lanolin an excellent emollient. With its high proportion of free alcohols, particularly cholesterol, and hydroxy acid esters, lanolin has the structural groups (—OH) to hydrogen bond water (to keep the skin moist) and to anchor within the skin (the fatty acid and ester hydrocarbon structures; see Figure 22–10).

Creams

Colloids are intermediate in size between small molecules and clumps of molecules sufficiently large to precipitate.

Creams are generally emulsions of either an oil-in-water type or a water-in-oil type. An ***emulsion*** is simply a colloidal suspension of one liquid in another. The oil-in-water emulsion has tiny droplets of an oily or waxy nature dispersed throughout a water solution (homogenized milk is an example). The water-in-oil emulsion has tiny droplets of a water solution dispersed throughout an oil (natural petroleum and melted butter are examples). An oil-in-water emulsion can be washed off the hands with tap water, while a water-in-oil one gives the hands a greasy, water-repellent surface.

Cold cream originally was a suspension of rose water in a mixture of almond oil and beeswax. Subsequently, other ingredients were added to

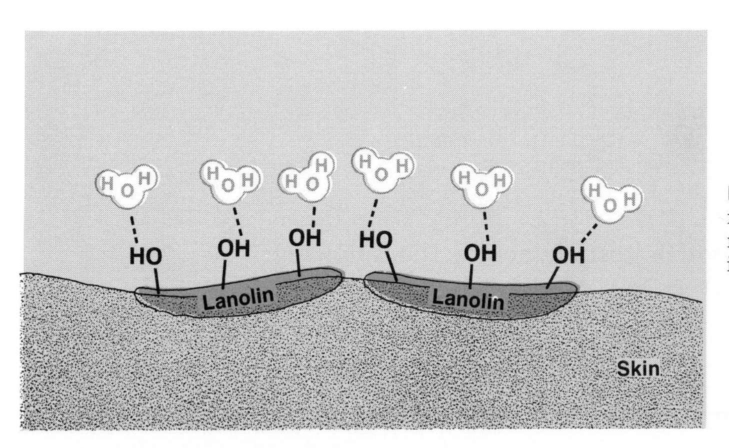

Figure 22–10 The hydroxyl groups of lanolin form hydrogen bonds with water and keep the skin moist. The fat parts of the molecule are "soluble" in the protein and fat layers of the skin.

get a more stable emulsion. An example of a modified cold cream composition is: almond oil, 35 percent; beeswax, 12 percent; lanolin, 15 percent; spermaceti (from whale oil), 8 percent; and strong rose water, 30 percent. Other oils can be substituted for some or all of the almond oil. Lanolin stabilizes the emulsion.

Creams add oil or fat content to surface skin.

Vanishing cream is a suspension of stearic acid in water, to which a stabilizer has been added to prevent the ingredients from separating. The stabilizer may be a soap, such as potassium stearate. These creams do not actually vanish; they merely spread as a smooth, thin covering over the skin.

The composition of rose water is given in Table 22-1.

Creams of various sorts may be used as the base for other cosmetic preparations; other ingredients are added, to give desired properties to the creams. As an example, hydrated aluminum chloride can be added to prepare a cream deodorant.

Lipstick

The skin on our lips is covered by a very thin corneal layer which is free of fat and is consequently dried out easily. A normal moisture content is maintained from the mouth. In addition to being a beauty aid, lipstick can be helpful under harsh conditions which tend to dry lip tissue.

Lipstick consists of a solution or suspension of coloring agents in a mixture of high molecular weight hydrocarbons or their derivatives, or both. The material must be soft enough to produce an even application when pressed on the lips, yet the film must not be too easily removed, nor may the coloring matter run. Lipstick is perfumed to give it a pleasant odor. The color usually comes from a dye or "lake" from the eosin group of dyes. A ***lake*** is a precipitate of a metal ion (Fe^{3+}, Ni^{2+}, Co^{3+}) with an organic dye. The metal ion enhances the color or changes the color of the dye.

A lake is a coloring agent made up of an organic dye adhering to an inorganic substance called a mordant.

Two suitable dyes, used in admixture and with their lakes, are dibromofluorescein (yellow-red) and tetrabromofluorescein (purple):

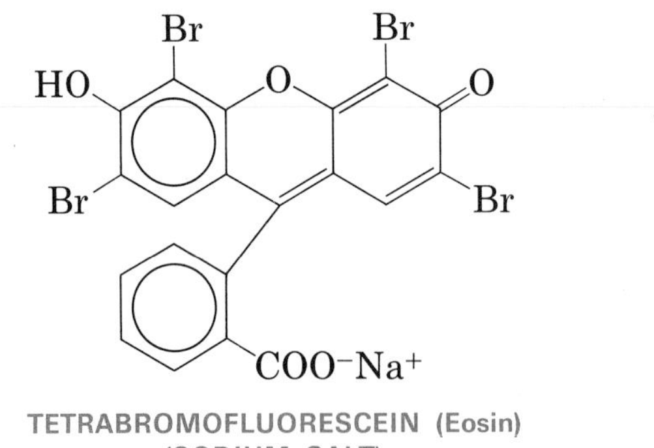

TETRABROMOFLUORESCEIN (Eosin)
(SODIUM SALT)

The ingredients in a typical formulation of lipstick include the following:

Dye	Furnishes color	4–8%
Castor oil, paraffins or fats	Dissolves dye	50%
Lanolin	Emollient	25%

List continued on following page

Carnauba wax Beeswax	Makes stick stiff by raising the melting point	18%
Perfume	Imparts pleasant odor	small amount

Carnauba wax and beeswax are high molecular weight esters.

In the manufacture of lipstick, the dye is first dispersed in the castor oil, and then the waxes, lanolin, and perfume are added, while the whole is heated and stirred to obtain a homogeneous mixture. The molten mass is then cast into suitable forms and subsequently inserted into holders, passed momentarily through a gas flame to obtain a smooth surface, and then packaged.

Suntan Lotions

Ultraviolet radiation tans skin.

One of the agents most harmful to skin is the short wavelength (ultraviolet) light from the sun. Various preparations have been used in the past to screen out all ultraviolet radiation and thus protect the skin when it must be exposed to the sun for long periods of time. Today, it is considered desirable to exclude the shorter, more harmful wavelengths, while transmitting enough less energetic, longer wavelength ultraviolet to permit gradual tanning.

The variety of suntan products ranges from lotions, which selectively filter out the higher energy ultraviolet rays of the sun, to preparations which essentially dye light-colored skin a tan color.

The lotions which filter out the ultraviolet rays are more accurately described as sunscreens, and their ingredients are often mixed with other materials, to give a lotion which both screens and tans. A common ingredient in preparations used to *prevent* sunburn is *p*-aminobenzoic acid.

p-AMINOBENZOIC ACID

Like most aromatic compounds, it absorbs strongly in the ultraviolet region of the spectrum (Figure 22–11).

The *p*-aminobenzoic acid is emulsified with a mixture of alcohols, an oil, and water by a high-molecular-weight fatty acid ester. A leading suntan lotion contains monoglyceryl *p*-aminobenzoate (3 percent), mineral oil (25 percent), sorbitan monostearate and polyoxyethylene sorbitan monostearate (10 percent; used as emulsifiers), and water (62 percent). Perfume is added to give the material a pleasant odor.

Newer suntan lotions contain 2-ethoxyethyl *p*-methoxycinnamate which absorbs the peak burning wavelength of ultraviolet radiation, at 308 nanometers. These lotions, like all the others, do not prevent sunburn completely. They merely prolong the possible length of exposure to the sun before severe sunburn occurs.

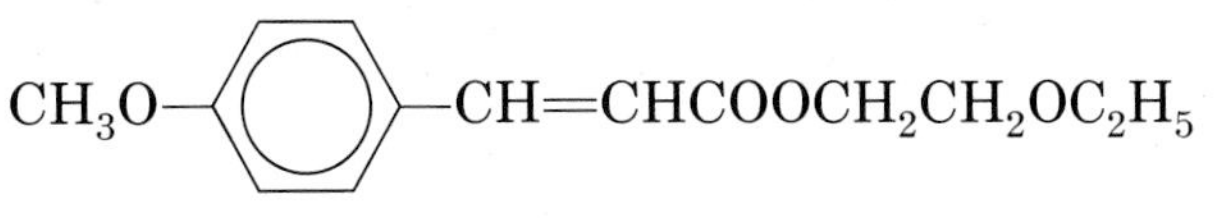

2-ETHOXYETHYL *p*-METHOXYCINNAMATE

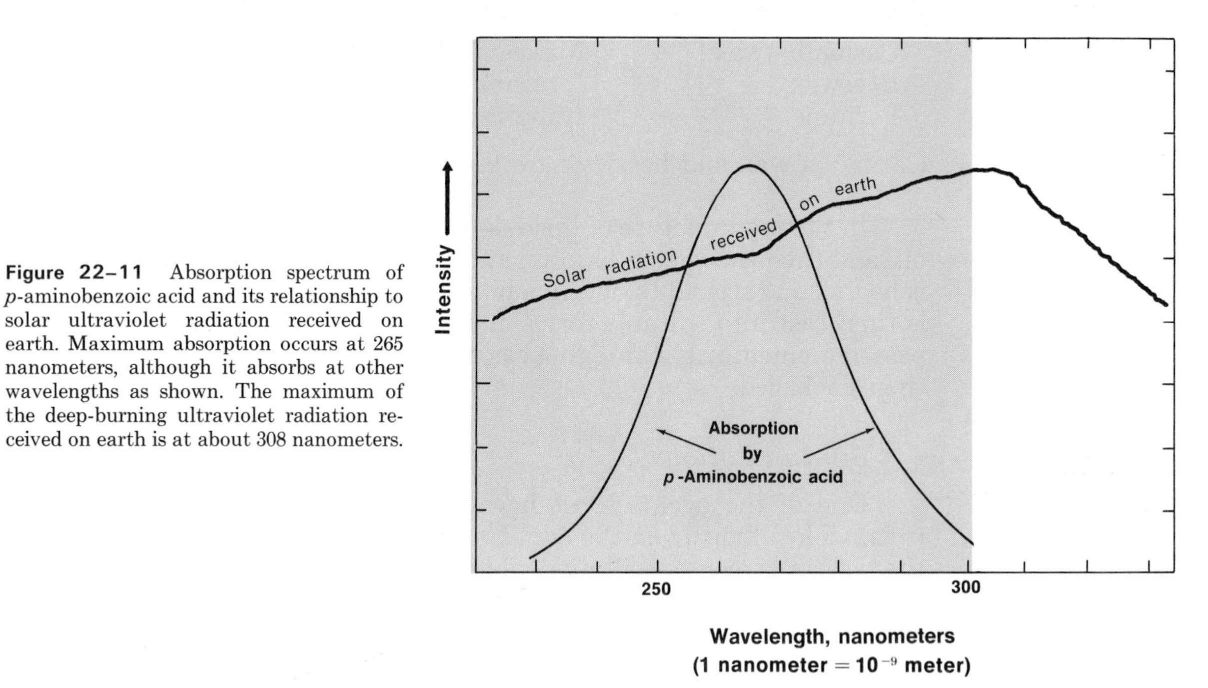

Figure 22–11 Absorption spectrum of *p*-aminobenzoic acid and its relationship to solar ultraviolet radiation received on earth. Maximum absorption occurs at 265 nanometers, although it absorbs at other wavelengths as shown. The maximum of the deep-burning ultraviolet radiation received on earth is at about 308 nanometers.

In tanning, the skin is stimulated to increase its production of the pigment ***melanin*** (see Hair Coloring and Bleaches in this chapter). At the same time, the skin thickens and becomes more resistant to deep burning.

Increased melanin protects sensitive lower layers of skin.

Preparations for the relief of the pain of sunburn are solutions of local anesthetics (such as benzocaine), plus other ingredients for color, odor, and softening of burned tissue. As you might expect, lanolin is especially suited for this purpose and it is often used. By softening the tissue, the emollient provides a ready access for the anesthetic to reach the nerve endings.

Face Powder

Face powder is used to give the skin a smooth appearance by covering up any oil secretions which would otherwise give it a shiny look. The powder must have some hiding ability, but if it is too opaque it will look too obvious. A powder which has the proper appearance, sticking properties, absorbance for oily skin secretions, and spreading ability usually requires several ingredients. A typical formula is the following:

Talc	65%	to which are added small amounts of perfume and coloring matter
Precipitated chalk	10%	
Zinc oxide	20%	
Zinc stearate	5%	

The absorbing properties of talc were discussed earlier. Precipitated chalk is $CaCO_3$, formed by precipitation from aqueous solutions of calcium chloride and sodium carbonate:

$$Ca^{2+} + 2Cl^- + 2Na^+ + CO_3^{2-} \longrightarrow \underline{CaCO_3} + 2Na^+ + 2Cl^-$$

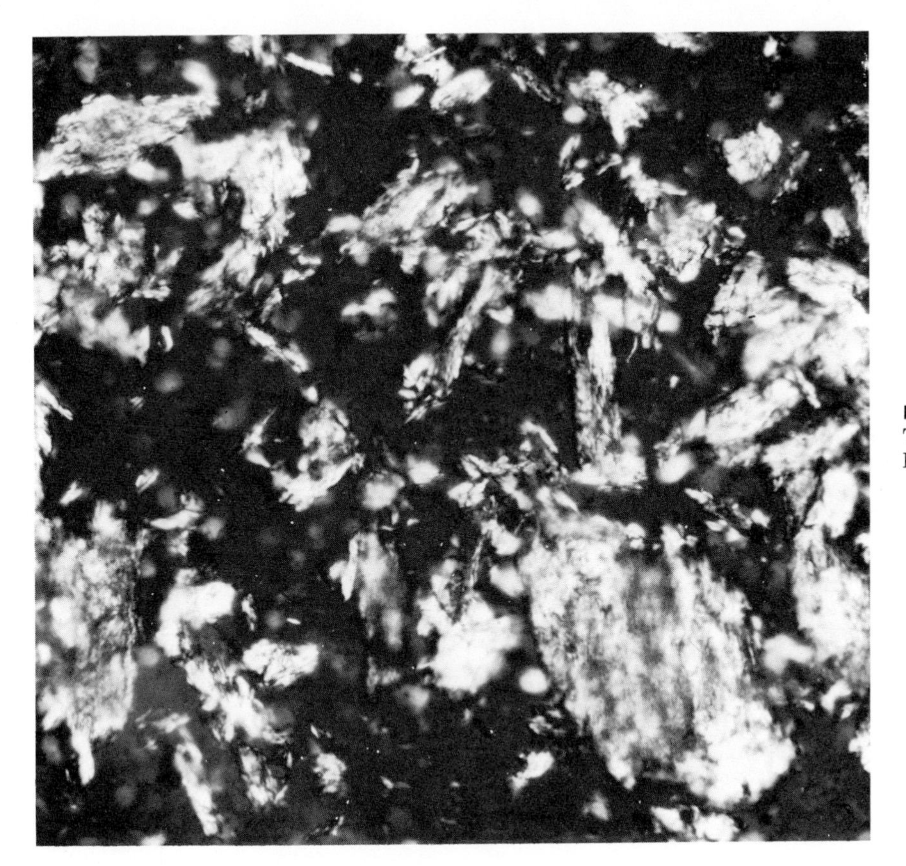

Figure 22–12 Photo of talc. The structure of talc is shown in Figure 22-9.

Zinc oxide has astringent properties; zinc stearate is a solid soap added for a binder.

Compact powders are formulations of a pigment blend, similar to face powder, with mineral oil (or lanolin) and organic hydroxy compounds as binders. They are pressed into cakes after thorough mixing.

Perfume

Perfumes are complex mixtures of odorous compounds.

A perfume is a material containing one or more volatile constituents which can produce a desired aroma. The sense of smell is quite complex and the nose is able to distinguish a truly amazing number of different molecules. The stereochemical theory of smell was described and illustrated in Figure 21–5. The chemistry of perfumes is quite complex, since it involves up to 5000 different natural or synthetic materials. A typical perfume will have at least three components of somewhat different volatility and molecular weights. (Recall that lower-molecular-weight compounds are generally more volatile.) The first, called the ***top note,*** is the most volatile and is the obvious odor when the perfume is first applied. The second, called the ***middle note,*** is less volatile and is generally a flower extract (violet, lilac, etc.). The last, or ***end note,*** is least volatile, and is usually a resin.

Most perfumes contain many components, and chemically they are often complex mixtures. As the analysis of natural perfume materials has progressed, the use of pure synthetic organic compounds to duplicate specific odors has become very common. An example is the isolation of civetone (see structure, below), a cyclic ketone from civet, a secretion of the civet cat of Ethiopia. It is highly valued for perfumes. The disgustingly obnoxious odor of civet becomes pleasant in extreme dilutions. The secretion, composed of free NH_3, resin, fat, and volatile oil, is located in a double pocket pouch under the skin of the animal's abdomen, with an opening near the tail. The animal uses its musk glands for scenting tree trunks, the ground, and similar places as a means of communication so that members of the species will be able to find each other at night in the forest. The musk is also used as a means of self-defense when the civet is attacked by dangerous carnivorous animals. The foul-smelling, burning secretion is discharged into the enemy's face; this momentarily stuns the attacker, giving the civet cat time to escape.

Many substances which have disagreeable odors at high concentrations have pleasant odors at low concentrations.

Civetone is now available in a synthetic form. It is prepared by forming 8-hexadecene-1,16-dicarboxylic acid into a ring. The thorium ion (Th^{4+}) catalyzes the closure of the ring.

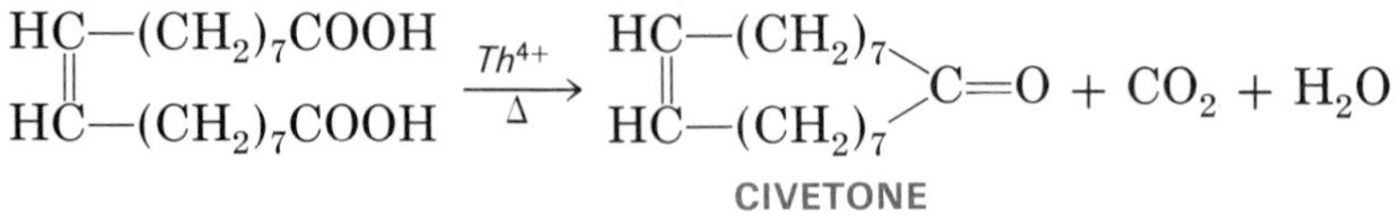

$$\begin{matrix} HC{-}(CH_2)_7COOH \\ \| \\ HC{-}(CH_2)_7COOH \end{matrix} \xrightarrow[\Delta]{Th^{4+}} \begin{matrix} HC{-}(CH_2)_7 \\ \| \\ HC{-}(CH_2)_7 \end{matrix}\!\!>\!C{=}O + CO_2 + H_2O$$

CIVETONE

Other compounds used in perfumes include high molecular weight alcohols and esters. An example is geraniol (b.p. 230°C), a principal component of Turkish geranium oil.

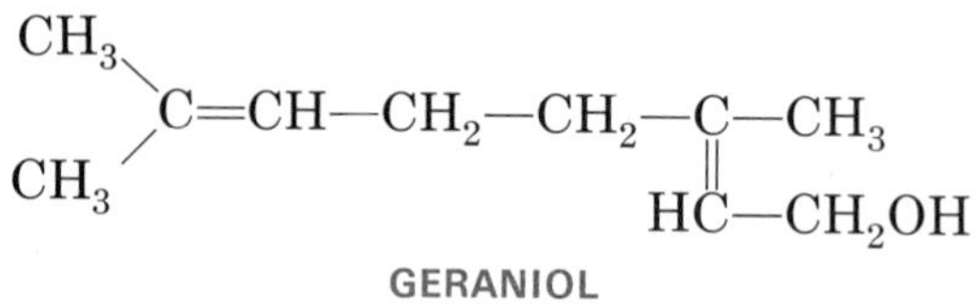

$$(CH_3)_2C{=}CH{-}CH_2{-}CH_2{-}\underset{\underset{HC-CH_2OH}{\|}}{C}{-}CH_3$$

GERANIOL

Esters of this alcohol are used to make synthetic rose aromas for perfumes. For example, the ester formed by reaction between geraniol and formic acid has a rose odor.

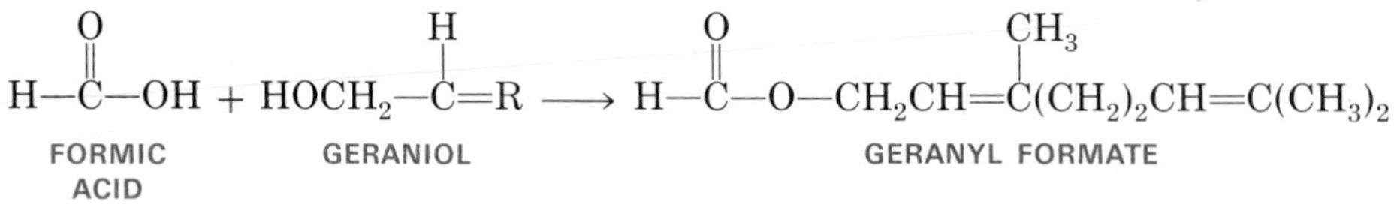

$$H{-}\overset{\overset{O}{\|}}{C}{-}OH + HOCH_2{-}\overset{\overset{H}{|}}{C}{=}R \longrightarrow H{-}\overset{\overset{O}{\|}}{C}{-}O{-}CH_2CH{=}\overset{\overset{CH_3}{|}}{C}(CH_2)_2CH{=}C(CH_3)_2$$

FORMIC ACID — GERANIOL — GERANYL FORMATE

Typical perfumes are 10 to 25 percent perfume essence and 75 to 90 percent alcohol. Perfumes are added to most cosmetics to give the product a desirable odor; they also mask the natural odor of other constituents. They are often mildly bactericidal and antiseptic. Some of the synthetic compounds used as fragrances in perfumes and colognes are given in Tables 22–1 and 22–2.

Ethyl alcohol is a major constituent of most perfumes.

Eye Makeup

There are several types of eye makeup: eyebrow pencils, mascara for eyelashes, and shading, among others. Their use varies with current fashion, though the variations can usually be traced back anywhere from

TABLE 22–1 Composition of a Typical Perfume

French Lilac-type Perfume

Material	*Source or formula*	*Percent by weight*
Jasmine, artificial	Mixture a (see below)	6.70
Pelargonic aldehyde (nonanal)	$CH_3(CH_2)_7CHO$	1.02
Terpineol	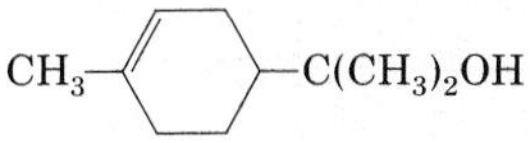 CH_3–(cyclohexene ring)–$C(CH_3)_2OH$	3.35
Musk ketone	Musk deer found principally in mountains of Northern India and Central Asia: $CH_3COC_6(C_4H_9)(CH_3)_2(NO_2)_2$	3.35
Hydroxycitronellal	$(CH_3)_2COH(CH_2)_3CH(CH_3)CH_2CHO$	67.00
Geraniol	$(CH_3)_2CCH(CH_2)_2C(CH_3){=}CHCH_2OH$	1.75
Phenylacetaldehyde, 50%	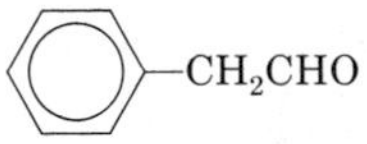 C_6H_5–CH_2CHO	1.75
β-Phenylethyl alcohol	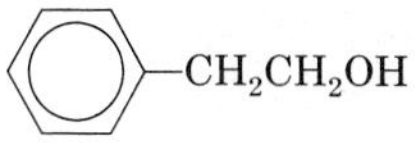 C_6H_5–CH_2CH_2OH	10.05
Anisic aldehyde	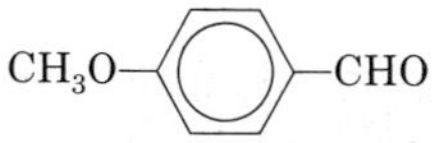 CH_3O–C_6H_4–CHO	0.66
Rose, artificial	Mixture b (see below)	1.02
Labdanum	Cistus bushes of Cyprus (Mixture c, see below)	3.35
		100.00

Mixture a

Material	Source or formula	Percent by weight
Benzyl acetate	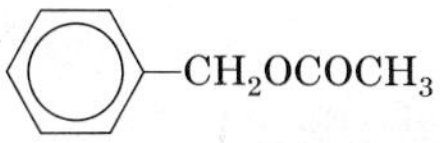 C_6H_5–CH_2OCOCH_3	65.0
Linalyl acetate	$CH_2{=}CHC(OCOCH_3)(CH_3)CH_2CH_2CH{=}C(CH_3)_2$	7.5
Linalool	$CH_2{=}CHCOH(CH_3)CH_2CH_2CH{=}C(CH_3)_2$	15.5
Benzyl alcohol	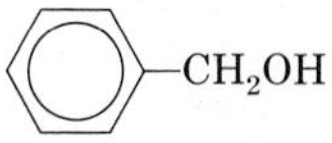 C_6H_5–CH_2OH	6.0
Indole	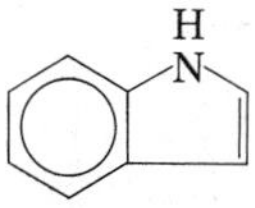 (indole structure, N–H)	2.5
Jasmone	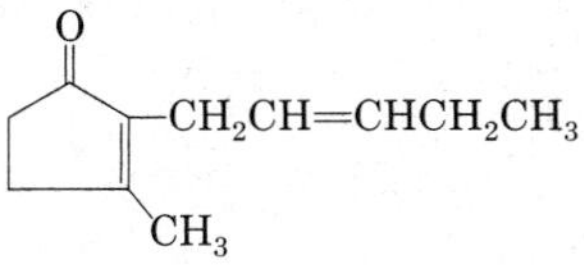 (cyclopentenone ring with =O, –$CH_2CH{=}CHCH_2CH_3$, and –CH_3)	3.0
Methyl anthranilate	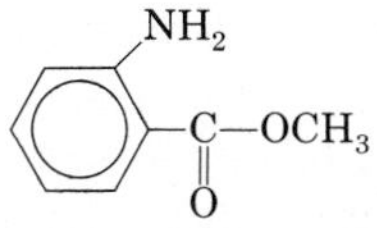 (C_6H_4 with NH_2)–$C(=O)$–OCH_3	0.5
		100.00

Mixture b

A mixture of geraniol, linalool, higher homologues of phenylethyl alcohol, citronellol, and esters of phenylethyl alcohol and phenylacetic acid.

Mixture c

A mixture of two ketones (acetophenone and 2,2,6-trimethylcyclohexanone) and other, still unidentified compounds.

TABLE 22-2 Composition of a Typical Cologne

	Eau de Cologne	
Material	*Source or formula*	*Amount*
Ethyl alcohol	C_2H_5OH	3 liters
Oil of bergamot	Peel of fruit of *Citrus aurantium.* Principal odorous constituent (36–45%) is linalyl acetate (above), but also contains other substances, such as 6% linalool, limonene, dipertene, and bergaptene	7 grams
Oil of lemon	Peel of lemons {90% limonene + terpene + phellandrene + pinene; 6% citral + citronellol + geranyl acetate + sesquiterpenes}	17 grams
Oil of neroli	Flowers of bitter orange, *Citrus aurantium,* subspecies, *sinensus* {90% limonene + citral + decylaldehyde + methylanthranilate + linalool + terpineol}	20 grams
Oil of rosemary	Herbs of the rosemary plant {10% borneol + 2.5% esters (bornyl actate) + camphor + eucalyptol + pinene + camphene}	7 grams

100 to 4000 years. Eye shadow, which currently is popular, was also very popular in ancient Egypt.

Eyebrow pencils are very much like lipstick, but they contain a different coloring matter. The coloring matter is a pigment such as lampblack; the other ingredients include fats, oils, petrolatum, and lanolin, blended to give the desired melting point, which may be raised by the addition of beeswax or paraffin. Petrolatum is a semisolid mixture of hydrocarbons (saturated, $C_{16}H_{34}$ to $C_{32}H_{66}$; and unsaturated, $C_{16}H_{32}$, etc.; melting point, 34 to 54°C). Brown pencils are made by using iron oxide pigments in place of lampblack.

Mascara is used to darken eyelashes and give them a longer appearance. The same colors as in eyebrow pencils are used, as well as other mineral coloring matters such as chromic oxide (dark green) and ultramarine (blue pigment of variable composition; probably a double silicate of sodium and aluminum silicate with some sodium sulfide). The coloring matter is suspended in a foundation that is a mixture of a soap, oils, fats, and waxes. The mascara may be water-soluble or water-resistant, depending upon the composition of the foundation. A typical formula consists of about 40 percent wax (beeswax, carnauba wax, and paraffin, adjusted for hardness), 50 percent soap, 5 percent lanolin and 5 percent coloring matter.

In the discussion of depilatories, the danger of the sulfide ion was emphasized. Through hydrolysis, the sulfide ion can produce a hydroxide ion, which functions as a corrosive poison. If very much mascara containing sodium sulfide gets into the eye, it can cause blindness.

$$S^{2-} + HOH \longrightarrow HS^- + OH^-$$

Take the case of *Lash-Lure,* a product sold many years ago to dye eyebrows and eyelashes. It was very toxic and the *Journal of the American Medical Association* reported at least 17 cases of *Lash-Lure* causing blindness. The active ingredient of *Lash-Lure* was a para-phenylenediamine derivative (see Hair Coloring and Bleaches). The blindness (or death in at least one case) was a disaster, but in addition, the severe pain was almost unbearable. Ulcers developed around the eyes: the cornea peeled off. The eyes drained constantly.

Para-phenylenediamine derivatives are still used in hair dyes. Extreme caution should be used when these dyes are applied. They should be kept away from the eyes and out of scrapes and cuts.

Titanium dioxide, a white powder, is also used as a base for many eye makeup preparations.

Eye shadow is a formulation of a coloring matter suspended in an oily-fatty-waxy base. A formula which has been used for this purpose is 60 percent petroleum jelly, 6 percent lanolin, 10 percent fats and waxes (approximately equal amounts of cocoa butter, beeswax, and spermaceti), and the balance zinc oxide (white) plus tinting or coloring dyes. Cocoa butter (melting point, 30 to 35°C) is composed of glycerides of stearic, palmitic, and lauric acids. It is obtained from natural products by compression of cacao seed. Spermaceti (melting point, 45°C) is chiefly cetyl palmitate, $CH_3(CH_2)_{14}COO(CH_2)_{15}CH_3$. It is taken from the solid fat from the head of the sperm whale.

In all these formulations, modern synthetic waxes, soaps, and fats of fixed and reproducible properties are being used successfully to obtain more standardized products. For example, the Carbowaxes, which are polyethylene glycols, are used in the formulation of protective hand creams, astringent cream, hair conditioners, and shaving creams. The molecular weights range from 1000 to 6000. They have low toxicity, are water soluble, and are easy to remove after using. Their preparation begins by adding water to ethylene oxide in the presence of sodium hydroxide. The reaction continues until the ethylene oxide is exhausted or the NaOH is neutralized.

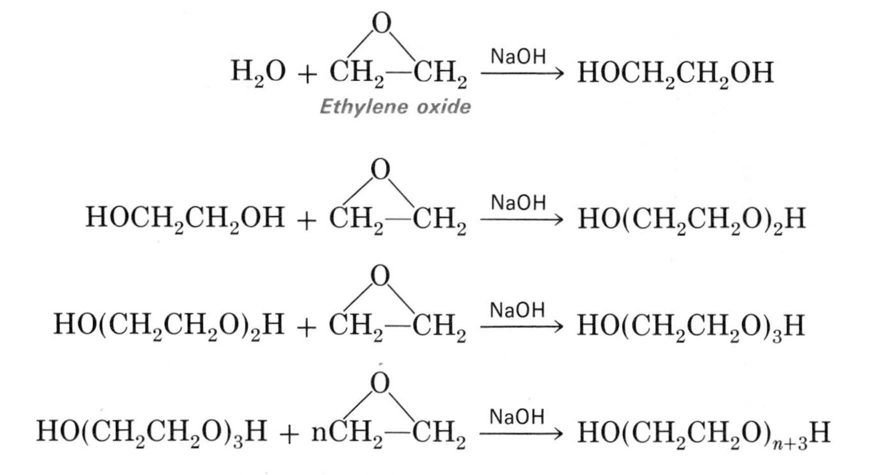

Nail Polish

Nail polish and hair sprays are formulated very much alike.

Nail polish is essentially a lacquer or varnish. It can be made of nitrocellulose, a plasticizer, a resin, a solvent, and perhaps a dye. The nitrocellulose can be replaced by another polymer molecule, which possesses similar qualities. The evaporation of the solvent leaves a film of nitrocellulose, plasticizer, resin, and dye. The nitrocellulose furnishes the shiny film; the plasticizer is added to make the film less brittle; and the resin is added to make the film adhere to the nail better and prevent flaking. Perfumes are added to cover the odor of the other constituents. A typical formulation is:

Nitrocellulose	15%
Acetone (solvent)	45%
Amyl acetate (solvent)	30%
Butyl stearate (plasticizer)	5%
Ester gum (resin)	5%

Perfumes and colors are added as needed.

Ester gum is a combination of esters—mainly glyceryl, methyl, and ethyl esters of rosin. Rosin is the resin remaining after distilling turpentine from pine exudate. It is 80 to 90 percent abietic acid. Rosin is slightly toxic to mucous membranes and slightly irritating to the skin. Its sticky nature is well known. The ester gum is prepared by heating rosin and the alcohol under pressure until the esterification occurs. The gums are soluble in nonpolar solvents.

Abietic acid

Nail polish removers are simply solvents which dissolve the film left by the nail polish. They consist largely of acetone or ethyl acetate, or both, to which small amounts of butyl stearate and diethylene glycol monomethyl ether have been added. However, some formulations contain combinations of amyl acetate, butyl acetate, ethyl acetate, benzene, olive oil, lanolin, and alcohol. Both nail polish and nail polish removers are very flammable, and care should be taken never to use them in the presence of open flames or lighted cigarettes.

$CH_2(OH)—CH_2OCH_2—CH_2(OCH_3)$

Diethylene glycol monomethyl ether

Cuticle softeners are primarily wetting agents and alkalies used to soften skin around the fingernail so it can be shaped as desired. The use of alkali to soften and swell protein is well known. A typical cuticle softener contains potassium hydroxide (3 percent), glycerol (12 percent), and water (85 percent). It may also contain sodium carbonate (an alkali), triethanol amine (a detergent, used as wetting agent), and trisodium phosphate (an alkali).

SELF-TEST 22-A

1. The surface pH of the human body is about ______.
2. Hair keratin contains considerable amounts of the sulfur-containing amino acid ______________.
3. Two types of bonds that hold strands of protein in place include ______________ and ______________.
4. Wet hair swells and stretches because of the breaking of ______________ bonds.
5. The peroxide oxidation of melanin produces a (colored/colorless) product.
6. A plasticizer causes a plastic material to be more ______________.
7. The propellants in most aerosol sprays are ______________.
8. Zinc peroxide acts in deodorants to oxidize amines and ______________ ______________.

9. Commercial lanolin comes from what animal? ____________

10. Vanishing cream is completely absorbed through the pores of the skin. (True/False)

CLEANSING AGENTS

Dirt has been defined as matter in the wrong place. Tomato catsup is esteemed as a palatable and nutritious food, but on your shirt it is dirt. There is a large number of cleansing, or ***surface-active,*** agents capable of removing the dirt without harm to the shirt (Figure 22–13). Indeed, radio and television advertising might lead us to believe that the soaps and detergents we have today are unique and vastly superior to the products of a year or a century ago. This is not always so. Soap, for example, has always been made by a time-tested recipe that dates back at least to the second century of the Christian era. Galen, the great Greek physician, mentions that soap was made from fat, ash lye, and lime. Moreover, Galen stated that soap not only served as a medicament but also removed dirt from the body and clothes.

Surface-active agents stabilize suspensions of nonpolar materials in polar solvents or vice versa. Examples include soaps, detergents, wetting agents, and foaming agents.

What *is* new is the greater purity of soap, the improvement in its cleaning action by numerous additives, and the advent of the relatively new synthetic detergents. The soap-making industry was revolutionized by two events. The first was the discovery of the process of making soda ash (Na_2CO_3; its water-softening and alkaline properties will be discussed later) from ordinary salt (NaCl) by Nicolas LeBlanc in 1791, and the second was the epoch-making work of the celebrated French chemist Chevreul, whose researches into the chemical constitution of the natural fats extended from 1813 to 1823. As a result of these two advances, the soap-makers of the 19th century were provided with ample quantities of sodium carbonate at a reasonable price and were armed with the knowledge of the true nature of fats and fatty acids. Accordingly, the soap industry made rapid progress; the products improved steadily and grew in diversity, until little by little the soap industry attained its present gigantic proportions.

Figure 22–13 Photomicrograph of clean cotton cloth (*left*) and soiled cotton cloth (*right*). The proper application of surface-active agents should return the soiled cloth to its original state.

Soaps

Glycerides are esters of glycerol and long-chain fatty acids derived from animal and vegetable fats and oils.

Soaps are salts of long-chain fatty acids (Chapter 14). The nature (length of the chain and number of double bonds) of the fatty acid chain is largely determined by the glyceride used, which ultimately determines the unique properties of the soaps. The principal glycerides used for soap-making are derived from both plants and animals:

(1) Tallow or animal fat from beef or mutton is primarily an ester of stearic acid [$CH_3(CH_2)_{16}COOH$]. It is usually mixed with coconut oil in making soap to prevent the product from being too hard.

(2) Coconut oil is a low melting solid. It is primarily an ester of lauric acid [$CH_3(CH_2)_{10}COOH$]. A soap made from coconut oil alone is very soluble in water and will lather even in sea water.

(3) Palm oil contains a very high concentration of free fatty acids, about 45 to 50 percent of which is oleic acid [$CH_3(CH_2)_7CH{=}CH(CH_2)_7COOH$]. It is an important constituent in toilet soaps.

(4) Olive oil is used in making Castile soap. It has a larger percentage (70 to 85 percent) of esters of oleic acid than palm oil.

(5) Bone grease is an animal fat of somewhat lower melting point than tallow, and it comes from a variety of sources. It is a relatively cheap source of fat. The esters of oleic acid (41 to 51 percent) are most prominent.

(6) Cottonseed oil is also a cheap source of glycerides for making soap. Its esters are mostly of linoleic acid [$CH_3(CH_2)_4(CH{=}CHCH_2)_2(CH_2)_6COOH$].

The length and degree of saturation of the fatty acid chain influence the solubility and hardness of the soap. A saturated, long-chain fatty acid makes a harder, more insoluble soap.

Salting out involves tying up the water molecules with small ions so that it can no longer dissolve the soap.

In large soap-making factories, after most of the glycerol is separated for sale, salt (NaCl) is added to facilitate the separation of the soap from the water present (Figure 22–14). The hot soap can then be poured into molds, cooled, and cut into bars, which contain about 30 percent water.

Fillers or Builders

A number of materials are added to soap powders for laundry purposes. These materials are often quite basic, and their addition gives the soap a greater detergent action. Commonly added materials include sodium carbonate, sodium phosphates, sodium polyphosphates, and sodium silicate. Rosin neutralized with sodium hydroxide is also commonly added to laundry soaps in large amounts. The rosin is mostly abietic acid. The neutralized acid has the nonpolar (hydrocarbon) part and polar end required for a soap. Such soaps are not to be recommended for use on the human skin. Phosphates, carbonates, and silicates hydrolyze to give OH^- ions, which react with grease to make soaps.

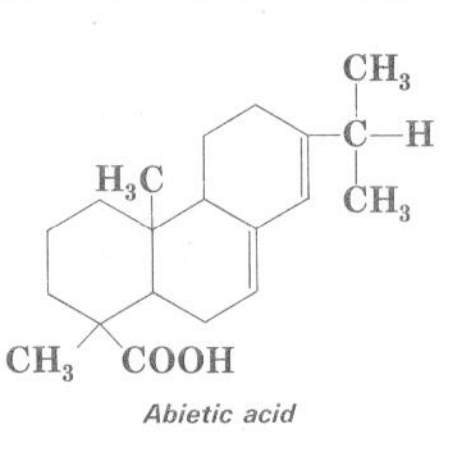

Abietic acid

$$PO_4^{3-} + H_2O \longrightarrow HPO_4^{2-} + OH^-$$
$$CO_3^{2-} + H_2O \longrightarrow HCO_3^- + OH^-$$
$$SiO_3^{2-} + H_2O \longrightarrow HSiO_3^- + OH^-$$

Types of Soaps

The variety of soaps available result from relatively minor variations in the soap-making process or in the ingredients added.

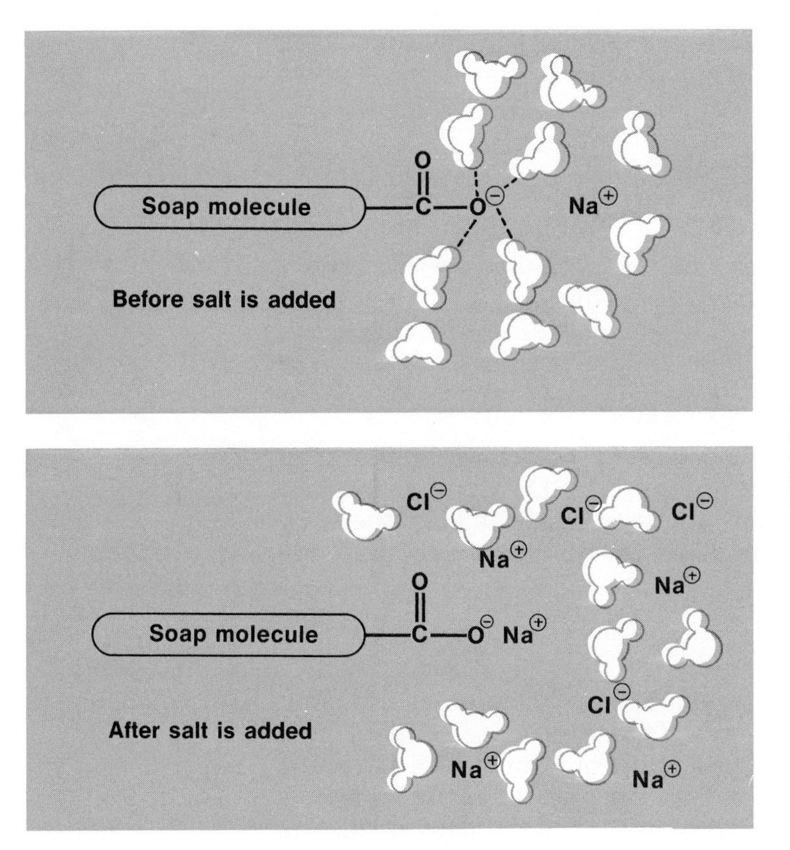

Figure 22–14 Salting out (shown diagrammatically). Without salt, there is sufficient water of hydration to keep the soap dissolved. When the salt is added, competition of the salt ions and negative soap ion for the polar water molecules decreases the hydration of the soap and makes it less soluble.

Toilet Soaps

Sodium—hard soap.

Potassium—soft soap.

Ammonium—liquid soap.

These generally have little or no filler and a minimal amount of free base, if any. Often much of the glycerol released in the saponification process is left in the soap. Perfumes, dyes, and medicinal agents may be added prior to casting the soap into a solid form. Floating soaps have air beaten into them as they solidify. A hard soap is obtained if there is a high percentage of a sodium salt of a relatively long-chain fatty acid, such as stearic acid, present. A soft or liquid soap is obtainable by saponification with potassium hydroxide, with the liquidity increasing as the chain length of the fatty acid decreases. Fatty acids with chains as short as C_{12} or shorter are not used because the resultant soaps irritate the skin. They are more volatile and create an odor problem. Fatty acids with chains longer than stearic acid tend to give very insoluble soaps.

Shampoos

These are often mixtures of several ingredients designed to satisfy a number of requirements. In addition to soaps, condensation products from diethanolamine and lauric acid are often used. These are essentially a type of detergent obtained by the following reaction:

$$\underset{\text{DIETHANOLAMINE}}{HN(CH_2CH_2OH)_2} + \underset{\text{LAURIC ACID}}{CH_3(CH_2)_{10}COOH} \xrightarrow{\Delta} \underset{\text{AN AMIDE DETERGENT}}{CH_3(CH_2)_{10}\overset{\overset{\displaystyle O}{\|}}{C}-\overset{\overset{\displaystyle H}{|}}{N}(CH_2CH_2OH)_2}$$

Some shampoos contain anionic detergents, which are less damaging to the eyes than cationic detergents. Sodium lauryl sulfate is an example of an anionic detergent.

$$CH_3(CH_2)_{11}OSO_3^-Na^+$$

SODIUM LAURYL SULFATE

The hair is more manageable and has a better sheen if all the shampoo is removed. An anionic detergent can be removed by rinsing with a cationic detergent (about a 1 percent solution), which neutralizes the anions and facilitates their removal. Caution should be exercised with rinses, since cationic detergents are damaging to the eyes. Types of cationic detergents are described in the next section.

Shampoos also contain compounds to prevent the calcium or magnesium in hard water from forming a precipitate; EDTA is often used for this purpose. Lanolin and mineral oil are often added to keep the scalp from drying out and scaling. Their presence is indicated by a cloudy appearance.

Other Soaps

Zinc stearate is only one of a large number of metallic soaps manufactured and used for special purposes. These are soaps of metal ions which are not used in regular consumer soaps. For example, lead stearate, a very poisonous soap, is used as a high pressure lubricant and as a catalyst to accelerate the drying of varnishes. Cadmium soaps are used in waterproofing and copper soaps in the manufacture of fungicides. In general, the soaps made with metallic ions with a charge of +2 or greater are very insoluble in water and are more like a grease (for which purpose some of them find considerable application).

Talcum Powders

Face powders and other talcum powders are generally composed of a relatively high percentage of talc, a naturally occurring hydrated magnesium silicate. The most common other ingredient is zinc stearate. While zinc stearate is a "soap," it is quite different from the common soaps in that it is water-repellent and is generally used as a fine, soft, bulky powder. Other materials, such as dyes, perfumes, and boric acid (mildly germicidal), are added to obtain the desired properties in the final product.

Synthetic Detergents

Synthetic detergents ("***syndets***") are derived from organic molecules which have been designed to have the same cleansing action, but less reaction than soaps with the cations found in hard water, such as Ca^{2+}, Mg^{2+} and Fe^{3+}. As a consequence, synthetic detergents are more effective in hard water than soap, which gives a precipitate in the presence of Ca^{2+}, Mg^{2+}, or Fe^{3+} ions. Since such precipitates have no cleansing action and tend to stick to laundry, their presence is very undesirable.

There is a synthetic detergent for almost every type of cleaning problem.

There is an enormous number of different synthetic detergents on the market. Their molecular structure consists of a long oil-soluble (hydrophobic) group and a water-soluble (hydrophilic) group. The hydrophilic groups include the sulfate ($—OSO_3—$), sulfonate ($—SO_3—$), hydroxyl (—OH), ammonium ($—NH_3^+$), and phosphate [$—OPO(OH)_2$] groups.

The early synthetic detergents were mostly sodium alkyl sulfate. An extension of the preparation of sodium lauryl sulfate, as presented in Chapter 14, is given here to illustrate the chemical processes involved. The principal starting material is a suitable vegetable oil, such as cottonseed oil or coconut oil. The first step is hydrogenation:

The preparation of all synthetic detergents begins with naturally occurring raw materials.

$$\begin{matrix} RCOOCH_2 \\ | \\ RCOOCH \\ | \\ RCOOCH_2 \end{matrix} + 6H_2 \xrightarrow{\text{Catalyst}} 3RCH_2OH + \begin{matrix} CH_2OH \\ | \\ CHOH \\ | \\ CH_2OH \end{matrix}$$

COCONUT OIL — HYDROGEN — [MAINLY LAURYL ALCOHOL, $CH_3(CH_2)_{11}OH$] — GLYCEROL

The second step involves esterification of the —OH group on the end of the lauryl alcohol hydrocarbon chain. This is accomplished by treating the lauryl alcohol with sulfuric acid:

$$CH_3(CH_2)_{11}OH + H_2SO_4 \longrightarrow CH_3(CH_2)_{11}OSO_3H + H_2O$$

LAURYL ALCOHOL — SULFURIC ACID — LAURYL HYDROGEN SULFATE

The final step involves neutralizing the acidic lauryl hydrogen sulfate with sodium hydroxide:

$$CH_3(CH_2)_{11}OSO_3H + NaOH \longrightarrow CH_3(CH_2)_{11}OSO_3Na + H_2O$$

SODIUM LAURYL SULFATE

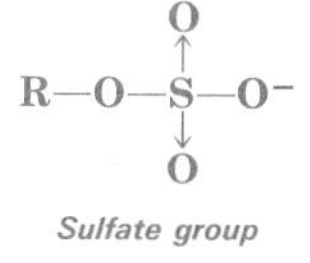

Sulfate group

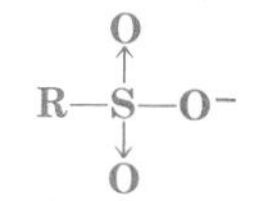

Sulfonate group

Other synthetic detergents (also called "surfactants," from ***surface-active agents***) are the alkylbenzenesulfonates. They are prepared by putting large alkyl groups on a benzene ring and then sulfonating the benzene ring with sulfuric acid or a related reagent. Before use, they are transformed into their sodium salts. The reactions involved are:

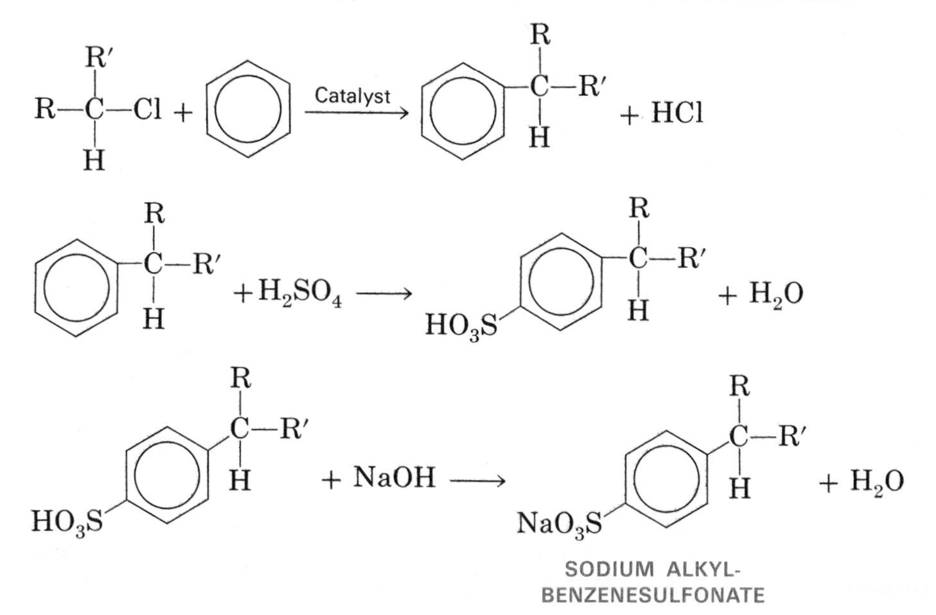

These molecules, like all others in this class, consist of a long hydrophobic chain and a highly polar group which interacts strongly with water.

In addition to the anionic (negatively charged) synthetic detergents already described, there are also detergents in which the polar group at the end of the hydrocarbon chain is positive or neutral.

Cationic (positively charged) detergents are almost all quaternary ammonium halides

$$R_1-\overset{\overset{\displaystyle R_2}{|}}{\underset{\underset{\displaystyle R_4}{|}}{N^+}}-R_3 \quad X^-$$

where one of the R groups is a long hydrocarbon chain and another frequently includes an —OH group. In these the water-soluble portion is positively charged; so they are sometimes called invert soaps (in soaps the water-soluble portion is negatively charged). They are prepared by treating the appropriate amine with an alkyl chloride:

$$R_1-\overset{\overset{\displaystyle R_2}{|}}{N}-R_3 + R_4Cl \longrightarrow R_1-\overset{\overset{\displaystyle R_2}{|}}{\underset{\underset{\displaystyle R_4}{|}}{N^+}}-R_3 \quad Cl^-$$

Cationic detergents act as disinfectants.

They frequently exhibit pronounced bactericidal qualities. Cationic detergents are incompatible with anionic detergents. When they are brought together, a high molecular-weight insoluble salt precipitates out, and this has none of the desired detergent properties of either starting material:

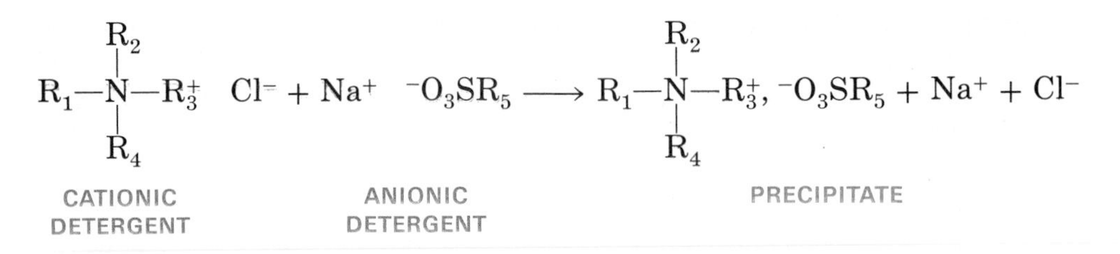

Nonionic detergents bear a polar, but not an ionic, grouping attached to a large organic grouping of low polarity. A typical example is a material prepared by the reaction of an acid with ethylene oxide:

$$RCOOH + (x+2)\underset{\diagdown O \diagup}{CH_2-CH_2} \longrightarrow R-\overset{\overset{\displaystyle O}{\|}}{C}-O-(CH_2)_2O(CH_2CH_2O)_xCH_2CH_2OH$$

In a typical nonionic detergent, $R = C_{12}H_{25}$ and $x = 2$. The large number of weakly polar C—O—C bonds has an effect similar to that of a single ionic group, and this end of the molecule provides the water solubility.

The nonionic detergents have several advantages over ionic detergents. Since they contain no ionic groups, they cannot form salts with calcium and magnesium ions and are, consequently, unaffected by hard water. For the same reason, they do not react with acids and may be used even in strong acid solutions.

In general, the nonionics foam less than ionic surface active agents, a property which is desirable where nonfoaming detergents are required, as in dishwashing. Nonionics do suffer from one drawback. They cannot be dried to solid powders. They are heavy liquids with melting points below room temperature. Consequently, nonionic detergents must be used in the form of water solutions which are far less convenient to handle than powdery materials.

Other Cleansers

Soaps containing pumice (finely powdered volcanic ash) will wash out ground-in dirt.

A very large number of special cleaners or cleansing agents is available. Simple abrasive cleansers contain a large percentage of an abrasive such as silica (SiO_2) or pumice (65 to 75 percent SiO_2, 10 to 20 percent Al_2O_3), a variable amount of soap, and generally some polyphosphates. They may also contain some synthetic detergent and a bleaching agent. All-purpose solid cleansers may contain one or more of a variety of salts which react with water to produce a basic solution: trisodium phosphate, sodium carbonate, sodium bicarbonate, sodium pyrophosphate, or sodium tripolyphosphate, plus a detergent and perhaps pine oil to give an attractive odor. Metal cleansers may contain strong acid or strong base to dissolve impurities. Many cleaning liquids contain organic solvents such as perchloroethylene, 1,1,1-trichloroethane, and the like. The vapors of these are quite toxic so the cleaners must be used in a ventilated area.

Whiter Whites

Bleaching agents are compounds which are used to remove color from textiles. Most commercial bleaches are oxidizing agents such as sodium hypochlorite. Optical brighteners are quite different, since they act by converting a portion of the invisible ultraviolet light, which impinges on them, into visible blue or blue-green light, which is emitted. Together or separately, these two classes of compounds find their way into commercial laundry and cleaning preparations, since they seem to be making clothes cleaner.

Chlorine produces hypochlorite when it reacts with water:

$$H_2O + Cl_2 \longrightarrow HOCl + HCl$$

In earlier times textiles were bleached by exposure to sunlight and air. In 1786, the French chemist Berthollet introduced bleaching with chlorine, and subsequently this process was carried out with sodium hypochlorite, an oxidizing agent prepared by passing chlorine into aqueous sodium hydroxide:

$$2Na^+ + 2OH^- + Cl_2 \longrightarrow \underset{\text{SODIUM HYPOCHLORITE}}{Na^+ + OCl^-} + Na^+ + Cl^- + H_2O$$

Shortly after this, hydrogen peroxide was introduced as a textile bleach. Later, a number of other compounds which contain oxidizing agents based on chlorine were developed and introduced.

One way to decolorize materials is to remove or immobilize those electrons in the material which are activated by visible light. The hypochlorite ion is capable of removing electrons from many colored materials.

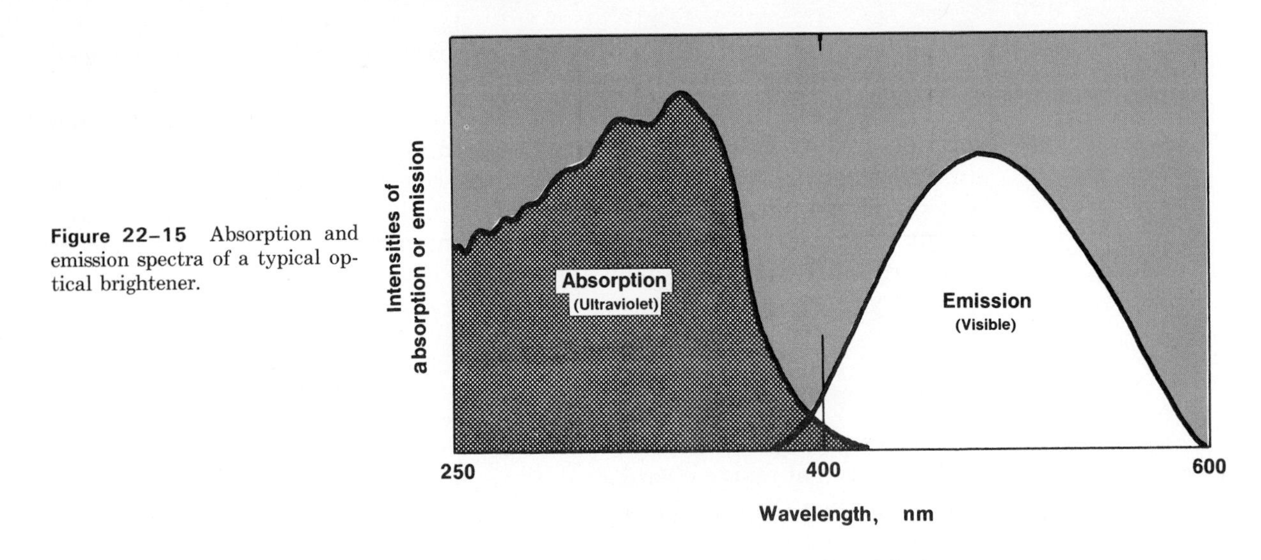

Figure 22–15 Absorption and emission spectra of a typical optical brightener.

In this process, the hypochlorite is reduced to chloride and hydroxide ions:

$$ClO^- + H_2O + 2e^- \longrightarrow Cl^- + 2OH^-$$

As stated above, optical brighteners are compounds which transform incident ultraviolet light into emitted visible light; this is a type of fluorescence. When optical brighteners are incorporated into textiles or paper, they make the material appear brighter and whiter (Figure 22–15).

A fluorescent material absorbs shorter wavelength light and emits light of a longer wavelength.

An example of such a brightener has this structure and its absorption and emission spectra are presented in outline form in Figures 22–15 and 22–16.

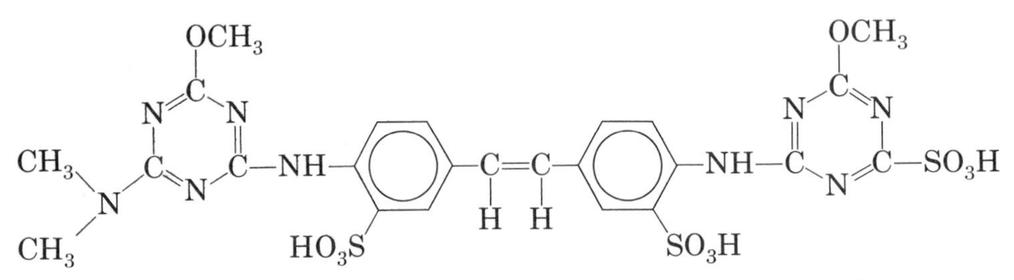

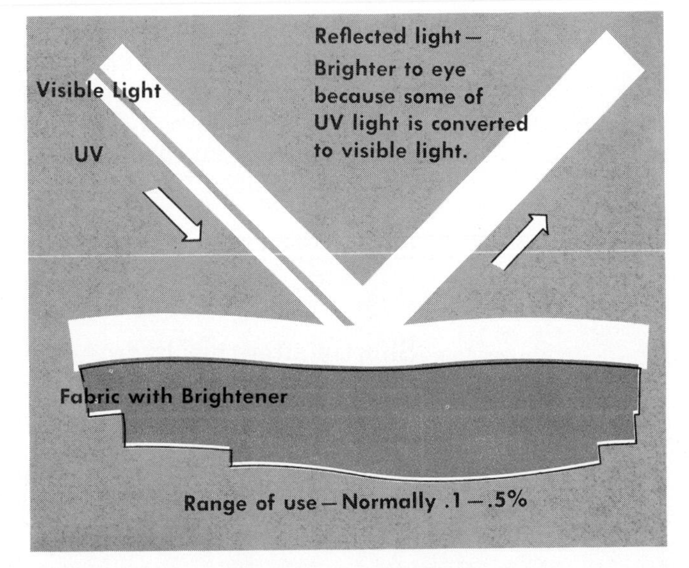

Figure 22–16 An optical brightener converts ultraviolet energy to visible light; hence, more light can be detected by the eye.

Spot and Stain Removers

To a large extent, stain removal procedures are based on solubility patterns or chemical reactions. Many stains, such as those due to chocolate or other fatty foods, can be removed by treatment with the typical dry-cleaning solvents such as tetrachloroethylene, $Cl_2C{=}CCl_2$.

Stain removers for the more resistant stains are almost always based upon a chemical reaction between the stain and the essential ingredients of the stain remover. A typical example is an iodine stain remover, which is simply a concentrated solution of sodium thiosulfate. The reaction here is

$$\underset{\text{IODINE}}{I_2} + 2Na_2S_2O_3 \longrightarrow \underbrace{2NaI + Na_2S_4O_6}_{\text{SOLUBLE IN WATER (COLORLESS)}}$$

Citric acid and tartaric acid will also remove iron stains and are less toxic than oxalic acid.

Iron stains are removed by treatment with oxalic acid, which forms a soluble coordination compound with the iron:

$$Fe_2O_3 + \underset{\text{OXALIC ACID}}{6H_2C_2O_4} \longrightarrow \underbrace{3H_2O + 2Fe(C_2O_4)_3^{3-} + 6H^+}_{\text{SOLUBLE IN WATER}}$$

Mildew stains can be removed by hydrogen peroxide or laundry bleach (sodium hypochlorite), which oxidizes the fungus responsible for the mildew. Blood stains on cotton can be removed by hypochlorite solution. Bleach should not be used on wool because it reacts chemically with the nitrogen atoms present in the peptide chains. The chemicals used to remove a few common stains are listed in Table 22-3.

Many stains can be removed by an appropriate solvent or chemical reagent.

TABLE 22-3 Some Common Stains And Stain Removers*

Stain	Stain Remover
Coffee	Sodium hypochlorite
Lipstick	Isopropyl alcohol, isoamyl acetate, Cellosolve ($HOCH_2CH_2OCH_2CH_3$), chloroform
Rust and ink	Oxalic acid, methyl alcohol, water
Airplane cement	50/50 amyl acetate and toluene or acetone
Asphalt	Benzene or carbon disulfide
Blood	Cold water, hydrogen peroxide
Berry, fruit	Hydrogen peroxide
Grass	50/50 amyl acetate and benzene or sodium hypochlorite or alcohol
Nail polish	Acetone
Mustard	Sodium hypochlorite or alcohol
Antiperspirants	Ammonium hydroxide
Perspiration	Ammonium hydroxide, hydrogen peroxide
Scorch	Hydrogen peroxide
Soft drinks	Sodium hypochlorite
Tobacco	Sodium hypochlorite

*Before any of these stain removers are used on clothing, the possibility of damage should be checked on a portion of the cloth that ordinarily is hidden.

SELF-TEST 22-B

1. A fat is a(n): (a) acid, (b) alcohol, (c) ester, (d) alkane, (e) olefin.
2. Vegetable oils can be used as well as animal oils to make soap. (True/False)
3. Is the hydrocarbon end of the soap molecule polar or nonpolar? ______________
4. Which is more soluble in water, calcium stearate or sodium stearate? ______________
5. Is a hydrocarbon chain hydrophobic or hydrophilic? ______________
6. Zinc stearate is an insoluble "soap" used in ______________.
7. Optical brighteners absorb ______________ light and change some of it to ______________ light.
8. Oxalic acid can be used to remove ______________ stains.

MATCHING SET

______	1. keratin	a.	salt of fatty acid
______	2. melanin	b.	reducing agent in wave lotion
______	3. sodium lauryl sulfate	c.	hair spray resin
______	4. polyvinylpyrrolidone	d.	holds moisture in skin
______	5. alcohol	e.	common alcohol in lanolin
______	6. hydrated aluminum chloride	f.	ultraviolet absorber in suntan lotion
______	7. soap	g.	alcohol produced in saponification of fat or oil
______	8. fat	h.	detergent builder
______	9. cholesterol	i.	radiates a different wavelength of light than that absorbed
______	10. thioglycolic acid	j.	a synthetic detergent
______	11. *p*-aminobenzoic acid	k.	laundry bleach
______	12. glycerol	l.	dark pigment
______	13. sodium tripolyphosphate	m.	deodorant component

_____ 14. whiteners — n. dehydrates skin microbes

_____ 15. sodium hypochlorite — o. skin and hair protein

QUESTIONS

1. You read in the newspaper about a new compound which will break disulfide bonds in proteins. What potential use might it have?

2. (a) What is the purpose of an emulsifier?
   (b) In which of the following cosmetics is an emulsifier important: suntan lotion, hair spray, cold cream?

3. Which one of each of the following pairs of properties would be appropriate for a hair spray propellant? Why?
   (a) high or low boiling point
   (b) soluble or insoluble in the active ingredients
   (c) capable or incapable of chemical reaction with the active ingredients
   (d) odorous or odorless
   (e) toxic or nontoxic

4. What is the structure of the monomer unit in polyvinylpyrrolidone?

5. What is the purpose of each of the following:
   (a) Freon in hair sprays?
   (b) Polyvinylpyrrolidone in hair sprays?
   (c) Aluminum chloride in deodorants?
   (d) *p*-Aminobenzoic acid in suntan lotion?

6. What specific substance is broken down during the bleaching of hair?

7. Use the structures of the constituents of lanolin to justify its ability to emulsify face creams.

8. Hydrogen bonding is a very handy theoretical tool. Name three applications of hydrogen bonding in consumer products.

9. If you were going to formulate a suntan lotion, what particular spectral property would you look for in choosing the active compound?

10. Why are detergents better cleansing agents than soaps in regions where the water supply contains calcium or magnesium salts?

11. Why is a soap from coconut oil more soluble in water than a soap made from palm oil?

12. Suggest ways of removing each of the following from clothing:
    (a) motor oil
    (b) iodine stain
    (c) lard
    (d) copper sulfate

13. Explain why vinegar is able to remove some stains which are soluble in weak acids.

14. Why was the discovery of cheap Na_2CO_3 important to the development of the soap industry?

15. Explain why Grandma's lye soap produced rough, red hands.

16. Explain how an optical brightener in a detergent works.

PROBLEMS

1. A typical hair spray for women *or* men has a formula:

resin (like PVP)	3.0%
plasticizer	0.2%
ethyl alcohol	26.0%
propellant (Freon 11/12)	70.8%
perfume (to suit application)	negligible
	100.0%

Costs on these ingredients vary with the number of cans of hair spray a manufacturer produces. Assume the resin costs \$0.90 per lb., the plasticizer, \$1.10 per lb., the ethyl alcohol, \$0.15 per lb., and the Freon 11/12 mixture \$0.45 per lb. Compute the costs of the ingredients of a 16-oz. (wt.) can of hair spray. *Ans.* \$0.39

2. A detergent label indicates that the material contains 20 per cent $Na_2CO_3 \cdot 10H_2O$. What is the percentage of water in the detergent from this source? *Ans.* 10.4% water

3. (a) Write an equation for the chemical reaction between the calcium ion (Ca^{2+}) and soap. *Ans.* (a) $2CH_3(CH_2)_{16}COO^- + Ca^{2+} \longrightarrow Ca[OOC(CH_2)_{16}CH_3]_2$

   (b) A washtub contains 40 liters of water. The water has run through a limestone region and picked up 0.01 mole calcium ion per liter. What is the maximum amount (in grams) of soap powder [$NaOCO(CH_2)_{16}CH_3$; molecular weight, 306 g/mole] required to precipitate the calcium ions? (b) 245 grams

SUGGESTIONS FOR FURTHER READING

Bennett, H., "The Chemical Formulary," Chemical Publishing Co., New York, 1933–1965. (Twelve volumes of formulas for making soaps, cosmetics, perfumes, and so forth.)

Bennett, H., "Chemical Specialties," Chemical Publishing Company, New York, 1969. (A detailed practical guide on how to set up a chemical specialties business to manufacture and sell cosmetics, herbicides, and so forth.)

"Enzymes in Detergents," *Chemistry,* Vol. 43, No. 2, p. 25 (1970).

Gleason, M. N., Gosselin, R. E., Hodge, H. C., and Smith, R. P., "Clinical Toxicology of Commercial Products," 3rd ed., The Williams & Wilkins Co., Baltimore, 1969. (Contains a wealth of information on the toxic aspects of various commercial products as well as their composition.)

Harry, R. G., "Modern Cosmetology," Chemical Publishing Co., New York, 1947. (A source of formulas and manufacturing techniques for almost all standard types of cosmetics.)

Jellinek, J. S., "Formulation and Function of Cosmetics," John Wiley and Sons, New York, 1970.

"The Smell of Detergents," *Chemistry,* Vol. 40, No. 5, p. 10 (1967).

CHAPTER 23

CONSUMER CHEMISTRY—AUTOMOTIVE PRODUCTS, PHOTOGRAPHY, AND PAINTS

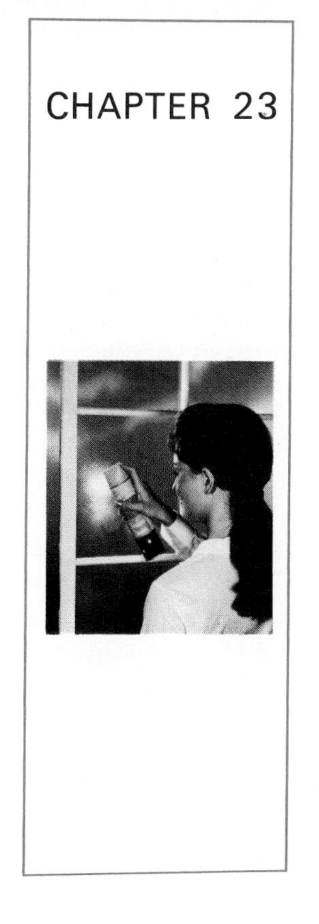

There is a simplifying thread running throughout consumer products that helps us understand how they all work: the molecular interpretation of the properties of materials.

As you study the chemistry of automotive products, photography, and paints in this chapter, you will see numerous examples of the chemical similarities among these three important categories of consumer products and also between these products and food, medicine, cosmetics, and cleansers. While each area is unique in the intended use of its chemicals and, to some degree, unique in its assortment of chemicals, the whole of consumer-product chemistry is blended together by the underlying thread of molecular interpretations of the actions of the various agents in the products. Antioxidants, wetting agents, emulsifying agents, dyes, perfumes, oxidizing agents, reducing agents, and chelating agents, to name only a few, are found in many consumer products. While the choice of the antioxidant in potato chips and motor oils varies, the mechanism of action of the two antioxidants will quite likely be very much alike. Each product usually claims its uniqueness by the *degree* to which its components accomplish their intended purpose and less often by the molecular mechanisms of its chemical reactants.

AUTOMOTIVE PRODUCTS

A significant amount of chemistry is involved in the production and operation of the automobile. Before an automobile can exist, metals must be won from their ores; plastics must be synthesized and fabricated;

paints have to be formulated; sulfuric acid for the battery must be made; rubber must be synthesized, formed into shape, and vulcanized; and glass has to be mixed, fired, molded, and cut. Many chemical reactions are required to power the automobile. The two most familiar are the combustion of the fuel and the electrochemistry of the battery. In its wake the automobile leaves its chemical exhaust to undergo a variety of chemical reactions, some of which lead to smog formation. The chemistry of fuels and exhaust gases has been discussed in previous chapters. In this section, we will discuss some of the chemicals that consumers buy for their automobiles.

Miscellaneous Gasoline Additives

The production of gasoline from petroleum and the antiknock properties of gasoline were discussed earlier. In addition to tetraethyl lead and/or aromatic compounds which reduce preignition or "knock" in an automobile engine, numerous other chemicals are added to gasoline to improve its properties.

Other chemicals are added to gasoline to prevent the ignition of new fuel by glowing particles from a previous ignition. These additives are called ***deposit modifiers.*** Phosphorus compounds such as tricresyl phosphate (one trade name is TCP) and, more recently, boron compounds, have been used for this purpose. These alter the composition of the deposits in the combustion chamber and make them less likely to glow. The phosphorus compounds also prevent spark plug deposits from becoming so electrically conductive that the charge leaks away instead of firing the plug. Boron compounds raise the octane number of the gasoline by preventing sulfur compounds from rendering tetraethyl lead ineffective.

$(CH_3-C_6H_4-O)_3PO$

Tricresyl phosphate

Antioxidants such as phenylenediamine, aminophenols, dibutyl-*p*-cresol, and ortho-alkylated phenols, are added to prevent the formation of peroxides that lead to knock and gum formation. About 2 or 3 pounds of these additives are added to every 1000 barrels of gasoline. The mechanism of antioxidation of automotive oils is very similar to the mechanism described for BHA and BHT in food additives (Chapter 21). Because copper ions catalyze gum formation, metal ion scavengers such as ethylenediamine are added to chelate the trace amounts of copper ions and render them ineffective. The copper gets into the gasoline from the copper tubing used for fuel lines and from brass parts of the engine.

Brass is an alloy of copper and zinc.

To inhibit water from corroding and rusting storage tanks, pipelines, tankers, and fuel systems of engines, ***antirust agents*** are added. Four compounds used to prevent corrosion are trimethyl phosphate, sodium and calcium sulfonates, and N,N′-di-sec-butyl-*p*-phenylenediamine. All these compounds have a polar or ionic end and a nonpolar end in the molecule. These agents coat metal surfaces with a very thin protective film that keeps water from contacting the surfaces, as shown in Figure 23–1. This also helps prevent gummy deposits in the carburetor and combats carburetor icing during cold weather.

Antiicing agents coat metal surfaces, as do antirust agents, and thus prevent ice particles from accumulating on surfaces, and/or depress the freezing point. The small ice particles pass harmlessly through the carburetor and into the engine where the heat converts them to water vapor that eventually exits with the exhaust gases. The freezing point

```
   OH OH
   |   |
H—C—C—H
   |   |
   H   H
```

Ethylene glycol

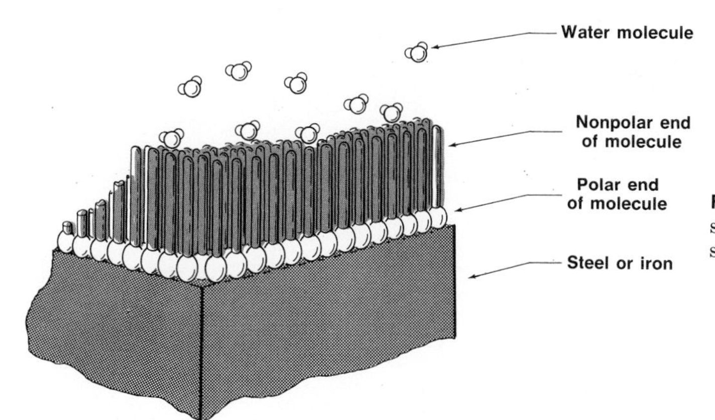

Figure 23–1 The action of surface active agents, such as rust inhibitors, mild antiwear agents, and some deicing agents.

depressants, which include alcohols and glycols, act in the same manner as the antifreeze in the engine's cooling system.

Detergents, which include alkylammonium dialkyl phosphates, are added to prevent the accumulation of high-boiling components on the walls of the carburetor. These deposits interfere with the air flow into the carburetor and cause rough idling, frequent stalls, poor performance, and increased fuel consumption. The effectiveness of these detergents stems from their surface active properties, as shown in Figure 23–1. The film of detergent provides a thin, nonpolar coating on the metal surfaces which prevents high molecular weight, nonpolar gums from forming thick deposits on the surfaces.

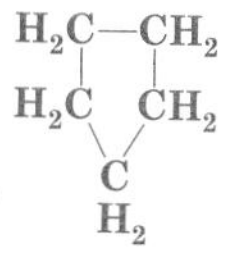

Cyclopentane

Upper cylinder lubricants are sometimes blended into gasoline. These are usually light mineral oils or low viscosity naphthenic distillates (such as cyclopentane) that dissolve away deposits in the intake system from the carburetor, cylinders, top piston rings, and valves. (Nonpolar deposits tend to dissolve in nonpolar solvents.)

At one time ***dyes*** were added to distinguish antiknock "ethyl" gasolines from nonantiknock gasolines. Always present in very small amounts, they are now used to distinguish grades and brands.

Lubricants and Greases

Lubricants have been used to separate moving surfaces, and thus minimize friction and wear, for a long time. Even before 1400 B.C., animal tallow was used to lubricate chariot wheels. Petroleum lubricating oils and greases came into widespread use after the famous Drake well was drilled at Titusville, Pennsylvania, in 1859. The use of additives in lubricants has progressed rapidly since about 1930; synthetic lubricants have been developed largely since World War II.

"Viscous" means resistant to flow, like molasses.

Lubricating oils from petroleum consist essentially of complex mixtures of hydrocarbon molecules. These generally range from low viscosity oils, having molecular weights as low as 250 amu to very viscous lubricants with molecular weights as high as about 1000 amu. The viscosity of an oil can often determine its use. For example, if the oil is too viscous, it offers too much resistance to the metal parts moving against each other. On the other hand, if the oil is not viscous enough, it will be squeezed out from between the metal surfaces, and consequently offer insufficient lubricating power. For these and other reasons, motor oil is often a

TABLE 23–1 Viscosity Data at −18°C and 99°C for the SAE Method of Rating Motor Oils

Motor Oil	Viscosity (SUS)* −18°C Min.	−18°C Max.	99°C Min.	99°C Max.
5W		4,000	39	
10W	6,000	12,000	39	
20W	12,000	48,000	39	
20			45	58
30			58	70
40			70	85
50			85	110

* SUS is the Saybolt Universal Second, which is the time in seconds required for 60 ml of oil to empty out of the cup in a Saybolt viscometer through a carefully specified capillary opening. Note the extremely shortened time for the outflowing of the hotter oil.

mixture of oils with varying viscosities. The common 10W-30 oil, for instance, combines the low-temperature viscosity of the Society of Automotive Engineers (SAE) 10W classification for easy low-temperature starting with SAE 30 high-temperature viscosity for better load capacity in bearings at the normal engine running temperature.

The SAE scale for rating motor oils is based on the viscosity of the oils. The viscosity criteria for the SAE scale are given in Table 23–1.

During distillation of petroleum crude oils, the lubricating crude oil fractions boil off after the lower boiling gasoline, kerosene, and fuel oils are removed. Most of the aromatic compounds are then removed from the lubrication oil fraction by solvent extraction to prevent the formation of sludge during high-temperature operation. Paraffin wax is then removed by low-temperature filtration to give the oil better flow characteristics. The final refining step is contact with an activated clay such as Fuller's earth, which absorbs many of the colored particles and provides an oil with a light color. Lubricating oils with the desired properties are then made by blending one or more refined stocks with the proper additives.

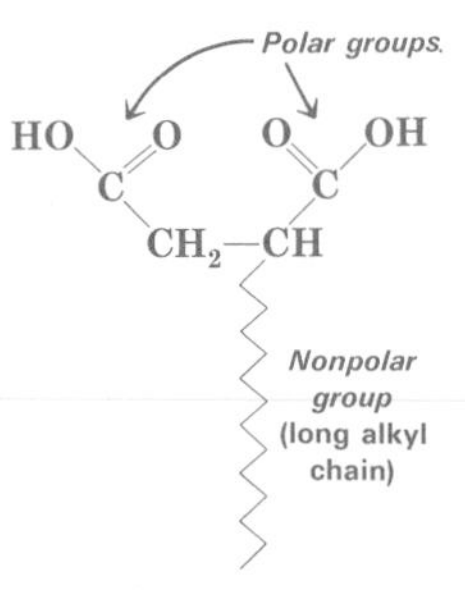

Alkylsuccinic acid

After the motor oil has been separated and refined, its usefulness is improved by the addition of substances such as antiwear agents, oxidation inhibitors, rust inhibitors, detergents, viscosity improvers, and foam inhibitors.

When two metal surfaces contact under heavy load and high temperature, as in the differentials of most cars, the friction produces intense heat which renders organic lubricant films ineffective. Under extreme conditions the metals can weld together. To combat this, *extreme pressure lubricants* were developed. These contain organic compounds as additives which react at the high contact temperatures to form high-melting inorganic lubricant films, such as lead sulfide and iron sulfide, on the metal surfaces; the presence of these films inhibits breakdown. These additives generally consist of sulfur, chlorine, phosphorus, and lead compounds which act either by providing layers that are hard and difficult to wear away or by serving as fluxing agents to contaminate the metal surface and prevent welding.

Under conditions of less severe friction, mildly polar organic acids,

Fluxing agents fuse or melt into the surface.

such as the alkylsuccinic type, and organic amines are often added as ***antiwear agents.*** These compounds provide an adherent, adsorbed film over metallic surfaces and reduce shearing of the metal. In somewhat more severe conditions about 1 percent tricresyl phosphate (TCP) or zinc dialkyldithiophosphate is widely used.

Rust inhibitors, such as the mild antiwear agents and antirust agents in gasoline, are preferentially adsorbed as a film on iron and steel surfaces to protect them from attack by moisture (Figure 23–1). If only a little water is present in a large amount of oil, mildly polar organic compounds such as alkylsuccinic acids and organic amines are often used. Where severe conditions are anticipated, more strongly adherent organic phosphates (TCP, for example), polyhydric alcohols, and sodium and calcium sulfonates are used.

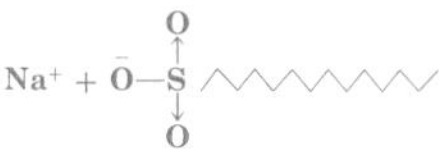

$Na^+ + \bar{O}-S(O)(O)-$ (alkyl chain)

A sodium sulfonate

Oil oxidation is thought to involve a chain reaction mechanism with hydroperoxide formation (—OOH) as the initiating process which eventually leads to the formation of organic acids and other products. ***Oxidation inhibitors*** appear to interrupt the chain reaction by tying up the hydroperoxide. This action delays the formation of sludge, varnish, and acids for extended operating periods and minimizes corrosion problems with the zinc-, cadmium-, and copper-containing alloys, which are corroded by organic acids in oxidized oils. Zinc, barium, and calcium thiophosphates are frequently used to prevent oxidation of the oils.

The prefix *thio* denotes sulfur replacing oxygen, as in sodium monothiophosphate, Na_3PSO_3, compared with sodium phosphate, Na_3PO_4.

Detergents are widely used in a 2 to 20 percent concentration in motor oils to prevent or remove deposits of oil-insoluble sludge, varnish, carbon, and lead compounds. The detergents are adsorbed on the insoluble particles, keeping them suspended in the oil so as to minimize deposits on rings, valves, and cylinder walls. The action is similar to the action of soap (or detergent) in removing grease from clothes or hands, as described in Chapter 22. A basic difference exists, however, because the polar end of the detergent molecule generally is attached to the particle, and its hydrocarbon end extends into the medium (oil). (In soapy water, the hydrocarbon end of the soap is in the oil or grease particle and the polar end is in the medium, water.) Barium and calcium sulfonates and phenoxides are used extensively as detergents in automotive motor oils.

Adsorb means to attach to the surface while absorb means to penetrate into the interior.

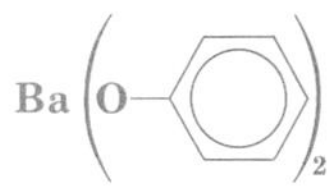

The viscosity of motor oils can be adjusted with additives. Polymethacrylate is added in small amounts (1 percent or less) to prevent wax, which condenses out at low temperatures, from forming a network of crystals that would immobilize the oil. The additive appears to adsorb on crystal faces, which prevents the interlocking crystal growth. These additives are ineffective in normally high viscosity oils. The viscosity of oils can be increased by adding linear polymers in the molecular weight range of about 5000 to 20,000 amu. The three types most commonly used are polyisobutylenes, polymethacrylates, and polyalkylstyrenes (Chapter 15). The entanglement of the long chains prevents easy flow of the oil. With use, the chains are broken into smaller fragments and the oil assumes its base viscosity.

$Ba(O-C_6H_5)_2$

Barium phenoxide

Severe churning and mixing of oil with air may cause foam and an oil overflow from the engine; failure of the machine may eventually ensue. Methyl silicone polymers (Chapter 15) in concentrations of only a few parts per million are effective for defoaming oil. Since the silicone additive is not completely soluble in the oil, it functions by forming minute droplets of low surface tension which aid in breaking up foam bubbles to release the trapped air (Figure 23–2).

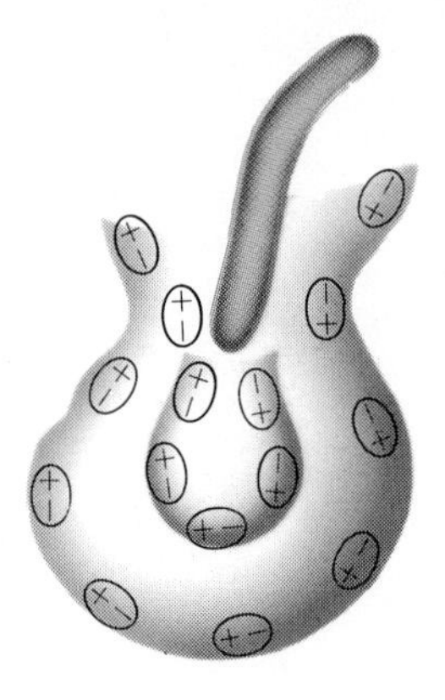

Figure 23–2 Possible defoaming action of substances like methylsilicone polymers in oil.

Greases are essentially lubricating oils thickened with a gelling agent such as fatty acid soaps of lithium, calcium, sodium, aluminum, or barium (Chapter 22). The fatty acids are usually oleic, palmitic, stearic, or other carboxylic acids derived from tallow, hydrogenated fish oil, castor oil, or, less often, wool grease and rosin. The soaps form a network of fibers that entrap the oil molecules within the interlacing fiber structure. Carbon black, silica gel, and clay are also used to thicken petroleum greases. Chemical additives similar to those used in lubricating oils and gasolines are added to greases to improve oxidation resistance, rust protection, and extreme pressure properties. Synthetic greases are being developed that deteriorate so slowly that longer intervals between grease jobs are now possible. Silicone greases have a useful life of up to 1000 hours at 450°F (232°C). Unfortunately, silicone greases provide relatively poor lubrication for gears and other sliding devices. Diester greases such as di(2-ethylhexyl) sebacate have found extensive use among synthetic greases. Lithium soaps dissolve well in the diester oil and form a grease with equal or better lubrication characteristics and a considerably longer useful life than petroleum greases. Blends of silicone oil and diester oil provide greases with good low-resistance lubricating power even at low temperatures (−73°C).

CH_3
$(CH_2)_3$
$CH—C_2H_5$
CH_2
O
C=O
$(CH_2)_8$ Diester
C=O
O
CH_2
$CH—C_2H_5$
$(CH_2)_3$
CH_3

Di(2-ethylhexyl) sebacate

Solid Lubricants

Perhaps the most tenacious and wear resistant solid lubricant is molybdenum disulfide, MoS_2. Like graphite, a very common solid lubricant, MoS_2 has a layered structure that enables one layer to float over another and provide lubrication (Figure 23–3). Strong bonds exist within

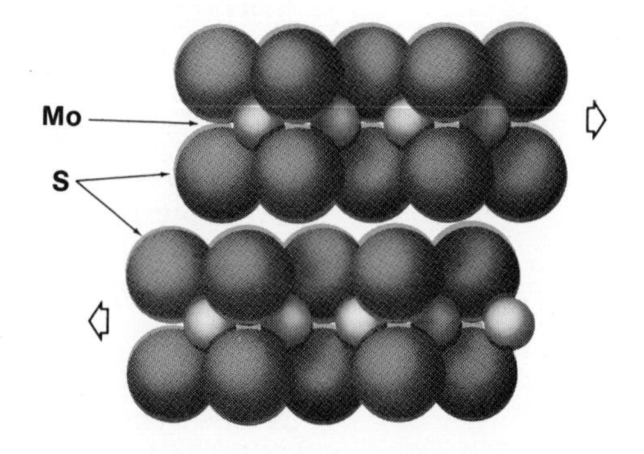

Figure 23–3 The structure of molybdenum disulfide, MoS_2. The small spheres represent molybdenum atoms. Note how the sulfides are adjacent to sulfides in every other layer. This allows one layer to slide over another, accounting for the slippery nature of MoS_2.

each S—Mo—S layer, while weak S—S bonds between the layers allow easy sliding of one layer over another. MoS_2 is extensively used in greases for automotive chassis lubrication. When used alone, MoS_2 suffers like any other solid lubricant from having poor wear resistance and being unable to heal any breaks in its surface coating.

Antifreeze

An antifreeze is a substance which is added to a liquid, usually water, to lower its freezing point. Although various substances have been used as antifreezes in the past, nearly all of the current market is supplied by ethylene glycol and methyl alcohol.

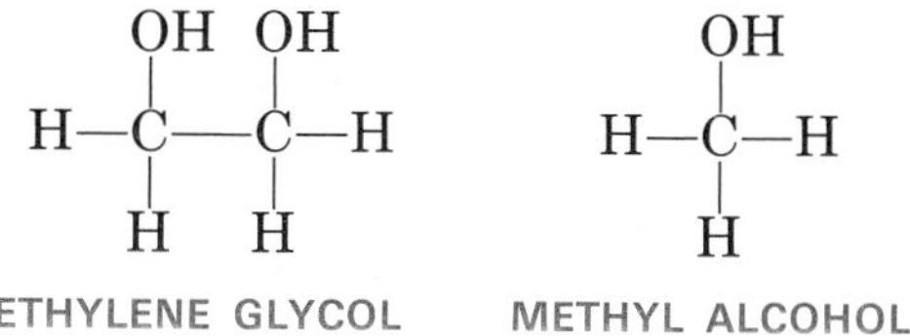

Ethylene comes from cracking petroleum. Ethylene glycol is prepared from ethylene, $CH_2{=}CH_2$, by an oxidation reaction

$$2CH_2{=}CH_2 + O_2 \longrightarrow 2\underset{\diagdown O \diagup}{CH_2{-}CH_2}$$

Ethylene oxide

followed by hydrolysis:

$$\underset{\diagdown O \diagup}{CH_2{-}CH_2} + H_2O \longrightarrow \underset{OH\quad OH}{CH_2{-}CH_2}$$

More than 95 percent of the antifreeze on the market is "permanent" antifreeze, having ethylene glycol as the major constituent. The largest use of antifreeze is in protection of water-cooled automobile and truck engines. Water has been selected as the coolant for these engines because of its universal availability, low cost, and good heat transfer properties: however, it has two serious disadvantages. First, it has a relatively high freezing point and second, under normal operating conditions, it is corrosive. Modern antifreeze mixtures effectively counteract these problems.

The temperature in the United States, except for Alaska, seldom, if ever, falls below $-40°F$ ($-40°C$). Both ethylene glycol and methyl alcohol can prevent water from freezing at these temperatures, as shown in the graph of Figure 23–4. Methyl alcohol more effectively lowers the

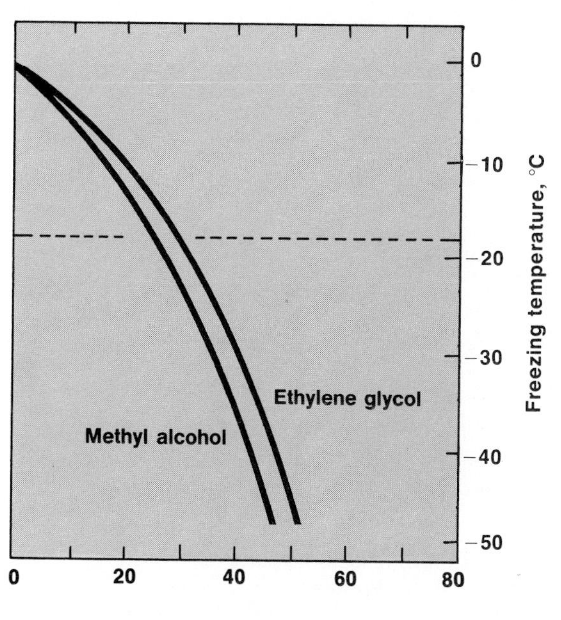

Figure 23–4 Freezing point depression of water by ethylene glycol and methyl alcohol.

freezing point, but because of its volatility and combustibility it is seldom used.

For ethylene glycol, methyl alcohol, and other nonelectrolytes, the depression of the freezing point in dilute solutions is proportional to the concentration of the solute and nearly independent of the nature of the solute. That is,

$$\Delta T_f = K_f m$$

where ΔT_f is the freezing point depression, K_f is the proportionality constant peculiar to each solvent (1.86°C for water), and m is the number of moles of solute per 1000 grams of water (molal concentration). In concentrated solutions this relationship can be used to obtain only a rough estimate of the freezing point.

Antifreeze protection charts always show the temperatures at which the first ice crystals form. Below this temperature the antifreeze solution turns to slush. If the slush is unable to circulate through the radiator, overheating, boiling and engine damage can result. For this reason it is best to add sufficient antifreeze to prevent the formation of the first ice crystals even at the lowest anticipated temperature (Figure 23–5).

The density of an antifreeze solution can be used to determine its freezing temperature.

A service station attendant measures the effectiveness of the antifreeze in your car by reading the position at which a hydrometer floats in a portion of a radiator solution. He is really measuring the solution's density, which varies with the amount of antifreeze present.

In 1960, automobile radiators were equipped with caps able to withstand pressures of 13 to 17 pounds per square inch; this change allowed a 20 to 28°C increase in maximum coolant temperatures. Thermostats now operate between 85 and 99°C, whereas in the past the range was 60 to 70°C. Ethylene glycol raises the boiling point of water as well as lowers its freezing point, as shown in Figure 23–6. Ethylene glycol reduces the vapor pressure of water, thus requiring a higher temperature for the solution to boil. Therefore, it is good policy to keep the antifreeze in the radiator all year. It prevents freezing in the winter and boiling over in the summer.

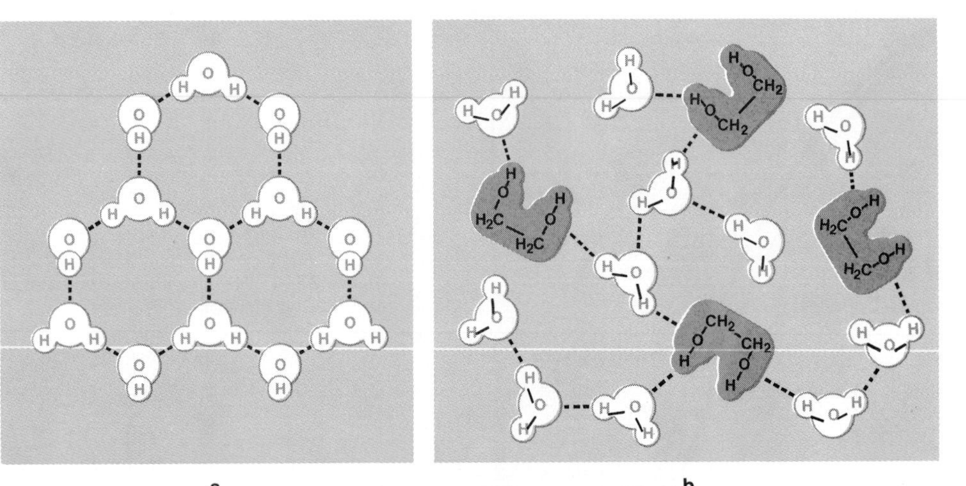

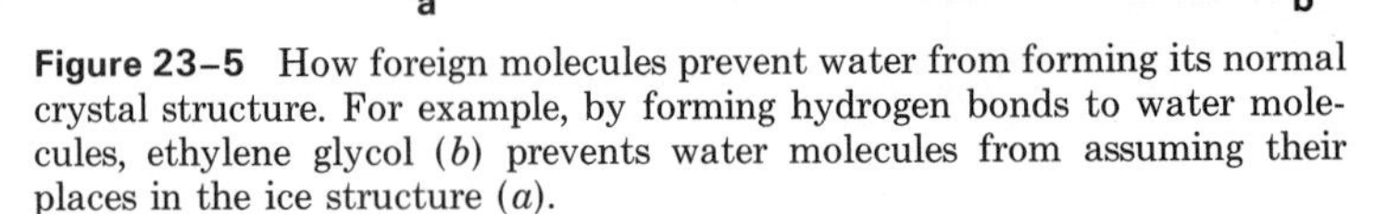

Figure 23–5 How foreign molecules prevent water from forming its normal crystal structure. For example, by forming hydrogen bonds to water molecules, ethylene glycol (*b*) prevents water molecules from assuming their places in the ice structure (*a*).

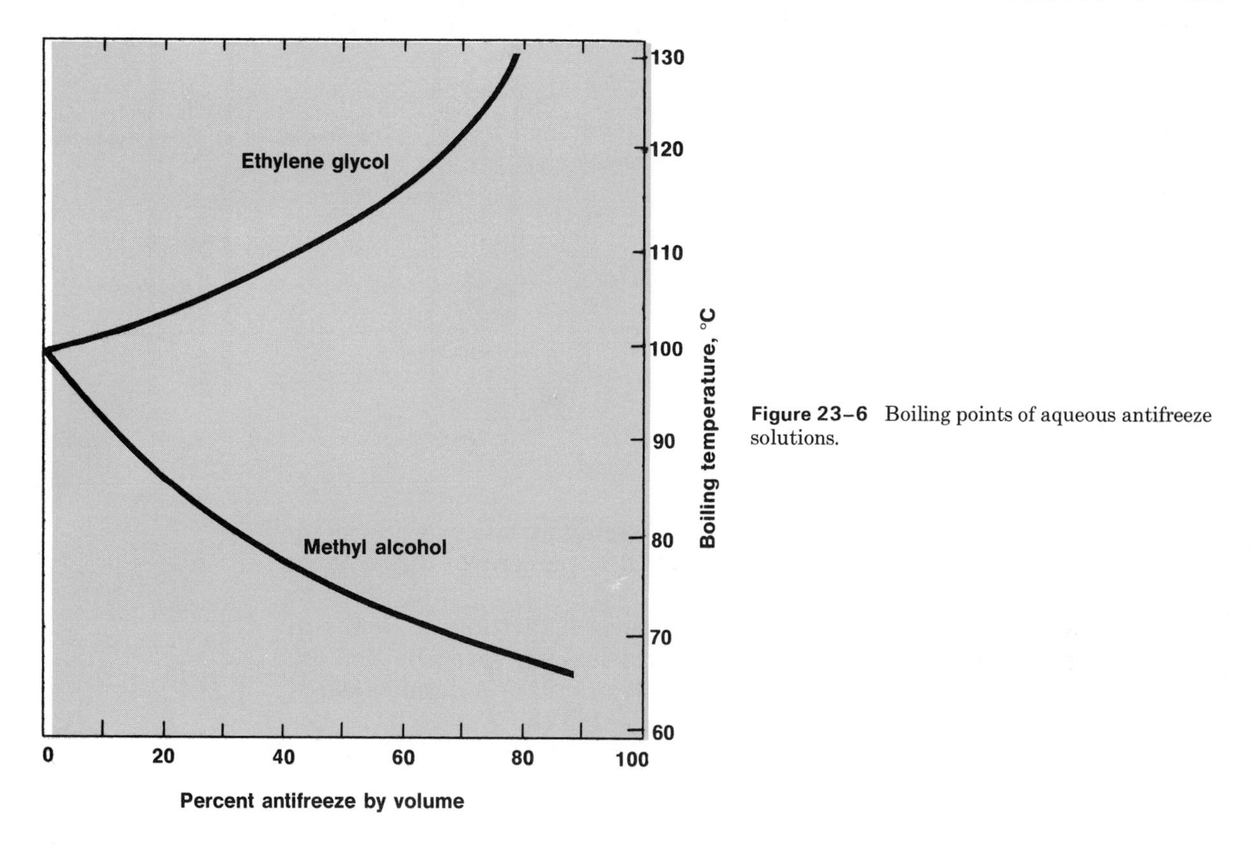

Figure 23–6 Boiling points of aqueous antifreeze solutions.

Commercial antifreeze contains various additives to prevent corrosion, leaks, damage to rubber, and foaming. A typical permanent antifreeze contains more than 95 percent ethylene glycol, several reducing agents to prevent corrosion, a substance to stop small leaks in the cooling system, and an antifoaming agent.

Good auto antifreeze formulations also contain rust inhibitors.

The prevention of corrosion is the second most important job of antifreeze. Metals in the cooling system which are subject to corrosion are copper, steel, cast iron, aluminum, solder (lead and tin), and brass (copper and zinc). The presence of oxygen, along with high temperatures, pressures, and flow rates, increases the possibility of general corrosion. Although corrosion can "eat" through the walls of the cooling system, the most general trouble is overheating caused by flakes of metal clogging the radiator. A large assortment of substances is used to inhibit corrosion. Some inhibit corrosion by acting as reducing agents (e.g., nitrites), some as ion scavengers (e.g., phosphates), and some as surface-active agents (detergents; e.g., triethanolamine). Most antifreezes contain two or more inhibitors for the different metals. All inhibitors are depleted with use.

Radiator sealants have been on the market for years. Modern sealants in antifreeze include asbestos fiber and polystyrene spheres. When a radiator springs a leak, the coolant penetrates the crack because the pressure inside the cooling system is greater than atmospheric pressure. Asbestos fibers are often too big to squeeze through and thus get caught in the crack, plugging the hole in the radiator. On the other hand, the larger polystyrene particles initially plug up most of the leak while the smaller ones build up behind them. The spheres fuse together under the pressure and temperature conditions that exist and form a solid plug in

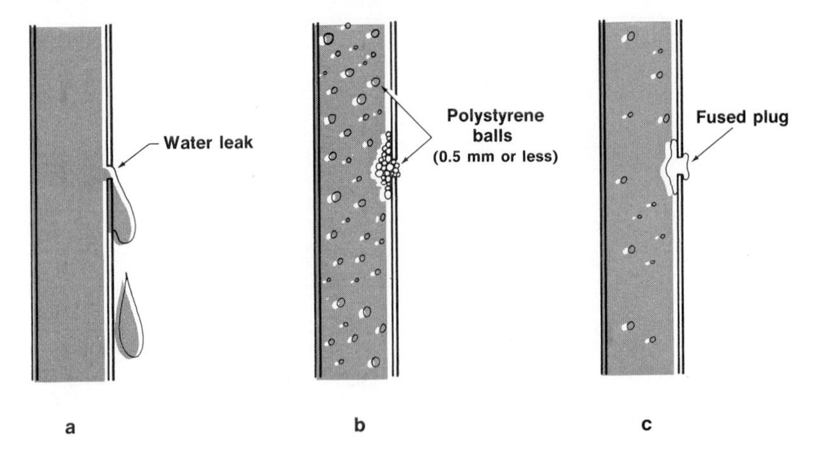

Figure 23–7 How polystyrene works to plug radiator leaks. Details are given in the text.

the crack (Figure 23–7). This is effective in stopping up holes or cracks up to 0.5 millimeter width, which includes about 90 percent of all radiator leaks.

Foaming is caused by one or more of several factors: air leaks in hoses, water pump, or radiator; exhaust gas leaking into the cooling system; failure to drain out cooling system cleansers; or extended use of antifreeze. Foaming can be corrected by tightening the system or by use of antifoam additives, such as silicones, polyglycols, mineral oils, high molecular weight alcohols, organic phosphates, alkyl lactates, castor oil soaps, and calcium acetate. These substances reduce the surface tension so bubbles cannot hold together (Figure 23–2).

With today's costs of antifreeze so inflated (approximately $5.00/gal. in 1975) due to increased demand and lower supply, extending the life of a charge of antifreeze is a prudent investment. Since corrosion begins in earnest when the pH of the coolant drops below 7, most commercial antifreeze contains an extra amount of alkali. As time passes this alkali is exhausted and corrosion begins. A check of your antifreeze's corrosion-fighting capability can be made with a piece of litmus paper. Pink means acid and trouble through corrosion. Adding a can of rust inhibitor (which contains an alkali) will adjust the pH back to a value above 7. A retest with litmus paper shows blue (basic). Your car is now protected as long as the mixture composition doesn't change due to boil-away or leaks.

Pink litmus—acidic. Blue litmus—basic.

Deicing fluids are a type of antifreeze, but they also will melt ice and frost. They are chiefly employed in removing ice and frost from parked aircraft and car windows. The glycol-alcohol type used on cars came on the market in 1959. The better formulations contain the following ingredients: ethylene or propylene glycol (for protection against refogging); one or more of the lower alcohols, such as 2-propanol, denatured alcohol, etc. (for low viscosity and good spray pattern); water (to minimize inside fogging by diminishing the evaporative-cooling effect of the formulation); nitrite or other inhibitor (to prevent corrosion of the container); and wetting agent, such as ethanolamine (to break surface tension so fluid will attack the ice readily).

Deicers lower the freezing point of water.

A simple alcohol-water mixture works satisfactorily for quick frost removal and to defog the inside of glass, but it evaporates quickly and refogging occurs unless glycol is present. Before a deicer formulation is applied, the snow should be removed and the ice should be scored to permit faster penetration.

Miscellaneous Automotive Products

Consumers spend millions of dollars each year on a large variety of specialized automotive products. The formulations of some typical examples of these products are given in Table 23-2.

TABLE 23-2 Typical Composition of Some Automotive Products

Engine and Motor Cleaners	
Ethylene dichloride	63%
p-Cresol	25%
Oleic acid	7.2%
Potassium (or sodium) hydroxide	1.4%
Water	3.0%

Radiator Cleaners	
(1) Oxalic acid	40%
Boric acid	60%
(2) Sodium carbonate	85%
Potassium dichromate	15%
May contain also:	
sulfamic acid	
1-butanol (n-butyl alcohol)	
2-propanol (isopropyl alcohol)	

Radiator Stop Leak	
Dextrin	5-10%
Cellulose gum	0.5%
Asbestos	5%
Sodium carbonate	0.8%
Isopropyl alcohol	10-15%
Water	to 100%

Shock Absorber Fluids	
(1) Petroleum ether	97%
Kerosene	3%
(2) Mineral oils	90-100%
Fatty oils	0-5%
May contain also:	
viscosity improvers (polymethacrylate esters)	0-5%
dyes	0-150 ppm

Automatic Transmission Fluids	
Mineral oils	75-100%
Oxidation inhibitors and detergents	0-20%
Viscosity improvers	0-5%
Polymethacrylate esters	
Polyisobutylenes	
Anti-wear agents	0-2%
Organic borates	
Antifoam agents	0-15%
Polysiloxanes	
Sealant	to 100%
Triaryl phosphate	
Dyes	0-200 ppm

Brake Fluids	
Lubricant	20-25%
Castor oil	
Butyl or glyceryl ether of polyoxyethylene propylene glycol	
Polypropylene glycol	
Solvent	80-85%
Methyl, ethyl and butyl ethers of ethylene glycol and related glycols	
May contain also:	
Inhibitors	
Amine soaps	
Potassium soaps	
Borax	
Antioxidants	
Hydroquinone	
Dyes	

SELF-TEST 23-A

1. Antirust agents such as trimethyl phosphate in gasoline prevent the formation of rust by: (a) reacting with oxygen; (b) coating metal surfaces; (c) converting rust (Fe_2O_3) into iron metal; (d) absorbing water. Select one answer.

2. Gum in the automobile engine is formed by ______________ of hydrocarbons which must have one or more ______________ bonds per molecule.

3. Since gums are nonpolar, a ______________ solvent such as cyclopentane is used to dissolve the gum in an engine.

4. Antioxidants such as aminophenols in gasoline prevent oxygen from reacting with gasoline prior to ignition by destroying ______________ ______________, which have unpaired electrons.

5. Lubricating oils with a rating of 10W-30: (a) have one type of molecule; (b) have the same viscosity at high and low temperatures; (c) are a mixture of oils with varying viscosities. Select one answer.

6. Molybdenum disulfide (MoS_2) is an excellent solid lubricant because it consists of ______________, which slide over each other.

7. The viscosity of oils and greases is increased by adding linear polymers because the chains of the polymers become ______________.

8. Automotive greases are essentially thickened (or gelled) motor oils. (True/False)

9. The development of what two kinds of greases makes it possible to go longer between automotive lubrication? ______________ and ______________ ______________

10. Ethylene glycol makes it more difficult for water to boil because the —OH groups on an ethylene glycol molecule form hydrogen ______________ with water molecules and detain them from escaping into the atmosphere.

11. Nitrite (NO_2^-) ions, being easily oxidizable, inhibit corrosion by being oxidized to ______________ ions and, at the same time, cause the reduction of metal ions to metal atoms and/or oxygen gas to oxide ions.

THE CHEMISTRY OF PHOTOGRAPHY

Black and White Photography

Many chemical substances are known to be photosensitive, that is, changeable in light; some of these, the halide (chloride, bromide, or iodide) salts of silver, have the useful ability of storing an image. The various processes known as ***photography*** are based upon this property.

Photochemistry deals with the chemical changes produced by absorbed radiant energy.

Photography as we know it today is the invention of no single individual, but rather the result of the efforts of many chemists, beginning as early as 1727, when J. H. Schulze observed that a mixture of silver nitrate and chalk darkened on exposure to light. These early images were not permanent, however, for there was no known way to rid the exposed material of unexposed silver compound.

In the early 1830's, a Frenchman, Louis Daguerre, discovered by accident that mercury vapor was capable of developing an image from a silver-plated copper sheet that had been sensitized by iodine vapor. The ***daguerreotype*** image was rendered permanent by washing the plate with hot concentrated salt solution. Daguerreotype portraits were very popular in the mid-nineteenth century.

Pronunciation of daguerreotype: (dä-′ger-ō-tīp)

In 1839 Daguerre demonstrated his photographic process in France. The process was improved by using sodium thiosulfate to wash off the unexposed silver salts, a chemical reaction that had been discovered by Sir J. W. F. Hershel twenty years earlier. By 1840, the daguerreotype process allowed portraits to be taken in less than a minute. Ten years later, every town of any size at all in the United States had a daguerreotype studio, The toxic nature of the mercury vapor helped this to be a short-lived commercial process.

In 1970, amateur photographers in America took some 4.75 billion still pictures and spent about $1.2 billion for film and processing.

In 1841, an Englishman, William Henry Fox-Talbot, announced the calotype process. The Fox-Talbot process (Figure 23–8) involved a paper made sensitive to light by silver iodide. The light-sensitive paper could be developed into an image with gallic acid in a development process essentially the same as is used today. This image turned out to an exact reverse of the original (the light areas were dark and the dark areas were light), which led to its name, the ***negative.*** When made with semitransparent paper, Talbot's negatives could be laid over another piece of photographic

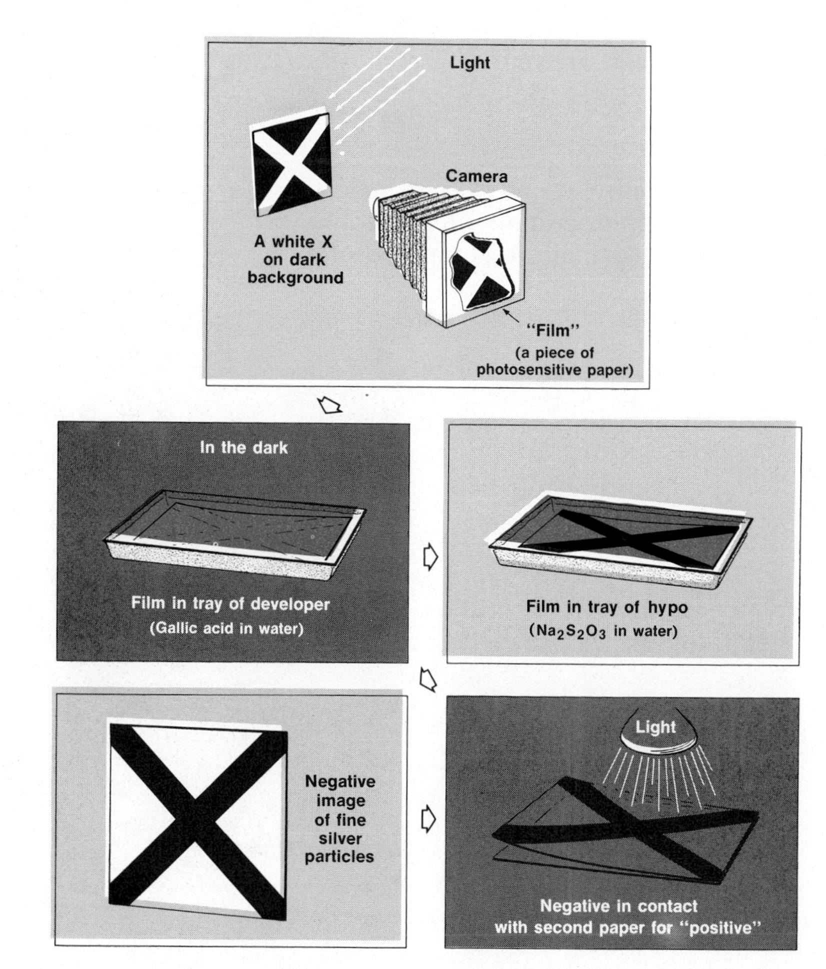

Figure 23–8 Fox-Talbot process.

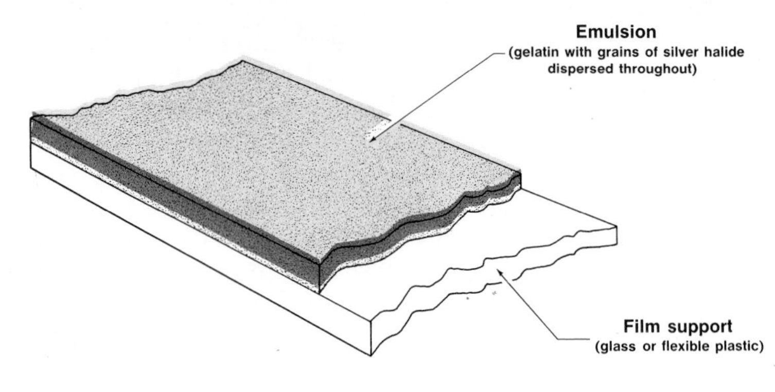

Figure 23–9 A modern photosensitive gelatin emulsion.

paper which, when exposed and developed, yielded a "positive," or direct copy of the original. Although the Talbot process required less time than the Daguerre process, the Talbot images were not sharp. It was obvious that some way of holding the silver halides on a transparent material would have to be devised.

At first, the silver salts were held on glass with the white of an egg. This provided sharp, though easily damaged, pictures. By 1871, the problem had been solved by an amateur photographer and physician, Dr. R. L. Maddox. He discovered a way to make a gelatin emulsion of silver salts and apply it to glass. In 1887, George Eastman introduced the Kodak, a camera using film made by attaching a gelatin emulsion to a plastic (cellulose nitrate) base (Figure 23–9). The camera could take 100 pictures and then camera and film had to be sent to Rochester, N.Y. for processing. The age of modern photography had arrived.

Cellulose acetate replaced easily combustible cellulose nitrate as the film support in 1951.

Photochemistry of Silver Salts

To understand the chemistry of photography, we must first look at the photochemistry of silver salts. A typical photographic film contains tiny crystallites called ***grains*** (Figure 23–10), which are composed of a

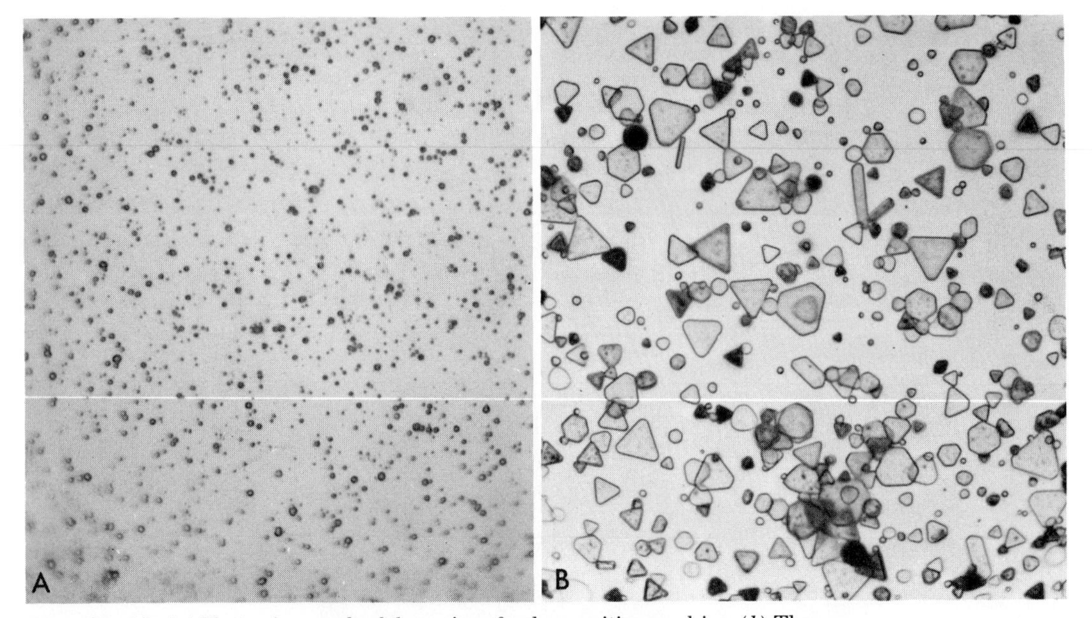

Figure 23–10 (*a*) Photomicrograph of the grains of a slow positive emulsion. (*b*) The grains of a high speed negative emulsion at the same magnification.

very slightly soluble silver halide, such as silver bromide, AgBr. The grains are formed when solutions of silver nitrate, $AgNO_3$, and potassium bromide, KBr, are mixed.

$$Ag^+ + NO_3 + K^+ + Br^- \longrightarrow \underline{AgBr} + K^+ + NO_3^-$$

The solid AgBr is then "ripened" to produce grains of the right size. After being washed, the grains are suspended in gelatin and the resulting gelatin emulsion is melted and applied as a coating on glass plates or plastic film.

When light of an appropriate wavelength strikes one of these grains, a series of reactions begins which leaves a small amount of free silver in the grain. Initially a free bromine atom is produced when the bromide ion absorbs the photon of light:

$$Ag^+Br^- \xrightarrow{\text{Light absorption}} Ag^+ + Br^0 + e^-$$

The silver ion-electron pair now may do one of two things: (1) it may recombine to form a free single silver atom:

$$Ag^+ + e^- \longrightarrow Ag^0$$

(but this is unlikely in a typical silver halide grain); or (2) the silver atom may combine with a neighboring silver ion to form an aggregate of silver atoms when finally neutralized by the available electrons in the grain:

In order for an exposed AgBr grain to be developable, it will need a minimum of four silver atoms as Ag_4^0.

$$Ag^0 + Ag^+ \longrightarrow Ag_2^+, \quad Ag_2^+ + e^- \longrightarrow Ag_2^0$$

$$Ag_2^+ + Ag^0 \longrightarrow Ag_3^+, \quad Ag_3^+ + e^- \longrightarrow Ag_3^0$$

$$Ag_3^+ + Ag^0 \longrightarrow Ag_4^+, \quad Ag_4^+ + e^- \longrightarrow Ag_4^0$$

It is the presence of this free silver in the exposed silver bromide grains that provides the ***latent image*** which is later brought out by the developer.

The free bromine atom, Br^0, tends to destroy the latent image by reactions which reduce either the number of free electrons

$$Br^0 + e^- + Ag^+ \longrightarrow Ag^+Br^-$$

The latent image is the "invisible developable image" stored in the silver halide grains.

or the size of the free silver particle

$$Br^0 + Ag_3^0 \longrightarrow Ag_2^0 + Ag^+Br^-$$

Hastening this process is the fact that the bromine atom, being neutral, essentially acts as a "positive hole" in the Ag^+Br^- lattice. Other neighboring bromide ions can and do give up electrons to this "hole," producing an effective migration of the image-destroying Br^0 throughout the grain (Figure 23–11).

A positive hole in the crystal is a neutral bromine atom which will accept one more electron to complete its electron pairing. $:\ddot{Br}\cdot + e^- \longrightarrow :\ddot{Br}:^-$

In order to overcome the problem of "migrating bromine atoms" within the exposed grain, chemical sensitizers are added. S. E. Sheppard discovered, in 1925, that these compounds were present in some gelatin emulsions as impurities, principally sulfur compounds. Now, compounds

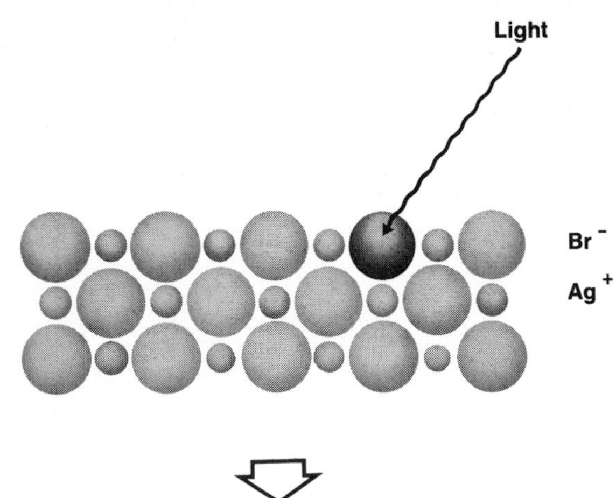

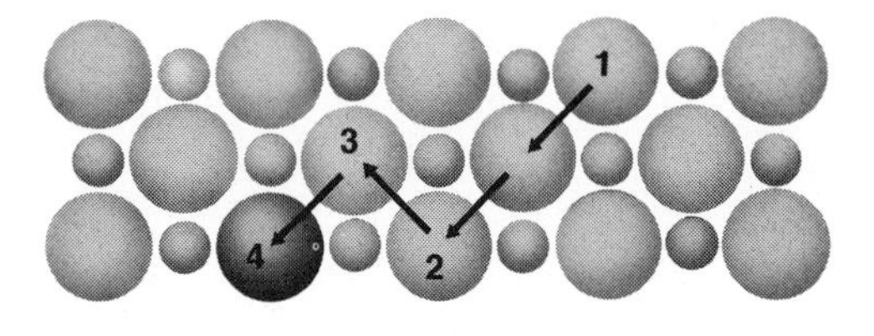

Figure 23–11 How light affects the silver bromide crystal. In the upper part of the illustration a photon strikes a bromide ion, producing a bromine atom and an electron. The bromine atom appears as a "positive hole" in the lattice of negative bromide ions. In the lower portion, the "hole" migrates by the movement of an electron from a neighboring bromide ion (1) to the "hole."
The "hole" is then located at the site from which the electron moved. Three more electron movements (2 to 1, 3 to 2, and 4 to 3) move the "hole" sequentially to lattice sites 2, 3, and finally 4.

such as silver sulfide (Ag_2S) are added to a photographic emulsion. It appears that the wandering "positive holes" (bromine atoms) react with the S^{2-} ions in the Ag_2S. The net reaction is the following:

$$2Br^0 + Ag_2S \longrightarrow 2Ag^+Br^- + \underset{\text{FREE SULFUR}}{S^0}$$

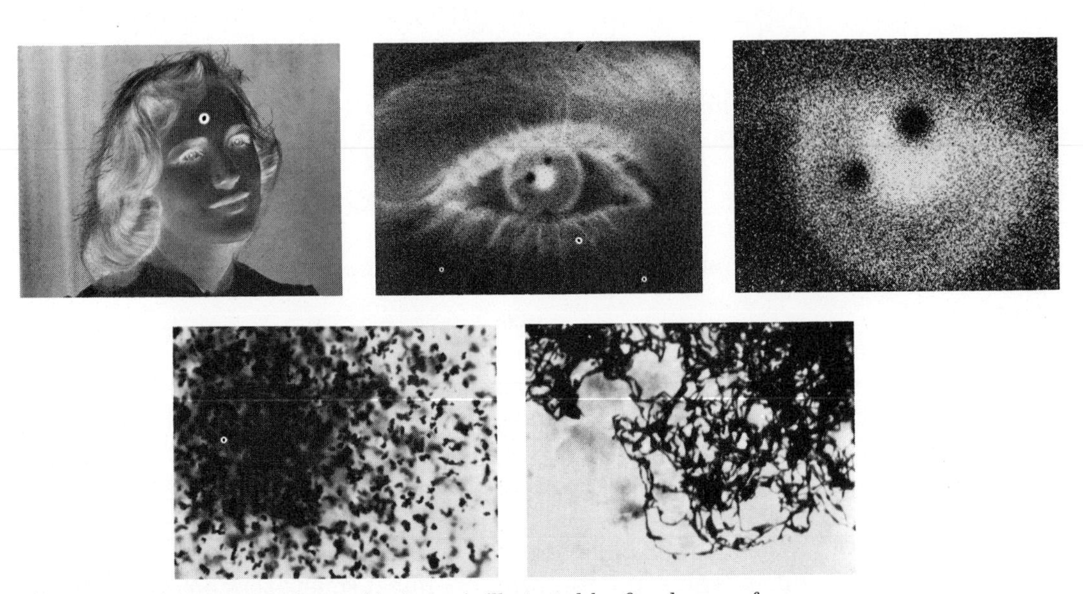

Figure 23–12 Grain in a photograhic image is illustrated by five degrees of magnification: (*a*) original size; (*b*) × 25; (*c*) × 250; (*d*) × 2500; and (*e*) × 25,000 (by electron microscope).

PLATE V

AIR POLLUTION

Air pollution in New York.

WATER POLLUTION

Industrial wastes pollute natural waters (Cumberland River, Nashville, Tennessee).

A foaming problem at the outlet of a sewer system (Cumberland River, Nashville, Tennessee).

Lower photos courtesy of Dr. David Wilson, Vanderbilt University.

PLATE VI

COLOR PHOTOGRAPH

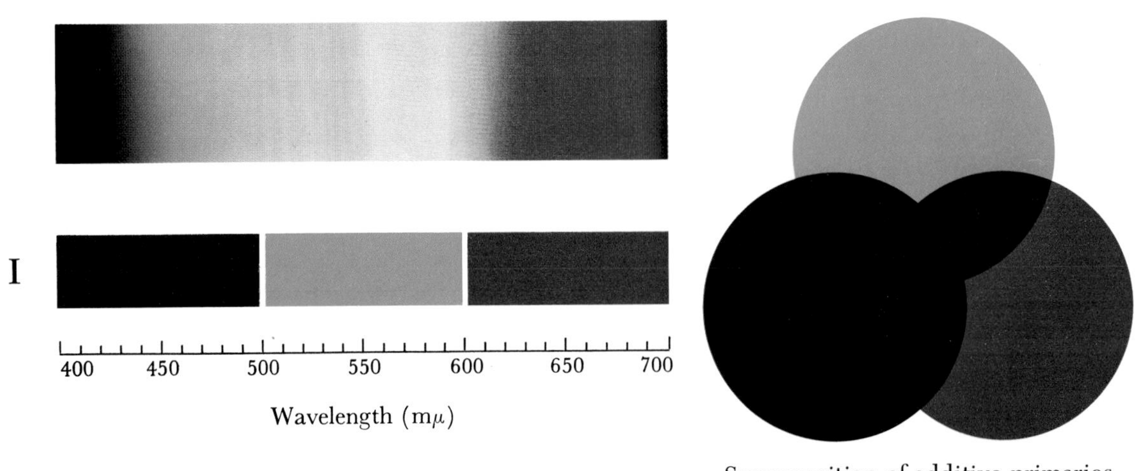

Superposition of additive primaries.

Visible spectrum and primaries. (I) Additive primaries: blue, green, red; (II) subtractive primaries: minus blue (yellow), minus green (magenta), minus red (cyan).

Superposition of subtractive primaries.

From "A Chemist's View of Color Photography," by Arnold Weissberger, *American Scientist*, 58(6), 1970; courtesy of Eastman Kodak Company, Rochester, N.Y.

PLATE VII

COLOR PHOTOGRAPHY

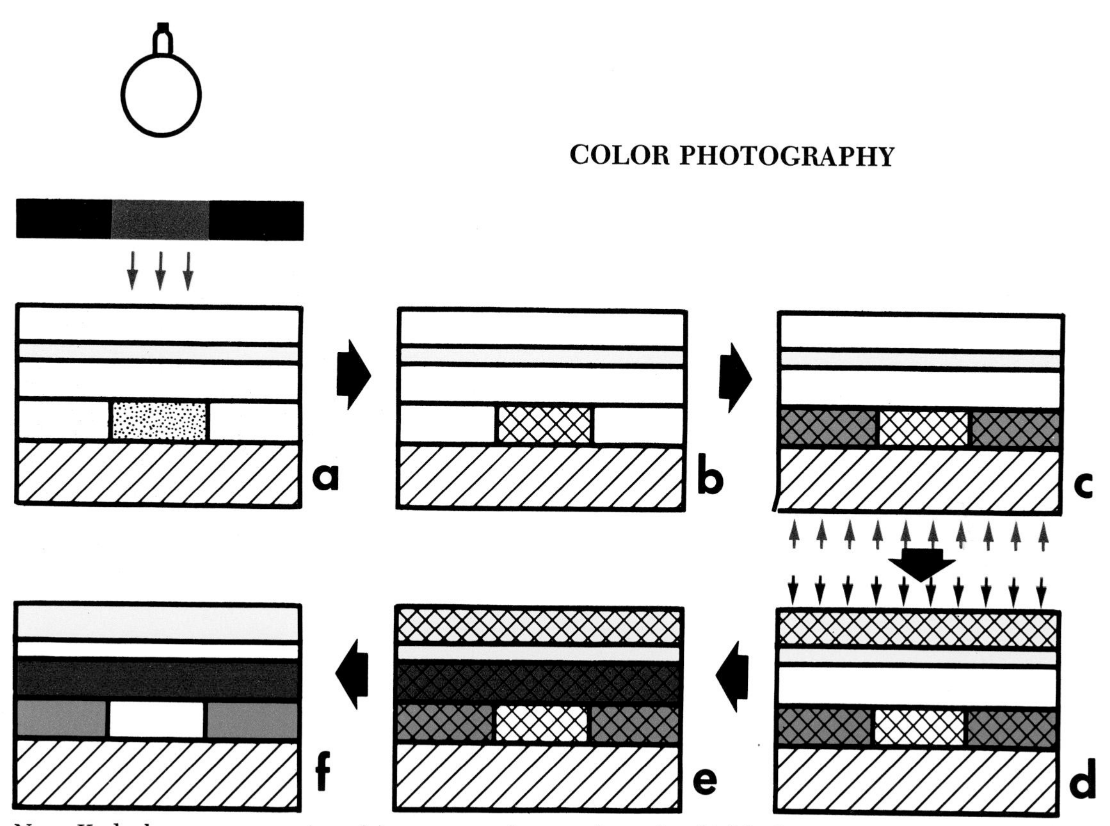

New Kodachrome processing: (a) exposure from red patch; (b) black-and-white development; (c) red exposure (through base) and cyan development using newly exposed silver halide; (d) blue exposure (from top) and yellow development using newly exposed silver halide; (e) fogging magenta development using residual silver halide (which is left only in green-sensitive layer); (f) silver and yellow filter layer removed by bleach and fix.

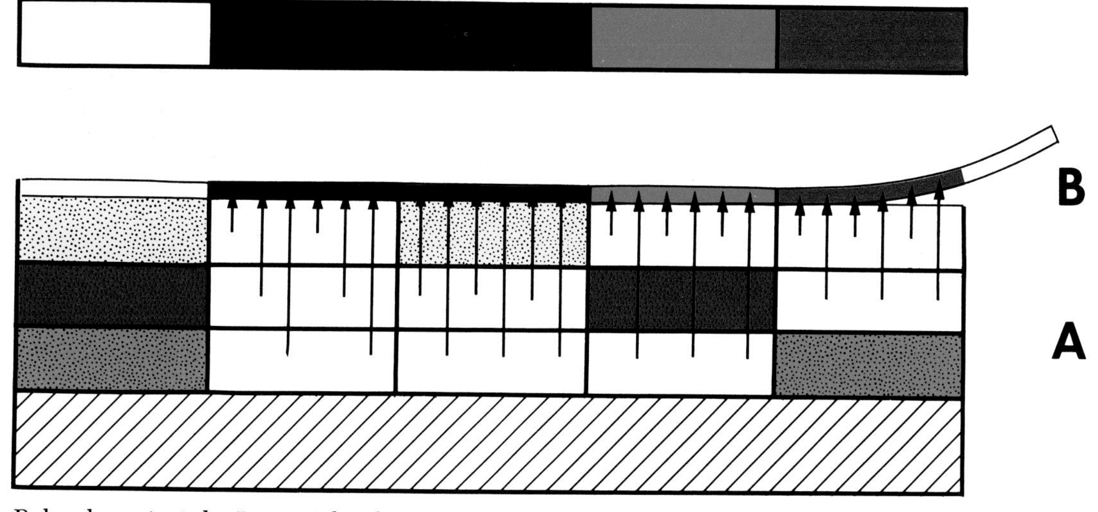

Polacolor principle. Layer A has been exposed and dyes have been made immobile by development. Mobile dyes have been transferred to receiving sheet B by contact.

From "A Chemist's View of Color Photography," by Arnold Weissberger, *American Scientist,* 58(6), 1970; courtesy of Eastman Kodak Company, Rochester, N.Y.

PLATE VIII

A.

B.

A. Cross dyeing of nylon. These carpet samples contain four different types of nylon. The top band is cationic dyeable, and the next three bands below have progressively increasing capacity for anionic dyes. All four types are dyed equally by disperse dyes. From right to left the strips were dyed with a red anionic dye, a blue cationic dye, a yellow disperse dye and a bath containing all three dyes. *B*. Cationic dyes on Dacron. A single cationic dye was used for two patches on the left of each group; the orange, purple, and green patches are dyed with a mixture of the two dyes at the left. (From Moore, J. A.: *Elementary Organic Chemistry*. Philadelphia, W. B. Saunders Co., 1974.)

Film sensitivity is rated on the American Standards Association (ASA) scale. The larger the number, the more sensitive the film is to light.

This mechanism renders a silver halide grain more sensitive than it would be ordinarily, since the latent image-destroying bromine atoms are used up in another process. Film sensitivity is also related to grain size. As the grain size in the emulsion increases, the effective light sensitivity of the film increases (up to a point; Figure 23–10). The reason for this is that the same number of silver atoms are needed to initiate reduction of the entire grain by the developer despite the grain size. (However, if the grain gets too large, the chances of four silver atoms forming a nucleus are not as great, so that a very large-grained emulsion would actually be less sensitive.)

Amplification of the Latent Image—Development

Silver halides are not the most photosensitive materials known. Why, then, are they effective image producers? The answer lies in the fact that the impact of a single photon on a silver halide grain produces a nucleus of at least four silver atoms, and this effect is amplified as much as a billion times by the action of a proper reducing agent ***(developer)*.**

When an exposed film is placed in developer, the grains that contain silver atom nuclei are reduced faster than those grains which do not. The more nuclei present in a given grain, the faster the reaction. The reduction reaction is

$$\underset{\text{IN GRAINS CONTAINING } Ag_4^0}{Ag^+} + e^- \longrightarrow Ag^0$$

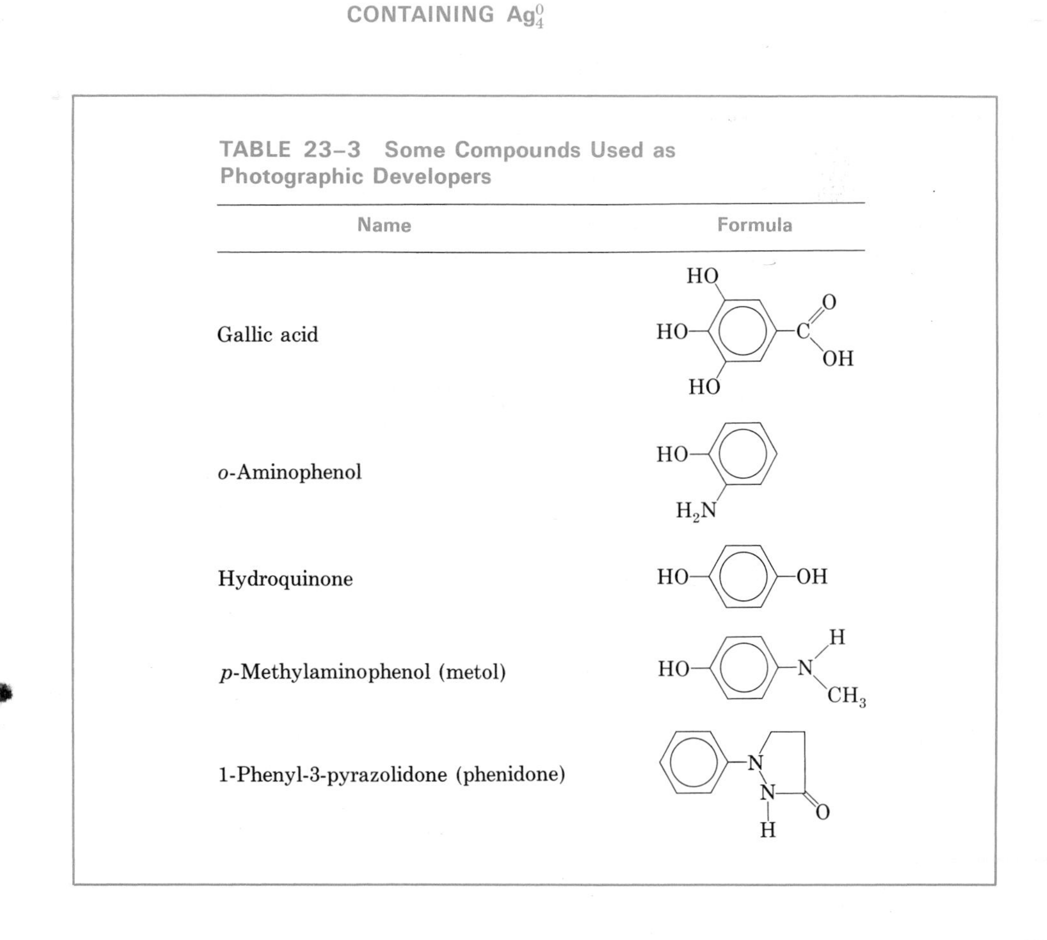

TABLE 23–3 Some Compounds Used as Photographic Developers

Name	Formula
Gallic acid	
o-Aminophenol	
Hydroquinone	
p-Methylaminophenol (metol)	
1-Phenyl-3-pyrazolidone (phenidone)	

Factors such as temperature, concentration of the developer, pH, and the total number of nuclei in each grain determine the extent of development and the intensity of free silver (blackness) deposited in the film emulsion.

The blackness on the negative is due to free silver atoms, Ag^0.

Not only must the developer be capable of reducing silver ions to free silver, but it must be selective enough not to reduce the unexposed grains, a process known as "fogging." Table 23–3 lists some substances that are used as developing agents.

Most developers used for black and white photography are composed of hydroquinone and metol or hydroquinone and phenidone. A typical developer consists of a developing agent (or two), a preservative to prevent air oxidation, and an alkaline buffer to prevent the actual reduction reaction from being retarded (Table 23–4). Other chemicals might be added but they are not absolutely necessary.

When hydroquinone acts as a developer, quinone is formed. Two protons are also produced for every two silver atoms:

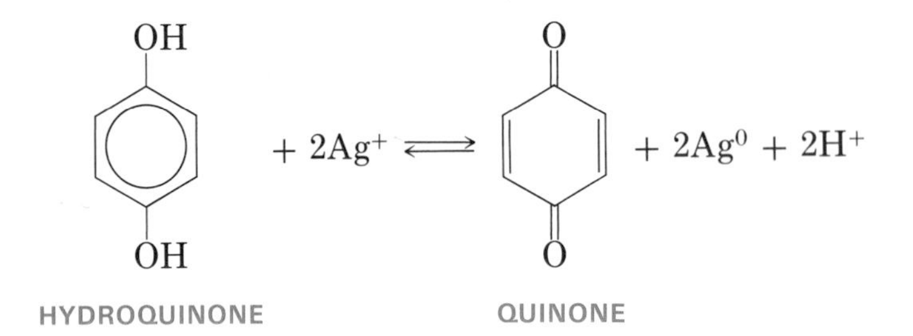

Since this reaction is reversible, a buildup of either protons or quinone would impede the development process. The sodium sulfite reacts with quinone and destroys its ability to revert back to hydroquinone.

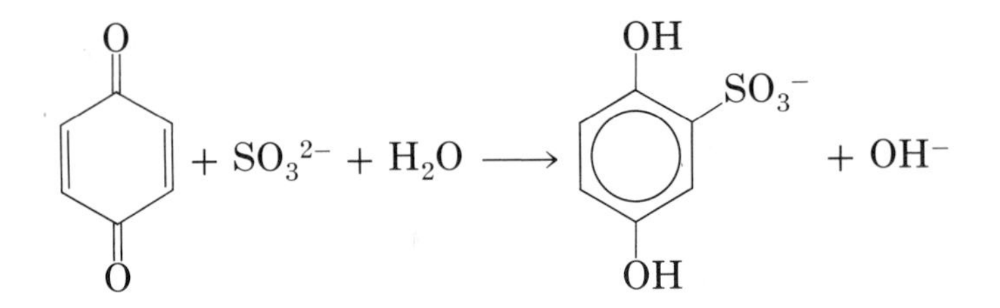

The protons are neutralized effectively by the hydroxide ions (OH^-).

$$H^+ + OH^- \longrightarrow H_2O$$

TABLE 23–4 Formula for a Typical Developer for Black and White Films

750 ml water at 50°C; dissolve in this water:	
Metol	2.0 g
Hydroquinone	5.0
Sodium sulfite	100.0
Borax, $Na_2B_4O_7 \cdot 10H_2O$	2.0
Add cold water to make 1000 ml of solution.	

When dissolved in water, a buffer such as borax yields the following products:

$$\underset{\text{BORAX}}{Na_2B_4O_7 \cdot 10H_2O} \longrightarrow \underset{\text{BORIC ACID}}{2H_3BO_3} + 2Na^+ + \underset{\text{BORATE ANION}}{2B(OH)_4^-} + 3H_2O$$

The presence of both the boric acid and the borate anion makes the solution a buffer solution (that is, it does not change pH easily). In the developer solution, some of the protons produced in the reduction reaction are neutralized by the borate anion:

$$H^+ + B(OH)_4^- \longrightarrow B(OH)_3 + H_2O$$

If development proceeds either too long or at a higher temperature than recommended, sufficient fogging occurs to render a negative useless. Since the rates of the development reactions increase with increasing temperature, the photographer usually controls the temperature of the development bath very carefully.

The development process is terminated by a ***stop bath.*** The stop bath usually contains a weak acid such as acetic acid which decreases the pH. The action of a stop bath is to build up the amount of hydrogen ions which effectively stops the hydroquinone $\longrightarrow$ quinone reaction.

Fixing

Solutions of sodium thiosulfate, $Na_2S_2O_3$, known as "hypo," were first used by Sir J. W. F. Hershel to "fix" negatives.

One of the principal problems in the early days of photography was the lack of permanence of the image. If development only produces free silver where the light intensity was greatest and nothing further is done to the negative, the undeveloped silver halide will be exposed the instant it is taken into the light. After that, almost any reducing agent will completely fog the negative. In order to overcome this problem, a suitable substance had to be found to remove the unreduced silver halides. The most commonly used ***fixing agent*** in black and white photography is the thiosulfate ion ($S_2O_3^{2-}$). Thiosulfate ions form stable complexes with silver ions in aqueous solution:

$$\underset{\text{INSOLUBLE SALT (UNDEVELOPED)}}{AgBr(s)} + \underset{\text{FROM "HYPO" SOLUTION}}{2S_2O_3^{2-}} \longrightarrow \underset{\text{WATER-SOLUBLE COMPLEX}}{Ag(S_2O_3)_2^{3-}} + Br^-$$

After fixing is complete, the film is washed sufficiently to remove the last traces of thiosulfate from the emulsion. Otherwise, the thiosulfate ion will decompose to produce free sulfur. The free sulfur will react with free silver to produce darkly colored silver sulfide and will mar the quality of the image.

$$S_2O_3^{2-} + 2H^+ \longrightarrow S^0 + SO_2 + H_2O$$

$$S + 2Ag \longrightarrow Ag_2S \text{ (brown)}$$

When certain types of black and white films are processed, as much as 60 to 80 percent of the original silver in the emulsion is removed by fixing. This silver can be recovered before the fixing bath is discarded, but the recovery only pays for itself when one has available large quantities of solution. For example, a photofinisher processing 100,000 rolls of black and white film a year would realize about $2900 of silver, with silver selling for $4.70 a troy ounce. Of course, eliminating this much silver from waste water makes sense also as an antipollution measure.

One troy ounce is about 31 grams.

Prints

A ***print*** is a reverse of a negative. Where the negative is dark, the print is light, and vice versa. The formation of the print is similar to the formation of the negative. Light-sensitive paper is exposed by light passed first through the negative and then onto the paper. The paper is then developed and fixed (Figure 23–13).

A fundamental difference between negative-making and print-making is the grain size of the emulsion. The silver halide grains in photographic papers are very small compared to the grains in film emulsions. Small grain size renders the paper less sensitive to light. In this way, more control can be exercised in the proper exposure of the paper by using longer exposure times. Silver chloride emulsions are often used in papers due to their lower overall sensitivity. The silver chloride is more stable relative to photodecomposition.

Photographic papers are much less light-sensitive than films.

Special precautions must be taken during the washing to eliminate the thiosulfate ions from the paper or they will decompose and stain the print with silver sulfide (see preceding section). A ***hypo eliminator,*** containing hydrogen peroxide, generally is used to oxidize the thiosulfate to the stable sulfate ion:

$$\underset{\text{THIOSULFATE ION}}{S_2O_3^{2-}} + \underset{\text{HYDROGEN PEROXIDE}}{4H_2O_2} \longrightarrow \underset{\text{SULFATE ION}}{2SO_4^{2-}} + 2H^+ + 3H_2O$$

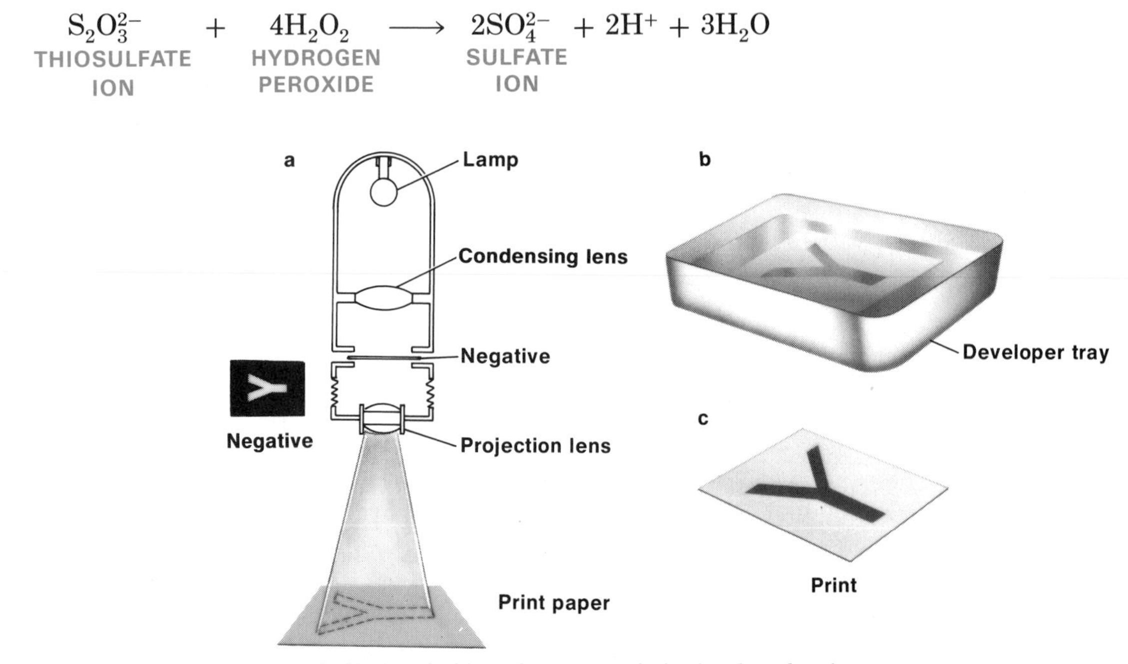

Figure 23–13 Black and white enlargement printing involves choosing a negative and placing it in an enlarger (*a*). The print paper is exposed for a length of time necessary to produce the desired print (usually by trial and error), and then placed in developer (*b*). After development, the print is fixed with hypo and washed thoroughly to produce the finished product (*c*), which is a direct opposite of the negative.

Instant Black and White Pictures

In 1947, Dr. Edwin H. Land introduced his invention of a process that produced a finished picture in one minute. Since its introduction, the Polaroid process has been popular for just this unique feature.

After exposure, a Polaroid film is brought into contact with a piece of receiver paper; at the same time a pod containing both a developer and a silver solvent is broken and spread over the film (Figure 23–14). As the developer reduces the exposed silver halide grains in the film emulsion, the silver solvent picks up the unexposed silver ions, which then diffuse across the boundary onto the receiver paper. There, in contact with minute grains of silver already in the paper, the developer reduces the silver in the hypo complex to free silver and a ***positive*** image. This positive image forms since black areas in the original object photographed

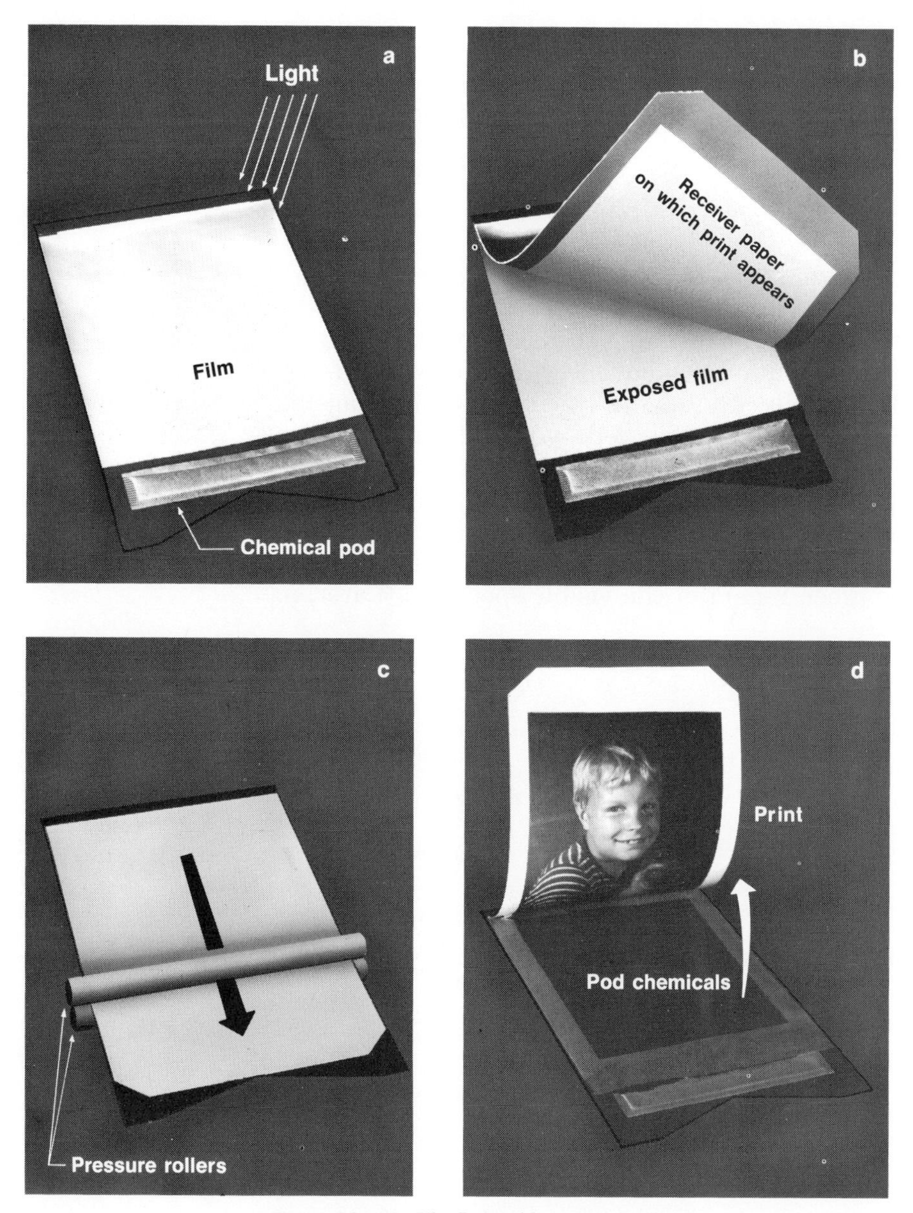

Figure 23–14 The Polaroid process for black and white photographs.

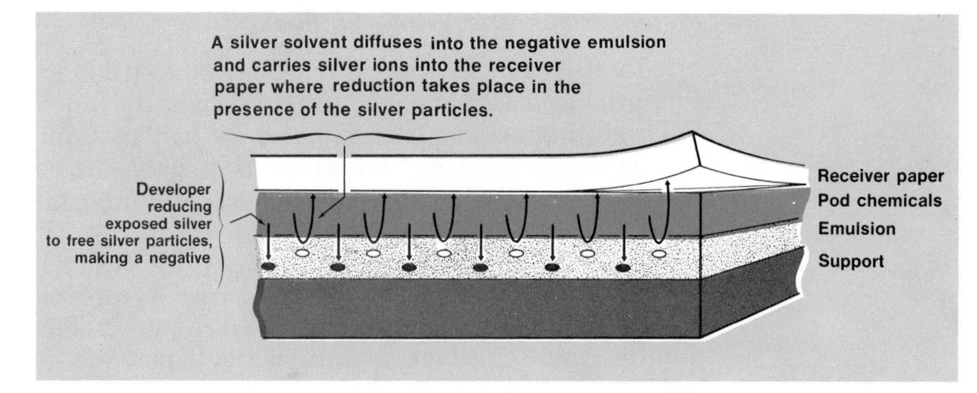

Figure 23–15 A schematic diagram of the chemistry of the Polaroid process.

exposed no silver grains in the emulsion, and it was this silver, as Ag^+ ions, which was then carried onto the receiver paper and reduced (Figure 23–15).

Spectral Sensitivity

Probably the most important ingredients in a black and white photographic emulsion, other than the silver halide salts themselves, are the spectral sensitizing dyes. Silver halides are most sensitive to blue light or higher energy electromagnetic radiation such as UV light (Figure 23–16). A film manufactured with only silver halides as the photosensitive agents will be only blue-sensitive and will not "see" reds, yellows, greens, and so on, as ordinary colors.

In 1873, while trying to eliminate light scattering problems in photographing the solar spectrum, W. H. Vogel, a German chemist, added a yellow dye to his emulsion. To his surprise he discovered that he could now record images in the green region of the visible spectrum. Later, in

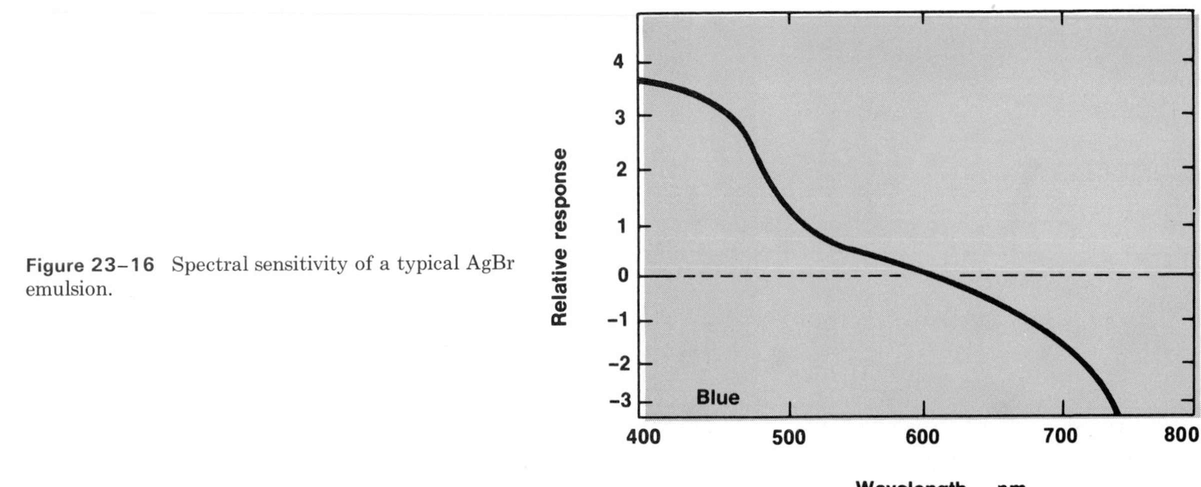

Figure 23–16 Spectral sensitivity of a typical AgBr emulsion.

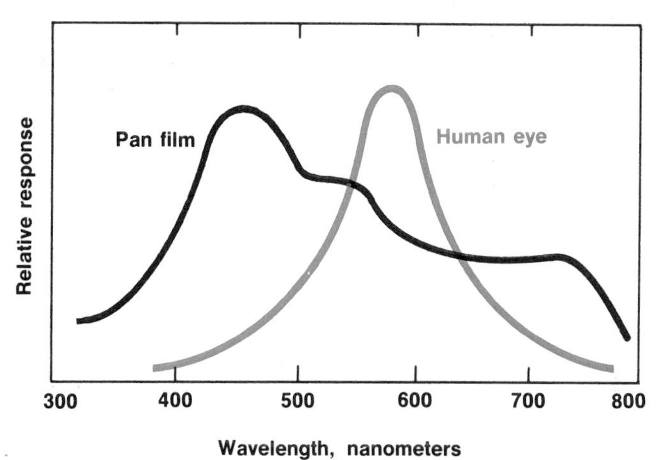

Figure 23–17 Spectral sensitivity of a panchromatic film compared to that of the human eye.

1904, another German, B. Homolka, discovered a dye, pinacyanol, which when added to a silver halide emulsion rendered it sensitive to the entire visible spectrum (Figure 23–18). Films of this type are called ***panchromatic*** or "pan" films.

By adding spectral sensitizing dyes, photographic emulsions can be made that are sensitive to selected regions of the spectrum with wavelengths from 100 nanometers (ultraviolet) to 1300 nanometers (infrared).

A panchromatic film is sensitive to the entire range of visible wavelengths.

The mechanism by which a dye molecule can impart spectral sensitivity to silver halide grains seems to involve initially the absorption of a photon of light (hν) by the dye molecule. Next, the excited molecule ejects an electron into the silver halide grain where a free silver atom is formed. The electron-deficient dye molecule then oxidizes the bromide ion, producing a bromine atom or positive hole:

$$\text{Dye} \xrightarrow{h\nu} \text{Dye}^+ + e^-$$

$$Ag^+ + e^- \longrightarrow Ag^0 \longrightarrow \text{silver nuclei}$$

$$\text{Dye}^+ + Br^- \longrightarrow \text{Dye} + Br^0 \text{ (positive hole)}$$

Thus, the process is effectively the same as a photon striking the bromide in the grain itself, the dye serving as a catalyst.

Color Photography

The chemistry of color photography, which dates back to 1861, is a good deal more complicated than that of black and white photography. James Clerk Maxwell, the famous English scientist, was the first to photograph an object in color. He used three exposures through three primary color filters and then resynthesized the color of the object by

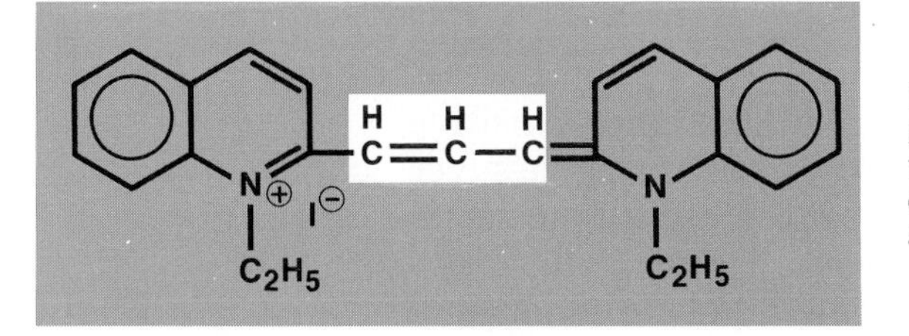

Figure 23–18 Pinacyanol, a cyanine dye. The conjugated group (double bond, single bond, etc.—one sequence in the white block) is the chromophore (Chapter 14). Other cyanine dyes have more CH groups, absorb longer wavelengths of light, and shift the film sensitivity toward the red.

projecting the image through the same filters. This experiment actually predated panchromatic film, but due to a peculiarity of his particular emulsion, it was essentially panchromatic nonetheless. The important point is that the results were consistent with the then emerging theory of color vision and thus led to other more significant results.

Additive and Subtractive Primary Colors

As early as 1611, De Dominis showed that the visible spectrum is composed of three fundamental colors: red, green, and blue (known as ***additive primaries***). This concept has since proved useful in the development of color vision theory and in color photography. After 1861 the idea slowly evolved that in order to reproduce color images, a film would have to be made with three different layers, each layer sensitive to one of the three primary colors. After the discovery of color-sensitizing dyes and panchromatic black and white film, several different techniques for color photography were developed, but not until 1935, when Kodachrome was placed on the market, did the products reach the consumer. The Kodachrome process produces transparencies (or slides) which are viewed as transmitted light.

If additive primary colors are used to form an image by superposition, such as we might expect in a color transparency, problems with light transmittance arise. Combinations of additive primary-color filters produce black. In order to overcome these problems and obtain a color slide (or color negative, for that matter) another system of primary-color filters was developed, known as ***subtractive primary*** colors. These colors are produced by dyes that absorb the additive primary colors. Thus a dye that absorbs red light transmits or reflects the remainder of the spectrum and appears greenish-blue (*cyan*). Absorption of blue light renders a dye yellow, and absorption of green light makes the dye appear bluish-red (*magenta*).

Blue light + Green light } Cyan
Green light + Red light } Yellow
Blue light + Red light } Magenta

When the proper mixture of subtractive primary dyes is formed in a photographic emulsion during the development process, an image is produced with the desired color. For example, a mixture of magenta and cyan dyes would appear blue, since the magenta dye absorbs green light and the cyan dye absorbs red light, leaving only blue to be transmitted out of the three components of white light.

In order to understand fully the next few paragraphs, turn to Color Plate IV and refer to it whenever needed.

The use of subtractive primaries in color photography was suggested as early as 1869, but it was much later before the chemistry was worked out in enough detail to yield good results. The problem is to get the right amount of the correct subtractive primary dye in the right place to reproduce the correct true-to-life color. White is produced by the absence of all three subtractive primaries, and black is produced by an equal balance of all three.

Color Film

Generally, a color film consists of a support and three color-sensitive emulsion layers. The blue-sensitive layer is usually on the top since silver halides are inherently blue-sensitive. Next, a yellow colored filter layer is added. This layer absorbs blue light and serves to protect the lower emulsion layers from blue light. A green-sensitive layer is added and

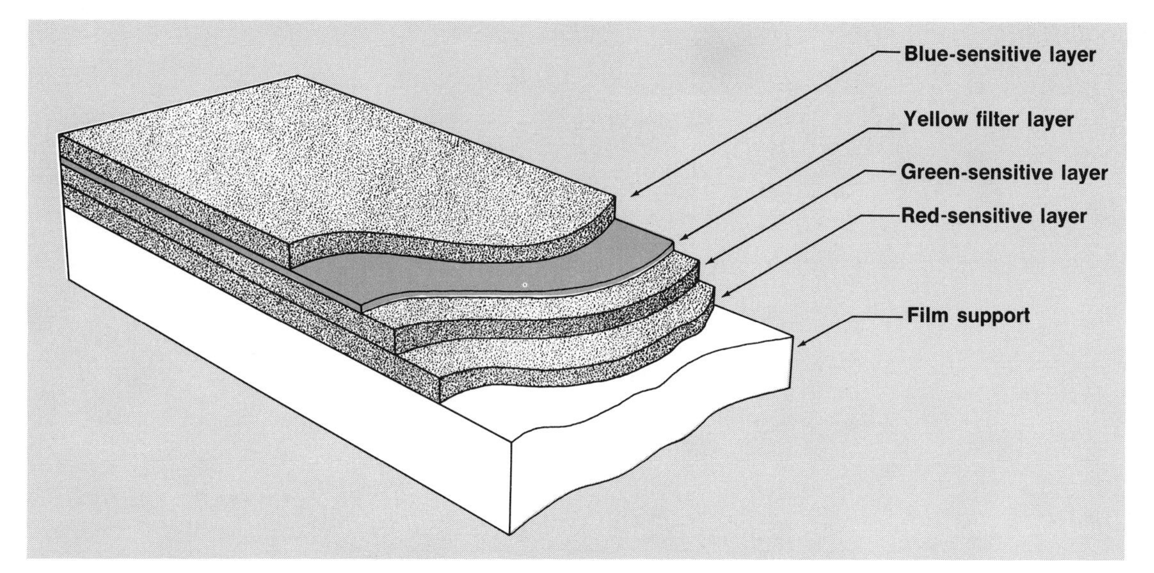

Figure 23–19 A typical arrangement of color-sensitive emulsion layers in a color film.

The thickness of the entire emulsion of color film is only about 0.0254 mm (0.001 inch) thick.

followed by a red-sensitive layer and the support (Figure 23–19). These layers are rendered color-sensitive by dyes similar to those in the cyanine class, which render black and white film panchromatic. It should be realized, however, that the color-sensitizing dyes are *not* generally involved in producing the final primary colors responsible for the color of the image. It is the final processing of the color film that yields the color image.

Color Development

Most color films are developed with the aid of a dye-forming color process first introduced by a German chemist, R. Fischer, in 1912. The basis for this process is the oxidation of the developer to a dye-forming substance, which is then allowed to react with a molecule called a ***coupler*** to form the dye.

In some color films such as Kodachrome II, the coupler is dissolved in the developer, and the two react together in the presence of the silver halide grain. In other color films, notably, Kodacolor, Ektachrome, and Anscochrome, the couplers are distributed evenly in the appropriate emulsion layers in which the desired dye is to be formed.

Color developers are generally substituted amines and as such are reducing agents. An example is N,N-diethyl-*p*-phenylenediamine:

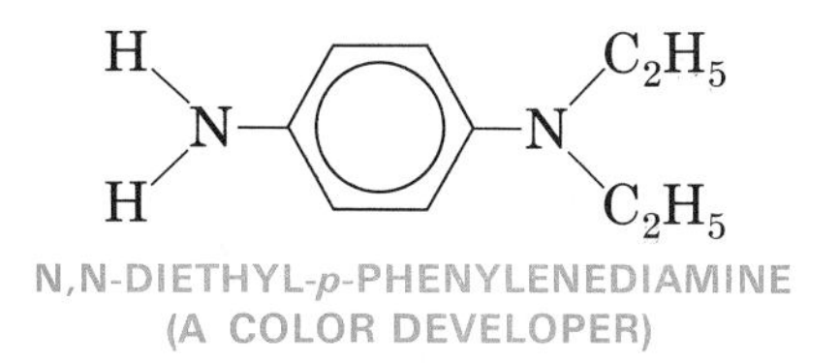

N,N-DIETHYL-*p*-PHENYLENEDIAMINE (A COLOR DEVELOPER)

To form a cyan dye during the development process, a phenol compound such as α-naphthol acts as a coupler:

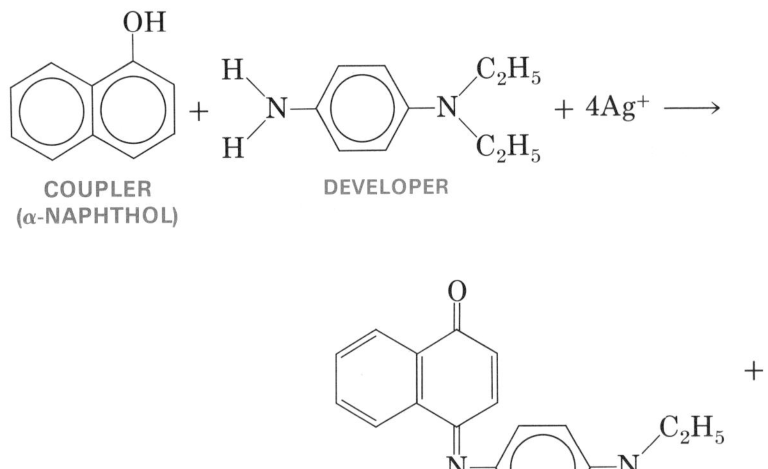

O
N
N
C_2H_5
C_2H_5

$+ 4Ag^0 + 4H^+$

A CYAN DYE (ABSORBS AT 630 NM)

Light with a wavelength of 630 nm is red.

Thus, in the development of an exposed silver halide grain in the red-sensitive emulsion layer, a small amount of cyan dye is produced. The free silver must be bleached out prior to finishing.

Minor changes in the structure of the developer are designed by the photographic chemist to vary solubility and speed of development. Perhaps the most bothersome aspect of these compounds is that they are ***allergenic*** and cause a skin irritation in most people similar to that caused by poison ivy. Amateur color developing is a rewarding pastime and has been made possible by specially designed developing agents which have low toxicity.

The Kodachrome Process

An interesting example of a widely used color photography system is the Kodachrome process of Eastman Kodak Company. The Kodachrome process is a ***reversal*** process; this means colors are reproduced in terms of their correct values and not their negative or complementary colors. The first developer in the Kodachrome process is a black and white developer. By careful temperature control, development of the exposed silver halide is made essentially complete.

The remaining unexposed silver halide in the three color-sensitive

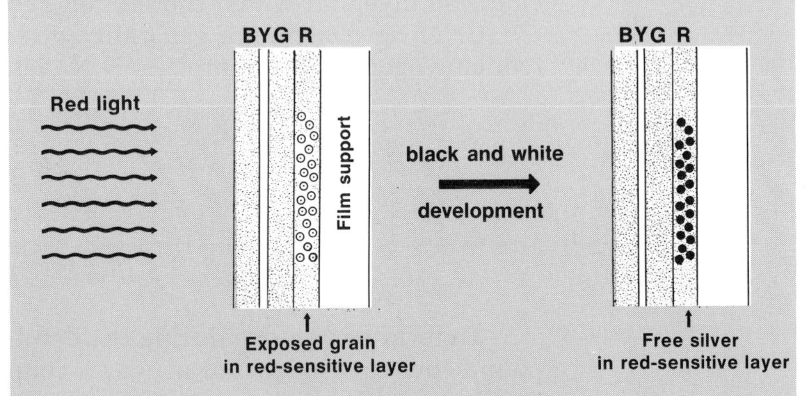

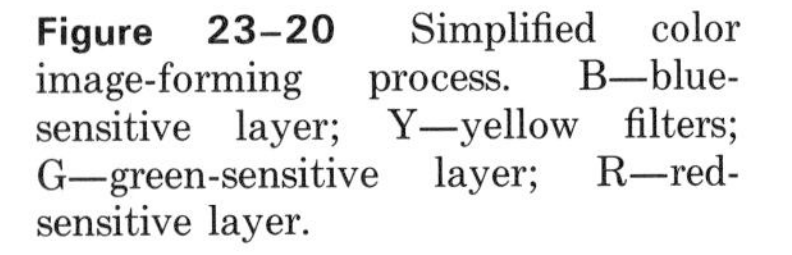
Figure 23–20 Simplified color image-forming process. B—blue-sensitive layer; Y—yellow filters; G—green-sensitive layer; R—red-sensitive layer.

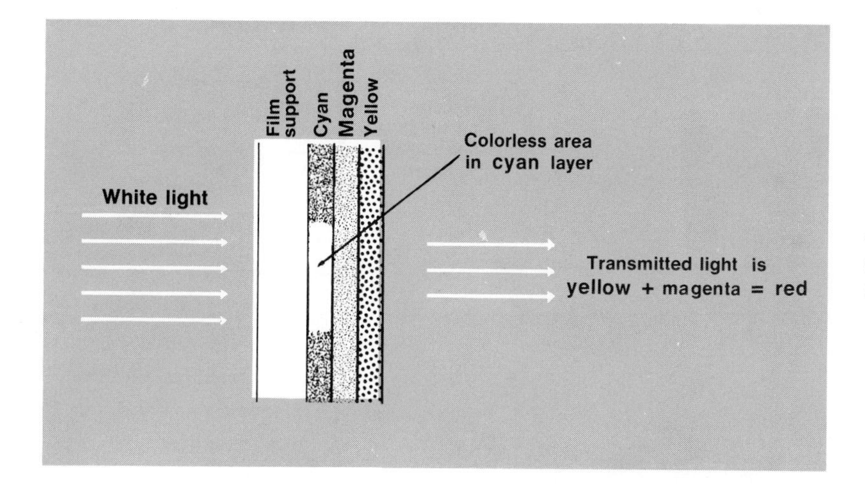

Figure 23–21 White light passes through a three-layer transparency. Since blue light and green light are absorbed, the transmitted light is red.

emulsions is a positive record of the original exposure. For example, refer to Color Plate IV and note that red light striking the film would, upon black and white development, leave free silver in the red-sensitive layer (Figure 23–20). Since no other color-sensitive layers were exposed by the original image, they contain no information. Now, selective reexposure and color development will produce free silver throughout the emulsion layers, along with the colored dyes, *except* where the red light originally struck the film. No dye forms there since the silver was previously reduced with a black and white developer.

All the silver is bleached out of color negative and reversal films during processing.

Next, all the silver in the three emulsion layers, as well as the yellow-colored protective layer, is bleached with an oxidant such as ferricyanide ion, $Fe(CN)_6^{3-}$:

$$Ag^0 + Fe(CN)_6^{3-} \longrightarrow Ag^+ + Fe(CN)_6^{4-}$$

Once oxidized, the silver is treated with hypo and washed from the emulsion. The resulting emulsion is colored, but transparent. Considering that red light originally exposed the film, we see that the transmitted light will appear red (Figure 23–21).

Color Pictures in a Minute

The Polaroid camera can produce an almost "instant" color picture. The chemical processes involved in making this type of color image (Polacolor) are similar to those mentioned earlier in this chapter, but there must be a delicate balance of properties of the developers, couplers, and dyes in order to obtain a good picture. Figure 23–22 shows the essential features of the Polacolor system. Prior to development, the receiver and photosensitive layers are separated. Light reaching the film strikes the blue-sensitive emulsion first, as in other color films. After exposure, development is initiated by pulling the film past pressure rollers, which also break the alkali-containing pod. The Polacolor dyes forming the negative image are "precipitated" in the photosensitive layer, which becomes a negative, while the dye-developer molecules which have not encountered exposed silver halide diffuse in the alkali up onto the receiver. There the dye-developer molecules react with the mordant to form the proper colors for a positive image.

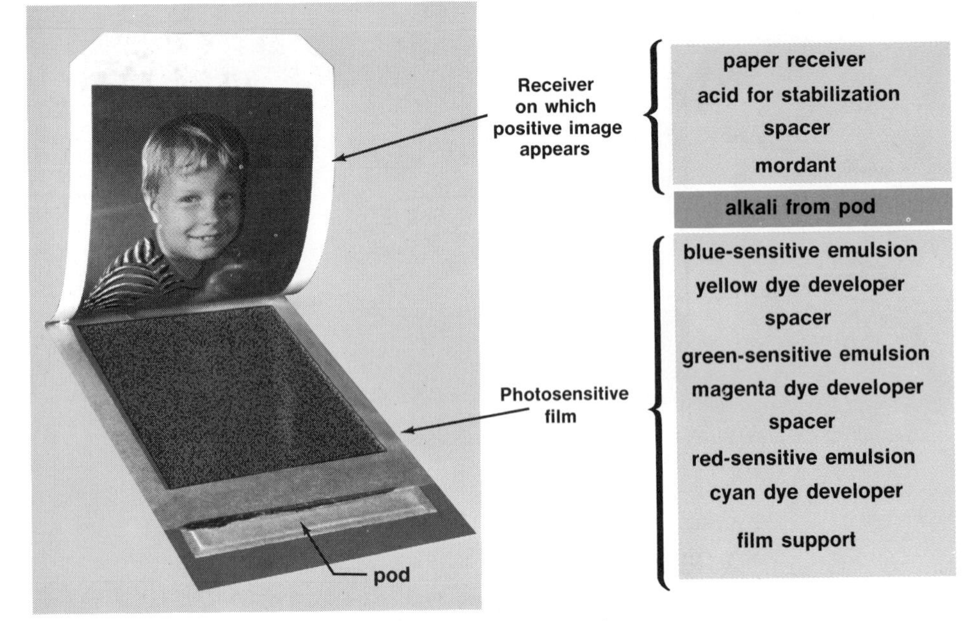

Figure 23–22 The Polacolor system.

The spacers serve to slow down the diffusion of the alkali into the photosensitive layer, allowing enough time for the developers to work. Even so, one minute is a relatively fast reaction time. Once the alkali has served its purpose of moving the dye-developer molecules onto the receiver, it is neutralized by the acid stabilizer. The finished Polacolor picture requires no further stabilization.

SELF-TEST 23-B

1. The branch of chemistry that deals with the changes in matter produced by absorbed radiant energy is ____________________.
2. What chemical composes the dark regions on a negative? ____________
3. What is the minimum number of silver atoms in an AgBr grain that is necessary for that grain to be developed? ________
4. In the development process, silver is (oxidized/reduced).
5. Complete the following equation for the fixing process.

   $Ag^+ + 2S_2O_3^{2-} \longrightarrow$ ____________________
6. Which are more sensitive to light, very small or somewhat larger grains of AgBr? ________________
7. Photographic developers are (oxidizing/reducing) agents.
8. Hypo is another name for __________ ____________.
9. Grain size is larger in photographic (negatives/prints).

10. (a) cyan + magenta = ____________

    (b) cyan + yellow = ____________

    (c) magenta + yellow = ____________

11. The primary additive colors are ____________, ____________, and ____________.

12. A Kodachrome color transparency has no silver in it. (True/False)

13. In the formation of a dye in color film, silver is (oxidized/reduced).

14. Film sensitive to all light in the visible range is known as ____________________ film.

THE CHEMISTRY OF PAINTS—PROTECTION AND BEAUTIFICATION

Most of us have taken up the brush, roller, or spray can at one time or another to apply a coating of what we call paint. We do this either to protect or beautify, or both. The things we paint are made of wood, metal, masonry, plastic, paper or ceramic. About $3 billion is spent on paints of all types in a typical year in the United States. In this section we will look at some of the chemistry of paints. From time to time we shall borrow on our knowledge of polymers from Chapter 15, since, as we will see, a paint forms a protective film which is much like a typical polymer film.

In 1970, nearly 830 million gallons of coatings were sold by U.S. companies. The $3 billion sales represents 5.5% of sales of all chemicals and allied products during 1970.

Figure 23–23 Photographers photographing a painter painting—in this picture the John Quincy Adams birthplace in Quincy, Massachusetts. The Sears Great American Home Series of advertisements illustrates the importance of surface coatings as preservatives. (Courtesy of Sears, Roebuck and Co., Chicago, Illinois.)

Basic Components of Paints

Paints are generally composed of two basic components: ***pigment*** and ***vehicle.*** The pigment supplies the desired color and, for exterior paints, provides a shield against the harsh ultraviolet radiation of the sun. The vehicle is the liquid part of the paint (in earlier years, linseed oil). Nowadays the vehicle often consists of a solvent and some sort of dissolved polymer, which act as a ***binder.***

A paint binder forms a molecular network to hold the pigment in place.

Unfortunately, most paint films break down in use, owing to attack by oxygen, sunlight, or pollutants, as well as through physical abuse. Much of the research on paints, and the interest the user has in paints, has to do with extending their life.

Film Formation—a Covering

In order to hold out moisture and air, and to hold the pigment in place, the binder of a paint must be capable of forming a solid film, which adheres to the painted object. Films can be formed by simply allowing the solvent to evaporate, or by more complicated processes in which polymerization reactions take place during the "drying" process.

For years, many farmers have used whitewash around the farm to brighten things up and protect wooden structures. Tom Sawyer's whitewash was a simple kind, composed of hydrated (slaked) lime [$Ca(OH)_2$] and water. (About 50 pounds of hydrated lime to 6 gallons of water will do the job.) After application, the lime slowly was converted to calcium carbonate (chalk) by carbon dioxide (CO_2) in the air. The resulting white color looked nice but protected the surface underneath very little.

$$\underset{\text{HYDRATED LIME}}{Ca(OH)_2} + CO_2(\text{in air}) \xrightarrow{\text{Slow}} \underset{\text{CHALK}}{CaCO_3} + H_2O$$

A film-forming, weather-resistant whitewash can be formed by adding hydrated lime to casein (milk protein; skimmed milk will do) and formaldehyde, HCHO. As we saw in Chapter 16, some proteins contain the amino acid lysine, which contains an amine group at the end of the hydrocarbon chain. In the presence of formaldehyde, the casein undergoes ***cross-linking*** (see Chapter 15) to produce a rigid film. The reaction is actually a condensation polymerization:

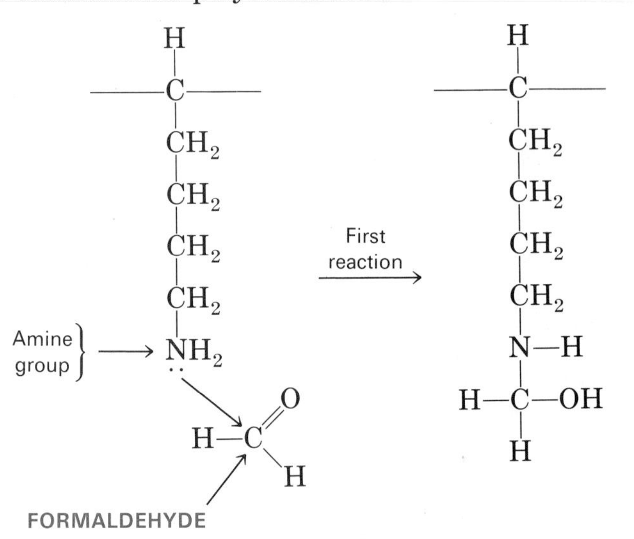

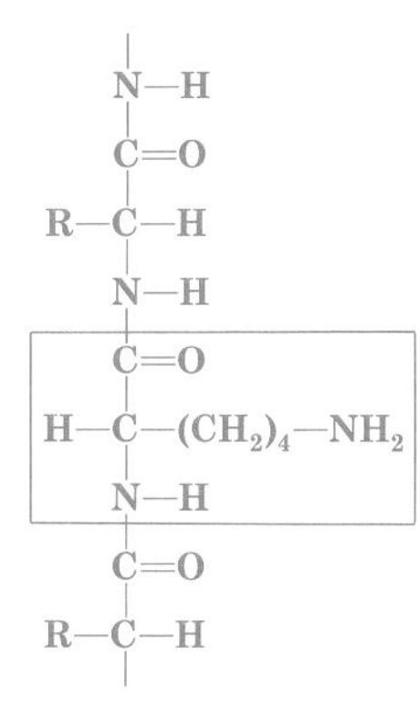

A portion of a protein chain with a lysine amino acid unit designated.

Then, cross-linking occurs between two chains:

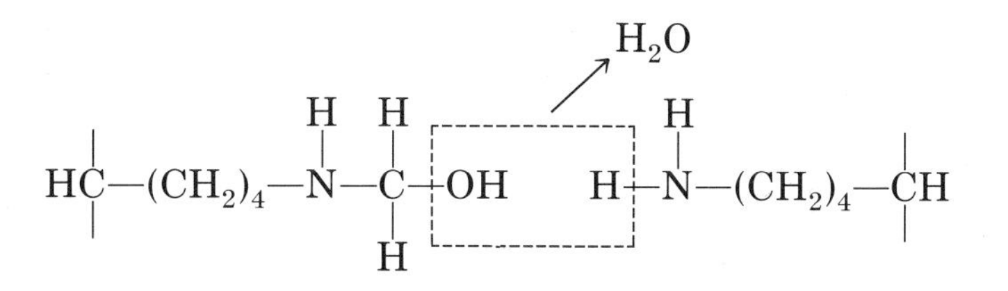

The resulting casein-formaldehyde polymer is a dense, three-dimensional web which holds the calcium carbonate very tightly and prevents it from washing off.

Oil-Based Paints

For outside painting, we used to use white lead [$Pb(OH)_2 \cdot 2PbCO_3$] dispersed in linseed oil and turpentine. This paint was used by the early Greeks and Romans. Of course, the lead is toxic, and lead-based paints are now banned in interior applications. In addition, lead reacts with hydrogen sulfide fumes present in industrial atmospheres to form black lead sulfide, PbS.

$$\underset{\text{(WHITE)}}{Pb(OH)_2} + \underset{\text{HYDROGEN SULFIDE}}{H_2S} \longrightarrow \underset{\text{(BLACK)}}{PbS} + 2H_2O$$

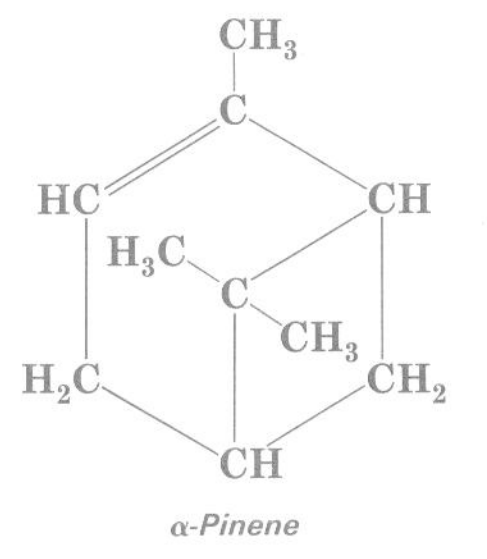

α-Pinene

The drying of an oil-based paint is a fairly complicated chemical process we shall look at later. However, the solvent mentioned above, turpentine, is a hydrocarbon mixture of two isomers of $C_{10}H_{16}$, α- and β-pinene—appropriately named since turpentine is obtained from pine trees. As a turpentine-based paint dries, the solvent simply evaporates, evolving a characteristic odor. Modern paints overcome the odor problem by using less obnoxious hydrocarbons such as mineral spirits, a petroleum fraction boiling between 93 and 150°C.

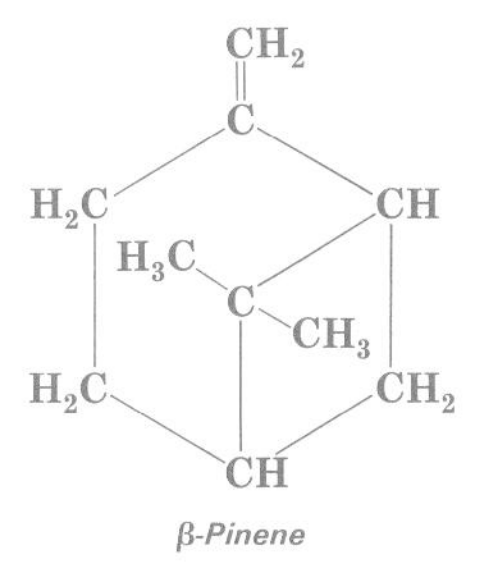

β-Pinene

The pigment of a modern oil-based paint contains titanium dioxide (TiO_2), "titanium calcium" (30 to 50 percent TiO_2 on the surface of calcium sulfate particles), zinc oxide (ZnO), and various extender pigments. Titanium dioxide has greater opacity and whiteness and is less toxic than white lead. The different crystalline forms of TiO_2 catalyze the decomposition of the binder (such as linseed oil) at different rates. The deterioration of the binder forms a white powder, which is easily rubbed off. This process is called ***chalking.*** The rate of chalking of the dried paint can be controlled with different crystalline forms of TiO_2. One form, rutile, causes slow chalking while another, anatase, promotes more rapid chalking. Controlled chalking is desirable because it gives a constantly renewed surface over a period of years.

Zinc oxide is a hardener in oil-based paints; zinc and other divalent ions (e.g., Ca^{2+}) form insoluble soaps with the fatty acids from the linseed oil, making the dried surface more durable. Extender pigments do not contribute much to opacity but are added to control the hardness of the paint film and to reduce the cost of the paint. If the relative amount of the total pigment is too low, the film is likely to be too soft. Talc [magnesium silicate, $Mg_3Si_4O_{10}(OH)_2$], limestone (calcium carbonate, $CaCO_3$)

Turpentine and mineral spirits help to lower the viscosity of oil paints and make them easier to spread.

and silica (silicon dioxide, SiO_2) are examples of extender pigments. A typical outdoor white oil-based paint could contain the following ingredients:

H_2S does not discolor ZnO- or TiO_2-containing paints. ZnS is white and TiO_2 does not react with H_2S.

	Percent by weight
Mineral spirits	34.2
Titanium dioxide (TiO_2)	29.0
Linseed oil	19.6
Calcium carbonate ($CaCO_3$)	12.0
Zinc oxide (ZnO)	2.2
Drier	1.7
Silica (SiO_2)	1.3

Interior and exterior oil paints are very similar. In interior paints titanium dioxide is added in greater amounts to give whiter whites and natural resins or greater concentrations of pigments are added to give a quicker drying film. Many additives are used in both exterior and interior oil-based paints to hasten the drying of the paint, provide better dispersion of the pigment, resist mildew and fungi formation, and prevent flooding, settling, skinning, yellowing, and destruction by ultraviolet light. For mildew resistance, mercury phenylacetate, $Hg(OCOCH_2C_6H_5)_2$, is used to the extent of about 0.5 percent by weight. The mercury ion can be the cause of slight darkening of the paint, since it reacts, like lead, with the pollutant hydrogen sulfide to form black mercuric sulfide.

Mineral spirits are petroleum fractions of moderate volatility.

$$Hg^{2+} + H_2S \longrightarrow \underset{\text{(BLACK)}}{HgS} + 2H^+$$

"Flooding" in paints results when the color components separate on drying and concentrate in streaks or patches in the surface of the film. Antiflooding agents are generally wetting agents, such as aluminum stearate or stearic acid. The polar end of the wetting agent bonds to the pigment particle, and the nonpolar hydrocarbon chain entangles itself compatibly in the oil film.

The Drying Mechanism

The drying of a modern oil-based paint involves much more than the evaporation of the solvent. A principal ingredient of these paints is a drying oil. Many drying oils are used such as soybean, castor, coconut, and linseed (obtained from the seed of the flax plant). All of these oils are esters of glycerin and various fatty acids, substances which were discussed in Chapter 14 in the making of soaps (Figure 23–24). The most important components of these oils in relation to their drying properties are the unsaturated fatty acids. The most important of these are oleic, linoleic, and linolenic acids.

Varnish contains a drying oil, rosin (see Chapter 22), and a thinner. Enamel is a pigmented varnish.

$$CH_3CH_2CH_2CH_2CH_2CH_2CH_2CH_2CH{=}CHCH_2CH_2CH_2CH_2CH_2CH_2CH_2\overset{O}{\overset{\|}{C}}OH$$

OLEIC ACID

$$CH_3CH_2CH_2CH_2CH_2CH{=}CHCH_2CH{=}CHCH_2CH_2CH_2CH_2CH_2CH_2CH_2\overset{O}{\overset{\|}{C}}OH$$

LINOLEIC ACID

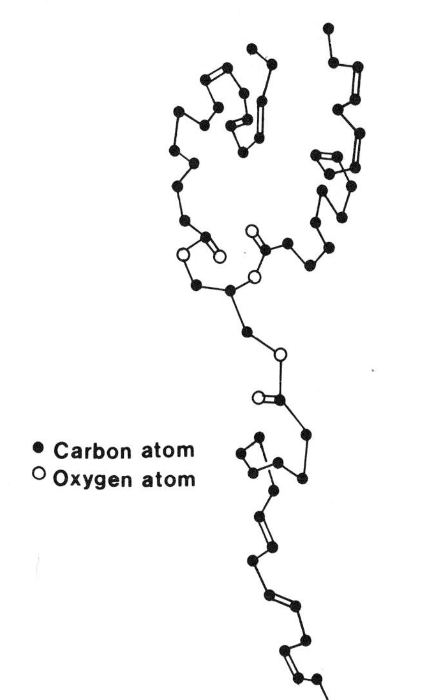

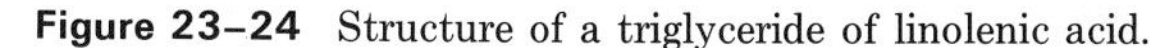

Figure 23–24 Structure of a triglyceride of linolenic acid.

$$CH_3CH_2CH{=}CHCH_2CH{=}CHCH_2CH{=}CHCH_2CH_2CH_2CH_2CH_2CH_2CH_2\overset{\overset{\displaystyle O}{\|}}{C}OH$$

LINOLENIC ACID

The most widely used drying oil is linseed oil, which contains varying proportions of the esters of these unsaturated acids. A typical linseed oil would yield, on hydrolysis, the following fatty acids:

4–7%	Palmitic acid (16 C atoms)
2–5%	Stearic acid (18 C atoms)
9–38%	Oleic acid (18 C atoms)
3–43%	Linoleic acid (18 C atoms)
25–58%	Linolenic acid (18 C atoms)

The attack of unsaturated fatty acids by oxygen causes fats and oils to turn rancid.

In one film-forming mechanism of drying, oxygen from the air attacks the unsaturated fatty acid chains at a carbon atom adjacent to the double bond, producing first a hydroperoxide radical ($\cdot OOH$) which then attaches itself to the fatty acid chain.

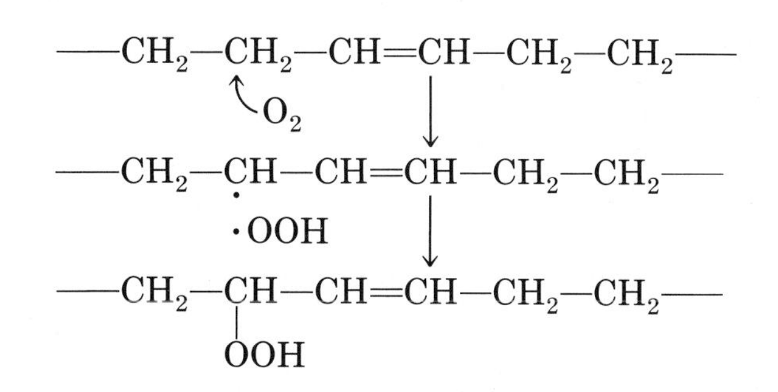

When a hydroperoxide group has attached itself to a fatty acid, it quickly forms a bridge to a neighboring fatty acid chain, again at a carbon atom adjacent to a double-bonded carbon atom. This cross-linking mechanism joins together two adjacent molecules by an ether linkage.

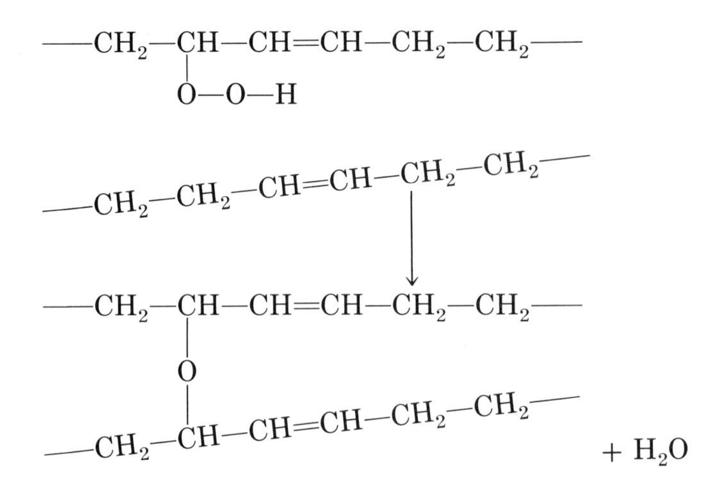

Another possible mechanism produces two ·OH radicals, which continue the process by other free-radical activation.

Free radical: molecular species with an unpaired electron.

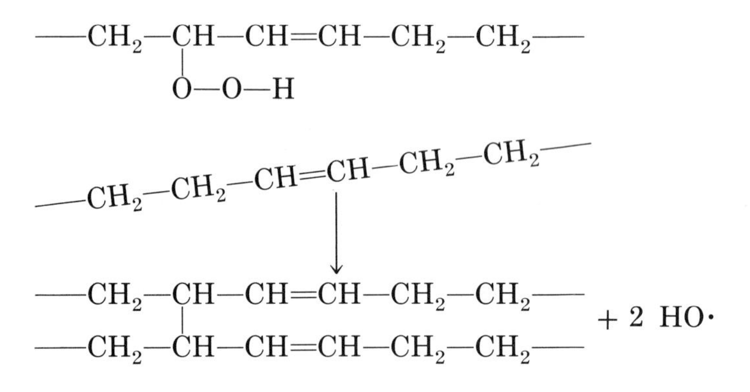

Both water and hydrogen peroxide are formed when oil paints dry. The hydrogen peroxide can be formed by the reaction of two hydroxy free radicals.

$$2\,HO\cdot \longrightarrow H_2O_2$$

The result of these cross-linking reactions taking place throughout the drying paint is the production of a web of linked molecules. Since the fatty acid molecules prior to drying were relatively small, some penetrated the openings in the surface being painted. After drying, good adhesion results. The solid nature of the film also holds the pigment particles in place.

To accelerate the decomposition of the hydroperoxide intermediates discussed above, metallic salts of fatty acids are added. One of the best metal ions is Pb^{2+}. This explains why lead salts were used for such a long time in paints until their toxic nature was discovered. Today salts of Zn, Co, Fe, Mn, and Ca are added to paints. These metal ions catalyze the decomposition of peroxides. They also precipitate fatty acids as insoluble soaps.

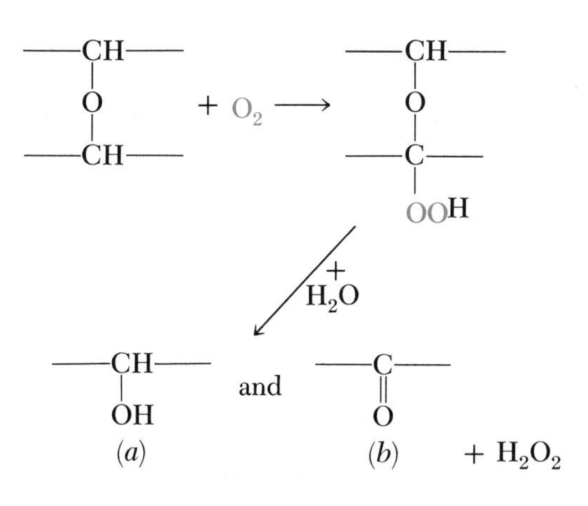

Figure 23–25 Breakdown mechanism for an ether linkage between two fatty acid chains. Oxidation first produces a hydroperoxy group. Water then reacts with this group producing (*a*) and (*b*).

Paint Film Breakdown—The Unwanted Kind

Although oxidation appears to cease when the paint film becomes solid, closer examination reveals that it continues at a slower rate over the lifetime of the film. For example, the oxidation of ether linkages can lead to a breakdown of the cross-linked polymer structure (Figure 23–25).

Ultraviolet radiation can cause paint to crack, fade, and depolymerize (Figure 23–26). Protection against ultraviolet rays can be afforded by: (1) pigmentation with suitable colors, which absorb the rays without being broken down (e.g., zinc oxide, titanium dioxide, carbon blacks, and iron oxides); (2) pigmentation with substances which reflect the rays (e.g., aluminum flakes); or (3) incorporation of small amounts of ultraviolet absorbers in the last coat (e.g., derivatives of benzotriazoles or benzophenone). Good absorbers have little color and absorb without being broken down themselves.

The yellowing of paint is observed often, particularly with interior paints. It has been confirmed experimentally that paints tend to yellow more in the dark and in the presence of moisture. Darkness favors the

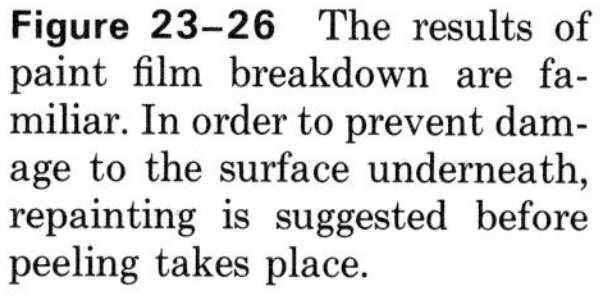
Figure 23–26 The results of paint film breakdown are familiar. In order to prevent damage to the surface underneath, repainting is suggested before peeling takes place.

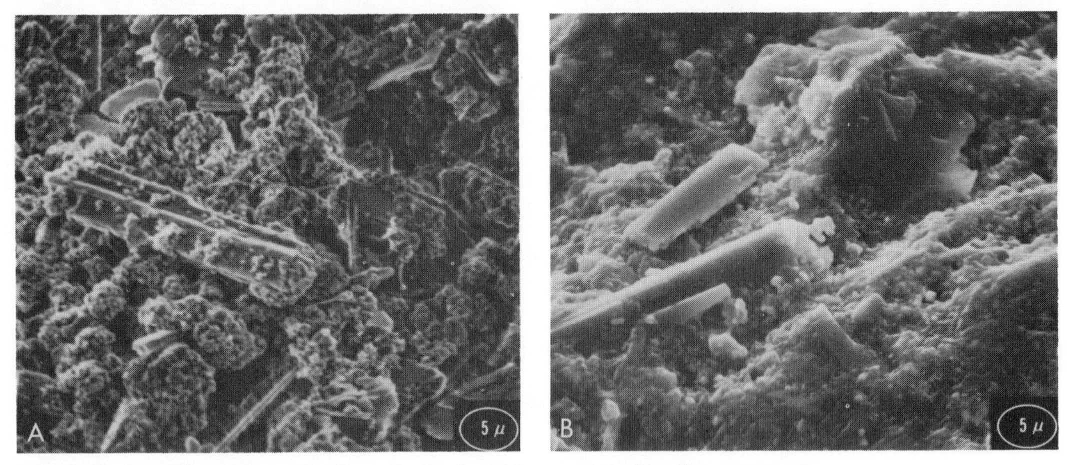

Figure 23–27 The surface of a linseed oil solvent base paint after four years of exterior exposure. *A*, before, and *B*, after washing with water and detergent to remove chalk; magnified 2000 times. (Courtesy of L. H. Princen, Oilseed Crops Laboratory, U.S. Department of Agriculture.)

condensation of atmospheric moisture on surfaces; light bleaches and tends to evaporate the moisture. The role (or roles) of moisture in the atmospheric yellowing of paints is not clearly understood. Water may hydrolyze linseed oil into acids that have stronger colors. It may catalyze the oxidation of the oils to other compounds that are highly colored.

Paint Film Breakdown—The Wanted Kind

Often it is necessary to remove paint for one reason or another. Films produced from drying oils are most easily removed by strong alkali such as solutions of lye (NaOH), or of trisodium phosphate (Na_3PO_4). Recalling what we learned in Chapter 14 about saponification of fats with lye to make water-soluble soaps, we should find it easy to understand why alkali literally eats up these paints. Drying oils are glycerides, and still contain ester linkages in the polymer film. Lye simply breaks down the polymer to a point at which it loses its structural integrity. When using lye or other strong alkali, lots of water should be used to wash the alkali away.

Latex Paints

A useful method of addition polymerization is called ***emulsion polymerization.*** This process is used to polymerize styrene and butadiene to make SBR rubber (see Chapter 15). By stabilizing the mixture of styrene and butadiene in water with a soap, the reaction heat is carried off easily by the water. In making tire rubber the emulsion must be broken down. If not destroyed, the emulsion can be used as a paint.

Latex paints have a styrene content as high as 85 percent with the butadiene about 15 percent. An advantage of these paints is that the polymer is already formed in a readily applicable form, the water emulsion (Figure 23–28).

Immediately after application with brush or roller, the water begins to evaporate. When only a part of the water is gone the emulsion breaks, and the remaining water quickly evaporates, leaving the film of paint.

The first commercial water-based Latex paint was Glidden's Spred Satin, introduced in 1948.

Water-based latex paints reduce fire hazards and air pollution associated with the handling and application of oil-based paints.

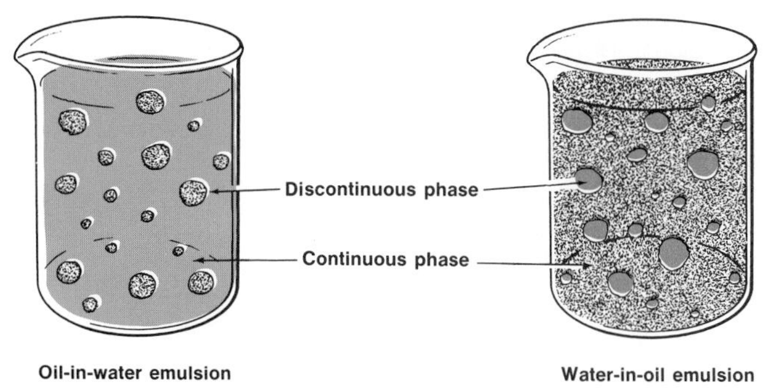

Figure 23–28 Two kinds of emulsions. An emulsion is composed of two immiscible liquids, one dispersed as tiny droplets in the other. An emulsifying agent is required to stabilize an emulsion.

Polymerization follows slowly. Drying times may be as short as 30 minutes. Since no hydrocarbon solvents are involved, odor is at a minimum, and there is no fire hazard. Interior painting with these paints allows quick occupancy after painting.

Since these paints are emulsions based on water, the painting tools can be easily cleaned with soap and water. This fact alone has had an important impact on their acceptance (Figure 23–29).

Arylic polymers have a sheen that allowed latex paint to compete in the exterior gloss market traditionally monopolized by oil-based coatings. Acrylics adhere well and control corrosion.

The styrene-butadiene resin is the least expensive material used, but it has a relatively long curing period, relatively poor adhesion, and a tendency to yellow with age. Polyvinylacetate is only a little more expensive and is an improvement over the styrene-butadiene resin. It quickly captured 50 percent of the latex market for interior paints. Another type with rapidly growing popularity, though about one third more expensive, are the acrylic resins and the "acrylic latex" paints. These are more washable and much more resistant to light damage. They are especially useful as exterior paints.

The fluoropolymers, similar to Teflon, are especially promising as surface coatings because of their great stability. Fluorine atoms are substituted for hydrogen atoms in the organic structure. Metals covered with polyvinylidene fluoride carry up to a 20-year guarantee against failure from exposure.

In the past few years, paint manufacturers have begun to blend linseed oil emulsions with latex emulsions in order to take advantage of the penetrating ability of the triglyceride molecules in the linseed oil.

Figure 23–29 Cleanup of latex paints is easy if you don't wait too long.

TABLE 23–5 Additives Used in Emulsion Paints

(1) Dispersing agents for pigments	Example: tetrasodium pyrophosphate ($Na_4P_2O_7$). Same principle of like-charged particles repelling as in oil-based paints.
(2) Protective colloids and thickeners	A thicker paint is slower to settle and drips and runs less. A protective colloid tends to stabilize the organic-water interface in the emulsion. Examples: sodium polyacrylates, carboxymethylcellulose, clays, gums. (Same mechanism as soap-dispersing oil in water; Chapter 14.)
(3) Defoamers	Foaming presents a serious problem if not corrected. The surface tension of the water must be decreased. Chemicals used: tri-n-butylphosphate, n-octyl alcohol and other higher alcohols.
(4) Coalescing agents	As the water evaporates and the paint dries, a coalescing agent is needed to stick the pigment particles together. As the resin film forms, the coalescing agent evaporates. Coalescing agents must volatilize very slowly. Examples: hexylene glycol and ethylene glycol.
(5) Freeze-Thaw Additives	Freezing will destroy the emulsion. Antifreezes such as ethylene glycol are used.
(6) pH Controllers	The effectiveness of the ionic or molecular form of the emulsifier depends on the acid or alkaline conditions (pH). The wrong pH will break down the emulsion. Most paints tend to be too acidic. Ammonia, NH_3, is added to neutralize the acid.

Presently, some "latex" paints contain as much as 75 percent linseed oil emulsion, but still have the desirable characteristics of latex. Table 23–5 gives a summary of various additives to emulsion paints and the rationale for their use.

When General Motors lacquered the 1923 Oakland with a nitrocellulose lacquer, the protective coatings industry first began its expansion into the use of a wide variety of materials instead of a few naturally occurring oils and minerals.

Baked-on Paints

If you have ever had a car repainted, perhaps you have had the opportunity to see the baking oven in which the paint is dried. Automobile finishes, and those on major appliances such as refrigerators, washing machines, and stoves, require very tough adherent paints in order to withstand the abuse they get (Figure 23–30).

Figure 23–30 The high temperatures in a drying oven cause numerous cross-linking reactions to take place, which increase the surface strength of an alkyd paint.

The chemistry of coatings such as these involves producing extensively cross-linked polymers. A popular type of baked-on paint is the so-called ***alkyd.*** The term alkyd comes from a mixture of ***al***cohol and ac***id.*** In Chapter 15, we saw that long-chain polyesters are produced by the reaction of difunctional (two-functional group) alcohols and acids. In alkyds, *poly*functional (more than one reactive group) alcohols and acids are used to form cross-linked films. Consider, for example, the possible reactions of the diacid, phthalic acid, and glycerol, a *tri*alcohol. Note the functional groups left unused.

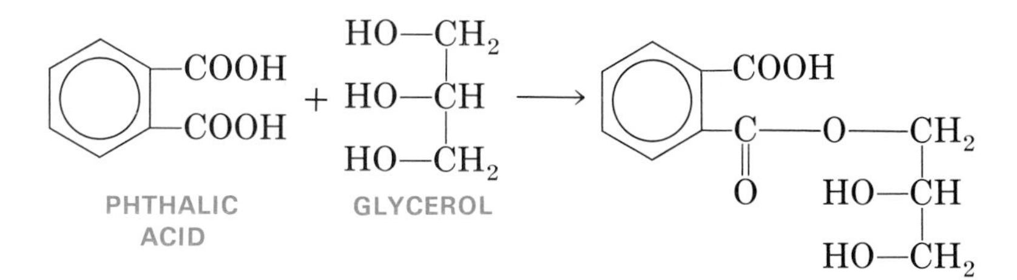

A second as well as a third molecule of phthalic acid can react with a molecule of glycerin.

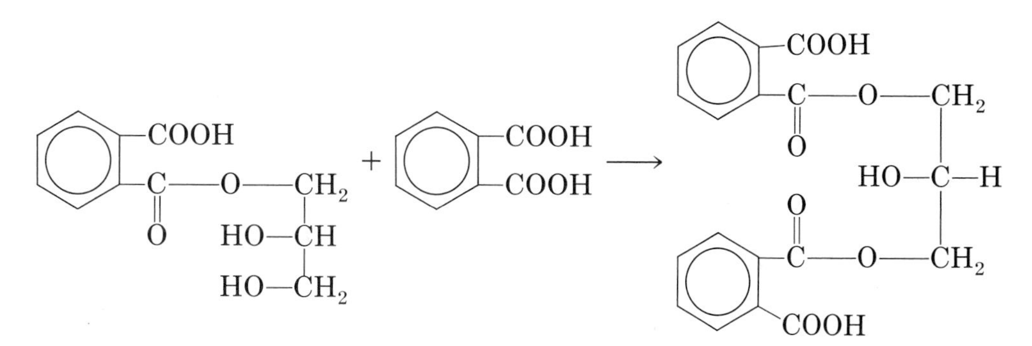

A resin is an amorphous solid or semisolid mixture of organic compounds.

With more reactants, extensive cross-linking becomes possible. The resulting polymer is called a ***resin.***

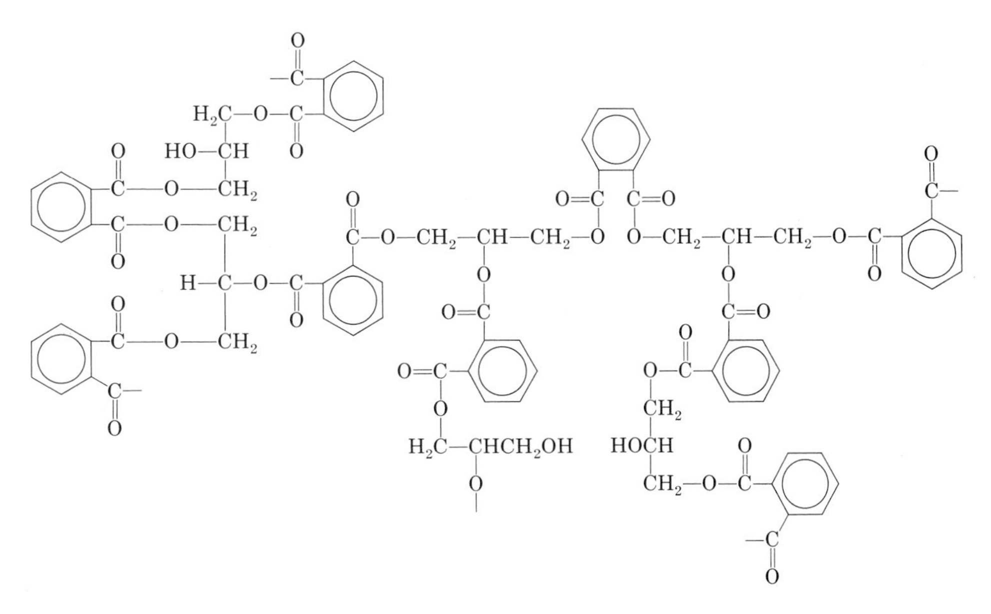

In commercial alkyd paints the resin is seldom as chemically simple as in this example. These resins are extremely tough. The cross-linking reactions are usually begun at the paint factory and carried to completion after the paint has been applied by heating the paint surface, a process known as baking.

Alkyds are used as binders in interior and exterior paints, enamels, lacquers, and varnishes for many uses.

Quite often a different type of resin is used along with an alkyd resin to give it more strength. This resin is one formed when urea reacts with formaldehyde. Recall a similar reaction in the casein-formaldehyde whitewash discussed earlier.

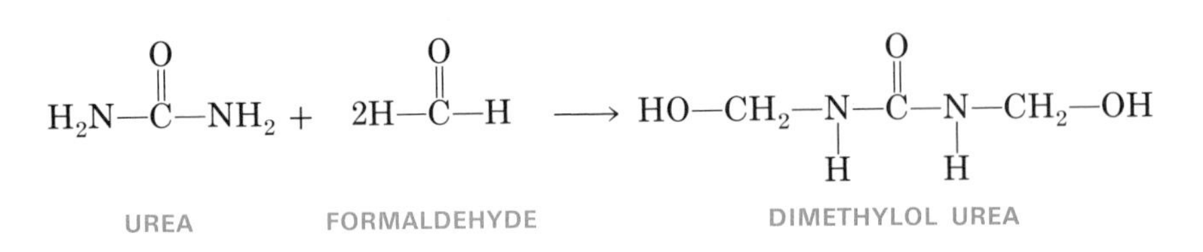

The dimethylol urea can then polymerize:

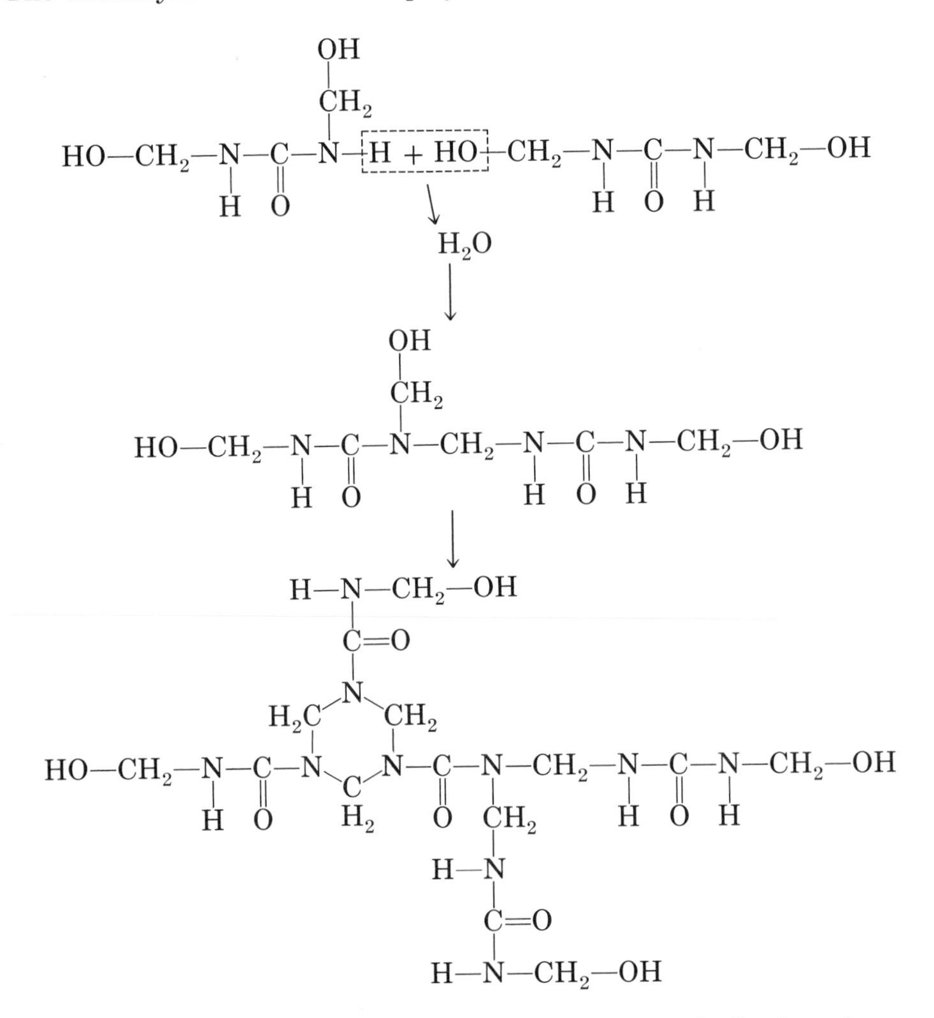

Many automobile finishes are combinations of alkyd and urea-formaldehyde resins which have been extensively cross-linked by a final baking at temperatures around 130°C for about 1 hour.

In applications like washing machine finishes, alkyds are not used as much as urea-formaldehyde resins since hot soap solution, being alkaline, attacks the ester linkages in the alkyd resin.

Paint chemistry is actually an extension of the chemistry of polymers, as we have seen. While it might be possible to make our own paints based on the information we have just studied, numerous other factors, some subtle, go into the making of a good paint. The chemistry is generally refined by a paint chemist so that the paint will protect well, look good, last long and be easy to apply. If it fails in one of these it will be replaced by another.

SELF-TEST 23-C

1. In paints, the components are dispersed in a liquid portion called the ________ and the color is supplied by the ________.
2. Whitewash is prepared by mixing water and ________.
3. White paint containing lead may be darkened because the lead reacts with the pollutant, ________, to form black PbS.
4. Modern white paints contain ________ as a pigment and eliminate the danger of ________ poisoning when white lead is used as the pigment.
5. Chalking is caused by deterioration of the ________.
6. The drying of oil-based paints involves the following three processes:
   (a) ________ of the turpentine or mineral spirits;
   (b) formation of hard soaps by the reaction of ________ ions with ________ acids;
   (c) ________ of fatty acids containing one or more carbon-carbon double bonds.
7. The drying oil used most often is ________ oil.
8. Linseed oil contains several types of triglycerides. (True/False)
9. A ________ is a network of organic molecules polymerized without a definite, uniform crystalline structure.
10. Water-based or "latex-type" paints are ________ because two immiscible liquids are stabilized.
11. Baked-on paints are usually alkyds. This means they are usually made from polyfunctional ________ and ________.

MATCHING SET

	Process		Explanation
______	1. gum formation in automobile	a.	layered crystalline structure which provides mild electrostatic repulsion between layers
______	2. antioxidant action, such as phenylenediamine in gasoline	b.	causes break in surface tension
______	3. antirust agents in gasoline, such as trimethyl phosphate	c.	chelates metals which catalyze gum formation
______	4. increased viscosity of oils	d.	interferes with hydrogen bonding of water
______	5. lubrication action of MoS_2	e.	polymerization of unsaturated hydrocarbon fuel components
______	6. action of ethylene glycol in permanent antifreeze	f.	polar end of molecule adheres to metal and blocks oxidation of the metal
______	7. defoaming action	g.	combination of reduction of light-sensitized silver grains and oxidation and coupling of dye molecules
______	8. formation of dark areas on a photographic negative	h.	addition of long-chain polymers which become entangled
______	9. light-sensitization of photographic film	i.	formation of soluble complex ion composed of unsensitized silver ions and thiosulfate
______	10. how hydroquinone develops a photographic film	j.	reduction of Ag^+
______	11. how a photographic film or negative is rendered stable to light and further reaction, i.e., fixed	k.	hydrogen ions are removed from —OH groups which oxidize the molecule and provides electrons for silver ion reduction
______	12. how colors are produced in a color film-developing process	l.	oxidation of Br^- to Br^0 which leads to Ag_4^0

QUESTIONS

1. Write a balanced equation for the complete combustion of 2,2,4-trimethylpentane.

2. What is meant by high detergency (HD) gasoline? Specifically, what do the detergents do for a gasoline engine?

3. What is the purpose of each of these substances in gasoline?

   (a) ethylenediamine
   (b) TCP
   (c) ethylene glycol
   (d) dyes

4. What is the meaning of 10W-30 oil?

5. How do the antiwear agents in motor oils work?

6. What is the difference in the composition of motor oils and vegetable oils, such as corn oil or palm oil?

7. What is the basic difference in the composition of motor oils and automotive greases?

8. Referring to the structure of ethylene glycol, explain why this compound is so soluble in water.

9. Why would it be undesirable to have a methanol-type antifreeze in your radiator during the summer?

10. Explain how a deicer melts ice.

11. Why is the methanol-type antifreeze called temporary and the ethylene glycol-type called permanent antifreeze?

12. Write a chemical equation for the rusting of iron in a radiator.

13. List three typical additives in antifreeze formulations and describe the action of each.

14. What is the purpose of sodium sulfite in a developer solution?

15. What would be the effect of low pH on a typical developer?

16. Explain the term ***fixing.*** What is it chemically and why is it important in photography?

17. What are the subtractive primary colors?

18. Explain this statement: "There is no silver in that color slide!"

19. Explain how a red dot would be photographed with a color film such as Kodachrome II.

20. What color is produced by the superposition of the following subtractive primary colors?

    (a) magenta and yellow
    (b) cyan and yellow
    (c) magenta and cyan
    (d) magenta, cyan, and yellow

21. Name three modes of film formation used in various paints and give an example of each.

22. Do modern latex paints always contain a rubber latex? Explain.

23. What is the chemical cause of chalking?

24. What agents in the environment destroy a paint?

25. Which white pigment is banned in interior paints? Explain.

26. If a white, outside paint becomes black with age, what is probably taking place?

27. What is the source of turpentine? How is it used in paints?

28. Could an oil-based paint "dry" in a vacuum? Explain.

29. Lye literally eats up an oil-based paint. What is the chemistry?

30. Would lye be effective on an alkyd paint? Explain.

31. Explain the drying of a latex paint.

32. Give two chief advantages of latex paints over oil-based paints.

33. Acrylics are a popular "latex-type" of paint today. The monomer of polyacrylates is

$$\begin{array}{c} CH_2{=}CH \\ \quad\;\; | \\ O{=}C{-}OR \end{array}$$

where R is CH_3 or C_2H_5. Draw a representative portion of polyacrylate.

PROBLEM

1. Suppose 186 grams of ethylene glycol (molecular weight, 62 amu) is added to 1000 grams of water. What is the freezing point of the mixture?

Ans. $-5.58°C$

SUGGESTIONS FOR FURTHER READING

"Chemistry in the Economy," American Chemical Society, Washington, D.C., (1973), Chapter 18 (Photographic Products).
Eaton, G. T., "Photographic Chemistry," Morgan and Morgan, New York, 1965. (A good, elementary monograph.)
Geller, I., "The World's First Oil Mine," *Chemistry,* Vol. 41, No. 8, p. 10 (1968).
Hahn, A. V., "The Petrochemical Industry," McGraw-Hill, New York, 1970.
Hillson, P., "Photography, A Study in Versatility," Doubleday and Co., Inc., Garden City, N.Y., 1969.
Keller, E., "Photography, Part I: Images in Silver," *Chemistry,* Vol. 43, No. 9, p. 6 (1970).
Keller, E., "Photography, Part II: Images in Color," *Chemistry,* Vol. 43, No. 11, p. 8 (1970).
Kirk, R. E., and D. F., Othmer, "Encyclopedia of Chemical Technology," Interscience Publishers, New York, 1967.
Miller, F. W., "Under the Hood," *Chemistry,* Vol. 44, No. 6, p. 12 (1971).
Newhall, B., "Latent Image—The Discovery of Photography," Doubleday and Co. Inc., Garden City, N.Y., 1967.
Schaar, B. E., "Chance Favors the Prepared Mind. Part Five—Photography," *Chemistry,* Vol. 39, No. 5, p. 34 (1966).
Weissberger, A., "A Chemist's View of Color Photography," *American Scientist,* Vol. 58, No. 6, p. 648 (1970).

CHAPTER 24

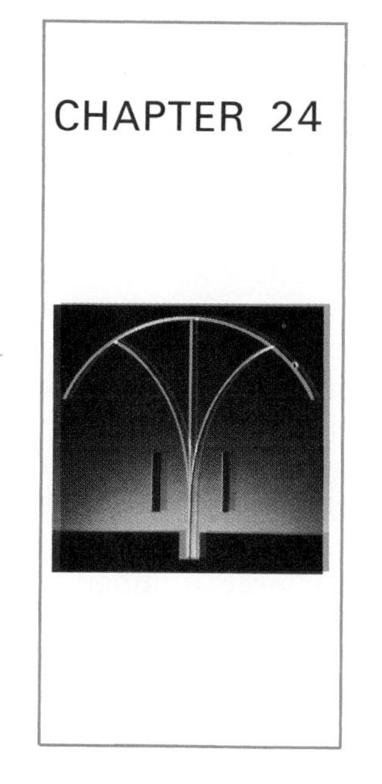

NUCLEAR CHANGES AND THEIR APPLICATIONS

The 19th century Daltonian atom pictured each element with its own type of characteristic, indestructible atom. The discovery of natural radioactivity led to the knowledge that some of the atoms are actually unstable and decompose spontaneously. Such discoveries revealed that the massive nuclei of some heavier elements decompose into less massive nuclei of lighter elements. Reactions involving changes in nuclear structure are called ***nuclear reactions.*** Nuclear reactions are fundamentally different from chemical reactions since the nucleus is undergoing change; as learned earlier, chemical changes involve only changes in the outer electronic structure of the atom.

Nuclear reactions are fundamentally different from chemical reactions.

Nuclear processes are especially important in terms of radiation-produced damage in the biosphere. The radiation accompanying most nuclear reactions is capable of bringing about chemical change since the energies available are sufficiently large to break chemical bonds. Nuclear reactions are a source of large amounts of energy (Chapter 25).

NUCLEAR PARTICLES AND REACTIONS

As in the case of atomic and molecular theory, we must depend on circumstantial evidence to establish the identity of the particles involved in nuclear reactions. The study is somewhat more difficult with nuclear reactions because many of the reactions of interest lead to products which can exist for only a very short period of time (sometimes as short as 10^{-8} second). For such nuclear reactions it is quite impossible to collect molar amounts of the reaction products, and thereby deduce characteristic properties of the submicroscopic particles involved. It is evident then that successful methods for the study of nuclear reactions must involve rapid observations and the ability to record these observations for later study.

Three methods for the detection of particles of nuclear reactions were briefly discussed in Chapter 6. Alpha, beta, and gamma rays from the spontaneous nuclear decays in uranium ore can be detected by the darkening produced on photographic film or the light produced on a phosphor (scintillation) screen. The scintillation screen is so sensitive that a single impact of a high energy atomic particle can be seen as a flash of light by observing the screen with a magnifying lens. A third method which can "see" only charged particles (not gamma rays) is the collector plate in a mass spectrometer. As the charged particles hit the plate, a current is induced, the size of which is a measure of the number of charged particles. Like the photographic exposure, a large number of particles is needed for detection in a mass spectrometer.

Three types of nuclear radiation:

alpha (α)—He^{2+} ions

beta (β)—electrons

gamma (γ)—electromagnetic radiation such as x-rays

Instruments that have been designed specifically for the detection of high-energy particles or photons from nuclear transitions include the Geiger counter and the scintillation counter. The ***Geiger counter*** is composed of a Geiger tube (Figure 24–1) and electrical equipment to amplify the current signal from the tube so it can be heard in the form of a click, seen as a flash of a light bulb, or stored for later study in a recording. The Geiger tube is made of a metal case with a window of thin mica (a mineral having suitable strength and transparency). Running through the center of the tube and insulated from the metal case is a wire charged at +1500 volts relative to the metal case at zero potential. The tube is filled with argon, an inert gas, at 0.05 atmosphere pressure. Under these conditions the charge is so great that the tube is on the verge of discharge. A slight increase in voltage will cause the argon atoms to ionize

$$Ar \longrightarrow Ar^+ + e^-$$

and the tube will discharge; the Ar^+ ions then rush to the negative

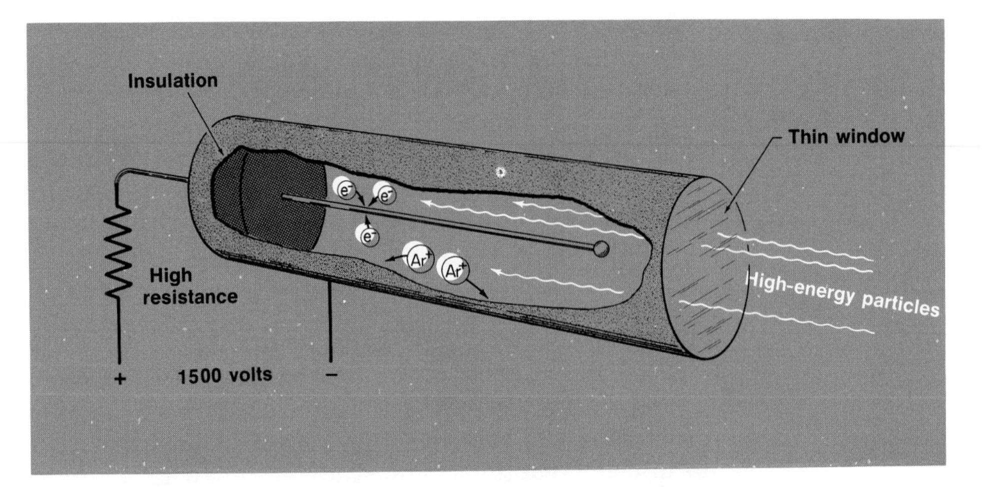

Figure 24–1 Schematic drawing of a Geiger tube. The voltage is adjusted to just below the discharge potential. Under these conditions high energy particles which penetrate the thin window cause the argon atoms to ionize. The result is a cascade effect, in which newly formed charged particles, accelerated by the electric field, produce more ions, and a massive and sudden discharge occurs. The high resistance in series with the high voltage electrodes prevents a large current for more than an instant, thus making the high voltage device safe to handle.

$e^- + Ar \longrightarrow Ar^+ + 2e^-$
Beta particle

electrode (the wall of the tube) and the free electrons will go to the positive electrode (the wire). If the voltage on the tube is adjusted to just below the discharge potential, a high-energy particle entering through the mica window will cause one or more argon atoms to ionize. The resulting charged particles cause other argon atoms to ionize in a cascade effect. Consequently there is a sudden and massive discharge, all produced by the single high-energy particle. The number of discharges per second can be measured by the current output of the tube. It is also possible that each pulse (discharge) can be counted with counting circuits. As a result, the Geiger counter can not only detect products of nuclear decay, but can actually measure their number as well. The Geiger counter is rugged and dependable, but it suffers from one disadvantage: in order to be detected, the particle must have sufficient energy to penetrate the mica window.

Light is emitted when high-energy particles strike a phosphor.

A number of interesting nuclear changes do not produce particles with sufficient energy for Geiger counter detection, so a more sensitive radiation detector is needed to study these cases. The ***scintillation counter*** is sensitive to such "soft-nuclear" emissions. The outside of a glass window in a scintillation tube is covered with a phosphor coating. The minute flash of light produced by a relatively low-energy nuclear emission passes through the glass to a very sensitive photoelectric detector. The resulting electric current is amplified and either displayed or recorded.

Scientists were not able to characterize the electron, alpha particle, beta particle, and positive ray particles until it was learned how to observe their behavior in electric and magnetic fields (Chapter 6). With the charge-to-mass ratios determined by Thomson and the charge per particle determined by Millikan, investigators were quickly able to study those particles which could be produced (by electrical discharge or natural radioactivity) in relatively large and steady streams. At this point, it was reasonable to assume that other nuclear particles escaped detection because they were produced in insufficient quantities to be detected. In

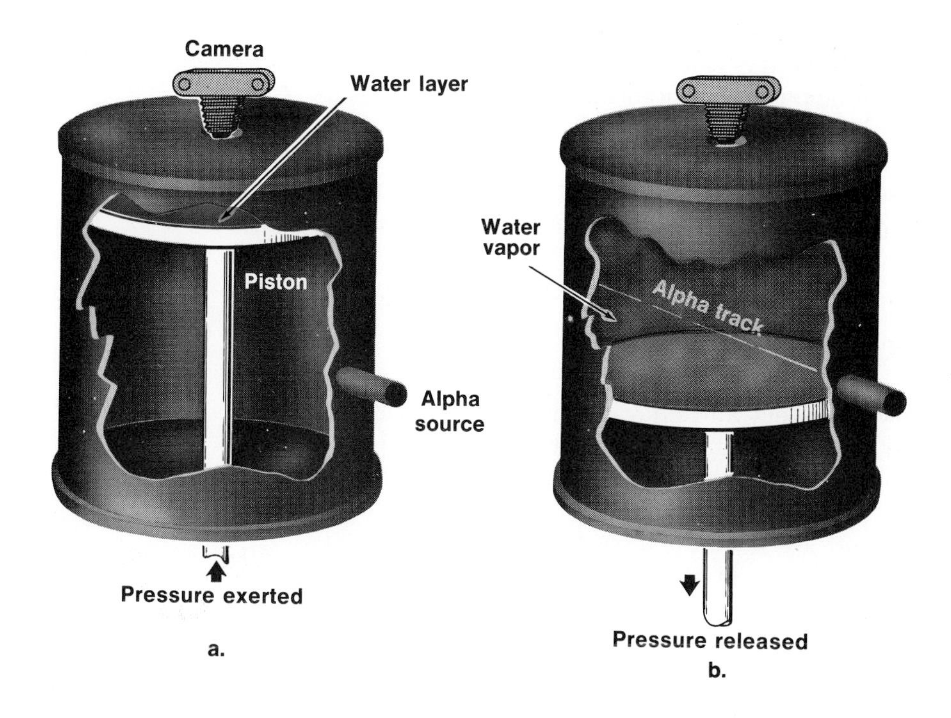

Figure 24–2 A simple Wilson cloud chamber.

Figure 24–3 Alpha tracks photographed in a cloud chamber. Although the alpha particles do have different energies, owing to collisions with other particles in escaping the source, the apparent difference in path lengths is exaggerated because they are not all moving in the same plane. Under the same conditions proton tracks from the reaction, ${}^{14}_{7}N + {}^{4}_{2}He \longrightarrow {}^{1}_{1}H + {}^{17}_{8}O$, would tend to be considerably longer. (From Wood, J. H., et al.: *Fundamentals of College Chemistry*. 1st ed. New York, Harper and Row, 1963.)

1911, C. T. R. Wilson invented the ***cloud chamber,*** which could actually "see" a single high-energy, nuclear particle in flight, the collision of such a particle with another nuclear particle, and the path of the products of such a reaction. A single nuclear event could be observed! Furthermore, when the cloud chamber is placed in a magnetic field, the charged particles will follow a curved path, and the individual particles, as a result, can be characterized.

The structure of a simple cloud chamber is illustrated in Figure 24–2. Pressure is exerted on a closed system containing a nuclear particle source (such as an alpha emitter), air, and a layer of water or other liquid, such as ethanol. The invisible alpha particles are constantly being emitted from the alpha source. If pressure is exerted on this system, the concentration of water vapor is increased in the air space around the alpha emitter. Now, if the pressure is suddenly reduced, the temperature of the air drops, and the air will contain more water vapor than it can normally hold. Consequently, there will be a strong tendency for the water vapor to condense (precipitate). In such a supersaturated system the water molecules readily condense on charged particles. Now, the alpha particle, because of its high energy, ionizes air particles in its path. As a result, there will be a visible path of condensed water (a cloud track, Figure 24–3) tracing the alpha particle pathway. Any charged particles with sufficient energy to ionize the molecules of the air can thus be observed. It is a relatively simple matter to photograph such cloud tracks and record nuclear events, such as a collision between an alpha particle and a nitrogen molecule, for later study.

The cloud track is somewhat analogous to the vapor trails of a high-flying jet airplane; even when the plane is too high to be seen itself, the condensed water from the exhaust clearly marks its pathway.

TRANSMUTATION IN NATURE

Armed with the ability to detect and characterize high energy particles in flight from nuclear reactions, nuclear scientists began collecting information concerning the nuclear reactions of naturally radioactive substances. There are many such reactions since all of the elements above bismuth in atomic number and a few below have one or more naturally occurring radioactive isotopes. Each of these isotopes is an emitter of alpha, beta, and/or gamma rays, and with each emission a nuclear reaction spontaneously occurs.

One of the dreams of the alchemists (1200–1700 A.D.) was to transmute base metals such as lead and iron into gold. The dream was discarded only after the acceptance of Dalton's indestructible atom.

The isotope of uranium with atomic mass 238 is an alpha emitter.

The atomic number of uranium is 92, which means that $^{238}_{92}U$ has 92 protons and 146 (i.e., 238 − 92) neutrons in the nucleus. When the $^{238}_{92}U$ nucleus gives off an alpha particle, made up of 2 protons and 2 neutrons, it necessarily loses 4 units of atomic mass and 2 units of atomic charge. The resulting nucleus has a mass of 234 and a nuclear charge of 90. Now, atoms containing 90 protons in the nucleus are atoms of thorium, not uranium. This spontaneous nuclear reaction then has changed an atom of one element into an atom of another element, and is an example of the ***transmutation*** of elements. The decomposition of the $^{238}_{92}U$ nucleus is stated briefly by the following nuclear equation:

$$^{238}_{92}U \longrightarrow ^{4}_{2}He + ^{234}_{90}Th$$

In this equation, the ***mass number*** of the particle is given by the superscript and the ***nuclear charge*** (atomic number) is given by the subscript. If the characterized alpha emission was not proof enough that this reaction occurs, additional evidence is supplied by the fact that $^{234}_{90}Th$ is always found with $^{238}_{92}U$ in natural ore deposits and almost always in just that concentration predicted by the rate of the reactions involved.

Mass number → $^{238}_{92}U$ ← Nuclear charge (atomic number or number of protons)

Thorium-234 is also radioactive. However, this nucleus is a beta emitter. This poses an interesting question: How can a nucleus containing protons and neutrons emit an electron? It has been established that an electron and a proton can combine outside the nucleus to form a neutron. Therefore, the reverse process is proposed to occur in the nucleus. A neutron decomposes, giving up an electron and changing itself into a proton:

$$^{1}_{0}n \longrightarrow ^{1}_{1}H + ^{0}_{-1}e + \text{Energy}$$

Since the mass of the electron is essentially zero compared to that of the proton and neutron, the nucleus would maintain essentially the same mass but it would now carry one more positive charge (a proton instead of one of the neutrons). This nucleus is no longer thorium since thorium has only 90 protons in the nucleus; it is now a nucleus of element 91, protactinium (Pa). The reaction is the following:

$$^{234}_{90}Th \longrightarrow ^{234}_{91}Pa + ^{0}_{-1}e + \text{Energy}$$

Gamma radiation may or may not be given off simultaneously with alpha or beta rays, depending on the particular nuclear reaction involved. Since gamma rays involve no charge and essentially no mass, it is evident that the emission of a gamma photon cannot alone account for a transmutation event.

No matter how much of a radioactive substance is present at the beginning, only half of it remains at the end of one half-life.

The uranium decay is extremely slow compared to the thorium decay. The rate of decay can be represented by a characteristic ***half-life.*** A half-life represents the period of time required for half of the radioactive material originally present to undergo transmutation. The half-life is independent of the amount and chemical form of radioactive material present and is determined only by the type of radioactive nucleus present in the sample. For example, in the reaction above the half-life of $^{234}_{90}Th$ is 24 days. This means that half of the thorium will be left after 24 days. In another 24 days half of the half ($\frac{1}{4}$) will be left. This process continues indefinitely with half of the $^{234}_{90}Th$ that exists decaying each 24 days.

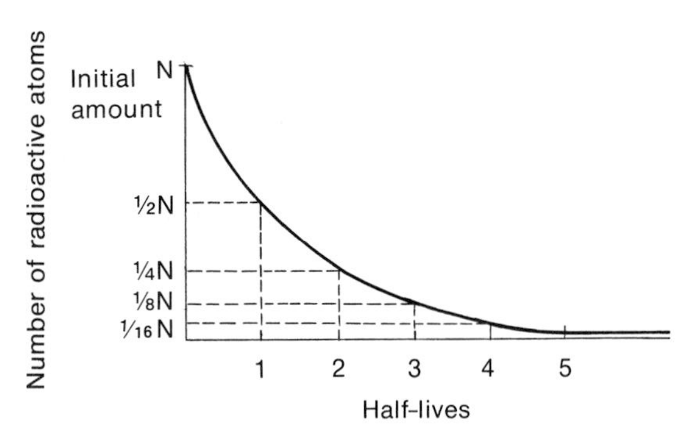

Figure 24–4 Half-life. The rate of decay for a radioactive atom depends in a very special way on the number of those atoms present. The rate is such that in a given period of time—the half-life for the species—half of the original number of atoms will be gone regardless of the number present at the start. In this graph the number of atoms remaining is plotted against time. At the end of one half-life period the original number, N, is reduced to $\frac{1}{2}$N. After two of these periods, the number is reduced to half of $\frac{1}{2}$N, or $\frac{1}{4}$N.

Figure 24–4 illustrates graphically how the concept of half-life works for a radioactive isotope. Some half-lives are extremely long and others are extremely short. The half-life for the $^{238}_{92}U$ alpha decay is 4.5 billion years. As one would expect, relatively large amounts of $^{238}_{92}U$ can be found in nature while only trace amounts of $^{234}_{90}Th$ occur.

The radioactive decay of $^{234}_{90}Th$ into $^{234}_{91}Pa$ is the second step in a series of nuclear decays that starts with $^{238}_{92}U$ and after 14 decays ends with a stable (nonradioactive isotope of lead, $^{206}_{82}Pb$. This decay series is called the ***uranium series*** (Figure 24–5). Two other natural decay series exist which are similar to the uranium series but they start out with a different parent isotope and proceed downward through a different set

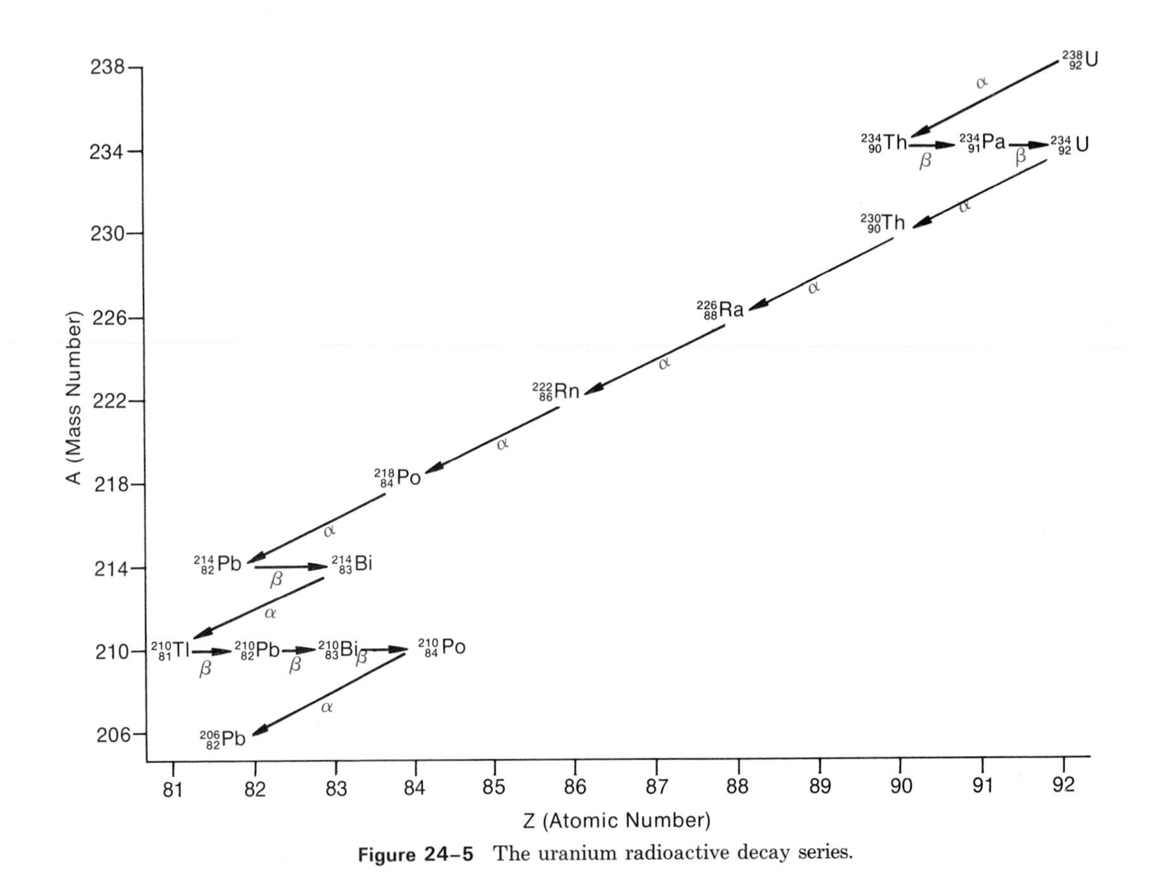

Figure 24–5 The uranium radioactive decay series.

TABLE 24–1 Half-Lives of the Naturally Occurring Radioactive Elements in the Uranium-238 ($^{238}_{92}U$) Series

Isotope	Type of Disintegration	Half-Life
U-238	α	4.5 billion years
Th-234	β	24.1 days
Pa-234	β	1.18 minutes
U-234	α	250,000 years
Th-230	α	80,000 years
Ra-226	α	1620 years
Rn-222	α	3.82 days
Po-218	α,β	3.05 minutes
Pb-214	β	26.8 minutes
Bi-214	α,β	19.7 minutes
Tl-210	β	1.32 minutes
Pb-210	β	22 years
Bi-210	β	5 days
Po-210	α	138 days
Pb-206	stable	

Each decay series ends with an isotope of lead as the final product.

of radioactive nuclei. The ***thorium series*** begins with $^{232}_{90}Th$ (a different isotope from the two thorium isotopes that occur in the uranium series) and ends with stable $^{208}_{82}Pb$. A third series, called the ***actinium series,*** begins with $^{235}_{92}U$ and ends with $^{207}_{82}Pb$. Most of the naturally occurring radioactive isotopes are members of one of these three decay series.

ARTIFICIAL NUCLEAR REACTIONS

After it was realized that the nuclei of some of the heavier isotopes were unstable, scientists wondered if nuclear reactions could be initiated with other nuclei that were apparently stable. In order to explain the stability of any nucleus, one must postulate the existence of short-range, attractive, nuclear forces between ***positive*** particles (protons) and/or ***neutral*** particles (neutrons). Such forces must be stronger than the electrostatic forces that would tend to make the positive particles fly apart. It seemed reasonable to believe that two nuclei might possibly react to form new nuclear species if they could be brought so close together that the short-range nuclear forces could be operative. In order to achieve this, the two nuclei would have to approach each other with sufficient kinetic energy to overcome the electrostatic repulsion and unite. In such a case, one could postulate an unstable ***compound nucleus*** that would emit particles, energy, or both, in seeking a stable structure.

Recall that like charges repel and unlike charges attract.

In 1919, Rutherford was successful in producing the first artificial nuclear change. He placed nitrogen gas in a cloud chamber and directed helium nuclei (alpha particles) into the box. The penetrating power of the alpha particles is determined by their kinetic energy, which in turn is determined by the reaction producing them. Since for a given source the alpha particles all have about the same energy, a photograph of such alpha tracks in a cloud chamber would show tracks of essentially the same length (Figure 24–3). Under the conditions of this experiment, Rutherford found some tracks that were much longer than the typical alpha track. Furthermore, these longer tracks did not appear to start at the origin of the alpha tracks but seemed to begin at the termination of an alpha track. When these tracks were studied in a magnetic field, their curvature indicated a particle with a charge-to-mass ratio identical to the value already established for the positive-ray particle, the proton. It was concluded then that the tracks were produced by high energy protons which, because of their smaller size and charge, are more penetrating than an alpha particle for a given amount of energy. All of the results of the experiment could be explained if one assumed the nuclear reaction to be:

$$^{14}_{7}N + ^{4}_{2}He \longrightarrow [^{18}_{9}F] \longrightarrow ^{17}_{8}O + ^{1}_{1}H$$

where $^{18}_{9}F$ is an unstable compound nucleus. Natural fluorine consists exclusively of the isotope $^{19}_{9}F$. Since both the $^{17}_{8}O$ and the hydrogen nuclei are stable, the products show no further tendency to undergo nuclear change.

Following Rutherford's original transmutation experiment, there was considerable interest in subjecting isotopes to high energy particles to discover new nuclear reactions. As you might guess, numerous reactions were found. For example, beryllium can be converted to carbon when subjected to an alpha ray bombardment:

$$^{9}_{4}Be + ^{4}_{2}He \longrightarrow ^{13}_{6}C \longrightarrow ^{12}_{6}C + \underset{\text{(NEUTRON)}}{^{1}_{0}n}$$

Although the $^{12}_{6}C$ produced in this reaction is stable, the neutron is given off with sufficient energy to provoke additional nuclear reactions in nuclei with which it collides. It was just this nuclear reaction that was used by James Chadwick in 1932 to finally prove the existence of the previously postulated neutron.

The neutron was discovered by this reaction by Chadwick in 1932.

Not all nuclear reactions produce stable isotopes as in the cases cited above. If $^{25}_{12}Mg$ is bombarded with an alpha source, a radioactive isotope of aluminum, $^{28}_{13}Al$, is produced which does not exist in nature:

$$^{25}_{12}Mg + ^{4}_{2}He \longrightarrow ^{29}_{14}Si \longrightarrow ^{28}_{13}Al^{*} + ^{1}_{1}H$$

The symbol * is often used to denote a radioactive isotope in nuclear equations. Radioactive isotopes, such as $^{28}_{13}Al$, have characteristic half-lives just as do the naturally occurring ones. The half-life of $^{28}_{13}Al$ is relatively short, only 2.3 minutes. The $^{28}_{13}Al$ nucleus emits a beta particle and becomes a stable isotope of silicon:

$$^{28}_{13}Al^{*} \longrightarrow ^{28}_{14}Si + ^{0}_{-1}e$$

Nitrogen can be bombarded with neutrons from the Chadwick reaction to produce radioactive $^{14}_{6}C$:

$$^{14}_{7}N + ^{1}_{0}n \longrightarrow ^{15}_{7}N \longrightarrow ^{14}_{6}C^{*} + ^{1}_{1}H$$

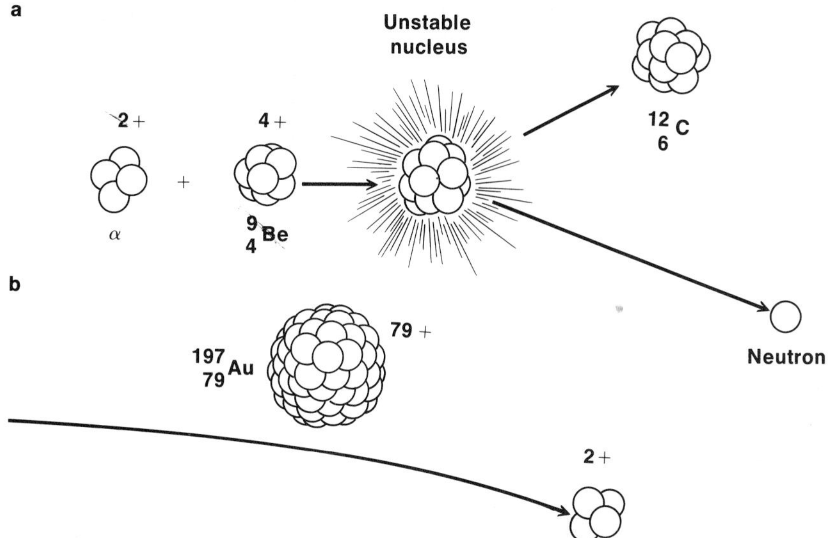

Figure 24–6 (*a*) An alpha particle can penetrate a beryllium (Be) nucleus since it has sufficient energy to overcome the repulsions of like charges and the nuclear stability. (*b*) In the case of Rutherford's gold-foil experiment (the experiment that suggested the nuclear atom), the alpha particles were not energetic enough to penetrate the gold (Au) nucleus and were deflected.

An interesting question arises as to why the alpha particles were scattered by the gold foil in Rutherford's gold-foil experiment (Chapter 6), and yet the same alpha source can produce a nuclear change with a smaller atom such as $^{9}_{4}Be$ (Figure 24–6). The answer lies in the fact that the charge on the gold nucleus is +79, whereas on the beryllium nucleus the charge is +4. Most of the alpha particles emitted from natural radioactive decay do not have enough energy to penetrate to a heavy, positively charged nucleus such as that of gold. Therefore, if artificial nuclear reactions are to be studied for the heavier elements, it is necessary to find ways of increasing the kinetic energy of the subatomic projectile particles.

SELF-TEST 24-A

1. The three types of natural radioactive emissions are ____________, ____________, and ____________ rays.

2. Of the three above, which one is not a particle in the conventional sense?

   ____

3. A nuclear radiation detection device utilizing a cloud of water vapor is called a ____________ ____________.

4. When a $^{87}_{35}Br$ nucleus emits a beta particle, the nuclear species which results is ____________.

5. When a $^{216}_{84}Po$ nucleus emits an alpha particle, the nuclear species that results is ____________.

6. The half-life of $^{44}_{19}K$ is 22 minutes. If a 1-gram sample of $^{44}_{19}K$ is taken, how much $^{44}_{19}K$ will remain after three half-lives (66 minutes)?

____

7. In the reaction below, what is the compound nucleus?

$$^{7}_{3}Li + ^{1}_{1}H \longrightarrow \_\_\_\_\_\_\_\_ \longrightarrow ^{7}_{4}Be + ^{1}_{0}n$$

8. The scientist who discovered the neutron was _______________ _______________. The process was _______________.

SOURCES OF HIGH-ENERGY PARTICLES

Monuments to man's recent technological skill are the intricate devices that have been developed to increase the kinetic energy of charged particles. Linear accelerators, cyclotrons, betatrons, synchro-cyclotrons, and synchrotrons are capable of accelerating electrons, protons, deuterons ($^{2}_{1}H$), alpha particles, and similarly charged particles to tremendous speeds. In the case of electron accelerators, the speed of the stream of electrons has been brought very close to the speed of light (186,000 miles per sec). Even with this costly hardware, there is still no way to increase directly the kinetic energy of uncharged particles like neutrons and the neutral atoms. The reasons for this will be evident after it is understood how these devices work.

The speed of light in metric units is 300,000 km/sec.

Although the electronic gear is quite elaborate and different for each apparatus, the basic principles of operation have been developed in earlier chapters and are comparatively simple: (1) opposite charges attract and (2) the path of a charged particle is curved as it passes through a magnetic field. When these effects are combined with the fact that electrical fields do not penetrate to the inside of a charged metal container (the fields are shielded by the "free" electrons in the metal), you have the fundamental facts necessary to explain how particle accelerators work.

Sufficient kinetic energy for a particle to penetrate the electrical fields of an atom and to enter a large nucleus cannot readily be gained by accelerating the particle in one step between two electrical poles. This would require an impossibly large potential difference of millions of volts. However, if the acceleration is done in several thousand steps, with readily obtainable potential differences of 2000 to 10,000 volts being applied to each step, sufficient energy can be imparted to the particle. The linear accelerator and the cyclotron illustrate the two primary ways the stepwise acceleration is accomplished.

The operation of the linear accelerator is less complicated than the cyclotron in that it does not require a magnetic field. The principle of operation is outlined in Figure 24–7. A source of electrons or protons is provided by ionization (as in a canal ray tube) at one end of the line of hollow metal cylinders and the target material is at the other end. The entire device is enclosed in a vacuum so that the accelerated particles may move in a straight path without the possibility of collisions with molecules in the air. Adjacent cylinders are charged oppositely so that when the charged particles traverse the space between the two cylinders,

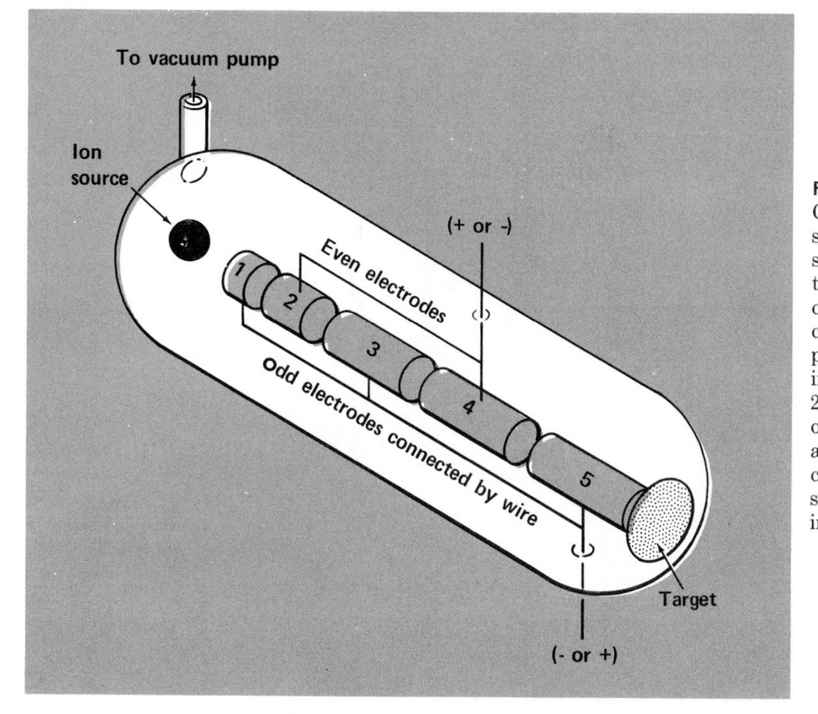

Figure 24–7 Linear accelerator diagram. Charged particles are produced at the ion source and are attracted toward an oppositely charged electrode, cylinder 1. When the charged particles are passing through cylinder 1, the charge on the cylinder is changed to the same sign as that on the particle being accelerated. At the same instant, the sign of the charge on cylinder 2 becomes opposite to that of the charge on the particle. Thus the particle is accelerated as it crosses the gap between cylinders. Note that the tubes become successively longer to accommodate increased speeds of the particle.

the particles are repelled by the previous tube and attracted by the upcoming one. When a given particle enters a cylinder, it is necessary to change suddenly the sign of the charge on that cylinder so that the particle will be repelled as it moves from the tube. Simultaneously, the signs of the charge on all the other cylinders are also changed. The result is that the particle is successively attracted to the upcoming cylinder and repelled by the previous cylinder. At each gap, then, the particles increase their speed. The cylinders are increasingly longer so that the increasingly greater velocities of the particles will not disrupt the timing of the changing of the charge on alternate cylinders. When the particles reach the target sample, they can enter the target nuclei if they have received sufficient energy to penetrate the fields. Since the nuclei of atoms are very small compared to the cross-sectional area of an atom, it is very difficult to hit a nucleus.

These accelerators are complicated, but their operating principles are simple.

The largest linear accelerator in the world went into operation in 1967 at Stanford, California. It is two miles long and accelerates electrons to energies of 20 to 40 billion electron volts (Bev). An electron volt is a unit of energy; it is the amount of energy gained by an electron when it is accelerated by an electric field, the electric potential across the field being one volt.

The cyclotron was developed in 1931, by Ernest O. Lawrence (element 103, a synthetic element, is named in his honor) and M. S. Livingston. In addition to the electronic gear and the huge magnets, the instrument consists of two hollow D-shaped metal containers enclosed in a vacuum as shown in Figure 24–8. Charged particles are formed near the center of the gap between the D's and begin their acceleration to high energy by being attracted into one of the D's. While the particles are in a D, the influence of the magnetic field causes their path to curve. By the time the particles have completed the semicircle, the signs on the D's have changed and the repulsion of the previous D and the attraction by

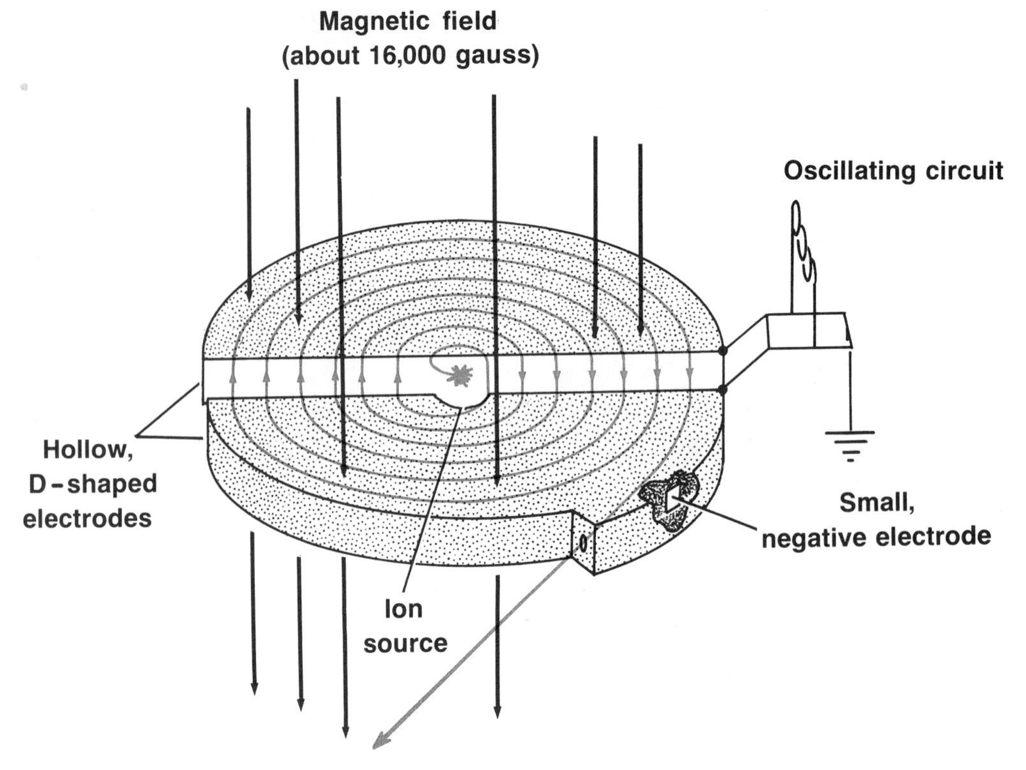

Figure 24–8 Schematic diagram illustrating the operation of the cyclotron. (Adapted from *College Chemistry,* 3rd ed., by Linus Pauling. San Francisco, W. H. Freeman and Company, Copyright © 1964.)

the upcoming one accelerate the particles across the gap between the D's. Each time the particles traverse the gap, they are accelerated to a greater speed. The increased speed causes the particles to move in a wider arc on each revolution. After many accelerations and revolutions, the arc is sufficiently large so the charged deflector plate can repel the particles through a window and onto the target sample. In some cyclotrons, the particles go too fast to be directed outward by the deflector plate so the target sample is placed inside one of the D's.

The earlier models of the cyclotron had D's with diameters of about 2 feet and accelerated protons to energies of about 0.5 ***million electron volts (Mev).*** Modern cyclotrons have diameters up to about 10 feet and accelerate protons to energies of 250 Mev. The voltage across the gap between D's may have any value between 10,000 and 200,000 volts.

The cyclotron is a source of relatively low-energy particles. More sophisticated accelerators are capable of much higher energies and include the betatron, synchro-cyclotron, and bevatron. These instruments are based on the same basic principles as the linear accelerator and the cyclotron, but are capable of producing particles with energies up to hundreds of billions of electron volts. In May, 1974, the world's largest particle accelerator was dedicated in Batavia, Illinois. This installation, called the Fermi National Accelerator Laboratory, after Enrico Fermi, cost $250 million to construct and has an annual budget under the direction of the Energy Research and Development Commission of approximately $36 million. The awesome power of the accelerator may be

Figure 24–9 Particle accelerator at the Fermi National Laboratory, near Batavia, Illinois. *A*, Aerial view of main accelerator. *B*, Magnets in interior of main accelerator, which is 4 miles in circumference and 1.27 miles in diameter. (Fermi Lab Photo.)

realized when it reaches its full potential of one million billion electron volts (1 million Bev).

When a beam of such high energy particles (electrons, protons, deuterons, alpha particles, or other ions) bombards nuclei, a wide variety of reactions take place. A massive nucleus, having captured a high energy particle, will usually emit one or more subatomic particles before reaching a stable state. If the bombarding particles have energies less than about 100 Mev the emitted particles can be described in terms of relatively simple subatomic particles such as protons, neutrons, electrons, positrons, neutrinos, and antineutrinos. However, when the bombarding

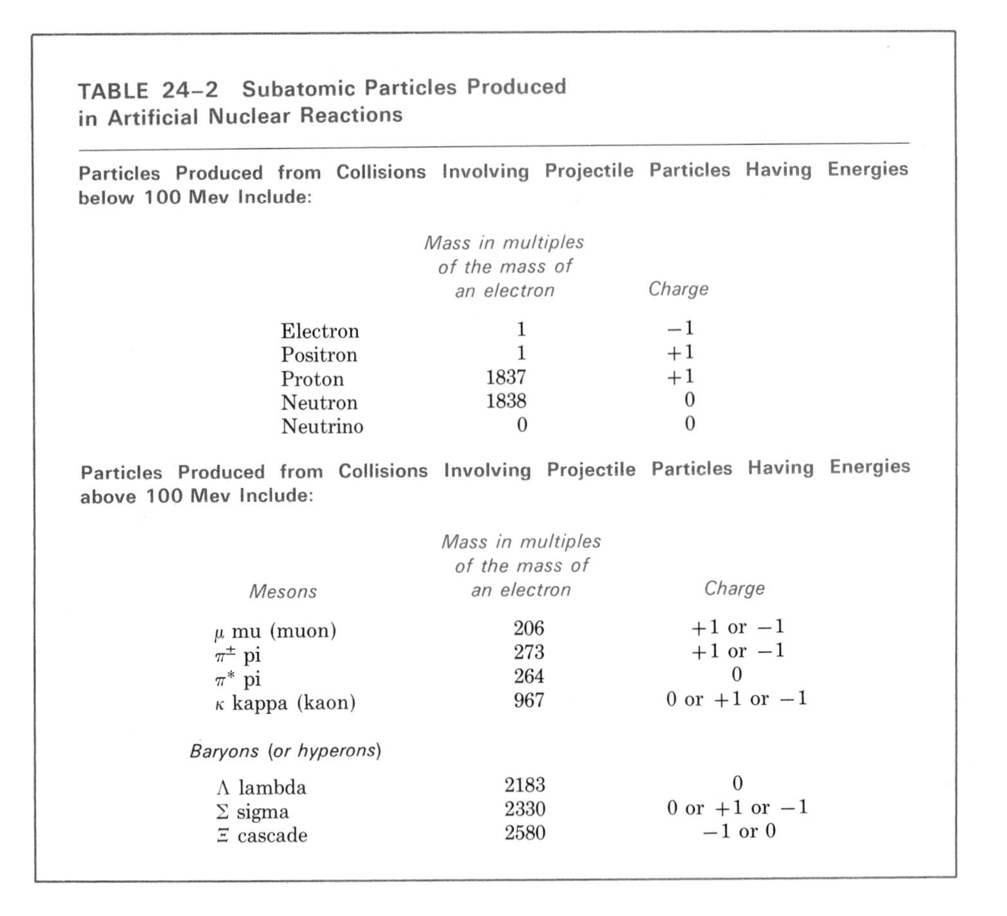

TABLE 24–2 Subatomic Particles Produced in Artificial Nuclear Reactions

Particles Produced from Collisions Involving Projectile Particles Having Energies below 100 Mev Include:

	Mass in multiples of the mass of an electron	*Charge*
Electron	1	−1
Positron	1	+1
Proton	1837	+1
Neutron	1838	0
Neutrino	0	0

Particles Produced from Collisions Involving Projectile Particles Having Energies above 100 Mev Include:

Mesons	*Mass in multiples of the mass of an electron*	*Charge*
μ mu (muon)	206	+1 or −1
$\pi^{\pm}$ pi	273	+1 or −1
π^{*} pi	264	0
κ kappa (kaon)	967	0 or +1 or −1
Baryons (or hyperons)		
Λ lambda	2183	0
Σ sigma	2330	0 or +1 or −1
Ξ cascade	2580	−1 or 0

particles have energies of thousands of Mev or greater, new and strange particles are created; some of these are briefly described in Table 24–2.

TRANSURANIUM ELEMENTS

The heaviest known element prior to 1940 was the element uranium. The invention of the cyclotron and other devices to obtain high energy particles made it possible to cause these particles to react with heavy nuclei to obtain even more massive nuclei. Thus, ***transuranium*** elements with atomic numbers greater than 92 were prepared.

In 1940, E. M. McMillan and P. H. Abelson, at the University of California, prepared element 93, the synthetic element neptunium (Np). The experiment involved directing a stream of high energy deuterons ($^{2}_{1}H$) onto a target of $^{238}_{92}U$. A deuteron is the nucleus of an isotope of hydrogen containing one neutron as well as one proton. The initial reaction involved the conversion of $^{238}_{92}U$ to $^{239}_{92}U$.

The first synthetic element was prepared in 1940.

$$^{238}_{92}U + ^{2}_{1}H \longrightarrow ^{239}_{92}U + ^{1}_{1}H$$

Uranium-239 has a half-life of 23.5 minutes and decays spontaneously to the element neptunium by the emission of beta particles.

$$^{239}_{92}U \longrightarrow ^{239}_{93}Np + ^{0}_{-1}e$$

Neptunium is also unstable, with a half-life of 2.33 days; it converts into a second new element, plutonium.

$$^{239}_{93}\text{Np} \longrightarrow {}^{239}_{94}\text{Pu} + {}^{0}_{-1}\text{e}$$

Plutonium is used to make atomic bombs.

Plutonium-239, like neptunium from which it is made, is radioactive, but its half-life is 24,100 years. Because of the relative values of the half-lives, very little neptunium could be accumulated, but the plutonium could be obtained in larger quantities. The $^{239}_{94}\text{Pu}$ is important as fissionable material since atomic bombs (see Chapter 25) can be made with it as well as with naturally occurring $^{235}_{92}\text{U}$. The names of these two elements were taken from the mythological names, Neptune and Pluto, in the same sequence as the planets Uranus, Neptune, and Pluto.

Although Neptune and Pluto are the last of the known planets in the solar system, their namesakes are not the last in the list of elements. The rush of transuranium experiments that followed produced additional elements: americium (Am), curium (Cm), berkelium (Bk), californium (Cf), einsteinium (Es), fermium (Fm), mendelevium (Md), nobelium (No), lawrencium (Lr), and elements 104, 105, and 106 (as yet unnamed). Obviously, the new elements were named after countries, states, cities, and people. Reactions employed in the production of the transuranium elements are given in Table 24–3. As accelerators with greater and greater energy capabilities are produced, even more nuclear reactions should be available for study.

TABLE 24–3 Nuclear reactions used to produce transuranium elements.*

Element	Atomic Number	Reaction
Neptunium, Np	93	$^{238}_{92}\text{U} + {}^{1}_{0}\text{n} \longrightarrow {}^{239}_{93}\text{Np} + {}^{0}_{-1}\text{e}$
Plutonium, Pu	94	$^{238}_{92}\text{U} + {}^{2}_{1}\text{H} \longrightarrow {}^{238}_{93}\text{Np} + 2{}^{1}_{0}\text{n}$
		$^{238}_{93}\text{Np} \longrightarrow {}^{238}_{94}\text{Pu} + {}^{0}_{-1}\text{e}$
Americium, Am	95	$^{239}_{94}\text{Pu} + {}^{1}_{0}\text{n} \longrightarrow {}^{240}_{95}\text{Am} + {}^{0}_{-1}\text{e}$
Curium, Cm	96	$^{239}_{94}\text{Pu} + {}^{4}_{2}\text{He} \longrightarrow {}^{242}_{96}\text{Cm} + {}^{1}_{0}\text{n}$
Berkelium, Bk	97	$^{241}_{95}\text{Am} + {}^{4}_{2}\text{He} \longrightarrow {}^{243}_{97}\text{Bk} + 2{}^{1}_{0}\text{n}$
Californium, Cf	98	$^{242}_{96}\text{Cm} + {}^{4}_{2}\text{He} \longrightarrow {}^{245}_{98}\text{Cf} + {}^{1}_{0}\text{n}$
Einsteinium, Es	99	$^{238}_{92}\text{U} + 15{}^{1}_{0}\text{n} \longrightarrow {}^{253}_{99}\text{Es} + 7{}^{0}_{-1}\text{e}$
Fermium, Fm	100	$^{238}_{92}\text{U} + 17{}^{1}_{0}\text{n} \longrightarrow {}^{255}_{100}\text{Fm} + 8{}^{0}_{-1}\text{e}$
Mendelevium, Md	101	$^{253}_{99}\text{Es} + {}^{4}_{2}\text{He} \longrightarrow {}^{256}_{101}\text{Mv} + {}^{1}_{0}\text{n}$
Nobelium, No	102	$^{246}_{96}\text{Cm} + {}^{12}_{6}\text{C} \longrightarrow {}^{254}_{102}\text{No} + 4{}^{1}_{0}\text{n}$
Lawrencium, Lr	103	$^{252}_{98}\text{Cf} + {}^{10}_{5}\text{B} \longrightarrow {}^{257}_{103}\text{Lr} + 5{}^{1}_{0}\text{n}$
name ?	104	$^{242}_{94}\text{Pu} + {}^{22}_{10}\text{Ne} \longrightarrow {}^{260}_{104}? + 4{}^{1}_{0}\text{n}$
name ?	105	$^{249}_{98}\text{Cf} + {}^{15}_{7}\text{N} \longrightarrow {}^{260}_{105}? + 4{}^{1}_{0}\text{n}$
name ?	106	$^{249}_{98}\text{Cf} + {}^{18}_{8}\text{O} \longrightarrow {}^{263}_{106}? + 4{}^{1}_{0}\text{n}$

*There is ample evidence to believe that elements 104, 105, and 106 have been produced, but at the time of the printing of this book, it was still not certain which research group (Russian or American) had made these elements first. Hence, the priority for naming the element was not established.

RADIOISOTOPE DATING OF THE UNIVERSE, MINERALS, AND ARTIFACTS

The concept of radioisotope half-life discussed earlier was almost immediately recognized as a useful tool for measuring the age of radioactive materials if reasonable assumptions were made. The assumptions for radioisotope dating are the following:

Assumptions are made in radio-dating.

1. The nuclear decay is independent of the past history of the isotope.
2. The decay is independent of the present chemical environment.
3. The rate of decay has always been constant.
4. There was definite initial isotope composition at the beginning of the radioactive decay process. (For example, no lead present initially in the uranium ore described below.)

Three methods of radioactive dating have proven widely applicable and are discussed here. They are ${}^{238}_{92}U/{}^{206}_{82}Pb$, ${}^{40}_{19}K/{}^{40}_{18}Ar$, and ${}^{14}_{6}C$ dating.

Uranium/Lead Dating

The decay scheme for natural ${}^{238}_{92}U$ is presented in Figure 24-4. Since the decay of ${}^{238}_{92}U$ eventually results in the stable ${}^{206}_{82}Pb$ isotope, an analytical determination of the relative amounts of these two isotopes can provide an estimate of the age of the rock formations in which they are found. This assumes, of course, that no ${}^{206}_{82}Pb$ was present in the sample at the initial time and that all of the ${}^{206}_{82}Pb$ present has appeared through this known process. An estimate of age is possible since the half-life for each decay reaction is known.

A related method was suggested as early as 1905 by Rutherford while lecturing at Yale University. He suggested that the helium resulting from alpha decays in the uranium series of decay reactions could be measured as an indication of age. About the same time, Bertram Boltwood suggested that the ${}^{238}_{92}U/{}^{206}_{82}Pb$ ratio could be measured as a criterion for dating rocks. Boltwood dated a sample of uraninite ore taken from Spruce Pine, North Carolina, as 510 million years old. Modern instrumentation using the same basic method on the same ore has yielded a date of 344 to 385 million years as its age.

The use of radioactive processes to determine the age of minerals was suggested by Rutherford in 1905.

To understand how this method works, consider the fact that 1.00 gram of ${}^{238}_{92}U$ in its half-life of 4.5 billion years would leave 0.50 grams of ${}^{238}_{92}U$ and in the process produce 0.43 gram of ${}^{206}_{82}Pb$. The time required for all of the other decay steps after the breakdown of ${}^{238}_{92}U$ is relatively short. Hence, the rate of lead formation is controlled by this *rate-determining* step. The amount of ${}^{206}_{82}Pb$ can be calculated by using the fact that one ${}^{238}_{92}U$ atom is converted into one ${}^{206}_{82}Pb$ atom and by using the conversion

$$(207.21 \text{ g Pb}/238.07 \text{ g U}) \times 0.50 \text{ g U}.$$

Now, if the ratio of 0.50 gram ${}^{238}_{92}U$ to 0.43 gram of ${}^{206}_{82}Pb$ is found in a uranium ore, it would follow that the rock is 4.5 billion years old.

Age determinations for various rocks taken from different parts of the world all indicate their ages to be in the neighborhood of 3 billion years. Some meteorites have been determined to be 4.5 billion years old.

TABLE 24–4 Radioisotope Dating of Lunar Samples from the Sea of Tranquility by Three Different Methods

			Age, Billions of Years	
Radioisotope	Half-life	Decay Product	*Crystalline sample*	*Lunar dust*
$^{232}_{90}Th$	13.9 billion years	$^{208}_{82}Pb$	3.6	4.5
$^{235}_{92}U$	0.71 billion years	$^{207}_{82}Pb$	3.9	4.7
$^{238}_{92}U$	4.5 billion years	$^{206}_{82}Pb$	3.8	4.7

As a consequence the age of the planets in the solar system is thought to be 4.5 billion years. Table 24–4 gives the ages of some lunar rocks and dust found at the Sea of Tranquility using $^{238}_{92}U$, $^{235}_{92}U$, and $^{232}_{90}Th$ radioisotope dating techniques.

Potassium/Argon Dating

The dating of mineral samples is possible due to the presence of a radioactive isotope of the element potassium, $^{40}_{19}K$. This isotope decays to a stable isotope of argon, $^{40}_{18}Ar$, by a process known as ***electron capture*** followed by gamma ray emission.

$$^{40}_{19}K + {}^{0}_{-1}e \longrightarrow {}^{40}_{18}Ar^* \longrightarrow {}^{40}_{18}Ar + {}^{0}_{0}\gamma$$

Electron capture is one method by which an unstable nucleus can decrease its atomic number by capturing an orbital electron close to the nucleus. The electron combines with a proton in the nucleus to form another neutron, thereby decreasing the atomic number by one unit. The $^{40}_{18}Ar^*$ species is unstable (denoted by the *) and will radiate energy in the form of a gamma ray in going to a lower energy state.

Argon is assumed to be trapped in the rock.

The potassium-argon method of age determination depends upon measuring the amount of $^{40}_{18}Ar$ trapped as a gas within the rock in which it was produced. Subsequently, the amount of $^{40}_{19}K$ in the rock is determined. The total amount of all isotopes of potassium is determined (usually by an emission spectrophotometric technique) and the amount of the $^{40}_{19}K$ isotope is measured. The amounts of $^{40}_{18}Ar$ and $^{40}_{19}K$ are determined by using a mass spectrograph. Once the amounts of $^{40}_{18}Ar$ and $^{40}_{19}K$ are known for a given sample, the age of the rock can be determined by using a graph such as Figure 24–10 or by using a rather complex mathematical equation (not given here) which was used to determine the graph.

Reliability of the potassium-argon dating method rests heavily upon the accuracy of measuring the very small amounts of $^{40}_{18}Ar$ and $^{40}_{19}K$ in a rock sample. $^{40}_{19}K$ not only decays by electron capture, but also by beta emission.

$$^{40}_{19}K \longrightarrow {}^{0}_{-1}e + {}^{40}_{20}Ca$$

For the calculation of age to be accurate, the two decay rates of $^{40}_{19}K$ must be known accurately. Repeated age determinations on different samples

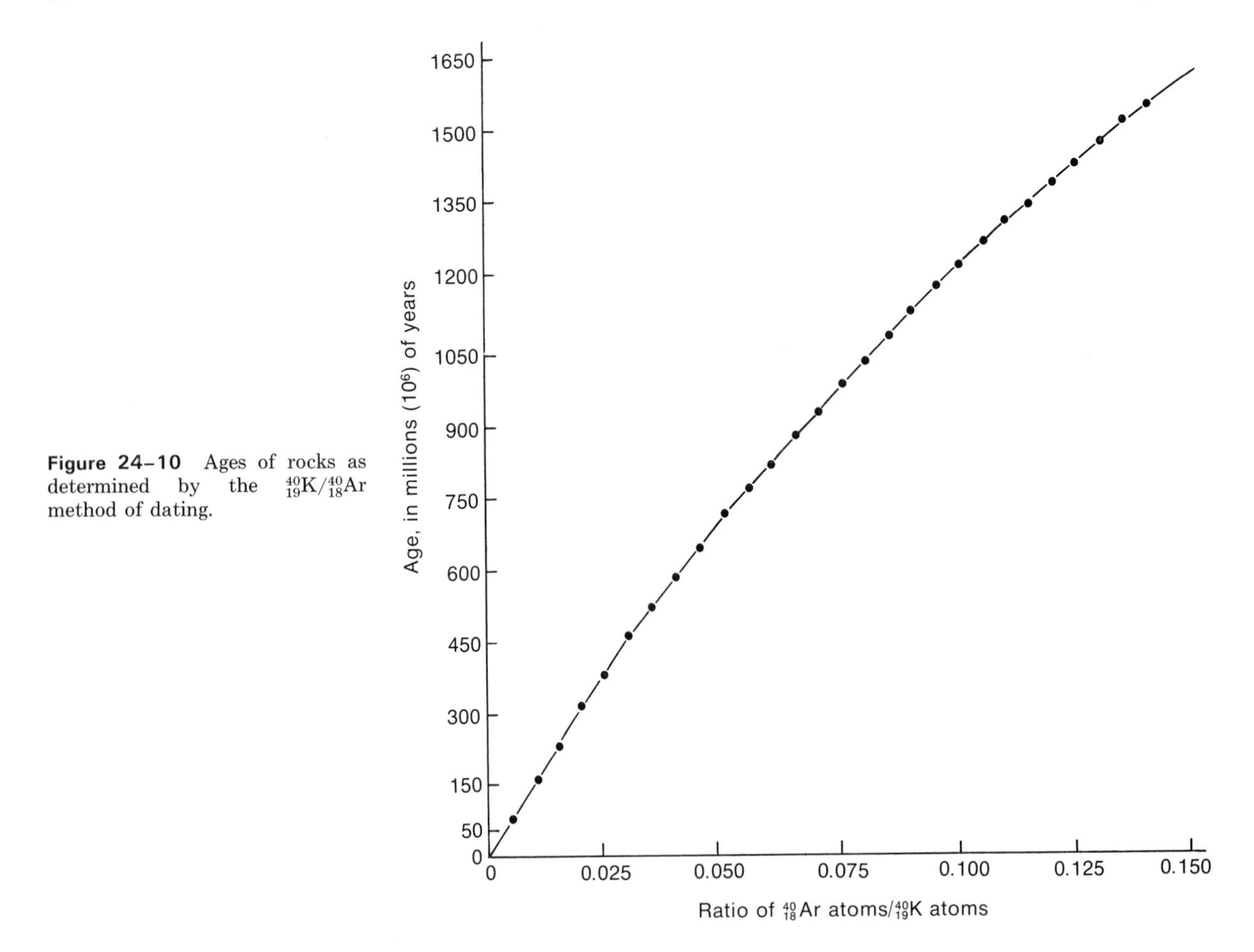

Figure 24–10 Ages of rocks as determined by the $^{40}_{19}K/^{40}_{18}Ar$ method of dating.

of the same material show that the method is about 98% accurate for ages about 3 million years and about 99% accurate for ages about 160 million years. This assumes that the rock sample contains all of the $^{40}_{18}Ar$ emitted by $^{40}_{19}K$ decay. The method, then, is limited to nonporous geologic materials which will retain argon completely. This rules out almost all sedimentary rocks.

Sedimentary rocks result from the compaction of sediments such as mud, sand, or precipitates.

Table 24–5 lists some of the results of using the potassium-argon method of dating geologic materials.

Unlike the radioactive $^{14}_{6}C$ dating procedure (discussed next) which depends upon the disappearance of $^{14}_{6}C$, the potassium-argon dating pro-

TABLE 24–5 Results of Several Potassium-Argon Age Determinations

Subject	Age
Rock from Olduvai Gorge, Tanzania (where L. S. B. Leakey found fossil remains of *Zinjanthropus*, an extinct primate)	1.75 million years
Quartz monzonite from Marysvale, Utah	26.0 million years
El Capitan, Yosemite, California	88.0 million years
Volcanic rock in central Arizona	1,800 million years

This method cannot be used on sedimentary rock.

cedure depends upon the appearance of the $^{40}_{18}Ar$ isotope. It is thus a clock, which is "set" at time zero when geologic conditions were such that gaseous argon could be trapped. This usually occurs when a molten rock formation solidifies. If events such as reheating take place during a rock's history, it is likely to lose some argon and, hence, appear too young (not enough $^{40}_{18}Ar$ will be found).

Carbon-14 Dating

Cosmic rays (interstellar radiation) are composed of many forms of very high-energy particles such as H^+ and He^{2+}. Many of these particles enter the earth's atmosphere every second. Cosmic rays undergo nuclear reactions with stable nuclei in the upper atmosphere to produce slow moving neutrons. These neutrons can react with $^{14}_{7}N$ nuclei present in nitrogen molecules in the upper atmosphere to produce a radioactive isotope of carbon, $^{14}_{6}C$, which has a half-life of 5730 years.

$$^{1}_{0}n + ^{14}_{7}N \longrightarrow ^{14}_{6}C^* + ^{1}_{1}H$$

The $^{14}_{6}C$ decays by a beta emission.

$$^{14}_{6}C^* \longrightarrow ^{0}_{-1}e + ^{14}_{7}N$$

Radioactive $^{14}_{6}C$ in the compound carbon dioxide mixes with the ordinary carbon dioxide in the atmosphere and, in turn, is incorporated into the structure of all living matter through natural food chains. Upon death of the organism, the intake of food ceases and the natural level of radioactive carbon present within the structure begins to decrease at the rate of 50 percent every 5730 years. It was the realization of this fact that led Professor Willard F. Libby, at the University of Chicago, to postulate that radioactive $^{14}_{6}C$ could be used to date ancient artifacts derived from living matter such as parchment, cloth, and wood carvings. The $^{14}_{6}C$ remaining in the artifact would have to be compared with the normal isotopic ratio of $^{12}_{6}C$, $^{13}_{6}C$, and $^{14}_{6}C$.

Professor Libby received a Nobel Prize in 1960 for $^{14}_{6}C$ dating.

TABLE 24–6 Comparison of Ages of Various Artifacts Over a Span of 3500 Years by Radiocarbon Dating and Other Methods such as Tree-ring Dating and Chronological Methods.*†

Material	Radiocarbon Age	Age by Another Method
Mammalian remains from middle of an Inca temple	450 ± 150 years	444 ± 25 years
Sequoia tree ring	930 ± 100	880 ± 15
Wood from Roman ship	2,030 ± 200	1,990 ± 3
Charcoal from Etruscan tomb	2,730 ± 240	2,600 ± 100
Wood from Egyptian tomb of Zoser	3,979 ± 350	4,650 ± 75

*These ages are given as the age in 1950.

†This table is taken from the "McGraw-Hill Encyclopedia of Science and Technology," Vol. 11. McGraw-Hill, New York, 1971, p. 291.

An important assumption is that the flow of $^{14}_{6}C$ into the biosphere is constant over time. Studies indicate that the radioactive carbon is indeed slowly mixed with its nonradioactive isotopes. The assumption that the rate of production of $^{14}_{6}C$ has been essentially constant over the past several thousand years is approximately true. Recently, growth rings on sequoia and bristlecone pine trees have been accurately measured and compared with $^{14}_{6}C$ dates. These experiments indicate that $^{14}_{6}C$ production has fluctuated, particularly during the first millenium B.C. In other words, there is not perfect agreement between the tree-ring age and the $^{14}_{6}C$ found in each ring. Nevertheless, radioactive $^{14}_{6}C$ dates do compare reasonably well with dates obtained from other methods when corrected for these discrepancies (Table 24–6).

The various dating methods complement one another in terms of the time spans for which they are useful.

RADIATION DAMAGE

We are constantly bombarded by radiation from a number of sources; this radiation includes cosmic rays, medical x-rays, radioactive fallout from countries which do nuclear testing, and naturally occurring radioisotopes which are widespread. Fortunately, most radiation damage is too slight to be noticed immediately, although its very presence should be regarded as one of the hazards of everyday life.

There is a normal background radiation from natural causes.

As we have seen earlier, a radioisotope will disintegrate into a stable species, or it will become part of a decay series. In a sample of radioactive matter large enough to measure, there will be many disintegrations over a given time if the half-life is short, or few disintegrations over the same time interval if the half-life is long. Normal background radiation on the human body is 2 or 3 disintegrations per second.

There are three principal factors which render a radioactive substance dangerous: (1) the number of disintegrations per second, (2) the half-life of the isotope, and (3) the type or energy of the radiation produced. In addition, radiation can be very damaging if the radioactive

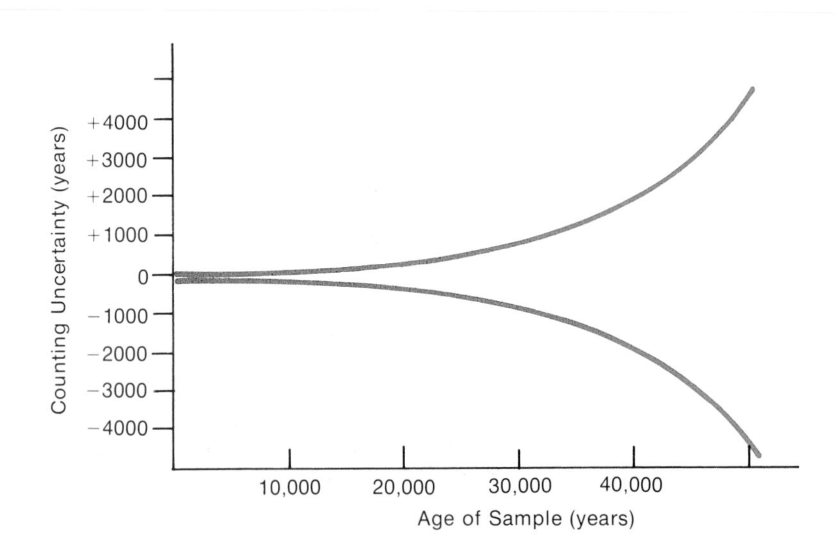

Figure 24–11 In radiocarbon dating, the counting uncertainty becomes increasingly important for older samples, resulting in very large uncertainties in ages beyond 30,000 years. Other factors, such as variabilities in cosmic rays over centuries, introduce still other uncertainties. (From *McGraw-Hill Encyclopedia of Science and Technology*. Vol. 11. New York, McGraw-Hill, 1971, p. 302.)

substance is of a chemical nature such that it can be incorporated into a food chain or otherwise enter a living organism.

Radioactive disintegrations are measured in ***curies*** (Ci; one Ci is 37 billion disintegrations per second). A more suitable unit is the microcurie (μCi) which is 37,000 disintegrations per second. One curie of a radioisotope is a potent sample if the energy per quantum is large enough to cause a biochemical change.

Wilhelm Roentgen discovered x-rays in 1895 and was awarded the Nobel Prize for this work in 1901.

The unit ***roentgen*** is used to measure the intensity of x-rays or gamma rays. One roentgen is the quantity of x- or gamma radiation delivered to 0.001293 gram of air such that the ions produced in the air carry 3.34×10^{-10} coulomb of charge. A single dental x-ray represents about one roentgen.

The three types of natural radioactive emissions differ in their penetrating ability (Figure 24–12), with gamma rays being by far the most penetrating.

Damage by radiation is due to ionization caused by the fast-moving particles colliding with matter, and by the excitation of matter by gamma and x-rays which in turn produce ionization. Neutrons are produced in nuclear explosions, in nuclear reactions, and by background cosmic radiation. A neutron does not produce ionization per se but instead imparts its kinetic energy to atoms, which in turn may ionize or break away from the atom to which they are bonded. Neutrons render many engineering materials such as plastics and metals structurally weak over long periods of time due to the decay caused by breaking chemical bonds.

Neutrons can damage metals causing structural failure. This is a severe problem in nuclear reactors.

Biological tissue is easily harmed by radiation. A flow of high-energy particles may cause destruction of a vital enzyme, hormone, or chromosome needed for life of a cell. The radiation may also produce free radicals

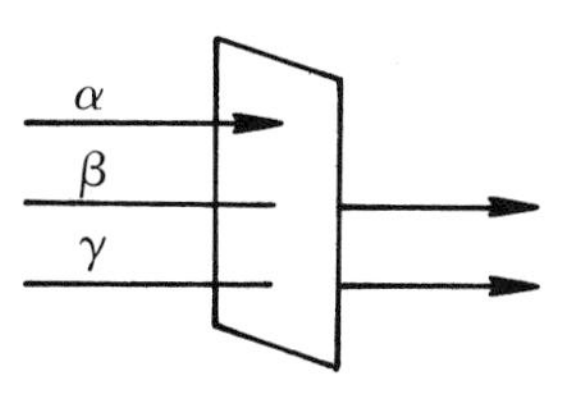

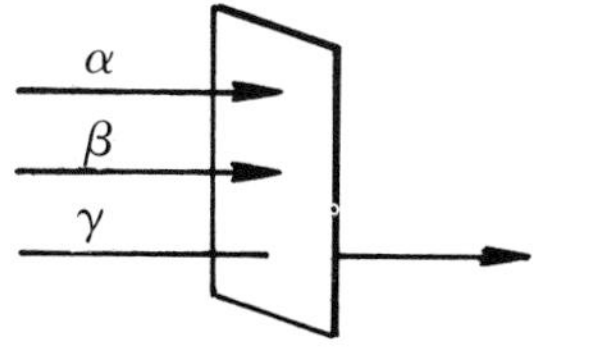

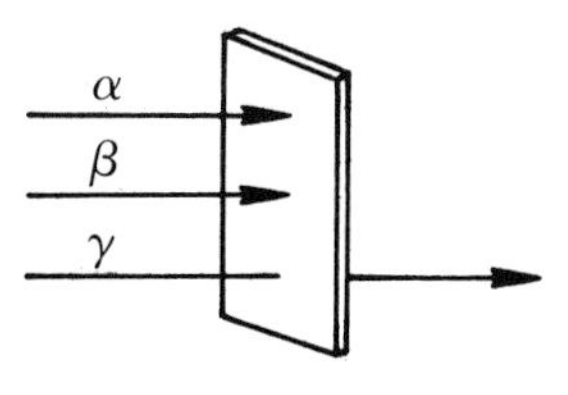

Figure 24–12 Penetrating ability of alpha (α), beta (β), and gamma (γ) radiation. Gamma rays even penetrate an 8-mm lead sheet. Skin will stop alpha rays but not beta rays.

which poison the cell. In general those cells which divide most rapidly are those most easily harmed by radiation (Figure 24–13).

Whole body radiation effects are divided into ***somatic effects,*** which are confined to the population exposed, and ***genetic effects,*** which are passed on to subsequent generations. A unit of measurement of radiation density is helpful in measuring the effect of radiation on tissue. The ***rad*** is defined as 100 ergs of energy imposed on a gram of tissue. Whole body doses of radiation of up to 150 rads produce scarcely any symptoms, whereas doses of 700 rads produce death. Intermediate doses produce vomiting, diarrhea, fatigue, and loss of hair. Often the somatic effects are delayed. Perhaps the best studied of the delayed effects are the incidences of cancer related to exposure to radiation. It has been estimated that 11% of all leukemia cases and about 10% of all forms of cancer are attributable to background radiations. Certainly an individual who is exposed to a higher than normal level of radiation over a considerable length of time increases the chances of cancer. The alteration of normal cells to cancerous cells caused by radiation is undoubtedly a series of changes, since in almost all cases the onset of cancer lags behind the exposure to radiation by an induction period of 5 to 20 years.

One *rad* is roughly the energy absorbed by tissue exposed to one roentgen of gamma rays.

1 joule = 10^7 ergs
1 calorie = 4.184×10^7 ergs

The genetic effects of radiation are the result of radiation damage to the germ cells of the testes (sperm) or the ovary (egg cells). Ionization caused by radiation passing through a germ cell may break a DNA strand or cause it to be altered in some other way. Replication of this altered DNA means transmission of a new message to successive generations, a ***mutation.*** Every type of laboratory animal upon which radiation dam-

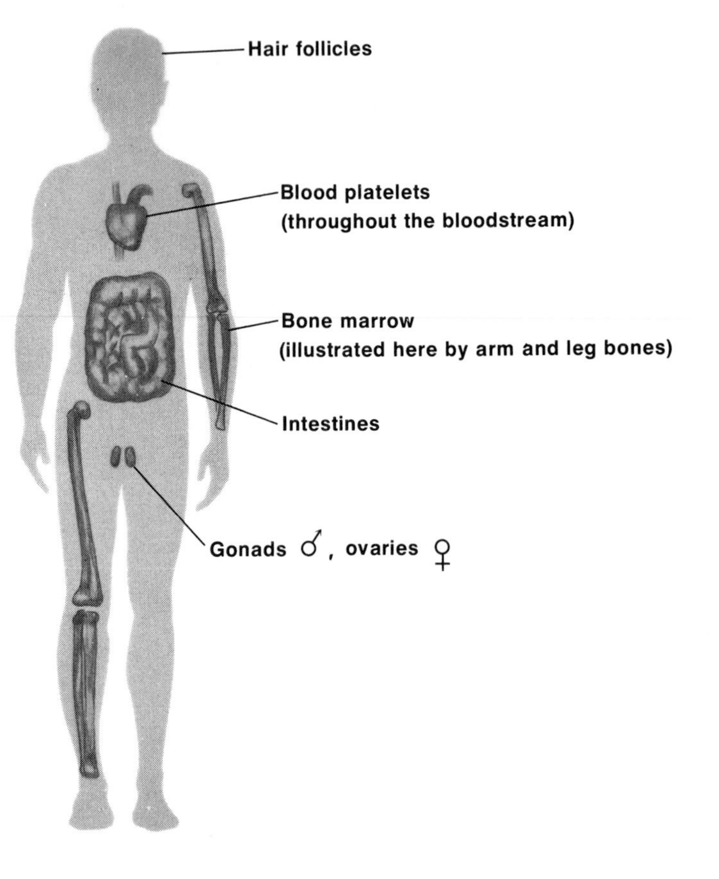

Figure 24–13 The fast-dividing cells within the body are the ones most harmed by radiation. These include cells in bone marrow, white cells, platelets of the blood, those lining the gastrointestinal tract, hair follicles, and gonads. In addition, the lymphocytes (cells producing the immune responses) are easily killed by radiation.

TABLE 24–7 Average dose of radiation to soft tissues and gonads from surroundings.*

Source	Dose to Gonads Per Year (rad)
Natural Background	
Cosmic rays	0.028
Local gamma rays	0.047
Radon in air	0.001
Potassium-40	0.019
Carbon-14	0.001
Other sources	0.002
Subtotal	0.098
Man-Made	
Medical X-rays	0.100
Luminous watch dials	0.001
Occupational exposure	0.002
Television sets	0.001
Fallout from weapons test	0.001
Subtotal	0.105
Total	0.203

*This table is taken from the "McGraw-Hill Encyclopedia of Science and Technology," Vol. 11. McGraw-Hill, New York, 1971, p. 250.

Medical X-rays account for about 50 percent of the average radiation dosage here.

age experiments have been performed has responded by increased incidence of mutation. Therefore, the necessity of protecting the population of childbearing age from radiation should be apparent. Theoretically, at least, one photon or one high-energy particle can ionize a chromosomal DNA structure and produce a genetic effect that will be carried for generations.

SELF-TEST 24-B

1. A particle having an energy of 1 Bev has an energy of ____________________ electron volts.
2. One type of particle accelerator which moves the charged particles in a circle is called a ____________________.
3. The first transuranium element to be "made by man" is ____________________.
4. The transuranium element of greatest atomic number to be "made by man" is element number ____________.
5. In $^{238}_{92}U$ dating of moon rocks, the final decay product measured is ____________.

6. Which type of rock cannot be dated by the potassium-argon method? (metamorphic, igneous, porous sedimentary)

7. The cause of radiocarbon in the atmosphere is ____________ ____________.

8. If the amount of radiocarbon in the atmosphere 2000 years ago had been less than is presently assumed, then an article from that time would appear (younger/older) as a result of radiocarbon determination.

9. The unit ____________ is used to measure the intensity of x-rays or gamma rays.

10. Which is the most penetrating type of radiation? (alpha, beta, gamma)

MATCHING SET

______	1. somatic effect	a.	intake stops when organism dies
______	2. one microcurie	b.	radiation effect on general population
______	3. $^{14}_{6}C$	c.	first suggested by Rutherford
______	4. $^{40}_{19}K$	d.	developed the cyclotron
______	5. genetic effect	e.	radiation damage to DNA
______	6. $^{238}_{92}U$ dating	f.	tissue easily damaged by radiation
______	7. element 106	g.	time required for half of the nuclei to disintegrate
______	8. E. O. Lawrence	h.	37,000 disintegrations per second
______	9. James Chadwick	i.	generates argon in a geologic clock
______	10. $^{218}_{84}Po$	j.	can detect ionizing radiation
______	11. half-life	k.	latest "man-made" element
______	12. C. T. R. Wilson	l.	developed cloud chamber
______	13. bone marrow	m.	discovered neutron
______	14. Geiger counter	n.	alpha decay product of $^{222}_{86}Rn$

QUESTIONS

1. Describe the operation of a Geiger counter.

2. Why is the cloud chamber (or the bubble chamber) more useful than other types of nuclear particle detectors?

3. What does the symbol $^{11}_{5}B$ mean?

4. In general, how have the synthetic transuranium elements been produced?

5. Complete or supply the following nuclear equations:

   (a) ${}^{1}_{1}H + {}^{35}_{17}Cl \longrightarrow {}^{4}_{2}He + ?$
   (b) beta emission of ${}^{60}_{27}Co$
   (c) alpha emission of ${}^{228}_{90}Th$
   (d) ${}^{1}_{0}n + {}^{60}_{28}Ni \longrightarrow ? + {}^{1}_{1}H$
   (e) ${}^{2}_{1}H + {}^{1}_{1}H \longrightarrow ? + {}^{0}_{0}\gamma$
   (f) ${}^{238}_{92}U + {}^{12}_{6}C \longrightarrow ? + 4{}^{1}_{0}n$

6. We speak of "seeing" a nuclear event with a cloud or bubble chamber. Explain why we are more truthful in using the quotation marks about the word *seeing.*

7. Look up the origin of the word *mutation* and explain why the word *transmutation* was an apt word choice to describe the changing of one element into another.

8. What is the difference between a thermal neutron and a high energy neutron resulting from cosmic radiation?

9. If a radium atom (${}^{226}_{88}Ra$) loses one alpha particle per atom, what element is formed? What is its atomic weight? What is its atomic number?

10. What are the important assumptions made in radiocarbon dating?

11. Name two methods by which the age of rocks can be determined. What assumptions are made in these methods?

12. What errors would be introduced into the age determination of a piece of granite if it had been reheated at some point during its existence?

13. What error would be introduced into the age determination of a tree ring if the amount of cosmic rays had been double their present value at the time the tree grew that ring?

14. What is meant by "delayed somatic effect?" Give an example.

PROBLEMS

1. The half-life of ${}^{218}Po$ is 3 minutes. How much of a 2-gram sample of this isotope remains after 15 minutes?

   Ans. 0.0625 g

2. Suppose you were given $1000.00 and told that you could spend one half of it the first year, one half of the balance the second year, and so on. If you spent the maximum allowed, at the end of what year would you have $31.25 of the original $1000.00 left? How is this problem analogous to radioactive dating techniques? Can you think of ways in which it is not?

   Ans. 5th year

3. A bacterial culture has a "doubling time" of 2 days. How many bacteria will be present in 2 weeks from 1000 initially present? How is "doubling time" related to half-life?

   Ans. 128,000

SUGGESTIONS FOR FURTHER READING

Clark, H. M., "The Origin of Nuclear Science," *Chemistry,* Vol. 40, No. 7, pp. 8–11 (1967).
Dalrymple, G. B., and Lamphere, M. A., "Potassium-Argon Dating," W. H. Freeman and Co., San Francisco, 1969.
"Dendrochronology," *Chemistry,* Vol. 43, No. 7, p. 26 (1970).
"Detecting Forgeries in Paintings," *Chemistry,* Vol. 41, No. 5, p. 5 (1968).
"Element 105," *Chemistry,* Vol. 43, No. 6, p. 20 (1970).
Faul, H., "Ages of Rocks, Planets, and Stars," McGraw-Hill Book Co., New York, 1966.
Flerov, G. N., and Zvara, I., "Synthesis of Transuranium Elements," *Science,* Journal, Vol. 4, No. 7, p. 63, July (1968).
"Fossil-Track Dating," *Chemistry,* Vol. 43, No. 7, p. 27 (1970).

Gordus, A. A., "Neutron Activation Analysis of Almost Any Old Thing," *Chemistry,* Vol. 41, No. 5, pp. 8–15 (1968).
Hudis, J., "Nuclear Reactions," *Chemistry,* Vol. 40, No. 7, pp. 20–24 (1967).
"Ionium Dating," *Chemistry,* Vol. 43, No. 7, p. 22 (1970).
Johnsen, R. H., "Radiation Chemistry," *Chemistry,* Vol. 40, No. 7, pp. 31–36 (1967).
Libby, W. F., "Radiocarbon Dating," University of Chicago Press, Chicago, 1955.
Maugh, T. H., II, "Element 106: Soviet and American Claims in Muted Conflict," *Science,* Vol. 186, p. 42 (1974).
"On Inhaling Plutonium: One Man's Long Story," *Science,* Vol. 185, p. 1028 (1974).
Overman, R. T., "Basic Concepts of Nuclear Chemistry," Reinhold Publishing Co., New York, 1963. (paperback)
"Radiocarbon Dating," *Chemistry,* Vol. 43, No. 7, p. 24 (1970).
Seaborg, G. T., "Man-Made Transuranium Elements," Prentice-Hall, Inc., New York, 1963. (paperback)
Seaborg, G. T., "Some Recollections of Early Nuclear Age Chemistry," *Journal of Chemical Education,* Vol. 45, p. 278 (1968).
"University of California's Berkeley Campus: 1941–42," *Chemistry,* Vol. 40, No. 3, pp. 25–26 (1967).

CHAPTER 25

ENERGY, MAN, AND SOCIETY

It is fitting as we draw this text to a close to consider one of the greatest unsolved problems of mankind. How will man meet his energy needs? The problem will have to be solved satisfactorily before man's other major problem—food supply—can be solved. In the United States, the average yearly consumption of energy is now in excess of 1.8×10^{16} kilocalories (energy units are given in Table 25-1) and this consumption has been increasing at the rate of about 3.5 percent per year over the last few years. The average consumption is therefore about 240,000 kilocalories per person per day for all forms of energy. This energy is currently being supplied by petroleum, coal, natural gas, water power, and nuclear sources (Figure 25-1). The predicted energy shortage, however, is now a reality. The power brownouts in the eastern United States in the late 1960's were a signal that something was amiss. In the early 1970's gasoline shortages were caused by inadequate refining capacity, low return on investment capital, and a Mid-East War. Problems of this type were

1.8×10^{16} kilocalories is the same as 7.1×10^{6} British thermal units (Btu) or 2.05×10^{13} kilowatt hours.

TABLE 25-1 Energy and Power Units

Energy—the Ability to do Work

1 calorie (cal) = 4.184 joule (J)*
1 kilowatt hour (kWh) = 8.604×10^5 calorie†
1 kilowatt hour (kWh) = 3.413×10^3 British thermal unit (Btu)

Power—the Rate of Doing Work

1 watt (W) = 1 joule per second
1 kilowatt (kW) = 10^3 W
1 megawatt (MW) = 10^3 kW = 10^6 W
1 watt = 1 volt × 1 ampere**

*The joule is the SI unit of energy.
†The electric company will connect your house for the maximum desired power, but they will sell you energy (not power) in kilowatt-hour units.
**One ampere flowing under an electrical potential of one volt requires one watt of power. A 100-watt bulb operating at 100 volts would draw 1 ampere of current.

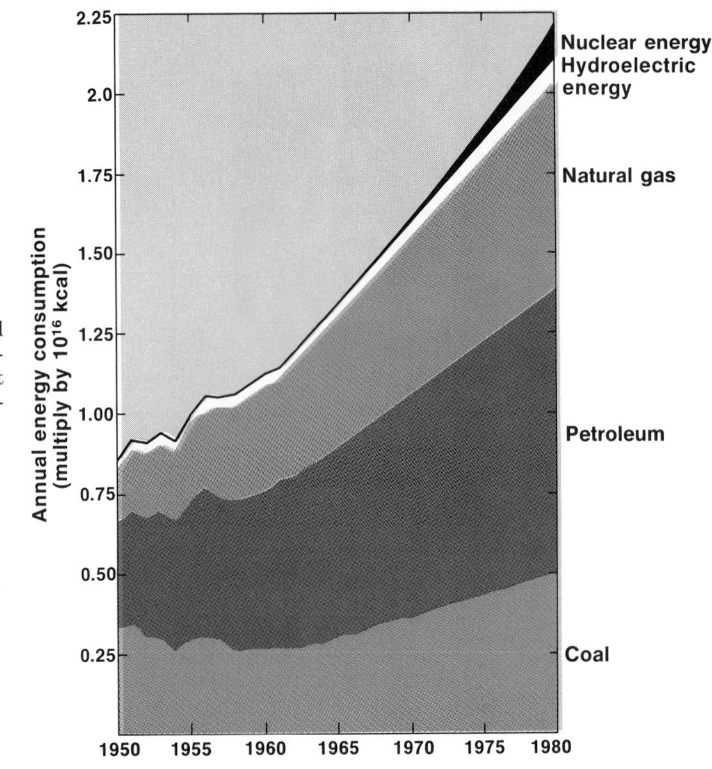

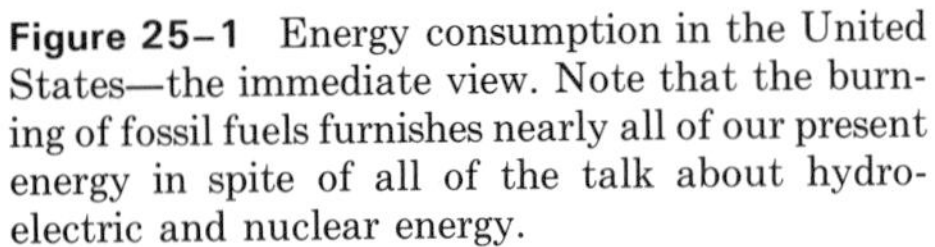

Figure 25–1 Energy consumption in the United States—the immediate view. Note that the burning of fossil fuels furnishes nearly all of our present energy in spite of all of the talk about hydroelectric and nuclear energy.

certain to occur because of two factors: (1) the world is using more energy while (2) the supplies of the type being used are becoming depleted.

As late as 1940, 80 percent of the world's population was living at the primitive or advanced agricultural level. Consequently, relatively little energy was being used. Today, the situation is in a rapid state of change. About 25 to 30 percent of the present world population lives in a very highly industrialized society and uses 80 percent of the world's energy (Figure 25-2). In the United States, 6 percent of the world's population uses 35 percent of the world's energy production in any given year. However, the relative use of energy in the United States is on the decline as compared with the rest of the world. This means that there is now taking place a serious world-wide competition for the types of energy sources currently available.

FOSSIL FUELS

Fossil fuels store the energy of the sun.

As indicated in Figures 25–1 and 25–3, nearly all of the present energy needs in the U.S. are being supplied by fossil fuels: petroleum, natural gas, and coal. This is also true on a worldwide basis. Although the chemistry and geology of fossil fuel formation is not thoroughly understood, it is generally agreed that buried plant material formed these fuels over millions of years. Perhaps the most significant point about fossil fuels is that modern man appears capable of using up in a few hundred years what it took nature millions of years to produce (Figure 25–3).

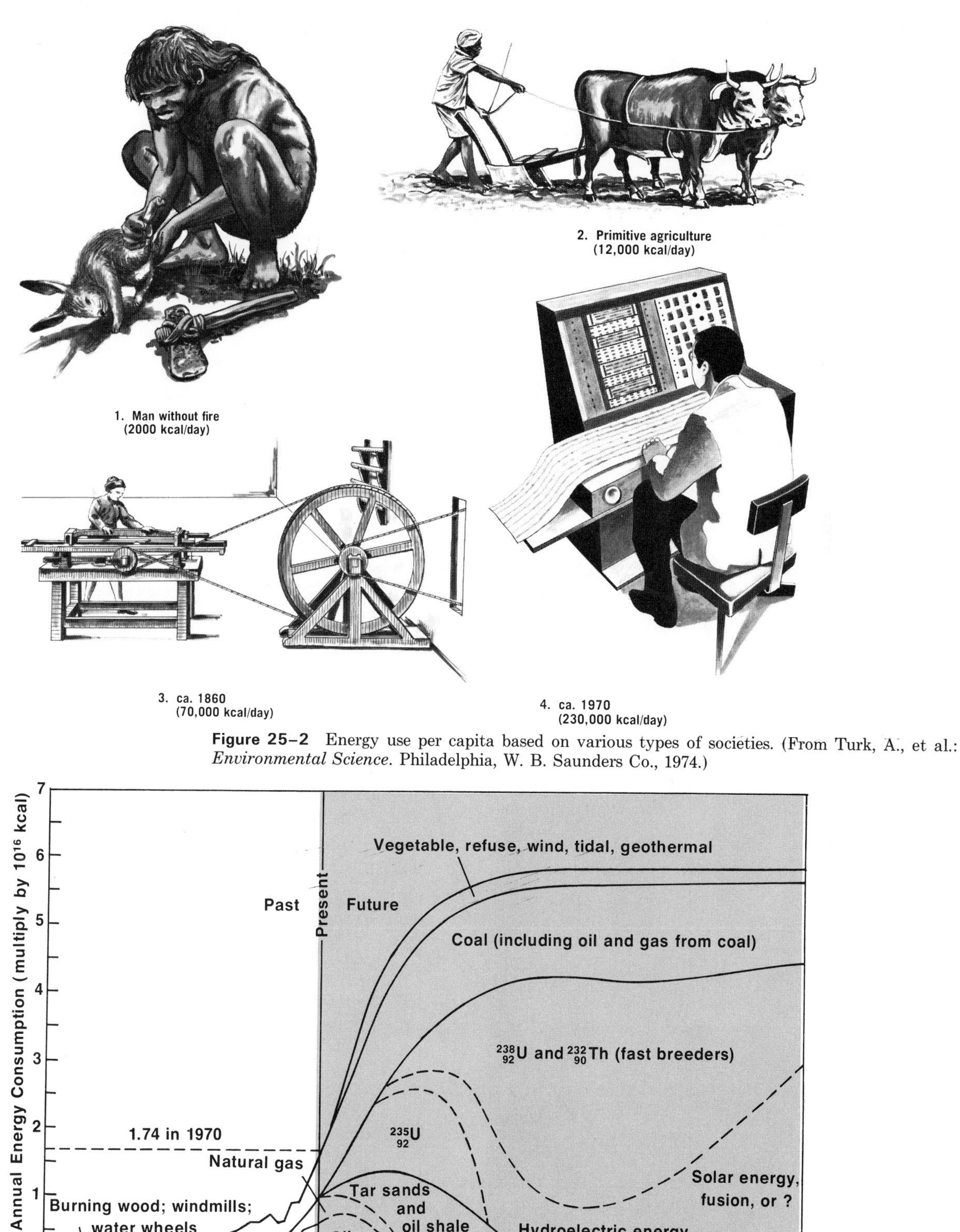

Figure 25–2 Energy use per capita based on various types of societies. (From Turk, A., et al.: *Environmental Science.* Philadelphia, W. B. Saunders Co., 1974.)

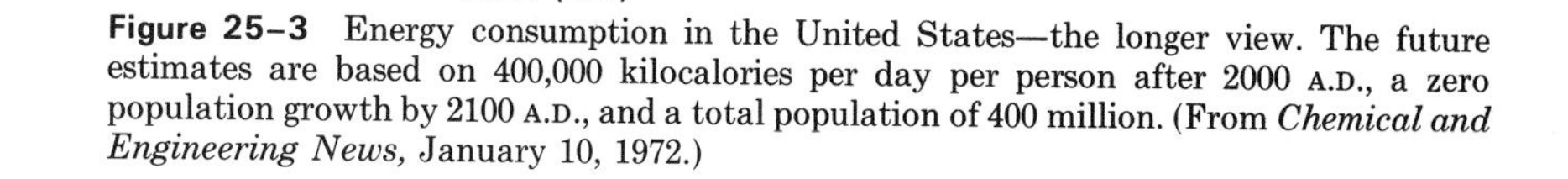

Figure 25–3 Energy consumption in the United States—the longer view. The future estimates are based on 400,000 kilocalories per day per person after 2000 A.D., a zero population growth by 2100 A.D., and a total population of 400 million. (From *Chemical and Engineering News,* January 10, 1972.)

TABLE 25–2 Energy Content of Fossil Fuels

Fuel	Reaction Equation	Heat Energy Evolved
Coal	$C + O_2 \longrightarrow CO_2$	94 kcal/mole; 7.8 kcal/g fuel
Natural Gas*	$CH_4 + 2O_2 \longrightarrow CO_2 + 2H_2O$	211 kcal/mole; 13.2 kcal/g fuel
Petroleum†	$2C_8H_{18} + 25O_2 \longrightarrow 16CO_2 + 18H_2O$	1,303 kcal/mole; 11.4 kcal/g fuel

* CH_4, methane, is the principal constituent in natural gas (up to 97%).
† C_8H_{18}, octane, is only one of many hydrocarbons present in petroleum.

When fossil fuels are burned, chemical energy stored long ago from sunlight via photosynthesis is released. The amount of energy varies with the type of fuel (Table 25–2). In each case the complete combustion products are carbon dioxide and water.

Petroleum

Vast deposits of petroleum were first discovered in the United States (Pennsylvania) in 1859 and in the Middle East (Iran) in 1908. Since that time petroleum has found wide use as an energy source, first as kerosene for lighting, then as gasoline for transportation, and more recently as fuel

One barrel of petroleum contains 42 gallons.

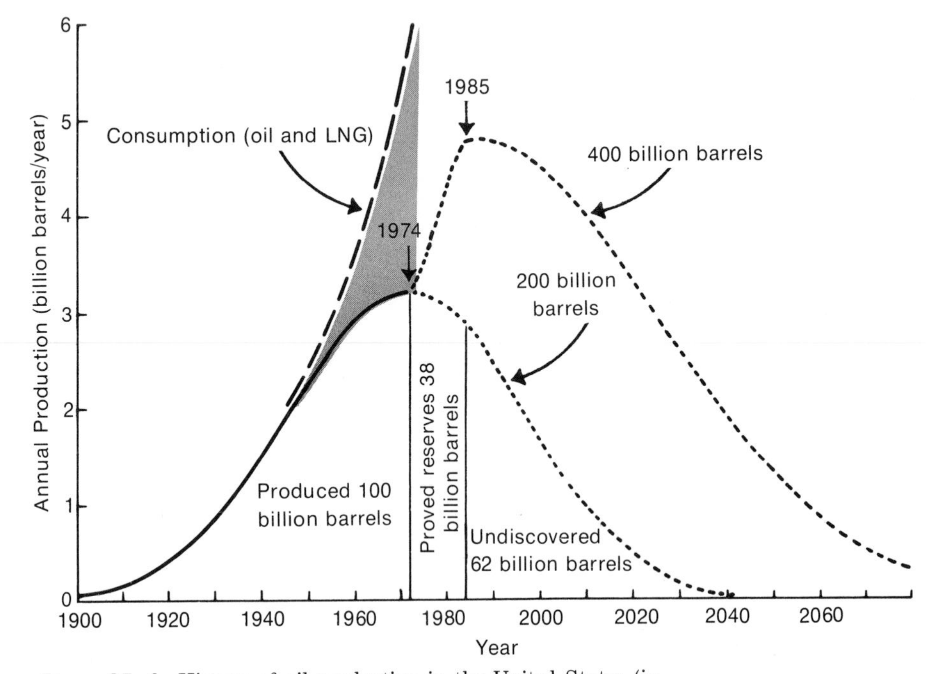

Figure 25–4 History of oil production in the United States (including Alaska). The dashed curve represents oil imports. The lower curve is based on known oil reserves, and the higher dotted curve is based on estimates of what is likely to be found with large deposits expected under the continental shelf. Regardless of how much is actually present, it is a finite amount and apparently will be consumed in a relatively short period of time, (NGL = natural gas liquefied). (From *Science,* April 19, 1974.)

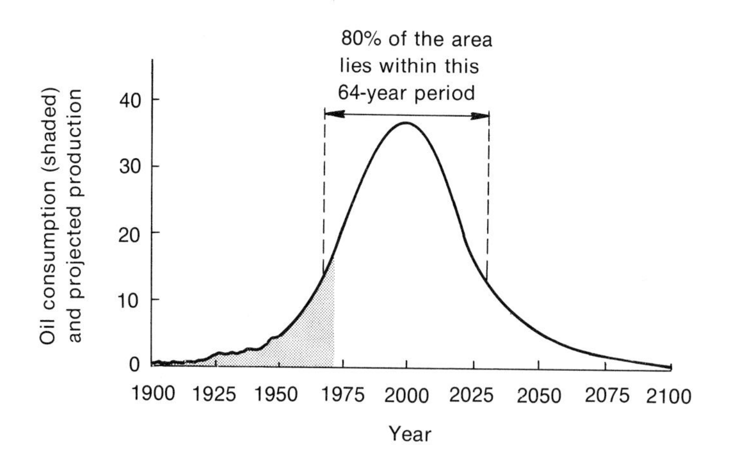

Figure 25–5 World oil use: past and projected.

oil to produce electricity. All of this has resulted in the use of over 5.4 billion barrels of oil per year in the U.S. in recent years.

It has been estimated that there were about 200 billion barrels of oil to be recovered in the United States (including Alaska) when oil production first began. Half of this amount has now been produced (see Figure 25–4). Numbers of this type are often quoted and a few qualifications should be noted. ***"Recoverable oil"*** is based on 30 percent recovery, i.e., pumping 30 percent from the ground and leaving 70 percent behind, since to recover more would cost more and hence raise the cost of the barrel of oil. It stands to reason that as supplies decrease and costs go up it will prove economically feasible to "recover" more of the oil.

The definition of recoverable oil is based on economics.

A second estimate of U.S. reserves of oil can be based on as yet undiscovered oil under the continental shelf. This estimate is 400 billion barrels, thus doubling the amount of U.S. oil. Either way, the consumption of petroleum by the United States, shown as the dashed line in Figure 25–4, is increasing at too fast a rate; obviously, this cannot continue. Figure 25–5 illustrates the situation world-wide. The assumption underlying the projection in Figure 25–5 is that the rest of the world will continue in its lower rate of energy consumption. This is not a feasible assumption. Based on the present trends, a reasonable projection is that 80 percent of the world's petroleum will be used in a 60- to 70-year period, ending at about the year 2025. Unless drastic changes are made at some point in the relatively near future, perhaps as early as the year 2000, petroleum will become an increasingly rare commodity.

As indicated earlier in this text, petroleum is the base for a very large petrochemical industry, including the plastics. A real chemical consideration that must come into focus in the future as the petroleum supply decreases is whether we want to burn petroleum for its energy content or save it as a starting material for petrochemical products.

Oil may become more valuable as a chemical raw material than as an energy source.

Coal

Unlike petroleum, where discoveries are being made almost daily, geologists believe that all the world's coal supplies have now been discovered. ***Minable coal*** is defined as 50 percent of all coal which is in a seam at least 12 inches thick and within 4000 feet of the surface. Estimates have been made as to the length of time this minable coal will supply our energy needs. Figure 25-6 shows that coal will be available as a

The largest supply of fossil fuel is in the form of coal.

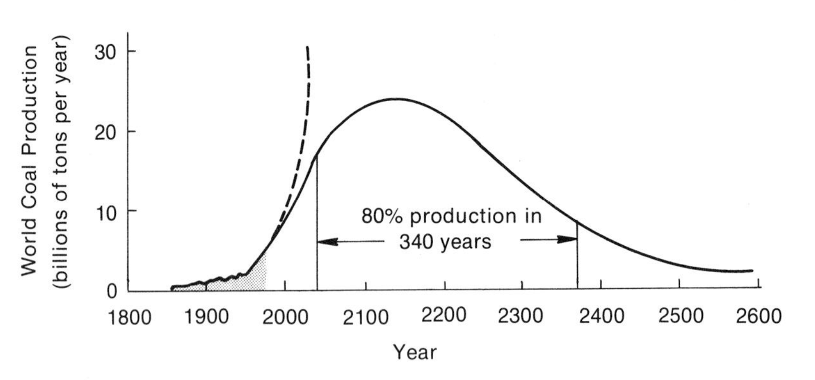

Figure 25–6 The coal mined to date (*shaded area*) represents only a small fraction of the minable coal. The rate of increase in coal consumption (*dashed line*) is 3.56 percent per year. It is obvious that such an exponential rise cannot continue long after the year 2000. At the present usage (held constant) coal would last for many hundreds of years.

fuel for a much longer period of time than oil. If new mining techniques are developed, then more of the deposited coal might be termed minable.

Two of coal's major drawbacks are that it is a relatively dirty fuel and difficult to handle. Coupled with the atmospheric pollution caused by sulfur-containing coal, these drawbacks were prime reasons for a major shift from coal to petroleum in electrical generating plants, particularly in the industrialized nations, late in the 1960s. The dangerous and unhealthful deep coal mining and the esthetically ugly strip mining contributed to the shift.

Coal Gasification

Gasification can make coal cleaner to burn and easier to handle from supplier to user.

Coal can be converted into a relatively clean-burning fuel by a process known as ***gasification*** (Figure 25–7). In this process, coal is made to react with either a limited supply of hot air or steam. In the first

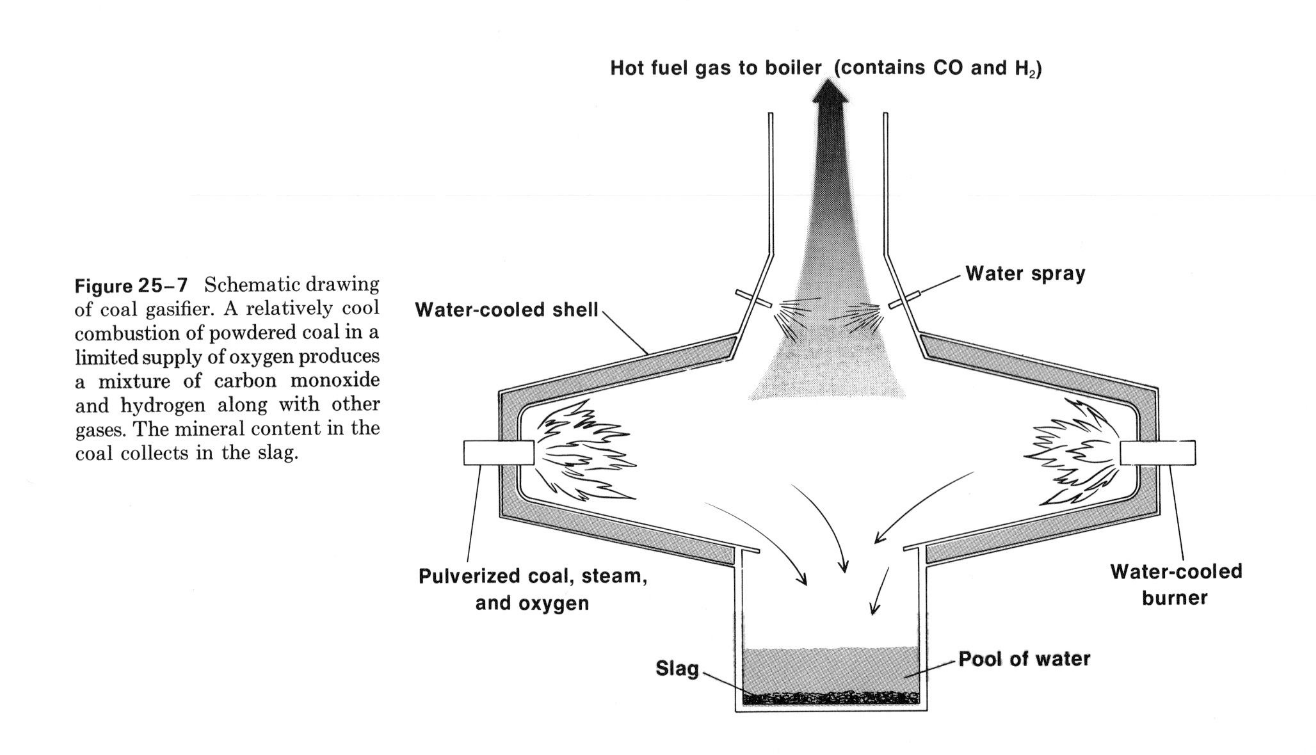

Figure 25–7 Schematic drawing of coal gasifier. A relatively cool combustion of powdered coal in a limited supply of oxygen produces a mixture of carbon monoxide and hydrogen along with other gases. The mineral content in the coal collects in the slag.

reaction the product is a gaseous mixture known as ***"power gas."***

$$\text{Coal} + \text{Air} \longrightarrow \underset{\text{POWER GAS}}{CO(g) + H_2(g) + N_2(g)}$$

Power gas contains up to 50 percent by volume nitrogen which does not burn and lowers the heating value of the gas considerably. In fact, power gas of this composition has only one sixth the heat content of methane.

If the coal is allowed to react with high-temperature steam a mixture of carbon monoxide and hydrogen (which contains no nitrogen) known as ***synthesis gas*** or ***coal gas*** is obtained. In either mixture, the CO and H_2 are burned in the air to produce heat.

Coal gas can be used like natural gas.

$$\underset{\text{COAL}}{C} + H_2O(g) \longrightarrow \underset{\text{SYNTHESIS GAS OR COAL GAS}}{CO(g) + H_2(g)}$$

The heat produced is about one third that of an equal volume of methane (natural gas).

$$2CO + O_2 \longrightarrow 2CO_2 + \text{Heat}$$
$$2H_2 + O_2 \longrightarrow 2H_2O + \text{Heat}$$

Other opportunities exist in the area of coal modification since liquid fuels can also be obtained from coal. Knowledgeable estimates indicate that between six and eight years will be needed to make such changes on an extensive scale.

Electricity Production

In 1970, about 25 percent of all the energy consumed in the United States was used for electricity production and 22 percent of the fossil fuels consumed were used for this purpose. Only 30 percent of the energy put into the generation of electricity is available at the point where the electricity is used (Figure 25–8). Part of the energy loss in electricity production is illustrated in a schematic diagram of a large modern fossil-fuel generating plant (Figure 25–9). For a 1,000-megawatt coal-burning plant, one hour of operation might look like this:

Electric generating plants yield about one-third of the fuel energy they consume in the form of electrical energy.

Coal consumed	696 tons producing 2.270 billion kilocalories
Smokestack heat loss	0.227 billion kilocalories
Heat loss in plant	0.106 billion kilocalories
Heat loss in evaporator to cool condenser	1.080 billion kilocalories
Electrical energy delivered to power lines	0.857 billion kilocalories
Percent of energy delivered as electricity	$\frac{0.857}{2.27} \times 100\% = 37.8\%$

There is a further energy loss in the power lines and the transformers

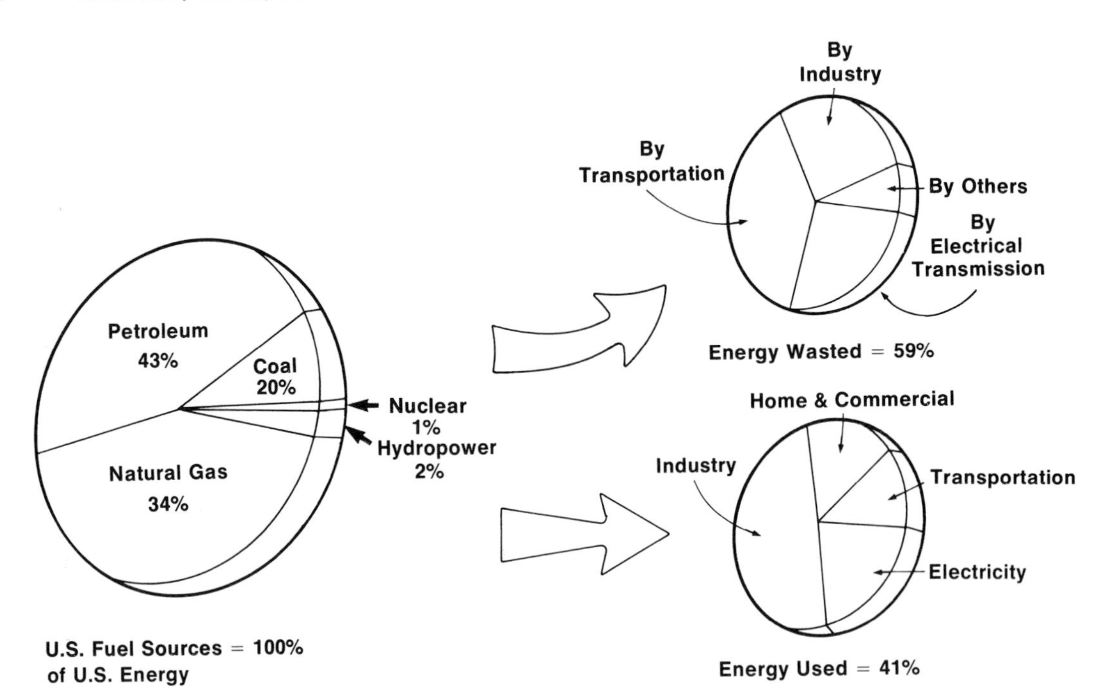

Figure 25–8 The annual fuel "pie" in the United States in the 1970's. Of the fuel sources left, this energy is more than half (59 percent) wasted. Only 41 percent of the fuel we pay for is actually used. This situation is due partly to the laws of thermodynamics and partly to wasteful practices, sloppy planning, and poor design of fuel-using devices.

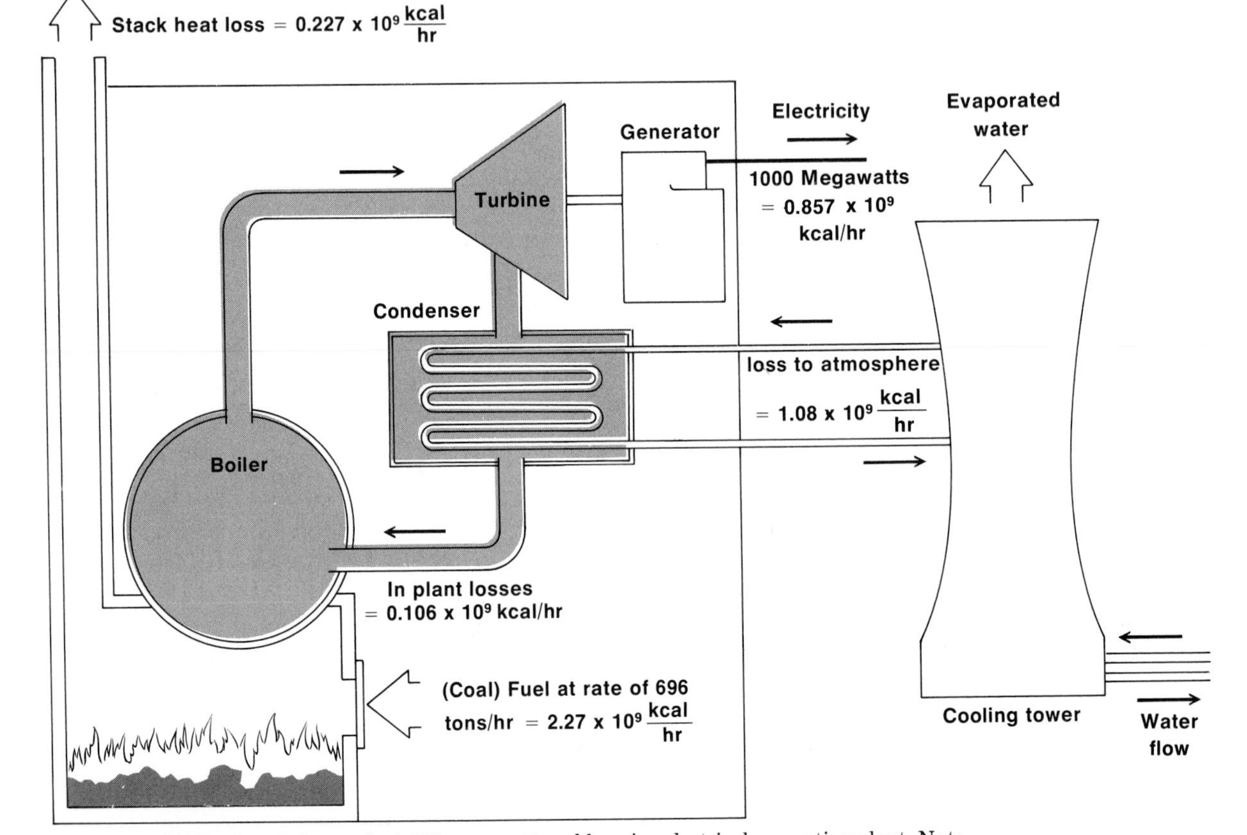

Figure 25–9 The heat balance of a 1,000-megawatt coal burning electrical generating plant. Note that the 696 tons of coal burned per hour furnish 2.27×10^9 kcal of heat energy, but only 0.857×10^9 kcal of energy is converted to electricity. Note also the large amounts of heat energy lost.

which lowers the useful output of the plant to 30 percent of the energy consumed. This is the ***efficiency*** figure for the overall operation. It is important to note that we pay for 300 kcal of heat energy in the form of coal or fuel oil but receive less than 100 kcal of energy in the form of electricity.

SELF-TEST 25-A

1. By 1980, approximately what percentage of the total U.S. energy consumption will be furnished by coal? Refer to Figure 25-1.__________
2. Which furnishes the most heat energy per gram—coal, petroleum, or natural gas?__________
3. How many gallons of oil are there in one barrel?__________
4. Using the recovery rate of 30 percent for oil in place, approximately how large a percentage of the oil in the United States has been thus far pumped from the earth (Figure 25-4)?__________
5. The composition of "power gas" obtained from coal gasification is: [(CO, H_2, N_2) or (CO, H_2)].
6. The typical efficiency of an electrical generating plant is about (100, 50, 33, 10) percent.

NUCLEAR ENERGY

Few things have captured the awe and imagination of mankind to quite the extent that atomic energy has in the past half-century. Vast amounts of energy are released in splitting heavy atomic nuclei, the ***fission process,*** and the same is also true in combining small atomic nuclei to make heavier ones, the ***fusion process.*** Consider the energy contrast between an oxidation of a fossil fuel and a nuclear fusion reaction. When one mole (6.02×10^{23} molecules, or 16 grams) of methane from natural gas is burned, over 200 kilocalories of heat are liberated:

$$CH_4 + 2O_2 \longrightarrow CO_2 + 2H_2O + 211 \text{ kilocalories (kcal)}$$

Energy changes associated with nuclear events may be many times larger than those associated with chemical events.

In contrast, a lithium nucleus can be made to react with a hydrogen nucleus to form two helium nuclei in a nuclear reaction. The energy released per mole of lithium in this reaction is 23,000,000 kilocalories. This means that 7 grams of lithium and 1 gram of hydrogen produce 100,000 times more energy through fusion of nuclei than 16 grams of methane and 32 grams of oxygen produce by exchanging electrons.

$${}^{7}_{3}Li + {}^{1}_{1}H \longrightarrow 2{}^{4}_{2}He + 23{,}000{,}000 \text{ kcal}$$

Realizing that nuclear changes could involve giant amounts of energy relative to chemical changes for a given amount of matter, Otto

Hahn, Fritz Strassman, Lise Meitner, and Otto Frisch discovered in 1938 that $^{235}_{92}U$ is fissionable. Subsequently the dream of controlled nuclear energy became a reality, followed by the bomb and nuclear power plants. In the 1950's it was hoped that nuclear energy would soon relieve the shortage of fossil fuels. To date this has not been accomplished, although the production of nuclear energy has grown very rapidly in recent years.

Fission Reactions

Fission is the breakup of heavy nuclei.

Fission can occur when a thermal neutron (with a kinetic energy about the same as that of gaseous molecules at ordinary temperatures) enters certain heavy nuclei with an odd number of neutrons ($^{235}_{92}U$, $^{233}_{92}U$, $^{239}_{94}Pu$). The splitting of the heavy nucleus produces two smaller nuclei, two or more neutrons (an average of 2.5 neutrons for $^{235}_{92}U$), and much energy. Typical nuclear fission reactions are written as follows:

$$^{235}_{92}U + ^{1}_{0}n \longrightarrow ^{141}_{56}Ba + ^{92}_{36}Kr + 3^{1}_{0}n + \text{Energy}$$

$$^{235}_{92}U + ^{1}_{0}n \longrightarrow ^{103}_{42}Mo + ^{131}_{50}Sn + 2^{1}_{0}n + \text{Energy}$$

Note that the same nucleus may split in more than one way. The fission products, such as $^{141}_{56}Ba$ and $^{92}_{36}Kr$, emit beta particles and gamma rays until stable isotopes are reached.

A low-energy neutron will disrupt some large nuclei.

The neutrons emitted can cause the fission of other heavy atoms. For example, the three neutrons emitted in the first reaction above could produce fission in three more uranium atoms, the nine neutrons emitted could produce nine more fissions, the 27 neutrons from these fissions could produce 81 neutrons, the 81 neutrons could produce 243, the 243 neutrons could produce 729, and so on. This process is called a ***"chain reaction"*** (Figure 25-10) and it occurs at a maximum rate when the uranium sample is large enough for most of the neutrons emitted to be captured by a nucleus before passing out of the sample. Sufficient sample in a certain volume to sustain a chain reaction is termed the ***critical mass.***

In the atomic bomb the critical mass is kept separated into several smaller subcritical masses until detonation, at which time the masses are

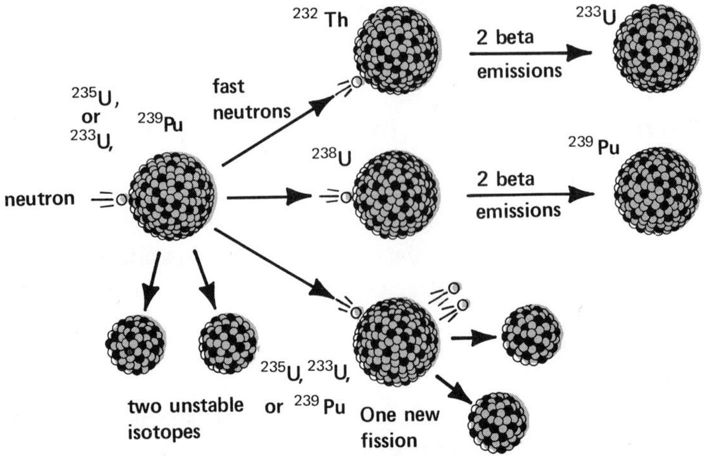

Figure 25–10 A chain reaction. A thermal neutron collides with a fissionable nucleus and the resulting reaction produces three additional neutrons. These neutrons can either convert nonfissionable nuclei, such as $^{232}_{90}Th$, to fissionable ones or cause additional fission reactions. If enough fissionable nuclei are present, a chain reaction will be sustained.

driven together by an implosive device. It is then that the tremendous energy is liberated and everything in the immediate vicinity is heated to temperatures of 5 to 10 million degrees. The sudden expansion of hot gases literally explodes everything nearby and scatters the radioactive fission fragments over a wide area. In addition to the movement of gases, there is the tremendous vaporizing heat that makes the atomic bomb so very devastating.

Separation of uranium isotopes had to preceed the control of atomic energy.

There is no danger of an atomic explosion in the uranium mineral deposits in the earth for two reasons. First, uranium is not found pure in nature—it is found only in compounds which in turn are mixed with other compounds. Second, less than 1 per cent of the uranium found in nature is fissionable $^{235}_{92}U$. The other 99 percent is $^{238}_{92}U$ which is not fissionable by thermal neutrons. In order to make nuclear bombs or nuclear fuel for electrical generation, a purification enrichment process must be carried out on the uranium isotopes, thus increasing the relative amount of $^{235}_{92}U$ atoms in a sample. Ordinary isotopic uranium such as found in ores is only 0.711 percent $^{235}_{92}U$.

Mass Defect—The Ultimate Nuclear Energy Source

What is the source of the tremendous energy of the fission process? It ultimately comes from mass being converted into energy, according to Einstein's famous equation, $E = mc^2$, where E is energy that results from the loss of an amount of mass, m, and c^2 is the speed of light squared. If separate neutrons, electrons, and protons are combined to form any particular atom, there is a loss of mass called the ***mass defect.*** For example, the calculated mass of one $^{4}_{2}He$ atom from the masses of the constituent particles is 4.032982 amu:

$$2 \times 1.007826 = 2.015652 \text{ amu, mass of two protons and two electrons}$$
$$2 \times 1.008665 = \underline{2.017330} \text{ amu, mass of two neutrons}$$
$$\text{Total} = 4.032982 \text{ amu, calculated mass of one } ^{4}_{2}He \text{ atom}$$

The mass that is lost leaves in the form of energy; $E = mc^2$.

Since the measured mass of a $^{4}_{2}He$ atom is 4.002604 amu, the mass defect is 0.030378 amu:

$$\begin{array}{r} 4.032982 \text{ amu} \\ - \underline{4.002604} \text{ amu} \\ 0.030378 \text{ amu, mass defect} \end{array}$$

Since the atom is more stable than the separated neutrons, protons, and electrons, the atom is in a lower energy state. Hence, the 0.030378 amu lost per atom would be released in the form of energy if the $^{4}_{2}He$ atom were made from separate protons, electrons, and neutrons. The energy equivalent of the mass defect is called the ***binding energy.*** The binding energy is analogous to the earlier concept of bond energy in that both are a measure of the energy necessary to separate the package (nucleus or molecule) into its parts.

Separated nuclear particles have more mass than when combined in a nucleus.

Atoms with atomic numbers between 30 and 63 have a greater mass defect per nuclear particle than very light elements or very heavy ones, as shown in Figure 25–11. This means the most stable nuclei are found in the atomic number range from 30 to 63.

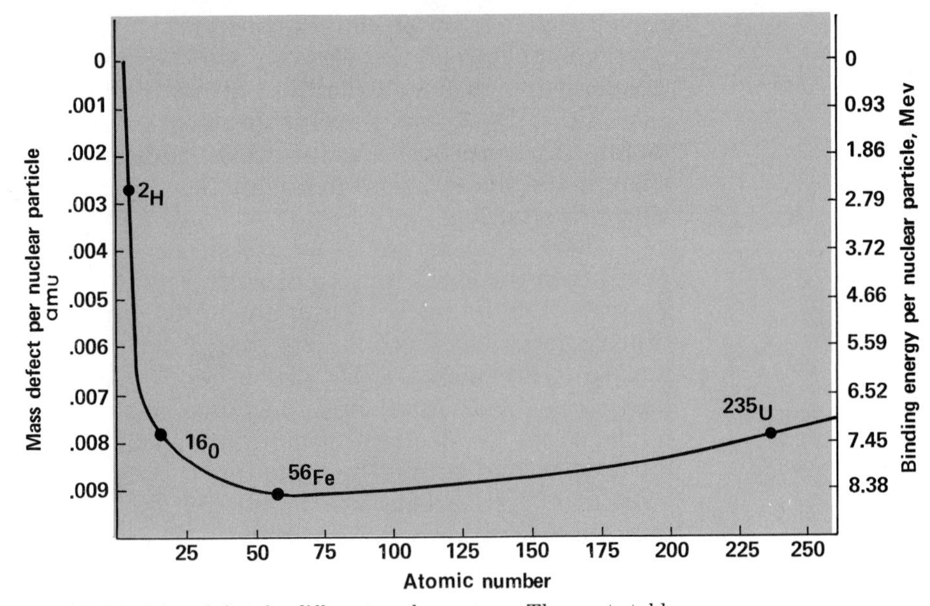

Figure 25–11 Mass defect for different nuclear masses. The most stable nuclei center around $^{56}_{26}Fe$, which has the largest mass defect per nuclear particle.

Because of the relative stabilities, it is in the intermediate range of atomic numbers that most of the products of nuclear fission are found. Therefore, when fission occurs and smaller, more stable nuclei result, these nuclei will contain less mass per nuclear particle. In the process, mass must be changed into energy. This energy gives the fission process its tremendous energy. It takes only about 1 kg of $^{235}_{92}U$ or $^{239}_{94}Pu$ undergoing fission to be equivalent to the energy released by 20,000 tons (20 kilotons) of ordinary explosives like TNT. The energy content in matter is further dramatized when it is realized that the atomic fragments from the 1 kg of nuclear fuel weigh 999 g, so only one tenth of one percent of the mass is actually converted to energy. The fission bombs dropped on Japan during World War II contained approximately this much fissionable material.

Intermediate-sized nuclei tend to have the greatest nuclear stability.

Controlled Nuclear Energy

The fission of a $^{235}_{92}U$ nucleus by a slow-moving neutron to produce smaller nuclei, extra neutrons, and large amounts of energy suggested to Enrico Fermi and others that the reaction could proceed at a moderate rate if the number of neutrons could be controlled. If a neutron control could be found, the concentration of neutrons could be maintained at a level sufficient to keep the fission process going but not high enough to allow an uncontrolled explosion. It would then be possible to drain the heat away from such a reactor on a continuing basis to do useful work. In 1942, Fermi, working at the University of Chicago, was successful in building the first atomic reactor, called an ***"atomic pile."***

An atomic reactor has a number of essential components. The charge material (fuel) must be fissionable or contain significant concentrations of a fissionable isotope such as $^{235}_{92}U$, $^{239}_{94}Pu$, or $^{233}_{92}U$. Ordinary uranium, which is mostly the nonfissionable $^{238}_{92}U$, can be used since it has a small con-

Atomic pile:
(1) carefully diluted fissionable material;
(2) moderator to control fission reaction;
(3) coolant to control heat;
(4) shielding to limit radiation.
The first one was piled together at the University of Chicago in 1942.

centration of the $^{235}_{92}U$ isotope. A moderator is required to slow the speed of the neutrons produced in the reactions without absorbing them. Graphite, water, and other substances have been used successfully as moderators. A substance that will absorb neutrons, such as cadmium or boron steel, is present in order to have a fine control over the neutron concentration. Shielding, to protect the workers from dangerous radiation, is an absolute necessity. Shielding tends to make reactors bulky installations. A heat-transfer fluid provides a large and even flow of heat away from the reaction center.

A schematic diagram of the X-10 graphite pile in Oak Ridge, Tennessee, is illustrated in Figure 25-12. Here the fuel is packed in openings in the graphite pile. Unlike the atomic bomb, the reaction cannot lead to an atomic explosion because less than the critical mass is assembled per unit volume. Even though the material of which the pile is made may absorb neutrons, the control rods offer a final and sensitive control to stop the fission reactions. With the control rods properly placed, the reaction can be maintained at various safe levels of activity.

Once the heat is produced in a nuclear reactor and safety measures are employed to protect against radiation, conventional technology allows this energy to be used to generate electricity, to power ships, or to operate any device that uses heat energy. A system for the nuclear production of electricity is illustrated in Figure 25-13.

What are the fuel requirements in nuclear fission energy production? In a typical fission event such as

When one gram of ^{235}U undergoes fission, it provides the same energy as burning about six tons of coal.

$$^{1}_{0}n + ^{235}_{92}U \longrightarrow ^{93}_{37}Rb + ^{141}_{55}Cs + 2^{1}_{0}n + 200 \text{ Mev},$$

the energy release, 200 Mev, is equivalent to 7.7×10^{-12} calories per atom

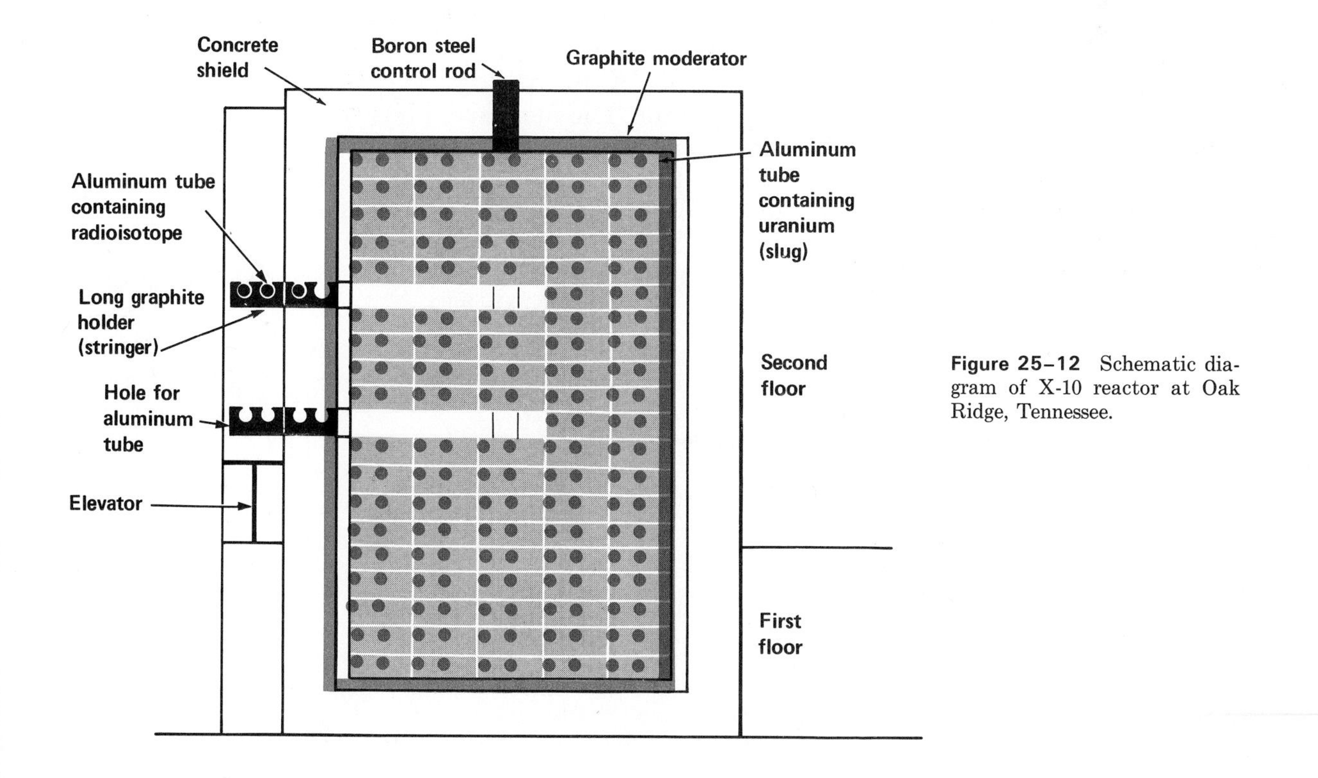

Figure 25-12 Schematic diagram of X-10 reactor at Oak Ridge, Tennessee.

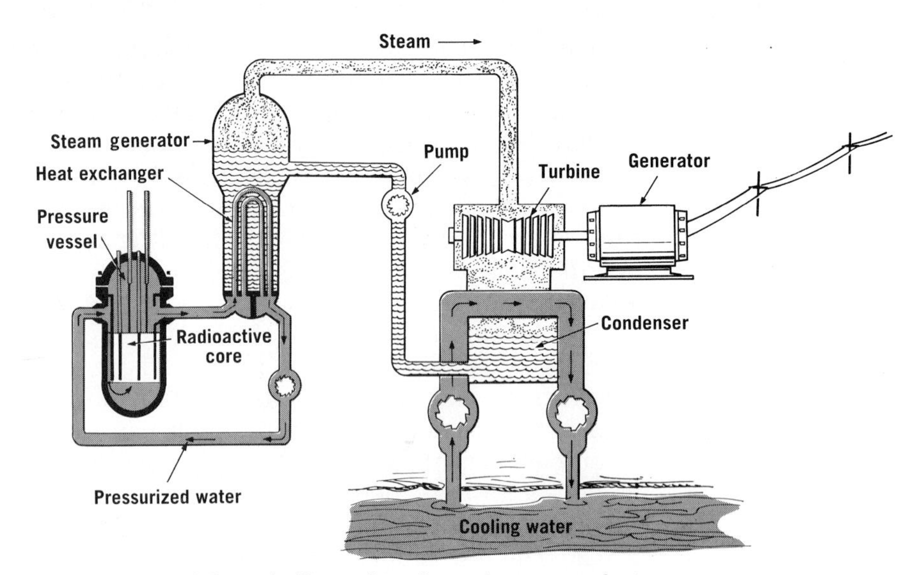

Figure 25–13 Schematic illustration of a nuclear power plant.

of $^{235}_{92}U$, or 4.64×10^9 kcal/mole. Since 1 gram of pure $^{235}_{92}U$ contains 2.56×10^{21} atoms, the total energy release for 1 gram of uranium-235 undergoing fission would be

$$2.56 \times 10^{21} \frac{\text{atoms}}{1\text{ g}} \times 7.7 \times 10^{-12} \frac{\text{cal}}{\text{atom}}$$

$$= 2.0 \times 10^{10} \frac{\text{cal}}{\text{g}}$$

This is the amount of energy which would be released if 5.95 tons of coal were burned, or if 13.7 barrels of oil were burned to produce heat to power a boiler. This means that about 3 kilograms of $^{235}_{92}U$ fuel per day would be required for a 1,000-megawatt electric generator. The fuel used, however, is not pure $^{235}_{92}U$, but ***enriched*** uranium containing up to 3 percent $^{235}_{92}U$. Considering the amount of $^{235}_{92}U$ needed in the future to meet the electrical demands of this nation (Table 25-3), $^{235}_{92}U$ will be in short supply in a few decades.

$^{235}_{92}U$ is in very short supply.

At $8 per ton, there appears to be enough uranium ore available to

TABLE 25–3 United States Uranium Requirements*

Year	Electric Generating Capacity in Megawatts (multiply by 10^3)		Uranium Fuel Consumed, Tons
	Total	*Nuclear*	
1970	300	6	—
1980	523	145	200,000
2000	1,550	735	1,600,000

* If breeder reactions are not used.

meet the electric generating needs until the last decade of this century; by then, however, the cheap ore will have run out and dramatic increases in electric rates will be necessary, brought on by fuel costs alone.

It is possible to convert the nonfissionable $^{238}_{92}U$ and $^{232}_{90}Th$ into fissionable fuels by using a ***breeder reactor.*** In such a reactor, a blanket of nonfissionable material is placed outside the fissioning $^{235}_{92}U$ fuel (Figure 25-14), which serves as the source of neutrons in the breeder reactions. The two breeder reaction sequences are the following:

$$^{238}_{92}U + ^{1}_{0}n \longrightarrow ^{239}_{92}U \xrightarrow{\beta} ^{239}_{93}Np \xrightarrow{\beta} ^{239}_{94}Pu$$

$$^{232}_{90}Th + ^{1}_{0}n \longrightarrow ^{233}_{90}Th \xrightarrow{\beta} ^{233}_{91}Pa \xrightarrow{\beta} ^{233}_{92}U$$

The products of the breeder reactions, $^{233}_{92}U$ and $^{239}_{94}Pu$, are both fissionable and neither are found in the earth's crust.

$^{239}_{94}Pu$ is toxic from a radiation as well as from a chemical point of view.

Breeder reactors present many technological problems, not the least of which is the potential of a disaster caused by mishandling of the $^{239}_{94}Pu$ isotope, which is extremely toxic and can also be fabricated into a fission bomb. Nevertheless, the expected benefit from the breeder program is massive amounts of energy. For example, if all the uranium used for electrical generation were used in breeder reactors, instead of running out of uranium fuel in several decades, the breeder fuels would supply the United States' electrical requirements for about 2600 years, assuming 1970 electricity-use levels were maintained, something euphemistically called ***Zero Energy Growth!*** If a breeder program cannot start in earnest soon enough, then the $^{235}_{92}U$ isotope will be used up only to generate electricity and another, as yet undiscovered, source of neutrons will be needed to convert the $^{238}_{92}U$ and $^{232}_{90}Th$ into fissionable fuels.

Radioactive isotopes are produced in all nuclear fission reactors. The radiation levels produced by these isotopes are dangerous to life and some of the half-lives involved extend into millions of years. We are, then, in the production of nuclear energy, building up large amounts of very dangerous materials. Serious objections have been raised to every suggested method of radioactive waste disposal. The radioactive waste is presently being held in underground depots and surface tanks. It is

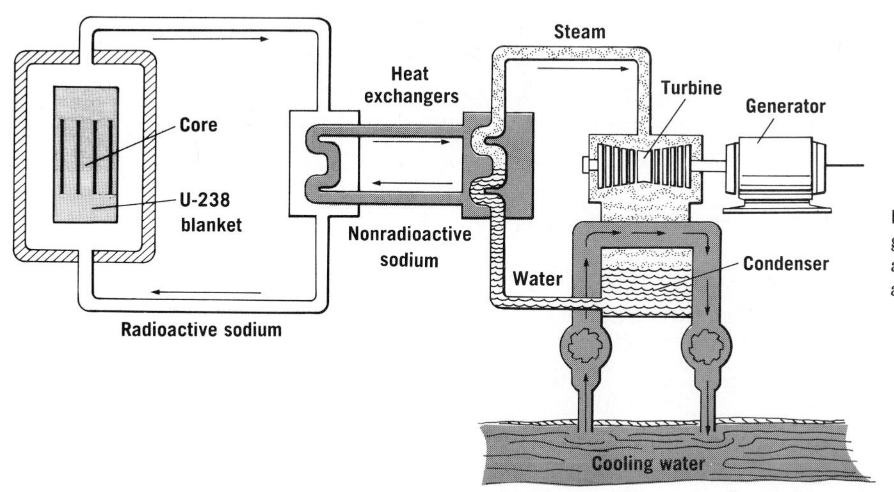

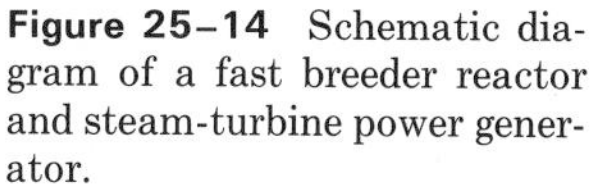
Figure 25–14 Schematic diagram of a fast breeder reactor and steam-turbine power generator.

argued by some that the exploitation of fission energy should await adequate methods for the disposal of these radioactive wastes.

Fusion Reactions

Fusion is the combination of very light nuclei.

When very light nuclei, such as H, He, and Li, are combined or ***fused*** to form an element of higher atomic number, energy must be given off consistent with the greater stability of the elements in this intermediate atomic number range (Figure 25–11). This energy, which comes from a decrease in mass, is the source of the energy released by the sun and by hydrogen bombs. Typical examples of fusion reactions are the following:

$$4\,{}^{1}_{1}H \longrightarrow {}^{4}_{2}He + 2\,{}^{0}_{+1}e + \text{Energy}$$

$${}^{2}_{1}H + {}^{2}_{1}H \longrightarrow {}^{3}_{2}He + {}^{1}_{0}n + 3.2 \text{ Mev}$$

$${}^{2}_{1}H + {}^{2}_{1}H \longrightarrow {}^{3}_{1}H + {}^{1}_{1}H + 4.0 \text{ Mev}$$

$${}^{3}_{1}H + {}^{2}_{1}H \longrightarrow {}^{4}_{2}He + {}^{1}_{0}n + 17.6 \text{ Mev}$$

${}^{2}_{1}H$ = Deuterium
${}^{3}_{1}H$ = Tritium
${}^{0}_{+1}e$ = Positron

The net reaction for the last three reactions given here is:

$$5\,{}^{2}_{1}H \longrightarrow {}^{4}_{2}He + {}^{3}_{2}He + {}^{1}_{1}H + 2\,{}^{1}_{0}n + 24.8 \text{ Mev}$$

or, for every five ${}^{2}_{1}H$ fused, 24.8 Mev of energy are released.

Materials for fusion reactions are available in enormous amounts.

Deuterium is a relatively abundant isotope—out of 6500 atoms of hydrogen in sea water, for example, one is a deuterium atom. What this means is that the oceans are a potential source of fantastic amounts of deuterium. There are 1.03×10^{22} atoms of deuterium in a single liter of seawater. In a single cubic kilometer of sea water, therefore, there would be enough deuterium atoms with enough potential energy to equal the burning of 1360 billion barrels of crude oil, and this is approximately the total amount of oil originally present in this planet.

High temperature and pressure are required for a fusion reaction to occur.

Fusion reactions take place rapidly only when the temperature is of the order of 100 million degrees or more. At these high temperatures atoms do not exist as such; instead, there is a ***plasma*** of nuclei and of electrons. In this plasma nuclei merge or combine. In order to achieve the high temperatures required for the fusion reaction of the hydrogen bomb, a fission bomb (atomic bomb) is first set off.

One type of hydrogen bomb depends on the production of tritium (${}^{3}_{1}H$) in the bomb. In this type lithium deuteride (${}^{6}_{3}Li{}^{2}_{1}H$, a solid salt) is placed around an ordinary ${}^{235}_{92}U$ or ${}^{239}_{94}Pu$ fission bomb. The fission is set off in the usual way. ${}^{6}_{3}Li$ absorbs some of the neutrons produced and splits into tritium, ${}^{3}_{1}H$, helium, ${}^{4}_{2}He$, and deuterium, ${}^{2}_{1}H$:

$${}^{6}_{3}Li + {}^{2}_{1}H + {}^{1}_{0}n \longrightarrow {}^{3}_{1}H + {}^{4}_{2}He + {}^{2}_{1}H$$

The temperature reached by the fission of ${}^{235}_{92}U$ or ${}^{239}_{94}Pu$ is sufficiently high to bring about the fusion of tritium and deuterium:

$${}^{3}_{1}H + {}^{2}_{1}H \longrightarrow {}^{4}_{2}He + {}^{1}_{0}n + 17.6 \text{ Mev}$$

A 20-megaton bomb usually contains about 300 lbs of lithium deuteride, as well as a considerable amount of plutonium and uranium.

Controlled Fusion

Containment is one of the biggest of the problems in developing controlled fusion.

There are three critical requirements for controlled fusion. First, the temperature must be high enough for ignition to occur. For the deuterium-tritium combination given earlier, a temperature of about 10^8 to 10^9 degrees is needed. Second, the plasma must be confined for a long enough time to release a net output of energy. Third, the energy must be recoverable in some usable form.

As yet the fusion reactions have not been "controlled." No physical container can contain the plasma without cooling it below the critical fusion temperature. Magnetic "bottles," enclosures in space bounded by a magnetic field, have confined the plasma, but not for long enough periods

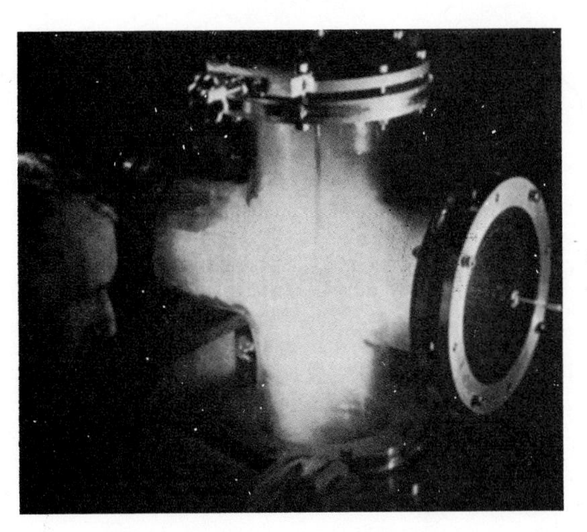

Figure 25–15 A controlled fusion experiment. (From *Chemical and Engineering News,* August 30, 1971, p. 5.)

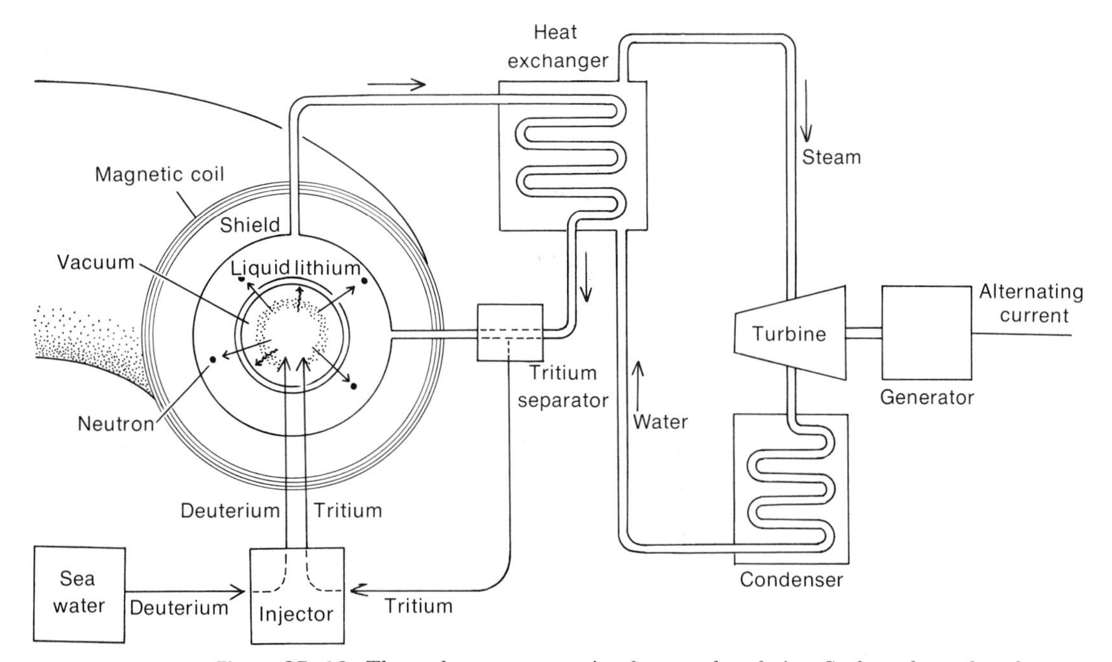

Figure 25–16 Thermal energy conversion from nuclear fusion. Such a scheme, based on the deuterium-tritium fuel cycle, relies on the energy of highly energetic neutrons. A liquid lithium shield absorbs these neutrons and is heated. The lithium then exchanges this heat with water to generate steam.

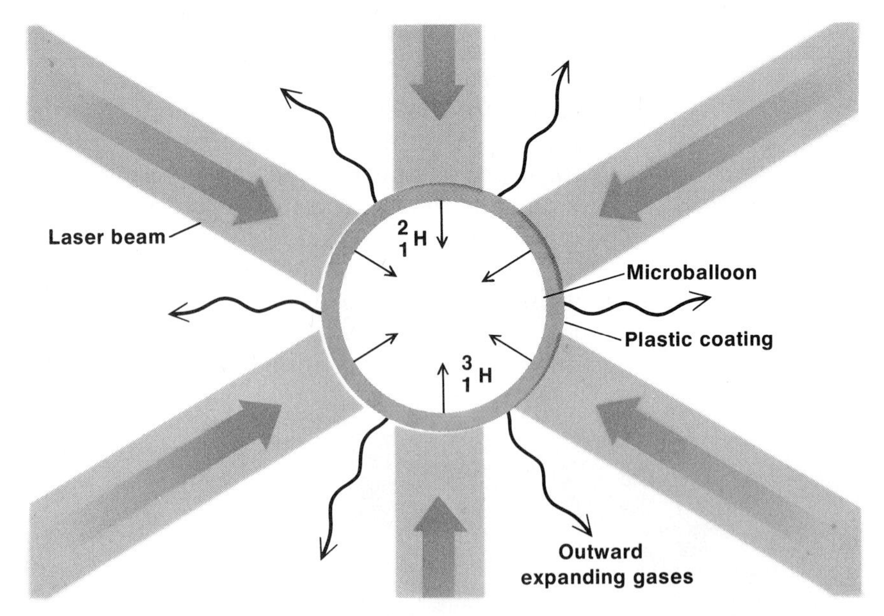

Figure 25-17 Focused laser light strikes the microballoon filled with deuterium and tritium, causing a plastic outer layer to burn off (or *ablate*). The outwardly expanding gases from the plastic material drive the glass sphere and its fuel contents inward. The high density and high temperatures produced result in fusion.

of time (Figure 25-15). Recent developments suggest that these "bottles" may soon hold the plasma long enough so that the fusion reaction can occur.

Thermal energy conversion (Figure 25-16) could be used to take the power from a deuterium-tritium fueled fusion reaction. Liquid lithium would be used to absorb kinetic energy of fast-moving neutrons and then exchange this heat with water to drive a steam turbine, thus producing electricity. This system is actually a breeder reactor as some of the lithium is converted into fuel—tritium, ^3_1H—by neutron absorption.

Controlled fusion might end many of the world's energy problems.

A newer confinement method is based on a laser system which simultaneously strikes tiny hollow glass spheres called ***microballoons*** which enclose the fuel, consisting of equal parts of deuterium and tritium gas, at high pressures (Figure 25-17).

There is hope that controlled fusion will probably be demonstrated by 1980 but it appears that it will only be after the turn of the century that fusion will furnish much of the world's energy needs.

Controlled fusion energy should result in a rather limited production of dangerous radioactivity. The lighter elements involved can be radioactive enough to be a serious hazard, but only for a short period of time; the half-lives of these isotopes are short. Storage and then return to the environment would be quite satisfactory.

SOLAR ENERGY

Solar energy is transmitted nuclear energy.

Earth's ultimate source of energy is the sun. The sun provides energy for photosynthesis, which gives us food to power our bodies, wood to burn, and the starting materials for coal and petroleum. In addition the sun provides the energy for the water cycle, which is the cause of our weather.

Although Earth receives only about three ten-millionths (0.0000003) of the total energy emitted from the sun, this amount of energy is enormous—about 2×10^{15} kcal/min, or 2.0 cal/cm^2/min. Due to reradiation from the atmosphere, and the absorption and scattering of radiant

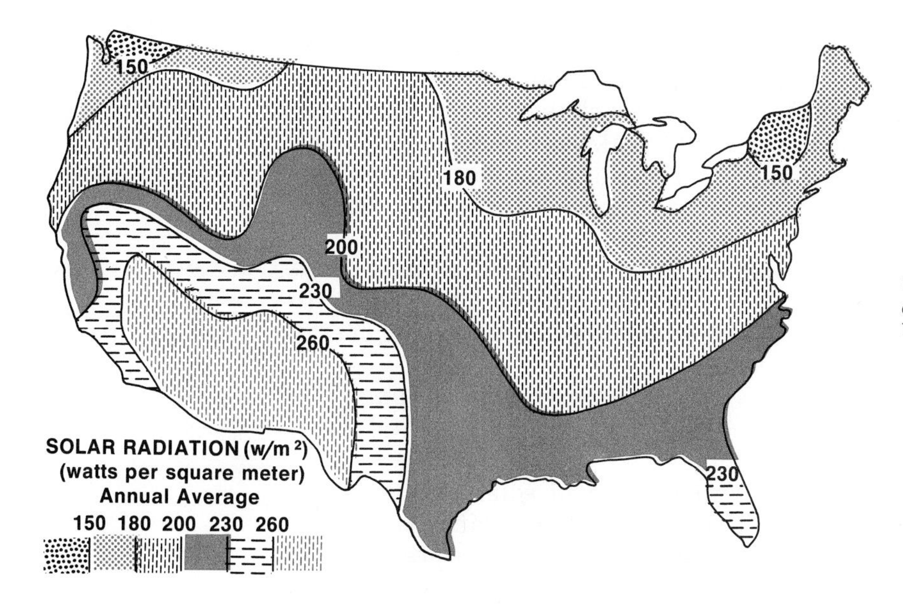

Figure 25–18 Annual rate of delivery of solar radiation in the United States.

energy by molecules in the lower portions of the atmosphere, the amount of radiation actually reaching the surface of this planet is about 1 cal/cm^2/min. The actual value depends on location, season, and weather conditions. Even this is a large amount of energy. For example, the roof of an averaged-sized house will receive about 10^8 calories/day when the radiation level is 1 cal/cm^2/min. This is equivalent to the heat energy derived from burning about 150 pounds of coal per day. This is also equal to 116 kilowatt hours per day of electrical energy—more than enough to heat an average American home on a winter day.

Only about 1 percent of the solar radiation used by plants in photosynthesis ends up as stored chemical energy such as foodstuffs. After fossil fuels are depleted, could photosynthesis be considered as a viable source of energy outside of food materials? Much of the energy used in photosynthesis is for "fixing" carbon in the form of cellulose and carbohydrates. Recently it has been shown that certain blue-green algae, *Anabeana cylindrica,* can convert sunlight and water into hydrogen and oxygen. A colony of such algae, coupled with a fuel cell utilizing hydrogen and oxygen, could be a source of electricity during sunlight hours (Figure 25–19).

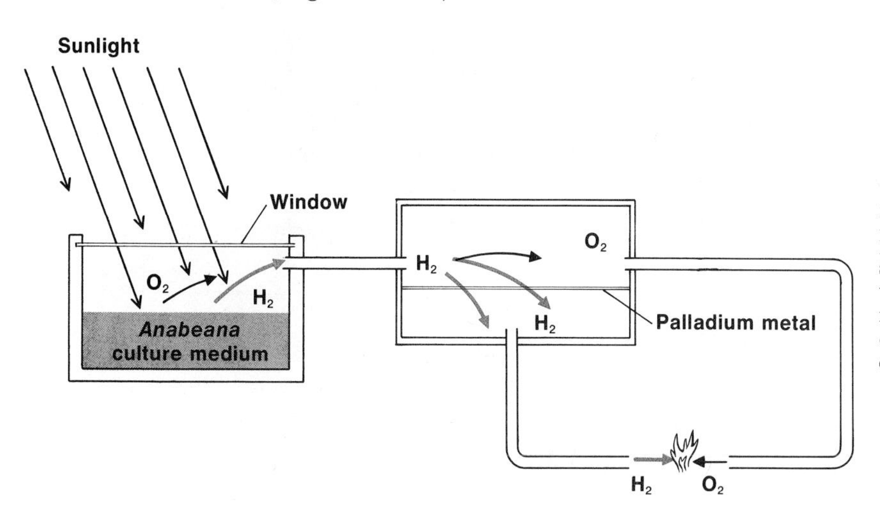

Figure 25–19 Schematic diagram of an electricity-producing photosynthesis process. H_2 and O_2 produced by the *Anabeana* are separated by palladium metal which is permeable toward H_2 but not O_2. The H_2 and O_2 are then combined in the fuel cell to produce electricity.

The attractiveness of solar energy is very great. For example, the world's present energy requirements per year are about 10^{18} kcal. An area of desert of about 28,000 square miles with little cloud cover or dust, near the equator, would receive about 1×10^{17} kcal/year of solar radiation. Such deserts exist in northern Chile, and could, if the need were great enough to justify the costs, supply a large portion of the world's energy needs.

The solar energy could heat water to steam, which in turn could generate electricity (Figure 25-20), which could electrolyze water to hydrogen and oxygen. The hydrogen could be piped to where the energy is needed and then converted to electricity.

Hydrogen can be burned in most devices which now burn natural gas.

Such an arrangement would give rise to a ***hydrogen economy,*** one which has many advantages over present energy sources such as fossil fuels.

Another approach to the direct utilization of solar energy is the ***solar battery,*** known as a photovoltaic device. The solar battery converts energy from the sun into electron flow. Solar batteries are about 13 to 14 percent efficient and are capable of generating electrical power from

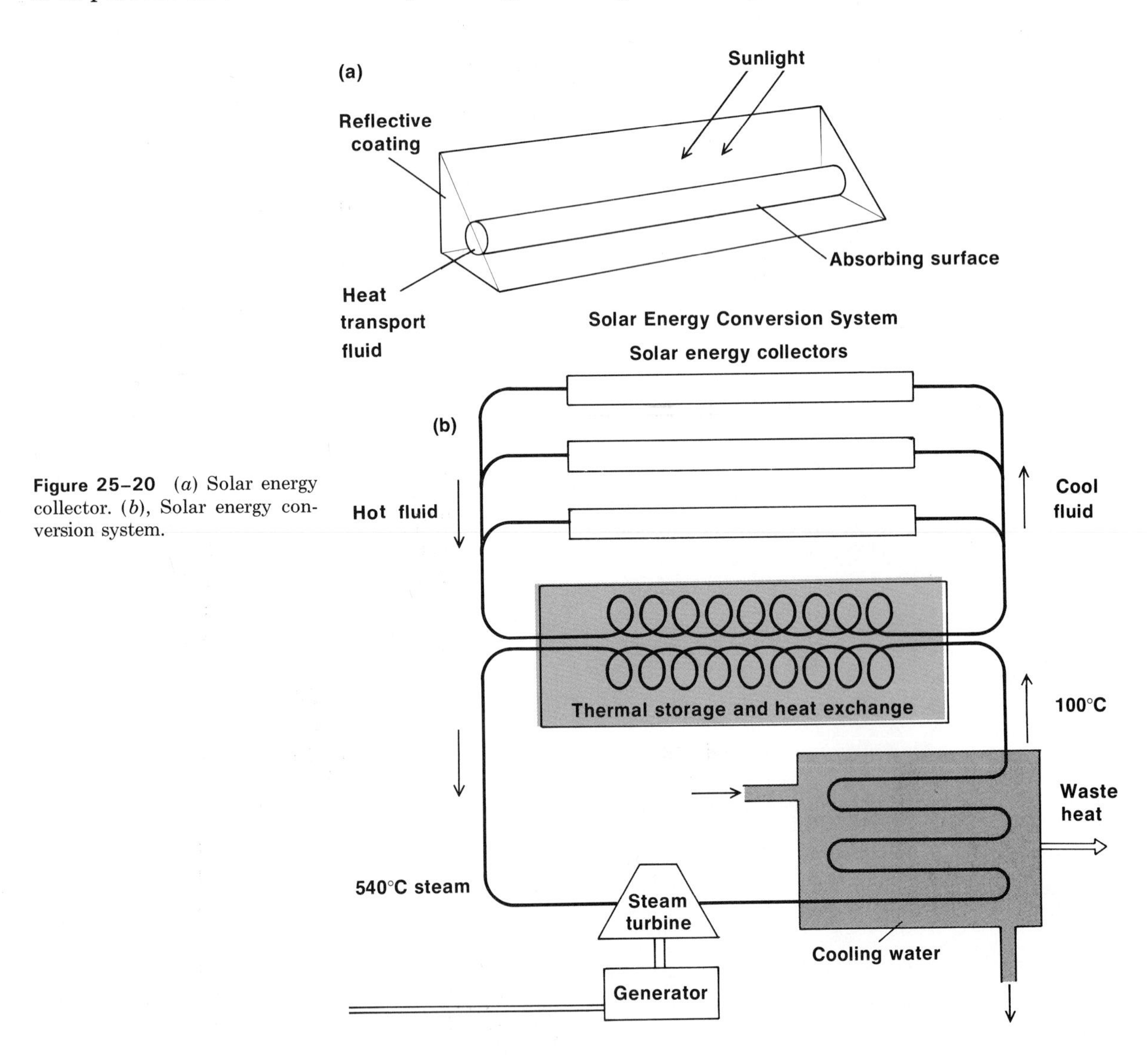

Figure 25-20 (*a*) Solar energy collector. (*b*), Solar energy conversion system.

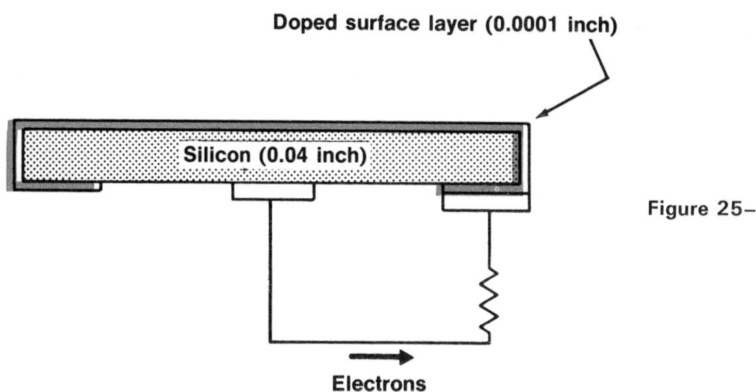

Figure 25–21 Silicon photodiode (Bell solar battery).

sunlight at the rate of at least 90 watts per square yard of illuminated surface. They are now used in space flight application and communication satellites, and in Israel, India, Pakistan, South Africa, and Azerbaijan SSR to obtain electric power.

The goal is to capture and use the solar energy "on the run" without upsetting the energy flow into (light side) and away from (dark side) Mother Earth.

One type of solar battery consists of two layers of almost pure silicon (Figure 25–21). The lower, thicker layer contains a trace of arsenic and the upper, thinner layer a trace of boron (Figure 25–22). Silicon has four valence electrons and forms a tetrahedral, diamond-like, crystalline structure. Each silicon atom is covalently bonded to four other silicon atoms. Arsenic has five valence electrons. When arsenic atoms are included in the silicon structure, only four of the five valence electrons of arsenic are used for bonding with four silicon atoms; one electron is relatively free to roam. Boron has three valence electrons. When boron atoms are included in the silicon structure, there is a deficiency of one electron around the boron atom; this creates "holes" in the boron-

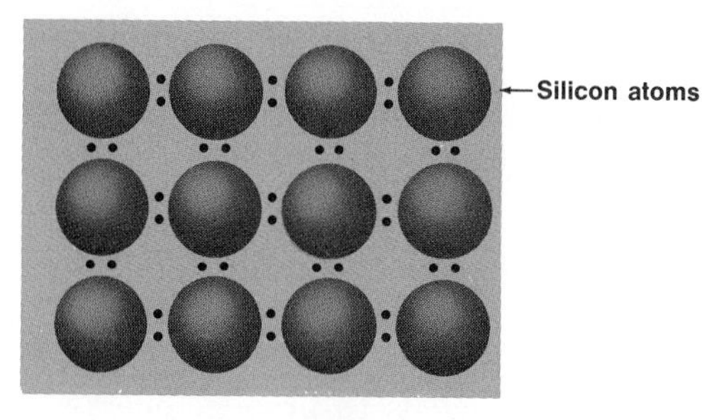

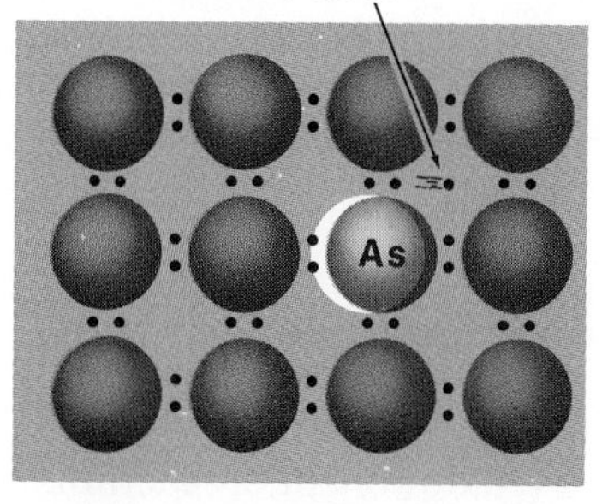

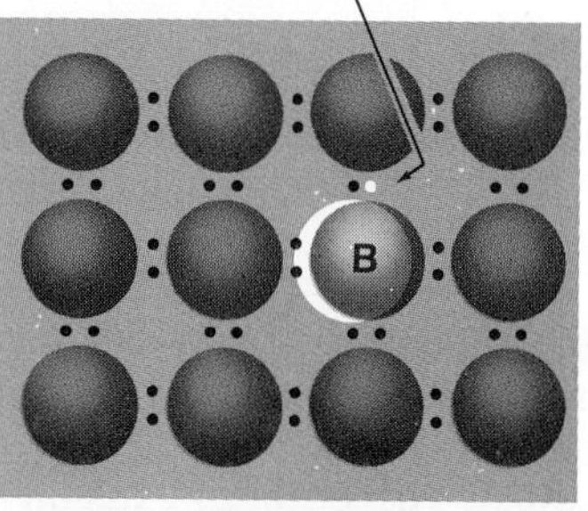

Figure 25–22 Schematic drawing of semiconductor crystal layers derived from silicon (after Masterton and Slowinski).

enriched layer. Even without sunlight, the "extra" electrons of arsenic diffuse into the holes in the boron. The driving force is the strong tendency to pair electrons. Externally applied potentials as high as 1000 volts are incapable of reversing the flow. The negative charge built up in the boron layer would hinder the flow of electrons into that layer and eventually stop the flow. The opposing factors—repulsion between free electrons and the drive to pair electrons—finally bring about an equilibrium.

When sunlight strikes the boron layer, the equilibrium is disturbed. If the wafer is connected as an ordinary battery would be, electrons flow from what appears to be the negative arsenic layer through the circuit and back into the boron layer (Figure 15–21). The fact that electrons enter the circuit from the arsenic layer can be explained if sunlight unpairs electrons in the boron layer, and the freed electrons are repelled to the arsenic layer and into the circuit. This pulls electrons into the boron layer to maintain the new balance between electron pairing and electron repulsion.

The advantage of the solar battery is that it has no moving parts, no liquids, and no corrosive chemicals—it just keeps on generating electricity indefinitely while exposed to sunlight. The drawbacks of the solar battery are the very large area required for large amounts of power, the high costs of the very pure materials (in 1974, costs of $6 to $25 per square foot were quoted), and the fact that they work only when the sun is shining. Since the first practical use of solar batteries in 1955 to power eight rural telephones in Georgia, they have undergone a great deal of development, and much more is expected because of their great potential use.

Solar batteries are still very expensive.

In theory, a number of huge energy sources are potentially available for our use. As yet we have not learned to control them sufficiently well to have cheap energy for all. However, progress is being made. It is evident that both chemical sources of energy and the fission of heavy nuclei are limited in their ability to supply effectively man's growing energy needs, and he must develop a wide variety of other energy sources.

SELF-TEST 25-B

1. The splitting of an unstable nucleus to produce energy is termed (fission /fusion).
2. A sufficiently large sample to sustain a chain reaction in a fission process is called the ______________ ______________.
3. Of the two major isotopes of uranium, which is fissionable, uranium-238 or uranium-235? ______________
4. When light nuclei combine to form heavy nuclei and energy, the process is (fission/fusion).
5. The major problem with obtaining fusion energy is (containment of reactants at high temperature/enough fuel/costs).

6. When and if fusion is used to produce energy in a breeder reaction, the fuel produced will be ________________.

7. When fission is used to produce energy in a breeder reaction the fuels produced will be ________________ and ________________.

MATCHING SET

______	1. user of 35 percent of world's energy	a.	1859
______	2. fossil fuels	b.	CO_2 and H_2O
______	3. combustion products of fossil fuels	c.	uranium-235 ($^{235}_{92}U$)
______	4. minable coal	d.	mass defect
______	5. synthesis gas	e.	plutonium-239 ($^{239}_{94}Pu$)
______	6. fissionable isotope	f.	United States
______	7. basis of nuclear energy	g.	$^{2}_{1}H$
______	8. product of a fission breeder reactor	h.	sea water
______	9. deuterium	i.	within 4000 feet of surface
______	10. date of petroleum discovery in U.S.	j.	microballoons
______	11. tritium	k.	coal, petroleum, natural gas
______	12. source of deuterium	l.	photosynthesis
______	13. used to confine fusion fuel	m.	CO and H_2
______	14. one use of solar radiation	n.	10 to 14 percent
______	15. approximate efficiency of a solar battery	o.	$^{3}_{1}H$

QUESTIONS

1. What is your attitude toward using up the fossil fuels within a few decades? Do we owe future generations a supply of these resources? Would you agree to give up air-conditioning, private cars, and power tools, to mention a few examples, and to limit heating and cooking if necessary to share these fuels with your grandchildren?

2. Which theoretically yields the greatest energy per mole?

   (a) the burning of gasoline
   (b) the fission of uranium-235
   (c) the fusion of iron-56

3. Which is the more efficient use of energy: burning coal in a house to heat it or heating the house electrically with energy produced in a coal-burning power plant?

4. Give three examples of systems which contain chemical energy that can be used as a source of heat energy.

5. Is the electrical energy where you live produced by burning fossil fuels? If not, what is the energy source? Are there pollution problems associated with the generation of the electrical power?

6. What produces the tremendous energy of a fission reaction?

7. What was the original source of energy which is tied up in fossil fuels?

8. Why is it difficult to fuse two $^{52}_{24}Cr$ nuclei? Explain.

9. What is meant by a chain reaction?

10. What major problem is associated with harnessing the energy from a fusion reaction?

11. Suggest several ways solar energy might be harnessed.

12. Name two other sources of energy not specifically mentioned in this chapter.

13. Explain how useful energy might be obtained from garbage.

14. Which is more fundamental—a supply of energy or a supply of food? Explain.

15. The energy consumption of the United States in 1970 was 2×10^{13} kilowatt hours. What is this amount of energy expressed in kilocalories? In Btu's?

16. Assume the world population to be 4.0 billion and calculate the earth's energy needs if everyone used as much energy as is used in the United States.

SUGGESTIONS FOR FURTHER READING

Angrist, S. W., and Hepler, L. G., "Order and Chaos," Basic Books, Inc., New York, 1967.
Ayres, E., and Scarlett, C. A., "Energy Sources—The Wealth of the World," McGraw-Hill Book Co., Inc., New York, 1952.
Beneman, J. R., and Weare, N. M., "Hydrogen Evolution by Nitrogen-Fixing *Anabeana cylindrica* Cultures," *Science,* Vol. 184, p. 174 (1974).
Chalmers, B., "Energy," Academic Press, New York, 1963.
Choppin, G. R., "Nuclear Fission," *Chemistry,* Vol. 40, No. 7, p. 25 (1967).
Daniels, F., "Direct Use of the Sun's Energy," Yale University Press, New Haven, 1964.
Eisenbud, Merril, "Environmental Radioactivity," 2nd ed., Academic Press, New York, 1973.
"Energy," a special issue of *Science,* Vol. 180, No. 4134 (1974).
"Energy and Power," *Scientific American,* Vol. 225, No. 3 (1971); (entire issue devoted to various facets of this important topic).
Goldsby, R. A., "Energy and Cells," MacMillan, New York, 1967.
Gregory, Dereck P., "The Hydrogen Economy," *Scientific American,* Vol. 228, No. 1, p. 13 (1973).
Hubbert, M. K., "The Energy Resources of the Earth," *Scientific American,* Vol. 225, No. 3, p. 60 (1971).
Mills, G. A., Johnson, H. R., and Perry, H., "Fuels Management in an Environmental Age," *Environmental Science and Technology,* Vol. 5, No. 1, p. 30 (1971).
"Nuclear Power and By-Product Plutonium," *Chemistry,* Vol. 40, No. 3, p. 7 (1967).
"Otto Hahn," *Chemistry,* Vol. 39, No. 12, p. 5 (1966).
Ross, F., "The World of Power and Energy, Lothrop, Lee and Shepard Co., New York, 1967.
Schur, S. H., "Energy," *Scientific American,* Vol. 209, No. 3, p. 111 (1963).
Shaheen, Baher I., "The Energy Crisis: Is It a Fabrication or a Miscalculation?," *Environmental Science and Technology* Vol. 8, No. 4, p. 316 (1974).
"Sun-Powered Furnace," *Chemistry,* Vol. 43, No. 11, p. 23 (1970).

APPENDIX A

THE INTERNATIONAL SYSTEM OF UNITS (SI)

Since 1960, a coherent system of units known as the Système International (SI system), bearing the authority of the International Bureau of Weights and Measures, has been in effect and is gaining acceptance among scientists. It is an extension of the metric system which began in 1790, with each physical quantity assigned an SI unit uniquely. An essential feature of both the older metric system and now the newer SI is a series of prefixes that indicate a power of ten multiple or submultiple of the unit.

I. Units of Length

The standard unit of length is the ***meter.*** It was originally meant to be one ten-millionth of the distance along a meridian from the North Pole to the equator. However, the lack of precise geographical information necessitated a better definition. For a number of years the meter was defined as the distance between two etched lines on a platinum-iridium

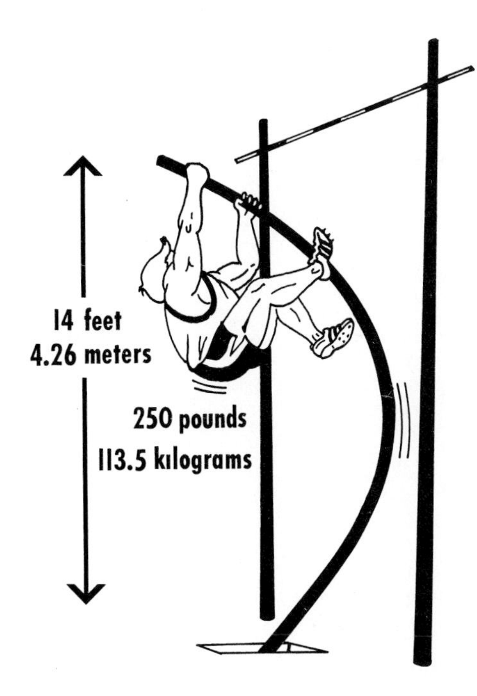

Figure A–1 The pole-vaulter is easily recognized as hefty when described by two hundred and fifty pounds and his jump something less than a record at fourteen feet. As Americans move closer to the use of the system of international measurements, the 113.5 kilograms and the 4.26 meters will produce similar conceptualizations related to previous experience.

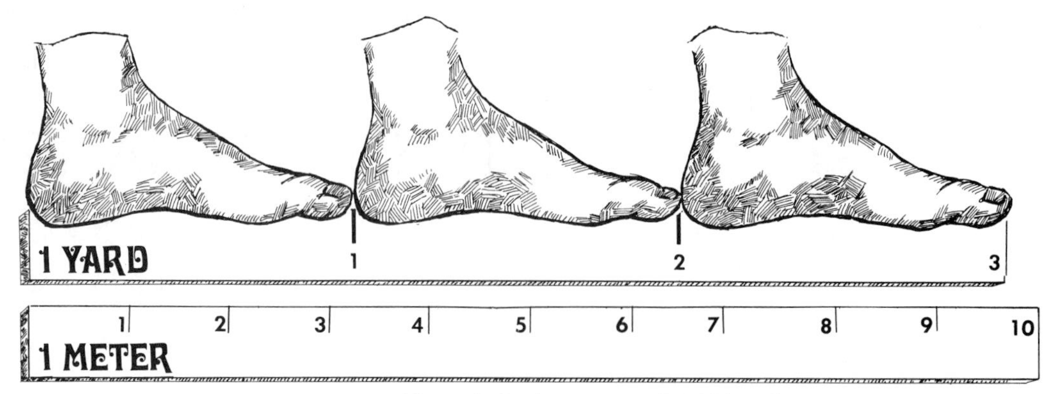

Figure A–2 A meter equals 1.094 yards.

bar kept at 0°C (32°F) in the International Bureau of Weights and Measures, at Sèvres, France. The inability to measure this distance as accurately as desired prompted a recent redefinition of the meter as being a length equal to 1,650,763.73 times the wavelength of the orange-red spectrographic line of ${}^{86}_{36}Kr$.

The meter (39.37 inches) is a convenient unit with which to measure the height of a basketball goal (3.05 meters), but it is unwieldy for measuring the parts of a watch or the distance between continents. For this reason, prefixes are defined in such a way that, when placed before the meter, they define distances convenient for man's purposes. Some of the prefixes with their meanings are:

nano—1/1,000,000,000 or 0.000000001
micro—1/1,000,000 or 0.000001
milli—1/1000 or 0.001
centi—1/100 or 0.01
deci—1/10 or 0.1
deca—10
hecto—100
kilo—1000
mega—1,000,000

The corresponding units of length with their abbreviations are the following:

nanometer (nm)—0.000000001 meter
micrometer (μm)—0.000001 meter
millimeter (mm)—0.001 meter
centimeter (cm)—0.01 meter
decimeter (dm)—0.1 meter
meter (m)—1 meter
decameter (dam)—10 meters
hectometer (hm)—100 meters
kilometer (km)—1000 meters
megameter (Mm)—1,000,000 meters

Since the prefixes are defined in terms of the decimal system the conversion from one metric length to another involves only shifting the decimal point. Mental calculations are quickly accomplished.

TABLE A–1 Conversion Factors*

Length:	1 inch (in)	=	2.54 centimeter (cm)
	1 yard (yd)	=	0.914 meter (m)
	1 mile (mi)	=	1.609 kilometers (km) (1,609 meters)
Volume:	1 ounce (oz)	=	29.57 milliliters (ml)
	1 quart (qt)	=	0.946 liter (l)
	1.06 quart (qt)	=	1 liter (l)
	1 gallon (gal)	=	3.78 liters (l)
Mass (weight)†:	1 ounce (oz)	=	28.35 grams (g)
	1 pound (lb)	=	453.6 grams (g)
	1 ton (tn)	=	907.2 kilograms (kg)

*Commonly used English units are used.

†Mass is a measure of the amount of matter, whereas weight is a measure of the attraction of the earth for an object at the earth's surface. The mass of a sample of matter is constant, but its weight varies with position and velocity. For example, the space traveler, having lost no mass, becomes weightless in earth orbit. Although mass and weight are basically different in meaning, they are often used interchangeably in the environment of the earth's surface.

How many centimeters are in a meter? Think: Since a centimeter is the one-hundredth part of a meter, there would be 100 centimeters in a meter.

Conversion of measurements from one system to the other is a common problem. Some commonly used English–SI equivalents (conversion factors) are given in Table A-1.

II. Units of Mass

The primary unit of mass is the ***kilogram*** (1000 grams). This unit is the mass of a platinum-iridium alloy sample deposited at the International Bureau of Weights and Measures. One pound contains a mass of 453.6 grams (a five-cent nickel coin contains about 5 grams).

Conveniently enough, the same prefixes defined in the discussion of length are used in units of mass, as well as in other units of measure.

III. Units of Volume

The SI unit of volume is the ***cubic meter*** (m^3). However, the volume capacity used most frequently in chemistry is the liter, which is defined as 1 cubic decimeter (1 dm^3). Since a decimeter is equal to 10 centimeters (cm), the cubic decimeter is equal to $(10\ cm)^3$ or 1000 cubic centimeters (cc). One cc, then, is equal to one milliliter (the thousandth part of a liter). The ml (or cc) is a common unit that is often used in the measurement of medicinal and laboratory quantities. There are then 1000 liters in a kiloliter or cubic meter.

IV. Units of Energy

The SI unit for energy is the ***joule*** (J), which is defined as the work performed by a force of one newton acting through a distance of one meter. A newton is defined as that force which produces an acceleration of one meter per second when applied to a mass of one kilogram. Conversion units for energy are:

$$1 \text{ calorie} = 4.184 \text{ joule}$$
$$1 \text{ kilowatt hour} = 3.5 \times 10^6 \text{ joule}$$

V. Other SI Units

Other SI units are listed below.

Time	second (s)
Temperature	Kelvin (K)
Electric current	ampere (A) = 1 coulomb per second
Amount of molecular substance	mole (mol) (6.023×10^{23} molecules)
Pressure	pascal (Pa) = newton per square meter
Power	watt (W) = 1 joule per second
Electric charge	coulomb (C) = 6.24196×10^{18} electronic charges = 1.036086×10^5 faradays

Further information on SI units can be obtained from "SI Metric Units—An Introduction," by H. F. R. Adams, McGraw-Hill Ryerson Ltd., Toronto, 1974.

APPENDIX B

TEMPERATURE SCALES

The system of measuring temperature which is used in scientific work is based on the ***centigrade*** or Celsius temperature scale. This temperature scale was defined by Anders Celsius, a Swedish astronomer, in 1742. The centigrade scale is based upon the expansion of a column of mercury which occurs when it is transferred from a cold standard temperature to a hot standard temperature. The cold standard temperature is the temperature of melting ice and is defined as 0°C. The hot standard temperature is the temperature at which water boils under standard conditions of pressure; it is defined as 100°C. The expansion of the mercury is assumed to be linear over this range, and the distance through which the mercury column expands between 0°C and 100°C is divided into 100 equal parts, each corresponding to one degree.

On the scale commonly used in the United States, the Fahrenheit scale (developed by and named after Gabriel Daniel Fahrenheit, a German-Dutch physicist of the early 18th century), the freezing mark is 32° and the boiling mark is 212°. Obviously, the ice is no hotter when a Fahrenheit thermometer is stuck into the system than when a centigrade thermometer is used. The marks of the tubes are simply different names for the same thing.

The ideal gas law predicts another temperature scale based on the temperature-volume behavior of a gas (Charles' law). Lord Kelvin reasoned that if an ideal gas lost $1/273$ of its volume during a temperature decrease from 0°C to −1°C, then at −273°C, theoretically at least, there should be no volume at all. All gases liquefy before this temperature is reached. However, this temperature, −273°C, appears to be a zero temperature defined by nature. Thus the Kelvin temperature scale was developed. Minus 273°C becomes 0 Kelvin and 0°C becomes 273 Kelvin. (More accurately, 273.15 K.) Note that current SI usage does not write °K, but K.

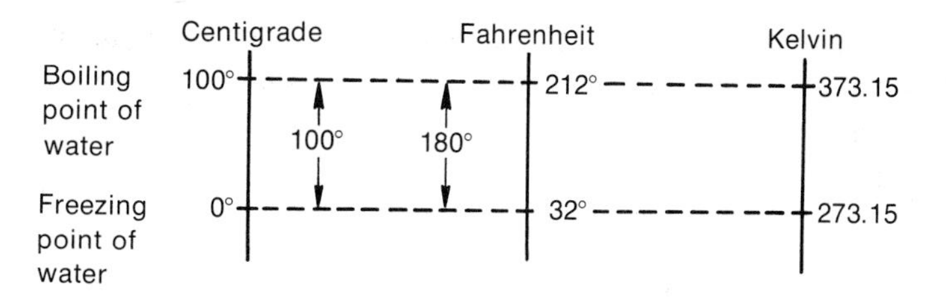

Because all these systems are now in common use, it is necessary at times to convert from one system to another.

Note that 100 centigrade degrees = 180 Fahrenheit degrees
or 1 C degree = 1.8 (or $^9/_5$) F degrees.

But if we wish to convert temperature from one scale to another, we must remember that 0°C is the same as 32°F. Therefore, 32 must be added to the calculated number of degrees Fahrenheit in order to revert to the start of the counting in the Fahrenheit system.

Therefore,

$$°F = {}^9/_5\,(°C) + 32$$

which can be arranged to

$$°C = {}^5/_9\,(°F - 32).$$

EXAMPLE 1

Convert 50°F to the corresponding centigrade temperature:

$$\begin{aligned} °C &= {}^5/_9\,(°F - 32) \\ &= {}^5/_9\,(50 - 32) \\ &= {}^5/_9\,(18) \\ &= 10°C \end{aligned}$$

EXAMPLE 2

Convert 25°C to the corresponding Fahrenheit temperature:

$$\begin{aligned} °F &= {}^9/_5\,(°C) + 32 \\ &= {}^9/_5\,(25) + 32 \\ &= 45 + 32 \\ &= 77°F \end{aligned}$$

To convert centigrade to Kelvin:

$$K = °C + 273$$

EXAMPLE 3

Convert 23°C to Kelvin:

$$\begin{aligned} K &= 23° + 273 \\ &= 296\ K \end{aligned}$$

Rather than memorizing formulas, many students prefer to figure out temperature conversions using common sense. A common-sense approach is shown in Table B–1.

TABLE B–1 Common-Sense Method for Converting Temperatures

Starting point: freezing point (f.p.) of water is 0°C = 32°F; boiling point (b.p.) of water is 100°C = 212°F

Other information needed: 100 C degrees = 180 F degrees

General Procedure	Example 1 (50°F = ?°C)	Example 2 (25°C = ?°F)	Example 3 (98.6°F = ?°C)
1. Select a reference point.	1. f.p. of water	1. f.p. of water	1. b.p. of water
2. Determine relationship of temperature of interest to reference point.	2. 50°F is 18 F degrees above f.p. of water	2. 25°C is 25 C degrees above f.p. of water	2. 98.6°F is 113.4 F degrees below b.p. of water
3. Convert the number of degrees.	3. 18 F degrees is $18\ F^\circ \times \frac{100\ C^\circ}{180\ F^\circ} =$ 10 C degrees	3. 25 C degrees is $25\ C^\circ \times \frac{180\ F^\circ}{100\ C^\circ} =$ 45 F degrees	3. 113.4 F degrees is $113.4\ F^\circ \times \frac{100\ C^\circ}{180\ F^\circ} =$ 63.0 C degrees
4. Express temperature, taking into account the selected reference point.	4. The temperature is 10 C degrees above the f.p. of water (which is 0°C), so the temperature is *10°C*	4. The temperature is 45 F degrees above the f.p. of water (which is 32°F), so the temperature is 45 + 32 = *77°F*	4. The temperature is 63.0 C degrees below the b.p. of water (which is 100.0°C), so the temperature is 100.0 − 63.0 = *37.0°C*

APPENDIX C

FACTOR-LABEL APPROACH TO CONVERSION PROBLEMS

For converting a measurement from one measurement system to another, the following method is straightforward and does not require the decision of whether to divide or multiply to obtain the proper answer. This method makes the decision for you, a decision which is sometimes difficult when dealing with new units with which you have had little experience.

EXAMPLE

How many liters are in 6 quarts?

Factor-Label Method of Solution:

1. Write down the unit to which you are converting. The question mark indicates the number of liters to be determined.

   1. ? liters =

2. On the right hand side of the equal sign, write down the quantity given. Write both the number and the name or label.

   2. ? liters = 6 quarts

3. Now look at the two units. Recall a conversion between these two units. These conversions must be learned or looked up.

   3. 1 liter = 1.06 quarts,
   1 liter per 1.06 quarts

   $$\text{or } \frac{1 \text{ liter}}{1.06 \text{ quarts}}$$

   $$\text{or } \frac{1.06 \text{ quarts}}{\text{liter}}$$

4. Write the conversion factor on the right hand side so that the unwanted units will cancel; that is, a unit in the numerator will cancel the same unit in the denominator, and only the unit you want will remain. Do not, of course, cancel the numbers, just the units.

   $$4.\ ? \text{ liters} = 6 \cancel{\text{quarts}} \times \frac{1 \text{ liter}}{1.06 \cancel{\text{quarts}}}$$

5. Do the indicated multiplication and division. The line, of course, means divided by. Check the units on both sides of the equation. The units should be the same and in the same position (numerator or denominator).

5. $? \text{ liters} = \dfrac{6}{1.06} \text{ liters}$
$= 5.66 \text{ liters}$

Now, suppose you needed to convert 3 pints to ml, and you do not know a conversion factor which will make the conversion in one step.

Proceed as before:

$$? \text{ ml} = 3 \text{ pints}$$

Note this is a volume conversion from the English system to the metric system as before. Recall the volume conversion factor which you know between these systems.

$$1 \text{ liter} = 1.06 \text{ quarts}$$

Convert in the system of the given quantity until you reach the unit in the conversion factor for this system. Then write down the conversion factor between systems so like units in separate factors will cancel.

$$? \text{ ml} = 3 \cancel{\text{ pints}} \times \frac{1 \cancel{\text{ quart}}}{2 \cancel{\text{ pints}}} \times \frac{1 \text{ liter}}{1.06 \cancel{\text{ quarts}}}$$

Cancel units and recall a conversion factor which converts the units you have left to the unit you want. Do the indicated multiplication and division.

$$\begin{aligned} ? \text{ ml} &= \frac{3 \times 1 \times \cancel{1 \text{ liter}}}{2 \times 1.06} \times \frac{1000 \text{ ml}}{\cancel{1 \text{ liter}}} \\ &= \frac{3 \times 1 \times 1 \times 1000}{2 \times 1.06 \times 1} \text{ ml} \\ &= 1420 \text{ ml} \end{aligned}$$

In brief, the method is very simple:

1. Write the units you want, an equals sign, and the quantity you have given.
2. Write conversion factors so the unwanted factors will cancel.
3. Keep in one system until you come to units in a familar intersystem conversion factor.
4. Do the indicated arithmetic and check the units.
5. Examine the size of the answer for reasonableness.

Note: This approach to problem solving has wide applicability to many other types of problems. It is especially useful in problems pertaining to weight relationships in chemical reactions. See Appendix D.

APPENDIX D

STOICHIOMETRIC CALCULATIONS

The bases for stoichiometric calculations were presented in Chapter 4. Problems of a more complex nature and a systematic approach to their solution are presented in the following examples. Finally, a list of exercise problems is given for further study.

EXAMPLE 1

Balanced Equations Express Particle Number Ratios

In the reaction of hydrogen with oxygen to form water, how many molecules of hydrogen are required to combine with 19 oxygen molecules?

Solution: A chemical equation can only be written for a reaction if the reactants and products are identified and the respective formulas determined. In this problem, the formulas are known and the unbalanced equation is:

$$H_2 + O_2 \longrightarrow H_2O.$$

It is evident that one molecule of oxygen contains enough oxygen for two water molecules and the equation, as written, does not account for what happens to the second oxygen atom. As it is, the equation is in conflict with the conservation of atoms in chemical changes. This conflict is easily corrected by balancing the equation:

$$2H_2 + O_2 \longrightarrow 2H_2O$$

Now, all atoms are accounted for in the equation and it is obvious that two hydrogen molecules are required for each oxygen molecule. In other words, two hydrogen molecules are equivalent to one oxygen molecule in their usage. This can be expressed as follows:

$$2 \text{ hydrogen molecules} \left\{\begin{matrix}\text{are}\\ \text{equivalent}\\ \text{to}\end{matrix}\right\} 1 \text{ oxygen molecule;}$$

or,

$$2H_2 \text{ molecules} \sim O_2 \text{ molecule;}$$

or,

$$\frac{2H_2 \text{ molecules}}{O_2 \text{ molecule}},$$

which can be read as two hydrogen molecules per one oxygen molecule.

Using now the factor-label approach developed in Appendix C, the solution is readily achieved.

$$? \ H_2 \text{ molecules} = 19 O_2 \text{ molecules}$$

$$? \ H_2 \text{ molecules} = 19 O_2 \cancel{\text{ molecules}} \times \frac{2H_2 \text{ molecules}}{O_2 \cancel{\text{ molecule}}}$$

$$= 38 H_2 \text{ molecules}$$

Note: The reader is likely to say at this point that the method is cumbersome and that he can quickly see the answer to be 38H_2 molecules without "the method." However, problems to follow are made much easier if a systematic method of approach is used.

EXAMPLE 2

Laboratory Mole Ratios Identical With Particle Number Ratios

How many moles of hydrogen molecules must be burned in oxygen (the reaction of example 1) to produce 15 moles of water molecules (about a glassful)?

Solution: The balanced equation,

$$2H_2 + O_2 \longrightarrow 2H_2O,$$

tells us that two molecules of hydrogen produce two molecules of water;

or,

$$2 \text{ molecules hydrogen} \sim 2 \text{ molecules water},$$

and therefore,

$$1 \text{ molecule hydrogen} \sim 1 \text{ molecule water}.$$

It is obvious then that the number of water molecules produced will be equal to the number of hydrogen molecules consumed regardless of the actual number involved. Therefore,

$$\left.\begin{matrix} 6.02 \times 10^{23} \text{ molecules} \\ \text{of hydrogen} \end{matrix}\right\} \sim \begin{matrix} 6.02 \times 10^{23} \text{ molecules} \\ \text{of water} \end{matrix}$$

Since 6.02×10^{23} is a number called the mole, it follows that one mole of hydrogen molecules will produce one mole of water molecules. The general conclusion, then, is the following: the ratio of particles in the balanced equation is the same as the ratio of moles in the laboratory. The solution to the problem logically follows:

$$\left.\begin{array}{l}\text{? moles of hydrogen}\\ \text{molecules}\end{array}\right\} = \left\{\begin{array}{l}15\ \cancel{\text{moles}}\\ \cancel{\text{water molecules}}\end{array}\right\} \times \frac{1\text{ mole hydrogen molecules}}{1\ \cancel{\text{mole water molecules}}}$$

$$= 15\text{ moles hydrogen molecules}$$

Note: Again the solution to the problem looks simple enough without resorting to the factor-label method. However in examples 3 and 4, the numbers become such that a quick mental solution is not readily achieved by most students.

EXAMPLE 3

Mole Weights Yield Weight Relationships

How many grams of oxygen are necessary to react with an excess of hydrogen to produce 270 grams of water?
Solution: From the balanced equation,

$$2H_2 + O_2 \longrightarrow 2H_2O,$$

the mole ratio between oxygen and water is immediately evident and is one mole of oxygen molecules per two moles of water molecules, or

$$\frac{1\text{ mole oxygen molecules}}{2\text{ moles water molecules}}.$$

This mole ratio can be changed into a weight ratio since the mole weight can be easily calculated from the atomic weights involved. One molecule of oxygen (O_2) weighs 32 amu (16 amu + 16 amu). Therefore a mole of oxygen molecules weighs 32 grams. Similarly, two moles of water weigh 36 grams [2(16 + 2 × 1)]. Therefore the weight ratio is:

$$\frac{1\ \cancel{\text{mole oxygen molecules}} \times \dfrac{32\text{ grams oxygen}}{\cancel{\text{mole oxygen molecules}}}}{2\ \cancel{\text{moles water molecules}} \times \dfrac{18\text{ grams water}}{\cancel{\text{mole water molecules}}}}$$

or,

$$\frac{32\text{ grams oxygen}}{36\text{ grams water}}.$$

This weight relationship is exactly the conversion factor needed to answer the original question:

$$\text{? grams oxygen} = 270\ \cancel{\text{grams water}} \times \frac{32\text{ grams oxygen}}{36\ \cancel{\text{grams water}}}$$

$$= 240\text{ grams oxygen}$$

Note: It should be observed that a weight relationship could be established between any two of the three pure substances involved in the reaction, regardless of whether they are reactants or products.

EXAMPLE 4

Gas Volumes Can Be Calculated by Knowing the Molar Volume

What volume of hydrogen measured at 0°C and 1 atmosphere of pressure is required to react with 240 grams of oxygen?

Solution: The mole ratio from the balanced equation,

$$2H_2 + O_2 \longrightarrow 2H_2O$$

is

$$\frac{\text{2 moles hydrogen molecules}}{\text{1 mole oxygen molecules}}.$$

When we remember that one mole of gas molecules (any gas) occupies 22.4 liters at "standard conditions" 0°C and 1 atmosphere), we can see that the mole ratio can be changed to a volume-mass ratio in the following way:

$$\frac{2\ \cancel{\text{moles hydrogen molecules}} \times \dfrac{\text{22.4 liters hydrogen}}{\cancel{\text{mole of hydrogen molecules}}}}{1\ \cancel{\text{mole oxygen molecules}} \times \dfrac{\text{32 grams oxygen}}{1\ \cancel{\text{mole oxygen molecules}}}}$$

or,

$$\frac{\text{2 (22.4) liters hydrogen}}{\text{32 grams oxygen}}.$$

The solution to the problem is as follows:

$$\begin{aligned} \text{? liters hydrogen} &= 240\ \cancel{\text{grams oxygen}}\ \frac{\text{2(22.4) liters hydrogen}}{32\ \cancel{\text{grams oxygen}}} \\ &= \text{336 liters hydrogen} \end{aligned}$$

The following examples are similar to the ones given above and a brief solution is given to each. After you have mastered the solutions to these problems, you should be able to work through the exercise list that follows.

EXAMPLE 5

How many molecules of water are produced in the decomposition of 8 molecules of table sugar? The unbalanced equation is as follows:

$$C_{12}H_{22}O_{11} \longrightarrow C + H_2O$$

Solution: Balance the equation,

$$C_{12}H_{22}O_{11} \longrightarrow 12C + 11H_2O$$

$$? \text{ molecules of water} = 8 \text{ molecules sugar} \times \frac{11 \text{ molecules water}}{1 \text{ molecule sugar}}$$
$$= 88 \text{ molecules water}$$

EXAMPLE 6

How many grams of mercuric oxide are necessary to produce 50 grams of oxygen? Mercuric oxide decomposes as follows:

$$2HgO \longrightarrow 2Hg + O_2$$

Solution:

(a) Weight of two moles of HgO = 2(201 + 16) = 2(217) = 434 g

(b) Weight of 1 mole of O_2 = 2(16) = 32 g

(c) $? \text{ g HgO} = 50 \text{ g oxygen} \times \frac{434 \text{ g mercuric oxide}}{32 \text{ g oxygen}}$
$= 678 \text{ g mercuric oxide}$

EXAMPLE 7

How many pounds of mercuric oxide are necessary to produce 50 pounds of oxygen by the reaction:

$$2HgO \longrightarrow 2Hg + O_2$$

Solution: Note that the problem is the same as Example 6 except for the units of chemicals. Also note that the conversion factor of Exercise 6,

$$\frac{434 \text{ g mercuric oxide}}{32 \text{ g oxygen}}$$

can be converted to any other units desired,

$$\frac{434 \text{ g mercuric oxide} \times \frac{1 \text{ pound}}{454 \text{ g}}}{32 \text{ g oxygen} \times \frac{1 \text{ pound}}{454 \text{ g}}} = \frac{434 \text{ pounds mercuric oxide}}{32 \text{ pounds oxygen}}$$

It is evident that the ratio, $\frac{434}{32}$, expresses the ratio between weights of mercuric oxide and oxygen in this reaction regardless of the units employed.

$$? \text{ pounds mercuric oxide} = 50 \text{ pounds oxygen} \times \frac{434 \text{ pounds mercuric oxide}}{32 \text{ pounds oxygen}}$$
$$= 678 \text{ pounds of mercuric oxide}$$

EXAMPLE 8

Carbon dioxide can be produced by the action of hydrochloric acid on calcium carbonate, the principal chemical in limestone. The balanced equation is as follows:

$$2HCl + CaCO_3 \longrightarrow CaCl_2 + H_2O + CO_2$$

What volume of carbon dioxide at 0°C and one atmosphere pressure can be produced from 75 grams of calcium carbonate, $CaCO_3$?

Solution:

(a) One mole of $CaCO_3$ produces one mole of carbon dioxide.

(b) Therefore 100 g of calcium carbonate (molecular weight = 100) produces 22.4 liters of carbon dioxide at STP.

(c) $? \text{ liters } CO_2 = 75 \cancel{\text{ g } CaCO_3} \times \frac{22.4 \text{ liters } CO_2}{100 \cancel{\text{ g } CaCO_3}} = 16.8 \text{ liters}$ of carbon dioxide

EXAMPLE 9

In the problem of Example 8, what volume of gaseous HCl at 0°C and 1 atmosphere would be required to prepare the acid solution?

$$? \text{ liters HCl} = 75 \cancel{\text{ g } CaCO_3} \times \frac{44.8 \text{ liters HCl}}{100 \cancel{\text{ g } CaCO_3}}$$

$$= 33.6 \text{ liters HCl}$$

or

$$? \text{ liters HCl} = 16.8 \cancel{\text{ liters } CO_2} \times \frac{44.8 \text{ liters HCl}}{22.4 \cancel{\text{ liters } CO_2}}$$

$$= 33.6 \text{ liters HCl}$$

PROBLEMS

1. What weight of oxygen is necessary to burn 28 g of methane, CH_4? The equation is the following:

$$CH_4 + 2O_2 \longrightarrow CO_2 + 2H_2O$$

Ans. 112 g oxygen

2. What volume would the oxygen in Problem 1 occupy at 0°C and 1 atmosphere pressure?

Ans. 78 liters

3. Potassium chlorate, $KClO_3$, releases oxygen when heated according to the equation:

$$2KClO_3 \longrightarrow 2KCl + 3O_2.$$

What weight of potassium chlorate is necessary to produce 1.43 grams of oxygen? What volume would this weight of oxygen occupy at 0°C and 1 atmosphere?

Ans. 3.65 g $KClO_3$

4. Fe_3O_4 is a magnetic oxide of iron. What weight of this oxide can be produced from 150 grams of iron?

Ans. 207 g oxide

5. Steam reacts with hot carbon to produce a fuel called water gas; it is a mixture of carbon monoxide and hydrogen. The equation is:

$$H_2O + C \longrightarrow CO + H_2$$

Ans. 60 g carbon

What weight of carbon is necessary to produce 10 grams of hydrogen by this reaction?

6. Iron oxide, Fe_2O_3, can be reduced to metallic iron by heating it with carbon.

$$2Fe_2O_3 + 3C \longrightarrow 4Fe + 3CO_2$$

Ans. 0.56 ton carbon

How many tons of carbon would be necessary to reduce 5 tons of the iron oxide in this reaction?

7. How many grams of hydrogen are necessary to reduce 1 pound (454 grams) of lead oxide (PbO) by the reaction:

Ans. 3.91 g hydrogen

$$PbO + H_2 \longrightarrow Pb + H_2O$$

8. How many liters of hydrogen at 0°C and one atmosphere pressure can be produced from one pound of water?

Ans. 565 l hydrogen

$$2H_2O \longrightarrow 2H_2 + O_2$$

9. Hydrogen can also be produced by the reaction of iron with steam.

$$4H_2O + 3Fe \longrightarrow 4H_2 + Fe_3O_4$$

Ans. 10.5 lb iron

What weight of iron would be needed to produce one-half pound of hydrogen?

10. Tin ore, containing SnO_2, can be reduced to tin by heating with carbon.

$$SnO_2 + C \longrightarrow Sn + CO_2$$

Ans. 79 tons tin

How many tons of tin can be produced from 100 tons of SnO_2?

ANSWERS TO SELF TESTS

CHAPTER 1

SELF-TEST 1-A

1. Any obvious mixtures containing two different kinds of things visible to the naked eye.
2. Operational definition.
3. Salt, sugar, silver, copper wires, baking soda, graphite, mercury, silica, sand, etc.
4. False.

SELF-TEST 1-B

1. Gasoline
2. Dry ice, iodine
3. Sublimation
4. Immiscible
5. Colors
6. Recrystallization
7. Gas; liquid

SELF-TEST 1-C

1. a. Metals.
   b. Nonmetals. Examples of ***metals:*** any element at or under Li, Be, Al, Ge, Sb, in the periodic table, or any of the metals between the Ti family and the Cu family. Examples of nonmetals: any element under He, F, or O in the periodic table plus H, B, C, Si, N, P, etc.
2. 106
3. False
4. Decay of food, digestion of food, any burning or corrosion process or any reaction of a metal and a nonmetal, etc., e.g. $Fe + S \longrightarrow FeS$.
5. Any transformation from solid to liquid or liquid to gas; any change which is only a change in physical appearance.
6. Substance
7. Submicroscopic
8. Macroscopic > microscopic > molecular
9. Theories, laws, facts
10. 1. One mole of sodium
    2. Two moles of sodium
    3. One mole of hydrochloric acid or HCl
    4. Yields or produces
    5. One mole of hydrogen gas molecules

MATCHING SET

1. e	4. a	7. c	10. i	13. k
2. h	5. b	8. d	11. m	14. l
3. g	6. j	9. f	12. o	15. n

CHAPTER 2

SELF-TEST 2-A

1. a. 2,1,2
   b. 1,2,1
   c. 4,3,2
2. a. SeO_2, SeO_3
   b. SeF_4, SeF_6
   c. H_2SeO_3, H_2SeO_4
3. a. increases
   b. decreases
   c. increases
4. 1/36 as much, or 3.8 kcalories
5. D_2O, DHO, H_2O

6. The mixture of the reactants will yield some of the products and a mixture of the products will yield some of the reactants.
7. The reaction proceeds in both directions.

MATCHING SET

1. h 3. e 5. c 7. a
2. f 4. b 6. d 8. g

CHAPTER 3

SELF-TEST 3-A

1. discontinuous
2. Leucippus
3. Aristotle and Plato
4. b
5. 2 atmospheres
6. 3 atmospheres
7. 4.75 (or 4.8) g
8. a. 303 K c. 373 K e. 327°C
   b. 248 K d. −23°C f. −269°C
9. 3 liters

SELF-TEST 3-B

1. destroyed, chemical
2. Lavoisier
3. compound
4. 75.1% Ag, 24.9% Cl
5. a. new because it has a different ratio by weight of oxygen to nitrogen
   b. 2:4:1

SELF-TEST 3-C

1. d
2. a. the same
   b. atoms
3. P_4, PF_3

MATCHING SET

1. d 4. j 7. a 10. i
2. f 5. b 8. c
3. h 6. g 9. e

CHAPTER 4

SELF-TEST 4-A

1. atomic mass unit
2. smaller than
3. 0.5, 20
4. carbon-12 isotope with atomic mass of 12

SELF-TEST 4-B

1. mole, Avogadro's number
2. K, 1 mole or 6.02×10^{23} atoms
   Kr, 83.8 g
   SO_2, 1 mole or 6.02×10^{23} molecules
   NO_2, 46 g
3. 180 amu
4. 20.183 g
5. a. CH_2O
   b. $C_2H_4O_2$
6. 0°C, 1 atm
7. 22.4

MATCHING SET

1. h 2. e 3. a 4. i 5. b 6. j 7. c 8. d 9. f 10. g

CHAPTER 5

SELF-TEST 5-A

1. (a) kinetic, (b) potential; 2. kinetic; 3. potential; 4. (a) mass, (b) velocity

SELF-TEST 5-B

1. potential, 2. Coulomb's, 3. constant, 4. kinetic, 5. equal to

SELF-TEST 5-C

1. change in temperature, change of phase, change in internal energy, physical change
2. heat of fusion
3. vaporization
4. loses (gives off, gives up)

MATCHING SET

1. l 2. k 3. a 4. f 5. i 6. b 7. d 8. c 9. e 10. j 11. g 12. h 13. m

CHAPTER 6

SELF-TEST 6-A

1. different, identical
2. electrons
3. equal, opposite
4. all
5. 1837
6. electron, proton
7. Thomson, Millikan

SELF-TEST 6-B

1. alpha, beta, gamma, gamma
2. protons, neutrons
3. small
4. nucleus
5. atomic numbers

6. protons, electrons (or vice versa)
7. 1837
8. uranium, thorium, radium

MATCHING SET

1. h
2. f
3. a
4. j
5. b
6. c
7. i
8. d
9. e
10. g

CHAPTER 7

SELF-TEST 7-A

1. 9p, 9e, 10n
2. 50p, 50e, 68n
3. waves, particles (either order)
4. spectrum
5. a. blue
   b. red
6. alpha particles
7. X-ray spectra

SELF-TEST 7-B

1. excited
2. a. n = 2 to n = 1
   b. n = 3 to n = 2
   c. n = 1 to n = 2
3. bright-line
4. quanta or photons
5. a, b, d
6. 10^{-7}, 10^{-9}
7. farther from, closer to

MATCHING SET

1. c
2. d
3. a
4. b

CHAPTER 8

SELF-TEST 8-A

1. standing
2. d and e
3. wave motion

SELF-TEST 8-B

1. ground
2. two
3. a. two; b. p-set; c. three.
4. $1s^2 2s^2 2p^6 3s^2 3p^6 4s^2 3d^{10} 4p^2$
5. n—state; ℓ—city; m—street; s—house number
6. two

SELF-TEST 8-C

1. a. Group I
   b. Group V
   c. Group I
2. s; p
3. a. no; b. no; c. yes

MATCHING SET

1. d
2. e
3. b
4. a
5. g
6. c
7. f
8. h

CHAPTER 9

SELF-TEST 9-A

1. ions
2. ionic lattice
3. one
4. helium, cesium
5. CaI_2
6. Cl^-
7. valence or bonding
8. losing
9. gaining
10. loses 1, gains 2, loses 2, loses 2, loses 3, gains 1

SELF-TEST 9-B

1. a. H_2 or Cl_2 or any other covalent bond between two identical atoms
   b. HCl, HF, or any other covalent bond between two different atoms
2. 6
3. 3
4. sodium
5. a. 8; b. octet; c. most of the time.
6. fluorine
7. O, N, or F

SELF-TEST 9-C

1. sigma, pi
2. sigma, pi
3. s, p
4. nonbonding, bonding
5. single
6. three
7. linear, trigonal planar, octahedral

MATCHING SET I

1. l
2. d
3. j
4. f
5. i
6. g
7. b
8. e
9. k
10. c
11. h
12. a
13. m

MATCHING SET II

1. d
2. g
3. h
4. f
5. a
6. e
7. b
8. c

MATCHING SET III

1. a
2. d
3. d
4. e
5. b

CHAPTER 10

SELF-TEST 10-A

1. electrolyte
2. 0.040 M
3. base, acid
4. amphiprotic
5. neutral
6. water
7. acid, base

SELF-TEST 10-B

1. a. 1×10^{-4}; b. 1×10^{-13}; c. 6; d. 1×10^{-8}
2. (10×140 g) or 1400 g
3. $OH^- + NH_4^+$
4. $H_3O^+ + Cl^-$
5. 2
6. yes

SELF-TEST 10-C

1. gains, reduced, loses, oxidized
2. oxidizing, reducing
3. reduced, cathode
4. no, yes, yes, no
5. reduction
6. oxidation

MATCHING SET

1. h
2. d
3. k
4. a
5. f
6. l
7. j
8. b
9. c
10. e
11. g
12. i

CHAPTER 11

SELF-TEST 11-A

1. experimental data
2. same
3. rate
4. change
5. radioactive
6. food

MATCHING SET

1. f
2. g
3. h
4. j
5. a
6. b
7. c
8. d
9. d
10. e
11. i

CHAPTER 12

SELF-TEST 12-A

1. oxygen
2. aluminum
3. Minnesota
4. limestone
5. copper
6. copper, aluminum, magnesium
7. slag
8. reduced
9. cathode
10. magnesium

SELF-TEST 12-B

1. oxygen
2. lead
3. ammonia
4. nitrogen, potassium, phosphorus
5. petroleum hydrocarbon
6. sulfuric
7. KCl

MATCHING SET

1. h
2. i
3. b
4. f or a
5. g
6. d
7. e
8. a
9. c

CHAPTER 13

SELF-TEST 13-A

1. organic
2. tetrahedral
3. hybridization
4. false
5. two (a or b or d, and e or f)
6. double bond
7. 2,4-dimethylhexane
8. —CH_3

SELF-TEST 13-B

1. four
2. rotation of plane polarized light
3. 32
4. delocalized
5. 12(6C, 6H)
6. 1,2,4-trimethylbenzene
7. coal, petroleum
8. true
9. substitution

MATCHING SET

1. f	4. c	7. b	10. l
2. h	5. g	8. i	11. e
3. a	6. k	9. d	12. j

CHAPTER 14

SELF-TEST 14-A

1. c
2. a. methanol
   b. ethanol
   c. formic acid or methanoic acid
   d. 2-butanol
   e. acetic acid or ethanoic acid
   f. ethylene glycol
   g. acetaldehyde
   h. 2-bromopropane
3. a. CH_3COOH
   b. CH_3COO^-, $Na^+ + H_2O$
   c. $CH_3CHCH_3 + H_2O$ (Br on the central C)
4. a. fermentation of starch
   b. hydration of ethylene
5. Stearic acid, palmitic acid, or oleic acid, among others.
6. a. alcohol
   b. carboxylic acid
   c. aldehyde
   d. ketone

SELF-TEST 14-B

1. a. $CH_3CH_2CH_2OCCH_2CH_3$ (C=O) (an ester) + H_2O

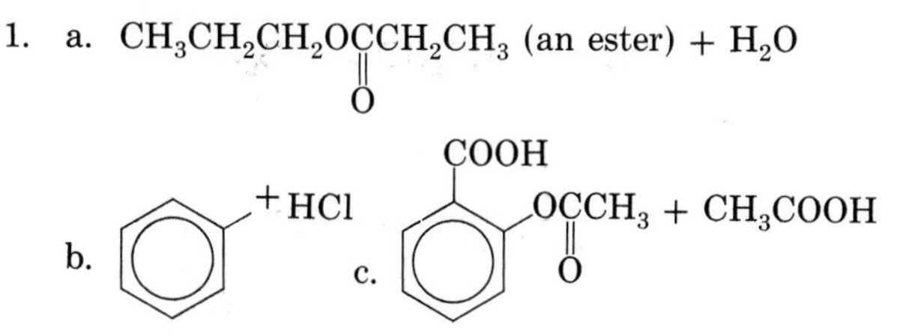

2. a. a group of atoms which absorbs light when present in molecules
   b. a derivative of sulfanilamide which can kill bacteria
   c. a compound used to impart a color to cloth or other material
   d. a salt of a long chain fatty acid

3. a. A fat is solid at room temperature, while an oil is liquid. The fat molecule has fewer double bonds.
   b. By hydrogenation.

MATCHING SET

1. h 4. j 7. b 10. g
2. e 5. a 8. f
3. i 6. d 9. c

CHAPTER 15

SELF-TEST 15-A

1. a. $H_2C{=}CH$ (with CH_3 on the CH) b. $HC{=}CH_2$ (with a benzene ring on the HC) c. $F_2C{=}CF_2$

2. isotactic, atactic, and syndiotactic (in any order)
3. free radical
4. isoprene
5. copolymer
6. polyamide or condensations
7. polyamine, polyacid (in any order), or a diamine and a dicarboxylic acid
8.

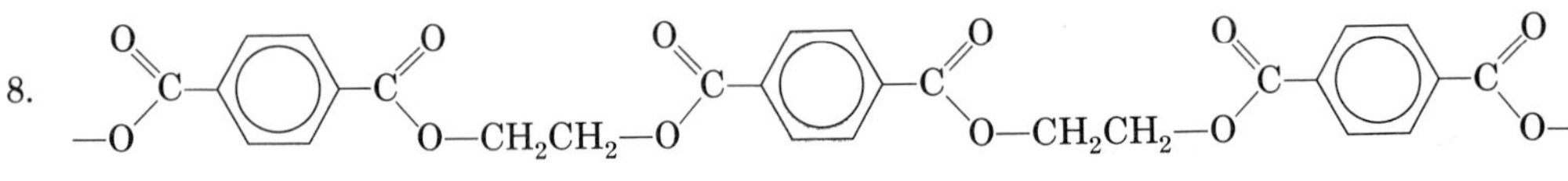

SELF-TEST 15-B

1. $-O-\underset{CH_3}{\overset{CH_3}{Si}}-\left(O-\underset{CH_3}{\overset{CH_3}{Si}}-\right)_n O-$

2. ultraviolet light
3. plasticizer
4. O (oxygen); it is a polymer held together by a network of Si—O bonds.

MATCHING SET

1. i 4. a 7. d 10. k
2. g 5. b 8. f 11. j
3. h 6. c 9. e

CHAPTER 16

SELF-TEST 16-A

1. carbon, hydrogen, and oxygen
2. monosaccharide
3. glucose, fructose
4. D-glucose
5. D-glucose
6. prostaglandins

SELF-TEST 16-B

1. amino acids
2. essential amino acids

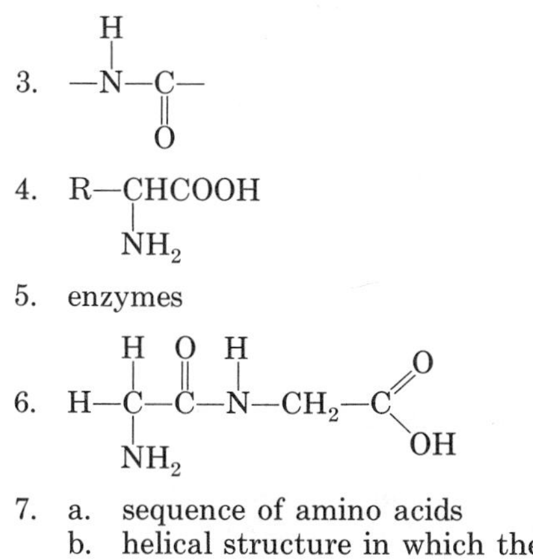

7. a. sequence of amino acids
   b. helical structure in which the amino acid chains are coiled
   c. the way in which the helical sections are themselves folded
8. a. 27
   b. 6

SELF-TEST 16-C

1. a. ribose
   b. deoxyribose
2. phosphoric acid, a sugar (ribose or deoxyribose), and a nitrogenous base.
3. double helix
4. True, when it invades a host cell
5. a. DNA
   b. tRNA
6. hydrogen

MATCHING SET

1. d
2. f
3. e
4. h
5. i
6. b
7. c
8. a
9. g
10. j
11. k

CHAPTER 17

SELF-TEST 17-A

1. catalyst
2. key, lock
3. niacin
4. enzyme
5. enzyme
6. coenzyme
7. niacin
8. active site

SELF-TEST 17-B

1. the sun
2. ATP
3. ADP, phosphate or phosphoric acid, energy
4. free
5. CO_2, H_2O, energy
6. NADPH
7. 3-phosphoglyceric acid
8. glucose

SELF-TEST 17-C

1. hydrolysis
2. lipids
3. emulsifying agents
4. ATP, lactic acid
5. CO_2, H_2O
6. coupled

MATCHING SET

1. i
2. f
3. a
4. h
5. g
6. c
7. e
8. b
9. j
10. d
11. k

CHAPTER 18

SELF-TEST 18-A

1. dehydration, hydrolysis
2. oxidizing
3. hemoglobin, oxygen
4. False
5. CN^-, cytochrome oxidase, oxygen
6. fluoroacetic acid
7. heavy metal poisoning or Lewisite, chelate
8. True

SELF-TEST 18-B

1. neurotoxins
2. synapses
3. acetylcholine
4. DNA, RNA
5. False
6. chimney sweeping
7. tetrahydrocannabinol
8. acetaldehyde
9. 4, 2
10. protein
11. complementary or light

MATCHING SET

1. b
2. g
3. i
4. j
5. m
6. e
7. h
8. f
9. c
10. l
11. n
12. a
13. d
14. k

CHAPTER 19

SELF-TEST 19-A

1. water
2. chlorination
3. soap
4. biochemical oxygen demand
5. oxygen, 1 liter
6. Limnology
7. phosphates

SELF-TEST 19-B

1. chlordan
2. toxic
3. electrostatic
4. calcium, magnesium
5. chlorine, fluorine
6. electric
7. two
8. more pure
9. settling, filtration
10. bacteria (small organisms), oxygen
11. soluble, organics

MATCHING SET

1. f
2. i
3. j
4. b
5. d
6. c
7. l
8. k
9. h
10. a
11. e
12. g

CHAPTER 20

SELF-TEST 20-A

1. parts per million; 10,000; 920
2. NO_2
3. CO, NO_2, SO_2 (any order)
4. troposphere
5. aerosols
6. surface area
7. electrostatic precipitator
8. NO_2
9. warm; cool
10. SO_2

SELF-TEST 20-B

1. SO_2
2. coal
3. first
4. another element
5. endothermic
6. oxides
7. False
8. False
9. limited
10. hemoglobin
11. False
12. benzo(α)pyrene (BaP)

SELF-TEST 20-C

1. eight
2. knocking
3. straight
4. cylinder
5. carburetor
6. tetraethyl lead
7. 93
8. yes
9. False
10. CO, hydrocarbons; nitrogen oxides
11. platinum (or palladium)

MATCHING SET

1. c
2. h
3. n
4. l
5. p
6. d
7. i
8. f
9. o
10. k
11. e
12. q
13. a
14. m
15. j
16. b

CHAPTER 21

SELF-TEST 21-A

1. synergism
2. salting, drying
3. False
4. cheese
5. more
6. sequestrant
7. False
8. flavor enhancer
9. dehydrating
10. free radicals

SELF-TEST 21-B

1. saccharin
2. qualitative
3. cyclamate
4. carbon dioxide
5. breaks
6. proteolytic
7. False
8. generally recognized as safe
9. active
10. acid, or hydrogen ion

SELF-TEST 21-C

1. hydrochloric acid
2. pain killer
3. morphine
4. acid; ester
5. fever
6. Alexander Fleming
7. True
8. sulfa drugs; antihistamines

MATCHING SET

1. l
2. h
3. n
4. a
5. p
6. b
7. c
8. e
9. d
10. g
11. f
12. i
13. j
14. k
15. m
16. o

CHAPTER 22

SELF-TEST 22-A

1. 4
2. cystine
3. disulfide, ionic
4. ionic
5. colorless
6. flexible
7. Freons (Freon 11 and Freon 12)
8. fatty acids
9. sheep
10. False

SELF-TEST 22-B

1. (c) ester
2. True
3. nonpolar
4. sodium stearate
5. hydrophobic
6. cosmetics (face powders)
7. short wavelength (or ultraviolet); long wavelength (or visible)
8. iron

MATCHING SET

1. o
2. l
3. j
4. c
5. n
6. m
7. a
8. d
9. e
10. b
11. f
12. g
13. h
14. i
15. k

CHAPTER 23

SELF-TEST 23-A

1. (b)
2. polymerization; double
3. nonpolar
4. free radicals
5. (c)
6. layers
7. entangled
8. True
9. silicone and diester greases with lithium soaps
10. bonds
11. nitrate

SELF-TEST 23-B

1. photochemistry
2. silver atoms, Ag^0
3. 4—Ag_4^0
4. reduced
5. $Ag(S_2O_3)_2^{3-}$
6. somewhat larger
7. reducing
8. sodium thiosulfate
9. negatives
10. (a) blue; (b) green; (c) red.
11. red, blue, green (any order)
12. True
13. reduced
14. panchromatic

SELF-TEST 23-C

1. vehicle; pigment
2. slaked lime [$Ca(OH)_2$]
3. H_2S
4. TiO_2; lead

5. binder
6. (a) evaporation;
   (b) metal, fatty;
   (c) polymerization.
7. linseed
8. True
9. resin
10. emulsions
11. acids, alcohols (either order)

MATCHING SET

1. e	4. h	7. b	10. k
2. c	5. a	8. j	11. i
3. f	6. d	9. l	12. g

CHAPTER 24

SELF-TEST 24-A

1. alpha, beta, gamma
2. gamma
3. cloud chamber
4. $^{87}_{36}Kr$
5. $^{212}_{82}Pb$
6. 0.125 g
7. $^{8}_{4}Be$
8. James Chadwick; beryllium bombardment by alpha rays.

SELF-TEST 24-B

1. one billion (10^9)
2. cyclotron
3. neptunium (Np)
4. 106
5. $^{206}_{82}Pb$
6. porous sedimentary
7. cosmic radiation
8. older
9. roentgen
10. gamma

MATCHING SET

1. b	5. e	9. m	13. f
2. h	6. c	10. n	14. j
3. a	7. k	11. g	
4. i	8. d	12. l	

CHAPTER 25

SELF-TEST 25-A

1. 20 percent
2. natural gas
3. 42
4. 50 percent
5. CO, H_2, N_2
6. 33 percent

SELF-TEST 25-B

1. fission
2. critical mass
3. uranium-235 ($^{235}_{92}U$)
4. fusion
5. containment of reactants at high temperature
6. $^{3}_{1}H$, tritium
7. plutonium-239, uranium-233 ($^{239}_{94}Pu$, $^{233}_{92}U$)

MATCHING SET

1. f	4. i	7. d	10. a	13. j
2. k	5. m	8. e	11. o	14. l
3. b	6. c	9. g	12. h	15. n

INDEX

Note: Numbers in *italic* indicate a figure; d following a page number indicates a definition; s indicates a molecular structure; and t indicates a table.

A

D

G

M

Q

R

U

V

Noble Gases

GROUPS

PERIODS

Period	IA	IIA	IIIB		IVB	VB	VIB	VIIB	VIII	VIII	VIII	IB	IIB	IIIA	IVA	VA	VIA	VIIA	O
1	1 H 1.00797																	1 H 1.00797	2 He 4.0026
2	3 Li 6.939	4 Be 9.0122												5 B 10.811	6 C 12.01115	7 N 14.0067	8 O 15.9994	9 F 18.9984	10 Ne 20.183
3	11 Na 22.9898	12 Mg 24.312												13 Al 26.9815	14 Si 28.086	15 P 30.9738	16 S 32.064	17 Cl 35.453	18 Ar 39.948
4	19 K 39.102	20 Ca 40.08	21 Sc 44.956		22 Ti 47.90	23 V 50.942	24 Cr 51.996	25 Mn 54.9380	26 Fe 55.847	27 Co 58.9332	28 Ni 58.71	29 Cu 63.54	30 Zn 65.37	31 Ga 69.72	32 Ge 72.59	33 As 74.9216	34 Se 78.96	35 Br 79.909	36 Kr 83.80
5	37 Rb 85.47	38 Sr 87.62	39 Y 88.905		40 Zr 91.22	41 Nb 92.906	42 Mo 95.94	43 Tc (99)	44 Ru 101.07	45 Rh 102.905	46 Pd 106.4	47 Ag 107.870	48 Cd 112.40	49 In 114.82	50 Sn 118.69	51 Sb 121.75	52 Te 127.60	53 I 126.9044	54 Xe 131.30
6	55 Cs 132.905	56 Ba 137.34	57 La 138.91	58 Ce → 71 Lu	72 Hf 178.49	73 Ta 180.948	74 W 183.85	75 Re 186.2	76 Os 190.2	77 Ir 192.2	78 Pt 195.09	79 Au 196.967	80 Hg 200.59	81 Tl 204.37	82 Pb 207.19	83 Bi 208.980	84 Po (210)	85 At (210)	86 Rn (222)
7	87 Fr (223)	88 Ra (226)	89 Ac (227)	90 Th → 103 Lr	104 *	105 *	106 *												

LANTHANIDE SERIES	58 Ce 140.12	59 Pr 140.907	60 Nd 144.24	61 Pm (147)	62 Sm 150.35	63 Eu 151.96	64 Gd 157.25	65 Tb 158.924	66 Dy 162.50	67 Ho 164.930	68 Er 167.26	69 Tm 168.934	70 Yb 173.04	71 Lu 174.97
ACTINIDE SERIES	90 Th 232.038	91 Pa (231)	92 U 238.03	93 Np (237)	94 Pu (242)	95 Am (243)	96 Cm (247)	97 Bk (249)	98 Cf (251)	99 Es (254)	100 Fm (253)	101 Md (256)	102 No (254)	103 Lr (257)